PRECALCULUS PLUS

RONALD D. FERGUSON

San Antonio College

WEST PUBLISHING COMPANY

Minneapolis/Saint Paul   New York   Los Angeles   San Francisco

DEDICATION

For your patience, love and understanding . . .
To Layne Ferguson, Julie Ferguson, and Lara Ferguson
. . . Thanks and love.

PRODUCTION CREDITS

Copyediting	Gnomi Gouldin, Phyllis Dunsay, Gary Phillips
Text Design	Geri Davis, Quadrata Inc.
Art Development	WPS, Inc.
Composition	Graphic World, Inc.
Cover Image/Design	Lois Stanfield, LightSource Images

Production, Printing and Binding by West Publishing Company

WEST'S COMMITMENT TO THE ENVIRONMENT

In 1906, West Publishing Company began recycling materials left over from the production of books. This began a tradition of efficient and responsible use of resources. Today, up to 95 percent of our legal books and 70 percent of our college and school texts are printed on recycled, acid-free stock. West also recycles nearly 22 million pounds of scrap paper annually—the equivalent of 181,717 trees. Since the 1960s, West has devised ways to capture and recycle waste inks, solvents, oils, and vapors created in the printing process. We also recycle plastics of all kinds, wood, glass, corrugated cardboard, and batteries, and have eliminated the use of Styrofoam book packaging. We at West are proud of the longevity and the scope of our commitment to the environment.

COPYRIGHT ©1995   By WEST PUBLISHING COMPANY
                           610 Opperman Drive
                           P.O. Box 64526
                           St. Paul, MN 55164-0526

Library of Congress Cataloging-in-Publication Data
Ferguson, Ronald D.
     Precalculus plus /   Ronald D. Ferguson.
       p.   cm.
     Includes index.
     ISBN 0-314-02835-8
     1. Functions.   I. Title.
QA331.F46   1994
512'.1—dc20   93-36017
                                                          CIP

British Library Cataloguing-in-Publication Data. A catalogue record for this book is available from the British Library.

TEXT IS PRINTED ON 10% POST CONSUMER RECYCLED PAPER

PRINTED WITH SOY INK

BRIEF CONTENTS

CONTENTS

PREFACE

PreCalculus Plus is for students preparing for calculus or introductory computer science. The text assumes a prerequisite of the equivalent of two years of high school algebra and includes all the standard topics of precalculus mathematics *plus* a few extras. Among these extras are extensive use of graphics and technology, previews of calculus concepts, and discussions of numerical methods from discrete mathematics.

PHILOSOPHY

A textbook is the consequence of many forces. Philosophy influences the presentation of topics as surely as tradition and market considerations. My goal in writing *PreCalculus Plus* was to implement the consensus topics of precalculus mathematics within the following framework:

A text should not break topics into isolated modules. This text endeavors to tie a current topic to previous topics. The connections usually take the form of some thematic element such as transformations.

Student comprehension improves when a concept is presented in more than one context. Although the principal discussion of a concept may occupy a single location in the text, the text explores many topics thematically. For example, linear models occur frequently throughout the text, as do quadratic functions, transformations, and the composition of functions.

Although Chapter 0 reviews prerequisite topics and much of Chapter 1 also may be review material, I believe a student benefits more from applying old skills in the context of new problems and concepts. For that reason, many exercises require the student to use prior concepts and skills.

Mathematics is a tool for modeling real-world situations. This text emphasizes that information can be expressed in numerous ways, including different algebraic forms, visually in the form of graphs, and even as a recursive process. As representations of information, each of these forms models the information. The text also discusses the limitations of models.

The transformation of information into different forms provides new insight into the information. For this reason a major theme of the text is transformations. From this perspective, the text presents algebra as a method for rewriting information. Analysis converts information from numerical to geometric representations and vice versa. Abstraction and idealization transform real-world problems into models. Technology provides a powerful tool for transforming the presentation of data.

The evolution of mathematical thought provides a model to improve student understanding of concepts. Etymology and historical anecdotes reveal mathematics as a human quest for knowledge rather than the arcane incantations of mathematical priests.

Concept development should exploit technology without inducing dependence on technology. Dependence should be avoided in two ways. First, the text should not predicate the development of a topic solely on an appeal to technology. Although calculator graphics are compelling, we should appeal to students' imaginations as well as their eyes. As a result, the text often uses the flaws in calculator graphics rather than the successes to illuminate a topic.

Second, the student must not become dependent on technology. The graphing calculator should be the student's tool, not the remote control to the student's brain. The text encourages, but does not require, that the student have graphing technology. Most examples include detailed graphs for the student without technology. Examples demonstrate methods for visualizing data that do not require technology, as well as examples that promote technology. Indeed, the text encourages the student to critically examine technology by providing assignments to convert graphs to charts, requesting annotations of graphs, and including examples of misrepresentation by technology.

A student without access to a graphing calculator or computer is given visual experience from the textbook. To improve student visualization, the text makes heavy use of computer-generated graphics. Be warned—exercises emphasizing the exploration of graphs are quite tedious without technology. A calculator icon marks these technology-intensive exercises.

The technological media must not become the message. In today's society, the availability of technology is commonplace. We should treat technology as if it were a compass, ruler, graph paper, or any other tool. This book tries not to let the student think that technology, rather than mathematics, is the goal of a discussion.

Technology is a valuable tool. On the train ride to knowledge, technology provides a window to view a concept. The train has other windows and multiple views. Without doubt, a good window seat can enhance almost any train ride. No matter what view the student finds most appealing, graphing technology is not the whole ride. Indeed, the text points out the dangers of applying a technological tool without first having a firm theoretical base.

As the computer provides a model for human thought, technology provides pedagogical models for learning and teaching. This text was designed to use technological approaches that offer insight to those without technology. For example, the text uses window views of computer screens for graphs. Other examples include transformations as a visual model for the composition of functions. Thus, the presentation should mesh nicely with such technological constructs as scrolling and setting ranges while serving equally well for students without calculators.

Structure of the Text

Outline of Topics

Chapter 1 reviews graphing techniques and introduces graphing calculators. The chapter presents graphs of lines, circles, and parabolas for calculator practice and to foreshadow functions in Chapter 2. Chapter 2 uses a window view for graphs of functions. Moreover, the chapter categorizes functions in terms of common properties such as increasing, decreasing, odd or even and so on. In addition, the algebra of functions helps prepare for the introduction of polynomials in Chapter 4. Compositions introduce the theme of transformations. Inverse functions set up logarithms in Chapter 6.

Chapter 3 introduces approximation methods, analysis of graphs, and graphical solutions of inequalities by examining familiar linear, quadratic, and absolute value functions. The chapter also introduces tracing functions, continued fractions, and fixed-point iteration. Subsequent chapters revisit these techniques.

Chapter 4 extends linear and quadratic functions to polynomials. The comparison of polynomials to integers leads to the introduction of rational and algebraic functions in Chapter 5. Chapter 5 reemphasizes earlier themes such as domain, continuity, and asymptotes as graphical aids.

The transcendental functions introduced in Chapter 6 develop from the natural base e. As a result, the development more closely resembles the approach of calculus texts rather than the arrangement usually found in a precalculus text. The chapter discusses the machine representation of a finite-precision decimal using the mantissa and characteristic. In addition, the concept of logarithmic transformations extends the theme of transformations.

Chapter 7 uses angular and linear velocity to introduce the use of radian measure. Early in chapter 7, the connection between angles of rotation and triangles is made. Hence, while the text emphasizes the unit circle principles, classic triangle trigonometry is interwoven in the development of concepts. Based on the unit circle, with occasional applications of technology, the chapter gives an early introduction to the graphs of circular functions. In addition, transformations tie the analysis of graphs to the composition of functions. Applications of circular functions include the modeling of wave forms. To take advantage of technology, graphical interpretations are used to teach the concepts of identities and conditional equations.

Chapter 8 develops the solution of right triangles and general triangles. The solution of conditional equations leads to inverse trigonometric equations. Other applications include polar coordinates, two-dimensional vectors, and complex numbers.

Chapter 9 presents classic analytical geometry of conic sections, culminating in the rotation of conic sections using trigonometric functions. The rotation of conics provides an additional example of transformations. Chapter 10 develops systems of linear and nonlinear equations and systems of inequalities. The linear systems of Chapter 10 set up the development of matrices and determinants in Chapter 11. In keeping with mathematical history and a theme of the text, transformations provide the definition for the multiplication of matrices.

Finally, Chapter 12 presents the discrete mathematics topics most commonly found in precalculus texts: sequences, series, the binomial theorem, and so on. Infinite-series examples explain the difficulties of representing infinite decimals in a finite-precision machine. As the last chapter, Chapter 12 often ties the last knot of a new concept to previous topics. For example, the discussion of formal recursion in the form of finite mathematical induction inspires a recursive definition for a polynomial.

STRUCTURE OF THE CHAPTERS

See the Summary of Features on page xxiii for specific examples of many of the following features.

QUOTES
Most sections begin with a quotation that reflects the nature of topics in the section, an historical insight, or a comment on the value of mathematics. These quotations affirm that humans are the source of our mathematical heritage.

SIDEBARS
In the margins of the text, a sidebar provides a mathematical note for student enrichment. Most sidebars fall into one or more of the following categories.

ETYMOLOGICAL.   Some sidebars trace the historical origin of a word. This etymology gives students insight into the meaning of a word and into the evolution of the concept the word represents.

HISTORICAL.   Historical notes provide a cultural heritage for mathematics. The evolution of mathematical thought may be more provocative than a simple statement of a modern view. Primary historical references in this book include *An Introduction to the History of Mathematics* by Howard Eves and Morris Kline's *Mathematical Thought from Ancient to Modern Times*.

ANECDOTAL.   Anecdotal sidebars consist of fictional dialogues to provide interactive student insight into a concept. Reading a fictional dialogue may even encourage students to discuss mathematics among themselves or with their instructor.

THEOREMS AND DEFINITIONS

The text highlights theorems and definitions. I do not expect students to become theorem-proving machines. However, reading, digesting, applying, or even formulating theorems and definitions encourages the thought process of students.

EXPOSITION

The exposition that is provided is conversational. The job of exposition is to introduce and explain. The mode of exposition may vary from simple declarations to exploratory queries. Sometimes the student is led by examples to a theorem or definition, so that the formal statement of the concept summarizes the process. At other times, a definition or theorem comes first so that the student can learn to read and interpret the principle before applying the concept to examples. While sentence structure and diction is not usually complex, developing students' thinking skills sometimes requires developing their vocabulary.

EXAMPLES/ILLUSTRATIONS

Much of the effectiveness of this text is in the use of Examples.

An Example consists of a series of related illustrations used to develop a concept. Most examples include a label identifying the concept developed. Clearly marked solutions and answers are detailed in each illustration. Examples often parallel exercises at the end of a section.

APPLICATIONS

Applications include a diversity of problems from fields such as engineering, biology, medicine, astronomy, architecture, and statistics. Applications offer opportunities to create mathematical models without too much distraction by artificial stories. Throughout the text, the use of numerical methods and finite precision decimals helps the student distinguish between exact representations and approximations. These discrete topics should benefit students using technology and students planning to study engineering or computer science.

FIGURES

For accurate representation, I generated figures for the text using *Graph-Windows* technology. The graphics were then imported into Aldus FreeHand for color enhancement and annotations. The annotations improve the clarity of the data.

Similarly, graphing exercises ask students to annotate the graphs they produce. This is especially important for graphical behavior that is poorly represented in a calculator display. Moreover, the student must then assume responsibility to do more than simply push a button and see a graph.

TECHNOLOGY

As more students acquire graphing calculators, it becomes more difficult to demand a specific brand and model for a class to use. If a student already has a graphing calculator, asking him or her to buy another for class uniformity is not likely to endear an instructor to the students. For that reason, graphing calculator examples are generic, in the sense that the text emphasizes the similarities among graphing calculators. Students using technology are reminded to set the window limits, enter a function, mind the mode, and execute the graph, zooming or scrolling as necessary. However, from time to time, some examples detail specific key strokes for the following calculators:

Casio fx7700-G

TI-81, TI-82, TI-85

Sharp EL9200/9300

There are a number of reasons for occasionally providing specific key strokes. First, it gives a student with one of these calculators quick access to an example without having to search a manual. It provides close comparison of key strokes to reinforce the similarities among calculators. Finally, it may inspire a student without a calculator to realize that a calculator is not only fast, but it is also easy to use. The last reason is important in a class that does not require calculators but does encourage their use.

Although several fine computer programs are available for producing graphs, for example Mathematica, Derive (see Available Supplements), MathCad, MicroCalc, etc., these programs are massive. I leave the discussion of these packages to the instructor and a user's manual. The computer graphics package discussed in this text is *GraphWindows*. Adopters of the textbook receive a limited license for the use of *GraphWindows*.

On the basis of ten years of experimenting with computer graphics to demonstrate concepts in precalculus and calculus, I wrote *GraphWindows* with several goals in mind. First among these goals was a gradual learning curve. Students should not have to learn specialized notation to store functions. Standard activities such as selecting a window, zooming, scrolling, and graphing should be intuitive. The display should be in color, so that functions are easy to identify. Compositions of user-defined functions should be natural to encourage experimentation with transformations.

Because EGA/VGA displays have higher resolution than current graphing calculators, *GraphWindows* provides sharper, more colorful classroom demonstrations. In particular, the graphs of three or more inequalities are much clearer in the blended colors produced by *GraphWindows* than on a graphing calculator. *GraphWindows* also supports polar and parametric representations.

Because the goal for *GraphWindows* is to allow a student to explore the two-dimensional graphs, the program is relatively small and should run on many older IBM compatible machines. If your department has a microcomputer lab and is reluctant to require that all students buy a graphing calculator, *GraphWindows* in a computer lab will allow students to experiment with graphing technology. The text includes examples of graphing using *GraphWindows* version 2.02. However, for details on the use of the latest version of *GraphWindows*, consult the *GraphWindows* user's manual.

EXERCISES

The exercises at the end of each section provide practice of concepts introduced in that section and review of related concepts. Red exercise numbers identify application exercises. Asterisks (*) mark those exercises requiring extra care. Exercise sets consist of parallel odd and even exercises. As a result, an instructor can assign all even, all odd, or perhaps multiples of three with the assurance that a student receives a full exposure to the concepts of the section. Appendix B provides answers to selected odd exercises. The problem sets are graded by level of difficulty in the sense that more difficult problems appear toward the end of each set of instructions, and in general toward the end of the exercises. While the exercises encourage a student to use technology, lack of a calculator usually will not keep a student from completing the exercises. You must take more care in making assignments from the problem sets.

PROBLEM SETS

Problem sets appear at the end of most sections. Problem sets consists of more challenging problems than in the exercises. As with the exercises, a red problem number identifies an application problem. Although the problem sets usually have parallel odd and even problems, such is not always the case. Often a sequence of several problems is necessary to complete the development of a concept. Depending on the section, many problem sets fall into one or more of the following categories.

TECHNOLOGY.   A graphing calculator icon marks problems that are difficult to do without technology. These problems may alternately be labeled as Problems for Technology. In general, technology problems overlap with problems in the Confirmation and Exploration categories.

CONFIRMATION.   Confirmation problems offer the student experiments with graphing technology that reinforce concepts from the section just completed. Based on these experiments, the confirmation problems may ask the student to summarize the concept.

EXPLORATION.   A binocular icon marks Exploration problems. Many Exploration problems preview concepts to be introduced in the next section. Some Exploration problems overlap the Preview of Calculus problems. Typically, these problems ask a student to make conjectures based on a sequence of graphic patterns. Usually the student must summarize, describe the new concept, or formulate a definition. Many of these summary problems overlap the Writing mathematics category.

WRITING MATHEMATICS.   A writing hand icon often labels a requirement to verbalize a concept. This task takes the form of formulating a definition, summarizing a result, or describing a concept. As a result, the problem encourages the student to look for patterns, make inferences, test conclusions, and communicate by writing. Constructing sentences to express what you have learned is a valuable opportunity to organize your thoughts.

CLOSER LOOKS.   A magnifying glass icon marks Closer Look problems. These problems explore concepts in greater detail or develop theory related to topics in the section. Many are more theoretical than usual.

PREVIEW OF CALCULUS.   Preview of Calculus problems give students a chance to connect precalculus concepts and techniques to elements of the calculus. Like Closer Look problems they are intended to challenge better students. The problems in Preview of Calculus include calculating slopes of secant lines, intuitive tangent lines, slopes of tangents to functions, and areas under curves. A blue-green problem number identifies these problems.

ENRICHMENT TOPICS.   Enrichment topics include concepts not normally found in precalculus texts. These topics include continued fractions, power series, radix notation, and finite-precision decimals. Other problems coach the student through the development of Cardan's solution of a cubic equation, the derivation of the quadratic formula using transformations, or the derivation of Hero's formula.

Choose carefully in making assignments from the problem sets.

SECTION SUMMARY

Every section includes a summary of introduced topics. Depending on time constraints, you may wish to spend more than one lecture on some sections. For example, you could follow an introductory lecture and assignment in the exercises with a second lecture and assignment in the problem sets. Sections are organized by convenient divisions for topics, not for presorted one hour lectures. The length of a section does not necessarily correspond to its importance. However, few sections demand more than one lecture.

KEY WORDS AND CONCEPTS

The end of each chapter lists key words and concepts to facilitate student review.

CHAPTER SUMMARY

Chapter summaries give concise review of major topics. These summaries repeat the major concepts from section summaries for easy reference.

REVIEW EXERCISES

Review exercises appear at the end of each chapter, keyed to the section where they were introduced. These exercises provide concise samples of exercises found within each section of a chapter. When a student encounters difficulty in reviewing these exercises, the section reference allows for quick access to the original material.

SIDELIGHTS

A sidelight appears at the end of each chapter. Sidelights are usually longer excursions into a historical reference, a short biography, or a discussion of the evolution of some other mathematical concept.

CHAPTER TESTS

A chapter test appears at the end of each chapter. Chapter tests are for student practice. Most instructors will not cover every chapter in a book nor give a test at the end of each chapter. For that reason, I made only a modest attempt at equalizing the size of chapters. I thought it was more important to have chapters meaningfully group topics. For this reason, a trigonometry chapter may be long whereas a chapter on algebraic functions is short.

AVAILABLE SUPPLEMENTS

SOLUTION MANUALS

The *Instructor's Solution Manual* contains solutions to even-numbered problems in the text. The *Student's Solution Manual* contains solutions to the odd-numbered problems. Sharon Edgmon of Bakersfield College prepared the solutions to all problems. At least one independent accuracy checker checked each solution.

TEST BANK

Norma James, New Mexico State University, wrote the Test Bank, which consists of four sample tests per chapter, two of which use multiple choice questions and two of which use open-ended questions.

GRAPHING CALCULATOR MANUAL

Samuel A. Lynch, Southwest Missouri State, prepared the *Graphing Calculator Manual* to accompany *PreCalculus Plus*. The manual introduces the use of the TI-81 calculator and illustrates commands and key strokes for the TI-81, often using examples from *PreCalculus Plus*. The manual points out differences between the TI-81 and TI-82. An appendix includes keystroke information for the Casio fx7700-G. The tutorial follows the same order of topics as *PreCalculus Plus* and includes graphing calculator exercises and answers.

EXPLORING PRECALCULUS WITH DERIVE

Exploring Precalculus with DERIVE is a workbook consisting of 28 explorations created by E. Hodes, P. Yuhn, and R.M. Mallen of Santa Barbara City College. Elizabeth Hodes edited the workbook, which coaches students through 28 precalculus experiments using DERIVE software (DERIVE, A Mathematical Assistant by Soft Warehouse, Inc.). Each exploration includes questions requiring the student to formulate and write conclusions about the exploration.

USER'S MANUAL

Designed as a reference for instructors, the *User's Manual for PreCalculus Plus* by Ronald D. Ferguson contains suggested course syllabi and lists of dependencies. To ease homework assignment the manual tabulates exercise page numbers with notes on the nature of the problems. For many sections, the manual suggests classroom technology demonstrations. These demonstrations may use technology for motivation or caution against difficulties faced in technology. The suggested demonstrations assume the instructor has access to graphing technology appropriate for classroom demonstrations. Some sections include a rationale for the approach to a concept. Rationales are most likely if the development of a topic differs from the standard treatment. Some rationales tie the new concept to a previous topic or foreshadow a topic developed in later sections. Some sections offer alternative approaches to a topic or particular cautions about an approach.

GraphWindows

GraphWindows, written by Ronald D. Ferguson, is available to book adopters. *GraphWindows* creates a gradual learning curve for technology by allowing students to enter formulas for functions much as they appear in textbooks. The program handles function composition, tracing, scrolling, and zooming in a natural manner to allow the student to concentrate on mathematics rather than technology. The screen design resembles a graphing calculator but takes advantage of the computer keyboard. The program operates in color under EGA/VGA on most IBM-compatible PCs.

Acknowledgments

Reviewers

I want to thank the following reviewers. In some cases, one reviewer's suggestion may conflict with another's. My job was to reconcile a multitude of suggestions. Therefore, where the text succeeds, much credit goes to the reviewers. Where it fails, I must blame myself.

Karen E. Barker, Indiana University at South Bend
Kathleen Bavelas, Manchester Technical Community College
June Bjercke, San Jacinto Community College
Joann Bossenbroek, Columbus State Community College
Gerald L. Bradley, Claremont McKenna College
Judy Cain, Tompkins Cortland Community College
Virginia L. Carson, DeKalb College
J. Curtis Chipman, Oakland University
John W. Coburn, St. Louis Community College at Florissant Valley
C. Patrick Collier, University of Wisconsin at Oshkosh
Ryness A. Doherty, Community College of Denver
Robert J. Fisher, Idaho State University
William R. Fuller, Purdue University
Dewey Furness, Ricks College
Paula Gnepp, Cleveland State University
William Grimes, Central Missouri State University
Frances Gulick, University of Maryland at College Park

Richard Hansen, De Anza College
Louise Hasty, Austin Community College
Edward Huffman, Southwest Missouri State University
David Johnson, Diablo Valley College
Donald R. LaTorre, Clemson University
Samuel A. Lynch, Southwest Missouri State University
Giles Wilson Maloof, Boise State University
Sheila D. McNicholas, University of Illinois at Chicago
Carolyn Meitler, Concordia University
Ann B. Megaw, University of Texas at Austin
Philip R. Montgomery, University of Kansas
Lynda Morton, University of Missouri at Columbia
John Oman, University of Wisconsin at Oshkosh
Jeanette Palmiter, Portland State University
Martin Peres, Broward Community College
Maria A. Reid, Borough of Manhattan Community College
Beverly Rich, Illinois State University
Joyce A. Riley, Louisburg University
Patricia Roecklein-Dalton, Montgomery College
Ross Rueger, College of the Sequoias
William Rundberg, College of San Mateo
Clarence H. Siemens, Solano College
Liz Sirjani, Washington State University
Laurence Small, L.A. Pierce College
Linda Smith, Tallahassee Community College
Tom Spradley, American River College
Louis A. Talman, Metropolitan State College of Denver
David Tartakoff, University of Illinois at Chicago
Lynn Tooley, Bellevue Community College
Anthony E. Vance, Austin Community College
Jan Vandever, South Dakota State University
David Van Langeveld, Clearfield High School
Joyce Kay Vetter, Indiana University/Purdue University
Carroll G. Wells, Western Kentucky University
Howard L. Wilson, Oregon State University
David Winslow, Louisiana State University
Gary L. Wood, Azusa Pacific University
Sandra Wray-McAfee, University of Michigan at Dearborn
Kemble Yates, Southern Oregon State College
Marvin Zeman, Southern Illinois University

(Reviewers of *GraphWindows* software)
Iris B. Fetta, Clemson University
Robert J. Lopez, Rose-Hulman Institute of Technology
Samuel A. Lynch, Southwest Missouri State University
William A. Rundberg, College of San Mateo
Laurence Small, L.A. Pierce College

Thanks also to Sharon Edgmon for her hard work on the solutions and to

Tanya Kalich and Susan Veyo for checking the accuracy of solutions. Special thanks go to my colleagues Tuvia Adar, Rita Bordano, Kenneth Reeves, Jack Shaw, and Reza Sojoodi who helped with solutions when time got short.

EDITORS

Editors wear many different hats. I have been fortunate to have an editor who provided not only support, but also gave me valuable feedback and encouragement even as he brought me back from flights of fancy. My thanks to Richard Mixter and Keith Dodson for their patience, persistence, and panache.

PRODUCTION STAFF

Finally, thanks to the production people at West, the copy editors, artists, and designers. And to Tom Modl, production editor, who kept it going and put it all together.

APPEAL FOR SUGGESTIONS AND COMMENTS

Textbooks evolve. This book is quite different from my original manuscript. Indeed, the changing target of technology has dictated constant revision and refinement during the design. Because production takes about a year, I rewrote not only according to what was current but toward what I imagined would be standard in a year. When the target moves, you must resight and hope you don't follow too much or lead too little. This is not a complaint or an excuse. I hope that the target will continue to move. Changing technology stimulates innovative teaching methods. So even as we go to press, I am thinking about revision. I would appreciate your comments. Please write. No text can be all things to all people. But I promise to consider every suggestion carefully and use it if I can.

Ronald D. Ferguson
San Antonio College

SUMMARY OF FEATURES WITH SELECTED EXAMPLES

This list contains samples only and is not intended as a comprehensive index of every example to a feature.

Feature	Description	Example	Location
Sidebars			
Etymological	The sidebars show the evolution of mathematical words and concepts.	"Pair"	Ch. 1-3
		Parameters	Ch. 3-1
		Quadratic	Ch. 3-2
		Logarithm	Ch. 6-2
		Mantissa	Ch. 6-3
		Sine	Ch. 7-2
Historical	This feature briefly explains the History or source of a topic. Alternative views and asides are also presented.	Pythagoras	Ch. 1-1
		Concavity	Ch. 1-5
		Equality	Ch. 2-1
		Absolute	Ch. 3-6
		Calculus	Ch. 4-2
		Root	Ch. 5-2
		e	Ch. 6-1
		Degree	Ch. 7-1
		Tangent & secant	Ch. 7-2
		Cardan	Ch. 8-7
		Transform	Ch. 9-5

Feature	Description	Example	Location
Applications	These problems are identified by a red problem number. Real world applications are found in Examples, Exercises, and Problem Sets.	Medicine	Ex. 3-1 #41
		Physics	Ex. 3-1 #44
		Windmills, Pendulums, & Business	Examp. 3-4
		Projectiles	P.S. 3-2 #1
		Torricelli	P.S. 3-2 #5
		Statistics	P.S. 3-2 #4
		Music	P.S. 3-2 #6
		Approximation	P.S. 3-3
		Precision	Ch. 3-7
		Radix	Ch. 4-1
		Half-life & biology	Examp. 6-1
		Inflation & Music	Ex. 6-3 #55
			Ex. 6-3 #57
		Nautical mile	P.S. 7-1
		Wave forms	Ch. 7-4
		Astronomy	Ex. 8-2 #87
		Astronomy	Ex. 8-4 #46
		Electronics	Ex. 8-4 #49
		Medic: ESL	Ch. 9-1
		Listening Stat	P.S. 9-4
		Electronics	Ex. 10-1 #34
		Economics	Ex. 10-1 #39
		Aspect ratio	Ex. 10-2 #35
Exercises	Parallel Odd and Even Exercises are designed to provide practice for topics in a section. Answers to selected Odd exercises appear in Appendix B. An instructor may assign all even, all odd, or multiples of three and still cover all topics in a chapter.	All Exercise Sets	See end of any section. E.g. Ch. 2-1 or Ch. 4-2, etc.
Problem Sets	Go beyond drill and practice of concepts. Included are the following categories.	All Problem Sets	See end of each section.

Feature	Description	Example	Location
Confirmation	Provides reinforcement of concepts in section usually with exploration using technology	Parabolas Sum functions Composition Identities	P.S. 1-5 P.S. 2-5 P.S. 2-6 P.S. 7-3
Exploration	Uses a binocular icon. Previews concepts of the next section, usually by experimenting with graphs. Sometimes it will include a preview of calculus. If a student is asked to describe or define a concept the Exploration may have a writing hand icon.	Slopes Lines Circles Parabolas Symmetry Increasing Composition Inverse Analytic Trig. Identity Conditional Graphing Sequences	P.S. 1-1 P.S. 1-2 P.S. 103 P.S. 1-4 P.S. 2-2 P.S. 2-3 P.S. 2-5 P.S. 2-6 P.S. 7-3 P.S. 7-5 P.S. 8-1 P.S. 12-1
Closer Looks	Uses a magnifying glass icon. These problems take a student into more detail or extend the development of a topic.	Symmetric Inverses Rat. Root 2 Vieta's Circular Funct.	P.S. 2-3 P.S. 2-8 P.S. 4-2 P.S. 4-4 P.S. 7-2

Technology	Uses a graphing calculator icon, and is often found in conjunction with one of the other types.	Circles (explore)
		Domain & Range
		Function sum
		Parametric:
		  parabolas, ellipses, &
		  hyperbolas
		Polar form of conics
			P.S. 1-3
			P.S. 2-1
			P.S. 2-4
			P.S. 9-2
			P.S. 9-3
			P.S. 9-4
			P.S. 9-5

Feature	Description	Example	Location
Preview of Calculus	The problems are identified by a blue problem number. This feature previews concepts and topics from the calculus.	Modeling rate of change	P.S. 2-1
			P.S. 2-4
		Increasing	
		Tangents to polynomials	P.S. 4-4
		Limits & continuity	Ch. 5-1
		Difference quotient	P.S. 5-2
		Secant lines	P.S. 6-1
		Area under curve	P.S. 6-2
		Trig substitution	P.S. 7-3
		Difference quotient	P.S. 7-5
			P.S. 8-2
			P.S. 9-2
		Tangents	
		Partial Fractions	P.S. 10-1
Enrichment	This feature presents topics not normally found in PreCalculus texts.	Division	P.S. 4-1
		Lagrange	P.S. 4-3
		Recursion	P.S. 4-5
		Involutions	P.S. 5-1
		Rule of 72	P.S. 6-1
		Hero's Form	P.S. 8-4
		Invariance	P.S. 9-5
		Eccentricity	P.S. 9-5
		Linear Programming	P.S. 10-3
		Recursive definitions	Ex. 12-4 #27

Visual Material	Figures and Displays		
Figures	Computer generated graphs are annotated using Aldus FreeHand. Students are encouraged to emulate these using graphing technology.	Fig. 1-4	Ch. 1-1
		Fig. 1-15	Ch. 1-3
		Fig. 2-21	Ch. 2-2
		Fig. 2-87	Ch. 2-5

Feature	Description	Example	Location
Technology Caution	Problems encountered when technology-displays are taken too literally.	Display 2-3 False harmonics	Ch. 2-2 Ch. 5-1 Examp. 7-9 P.S. 7-4
Sidelights **Graphing Technology**	Longer than Sidebars. Usually contain - historical or biographical information or anecdotal development of a topic. Examples of applied technology are integrated into the text.	See end of chapters 1 through 12.	At end of every chapter.
Key strokes	Specific key strokes for various graphing technologies to encourage students to use technology.	All Ex. 2-13Ill. 3 Examp. 7-9 Examp. 8-11ill5 Parametric Matrices	Ch. 1-2 Ch. 2-5 Ch. 4-5 Ch. 7-4 Ch. 8-5 P.S. 9-1 P.S. 11-1
Generic	Examples of concepts reinforced through implementation on technology	Composition Inverse Iteration	Ch. 2-6 Ex. 2-15Ill. 5 Ex. 8-4Ill. 7

PRELIMINARIES

0 INTRODUCTION

. . . the enormous usefulness of mathematics in the natural sciences is something bordering on the mysterious and there is no rational explanation for it. . . . the miracle of the appropriateness of the language of mathematics for the formulation of the laws of physics is a wonderful gift which we neither understand nor deserve.
—E.P. WIGNER

Words are the wheels of communication. Definitions steer our words. Many words have both common and formal definitions. Some definitions are short, other words take an entire textbook to digest. The history of a word often provides insight into current usage. Consider some words commonly used in science and mathematics.

Geometry translates literally as "earth measure." One characteristic found in our world is shape. Shapes are the essence of geometry. Geometry is a visually oriented science with real world applications.

If geometry examines shapes, algebra must be the study of numbers. Many texts describe algebra as generalized arithmetic. Such a view indicates the importance that numbers have in algebra. In this text, *algebra* is the science of rewriting (transforming) information. Algebra supplies the methods to alter the form of information, without changing the meaning. Thus, rewriting $2 + 3$ as 5 is algebra. So is rewriting $2 + 3$ as $3 + 2$.

Galileo claimed that without measurement, there is no science. As a result, numbers dominate most sciences. The numbers used for measurements are scalar numbers. You find scalars on rulers, thermometers and so forth. The scalar numbers we study here are the real numbers. The most common algebra is the algebra of real numbers.

Many computers can display numerical information visually. The visualization of numeric relationships provides new insight into a problem. The connection of algebraic (manipulative) techniques to visualization (geometry) is called *analysis*. Chapter 1 develops some graphing techniques as an introduction to analysis.

Science is from the Latin for "knowledge" or "to know." *Mathematics*, a Greek word, means "to learn." Mathematics defines the methods by which we collect bits of knowledge.

Algebra—Al-jebr w' almaqabala was an algebra book written in the ninth century by Al-Khowarizmi. The title means "restoration and reduction." (Restoration was the transposition of negative terms to the other side of an equation. Reduction was the uniting of similar terms.) *Algebra* is a Latin derivation of the first word in the title of the book.

Precalculus mathematics refers to the algebra, geometry and analysis topics useful to students who take calculus. Precalculus topics include real numbers, functions, and ideas such as graphs and continuity.

A model is a structure intended to represent or idealize a situation. One purpose of building a model is to predict the consequences of actions before undertaking the action.

A mathematical model usually begins with definitions and the assumption of basic principles. Predictions are based on the logical consequences of definitions and axioms, known as *theorems*. A mathematical system is composed of definitions, assumed principles and theorems.

Some ambiguity surrounds the idea and purpose of a review. The reviews in this text emphasize those topics considered prerequisite to the remaining chapters. The text makes no attempt to discuss review topics in detail. Such a review serves at least two useful purposes: (1) to establish agreeable terminology for previously learned information; and (2) to indicate prerequisite knowledge. As an extra, reviews can stimulate the memory of a conscientious student.

The review begins with real numbers in the next section. The remaining sections highlight topics from prerequisite courses. These sections are not comprehensive, nor are they intended to be more than brief essays that a student might read as needed. For that reason, some overlap whereas others are separate. In any case, although reading them in sequence may be helpful, it is not necessary.

1 SYSTEM OF REAL NUMBERS
The best review of arithmetic consists in the study of algebra.
—F. CAJORI

A **set** is a collection of objects called *elements*. A simple method to describe a set is to list the elements within a pair of braces. For example, the set of vowels is given by {a, e, i, o, u}. A subset is a set in which all of the elements belong to a larger set. The set of vowels is a subset of the set of letters of the alphabet.

One of the more useful collections is the set of **real numbers.** The set of real numbers consists of several subsets of useful numbers. Table 0-1 categorizes these subsets of the real numbers.

A **constant** is a symbol that represents exactly one value. Therefore, each of -2, 5.81, $\frac{3}{4}$ and π is a real constant. A **variable** is a symbol that represents any element from a collection known as the *domain* of the variable. A variable represents any element in the domain in much the same way the pronoun *she* represents any person from the collection of all females. The domain of *she* is females; the domain of *he* is males. Similarly, we could let the variable n represent any counting number in order to discuss properties of counting numbers only.

A real variable is a variable for the set of real numbers. Thus, using x as a variable to describe a characteristic of a real number describes the characteristic for every real number. Unless otherwise stated, we assume a variable in this text is a real variable.

Table 0-1
Subsets of Real Numbers

Category	Description	Example
Counting Numbers	Numbers used for counting, it measures number of items. Also called natural numbers	$\{1, 2, 3, 4, 5, \ldots\}$
Integers	Signed "whole" numbers indicating both direction and magnitude, it includes the counting numbers, 0, and negatives of the counting numbers	$\{\ldots, -4, -3, -2, -1, 0,$ $1, 2, 3, 4, \ldots\}$
Rational Numbers	Measures fractional parts, includes any number expressible as a quotient of two integers	$\left\{\ldots, -\frac{1}{2}, \frac{5}{3}, -\frac{3}{201}, \frac{4}{13}, 0, 5,\right.$ $\left. -13.2, 11.345, 2.3\overline{7}, \ldots\right\}$
Irrational Numbers	Any real number that is not rational; rational numbers and irrational numbers complete the real numbers	$\left\{\ldots, \pi, \frac{3\pi}{2}, e, \pi^2,\right.$ $\sqrt{5}, \sqrt[3]{4}, -\sqrt{17}, \sqrt{\pi},$ $\left.\frac{3 + \sqrt{5}}{\sqrt{2}}, \ldots\right\}$

Display 0-1 indicates a relationship among counting numbers, integers, rational numbers, irrational numbers and real numbers.

S I D E B A R

Expressions containing variables, such as $x + 5 = 1$, are sometimes true and sometimes false, depending on the value that is substituted for x. Pronouns play a similar role in everyday language. For example, if you make the statement "He is my dad," referring to a group of men, the statement will be true only for the one person in the group who actually is your father. The other men would be incorrect substitutions for the pronoun "he." In the same way, $x = -4$ would be the only correct substitution for x in $x + 5 = 1$.

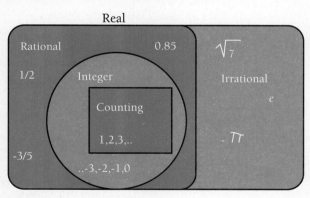

DISPLAY 0-1

Note that if x represents a counting number, then x is also an integer, a rational number and a real number, but x is not irrational.

Many manipulative properties characterize real numbers. These properties are a result of operations on real numbers. The operations of addition, subtraction, multiplication and division each take two real numbers and produce a real number result. Table 0-2 lists the properties that characterize the addition and multiplication of real numbers.

Other sets of numbers also obey these properties. Among these are rational numbers and complex numbers. Section 0-7 distinguishes complex

Table 0-2
Properties for Adding and Multiplying Real Numbers

Suppose that a, b and c are real numbers.

Add	Multiply	Property
$a + b$ is real	ab is real	Closure
$a + b = b + a$	$ab = ba$	Commutative
$a + (b + c) = (a + b) + c$	$a(bc) = (ab)c$	Associative
For all x, $x + 0 = x$	For all x, $1x = x$	Identity
For every real x, there is a real $-x$ such that $x + (-x) = 0$	For every real $x \neq 0$, there is a real $1/x$ such that $(x)(1/x) = 1$	Inverse
$a(b + c) = (ab) + (ac)$		Distributive

numbers from real numbers. These properties characterize the addition and multiplication of real numbers. The following definitions include subtraction and division by converting them to addition and multiplication, respectively.

DEFINITION 0-1
Subtraction

$$a - b = a + (-b)$$

DEFINITION 0-2
Division

$$a \div b = a(1/b), \quad \text{where } b \neq 0$$

The advantages of converting subtraction to addition or division to multiplication become obvious with the following examples:

$$[10 + (-6)] + (-2) = 10 + [(-6) + (-2)] \qquad \text{(associative)}$$
But $$[10 - 6] - 2 \neq 10 - [6 - 2] \qquad \text{(not associative)}$$
Also $$24\left(\tfrac{1}{12}\right) = \left(\tfrac{1}{12}\right)24 \qquad \text{(commutative)}$$
but $$24 \div 12 \neq 12 \div 24 \qquad \text{(not commutative)}$$

Because addition is commutative and associative, we abbreviate some expressions without ambiguity:

$$[(23 + (-2)] + [7 + (-8)] = 23 + (-2) + 7 + (-8)$$

No matter the order of the indicated additions, the answer is 20.
Similar claims cannot be made for subtraction:

$$23 - 10 - 4 = 9$$

Subtract from left to right to obtain 9. If the last subtraction is done first, the result is 17. A sequence of two or more subtractions is ambiguous. We must either agree on the order for performing subtraction or always punctuate subtractions with parentheses, and even brackets, to eliminate ambiguity. A similar difficulty occurs with repeated divisions. The next section resolves these ambiguities.

These last examples contained arithmetic expressions. An arithmetic expression consists of one or more constants joined by arithmetic operations. Thus, $15 - 12 + 3$ and $2\pi + 6$ are both arithmetic expressions. Extending this idea to include variables produces an **algebraic expression.** For example, $3x + 5(2 - y)$ is an algebraic expression with variables x and y.

The arithmetic expression $15 - 12 + 3$ may be simplified to 6 by performing the indicated arithmetic. Then we say the expression was *evaluated.* Note that no further simplification is available for the expression $2\pi + 6$.

The algebraic expression $3x + 5(2 - y)$ cannot be evaluated unless specific values for x and y are given. Let $x = 2$ and $y = 3$. Substitute 2 for x and 3 for y by replacing the variables with the given constants in the expression to obtain $3(2) + 5[2 - (3)]$. Perform the indicated arithmetic and the expression will become 1. Choose different values for x and y and the value of the expression may be different. The domain of an algebraic expression is the "largest" subset of the domain(s) of the variable(s) for which evaluation produces real values.

An equation is a statement of equality of two expressions. If at least one of the expressions contains a variable, the equation is an algebraic equation. Thus, $3(2) + 5[2 - (3)] = 1$ is an equation and $3x + 5y = 10$ is an algebraic equation. The expression to the left of the equals sign is the *left side* of the equation; the expression to the right, is the *right side*.

An equation is a complete sentence in which the verb is *equals*. As with any sentence an equation may be true or false. For example, the equation $3 + 2 = 4$ is false. A false equation is known as a *contradiction.* The algebraic equation $x + x = 2x$ is true for any real number substituted for x. Equations that are true for all values in the domains of both sides of the equation are known as *identities*. The equation $3x/x = 3$ is an identity where 0 is not in the domain of the lefthand expression.

Finally, the equation $x + 1 = 4$ is true when 3 is substituted for x and false for every other value of x. Algebraic equations true for some values of their variable(s) and false for others are known as *conditional equations*. Those values from the domain(s) of the variable(s) that produce a true statement when substituted for the variable(s) are said to *satisfy* the equation. The collection of all domain values that satisfy a given equation form the solution(s) or roots of the equation. For example the solution of $4x - 3 = x$ is $\{1\}$. To *solve* an equation means to determine the solution of the equation. If the solution is listed in a set, the set is the solution set.

Although the real number properties describe the simplest behavior of real number arithmetic, they offer no clue to manipulate statements using "equals" as a verb. The properties of equality (Table 0-3) fill this gap.

Table 0-3
Properties of Equality

If a, b and c are real numbers then	
$a = a$	Reflexive property of equality
If $a = b$ then $b = a$	Symmetric property of equality
If $a = b$ and $b = c$ then $a = c$	Transitive property of equality
If $a = b$ then $a + c = b + c$	Addition property of equality
If $a = b$ then $ac = bc$	Multiplication property of equality

Other characteristics of the real numbers follow as theorems from these properties.

Theorems usually come in one of the following standard formats.

Theorem Format

> Hypothesis implies Conclusion *or* If Hypothesis, then Conclusion.

The *hypothesis* states conditions required to apply a theorem. Sometimes hypotheses are not explicitly stated. For example, implicit in the multiplicative property of 0 ($0x = 0$) is that x is a real number. We restate the theorem in a standard format as "For any real number x, $0x = 0$," or "if x is a real number, then $0x = 0$." The *conclusion* of a theorem indicates the actions possible from a true hypothesis. One type of proof consists of forming a logical connection between the hypothesis and the conclusion. This connection is often a series of statements each justified by a previous property, definition or theorem. The sequence of statements forms an argument that bridges the gap between the hypothesis and conclusion: if the hypothesis is true then the conclusion must follow.

Sample Theorem 1

> If a and b are real constants then a solution of the equation $a + x = b$ is $x = b + (-a)$.

Sample Proof 1: The simplest type of proof is a verification. Let x be a real number. To verify the solution substitute $b + (-a)$ into the given equation:

$$a + x = b \qquad \text{(given)}$$
$$x + a = b \qquad \text{(commutative +)}$$
$$[(b + (-a)] + a = b \qquad \text{(substitute)}$$
$$b + (-a + a) = b \qquad \text{(associative)}$$
$$b + 0 = b \qquad \text{(additive inverse)}$$
$$b = b \qquad \text{(addition of 0)}$$

The last statement is true and so verifies the theorem.

Now we return to the multiplication property of 0.

Sample Theorem 2
Multiplication Property of 0.

> If x is a real number, then $0x = 0$.

Sample Proof 2: We begin with the hypothesis and "work our way" to the conclusion. Since we are given only that x is a real number, we have few clues for where to start. Because the theorem is about the number 0, begin with a known property of 0.

$$1 + 0 = 1 \qquad \text{(additive identity)}$$
$$x(1 + 0) = x1 \qquad \text{(multiplicative property of =)}$$
$$x + x0 = x \qquad \text{(distributive property)}$$

Now solve for $x0$:

$$x0 = x + (-x) \qquad \text{(see Sample Theorem 1)}$$
$$x0 = 0 \qquad \text{(additive inverse)}$$
$$0x = 0 \qquad \text{(commutative multiply)}$$

This completes the proof. We have constructed a sequence of statements that lead from the hypothesis to the conclusion. Each statement is justified by a previous axiom or theorem.

Sample Theorem 3

> For every real number x, $-1x = -x$.

Sample Proof: There is not much hypothesis here. From the conclusion, the theorem must be about the number -1. Therefore, begin with what we know about -1. (Let x be any real number.)

$$-1 + 1 = 0 \qquad \text{(additive inverse)}$$
$$x(-1 + 1) = x(0) \qquad \text{(multiplication property of =)}$$
$$x(-1) + x(1) = x(0) \qquad \text{(distributive property)}$$
$$-1x + 1x = 0x \qquad \text{(commutative property of multiplication)}$$
$$-1x + x = 0x \qquad \text{(multiplicative identity)}$$
$$-1x + x = 0 \qquad \text{(Sample Theorem 2)}$$

But
$$-x + x = 0 \qquad \text{(additive inverse)}$$

Thus,
$$-x + x = -1x + x \qquad \text{(transitive and symmetric)}$$
$$x + (-x) = -1x + x \qquad \text{(commutative for addition)}$$
$$-x = (-1x + x) + (-x) \qquad \text{(Sample Theorem 1)}$$
$$-x = -1x + [x + (-x)] \qquad \text{(associative for addition)}$$
$$-x = -1x + 0 \qquad \text{(additive inverse)}$$
$$-x = -1x \qquad \text{(additive identity)}$$

Real numbers are **ordered**. By this we mean that if a and b are distinct real numbers, then either a is larger than b or b is larger than a. If x is larger than 0 then x is a positive number. If 0 is larger than x, then x is negative. Section 0-6 details many well-known properties of positive and negative numbers. For now, the order property of real numbers promotes the organization of real numbers using a real number line. A real **number line** forms a one-to-one correspondence between the real numbers and the points of a line. Each point on the line represents a unique real number as the **graph** of that number. In turn, each real number locates a unique point on the line by providing the **coordinate** of that point. By constructing the number line so that the coordinates are in order, the line becomes a measuring device similar to a ruler. The number line shown in Figure 0-1 illustrates how order enhances the scalar nature of real numbers.

Order provides direction to real numbers. A negative sign indicates a measurement to the left of 0; a positive number indicates a measure to the right.

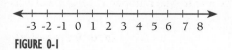

FIGURE 0-1

SUMMARY

The properties of real numbers are listed below. Suppose that a, b, and c are real numbers.

ADD	MULTIPLY	PROPERTY
$a + b$ is real	ab is real	Closure
$a + b = b + a$	$ab = ba$	Commutative
$a + (b + c) = (a + b) + c$	$a(bc) = (ab)c$	Associative
For all x, $x + 0 = x$	For all x, $1x = x$	Identity
For every real x, there is a real $-x$ such that $x + (-x) = 0$	For every real $x \neq 0$, there is a real $1/x$ such that $x(1/x) = 1$	Inverse
$a(b + c) = (ab) + (ac)$		Distributive

Subtraction: $a - b = a + (-b)$

Division: $a \div b = a(1/b)$

If x is larger than 0, then x is positive. If x is less than 0, then x is negative. The real number line pairs real numbers with points on a line so that larger numbers are to the right of smaller numbers.

0-1   EXERCISES

Identify the axiom illustrated by each of the following.

1. $5(3) = 3(5)$

2. $3 + (5 + 4) = (3 + 5) + 4$

3. If $5(3) = 15$, then $15 = 5(3)$.

4. $3 + (5 + 4) = 3 + (4 + 5)$

5. $3(5 + 4) = 3(4 + 5)$

6. $3(5 + 4) = 15 + 12$

7. If $5 + 4 = 9$, then $3(5 + 4) = 27$.

8. If $(5 + 4) = 9$, then $(5 + 4) + 3 = 9 + 3$.

9. $4[7(8)] = [4(7)]8$

10. $1(10 + 2) = 10 + 2$

11. $15 + 0 = 15$

12. $12 + [4 + (-4)] = 12 + 0$

13. $3[4(\frac{1}{4})] = 3(1)$

14. $4(6 + 3) = (6 + 3)4$

15. $4(6 + 3) = 4(3 + 6)$

16. $4(6 + 3) = 24 + 12$

17. $4 + (6 + 3) = (4 + 6) + 3$

18. If $4 + (6 + 3) = 13$ and $13 = (4 + 6) + 3$, then $4 + (6 + 3) = (4 + 6) + 3$.

19. $5 = 5$

Construct a number line and indicate the approximate location of each of the following real numbers (use a calculator if necessary).

20. -3 **21.** 7 **22.** $\frac{5}{4}$

23. $-\frac{9}{5}$ **24.** -3.45 **25.** 4.12

26. $\sqrt{3}$ **27.** $\sqrt{5}$ **28.** $\frac{\pi}{2}$

29. $\pi - 1$ **30.** $\pi + 2$ **31.** $\frac{\pi}{3}$

32. $\sqrt{\pi}$ **33.** $\sqrt[3]{\pi}$ **34.** $1 + \sqrt{2}$

38. $1 + \sqrt{2} + 1 - \sqrt{2}$

39. $(1 + \sqrt{3})(1 - \sqrt{3})$

40. $(1 + \sqrt{2})(1 - \sqrt{2})$

41. $1 + \sqrt{3} + 1 - \sqrt{3}$

42. $\dfrac{1 - \sqrt{2}}{2}$

43. $\dfrac{1 + \sqrt{3}}{2}$

44. $\sqrt{\dfrac{1 + \sqrt{2}}{2}}$

45. $\sqrt{\dfrac{1 - \sqrt{3}}{2}}$

46. $\dfrac{2\pi}{3}$

47. $\dfrac{\pi}{6}$

48. $\dfrac{3\pi}{4}$

49. $\dfrac{-\pi}{4}$

Use substitution to determine which of the following are solutions to the given equations.

50. $3x - 5 = 13$, try -6, -3, 0, 3, 6

51. $2x + 5 = 11$, try -6, -3, 0, 3, 6

52. $5 - 3x = 14$, try -6, -3, 0, 3, 6

53. $3 - 5x = 33$, try -6, -3, 0, 3, 6

54. $xx - 3 = 13$, try -4, -2, 0, 2, 4

55. $xx + 5 = 9$, try -4, -2, 0, 2, 4

56. $xx - x = 6$, try -4, -2, 0, 1, 3

57. $xx - 3x = 4$, try -1, -3, 0, 2, 4

0-1*   PROBLEM SET

Suppose that each transaction in a check ledger consists of a pair of numbers written in the form (Deposit, Withdrawal). Define the *net value* of the transaction as Deposit − Withdrawal. For the transaction ($100, $25) the net value is $75. For the transaction ($50, $137) the net value is -$87. A *positive cash flow* occurs when Deposit > Withdrawal. A *negative cash flow* occurs when Deposit < Withdrawal.

1. Give two examples of transactions for which the net value is $0.

2. Give two examples of transactions for which the net value is $1.

3. Give the net value of ($4, $9)

4. Give the net value of ($5, $13).

5. Give the net value of ($13, $13).

6. Give the net value of ($7, $7).

7. Give the net value of ($13, $5).

8. Give the net value of ($9, $4).

9. Verify that transaction ($S + d, $M + d) has the same net value as ($S, $M).

10. Determine whether ($0, $M) represents a positive or negative cash flow.

11. Determine whether ($S, $0) represents a positive or negative cash flow.

12. Suppose that ($S, $M) and ($T, $R) are transactions. Combine these into a single transaction by defining the sum as ($S, $M) + ($T, R) = ($S + T, $M + R). Confirm the reasonableness of this definition by determining the net value of these transactions.

13. See Problem 12. Use the sum definition for ($100, $0) + ($200, $0). Interpret the results in terms of the cash flow of the two transactions and their sum.

14. See Problem 12. Use the sum definition for ($0, $500) + ($0, $300). Interpret the results in terms of the cash flow of the two transactions and their sum.

15. See Problem 12. Use the sum definition for ($500, $0) + ($0, $300). Interpret the results in terms of the cash flow of the two transactions and their sum.

16. See Problem 12. Use the sum definition for ($0, $700) and ($200, $0). Interpret the results in terms of the cash flow of the two transactions and their sum.

17. Define a "multiplication" transaction as ($S, $M) × ($T, $R) = ($ST + MR, $SR + MT). Rewrite each of these transactions in terms of its net value.

18. See Problem 17. Try multiplication of ($10, $0) × ($0, $5). Interpret the results in terms of the cash flow of the two transactions and their product.

19. See Problem 17. Try multiplication of ($0, $15) × ($8, $0). Interpret the results in terms of the cash flow of the two transactions and their product.

20. See Problem 17. Try multiplication of ($20, $0) × ($12, $0). Interpret the results in terms of the cash flow of the two transactions and their product.

---

*The handwriting icon indicates that for these problems you should draw conclusions or express concepts in your own words. The magnifying glass icon indicates that these problems develop a concept in more detail or are more theoretical. Problem numbers in this color indicate an application. Problem numbers in this color indicate a preview of calculus. Problems marked with an asterisk(*) are more difficult or require extra care.

21. See Problem 17. Try $(0, M) \times (0, R)$. Interpret the results in terms of the cash flow of the two numbers and their product.

22. See Problems 12 through 21. Determine $(\$5, \$0) \times (\$0, \$3)$.

23. See Problems 12 through 21. Determine $(\$4, \$0) \times (\$5, \$0)$.

24. See Problems 12 through 21. Determine $(\$4, \$0) + (\$5, \$0)$.

25. See Problems 12 through 21. Determine $(\$6, \$0) + (\$0, \$2)$.

26. Which answers in Problems 22 through 25 are negative cash flows?

27. One formal definition of an *integer* is that an integer is an ordered pair of counting numbers. Based on your experience with transactions write a formal definition for an *integer*. Also define integer addition and multiplication.

28. See Problem 27. Define integer subtraction.

2	**CALCULATORS AND THE ORDER OF OPERATIONS**

Do not imagine that mathematics is hard and crabbed, and repulsive to common sense. It is merely the etherealization of common sense.
—W. THOMSON (LORD KELVIN)

Operations—The four fundamental operations of arithmetic have been known by various names. The thirteenth century referred to them as *species*. Ramus used *parts* in 1569, and the sixteenth century discussed *acts* and *passions*. Clabius introduced the word *operation* in 1608.

Add is from the Latin *adere* meaning "to put to." Widman first used plus and minus signs in 1489. Most historians believe the symbol + derives from the Latin word for *and*, et. *Multiply* comes from the Latin *multiplicare*, a root word meaning "having many folds." In 1698 Leibniz wrote Bernoulli, "I do not like × as a symbol for multiplication, as it is easily confounded with *x*, . . . often I simply relate two quantities by an interposed dot."

Arithmetic is easier with a calculator. Hand-held calculators often specialize in some application. Business calculators provide users with results based upon built-in formulas for specialized business needs. Some calculators feature statistical functions. Other calculators are "programmable." Scientific calculators vary among manufacturers. But most scientific calculators extend the usual arithmetic keys found on every calculator to include specialized function keys such as Ln, Log, e^x, 10^x, Sin, Cos, Tan. Most calculators have a memory key to store values.

Memory and function keys are invaluable to scientists and engineers, hence the name *scientific calculator*. We represent the calculator key labeled $f1$ with the notation $\boxed{F1}$. For example, on most calculators $\boxed{+/-}$ represents the function key for changing the sign of a number.

The operation of scientific calculators varies from model to model. Most come with detailed instruction books. A calculator's logic controls the entry of data. Two popular logics dominate hand-held calculators: algebraic logic and reverse polish notation (RPN). Although RPN is an efficient method for entering a problem, the user must organize the problem for entry and track pending operations.

An advantage of algebraic logic is that data entry resembles the order in which expressions are typed in algebra textbooks. Scientific calculators even have keys for lefthand and righthand parentheses, $\boxed{(}$ and $\boxed{)}$.

Once a calculator is on, pressing digit keys enters a number into the calculator's display and memory. After entering a number, execute a function (operation) by pressing the appropriate function key. Function keys represent two types of functions, monadic and dyadic. A *monadic* function operates on one input value (the argument). For example, the sign change key $\boxed{+/-}$ operates on one value. Enter a number, say 17. (Press All Clear $\boxed{AC}$ then $\boxed{1}$ $\boxed{7}$, and 17 will be displayed). Now press the $\boxed{+/-}$ function key, and the displayed 17 becomes -17. Press $\boxed{+/-}$ again and the -17 becomes 17. A monadic function key takes a displayed value and converts it to a new value. The value displayed depends upon the function key pressed.

Consider the reciprocal function key $\boxed{1/x}$. Pressing this key produces the multiplicative inverse of the current display value. For example, press $\boxed{AC}$, then enter 5 into the display. Now pressing $\boxed{1/x}$ causes the 5 to be replaced with its reciprocal, 0.2 ($\frac{1}{5}$ = 0.2). Other monadic keys include $\boxed{\sqrt{x}}$, $\boxed{LOG}$, $\boxed{LN}$, $\boxed{SIN}$ and so on. Do not worry about the exact purpose of these keys at this time, remember only that they are monadic and usually act upon the number in the display.

A *dyadic* function requires two arguments. Examples of dyadic functions include $\boxed{+}$, $\boxed{-}$, $\boxed{\times}$, $\boxed{\div}$, $\boxed{x^y}$. For algebraic entry, a dyadic function comes between the two arguments. Because the display shows only one number at a time, pressing a dyadic function key causes the calculator to wait for the next entered number as a second argument. For example, clear the display, enter 37 and press the $\boxed{+}$ key. Nothing seems to happen. This is because $\boxed{+}$ is dyadic and requires a second value. Without clearing the display, enter 13. Note that the 37 disappears from the display as you enter 13. But the 37 is not forgotten. The calculator remembers the 37 as well as the displayed 13. Pressing another function key usually completes a pending dyadic operation. Pressing the $\boxed{=}$ key completes all pending operations and functions and displays the final result. Press $\boxed{=}$, and the calculator displays 50.

In evaluating expressions with a graphing calculator, the display usually shows a complete entry expression and does the evaluation only when instructed. Graphing calculators do not usually display intermediate calculations. Graphing calculators are discussed in detail in Chapter 1.

Suppose we must enter a complicated arithmetic expression into a calculator. An algebraic logic calculator does arithmetic by making some operations of higher priority than others. For us to use such a calculator we must understand the logic that controls the calculator's operation. An algebraic logic calculator's priorities are controlled by the following principals known as the **order of operations** (Table 0-4).

TABLE 0-4
Order of Operations

G	**Grouping symbols:** Operations enclosed in parentheses are done first. If nested, they are done from innermost to outermost.
E	Exponentiation and radicals. *Followed* by negation and all other monadic functions.
M	Multiplication and division. Done left to right.
A	Addition and subtraction. Done left to right.

We have not yet considered exponentiation. The following definition provides a limited view of exponentiation. The next section expands the definition of exponents and defines radical notation. Chapter 6 provides a more complete discussion.

DEFINITION 0-3
Exponentiation

The exponential expression x^n means that x is to be used as a factor n times. That is,

$$x^n = \underbrace{xxxx \ldots x}_{n \text{ factors of } x},$$

where the base x is a real number and the **exponent** n is a counting number.

As a result, $3^4 = 3 \times 3 \times 3 \times 3$. The value of 3^4 is 81.

Based on the priorities described in the acronym GEMA, the expression $5 + 3 \times 2^4 + 1$ is not ambiguous. The order is exponentiation ($2^4 = 16$), then multiplicaton ($3 \times 16 = 48$), then addition ($5 + 48 + 1 = 54$). Any other order requires explicit punctuation to indicate our preference. *Parentheses* (and brackets) are algebra's punctuation marks. To indicate that the addition of 5 and 3 precedes the other operations, write $(5 + 3) \times 2^4 + 1$. Evaluating this expression produces 129.

Negation comes *after* exponentiation. There is no universal agreement on this choice, but for algebra we believe it to be reasonable. In general, calculators complete monadic functions such as negation before dyadic functions such as addition. Two operations of the same precedence level, for example, multiplication and division, are done from left to right.

The instruction to simplify an algebraic expression means to carry out, in so far as is possible, the order of operations according to GEMA. For now, consider the simplification of numeric expressions.

EXAMPLE 0-1

Simplification of Numerical Expressions by GEMA. Simplify the following expressions (Check by calculator. See the next example):

Illustration 1:

$$\frac{(5 - 7)^4}{8} + 1$$

Solution:

$$\frac{(5 - 7)^4}{8} + 1 = \frac{(5 - 7)^4}{8} + 1$$

$$= \frac{(-2)^4}{8} + 1 \qquad \text{(G)}$$

$$= \frac{16}{8} + 1 \qquad \text{(E)}$$

$$= 2 + 1 \qquad \text{(M\&D)}$$

$$= 3 \qquad \text{(A\&S)}$$

Illustration 2:

$$-3^2 + 5(-2)^4$$

Solution:

$$-3^2 + 5(-2)^4 = -3^2 + 5(-2)^4$$
$$= -9 + 5(16) \quad \text{(E)}$$
$$= -9 + 80 \quad \text{(M)}$$
$$= 71 \quad \text{(A)}$$

Illustration 3:

$$-5 - 8 + 3 - 4(40) \div 10 \div 4$$

Solution:

$$-5 - 8 + 3 - 4(40) \div 10 \div 4 = -5 - 8 + 3 - 4(40) \div 10 \div 4$$
$$= -5 - 8 + 3 - 160 \div 10 \div 4$$
$$= -5 - 8 + 3 - 16 \div 4$$
$$= -5 - 8 + 3 - 4$$
$$= -13 + 3 - 4$$
$$= -14$$

Scientific Calculators. Use a scientific calculator to approximate the following.

EXAMPLE 0-2

Illustration 1:

$$-3^2 + 5(-2)^4$$

(Compare to Example 0-1, Illustration 2)

Solution: Caution! The sign change key $\boxed{+/-}$ on some algebraic logic calculators is equal in priority to the square key $\boxed{x^2}$ and of higher priority than the exponent key $\boxed{x^y}$. If you enter $\boxed{3}\boxed{+/-}\boxed{x^y}\boxed{+}\boxed{5}\boxed{\times}\boxed{2}\boxed{+/-}\boxed{x^y}\boxed{4}\boxed{=}$ the display may be 89 not 71. This is usually a problem only with leading negative signs. Try $\boxed{3}\boxed{x^2}\boxed{+/-}$ to correct the difficulty, or change the problem to $0 - 3^2 + 5(-2)^4$. The answer will be 71 as before. Whether by calculator or not, in algebra *exponents come before negation*. Thus, $-3^2 = -9$. To override the priority use parenthesis: $(-3)^2 = 9$.

Illustration 2:

$$[\sqrt{23} \, (17.34)/0.007)]^{1.3}$$

Solution:

Enter	Press	Display	Status Note
	$\boxed{(}$	0	
23		23	
	$\boxed{\sqrt{x}}$	4.7958315	Monadic
	$\boxed{\times}$	4.7958315	Dyadic

Continued

Enter	Press	Display	Status Note
17.34		17.34	
	÷	83.159718	Complete pending ×
0.007		0.007	
	)	11879.9598	Complete pending ÷
	x^y	11879.9598	Dyadic
1.3		1.3	
	=	198271.121	Complete multiply

Note: A calculator provides only finite precision decimal solutions. An exact decimal representation of a number like $\sqrt{23}$ is not possible. For precision, most scientific calculators store two more decimals than they display. For example, calculators with displays of eight digits store ten decimal places.

Caution: Division poses entry problems on a scientific calculator. Many students attempt to enter 8/[4(2)] as 8 ÷ 4 × 2 = . This produces the incorrect result of 4. Since both the 4 and the 2 are divisors, the correct entry is 8 ÷ 4 ÷ 2 = . Parentheses offer another solution: 8 ÷ (4 × 2) = .

Illustration 3:   Determine the value of $4M^3 - 5M^2 + 3M + 7$, where $M = 2$.

Solution: Store 2 (to substitute for M). Enter the formula as written.

Enter	Press	Display	Status
2	STO	2	$M = 2$
4	×	4	4
	RCL	2	$4M$

Continued

Enter	Press	Display	Status
	x^y	2	$4M$
3		3	$4M^3$
	$-$	32	$4M^3 -$
5	$\times$	5	$4M^3 - 5$
	RCL	2	$4M^3 - 5M$
	x^y	2	$4M^3 - 5M^2$
	$+$	12	$4M^3 - 5M^2 +$
3	$\times$	3	$4M^3 - 5M^2 + 3$
	RCL	2	$4M^3 - 5M^2 + 3M$
	$+$	18	$4M^3 - 5M^2 + 3M +$
7	$=$	25	$4M^3 - 5M^2 + 3M + 7$

The displayed value is 25.   ▬

Because scientists and engineers often calculate with extremely large or exceptionally small numbers, most scientific calculators accept and display numbers in scientific notation. Scientific notation expresses a number between 1 and 10 times a power of 10. In scientific notation a calculator displays as many digits as possible (about five) and then an exponent for base 10. Some scientific calculators display an **E** between the digits and the exponent, some simply inset a space. A calculator usually could not display 5,637,296,000,000,000 because there are too many digits. See Display 0-2 for two examples for equivalents of the scientific notation form 5.637296×10^{15}. Consult your user's manual for instructions on entry in scientific notation.

Calculators speed up numerical problem solving. Once the calculator lifts the burden of tedious arithmetic, solutions to many problems flow more smoothly. Caution is important. A calculator provides useful answers if the user understands the principles controlling data entry. For an algebraic logic calculator, this means data entry according to the order of operations.

DISPLAY 0-2

0-2   E X E R C I S E S

Simplify each of the following numeric expressions.

1. $3(5 - 9) \div 2 + 1$

2. $-5^2 + 3(4 - 8) + 1$

3. $5 - 2^3 - 10 \div 5 + 5$

4. $3(4 + 2^3 - 1) + 2$

5. $-5(-3)^2(2) + (-2)^3$

6. $3\{5 - [4 - 2(1 + 3)]\} + 1$

7. $-(-2)^3 - (-1)^{10}$

8. $3 - (-2)^4 - (-3)^2$

9. $4 - 5\{2 - [3(-1) - 2]\} - 1$

10. $64 \div 8 \div 4 \div 2$

11. $(3^{-1} + 3)^{-1}$

12. $(2 + 2^{-1})^{-1}$

13. $3^1 + 3^0 + 3^{-1} + 3^{-2}$

14. $2^2 + 2^0 + 2^{-2} + 2^{-4}$

15. $\dfrac{3^2 - 2^2}{3 - 2}$

16. $\dfrac{4^2 - 3^2}{4 - 3}$

17. $\dfrac{3}{4} + \dfrac{4}{3}$

18. $\dfrac{2}{3} + \dfrac{3}{2}$

19. $\dfrac{\dfrac{3}{4} + \dfrac{4}{3}}{\dfrac{4}{3} - \dfrac{1}{2}}$

20. $\dfrac{\dfrac{2}{3} - \dfrac{3}{2}}{\dfrac{1}{3} + \dfrac{3}{2}}$

21. $3.2(5.1 - 2^2)^2$

22. $2.3(7.8 - 3^2)^2$

23. $5.73 - 2.1(4.3)$

24. $1.9(3.1) - 6.39$

25. $\dfrac{3.1^2 - 3^2}{3.1 - 3}$

26. $\dfrac{2.7^2 - 2.6^2}{2.7 - 2.6}$

27. $\dfrac{3.15^2 - 3.14^2}{3.15 - 3.14}$

28. $\dfrac{2.71^2 - 2.7^2}{2.71 - 2.7}$

29. $\dfrac{(2.6 + 0.01)^2 - 2.6^2}{0.01}$

30. $\dfrac{(3.5 + 0.01)^2 - 3.5^2}{0.01}$

Approximate the following with a scientific calculator.

31. $(2.37)^5 \sqrt{2}$

32. $\sqrt{3}(5.61)^3$

33. $\dfrac{(15 + 7.3)^{1.5}}{7.2}$

34. $\dfrac{(13 - 2.6)^{1.4}}{5}$

35. $\sqrt{7 + \sqrt[3]{5}}$

36. $\sqrt[3]{11 + \sqrt{6}}$

37. $\dfrac{25.3\sqrt{17^3}}{3.16 + 105.7}$

38. $\dfrac{(25.4 - 2.13479)^2}{16.2\sqrt[3]{5}}$

39. $\dfrac{1 + \dfrac{1}{3}}{1 - \sqrt[3]{5}}$

40. $\dfrac{2 - \dfrac{3}{7}}{1 + \sqrt[5]{3}}$

Approximate the value of each of the following, where $M = 3$.

41. $2M^3 - 5M^2 + 3M + 7$

42. $5M^3 - 3M^2 + 7M - 2$

43. $\dfrac{3M - 5}{2M + 7}$

44. $\dfrac{6M + 7}{M - 3}$

45. $\sqrt{7M^2 + 7M}$

46. $\sqrt{5\,M^3 + M}$

47. $M^{-1} + M^{0.5}$

48. $M^{0.4} - M^{-2}$

*49. $\sqrt{M^2 - 4M}$

*50. $\sqrt{5 - M^{\,2}}$

Convert each of the following numbers to scientific notation.

51. 3,650,200

52. 1,290,000,000

53. 0.000384

54. -0.0000093

55. -456,700,000

56. 0.034

Rewrite each of the following without scientific notation.

57. 1.34E4

58. -2.09E5

59. -1.34E-5

60. 2.09E-4

61. 1.01E12

62. 9.003E-11

63. Suppose 100 pages of a book are $\frac{3}{8}$" thick and each cover is $\frac{1}{4}$" thick. If the book has 450 pages, how thick is it?

64. See Problem 63. If two identical books are placed on a shelf, what is the distance from the last page of the first book to the first page of the last book?

65. Suppose 100 pages of a book are 0.375" and the cover of the book is 0.20". How thick is the book if it has 500 pages?

66. See Problem 65. If two identical books are placed on a shelf, what is the distance from the first page of the first book to the last page of the last book?

67. The speed of light is 186,282 miles/sec. If the moon is 240,000 miles from the earth, how long does it take a laser beam to travel to the moon and back?

68. The speed of light is 299,800 km/sec. If a laser beam takes 2.6 seconds to travel to the moon and back, what is the moon's distance from the earth?

69. President Reagan once compared the national debt to a stack of one dollar bills. Suppose that a stack of 1000 one dollar bills is 4" tall. If there are 5280 feet in a mile, what is the value of a mile-high stack of one dollar bills?

70. See Problem 69. If the national debt is $3 trillion, how tall a stack of one dollar bills would it take to pay off the debt?

---

3 EXPONENTS AND RADICALS

We should endeavor to make everything as simple as possible but not simpler.
—ALBERT EINSTEIN

Often, there are alternatives to accomplish the standard order of operations, GEMA. The instruction to simplify an algebraic expression means to carry out as much of the order of operations as is feasible. Sections 0-3, 0-4 and 0-5 explore properties of real number arithmetic that allow us to accomplish steps in simplification including expressions with mixed operations or variables.

Exponentiation is not mentioned in the axioms for real numbers. This section explores some properties of exponents and alternate forms of exponent notation. Then we formulate a step-by-step method for simplifying expressions involving an exponent or radical. The following theorem summarizes properties of exponents.

<table>
<tr>
<td>

Theorem 0-1
Properties of Exponents

</td>
<td>

Suppose that m, n and p are real numbers, and b and c are positive real numbers, then

1. $b^m b^n = b^{m+n}$

2. $\dfrac{b^m}{b^n} = b^{m-n}, \quad b \neq 0$

3. $b^0 = 1, \quad b \neq 0$

4. $b^{-n} = \dfrac{1}{b^n}, \quad b \neq 0$

5. $(b^m c^n)^p = b^{mp} c^{np}$ (distributive property of exponents)

</td>
</tr>
</table>

Most of the statements in Theorem 0-1 are easy to justify for counting number exponents. Formal proofs rely upon a more general definition of exponent. Parts 3 and 4 of the theorem extend the concept of exponents to integers $\{\ldots -3, -2, -1, 0, 1, 2, 3, \ldots\}$. One goal in this text is to extend exponents to include rational and irrational numbers. We begin with rational exponents here and approach irrational exponents in Chapter 6.

First we assign meaning to an exponent that is a reciprocal of a counting number. For example, choose the reciprocal of 5, $\frac{1}{5}$. What does $32^{1/5}$ mean? Factor 32 into 2^5 to obtain

$$32^{1/5} = (2^5)^{1/5}$$
$$= 2$$

Therefore $32^{1/5} = 2$, because $2^5 = 32$. We expect $8^{1/3} = 2$, because $2^3 = 8$. Also, $16^{1/2} = 4$, because $4^2 = 16$. Finally, $(-27)^{1/3} = -3$, because $(-3)^3 = -27$. However, negative bases pose difficulties for fractional exponents with even denominators: $(-25)^{1/2}$ is not defined. Suppose x were the answer. Then $x^2 = -25$, not a possibility with real numbers. This discussion inspires the following definition.

<table>
<tr>
<td>

DEFINITION 0-4
Principal nth Root

</td>
<td>

The principal nth root of a nonnegative real number b is a nonnegative real number x given by $b^{1/n} = x$, where n is a counting number greater than 1 and $x^n = b$.

</td>
</tr>
</table>

Hence, the principal 5th root of 32 is given by $32^{1/5} = 2$. To avoid fractional exponents we introduce an equivalent notation known as **radical notation**.

<table>
<tr>
<td>

DEFINITION 0-5
Radical Notation

</td>
<td>

$\sqrt[n]{b}$ is the radical form of $b^{1/n}$, where b is not negative if n is even, and n is a counting number larger than 1. Here, n is the index of the radical, $\sqrt{}$ is the radical sign and b is the radicand.

</td>
</tr>
</table>

Rewriting the exponential expression $32^{1/5} = 2$ into radical form, we obtain $\sqrt[5]{32} = 2$. Read this notation as "the principal 5th root of 32 is 2."

Based on the preceding definitions we can extend the definition of exponents to include any rational number.

DEFINITION 0-6 **Rational Exponents**	Now suppose that b is a positive real number and m and n are integers, $(n \neq 0)$, then $b^{m/n} = (b^m)^{1/n}$ or $b^{m/n} = (b^{1/n})^m$. If n is an odd number $(n = 2k + 1$, where k is an integer), then the base b may be any real number.

As a direct result of the definition, $b^{m/n} = \sqrt[n]{b^m} = (\sqrt[n]{b})^m$, where n is a counting number larger than 1. If n is an even counting number, then b must not be negative. As a result we can rewrite the exponential expression $x^{3/4}$ into the radical form $\sqrt[4]{x^3}$. Similarly, the radical expression $\sqrt[3]{xy^2}$ into the exponential form $(xy^2)^{1/3} = x^{1/3}y^{2/3}$.

All the properties of exponents from Theorem 0-1 continue to hold for rational exponents. Note that rewriting $b^{2/3}b^{4/7}$ as $b^{2/3+4/7}$ requires a common denominator to obtain $b^{26/21}$. Arithmetic with rational numbers is usually more difficult than integer arithmetic. Each notation has its advantages and disadvantages.

The properties of exponents carry over immediately to radical notation.

Theorem 0-2 **Properties of Radicals**	Suppose that b and c are nonnegative real numbers when n is even, then 1. $\sqrt[n]{bc} = \sqrt[n]{b}\,\sqrt[n]{c}$ 2. $\sqrt[n]{\dfrac{b}{c}} = \dfrac{\sqrt[n]{b}}{\sqrt[n]{c}}, c \neq 0$ 3. $\sqrt[mn]{b^m} = \sqrt[n]{b}$ 4. $\sqrt[n]{b^m} = b^q \sqrt[n]{b^r}$, where q is the integer quotient of m divided by n and r is the integer remainder of m divided by n: $m = nq + r$. 5. $\sqrt[n]{-1} = \begin{cases} -1 & \text{if } n \text{ is odd} \\ \text{undefined} & \text{if } n \text{ is even} \end{cases}$ 6. $\sqrt[n]{b^n} = \begin{cases} b & \text{if } n \text{ is odd} \\ \|b\| & \text{if } n \text{ is even} \end{cases}$ 7. $\sqrt[n]{b^0} = \sqrt[n]{1} = 1 \quad$ if $b \neq 0$ 8. $\sqrt[n]{b}\,\sqrt[m]{b} = \sqrt[mn]{b^{m+n}}$

We define the **simplification** of an expression in the following summary.

S U M M A R Y

To simplify an expression means to accomplish the following steps.

G   Remove grouping symbols.   There are three cases:

1. Multiply across addition: $a(b + c) = ab + ac$
2. Exponent across multiply: $(b^m c^n)^p = b^{mp} c^{np}$
3. Exponents across addition: $(b^m + c^n)^p = ?$

Note: Exponents *do not* distribute across additions. The general technique for removing parentheses in this type of problem is complex. We delay its discussion until Chapter 12. In the meantime, we will not require simplification of expressions of the form in case 3 for exponents larger than 3. To simplify one of these expressions, rewrite it as a product.

Example, $(a + b)^2 = (a + b)(a + b)$

E   Simplify products and quotients using the rules of exponents and radicals.

Exponents:

1. Eliminate negative exponents (Theorem 0-1, case 4) and zero exponents (Theorem 0-1, case 3)
2. Eliminate duplicate bases in products (Theorem 0-1, case 1) and quotients (Theorem 0-1, case 2)

Radicals:

1. Radicands are factored and simplified according to the rules of exponents.
2. No rationals appear in the radicand.
3. No radicals appear in any denominator.
4. All radicand exponents are less than the index.
5. The index should be as small as possible.
6. There should be as few radicals as possible.

(Note: sometimes radical steps 5 and 6 are in conflict. Sometimes we may be unable to accomplish radical step 3.)

(Note: it is permissible to rewrite exponential expressions in radical form and vice versa to facilitate your simplification. The usual convention is that expressions in exponential form are simplified to exponential form unless there is an explicit request to the contrary; similarly for radical expressions.)

M   Complete any familiar multiplication (or division).

A   Combine like terms. (Note: as a special case *like terms* include rational expressions for which you obtain a common denominator. See Section 0-5.

This section investigates simplification using step G2 and step E. The next section examines the remainder of the simplification steps.

EXAMPLE   0-3

Simplification of Exponential Expressions. Simplify the following expressions. Assume that the variable bases are positive.

Illustration 1:

$$x^{-3}y^4x^2y^{-2}$$

Solution:

$$x^{-3}y^4x^2y^{-2} = \frac{y^4x^2}{x^3y^2} \quad \text{(eliminate negative exponents)}$$

$$= \frac{y^2}{x} \quad \text{(eliminate duplicate bases)}$$

Illustration 2:

$$\frac{x^{-1/2}y^{2/3}}{x^{1/2}y^{1/2}}$$

Solution:

$$\frac{x^{-1/2}y^{2/3}}{x^{1/2}y^{1/2}} = \frac{y^{2/3}}{x^{1/2}x^{1/2}y^{1/2}} \quad \text{(eliminate negative exponents)}$$

$$= \frac{y^{2/3-1/2}}{x^{1/2+1/2}} \quad \text{(eliminate duplicate bases)}$$

$$= \frac{y^{1/6}}{x}$$

Illustration 3:

$$\left(\frac{x^{3/4}y^{-1/2}}{x^{1/5}y^{1/3}}\right)^2$$

Solution:

$$\left(\frac{x^{3/4}y^{-1/2}}{x^{1/5}y^{1/3}}\right)^2 = \frac{x^{3/2}y^{-1}}{x^{2/5}y^{2/3}} \quad \text{(remove grouping symbols)}$$

$$= \frac{x^{3/2}}{x^{2/5}y^{2/3}y} \quad \text{(eliminate negative exponents)}$$

$$= \frac{x^{11/10}}{y^{5/3}} \quad \text{(eliminate duplicate bases)}$$

Illustration 4:

$$xy^{-1/2}(x^{1/2}y^{1/2} - xy^{-3/2}) + y(x^3y^{-1})^2$$

Solution:

$$xy^{-1/2}(x^{1/2}y^{1/2} - xy^{-3/2}) + y(x^3y^{-1})^2 \qquad \text{(eliminate grouping symbols)}$$
$$= xy^{-1/2}(x^{1/2}y^{1/2} - xy^{-3/2}) + y(x^3y^{-1})^2$$
$$= x^{3/2}y^0 - x^2y^{-2} + yx^6y^{-2}$$
$$= x^{3/2} - \frac{x^2}{y^2} + \frac{x^6}{y}$$
$$= \frac{x^{3/2}y^2}{y^2} - \frac{x^2}{y^2} + \frac{x^6y}{y^2}$$
$$= \frac{x^{3/2}y^2 - x^2 + x^6y}{y^2}$$

∎

EXAMPLE 0-4

Simplifying Radical Expressions. Simplify each of the following.

Illustration 1:

$$\sqrt{125x^5}$$

Solution:

$$\sqrt{125x^5} = \sqrt{5^3x^5} \qquad \text{(factor radicand)}$$
$$= 5x^2\sqrt{5x} \qquad \text{(extract roots)}$$

Illustration 2:

$$\frac{3}{\sqrt{6}}$$

Solution:

$$\frac{3}{\sqrt{6}} = \frac{3}{\sqrt{6}}\frac{\sqrt{6}}{\sqrt{6}} \qquad \text{(rationalize denominators)}$$
$$= \frac{3\sqrt{6}}{\sqrt{6^2}}$$
$$= \frac{3\sqrt{6}}{6}$$
$$= \frac{\sqrt{6}}{2}$$

Illustration 3:

$$\sqrt[6]{x^3}$$

Solution:

$$\sqrt[6]{x^3} = x^{3/6} \qquad \text{(reduce to lowest terms)}$$
$$= x^{1/2}$$
$$= \sqrt{x}$$

Illustration 4:

$$\sqrt[3]{\frac{48x^5y^2}{z^8w}}$$ (rationalize denominator)

Solution:

$$\sqrt[3]{\frac{48x^5y^2}{z^8w}} = \sqrt[3]{\frac{2^4 3x^5y^2}{z^8w}}$$ [1]

$$= \sqrt[3]{\frac{2^4 3x^5y^2 \, zw^2}{z^8w \, zw^2}}$$ [2]

$$= \sqrt[3]{\frac{2^4 3x^5y^2 zw^2}{z^9 w^3}}$$ [3]

$$= \frac{2x}{z^3 w} \sqrt[3]{\frac{2(3)x^2 y^2 zw^2}{1}}$$ [4]

$$= \frac{2x}{z^3 w} \sqrt[3]{6x^2 y^2 zw^2}$$ [5]

Note: In step 2 of the solution we multiplied the numerator and denominator in the radicand by $zw^2/(zw^2)$. This prepared the exponents of the denominator for division by the index 3. From step 3 to step 4 we divide each exponent in the radicand by the index 3. The quotient is the exponent for that base outside the radical, the remainder is the exponent for that base in the radicand.

Illustration 5:

$$\sqrt{32x^5} + 3x\sqrt{8x^3} - 5\sqrt{18x^9}$$

Solution:

$$\sqrt{32x^5} + 3x\sqrt{8x^3} - 5\sqrt{18x^9}$$ (simplify each radical)
$$= \sqrt{2^5 x^5} + 3x\sqrt{2^3 x^3} - 5\sqrt{2(3^2)x^9}$$
$$= 4x^2\sqrt{2x} + 6x^2\sqrt{2x} - 15x^4\sqrt{2x}$$
$$= 10x^2\sqrt{2x} - 15x^4\sqrt{2x}$$

Note: The first two radicals are like terms that combine into a single term. The last term is not literally like the other term. No further combination is possible. Although we prefer as few radicals as possible in an expression, do not be tempted to use the distributive property to remove factors common to both terms. Then you would write

$$10x^2\sqrt{2x} - 15x^4\sqrt{2x} = 5x^2\sqrt{2x}(2 - 3x^2)$$

Although this form of the expression may be useful, it is not simplified because of the parentheses. Section 0-4 treats $5x^2\sqrt{2x}(2 - 3x^2)$ as the factored form of the expression.

0-3   E X E R C I S E S

Simplify the following.

1. $xxxyy$ **2.** $xxyyxxy$

3. xy^5x^3 **4.** $x^4y^2x^5$

5. $x^{1/3}x^{1/2}$ **6.** $y^{1/4}y^{2/3}$

7. $x^4y^{-3}x^3y^{-7}$ **8.** $x^{-3}y^2x^5y^{-7}$

9. $y^{-2/3}y^{1/4}y^{-1/6}$ **10.** $x^{-1/2}x^{1/3}x^{-1/6}$

11. $(x^{-3}y)^{-2}x^4y^5$ **12.** $(x^{-4}y)^{-6}y^3x^5$

13. $x^3y^{-2}(x^2y^{-3})^{-3}$ **14.** $x^2y^3(x^{-4}y)^{-3}$

15. $x^{-1/2}y^{1/2}(x^{-1/2}y^{1/2} + x^{1/2}y^{-1/2})$

16. $x^{2/3}y^{-1/3}(x^{-2/3}y^{1/3} - x^{1/3}y^{-2/3})$

17. $\sqrt{25}$ **18.** $\sqrt{36}$

19. $\sqrt[3]{-27}$ **20.** $\sqrt[3]{-8}$

21. $\sqrt{x^2y^4z^6}$ **22.** $\sqrt{\dfrac{x^2y^4}{z^6}}$

23. $\sqrt{\dfrac{xy^3}{z^4}}$ **24.** $\sqrt{\dfrac{x^3y^4}{z^6}}$

25. $\sqrt{\dfrac{xy^3}{z^5}}$ **26.** $\sqrt{\dfrac{x^3y^2}{z^3}}$

27. $\dfrac{1}{\sqrt{3}}$ **28.** $\dfrac{1}{\sqrt{2}}$

29. $\dfrac{3x}{\sqrt{x}}$ **30.** $\dfrac{5y}{\sqrt{y^3}}$

31. $5\sqrt{x} + 7\sqrt{x} - 2\sqrt{y}$ **32.** $4\sqrt{x} + 3\sqrt{y} - 7\sqrt{x}$

33. $7\sqrt{x} + 5\sqrt[3]{x^2} + 2\sqrt{x} + \sqrt[3]{x^2}$

34. $5\sqrt{x} - 3\sqrt[3]{x^2} - 6\sqrt{x} + 2\sqrt[3]{x^2}$

35. $\dfrac{3}{\sqrt{x}} - \dfrac{5}{\sqrt{x}}$ **36.** $\dfrac{7}{\sqrt{x}} + \dfrac{4}{\sqrt{x}}$

37. $\dfrac{3}{x} - \dfrac{5}{\sqrt{x}}$ **38.** $\dfrac{7}{x} + \dfrac{4}{\sqrt{x}}$

39. $\dfrac{5}{\sqrt{3} - \sqrt{2}}$ **40.** $\dfrac{7}{\sqrt{5} + \sqrt{2}}$

41. $\dfrac{x-2}{\sqrt{x} - \sqrt{2}}$ **42.** $\dfrac{x-3}{\sqrt{x} - \sqrt{3}}$

43. $\sqrt{3^25^4}$ **44.** $\sqrt{2^43^6}$

45. $\sqrt[3]{2^35^3}$ **46.** $\sqrt[3]{-1(3^6)}$

47. $\sqrt[3]{27x^9y^{12}}$ **48.** $\sqrt{121x^{10}y^{11}}$

49. $\sqrt{\dfrac{81x^6y^4}{z^3}}$ **50.** $\sqrt[3]{\dfrac{-64x^7z^6}{y^4}}$

Identify each of the following statements as true or false.

51. $\sqrt{25 - 16} = 5 - 4$ **52.** $\sqrt{25 + 144} = 5 + 12$

53. $\sqrt{\dfrac{25}{16}} = \dfrac{5}{4}$ **54.** $\sqrt{\dfrac{25}{144}} = \dfrac{5}{12}$

55. $\sqrt{12 - 144} = 5 - 12$ **56.** $\sqrt{16(25)} = 4(5)$

57. $-\sqrt{9} = -3$ **58.** $-\sqrt[3]{8} = -2$

59. $\sqrt{-9} = -3$ **60.** $\sqrt[3]{-8} = -2$

61. $\sqrt{(-3)^2} = -3$ **62.** $\sqrt[3]{(-2)^6} = (-2)^2$

63. $\sqrt{(-3)^2} = |-3|$ **64.** $\sqrt[3]{(-2)^3} = |-2|$

65. $\sqrt{16} = 8$ **66.** $\sqrt[3]{27} = 3$

67. $\sqrt{100} = \pm10$ **68.** $\sqrt[3]{81} = 27$

69. $\sqrt[5]{(-2)^5} = |-2|$ **70.** $\sqrt{81} = \pm9$

Rewrite each of the following rational expressions so that no radical expression appears in the numerator. For example,

$$\frac{\sqrt{3} - 1}{2} = \frac{\sqrt{3} - 1}{2} \cdot \frac{\sqrt{3} + 1}{\sqrt{3} + 1}$$

$$= \frac{3 - 1}{2(\sqrt{3} + 1)}$$

$$= \frac{1}{\sqrt{3} + 1}$$

71. $\dfrac{\sqrt{7} - 2}{5}$ **72.** $\dfrac{3 - \sqrt{5}}{4}$

73. $\dfrac{\sqrt{5} + \sqrt{2}}{\sqrt{5} - \sqrt{2}}$ **74.** $\dfrac{\sqrt{3} - \sqrt{7}}{\sqrt{3} + \sqrt{7}}$

75. $\dfrac{\sqrt{x} - \sqrt{3}}{x - 3}$ **76.** $\dfrac{\sqrt{x} - \sqrt{2}}{x - 2}$

77. $\dfrac{\sqrt{x + h} - \sqrt{x}}{h}$ **78.** $\dfrac{\sqrt{x + \Delta} - \sqrt{\Delta}}{\Delta}$

79. $\dfrac{\sqrt[3]{x} - \sqrt[3]{2}}{x - 2}$ $\left(\text{Hint: multiply by } \dfrac{\sqrt[3]{x^2} + \sqrt[3]{2x} + \sqrt[3]{4}}{\sqrt[3]{x^2} + \sqrt[3]{2x} + \sqrt[3]{4}}\right)$

80. $\dfrac{\sqrt[3]{x} + 1}{x + 1}$ $\left(\text{Hint: multiply by } \dfrac{\sqrt[3]{x^2} - \sqrt[3]{x} + 1}{\sqrt[3]{x^2} - \sqrt[3]{x} + 1}\right)$

Construct a real number line and indicate the approximate location of each of the following.

81. $\sqrt{5}$ **82.** $\sqrt{3}$

83. $\dfrac{3+\sqrt{5}}{2}$ **84.** $\dfrac{1-\sqrt{3}}{2}$

85. $\dfrac{3-\sqrt{5}}{2}$ **86.** $\dfrac{1+\sqrt{3}}{2}$

87. $\dfrac{3+\sqrt{5}}{2}+\dfrac{3-\sqrt{5}}{2}$ **88.** $\dfrac{1-\sqrt{3}}{2}+\dfrac{1+\sqrt{3}}{2}$

89. $\left(\dfrac{3+\sqrt{5}}{2}\right)\left(\dfrac{3-\sqrt{5}}{2}\right)$ **90.** $\left(\dfrac{1-\sqrt{3}}{2}\right)\left(\dfrac{1+\sqrt{3}}{2}\right)$

Rewrite each of the following using radical notation. Assume all variables represent positive numbers.

91. $x^{2/3}$ **92.** $x^{3/4}$
93. $3x^{1/2}$ **94.** $(5x)^{1/3}$
95. $(3x)^{1/2}$ **96.** $5x^{1/3}$
97. $x^{2/3}y^{1/3}$ **98.** $x^{1/4}y^{3/4}$
99. $x^{2/3}+y^{-1/3}$ **100.** $x^{-1/4}-y^{3/4}$

Rewrite each of the following using exponential notation. Assume all variables represent positive numbers.

101. $\sqrt{x}$ **102.** $\sqrt[3]{x}$
103. $\sqrt[4]{x^3y}$ **104.** $\sqrt[3]{xy^2}$
105. $\sqrt[3]{\sqrt{x}}$ **106.** $\sqrt{\sqrt[3]{x^2}}$
107. $\sqrt[6]{x^3}$ **108.** $\sqrt[4]{x^2}$
109. $\sqrt{1+\sqrt{x}}$ **110.** $\sqrt[3]{1-\sqrt{x}}$

---

4   POLYNOMIAL EXPRESSIONS: SIMPLIFYING AND FACTORING

Integral numbers are the fountainhead of all mathematics.
—H. MINKOWSKI

A *monomial* is a constant, a variable, or the product of a constant and one or more variables. As a result, exponents for variables can be only counting numbers. If a monomial includes a variable, the constant factor is the *coefficient* of the monomial. The *degree* of a monomial is the sum of the exponents of the variables. For example, $-2\sqrt{3}\,x^2y^5w$ is an eighth degree monomial with coefficient of $-2\sqrt{3}$.

Two monomials are like monomials if each has the same variables, with the same exponents. For example, $5x^2y$ and $-1.7x^2y$ are like monomials, but $5.34x^2y$ and $17xy^2$ are not like monomials. By using the distributive property the sum or difference of two like monomials combine into a single monomial. For example,

$$5x^2y - 1.7x^2y = (5 - 1.7)x^2y = 3.3x^2y$$

Each monomial in a sum is a term of the sum. Two unlike terms cannot combine into a single term. Thus, an expression $5.34x^2y + 7xy^2$ has two unlike terms. Such expressions are known as *binomials*. Similarly, the sum of three unlike monomials is a *trinomial*. In general, a monomial or an expression consisting of the sum or difference of two or more unlike monomials is known as a *multinomial*. A special case of the multinomial occurs when the multinomial expression contains at most one variable. Then the multinomial is known as a **polynomial expression**. Hence, 5, $3x + 2$, $5x^2 - 4x + 7$, and x^5 are polynomials in x. However, $x + y$ is a multinomial of two variables, not a polynomial.

Note the three term polynomial $-4.3x^5 + 7x^2 - 2$. The degree of the first term is 5. The degree of the second term is 2. The constant term -2 contains no variable. The degree of a polynomial is the highest of the degrees of all the terms. Therefore, the degree of the given trinomial is 5. It is customary to list the terms of a polynomial in order of descending degrees. As a result, the highest degree term appears first in this standard form. The coefficient of the highest degree term is then called the *leading coefficient* of the polynomial. For the given polynomial, the leading coefficient is -4.3. Because there is only one variable in a polynomial, terms may be uniquely identified by their degree. Note that the coefficient of the second degree term is 7. First-degree polynomials, such as $3x - 4$ are also known as *linear* polynomials. Second-degree polynomials, such as $5x^2 - 2x$ are also known as *quadratic* polynomials.

Two polynomials are equal if they are the same degree and the coefficient of each term in one polynomial equals the coefficient of the like degreed term in the other polynomial. One characteristic of a simplified polynomial is that all like terms are combined. In the order of operations (GEMA), combination of like terms is an addition and thus occurs as the last step.

EXAMPLE 0-5

Simplifying Polynomials. Simplify each of the following polynomial expressions.

Illustration 1:

$$3x^2 - 5x + 17 - x^2 + 2x - 4$$

Solution:

$$3x^2 - 5x + 17 - x^2 + 2x - 4 = 3x^2 - x^2 - 5x + 2x + 17 - 4$$
$$= 2x^2 - 3x + 13$$

Illustration 2:

$$5(3x^3 - 7x^2 + 14.5) - x(2x^2 - 5)$$

Solution:

$$5(3x^3 - 7x^2 + 14.5) - x(2x^2 - 5) = 15x^3 - 35x^2 + 72.5 - 2x^3 + 5x$$
$$= 13x^3 - 35x^2 + 5x + 72.5$$

Illustration 3:   A Binomial Times a Trinomial

$$(3x - 2)(x^2 - 4x + 5)$$

Solution: Apply the distributive property:

$$(3x - 2)(x^2 - 4x + 5) = \qquad\qquad (3x - 2)(x^2 - 4x + 5)$$

$$= 3x(x^2) + 3x(-4x) + 3x(5) + (-2)(x^2) + (-2)(-4x) + (-2)5$$
$$= 3x^3 - 12x^2 + 15x - 2x^2 + 8x - 10$$
$$= 3x^3 - 14x^2 + 23x - 10 \qquad\blacksquare$$

Multiplication of polynomials can be tedious. The following special forms help speed the multiplication of some polynomials.

Special Products	
Clues	**Formula**
Two binomials formed by sum and difference of same terms	$(x + a)(x - a) = x^2 - a^2$
Square of a binomial	$(x + a)^2 = x^2 + 2ax + a^2$ $(x - a)^2 = x^2 - 2ax + a^2$
Product of two binomials	$(ax + b)(cx + d)$
Use FOIL: F: Product of first terms O: Product of outside terms I: Product of inside terms L: Product of last terms	$ax(cx) = acx^2$ $ax(d) = adx$ $b(cx) = bcx$ $b(d) = bd$
Combine like terms	$acx^2 + (ad + bc)x + bd$

EXAMPLE 0-6

Using Special Forms to Simplify Polynomials. Simplify each of the following polynomial expressions.

Illustration 1:

Binomial Sum and Difference

$$(3x - 5)(3x + 5)$$

Solution:

$$(3x - 5)(3x + 5) = 9x^2 - 25$$

Note: The resulting polynomial is known as a **difference of** two **squares**. In general, a difference of two squares is a binomial of the form $a^2 - b^2$.

Illustration 2:

Square of a Binomial

$$(4x - 3)^2$$

Solution:

$$(4x - 3)^2 = (4x)^2 - 2(3)(4x) + 3^2$$
$$= 16x^2 - 24x + 9$$

Note: The resulting polynomial is known as a **perfect square** trinomial. In general, a perfect square trinomial takes the form $a^2x^2 + 2abx + b^2$.

Illustration 3:

Square of a Binomial

$$(2x + 5)^2$$

Solution:

$$(2x + 5)^2 = (2x)^2 + 2(2)(5x) + 5^2$$
$$(2x + 5)^2 = 4x^2 + 20x + 25 \qquad \text{(perfect square trinomial)}$$

Illustration 4: Product of Two Binomials

$$(3x - 4)(2x + 5)$$

Solution 1: Apply Distributive Property

$$(3x - 4)(2x + 5) = 3x(2x) + 3x(5) - 4(2x) - 4(5)$$
$$= 6x^2 + 15x - 8x - 20$$
$$= 6x^2 + 7x - 20$$

Solution 2: Foil

First terms product $= 3x(2x) = 6x^2$
Outside terms product +
Inside terms product $= 3x(5) - 4(2x) = 7x$
Last terms product $= -4(5) = -20$
$(3x - 4)(2x + 5) \qquad = 6x^2 + 7x - 20$

$$(x - 3)(x^2 + 3x + 9)$$

Illustration 5: Simplify:

$$(x - 3)(x^2 + 3x + 9) = x^3 + 3x^2 + 9x - 3x^2 - 9x - 27$$
$$= x^3 - 27$$

Note: The resulting polynomial is known as a **difference of** two **cubes**. In general, a difference of two cubes is a binomial of the form $a^3 - b^3$. A similar polynomial known as the **sum of** two **cubes** takes the form $a^3 + b^3$. ▬

Simplified form expresses the idea that an expression be written to the GEMA order of operation specifications. *Simplified* implies that the resulting expression should be more convenient. However, there are other useful forms for expressions. One such form is the factored form.

Just as the components of addition are terms, the component of a multiplication are **factors**. A *product* is the result of multiplying factors. An expression is in *factored form* if it written so that the last indicated operation is multiplication.

A familiar example of both simplified and factored formats of the same expression is the distributive property of multiplication across addition:

$$a(b + c) = ab + ac$$

The righthand side of the equation is in simplified form: replace a, b and c with specific values, then multiply for ab and ac. The last arithmetic operation indicated in evaluating the righthand side is addition.

Look at the lefthand side of the equation. Substitute values for a, b and c. The parentheses indicate that addition is done first so that the last arithmetic operation in evaluating the expression on the left is multiplication. The left-hand side of the equation is factored. The component factors are a and $(b + c)$.

The distributive property is the primary tool for simplifying and factoring. Although factoring does not undo simplification and simplifying does not destroy factorization—notice that x^2y is both simplified and factored—often the forms are different.

In a previous course you may have factored 120 into prime numbers. Usually, skill in factoring relates directly to skill in multiplying. Because simplifying resembles the opposite of factoring, the previous examples suggests methods for factoring special expressions. ▬

Factoring Polynomials. Factor the given expression.

EXAMPLE 0-7

Illustration 1:

Distributive Property

$$5x^3 - 15x^2 - 35x$$

Solution: Factor

$$5x^3 - 15x^2 - 35x = 5xx^2 - 5x3x - 5x7$$
$$5x^3 - 15x^2 - 35x = 5x(x^2 - 3x - 7)$$

Note: Factoring by using the distributive property is known as removing a **common factor**.

Illustration 2:

Difference of Two Squares

$$x^2 - 25$$

Solution: Factor

$$x^2 - 25 = x^2 - 5^2$$
$$= (x - 5)(x + 5)$$

Illustration 3:

Difference of Two Squares

$$(x - 3)^2 - 25$$

Solution: Factor

$$(x - 3)^2 - 25 = (x - 3)^2 - 5^2$$
$$= [(x - 3) - 5][(x - 3) + 5]$$
$$= [x - 8][x + 2]$$

Illustration 4:

Difference of Two Squares

$$4x^2 - 7$$

Solution: Factor

$$4^2 - 7 = (2x)^2 - (\sqrt{7})^2$$
$$= (2x - \sqrt{7})(2x + \sqrt{7})$$

Illustration 5:   Difference of Two Squares

$$(2x - 3)^2 - 25$$

Solution: Factor

$$(2x - 3)^2 - 25 = (2x - 3)^2 - 5^2$$
$$= [(2x - 3) - 5][(2x - 3) + 5]$$
$$= [2x - 8][2x + 2]$$
$$= 2[x - 4]\,2[x + 1] \qquad \text{(common factor)}$$
$$= 4[x - 4][x + 1]$$

Illustration 6:   Difference of Two Cubes

$$x^3 - 27 \qquad\qquad \text{(See Example 0-6, Illustration 5)}$$

Solution: Factor

$$x^3 - 27 = (x - 3)(x^2 + 3x + 9)$$

Note: To check the result, simplify $(x - 3)(x^2 + 3x + 9)$.

Illustration 7:   Perfect Square Trinomials
Factor

$$x^2 - 6x + 9 = (x - 3)^2$$

Note: To recognize a perfect square in trinomial form, note that the coefficient of x^2 is 1 and that the square of half the coefficient of the x term equals the third term: $(-2b/2)^2 = b^2$. For this illustration $(-6/2)^2 = 9$.

Illustration 8:   Factor

$$2x^2 - 12x - 32$$

Solution 1: Remove the common factor 2.

$$2x^2 - 12x - 32 = 2(x^2 - 6x - 16)$$

Is $x^2 - 6x - 16$ prime or will it factor? Based on Example 6, we suspect that $x^2 - 6x - 16 = (x + a)(x + b)$. Now guess values for a and b, then check the guess. Because

$$(x + a)(x + b) = x^2 + (a + b)x + ab$$

then $x^2 - 6x - 16 = x^2 + (a + b)x + ab$.

We have powerful clues: $a + b = -6$ and $ab = -16$. One possibility is $a = -8$ and $b = 2$. Because $(x - 8)(x + 2) = x^2 - 6x - 16$, we have the following result:

$$2x^2 - 12x - 32 = 2(x^2 - 6x - 16)$$
$$= 2(x - 8)(x + 2)$$

Solution 2: Solution 2 uses the experience gained from Illustrations 3 and 7. Expression $x^2 - 6x - 16$ would be easy to factor if it were a perfect square trinomial or a difference of two squares. But a polynomial beginning with $x^2 - 6x$ must end with $+9$ to be a perfect square. Recall that the check for a perfect square is that the last term is the square of half the coefficient of the middle term x. Because $(-6/2)^2 = 9$, the $+9$ is required to complete the square for $x^2 - 6x$. However, we do not wish to change the value of the original expression, just its form. To accomplish this, add 0 to the expression in the form $+9 - 9$. The following steps illustrate the process:

$$x^2 - 6x - 16 = x^2 - 6x + \underline{0} - 16$$
$$= x^2 - 6x + \underline{9 - 9} - 16$$
$$= x^2 - 6x + 9 - (9 + 16)$$
$$= x^2 - 6x + 9 - 25$$

The first three terms are a perfect square trinomial (see Illustration 7). Factor the perfect square to obtain

$$x^2 - 6x - 16 = (x - 3)^2 - 25$$

The resulting expression is a difference of two squares (see Illustration 3), which factors immediately:

$$x^2 - 6x - 16 = [(x - 3) + 5][(x - 3) - 5]$$
$$= (x + 2)(x - 8)$$

As a result,

$$2x^2 - 12x - 32 = 2(x + 2)(x - 8)$$

Note: This process of converting a quadratic trinomial to a difference of two squares is known as **completing the square**.

Illustration 9:

Quadratic Trinomial

$$2x^2 + 11x - 21$$

Solution: There is no common integer factor. From Illustration 8, we suspect that a quadratic trinomial would factor into two binomials:

$$2x^2 + 11x - 21 = (ax + b)(cx + d)$$

Use FOIL to simplify $(ax + b)(cs + d)$ to obtain

$$2x^2 + 11x - 21 = abx^2 + (ad + bc)x + bd$$

Then $ab = 2$, $(ad + bc) = 11$ and $bd = -21$. For feasibility restrict c and d to integers and a and b to positive integers. Then either $a = 2$ and $b = 1$ or vice versa. Because $bd = -21$ the signs of b and d are opposite and the absolute value of choices for b and d are as follows:

b	1	3	7	21
d	21	7	3	1

Any of these choices will produce the correct constant term. To verify which, if any, is correct, we must check whether the combination produces the middle term $11x$. The process is known as *trial and error*.

Factors: $(2x + b)(x + d)$	Trial		Test	Conclusion
Use $b = 1, d = -21$	$(2x + 1)(x - 21)$	$=$	$2x^2 - 41x - 21$	Error
Note $b = -21, d = 1$	$(2x - 21)(x + 1)$	$=$	$2x^2 - 19x - 21$	Error
Use $b = 3, d = -7$	$(2x + 3)(x - 7)$	$=$	$2x^2 - 11x - 21$	Close

The desired middle term is $11x$, not $-11x$. Adjust the signs of b and d:

Use $b = -3, d = 7$ $(2x - 3)(x + 7) = 2x^2 + 11x - 21$ Correct

Then $2x^2 + 11x - 21 = (2x - 3)(x + 7)$.

Illustration 10:   Common Factor

$$20x^3y + 110x^2y - 210xy$$

Solution: Remove the common factor $10xy$:

$$20x^3y + 110x^2y - 210xy = 10xy(2x^2 + 11x - 21)$$

Note that $(2x^2 + 11x - 21)$ is the expression from Illustration 9. Replace it with the factors from that illustration. When each factor is **prime** and will not factor further the expression is completely factored. For polynomials this *usually* means the factors are constant, linear polynomials with integer coefficients or quadratic polynomials with integer coefficients. Therefore,

$$20x^3y + 110x^2y - 210xy = 10xy(2x - 3)(x + 7)$$
$$= 2(5)xy(2x - 3)(x + 7)$$

The component factors are $2, 5, x, y, (2x - 3)$ and $(x + 7)$.   ▬

Illustration 8 of Example 0-7 demonstrated factoring by completing the square. Whereas the simplification process uses GEMA as a step-by-step procedure to accomplish its goal, factoring is more an art than a science. For that reason we propose a strategy rather than a method for factoring expressions. The strategy may not always be successful.

S U M M A R Y   **Factoring Strategy**

To factor an expression, try the following sequence of activities.

1. Remove common factors

Identify any common factors in the expression. If no other common factors are apparent, 1 or -1 will work. "Factor out" the common factor by writing it down once. In parentheses place what remains of the terms of the original expression with the common factor removed from each term. For example, in $ab + ac$, the factors of the first term are a and b. The factors of the second term are a and c. Clearly a is a factor common to both terms. Write down a times parentheses. Inside the parentheses write $b + c$, which is $ab + ac$ with a removed:

$$ab + ac = a(b + c)$$

2. Identify special forms

Count the number of terms in the expression.

a. If the expression has two terms

If the expression resembles one of the following simplified forms, use the equivalent factored format as a pattern to factor your expression.

$$a^2 - b^2 = (a + b)(a - b) \qquad \text{(difference of two squares)}$$
$$a^2 + b^2 \qquad \text{(does not factor using real numbers)}$$
$$a^3 - b^3 = (a - b)(a^2 + ab + b^2) \quad \text{(difference of two cubes)}$$
$$a^3 + b^3 = (a + b)(a^2 - ab + b^2) \qquad \text{(sum of two cubes)}$$

b. If the expression has three terms   (quadratic trinomial)

If the expression has three terms and can be written to resemble one of the following, apply the format for factoring to your expression:

$$x^2 - 2bx + b^2 = (x - b)^2 \quad \text{(see Example 0-7, Illustration 7)}$$
$$x^2 + 2bx + b^2 = (x + b)^2 \qquad \text{(perfect square)}$$

Method 1:

$$ax^2 + bx + c = (kx + d)(jx + e) \quad \text{(test trial values for } k, d, j \text{ and } e)$$

Method 2: Complete the square.

c. If the expression has more than three terms, factor by grouping

Experimentally group two or three of the terms together. See if you can factor just this group of terms. If not try another group. If you manage to factor a group of terms, reexamine the problem in terms of steps 1 and 2.

Continued

Consider, for example, $ax + ay + cx + cy$. The expression has four terms. Group the first two terms and last two terms: $(ax + ay) + (cx + cy)$. Factor these groups separately: $a(x + y) + c(x + y)$. Reexamine the results. We now have only two terms and they share a common factor of $(x + y)$. Remove this common factor:

$$ax + ay + cx + cy = (x + y)(a + c)$$

3. **Repeat steps on component factors**
   If steps 1 and 2 do not work on the expression, assume the expression is prime. Express this by surrounding the expression with parentheses, and preface it with an optional factor of 1. If any of steps 1 or 2 resulted in factors, examine each factor separately as if it were a separate expression. Repeat steps 1, 2 and 3 for each factor. Quit when all factors are prime. Check by multiplying factors.

EXAMPLE 0-8

Applying the Factoring Strategy for Factoring Expressions. Factor each of the following completely.

Illustration 1:

Difference of Two Squares Form

$$80x^4 - 45w^4$$

Solution:

$$
\begin{aligned}
80x^4 - 45w^4 &= 5(16x^4) - 5(9w^4) && \text{(1. Common factor)} \\
&= 5(16x^4 - 9w^4) && \text{(2. Difference of two squares)} \\
&= 5(4x^2 - 3w^2)(4x^2 + 3w^2) && \text{(3. Again!)} \\
&= 5(2x - \sqrt{3}w)(2x + \sqrt{3}w)(4x^2 + 3w^2) && \text{(Optional?)}
\end{aligned}
$$

Note: If the instructions to factor included the restriction of using integers, we would not have broken $4x^2 - 3w^2$ into factors that included irrational numbers. Step 3 would end the solution.

Illustration 2:

Quadratic Trinomial

$$3x^2 - 10x - 8$$

Solution: You already know how to factor this expression. For many problems, testing trial factors is fast and efficient. However, we shall demonstrate factoring by completing the square.

$$3x^2 - 10x - 8 = 3(x^2 - \tfrac{10}{3}x - \tfrac{8}{3}) \qquad\qquad (\text{note } \tfrac{1}{2}(\tfrac{10}{3}) = \tfrac{5}{3})$$

$$= 3(x^2 - \tfrac{10}{3}x + (\tfrac{5}{3})^2 - (\tfrac{5}{3})^2 - \tfrac{8}{3})$$

$$= 3(x^2 - \tfrac{10}{3}x + \tfrac{25}{9} - \tfrac{25}{9} - \tfrac{8}{3})$$

$$= 3[(x - \tfrac{5}{3})^2 - \tfrac{25}{9} - \tfrac{24}{9}]$$

$$= 3[(x - \tfrac{5}{3})^2 - \tfrac{49}{9}]$$

$$= 3[(x - \tfrac{5}{3}) - \tfrac{7}{3}][(x - \tfrac{5}{3}) + \tfrac{7}{3}]$$

$$= 3[x - \tfrac{12}{3}][x + \tfrac{2}{3}]$$

$$= 3(x - 4)(x + \tfrac{2}{3}) \qquad\qquad (\text{includes rational numbers})$$

$$= (x - 4)(3x + 2) \qquad (\text{integers only: } 3(x + \tfrac{2}{3}) = (3x + 2))$$

Illustration 3:

Quadratic Trinomial

$$2x^2 + 11x + 18$$

Solution:

$$2x^2 + 11x + 18 = 2(x^2 + \tfrac{11}{2}x + 9) \qquad\qquad (\text{Note } \tfrac{1}{2}(\tfrac{11}{2}) = \tfrac{11}{4})$$

$$= 2[x^2 + \tfrac{11}{2}x + (\tfrac{11}{4})^2 - (\tfrac{11}{4})^2 + 9]$$

$$= 2(x^2 + \tfrac{11}{2}x + \tfrac{121}{16} - \tfrac{121}{16} + 9)$$

$$= 2[(x - \tfrac{11}{4})^2 - \tfrac{121}{16} + \tfrac{144}{16}]$$

$$= 2[(x - \tfrac{11}{4})^2 + \tfrac{23}{16}]$$

This expression will not factor using real numbers. Note that the form is a *sum* of two squares, not a *difference*. The expression is said to be *prime over the integers*. The best we can do is $2x^2 + 11x + 18 = 1(2x^2 + 11x + 18)$. ▬

One advantage of factored expressions results from the zero factor theorem.

Theorem 0-3
Zero Factor Theorem

If a product is 0, then at least one of the factors is 0.

For example if $rs = 0$, then $r = 0$ or $s = 0$. If $(x - 3)(x + 4)(x - 7) = 0$ then $x - 3 = 0$ or $x + 4 = 0$ or $x - 7 = 0$, in which case $x = 3$ or $x = -4$ or $x = 7$. For an equation given by a polynomial equal to 0, factoring the polynomial provides a method for solving the equation. Consider the equation given by

$$2x^2 - 5x = x^2 - 2x + 10$$

To apply the zero factor theorem, first rewrite the equation so that the righthand side is 0, then factor the lefthand side:

$$2x^2 - 5x = x^2 - 2x + 10$$
$$2x^2 - 5x - x^2 + 2x - 10 = x^2 - 2x + 10 - x^2 + 2x - 10$$
$$x^2 - 3x - 10 = 0$$
$$(x - 5)(x + 2) = 0 \qquad \text{(apply the zero factor theorem)}$$
$$x - 5 = 0 \quad \text{or} \quad x + 2 = 0$$
$$x - 5 + 5 = 0 + 5 \quad \text{or} \quad x + 2 - 2 = 0 - 2$$
$$x = 5 \quad \text{or} \quad x = -2$$

Either solution will check in the original equation: {5, -2} is the solution.

0-4   EXERCISES

Simplify the following.

1. $3(x + 5) - 7(2 - x)$
2. $4(x + 3) - 2(5x - 4)$
3. $x(x + 7) - 2(x + 3)$
4. $x(x + 3) + 2(x - 5)$
5. $x(x + 3) - 5(x + 3)$
6. $x(x + 4) - 3(x + 4)$
7. $(x - 5)(x + 3)$
8. $(x - 3)(x + 4)$
9. $(x + 3)(x^2 - 5x + 2)$
10. $(x + 4)(x^2 + 3x - 7)$
11. $(2x + 7)(2x - 7)$
12. $(3x - 4)(3x + 4)$
13. $(3x - 4)^2$
14. $(2x + 7)^2$
15. $(x - 3)(x^2 + 3x + 9)$
16. $(2x + 1)(4x^2 - 2x + 1)$
17. $(2x - 5)^2$
18. $(5x - 2)(5x + 2)$
19. $(5x - 2)(2x + 5)$
20. $(3x - 4)(2x + 7)$

Factor each of the following.

21. $5x - 15$
22. $3x - 15$
23. $x^2 - 16$
24. $x^2 - 25$
25. $x^2 + 25$
26. $x^2 + 16$
27. $x^2 - 7$
28. $x^2 - 8$
29. $(x + 3)^2 - 25$
30. $(x - 2)^2 - 16$
31. $3x^2 - 27$
32. $2x^2 - 8$
33. $3(x - 2)^2 - 27$
34. $2(x - 1)^2 - 8$
35. $x^3 - 27$
36. $x^3 + 8$
37. $(x + 5)^2 - 36$
38. $(x - 5)^2 - 16$
39. $(x - 5)^2 - 8$
40. $(x + 5)^2 - 7$
41. $x(x + 5) - 3(x + 5)$
42. $x(x - 2) + 5(x - 2)$
43. $x^2 - 5x + 6$
44. $x^2 - 5x - 6$
45. $x^2 - x - 6$
46. $x^2 + 5x + 6$
47. $x^2 - 7x + 12$
48. $x^2 + 7x + 12$
49. $x^2 - x - 12$
50. $x^2 + x - 12$
51. $x^2 + 4x - 12$
52. $x^2 - 4x - 12$
53. $x^2 - 11x - 12$
54. $x^2 + 11x - 12$
55. $x^2 - 6x - 12$
56. $x^2 + 6x - 12$
57. $2x^2 + 14x + 24$
58. $2x^2 - 14x + 24$
59. $2x^2 - 13x - 24$
60. $2x^2 - 19x + 24$
61. $2x^2 + 2x - 24$
62. $2x^2 + 8x - 12$
63. $x^2 - 8x + 1$
64. $x^2 - 6x + 2$
65. $x^2 + 4x - 7$
66. $x^2 + 10x - 5$
67. $x^2 + 5x - 1$
68. $x^2 - 3x + 1$
69. $x^2 - 2x + 7$
70. $x^2 - 4x + 9$
71. $x^3 - 1$
72. $x^3 + 1$
73. $27x^3 + 8$
74. $8x^3 - 27$
75. $x^6 + 1$
76. $x^6 - 1$
77. $x^6 - 64$
78. $64x^6 - 1$
79. $3ax + 6ay - 5bx - 10by$
80. $20xy - 6az - 10xz + 12ay$
81. $2x(x + 7) - 3(x + 7)$
82. $3x(x - 2) + 5(x - 2)$
83. $(x + 1)^2 - 3(x + 1) + 2$
84. $(x - 3)^2 - 5(x - 3) + 4$
85. $x^3 - 5x^2 + 4x$
86. $x^4 - 3x^3 + 2x^2$
87. $x^4 - 5x^2 + 4$
88. $x^4 - 2x^2 + 1$
89. $y^2 - x^2 + 6x - 9$
90. $y^2 - x^2 - 6x - 9$

For each of the following, (a) factor the expression and (b) indicate the values of x for which the expression equals 0.

91. $x^2 - 7x + 12$
92. $x^2 - 8x + 12$
93. $x^2(x + 1) - 4(x + 1)$
94. $(x + 1)^2 - 7(x + 1) + 12$

95. $3x^2(x^2 - \frac{5}{3}) + 2xx^3$ **96.** $(x - 2)x^3 - (x - 2)x$

97. $(2x + 1)^2 - 8(2x + 1) + 12$

98. $x^2(3x^2) + (x^3 - \frac{5}{2})2x$

99. $(x^2 - 3x + 2)(x^2 + 4) - 13(x^2 - 3x + 2)$

100. $(x^2 - 9)x^2 - 5x(x^2 - 9) + 6(x^2 - 9)$

Factor each of the following as suggested (cf = common factor).

101. $x^{-1/2} + x^{1/2}$ (Hint: cf = $x^{-1/2}$)

102. $\sqrt{x} + \dfrac{1}{\sqrt{x}}$ $\left(\text{Hint: cf} = \dfrac{1}{\sqrt{x}}\right)$

103. $\sqrt[3]{x^2} + \sqrt[3]{x}$ (Hint: cf = $\sqrt[3]{x}$)

104. $x^{2/3} - x^{1/3}$ (Hint: cf = $x^{2/3}$)

105. $x^{2/3} - x^{1/3} - 2$ (Hint: let $y = x^{1/3}$)

106. $\sqrt[3]{x^2} + \sqrt[3]{x} - 2$ (Hint: let $y = \sqrt[3]{x}$)

107. $x^{-1/3} + x^{2/3}$ **108.** $x^{3/4} - x^{-1/4}$

109. $x^{1/2}y^{-1/2} + x^{-1/2}y^{1/2}$ **110.** $x^{1/3}y^{1/5} - x^{-1/3}y^{6/5}$

111. Note that $1004(996) = (1000 + 4)(1000 - 4)$. Discuss the advantages of the second form for doing calculations. What is the basis for this "shortcut"?

112. Note that $109^2 = (100 + 9)^2$. Discuss the advantages of the second form for doing calculations. What is the basis for this "shortcut"?

5   SIMPLIFYING RATIONAL EXPRESSIONS

Mighty are numbers, joined with art restless.
—EURIPIDES

In the previous section we avoided the division of polynomials. The quotient of polynomials is not always a polynomial much in the same manner that the quotient of integers is not always an integer. Recall that a rational number is the ratio of two integers. A rational expression is the algebraic parallel of a rational number.

A **rational expression** is an expression of the form N/D where N and D are polynomial expressions and $D \neq 0$. N is the **numerator** of the rational expression and D is the **denominator**. The domain of a rational expression is the set of all real numbers for which the denominator is not 0.

As with rational numbers, arithmetic for rational expression includes factoring, reducing to lowest terms, multiplying, dividing and determining common denominators for use in addition and subtraction.

DEFINITION 0-7
Reduced to Lowest Terms

A rational expression is reduced to lowest terms, if the numerator and denominator share no common factors other than 1 or -1.

For example, $5(x - 3)/[(x + 7)(x - 3)]$ is not reduced to lowest terms because the numerator and denominator each contain the factor $(x - 3)$. Also the domain of the rational expression consists of all real numbers except -7 and 3. To reduce the expression to lowest terms we divide out (cancel) the common factors. Thus,

$$\frac{5(x - 3)}{(x + 7)(x - 3)} = \frac{5}{(x + 7)}, \quad \text{where } x \neq -7 \text{ and } x \neq 3$$

DEFINITION 0-8
Multiplication and Division

If N/D and P/Q are rational expressions then

$$\left(\frac{N}{D}\right)\left(\frac{P}{Q}\right) = \frac{NP}{DQ}$$

$$\frac{N}{D} \div \frac{P}{Q} = \left(\frac{N}{D}\right)\left(\frac{Q}{P}\right) \qquad P \neq 0$$

DEFINITION 0-9
Least Common Multiple

The least common multiple of polynomial expressions D and Q is the smallest degree polynomial expression L such that D is a factor of L and Q is a factor of L.

Notes on the least common multiple:

1. Because D is a factor of L, there is a polynomial expression d such that $dD = L$. We say that D divides L and the quotient is d.

2. Because Q divides L, there is a polynomial expression q such that $qQ = L$. We say that Q divides L and the quotient is q.

DEFINITION 0-10
Addition of Rational Expressions

$$\frac{N}{D} + \frac{P}{Q} = \frac{N}{D}\frac{d}{d} + \frac{P}{Q}\frac{q}{q} = \frac{Nd + Pq}{L}$$

where N/D and P/Q are rational expressions and L is the least common multiple of D and Q. Also $d = L/D$ and $q = L/Q$. L is the **least common denominator** (LCD) of the two rational expressions.

EXAMPLE 0-9

Algebra of Rational Expressions. Simplify each of the following rational expressions.

Illustration 1:

Product of Two Rational Expressions

$$\left(\frac{x^2 - 25}{x^2 - 7x + 12}\right)\left(\frac{3x^3 - 27x}{5x + 25}\right)$$

$$= \left(\frac{(x + 5)(x - 5)}{(x - 3)(x - 4)}\right)\left(\frac{3x(x + 3)(x - 3)}{5(x + 5)}\right) \qquad \text{(factor)}$$

$$= \frac{(x + 5)(x - 5)3x(x + 3)(x - 3)}{(x - 3)(x - 4)5(x + 5)} \qquad \text{(multiply)}$$

$$= \frac{3x(x + 3)(x - 5)}{5(x - 4)}, \qquad x \neq 3, \quad x \neq 4 \text{ and } x \neq \text{-5 (reduce)}$$

Illustration 2:   Quotient of Two Rational Expressions

$$\frac{x^2 - 16}{3x + 9} \div \frac{x^2 - x - 20}{x^2 - 9}$$

$$= \frac{(x-4)(x+4)}{3(x+3)} \div \frac{(x-5)(x+4)}{(x+3)(x-3)} \qquad \text{(factor)}$$

$$= \frac{(x-4)(x+4)}{3(x+3)} \; \frac{(x+3)(x-3)}{(x-5)(x+4)} \qquad \text{(convert to multiply)}$$

$$= \frac{(x-4)(x+4)(x+3)(x-3)}{3(x+3)(x-5)(x+4)} \qquad \text{(multiply)}$$

$$= \frac{(x-4)(x-3)}{3(x-5)}, \quad x \neq -3, \;\; x \neq 3, \;\; x \neq 5 \;\; \text{and} \;\; x \neq -4 \text{(reduce)}$$

Illustration 3:   Addition of Rational Expressions: Same Denominator

$$\frac{3x - 7}{x^2 - 5x + 6} + \frac{4 - x}{x^2 - 5x + 6} + \frac{1 - x}{x^2 - 5x + 6}$$

$$= \frac{3x - 7}{(x-3)(x-2)} + \frac{4 - x}{(x-3)(x-2)} + \frac{1 - x}{(x-3)(x-2)}$$

$$= \frac{(3x - 7) + (4 - x) + (1 - x)}{(x-3)(x-2)} \qquad \text{(add)}$$

$$= \frac{x - 2}{(x-3)(x-2)} \qquad \text{(simplify numerator)}$$

$$= \frac{(x - 2)}{(x-3)(x-2)} \qquad \text{(factor numerator)}$$

$$= \frac{1}{x - 3}, \quad x \neq 2 \;\; \text{and} \;\; x \neq 3 \qquad \text{(reduce)}$$

Illustration 4:   Subtraction of Rational Expressions: Same Denominator

$$\frac{2x - 3}{(x+2)(x-5)} - \frac{x - 5}{(x+2)(x-5)}$$

$$= \frac{(2x - 3)}{(x+2)(x-5)} - \frac{(x - 5)}{(x+2)(x-5)} \qquad \text{(factor)}$$

$$= \frac{(2x - 3) - (x - 5)}{(x+2)(x-5)} \qquad \text{(add)}$$

$$= \frac{2x - 3 - x + 5}{(x+2)(x-5)} \qquad \text{(simplify the numerator)}$$

$$= \frac{(x + 2)}{(x+2)(x-5)} \qquad \text{(factor the numerator)}$$

$$= \frac{1}{(x - 5)}, \quad x \neq -2, \;\; x \neq 5 \qquad \text{(reduce)}$$

Illustration 5:   Addition and Subtraction of Rational Expressions: Different Denominator

$$\frac{3}{x-4} - \frac{1}{x+3} - \frac{x+10}{x^2-x-12}$$

$$= \frac{3}{(x-4)} - \frac{1}{(x+3)} - \frac{(x+10)}{(x-4)(x+3)} \qquad \text{(factor)}$$

$$= \frac{3}{(x-4)} - \frac{1}{(x+3)} - \frac{(x+10)}{(x-4)(x+3)} \qquad (\text{LCD} = (x-4)(x+3))$$

$$= \frac{3(x+3)}{(x-4)(x+3)} - \frac{1(x-4)}{(x+3)(x-4)} - \frac{(x+10)}{(x-4)(x+3)} \qquad \text{(rewrite)}$$

$$= \frac{3(x+3) - 1(x-4) - (x+10)}{(x-4)(x+3)} \qquad \text{(add)}$$

$$= \frac{3x+9 - x+4 - x-10}{(x-4)(x+3)} \qquad \text{(simplify the numerator)}$$

$$= \frac{(x+3)}{(x-4)(x+3)} \qquad \text{(factor the numerator)}$$

$$= \frac{1}{x-4}, \qquad x \neq 4, \quad x \neq -3 \qquad \text{(reduce)}$$

Illustration 6:   Combined Operations

$$\frac{x}{x-4} - \frac{2x+2}{x-3} \div \frac{x^2-3x-4}{2x-6}$$

$$= \frac{x}{(x-4)} - \frac{2(x+1)}{(x-3)} \div \frac{(x-4)(x+1)}{2(x-3)} \qquad \text{(factor)}$$

$$= \frac{x}{(x-4)} - \frac{2(x+1)}{(x-3)} \frac{2(x-3)}{(x-4)(x+1)} \qquad \text{(convert division to multiplication)}$$

$$= \frac{x}{(x-4)} - \frac{2(x+1)2(x-3)}{(x-3)(x-4)(x+1)} \qquad \text{(multiply before subtraction)}$$

$$= \frac{x}{(x-4)} - \frac{4}{(x-4)} \qquad \text{(subtract, already has LCD)}$$

$$= \frac{x-4}{(x-4)} \qquad \text{(simplify numerator)}$$

$$= \frac{(x-4)}{(x-4)} \qquad \text{(factor numerator)}$$

$$= 1, \qquad x \neq 4, \quad x \neq 3, \quad x \neq -1 \qquad \text{(reduce)}$$

Illustration 7:   Combined Operations

$$\left(\frac{x}{x-4} - \frac{2x+2}{x-3}\right) \div \frac{x^2 - 3x - 4}{2x - 6}$$

$$= \left[\frac{x}{(x-4)} - \frac{2(x+1)}{(x-3)}\right] \div \frac{(x-4)(x+1)}{2(x-3)}$$

$$= \left[\frac{x}{(x-4)} - \frac{2(x+1)}{(x-3)}\right] \frac{2(x-3)}{(x-4)(x+1)} \qquad \text{(convert division to multiplication)}$$

$$= \left[\frac{x(x-3)}{(x-4)(x-3)} - \frac{2(x+1)(x-4)}{(x-4)(x-3)}\right] \frac{2(x-3)}{(x-4)(x+1)} \qquad \text{(get LCD} = (x-4)(x-3))$$

$$= \left[\frac{x(x-3) - 2(x+1)(x-4)}{(x-4)(x-3)}\right] \frac{2(x-3)}{(x-4)(x+1)} \qquad \text{(subtract)}$$

$$= \left[\frac{x^2 - 3x - 2(x^2 - 3x - 4)}{(x-4)(x-3)}\right] \frac{2(x-3)}{(x-4)(x+1)} \qquad \text{(simplify numerator)}$$

$$= \left[\frac{x^2 - 3x - 2x^2 + 6x + 8}{(x-4)(x-3)}\right] \frac{2(x-3)}{(x-4)(x+1)}$$

$$= \left[\frac{-x^2 + 3x + 8}{(x-4)(x-3)}\right] \frac{2(x-3)}{(x-4)(x+1)}$$

$$= \left[\frac{-(x^2 - 3x - 8)}{(x-4)(x-3)}\right] \frac{2(x-3)}{(x-4)(x+1)} \qquad \text{(factor numerator)}$$

$$= \frac{-(x^2 - 3x - 8)2(x-3)}{(x-4)(x-3)(x-4)(x+1)} \qquad \text{(multiply)}$$

$$= \frac{-2(x^2 - 3x - 8)}{(x-4)^2(x+1)}, \qquad x \neq 4, \quad x \neq -1, \quad x \neq 3 \qquad \text{(reduce)}$$

Note: The expression

$$\left(\frac{x}{x-4} - \frac{2x+2}{x-3}\right) \div \frac{x^2 - 3x - 4}{2x - 6} \quad \text{could be written as} \quad \frac{\left(\dfrac{x}{x-4} - \dfrac{2x+2}{x-3}\right)}{\dfrac{x^2 - 3x - 4}{2x - 6}}$$

However, this form is not a rational expression because the numerator and denominator are not polynomials. In general, ratios of algebraic expressions are known as *algebraic fractions*. Algebraic fractions include rational expressions. In particular, a compound fraction is an algebraic fraction in which the numerator and denominator are sums or differences of rational expressions. A compound fraction usually simplifies to a rational expression.

Illustration 8:

Simplifying a Compound Fraction

$$\frac{\dfrac{9}{x^2} - 1}{\dfrac{1}{x} + \dfrac{1}{3}}$$

Solution: Convert to rational expressions.

$$\frac{\dfrac{9}{x^2} - 1}{\dfrac{1}{x} + \dfrac{1}{3}} = \frac{\dfrac{9}{x^2} - \dfrac{x^2}{x^2}}{\dfrac{3}{3x} + \dfrac{x}{3x}} \qquad \text{(combine terms in the numerator)}$$

$$= \frac{\dfrac{9 - x^2}{x^2}}{\dfrac{3 + x}{3x}} \qquad \text{(combine terms in the denominator)}$$

$$= \frac{9 - x^2}{x^2} \div \frac{3 + x}{3x} \qquad \text{(rewrite as a division)}$$

$$= \frac{(3 + x)(3 - x)}{x^2} \times \frac{3x}{(3 + x)} \qquad \begin{array}{l}\text{(factor and convert}\\\text{to multiplication)}\end{array}$$

$$= \frac{3(3 - x)}{x}, \qquad x \neq 0, \quad x \neq \text{-}3 \qquad \text{(multiply and reduce)}$$

Solution 2: Use LCD of "small" denominators.

$$\frac{\dfrac{9}{x^2} - 1}{\dfrac{1}{x} + \dfrac{1}{3}} = \left(\frac{\dfrac{9}{x^2} - \dfrac{1}{1}}{\dfrac{1}{x} + \dfrac{1}{3}}\right)\left(\frac{\dfrac{3x^2}{1}}{\dfrac{3x^2}{1}}\right) \qquad \begin{array}{r}[3x^2/(3x^2) = 1]\\ \text{(LCD of small denominators is } 3x^2)\end{array}$$

$$= \frac{\left[\dfrac{9(3x^2)}{x^2} - \dfrac{1(3x^2)}{1}\right]}{\left[\dfrac{1(3x^2)}{x} + \dfrac{1(3x^2)}{3}\right]} \qquad \text{(multiply)}$$

$$= \frac{9(3) - 3x^2}{3x + x^2}, \qquad x \neq 0 \qquad \text{(reduce each "small" fraction)}$$

$$= \frac{3(3 - x)(3 + x)}{x(3 + x)} \qquad \text{(factor and reduce again)}$$

$$= \frac{3(3 - x)}{x}, \qquad x \neq 0, \quad x \neq 3$$

Illustration 9:   Simplifying an Algebraic Fraction

$$\frac{x^2y}{w(x+2)^2} - \frac{3-x}{w^3(x+2)} = \frac{x^2y}{w(x+2)^2} - \frac{3-x}{w^3(x+2)} \quad (\text{LCD} = w^3(x+2)^2)$$

$$= \frac{x^2y}{w(x+2)^2}\frac{w^2}{w^2} - \frac{3-x}{w^3(x+2)}\frac{(x+2)}{(x+2)}$$

$$= \frac{x^2yw^2}{w^3(x+2)^2} - \frac{(3-x)(x+2)}{w^3(x+2)^2}$$

$$= \frac{x^2yw^2 - (3-x)(x+2)}{w^3(x+2)^2}$$

$$= \frac{x^2yw^2 + x^2 - x - 6}{w^3(x+2)^2}$$

0-5   EXERCISES

Simplify each of the following.

1. $\dfrac{16}{36}$

2. $\dfrac{-24}{36}$

3. $\dfrac{x^2-9}{2x+6}$

4. $\dfrac{x^2-4}{3x+6}$

5. $\dfrac{x^2-7x+12}{x^2-16}$

6. $\dfrac{x^2-9}{x^2-2x-3}$

7. $\dfrac{3x+6}{x2-9}\dfrac{x^2+4x+3}{x+2}$

8. $\dfrac{x^2-2x-15}{2x+8}\dfrac{x^2-16}{x^2-x-12}$

9. $\dfrac{5x-10}{} \div \dfrac{x^2-25}{x^2+9x+20}$

10. $\dfrac{3x^2+6x}{x^2-9} \div \dfrac{x^2-4}{x^2+x-6}$

11. $\dfrac{3}{x}+\dfrac{1}{x^2}$

12. $\dfrac{5}{3x}-\dfrac{2}{x^2}$

13. $\dfrac{2}{x-1}-\dfrac{3}{x+1}$

14. $\dfrac{x}{x-2}+\dfrac{3}{x+2}$

15. $\dfrac{2}{x-3}-\dfrac{1}{x+3}-\dfrac{6}{x^2-9}$

16. $\dfrac{3}{x+2}-\dfrac{2}{x-2}+\dfrac{8}{x^2-4}$

17. $\dfrac{x}{x+2}+\dfrac{3}{x-4}\div\dfrac{3x+6}{2x-8}$

18. $\dfrac{2x}{x-3}-\dfrac{3}{x+5}\div\dfrac{x-3}{2x+10}$

19. $\left(\dfrac{x}{x+2}+\dfrac{3}{x-4}\right)\div\dfrac{3x+6}{2x-8}$

20. $\left(\dfrac{2x}{x-3}-\dfrac{3}{x+5}\right)\div\dfrac{x-3}{2x+10}$

21. $\dfrac{\dfrac{x}{3}-\dfrac{3}{x}}{1+\dfrac{3}{x}}$

22. $\dfrac{\dfrac{x^2}{16}-1}{\dfrac{x}{4}+1}$

23. $\dfrac{\dfrac{x}{25}-\dfrac{1}{x}}{\dfrac{1}{x}+\dfrac{1}{5}}$

24. $\dfrac{\dfrac{5}{x}-\dfrac{x}{5}}{\dfrac{1}{5}-\dfrac{1}{x}}$

25. $7+\dfrac{\dfrac{x}{8}-\dfrac{2}{x}}{\dfrac{1}{4x}}$

26. $1-\dfrac{\dfrac{x^2}{9}-\dfrac{1}{4}}{\dfrac{2x+3}{36}}$

27. $\dfrac{\dfrac{x}{3} - \dfrac{3}{x}}{1 + \dfrac{3}{x}} - \dfrac{\dfrac{x}{25} - \dfrac{1}{x}}{\dfrac{1}{x} + \dfrac{1}{5}}$

28. $\dfrac{\dfrac{x^2}{16} - 1}{\dfrac{x}{4} + 1} + \dfrac{\dfrac{5}{x} - \dfrac{x}{5}}{\dfrac{1}{5} - \dfrac{1}{x}}$

29. $\dfrac{\dfrac{5}{x} - \dfrac{x}{5}}{\dfrac{1}{5} - \dfrac{1}{x}} \div \dfrac{\dfrac{x}{25} - \dfrac{1}{x}}{\dfrac{1}{x} + \dfrac{1}{5}}$

30. $\dfrac{\dfrac{x}{3} - \dfrac{3}{x}}{1 + \dfrac{3}{x}} \div \dfrac{\dfrac{x^2}{16} - 1}{\dfrac{x}{4} + 1}$

31. $\dfrac{x^2 - 3^2}{x - 3}$

32. $\dfrac{x^2 - 2^2}{x - 2}$

33. $\dfrac{(x + h)^2 - x^2}{h}$

34. $\dfrac{(x + h)^2 - x^2}{-h}$

35. $\dfrac{x^3 - 3^3}{x - 3}$

36. $\dfrac{x^3 - 2^3}{x - 2}$

37. $\dfrac{(x - h)^3 - x^3}{-h}$

38. $\dfrac{(x + h)^3 - x^3}{h}$

39. $\dfrac{x^4 - 3^4}{x - 3}$

40. $\dfrac{x^4 - 2^4}{x - 2}$

6

INEQUALITIES AND INTERVAL NOTATION

Every mathematical book that is worth the reading must be read "backwards and forwards," if I may use the expression. I would modify Lagrange's advice a little and say, "Go on, but often return to strengthen your faith." When you come on a hard or dreary passage, pass it over; and come back to it after you have seen its importance or found the need for it further on.
—GEORGE CHRYSTAL

What do we mean by *order*? With counting numbers, we mean the next number or the previous number. The words *successor* and *predecessor* identify these: the successor of 37 is 38; the predecessor of 5307 is 5306. We cannot be so specific with rational numbers.

There are rational numbers after $\frac{2}{3}$ but not a next rational number after $\frac{2}{3}$. Choose any number, no matter how close, and it will not be the next rational number. There is always another rational number "between" any two given rational numbers. Between $\frac{2}{3}$ and the number chosen is another rational number, for example, the average of the two numbers.

This section gives precision, both by algebraic and geometric approaches, to the concept of order of numbers. You are already familiar with positive real numbers. We begin with the properties that describe the set of positive real numbers.

Axiom—from the Greek *axioma* meaning authoritative sentence. Generally an axiom is a statement or property universally accepted as true.

Axiom 0-1
Closure Properties of Positive Numbers

Suppose that a and b are positive real numbers, then $a + b$ is a positive real number and ab is a positive real number. The multiplicative identity, 1, is a positive real number.

Axiom 0-2
The Trichotomy Property

> For any real number x, only one of the following statements is true:
>
> **1.** x is positive
> **2.** x is 0
> **3.** $-x$ is positive (then x is a negative real number).
>
> This is called the **trichotomy property**.

These order properties lead to the definition of order relations.

DEFINITION 0-11
Greater Than

> Suppose that a and b are real numbers, then $a > b$ means that $a - b$ is positive; $a > b$ is read "a is greater than b."

DEFINITION 0-12
Less Than

> Suppose that a and b are real numbers, then $a < b$ means that $b > a$; $a < b$ is read "a is less than b."

The order relations *greater than* and *less than* combine with the axioms for real numbers to provide greater insight into the real numbers. Any collection of numbers that satisfies the preceding order properties is said to be *ordered*. The rational numbers are ordered. The real numbers are ordered. Sentences that use $>$ or $<$ as a verb are **inequalities.**

Theorem 0-4
Positive Numbers

> If $x > 0$, then x is positive. If x is positive, then $x > 0$.

Proof: Suppose x is positive, then $x - 0$ is positive and so $x > 0$. By definition, if $x > 0$, then $x - 0$ is positive. Because $x - 0 = x$, then x is positive. ▬

Theorem 0-5
Negative Numbers

> If $x < 0$, then x is negative. If x is negative, then $x < 0$.

The proof of Theorem 0-5 is similar to the proof of Theorem 0-4. The successor of a counting number n is $n + 1$. For example, the successor of 17 is $17 + 1$ or 18. Recall that 1 is a positive real number. Because positive numbers are closed for addition and 1 is positive, then the successor of 1, $1 + 1$ or 2, is positive. Also $2 > 1$, because $2 - 1 = 1$.

Similarly, 3 is positive and $3 > 2$. Perhaps the parallel in integers seems more challenging. Can you show that $1 > 0$? From there can you show that $0 > -1$?

Theorem 0-6
Relation of a Successor to a Number

> For any integer x, $x + 1 > x$.

The preceding discussion and theorem provide some insight for a graphical representation of the order relations of real numbers. The physical **distance** used to mark x and its relation to $x + 1$ on a number line is the *scale* of the graph. The following summarizes the concept of a number line as introduced in Section 0-1.

Number Line Graphing Convention

A real number line is a graphic representation for real numbers that uses the following conventions:

1. Every real number addresses a unique point of the line. The point is the graph of the number. The real number is the coordinate of the point.
2. Every point of the line is the image of a unique real number.
3. If the real number line is horizontal, then the positive real numbers address points to the right of the graph of 0. Graphs of negative numbers are to the left of the graph of 0. For a vertical number line, graphs of positive numbers are above the graph of 0, negative numbers are below. The graph of 0 is called the *origin* of the number line.
4. For a horizontal number line, if a and b are real numbers and $a < b$, then graph b to the right of a. Similarly, for a vertical line, graph b above a.

Part 3 of the graphing convention gives an interpretation of positive versus negative numbers. The sign (positive or negative) of a real number indicates direction (right or left, up or down). The remaining numerical information represents distance from the origin. Thus -7 is 7 units to the left of the origin, and 12.3 is 12.3 units to the right of the origin. The concept of the distance of a real number as distinct from its sign constitutes its **absolute value.** Because the description of absolute value is geometric and distance should never be negative, refer to the trichotomy property for the inspiration of the following algebraic definition of absolute value.

DEFINITION 0-13
Absolute Value

The absolute value of a real number x symbolized by $|x|$ is given by

$$|x| = \begin{cases} x, & \text{if } x \text{ is positive or } 0 \\ -x, & \text{if } x \text{ is negative} \end{cases}$$

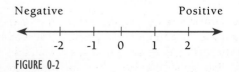

Negative   Positive

-2   -1   0   1   2

FIGURE 0-2

Remember, the graphing convention is a tradition. As a tradition, it provides us with a common means of communication. But it is not a hard, fast rule. Moreover, even within the scope of these conventions, room is left for creative variation. See Figure 0-2 for a common real number line representation.

Figure 0-2 labels the graphs for only a few integers. What of the graphs of the rational numbers? What of the graphs of the real numbers? First we must have some insight into what order means for these numbers.

DEFINITION 0-14
Order for Rational Numbers

Suppose that p/q and n/d are rational numbers with $q > 0$ and $d > 0$, then

$$\frac{p}{q} > \frac{n}{d}, \text{ if and only if } pd > qn; \frac{p}{q} < \frac{n}{d} \text{ means } \frac{n}{d} > \frac{p}{q}$$

Notice how the definition of order for rational numbers refers back to the order relation of integers. Mathematics builds new concepts out of prior concepts. Now $\frac{3}{5} > \frac{7}{12}$ because $3(12) > 5(7)$. Back to the number line. Note that $\frac{3}{5} < 1$, because $\frac{3}{5} < \frac{1}{1}$ and $3 < 5$. This leads to the following theorem.

Theorem 0-7
Rationals Between 0 and 1

Suppose that p/q is a rational number and that $p > 0$ and $q > 0$ and $p < q$, then $p/q > 0$ and $p/q < 1$.

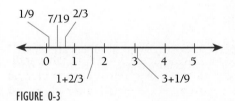

FIGURE 0-3

Begin by graphing numbers like $\frac{2}{3}$, $\frac{7}{19}$ and $\frac{1}{9}$. Theorem 0-6 suggests the location of the points for $1 + \frac{2}{3}$ or $3 + \frac{1}{9}$. See Figure 0-3.

The distance between any integer and its successor is uniform. This is not the only way to scale a graph, but it is the most common. For specific applications, attach units of measure to the integers. These units of measure could be inches, meters, degrees, grams, or other types of measure in which order plays a role. For a general discussion, do not specify a unit of measure. When appropriate, we attach units of measure for better insight into a problem.

As yet we have no easy scheme for locating irrational numbers on the number line. The Pythagorean theorem allows us to locate the graph of numbers like $\sqrt{5}$. Mark the point representing the graph of 2. At that point construct a line perpendicular to the number line. From 2 along this perpendicular, copy the distance from 0 to 1. Label the point on the perpendicular 1 unit from 2 as P. Now draw a line from P to the origin. See the resulting right triangle in Figure 0-4.

By the **Pythagorean theorem**, the length of the hypotenuse of this triangle is $\sqrt{5}$. Copy this distance on the number line from the origin to the right. Label the righthand endpoint $\sqrt{5}$. In a similar manner copy the distance from the origin to the left and create the graph of $-\sqrt{5}$.

As an alternative to geometric constructions, approximate irrational numbers on a calculator to estimate the location of the graphs. Because graphs are inaccurate, the approximation should be adequate. By calculator, $\sqrt{5}$ is about 2.236. But 2.236 is indistinguishable from $\sqrt{5}$ as scaled in Figure 0-4.

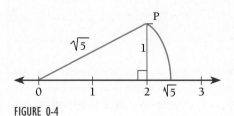

FIGURE 0-4

Look back at Theorem 0-7. One use of an inequality is the description of an interval of numbers. An **interval** is a set of numbers that satisfy an inequality. For a noun to satisfy a sentence containing a pronoun means that replacing the pronoun with the noun produces a true sentence. Similarly for a real number to satisfy an inequality containing a variable means that substituting the number for the variable produces a true inequality. As an example, note that many real numbers satisfy the inequality

$$x + 1 > 17$$

Among these numbers are 327 and 47.093. Numbers that do not satisfy the inequality include 14 and -5000.

Like equations, we rewrite inequalities to gain insight into their meaning. Compare the properties of equality to the next theorems to discover the similarities and differences for equalities and inequalities. Each theorem is written for "greater than." A similar theorem for "less than" follows immediately from the definition.

Theorem 0-8
Transitive Property of Inequalities

If $a > b$ and $b > c$, then $a > c$, where a, b and c are real numbers.

Proof: We begin with the hypothesis and work our way to the conclusion.

$a > b$ means $a - b$ is positive.

$b > c$ means $b - c$ is positive.

But $(a - b) + (b - c)$ is positive because positive real numbers are closed for addition. Now $(a - b) + (b - c) = a - c$. Because $a - c$ is positive, by definition we have $a > c$. ▬

Theorem 0-9
Addition and Multiplication Properties

If $a > b$ (a, b and c are real numbers)
then $a + c > b + c$ (addition property)
Moreover, if c is positive, then   (multiplication properties)
$ac > bc$
If c is negative, then
$ac < bc$

Proof: The proofs of the parts of the theorem are similar. We choose the last conclusion as a demonstration and leave the other two parts as exercises.

Suppose that $a > b$ and c is negative (hypothesis). By definition, $a > b$ means that $a - b$ is positive. By the trichotomy property, if c is negative, then $-c$ is positive. Then $-c(a - b)$ is positive because positive real numbers are closed under multiplication.

$$-c(a - b) = -ac + bc \qquad \text{(see Exercise 0-2)}$$

Because $-ac + bc$ is positive, rewrite it as $bc - ac$ and obtain $bc > ac$. By the definition of less than, reexpress this as $ac < bc$. ▬

Graphs of Inequalities. Graph each of the following inequalities.

EXAMPLE 0-10

Simple Inequality

$$x > 5$$

Solution: By graphing convention, any number to the right of 5 should satisfy $x > 5$. There are infinitely many such numbers. Some of these numbers are close to 5, and their graph is indistinguishable from the graph of 5. In Figure 0-5, we shade the number line to the right of 5 to represent the values of x, but to indicate that 5 is excluded, we place a small circle about the point labeled 5.

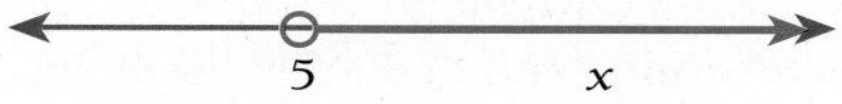

FIGURE 0-5

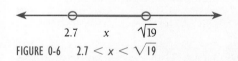

Compound Inequality

$$x > 2.7 \text{ and } x < \sqrt{19}$$

Solution: This is a *compound* sentence formed from two simple sentences, $x > 2.7$, $x < \sqrt{19}$. Each of these is easy to graph separately. What requires discussion is the nature of the conjunction *and*. In mathematics, a compound sentence using *and* is true if and only if each component simple sentence is true. Therefore, 3 satisfies both parts of the compound sentence, but 25 and -2 do not. The graphical equivalent of this is the collection of points on the number line where the component graphs overlap. The shaded region of Figure 0-6 represents the graph of this compound sentence. This shaded region is the overlap of the solution of $x > 2.7$ with the solution of $x < \sqrt{19}$.

FIGURE 0-6 $2.7 < x < \sqrt{19}$

Note: Figure 0-6 indicates a natural abbreviation for compound statements using the conjuction *and*. Abbreviate $a < x$ and $x < b$ as $a < x < b$. Read the abbreviation "a is less than x and x is less than b. The word *and* is implied by the "run-on" sentence. We say that x is *between* a and b.

$$x > \tfrac{9}{2} \text{ or } x < -2$$

Solution: This is a compound sentence joined by the disjunction *or*. Compound sentences using *or* are true if at least one of the simple sentence components is true. The graphical equivalent is to join the shaded regions of the component solutions into the final solution. Therefore, any shaded region is part of the entire solution.
The shaded regions in Figure 0-7 represent the solution of $x > \tfrac{9}{2}$ or $x < -2$.

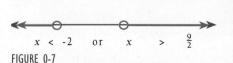

FIGURE 0-7

Note: These last two illustrations reinforce a fundamental strategy of mathematics: break a complex problem into its component parts, analyze the component parts, then reassemble the component solutions into a whole solution. The distinction between these last illustrations is the nature of the reassembly as implied by the words *and* and *or*. Although $a < b$ and $b < c$ abbreviates as $a < b < c$, there is no equivalent abbreviation for $a < b$ or $b > c$. The problem lies with the disjunction *or*. Except as follows, there are few suitable abbreviations for a compound inequality using the disjunction *or*. In any case, *never* abbreviate $a < x$ or $x < b$ as $a < x < b$. This abbreviation is for the conjunction *and*.

$$x > 5 \quad \text{or} \quad x = 5$$

Solution: Refer to Illustration 1. The graph of $x > 5$ is unchanged. Join it to the graph of $x = 5$. The graph of 5 is the omitted point circled in Illustration 1. Shading the circle results in a graph representing the compound sentence (Figure 0-8).

Note: To conserve space, abbreviate $x > 5 \quad or \quad x = 5$ as $x \geq 5$. Read the abbreviation as "x is greater than *or* equal to 5." Similarly, read $x \leq \sqrt{3}$ as "x is less than *or* equal to the square root of 3." The symbols $\geq$ and $\leq$ are two of the few abbreviations for the disjunction *or*. A more compact notation for inequalities is interval notation.

Good symbols make concepts easier to understand. Consider international highway signs. Even if the language is foreign, with clear symbols you grasp the meaning. **Interval notation** is a symbolism designed to improve insight into graphic representations of inequalities.

Reference to endpoints provides the underlying principle for abbreviating an inequality with interval notation. However, the notation should distinguish between $a < x < b$ and $a \leq x \leq b$. Both inequalities represent unbroken intervals of numbers. One final goal is to represent unbroken intervals that have only one endpoint. For example, consider the interval given by $x \geq 5$. For this last goal, we need a symbol to represent "no endpoint." The classic symbol for this concept is ∞ read "infinity." Use ∞ to indicate no righthand endpoint and $-\infty$ to represent no lefthand endpoint.

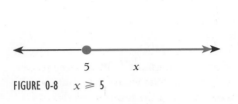

Illustration 4:

FIGURE 0-8 $x \geq 5$

S I D E B A R
Loveknot, Lazy Eight, the Worm Ouroborous, the ancient symbol for a snake devouring its own tail: all of these are symbolized by ∞. All of these imply the meaning for this ancient symbol for infinity.

DEFINITION 0-15
Interval Notation

INEQUALITY	INTERVAL NOTATION	NAME
$a < x < b$	(a, b)	Open interval
$x < b$	$(-\infty, b)$	Open interval
$a < x$	(a, ∞)	Open interval
$a \leq x \leq b$	$[a, b]$	Closed interval
$x \leq b$	$(-\infty, b]$	Closed interval
$a \leq x$	$[a, \infty)$	Closed interval
$a < x \leq b$	$(a, b]$	Half open, half closed
$a \leq x < b$	$[a, b)$	Half closed, half open

Admire the simplicity of the notation. The lefthand endpoint is to the left, the righthand endpoint is to the right. A bracket, [or], indicates the endpoint is included in the graph. A parenthesis, (or), indicates the endpoint is excluded from the graph. *Closed* refers to including the endpoint(s). *Open* excludes the endpoint. Note that ∞ is not a number, has no graph and is never included in an interval: it makes no sense to include ∞ as an endpoint; we use ∞ to mean there is no endpoint.

Refer to Example 0-10. The following express each of the inequalities in interval notation:

$$x > 5 \qquad \rightarrow \qquad (5, \infty)$$

$$2.7 < x < \sqrt{19} \qquad \rightarrow \qquad (2.7, \sqrt{19})$$

$$x > \tfrac{9}{2} \text{ or } x < -2 \qquad \rightarrow \qquad \left(\tfrac{9}{2}, \infty\right) \text{ or } (-\infty, -2)$$

$$x \geq 5 \qquad \rightarrow \qquad [5, \infty)$$

Moreover, as shown graphically in Figure 0-9,

$-20 < x \leq 10$ or $x > 50$ is $(-20, 10]$ or $(50, \infty)$.

Note: In set theory, the symbol $\cap$ is used to indicate the common part or **intersection** of two sets. The symbol $\cap$ is logically equivalent to the conjunction *and*. Similarly, the symbol $\cup$ indicates the joining or **union** of two sets. Therefore, $\cup$ can replace the disjunction *or* when discussing the union of two intervals of numbers.

As a result, $(-20, 10]$ or $(50, \infty)$ becomes $(-20, 10] \cup (50, \infty)$. Moreover, use the set notation symbol $\in$ for "is an element of," to indicate that a number lies in an interval. For example, $3 \in (-1, 5]$, because $-1 < 3 \leq 5$. Similarly, write the compound statement $-5 < x < -1$ or $7 \leq x$ as $x \in (-5, -1) \cup [7, \infty)$.

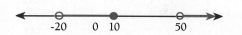

FIGURE 0-9 $(-20, 10]$ or $(50, \infty)$

0-6   EXERCISES

Determine the indicated absolute value.

1. $|-5.7|$
2. $|0 - 1.3|$
3. $|13 - 4|$
4. $|4 - 13|$
5. $|13 - (-4)|$
6. $|-13 - 4|$
7. $|3 - \pi|$
8. $|\pi - 2|$
9. Is $|-x| = |x|$ an identity? Explain your answer.
10. Is $|a - b| = |b - a|$ an identity? Explain your answer.

Identify each of the following as true or false.

11. $\frac{11}{13} < \frac{10}{12}$
12. $-\frac{12}{13} < -\frac{10}{11}$
13. $\sqrt[3]{4} > \sqrt[2]{3}$
14. $\sqrt{5} > \sqrt[3]{9}$
15. $1.4 < \sqrt{2} < 1.5$
16. $1.7 < \sqrt{3} < 1.8$
17. If $-x > 2$, then $x < 0$.
18. If $\frac{1}{x} < \frac{1}{2}$, then $x > 2$.
19. If $x > 5$, then $\frac{1}{x} < \frac{1}{5}$.
20. If x is real, then $-x < 0$.

For each of the following, (a) graph the solution on a number line and (b) write the solution using interval notation. Use $\cap$ or $\cup$ to indicate intersection or union, respectively, of the intervals.

21. $x < 7$
22. $x > -2$
23. $x \geq -5$
24. $x \leq -1$
25. $x > 2$ and $x \leq 8$
26. $x \geq -3$ and $x \leq 4$
27. $x \leq -1$ or $x > 2$
28. $x < 1$ or $x > 4$
29. $x \leq 2$ and $x > 0$
30. $x \leq -3$ or $x > -5$
31. $-2 < x < 5$
32. $-1.3 < x < 7$
33. $-1 \leq x \leq 2$ or $x > 5$
34. $x < -3$ or $0 < x \leq 7$
35. $-1 \leq x \leq 3$ or $5 < x \leq 7$
36. $-5 \leq x \leq 0$ or $1 < x \leq 3$
37. $-1 \leq x \leq 3$ and $x > 0$
38. $-5 \leq x \leq 0$ and $x < -2$
39. $-1 \leq x \leq 3$ and $5 < x \leq 7$
40. $-5 \leq x \leq 0$ and $1 < x \leq 3$

Graph each of the following.

41. $(-3, 4)$
42. $(3, 5]$
43. $[-2, 1.5)$
44. $[-1.1, 3.2]$
45. $[-3, \pi) \cap (0, \infty)$
46. $(-\infty, 4] \cap (-5, \infty)$
47. $(-\infty, 4] \cup (-5, 3]$
48. $[-3, \pi) \cup (0, \infty)$
49. $(-\infty, 4] \cap (-5, 3]$
50. $[-3, \pi) \cap (0, \infty)$

Solve each of the following. Write the solution using interval notation.

51. $3x + 2 > 5$

52. $3x - 2 < 7$

53. $-2x + 1 < 5$

54. $-2x + 7 > 2$

55. $3x - 1 \geq x - 2$

56. $4x + 2 \leq 3 - x$

57. $x + 7 \leq 5x - 5$

58. $2x - 1 \geq 4 + 3x$

59. $2x < \pi$

60. $6x > 5\pi$

61. $-2\pi \leq 2x \leq 2\pi$

62. $-2\pi \leq 3x \leq 2\pi$

63. $-2\pi \leq 2x - \pi \leq 2\pi$

64. $-2\pi \leq 3x - \pi \leq 2\pi$

65. $-2\pi \leq 2x - \dfrac{\pi}{2} \leq 2\pi$

66. $-2\pi \leq 3x - \dfrac{\pi}{2} \leq 2\pi$

67. $-2\pi \leq 5x - \dfrac{3\pi}{4} \leq 2\pi$

68. $-2\pi \leq 4x - \dfrac{2\pi}{3} \leq 2\pi$

69. $-2\pi \leq 1 - \dfrac{x}{2} \leq 2\pi$

70. $-2\pi \leq 2 - \dfrac{x}{3} \leq 2\pi$

71. Some texts define $|x|$ as $|x| = \sqrt{x^2}$. Verify the workings of this definition by evaluating $\sqrt{x^2}$ for $x = 5$, $x = 0$ and $x = -3$.

72. See Problem 71. What does this alternative definition of absolute value indicate as a geometric interpretation of $\sqrt{}$?

73. Show that $|x| < 3$ is equivalent to $-3 < x < 3$.

74. Show that $|x| \leq 4$ is equivalent to $-4 \leq x \leq 4$.

75. See Problem 73. Write $|x| < 6$ as an inequality without using the absolute value symbols.

76. See Problem 74. Write $|x| \leq 7$ as an inequality without using the absolute value symbols.

77. See Problem 73. Show that $|x - 2| < 4$ is equivalent to $-4 < x - 2 < 4$.

78. See Problem 74. Show that $|x + 3| \leq 5$ is equivalent to $-5 \leq x + 3 \leq 5$.

79. See Problem 77. Show that $|x - 3| < 4$ represents the interval $(-1, 7)$. Note that this interval is centered at 3 and has a width of $2(4)$.

80. See Problems 78 and 79. Show that $|x + 4| \leq 7$ represents the interval centered at -4 with a width of $2(7)$. This is the interval $[-11, 3]$.

Express the following in interval notation. See Problems 73–80.

81. $|x| < 4$

82. $|x| < 5$

83. $|x - 1| < 3$

84. $|x - 3| < 1$

85. $|x - 5| \leq 2$

86. $|x - 4| \leq 3$

87. $|x + 6| \leq 5$

88. $|x + 2| \leq 8$

89. $|3 - x| < 10$

90. $|5 - x| \leq 11$

Refer to problems 73–90. Rewrite each of the following as an inequality using absolute value notation. For example, if $-2 < x < 12$, then the midpoint of the interval is $(-2 + 12)/2$ or 5. The distance from the midpoint to either endpoint is 7 ($12 - 5 = 7$). Expressing this information with absolute value we have $|x - 5| < 7$.

91. $-5 < x < 5$

92. $-4 \leq x \leq 4$

93. $2 \leq x \leq 4$

94. $-2 < x < 4$

95. $-4 < x < 14$

96. $-5 < x < -1$

97. $0 \leq x \leq 10$

98. $2 < x < 20$

99. $-1 < x < 8$

100. $-4 \leq x \leq 15$

7   *COMPLEX NUMBERS

. . . the imaginary, this bosom-child of complex mysticism.
—EUGEN DUHRING

You are familiar with the construction of rational numbers by indicating a quotient of integers: a pair of integers makes up one rational number. The notation for rational numbers distinguishes the separate parts of the rational number, the numerator and denominator. Problem set 0-1 indicated how to define integers as a pair of counting numbers. This is less familiar to most people, so we are not as quick to identify the separate parts called the *subtrahend* and *minuend*. Nonetheless, these constructions reveal how mathematicians build new concepts on simpler ideas.

Counting numbers are not closed under subtraction. This provided a reason to develop integers. However, integers are not closed under division. The desire for closure under division motivates the development of rational numbers. Recall that $\sqrt{2}$ is not a rational number and so we have irrational numbers. Finally, $\sqrt{-1}$ is not a real number. We have no closure for taking square roots of negative real numbers. This section develops the complex numbers. Complex numbers provide answers for operations such as taking the square root of -1: $\sqrt{-1}$.

To begin, we adopt the symbol i to represent $\sqrt{-1}$. By the definition of square root, this means that $i^2 = -1$. The following theorem, gives some insight into the nature of i and the numbers that develop from it.

Theorem 0-10
Property of i

> If i is a number such that $i^2 = -1$, then i is not a real number.

Proof: By the trichotomy property every real number is either positive, negative or 0. Test i against each of these cases.

Suppose i is positive. Then, $i > 0$. Multiply each side of the inequality by i. Because i is assumed to be positive, by the multiplication property of inequalities, $ii > i0$. Then $i^2 > 0$. But $i^2 = -1$. This makes $-1 > 0$, a contradiction of the known relation of -1 and 0. Therefore i cannot be positive.

Suppose that i is negative. Then, $i < 0$. Multiply each side of the inequality by i. Because i is assumed to be negative, multiplication by a negative number reverses the direction of the inequality and we have $ii > i0$. However, $i^2 > 0$. This is the same contradiction as before.

The final case is to assume that $i = 0$. Multiplying both sides by i one last time, we obtain $i^2 = 0$. A clear contradiction, since $-1 \neq 0$.

Because i can be neither positive, negative, nor 0, i cannot be a real number.

Because $i^2 = -1$, then $i^4 = 1$. These facts allow us to reduce any integral power of i to 1, -1, i or $-i$. For example, $i^3 = ii^2$, then $i^3 = -i$. Similarly, $i^{23} = i^{4(5)}i^3$. Because $i^4 = 1$, then $i^{4(5)}i^3 = 1^5 i^3$. Thus, $i^{23} = -i$. A simple rule is to divide the exponent of i by 4 and discard the quotient, the remainder of the division is the new exponent for i. Consider i^{42}. Divide 42 by 4. Discard the quotient of 10. The remainder is 2 and this is the new exponent for i: $i^{42} = i^2$. Thus, $i^{42} = -1$.

Just as a fraction bar allows us to identify the numerator and denominator of a rational number, we use i to distinguish the two real numbers used to build a complex number.

DEFINITION 0-16
Complex Numbers

> A **complex number** z is any number expressible in the form
>
> $$z = a + bi$$
>
> where a and b are real numbers and $i^2 = -1$.
> The number a is the **real part** of z, and b is the **imaginary part** of z.

Just as a rational number has a numerator and a denominator, so a complex number has a real part and an imaginary part. In the case that $b = 0$, the complex number has no imaginary part and is strictly real. With this interpretation, Display 0-3 indicates the relationship among counting numbers, integers, rational numbers, irrational numbers, real numbers and, at last, complex numbers.

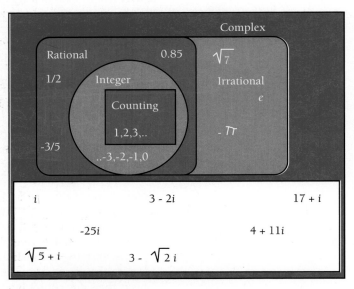

DISPLAY 0-3

Complex numbers eliminate the restriction imposed on square roots. Although the square root of a negative number is not real, it is complex. Consider $\sqrt{-36} = \sqrt{36}\ \sqrt{-1} = 6i$. Similarly, $\sqrt{-50} = \sqrt{25(2)}\ \sqrt{-1} = 5\sqrt{2}\ i$. Be careful! $\sqrt{2i} \neq \sqrt{2}\ i$. Although $\sqrt{-1} = i$, the value of $\sqrt{i}$ is not clear.

DEFINITION 0-17
Equality of Complex Numbers

Two complex numbers are equal, if their real parts and their imaginary parts are respectively equal.

The arithmetic for rational numbers is defined in terms of the arithmetic for the component parts of a rational number, integers. As a result, rational numbers possessed all the properties of integers and added a new property, multiplicative inverses. Rational numbers conform to the axioms for real numbers. With appropriate definitions for addition and multiplication, the complex numbers should inherit these properties from the real numbers. However, based on what we know of i, at least one property of real numbers is lost in the expansion to complex numbers. Complex numbers are not ordered. Greater than and less than have no meaning for complex numbers. Within this context, we say that real numbers form the "largest" *ordered* field. Extensions beyond complex numbers usually result in some properties being lost.

DEFINITION 0-18
Addition of Complex Numbers

> If $z = a + bi$ and $w = c + di$ are complex numbers, then
> $$z + w = (a + c) + (b + d)i$$

The addition of complex numbers resembles combining like terms when simplifying an expression. Rather than memorize this definition, assume that i marks the imaginary part of a complex number as a different term than the real part. Then given $z = 3 - 4i$ and $w = 2 + 5i$, to determine $w - z$ consider the problem a simplification:

$$w - z = (2 + 5i) - (3 - 4i)$$
$$= 2 + 5i - 3 + 4i$$
$$= 2 - 3 + (5 + 4)i$$
$$= -1 + 9i$$

The preceding example suggests a definition for $-z$ and for subtraction of complex numbers.

A similar approach works for the product of z and w:

$$wz = (2 + 5i)(3 - 4i)$$
$$= 2(3) - 2(4i) + 3(5i) - 5i(4i)$$
$$= 6 - 8i + 15i - 20i^2$$
$$= 6 - 8i + 15i - 20(-1) \qquad (i^2 = -1)$$
$$= 6 + 20 - 8i + 15i$$
$$= 26 + 7i$$

Rather than formally define multiplication of complex numbers, assume the properties for multiplication of binomials holds. The problem set will give you the opportunity to formulate a definition for multiplication of complex numbers.

To define the quotient of complex numbers is not so obvious. Division requires the existence of a multiplicative inverse. For consistency, use $1 = 1 + 0i$ as the multiplicative identity and $0 = 0 + 0i$ as the additive identity. We begin with the definition of *complex conjugate*.

DEFINITION 0-19
Complex Conjugate

> The complex **conjugate** of a complex number, $z = a + bi$, is $\bar{z} = a - bi$.

Simple arithmetic should convince you that $z\bar{z} = a^2 + b^2$. The product of complex conjugates is a real number!

Theorem 0-11
Multiplicative Inverse

> Suppose that z is a complex number, $z \neq 0$, then the multiplicative **inverse** of z is
> $$z^{-1} = \frac{\bar{z}}{z\bar{z}}$$

The proof of Theorem 0-11 is straightforward. First verify that $\overline{z}/(z\overline{z})$ is a complex number, then show that $z\,[\overline{z}/(z\overline{z})] = 1$.

<table>
<tr><td>DEFINITION 0-20
Division of Complex Numbers</td><td>Suppose w and z are complex numbers, $z \neq 0$, then

$$\frac{w}{z} = w\,\frac{\overline{z}}{z\overline{z}}$$
$$= \frac{w\overline{z}}{z\overline{z}}$$</td></tr>
</table>

To divide complex numbers expressed in fraction form, multiply the numerator and denominator by the conjugate of the denominator:

$$\frac{2 + i}{3 + 4i} = \frac{2 + i}{3 + 4i}\,\frac{3 - 4i}{3 - 4i}$$

$$= \frac{(2 + i)(3 - 4i)}{(3 + 4i)(3 - 4i)}$$

$$= \frac{10 - 5i}{25}$$

$$= \frac{2}{5} - \frac{1}{5}i$$

Chapter 8 adds details to the discussion of complex numbers.

SUMMARY

A complex number is $z = a + bi$, where a, b are real and $i^2 = -1$. The complex conjugate of z is $\overline{z}$, where $\overline{z} = a - bi$.

Suppose that z and w are complex numbers, $z = a + bi$ and $w = c + di$, then

1. $z + w = (a + c) + (b + d)i$
2. $z - w = (a - c) + (b - d)i$
3. $zw = (ac - bd) + (ad + bc)i$
4. $w \div z = \dfrac{w\overline{z}}{z\overline{z}}$

The arithmetic of complex numbers conforms to the field properties for real numbers. But complex numbers are not ordered. Two complex numbers are either equal or not equal: one is not "larger than" the other.

0-7 EXERCISES

Simplify the following.

1. i^2
2. i^3
3. i^4
4. i^5
5. i^6
6. i^7
7. i^8
8. i^9
9. i^{103}
10. i^{101}
11. i^{98}
12. i^{207}
13. i^{17}
14. i^{26}
15. i^{-1}
16. i^{-2}
17. i^{-4}
18. i^{-3}
19. i^{-105}
20. i^{-52}
21. $(3 + 2i) - (2 - i)$
22. $(-2 + 5i) - (3 - 7i)$
23. $(3 + 2i)(3 - 2i)$
24. $(-2 + 5i)(-2 - 5i)$
25. $\dfrac{3 + 2i}{3 - 2i}$
26. $\dfrac{-2 + 5i}{-2 - 5i}$
27. $\dfrac{3 - 2i}{3 + 2i}$
28. $\dfrac{-2 - 5i}{-2 + 5i}$
29. $(5 + 2i)\dfrac{3 - i}{1 + i}$
30. $\dfrac{(2 - 4i)(3 + 2i)}{1 - i}$

Rewrite the following in the form $a + bi$, where a and b are real and $i^2 = -1$.

31. $\sqrt{-4}$
32. $\sqrt{-9}$
33. $\sqrt{-18}$
34. $\sqrt{-28}$
35. $\sqrt{3} - \sqrt{-25}$
36. $\sqrt{7} + \sqrt{-36}$
37. $\dfrac{-10 - \sqrt{-72}}{2}$
38. $\dfrac{-15 + \sqrt{-75}}{5}$
39. $\dfrac{6 + \sqrt{-28}}{4}$
40. $\dfrac{8 - \sqrt{-20}}{6}$

For each of the following complex numbers z, determine (a) the conjugate of z, $\bar{z}$, (b) $z + \bar{z}$ and (c) $z\bar{z}$.

41. $z = 3 - 4i$
42. $z = 3 + 4i$
43. $z = 2 + 3i$
44. $z = 2 - 3i$
45. $z = 5i$
46. $z = 7$
47. $z = -5$
48. $z = -3i$
49. $z = \dfrac{-3 + i\sqrt{5}}{2}$
50. $z = \dfrac{5 - i\sqrt{3}}{2}$

For each of the following, Problems 51–70 assume that z and $\bar{z}$ are complex numbers, where $z = a + bi$ and $\bar{z} = a - bi$. Similarly, $w = c + di$ and $\bar{w} = c - di$.

51. Show that $z + \bar{z}$ is a real number.
52. Show that $z - \bar{z}$ has a real part of 0.
53. Show that $z\bar{z}$ is a real number.
54. The magnitude of z, denoted by $|z|$, is defined as $|z| = \sqrt{a^2 + b^2}$. Show that $|z| = \sqrt{z\bar{z}}$.
55. Let $u = z + w$. Determine u.
56. Let $v = zw$. Determine v.
57. See Problem 55. Determine $\bar{u}$. (Note: $\bar{u}$ represents $\overline{z + w}$)
58. See Problem 56. Determine $\bar{v}$. (Note: $\bar{v}$ represents $\overline{zw}$)
59. Determine the sum of $\bar{z} + \bar{w}$.
60. Determine the product of $\bar{z}\,\bar{w}$.
61. See Problems 55, 57 and 59. Verify that $\overline{z + w} = \bar{z} + \bar{w}$.
62. See Problems 56, 58 and 60. Verify that $\overline{zw} = \bar{z}\,\bar{w}$.
63. Verify that $\overline{z - w} = \bar{z} - \bar{w}$.
64. Verify that the conjugate of z/w is $\bar{z}/\bar{w}$.
65. Suppose we square z and then take the conjugate. Will the answer be the same as taking the conjugate first and then squaring?
66. Show that the conjugate of $\bar{z}$ is z.
67. Show that the conjugate of $-z$ is the negative of $\bar{z}$.
68. Verify that $1/z = \bar{z}/(z\bar{z})$ is a complex number.
69. See Problem 54. Suppose $u = z/|z|$. Determine $|u|$.
70. See Problem 69. Suppose $v = \bar{z}/|z|$. Determine $|v|$.
71. If $w = \dfrac{\sqrt{2} - i\sqrt{2}}{2}$, determine w^2.
72. If $u = \dfrac{-\sqrt{2} + i\sqrt{2}}{2}$, determine u^2.
73. Recall that $\sqrt{a} = b$ if $b^2 = a$. Verify that $\sqrt{-2 + 2i\sqrt{3}} = 1 + i\sqrt{3}$.
74. See Problem 73. Verify that $\sqrt{2 - 2i\sqrt{3}} = \sqrt{3} - i$.
75. See Problem 73. Verify $\sqrt{i} = \dfrac{\sqrt{2} + i\sqrt{2}}{2}$.
76. See Problem 73. Verify $\sqrt{-i} = \dfrac{\sqrt{2} - i\sqrt{2}}{2}$.
77. Two complex numbers are equal if and only if their real parts and imaginary parts are respectively equal. Suppose x and y are real numbers. Determine x and y if $3x + 2i = 12 + 4yi$.
78. See Problem 77. Determine real values for x and y if $x + (x + y)i = 5 + 7i$.

79. See Problem 77. Determine values for x and y if $2x + (y - x)i = 8 - 3i$.

80. See Problem 77. Determine values for x and y if $x + yi + (y - x)i = 3 + i$.

81. See Problems 1–20. Prove that $i^{4n} = 1$.

82. See Problems 1–20. Prove that $i^{4n+2} = -1$.

83. See Problems 1–20. Prove that $i^{4n+3} = -i$.

84. See Problems 1–20. Prove that $i^{4n+1} = i$.

CHAPTER SUMMARY

Real numbers and the operations of addition and multiplication obey the laws of the field properties.

Subtraction: $a - b = a + (-b)$

Division: $a \div b = a\, \dfrac{1}{b}$

Expressions are simplified according to the field properties and the order of operations.

G	Grouping symbols. Operations enclosed in parentheses are done first. If nested, done from innermost to outermost.
E	Exponents and radicals. Followed by negation.
M	Multiplication and division. Done left to right.
A	Addition and subtraction. Done left to right.

To factor an expression, try the following sequence of activities.

1. Factor out any common factors.
2. If there are no common factors, then count the number of terms:
If the expression has two terms:

$a^2 - b^2 = (a + b)(a - b)$ (difference of two squares)
$a^3 - b^3 = (a - b)(a^2 + ab + b^2)$ (difference of two cubes)
$a^3 + b^3 = (a + b)(a^2 - ab + b^2)$ (sum of two cubes)

If the expression has three terms: $ax^2 + bx + c$. Use trial and error or complete the square.
If the expression has more than three terms, experimentally group two or three of the terms together. See if you can factor just this group of terms. If not try another group. If you manage to factor a group of terms, reexamine the problem in terms of steps 1 and 2.

3. If steps 1 and 2 did not work on the expression, assume the expression is prime and cannot be factored. Repeat steps 1, 2 and 3 for each factor. Quit when all factors are prime.

A complex number $z = a + bi$, where a, b are real and $i^2 = -1$.

The complex conjugate of z is $\bar{z}$, where $\bar{z} = a - bi$.

The arithmetic of complex numbers conforms to the field axioms, but complex numbers are not ordered.

KEY WORDS AND CONCEPTS

Absolute value
Algebraic expressions
Common factor
Completing the square
Complex number
   Imaginary part
   Real part
Conjugate
Constant
Counting numbers
Coordinate
Denominator
Difference of cubes
Difference of squares
Exponent

Factor
Graph
Grouping symbols
Index of a radical
Inequalities
Integers
Intersection
Interval
Interval notation
Inverse
Irrational numbers
Least Common Denominator
Number line
Numerator
Order

Order of operations (GEMA)
Perfect square
Polynomial expressions
Prime
Pythagorean Theorem
Radical
Rational expression
Rational numbers
Real numbers
Sets
Simplification
Sum of cubes
Trichotomy property
Union
Variable

0 REVIEW EXERCISES

SECTION 0-1

Identify the field property or property of equality illustrated by each of the following.

1. $7 + (6 + 5) = (7 + 6) + 5$

2. $7 + (6 + 5) = 7 + (6 + 5)$

3. $2(3 + 4) = 6 + 8$

4. $5(6 + 4) = (6 + 4)5$

5. $5 + (6 + 4) = (6 + 4) + 5$

Construct a number line and indicate the approximate location of each of the following real numbers (use a calculator if necessary).

6. $\dfrac{-3}{5}$ **7.** 2.8 **8.** $\sqrt{6}$

9. $\pi - 2$ **10.** $\dfrac{2\pi}{5}$

SECTION 0-2

Simplify each of the following numeric expressions.

11. $4 - 3^2 - 12 \div 4 + 2$

12. $3(2 + 2^4 - 3) + 1$

13. $-(-3)^3 - (-1)^{11}$

14. $5 - (-2)^3 - (-3)^2$

15. $3(4^{-1})^{-1}$

16. $(3^{-1} + 2^{-1})^{-1}$

17. $\dfrac{5^2 - 2^2}{5 - 2}$ **18.** $\dfrac{14^2 - 3^2}{14 - 3}$

19. $\dfrac{\frac{3}{4} + \frac{4}{3}}{\frac{4}{3} - \frac{1}{2}}$

20 $\dfrac{\frac{2}{3} - \frac{3}{2}}{\frac{1}{3} + \frac{3}{2}}$

21. $\dfrac{5.1^2 - 5^2}{5.1 - 5}$

22. $\dfrac{1.2^2 - 1.1^2}{1.2 - 1.1}$

23. $\dfrac{(2 + 0.01)^2 - 2^2}{0.01}$

24. $\dfrac{(3.5 + 0.1)^2 - 3.5^2}{0.1}$

SECTION 0-3

Simplify the following.

25. $x^2 y^5 x^7 y^4$

26. $x^{-1/3} y^{1/2} (x^{-2/3} y^{1/2} + x^{1/3} y^{-1/2})$

27. $\sqrt{\dfrac{25}{16}}$

28. $\sqrt{x^4 y^2 z^8}$

29. $\sqrt{\dfrac{x^2 y^5}{z^3}}$

30. $\dfrac{3y}{\sqrt{y^5}}$

31. $4\sqrt{x} + 2\sqrt{x} - 3\sqrt{y}$

32. $\dfrac{3}{x} - \dfrac{2}{\sqrt{x}}$

33. $\dfrac{x - 7}{\sqrt{x} - \sqrt{7}}$

34. $\sqrt[5]{-1(3^7)}$

35. $\sqrt{\dfrac{49 x^6 y^4}{z^5}}$

36. $\sqrt[3]{\dfrac{-16 x^5 z^8}{y^2}}$

Rewrite each of the following rational expressions so that no radical expression appears in the numerator.

37. $\dfrac{\sqrt{x} - \sqrt{5}}{x - 5}$

38. $\dfrac{\sqrt[3]{x} - \sqrt[3]{5}}{x - 5}$

Construct a real number line and indicate the approximate location of each of the following.

39. $\dfrac{5 - \sqrt{7}}{2}$

40. $\dfrac{3 + \sqrt{2}}{2}$

Rewrite each of the following using radical notation. Assume all variables represent positive numbers.

41. $x^{3/4}$

42. $x^{-1/5} - y^{1/4}$

Rewrite each of the following using exponential notation. Assume all variables represent positive numbers.

43. $\sqrt[3]{x^2}$

44. $\sqrt[4]{\sqrt[3]{x}}$

45. $\sqrt[5]{x^2}$

46. $\sqrt{1 - \sqrt{x}}$

SECTION 0-4

Simplify each of the following.

47. $4(x + 3) - 5(2 - x)$

48. $6(x + 2) - 3(5x - 4)$

49. $(x - 8)(x + 7)$

50. $(x - 2)(x^2 - 3x + 1)$

Factor each of the following.

51. $5x - 25$

52. $x^2 - 12$

53. $(x - 5)^2 - 36$

54. $x^3 + 125$

55. $(x + 8)^2 - 49$

56. $x(x - 3) + 5(x - 3)$

57. $x^2 - 7x + 6$

58. $x^2 - x - 12$

59. $x^2 + x - 12$

60. $x^2 + 11x - 12$

61. $2x^2 - 14x + 24$

62. $x^2 - 6x + 1$

63. $x^2 + 4x - 2$

64. $x^3 - 1$

65. $8x^3 + 27$

66. $x^6 - 64$

67. $10xy - 4az - 10xz + 4ay$

68. $5x(x + 4) - 3(x + 4)$

69. $(x - 1)^2 - 5(x - 1) + 4$

70. $x^4 - 5x^3 + 4x^2$

71. $x^4 - 5x^2 + 4$

72. $y^2 - x^2 - 6y + 9$

For each of the following, (a) factor the expression and (b) indicate the values of x for which the expression equals 0.

73. $x^2 - 8x + 12$

74. $(x - 2)^2 - 7(x - 2) + 12$

75. $(x^2 - x - 2)(x^2 + 1) - 10(x^2 - x - 2)$

Factor each of the following.

76. $x^{-1/3} + x^{2/3}$

77. $\dfrac{\sqrt{x}}{x} + \dfrac{x}{\sqrt{x}}$

78. $x^{2/3} - 3x^{1/3} + 2$

SECTION 0-5

Simplify each of the following.

79. $\dfrac{x^2 - 4}{3x + 6}$

80. $\dfrac{x^2 - 4}{x^2 - 4x + 4}$

81. $\dfrac{3x + 9}{x^2 - 4} \cdot \dfrac{x^2 - x - 2}{x + 3}$

82. $\dfrac{5x^2 + 10x}{x^2 - 16} \div \dfrac{x^2 - 4}{x^2 + 2x - 8}$

83. $\dfrac{3}{x^2} + \dfrac{1}{x}$

84. $\dfrac{x}{x - 3} + \dfrac{2}{x + 2}$

85. $\dfrac{3}{x - 3} - \dfrac{1}{x + 3} - \dfrac{6}{x^2 - 9}$

86. $\dfrac{3x}{x-2} - \dfrac{3}{x+5} \div \dfrac{x-2}{2x+10}$

87. $\left(\dfrac{x}{x-2} + \dfrac{3}{x-3}\right) \div \dfrac{3x+6}{2x-6}$

88. $\dfrac{\dfrac{x^2}{4} - 4}{\dfrac{x}{2} + 2}$

89. $\dfrac{\dfrac{x}{5} - \dfrac{5}{x}}{\dfrac{1}{x} - \dfrac{1}{5}}$

90. $2 - \dfrac{\dfrac{x^2}{9} - \dfrac{1}{4}}{\dfrac{2x+3}{18}}$

91. $\dfrac{\dfrac{x}{3} - \dfrac{3}{x}}{1 - \dfrac{3}{x}} - \dfrac{x - \dfrac{25}{x}}{\dfrac{1}{x} + \dfrac{1}{5}}$

92. $\dfrac{\dfrac{x}{3} - \dfrac{3}{x}}{1 - \dfrac{3}{x}} \div \dfrac{\dfrac{x^2}{16} - 1}{\dfrac{x}{2} + 2}$

93. $\dfrac{x^2 - 5^2}{x - 5}$

94. $\dfrac{(x+h)^2 - x^2}{h}$

95. $\dfrac{x^3 - 1^3}{x - 1}$

96. $\dfrac{(x+h)^3 - x^3}{h}$

SECTION 0-6

Identify each of the following as true or false.

97. $\sqrt{3} > \sqrt[3]{7}$

98. $2 < \sqrt{5} < 2.5$

99. If $\dfrac{1}{x} < \dfrac{1}{10}$, then $x > 10$.

100. If $x > 9$, then $\dfrac{1}{x} < \dfrac{1}{9}$.

For each of the following, (a) graph the solution on a number line and (b) write the solution using interval notation.

101. $x < -3$

102. $x < 0$ or $x > 5$

103. $x \le 5$ and $x > -1$

104. $-\pi < x < 10$

105. $-2 \le x \le 3$ or $x > 4$

106. $-4 \le x \le 0$ and $-1 < x \le 2$

Graph each of the following.

107. $(-2, \pi)$

108. $[-1, \pi) \cup (5, \infty)$

109. $(-\infty, -1] \cap (-3, 2]$

110. $[-1, \pi) \cap (0, \infty)$

Solve each of the following. Write the solution using interval notation.

111. $2x + 3 > 5$

112. $x - 7 \ge 4 + 3x$

113. $3x < \pi$

114. $-2\pi \le 6x \le 2\pi$

115. $-2\pi \le 3x - \pi \le 2\pi$

116. $-2\pi \le 1 - \dfrac{x}{3} \le 2\pi$

Express the following in interval notation.

117. $|x| < 2$

118. $|x - 4| < 5$

119. $|x - 21| \le 7$

120. $|2 - x| \le 8$

Rewrite each of the following as an inequality using absolute value notation.

121. $-15 < x < 15$

122. $-3 < x < 7$

123. $-6 < x < 14$

124. $-6 \le x \le 15$

SECTION 0-7

Simplify the following.

125. i^2

126. i^9

127. i^{15}

128. i^{31}

129. i^{153}

130. i^{240}

131. i^{-17}

132. i^{-3}

133. i^{-8}

134. i^{-57}

135. $(5 + 2i) - (4 - i)$

136. $(3 + 2i)(4 - 5i)$

137. $\dfrac{5 + 2i}{5 - 2i}$

138. $\dfrac{4 - 5i}{3 - 5i}$

139. $(5 + 2i)\dfrac{3 - 2i}{2 + i}$

140. $\dfrac{(3 - 2i)\,3 + 2i}{2 - i}$

Rewrite the following in the form $a + bi$, where a and b are real and $i^2 = -1$.

141. $\sqrt{-49}$

142. $\sqrt{-72}$

143. $\sqrt{13} - \sqrt{-125}$

144. $\dfrac{10 - \sqrt{-50}}{5}$

145. $\dfrac{-6 + \sqrt{-12}}{4}$

146. $\dfrac{18 - \sqrt{-18}}{6}$

For each of the following complex numbers z, determine (a) the conjugate of z, $\bar{z}$, (b) $z + \bar{z}$ and (c) $z\bar{z}$.

147. $z = 5 - 12i$

148. $z = 3 - 2i$

149. $z = 4$

150. $z = -5i$

151. $z = \dfrac{-5 + i\sqrt{7}}{2}$

152. $z = \dfrac{1 - i\sqrt{2}}{2}$

Algorithms

Before Mohammed, before the rise of Western civilization, people in the Middle East wrote numbers as words. Although Hindus and Greeks influenced the Middle Eastern culture of the tenth and eleventh centuries, modern day numerals first came to civilization through the work of al-Khowarizmi. Even today, we still call them Arabic numerals.

ALGEBRA

The word *algebra* comes from the title of al-Khowarizmi's treatise, *Hisab al-jabr w'al-muqabalah*. The title translates as "the science of transposition and simplification." The word *al-jabr* evolved, through the Latin, into our modern word *algebra*. From common usage, *al-jabr* meant "reunite." The Moors of Spain referred to a reuniter of broken bones as an *algebrista*. Healers have not always been called as they are now.

ALGORITHM

The contents of al-Khowarizmi's treatise were not elegant mathematics in the modern sense. Rather they consisted of methods of obtaining arithmetical answers to problems. An 1857 Latin translation of the work referred to the author as Algoritmi. Eventually references to the name evolved to the word *algorithm*, a word synonymous with a method for solving a problem.

Although *algorithm* originally meant any method by which an answer could be obtained, computer scientists refined the idea of the algorithm. Given a solvable problem and a device to solve the problem, an algorithm is a precise, step-by-step description of the method for solving the problem. The description of the method of solution is in a language appropriate for the device used in the solution.

CHARACTERISTICS OF ALGORITHMS

Algorithms have the following characteristics.

1. The description of the algorithm results in a finite sequence of activities.
2. The first activity in the sequence of activities is unique.
3. Every activity (except the last) has a unique successor activity.
4. The last activity is either the solution to the problem or a conclusion that the problem has no feasible solution.

Other descriptions of an algorithm are more poetic.

Hermes (1965): "a general procedure such that for any appropriate question the answer can be obtained by the use of a simple computation according to a specified method."
Minsky (1967): "an effective procedure is a set of rules that tells us, from moment to moment, precisely how to behave."

In Chapter 12, we introduce the idea of finite mathematical induction. In many respects the description of the steps required for an algorithm resemble the principle of finite mathematical induction. At least three descriptions of mathematical statements are relevant to algorithms.

EXISTENCE STATEMENTS

A statement may specify that a problem has a solution, but give no hint how to determine the answer. Such statements are existence statements. For example we know that every equation of the form

$$ax^5 + bx^4 + cx^3 + dx^2 + ex^1 + f = 0 \qquad (a \neq 0, \quad a, b, c, d, e, f \in \text{Real})$$

has at least one real solution. We may not be able to determine that exact solution in every case.

OBJECT-ORIENTED STATEMENTS

A statement may be object oriented. Often the equality relationship plays a role in such statements; for example, $A + B = B + A$. In object-oriented statements, we simply claim that two objects represent the same thing: they are equal. Substitution is a common use for such information. For example, determine the value of $x^2 + 3x$, when $x = 2$. Computer science describes binding the value of 2 to x for substitution as specifying an "instance" (instantiation) of x. An English teacher might indicate that x is a pronoun and that the noun 2 is given as the antecedent.

PROCEDURAL STATEMENTS

Finally, statements can describe procedures for obtaining solutions to problems. These procedures are step-by-step activities for obtaining solutions. A procedure is a modern day instance of an algorithm. Where appropriate, this text will describe the process of solving problems as step-by-step procedures. Whether the description is informal or detailed, the intent is the same, to organize the process into a scheme for problem solving. The student retains the responsibility of problem recognition and the creative application of an appropriate problem solving process.

0 CHAPTER TEST

1. Simplify
  a. $-5^2 - 2^2 + (-3)^2$
  b. $(x - 3)(x + 2) + (x - 3)(x + 2)$

2. Prove: if $a > 3$ and $b > 4$, then $a + b > 7$.

3. Simplify $\sqrt[3]{\dfrac{81x^5}{yz^5}}$

4. Rewrite $x^{2/3}y^{1/3} - x^{1/2}$ using radicals.

5. Factor
  a. $(x + 3)^2 - 7(x + 3) + 12$
  b. $(2x - 1)^2 - 16$
  c. $x^2 + x + 1$

6. Simplify
  a. $\dfrac{2x}{x - 1} - \dfrac{3}{x + 2} \div \dfrac{3x - 3}{2x + 4}$

  b. $\dfrac{\dfrac{x}{12} - \dfrac{3}{x}}{1 - \dfrac{6}{x}}$

  c. $\dfrac{x^2 - 1.5^2}{x - 1.5}$

7. Express $|x - 4| < 11$ in interval notation.

8. Solve and express the answer in interval notation
$$-2\pi < 2x + \pi < 2\pi.$$

9. Simplify $\dfrac{5 - 2i}{3 + 4i}$.

10. If $z = \dfrac{3 + i\sqrt{5}}{2}$, determine $z\bar{z}$.

GRAPHS

This chapter introduces the relationship of algebra to graphs. The graphs we shall study have familiar shapes: the line, the parabola and the circle. Each graph will help introduce the concept of analysis. One of the great concepts of mathematics is that geometric patterns correspond to algebraic patterns. *Analysis* is the study of algebraic patterns and their graphs.

To explore analysis, we shall discuss the use of technology. Modern graphing calculators can do much of the work of producing a graph. Personal computers and software also have graphing capabilities. Section 1-2 introduces these technological aids. Graphing technology will give you novel and useful insights into the connection of algebra and geometry.

1   THE PLANE

There is now flourishing a certain kind of arithmetic called algebra, which endeavors to accomplish in regard to numbers what the ancients achieved in respect to geometrical figure.
—RENÉ DESCARTES (1596–1650)

Geometry reduces data to a visual level. Algebra more often attacks problems from a numerical viewpoint. We see planes, angles and lines in nature. These are visualizations of qualities. Distractions, such as color, sound and texture influence our insight into these qualities: our visualization is imprecise. We see graphs of economic trends, or performance curves for stereo equipment or perhaps even bar graphs to represent the various component of personality. But understanding economic trends, engineering formulas and psychological measurements benefits from the precision of numbers. Using the precision of algebra to examine geometrical information is the basis for analytical geometry.

Two intersecting lines define a plane. Therefore, as a single number locates a point on a number line, a pair of numbers can address a point in a plane. Several methods exist for locating points in a plane by using a pair of numbers. The most common method for locating points in the plane using pairs of numbers is the **Cartesian coordinate system.**

A Cartesian coordinate system consists of two number lines constructed at right angles. One of the lines, the *x*-**axis,** is drawn horizontally. A value of *x* is located by going left or right along the *x* axis. The second line, called the *y*-**axis,** is a vertical line. A value of *y* is located by going up or down along the *y* axis. These intersecting axes determine a unique plane. Locating a point in this plane requires its address (or location) as an **ordered pair** of real numbers, symbolized by (*x, y*). The first component of the pair is the horizontal axis address. The second component is the vertical axis address.

Graph is from the Greek *graphikos* meaning "written or recorded."

As with a number line, the located point is the graph of the ordered pair of numbers. The ordered pair of numbers are the **coordinates** of the point. To plot a point means to mark the location of the point (usually with a "dot") based on its coordinates. Similarly, we represent a collection of points by plotting each point in the same coordinate system. The resulting picture is the *graph* of the collection of points.

Because an infinite plane extends beyond our view, we must be satisfied with viewing a portion of the plane. A graph may also be infinite in nature and beyond our ability to view all at once. Instead, we see a portion of the graph. To organize our view, imagine a **window,** showing us a limited portion of the plane. See Figure 1-1. The graphs displayed in Cartesian coordinates are "snapshots," capturing part of the complete picture, much as a photo*graph* freezes a moment in time.

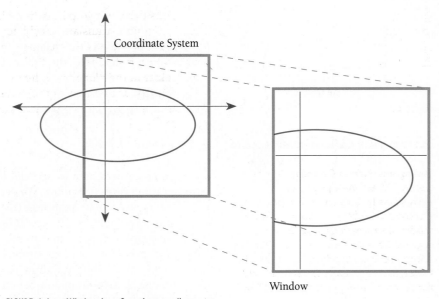

Coordinate System

Window

FIGURE I-I   Window into Cartesian coordinates

Illustration 1:

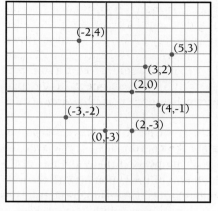

FIGURE 1-2   Plotted points

Figure 1-2 shows a graph of the points represented by the following ordered pairs of numbers:

$$A = (2, -3), \quad B = (-2, 4), \quad C = (3, 2), \quad D = (-3, -2)$$
$$E = (4, -1), \quad F = (5, 3), \quad G = (2, 0), \quad H = (0, -3).$$

Graphs express relationships, usually y-axis measurements to x-axis measurements. For example, consider a system in which the x-axis represents elapsed time and the y-axis represents distance from home. Then (3 hours, 130 miles) contains the information that three hours into a trip the distance from home is 130 miles. Similarly, (5 hours, 250 miles) indicates that five hours into the trip the distance home is 250 miles. Interpret each hour of the trip as shown in Display 1-1. For example, between hour 5 and hour 6, perhaps the driver started back home.

Because the coordinate axes are at right angles, another name for Cartesian coordinates is *rectangular coordinates*. Because the axes are at right angles, the Pythagorean Theorem applies.

Pythagorean Theorem

> Given a right triangle with hypotenuse of length c and sides of lengths a and b, then
>
> $$a^2 + b^2 = c^2$$
>
> also see Display 1-2.

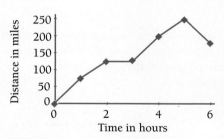

Display 1-1

Consider any two points in a plane. Several geometric concepts concerning two points translate nicely to algebraic equivalents. One of the most important of these is the distance between the two points. Since the x- and y-axes form a right angle, the **Pythagorean theorem** gives the distance between two points in the plane from their coordinates.

Our first application of the Pythagorean theorem is to obtain a formula for the distance between two points in the plane. We use subscripts as in x_1 and x_2

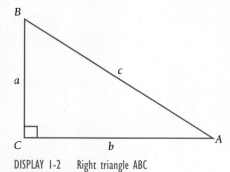

DISPLAY 1-2   Right triangle ABC

to distinguish points. See Figure 1-3. Because the line segments connecting (x_1, y_1), (x_2, y_2) and (x_2, y_1) form a right triangle, then

$$d^2 = |x_2 - x_1|^2 + |y_2 - y_1|^2$$

The square of any real number is nonnegative, making the absolute values unnecessary. Because distance is never negative, take the positive square root and obtain the **distance formula**:

Distance Formula

> The distance between (x_1, y_1) and (x_2, y_2) is
> $$d = \sqrt{(x_2 - x_1)^2 + (y_2 - y_1)^2}$$

Another geometric concept expressible in algebraic terms is the midpoint between two given points. Refer to Figure 1-3. Connect the midpoints of the sides of the triangle. Also connect the midpoint of the hypotenuse to the midpoints of the sides. The similar right triangles formed show that the coordinates of the midpoint of the line segment connecting $P_1 (x_1, y_1)$ with $P_2 (x_2, y_2)$ are

Midpoint Formula

> $$\text{Midpoint } (P_1 P_2) = \left(\frac{x_1 + x_2}{2}, \frac{y_1 + y_2}{2} \right)$$

Interpret this **midpoint formula** as the *average* of the coordinates. You can verify the formula by use of the distance formula. Show that the distances from each end point to the midpoint are equal and that the sum of these distances equals the distance between the endpoints.

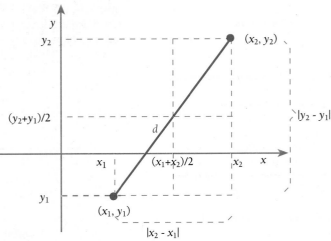

FIGURE 1-3   Distance between two points

For each of the following pairs of points, determine (a) the distance between the points (geometrically, the length of the line segment connecting the two points), and (b) the coordinates of the midpoint of the line segment joining the two points.

EXAMPLE 1-2

Illustration 1:

Figure 1-4 plots the points (5, 3) and (2, -1).

$$\text{a. Distance } d = \sqrt{(5-2)^2 + (3-(-1))^2}$$
$$= \sqrt{9+16}$$
$$= \sqrt{25}$$
$$= 5$$

$$\text{b. Midpoint: } \left(\frac{5+2}{2}, \frac{3+(-1)}{2}\right) = \left(\frac{7}{2}, 1\right)$$

Illustration 2:

Similarly for (5, -2) and (0, 10).

$$\text{a. Distance: } d = \sqrt{(5-0)^2 + (-2-10)^2} = 13$$

$$\text{b. Midpoint: } \left(\frac{5+0}{2}, \frac{-2+10}{2}\right) = (2.5, 4)$$

Illustration 3:

Copper City is 3 km due east and 2 km due south of a weather radar station (3 km, -2 km). Silver City is 10 km due west and 5 km due north of the same station (-10 km, 5 km). Determine the distance between the two cities and locate a point halfway between them. See Figure 1-5.

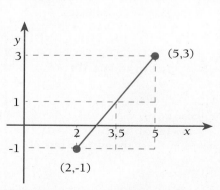

FIGURE 1-4   Length and midpoint

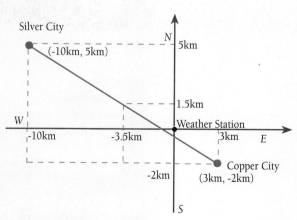

FIGURE 1-5   Distance between cities

Solution: Let the station be the origin for a Cartesian coordinate system:

$$\text{Distance} = \sqrt{[3\text{ km} - (-10\text{ km})]^2 + (-2\text{ km} - 5\text{ km})^2}$$
$$= \sqrt{169\text{ km}^2 + 49\text{ km}^2}$$
$$= \sqrt{218\text{ km}^2}$$
$$= \sqrt{218}\ \sqrt{\text{km}^2}$$
$$= \sqrt{218}\ \text{km}$$
$$\cong 14.8\text{ km}$$
$$\text{Midpoint} = \left[\frac{3\text{ km} + (-10\text{ km})}{2}, \ \frac{-2\text{ km} + 5\text{ km}}{2} \right]$$
$$= (-3.5\text{ km}, 1.5\text{ km}).$$

The midpoint is 3.5 km west and 1.5 km north of the weather station.

Illustration 4:   Determine the coordinates of a point on the x-axis 5 units from $(0, 3)$ on the y-axis.

Solution: Let $(x, 0)$ be the required point on the x-axis. Then, the distance from $(0, 3)$ to $(x, 0)$ is 5 (see Figure 1-6).

$$\sqrt{(x - 0)^2 + (0 - 3)^2} = 5$$
$$\sqrt{x^2 + 9} = 5$$
$$x^2 + 9 = 25$$
$$x^2 = 16$$
$$x = 4 \text{ or } x = -4.$$

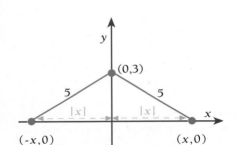

FIGURE 1-6

Change is a property of natural and social life: few things remain constant. Points on a plane may represent geographical position, or gross income in a given year or any number of items subject to change. Often we draw a graph to gain insight into that change. If we can understand the nature of change, then perhaps we can predict what changes are yet to happen.

Consider two points (observations) on a coordinate plane. Suppose that we shift our attention from one of two points to the other. What kind of changes were necessary to redirect our attention? The distance between two points is one way to measure that change between the two points. Another way is to examine the individual changes in the x and y coordinates.

DEFINITION 1-1
Average Rate of Change

Suppose that (x_1, y_1) and (x_2, y_2) are two points in a Cartesian coordinate system.
Then the change in x from x_1 to x_2 is $\Delta x = x_2 - x_1$.
Similarly, the change in y from y_1 to y_2 is $\Delta y = y_2 - y_1$.
(The Greek letter delta, Δ, is used here to suggest "difference.") To compare the change in y with the change in x, use the **average rate of change**:

$$\frac{\Delta y}{\Delta x} = \frac{y_2 - y_1}{x_2 - x_1}$$

EXAMPLE 1-3

Illustration 1:

Illustration 2:

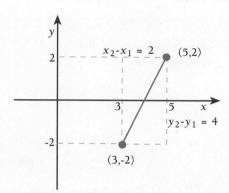

FIGURE 1-7   Average rate of change

Illustration 3:

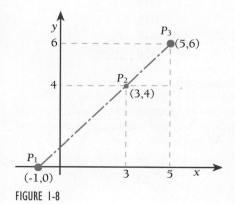

FIGURE 1-8

Calculate the average rate of change for the points indicated in Figure 1-7.

$$\frac{\Delta y}{\Delta x} = \frac{y_2 - y_1}{x_2 - x_1}$$

$$= \frac{2 - (-2)}{5 - 3}$$

$$= \frac{4}{2}$$

$$= 2$$

Henry Morita records his travels on a graph using the time t and the distance d he traveled from San Francisco. He records the information as the point (t, d). If after 1 hour, he is 50 km from San Francisco and after 3 hours, 200 km from San Francisco, then he records the points (1 hr, 50 km) and (3 hr, 200 km). What is his average rate of change for these two points?

$$\frac{\Delta y}{\Delta x} = \frac{200 \text{ km} - 50 \text{ km}}{3 \text{ hr} - 1 \text{ hr}}$$

$$= \frac{150 \text{ km}}{2 \text{ hr}}$$

$$= 75 \text{ km/hr}$$

Note: Here, we express speed as an average rate of change.

One application of average rate of change is to determine whether three points lie in the same straight line. Three or more points are **collinear points** if it is possible to construct a single straight line that contains all of the points.

Determine whether P_1 (3, 4), P_2 (-1, 0) and P_3 (5, 6) are collinear.

Solution 1: From geometry, the points are collinear if the average rate of change from P_1 and P_2 is the average rate of change from P_2 to P_3 (see Figure 1-8):

$$\frac{\Delta y}{\Delta x} = \frac{4 - 0}{3 - (-1)} \quad [P_1 P_2]$$

$$= 1$$

$$\text{Also, } \frac{\Delta y}{\Delta x} = \frac{0 - 6}{-1 - 5} \quad [P_2 P_3]$$

$$= 1$$

Solution 2: Consider the three distances measured between all possible pairs of points. Three points are collinear if the sum of the two shorter distances *equals* the longer distance (this is a consequence of the axiom that the shortest distance between two points lies along a straight line). (see Figure 1-8).

Collinear Points

Points		Distances	
$P_1 P_2$	$=$	$\sqrt{[3-(-1)]^2 + (4-0)^2}$	$= 4\sqrt{2}$
$P_2 P_3$	$=$	$\sqrt{(4-6)^2 + (3-5)^2}$	$= 2\sqrt{2}$
$P_1 P_3$	$=$	$\sqrt{(6-0)^2 + [5-(-1)]^2}$	$= 6\sqrt{2}$

Because $P_1 P_2 + P_2 P_3 = P_1 P_3$ the points are collinear. Be cautious. Remember, $\sqrt{14} + \sqrt{18} \neq \sqrt{32}$!

Later sections of this chapter use the distance formula, midpoint formula and the concept of average rate of change to analyze some special types of graphs. These graphs are the line, the circle and the parabola.

SUMMARY

For two points in a plane with coordinates (x_1, y_1) and (x_2, y_2)

Distance $= \sqrt{(x_2 - x_1)^2 + (y_2 - y_1)^2}$

Midpoint $= \left(\dfrac{x_1 + x_2}{2}, \dfrac{y_1 + y_2}{2} \right)$.

Average rate of change: $\dfrac{\Delta y}{\Delta x} = \dfrac{y_2 - y_1}{x_2 - x_1}$

1-1   EXERCISES

Plot the points whose locations are given by the following coordinates.

1. $(3, 5)$
2. $(5, 3)$
3. $(-2, 4)$
4. $(4, -2)$
5. $(-5, -3)$
6. $(-3, -5)$
7. $(4, 0)$
8. $(0, 3)$
9. $(0, -2)$
10. $(-5, 0)$
11. $(5, -2)$
12. $(-2, 5)$

13. $(3, 3)$
14. $(-2, -2)$
15. $(-3, -3)$
16. $(2, 2)$
17. $(2, -3)$
18. $(-5, 1)$
19. $(2, 3)$
20. $(5, 1)$

For the given coordinates of pairs of points determine (a) the distance between the points, (b) the coordinates of the midpoint of the segment joining the points, (c) the average rate of change between the points, and (d) graph the points and label the graph with the results of a, b, and c.

21. $(-2, 3), (10, -2)$
22. $(5, -2), (2, 2)$
23. $(-1, 7), (3, 4)$
24. $(5, -1), (-7, 4)$

Problem numbers in this color indicate an application. Problem numbers in this color indicate a preview of calculus. **Problems marked with an asterisk(*) are more difficult or require extra care.**

25. (0, 0), (1, -1)

27. (-2, 0), 3, -2)

29. (4a, 0), (a, 3b), $a \neq 0$

26. (-1, 1), (0, 0)

28. (5, 1), (-2, 3)

30. (5b, 2a), (2b, -2a), $b \neq 0$

Suppose that the following sets of triple points represent the vertices of a triangle. Determine whether each triangle is scalene (no equal sides), isosceles (exactly two equal sides) or equilateral (all three sides equal). Sketch each triangle.

31. (0, 5), (-5,0), (0, 0)

33. (0, 0), (0, -6), $(3\sqrt{3}, 3)$

35. $(\sqrt{2}, -\sqrt{2})$, (0, 0), (2, 0)

37. (1, 2), (3, 4), (4, 0)

39. (0.5, -3), (1.5, 0), (1, 1)

32. (0, 0), (4, 0), $(2, 2\sqrt{3})$

34. (1, 1), (-1, -1), (1, -1)

36. (3, 0), (1, 0), $(3, \sqrt{3})$

38. (-1, 2), (2, -1), (1, 1)

40. (-0.3, 11), (1.7, 1), (2, 2)

Determine which of the following sets of points are collinear.

41. (1, 2), (3, 4), (5, 6)

43. (-1, -3), (0, 0), (2, 6)

45. (1, 4), (3, 1), (-2, 5)

47. (0.5, 1), (0.3, 0.2), (10, 11)

42. (0, 0), (1, 2), (3, 6)

44. (-2, 4), (1, 1), (4, 4)

46. (5, 5), (2, 1), (-1, 3)

48. (2, 0.4), (-1, 0.9), (-2, 20)

49. (a, a^2), (b, b^2), $(0, -ab)$

50. $\left(b, \dfrac{1}{b}\right), \left(a, \dfrac{1}{a}\right), \left(\dfrac{1}{b} + 1, a\right), b \neq 0$

51. Tom's house is 300 feet due west and 400 feet due north of a fire plug. How far is his house from the plug?

52. A carpenter lays out an 8 ft stud and a 6 ft beam at a right angle and end to end. What is the distance between the other two endpoints?

53. Are the points (1, -2), (5, 2), (-3, 2) the vertices of a right triangle? (According to the Pythagorean theorem, a triangle with sides of length *a*, *b* and *c* (*c* is longest) is a right triangle if $a^2 + b^2 = c^2$.)

54. Are the points (3, -2), (1, 2), (-5, -6) the vertices of a right triangle?

55. Are the points (1, 1), (3, 3), (-1, 3), (1, 5) the vertices of a rectangle?

56. Are (1, 3), (3, 1), (-3, -1), (-1, -3) the vertices of a rectangle?

57. Given the vertices of a triangle, P_1 (0, 0), P_2 (3, 4) and P_3 (-1, 5), graph the vertices and connect them to form the triangle. Determine the coordinates of the midpoints of

the sides of the triangle. Label the midpoint P_1P_2 as M_1. Label the midpoint of P_1P_3 and M_2. Draw a line segment to connect M_1 and M_2. Show that the length of M_1M_2 is half the length of P_2P_3.

58. Given the vertices of a triangle, P_1 (0, 0), P_2 (-3, 5) and P_3 (2, 7), graph the vertices and connect them to form the triangle. Determine the coordinates of the midpoints of the sides of the triangle. Label the midpoint of P_1P_2 as M_1. Label the midpoint of P_1P_3 and M_2. Draw a line segment to connect M_1 with M_2. Show that $\Delta y/\Delta x$ from P_2 to P_3 equals $\Delta y/\Delta x$ from M_1 to M_2.

59. Determine x for (x, 7), (3, 2) so that $\Delta y/\Delta x = 0.5$.

60. Determine y for (3, y), (-2, 5) so that $\Delta y/\Delta x = 1.5$.

Determine the average rate of change of *y* with respect *x* for the following pairs of points. Leave the units of measure in the formula for $\Delta y/\Delta x$.

61. (3 sec., 157 ft.), (1 sec., 95 ft.)

62. (8 gals., 235 mi.), (6 gal., 175 mi.)

63. ($300, $15), ($200, $10)

64. ($1600, $200), ($1500, $220)

65. (3 sec., 15 m./sec.), (4 sec., 20.1 m./sec.)

66. (5 ohms, 200 volts), (25 ohms, 300 volts)

67. (5 amp., 100 volts), (10 amp., 150 volts)

68. (4 sec., 64 ft./sec.), (2 sec., 32 ft./sec.)

*69. (5 milli sec., 10 flops), (1000 milli sec., 2000 flops)

*70. (3 microsec., 5 adds), (1000 microsec., 6000 adds)

71. Verify the distance formula by using the Pythagorean theorem.

72. Show that the distance from $(0.5(x_1 + x_2), 0.5(y_1 + y_2))$ to (x_1, y_1) is half the distance from (x_1, y_1) to (x_2, y_2).

73. If the average rate of change between (x_1, y_1) and (x_2, y_2) is equal to the average rate of change between (x_1, y_1) to (x_3, y_3) show that the three points lie on the same straight line.

74. Prove that, on a straight line, the average rate of change between any two points of the line is equal to the average rate of change between any two other points of the line.

75. Show that the distance between two points is given by $$d = \sqrt{(\Delta x)^2 + (\Delta y)^2}$$

76. See Problem 75. Show that $d = \sqrt{1 + m^2}\,|\Delta x|$, where $m = \Delta y/\Delta x$

I-I*   P R O B L E M   S E T

1. Suppose a circle of radius r about the origin $(0, 0)$ consists of all points (x, y) such that the *distance* of (x, y) to the origin is r. Prove that these points satisfy the equation $x^2 + y^2 = r^2$.

2. If a circle of radius r about the point (h, k) consists of all points (x, y) such that the *distance* of (x, y) to the (h, k) is r, prove that these points satisfy the equation $(x - h)^2 + (y - k)^2 = r^2$.

3. Prove that if $(0, 0)$ and $(a, 0)$ and (b, h) are the vertices of two adjacent sides of a parallelogram, then the fourth vertex is at $(a + b, h)$.

4. See Problem 3. Prove that $(0, 0)$, $(a, 0)$, (b, h) and $(a + b, h)$ are the vertices of a parallelogram.

5. Suppose that (a, b), (c, d) and (x, y) are the vertices of a right triangle with the right angle at (x, y) (where *none* of the x coordinates are equal and *none* of the y coordinates are equal). Prove that $(y - b)(y - d) + (x - a)(x - c) = 0$.

6. See Problem 5. Prove that $\dfrac{y - b}{x - a} = -\dfrac{x - c}{y - d}$.

7. See Problem 5. Suppose that m is the average rate of change from (a, b) to (x, y) and n is the average rate of change from (c, d) to (x, y). Prove that $mn = -1$.

9. See Problems 5 and 7. Prove that if $mn = -1$ then the points (a, b), (c, d), (x, y) are vertices of a right triangle.

*The binoculars logo indicates problems that *explore* topics found in future sections of the book or in calculus.

---

2   *GRAPHING TECHNOLOGY

A rule to trick the arithmetic.
—R. KIPLING

DISPLAY 1-3   Sharp EL9300

We use the notation F1 to represent the key on a calculator labeled *f*1. For example, on most calculators ⊠ represents the function key for multiplication. Multiplication is a *dyadic* operation, by which we mean that multiplication operates upon two numbers. This is in contrast to a *monadic* operation such as square root that operates on one number. This section examines the operation of calculators designed not only as powerful arithmetic devices, but also to display numerical patterns visually. These calculators are part of graphing technology.

Like scientific calculators, graphing calculators also calculate. But one look at their display panel suggests more than just calculation. The power of the graphing calculator is that it can represent numerical patterns visually.

In this section we introduce graphing calculators with emphasis on the Casio fx7700G, Sharp EL9200/EL9300 and Texas Instruments TI-81, TI-82 and TI-85. Other models from these manufacturers operate similarly. As this book was written, four manufacturers provided the bulk of graphing calculators to the education market: Casio, Sharp, Texas Instruments and Hewlett Packard. All these calculators provide excellent graphics. All share common concepts. Each has its advantages and disadvantages. Key stroke examples for the Casio, Sharp and Texas Instrument calculators were chosen because each is widely available, economical, and easy to use. The Hewlett Packard 48 series are powerful machines, as reflected by both their price and performance.

Most comments about Sharp or Texas Instrument calculators apply to the Casio and vice versa. In general, these comments also apply to other graphing

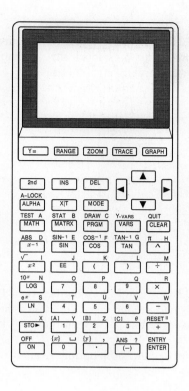

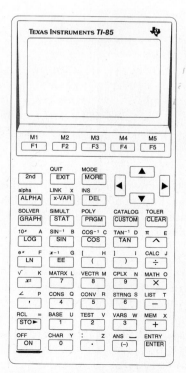

DISPLAY 1-4   TI-81 (above) and TI-85

calculators. For example, the TI-85 provides the power of the TI-81 plus additional features. Even though some TI-85 keys are different, menus on the screen coach you through key strokes similar to the TI-81. The TI-82 offers substantial improvements over the TI-81 while maintaining most of the same keys. However, for the purpose of this text, the primary difference is that the RANGE key has been replaced with the Window key. Although the TI-82 exhibits major improvements in its menus, other key changes are minor. Therefore we shall treat the TI-82 as identical to the TI-81. The Casio fx-6300G is an economical cousin to the fx-7700G. For our purpose, the Sharp EL9200 is identical to the Sharp EL9300.

With such diversity among the available machinery, we cannot detail the operation of each. The authority for the use of your graphing calculator is your user's manual. *The keystrokes in this text are simply to encourage your use of technology.* Neither this discussion nor any examples in this text replace a good user's manual.

On the other hand, user's manuals do not point out the similarities (or differences) among models, which this text will sometimes do. The sample key strokes in this text were not chosen to promote a particular brand name. Rather these examples are to encourage a student who has not used a graphing calculator to take an "I can do that" attitude.

Display 1-3 shows the key layout of the Sharp EL9300. Display 1-4 compares two Texas Instrument keyboards. Many keys perform similar actions among all the models. In the Casio fx7700G, the AC (All Clear) function key turns on the calculator and clears the display. (To conserve battery power, most calculator models automatically shut off if no key is pressed for several minutes. The contents of the calculator memory are not lost.) Thus, to turn on the Casio, press the key labeled AC. If necessary press AC again to clear the screen. The Texas Instruments and Sharp calculators have separate ON/OFF and CLEAR keys.

Each calculator has a method to turn the display on or off. In what other manners are they similar? First we present an overview of their common features, then we examine in detail some of their differences.

Each calculator features a *mode* option. Mode determines whether the calculator does calculations or displays graphs. Mode also controls the type of graphical display used and the manner for graphing formulas.

The *range* option allows you to select the display window for the graph. Set the left and right borders of the window using xMin and xMax. The bottom and top of the window follow from yMin and yMax. You may also space *tics* displayed on the x-axis and y-axis using xScl and yScl.

Each calculator provides a method for entering and saving formulas. You may retrieve a stored formula and *edit* it. From the stored formulas you may *select* those to display and instruct the calculator to graph them.

Graphing a formula on a calculator involves the following steps:

1. Select the *mode* of calculator operation.
2. Select the *range* for the graphing window.
3. *Enter* and store one or more formulas for graphing.
4. *Select* the formulas to graph.
5. Instruct the calculator to *display* the graph.

Now for some details and differences.

The Casio and Texas Instrument calculators have a MODE key. The equivalent Sharp key is SET-UP. The purpose of the key is to establish the conditions under which the machine will operate. For example, press the Casio MODE key and the following menu appears:

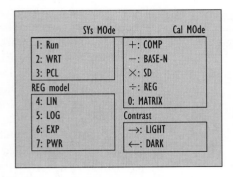

The first rule of pushing keys on a calculator is that keys must be pressed in sequence. Release the first key before pressing a second key. Do not press keys simultaneously. As an example on the Casio, follow MODE with the SHIFT key and an alternate menu follows.

In most cases the TI-85 MODE key simply extends the options available on the TI-81. For the TI-81 MODE key, we have

where the current default is in reverse video. A flashing bar (the *cursor*) shows where changes will be made. The arrow keys move the cursor. From the TI menu you should select Norm, Float 5, Rad, Function, Connected, Sequence, Grid Off, and Rect for the graphics introduced here.

The set-up menu for the Sharp is as follows:

The current selection is in reverse video. Choice *A* displays the current defaults. The up/down arrow keys move the selector from *A* through *H*. As each choice is highlighted, the boxed menu to the right changes to reflect new choices. For example, move the highlight to choice *B* or press the key labeled with a blue *B* [COS] key pad and the menu changes.

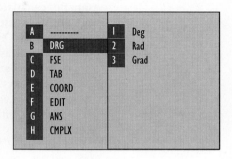

Use the righthand arrow key to move to the boxed choices then use the down arrow key to highlight *Rad*. Finally press [ENTER]. Even quicker, choose **Rad** by pressing [2].

Let us try some arithmetic. Select the MATH mode. For the Sharp, choose the key with the *icon* for arithmetic symbols [⊞] in the upper left corner. On the TI-81, choose [MATH]. On the Casio choose [MODE][+].

Clear your screen. We are ready to do calculations. Enter the expression $3 + 4 \times 5$ by pressing the key sequence [3][+][4][×][5]. Notice that $3 + 4 \times 5$ appears at the top of the screen. If you make a mistake typing, press the lefthand arrow key [←] to back up and type over your error. Use the backspace key [BS] on the Sharp. Once the formula is correct press [ENTER] ([EXE] on the Casio). Immediately, 23 appears on the next line at the righthand side of the display.

Now enter $15 \times 23 + -1$ [ENTER]. Be careful with the minus in the expression. The Casio calculator can distinguish when the [−] key means negative and when it means subtract. On the Sharp and TI calculators there are separate keys for subtraction [−] and negation [(-)]. *These keys are not interchangeable!*

Next try $3 + \times 5$ [ENTER]. You will receive a syntax error message. *Syntax* refers to the grammatically correct method for constructing an expression. Clear the screen to eliminate the error message.

Correct syntax varies among calculators. For example the Casio interprets $3 \times +5$ as multiply 3 times positive 5. But the Sharp and TI find this entry unacceptable and generate a syntax error.

Because of the limited number of keys available on each calculator, keys usually have several purposes. Labels above each key indicate the secondary tasks of the key. For example the TI-81 negative key [(-)] has two labels above it: ANS and ?. The Sharp [ENTER] also has an ANS label above it. Above the Casio fx-7700G [1] are the labels DRG and U. All four calculators have an [ALPHA] key. The purpose of this key is for access to the letters of the alphabet. To enter an *R* on a Sharp or TI-85 press the sequence [ALPHA][5] because the *R* is above the [5] key. On the Casio *R* is above [6]. On the TI-81, the *R* is above [×]. The purpose of the [ALPHA] key is to alter the action of a function key.

Another key that alters the action of the keys on the Sharp is the $\boxed{\text{2ndF}}$. This key is $\boxed{\text{2nd}}$ on the Texas Instrument calculators. The equivalent Casio key is $\boxed{\text{SHIFT}}$.

All these calculators have access to the variable ANS. Each time a calculation is done the resulting value is stored under ANS. Entering ANS into an expression is equivalent to inserting the results of the last calculation. For example, calculate 5 + 4, then ANS contains the result 9. If we enter 16 + ANS, the calculator displays the result 25 and stores 25 in ANS (replacing the previous value). For access to ANS on a TI press $\boxed{\text{2nd}}$ $\boxed{(-)}$. On the Sharp use $\boxed{\text{2nd}}$ $\boxed{\text{ENTER}}$. The Casio has an $\boxed{\text{ANS}}$ key.

Most calculators can display about seven or eight lines of sixteen to twenty characters. More information on the entry and editing of expressions can be found in your calculator user's manual. We shall not spend much time in this section with mundane calculations. What we want to introduce here is graphing.

Access your MODE menu. Be sure the default selections include REC-TANGULAR (RECT) and FUNCTION (FUNC). For example, to select a rectangular coordinate graphing mode on the Casio press the key sequence $\boxed{\text{MODE}}$ $\boxed{\text{SHIFT}}$. Under the Graph type menu choose + : REC for rectangular coordinations.

On the TI-85, press $\boxed{\text{2nd}}$ $\boxed{\text{MORE}}$. The current defaults are shaded. To change a default, use the arrow keys to move the cursor to your choice, then press $\boxed{\text{ENTER}}$. Press $\boxed{\text{EXIT}}$ when you have set all defaults. For the Sharp select $\boxed{\text{SET-UP}}$ $\boxed{\text{E}}$ $\boxed{1}$ to choose COORDinate XY.

Once we are in the correct mode, there are two things we need to do to assure a good graph: first, we must choose a *graphing window*, and second, we must indicate the relationship of *x* and *y* values. To choose a graphing window press the $\boxed{\text{RANGE}}$ key. The Sharp, Casio and TI-81 have a $\boxed{\text{RANGE}}$ key. On the TI-85, press $\boxed{\text{GRAPH}}$ first. The $\boxed{\text{F1}}$ key will be labeled Range. Select $\boxed{\text{RANGE}}$. The following screen will appear:

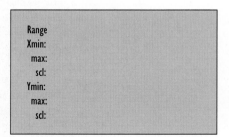

Next to Xmin: through scl: will be some values. Next to the Xmin: should be a blinking cursor. The cursor indicates where the next key stroke will display. For Xmin: enter -10 $\boxed{\text{ENTER}}$. The cursor moves to max:. Now enter 10 $\boxed{\text{ENTER}}$. Next to scl:, enter 1 $\boxed{\text{ENTER}}$. This will set the scale for "tic" marks on the *x*-axis to be 1 unit apart.

Note: on Sharp and TI calculators use the negation key $\boxed{(-)}$ to enter the range values. *The subtraction key is not the sign change key*. To exit Range on the TI-81 (window on the TI-82), press GRAPH or y=. On the TI-85 press EXIT. On the Sharp choose QUIT . When you press EXE for the last screen entry in the Casio, a second screen will appear. Do not change any values on this screen. The second screen has nothing to do with rectangular coordinates. Simply press EXE repeatedly until the second screen disappears.

Complete the screen as follows:

```
Range
Xmin:  -10.
  max:  10.
  scl:   1.
Ymin:  -10.
  max:  10.
  scl:   1.
INIT
```

We refer to this range setting as the *default window*.

Now we must enter an expression that gives *y* as a formula using *x*. For the Casio, select the GRAPH button. The word *Graph* followed by *y* = appears. Locate the key labeled XθT or XVAR. This is the variable key. Press the key and *x* will appear on the screen. To square *x* press the square button, SHIFT x². The Casio will store and recall up to six formulas under the names f1, f2, f3, f4, f5 and f6. See your user's manual.

On the TI-81, you press y= to enter a formula. The TI-81 will store up to four rules of pairing. These are labeled Y1 =, Y2 =, Y3 = and Y4 =. Gain access to any one of these, then enter the formula. The F1 button becomes y= on the TI-85. (The TI-82 allows up to ten formulas; the TI-85 allows up to ninty-nine formulas.) The TI recognizes implied multiplication but not inplied negation. Unlike the Casio, the TI does not graph a formula upon pressing EXE (ENTER) on the TI), rather you must mark the formulas you want graphed by moving the cursor to the equals and pressing ENTER . Selected formulas display the equals in reverse video. You may unmark a formula similarly. All of the marked functions will be graphed when you press the GRAPH key or take any other action that requires complete updating of the graph screen. y= XT x² GRAPH. You should produce the same graph as the Casio, assuming no other formula is selected except Y1 =.

Choose the graph icon and MENU for access to formulas in the Sharp. Selection of formulas for graphing is like the TI. An outstanding attribute of the Sharp calculators is that formulas may be entered as they appear in textbooks. The Sharp allows storage of up to ninty-nine formulas with up to four *active* formulas.

····································

EXAMPLE 1-4

Graph the following on the same coordinate system:

1. $y = 5 - 2x$ **2.** $y = x^2 - 3$ **3.** $y = (5 - 2x)^2 - 3$

Preliminary: Check your mode to be sure the defaults are correct. Set the range to the default window. Be sure the screen is clear.

Illustration 1:

$y = 5 - 2x$ (for the Casio fx-7700G)

Solution: Press $\boxed{\text{GRAPH}}$ $\boxed{5}$ $\boxed{-}$ $\boxed{2}$ $\boxed{\text{X}\Theta\text{T}}$ $\boxed{\text{EXE}}$. (See Figure 1-9.)

Notice that no $\times$ is needed between the 2 and x. These graphing calculators understand implied multiplication. On the Casio, press $\boxed{\text{G}\leftrightarrow\text{T}}$ (switch between graph and text). So long as we do not press the clear screen key CLS $\boxed{\text{F5}}$ $\boxed{\text{EXE}}$ the previous graph will remain on the screen.

Note: On the TI or Sharp, enter each graph separately under Y1=, Y2=, Y3=, Y4= and be sure to select each formula you want graphed.

Illustration 2:

$y = x^2 - 3$ (for the TI-85)

Solution: Press $\boxed{\text{GRAPH}}$ $\boxed{\text{F1}}$ (Display : y1=)
   Press $\boxed{\text{xVAR}}$ $\boxed{x^2}$ $\boxed{-}$ $\boxed{3}$ $\boxed{\text{ENTER}}$ $\boxed{\text{EXIT}}$ $\boxed{\text{F5}}$ (See Figure 1-10.)
Note: to deselect the graph, move the cursor to Y1= and press $\boxed{\text{CLEAR}}$.

Illustration 3:

$y = (5 - 2x)^2 - 3$ (for the Sharp)

Solution: Press $\boxed{\text{/\textbackslash/}}$ $\boxed{(}$ $\boxed{5}$ $\boxed{-}$ $\boxed{2}$ $\boxed{\text{X}\Theta\text{T}}$ $\boxed{)}$ $\boxed{\text{a}^\text{b}}$ $\boxed{2}$ $\boxed{\rightarrow}$ $\boxed{-}$ $\boxed{3}$ $\boxed{\text{/\textbackslash/}}$. (See Figure 1-11.)

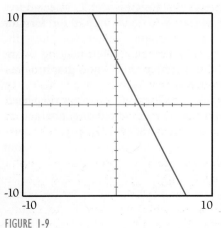

FIGURE 1-9

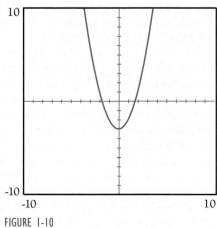

FIGURE 1-10

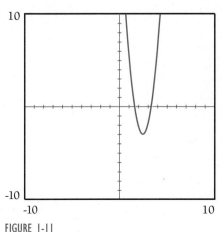

FIGURE 1-11

Although the graphs discussed came from pairs of numbers, we were not able to see the coordinates of each point as it was plotted. To do this we need to trace the x values of the graph. First clear previous graphs and formulas from the calculator's memory. Enter sin x as a formula for y1=. In Chapter 7 we will begin our study of Sin, but with the aid of the graphing calculator we can preview its graph.

Now press the Trace key. (On the Sharp, press an arrow key while the graph is displayed.) An x-shaped flashing cursor will appear on the graph. At the bottom of the screen we see the coordinates for the cursor. Pressing ⬅ decreases the value of x. Pressing ➡ increases the value of x. As you press ➡ , note that the cursor moves to the right, but clings to the graph as a path. Simultaneously the values of x and y change to reflect the current position of the cursor on the graph.

Move the cursor until x is very close to 3. Note that x skips values as we press the arrow key. This is because the cursor is jumping from point to point on the graph. The screen resolution is not fine enough to "tune in" to finer x values. However, we can *zoom in* on a region of the graph. Zooming in gives us better detail by magnifying the graph.

To zoom-in press the Zoom key. A menu should appear. Consult your user's manual concerning the various zoom options.

Numerous graphing programs for personal computers are available to improve your exploration of analysis. Among these are CalcPad, MicroCalc and MathCad. As with calculators, each program has its strong points and weaknesses.

GRAPHWINDOWS

GraphWindows is a copyrighted program written by the author of this book to work a computer like a graphing calculator. The publisher provides copies of *GraphWindows* to adopters of this text. Like a graphing calculator, *GraphWindows* allows you to enter an algebraic formula that expresses y in terms of x, then displays a graph of that relationship. *GraphWindows* is designed for an EGA/VGA video display, but it should run on any IBM compatible that supports at least the CGA graphics standard and 512 kb memory. Check with your instructor on the availability of *GraphWindows* and how to start the *GraphWindows* system.

Once *GraphWindows* is running on your computer, you may do many of the same actions available on the Casio, Sharp and Texas Instruments graphing calculators. You may enter formulas that relate y and x. You may choose a window for viewing the graph. Once *GraphWindows* has displayed the graph, you may zoom in to view graphic details or zoom out to get an overview. *GraphWindows* includes a Help system that can help you experiment with graphics in other ways.

You may store and use formulas simultaneously in graph windows, up to six active under the function keys labeled F1 , F2 , F3 , F4 , f and g . To enter a new formula or replace a previous formula for a particular key, hold down the Alt key and press the function key for which you want to store a formula For example, to store a formula under F2, press Alt + F2 . An entry box similar to that in Display 1-5 will pop up on the screen. You need not switch back and forth between the entry screen and the graphics screen.

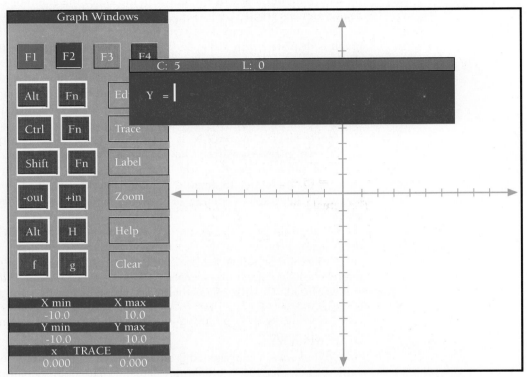

DISPLAY 1-5   Dialog box

The vertical bar, |, following the Y = is the cursor, where you will type your formula. The Y = is not displayed by *GraphWindows*. You may enter Y = yourself or leave it out, it is optional. Like the TI, Sharp and Casio, *GraphWindows* understands that placing numbers or variables and so forth next to each other, with no symbol in between, implies multiplication. Like the Casio, *GraphWindows* also distinguishes between subtraction and implied negation. Similar to the Sharp, *GraphWindows* allows you to write formulas pretty much as you write them in algebra.

If you feel more secure by typing in a multiplication symbol, remember that standard computer keyboards do not have a ⊠ key. Use an asterisk, *, if you want *explicit* multiplication. Although *GraphWindows* understands implicit negation, for *explicit negation* (rather than subtraction) you may use the tilde, ~. Like the Sharp, enter exponents in the standard algebraic manner, in superscripts. The up arrow key ↑ begins an exponent by moving the cursor up half a line. Type in your exponent expression. To move back down a half line, use the down arrow key ↓ . Consider the following examples.

EXAMPLE 1-5

Illustration 1:

These are the same as the formulas Example 1-4. Graph the following on the same coordinate system:

1. $y = 5 - 2x$ **2.** $y = x^2 - 3$ **3.** $y = (5 - 2x)^2 - 3$

Solution: Store $y = 5 - 2x$ by pressing [Alt] + [F1] . In the dialog box type

$$5 - 2x$$

Press [ENTER] .

Store $y = x^2 - 3$ by pressing [Alt] + [F2] . In the dialog box type

$$Y = x^2 - 3 \quad (\text{Remember, Y} = \text{is optional})$$

Press [ENTER] . The actual key strokes for entering this formula are

[y] [=] [x] [↑] [2] [↓] [−] [3] [ENTER]

Store $y = (5 - 2x)^2 - 3$ by pressing [Alt] + [F3] .

The actual key strokes for entering this formula are

[(] [5] [−] [2] [X] [)] [↑] [2] [↓] [−] [3] [ENTER]

To view any of these graphs, press the corresponding function key. For example, to graph $y = x^2 - 3$, press [F2] . If you want each graph on a separate screen, press [Alt] + [C] to clear the screen. If you do not clear the screen and press [F1] then the graph of $y = 5 - 2x$ will be drawn on top of the graph for $y = x^2 - 3$. See Figures 1-9 through 1-11 for the graphs of F1, F2 and F3.

Illustration 2:

Graph $y = e^{-x^2}$.

Solution: Store the formula under F4 by pressing [Alt] + [F4] . Enter the formula with the following keystrokes: [e] [↑] [−] [x] [↑] [2] [ENTER] . Press [Alt] + [C] to clear the screen. Then press [F4] to graph the formula. See Figure 1-12.

Because we have not chosen a window for this graph, it is drawn in the default window. In the default window, the x values range from Xmin = -10 to Xmax = 10, and the y values range from Ymin = -10 to Ymax = 10. This is "too far back" for a good view of F4. Zoom in for a better look.

To zoom in, press [Alt] + [+] . To zoom out, press [Alt] + [−] . Zoom the window in or out based on a predetermined scaling factor. To return to the default window size press [HOME] . The arrow keys [↑] [←] [→] [↓] allow you to scroll up, left, right or down with the graphing window. [Alt] + [A] toggles the xy axes off and on.

At this point we would like to paste a label on the graph. Press [SHIFT] + [F4] . In the upper lefthand corner of the graph appears a blurred

$$y = e^{-x^2}$$

Use arrow keys to reposition the label on the graph. When the label is where you want it, press [ENTER] . Pressing [ESC] cancels the pasting. If you have a graphics screen dump for your computer, you can usually send a copy of the graph to the printer by pressing [SHIFT] + [PrintScreen] .

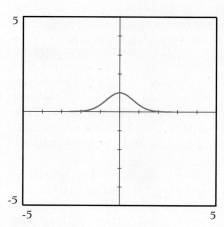

FIGURE 1-12   F4

Illustration 3:

Graph $y = -5 \sin x$.

Solution: Unlike graphing calculators, *GraphWindows* has no single key for Sin. You must type the name in yourself. (Normally, you should expect to type all the characters associated with these special names.) Therefore, to enter this formula, type it exactly as it appears. The actual key strokes are

[−] [5] [S] [i] [n] [Space] [x] [ENTER]

Notice the [Space] key to insert a space between Sin and x. The space is for readability and is not required. Sin is a monadic operator and *GraphWindows* expects an argument to follow Sin, the x positioned next to Sin. Once more, a space between the 5 and Sin is optional. But no operation symbol between the two indicates an implied multiplication. *GraphWindows* automatically assumes negation on the 5 and multiplication between the 5 and Sin. ■

Most of the graphs in this text were produced by graphing technology. If you have access to a graphing calculator or computer, you are encouraged to use technology to explore the examples and exercises in each section. The exercises at the end of this section include formulas to explore from future sections.

SUMMARY   Calculators

- The usual order of operations for graphing calculators is as follows:

  Grouping symbol   [(] [)]
  Monadic function keys   [(−)] [Ln] [Sin] [eˣ] [etc]
  Dyadic function keys (the usual order)
      exponents   [xʸ]
      multiply/divide (in order of entry)   [×] [÷]
      add/subtract (in order of entry)   [+] [−]

- Most graphing calculators have the ability to do standard computation like scientific calculators. Graphing calculators also have modes for displaying relationships between x and y variables visually.

- To graph an expression where y is given as a formula using x,

  1. Be sure that the calculator is in the correct mode (usually rectangular (XY), function, connected).

  2. Set the range to display an appropriate window for the graph.

  3. Enter the formula for y= using correct syntax.

  4. Select the formula for graphing.

  5. Execute the graph command.

- The Trace command is used to locate particular plotted points of the graph. If the Trace command is too coarse in choosing points of the graph, we can zoom in on a region of the graph by marking a BOX about the portion of the graph we wish to examine.

CALCULATOR	KEYS			
ACTION	CASIO FX7700G	SHARP EL-9300	TI-81/TI-82	TI-85
Set mode	[MODE] [SHIFT]	[SET-UP]	[MODE]	[2nd] [MODE]
Rectangular	[+]	[E] [1]	[↓↔] … [ENTER]	[↓↔] … [ENTER]
Polar	[−]	[E] [2]	[↓↔] … [ENTER]	[↓↔] … [ENTER]
Parametric	[×]	[E] [3]	[↓↔] … [ENTER]	[↓↔] … [ENTER]
Function			[↓↔] … [ENTER]	[↓↔] … [ENTER]
Radian	[SHIFT] [1] [F2] [EXE]	[B] [2]	[↓↔] … [ENTER]	[↓↔] … [ENTER]
Degree	[SHIFT] [1] [F1] [EXE]	[B] [1]	[↓↔] … [ENTER]	[↓↔] … [ENTER]
Connected	[5]	[graph] [MENU] [C] [1]	[↓↔] … [ENTER]	[↓↔] … [ENTER]
Dot	[6]	[graph] [MENU] [C] [2]	[↓↔] … [ENTER]	[↓↔] … [ENTER]
Sequence		[graph] [MENU] [D] [1]	[↓↔] … [ENTER]	[↓↔] … [ENTER]
Simultaneous		[graph] [MENU] [D] [2]	[↓↔] … [ENTER]	[↓↔] … [ENTER]
Grid			[↓↔] … [ENTER]	[↓↔] … [ENTER]
Exit		[QUIT]	[2nd] [CLEAR]	[2nd] [EXIT]
Calculation	[MODE] [+]	[calc]		
Set Range	[RANGE]	[RANGE]	[RANGE] / [Window]	[GRAPH] [F2]
Exit	[EXE] … [EXE]	[QUIT]	[2nd] [CLEAR]	[2nd] [EXIT]
Clear Screen	[F5] [EXE]	[CL]	[CLEAR]	[CLEAR]
Store Formula	[SHIFT] [0] … [F1] [1…6] [AC]	[graph] [MENU] [A] [1…4]	[y=] … [ENTER]	[GRAPH] [F2]
Select	[SHIFT] [0] [F3] [1…6]	[graph] [MENU] [A] [1…4]	[y=] [<] [ENTER]	[<] [CLEAR]
Deselect		[graph] [MENU] [A] [1…4]	[y=] [<] [ENTER]	[CLEAR]
Edit	[SHIFT] [0] [F2] [1…6]	[graph] [MENU] [A] [1…4]		
Draw Graph	[GRAPH] … [EXE]	[graph]	[GRAPH]	[GRAPH] [F5]
Zoom In	[F2] [F3]	[ZOOM]	[ZOOM] [2]	[GRAPH] [F3]
Zoom Out	[F2] [F4]	[ZOOM]	[ZOOM] [1]	[GRAPH] [F3]
Trace	[F1]	[<] or [>]	[TRACE]	[GRAPH] [F4]
System Reset	[RESET] (On Back) [F1]	[RESET] (On Back)	[2nd] [+] [2]	[2nd] [+] [F3]

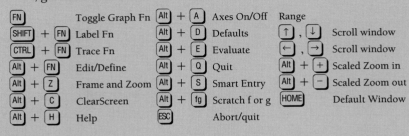

1-2* **E X E R C I S E S**

Approximate the following with a graphing calculator.

1. $(2.37)^5\sqrt{2}$

2. $\sqrt{3}(5.61)^3$

3. $\dfrac{(15 + 7.3)^{1.5}}{7.2}$

4. $\dfrac{(13 - 2.6)^{1.4}}{5}$

5. $\sqrt{7 + \sqrt[3]{5}}$

6. $\sqrt[3]{11 + \sqrt{6}}$

7. $\dfrac{25.3\sqrt{17^3}}{3.16 + 105.7}$

8. $\dfrac{(25.4 - 2.13479)^2}{16.2\sqrt[3]{5}}$

9. $\dfrac{1 + \dfrac{1}{3}}{1 - \sqrt[3]{5}}$

10. $\dfrac{2 - \dfrac{3}{7}}{1 + \sqrt[5]{3}}$

Use graphing technology to sketch the graphs of the following. For each new graph, reset the Range to the default window: $X\min = -10$, $X\max = 10$, $Y\min = -10$, $Y\max = 10$. (Check with your instructor for a method to produce any required "hard copies" of the graphs.)

11. $y = 2x$

12. $y = 3x$

13. $y = -2x$

14. $y = -3x$

15. $y = x^2$

16. $y = -2x^2$

17. $y = -x^2$

18. $y = 2x^2$

19. $y = 2x - 5$

20. $y = 3x + 2$

21. $y = 5 - 2x$

22. $y = 2 - 3x$

23. $y = x^2 - 3$

24. $y = 5 - 2x^2$

25. $y = x^2 - 2x - 6$

26. $y = 2x^2 - 5x - 3$

27. $y = 2x - 1$

28. $y = 1 - 2x$

29. $y = (2x - 1)^2$

30. $y = (1 - 2x)^2$

31. $y = \sin x$

32. $y = \cos x$

33. $y = \log x$

34. $y = \ln x$

35. $y = \tan x$

36. $y = \sqrt{x}$

37. $y = 1/\sin x$

38. $y = 1/\cos x$

39. $y = 10^x$

40. $y = e^x$

41. $y = 1/\tan x$

42. $y = x^2$

---

*The calculator logo indicates exercises in which a graphing calculator or GraphWindows can be used.

Sketch the graphs of the following pairs so that each pair is on a separate coordinate system. For each pair, compare the shapes.

43. $y = \cos x$
$y = \sin(x - \pi/2)$

44. $y = \sin x$
$y = \cos(x - \pi/2)$

45. $y = \ln x$
$y = e^x$

46. $y = \log x$
$y = 10^x$

47. $y = \sqrt{x}$
$y = x^2$

48. $y = \tan x$
$y = \sin x/\cos x$

*49. $y = \tan x$
$y = \dfrac{\sin x}{\cos x} + 1$

*50. $y = \log x$
$y = \dfrac{\ln x}{\ln 10} + 1$

1-2 PROBLEM SET

Use graphing technology to graph each of the following collections of equations in the same window. These exercises explore the graphs that will be discussed in Section 1-3.

1. **(a)** $y = x, y = 2x, y = 3x, y = 4x, y = 5x$
   (b) Describe the effect of a in the graph of $y = ax$.

2. **(a)** $y = -x, y = -2x, y = -3x, y = -4x, y = -5x$
   (b) Describe the effect of a in the graph of $y = -ax$.

3. **(a)** $y = x, y = -x, y = 2x, y = -2x$
   (b) Compare the graphs of $y = ax$ and $y = -ax$.

4. **(a)** $y = x, y = x/2, y = x/3, y = x/4, y = x/5$
   (b) Describe the effect of d in the graph of $y = x/d$.

5. **(a)** $y = -1x/(-2), y = 5x/3, y = 3x/5, y = 2x/9, y = 5x/9$
   (b) Describe the effect of a and d in the graph of $y = ax/d$.

6. **(a)** $y = x, y = x + 2, y = x - 3, y = x + 5, y = x - 4$
   (b) Describe the effect of b in the graph of $y = x + b$.

7. **(a)** $y = x + 5, y = 2x + 5, y = x + 4, y = -3x + 4$
   (b) Describe the effect of a and b in the graph of $y = ax + b$.

---

3 LINES

Algebra is but written geometry and geometry is but figured algebra.
—SOPHIE GERMAIN

Pair—from the Latin for "equal." See also the root for the words *par* (golf) and com*pare*.

Pairing of objects is a fundamental human activity. Pairing provides insight that leads to learning. Graphing is a visual method for examining trends in pairings, whether that pairing represents (increase in national debt, divorces per year) or (number of cigarettes smoked, incidence of lung cancer) or some other pairing. Pairings establish relations. Ordered pairs of numbers represent points on the plane and vice versa.

You are already familiar with collections of points that form familiar patterns on a plane. These patterns have names like *circle, line,* and *rectangle.*

In general, any collection of ordered pairs is a **relation**. Because a collection of ordered pairs of real numbers have a graphical representation in the Cartesian coordinate system, we refer to these two-dimensional graphs as *relations* although they simply represent the relations. An equation or inequality between two variables (say, x and y) algebraically describes points on a plane and thus represents a relation. We expect that algebraic patterns given by equations produce geometric patterns when graphed. One goal of analysis is the recognition of equation formats that describe well-known geometrical patterns.

Tom McKenzie is investigating the relationship of economic hardship to academic success. Tom has developed his own scales for measuring economic hardship and academic success. He collects data from students and plots the two measurements against each other on a Cartesian coordinate system. Over the years, Tom has noted a pattern to his "scatter diagram": the points lie "near" a straight line. If Tom is correct, he may write a formula that attempts to relate hardship to success. Such a formula might be used to predict the impact of economic hard times on academic success. The ability to predict is essential for planning and developing programs to enhance success.

For example, consider the algebraic patterns represented by the following equations:

$$5x + 2y = 17$$
$$x^2 - xy + y = 3$$
$$\frac{5x}{y} = 22$$
$$y^2 - x^3 = 0$$

It is usually not difficult to rewrite an equation so that all the variables are on the left side of the equation. Specifically, suppose that we have an equation containing x or y. Some people refer to such equations as *formulas*. For that reason, we use the notation $F(x, y) = c$ to represent the idea of a relationship between x and y. The letter c represents some constant value in the equation. At this point $F(x, y)$ represents an abbreviation of the "formula of two variables x and y."

Consider the equation $5x + 2y = 17$. Abbreviate $5x + 2y$ as $F(x, y)$. Then $F(x, y) = 17$. The notation $F(3, 1)$ abbreviates the statement "the formula F with x replaced by 3 and y replaced by 1." The equation is true for $F(3, 1)$:

$$F(3, 1) = 17$$
since $5(3) + 2(1) = 17$

We say the point in the plane $(3, 1)$ *satisfies* the equation $5x + 2y = 17$. Similarly, $(-1, 11)$ satisfies the equation. The point $(2, 5)$ does not satisfy the equation because $5(2) + 2(5) = 17$ is false.

DEFINITION 1-2
Graph of a Relation

> The **graph** of the relation $F(x, y) = c$ is the collection of all points (x, y) in the plane that satisfy the equation $F(x, y) = c$.

This section examines linear relations. The graph of a linear relation is a straight line.

The straight **line** is one of the simplest graphical relations. Graphs of lines fall into three classes: those perpendicular to the y-axis, those perpendicular to the x-axis and those perpendicular to neither.

A line perpendicular to the y-axis is a **horizontal line.** All the points on a horizontal line have the same second component (Figure 1-13). For a horizontal line the y value is constant. Therefore, an algebraic description of the points forming the displayed horizontal line is $y = 4$. In general, the graph of an equation of the form $y = c$ is a horizontal line.

Lines perpendicular to the x-axis are **vertical lines.** All the points on a vertical line have a constant first component (Figure 1-14). An equation to represent the given vertical line is $x = -3$. In general, the graph of an equation of the form $x = c$ is a vertical line.

An **oblique line** is perpendicular to neither axis. The equation of an oblique line is more complex than for vertical or horizontal lines. To discuss oblique lines, first recall the concept of average rate of change.

Figure 1-15 shows several triangles with sides labeled to represent the average rate of change of y with respect to x, $\Delta y / \Delta x$. Because the triangles are similar, the ratio $\Delta y / \Delta x$ remains constant for a given straight line no matter how points are chosen from the line. The **slope** of an oblique line is the constant average rate of change between points of the line.

DEFINITION 1-3
Slope

The slope m of a line is the average rate of change of y with respect to x for the line:

$$m = \frac{\Delta y}{\Delta x}$$

The slope of a horizontal line is 0. Vertical lines have no slope.

Consider the set of all points (x, y) for which the average rate of change from (x, y) to the fixed point (x_1, y_1) is the constant m. By definition of the rate of change, these points satisfy

$$\frac{y - y_1}{x - x_1} = m$$

Multiply both sides of the equation by $(x - x_1)$ to obtain Theorem 1-1, for a **linear equation**.

Theorem 1-1
Linear Equation

An equation representing the line through a given point (x_1, y_1) with a given slope $m = \Delta y / \Delta x$, where $\Delta x \neq 0$, is

$$(y - y_1) = m\,(x - x_1)$$

Replace m with $\Delta y / \Delta x$ and multiply both sides of the equation by Δx for a more general form of the equation: $\Delta x(y - y_1) = \Delta y(x - x_1)$.

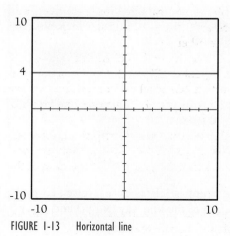

FIGURE 1-13   Horizontal line

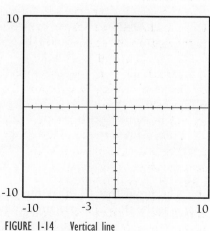

FIGURE 1-14   Vertical line

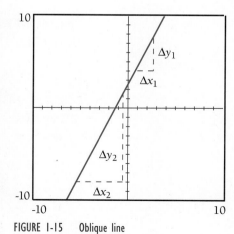

FIGURE 1-15   Oblique line

Algebraic manipulation of equations representing vertical lines or lines like those of Theorem 1-1 produces the general form of the equation of a line: $ax + by = c$.

The letters x and y represent real variables, and a and b represent real constants that are not both zero. The equation is of the first degree. In the plane, the graphs of such equations are straight lines. Equations expressible in this general format are **linear equations** in two variables.

The **x-intercept** of a line is the point (if any) where the line intersects the x-axis. Because all points on the x-axis have a y component of 0, so does the x-intercept of a line. Similarly, the **y-intercept** of a line is the point of its graph (if any) at which the x component is 0. Every oblique line intersects both the x-axis and the y-axis.

A linear equation may be written in forms other than the general form. Some forms for linear equations provide extra insight into the lines they represent. Using algebra to manipulate a linear equation into one of these forms gives insight in visualizing the graph. Analysis uses algebra to improve insight into geometry.

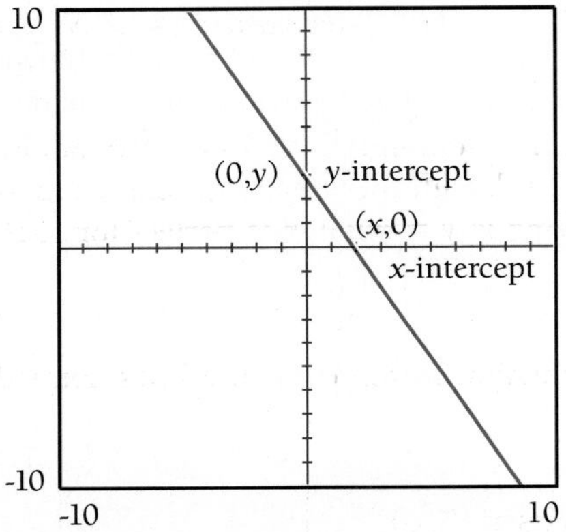

For example, $3x + 2y = 6$ is linear. The graph is a line containing infinitely many points. We cannot individually plot the infinite points of the graph. However, the graph becomes obvious from any two points that satisfy the equation. Rewriting the equation may help identify two such points. Try rewriting the equation by dividing by 6:

$$\frac{x}{2} + \frac{y}{3} = 1$$

Now the intercepts are easy to calculate. First, replace x with 0. Then $y/3 = 1$ so that $y = 3$. The y-intercept is $(0, 3)$. Similarly, $y = 0$ establishes $(2, 0)$ as the x-intercept. The result is shown in Figure 1-16. The steps in the preceding discussion establish the theorems for the **intercept form of a line**. Other formats for linear equations also yield graphical information, such as that for the **slope-intercept form of a line**.

FIGURE I-16   Intercepts of a line

Theorem I-2 Intercept Form	The intercept form of a linear equation is $x/a + y/b = 1$ where the x-intercept is $(a, 0)$ and the y-intercept is $(0, b)$.

Theorem I-3 Slope-Intercept Form	The slope-intercept form of a linear equation is $y = mx + b$ where the y-intercept is $(0, b)$ and the slope is m.

The proof that $(0, b)$ is the y-intercept is left as an exercise. To prove that m is the slope, choose any two points of the line, say (x_1, y_1) and (x_2, y_2).

Then these points will satisfy the equation. Thus, $y_1 = mx_1 + b$ and $y_2 = mx_2 + b$. Calculate the slope as the average rate of change for y compared with x.

$$\frac{\Delta y}{\Delta x} = \frac{y_2 - y_1}{x_2 - x_1} \qquad \text{(definition of slope)}$$

$$= \frac{(mx_2 + b) - (mx_1 + b)}{x_2 - x_1} \qquad \text{(substitute for y)}$$

$$= \frac{mx_2 - mx_1}{x_2 - x_1} \qquad \text{(simplify)}$$

$$= \frac{m(x_2 - x_1)}{x_2 - x_1} \qquad \text{(factor and reduce)}$$

$$= m \qquad \text{(because } x_2 \neq x_1. \text{ Why?)}$$

Equations solved for y, such as in the slope-intercept form, are among the easiest to graph using technology. The default for most graphing technology is to define $y =$ as a formula involving x: $y = mx + b$ is perfect for such entry.

EXAMPLE 1-6

Writing Linear Equations. Write a linear equation to represent the lines defined by the following information.

Illustration 1:

Represent a line perpendicular to the x-axis at $(5, 0)$.

Solution 1: $x = 5$, because the x coordinate is constant for a vertical line.

Solution 2: The line also passes through the point $(5, 1)$. Thus, $\Delta y = 1$ and $\Delta x = 0$. Based on the general form following Theorem 1-1,

$$\Delta x(y - y_0) = \Delta y(x - x_0)$$
$$0(y - 0) = 1(x - 5)$$
$$0 = x - 5$$
$$x = 5$$

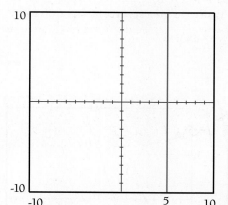

FIGURE 1-17 $x = 5$

Note: Because $\Delta x = 0$, the slope m for this line is undefined. The slope is not defined for vertical lines. This would not be suitable to use with graphing technology when the formula must be entered in the form $y =$. There is no y in this representation of a vertical line (Figure 1-17).

Illustration 2:

Represent a line perpendicular to the y-axis at $(0, -4)$.

Solution 1: $y = -4$, because the y coordinate is constant for a horizontal line.

Solution 2: The line also contains $(1, -4)$. Then $\Delta y = 0$ and $\Delta x = 1$. Thus,

$$m = \frac{0}{1} = 0$$

$$y - (-4) = 0(x - 0)$$

$$y = -4$$

Because $\Delta y = 0$, the slope for this line is $m = 0$. The slope of a horizontal line is 0 (Figure 1-18).

Illustration 3:   Represent a line through $(5, 1)$ with slope $m = -\frac{2}{3}$.

Solution: Because $m = -\frac{2}{3}$,

$$(y - 1) = -\frac{2}{3}(x - 5)$$

$$3y - 3 = -2x + 10$$

General form: $2x + 3y = 13$ (see Figure 1-19)

Intercept form: $\dfrac{x}{13/2} + \dfrac{y}{13/3} = 1$

Slope-intercept form: $y = -\frac{2}{3}x + \frac{13}{3}$

This last form is appropriate for use with graphing technology. For that reason we limit our answers to this form.

Illustration 4:   Represent a line through $(0, 5)$ with slope $m = 3$.

$$(y - 5) = 3(x - 0)$$

$$-3x + y = 5$$

Slope-intercept form: $y = 3x + 5$

See Figure 1-20.   —

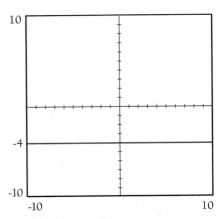

FIGURE 1-18 $y = -4$

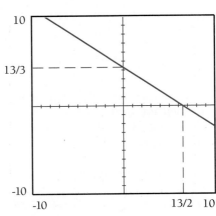

FIGURE 1-19 $2x + 3y = 13$

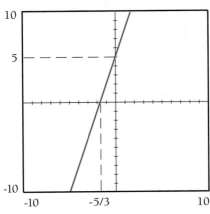

FIGURE 1-20 $y = 3x + 5$

For any given linear equation in two variables, the graph is a line. Once we know two points of the line, the remaining points of the line are obvious. Simply sketch a straight line segment through the two points to represent the entire graph. Even quicker, use graphing technology.

Graphing Linear Equations in the xy Plane. Sketch the graph of the following linear equations.

EXAMPLE 1-7

Illustration 1:

$$x = 3$$

Solution: Choose any two points with x coordinates of 3, say, (3, 2) and (3, -1). Draw a line through the points (Figure 1-21).

Illustration 2:

$$y = 2$$

Solution: Choose points, say (3, 2) and (1, 2), each of which has a y coordinate equal to 2. Draw a line through the points, as in Figure 1-22.

Note: Sometimes a linear equation has only one explicit variable. The graph of this equation in the plane is one of the following lines:

(a) Horizontal $(y = k)$ [through $(0, k)$]   or
(b) Vertical $(x = h)$ [through $(h, 0)$]

Linear equations expressible in the form $ax + by = c$, where $a \neq 0$ and $b \neq 0$, represent oblique lines.

Illustration 3:

Represent the line where $3x - 2y = 6$.

Solution 1: The graph is an oblique line. We need two points. Choose arbitrary values for x and substitute these into the equation to determine corresponding

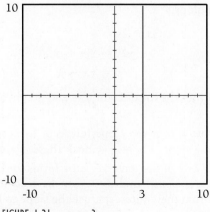

FIGURE 1-21 $x = 3$

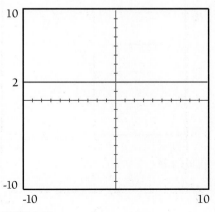

FIGURE 1-22 $y = 2$

values for y. Choose a minimum of three values for x to provide a check against arithmetic error: the calculated points must lie on a line.

Let $\qquad x = 1,$

$$3(1) - 2y = 6$$
$$-2y = 3$$
$$y = -\tfrac{3}{2}$$

Thus, $(1, -\tfrac{3}{2})$ is a point on the line.

Let $\qquad x = 2,$

$$3(2) - 2y = 6$$
$$-2y = 0$$
$$y = 0$$

The x-intercept is $(2, 0)$. Is this just luck?

Let $\qquad x = 0$ (try for the y-intercept),

$$3(0) - 2y = 6$$
$$-2y = 6$$
$$y = -3$$

The y-intercept is $(0, -3)$. We need not restrict ourselves to choosing only x values.

Let $\qquad y = 0$ (try for x-intercept),

$$3x - 2(0) = 6$$
$$3x = 6$$
$$x = 2$$

Again the x-intercept is $(2, 0)$. See Figure 1-23.

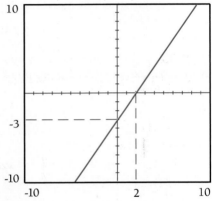

FIGURE 1-23 $3x - 2y = 6$

The easiest arithmetic occurs in substituting 0 for one of the variables. These values in turn produce the coordinates of an intercept.

Solution 2: Rewrite the equation in slope-intercept form and use graphing technology.

$$3x - 2y = 6$$
$$-2y = -3x + 6$$
$$y = \tfrac{3}{2}x - 3$$

Enter the formula for $y =$ and graph. See Figure 1-24.

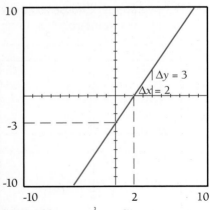

FIGURE 1-24 $y = \tfrac{1}{2}x - 3$

Note: From the slope-intercept form read that the y-intercept is $(0, -3)$ and the slope is $m = \tfrac{3}{2}$. Check these against the graph. The slope-intercept form is one of the most versatile forms for linear equations. With it, the y-intercept (first point) and slope can be read from the equation. Plot the y-intercept. From the y-intercept, use the slope to locate a second point.

Illustration 4:
$$y = -\tfrac{3}{5}x + 1$$

The equation is in slope-intercept form. The y-intercept is $(0, 1)$. The slope is $m = -\tfrac{3}{5}$. Let $\Delta y = -3$ and $\Delta x = 5$ for plotting a second point from $(0, 1)$. Finally, calculate the x-intercept by substituting 0 for y. Or with graphing technology obtain Figure 1-25.

$$0 = -\tfrac{3}{5}x + 1$$

where $x = \tfrac{5}{3}$; that is, the x-intercept is $\tfrac{5}{3}$. Check this with the Trace feature of your technology.

Illustration 5:

Represent the line where $3y - 6x = 2$.

Solution: Solve for y to rewrite the equation in slope-intercept form:

$$y = 2x + \tfrac{2}{3} \qquad\qquad\qquad\text{(Figure 1-26)}$$

The slope is $m = \tfrac{2}{1}$. The y-intercept is $(0, \tfrac{2}{3})$. Obtain the x-intercept by substituting 0 for y. To graph by hand, the original equation is easier than the slope-intercept form:

$$3(0) - 6x = 2$$
$$x = -\tfrac{1}{3}$$

Thus, $(-\tfrac{1}{3}, 0)$ is the x-intercept. ∎

Annotations are notes attached to a document to help explain it. Although technology produced Figure 1-26, the labels for the x-intercept and y-intercept are annotations. Such annotations do not appear automatically with graphing technology. Usually they are handwritten onto the graph.

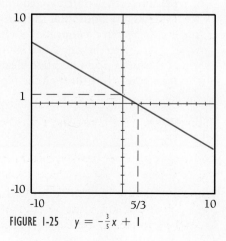

FIGURE 1-25　$y = -\tfrac{3}{5}x + 1$

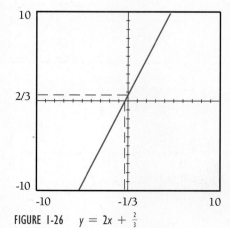

FIGURE 1-26　$y = 2x + \tfrac{2}{3}$

SUMMARY

To write an equation for a line

- Given two points (x_1, y_1) and (x_2, y_2), an equation for the line through these two points is

$$\Delta x(y - y_1) = \Delta y(x - x_1)$$

where $\Delta x = x_2 - x_1$ and $\Delta y = y_2 - y_1$.

- Given a point (x_1, y_1) and a slope $m = \Delta y/\Delta x$ for a nonvertical line, an equation for the line is $y = m(x - x_1) + y_1$.

Formats for an equation representing a line:

- The general form of a line is $ax + by = c$.

    (a) $a \neq 0$ and $b \neq 0$ intersects both axes. Determine two points of the line, and sketch a line through them.
    (b) $a \neq 0$ and $b = 0$ ($x = h$). Draw a vertical line through $(h, 0)$.
    (c) $a = 0$ and $b \neq 0$ ($y = k$). Draw a horizontal line through $(0, k)$.

- The intercept form of a line is $x/a + y/b = 1$. Locate the x-intercept $(a, 0)$ and the y-intercept $(0, b)$ and sketch a line through them.
- The slope-intercept form of a line is $y = mx + b$ (best for graphing). Use graphing technology or

    (a) Identify the y-intercept $(0, b)$.

    (b) Identify the slope $m = \Delta y/\Delta x$. Select appropriate values for Δy and Δx.
    (c) Plot the y-intercept $(0, b)$.
    (d) From the y-intercept count horizontally Δx units and vertically Δy units; plot the point at the resulting location. (If Δx is negative, count to the left $|\Delta x|$ units. If Δx is positive, count to the right. In a similar manner for Δy, positive means count up and negative means count down.
    (e) Draw a line through the plotted points.

- Use algebraic manipulation to convert an equation among these forms.

1-3   E X E R C I S E S

Write a linear equation for the line through the following pairs of points.

1. $(3, 5), (3, 7)$ **2.** $(5, 4), (5, -1)$

3. $(-4, 2), (-4, -3)$ **4.** $(-3, -2), (-3, -1)$

5. $(5, 1), (2, 1)$ **6.** $(-4, 3), (-1, 3)$

7. $(2, -3), (4, -3)$ **8.** $(5, -1), (17, -1)$

9. $(3, -4), (-5, 4)$ **10.** $(7, 1), (-3, 11)$

11. $(4, 7), (3, -2)$ **12.** $(-2, 3), (-1, -1)$

13. $(5, 3), (-2, 4)$ **14.** $(-2, 1), (7, -3)$

15. $(\frac{1}{2}, -\frac{1}{3}), (2, -3)$ **16.** $(0.4, -1.3), (7, -1)$

17. $(0, 0), (k, km), km \neq 0$ **18.** $(a, a^2), (b, b^2), a^2 \neq b^2$

19. $(a, b^2), (b, a^2), a^2 \neq b^2$ **20.** $(a, b), (a + 1, b + m),$ $m \neq 0$

Write a linear equation representing the line through each of the following points and with the indicated slope:

21. $m = 2, (0, 3)$ **22.** $m = 3, (0, -1)$

23. $m = 3, (0, -2)$ **24.** $m = 2, (0, 4)$

25. $m = \frac{1}{2}, (-1, 3)$ **26.** $m = -\frac{3}{4}, (8, -2)$

27. $m = -\frac{3}{5}, (2, -4)$ **28.** $m = \frac{1}{3}, (3, -5)$

29. $m = 0, (1, 2)$ **30.** $m = 0, (3, -5)$

    (a) Rewrite each of the following linear equations in the slope intercept form by solving the equation for y.

    (b) Determine the slope of the line for each equation.

    (c) Give the x-intercept (if any) for each of the lines.

    (d) Give the y-intercept (if any) for each of the lines.

    (e) Use graphing technology to graph each one.

31. $2y + 3x = 8$ **32.** $3y + 6x = 9$

33. $3y - 9x = 6$ **34.** $2y - 8x = 6$

35. $6x + 3y = 5$ **36.** $3x + 6y = 4$

37. $5x - 2y = 3$ **38.** $3x - 4y = 2$

39. $\frac{1}{2}x - \frac{2}{3}y = \frac{1}{4}$ **40.** $\frac{3}{5}x + \frac{1}{2}y = \frac{1}{3}$

Graph each of the following.

41. $2y + 6x = 8$ **42.** $3y + 6x = 9$

43. $\dfrac{x}{2} + \dfrac{y}{3} = 1$ **44.** $\dfrac{x}{3} + \dfrac{y}{2} = 1$

45. $y = \frac{1}{3}x - 3$ **46.** $y = -\frac{2}{3}x + 5$

47. $y = -\frac{3}{5}x + 7$ **48.** $y = \frac{3}{4}x - 2$

49. $y = 5x - 2$ **50.** $y = -3x + 5$

51. $x = 5$ **52.** $y = -2$

53. $y = -3$ **54.** $x = 4$

55. $6x + 3y = 5$ **56.** $3x + 6y = 4$

57. $5x - 2y = 3$ **58.** $3x - 4y = 2$

59. $\frac{1}{2}x - \frac{2}{3}y = \frac{1}{4}$ **60.** $\frac{3}{5}x + \frac{1}{2}y = \frac{1}{3}$

For each of the following lines, identify the coordinates of the intercepts. Determine whether the slope is positive, negative, 0 or undefined. Refer to the Summary to write an equation to represent each line.

61. 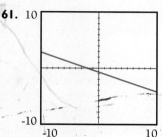   **62.**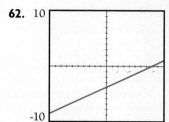

63. 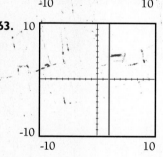   **64.**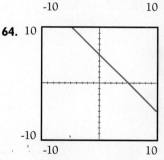

65. 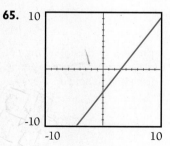   **66.**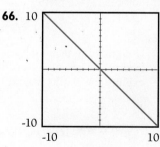

67. 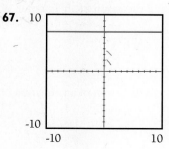   **68.**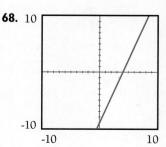

69.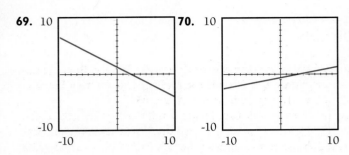

70.

71. Prove that $(0, b)$ is the y-intercept of $y = mx + b$.

72. Prove that $(0, b)$ is the y-intercept of $x/a + y/b = 1$.

73. Prove that $(a, 0)$ is the x-intercept of $x/a + y/b = 1$.

74. Prove that a vertical line has no slope.

75. Prove that, if $y = mx + b$ is parallel to $y = nx + c$, then $m = n$.

Note: two lines in a plane are parallel if they do not intersect. Two lines are perpendicular if they intersect at a right angle. If two sides of a triangle coincide respectively with two perpendicular lines, then the triangle is a right triangle.

76. Prove that $y = mx + b$ and $y = mx + c$, $b \neq c$, are parallel lines.

*77. Prove that $y = mx$ and $y = (-1/m)x$ are perpendicular lines.

*78. Prove that, if $y = mx$ and $y = nx$ are perpendicular lines, then $mn = -1$.

Data for the Municipal Development Corporation (MDC Fund).

YEAR t	1988	1989	1990	1991	1992
Fund $	21000	30000	39000	50000	65000

79. Graph the data for the MDC Fund using points of the form $(t, \$)$.

Hint: choose a window where $1987 < t < 1993$. Scale the $ axis in thousands of dollars.

80. Examine the graph in Problem 79. Do the points appear to be "close" to a line?

81. Determine the average change per year (slope) in fund assets by using the first and last year of the data.

82. Use the first and last years of the data to calculate the midpoint of a line segment connecting them. The resulting year should be 1990. How close is the midpoint fund value to the listed value for 1990?

I-3   **P R O B L E M   S E T**

Use graphing technology to graph each of the following collections of equations in the same window. These exercises explore the graphs that will be discussed in Section 1-4.

1. (a) $y = \sqrt{4 - x^2}$, $y = \sqrt{9 - x^2}$,

   $y = \sqrt{16 - x^2}$, $y = \sqrt{25 - x^2}$

   (b) Describe the effect of a in the graph of an equation of the form $y = \sqrt{a^2 - x^2}$.

2. (a) $y = -\sqrt{4 - x^2}$, $y = -\sqrt{9 - x^2}$,

   $y = -\sqrt{16 - x^2}$, $y = -\sqrt{25 - x^2}$

   (b) Describe the effect of a in the graph of an equation of the form $y = -\sqrt{a^2 - x^2}$.

3. (a) $y = \sqrt{9 - x^2}$, $y = -\sqrt{9 - x^2}$,

   $y = \sqrt{16 - x^2}$, $y = -\sqrt{16 - x^2}$

   (b) Describe the shape of graph formed by the pair of equations $y = \sqrt{a^2 - x^2}$ and $y = -\sqrt{a^2 - x^2}$.

4. (a) $y = 3 - \sqrt{9 - (x - 5)^2}$, $y = 3 - \sqrt{9 - (x - 5)^2}$

   (b) Describe the role of h and k in the graph of $y = k - \sqrt{r^2 - (x - h)^2}$, $y = k - \sqrt{r^2 - (x - h)^2}$

| 4 | CIRCLES AND TRANSLATIONS |

Section 1-3 used the concept of average rate of change to develop linear equations. In this section, the distance formula is the tool used to derive a formula for one of the most familiar of all graphic shapes, the circle.

To draw a circle in a plane, you first locate a reference point called the **center** of the circle. This point provides the primary location information about the circle. Although the center is not an actual point of the circle, the center positions the circle in the plane.

To complete the description of a circle about a chosen center, we must decide how large the circle is to be. The size of a circle is measured by its radius. The **radius** of a circle measures the distance from the center of the circle to any point on the circle. Because the radius measures distance, it may not be negative. The following definition summarizes the physical attributes of a circle.

DEFINITION 1-4
Circle

> A **circle** is the set of points in a plane whose distance from a given point in the plane is constant. The given point is the center of the circle. The constant distance is the radius.

The distance formula provides an immediate conversion of the definition of a circle to a representative equation in the Cartesian coordinate system. Let (x, y) represent any point of the circle and (h, k) be the center. With r as a radius, apply the distance formula:

$$\sqrt{(x - h)^2 + (y - k)^2} = r$$

Because the radicand and the radius represent nonnegative numbers, square both sides of the equation to avoid radicals.

$$(x - h)^2 + (y - k)^2 = r^2$$

This is the general form of the equation for a circle. An equation of this form not only indicates that the graph will be a circle, but also specifies the center (h, k) and the radius, r. The equation representing a circle is of the second degree (**quadratic**; see Section 0-4) in both variables. Equations for lines were of the first degree (linear) in both variables. Because the general equation for a circle is not solved for y, the format is not useful with graphing technology.

EXAMPLE 1-8

Illustration 1:

Graphing Circles From Their Equations. Identify the center and radius for the circles represented by each of the following equations. Use this information to sketch the graph.

$$(x - 3)^2 + (y + 2)^2 = 25$$

Solution: The equation is in graphing format. Because $r^2 = 25$, and the radius represents a distance, choose the positive square root of 25 as the radius:

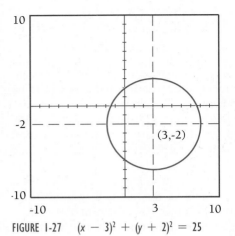

FIGURE 1-27 $(x - 3)^2 + (y + 2)^2 = 25$

$r = 5$. An easy way to get the center is to note that in the graphing form the center occurs where the "squared" terms equal 0. Therefore, one coordinate of the center is given by

$$(x - 3)^2 = 0$$
$$x - 3 = 0$$
$$x = 3$$

Similarly, for the y coordinate,

$$(y + 2)^2 = 0$$
$$y + 2 = 0$$
$$y = -2$$

Therefore the center is (3, -2) and the radius is 5 (Figure 1-27). With practice, you can "read" the coordinates of the center as easily as the radius.

A word of warning concerning graphing technology to produce the graphs of circles is in order: the window displaying the graph may produce some distortion. The distortion is the result of differences in scale on the x- and y-axes. As a result, circles may not appear "round."

Illustration 2:

$$(x + 7)^2 = 16 - y^2$$

Solution: The equation is not in the graphing format for a circle. We cannot be sure the equation represents a circle until we "fit it" to the graphing format. Add y^2 to both sides of the equation and obtain

$$(x + 7)^2 + y^2 = 16$$

The radius is 4 and the center is (-7, 0), see Figure 1-28.

Illustration 3:

None of the following equations represents a circle:

$$(x + 7)^2 + y = 16$$
$$(x + 7) + y^2 = 16$$
$$(x + 7) + y = 16 \qquad \text{(a line)}$$
$$(x + 7)^2 - y^2 = 16$$
$$(x + 7)^2 = y^2 + 16$$

Illustration 4:

$$x^2 - 4x + y^2 - 7y - \tfrac{3}{4} = 0$$

Solution: Although the equation is not in the graphing form for a circle, it might represent a circle because of the positive quadratic terms on the lefthand side of the equation. We will try to put it into the graphing form. The presence of the linear terms suggests completing the square.

$$x^2 - 4x + y^2 - 7y - \tfrac{3}{4} = 0$$

$$x^2 - 4x + \underline{\hspace{1cm}} + y^2 - 7y + \underline{\hspace{1cm}} = \tfrac{3}{4}$$

$$(x^2 - 4x + 2^2) + (y^2 - 7y + \left(\tfrac{7}{2}\right)^2) = \tfrac{3}{4} + 2^2 + \left(\tfrac{7}{2}\right)^2$$

$$(x - 2)^2 + \left(y - \tfrac{7}{2}\right)^2 = \tfrac{68}{4}$$

The radius of the circle is $\sqrt{\frac{68}{4}}$ (or $\sqrt{17}$) and the center is $\left(2, \frac{7}{2}\right)$. Locate the center at (2, 3.5). A calculator approximates the radius as 4.12310563. Graphs are approximations at best. Use the approximations to sketch the graph, then annotate the graph with precise values. See Figure 1-29.

Illustration 5:

Writing the Equation for a Circle. Write an equation for a circle with radius 7 and center at (-4, 1).

Solution: Use the graphing form of a circle and substitute the values from the radius and center:

$$[x - (-4)]^2 + (y - 1)^2 = 7^2$$
$$(x + 4)^2 + (y - 1)^2 = 49$$

See Figure 1-30. ■

Other parts of a circle have names. An *arc* is any connected portion of the curve that forms the circle. The *circumference* of the circle is the length of the complete arc of the circle, its perimeter.

The *diameter* of a circle is twice the length of the radius. As circumference measures the distance "around" a circle, diameter measure the "width" across a circle.

One of the best known geometric relationships is the relation of the circumference of a circle to its diameter. Since all circles are similar, in that they all have the same shape, the corresponding parts of any two circles are in proportion. As a result, for any size circle, the ratio of the circumference to the diameter is constant. This constant value has been approximated as 3.141592654 . . . Because the constant's representation is an infinite decimal with no repeating pattern of digits, we refer to it by the Greek letter π (pronounced "pie" as in apple pie). Now we have

$$\frac{C}{D} = \pi$$

so that $C = \pi D$

where C is the circumference and D is the diameter of a circle.

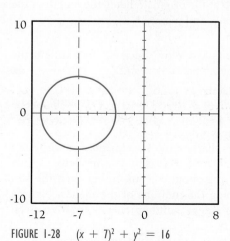

FIGURE 1-28 $(x + 7)^2 + y^2 = 16$

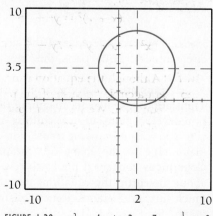
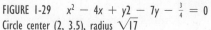

FIGURE 1-29 $x^2 - 4x + y2 - 7y - \frac{3}{4} = 0$
Circle center (2, 3.5), radius $\sqrt{17}$

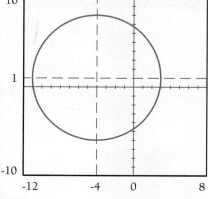

FIGURE 1-30 $(x + 4)^2 + (y - 1)^2 = 49$

EXAMPLE 1-9

Illustration 1:

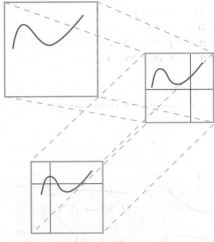

FIGURE 1-31   Translation

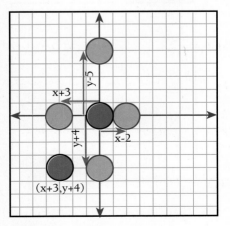

FIGURE 1-32   Multiple translations

Calculations Connected with Circles

Write an equation representing a circle with a diameter whose endpoints are $(2, -3)$ and $(8, 5)$.

Solution: We need to know the center and radius to write the equation. The center is at the midpoint of the diameter. From Section 1-1, use the midpoint formula to get the coordinates of the center:

$$\left(\frac{2+8}{2}, \frac{-3+5}{2}\right) = (5, 1).$$

The radius is the distance from the center to either endpoint. Apply the distance formula:

$$r = \sqrt{(8-5)^2 + (5-1)^2}$$
$$r = 5$$

Now substitute these values into the graphing format for a circle:

$$(x-5)^2 + (y-1)^2 = 5^2$$

You may check this solution by substituting the endpoints into the equation to verify that they satisfy the equation.

Illustration 2:

See Illustration 1. Determine the circumference of the circle.

Solution: Because the radius of the circle was 5, the circumference is

$$C = 2\pi r \qquad\qquad (D = 2r)$$
$$= 2\pi(5)$$
$$= 10\pi \qquad\qquad (\text{approximately } 31.4159\ldots)$$

Illustration 3:

Consider the circle of radius 1, centered at the origin. What is the length of the arc of the circle "cut off" by the positive x- and y-axes?

Solution: The circumference of the circle is 2π. Why? Because the axes cut the circle into four equal parts, the length must be $2\pi/4 = \pi/2$. ▬

A **translation** of a graph repositions a graph in the plane (Figure 1-31). Translation is similar to imagining the graph at a different location. This section develops translation as shifting our window viewpoint.

One reason for using translations is that it reinforces formula patterns as a major clue to the shape of a graph. Another reason is simplicity. When Copernicus conceived the earth revolving about the sun rather than the sun about the earth, the equations describing the motions of the planets became much simpler. A translation establishes a new "window" viewpoint for a given graph. See Figure 1-31.

A translation is a special case of a more general concept known as a *trans-*

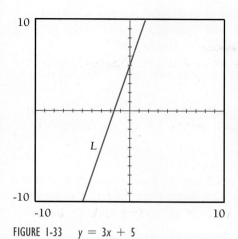

FIGURE 1-33 $y = 3x + 5$

formation. Chapter 2 introduces transformations. For now consider the graph in Figure 1-32.

Observe that, in Figure 1-32, the same shape is in different positions. Translation preserves the graph's shape while relocating it. We translate the graph given by $F(x, y) = c$ by altering the equation. For example, replace x with $x - 2$: $F(x - 2, y) = c$. To satisfy the new equation the value substituted for x must be 2 more than x in the original equation. As a result, the new graph appears 2 units to the right of the original. Replacing x with $x + 3$ causes the x values to appear 3 units sooner and relocates the graph 3 to the left. Similarly, replace y with $y + 4$ to shift the graph down 4 units. Replace y with $y - 5$ to move the graph up 5 units.

Refer to the points marked in Figure 1-32. For each point, the change in coordinate value is constant under a given translation. The definition of *translation* summarizes these observations.

DEFINITION 1-5
Translation

Replacing x with $(x - h)$ in $F(x, y) = c$ produces an equation whose graph is the same shape as the original graph, but translated horizontally h units.

Replacing y with $(y - k)$ in $F(x, y) = c$ produces an equation whose graph is the same shape as the original graph, but translated vertically k units.

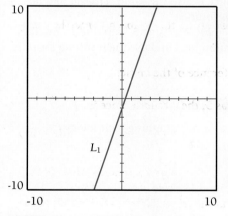

FIGURE 1-34 $y = 3(x - 2) + 5$

Note: do not let the subtraction in the definition distract you. For example to shift a graph 4 units *right* requires a substitution in which 4 replaces h: $(x - 4)$. To shift a graph 5 units *down* requires a substitution of -5 for k: $[y - (-5)] = (y + 5)$. Also, $(x + 2) = [x - (-2)]$ and thus translates a graph 2 units *left*. Finally $(y - 7)$ translates a graph 7 units *up*. In each case, the translation is in the opposite direction than you might first expect.

EXAMPLE   1 - 10

Illustration 1:

Graphing Translations. Graph each of the following. Translate the graph as indicated. Graph the translation.

$$y = 3x + 5 \qquad\qquad (L)$$

Label the line (L) for easy reference (Figure 1-33). Translate (L) 2 units horizontally. Name the translation (L_1).

Solution: Replace x with $(x - 2)$ to obtain (L_1). See Figure 1-34.

$$y = 3(x - 2) + 5 \qquad\qquad (L_1)$$
$$y = 3x - 1 \qquad\qquad (L_1)$$

Note: Another way to visualize translation is to imagine that the graphical window captures the shape of the graph and then superimposes that shape at another position in the coordinate system, as in Figure 1-35.

Illustration 2:

Refer to Illustration 1. Translate (L) -6 units vertically. Call this translation (L_2).

Solution: Replace y with $[y - (-6)]$ to obtain (L_2), see Figure 1-36.

$$[y - (-6)] = 3x + 5 \qquad (L_2)$$
$$y = 3x - 1 \qquad (L_2)$$

Note: Sometimes it is difficult to distinguish among translations. Note that the horizontal translation L_1 in Illustration 1 produces the same graph as the vertical translation L_2 in Illustration 2.

Illustration 3:

Consider a circle at the origin $(0, 0)$ with radius r. The equation is $x^2 + y^2 = r^2$. See Figure 1-37. Translate the circle so that the center is relocated to (h, k). Substitute $x - h$ for x and $y - k$ for k.

$$(x - h)^2 + (y - k)^2 = r^2$$

Although the translation relocates the graph of the circle, the radius remains unchanged. As a result, the circumference is also the same. *Translations do not alter the size or shape of a figure, only the location.*

How can we graph $(x - h)^2 + (y - k)^2 = r^2$ using graphing technology? Our current method is to solve the equation for y and use the resulting formula.

$$(y - k)^2 = r^2 - (x - h)^2$$
$$y - k = \pm \sqrt{r^2 - (x - h)^2}$$
$$y = k \pm \sqrt{r^2 - (x - h)^2}$$

This last equation implies two formulas for our graphing technology:

$$y = k + \sqrt{r^2 - (x - h)^2}$$
$$y = k - \sqrt{r^2 - (x - h)^2}$$

Graph both in the same window to obtain one circle.

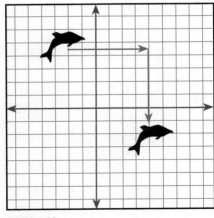

FIGURE 1-35

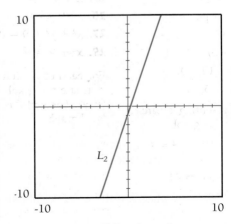

FIGURE 1-36 $y - (-6) = 3x + 5$

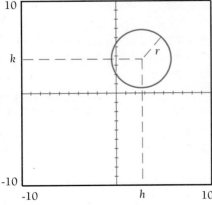

FIGURE 1-37 $(x - h)^2 + (y - k)^2 = r^2$

- The graphing format for an equation representing a circle is given by

$$(x - h)^2 + (y - k)^2 = r^2, \qquad r \geq 0,$$

where r is the radius of the circle and the center is at (h, k).
- The diameter of a circle is any line segment through the center of the circle with endpoints on the circle.
- The circumference of a circle measures the length of the arc that forms the circle and is given by $2\pi r$, where π is approximately 3.141592654.
- Suppose that $F(x, y) = c$ is a formula in x and y that represents the graph of curve in the Cartesian coordinate system. Then substituting $x - h$ for x and $y - k$ for y in the equation produces a translation of the curve. The translated curve has the same shape and is the same size as the original curve but is relocated horizontally by h units and relocated vertically by k units.

1-4   EXERCISES

Write an equation to represent a circle in the Cartesian coordinate system described by each of the following:

1. center (5, 2), radius 3
2. center (3, 4), diameter 4
3. center (-3, 1), circumference 10.
4. center (4, -5), circumference 10π.
5. center (5, 0) and passing through the origin.
6. center (0, -4) and passing through the origin.
7. center (0, 0) and passing through (3, -4).
8. center (0, 0) and passing through (-5, 12).
9. With diameter endpoints at (-7, 17) and (3, -7).
10. With diameter endpoints at (-1, -7) and (5, 1).

For each of the following equations representing a circle, determine the center and radius and sketch the graph.

11. $x^2 + (y - 2)^2 = 9$
12. $(x - 3)^2 + y^2 = 4$
13. $(x + 3)^2 + (y - 5)^2 = 36$
14. $(x - 4)^2 + (y + 2)^2 = 25$
15. $x^2 + y^2 + 10y + 19 = 0$
16. $x^2 + 8x + y^2 - 9 = 0$
17. $x^2 + y^2 - 6y + 4x = 3$
18. $x^2 + y^2 + 10x - 6y = 2$

19. $x^2 + y^2 + 5y - 3x = 1$
20. $x^2 + y^2 - 3y + 7x = 0$

For each of the following equations identify the graph as a circle, a line or neither. If a circle or line, sketch the graph. Use graphing technology.

21. $x^2 + y^2 = 9$
22. $x + y = 4$
23. $x + y = 9$
24. $x + y^2 = 4$
25. $x^2 + y = 9$
26. $x^2 + y^2 = 4$
27. $x^2 + y^2 + 9 = 0$
28. $x + y + 4 = 0$
29. $x^2 + 9 = y^2$
30. $x^2 = 9 + y^2$

For each of the following equations, sketch the graph, then translate the graph with the substitution $(x - 3, y + 1)$ for (x, y). Then on the same coordinate system sketch the translated graph.

31. $x^2 + y^2 = 4$
32. $x^2 + y^2 = 9$
33. $x + y = 4$
34. $x + y = 9$
35. $x^2 + 6x + y^2 + 5 = 0$
36. $x^2 + y^2 - 2y = 8$
37. $x^2 + y^2 - 2y = 3$
38. $x^2 + 6x + y^2 = 6$
39. $x^2 + y^2 + 6x - 2y + 9 = 0$
40. $x^2 + y^2 - 2y + 6x + 4 = 0$

41. Refer to Problems 39 and 40. Are the resulting equations "simpler" or more complex?

42. Refer to Problem 41. Discuss any advantages or disadvantages to translating.

For each of the following pairs of equations, determine substitutions that transform equation *a* into equation *b*. Also indicate the effect the translation will have on the graph of *a*. Graphing technology should help.

43. (a) $y = 2x$ **(b)** $y + 3 = 2(x - 5)$

44. (a) $y = 3x$ **(b)** $y - 2 = 3x + 6$
45. (a) $x^2 + y^2 = 16$ **(b)** $(x - 1)^2 + (y + 3)^2 = 16$
46. (a) $y = 5x^2$ **(b)** $y = 5(x + 2)^2 - 3$
47. (a) $y = 2(x - 3)^2 + 1$ **(b)** $y = 2x^2$
48. (a) $x^2 + (y - 2)^2 = 9$ **(b)** $(x - 5)^2 + y^2 = 9$
49. (a) $y = (x + 3)^2 - 2$ **(b)** $y = x^2 + 6x + 5$
50. (a) $y^2 = x - 4$ **(b)** $y^2 - 4y = x + 1$

1-3   PROBLEM SET

Use graphing technology to graph each of the following collections of equations in the same window. These exercises explore the graphs that will be discussed in Section 1-5.

1. (a) $y = x^2, y = 2x^2, y = 3x^2, y = 4x^2$

   (b) Describe the effect of *a* in the graph of an equation of the form $y = ax^2$

2. (a) $y = -x^2, y = -2x^2, y = -3x^2, y = -4x^2$

   (b) Describe the effect of *a* in the graph of an equation of the form $y = -ax^2$

3. (a) $y = x^2, y = -x^2, y = -2x^2, y = 3x^2$

   (b) Describe the effect of the sign of *a* on the graph of an equation of the form $y = ax^2$

4. (a) $y = x^2, y = x^2 + 1, y = x^2 + 2, y = x^2 - 3, y = x^2 - 4$

   (b) Describe the effect of *k* on the graph of an equation of the form $y = x^2 + k$

5. (a) $y = x^2, y = (x - 2)^2, y = (x + 3)^2, y = (x - 5)^2$

   (b) Describe the effect of *h* on the graph of an equation of the form $y = (x - h)^2$

6. (a) $y = (x - 1)^2 + 3, y = (x + 3)^2 - 1, y = (x - 2)^2 + 5, y = (x + 5)^2 - 2$

   (b) Describe the effect of *h* and *k* on the graph of an equation of the form $y = (x - h)^2 + k$

7. (a) $y = 3(x - 1)^2 + 3, y = -2(x + 3)^2 - 1, y = -2(x - 2)^2 + 5, y = 3(x + 5)^2 - 2$

   (b) Describe the effect of *a*, *h* and *k* on the graph of an equation of the form $y = a(x - h)^2 + k$

---

5   THE PARABOLA

One cannot escape the feeling that these mathematical formulas have an independent existence and an intelligence of their own, that they are wiser than we are, wiser even than their discoverers, that we get more out of them than was originally put into them.
—HEINRICH HERTZ

Consider an equation that is not linear. The shape of the graph will not be a straight line. What will be the shape of the graph?

   The last section explored the graphs of circles. In an equation for a circle, *both* the variables are quadratic. For example, $x^2 + y^2 = 1$ represented a circle

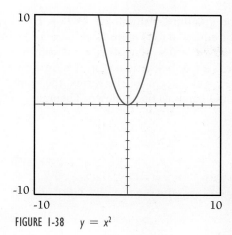

FIGURE 1-38 $y = x^2$

at the origin $(0, 0)$ with a radius of 1. This section examines equations that fall between a first degree equation (a line) and an equation in which both variables are quadratic. These equations are quadratic in one variable and linear in the other. As a first example, consider the equation $y = x^2$.

What is the graph of $y = x^2$? Graphing technology provides the easiest way to explore the graph. Alternately, select x values, calculate corresponding y values to satisfy the equation, then plot these points. From sufficient points infer a graphical pattern. From the pattern, conjecture the rest of the graph. Graphing technology provides the shape in Figure 1-38.

A goal of analysis is to learn to *read* each change in an equation as it affects the shape or position of a graph. Examine the x^2 term of the equation. Squaring a real number results in a nonnegative real number. Therefore, y will never be negative. If $x = 2$ the value for y is 4; the same y value produced by $x = -2$. Similarly, the equation duplicates y values for each value of x and $-x$ except when $x = 0$. The point where $x = 0$ is an interesting point of the graph: the y value produced is unique, and no other x value can duplicate it. For negative values of x, we obtain one set of y values. The positive values of x produce duplicates of these y values. Viewed from left to right, the graph changes direction at $x = 0$ to pass through the duplicated y values.

The shape of the graph in Figure 1-38 is a parabola. The point $(0, 0)$ where the graph changes direction is the *vertex* of the parabola. The following formal definition of a *parabola* echoes a method for mechanically drawing a parabola based on properties long associated with the parabolic shape. Because many "reflection" applications of parabolas make reference to a *focus*, the definition is traditional.

DEFINITION 1-6
Parabola

A **parabola** is the set of points in a plane equally distant from a line and a fixed point not on the line.

The line is the *directrix* of the parabola. The fixed point is the *focus*. The *vertex* of the parabola is the midpoint of a line segment from the focus perpendicular to the directrix.

S I D E B A R

Marilyn Bronowski decides to build her own satellite antenna. A receiving horn for a high-gain linear amplifier gathers signals from the satellite. The problem is that satellite signals are weak. Marilyn needs to collect signals over a large area and concentrate them on the receiving horn. From her study of geometry, Marilyn knows that parabolas possess the ideal reflection properties for her purpose. Therefore she constructs a surface based upon the shape of a parabola.

Consider a focus $P = (0, c)$ and a directrix $y = -c$. Note that $(0, 0)$ will be the vertex of the parabola. The vertex must be a point of the parabola. Why? (See Figure 1-39.) Calculate the distances from (x, y) to the focus and from (x, y) to the directrix. Set these two distances equal, square both sides and obtain the equation for a parabola:

$$\sqrt{(x - 0)^2 + (y - c)^2} = \sqrt{(x - x)^2 + [y - (-c)]^2}$$
$$x^2 + y^2 - 2cy + c^2 = y^2 + 2cy + c^2$$
$$-4cy = -x^2$$
$$4cy = x^2$$

To obtain a form convenient for graphing technology, let $a = 1/(4c)$ and obtain $y = ax^2$. Compare this to the equation for Figure 1-38.

Notice that if $a > 0$, and $x \neq 0$ then $ax^2 > 0$ so that all $y > 0$. Except for the vertex, the graph is above the x-axis. A parabola oriented in this direction opens up. If $a < 0$, then $ax^2 < 0$ when $x \neq 0$. Except for the vertex, the graph is below the x-axis. In this orientation the parabola opens down. Theorem 1-4 summarizes.

Theorem 1-4

> The graph of $y = ax^2$ is a parabola with vertex at $(0, 0)$. If a is a positive number, then the parabola "opens up." If a is a negative number, then the parabola "opens down."

Concavity— Some texts refer to parabolas that open up as being "concave up." Parabolas that open down are "concave down." Concavity is also common terminology in calculus texts.

Symmetry is a useful property in graphing parabolas. Squaring corresponding positive and negative x values produces duplicate y values. In the equation $y = x^2$, squaring -3 and 3 produced duplicate y values of 9. In effect, the graph duplicated itself on either side of the y-axis. *Symmetry* is the term for such duplication.

Based on circles and parabolas, the graphs of second-degree equations change direction where the square of a variable expression is 0. Translate axes on $x^2 + y^2 = r^2$ or $y = x^2$ to verify this conjecture.

For a parabola with a vertex off the origin, translate the graph by replacing x with $(x - h)$ and y with $(y - k)$. The resulting parabola has a vertex located at (h, k): $y - k = a(x - h)^2$, see Figure 1-40.

Note the influence of the squared term. Duplicate values for y occur wherever $(x - h)^2 \neq 0$. The vertex must occur where $(x - h)^2 = 0$. Therefore, $x = h$. Substitute h for x in the equation, solve for y and obtain the vertex in the following theorem.

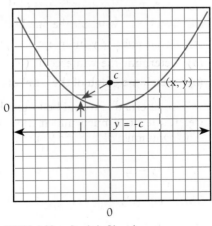

FIGURE 1-39   Parabola Directrix

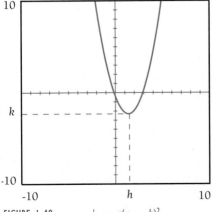

FIGURE 1-40 $y - k = a(x - h)^2$

Theorem 1-5
Graphing Form of a Parabola

> The graph of an equation of the form
>
> $$y = a(x - h)^2 + k$$
>
> is a parabola with vertex at (h, k).
>
> If a is positive, then $y \geq k$ so that the graph "opens up."
> If a is negative, then $y \leq k$ so that the graph "opens down."

The following definition generalizes the concept of intercepts.

DEFINITION 1-7
Intercepts

> Suppose that F is a relation graphed in the Cartesian coordinate system. The y-intercepts of F are those points of F (if any) for which the x coordinates are 0. Geometrically, these are the points where the graph intersects the y-axis. Similarly, if the graph intersects the x-axis, the x-intercepts of F are those points of F for which y is 0.

An intercept occurs when one variable assumes a value of 0. For that reason, some texts refer to an intercept by listing only the nonzero component of the point. For example, suppose that $(0, 3)$ is a y-intercept of F; we might write y-intercept = 3. If F has x-intercept = -2 then $(-2, 0)$ is a point of the graph F.

In summary, suppose that the points of F satisfy the equation $F(x, y) = c$. The y-intercepts of F are the solutions of $F(0, y) = c$. The x-intercepts are the solutions of $F(x, 0) = c$.

EXAMPLE 1-11

Graphing and Analyzing Parabolas. For each of the following equations in two variables, show that the graph is a parabola by rewriting the equation into the graphing format, $y = a(x - h)^2 + k$. Also,

(a) Determine the coordinates of the vertex.

(b) Determine whether the parabola opens up or down.

(c) Determine the y-intercept.

(d) Determine the x-intercepts (if any).

(e) Sketch the graph of the parabola.

An alternative would be to graph each of the following using technology. Determine how much information you can obtain from the graph and how much you must determine from the *format* of the equation.

Illustration 1:

$$y = 2(x + 3)^2 - 5$$

Solution: The equation is in graphing format. Thus, $a = 2$, $h = -3$, $k = -5$.

(a) Vertex: $(-3, -5)$

(b) Opens up because $a = 2$ is positive.

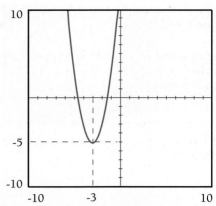

FIGURE 1-41 $y = 2(x + 3)^2 - 5$

(c) The y-intercept requires that $x = 0$:

$$y = 2(0 + 3)^2 - 5$$
$$y = 13$$

y-intercept: $(0, 13)$

(d) The x-intercept(s) require that $y = 0$

$$0 = 2(x + 3)^2 - 5$$
$$0 = (x + 3)^2 - 2.5 \qquad \text{(difference of squares)}$$
$$0 = [(x + 3) + \sqrt{2.5}][(x + 3) - \sqrt{2.5}] \qquad \text{(factor)}$$
$$(x + 3) + \sqrt{2.5} = 0 \quad \text{or} \quad (x + 3) - \sqrt{2.5} = 0 \qquad \text{(solve)}$$
$$x = -3 - \sqrt{2.5} \quad \text{or} \quad x = -3 + \sqrt{2.5}$$

The x-intercepts are $(-3 - \sqrt{2.5}, 0)$, $(-3 + \sqrt{2.5}, 0)$

(e) For the graph, locate the vertex. Use a calculator to approximate the location of the intercepts:

$$x\text{-intercepts} \cong (-4.58, 0), \quad (-1.42, 0).$$

Since the shape of a parabola is well-known, it is easy to imagine the remainder of the graph, shown in Figure 1-41.

Illustration 2:

$$y = -3x^2 + 6x - 9$$

Solution: Complete the square to obtain an equation in the form of Theorem 1-5.

$$y = -3(x^2 - 2x + \underline{\quad\quad}) - 9$$
$$y = -3(x^2 - 2x + 1) - 9 + 3$$
$$y = -3(x - 1)^2 - 6.$$

(a) Vertex $(1, -6)$

(b) Parabola opens down since $a = -3$.

(c) Substitute 0 for x and obtain

$$y = -3(0 - 1)^2 - 6$$
$$y = -9$$

The y-intercept is $(0, -9)$.

(d) Substitute 0 for y to obtain the x-intercept:

$$0 = -3(x - 1)^2 - 6$$
$$-2 = (x - 1)^2$$

Impossible! A real number is never negative when squared. This parabola has no x-intercepts. Since the vertex is below the x-axis and the parabola opens down, there are no x-intercepts.

(e) Substitute another value for x, say $x = 2$:

$$y = -3(2 - 1)^2 - 6$$
$$y = -9$$

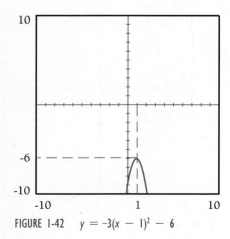

FIGURE 1-42 $y = -3(x - 1)^2 - 6$

Note that the vertex is the highest point on the graph: thus the *maximum y* value is -6 and occurs when *x* is 1 (Figure 1-42). ▬

Reversing the roles of *x* and *y* in an equation for a parabola gives an equation for a parabola that opens right or left, instead of up or down. Consider $x = a(y - k)^2 + h$ as you read the following example.

EXAMPLE 1-12

Horizontal Parabolas. Locate the vertex and sketch the graph for each of the following. Caution, some of these equations are in an inconvenient form for graphing technology. They are solved for *x*, not *y*.

Illustration 1:

$$x = 3(y - 2)^2 - 3$$

Solution: The squared term is the key to analyzing the graph. Except for $y = 2$, $(y - 2)^2 > 0$ produces duplicate values for *x* so that the vertex is at $(-3, 2)$ and the graph opens in a positive direction: to the right. See Figure 1-43.

Illustration 2:

$$x = -2(y + 1)^2 + 2$$

Solution: The vertex is $(2, -1)$. Since $a = -2$, the graph opens to the left. See Figure 1-44.

Illustration 3:

$$x = 2y^2 + 8y - 2$$

Solution: The presence of a *y* term complicates the influence of y^2 on the graph. Complete the square in *y* so that all *y* values are in a single squared term.

$$x = 2(y^2 + 4y + \underline{\quad}) - 2$$
$$x = 2(y^2 + 4y + 4) - 2 - 8$$
$$x = 2(y + 2)^2 - 10$$

The vertex is $(-10, -2)$ and the parabola opens to the right. See Figure 1-45.

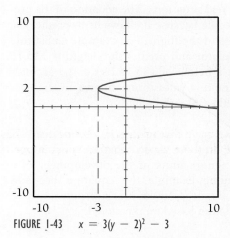

FIGURE 1-43 $x = 3(y - 2)^2 - 3$

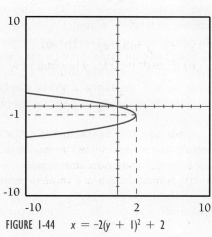

FIGURE 1-44 $x = -2(y + 1)^2 + 2$

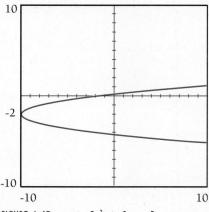

FIGURE 1-45 $x = 2y^2 + 8y - 2$

Illustration 4:

The path of a projectile under the influence of gravity resembles a parabola. Experimentally, Adrian's physics class collects data on a projectile. After measuring the height h of the projectile t seconds into the flight, the class records the following information (plotted in Figure 1-46):

height, h	0	80	128	144	128	80	0
time, t	0	1	2	3	4	5	6

The class notes the duplicate values for height (once on the way up, again on the way down). Adrian estimates that the maximum height is 144 feet and occurs 3 seconds into the flight. Based on these observations, what equation *models* the flight?

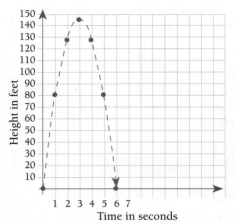

FIGURE 1-46   Plotted points of a projectile

Solution: The equation should represent a parabola. Because the vertex occurs at the point (3 seconds, 144 feet), the equation is expressible as

$$\text{height} = A(\text{time} - 3 \text{ sec.})^2 + 144 \text{ feet}$$

The only value remaining is A. The other observations must also satisfy the equation, substitute one point and solve for A. The point (0 sec., 0 feet) offers the easiest arithmetic.

$$0 \text{ feet} = A(0 \text{ sec.} - 3 \text{ sec.})^2 + 144 \text{ feet}$$
$$-144 \text{ feet} = A(9 \text{ sec.}^2)$$
$$-16 \text{ feet/sec.}^2 = A$$

Therefore,

$$\text{height} = \frac{-16 \text{ feet}}{\text{sec.}^2} (\text{time} - 3 \text{ sec.})^2 + 144 \text{ feet}$$

Note: A *model* predicts characteristics of real situations. The equation in the preceding example is a model of the projectile because it allows us to predict the height of the projectile at various times in the flight. For example, substitute 2.3 seconds for time, apply your calculator and predict the height = 136.16 feet. With the model's aid, we predict a height of 136.16 feet at 2.3 seconds after launch. Similarly, a model airplane in a wind tunnel allows an engineer to estimate airflow across the wings of the full-sized plane. ▬

A model is an *abstraction*. By this we mean that models ignore the distracting details of a problem to concentrate on those aspects that are most important. For that reason, we sometimes ignore units in the development of a mathematical model. For the preceding example the equation becomes $h = -16(t - 3)^2 + 144$. In these cases, it is the model user's responsibility to reintroduce units when making a prediction based on the model.

For equations expressible in the graphing form

$$y = a(x - h)^2 + k \quad \text{or} \quad x = a(y - k)^2 + h$$

• The graph is a parabola.
• If a is positive the graph opens in a positive direction; a *minimum* value for the first-degree variable occurs at the vertex. If a is negative the graph opens in a negative direction; a *maximum* value for the first-degree variable occurs at the vertex.
• The vertex of the parabola is (h, k).
• If the equation is in the standard form

$$y = ax^2 + bx + c \text{ or}$$
$$x = ay^2 + by + c$$

completing the square converts the equation to graphing form.

1-5 E X E R C I S E S

For each of the following equations, show that the graph of the equation is a parabola by rewriting the equation in graphing form. Give the direction the parabola opens, the coordinates of the vertex and whether the vertex is a maximum or minimum. Sketch the graph of the parabola. Use graphing technology whenever possible.

1. $y = (x - 1)^2 + 3$
2. $y = (x + 3)^2 - 1$
3. $y = -2(x + 3)^2 + 8$
4. $y = -3(x - 2)^2 - 3$
5. $y = -(x + 5)^2$
6. $y = 2x^2 - 3$
7. $y = 3x^2 + 2$
8. $y = 2(x - 2)^2$
9. $y = 0.5(x - 2)^2 + 1.2$
10. $y = -\frac{1}{3}(x + \frac{1}{2})^2 - 1$

11. $y = x^2 + 6x + 5$
12. $y = x^2 - 4x + 3$
13. $y = x^2 - 8x$
14. $y = x^2 + 10x$
15. $y = -2x^2 + 6x - 1$
16. $y = -x^2 - 8x + 3$
17. $y = 2x^2 + 5x + 4$
18. $y = 3x^2 - 5x + 3$
19. $y = -3x^2 + 7$
20. $y = -2x^2 + 6$
21. $x = (y + 3)^2 - 1$
22. $x = (y - 1)^2 + 3$
23. $x = -3(y - 2)^2 + 4$
24. $x = -2(y + 3)^2 + 8$

25. $x = 2y^2 - 3$
26. $x = -(y + 5)^2$
27. $x = 2(y - 2)^2$
28. $x = 3y^2 + 2$
29. $x = -\frac{1}{3}(y + \frac{1}{2})^2 - 1$
30. $x = 0.4(y - 2)^2 + 1.3$
31. $x = y^2 - 4y + 3$
32. $x = y^2 + 6y + 5$
33. $x = y^2 + 10y$
34. $x = y^2 - 8y$
35. $x = -y^2 - 8y + 3$
36. $x = -2y^2 + 6y - 1$
37. $x = 3y^2 - 5y + 3$
38. $x = 2y^2 + 5y + 4$
39. $x = -2y^2 + 6$
40. $x = -3y^2 + 7$

Graph each of the following. Annotate the vertex and intercepts.

41. $y = (x + 1)^2 - 3$
42. $x = -(y + 2)^2 - 1$
43. $y = (x + 1) - 3$
44. $x = -(y + 2) - 1$
45. $x = -(y - 1)^2 + 2$
46. $y = 2(x - 1)^2 + 3$
47. $x = -(y - 1) + 2$
48. $y = 2(x - 1) + 3$
49. $y = 8x + 12$
50. $x = 8y + 12$
51. $y = x^2 + 8x + 12$
52. $y = y^2 + 8y + 12$
53. $x = y^2 - 6y + 9$
54. $y = x^2 - 6x + 9$

55. $y + 6x = x^2 + 9$

56. $x + 6y = y^2 + 9$

57. $x + 6y = 9$

58. $y + 6x = 9$

59. $6x + 3y = 3x^2 + 12$

60. $6y + 3x = 3y^2 + 12$

61. Use $x = -c$ as a directrix and $(c, 0)$ as a focus to derive a graphing form for a parabola that opens horizontally.

62. Refer to Problem 61. Which direction does the parabola open if c is negative?

63. Refer to Problem 61. Which direction does the parabola open is c is positive?

64. Refer to the derivation of Theorem 1-4. Show that if $y = ax^2$, then the focus is $(0, c)$, where $c = 1/(4a)$.

65. Refer to the derivation of Theorem 1-4. Show that if $y = ax^2$, then the directrix is $y = -c$, where $c = 1/(4a)$.

66. Refer to Problem 64. Show that the focus of $y = a(x - h)^2 + k$ is $(h + c, k)$, where $c = 1/(4a)$.

67. Refer to Problem 65. Show that the directrix of $y = a(x - h)^2 + k$ is $y = h - c$, where $c = 1/(4a)$.

68. Show that the x-intercepts of $y = ax^2 - b$ are $(\sqrt{b/a}, 0)$ and $(-\sqrt{b/a}, 0)$ where a and b are both positive.

69. Refer to Problem 68. Show that the x-intercepts of $y = a(x - h)^2 - b$ are $(h + \sqrt{b/a}, 0)$ and $(h - \sqrt{b/a}, 0)$, where both a and b are positive.

70. Show that $y = ax^2 + bx + c$ can be rewritten in the form $y = a(x - h)^2 + k$, where $h = -b/(2a)$ and $k = -(b^2 - 4ac)/(4a)$.

71. Refer to Problems 69 and 70. Show that the x-intercepts of $y = ax^2 + bx + c$ are $((-b + \sqrt{b^2 - 4ac})/(2a), 0)$ and $((-b - \sqrt{b^2 - 4ac})/(2a), 0)$, where $b^2 - 4ac > 0$ and $a \neq 0$.

 In calculus it can be shown that the slope of a line tangent to the parabola $y = x^2$ at the point (x_0, y_0) is $m = 2x_0$. See Figure 1-47.

72. Determine the slope of the tangent line to $y = x^2$ at $x_0 = 2$. Is the slope positive or negative?

73. Determine the slope of the tangent line to $y = x^2$ at $x_0 = -1$. Is the slope positive or negative?

74. Sketch the graph of $y = x^2$ and the tangent line at $x_0 = 2$. See Problem 72.

75. Sketch the graph $y = x^2$ and the tangent line at $x_0 = -1$. See Problem 73.

76. Determine the slope of the tangent line to $y = x^2$ at the vertex. Sketch the graph of the parabola and the tangent at the vertex.

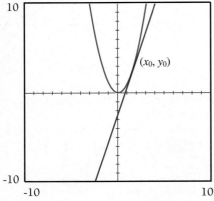

FIGURE 1-47   Tangent to $y = x^2$ at x_0

 PROBLEM SET

One quality associated with a line is the that of being *straight*. The following sequence of problems should confirm your intuition that no part of a parabola is straight. A line segment consists of two points of a line and the points of the line "between" them. In a similar manner, a *parabolic segment* is two points of a parabola and the points of the parabola "between" them.

1. Recall that if (x_1, y_1) is a point on the parabola given by $y = x^2$, then $y_1 = (x_1)^2$. What would it mean for (x_2, y_2) to be a point of $y = x^2$? Use these results to show that $y_1 + y_2 = (x_1)^2 + (x_2)^2$ and $2y_1 + 2y_2 = 2(x_1)^2 + 2(x_2)^2$.

2. Recall that $((x_1 + x_2)/2, (y_1 + y_2)/2)$ represents the midpoint on the line segment connecting two points (x_1, y_1) and (x_2, y_2). What would it mean for $((x_1 + x_2)/2, (y_1 + y_2)/2)$ to be a point of $y = x^2$? Use this result to show that $2y_1 + 2y_2 = (x_1)^2 + 2x_1x_2 + (x_2)^2$.

3. Use the results of Problems 1 and 2 to show that $(x_1)^2 - 2x_1x_2 + (x_2)^2 = 0$ and $(x_1 - x_2)^2 = 0$.

4. Refer to Problem 3. Can x_1 and x_2 be different? Why? If $x_1 = x_2$ what can we conclude about y_1 and y_2? Explain.

5. Refer to Problems 1–4. Under what conditions can two points and the midpoint of the segment connecting them all be points of the parabola $y = x^2$?

6. Argue that no parabolic segment of $y = x^2$ can be a line segment.

---

*The magnifying glass icon indicates that these problems develop a concept in more detail or are more theoretical.

PROBLEMS FOR TECHNOLOGY

7. Use graphing technology to scroll about the graph of $y = x^2$. Particularly examine the graph well away from $(0, 0)$. Does any segment of the graph appear to be "straight"?

8. Use graphing technology to graph $y = x^2$. Experiment by zooming in on several parabolic segments. Is it possible to zoom in until a parabolic segment appears "straight." If so, is the segment really straight? Discuss the dangers of drawing conclusions based on graphs alone.

CHAPTER SUMMARY

- For two points in a plane with coordinates (x_1, y_1) and (x_2, y_2)

$$\text{Distance} = \sqrt{(x_2 - x_1)^2 + (y_2 - y_1)^2}$$

$$\text{Midpoint} = \left(\frac{x_1 + x_2}{2}, \frac{y_1 + y_2}{2}\right).$$

Average rate of change: $\dfrac{\Delta y}{\Delta x} = \dfrac{y_2 - y_1}{x_2 - x_1}$

- Given a point (x_1, y_1) and a slope $m = \Delta y/\Delta x$ for a nonvertical line, an equation for the line is $y = m(x - x_1) + y_1$
- The slope-intercept form of a line is $y = mx + b$, where m is the slope and $(0, b)$ is the y-intercept.
- The graph of equation of the form

$$(x - h)^2 + (y - k)^2 = r^2, \ r \geq 0,$$

is a circle of radius r and center at (h, k).
- Translate the graph of $F(x, y) = c$ by substituting $x - h$ for x and $y - k$ for y in the equation. The translation preserves the shape and size of the original curve but relocates the graph horizontally by h units and vertically by k units.
- The graph of an equation of the form $y = a(x - h)^2 + k$ or $x = a(y - k)^2 + h$ is a parabola.

If a is positive the graph opens in a positive direction. A minimum value for the first-degree variable occurs at the vertex.

If a is negative the graph opens in a negative direction. A maximum value for the first-degree variable occurs at the vertex.

The vertex of the parabola is (h, k).

Completing the square converts an equation of the form $y = ax^2 + bx + c$ or $x = ay^2 + by + c$ to graphing form.

KEY   WORDS   AND   CONCEPTS

Average rate of change
Cartesian coordinate system
Center
Circle
Collinear points
Coordinates
Distance formula
Graph
Horizontal line
Intercept form of a line

Line
Linear equation
Midpoint formula
Oblique Line
Ordered pair
Parabola
Pythagorean theorem
Quadratic
Radius
Relation

Slope
Slope-intercept form of a line
Translation
Vertical line
Window
x-axis
x-intercept
y-axis
y-intercept

REVIEW   EXERCISES

SECTION 1-1

Plot the points whose locations are given by the following coordinates:

1. (4, 7)

2. (4, -7)

3. (-5, -6)

4. (0, 5)

5. (0, -5)

6. (5, 0)

7. (2, 3)

8. (3, 2)

9. (2, -3)

10. (5, 1)

For the following pairs of coordinates representing points determine (a) the distance between the two points, (b) the coordinates of the midpoint of the segment joining the points, (c) the average rate of change between the points. Graph the points and label the graph with the answers from parts a, b, and c.

11. (-3, 8), (11, -4)

12. (8, -5), (-8, 7)

13. (0, 0), (4, -3)

14. (5, 1), (-4, 8)

15. (5a, 0), (a, 3b), $a \neq 0$

16. (7b, 2a), (3b, -3a), $b \neq 0$

Suppose that each of the following collections of points represents the vertices of a triangle. Determine whether each triangle is scalene (no equal sides), isosceles (exactly two equal sides) or equilateral (all three sides equal). Sketch each triangle.

17. (0, 7), (-7, 0), (0, 0)

18. (1, 1), (-1, -1), (3, -3)

19. ($\sqrt{2}$, -$\sqrt{2}$), (0, 0), (1, 0)

20. (-1, 2), (2, -1), (0, 0)

21. (0.5, -3), (4.5, 0), (2, 1)

22. (-0.3, 11), (2.7, 0), (2, 1)

Determine which of the following sets of points are collinear:

23. (1, 3), (3, 5), (5, 7)

24. (-2, 3), (1, 4), (4, 5)

25. (1, 5), (3, 1), (-2, 7)

26. (2, -4), (-1, .3), (-2, 10)

27. (a, a^2), (b, -b^2), (0, ab), $b \neq a$, $b \neq -a$

28. $\left(b, \dfrac{1}{b}\right), \left(a, \dfrac{1}{a}\right), \left(\dfrac{1}{ab}, 0\right), b \neq 0, a \neq 0$

29. Are the points (4, -4), (1, 0), (9, 6) the vertices of a right triangle?

30. Are the points (4, -4), (9, 6), (1, 0), (8, 0) the vertices of a rectangle?

31. Determine x for (x, 5), (2, -3) so that $\Delta y / \Delta x = 1.5$.

Determine the average rate of change of y with respect x for each of the following pairs of points. Leave the units of measure in the formula for $\Delta y / \Delta x$.

32. (3 seconds, 17 feet), (1 second, 5 feet)

33. (10kg, $100), (20kg, $300)

34. (2 seconds, 25 meters/second), (4 seconds, 15 meters/second)

35. (5 amp., 110 volts), (6 amp., 130 volts)

36. (5 seconds, 100 milliflops), (8 seconds, 200 flops) [milli = 0.001]

SECTION 1-2

Use graphing technology to graph the following.

37. $y = -2x$ **38.** $y = 5x$

39. $y = 2x^2$ **40.** $y = -3x^2$

41. $y = 3x - 4$ **42.** $y = 1 - 3x^2$

43. $y = 2x^2 - 3$ **44.** $y = 1 - 3x$

45. $y = 3x - 1$ **46.** $y = (1 - 3x)^2$

47. $y = 1 + \sin x$ **48.** $y = \ln|x|$

49. $y = 1 - \tan x$ **50.** $y = \cos^{-1} x$

51. $y = 10^x - 5$ **52.** $y = \ln(x^2)$

Graph the following so that each is on a separate coordinate system. (Set your calculator to Rad mode). Compare the shapes in each pair:

53. $y = 1 + \sin x$ **54.** $y = \cos(2x)$
$\quad\;\; y = \sin^{-1}(x - 1)$ $\quad\;\; y = \cos^{-1}\left(\dfrac{x}{2}\right)$

55. $y = \sqrt{|x|}$ **56.** $y = 1 + \tan x$
$\quad\;\; y = x^2$ $\quad\;\; y = 1 + \tan^{-1} x$

SECTION 1-3

Write a linear equation for the lines through each of the following pairs of points:

57. $(4, 7), (3, 7)$ **58.** $(-3, -5), (-3, -1)$

59. $(5, 5), (5, 1)$ **60.** $(-5, 3), (-2, 3)$

61. $(3, -3), (-3, 3)$ **62.** $(7, 4), (-2, 11)$

63. $(4, 7), (9, -12)$ **64.** $(-4, 5), (7, -13)$

65. $\left(\frac{1}{2}, -\frac{1}{3}\right), (5, -2)$ **66.** $(b, a^2), (a, b^2), a \neq b$

Write a linear equation representing the line through each of the following points and with the indicated slope:

67. $m = -2, (1, 3)$ **68.** $m = 3, (0, 5)$

69. $m = \frac{1}{2}, (-2, 5)$ **70.** $m = 0, (4, -1)$

(a) Rewrite each of the following linear equations in the slope-intercept form by solving the equation for y. (b) Determine the slope of the line for each equation. (c) Give the x-intercept for each line. (d) Give the y-intercept for each line.

71. $3y + 4x = 12$ **72.** $2y - 3x = 6$

73. $2x - 3y = 5$ **74.** $3x - 4y = 5$

75. $\frac{1}{2}x + \frac{2}{3}y = \frac{1}{4}$ **76.** $\frac{3}{5}x - \frac{1}{15}y = \frac{1}{3}$

Graph each of the following:

77. $3y + 6x = 12$ **78.** $\dfrac{x}{5} - \dfrac{y}{2} = 1$

79. $y = \frac{2}{3}x - 1$ **80.** $y = -\frac{3}{4} + 2$

81. $y = -\frac{3}{5}x + 3$ **82.** $y = -2x + 3$

83. $x = 4.7$ **84.** $y = -5$

85. $y = -2.5$ **86.** $2x + 5y = 4$

87. $5x - 2y = 10$ **88.** $\frac{3}{5}x + \frac{1}{2}y = \frac{1}{3}$

SECTION 1-4

Write an equation to represent the circle described by each of the following:

89. center $(3, 4)$, radius 5.

90. center $(6, 0)$ and passing through the origin.

91. center $(0, 0)$ and passing through $(5, -12)$.

92. With diameter endpoints at $(7, -15)$ and $(-3, 9)$.

Graph each of the following circles. Label the center and radius.

93. $x^2 + (y - 3)^2 = 25$

94. $(x - 3)^2 + (y + 4)^2 = 16$

95. $x^2 + y^2 + 8y - 9 = 0$

96. $x^2 + y^2 - 4y + 5x = 0$

Graph the following. Translate the graph by substituting $(x - 5, y + 2)$ for (x, y). On the same coordinate system graph the translated graph.

97. $x^2 + y^2 = 16$ **98.** $x + y = 9$

99. $x^2 + 4x + y^2 - 5 = 0$

100. $x^2 + y^2 - 6y + 4x + 4 = 0$

SECTION 1-5

Show that the graph of each of the following equations is a parabola by rewriting the equation in graphing form. Give the direction the parabola opens, the coordinates of the vertex and whether the vertex is a maximum or minimum. Graph the parabola and annotate this information.

101. $y = (x - 2)^2 + 5$ **102.** $y = -3(x - 5)^2 - 2$

103. $y = -(x + 3)^2$ **104.** $y = 2(x - 5)^2$

105. $y = 0.5(x - 1)^2 + 1.2$ **106.** $y = x^2 - 6x + 3$

107. $y = x^2 - 6x$ **108.** $y = -x^2 - 5x + 1$

109. $y = 2x^2 + 5x + 1$ **110.** $y = -2x^2 + 6$

111. $x = (y + 2)^2 - 3$ **112.** $x = -2(y + 3)^2 + 8$

113. $x = 3y^2 - 2$ **114.** $x = 3y^2 - 2$

115. $x = -\frac{1}{3}\left(y - \frac{1}{2}\right)^2 + 2$ **116.** $x = y^2 - 6y + 2$

117. $x = y^2 + 18y$ **118.** $x = -2y^2 + 6y - 2$

119. $x = 2y^2 - 5y + 3$ **120.** $x = -3y^2 + 4$

Graph each of the following:

121. $y = (x - 1)^2 - 4$

122. $x = -(y + 1) - 4$

127. $x = y^2 - 6y + 9$

128. $6y = y^2 - 9 + x^2$

123. $x^2 = -(y - 1)^2 + 4$

124. $y = 2(x - 1) + 4$

129. $x^2 + 6y = 9$

130. $6y + 2x = 2y^2 + 12$

125. $y = 8x + 16$

126. $x = y^2 + 8y + 16$

SIDELIGHT

Windows

Window—literally, "wind eye." A window is an opening for letting in light or air or for looking through.

In Robert Heinlein's novel *The Door into Summer*, hydraulic pressure forces the hero's cat to seek the outdoors in the midst of winter. The cat steadfastly marches from door to door to request an exit. As each exit door opens to reveal the snow and cold, the cat retreats to lead the hero to the next door. The ritual continues through every door in the house: no one can persuade the cat that no door leads to summer.

One advantage of algebra is that rewriting an expression gives us new insight, a different view of the information represented by the expression. Massaging the data into different forms broadens our understanding.

Similarly for visual information, all windows in a house do not open onto the same scene. Some views are more informative. Some windows are more likely to inform you of the arrival of the mail carrier, others convey messages about plants and animals. Although you likely have some favorite window, a "best view" depends upon what you wish to see.

We extend the window metaphor to graphs. Theoretically, many graphs consume an infinite amount of space. Physically, the human eye would be overloaded with an infinite panorama. Instead, the eye concentrates on a smaller region, examines the region for interesting details, and then moves to examine a new region. If the eye discovers patterns in the graph, the mind becomes convinced that it understands even the hidden portion of the infinite graph. The "window" into the total graph satisfies our curiosity.

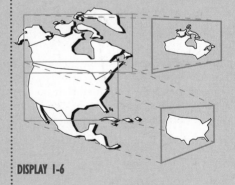

DISPLAY 1-6

Consider the graphic panorama in Display 1-6 and the selected windows to concentrate our attention on some details of the total scene. Notice how relocating a window can provide different views of the panorama. *Translation* is the mathematical term for sliding a window about to obtain a more advantageous view.

A window can be no larger than its house. Similarly, printing restricts window sizes to a textbook page. Because window construction has physical limits, let us increase the versatility of windows in another way. Imagine that the panes in our windows have the cameralike ability to magnify or reduce a transmitted image. Like fun house mirrors, the panes are even versatile enough to magnify in one direction while reducing in another direction. Of course, then the image becomes distorted. Consider the windows in Display 1-7, all displaying parts of a graph.

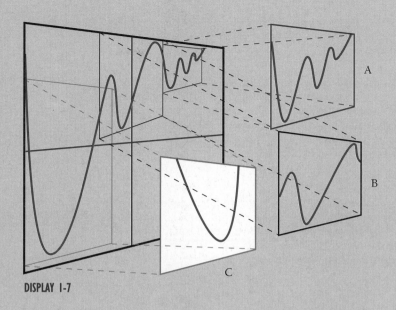

DISPLAY 1-7

Physically, all the windows in Display 1-7 are the same size. However their magnification properties allow them to transmit different amounts of information from the same graph. Window A magnifies some details in a small but interesting portion of the graph. Window B encompasses more of the total graph. Imagine that Window A "zooms" in on the details of the graph. Window B zooms out from the graph to provide a broader view.

Window C zooms in horizontally while zooming out vertically. Notice the distortion when the zooms are not uniform in both directions. In every case, you must pay close attention to the scales that indicate whether a view highlights details or expands on broad patterns. The aspect ratio measures the distortion in graphs due to unequal horizontal and vertical scaling. In summary, we translate a window to a new location. Transformation not only translates but also rescales the window to improve our view.

Finally, like a camera, it is possible to rotate a window to reorient our view. See Display 1-8.

Throughout this text, we use windows to provide a view into a graph. Often, however, we simply refer to a given window view as a graph. You must visualize the graph beyond the given window view.

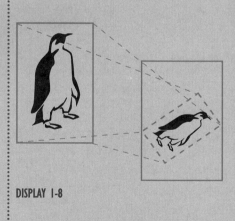

DISPLAY 1-8

CHAPTER TEST

1. For the pairs of points $(-5, 4)$, $(11, -12)$ determine (a) the distance between the two points, (b) the coordinates of the midpoint of the segment joining the points, and (c) the average rate of change between the points. Graph the two points and appropriately label the graph with the answers from parts a, b, and c.

   Are the points $(0, 3)$, $(4, 0)$, $(-3, -1)$, $(0, -3)$ the vertices of a rectangle?

2. (a) Rewrite each of the following linear equations in the slope-intercept form by solving the equation for y. (b) Determine the slope of the line for each equation. (c) Give the x-intercept for each of the lines. (d) Give the y-intercept for each of the lines.

   $$\frac{y}{3} + \frac{x}{4} = 1$$

   $$3y - 6x = 18$$

3. Write a linear equation for the lines described by going
   (a) through $(2, 5)$, with slope 3.
   (b) through $(1, 4)$ and $(3, -2)$.

4. (a) Write and equation for the circle with center at $(2, -3)$ and radius 7.

   (b) Write an equation for the parabola with vertex at $(1, 0)$ and y-intercept of $(0, 4)$.

Identify and sketch the graph of each of the following:

5. (a) $y = (x - 1)^2 - 5$ (b) $y - 3 = 6x - 1$

6. (a) $y = \dfrac{x - 2}{3}$ (b) $x^2 = 7 - (y + 2)^2$

7. $16x^2 + 16y^2 = 400$ 8. $(x - 1) - (y - 2)^2 = -16$

9. $x^2 + y^2 = -2$ 10. $x^2 + y^2 - 8x + 4y = 5$

FUNCTIONS

This chapter explores a special kind of relation called a *function*. As before, we examine this idea from the algebraic (numerical) and geometric (visual) viewpoints. Functions are a fundamental component for the study of calculus and play an important role in computer programming.

1 INTRODUCTION

The Modern Theory of Functions—that stateliest of all the pure creations of the human intellect.
—C. J. KEYSER

As discussed in the last chapter, pairing is a natural process in human thought. Our minds pair what we know with what is new and what we understand with what we want to understand. Paired items range from cause paired with effect to interest rates paired with inflation. Thus far, the pairings have been simple pairings of real numbers. In particular, (x, y) denoted an ordered pair of real numbers.

Patterns are often the only way to make "sense" from a collection of ordered pairs of real numbers. Chapter 1 indicated that some geometric patterns have corresponding algebraic patterns expressible as a formula of x and y. Other pairings have no formula representation. Consider the number of flips of a coin paired with the number of heads observed. If a formula model could predict the number of heads on a given sequence of coin flips, gambling would not depend on chance.

A function f is a collection of ordered pairs (x, y) such that no two distinct pairs of f have the same x value, in which case we say that y is a function of x.

Suppose f is a function. Denote the ordered pairs of f as (x, y). The collection of x values for f is the **domain** of f, denoted D_f. If the domain is not expressly given, assume the domain consists of all real numbers for which the rule of pairing produces a real y value. These assumed values form the **implied domain** of the function.

The **range** of f is the collection of associated y values. Denote the range by R_f. A **real function** is a function in which the domain and range are subsets of the set of real numbers.

EXAMPLE 2-1

Identifying Functions and Domains

(a) Indicate whether each of the following rules of pairing represent y as a function of x. If not a function, list some domain component (x value) paired with more than one range component (y value).

(b) If the pairing is a function, determine the implied domain.

Illustration 1:

$$y = 3x$$

Solution:

(a) This rule of pairing represents a function because any real value substituted for x produces exactly one real value for y. Standard arithmetic operations, addition, subtraction, multiplication and division (except by 0), each produces one answer. Therefore, formulas based on elementary arithmetic calculations using x or constants to produce y represent functions.

(b) Each real value of x produces one real value for y. There are no unacceptable values for x. Therefore, assume that $D_f =$ Reals. See Figure 2-1.

Illustration 2:

$$y^2 = x$$

Solution:

(a) The variable y is not a function of x. Substitute 9 for x, and y could assume values of 3 or -3. Thus the x value 9 produces two different y values. Substitute any positive number a for x. Then $y = \sqrt{a}$ or $y = -\sqrt{a}$. However, we need only one example to demonstrate that $y^2 = x$ does not represent a function. Such an example is known as a *counterexample*. Substituting 9 for x provides a *counterexample* to the claim that $y^2 = x$ represents a function. See Figure 2-2.

(b) Not applicable because the equation does not represent a function.

Illustration 3:

$$y = \frac{3}{x + 1}$$

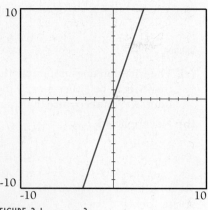

FIGURE 2-1 $y = 3x$

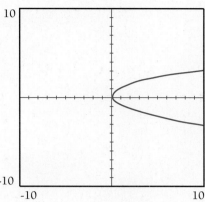

FIGURE 2-2 $y^2 = x$

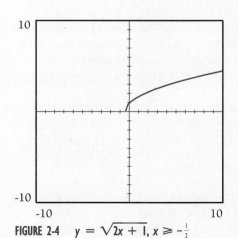

FIGURE 2-3 $y = \dfrac{3}{x+1}, x = -1$

Solution:

(a) Except for $x = -1$, arithmetic with each real value of x produces exactly one corresponding y value. Therefore y is a function of x.

(b) We cannot allow any denominator to become 0 (see Figure 2-3):

$$y = \frac{3}{x+1}, x + 1 \neq 0, x \neq -1$$

$$D_f = (-\infty, -1) \cup (-1, +\infty).$$

Illustration 4:

$$y = \sqrt{2x + 1}$$

Solution:

(a) The pairing is a function. Calculate y from x using multiplication, addition and taking the square root. Standard arithmetic produces a unique values for y for each value of x.

(b) We must take care with a radical having an even index. The radicand cannot be negative. Therefore

$$y = \sqrt{2x + 1}, \quad 2x + 1 \geq 0, \quad x \geq -\tfrac{1}{2}.$$

The domain is $[-\tfrac{1}{2}, +\infty)$. See Figure 2-4.

FIGURE 2-4 $y = \sqrt{2x + 1}, x \geq -\tfrac{1}{2}$

Illustration 5:
$$x^2 + y^2 = 9$$

Solution:

(a) The equation does not imply that y is a function of x. If x is chosen to be 0, then y may be either 3 or -3. Both (0, 3) and (0, -3) are pairs in this relation. Do you recognize the equation for a circle? See Figure 2-5.

(b) Not applicable.

Illustration 6:
$$y = \frac{x}{\sqrt{3 - x}}$$

Solution:

(a) The equation represents y as a function of x because arithmetic on x produces unique y values.

(b) Denominators may not be 0 and the radicand must not be negative. Hence, the domain is $(-\infty, 3)$.

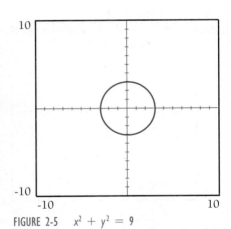

FIGURE 2-5 $x^2 + y^2 = 9$

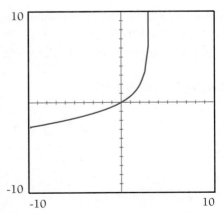

FIGURE 2-6 $y = \dfrac{x}{\sqrt{3 - x}}, 3 > x$

In the preceding illustration, the test whether a relation is a function is to decide if each domain value substituted for x produces a unique corresponding value for y. The x values are chosen first. As a result, x is the **independent variable**. Any value from the domain is available to substitute for the independent variable.

From the formula, the calculated value for y results from the value selected for x. Because the value of y *depends* upon the selected x value, y is the **dependent variable**. Solving for the dependent variable makes it easier to determine whether the equation represents a formula for a function.

Although variables do not have to be x or y in a function, this is a common tradition: we say that y is a function of x. This means that you determine y from some rule of pairing (perhaps an equation) using x. You could rewrite $3s + 4t = 5$ so that s is expressed as a function of t. Then t would be the independent variable and s would be the variable dependent on t. Consider that $x = 3 + 5y$ suggests that x is a function of y.

An equation solved for the dependent variable y in terms of x is an *explicit* rule for the function. For example, $y = 5x^2 - 3x + 7$ is a formula that defines y explicitly in terms of x. Explicit rules make functions easier to identify. Substituting a given value for x in an explicit rule produces the unique, corresponding y value.

It is now apparent why graphing technology labels one of the graphing modes as *function*. In the function mode, you define $y =$ using a formula involving x. Therefore, function mode requires formulas solved for the dependent variable y. Because the emphasis in this text is on functions, from this point forward your graphing technology becomes more useful.

If a formula does not explicitly express y as a function of x, we still may be able to solve for the dependent variable. Unsolved rules of pairing do not explicitly define a function. Rather the rule hints that we may be able to obtain a function from the formula. In this case, we say that the rule of pairing *implies* the existence of a function, and that function is called an **implied function**.

For example, consider the formula $5R - 3S = 15$. Determine whether R is a function of S. The instruction implies that S should be the independent variable, R should be the dependent variable and that we should solve the equation for R in terms of S: $R = \frac{3}{5}S + 3$. To determine whether the equation implied that S is a function of R, reverse the roles of S and R to obtain $S = -\frac{5}{3}R - 5$.

Consider the circle given by $x^2 + y^2 = 9$ (Figure 2-7). Substitute 0 for x and y could be 3 or -3: y is not a function of x. Similarly, substitute 0 for y and x is 3 or -3: x is not a function of y.

Solving for y reinforces this belief:

$$y^2 = 9 - x^2$$
$$y = \pm\sqrt{9 - x^2}$$

However, the solution implies at least two functions:

$$y = \sqrt{9 - x^2}, \quad y = -\sqrt{9 - x^2}$$

The graphs associated with these functions are the top half of the circle and the bottom half of the circle respectively. See Figure 2-8.

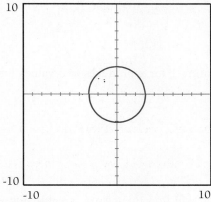

FIGURE 2-7 $x^2 + y^2 = 9$

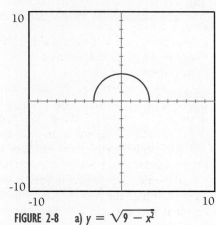 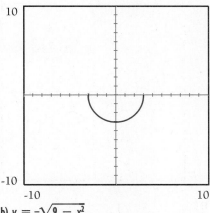

FIGURE 2-8   a) $y = \sqrt{9 - x^2}$ b) $y = -\sqrt{9 - x^2}$

Identifying Functions from Implicit Rules of Pairing

EXAMPLE 2-2

(a) Solve the following equations for the dependent variable y.

(b) Determine whether the resulting equation represents the rule of pairing for a function.

(c) If a function, give the domain implied by the rule of pairing.

Illustration 1:

$$xy = 10y + 1$$

Solution:

(a) $xy - 10y = 1$

$(x - 10)y = 1$

$$y = \frac{1}{x - 10}, \qquad x \neq 10$$

(b) A function.

(c) Domain: $(-\infty, 10) \cup (10, \infty)$.

Illustration 2:

$$y^2 - x = 3$$

Solution:

(a) $y^2 = x + 3$
$$y = \pm\sqrt{x + 3}$$

(b) Not a function. Notice that substituting 1 for x produces two values for y: 2, -2.

(c) Not applicable.

Illustration 3:

$$3x + 5y = 6 + 5y$$

Solution:

(a) Try to solve this equation for y. You will find the task impossible: the y terms disappear. This equation cannot represent y as a function of x.

Illustration 4:

$$3R + 4S = 10 \qquad\qquad \text{(express } R \text{ as a function of } S\text{)}$$

Solution:

(a) Solving for R: $R = (-4S + 10)/3$.

(b) R is a function of S.

(c) The domain for S is all real numbers.

The function notation is an outgrowth of the desire that formulas representing functions should be solved for the dependent variable.

S   I   D   E   B   A   R

Functions as a Black Box

Sometimes it is useful to think of functions as a "black box." By this, we mean that you may visualize a function as an abstract machine that manufactures y values (the result) based on the input of x values (the raw materials). The major requirement is that for a given raw material the product must be unique. This approach to functions reflects a computer science philosophy of programming that uses data abstraction and object manipulation. See Display 2-1.

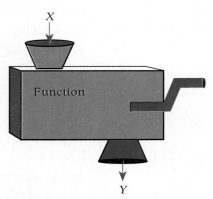

DISPLAY 2-1   Black box function machine

DEFINITION 2-2
Function Notation

> Let f be a function that pairs x values with unique y values. Then the **function notation** to abbreviate f pairs x with y is $f(x) = y$. Read the notation any one of the following ways:
>
>   **1.** f of x is y.
>   **2.** f at x is y.
>   **3.** f evaluated at x is y.
>   **4.** f pairs x with y.

The symbol $f(x)$ represents the dependent variable y. This notation simultaneously identifies the name of the function f, while specifying that the domain component x produces y.

EXAMPLE 2-3

Function Notation. In the following, use function notation to express y as a function of x.

Illustration 1:

$y = 5x + 7$ represents function f.

Solution:
Because $f(x) = y$, then $f(x) = 5x + 7$.

Illustration 2:

$y = x^2 - 8$ represents function g.

Solution:
$g(x) = y$; therefore, $g(x) = x^2 - 8$.

Illustration 3:

$xy = \sqrt{x + 5}$ implies a function h.

Solution:
Solve for the dependent variable y.

$$y = \frac{\sqrt{x + 5}}{x}$$

Domain: $x \neq 0$ and $x + 5 \geq 0$ $x \neq 0$ and $x \geq -5$.

Because $h(x) = y$, we have

$$h(x) = \frac{\sqrt{x + 5}}{x}, \qquad D_h = [-5, 0) \cup (0, +\infty).$$

Suppose function notation for a function f provides us with a formula for y. Then given a domain value for f we can calculate the corresponding range component. Evaluating the function f is the process of determining the y value f pairs with a given x value. The advantages of function notation become apparent during evaluation. Consider the instruction "if $y = x^2 + 3$, determine y given that $x = 2$." The same instruction becomes "if $f(x) = x^2 + 3$, evaluate $f(2)$." The symbol, $f(2)$, indicates the name of the function, f, and the value to substitute in the rule of pairing for x, 2.

Function Evaluation.   Evaluate the given function as indicated.

EXAMPLE 2-4

Illustration 1:

If $f(x) = x^2 + 3$ (Figure 2-9), determine (a) $f(2)$, (b) $f(3)$, and (c) $f(-2)$.

Solution:

(a) $f(x) = x^2 + 3$
$f(2) = (2)^2 + 3$
$f(2) = 7$

(b) $f(x) = x^2 + 3$
$f(3) = (3)^2 + 3$
$f(3) = 12$

(c) $f(x) = x^2 + 3$
$f(-2) = (-2)^2 + 3$
$f(-2) = 7$

Illustration 2:

If $g(x) = \sqrt{x + 1}$ (Figure 2-10), determine (a) $g(2)$, (b) $g(3.2)$, (c) $g(-2)$.

Solution:

(a) $g(2) = \sqrt{2 + 1}$
$= \sqrt{3}$

(b) $g(3.2) = \sqrt{3.2 + 1}$
$= \sqrt{4.2}$

(c) $g(-2) = \sqrt{-2 + 1}$
$= \sqrt{-1}$

Not allowed! The domain of g does not include -2. The implied domain is $Dg = [-1, +\infty)$.

Illustration 3:

If $h(x) = 3/(x - 2)$ (Figure 2-11), determine (a) $h(5)$, (b) $h(-3.2)$, (c) $h(2)$.

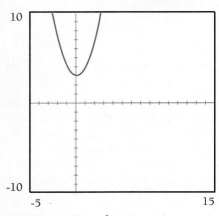

FIGURE 2-9 $f(x) = x^2 + 3$

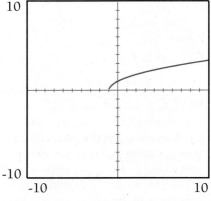

FIGURE 2-10 $g(x) = \sqrt{x + 1}$

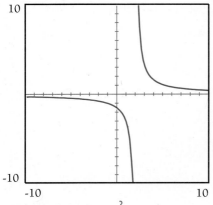

FIGURE 2-11 $h(x) = \dfrac{3}{x - 2}$

SIDEBAR

Assignment versus Equality

Most computer languages use some symbol to indicate that a value is to be stored in variable. Among the better assignment symbols to indicate the storage of a value is ← from APL: $x \leftarrow 5$, which stores 5 in the variable x. Similarly, $x \leftarrow x + 7$ replaces the value of x with the current value of x (5 from the previous assignment) plus 7. Thus, the new value of x is 12.

Other languages use different assignment statements. For example, Pascal uses $x := x + 7$ whereas BASIC uses $x = x + 7$. Especially in BASIC, we must not confuse assignment (replacement) with equality. Even in a mathematical context we sometimes use equals to represent replacement: $f(x) = f(x) + 7$. This statement makes no sense as an equation, but it can represent the replacement of $f(x)$ with $f(x) + 7$. We shall use a similar notation later when we discuss recursion.

Solution:

(a) $h(5) = \dfrac{3}{5 - 2}$

$= 1$

(b) $h(-3.2) = \dfrac{3}{-3.2 - 2}$

Without a calculator, algebraic manipulation allows us to obtain

$$h(-3.2) = -\frac{15}{26}$$

Direct entry into a calculator will approximate $h(-3.2)$. But then we must use $\cong$ to represent "is approximately." As a result,

$$h(-3.2) \cong -0.576231$$

In Section 3-7 we shall discuss what we mean by *approximately*.

(c) 2 is not in the domain of h; $h(2)$ does not exist.

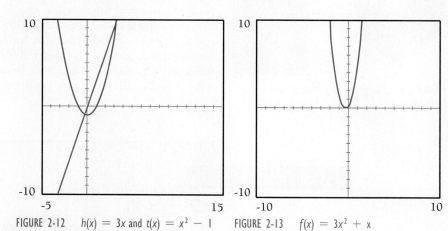

FIGURE 2-12 $h(x) = 3x$ and $t(x) = x^2 - 1$ FIGURE 2-13 $f(x) = 3x^2 + x$

Illustration 4:

If $k(x) = 3x$ and $t(x) = x^2 - 1$ (Figure 2-12), determine (a) $k(5)$, (b) $t(-2)$, (c) $k(3) + t(3)$.

Solution:

(a) $k(5) = 3(5)$

$= 15$

(b) $t(-2) = (-2)^2 - 1$

$= 3$

(c) $k(x) + t(x) = (3x) + (x^2 - 1)$

$k(3) + t(3) = 3(3) + (3)^2 - 1$

$= 17$

Illustration 5:

If $f(x) = 3x^2 + x$ (Figure 2-13), determine (a) $f(z)$, (b) $f(x + h)$, (c) $f(2z)$, (d) $f(w^2)$.

Solution:

(a) $f(z) = 3(z)^2 + (z)$

$\qquad = 3z^2 + z$

(b) $f(x + h) = 3(x + h)^2 + (x + h)$

$\qquad = 3(x^2 + 2hx + h^2) + (x + h)$

$\qquad = 3x^2 + 6hx + 3h^2 + x + h$

(c) $f(2z) = 3(2z)^2 + (2z)$

$\qquad = 12z^2 + 2z$

(d) $f(w^2) = 3(w^2)^2 + (w^2)$

$\qquad = 3w^4 + w^2$

The idea of a function is among the most elegant concepts in mathematics. But the power of functions does not end with mathematics. Computer science uses functions to do repetitive, specific tasks, such as selecting the smallest number from a list. The key to functions in computer science is that each execution of the function should return a single range component. The programming maxim is that a function should do only one job—but do it well.

Computer science provides further insight into functions. Think of a function as a process. We enter information, domain components, into the function. The function process operates to transform the information into data. Finally, the function produces the resulting range value data. See Figure 2-14.

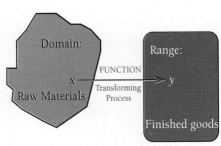

FIGURE 2-14   Diagram of function as transformation

S U M M A R Y

- A function f pairs x values from its domain with unique y values in its range. The notation $f(x)$ symbolizes the y value paired with the given x value.
- If the domain of a function is not explicitly stated, assume the implied domain to be any real number x for which the rule of pairing produces a unique real value for y. Functions where the domain and range are subsets of the real numbers are real functions.
- There are two areas of concern for implied domains.

    (a) Rational rule of pairing: $f(x) = N(x)/D(x)$, D_f is real numbers where $D(x) \neq 0$.

    (b) Rules of pairing with even indexed radicals: $f(x) = \sqrt[2n]{R(x)}$, where n is a counting number. D_f is all real numbers such that $R(x) \geq 0$.

- To evaluate a function f for a given value x means to determine the y value, represented by $f(x)$, which f pairs with x. Evaluate a function by replacing x with the given value everywhere in the rule of pairing, then do the arithmetic.

2-1*   EXERCISES

Identify which of the following expresses y as a function of x.

1. $y = x$ **2.** $y = 2x$

3. $y^2 = x - 1$ **4.** $y = x^2 - 1$

5. $y = x^3$ **6.** $y^2 = x^3$

7. $x^2 + y^2 = 4$ **8.** $x^2 - y^2 = 4$

9. $y = x^2 + 6x + 5$ **10.** $x = y^2 + 6y + 5$

Solve each of the following for the dependent variable y. Name each function f and express y as a function of x.

11. $3x + 2y = 5$ **12.** $2y - xy = 7$

13. $3y + x^2 = 2$ **14.** $x^2 - 5y = 7$

15. $xy = 10$ **16.** $xy = -5$

17. $3xy + y = x$ **18.** $5y - 2xy = 5$

19. $x^2 y - 5 = xy$ **20.** $xy + x = 2x^2 y$

State the *implied* domain for each of the following functions. Use graphing technology to determine the influence of the domain of the graph.

21. $f(x) = 3x + 1$ **22.** $g(x) = 1 - 5x$

23. $h(x) = x^2 + 6x$ **24.** $k(x) = 2x^2 - 7$

25. $t(x) = \sqrt{x}$ **26.** $P(x) = \sqrt{x - 2}$

27. $j(x) = \sqrt[5]{x + 1}$ **28.** $R(x) = \sqrt[3]{x}$

29. $t(x) = \sqrt{2 - 3x}$ **30.** $P(x) = \sqrt{3 - 5x}$

31. $g(x) = \dfrac{5}{x - 2}$ **32.** $f(x) = \dfrac{3}{x + 5}$

33. $k(x) = \dfrac{3x}{x^2 - 9}$ **34.** $h(x) = \dfrac{x}{x^2 - 4}$

35. $p(x) = \dfrac{x}{\sqrt{x + 7}}$ **36.** $s(x) = \dfrac{3}{\sqrt{x - 2}}$

37. $w(x) = \dfrac{\sqrt{x}}{x + 2}$ **38.** $q(x) = \dfrac{\sqrt{x}}{x + 1}$

39. $f(x) = \sqrt{x} + \dfrac{1}{x}$ **40.** $g(x) = \sqrt{x} - \dfrac{1}{x}$

Evaluate $f(x) = x^2 + 1/x$ as indicated.

41. $f(4)$ **42.** $f(-3)$

43. $f(2)$ **44.** $f(-1)$

45. $f(0)$ **46.** $f(-0)$

47. $f(-2)$ **48.** $f(1)$

49. $f(-4)$ **50.** $f(3)$

Evaluate $g(x) = x^3 - x$ as indicated.

51. $g(-3)$ **52.** $g(4)$

53. $g(-1)$ **54.** $g(2)$

55. $g(0.5)$ **56.** $g(0.25)$

57. $g(a)$ **58.** $g(-b)$

59. $g(a + b)$ **60.** $g(a - b)$

Determine $f(5)$ for each of the following functions f.

61. $f(x) = 3x + 1$ **62.** $f(x) = 2x - 5$

63. $f(x) = \sqrt{x + 4}$ **64.** $f(x) = \sqrt{x - 1}$

65. $f(x) = \dfrac{x}{x + 2}$ **66.** $f(x) = \dfrac{x + 2}{x}$

67. $f(x) = \sqrt{x - 1} + \dfrac{10}{x}$

68. $f(x) = \sqrt{x + 4} + \dfrac{x + 5}{x}$

69. $f(x) = \dfrac{2x - 1}{\sqrt{x}}$ **70.** $f(x) = \dfrac{\sqrt{x}}{3x + 1}$

71. $f(x) = \sqrt[4]{72}$ **72.** $f(x) = -3$

Evaluate $f(x) = 5x - x^2$ as indicated.

73. $f(w)$ **74.** $f(p)$ **75.** $f(z + 3)$

76. $f(q + 2)$ **77.** $f(x + t)$ **78.** $f(x + h)$

79. $f(3t)$ **80.** $f(5t)$ **81.** $f(-x)$

82. $f(-z)$ **83.** $f(x^2)$ **84.** $f(x^3)$

85. $f(x + h) - f(x)$ **86.** $f(b) - f(a)$

87. $\dfrac{f(x + h) - f(x)}{h}$ **88.** $\dfrac{f(b) - f(a)}{b - a}$

89. $\dfrac{f(x) - f(x_0)}{x - x_0}$ **90.** $\dfrac{f(x + \Delta x) - f(x)}{\Delta x}$,

where Δx represents a single symbol

For each of the following functions f, evaluate $f(-x)$.

91. $f(x) = 5x$ **92.** $f(x) = x^2 + 3$

93. $f(x) = x^4 - 3x^2 + 5$ **94.** $f(x) = x^3 - 7x$

95. $f(x) = x^{15}$ **96.** $f(x) = x^{28}$

Problem numbers in this color indicate an application. Problem numbers in this color indicate a preview of calculus. **Problems marked with an asterisk(*) are more difficult or require extra care.**

For each of the following functions f (Problems 97–106), determine $[f(x + h) - f(x)]/h$.

97. $f(x) = x^2 + 3$

98. $f(x) = 5 - x^2$

99. $f(x) = 2x - 3x^2$

100. $f(x) = x^2 - 4x$

101. $f(x) = 5x - 2$

102. $f(x) = 3 - 4x$

103. $f(x) = x^3$

104. $f(x) = 2x^3$

105. $f(x) = \sqrt{x}$

106. $f(x) = \sqrt{x} + 3$

107. For $5z - 2w = 6$, write w as a function of z.

108. For $3t - 6d = 8$, write d as a function of t.

109. For $5z - 2w = 6$, write z as a function of w.

110. For $3t - 6d = 8$, write t as a function of d.

111. For $3x + 2y = 9$, write x as a function of y.

112. For $5x - 3y = 11$, write x as a function of y.

113. For $d + 16t^2 - 120t = 0$, write d as a function of t.

114. For $s + 7.5t^2 + 200t = 0$, write s as a function of t.

*115. For $d + 16t^2 - 120t = 0$, write t as a function of d.

*116. For $s + 7.5t^2 + 200t = 0$, write t as a function of s.

117. Suppose that candy sells for $2.25 a piece. Then the total cost is a function of the number of pieces (n) sold. Call this function c and write a rule of pairing for c.

118. The revenue acquired from a bicycle rented at $7.50/hour is a function of the number (n) of hours rented. Call this function R and write a rule of pairing for R.

Problem numbers in this color indicate an application. Problem numbers in this color indicate a preview of calculus. Problems marked with an asterisk(*) are more difficult or require extra care.

2-1*   PROBLEM SET

1. (a) Ted stands 5 feet away from a lamppost 12 feet tall. Make a diagram to represent the lamppost, Ted and a beam of light from the lamp post that defines the limit of Ted's shadow. Label the length of Ted's shadow as y. If Ted is 6 feet tall, use similar triangles to determine the length of Ted's shadow.

(b) Now suppose that Ted stands x feet away from the lamppost. Use similar triangles to express the length of Ted's shadow as a function of x.

(c) Suppose that Ted moves to the base of the lamppost ($x = 0$). He starts a stopwatch and begins to walk away from the lamppost at the rate of 2 feet/second. Express Ted's distance x from the base of the lamp as a function of the time t he has walked. (This is a model.)

(d) Relabel the diagram from part a so that Ted's distance from the lamppost uses t (time) instead of x (feet). Hint: see parts b and c.

(e) See parts c and d. Express the length of Ted's shadow as a function of the time t that Ted has walked away from the base of the lamppost.

(f) The preceding functions model an event. Describe, in words, what happens to the length of Ted's shadow as Ted increases his distance from the lamppost.

Describe how the time that Ted walks affects the length of his shadow.

(g) What kind of values are legitimate for t?

(h) What kind of results are legitimate for y?

2. (a) Mary wishes to wind a line around a cylinder. From geometry the circumference of the cylinder is 2π times the radius. Draw a diagram representing the cylinder. If the cylinder radius is 10 cm, determine the circumference.

(b) Suppose that as the line winds about the cylinder it "stacks up." If the line is 0.1 cm in diameter, then after one revolution, the radius of the cylinder would increase to (approximately) 10.1 cm. After two revolutions, the radius would be 10.2 cm. If the cylinder completes x revolutions express the circumference c of the cylinder as a function of x.

(c) Approximate the length of line y wound on the cylinder as a function of the number of revolutions.

(d) What are the limitations on values for x? How would you interpret the results for $x = 3.5$ revolutions?

(e) Suppose that the cylinder rotates at a uniform 10 revolutions per minute (10 rpm). Express x as a function of time t.

(f) See parts c and e. Express y as a function of t.

3. (a) Suppose a rope of uniform density weighs 1000 grams per meter. If 3.5 meters of a 10 meter rope is dangled

*The handwriting icon indicates that for these problems you should draw conclusions or express concepts in your own words.·

over a table edge, determine the weight of the segment of dangling rope.

(b) Draw a diagram representing the rope dangling over the edge of a table. Label the amount of rope supported by the table as x. Express the amount of rope dangling off the table as a function of x. What are the allowable values for x?

(c) Express the weight of the dangling rope as a function of x.

(d) If the rope slides from the table at a uniform rate of 10 cm per minute, express the weight of the dangling segment as a function of time t. What is the domain of this function?

EXPLORATION

The following problems anticipate concepts introduced in the next section. Use graphing technology to graph each of the following in the default window $[-10, 10]$, $[-10, 10]$. Estimate the domain and range from the graph.

4. $y = x - 3$
6. $y = x^2 - 3$
5. $y = 2x - 1$
7. $y = 5 - x^2$

8. $y = x^3$
10. $y = x^4 - x^2$
*12. $y = \sin x$
*14. $y = e^x$
*16. $y = 1/x$
9. $y = x - x^3$
11. $y = x^4$
*13. $y = \cos x$
*15. $y = \ln x$
*17. $y = \sqrt{x}$

*The calculator logo indicates exercises in which a graphing calculator or GraphWindows can be used.

*The binoculars logo indicates problems that *explore* topics found in future sections of the book or in calculus.

2   GRAPHS OF FUNCTIONS

Mathematics is on the artistic side a creation of new rhythms, orders, designs, harmonies, and on the knowledge side, is a systematic study of various rhythms, orders, designs and harmonies.
—WILLIAM L. SCHAAFF

Graph is from the Greek meaning "written." Therefore, *photograph* means "written with light." Graphs give visual insight into relations. There are several graphic representations for functions.

Figure 2-15 gives one representation of a function f. This diagram shows that f pairs objects from its domain with objects in its range. No other information is included: not the nature of the domain, not the nature of the range, not a hint about f. If we knew something specific about f, for example $f(1) = 3, f(2) = 7, f(3) = 4$, and $f(4) = 0$, we could collect the data in tabular form.

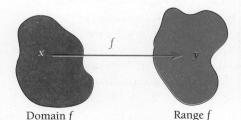

FIGURE 2-15   f: Domain → Range

x	→	$f(x)$
1	→	3
2	→	7
3	→	4
4	→	0

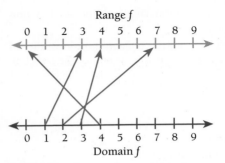

Range f

Domain f

FIGURE 2-16 $f{:}x \longrightarrow y$. Arrow representation

This leads to a more informative picture of f. Figure 2-16 shows an arrow representation of the given pairs of f.

If the domain of f is large, the arrow representation of f will become cumbersome. Are there graphic methods friendlier to the user?

Recall that $f(x) = y$ means the function f pairs x with y. As in Chapter 1, represent (x, y) as a point in the plane. Then, for $f(1) = 3$ use the point $(1, 3)$ as a point of the graph of f. More briefly, let $f(2) = 7$ mean that $(2, 7)$ is a point of the graph of f. If we think of f and the graph of f as identical, then $f(3) = 4$ means that $(3, 4)$ is a point of f. Similarly, $f(4) = 0$ means that $(4, 0)$ is a point of f. Recall that $(4, 0)$ lies on the x-axis so that f crosses the x-axis at 4.

DEFINITION 2-3
Graph of a Function

The **graph of a function** f consists of all points (x, y) such that $f(x) = y$.

EXAMPLE 2-5

Sketch the graph for each of the following functions.

Illustration 1:

Suppose that f is a function whose only points are as follows:

$$f(0) = 3, \qquad f(1) = 2, \qquad f(3) = 4, \qquad f(4) = -1$$

Solution: See Figure 2-17.

These points are equivalent to the ordered pairs $(0, 3)$, $(1, 2)$, $(3, 4)$ $(4, -1)$. Figure 2-17 shows the graph of f. Note that the graph consists of four distinct points. Such graphs are discrete graphs. By *discrete* we mean that the points are "isolated" or "distinct." The graph for f contains four distinct points: f is discrete.

Illustration 2:

Suppose $f(n) = 1/n$, where $n \in \{1, 2, 3, 4, \ldots\}$

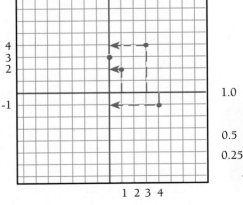

FIGURE 2-17 f

FIGURE 2-18 $f(n) = \dfrac{1}{n}, n \in \{1, 2, 3, 4 \ldots\}$

Solution:

$$f = \{(1, 1), (2, \tfrac{1}{2}), (3, \tfrac{1}{3}), (4, \tfrac{1}{4}), \ldots\}$$

The graph of f contains infinitely many points. Each point of f is isolated. The graph of f is discrete. See Figure 2-18.

Illustration 3:

$$g(x) = 3x, \text{ where } x \in (-\infty, \infty)$$

Solution: The formula for the function g is $y = 3x$. Therefore, g is a linear function. The graph of g is the line shown in Figure 2-19.

The points of g are not distinct but connected. Functions with such graphs are **continuous** functions. The graph of the continuous function g consists of infinitely many points, as would seem correct intuitively. Because there are no "gaps" between the points, sketch a continuous graph without lifting your pencil from the paper. Because of the separation of points, we must lift our pencil when graphing a discrete function.

Illustration 4:

$$h(x) = x^2$$

(Figure 2-20)

Solution: The function h has rule of pairing $y = x^2$. Hence h is a quadratic function. The graph of h is a parabola. Enter $y = x^2$ to produce the graph of h using technology. The graph of h is continuous.

Note: By custom, we name functions according to the nature of their formulas: *linear* and *quadratic* functions in the preceding illustrations, and because of the indicated division, a *rational* function in the following one.

Illustration 5:

$$k(x) = \frac{x}{x + 1}$$

Solution: The domain of k is the collection of all real numbers except -1. We have not previously discussed the graph of a function like k. One method to produce the graph is to plot numerous points of the graph and imagine the graph of k as shown in Figure 2-21.

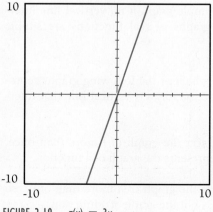

FIGURE 2-19 $g(x) = 3x$

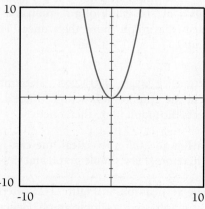

FIGURE 2-20 $h(x) = x^2$

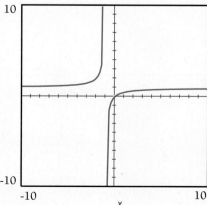

FIGURE 2-21 $k(x) = \dfrac{x}{x + 1}$

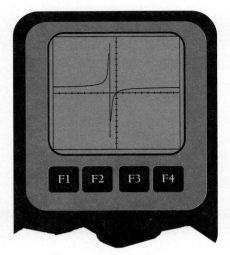

DISPLAY 2-2   Graphing calculator with display

A better method is to use graphing technology. First set the range to the default window: Xmin = -10, Xmax = 10, Xscl = 1, Ymin = -10, Ymax = 10, Yscl = 1. Be sure you are in the rectangular, function mode and that points will be connected. ▬

If you use technology to graph h, the graph may not look exactly like Figure 2-21 (see Display 2-2). An *extra* vertical line may connect the two separate pieces of the graph or the graph might not extend to the boundaries of your display. This is due to the methods used to display continuous graphs on discrete devices. Try zooming in on the interval containing $x = 1$ (see Section 1-2) or increasing the number of plotted points to improve the graph. A less satisfactory method is to switch from the connect mode (connected) to the plot mode (dot). In the dot mode adjacent plotted points remain unconnected and the graph does not appear continuous.

If you recognize patterns in the equations representing functions, then graphing those functions is easier: you know what "shape" to expect. Plotting individual points is always available when all else fails. Graphing technology produces graphs through a process that plots many individual points. Because graphing technology plots points, it produces discrete representations of graphs that are continuous. As a result, graphing technology may misrepresent a graphed function. In Display 2-2 the graphing calculator portrays a function that is more accurately graphed in Figure 2-21. The calculator did not recognize the behavior of the graph between closely plotted points.

Even with graphing technology, you have the responsibility of recognizing whether a graph is correct. The remainder of Chapter 2 explores clues to help you anticipate the graphs of functions based on their formulas. Graphing technology can help you explore the behavior of the example functions. We begin with the following theorem for the **vertical line test**.

Theorem 2-1
Vertical Line Test

> If f is a real function, no vertical line may intersect the graph of f more than once.

The proof of Theorem 2-1 follows immediately from the definition of *function*. The graph of a relation R is a simple graph if no vertical line intersects the graph more than once. The graphs of real functions are simple graphs.

EXAMPLE 2-6

Recognizing Graphs of Functions.   Determine which of the following graphs represent the graph of a function. If not a function, draw a vertical line that intersects the graph more than once.

Illustration 1:

In Figure 2-22 no vertical line can intersect the graph of f more than once. Therefore, f is a simple graph and thus represents the graph of a function.

Illustration 2:

The vertical line in Figure 2-23 intersects the graph of h more than once, so the graph is not a simple graph. Hence h is not the graph of a function.

Illustration 3:

The graph of t in Figure 2-24 is a simple graph. Therefore, t is a function.

Illustration 4:

The graph in Figure 2-25 of q is not a simple graph. The y-axis is a vertical line that intersects q more than once. As a result q is not a function.

Because of Theorem 2-1, the graph of a function will not backtrack horizontally as in Illustration 4. If we plot sample points of a function, the remainder of the curve representing the function "proceeds" through the plotted points from *left to right*.

The range and domain of a function have visual analogies. A sunrise shadow of the graph projected onto the y-axis represents the function's range: a "height" for the graph. The domain is like the "width" of the graph: imagine a shadowlike projection of the graph onto the x-axis. This projection is the function's domain. Imagining geometric projections of the function onto the x-axis and y-axis provides a method to represent the domain and range using interval notation.

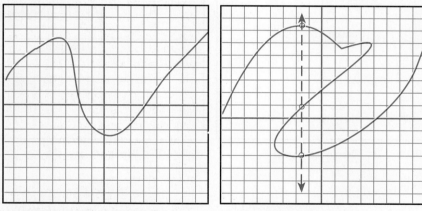

FIGURE 2-22   A simple graph f FIGURE 2-23   Vertical line test: h

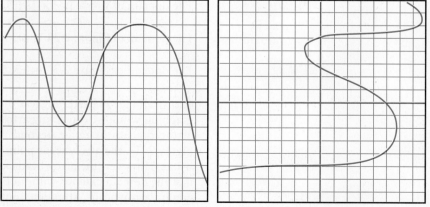

FIGURE 2-24 t FIGURE 2-25 q

EXAMPLE 2-7

Illustration 1:

Illustration 2:

Illustration 3:

Use interval notation to express the domain and range of the following functions.

Figure 2-26 displays the projections of f shaded upon the x- and y-axes. The domain of f is $(-\infty, +\infty)$, the range of f is $(-\infty, +\infty)$.

The domain of h is $(-\infty, +\infty)$, the range of h is $[0, +\infty)$; see Figure 2-27.

The domain of g is $(-\infty, +\infty)$, the range of g is $(-\infty, 3]$; see Figure 2-28.

Note: The rule of pairing for g is $g(x) = -(x - 2)^2 + 3$. The domain of g is obvious from the rule of pairing. We may be able to guess the range from the graph of g.

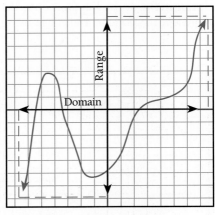

FIGURE 2-26   Domain and range of f

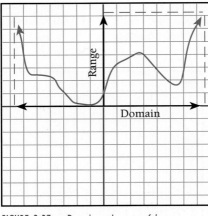

FIGURE 2-27   Domain and range of h

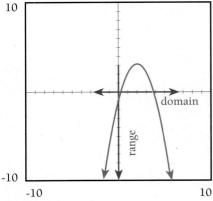

FIGURE 2-28   Domain and range of g,
$g(x) = -(x - 2)^2 + 3$

Illustration 4:

Illustration 5:

Illustration 6:

Illustration 7:

The domain of $xy = 1$ is $(-\infty, 0) \cup (0, +\infty)$, the range is $(-\infty, 0) \cup (0, +\infty)$; see Figure 2-29.

The domain of $y = \sqrt{9 - x^2}$ is $[-3, 3]$, the range is $[0, 3]$; see Figure 2-30.

The domain of Figure 2-31 is $[0, +\infty)$, the range is $(-\infty, 0]$.

The domain of Figure 2-32 is $[-3, 5)$, the range is $[-3, 3]$.

Note: Graph arrowheads suggest the extension of function pattern beyond the graphing window. Projecting arrowheads onto an axis suggests the extension of the domain or range beyond the window. This is not a general rule. In Figure 2-29, the arrowheads pointing toward $x = 0$ do not project well on the x-axis. However, the arrowheads at the far left and far right correctly indicate that the domain extends without bounds. Because we cannot see the complete graph, arrowheads stimulate our imgination beyond the bounds of the window.

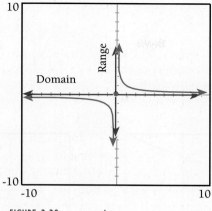

FIGURE 2-29 $xy = 1$

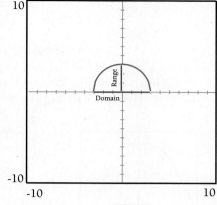

FIGURE 2-30 $y = \sqrt{9 - x^2}$

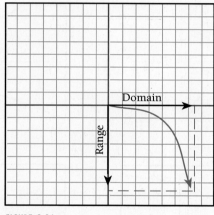

FIGURE 2-31

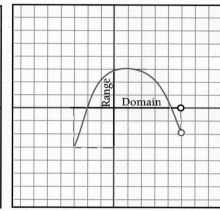

FIGURE 2-32

Summary

Simple graphs represent real functions. To test whether a graph is simple, use the vertical line test: no vertical line can intersect a simple graph more than once. Project the graph of the function onto the x-axis to represent the domain of the function. Project the graph of the function onto the y-axis for the range of a function.

2-2   E X E R C I S E S

Determine whether each of the graphs, 1–20, represents the graph of a function. If not a function draw a vertical line that intersects the graph more than once.

I.
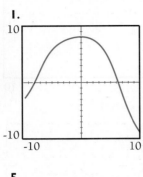

2.
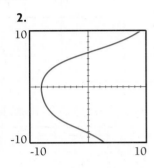

3.
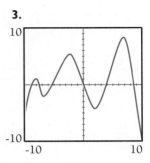

4.
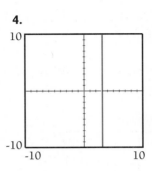

5.
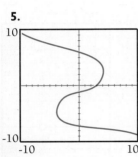

6.
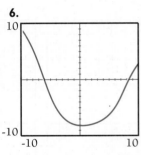

7.
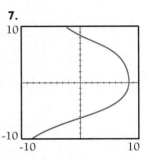

8.
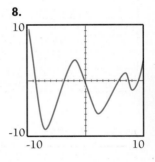

9.
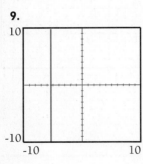

10.
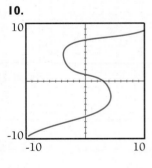

11.
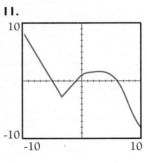

12.
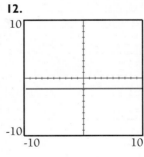

13.
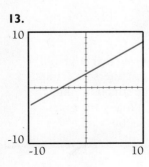

14.
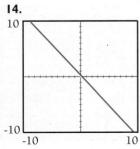

15.
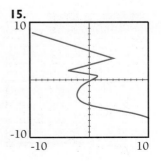

16.
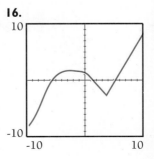

17.

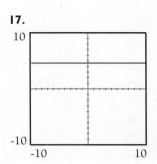

18.

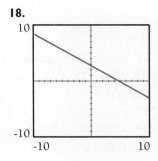

19.

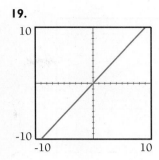

20.

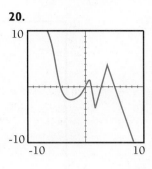

Give the domain and range for each of the graphs (21–30.)

21.

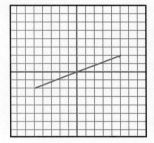

22.

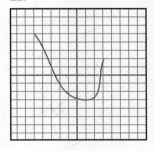

23.

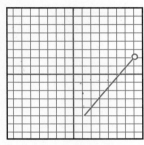

24.

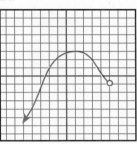

25.

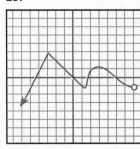

26.

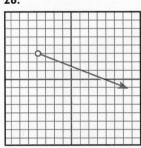

27.

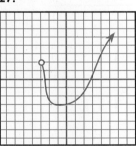

28.

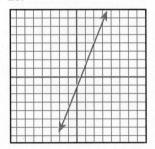

29.

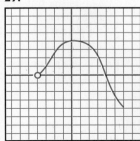

30.

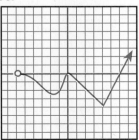

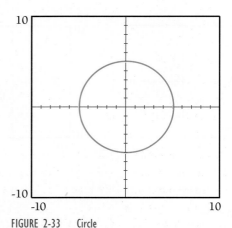

FIGURE 2-33   Circle

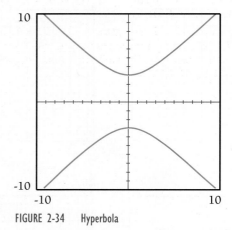

FIGURE 2-34   Hyperbola

Graph the following functions. Indicate the domain of each function.

31. $f(x) = 3x - 1$ **32.** $g(x) = 2x + 1$

33. $k(x) = 5 - 0.5x$ **34.** $j(x) = -3 + 0.3x$

35. $q(x) = x^2$ **36.** $f_2(x) = -x^2$

37. $p(x) = (x - 3)^2 + 1$ **38.** $z(x) = (x + 2)^2 - 1$

39. $f_3(x) = 3 - x^2$ **40.** $f_4(x) = x^2 - 3$

41. $c(x) = x^3$ **42.** $m(x) = -x^3$

43. $f_5(x) = \sqrt{x}$ **44.** $n(x) = -\sqrt{x}$

45. $R(t) = -\dfrac{1}{t}$ **46.** $s(t) = \dfrac{1}{t}$

47. $A(x) = |x|$ **48.** $B(x) = |x| + 1$

49. $f^*(u) = u\sqrt{9 - u^2}$ **50.** $g^*(u) = u\sqrt{4 - u^2}$

51. Prove Theorem 2-1.

52. Suppose that a horizontal line intersects the graph of g more than once. Can g be a function? Explain.

53. Consider the graph in Figure 2-33, which is not the graph of a function. It is possible to erase a portion of the graph so that the resulting graph represents a function with the same domain as the original graph. How many different such functions can be formed by judicious erasure?

54. See Problem 53. Answer the same question for the graph shown in Figure 2-34.

 2-2   P R O B L E M   S E T

1. (a) Marilyn collects data during an experiment to calibrate a spring in a spring scale. She hangs a weight from the spring and records as an ordered pair the weight and the "stretch" in the spring: (10 kg, 15 cm), (5 kg, 7 cm), (16 kg, 24 cm), (9 kg, 14 cm), (20 kg, 30 cm), (21 kg, 30 cm), (10 kg, 16 cm), (2 kg, 3 cm), (13 kg, 20 cm), (14 kg, 21 cm). Plot these points on a Cartesian coordinate using the weight in kilograms on the x-axis and the stretch in centimeters on the y-axis.

(b) Calculate the "average" (mean) of all weights used during the experiment. Similarly, calculate the mean of the stretch. Plot this "average" point of the experiment with the other points.

(c) Are these points of a function? Use a straight edge to "fit" some lines to the plotted data. Be sure each line goes through the average point. Are there any other points the line should pass through? Remember that (0 kg, 0 cm) is an unlisted point of this experiment: if no weight is hung the spring is not stretched. Can any line pass through all the data points?

(d) Choose a line that seems to fit the data best. Write an equation for this line. Now express the weight suspended in the experiment as a function of the corresponding stretch in the spring.

(e) What is the domain of your function? Use your function to predict the weight of an object that stretches the spring 25 cm.

EXPLORATION

The following problems anticipate concepts introduced in the next section. Use graphing technology to graph each of the following in the graphing window $x \in [-3, 3]$, $y \in [-20, 20]$:

2. (a) $y = x$

(b) $y = x^3$

(c) $y = x^5$

(d) $y = 2x^3 - 4x$

(e) $y = x^5 - x^3 + x$

(f) Discuss any common pattern to the rules of pairing.

(g) Discuss any common pattern to the graphs.

(h) Does $y = x^2$ have the same common pattern?

3. (a) $y = x^2$

(b) $y = x^2 + 2$

(c) $y = x^4$

(d) $y = x^4 - x^2 - 3$

(e) $y = x^2 - x^4$

(f) Discuss any common pattern to the rules of pairing.

(g) Discuss any common pattern to the graphs.

4. (a) $y = x + x^2$

(b) $y = 1 + x$

(c) $y = x^2 - 3x$

(d) $y = x^4 - x^3 + x^2 - x + 1$

(e) $y = x^5 - x^4$

(f) Discuss any common pattern to the rules of pairing.

(g) Discuss any common pattern to the graphs.

(h) Compare the results of Problems 2, 3 and 4.

5. Graph $f(x) = x + 3$ and $g(x) = (x^2 - 2x - 15)/(x - 5)$ in separate windows. Are the graphs the same? Are the functions the same? Justify your answers. For each graph, zoom in on $(5, f(5))$

6. Graph $f(x) = x - 2$ and $g(x) = (x^2 - 5x + 6)/(x - 3)$ in separate windows. Are the graphs the same? Are the functions the same? Justify your answers. For each graph, zoom in on $(3, f(3))$.

---

3   ODD OR EVEN FUNCTIONS: SYMMETRY

An idea reaches its full usefulness only when one understands it so well that one believes that one has always possessed it and becomes incapable of seeing it as anything but a trivial and immediate remark.
—HENRI LEBESQUE

Symmetry is often observed in the physical world, both at the macroscopic and microscopic scales. In the field of nuclear physics, it can even be observed at the submicroscopic scale. Electrons and positrons, neutrinos and antineutrinos, and protons and antiprotons are just a few examples of particle-antiparticle pairs of the subatomic world that exhibit symmetric characteristics.

Examine the butterfly wings in Display 2-3. The butterfly's wings are not identical. They are approximately mirror images. Mirror symmetry relates an object to its mirror reflection. Mirrors provide one of the most obvious devices for creating symmetrical displays. Symmetry helps us to describe the world about us.

Because graphs are models for many situations in nature, it is appropriate that symmetry be a property of some graphs. There are many different types of symmetry. We shall concentrate on three of the most commonly discussed kinds of symmetry. Windows provide the vehicle for a visual discussion.

DISPLAY 2-3   A butterfly

Moreover, one use of symmetry is to improve the sketching of graphs. If a function displays symmetry, fewer points need be plotted to sketch the remainder of the graph. See Figure 2-35.

DEFINITION 2-4
Symmetry

Visual definitions:

1. A graph f is symmetric about the x-axis if for any window of the graph with horizontal bounds equal distance from the x-axis, "folding" the window along the x-axis causes the graph to coincide with itself. The x-axis serves as the mirror for the mirror image symmetry of the portion of f above the x-axis to the portion of f below the x-axis (Display 2-4).

2. A graph f is symmetric about the y-axis if for any window with vertical bounds equal distance from the y-axis, "folding" the window of the graph along the y-axis causes the graph to coincide with itself. The y-axis provides mirror image symmetry for f (Display 2-5).

3. A graph f is symmetric about the origin if for any window of the graph with horizontal boundaries equal distance from the origin and vertical boundaries equal distance from the origin, "folding" the window along both the x-axis and the y-axis causes the graph to coincide with itself. Alternately, the origin acts like camera lens to inverte and reverse the image across the origin to opposite quadrants of the plane (Display 2-6). This is **origin symmetry**.

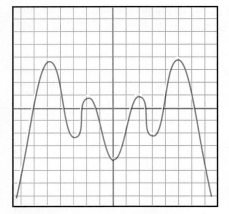

FIGURE 2-35   Window for symmetry

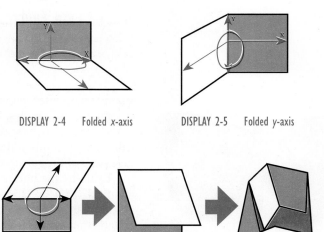

DISPLAY 2-4   Folded x-axis   DISPLAY 2-5   Folded y-axis

DISPLAY 2-6   Folded both axes

EXAMPLE 2-8

Visual Symmetry. Determine any symmetry about the x-axis, y-axis or origin exhibited by the following graphs.

Illustration 1:   The graph in Figure 2-36 is symmetric about the y-axis. Imagine folds along the x-axis and y-axis separately and then along both axes. Only the fold along the y-axis causes the graph to exactly overlap itself.

Illustration 2:   The graph in Figure 2-37 is symmetric about the x-axis. An imagined fold along the x-axis causes the graph to overlap itself. Folds along the y-axis or along both axes do not cause an overlap. Notice that this graph *does not* represent a function.

Illustration 3:   The graph in Figure 2-38 is symmetric about the x-axis, y-axis and origin. The graph overlaps itself when folded along the x-axis, the y-axis and when folded along both axes.

Illustration 4:   The graph in Figure 2-39 is symmetric about the origin. The graph does not overlap when folded about the x-axis or the y-axis. But when folded about both axes the graph overlaps itself.

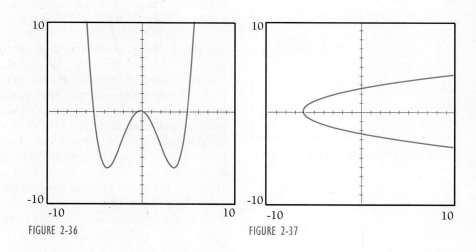

FIGURE 2-36   FIGURE 2-37

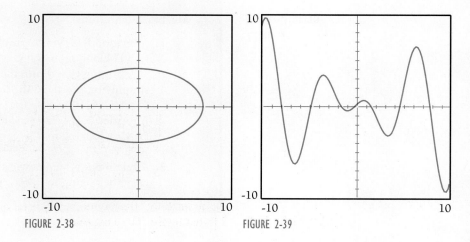

FIGURE 2-38   FIGURE 2-39

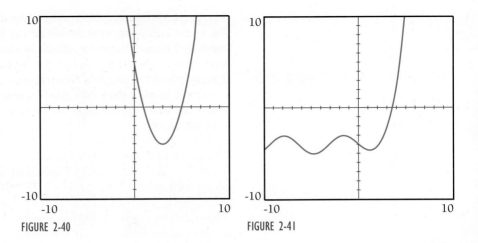

FIGURE 2-40   FIGURE 2-41

Illustration 5:   The graph in Figure 2-40 is symmetric, but not about axes or origin. The graph has some type of symmetry, but is not symmetric about the *x*-axis, *y*-axis or origin.

Illustration 6:   The graph in Figure 2-41 has no symmetry.   ▄

Were we correct in the preceding illustrations? Did the folded graphs in each window really coincide? Since graphs are visual and, by that nature, inexact, perhaps the overlap was only close. Let us reformulate the definitions of *x*-axis, *y*-axis and origin symmetries algebraically. Recall that a relation *R* is a collection of ordered pairs of real numbers. Because collections of ordered pairs of real numbers correspond to the points in the plane, if (x, y) belongs to *R*, denoted by $(x, y) \in R$, then (x, y) represents a point in the graph of *R*. However, the use of **relation** emphasizes ordered pairs of numbers rather than points in a plane. Thus, if we want to reformulate the definition of symmetry with algebraic precision we refer to relations.

DEFINITION 2-5
Symmetry

Algebraic definitions:

1. A relation *R* is symmetric about the *x*-axis if for every point (x, y) belonging to *R*, $(x, -y)$ also belongs to *R*. Therefore, substituting $(x, -y)$ into the rule of pairing for *R* and simplifying produces the original rule of pairing, see Figure 2-42.
2. A relation *R* is symmetric about the *y*-axis if for every point (x, y) belonging to *R*, $(-x, y)$ also belongs to *R*, see Figure 2-43.
3. A relation *R* is symmetric about the origin if for every point (x, y) belonging to *R*, $(-x, -y)$ also belongs to *R*; see Figure 2-44.

Suppose that an equation in x and y describes a relation. Apply the definition for y-axis symmetry by replacing (x, y) in the equation with $(-x, y)$. Simplify the resulting equation. If the simplified equation is equivalent to the original rule of pairing, the graph is symmetric about the y-axis. A similar method works for origin symmetry.

The following theorem reduces the arithmetic required to check relations for symmetry.

Theorem 2-2
Symmetry

> If a relation R is symmetric about any two of the following three, x-axis, y-axis, origin, then R must be symmetric about the third.

Partial Proof: Suppose R is symmetric about the x-axis and y-axis, show that R is symmetric about the origin. Let $(x, y) \in R$. Because R is symmetric about the x-axis then $(-x, y) \in R$. But R is symmetric about the y-axis, so that $(-x, -y) \in R$. From $(x, y) \in R$ we have obtained $(-x, -y) \in R$, which satisfies the definition of symmetric about the origin.

The proofs that x-axis and origin symmetry imply y-axis symmetry, and that y-axis and origin symmetry imply x-axis symmetry are similar and we leave them as exercises.

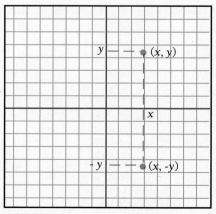

FIGURE 2-42 (x, y), $(x, -y)$

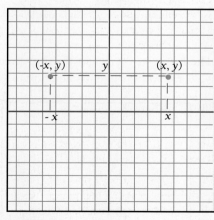

FIGURE 2-43 (x, y), $(-x, y)$

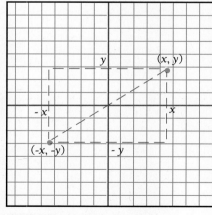

FIGURE 2-44 (x, y), $(-x, -y)$

EXAMPLE 2-9

Illustration 1:

Algebraic Symmetry. Determine the x-axis, y-axis or origin symmetry of the following relations.

$$y = x^2$$

Original equation $y = x^2$	x-axis test: substitute $(x, {}^-y)$	y-axis test: substitute $({}^-x, y)$	Origin test: substitute $({}^-x, {}^-y)$
$y = x^2$	$({}^-y) = (x)^2$	$(y) = ({}^-x)^2$	$({}^-y) = ({}^-x)^2$
Simplify	${}^-y = x^2$	$y = x^2$	${}^-y = x^2$
Compare to original	Different	Same	Different
Conclusion		Symmetric	

The graph of $y = x^2$ in Figure 2-45 is symmetric about the y-axis.

Note: After checking x-axis and y-axis symmetry in Figure 2-45, we need not check for symmetry about the origin because it is impossible for a relation to be symmetric about exactly two of the three. See Theorem 2-2.

$$x^2 + y^2 = 4$$

Illustration 2:

Original equation $x^2 + y^2 = 4$	x-axis test: substitute $(x, {}^-y)$	y-axis test: substitute $({}^-x, y)$	Origin test: substitute $({}^-x, {}^-y)$
$x^2 + y^2 = 4$	$(x)^2 + ({}^-y)^2 = 4$	$({}^-x)^2 + (y)^2 = 4$	$({}^-x)^2 + ({}^-y)^2 = 4$
Simplify	$x^2 + y^2 = 4$	$x^2 + y^2 = 4$	$x^2 + y^2 = 4$
Compare to original	Same	Same	Same
Conclusion	Symmetric	Symmetric	Symmetric

The result in the origin test was predictable by Theorem 2-2. The graph of $x^2 + y^2 = 4$ is symmetric about the x-axis, y-axis and origin.

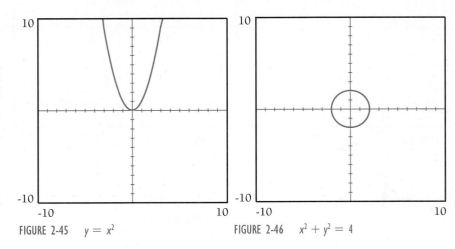

FIGURE 2-45 $y = x^2$ FIGURE 2-46 $x^2 + y^2 = 4$

Illustration 3:

$$y = \frac{1}{x}$$

Original equation $y = \dfrac{1}{x}$	x-axis test: substitute $(x, -y)$	y-axis test: substitute $(-x, y)$	Origin test: substitute $(-x, -y)$
$y = \dfrac{1}{x}$	$-y = \dfrac{1}{x}$	$y = \dfrac{1}{-x}$	$-y = \dfrac{1}{-x}$
Simplify	$y = -\dfrac{1}{x}$	$y = -\dfrac{1}{x}$	$y = \dfrac{1}{x}$
Compare to original Conclusion	Different	Different	Same Symmetric

The graph of $y = 1/x$ in Figure 2-47 is symmetric about the origin.

Illustration 4:

$$y = x + 1$$

Original equation $y = x + 1$	x-axis test: substitute $(x, -y)$	y-axis test: substitute $(-x, y)$	Origin test: substitute $(-x, -y)$
$y = x + 1$	$-y = x + 1$	$y = -x + 1$	$-y = -x + 1$
Simplify	$y = -x - 1$	$y = -x + 1$	$y = x - 1$
Compare to original	Different	Different	Different
Conclusion			

The graph of $y = x + 1$ in Figure 2-48 is not symmetric about the x-axis, y-axis or origin.

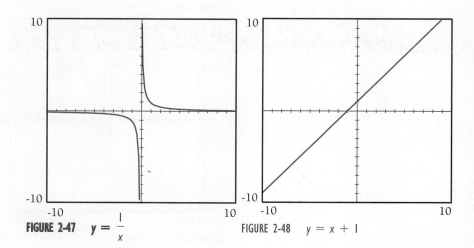

FIGURE 2-47 $y = \dfrac{1}{x}$ FIGURE 2-48 $y = x + 1$

Illustration 5:

$$y^2 = x^3$$

Original equation $y^2 = x^3$	x-axis test: substitute $(x, -y)$	y-axis test: substitute $(-x, y)$	Origin test: substitute $(-x, -y)$
$y^2 = x^3$	$(-y)^2 = x^3$	$y^2 = (-x)^3$	$(-y)^2 = (-x)^3$
Simplify	$y^2 = x^3$	$y^2 = -x^3$	$y^2 = -x^3$
Compare to original	Same	Different	Different
Conclusion	Symmetric		

The graph of $y^2 = x^3$ in Figure 2-49 is symmetric to the x-axis.   ▬

What kind of symmetry do we expect in functions? The following definition and theorems provide both terminology and a simple method for checking symmetry in functions.

DEFINITION 2-6
Odd and Even Functions

> **1.** A function f is *odd* if $f(-x) = -f(x)$ for every x in the domain of f.
> **2.** A function f is *even* if $f(-x) = f(x)$ for every x in the domain of f.

Theorem 2-3
Odd Function Symmetry

> A function f is *odd* if and only if f is symmetric about the origin.

Partial Proof: Suppose that f is odd (see Figure 2-50). By definition, $f(-x) = -f(x)$ for every x in the domain of f. Suppose x is in the domain of f, then $(x, f(x))$ is a point of the graph. Note that $(-x, f(-x))$ is a point of the graph because $f(-x)$ exists in the hypothesis. Substitute $-f(x)$ for $f(-x)$ and $(-x, -f(x))$ is a point of the graph as required for symmetry about the origin.

The remainder of the proof and the proof of the next theorem are left as exercises.

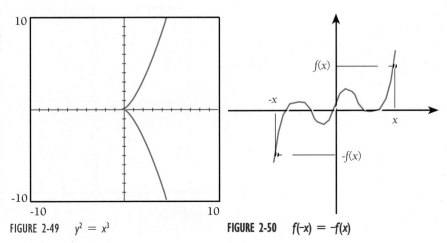

FIGURE 2-49 $y^2 = x^3$ **FIGURE 2-50** $f(-x) = -f(x)$

Theorem 2-4
Even Function Symmetry

A function f is *even* if and only if f is symmetric about the *y*-axis.

Check a function for symmetry about the origin or *y*-axis by determining whether the function is odd or even. Substituting $-x$ for x in the rule of pairing for a function provides an easy odd or even (or neither) check. If the original rule of pairing results, the function is even. If the negative of the original rule of pairing results, the function is odd. If neither the same formula nor the negative results, then the function is neither even nor odd.

EXAMPLE 2-10

Symmetry and Odd and Even Functions. Determine whether each of the following functions is odd, even or neither. Use this information to predict the symmetry of the functions graph about the *y*-axis or origin.

Illustration 1:

$$f(x) = x^2 - 3$$

Solution: $f(-x) = (-x)^2 - 3$
$$= x^2 - 3 \qquad \text{(same as } f(x)\text{)}$$
$$f(-x) = f(x) \qquad \text{(even)}$$

The graph of f in Figure 2-51 is symmetric about the *y*-axis.

Illustration 2:

$$g(x) = x^3 - 3x$$

Solution: $g(-x) = (-x)^3 - 3(-x)$
$$= -x^3 + 3x \qquad \text{(opposite of } g(x)\text{)}$$
$$= -(x^3 - 3x)$$
$$g(-x) = -g(x)$$

The function g is odd, and the graph of g in Figure 2-52 is symmetric about the origin.

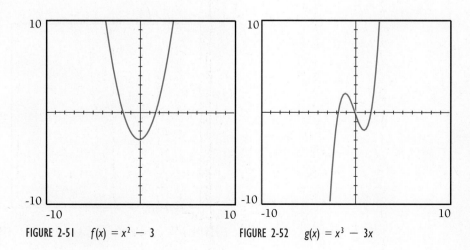

FIGURE 2-51 $f(x) = x^2 - 3$ FIGURE 2-52 $g(x) = x^3 - 3x$

Illustration 3:

$$h(x) = x^2 + 2x$$

Solution: $h(-x) = (-x)^2 + 2(-x)$
$$= x^2 - 2x \qquad\qquad\text{(neither } h(x) \text{ nor } -h(x))$$

The function h is neither odd nor even, and the graph of h in Figure 2-53 is symmetric about neither the origin nor the y-axis.   ▬

How can we apply symmetry to sketching graphs? Even without technology, symmetric graphs require fewer plotted points. Refer to Figures 2-51 and 2-52. If the shape of the graph to the right of the y-axis is known, then the rest of the graph is apparent from symmetry.

Consider $j(x) = x^4 - x^2$. A quick examination of $j(-x)$ reveals that j is even and thus symmetric about the y-axis. Plotting selected points of j for values of $x \geq 0$, we obtain Figure 2-54. From these points we determine part of the graph of j as Figure 2-55 shows. Because of symmetry, the remainder of the graph of j is the mirror image reflected about the y-axis as shown in Figure 2-56.

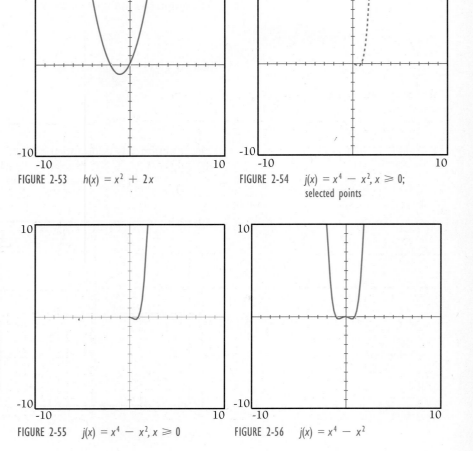

FIGURE 2-53 $h(x) = x^2 + 2x$

FIGURE 2-54 $j(x) = x^4 - x^2, x \geq 0$; selected points

FIGURE 2-55 $j(x) = x^4 - x^2, x \geq 0$

FIGURE 2-56 $j(x) = x^4 - x^2$

A similar process works for $h(x) = x^3 - 5x$, an odd function. Examine the rules of pairing for j and h and guess the inspiration for the names odd and even.

SUMMARY

- A given relation can be symmetric about none, one or all three of the x-axis, y-axis and origin.
- Imagine a graph folded along an axis for a visual method of checking symmetry of the graph about that axis. To algebraically check a function for symmetry, evaluate the function at $-x$. If you obtain the original rule of pairing then the function is even. If you obtain the negative of the original rule of pairing, the function is odd. Odd functions are symmetric about the origin, whereas even functions are symmetric about the y-axis.
- Symmetry of functions is a valuable aid in graphing.

2-3   EXERCISES

Identify the x-axis, y-axis or origin symmetry for graphs 1–20.

1.

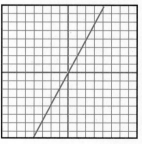

2.

3.

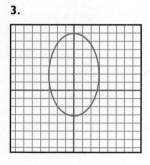

4.

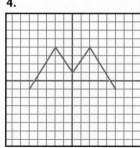

5.

6.

7.

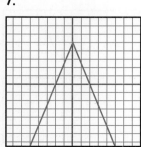

8.

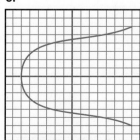

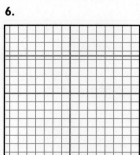

9.   **10.**   **11.**   **12.**

13.   **14.**   **15.**   **16.**

17.   **18.**   **19.**   **20.**

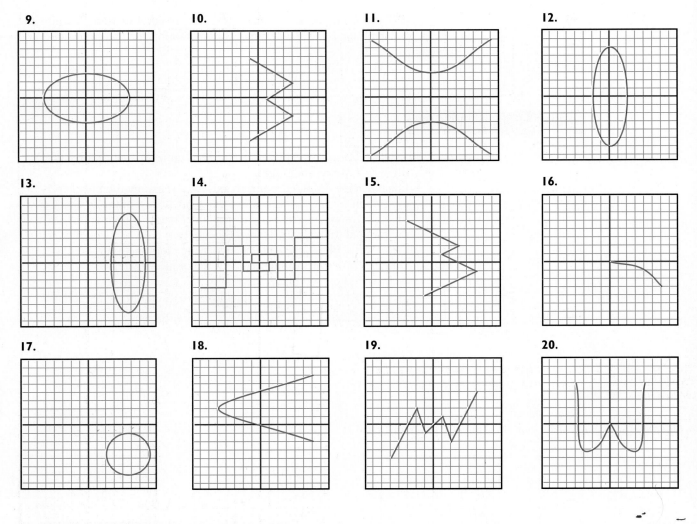

For each partial graphs 21–30, drawn for the positive portion of its domain, draw the remainder of the graph, assuming (a) the graph is symmetric about the *y*-axis and (b) the graph is symmetric about the origin.

21.   **22.**   **23.**   **24.**

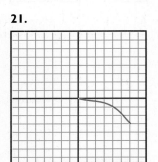
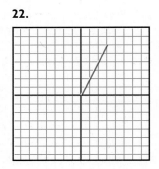
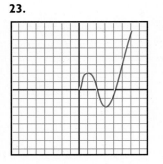
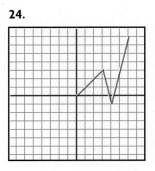

25.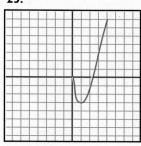

26.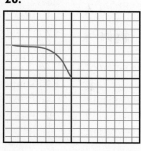

27.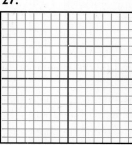

28.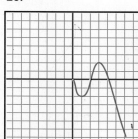

29.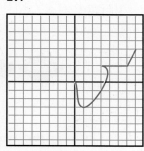

30.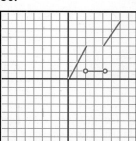

For each of the following relations, determine any symmetry about the *x*-axis, *y* axis, or origin. (Check your analysis using technology.)

31. $y = 3x$

32. $y = -2x$

33. $y = x + 1$

34. $y = 1 - x$

35. $y = \dfrac{1}{x^2}$

36. $y = \dfrac{-1}{x^2}$

37. $y = x^3$

38. $y^3 = x$

39. $y = x^4$

40. $y = x^2$

41. $y = x^3 - x$

42. $y = x^3 + x$

43. $y = x^2 + x$

44. $y = x^2 - x$

45. $x^2 + y^2 = 16$

46. $x^2 + y^2 = 25$

47. $x = y^2 + 3$

48. $x = 3 - y^2$

49. $x^2 + y^2 = 9$

50. $x^2 + y^2 = 16$

Check each of the following functions to determine (a) whether the function is odd, even or neither and (b) whether the function is symmetric about the origin or *y*-axis. Confirm your analysis with technology.

51. $f(x) = x^2$

52. $h(x) = -x^4$

53. $g(x) = x^3$

54. $k(x) = x$

55. $t(x) = x^4 - x^2 + 1$

56. $p(x) = x^3 + 3x$

57. $s(x) = 3x^5 - x$

58. $w(x) = 2x^6 + x^2 + 3$

59. $q(x) = x^4 + x^3$

60. $r(x) = x^3 + x^2 + 3$

61. $c(x) = 5$

62. $B(x) = -3$

63. $L(x) = 3x + 2$

64. $M(x) = 5x - 1$

65. $A(x) = |x|$

66. $B(x) = |x| + 1$

67. $T(x) = \dfrac{x^2 + 1}{x^4}$

68. $S(x) = \dfrac{x^3 + 1}{x^5}$

69. $W(x) = \dfrac{x^5 + x^3}{x}$

70. $P(x) = x^2(x^4)$

71. If $f(x) = x^{2n+1}$ and n is an integer, show that f is odd.

72. If $f(x) = x^{2n}$ and n is an integer, show that f is even.

73. Refer to Problem 71. Conjecture how odd functions got their name.

74. Refer to Problem 72. Conjecture how even functions got their name.

75. Suppose we define Signum$(x) = |x|/x$ if x is not 0, and Signum$(0) = 0$. Determine whether Signum is odd, even or neither.

76. Refer to Problem 75. Plot the points and use the symmetry, if any, to sketch the graph of the Signum function.

77. Sketch the graph of $f(x) = |x^2 - 4|$. Plot points and use symmetry, if appropriate.

78. Sketch the graph of $g(x) = |x| - 2$. Plot points and use symmetry, if appropriate.

79. Sketch the graph of $h(x) = x^3 - x$. Plot points and use symmetry, if appropriate.

80. Sketch the graph of $j(x) = \sqrt{x}$. Plot points and use symmetry, if appropriate.

81. Attack or defend this statement: a function f cannot be both odd and even.

82. Attack or defend this statement: no function is symmetric to the x-axis.

83. If a relation is symmetric about the x-axis and origin, prove it is symmetric about the y-axis.

84. If a relation is symmetric about the y-axis and origin, prove it is symmetric about the x-axis.

85. Prove Theorem 2-2.

86. If f is a real function symmetric about the origin, prove that f is odd (see Theorem 2-3).

87. Prove Theorem 2-4.

88. Suppose that f is a function with domain $(-a, a)$, $a > 0$. Prove that $g(x) = [f(x) - f(-x)]/2$ is odd.

89. Suppose that f is a function with domain $(-a, a)$, $a > 0$. Prove that $h(x) = [f(x) + f(-x)]/2$ is even.

90. Prove that any function f with domain $(-a, a)$, $a > 0$ can be written as the sum of an odd function plus an even function. Hint: see Problems 88 and 89.

2-3   PROBLEM SET

1. **(a)** Make up a visual definition to describe a graph symmetric about the line represented by $y = x$.

 (b) Restate your definition in algebraic terms.

 (c) Prove that the graph of $x^2 + y^2 = 9$ is symmetric about the line $y = x$.

2. **(a)** Make up a visual definition to describe a graph symmetric about the line $y = -x$.

 (b) Restate our definition in algebraic terms.

 (c) Prove that $y = 2x + 1$ is not symmetric about $y = -x$.

3. **(a)** Make up a definition to describe a graph symmetric about the vertical line $x = a$. Can such a graph represent a function? Under what circumstances?

 (b) Prove that $y = x^2 - 6x + 9$ is symmetric about $x = 3$.

4. **(a)** Make up a definition to describe a graph symmetric about the horizontal line $y = b$. Can such a graph

represent a function? Under what circumstances?

 (b) Give an example of a relation symmetric about $y = 4$.

5. **(a)** Suppose that (a, b) is a point in the plane. Use the definition of symmetric about the origin as an example to define visually a graph symmetrical about (a, b).

 (b) Restate your definition in algebraic terms.

 (c) Prove that $(x - a)^2 + (y - b)^2 = r^2$ is symmetric about (a, b).

 (d) Prove that $(x - a)^2 - (y - b)^2 = r^2$ is symmetric about (a, b).

 (e) Give a graphical example of a function symmetric about (a, b).

 (f) Give a nonfunction example of a graph symmetric about (a, b).

EXPLORATION

The following problems use graphing technology to anticipate concepts introduced in the next section. Graph the given functions.

6. **(a)** $y = 3x$ **(b)** $y = 2x$

 (c) $y = 0.4x$ **(d)** $y = x^3/5$

 (e) $y = \ln x$

 (f) Each of these functions *increases*. Describe an increasing function. Formulate a visual definition for *increasing*.

7. **(a)** $y = -0.7x$ **(b)** $y = -3x$

 (c) $y = -x^3/4$ **(d)** $y = e^{-x}$

 (e) $y = -x$

 (f) Each of these functions *decreases*. Describe a decreasing function. Formulate a visual definition for *decreasing*.

8. **(a)** $y = x^2$ **(b)** $y = -e^{-x^2}$

 (c) $y = |x|$ **(d)** $y = \ln|x|$

 (e) $y = |x| - 3$

(f) Each of these functions decreases where x is negative and increases where x is positive. Describe a function increasing (or decreasing) over an x interval.

(g) Formulate a visual definition for a function increasing on a given x interval.

9. Describe the intervals (if any) over which the following functions are increasing and the intervals (if any) over which they are decreasing.

(a) $y = x^2 - 3$ **(b)** $y = |x - 2|$

(c) $y = |x^2 - 9|$ **(d)** $y = e^x$

(e) $y = \ln(1/x)$

10. Sketch each of the following in the window $x \in (-20, 20)$, $y \in (-5, 5)$, then scroll the graph to the right or left 10 units.

(a) $y = \sin(\pi x)$

(b) $y = \cos(\pi x)$

(c) $y = \sin(\pi x) + \cos(\pi x)$

(d) $y = 2\sin(\pi x) - \cos(\pi x)$

(e) $y = \tan(\pi x/2)$

(f) Each of these functions is *periodic*. Describe a periodic function.

(g) Each of the functions completes one cycle for $x \in (0, 2)$. Formulate a visual definition for a periodic function.

4   INCREASING, DECREASING AND PERIODIC FUNCTIONS

To think the thinkable—that is the Mathematician's aim.
—C. L. KEYSER

In the last section, the properties of odd and even functions provided insight into the graphs of functions. The more clues obtained from the equation for a function, the better we can visualize its graph. Whether with technology or plotting points, visualization skills improve our understanding of functions.

This is a key feature of analysis: the interpretation of the algebraic properties of a function's formula as graphical attributes. By anticipating the shape of a graph, we reduce points plotted before visualizing the complete graph. This section explores functional properties helpful in sketching graphs. The first such properties are increasing and decreasing.

Because graphs of functions do not backtrack horizontally, we view them like we read sentences—from left to right. As a natural consequence, we examine a simple graph from left to right to determine where it is increasing and where it is decreasing. Consider the function f represented by the graph in Figure 2-57.

As seen from left to right, f is sometimes increasing and sometimes decreasing. To indicate where a function is increasing or decreasing, refer to *intervals* within the domain of the function. We do not list the y values through which the function increases or decreases.

In Figure 2-57, f is increasing on the following intervals: $[a, b]$, $[c, d]$ and $(d, e]$. Note particularly the point where $x = d$. The function increases on the interval from c to d, up to and including d. On the interval from d to e, d marks the beginning of the increase, but the functional values immediately to the right of d are less than $f(d)$. Observe that f is not increasing on $[c, e]$ despite the fact that $[c, e] = [c, d] \cup (d, e]$. Finally, f is decreasing on $[b, c]$. Intuition drives our idea of increasing and decreasing. Here, visualization is not difficult. Examining the ideas algebraically requires more thought.

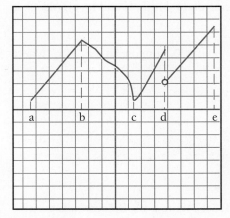

FIGURE 2-57 f

DEFINITION 2-7
Increasing and Decreasing Functions

1. A function f is **increasing** on an interval I in the domain$_f$ if, for every x_1 and x_2 in I where $x_1 < x_2$ (visualize left to right), $f(x_1) < f(x_2)$.

2. A function f is **decreasing** on an interval I in the domain$_f$ if, for every x_1 and x_2 in I where $x_1 < x_2$ (left to right), $f(x_1) > f(x_2)$.

E X A M P L E   2 - 1 1

Increasing and Decreasing Functions. Determine the intervals over which each of the following functions is increasing and over which each is decreasing.

Illustration 1:

The function g in Figure 2-58 is decreasing on (a, b).

Illustration 2:

The function h in Figure 2-59 is neither increasing nor decreasing on any interval. The function h is a *constant* function.

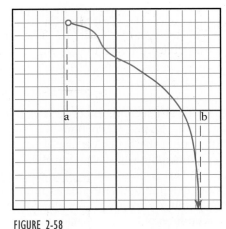

FIGURE 2-58

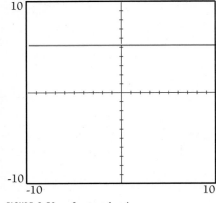

FIGURE 2-59   Constant function

Illustration 3:

The function k in Figure 2-60 is increasing on $[a, b)$, $[b, c]$ and $[d, e]$. Also, k is decreasing on $[c, d]$ and $[e, f)$.

Illustration 4:

Consider a linear function $f(x) = mx + b$, $m > 0$. Suppose that $x_1 < x_2$, where $x_1, x_2 \in R$.

Then, $mx_1 < mx_2$

$mx_1 + b < mx_2 + b$

$f(x_1) < f(x_2)$

Therefore, f is increasing on $(-\infty, +\infty)$.

Illustration 5:

$f(x) = 3x + 5$

Because f is linear, refer to Illustration 4. See Figure 2-61. Because m is positive ($m = 3$), f is increasing on $(-\infty, +\infty)$.

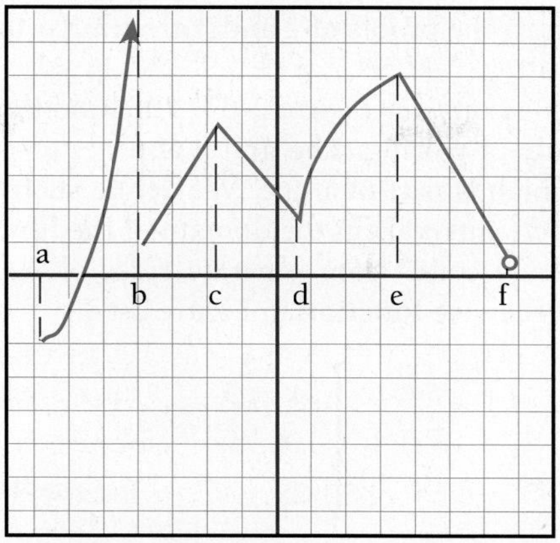

FIGURE 2-60

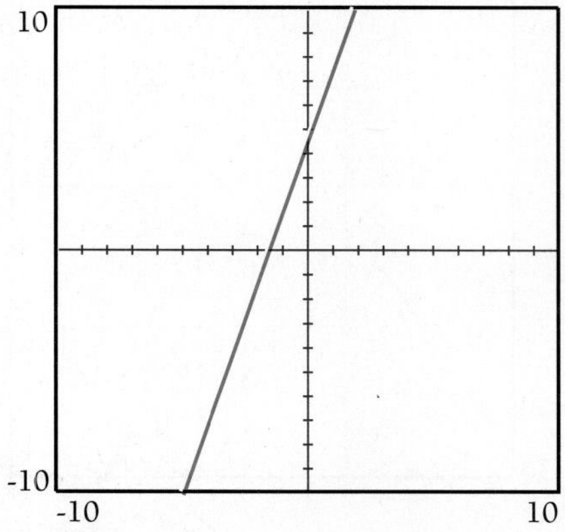

FIGURE 2-61 $f(x) = 3x + 5$

Illustration 6:

$$f(x) = x^2 - 2x + 3$$

The function f is a quadratic function. The graph in Figure 2-62 is a parabola. Also, f decreases on $(-\infty, 1]$ and increases on $[1, +\infty)$. ▬

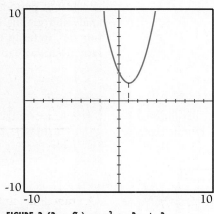

FIGURE 2-62 $f(x) = x^2 - 2x + 3$

Illustration 4 of Example 2-11 forms part of the proof of the next theorem.

Theorem 2-5
Increasing, Decreasing Linear Functions

Suppose that f is a linear function given by

$$f(x) = mx + b.   \text{Then,}$$

1. If $m > 0$, f is increasing on $(-\infty, +\infty)$.
2. If $m < 0$, f is decreasing on $(-\infty, +\infty)$.
3. If $m = 0$, f is a constant function.

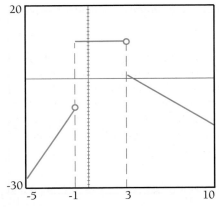

FIGURE 2-63   Piecewise graph

Proof: The proofs of 2 and 3 are similar to Illustration 4, Example 2-11. ▬

Consider Figure 2-63. The function shown is too complicated for a single rule of pairing. The graph appears to consist of three distinct pieces, each of which is part of a line. We define such functions piecewise across several discrete intervals in the domain of the function. The range values for each subinterval come from separate rules of pairing. Consider the description of the **piecewise function** in Figure 2-63.

$$f(x) = \begin{cases} 5x - 3 & \text{for } x < -1 \\ 10 & \text{for } -1 \leq x < 3 \\ -2x + 7 & \text{for } x \geq 3 \end{cases}$$

The function f uses three rules of pairing. On $(-\infty, -1)$ use the rule $y = 5x - 3$. From Theorem 2-5, f is increasing on $(-\infty, -1)$. Similarly, $y = 10$ indicates that f is constant on $[-1, 3)$. Finally, f decreases on $[3, \infty)$ because $y = -2x + 7$ has a negative slope.

One of the remarkable things about Halley's comet is that it returns on schedule. In much the same way, but on a smaller time scale, we thrill to the eruption of Old Faithful because we know *when* to expect it. These and other regular, repeating natural occurrences are periodic.

Just as periodicity helps us anticipate natural events, periodicity assists in analyzing graphs of functions. We would like to define the periodicity of a function in terms of consistent, repetitive patterns. Windows supply a particularly useful way to imagine periodicity.

DEFINITION 2-8
Periodic Function

> A function f has a *period k* if $f(x + k) = f(x)$ $(k > 0)$ for every x in the domain of f. If f has a period of k, then f is **periodic**. The smallest k such that k is a period of f is the *fundamental period* of f.

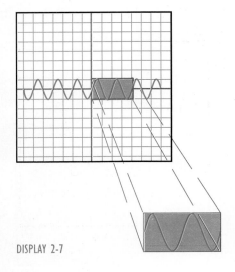

DISPLAY 2-7

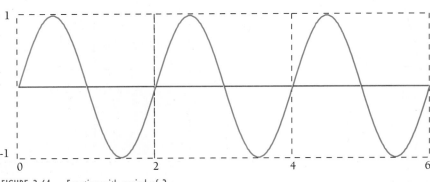

FIGURE 2-64   Function with period of 2

EXAMPLE 2-12

Based on the window displayed, determine the fundamental period, if any, for the following functions.

Illustration 1:

Note in Figure 2-64 the repetition of pattern in the graph every two units. A repetition also occurs every four units, every six units, and so on. Then, 2, 4, 6, 8, and so on represent periods of the function f. The fundamental period of f is 2. The custom is to list only the fundamental period when asked for the period. Imagine a window that captures a snapshot of the graph across one fundamental period. Carefully pasting several copies of the snapshot end to end (concatenating) extends the view of the graph. See Display 2-7.

Illustration 2:

Figure 2-65 displays two cycles of g. The period of g is four. Note the duplicate snapshots in the windows.

Illustration 3:

The function h in Figure 2-66 is not periodic.

Illustration 4:

$$E(x) = \begin{cases} 1 & \text{if the largest integer} \leq x \text{ is odd} \\ 0 & \text{if the largest integer} \leq x \text{ is even} \end{cases}$$

Note that $E(3) = 1$, $E(5.4) = 1$, $E(-19.375) = 0$, $E(\pi) = 1$, $E(0.2) = 0$, $E(-258.387654 \ldots) = 1$. The graph in Figure 2-67 suggests the period is two. Argue that two is the period as follows:

Consider $E(x)$ and $E(x + 2)$. If the largest integer $\leq x$ is odd, then so is the largest integer $\leq x + 2$.

$$E(x) = 1 = E(x + 2).$$

If the largest integer $\leq x$ is even, then so is the largest integer $\leq x + 2$. Therefore,

$$E(x) = 0 = E(x + 2)$$

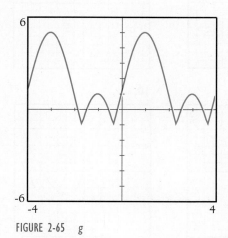

FIGURE 2-65   g

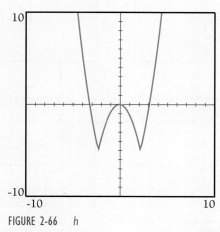

FIGURE 2-66   h

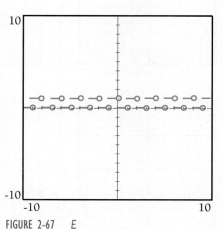

FIGURE 2-67   E

S U M M A R Y

• A function *f* is *increasing* on an interval *I* if the *y* values get larger as we examine the graph from left to right on the interval. A function is *decreasing* on an interval if the *y* values get smaller as we examine the graph from left to right on the interval.

• A function *f* is *periodic* with fundamental period *k* ($k > 0$), if *k* is the smallest number such that $f(x) = f(x + k)$ for every *x* (and $x + k$) in the domain of *f*. Joining a series of identical snapshots of one period of a function produces a larger window into the graph.

2-4   E X E R C I S E S

Use interval notation to indicate where the following functions are increasing or decreasing in graphs 1–10.

1.
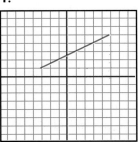

2.
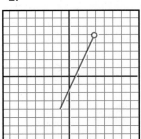

3.
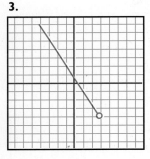

4.
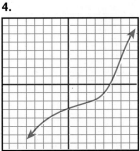

5.
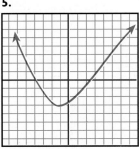

6.
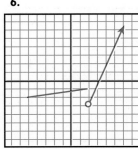

7.
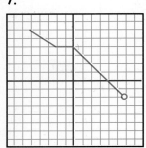

8.
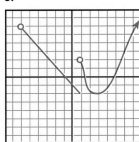

9.
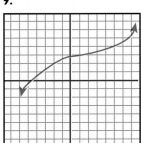

10.
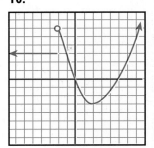

Indicate whether each of the following linear functions is increasing, decreasing or neither.

11. $f(x) = 5x - 3$

12. $g(x) = 5 - 3x$

13. $k(x) = 0.5 - 2x$

14. $t(x) = 3 + 0.25x$

15. $w(x) = -5x - 3$

16. $q(x) = 7x + 3$

17. $p(x) = \dfrac{3x}{5} + \dfrac{2}{4}$

18. $s(x) = \dfrac{2x - 1}{5}$

19. $m(x) = \dfrac{5x - 3}{2}$

20. $n(x) = \dfrac{1}{3} - \dfrac{x}{2}$

Use interval notation to indicate where the following functions are increasing, decreasing or constant.

21. $f_0(x) = \begin{cases} 3x & \text{if } x \le 0 \\ -x & \text{if } x > 0 \end{cases}$

22. $f_1(x) = \begin{cases} -5x & \text{if } x < 0 \\ 2x & \text{if } x \ge 0 \end{cases}$

23. $f_2(x) = \begin{cases} x + 2 & \text{if } x < 3 \\ x - 4 & \text{if } x \ge 3 \end{cases}$

24. $f_3(x) = \begin{cases} 7x - 5 & \text{if } x \le -2 \\ x + 6 & \text{if } x > -2 \end{cases}$

25. $f_5(x) = \begin{cases} 2x & \text{if } x \le -3 \\ -3 & \text{if } -3 < x \le 4 \\ 5 & \text{if } 4 < x \end{cases}$

26. $f_4(x) = \begin{cases} -x & \text{if } x < -2 \\ 4x & \text{if } -2 \le x \le 5 \\ x - 2 & \text{if } 5 < x \end{cases}$

27. $f_6(x) = \begin{cases} 3 - x^2 & \text{if } x \le -3 \\ x - 3 & \text{if } -3 < x \end{cases}$

28. $f_7(x) = \begin{cases} x - 1 & \text{if } x < 2 \\ 1 - x^2 & \text{if } 2 \le x \end{cases}$

29. $f_8(x) = \begin{cases} 10 & \text{if } x \le 3 \\ x + 7 & \text{if } x > 3 \end{cases}$

30. $f_9(x) = \begin{cases} 3x - 1 & \text{if } x \le 4 \\ 11 & \text{if } x > 4 \end{cases}$

31. $f_5(x) = \begin{cases} x^2 & \text{if } x \le -3 \\ -3x & \text{if } -3 < x \le 2 \\ 6 & \text{if } 2 < x \end{cases}$

32. $f_4(x) = \begin{cases} -x & \text{if } x < -5 \\ 4 + 3x & \text{if } -5 \le x \le 5 \\ x^2 - 2 & \text{if } 5 < x \end{cases}$

 Determine whether each of the functions in graphs 33–42 is periodic. If periodic, give the fundamental period.

33.

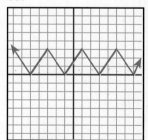

34.

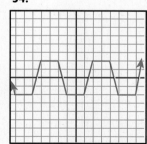

35.

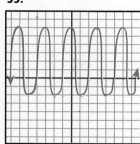

36.

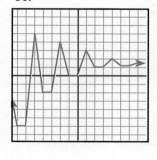

37.

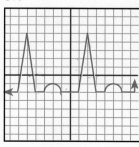

38.

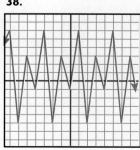

39.

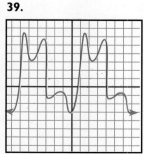

40.

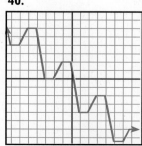

41.

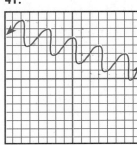

42.

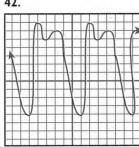

43. Prove Theorem 2-5, part 1.

44. Prove Theorem 2-5, part 2.

45. Prove Theorem 2-5, part 3.

46. Prove $f(x) = x^2$ is increasing in $[0, +\infty)$.

47. Give an example of some natural periodic occurrence.

48. Give an example of a business situation in which information about increases or decreases might be important.

    2-4   P R O B L E M   S E T

1. In calculus, a function f is increasing at a point $(x, f(x))$, if the line tangent to f at $(x, f(x))$ is increasing. Draw the graph of a parabolic function f. Sketch several lines so that they appear tangent to f to convince yourself of the reasonableness of this test for increasing.

2. Refer to Problem 1. Formulate a similar test to determine whether a function is decreasing at a point.

3. In calculus it can be shown that for the function $f(x) = x^2 - 6x + 8$, the slope of the tangent line at any point

$(x, f(x))$ is $m = 2x - 6$. Solve the inequality $2x - 6 > 0$ to determine where the slope of the tangent line is positive. Interpret this result in terms of the graph of f.

4. Refer to Problem 3. Solve $2x - 6 < 0$. Interpret the results in terms of the graph of f. What is the significance of the solution to $2x - 6 = 0$?

EXPLORATION

The following problems anticipate concepts introduced in the next section. Graph each of the following collection of functions in the same window. In each case, discuss the graphical and algebraic relations of h to f and g.

5. $f(x) = x + 3$, $g(x) = x + 1$, $h(x) = (x + 3) - (x + 1)$

6. $f(x) = 3x + 2$, $g(x) = x - 4$, $h(x) = (3x + 2) - (x - 4)$

7. $f(x) = x^2$, $g(x) = 2x$, $h(x) = x^2 - 2x$

8. $f(x) = x^2$, $g(x) = 3$, $h(x) = x^2 + 3$

9. $f(x) = |x|$, $g(x) = x$, $h(x) = x + |x|$

10. $f(x) = |x|$, $g(x) = x$, $h(x) = x|x|$

11. $f(x) = |x|$, $g(x) = x$, $h(x) = \dfrac{x}{|x|}$

12. $f(x) = x^2 - 4$, $g(x) = x$, $h(x) = x(x^2 - 4)$

13. $f(x) = x - 5$, $g(x) = x + 2$, $h(x) = (x - 5)(x + 2)$

14. $f(x) = x - 5$, $g(x) = x + 2$, $h(x) = \dfrac{x - 5}{x + 2}$

5   ALGEBRA OF FUNCTIONS

The Mathematician, carried along on his flood of symbols, dealing apparently with purely formal truths, may still reach results of endless importance for our description of the physical universe.
—KARL PEARSON

Formulas for functions sometimes get complicated. These complications interfere with our analysis. One method for solving a problem is to break the problem into simpler components. You used such an approach when you factored numerators and denominators of fractions to reduce the fractions to lowest terms. To become good at factoring you first had to become good at multiplication.

This section introduces methods to construct new functions from given functions in a manner similar to arithmetic. With sufficient practice at forming new functions from old, we can reverse the process to break complicated functions into simpler components. The **algebra of functions** is the process of constructing new functions from given functions.

DEFINITION 2-9
Algebra of Functions

Suppose that f and g are functions and x is any number in the domains of f and g.

1. The *sum* $f + g$ is a function defined as $(f + g)(x) = f(x) + g(x)$.
2. The *difference* $f - g$ is a function defined as $(f - g)(x) = f(x) - g(x)$.
3. The *product* fg is a function defined as $(fg)(x) = f(x)g(x)$.
4. The *difference* f/g is a function defined as $(f/g)(x) = f(x)/g(x)$, $g(x) \neq 0$.

EXAMPLE 2-13

Constructing Functions Using the Algebra of Functions. For each of the following pairs of functions f and g form the rules of pairing for (a) $f + g$, (b) $f - g$, (c) fg, (d) f/g and give the domain of each.

Illustration 1:

$$f(x) = 3x + 1, \qquad g(x) = x - 2$$

(a) $(f + g)(x) = f(x) + g(x)$
$$= (3x + 1) + (x - 2)$$
$$= 4x - 1$$

The domain of $f + g$ is $(-\infty, \infty)$.

(b) $(f - g)(x) = f(x) - g(x)$
$$= (3x + 1) - (x - 2)$$
$$= 2x + 3$$

The domain of $f - g$ is $(-\infty, \infty)$.

(c) $(fg)(x) = (3x + 1)(x - 2)$
$$= 3x^2 - 5x - 2$$

The domain of fg is $(-\infty, \infty)$.

(d) $(f/g)(x) = \dfrac{f(x)}{g(x)}, g(x) \neq 0.$

$$= \frac{3x + 1}{x - 2}, \quad x - 2 \neq 0.$$

The domain of f/g is $(-\infty, 2) \cup (2, \infty)$.

Illustration 2: $f(x) = \dfrac{3}{x + 2}, g(x) = \sqrt{x + 3}$

In Figure 2-68, the domain of f is $x \neq -2$ or $(-\infty, -2) \cup (-2, \infty)$.

In Figure 2-69, the domain of g is $x \geq -3$ or $[-3, \infty)$.

The values of x in both domains are $[-3, -2) \cup (-2, +\infty)$. See Figure 2-70.

(a) $(f + g)(x) = \dfrac{3}{x + 2} + \sqrt{x + 3}$

The domain of $f + g$ is $[-3, -2) \cup (-2, +\infty)$.

(b) $(f - g)(x) = \dfrac{3}{x + 2} - \sqrt{x + 3}$

The domain of $f - g$ is $[-3, -2) \cup (-2, +\infty)$.

(c) $(fg)(x) = \dfrac{3}{x + 2} \sqrt{x + 3}$

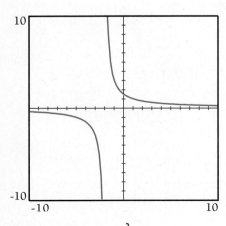

FIGURE 2-68 $f(x) = \dfrac{3}{x + 2}$

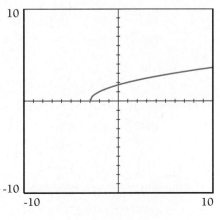

FIGURE 2-69 $g(x) = \sqrt{x + 3}$

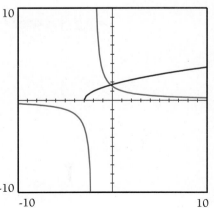

FIGURE 2-70 $f(x) = \dfrac{3}{x + 2}, g(x) = \sqrt{x + 3}$

The domain of fg is $[-3, -2) \cup (-2, +\infty)$.

(d) $(f/g)(x) = \dfrac{3}{(x + 2)\sqrt{x + 3}}, \qquad g(x) \neq 0$

$$x + 3 \neq 0$$

The domain of f/g is $[-3, -2) \cup (-2, +\infty)$, $x \neq -3$. Thus, the domain of f/g is $(-3, -2) \cup (-2, \infty)$.

Illustration 3:

Most graphing technology supports the algebra of functions. For example, suppose you have stored $Y1 = 3 + X$ and $Y2 = X^2$ using your technology. Then you can store $Y1 + Y2$ under $Y3$. Methods for entering $Y1 + Y2$ to define $Y3$ vary. For the TI-81, enter $Y1$ and $Y2$ using the [2nd] [Y vars] menu. For the Sharp use [MATH] [E] for access to $Y1$ and $Y2$. For *GraphWindows* type $F1(x) + F2(x)$. Entry of $Y1/Y2$ or $Y1*Y2$ is similar. Consult your user's manual for more details. ■

Many times a number is easier to use in factored form or as a sum of terms. Similarly, functions may be more easily analyzed if broken into simpler, component functions. For example, $f(x) = x^2 + 5x + 6$ could be thought of as

$f(x) = g(x) + h(x)$,   where $g(x) = x^2 + 3x$ and $h(x) = 2x + 6$

or $f(x) = k(x) + m(x)$,   where $k(x) = x^2 + 6$ and $m(x) = 5x$

or $f(x) = t(x)r(x)$,   where $t(x) = x + 3$ and $r(x) = x + 2$

There are infinitely many variations on this theme. But skill in rewriting a complicated function in terms of simpler, component functions requires an ability to do the algebra of functions. The next several chapters examine classes of functions constructed by the algebra of functions.

SUMMARY

For two functions f and g with overlapping domains, $f + g$, $f - g$, fg and f/g are functions with rules of pairing defined as follows:

(a) $(f + g)(x) = f(x) + g(x)$
(b) $(f - g)(x) = f(x) - g(x)$
(c) $(fg)(x) = f(x)g(x)$
(d) $(f/g)(x) = f(x)/g(x), \qquad g(x) \neq 0$

The domain for each of these functions consists of the real numbers shared by the domains of f and g, except that f/g must also exclude any numbers x for which $g(x) = 0$.

2-5   E X E R C I S E S

For each of the following pairs of functions f and g, form the rules of pairing for $f + g$, $f - g$, fg and f/g. Also, give the domain for each of these new functions.

1. $f(x) = 2x$, $g(x) = 3$

2. $g(x) = -x$, $f(x) = 5$

3. $f(x) = 3$, $g(x) = 2x$

4. $g(x) = 5$, $f(x) = -x$

5. $g(x) = \sqrt{x}$, $f(x) = \dfrac{1}{x}$

6. $f(x) = \sqrt{x}$, $g(x) = \dfrac{1}{x}$

7. $f(x) = x + 3$, $g(x) = x - 2$

8. $f(x) = x + 1$, $g(x) = x - 3$

9. $f(x) = 2x + 1$, $g(x) = x + 5$

10. $f(x) = 3x - 1$, $g(x) = x + 4$

11. $f(x) = x^2$, $g(x) = x^4$

12. $f(x) = x^3$, $g(x) = x$

13. $f(x) = x^3$, $g(x) = x^5$

14. $f(x) = 5$, $g(x) = x^2$

15. $f(x) = x^2$, $g(x) = 3x - 2$

16. $f(x) = x^3$, $g(x) = 2x - 3$

17. $f(x) = x^2 - 1$, $g(x) = 9 - x^2$

18. $f(x) = 4 - x^2$, $g(x) = x^2 - 9$

19. $f(x) = \sqrt{x}$, $g(x) = \sqrt{2x + 3}$

20. $f(x) = \sqrt{x + 1}$, $g(x) = \sqrt{3x - 2}$

For each of the following functions f, (a) determine rules of pairing for two functions g and h such that $f = g + h$, $g(x) \neq$ 0 and $h(x) \neq 0$; (b) determine rules of pairing for two functions m and n such that $f = mn$, $m(x) \neq 1$, $n(x) \neq 1$.

21. $f(x) = 2x + 4$

22. $f(x) = 3x - 6$

23. $f(x) = x^2 - 7x$

24. $f(x) = x^2 + 3x$

25. $f(x) = x^2 + 5x + 6$

26. $f(x) = x^2 - 5x + 6$

27. $f(x) = x^2 - 7x + 12$

28. $f(x) = x^2 + 7x + 12$

29. $f(x) = x^2 - x - 12$

30. $f(x) = x^2 + x - 12$

31. $f(x) = x^2 + 4x - 12$

32. $f(x) = x^2 - 4x - 12$

33. $f(x) = x^2 + 8x + 12$

34. $f(x) = x^2 - 8x + 12$

35. $f(x) = x^2 - 9$

36. $f(x) = x^2 - 16$

37. $f(x) = 3x^2 + 2x - 5$

38. $f(x) = 2x^2 - 5x - 7$

39. $f(x) = x^3 + 1$

40. $f(x) = x^3 - 1$

41. If f and g are odd functions, prove $f + g$ is odd.

42. If f and g are even functions, prove $f + g$ is even.

43. If f and g are even functions, prove fg is even.

44. If f and g are odd functions, prove fg is even.

45. Give an example to show that if f and g are odd, then f/g does not have to be odd.

46. If f is odd and g even, what kind of conclusion can be drawn about $f + g$? About fg?

2-5   P R O B L E M   S E T

For each of the following functions, f and g, use technology to examine the graph of f, g and $f + g$. In each case describe an effect each f and g has on the graph of $f + g$.

1. $f(x) = 2$, $g(x) = 3$

2. $f(x) = 4$, $g(x) = -1$

3. $f(x) = x$, $g(x) = -2$

4. $f(x) = x$, $g(x) = 1$

5. $f(x) = x$, $g(x) = 2x$

6. $f(x) = -x$, $g(x) = 2x$

7. $f(x) = x + 3$, $g(x) = 2 - x$

8. $f(x) = x + 1$, $g(x) = x + 2$

9. $f(x) = x^2$, $g(x) = 2x + 1$

10. $f(x) = x^2$, $g(x) = -2x - 3$

For each of the following functions, f, g and h, use technology to examine the graph of f, g and h. In each case describe an effect of f and g on the graph of h.

11. $f(x) = 8$, $g(x) = 2$, $h = f/g$

12. $f(x) = 3$, $g(x) = x$, $h = fg$

13. $f(x) = 2$, $g(x) = x$, $h = f/g$

14. $f(x) = x - 2$, $g(x) = x$, $h = g/f$

15. $f(x) = x - 2$, $g(x) = x + 3$, $h = fg$

16. $f(x) = x - 2$, $g(x) = x + 3$, $h = f/g$

17. $f(x) = x - 2$, $g(x) = x + 3$, $h = g/f$

18. $f(x) = x^2 - 4$, $g(x) = x + 2$, $h = f/g$

19. $f(x) = x^2 - 4$, $g(x) = x + 2$, $h = g/f$

20. $f(x) = x^2 - 4$, $g(x) = x + 2$, $h = f - g$

EXPLORATION

The following problems anticipate concepts introduced in the next section. Graph each of the following collection of functions in the same window. In each case describe a graphical (or algebraic) relation of h to f and g.

21. $f(x) = x^2, g(x) = x + 1, h(x) = (x + 1)^2$
22. $f(x) = x^2, g(x) = x - 4, h(x) = (x - 4)^2$
23. $f(x) = 3x - 2, g(x) = x^2, h(x) = 3x^2 - 2$
24. $f(x) = x^2 + x, g(x) = x - 5, h(x) = (x-5)^2 + (x-5)$

25. $f(x) = |x|, g(x) = x - 3, h(x) = |x - 3|$
26. $f(x) = |x|, g(x) = x^2 - 9, h(x) = |x^2 - 9|$
27. $f(x) = \sqrt{x}, g(x) = x + 3, h(x) = \sqrt{x + 3}$
28. $f(x) = 4 - x^2, g(x) = \sqrt{x}, h(x) = \sqrt{4 - x^2}$
29. $f(x) = \sin x, g(x) = x + 2, h(x) = \sin(x + 2)$
30. $f(x) = e^x, g(x) = x - 3, h(x) = e^{x-3}$

6   COMPOSITES OF FUNCTIONS

Algebra is generous, she often gives more than is asked of her.
—D'ALEMBERT

S I D E B A R

"Let me be sure I understand," Eddie said anxiously to the tax consultant. "I not only have to pay $743 in taxes, I also have to pay 5 percent tax on my taxes?"

The consultant nodded. "It's the tax surcharge. Congress says it's temporary. But . . ."

"I can't believe I have to pay a tax on the taxes I pay. Wait a minute. The $743 was because I am in the 15 percent tax bracket. What does the 5 percent surtax do to my tax bracket?"

"To determine that," said the consultant, "we have to compose the two rates into one."

There is yet another way to form a function from two given functions. Imagine a manufacturing process takes liquid plastic and molds it into balls. A second process takes molded balls of plastic and polishes and drills them to make bowling balls. Suppose we use the output for one function as the input for the other function. Combining the processes into a single process takes liquid plastic for input and produces finished bowling balls for output. This combined process is **composition**.

DEFINITION 2-10
Composition

The composite function $f \circ g$ of two functions f and g is defined by
$$f \circ g\,(x) = f[g(x)]$$
The domain of $f \circ g$ consists of all real numbers x from the domain of g such that $g(x)$ is in the domain of f.

Visually, we have the diagram of $f \circ g$ in Figure 2-71.

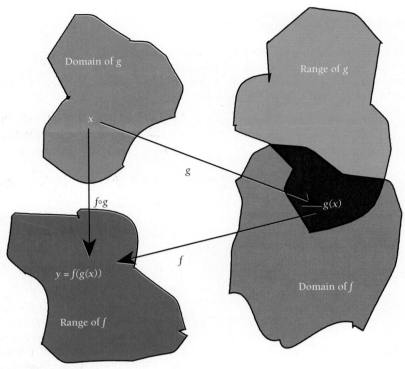

FIGURE 2-71   Diagram for $f \circ g$

Composites of Functions. For the following pairs of functions determine the indicated composition evaluated at 3 and x. Also determine the domain of the composition.

EXAMPLE 2-14

Illustration 1:

$$f(x) = 3x, \qquad g(x) = x^2 + 1$$
$$f \circ g\,(3) = f[g(3)]$$
$$= f[3^2 + 1]$$
$$= f[10]$$
$$= 3(10)$$
$$f \circ g(3) = 30$$
$$f \circ g(x) = f[g(x)]$$
$$= f[x^2 + 1]$$
$$f \circ g(x) = 3(x^2 + 1)$$
$$= 3x^2 + 3$$

The domain of $f \circ g$ is a restriction on the domain of g, all real numbers. The fact that the range values of g, $x^2 + 1$, must lie in the domain of f imposes the restriction. But the domain of f is all real numbers and $x^2 + 1$ is a real number. Therefore no restriction is necessary. The domain of $f \circ g$ is all real numbers.

Alternately, we could have determined the domain of $f \circ g$ directly from the rule of pairing for $f \circ g(x)$.

Illustration 2:

$$f(x) = 3x, \qquad g(x) = x^2 + 1$$
$$g \circ f(3) = 82$$
$$g \circ f(x) = 9x^2 + 1$$

The domain of $g \circ f$ is $(-\infty, \infty)$.

Note: These are the same functions as Illustration 1 with the roles of f and g reversed. Note that the resulting compositions are not equal. In general, $f \circ g \neq g \circ f$.

Illustration 3:

$$f(x) = 5x - 1, \qquad g(x) = \sqrt{x + 1}$$
$$\begin{aligned} f \circ g(3) &= f[g(3)] \\ &= f[2] \\ &= 9 \\ f \circ g(x) &= f[g(x)] \\ &= f[\sqrt{x + 1}] \\ &= 5\sqrt{x + 1} - 1 \end{aligned}$$

Begin with the domain of g, $x \geqslant -1$. Now the range value of g, $\sqrt{x + 1}$, must be in the domain of f. Because the domain of f is all real numbers and $\sqrt{x + 1}$ is real, no additional restrictions are necessary. The domain of $f \circ g$ is all real x such that $x \geqslant -1$. Notice that the formula for $f \circ g$ also implies the domain for $f \circ g$.

Illustration 4:

$$g(x) = \sqrt{x + 1}, \qquad f(x) = 5x - 1$$
$$\begin{aligned} g \circ f(3) &= g[f(3)] \\ &= g[14] \\ g \circ f(3) &= \sqrt{15} \\ g \circ f(x) &= g[f(x)] \\ &= g[5x - 1] \\ &= \sqrt{(5x - 1) + 1} \\ g \circ f(x) &= \sqrt{5x} \end{aligned}$$

For the domain of $g \circ f$, begin with the domain of f, all real numbers. The range values of f, $5x - 1$ must lie in the domain of g. The domain of g is $x \geqslant -1$. Therefore, $5x - 1 \geqslant -1$ is necessary. $5x \geqslant 0$. And, $x \geqslant 0$, is the domain of $g \circ f$. Alternately, examine the rule of pairing for $g \circ f$:

$$g \circ f(x) = \sqrt{5x}$$

From this, determine the domain is $x \geqslant 0$.

Illustration 5:

$$f(x) = x, \qquad g(x) = 3x^2 - 5$$
$$\begin{aligned} f \circ g(3) &= f[g(3)] \\ &= f[22] \\ &= 22 \end{aligned}$$

$$f \circ g(x) = f[g(x)]$$
$$= f[3x^2 - 5]$$
$$= 3x^2 - 5$$

Domain of $f \circ g$ is all real numbers.

Note: From Section 0-1, the additive identity is 0. The multiplicative identity is 1. In the preceding example, $f(x) = x$ acts like an identity for composition, $g \circ f = g$ and $f \circ g = g$. Therefore, denote I, $(I(x) = x)$, as the **compositional identity**.

Illustration 6:

$$f(x) = \tfrac{1}{2}x - 3, \qquad g(x) = 2x + 6$$
$$f \circ g(3) = f[g(3)]$$
$$= f[12]$$
$$= 3$$
$$f \circ g(x) = f[2x + 6]$$
$$= \tfrac{1}{2}(2x + 6) - 3$$
$$= x$$

Domain of $f \circ g$ is all real numbers.

Note: The composition of f and g produced the compositional identity $[I(x) = x]$. The rational numbers $\frac{2}{3}$ and $\frac{3}{2}$ are *multiplicative inverses,* because $\frac{2}{3} \cdot \frac{3}{2} = 1$, the *multiplicative identity.* For a similar reason we say that f and g are **compositional inverses**.   ▬

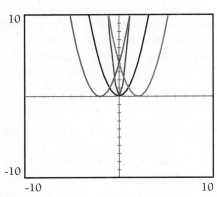

FIGURE 2-72 $f(x) = x^2$ with compositions

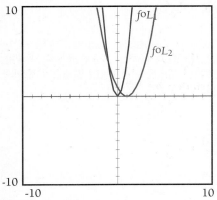

FIGURE 2-73 $f \circ L_1 = (2x)^2, f \circ L_2 = (x - 1)^2$

What about the graph of $f \circ g$? Let $f(x) = x^2$, see Figure 2-72. The graph of f is a parabola. As an experiment, use graphing technology to compare the graph of f with the graph of $f \circ g$ where $g(x) = x - 2$. Try $g(x) = x + 5$ and $g(x) = -3x$. Now use the same functions and graph $g \circ f$. In each case, the graph of the composition is a parabola. The characteristic shape of the original graph does not change.

These compositions illustrate an elegant idea: the idea of a **transformation**. Suppose that f is a function. A happy circumstance would be that the graph of f is well known. By well known, we mean that the rule of pairing of f is easy to recognize and the associated graph has a familiar shape such as $f(x) = x^2$ in Figure 2-72. If we compose f with linear functions, say $L_1(x) = 2x$ and $L_2(x) = x - 1$, the resulting functions graphs would be as shown in Figure 2-73.

The characteristic shape of the graph of f does not change with these compositions. Rather the shape is shifted, stretched (compressed) or perhaps reflected.

In general, a transformation L on the function f is the composition of f and L where L is a first-degree function given by $L(x) = mx + b, m \neq 0$.

A transformation L applies to f in one of two ways:

$$L \circ f(x) = mf(x) + b \quad \text{or} \quad f \circ L(x) = f(mx + b)$$

S I D E B A R

Suppose that $L(x) = x + b$ represents a transformation on a function f, then L is called a *linear transformation*. In a later chapter we extend the idea of transformations. For now, the major concept connected with a transformation is that a transformation "stretches" or "reflects" a graph or "shifts" the graph to a new location, but the "shape" of the graph remains intact. The kind of transformations introduced in this section are affine transformations.

Experiment with your graphing technology to confirm that the results of the following table provide graphical interpretations of various L transformations.

Original graph f

Transformation		Composition	Resulting graph
$f \circ L$	$L(x) = x + b$	$f(x + b)$	"shift" left or right
$L \circ f$	$L(x) = x + b$	$f(x) + b$	"shift" up or down
$f \circ L$	$L(x) = mx$	$f(mx)$	"stretch" horizontally
$L \circ f$	$L(x) = mx$	$mf(x)$	"stretch" vertically
$f \circ L$	$L(x) = -x$	$f(-x)$	reflect left to right
$L \circ f$	$L(x) = -x$	$-f(x)$	reflect top to bottom

Imagine that the transformation L repositions the graph of the function f within our window view, or perhaps the transformation stretches or reflects the image of f within the window. Such an approach can help our examination of graphs. For example, if the shape of $f(x) = x^2$ is a parabola, then the possible linear transformations on $f(x)$, $m_1 x^2 + b_1$ and $(m_2 x + b_2)^2$ should simplify to the form $ax^2 + bx + c$. As a result we expect the resulting graph to be a parabola, translated to another position in the plane and perhaps reflected or rescaled along the two axes. The key is that the shape is still a parabola.

Transformations

Consider the parabola representing the graph of $f(x) = x^2$ in Figure 2-74; the window is $-3 \leq x \leq 3$ and $0 \leq y \leq 9$. Now examine $g(x) = -2(x - 1)^2 + 3$. Although the graph of g is known from Chapter 1, we shall approach the graph by use of transformations. First rewrite g in terms of f.

$$g(x) = -2f(x - 1) + 3$$

One transformation is on the domain of f: $L_1(x) = x - 1$.
A second transformation is on the range of f: $L_2(x) = -2x + 3$.

$$g(x) = L_2 \circ [f \circ L_1(x)]$$

To determine a suitable window to display g we must work backward from the domain of f to the domain of g. Then we must build up from the range of f to the range of g. The domain window for f is

$$-3 \leq x \leq 3$$

Replace x with L_1: $-3 \leq x - 1 \leq 3$.
Solve for x: $-2 \leq x \leq 4$ (window for g).
Range window for f: $0 \leq f(x) \leq 9$.
Build L_2:

$$0 \geq -2f(x) \geq -18 \qquad \text{(invert graph)}$$
$$3 \geq -2f(x) + 3 \leq -15$$

EXAMPLE 2-15

Illustration 1:

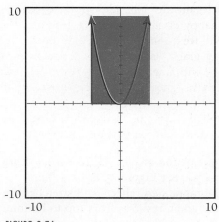

FIGURE 2-74

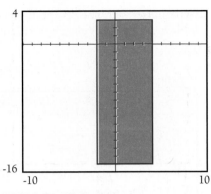

FIGURE 2-75   Window for g   FIGURE 2-76 $g(x) = -2(x - 1)^2 + 3$

Obtain $g(x)$: $3 \geq g(x) \geq -15$ (window for g).

Now construct a window based on $-2 \leq x \leq 4$ and $3 \geq g(x) \geq -15$. Note the inversion of the window based on the multiplication by -3. See Figure 2-75.

Copy the shape of f into the window for g, inverting the graph from top to bottom. Note the extension of the graph beyond this window in Figure 2-76. Verify that the same graph results from the methods of Chapter 1.

Illustration 2:   Suppose that S is a periodic function, with domain of all real numbers and a range of $[-1, 1]$. A window containing one period, $[0, 2\pi]$, for S is shown in Figure 2-77.

To sketch the graph of $y = -3S(2x + \pi) + 1$, first decompose the function to determine the appropriate transformations on one window.

Domain: Start with the restrictions on the argument of S and reduce it to a statement about x:

$$0 \leq \text{domain component for } S \leq 2\pi$$
$$0 \leq \quad 2x + \pi \quad \leq 2\pi$$
$$-\frac{\pi}{2} \leq \quad x \quad \leq \frac{\pi}{2}$$

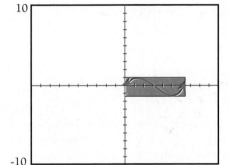

FIGURE 2-77   S

Range: Start with the range of S and build it into $-3S(x) + 1$:

$$-1 \leq \text{range component for } S \leq 1$$
$$-1 \leq \quad S(x) \quad \leq 1$$
$$3 \geq \quad -3S(x) \quad \geq -3$$
$$4 \geq \quad -3S(x) + 1 \quad \geq -2$$
$$4 \geq \quad y \quad \geq -2$$

The bounds for a suitable window for one period of $y = -3S(2x + \pi) + 1$ are given by $x = -\pi/2$, $x = \pi/2$, $y = 4$ and $y = -2$. See Figure 2-78.

Note that multiplication by -3 in the range step reflects and stretches the graph across the x-axis. The image of the graph reverses from top to bottom. The range of the graph triples. Now fit the original shape of the graph into the

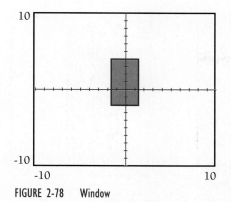

FIGURE 2-78   Window

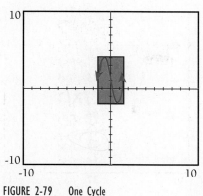

FIGURE 2-79   One Cycle

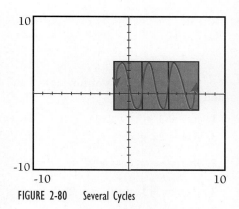

FIGURE 2-80   Several Cycles

new window (Figure 2-79). Append several copies of the window end to end for an enhanced, more general view of the graph (Figure 2-80).

Note: Experiment with these transformations using a graphing calculator. Set up a range for x from 0 to 6.5 and for y from -4 to 4. Use [SIN] for S. For a Sharp or TI calculator, check the defaults. Set Rad as the default. Under $Y1 =$ store $2x + \pi$. Store sin $Y1$ under $Y2 =$. Store $-3Y2 + 1$ under $Y3$. Graph, then compare the graphs of $Y1$, $Y2$ and $Y3$.

A similar method works with *GraphWindows* to store $-3x + 1$ under [Alt][F1], Sin x under [Alt][F2] and $2x + \pi$ under [Alt][F3]. Finally, under [Alt][F4] store $F1(F2(F3(X)))$. Graph [F4].

Later chapters use transformations for analyzing the graphs of functions. For now, we explore the changes to the graph of a function f when it is composed with a first-degree function L.

SUMMARY

- The composite function $f \circ g$ of two functions f and g is defined by $f \circ g(x) = f[g(x)]$. The domain of $f \circ g$ is all real numbers x such that x is in the domain of g and $g(x)$ is in the domain of f.
- $I(x) = x$ is the compositional identity because, for any function f, $f \circ I = f$ and $I \circ f = f$.
- In general, $f \circ g \neq g \circ f$.
- A transformation of a real function f is given by $L_1 \circ (f \circ L_2)$, where L_1 and L_2 are first-degree functions. A transformation of f does not alter the fundamental shape of the graph of f. However, the transformation may shift the graph of f right, left, up or down according to the intercepts of L_1 and L_2. Also, the transformation may stretch the graph of f vertically or horizontally based on the slopes of L_1 and L_2.

2-6   E X E R C I S E S

For each of the pairs of functions f and g, evaluate $f \circ g(2)$ and $g \circ f(2)$.

1. $f(x) = 3, g(x) = 2x$

2. $f(x) = 5, g(x) = -x$

3. $f(x) = 2x, g(x) = 3$

4. $f(x) = -x, g(x) = 5$

5. $f(x) = \dfrac{1}{x}, g(x) = \sqrt{x}$

6. $f(x) = \sqrt{x}, g(x) = \dfrac{1}{x}$

7. $f(x) = x + 3, g(x) = x - 2$

8. $f(x) = x + 1, g(x) = x - 3$

9. $f(x) = 2x + 1, g(x) = x + 5$

10. $f(x) = 3x - 1, g(x) = x + 4$

11. $f(x) = x^2, g(x) = x^4$

12. $f(x) = x^3, g(x) = x$

13. $f(x) = x^3, g(x) = x^5$

14. $f(x) = 5, g(x) = x^2$

15. $f(x) = x^2, g(x) = 3x - 2$

16. $f(x) = x^3, g(x) = 2x - 3$

17. $f(x) = x^2 - 1, g(x) = 9 - x^2$

18. $f(x) = 4 - x^2, g(x) = x^2 - 9$

19. $f(x) = \sqrt{x}, g(x) = \sqrt{2x + 3}$

20. $f(x) = \sqrt{x + 1}, g(x) = \sqrt{3x - 2}$

21. $f(x) = \dfrac{1}{x}, g(x) = \dfrac{1}{x}$

22. $f(x) = -\dfrac{1}{x}, g(x) = -\dfrac{1}{x}$

23. $f(x) = 0.2x - 1, g(x) = 5x + 5$

24. $f(x) = 1.5x + 2, g(x) = \dfrac{2}{3}x - \dfrac{4}{3}$

25. $f(x) = x^2 + 1, g(x) = \sqrt{x - 1}$

26. $f(x) = \sqrt{x + 1}, g(x) = x^2 - 1$

*27. $f(x) = \sqrt{x}, g(x) = 1 - x$

*28. $f(x) = \sqrt{x - 1}, g(x) = -x$

*29. $f(x) = \dfrac{1}{x}, g(x) = x - 2$

*30. $f(x) = \dfrac{2}{x + 1}, g(x) = 1 - x$

For each of the following pairs of functions f and g, determine a formula for $f \circ g$ and the domain of $f \circ g$.

31. $f(x) = 3, g(x) = 2x$

32. $f(x) = 5, g(x) = -x$

33. $f(x) = \dfrac{1}{x}, g(x) = \sqrt{x}$

34. $f(x) = \sqrt{x}, g(x) = \dfrac{1}{x}$

35. $f(x) = 2x + 1, g(x) = x + 5$

36. $f(x) = 3x - 1, g(x) = x + 4$

37. $f(x) = x^2, g(x) = x^4$

38. $f(x) = x^3, g(x) = x^3$

39. $f(x) = x^2 - 1, g(x) = 9 - x^2$

40. $f(x) = 4 - x^2, g(x) = x^2 - 9$

41. $f(x) = \dfrac{1}{x}, g(x) = \dfrac{1}{x}$

42. $f(x) = -\dfrac{1}{x}, g(x) = -\dfrac{1}{x}$

43. $f(x) = \dfrac{1}{5}x - 1, g(x) = \sqrt{x + 1}$

44. $f(x) = 1.5x + 2, g(x) = \dfrac{2}{3}x - \dfrac{4}{3}$

45. $f(x) = x^2 + 1, g(x) = \sqrt{x - 1}$

46. $f(x) = \sqrt{x + 1}, g(x) = x^2 - 1$

*47. $f(x) = \sqrt{x}, g(x) = 1 - x$

*48. $f(x) = \sqrt{x - 1}, g(x) = 2x$

*49. $f(x) = \dfrac{1}{x}, g(x) = x - 2$

*50. $f(x) = \dfrac{2}{x + 1}, g(x) = 1 - x$

Suppose that S has the graph shown in the window of Figure 2-81. Sketch the graph of each of the following transformations of S. (Use graphing technology. The graph of S is identical to the graph of sin.)

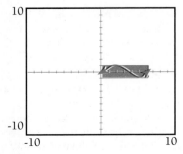

FIGURE 2-81   S

51. $f(x) = S(3x)$ **52.** $f(x) = S(x + 3)$
53. $f(x) = S(x - 3)$ **54.** $f(x) = S(-3x)$
55. $f(x) = -2S(x)$ **56.** $f(x) = S(x) - 2$
57. $f(x) = S(x) + 2$ **58.** $f(x) = 2S(x)$
59. $f(x) = 3S(x) + 2$ **60.** $f(x) = S(3x + 2)$
61. $f(x) = S(2x - 3)$ **62.** $f(x) = 2S(x) - 3$
63. $f(x) = 2S(x - 3)$ **64.** $f(x) = S(2x) - 3$
65. $f(x) = -2S(x + 1) - 3$ **66.** $f(x) = -3S(x - 2) + 3$
67. $f(x) = 3S(-2x + 1)$ **68.** $f(x) = 2S(-3x + 1)$
69. $f(x) = -2S(-3x + 4) - 1$
70. $f(x) = -3S(-2x - 5) + 1$

Attack or defend the following statements.

71. The composition of two odd functions is odd.

72. The composition of two even functions is even.

73. The composition of an odd function and an even function is even.

74. The composition of an even function and an odd function is odd.

75. If $f(x) = x^2$ and $g(x) = \sqrt{x}$, then $f \circ g(x) = g \circ f(x)$.

76. If $f(x) = x^2$ and $g(x) = \sqrt{x}$, and $x > 0$, then $f \circ g(x) = g \circ f(x)$.

77. If $f(x) = x^3$ and $g(x) = \sqrt[3]{x}$ and $x < 0$, then $g \circ f(x) = f \circ g(x)$.

78. If $f(x) = x^3$ and $g(x) = \sqrt[3]{x}$, then $g \circ f(x) = f \circ g(x)$.

79. If $f \circ f(x) = x$, then $f(x) = x$.

80. If f increases on I and g is increasing on I, then $f \circ g$ is increasing on I.

81. If f decreases on I and g is increasing on I, then $f \circ g$ is decreasing on I.

2-6 PROBLEM SET

Problem 1 leads you through a manual construction of the graph of $f \circ g$ from the graphs of f and g. The function I is the line given by the equation $y = x$. See Display 2-7.

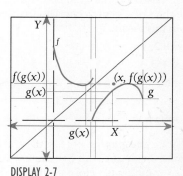

DISPLAY 2-7

1. (a) Reproduce the graphs of the functions f, g and I shown in Display 2-7. Label point x from the domain as g as shown. Sketch a dashed vertical through x so that it intersects the graph of g. Label the point of intersection $(x, g(x))$.

(b) Sketch a dashed horizontal line through $(x, g(x))$. Note that the horizontal line intersects the y-axis at the point $(0, g(x))$. Label the y-intercept. The horizontal line also intercepts the identity function $I(x) = x$. Recall that the coordinates of any point on the identity are equal. Therefore the point where the hor-

izontal line intersects the identity is $(g(x), g(x))$. Label this point.

(c) Extend a dashed vertical line through $(g(x), g(x))$. The vertical line intersects the x-axis at $(g(x), 0)$. Label this point. The vertical line also intersects the graph of f. Because we have located $g(x)$ on the x-axis in the domain of f, the coordinate of the point on f is $(g(x), f(g(x)))$. Label the point on f.

(d) Sketch a horizontal dashed line through the point labeled $(g(x), f(g(x)))$. The horizontal line intercepts the y-axis at $f(g(x))$. Because $f \circ g(x) = f(g(x))$, label the y-intercept as $f \circ g(x)$.

(e) Notice that the horizontal line with y-intercept $f \circ g(x)$ intersects the vertical line through $(x, 0)$. Therefore the coordinate of this point of intersection is $(x, f \circ g(x))$, which is a point of $f \circ g$. Label this point.

(f) Parts a–e provide an algorithm for constructing the graph of $f \circ g$ from the graphs of f and g. Start with a domain component x of g. Go vertically to g. Go horizontally to I. Go vertically to f. Go horizontally until positioned above the original x. You are now at a point for $f \circ g$. Use this algorithm to plot ten points of $f \circ g$.

(g) Discuss any difficulties that arise in the algorithm in part f.

(h) Use the points plotted in part g to sketch the remainder of the graph.

2. Start with a fresh copy of Display 2-7. Now draw the graph of $g \circ f$.

3. Start with fresh graph paper. Sketch the graph of $f(x) = x^2$ and $g(x) = x - 2$ on the same coordinate system.

4. See 3. Use the same method to construct the graph of $g \circ f$.

Decomposition of functions

In algebra, once you learn to multiply well, you reverse this skill by learning to factor. In calculus, once you have mastered composition of functions, you learn to break a complicated function h into two functions f and g so that $f \circ g = h$. For example, if $h(x) = 3(2x - 1)^2 - 5$ then let $f(x) = 3x^2 - 5$ and $g(x) = 2x - 1$. Now $f \circ g = h$. For each of the following functions h, determine f and g so that $h = f \circ g$.

5. $h(x) = 2(x + 3)^2 - 5$ **6.** $h(x) = 5(x - 4)^2 + 2$

7. $h(x) = (x + 1)^2 - 3(x + 1) + 2$

8. $h(x) = (5x)^2 - 3(5x) + 1$

9. $h(x) = \sqrt{3x - 1}$ **10.** $h(x) = \sqrt[3]{2 - 5x}$

11. $h(x) = \dfrac{3}{(5x - 2)^2}$ **12.** $h(x) = \dfrac{-2}{(2 - 5x)^3}$

13. $h(x) = (x^2 + 1)^{23}$ **14.** $h(x) = 3(x^3 - 2)^{-17}$

Problems for graphing technology

Use graphing technology to graph each of the following groups of functions. In each case, one of the functions is a transformation of a shape based on composition with the other functions. These functions were chosen for easy identification of the effect of the transformation.

15. Compare the y-intercepts on each group of graphs in a through e. Based on the graphs, identify the functions that have the same y-intercept. Discuss the role of composition in this result.

    (a) $y = 2x + 3, y = x^2, y - 2(x)^2 + 3$

    (b) $y = \sin(x), y = 3\sin(x) - 1, y = 3x - 1$

    (c) $y = 3x - 5, y = x^2, y = 3(x)^2 - 5$

    (d) $y = 2\sin(3x - 1) + 5, y = 2x + 5, y = 3x - 1$

    (e) $y = 3\sin(2x + 5) - 1, y = 2x + 5, y = 3x - 1$

16. Compare the x-intercepts of each group of graphs in a through e. Based on the graphs, identify the functions that have the same x-intercept. Discuss the role of composition in this result.

    (a) $y = 2x + 3, y = x^2, y = (2x + 3)^2$

    (b) $y = \sin(x), y = \sin(5x - 2), y = 5x - 2$

    (c) $y = 3x - 5, y = x^2, y = (3x - 5)^2$

    (d) $y = 5\sin(4x + 3) - 2, y = 5x - 2, y = 4x + 3$

    (e) $y = 4\sin(5x - 2) + 3, y = 5x - 2, y = 4x + 3$

17. Compare the x-intercepts of each group of graphs in a through e. Based on the graphs, identify the functions that have the same x-intercept. Discuss the role of composition in this result.

    (a) $y = x - 3, y = -x^3, y = -(x - 3)^3$

    (b) $y = x + 2, y = x^3, y = (x + 2)^3$

    (c) $y = \sin(x), y = \sin(4x + 3), y = 4x + 3$

    (d) $y = \sin(x), y = \sin(3x - 1), y = 3x - 1$

    (e) $y = \sin(x), y = \sin(2x + 5), y = 2x + 5$

18. Compare the y-intercepts of each group of graphs in a through e. Based on the graphs, identify the functions that have the same y-intercept. Discuss the role of composition in this result.

    (a) $y = x + 2, y = x^3, y = (x)^3 + 2$

    (b) $y = x - 3, y = -x^3, y = (-x^3) - 3$

    (c) $y = \sin(x), y = 5\sin(x) - 2, y = 5x - 2$

    (d) $y = \sin(x), y = 4\sin(x) + 3, y = 4x + 3$

    (e) $y = \sin(x), y = -2\sin(x) + 3, y = -2x + 3$

EXPLORATION

The following problems anticipate concepts from the next section. Graph each of the following collection of functions in the same standard window. In each case describe a graphical (or algebraic) relation of h to f and g.

19. $f(x) = 2x + 4, g(x) = 0.5x - 2, h(x) = f \circ g(x)$

20. $f(x) = 3x + 6, g(x) = x/3 - 2, h(x) = f \circ g(x)$

21. $f(x) = x^2, g(x) = \sqrt{x}, h(x) = \sqrt{x^2}$

22. $f(x) = x^3, g(x) = \sqrt[3]{x}, h(x) = \sqrt[3]{x^3}$

23. $f(x) = \dfrac{1}{x}, g(x) = f(x), h(x) = f \circ g(x)$

24. $f(x) = e^x, g(x) = \ln x, h(x) = e^{\ln x}$

25. $f(x) = e^x, g(x) = \ln x, h(x) = \ln(e^x)$

26. $f(x) = \dfrac{2}{x - 3}, g(x) = \dfrac{2 + 3x}{x}, h(x) = f \circ g(x)$

7 INVERSE FUNCTIONS

How can it be that mathematics, being after all a product of human thought independent of experience, is so admirably adapted to the objects of reality?
—ALBERT EINSTEIN

S I D E B A R

Many computer programs have features and converse features. For example, insert a blank line and delete a blank line undo each other and are thus converses. Similarly, up a page and down a page are converses.

Suppose a function describes the amount of interest I, due on a loan of $100 at 10 percent interest after t years. Here I is a function of t: $I = f(t)$. In particular, $I = 0.10(\$100)t$.

Suppose that we pay $25 interest at the end of the loan. A reasonable question is, "How many years did we keep the loan?" The question seems backward to the original. We are asking if t could be a function of I. The generalized concept reversing the roles of domain and range is an inverse.

DEFINITION 2-11
Inverse Functions

> Suppose that f is a function $\qquad f = \{(x, f(x)): x \in D_f\}$.
> then the *converse* of f is the relation $\qquad f^* = \{(x, y): f(y) = x\}$.
> If f^* is a function then we reserve the special notation f^{-1} for f^* and we say that f and f^{-1} are **inverse functions**.

The preceding definition implies that the domain of f is equal to the range of f^*. Similarly the range of f equals the domain of f^*. Visually, interchanging the roles of the domain and range to construct f^* from f is equivalent to reflecting the graph about the line $y = x$. Compare the definition of inverse functions with the discussion of compositional inverses in the preceding chapter. These two concepts are equivalent: f and g are compositional inverses if and only if $g = f^{-1}$ (and $f = g^{-1}$). The proof is left as an exercise.

What conditions guarantee that a given function f has an inverse function f^{-1}? In Figure 2-82, vertical lines—used to test whether a graph is a function—become horizontal lines when reflected. Similarly, imagined horizontal lines will become vertical after the reflection. Because the new vertical lines test whether the new relation is a function, we conclude as follows.

A graph is the graph of a function if every vertical line intersects the graph in at most one point; that is, it passes the **vertical line test**. Moreover, if every horizontal line intersects the graph in at most one point then the *reflection* of the graph about the line $y = x$ is a function; that is, it passes the **horizontal line test**.

The horizontal and vertical line tests are visual checks. A function that satisfies these tests is a **one-to-one function**. The following definition provides an equivalent, algebraic test for one-to-one functions.

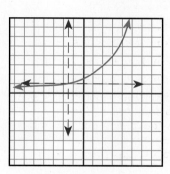

f

f^{-1}

FIGURE 2-82

DEFINITION 2-12
One-to-One Function

> A relation t is one-to-one if t pairs each domain component with exactly one range component *and* t pairs each range component with exactly one domain component.

Every one-to-one relation is a function and so is the converse of the relation.

EXAMPLE 2-16

One-to-One Functions and Inverse Functions. For each of the following relations g, determine whether (a) the relation is a function, (b) the converse of g, g^*, is a function, *and* (c) g is one to one.

Illustration 1:

See Figure 2-83.

Solution: Examine possible vertical and horizontal lines.

Answer: The graph represents a function, but it is not one to one.

Illustration 2:

See Figures 2-84 and 2-85.

Answer: The graph of Figure 2-84 does not represent a function. The graph of the converse in Figure 2-85 does not represent a function. Neither graph is one-to-one.

Illustration 3:

See Figures 2-86 and 2-87.

Answer: The graph in Figure 2-86 is one to one. Both the graph and its converse represent functions.

Illustration 4:

$$f = \{(3, 2), (4, 7), (6, 2)\}$$

Answer: The set f is a function. See Figure 2-88. But f is not one to one: the f range component 2 is paired with two different x components. Hence f^* is not a function.

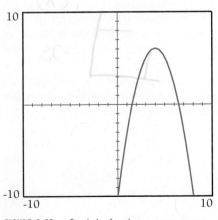

FIGURE 2-83   Parabola: function, not one to one

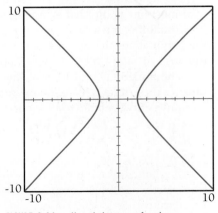

FIGURE 2-84   Hyperbola: not a function

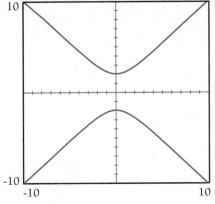

FIGURE 2-85   The converse of Figure 2-84

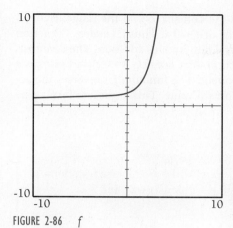

FIGURE 2-86   f

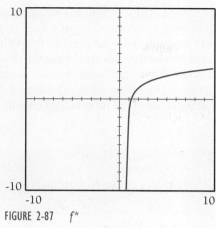

FIGURE 2-87   f*

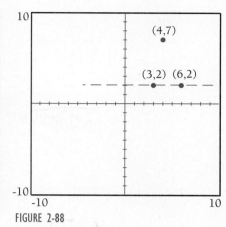

FIGURE 2-88

Illustration 5:

$$y = x^2 + 1$$

Solution: See Figure 2-89. The equation expresses y explicitly in terms of x, therefore y is a function of x:

$$y = f(x)$$

Also, recall that the graph of $y = x^2 + 1$ is a parabola that opens up. By examining the graph or by noting that x cannot be explicitly solved for in terms of y, conclude that x is *not* a function of y.

Answer: The function f is not one to one.

Figure 2-90 diagrams the complementary action of a function and its inverse.

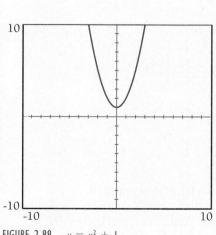

FIGURE 2-89 $y = x^2 + 1$

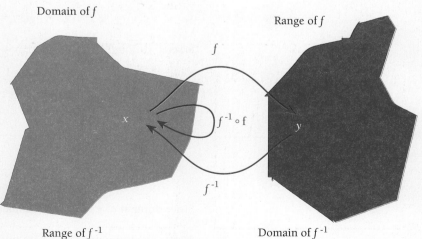

FIGURE 2-90 $f^{-1} \circ f$

Note that the function f maps x in the domain to y in the range. Here f^{-1} takes that y value and maps it back to the original x. Thus f^{-1} *undoes* the action of f in much the same manner that subtraction undoes addition. The composition $f^{-1} \circ f$ expresses this sequence of action. As a result, $f^{-1} \circ f(x) = x$ returns a range value equal to the domain component. The function f^{-1} reverses the activity of f to return us to our starting domain value. This argument forms part of the proof of the following theorem:

Theorem 2-6
Inverse Functions

> Suppose that f and g are functions, then f and g are inverse functions ($f^{-1} = g$, $g^{-1} = f$), if and only if $f \circ g(t) = t$ for all $t \in D_g$ and $g \circ f(v) = v$ for all $v \in D_f$.

EXAMPLE 2-17

Identifying Inverse Functions. Determine whether the following are pairs of inverse functions.

Illustration 1:

$$f(x) = 3x - 2, \qquad g(x) = \tfrac{1}{3}x + \tfrac{2}{3}$$

Solution:

$$\begin{aligned} f \circ g(x) &= f[g(x)] \\ &= f[\tfrac{1}{3}x + \tfrac{2}{3}] \\ &= 3[\tfrac{1}{3}x + \tfrac{2}{3}] - 2 \\ &= x + 2 - 2 \\ &= x \end{aligned}$$

Similarly, $g \circ f(x) = x$, see Figure 2-91.

Answer: The functions f and g are inverse functions: $f = g^{-1}$.

Illustration 2:

$$f(x) = \sqrt{x - 1}, \qquad g(x) = x^2 + 1$$

Solution: Although it is correct that

$$\begin{aligned} g \circ f(x) &= (\sqrt{x - 1})^2 + 1 \\ &= x, \qquad \text{since } x \geq 1 \text{ for domain of } f \end{aligned}$$

Note that

$$\begin{aligned} f \circ g(x) &= \sqrt{x^2 + 1 - 1} \\ &= \sqrt{x^2} \\ &= |x| \end{aligned}$$

Since domain of g is all real numbers, then $|x| \neq x$, as required.

Answer: The functions f and g are not inverse functions. See Figure 2-92.

Illustration 3:

$$f(x) = \sqrt{x - 1}, \quad \text{and} \quad g(x) = x^2 + 1, \quad x \geq 0$$

Solution:

$$g \circ f(x) = (\sqrt{x - 1})^2 + 1$$
$$= x, \quad x \geq 1$$
$$f \circ g(x) = \sqrt{x^2 + 1 - 1}$$
$$= \sqrt{x^2}$$
$$= |x|$$
$$= x, \quad \text{since } x \geq 0$$

Answer: The functions f and g are inverse functions. See Figure 2-93. Note the similarities and differences in Illustrations 2 and 3.

Illustration 4:

$$f = \{(1, 2), (5, -7), (3, 8)\}, \quad g = \{(2, 1), (-7, 5) (8, 3)\}$$

Solution: Notice that
$$f \circ g(2) = f[g(2)]$$
$$= f[1]$$
$$= 2$$

We need not check every possible form of $f \circ g$ and $g \circ f$ to verify that f and g are inverses. Note that the pairs for f are the same as the pairs for g and that the roles of the domain and range have been interchanged.

Illustration 5:

Many calculators feature inverse function keys. The inverse functions usually execute by first pressing a special inverse key. For example suppose that a calculator has a "square" $\boxed{x^2}$ function key and an inverse $\boxed{\text{INV}}$ function key. Then the following sequence will produce the square root of 16: square root is the inverse of square (see Illustration 3).

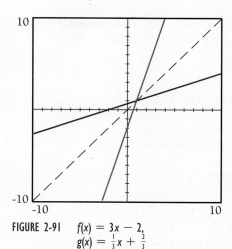

FIGURE 2-91 $f(x) = 3x - 2,$
$g(x) = \frac{1}{3}x + \frac{2}{3}$

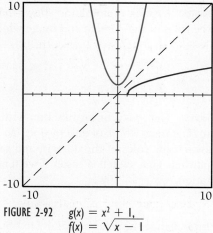

FIGURE 2-92 $g(x) = x^2 + 1,$
$f(x) = \sqrt{x - 1}$

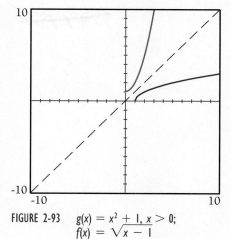

FIGURE 2-93 $g(x) = x^2 + 1, x > 0;$
$f(x) = \sqrt{x - 1}$

The key sequence $\boxed{1}\boxed{6}\boxed{\text{INV}}\boxed{x^2}\boxed{=}$ displays a 4. On the other hand, $\boxed{4}\boxed{x^2}\boxed{=}$ displays 16. The Casio FX-7700G has a $\boxed{\text{SHIFT}}$ key instead of an $\boxed{\text{INV}}$ key. In these cases the $\boxed{\text{SHIFT}}$ key may do more than produce the inverse of a given function key. In the Casio, the key square root sequence is $\boxed{\sqrt{}}\boxed{1}\boxed{6}\boxed{\text{EXE}}$. Whereas, $\boxed{4}\boxed{\text{SHIFT}}\boxed{\sqrt{x}}\boxed{\text{EXE}}$ produces 16. The TI calculators mark this special key as $\boxed{\text{2nd}}$. The Sharp uses $\boxed{\text{2ndF}}$.

In Example 2-17, you may have noticed that increasing functions had inverses that also were increasing. The following theorems summarize relationships between increasing (or decreasing) and inverse functions. We leave the proofs as exercises.

Theorem 2-7
Increasing or Decreasing with One to One

If a function f is increasing (or decreasing) on its domain, then f is one to one.

Corollary

A **linear function** with a nonzero slope is one to one.

Theorem 2-8
Increasing Functions and Inverses

If a function f is increasing on its domain, then f^{-1} is increasing on its domain.

EXAMPLE 2-18

One-to-One Functions and Inverse Functions. Determine which of the following functions are one to one. If one to one, determine a rule of pairing for the inverse of the function.

Illustration 1:

$$f = \{(1, 3), (-2, 4), (5, 7)\}$$

Solution: A quick examination shows that f is one to one. To construct f^{-1} switch the roles of the domain and range while maintaining the same pairings:

$$f^{-1} = \{(3, 1), (4, -2), (7, 5)\}$$

Illustration 2:

$$f(x) = 4x - 3$$

Solution: Here f is a linear function with a positive slope $m = 4$. By the corollary to Theorem 2-7, f is one to one. To determine a rule of pairing for f^{-1}, we switch the roles of the domain and range while maintaining the same rule of pairing. That switch is easier if we temporarily abandon standard function notation.

$$f: y = 4x - 3$$

Interchange the roles of x and y to obtain f^{-1}.

$$f^{-1}: x = 4y - 3$$

We prefer to have our rule of pairing solved for y.

$$f^{-1}: x + 3 = 4y$$

$$\frac{x}{4} + \frac{3}{4} = y$$

Now back to function notation.

$$f^{-1}(x) = \frac{x}{4} + \frac{3}{4}$$

Check:

$$f \circ f^{-1}(x) = f\left(\frac{x}{4} + \frac{3}{4} \right)$$

$$= 4\left(\frac{x}{4} + \frac{3}{4} \right) - 3$$

$$= x + 3 - 3$$

$$= x \qquad\qquad \text{(the identity, as expected)}$$

Illustration 3:
$$f(x) = x^2$$

Solution: Even functions are symmetric to the y-axis. Therefore, f cannot be one to one because f is even (f is a parabola that opens upward).

Illustration 4:
$$f(x) = \frac{2x + 3}{x - 1} \qquad (x \neq 1)$$

Solution: It is not immediately clear whether f is one to one. We shall illustrate an algebraic method for determining an inverse. This attempt to determine the inverse allows us to judge whether the function is one to one.

$$f: y = \frac{2x + 3}{x - 1}$$

Switch x and y in the rule of pairing.

$$x = \frac{2y + 3}{y - 1}$$

Judge whether this rule of pairing defines a function by solving for y. Multiply by $(y - 1)$.

$$x(y - 1) = 2y + 3$$

$$xy - x = 2y + 3$$

$$xy - 2y = x + 3$$

$$y(x - 2) = x + 3$$

Then, if $x \neq 2$, $y = \dfrac{x + 3}{x - 2}$

Because we were able to solve for y, we conclude that the rule of pairing implied a unique function, f^{-1}.

$$f^{-1}(x) = \frac{x+3}{x-2}$$

The existence of f^{-1} indicates that f is a one-to-one function.

Note: Because the domain of f^{-1} is all real x except that $x \neq 2$, then the range of f is all real y except that $y \neq 2$. The range of f^{-1} is restricted to $x \neq 1$. Why?

S U M M A R Y

For a relation f,

- The converse of f, f^*, comes from replacing (x, y) with (y, x) in the rule of pairing for f. $D_f = R_{f^*}$ and $R_f = D_{f^*}$.
- The graphs of f and f^* are symmetric about the line $y = x$.

For one-to-one functions f and g:

- The inverse of f, f^{-1}, comes from replacing (x, y) with (y, x) as in forming the converse of a relation.
- The graphs of f and f^{-1} are symmetric about the line $y = x$.

$$f \circ f^{-1}(t) = t, \qquad t \in D_{f^{-1}}$$
$$\text{and } f^{-1} \circ f(r) = r, \qquad r \in D_f.$$

- f and g are inverse functions if and only if

$$f \circ g(x) = x \text{ and } g \circ f(x) = x$$

for all values of x in their respective domains.

2-7   E X E R C I S E S

For each of the relations represented by graphs 1–10, determine whether (a) the relation is a function, (b) the relation is one to one. If one to one, graph the inverse function.

1.

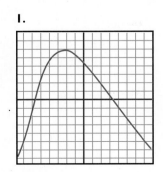

2.

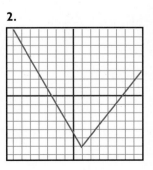

3.
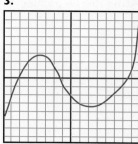

4.
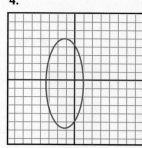

5.
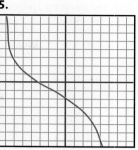

6.
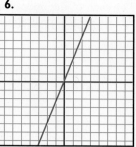

7.
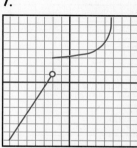

8.
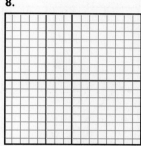

9.
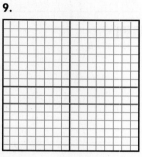

10.
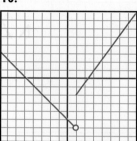

For each of the following functions determine whether the function is one to one. If one to one, determine the inverse of the function.

11. $f(x) = 5x - 3$

12. $g(x) = 3x - 5$

13. $h(x) = -\frac{2}{3}x + 2$

14. $k(x) = 0.3x - 1.5$

15. $j(x) = 1.4x - 3.7$

16. $t(x) = \frac{3}{7}x - 3$

17. $w(r) = r^2 - 2$

18. $v(s) = s^2 - 2, \ s > 0$

19. $q(u) = 2u - u^2, u < 1$

20. $z(w) = w^2 + 6w$

21. $f(x) = \dfrac{1}{x}$

22. $g(x) = -\dfrac{1}{x}$

23. $g(x) = \dfrac{x + 1}{x + 2}$

24. $f(x) = \dfrac{x - 2}{x + 3}$

25. $k(x) = \sqrt{x}$

26. $j(x) = \sqrt[3]{x}$

27. $p(r) = \sqrt[5]{r}$

28. $s(t) = \sqrt[4]{t}$

29. $z(x) = \sqrt{2x + 3}$

30. $w(x) = \sqrt{2 - x}$

Use composition to determine which of the following pairs of functions are inverse functions.

31. $f(x) = 5x - 3, g(x) = \frac{1}{5}x + \frac{3}{5}$

32. $f(x) = \frac{2}{3}x - 1, g(x) = \frac{3}{2}x + 3$

33. $h(x) = 2.5x + 2.5, j(x) = 0.4x - 4$

34. $h(x) = -x + 3, j(x) = 3 - x$

35. $f(x) = \dfrac{1}{x - 1}, g(x) = \dfrac{1}{x - 1}$

36. $f(x) = -\dfrac{1}{x}, g(x) = -\dfrac{1}{x}$

37. $f(x) = x^2, g(x) = \sqrt{x}$

38. $f(x) = x^3, x < 0, g(x) = \sqrt[3]{x}$

39. $f(x) = x^2, x > 0, \quad g(x) = \sqrt{x}$

40. $f(x) = x^3, g(x) = \sqrt[3]{x}$

For each of the one-to-one functions represented by graphs 41–50, sketch the graph of the inverse function.

41.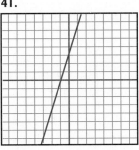

42.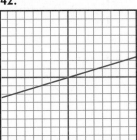

43.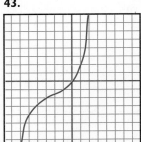

44.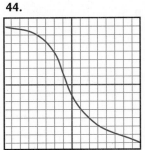

45.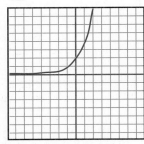

46.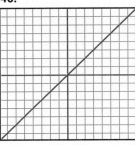

47.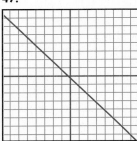

48.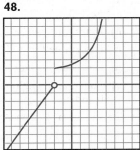

49.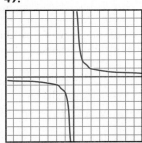

50.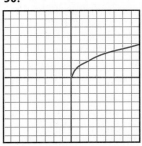

Prove the following theorems.

51. Every increasing function is one to one.

52. Every decreasing function is one to one.

53. No even function with domain $(-a, a)$, $a > 0$, is one to one.

54. If a function is one to one, it is not necessarily strictly increasing or strictly decreasing in its domain. (Give a counterexample.)

55. At least one odd function is one to one.

56. At least one odd function is not one to one.

57. $I(x) = x$ is its own inverse.

58. $H(x) = \dfrac{1}{x}$ is its own inverse.

59. $A(x) = |x|$ has no inverse.

60. $t(x) = x^{10} + x^6 + x^2 + 10$ has no inverse.

61. The inverse of an increasing function is increasing.

62. The inverse of a decreasing function is decreasing.

63. A constant function is not one to one.

64. If f is one to one, then $(f^{-1})^{-1} = f$.

2-7 P R O B L E M   S E T

1. (a) Reproduce the graph of a function f shown in Display 2-8. On the same copy sketch the graph of $I(x) = x$. Now choose any point of f and label it as $(x, f(x))$. Construct a horizontal line through $(x, f(x))$. Label the point where it intersects $I(x) = x$ as $(f(x), f(x))$.

(b) Construct a vertical line through $(x, f(x))$. Label the point where it intersects $I(x) = x$ as (x, x).

(c) Construct a vertical line through $(f(x), f(x))$ and a horizontal line through (x, x). These two new lines intersect at $(f(x), x)$. Label this point.

(d) Connect $(x, f(x))$ and $(f(x), x)$ with a line segment. This segment intersects $I(x) = x$ as some point, call it (a, a). Label this point.

(e) Prove that the distance from $(x, f(x))$ to (a, a) equals the distance from (a, a) to $(f(x), x)$. Mark these equal distances.

(f) Prove that the distance from $(x, f(x))$ to (x, x) equals the distance from (x, x) to $(f(x), x)$. Mark these equal distances.

(g) From geometry, if a line passes through two points equidistant from the endpoints of a segment, then the line is perpendicular to the segment. Prove that $I(x) = x$ is perpendicular to the segment from $(x, f(x))$ to $(f(x), x)$.

2. Use the results of Problem 1 for an algorithm to construct the graph of the converse of f. Use your algorithm to sketch the graph of f^*. Note that f and f^* are mirror images about the line $y = x$.

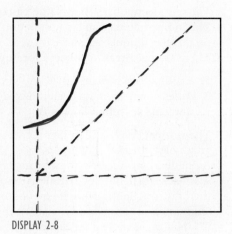

DISPLAY 2-8

3. (a) Determine the inverses of $f(x) = 2x - 4$ and $g(x) = 5x + 3$.

(b) Determine $f \circ g(x)$ and $g \circ f(x)$.

(c) Determine $f^{-1} \circ g^{-1}(x)$ and $g^{-1} \circ f^{-1}(x)$.

(d) Determine $(f \circ g)^{-1}(x)$ and $(g \circ f)^{-1}(x)$.

(e) Formulate relationships among the results of parts a, b and c.

4. See Problem 3. Repeat parts b, c, d and e, for $f(x) = 2 - x$ and $g(x) = \sqrt[3]{x}$.

PROBLEMS FOR GRAPHING TECHNOLOGY

Use graphing technology to graph each of the following pairs of functions in the same window. From the graphs of each pair, conjecture whether the pair of functions are inverses.

5. $y = 5x - 3$, $y = \frac{1}{5}x + \frac{3}{5}$

6. $y = \frac{2}{3}x + 1$, $y = \frac{3}{2}x - \frac{3}{2}$

7. $y = x^3$, $y = \sqrt[3]{x}$

8. $y = x^2$, $y = \sqrt{x}$

9. $y = x^4$, $y = \sqrt[4]{x}$

10. $y = x^5$, $y = \sqrt[5]{x}$

11. $y = x^3 + 2$, $y = \sqrt[3]{x - 2}$

12. $y = x^5 - 3$, $y = \sqrt[5]{x + 3}$

13. $y = \sin^{-1}x$, $y = \sin x$

14. $y = e^x$, $y = \ln x$

CHAPTER SUMMARY

- A function f pairs x values from the domain of f with unique y values from the range of the function. Thus $f(x)$ symbolizes the y value paired with the given x value.
- The implied domain of a real function f is assumed to be all real numbers x for which the rule of pairing produces a unique real value for y.
- There are two areas of concern for implied domains.

  1. Rational rule of pairing: $f(x) = N(x)/D(x)$, D_f is real numbers, where $D(x) \neq 0$.
  2. Rules of pairing with even indexed radicals: $f(x) = \sqrt[2n]{R(x)}$, where n is a counting number, D_f is all real numbers such that $R(x) \geq 0$.

- The vertical line test identifies a given graph as the graph of a function. No vertical line intersects the graph of a function more than once.
- A relation may not be symmetrical about exactly two of the x-axis, y-axis or origin.
- To check a function for symmetry, evaluate the function at $-x$. If you obtain the original rule of pairing then the function is even. If you obtain the negative of the original rule of pairing, the function is odd. Odd functions are symmetric about the origin, whereas even functions are symmetric about the y-axis.
- A function f is increasing on an interval I if the y values get larger as we examine the graph from left to right on the interval. A function is decreasing on an interval if the y values get smaller as the x values increase.
- A function f is periodic with fundamental period $k(k > 0)$, if k is the smallest number such that $f(x) = f(x + k)$ for every x (and $x + k$) in the domain of f.
- For two functions f and g with overlapping domains, $f + g, f - g, fg$ and f/g are functions with rules of pairing defined as follows:

$$(f + g)(x) = f(x) + g(x)$$
$$(f - g)(x) = f(x) - g(x)$$
$$(fg)(x) = f(x)g(x)$$
$$(f/g)(x) = f(x)/g(x), \qquad g(x) \neq 0$$

- The domain for each of these functions is $D_f \cap D_g$, except that f/g must also exclude any numbers x for which $g(x) \neq 0$.
- The domain of the composition $f \circ g$ is the real values x such that x is in the domain of g and $g(x)$ is in the domain of f. The rule of pairing for $f \circ g$ is

$$f \circ g(x) = f[g(x)]$$

$I(x) = x$ is the compositional identity: $f \circ I = f$ and $I \circ f = f$.
In general, $f \circ g \neq g \circ f$.
- A transformation of a real function f is a function given by $L_1 \circ (f \circ L_2)$, where L_1 and L_2 are first-degree functions. A transformation does not alter the fundamental shape of f. However, the transformation may shift the graph of f right, left, up or down according to the intercepts of L_1 and L_2. Also, the transformation may stretch the graph of f vertically or horizontally based on the slopes of L_1 and L_2.
- For a relation f,

  1. The converse of f, f^*, comes from replacing (x, y) with (y, x) in the rule of pairing for f, $D_f = R_{f*}$ and $R_f = D_{f*}$.

  2. The graphs of f and f^* are symmetric about the line $y = x$.

- For one-to-one functions f and g,

  1. The inverse of f, f^{-1}, is the converse of f.

  2. $f \circ f^{-1}(t) = t$, $t \in D_f{}^{-1}$,

  and $f^{-1} \circ f(r) = r$, $r \in D_f$.

  3. f and g are inverse functions if and only if

    $f \circ g(x) = x$ and $g \circ f(x) = x$

    for all values of x in their respective domains.

KEY WORDS AND CONCEPTS

Algebra of functions
Composition
Compositional identity
Compositional inverses
Continuous
Decreasing function
Dependent variable
Domain
Even function symmetry
Function
Function evaluation

Function notation
Graph of a function
Horizontal line test
Identity
Implied domain
Implied functions
Increasing function
Independent variable
Inverse function
Linear function
Odd function symmetry

One-to-one function
Origin symmetry
Periodic function
Piecewise function
Range
Real function
Relation
Transformation
Vertical line test

2 REVIEW EXERCISES

SECTION I

Identify which of the following expresses y as a function of x.

1. $y = 3x - 5$

2. $y = 5 - x^2$

3. $y = 2x^3$

4. $x^2 + y^2 = 4$

5. $y = x^2 + 8x + 1$

6. $x - 2 = y^2 + 4y$

Solve each of the following for the dependent variable y, and then express y as a function of x. Name each function f.

7. $5x + 3y = 8$

8. $x^2 - 2y = 4$

9. $xy = -2$

10. $9y - 3xy = 7$

11. $x^2y - 3 = -xy$

12. $xy - x = 3x^2y$

State the domain implied by the rule of pairing for each of the following functions.

13. $f(x) = 5x - 4$

14. $k(x) = 3x^2 - 4$

15. $t(x) = \sqrt{x + 2}$

16. $R(x) = \sqrt[5]{x}$

17. $t(x) = \sqrt{4 - 2x}$

18. $P(x) = \sqrt{5 - 7x}$

19. $g(x) = \dfrac{x}{x + 2}$

20. $h(x) = \dfrac{3x}{x^2 - 9}$

21. $p(x) = \dfrac{x + 1}{\sqrt{x - 6}}$

22. $q(x) = \dfrac{\sqrt{2 + x}}{x - 3}$

23. $f(x) = \sqrt{x+1} + \dfrac{5}{x}$ **24.** $g(x) = \sqrt{x} - \dfrac{1}{x-2}$

Evaluate $f(x) = 2x^2 - x$ as indicated.

25. $f(4)$ **26.** $f(-1)$
27. $f(0)$ **28.** $f(1)$
29. $f(-4)$ **30.** $f(3)$

Evaluate $g(x) = 2x^3$ as indicated.

31. $g(-3)$ **32.** $g(2)$
33. $g(0.5)$ **34.** $g(-b)$
35. $g(a+b)$ **36.** $g(a-b)$

Determine $f(3)$ for each of the following functions f.

37. $f(x) = 3x - 9$ **38.** $f(x) = \sqrt{4x+4}$

39. $f(x) = \dfrac{x}{x+3}$

40. $f(x) = \sqrt{x+6} + \dfrac{x+5}{x}$

41. $f(x) = \dfrac{2x-3}{\sqrt{x+1}}$ **42.** $f(x) = -3$

Evaluate $f(x) = 4x - x^2$ as indicated.

43. $f(w)$ **44.** $f(q+2)$
45. $f(x+t)$ **46.** $f(5t)$
47. $f(-x)$ **48.** $f(x^3)$

49. $f(x+h) - f(x)$ **50.** $\dfrac{f(b) - f(a)}{b-a}$

51. $\dfrac{f(x) - f(x_0)}{x - x_0}$ **52.** $\dfrac{f(x+h) - f(x)}{h}$

For each of the following functions f, evaluate $f(-x)$.

53. $f(x) = 2x^2 - 5$ **54.** $f(x) = \dfrac{x^3}{2}$

55. $f(x) = x^5 - x$ **56.** $f(x) = -3x^8 + x^2$
57. For $7z - 3w = 4$, write w as a function of z.
58. For $7z - 3w = 4$, write z as a function of w.
59. For $5x + 9y = 8$, write x as a function of y.
60. For $d + 16t^2 - 100t - 27 = 0$, write d as a function of t.

SECTION 2

Graph each of the following functions.

61. $f(x) = 3 - x$ **62.** $j(x) = -0.4x + 2$
63. $q(x) = -x^2$ **64.** $z(x) = (x+1)^2 - 3$

65. $f_3(x) = 5 + x^2$ **66.** $m(x) = -2x^3$

67. $f_5(x) = \sqrt{x+3}$ **68.** $s(t) = -\dfrac{1}{t}$

69. $A(x) = |x-2|$ **70.** $g(u) = \dfrac{u}{3}\sqrt{9 - u^2}$

SECTION 3

For each of the following relations, determine any symmetry to the x-axis, y-axis or origin. Use graphing technology.

71. $y = -2x$ **72.** $y = 3 + x$

73. $y = \dfrac{1}{x^3}$ **74.** $y = \sqrt[3]{x}$

75. $y = x^2$ **76.** $y = x^3 - 2x$
77. $y = x^2 + 4$ **78.** $x^2 + y^2 = 9$
79. $x = y^2 + 3$ **80.** $x - y = 9$

Check each of the following functions to determine (a) whether the function is odd, even or neither, and (b) whether the function is symmetric to the origin or y-axis. Use graphing technology to confirm your analysis.

81. $f(x) = 5x^2$ **82.** $k(x) = -2x$
83. $t(x) = x^4 - 3x^2 + 5$ **84.** $w(x) = -x^3 + x + 3$
85. $q(x) = x + x^3$ **86.** $B(x) = -3$
87. $L(x) = 5x + 2$ **88.** $B(x) = |2x| + 1$

89. $T(x) = \dfrac{x^4 - 3}{x^2}$ **90.** $P(x) = x^3(x^4)$

SECTION 4

Indicate whether each of the following linear functions is increasing, decreasing or neither.

91. $f(x) = 0.5x - 3$ **92.** $t(x) = 3 + 5x$

93. $w(x) = 3 - 4x$ **94.** $n(x) = \dfrac{2}{3} - \dfrac{x}{4}$

Determine where the following functions are increasing, decreasing or constant.

95. $f_0(x) = \begin{cases} -3x & \text{if } x \leq 1 \\ 2x & \text{if } x > 1 \end{cases}$

96. $f_3(x) = \begin{cases} 7x - 5 & \text{if } x \leq -3 \\ -x + 5 & \text{if } x > -3 \end{cases}$

97. $f_5(x) = \begin{cases} 3x & \text{if } x \leq -3 \\ 4 & \text{if } -3 < x \leq 4 \\ 5 - x & \text{if } 4 < x \end{cases}$

98. $f_7(x) = \begin{cases} 2x - 7 & \text{if } x < -1 \\ x - 8 & \text{if } -1 \leq x < 4 \\ 7 - 3x & \text{if } 4 \leq x \end{cases}$

99. $f_8(x) = \begin{cases} 10 & \text{if } x \leq 1 \\ 2x + 3 & \text{if } x > 1 \end{cases}$

100. $f_9(x) = \begin{cases} 3x - 5 & \text{if } x \leq 2 \\ -11 & \text{if } x > 2 \end{cases}$

SECTION 5

For each of the following pairs of functions f and g, form the rules of pairing for $f + g, f - g, fg$ and f/g. Also, give the domain for each of these new functions.

101. $f(x) = 3x, g(x) = 2$ **102.** $g(x) = 4, f(x) = -x$

103. $g(x) = \sqrt{x - 1}, f(x) = \dfrac{1}{x}$

104. $f(x) = 2x + 1, g(x) = x + 3$

105. $f(x) = 3x + 1, g(x) = x - 5$

106. $f(x) = 2x^3, g(x) = x$

107. $f(x) = x^3, g(x) = x$ **108.** $f(x) = 5, g(x) = x^2$

109. $f(x) = x^2, g(x) = 2x - 3$

110. $f(x) = 1 - x^2, g(x) = x^2 - 4$

111. $f(x) = \dfrac{x^2 - 1}{3}, g(x) = \sqrt{3x + 1}$

112. $f(x) = \sqrt{x - 1}, g(x) = x^2 + 1$

For each of the following functions f, (a) determine rules of pairing for two functions g and h such that $f = g + h, g(x) \neq 0$ and $h(x) \neq 0$; (b) determine rules of pairing for two functions m and n such that $f = mn, m(x) \neq 1, n(x) \neq 1$.

113. $f(x) = 3x + 9$ **114.** $f(x) = x^2 - 5x$

115. $f(x) = 2x^2 - 10x + 12$ **116.** $f(x) = x^2 + 8x + 12$

117. $f(x) = x^2 + x - 12$ **118.** $f(x) = x^2 - 4x - 12$

119. $f(x) = x^2 + 7x + 12$ **120.** $f(x) = x^2 - 25$

121. $f(x) = 3x^2 + 7x - 6$ **122.** $f(x) = x^3 + 8$

For each of the pairs of functions f and g, determine $f \circ g(3)$ and $g \circ f(3)$.

123. $f(x) = 5, g(x) = 3x$ **124.** $f(x) = -2x, g(x) = 4$

125. $f(x) = \dfrac{1}{x}, g(x) = \sqrt{x}$

126. $f(x) = x - 1, g(x) = x + 2$

127. $f(x) = 2x - 5, g(x) = x + 3$

128. $f(x) = x^3, g(x) = x - 1$

129. $f(x) = x^3, g(x) = x - 2$

130. $f(x) = x^3, g(x) = x + 1$

131. $f(x) = x^2 - 3, g(x) = 9 - x^2$

132. $f(x) = \sqrt{x + 1}, g(x) = \sqrt{2x - 2}$

133. $f(x) = \dfrac{1}{x}, g(x) = \dfrac{1}{x}$

134. $f(x) = -\dfrac{5}{x}, g(x) = -\dfrac{5}{x}$

135. $f(x) = 0.2x - 3, g(x) = 5x + 15$

136. $f(x) = \sqrt{x + 5}, g(x) = x^2 - 5$

137. $f(x) = \sqrt{x + 1}, g(x) = 1 - x$

138. $f(x) = \dfrac{2}{x + 1}, g(x) = 1 - x$

SECTION 6

For each of the following pairs of functions f and g, determine (a) a rule of pairing for $f \circ g$ and (b) the domain of $f \circ g$.

139. $f(x) = 5, g(x) = -2x$

140. $f(x) = \sqrt{x}, g(x) = -\dfrac{1}{x}$

141. $f(x) = 3x - 2, g(x) = x + 5$

142. $f(x) = x^3, g(x) = \dfrac{1}{x^3}$

143. $f(x) = x^2 - 1, g(x) = 1 + x^2$

144. $f(x) = -\dfrac{5}{x}, g(x) = -\dfrac{5}{x}$

145. $f(x) = \frac{1}{5}x - 1, g(x) = \sqrt{5x + 1}$

146. $f(x) = \sqrt{x + 7}, g(x) = x^2 - 7$

147. $f(x) = \sqrt{1 - x}, g(x) = 1 - x$

148. $f(x) = \dfrac{2}{x + 1}, g(x) = 2x - 1$

Suppose that S has the graph shown in the window of Figure 2-77. Sketch the graph of *each of* the following transformations of S. (Use graphing technology. The graph of S is identical to the graph of sin.)

149. $f(x) = S(5x)$ **150.** $f(x) = S(\pi x)$

151. $f(x) = -4S(x)$ **152.** $f(x) = 4S(x)$

153. $f(x) = 5S(x) + 2$ **154.** $f(x) = 2S(x) - 5$

155. $f(x) = 5S(x - 3)$

156. $f(x) = -2S(x - 3) + 1$

157. $f(x) = 4S(3x - 5)$

158. $f(x) = -4S(-3x - 5) + 2$

SECTION 7

For each of the following functions determine whether the function is one to one. If the function is one to one, determine the inverse of the function.

159. $f(x) = 50x - 30$

160. $k(x) = 0.03x - 0.05$

161. $j(x) = \frac{2}{3}x - \frac{3}{5}$

162. $v(s) = 2 - s^2,\ s > 0$

163. $q(u) = 4u - u^2, u < 2$

164. $g(x) = -\dfrac{3}{x}$

165. $g(x) = \dfrac{x}{x-1}$

166. $j(x) = \sqrt[3]{x - 2}$

167. $p(r) = \sqrt{r}$

168. $w(x) = \dfrac{1}{\sqrt{2 - x}}$

Use composition to determine which of the following pairs of functions are inverse functions.

169. $f(x) = 4x - 7, g(x) = \frac{1}{4}x + \frac{7}{4}$

170. $f(x) = 0.5x + 3.5, g(x) = 2x - 7$

171. $f(x) = \dfrac{2}{x - 3}, g(x) = \dfrac{3x + 2}{x}$

172. $f(x) = x^2, g(x) = \sqrt{x}$

173. $f(x) = x^2 + 1, x > 0, g(x) = \sqrt{x - 1}$

SIDELIGHT

History of Functions

The flower of modern mathematical thought—the notion of a function.
—Thomas J. McCormack

Like most ideas, the concept of function evolved over a long period, changing with the needs of the time. Many mathematicians applied the whetstone of their intellect to sharpen and refine the idea of function into its present usage.

G. W. Leibniz (1646–1716) introduced the name function in 1694. Late in *Historia* (1714), Leibniz used the word *function* to mean any quantities that depend on a variable. Leibniz used the term to indicate any quantity connected with the graph of a curve. Some twenty-four years later, Johann Bernoulli (1667–1748) pioneered the idea that a function was an expression that incorporated a variable and some constants.

Leonhard Euler (1707–1783) refined the concept once more. Euler reserved the word *function* for equations involving variables and constants. The Euler viewpoint of function dominated until Joseph Fourier (1768–1830).

The investigations of Fourier into heat flow mechanics motivated the evolution of function into its modern-day usage. The problems Fourier studied presented relationships between variables that were more general than any previously considered. As a result, Lejeune Dirichlet (1805–1859) expanded the definition of function to include the formulas of Fourier. The Dirichlet definition of function is the one in common usage today and is equivalent to the definition given in Section 2-1. Approximately 150 years of evolution from Leibniz to Dirichlet has modified the idea of function into a modern tool of analysis.

Modern set theory expands Dirichlet's generalization. The function concept grew from a simple idea of pairings into an approach that permeates much of modern mathematics and science. The function has evolved into an important tool in our efforts to understand nature.

2 CHAPTER TEST

1. Which of the following represent functions and which represent one-to-one functions?

  (a) $x + y = 2$

  (b) $x^2 + y^2 = 4$

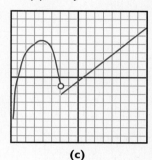   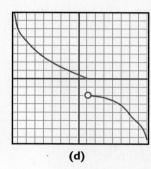

          (c)   **(d)**

2. If $g(x) = \dfrac{x^2}{x^2 + 1}$, determine

  (a) $g(-2)$

  (b) $g(x + h)$

  (c) $\dfrac{g(x + h) - g(x)}{h}$

  (d) Is g odd, even or neither?

  (e) Is g symmetric to the y-axis, x-axis, origin or none of these?

3. Sketch the graph of $f(x) = x^2 - 4x - 6$. If f is symmetric to a line, draw the axis of symmetry.

4. Give the domain and range of $f(x) = \sqrt{x^2 - 4x - 6}$

5. For $5r + 3s^2 - 2 = 0$, express r explicitly as a function of s.

6. Suppose that $f(x) = x^2 + 1$ and $g(x) = \sqrt{x + 2}$. Determine the domains and rules of pairing for each of the following:

  (a) $f + g$

  (b) $f - g$

  (c) gf

  (d) f/g

7. For $f(x) = x^2 - 1$ and $g(x) = \sqrt{x + 1}$, determine

  (a) $f \circ g(x)$

  (b) $D_{f \circ g}$

  (c) $g \circ f(x)$

  (d) $D_{g \circ f}$

8. **(a)** Determine $f^{-1}(x)$ for $f(x) = -\frac{2}{3}x - 5$.

  (b) Sketch the graphs of f and f^{-1} on the same coordinate system.

9. If $f(x) = x^3 - 1$ and $g(x) = \sqrt[3]{x + 1}$, does $f^{-1} = g$? Justify your answer.

10. Prove that, if f is a decreasing function, then f^{-1} exists and is a decreasing function.

LINEAR AND QUADRATIC FUNCTIONS

This chapter examines some special functions. Among these are linear, quadratic and absolute value functions. In later chapters these functions, along with others, provide the blocks for building more intricate functions.

1 LINEAR FUNCTIONS

The essence of mathematics lies in its freedom.
—GEORGE CANTOR

We have investigated general functional patterns such as symmetry, increasing and periodicity. We turn now to categorizing functions. The classification of a function provides a powerful clue to the shape of the graph. Indeed, you should recognize the functions of this section as old friends.

The line is among the simplest graphic shapes. Our first category of function is the linear function, or first-degree function.

DEFINITION 3-1
First Degree Function

> A first-degree function f is a function defined by
>
> $$f(x) = mx + b \quad (m \neq 0, \ b, \ x \in \text{Real})$$

Note: We classify functions by the role of the variable in the rule of pairing. In definition 3-1, the exponent for x was 1, hence the name *first degree*. Because the graph of $y = mx + b$ is a straight line, a first-degree function is a **linear function.** A **constant function**, such as $g(x) = c$, is not first degree. (Functions were discussed in Chapter 2). Because the graph of g is a horizontal line, g is also linear.

EXAMPLE 3-1

Analyzing and Graphing Linear Functions. Identify each of the following functions as linear or not. If linear then determine the slope, x-intercept and y-intercept. Graph the function using graphing technology.

Illustration 1:

$$f(x) = \tfrac{2}{3}x - 5$$

Solution: First-degree function, linear. See Figure 3-1.

Slope: $m = \tfrac{2}{3}$; y-intercept: $(0, -5)$

Note: The **x-intercept** of a function f is the point where the graph of f intersects the x-axis. Constant functions do not intercept the x-axis in a single point, but any first-degree function intercepts the x-axis exactly once. To determine that unique point, set $f(x)$ equal to 0 and solve for x.

$$0 = \tfrac{2}{3}x - 5$$

$$5 = \tfrac{2}{3}x$$

$$\tfrac{15}{2} = x$$

The x-intercept for f is $\left(\tfrac{15}{2}, 0\right)$.

Use the graphing technology trace feature with f to position the cursor at $(3, -3)$. Move the cursor 3 units left. The x coordinate is 0. Examine the change in the y coordinate to verify that the slope is $\tfrac{2}{3}$. Also confirm that the x-intercept is 7.5.

Illustration 2:

$$g(x) = -5x + \tfrac{2}{3}$$

Solution: First-degree function, linear. See Figure 3-2.

Slope: $m = -\tfrac{5}{1}$, y-intercept: $\left(0, \tfrac{2}{3}\right)$, x-intercept: $\left(\tfrac{2}{15}, 0\right)$

Check these using Trace with your graphing technology.

Illustration 3:

$$h(x) = -2$$

Solution: Constant function, linear. See Figure 3-3.

Slope: $m = 0$, y-intercept: $(0, -2)$

No x-intercept.

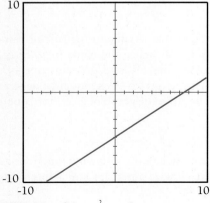

FIGURE 3-1 $f(x) = \tfrac{2}{3}x - 5$

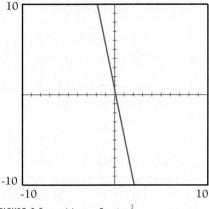

FIGURE 3-2 $g(x) = -5x + \tfrac{2}{3}$

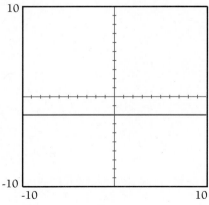

FIGURE 3-3 $h(x) = -2$

Illustration 4:

$$k(x) = x^2$$

Because the rule of pairing is second degree, k is not linear. ▬

The average rate of change of y compared to x measures the **slope of** a linear function and is given by

$$\frac{\Delta y}{\Delta x} = \frac{y_2 - y_1}{x_2 - x_1}$$

$$\frac{\Delta y}{\Delta x} = \frac{f(x_2) - f(x_1)}{x_2 - x_1}$$

Calculus refers to this expression as the *difference quotient*. For any two points of linear function f, where $f(x) = mx + b$, the difference quotient f is m. Because $f(0) = b$, the y-intercept of f is $(0, b)$.

EXAMPLE 3-2

Evaluating the Difference Quotient. Evaluate the difference quotient for the following pairs of points.

Illustration 1:

$$k(x) = 5x - 1$$
$$(6079.3, k(6079.3)), \quad (-3.892, k(3.892))$$

Solution: No calculation is necessary because k is linear (Figure 3-4). The slope is 5 for *any* pair of points of k.

Illustration 2:

$$g(x) = x^2$$
$$(1, g(1)), \quad (2, g(2))$$

Solution: See Figure 3-5.

$$\frac{g(2) - g(1)}{2 - 1} = \frac{2^2 - 1^2}{2 - 1} = 3$$

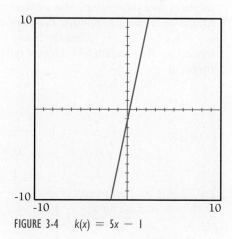

FIGURE 3-4 $k(x) = 5x - 1$

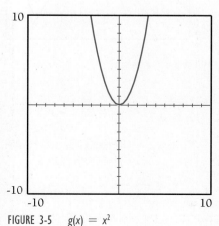

FIGURE 3-5 $g(x) = x^2$

Illustration 3:

$$g(x) = x^2$$
$$(2, g(2)), \ (3, g(3))$$

Solution:

$$\frac{g(3) - g(2)}{3 - 2} = \frac{3^2 - 2^2}{3 - 2} = 5$$

Note: From Illustrations 2 and 3, g does *not* have the same slope between any two pairs of points. The average change is "steeper" between points at $x = 2$ and $x = 3$ than between $x = 1$ and $x = 2$. See Figure 3-5. The graph of g is not a straight line.   ▬

The slope of a linear function provides insight into its graph. Examine the illustrations of Example 3-1. What is the relation of a line's slope to whether the function is increasing or decreasing? For easy reference we repeat Theorem 2-5 as Theorem 3-1.

Theorem 3-1
Increasing and Decreasing

Suppose that f is a linear function given by $f(x) = mx + b$.
  1. If $m > 0$, then f is increasing on $(-\infty, \infty)$;
  2. If $m < 0$, then f is decreasing on $(-\infty, \infty)$;
  3. If $m = 0$, then f is constant on $(-\infty, \infty)$.

FIGURE 3-6 $y = mx + 3$

Consider two linear functions in the same plane:

$$f_1(x) = m_1 x + b_1$$
$$f_2(x) = m_2 x + b_2$$

If $b_1 = b_2$ then f_1 and f_2 share the same y-intercept. Imagine all linear functions with a common y-intercept, say $(0, 3)$. See Figure 3-6.

There are infinitely many such linear functions. We specified one constant, $b = 3$, but b alone does not identify a particular linear function. A constant used in this manner is a **parameter**. To produce a specific linear function requires two parameters: the y-intercept b and the slope m. Imagine all linear functions with a given slope as a parameter, say $m = 2$ (Figure 3-7).

Again there are infinitely many such linear functions; each line in this family is a line **parallel** to the others. These observations lead to the following theorem.

Theorem 3-2
Perpendicular and Parallel

Suppose f and g are two linear functions,

$$f(x) = mx + b \quad \text{and} \quad g(x) = nx + c$$

  1. Then $f = g$, if and only if $m = n$ and $b = c$ (the graphs of f and g are the same line).
  2. If $m = n$ and $b \neq c$, then f and g are parallel lines.

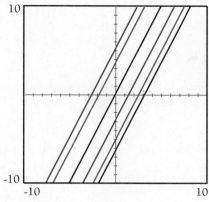

FIGURE 3-7 $f(x) = 2x + b$

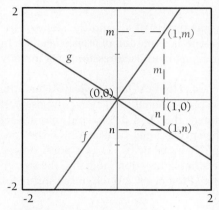

FIGURE 3-8   Pythagorean theorem

3. If $m \neq n$, then f and g intersect at a single point.

4. Suppose $m \neq 0$ and $n \neq 0$, f and g are perpendicular if and only if $mn = -1$.

Proof: Part 1, where the slopes and intercepts are both equal, is a summary of the parameter discussion. Equal slopes with unequal intercepts is left as an exercise. Part 3 is a direct result of Part 2 and is also an exercise.

For Part 4, consider f and g intersecting at $(0, 0)$ as shown in Figure 3-8. The slopes of the two lines are not equal and neither slope is 0. (In fact $m > 0$ and $n < 0$.) Therefore neither line is horizontal. On the x-axis, one unit to the right of the point of intersection, erect a perpendicular line. Because the slopes are m and n for f and g, respectively, a right triangle is formed as shown in Figure 3-8. By the Pythagorean theorem, the hypotenuse on f will be $\sqrt{1 + m^2}$. Similarly, the hypotenuse on g is $\sqrt{1 + n^2}$. Suppose that f and g are **perpendicular**; that is, they form a right angle. The hypotenuse of the large right triangle with sides on f and g is of length $\sqrt{(1-1)^2 + (m-n)^2} = \sqrt{(m-n)^2}$. Apply the Pythagorean theorem again:

$$1 + m^2 + 1 + n^2 = \sqrt{(m-n)^2}^2 \qquad \text{(because } (m-n)^2$$
$$2 + m^2 + n^2 = m^2 - 2mn + n^2 \qquad = m^2 - 2mn + n^2)$$
$$2 = -2mn$$

Neither m nor n is 0, therefore $mn = -1$.

These steps are reversible, so that if $mn = -1$, then f and g intersect at right angles. If we translate the two lines away from the origin, substitute $x - r$ for x and $y - s$ for y, neither the slope of the lines nor the angle between them changes. Therefore the theorem still applies.

Write a formula for the following linear functions.

EXAMPLE 3-3

Illustration 1:

See Figure 3-9. The linear function f through $(2, 5)$ and parallel to $g(x) = 3x + 1$.

Solution: The slope of f equals the slope of g because f is parallel to g. Therefore $f(x) = 3x + b$. All that remains is the parameter b. Because f passes through $(2, 5)$, substitute 2 for x and 5 for $f(x)$ in the formula

$$f(2) = 5$$
$$5 = 3(2) + b$$

Solve for b: $-1 = b$

$f(x) = 3x - 1$ is the *required* linear function.

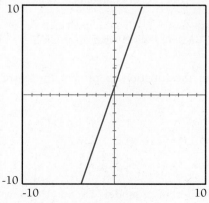

FIGURE 3-9 $g(x) = 3x + 1$

Illustration 2:   The linear function f with y-intercept of 7 and perpendicular to

$$g(x) = \tfrac{2}{3}x - 4$$

Solution: The y-intercept b is given. See Figure 3-10. Because f is perpendicular to g, the slope of f is the negative reciprocal of the slope of g:

$$m = \frac{-1}{\frac{2}{3}}$$

$$= -\tfrac{3}{2}$$

The formula for f is $f(x) = -\tfrac{3}{2}x + 7$.

Illustration 3:   Hooke's Law: If the y-intercept of a linear function is $(0, 0)$ and the slope of the line is *not* 0, we often refer to the line's slope as a *constant of proportionality*. In many applications we indicate such functions by the expression "y is directly proportional to x."

A spring is given to a physics class. Hooke's Law indicates that the "stretch" of a spring y, in centimeters, is proportional to the force imposed on the spring, x, in kilograms. By experiment, the class determines that a 2 kg mass hung from the spring stretches the spring 5 cm. (a) Determine the constant of proportionality for this spring. (Remember $(0, 0)$ is a point on the line: (no weight, no stretch).) (b) Write a linear function modeling the stretch of the spring as a function of the mass. (c) Predict the stretch in the spring for a 12 kg weight.

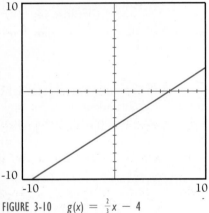

FIGURE 3-10 $g(x) = \tfrac{2}{3}x - 4$

Solution:

We require	$f(x) = Mx$	(general model from Hooke's Law)
Since	$f(2\text{ kg}) = 5\text{ cm}$	(specific spring, by experiment)
then	$5\text{ cm} = M2\text{ kg}$	

$$M = \frac{5\text{ cm}}{2\text{ kg}} \quad \text{(the rate of change is 2.5 cm/kg)}$$

The function is $f(x) = \dfrac{2.5\text{ cm}}{\text{kg}}x$ (specific model, Figure 3-11)

Suppose that $x = 12$ kg. Substitute into the function to predict the corresponding $f(12\text{ kg})$ stretch:

$$f(12\text{ kg}) = \frac{2.5\text{ cm}}{\text{kg}}(12\text{ kg})$$

$$f(12\text{ kg}) = 30\text{ cm}$$

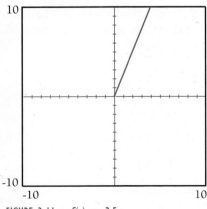

FIGURE 3-11 $f(x) = 2.5x$

Note: The function $f(x) = 2.5$ (cm/kg) x represents a **model**. The model allows us to predict the spring's behavior under certain circumstances. However, a model is not reality. Substituting 1000 kg for x may not produce meaningful results: the limit of elasticity of the spring will long since have been exceeded. Models help us predict reality, but they have their limits. ▬

SUMMARY

In computer science, *parameters* are the values passed to a program (or subprogram) to complete the information needed to execute the program. In that respect, parameters play a role similar to arguments. In mathematics, parameters are usually constants required to complete the description of a specific problem. The distinction between parameters and variables is a matter of emphasis.

- The graph of a linear function f given by $f(x) = mx + b$ is a straight line. The domain of f is the set of real numbers. The slope of the line is m. The y-intercept of the graph is $(0, b)$.
- The graphs of two linear functions are parallel if they have the same slope. The graphs are perpendicular if the product of their slopes is -1.
- *Parameter* translates literally as "along-side measurement." Parameters specify information essential to describe a specific linear function. Linear functions require two parameters; for example, the slope and the y-intercept, although alternate parameters are possible.

3-1   EXERCISES

For the following linear functions, determine the slope and y-intercept. Determine whether the function is increasing, decreasing or neither. Graph the function.

1. $f(x) = -3$

2. $g(x) = 5$

3. $h(x) = 3x$

4. $j(x) = -2x$

5. $k(x) = -3x + 4$

6. $w(x) = 4x - 5$

7. $q(x) = 0.5x - 2$

8. $t(x) = -0.3x + 2$

9. $v(x) = 4 - \frac{1}{2}x$

10. $u(x) = \frac{2}{3}x - 1$

Write a formula for the linear function to satisfy the following conditions.

11. Through $(0, 5)$ with slope of -4.

12. Through $(0, -3)$ with slope of 2.

13. Through $(0, -2)$ and perpendicular to $g(x) = -\frac{2}{3}x + 1$.

14. Through $(0, \frac{1}{2})$ and perpendicular to $h(x) = \frac{3}{4}x - 2$.

15. Through $(2, 5)$ and with slope of $\frac{1}{2}$.

16. Through $(-1, 4)$ and with slope of $\frac{2}{3}$.

17. Through $(2, -1)$ and $(1, 5)$.

18. Through $(-5, 2)$ and $(1, 1)$.

19. Parallel to $j(x) = 2x - 1$ and with the same x-intercept as $k(x) = -3x + 5$.

20. With the same x-intercept as $j(x) = 2x - 1$ and parallel to $k(x) = -3x + 5$.

For the following pairs of linear functions, determine whether the graphs intersect in a single point, are parallel or are the same line. For intersecting lines indicate whether the lines are perpendicular.

21. $f(x) = 2x - 5$ $g(x) = 5x - 2$

22. $f(x) = \frac{1}{2}x + 3$ $g(x) = \frac{1}{2}x - 3$

23. $f(x) = \frac{1}{2}x + 3$ $g(x) = -2x + 3$

24. $f(x) = 4x + 5$ $g(x) = 2x + 5$

25. $f(x) = 7 - x$ $g(x) = x + 7$

26. $f(x) = \frac{3}{4}x - 2$ $g(x) = \frac{3}{4}x + 7$

27. $f(x) = \frac{3}{4}x + 2$ $g(x) = -\frac{3}{4}x + 7$

28. $f(x) = -\frac{3}{4}x + 2$ $g(x) = \frac{4}{3}x - 7$

29. $f(x) = -\frac{3}{4}x - 2$ $g(x) = -\frac{4}{3}x + 7$

30. $f(x) = -5$ $g(x) = \frac{1}{5}$

Evaluate the difference quotient for the indicated points of the following.

31. $g(x) = 2x - 5$ for $(3, g(3))$ and $(-5, g(-5))$.

32. $h(x) = \frac{1}{2}x$ for $(-2, h(-2))$ and $(1, h(1))$.

33. $j(x) = x + 5$ for $(3678, j(3678))$ and $(-200.3, j(-200.3))$.

34. $t(x) = -x + 3$ for $(6005, t(6005))$ and $(-0.004, t(-0.004))$.

35. $f(x) = x^2$ for $(2, f(2))$ and $(5, f(5))$.

36. $f(x) = x^2$ for $(-1, f(-1))$ and $(4, f(4))$.

37. $f(x) = mx + b$ for $(c, f(c))$ and $(d, f(d))$.

38. $f(x) = mx + b$ for $(x, f(x))$ and $(x + h, f(x + h))$.

39. $f(x) = x^2$ for $(x, f(x))$ and $(a, f(a))$.

40. $f(x) = x^2$ for $(x, f(x))$ and $x + h, f(x + h)$.

41. Suppose the dosage y of the drug chloramphenicol is proportional to the patient's body weight x. (a) If a 50 kg patient requires 2500 mg of the drug, determine the constant of proportionality. (b) Write a linear equation relating x to y. (c) Determine the best dosage for a 75 kg patient.

42. Suppose the average number of transmission errors y in a large file is proportional to the number of bits per second (BPS) x transmitted. (a) Determine the constant of proportionality if 1200 BPS produces 2 errors. (b) Write a linear equation relating x to y. (c) Predict the expected number of errors when transmitting at 9400 BPS.

43. Suppose the most pleasing shape for a window occurs when the width y is proportional to the height x of the window. (a) If you select a 30 cm wide by 50 cm high window as most pleasing, determine the constant of proportionality. (b) Write a linear equation relating x to y for this choice. (c) Predict the width of a 125 cm high window so that it will be judged most pleasing.

44. Suppose the speed y of a sailboat at "broad reach" is approximately proportional to the area x of sail unfurled (where wind is of constant velocity). (a) If 20 square meters of sail produce 10 knots, determine the constant of proportionality. (b) Write a linear equation relating x and y. (c) If unfurling a jib sail brings the total sail area to 28 square meters, predict the boat's speed.

45. The length of an item in centimeters is proportional to the length measured in inches. (a) If a 50 inch item measures 127 centimeters on a metric scale, determine the constant of proportionality of centimeters to inches. (b) Write a formula for converting inches to centimeters. (c) Convert 40 inches to centimeters. (d) Convert 40 centimeters to inches.

46. The volume of a liquid in liters is proportional to its volume in gallons. (a) Supposing that 5 gallons has the same volume as 19 liters, determine the constant of proportionality for liters to gallons. (b) Write a formula to convert liters to gallons. (c) What is the liter equivalent of 8 gallons? (d) What is the gallon equivalent of 8 liters?

47. Prove Case 2 of Theorem 3-2.

48. Prove Case 3 of Theorem 3-2.

49. See Figure 3-8. Prove that the length of the segment from $(1, 0)$ to $(1, n)$ is $-n$.

50. See Problem 49. Show that the triangle with vertices at $(0, 0)$, $(1, 0)$ and $(1, m)$ is similar to the triangle with vertices at $(0, 0)$, $(1, 0)$ and $(1, n)$. Hint: The angles of these triangles at $(0, 0)$ are complementary. Why?

51. See Problems 49 and 50. Prove that $-n/1 = 1/m$.

52. See Problem 51. Prove part 4 of Theorem 3-2 (i.e., prove $mn = -1$).

3-1   PROBLEM SET

PROBLEMS FOR GRAPHING TECHNOLOGY

Predict whether the graph increases or decreases in the standard window: ($x \in [-10, 10]$, $y \in [-10, 10]$). Graph and use the trace feature to approximate the x and y intercepts.

1. $y = 3 - 2x$

2. $y = 4x - 5$

3. $y = 3x - 2$

4. $y = 4 - 5x$

5. $y = \pi x - 1$

6. $y = x - \pi$

7. $3y = 2x - 7$

8. $5x - 3y = 4$

9. $y = 4 - 3.7x$

10. $y = 2.7x - 3$

EXPLORATION

The following problems anticipate concepts to be introduced in the next section. Graph the following. Identify an x interval over which the graph is increasing and an x interval over which the graph is decreasing.

11. (a) $y = x^2$

(b) $y = x^2 - 5$

(c) $y = x^2 + 3$

(d) $y = 3x^2$

(e) $y = 3x^2 - 2$

(f) The graph of each of parts a through e is a parabola. Each graph is said to *open up*. Describe what is meant by *opening up*.

(g) The graph of each of parts a through e decreases over an interval and then increases over an interval. The point at which the graph changes from increasing to decreasing is the *vertex* of the parabola. Give the coordinates of the vertex of each parabola.

12. (a) $y = -x^2$

  (b) $y = 5 - x^2$

  (c) $y = -x^2 + 3$

  (d) $y = -2x^2$

  (e) $y = 4 - 3x^2$

  (f) The graph of each of parts a through e is a parabola. Each graph is said to *open down*. Describe what is meant by *opening down*.

(g) The graph of each of parts a through e increases over an interval and then decreases over an interval. The point at which the graph changes from decreasing to increasing is the vertex of the parabola. Give the coordinates of the vertex of each parabola.

13. See Problems 11 and 12. Indicate whether the graphs of the following represent parabolas that open up or down. Give the interval over which the graph is increasing and the interval over which the graph is decreasing. Give the coordinates of the vertex (use the trace feature, if necessary).

  (a) $y = x^2 - 4x$ **(b)** $y = 4x - x^2$

  (c) $y = 3x^2 - 5x - 2$ **(d)** $y = 2x^2 - 3x - 2$

  (e) $y = -x^2 + 5x + 2$ **(f)** $y = -x^2 - 3x + 1$

14. See Problem 13. Use the trace feature on the graphs of parts a through f to determine the x-intercepts of the graphs.

---

2 **QUADRATIC FUNCTIONS**

The greatest mathematicians, as Archimedes, Newton, and Gauss, always united theory and applications in equal measure.
—FELIX KLEIN

Quadratic is from the Latin "quadratum" meaning squared. The primary characteristic of a quadratic equation is the presence of a squared term as the highest degree term.

The next type of function we consider is the quadratic function. **Quadratic function** is another name for a second-degree function. In general, a quadratic function f is defined by

$$f(x) = ax^2 + bx + c, \qquad a \neq 0$$

You may remember from Chapter 1 that the graph of a quadratic function is a **parabola**. However, not all parabolas are graphs of functions. Parabolas that open left or right fail the vertical line test for functions. The graph of a quadratic function is a parabola that opens up or down. As a result, each quadratic function has either a *maximum* value (opens down) or a *minimum* value (opens up) at its vertex. See Figure 3-12.

The parabolic shape is common. We find it in the path of a projectile or the arch of a bridge. In technology, the reflectors of automobile headlights and satellite dishes are parabolic in shape. You may imagine other possible applications. Despite the shape of the graph, quadratic functions have applications that may not, at first, appear to be related to parabolas. This section examines some uses of quadratic functions.

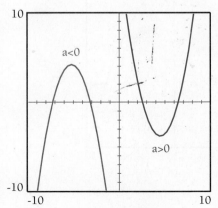

FIGURE 3-12 Quadratic function: $y = ax^2 + bx + c$ ($a > 0$: opens up; $a < 0$: opens down)

EXAMPLE 3-4

Illustration 1:

Consider a pen for chickens. What size should we build a rectangular pen, if the pen (a) is bounded on one side by the barn, (b) uses exactly 100 meters of chicken wire and (c) is to enclose the maximum area? Because an area is measured in square units, we anticipate that our formula will include a quadratic term.

Solution: Let $x =$ the width of the rectangle and, because area is a function of the width, $A(x) =$ the area of the rectangle. For all such possible rectangles (Figure 3-13),

$$A(x) = x(100 - 2x)$$
$$A(x) = 100x - 2x^2$$

See Figure 3-14. Plotting the area of the pen against the width of the pen produces a parabolic graph that opens downward. Therefore, the maximum value for the area $A(x)$ in terms of the width x of the rectangle occurs at the vertex of the parabola. **Complete the square** to determine the coordinates of the vertex.

$$A(x) = -2\ (x^2 - 50x + \underline{\hspace{1cm}}\ )$$
$$= -2(x^2 - 50x + 625) + 2(625)$$
$$= -2(x^2 - 50x + 625) + 1250$$
$$= -2(x - 25)^2 + 1250$$

The vertex is given by (width, area) = (25, 1250). $A(25) = 1250$. So, the area is a maximum of 1250 square meters when the width of the pen is 25 meters. With a width of 25 meters, the length must be 50 meters.

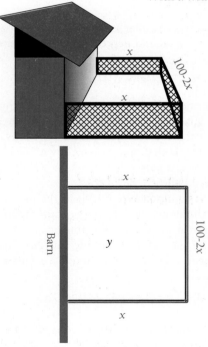

FIGURE 3-13   Rectangle x by $100 - 2x$

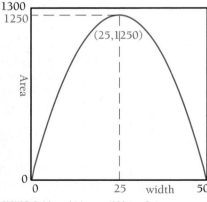

FIGURE 3-14 $A(x) = x(100 - 2x)$

Illustration 2:

The statement *y varies as the square of x* means that $y = kx^2$, where k is the constant of variation. As an example, consider the design of a windmill. Many factors affect the performance of a windmill: blade design, blade surface area, diameter of the turning circle, wind speed, and wind direction. To construct a mathematical model, we ignore the influence of most aspects to concentrate on one, the diameter of the turning circle. We consider the other features to be constant and lump them into a constant factor while we examine how changes in windmill diameter produce changes in power. The power generated by a windmill varies directly as the square of the length of the windmill blades. Suppose that a 10 knot wind produces 5 horsepower on a windmill with blade length of 2 meters. If the wind speed (and other factors) remains constant, determine the horsepower produced by changing the blade length to 3 meters.

Solution: Let y be the power in horsepower, x be the blade length in meters. To shorten our computations, we leave out the units when solving for k.

$$y = kx^2$$
$$5 = k(2)^2$$
$$5 = 4k$$
$$1.25 = k \qquad \text{(hp/m}^2\text{)}$$

Therefore, $y = 1.25x^2$

Let $x = 3$, $y = 1.25(3)^2$

$$y = 11.25 \qquad \text{(horsepower)}$$

Illustration 3:

The length of a pendulum varies directly with the square of the time that the pendulum takes for one complete cycle. Suppose that a given pendulum is 1 meter long and takes 2 seconds to complete a swing. Write a formula that expresses the relation between the length of the pendulum and the time of its swing. Also, determine the time of the pendulum's swing when shortened to 90 cm.

Solution: Let T be the time, in seconds, of the swing, and L the length of the pendulum in cm. Then,

$$L = kT^2$$
$$100 \text{ cm} = k(2 \text{ sec})^2$$
$$25 \text{ cm/sec}^2 = k$$

Now, $L = (25 \text{ cm/sec}^2)T^2$

Let $L = 90$ cm, then $90 \text{ cm} = (25 \text{ cm/sec}^2)T^2$

$$T^2 = 3.6 \text{ sec}^2$$
$$T = \sqrt{3.6} \text{ sec}$$

By calculator, $T \cong 1.897366596$ sec

Note: We have ignored friction and other factors affecting the pendulum. Moreover, we do not know the *accuracy* of the original measurements. Therefore nine-digit *precision* in the answer may not be justified.

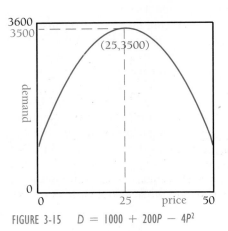

Illustration 4:

FIGURE 3-15 $D = 1000 + 200P - 4P^2$

Suppose the demand D (in thousands) for a product is a function of the price P (in dollars). A manufacturer gathers (price, demand) statistics and estimates the demand for a product is given by $D(P) = 1000 + 200P - 4P^2$ (Figure 3-15). Estimate the average rate of change of the demand as the price is increased from \$25 to \$26. This is often called the *marginal demand*. Also determine the average rate of change of the demand as the price is increased from \$$x$ to \$$(x + h)$. For clarity, we will omit the units.

Solution 1:

$$D(25) = 3500, \qquad D(26) = 3496 \qquad \text{Hence}$$

$$\frac{\Delta D}{\Delta P} = \frac{3496 - 3500}{26 - 25}$$

$$\frac{\Delta D}{\Delta P} = \frac{-4}{1} \qquad\qquad \left(\frac{-4 \text{ thousand}}{\$1} \right)$$

Solution 2:

$$D(x) = 1000 + 200x - 4x^2$$

$$D(x + h) = 1000 + 200(x + h) - 4(x + h)^2$$

$$= 1000 + 200x + 200h - 4x^2 - 8hx - 4h^2$$

$$\text{Then } \Delta D = D(x + h) - D(x)$$

$$= (1000 + 200x + 200h - 4x^2 - 8hx - 4h^2)$$
$$- (1000 + 200x - 4x^2)$$

$$= 200h - 8hx - 4h^2$$

$$\text{and } \frac{\Delta D}{\Delta P} = \frac{200h - 8hx - 4h^2}{h}$$

$$= 200 - 8x - 4h$$

Note: Replace x with 25 and h with 1 (since $25 + 1 = 26$) and obtain the same results as in Solution 1.

SUMMARY

• The graph of a quadratic function f given by

$$f(x) = ax^2 + bx + c, \qquad a \neq 0$$

is a parabola. Completing the square converts a quadratic function in standard form to graphing form:

$$f(x) = a(x - h)^2 + k, \qquad a \neq 0$$

3-2 EXERCISES

Graph of the following quadratic functions. Label the coordinates of the vertex.

1. $f(x) = x^2$
2. $f(x) = -x^2$
3. $f(x) = x^2 - 3$
4. $f(x) = x^2 + 5$
5. $f(x) = x^2 + 2x$
6. $f(x) = x^2 - 4x$
7. $f(x) = 1 - x^2$
8. $f(x) = 4 + x^2$
9. $f(x) = (x + 1)^2 - 2$
10. $f(x) = (x - 2)^2 + 1$
11. $f(x) = (x - 4)^2 + 3$
12. $f(x) = (x + 3)^2 - 4$
13. $f(x) = -2(x + 1)^2 + 3$
14. $f(x) = -3(x - 2)^2 + 1$
15. $f(x) = x^2 + 4x - 1$
16. $f(x) = x^2 - 6x + 1$
17. $f(x) = 2x^2 - 12x + 4$
18. $f(x) = 3x^2 + 6x - 9$
19. $f(x) = x^2 + x + 5$
20. $f(x) = x^2 - 3x - 5$

Recall that the difference quotient measures the average rate of change through two given points. Suppose that the given points are $(x, f(x))$ and $(x + h, f(x + h))$, then the difference quotient becomes $[f(x + h) - f(x)]/h$. For each of the following functions f, determine (a) $f(x + h)$, (b) $f(x + h) - f(x)$, and (c) $[f(x + h) - f(x)]/h$.

21. $f(x) = x^2$
22. $f(x) = -x^2$
23. $f(x) = x^2 - 3$
24. $f(x) = x^2 + 5$
25. $f(x) = x^2 + 2x$
26. $f(x) = x^2 - 4x$
27. $f(x) = 1 - x^2$
28. $f(x) = 4 + x^2$
29. $f(x) = (x + 1)^2 - 2$
30. $f(x) = (x - 2)^2 + 1$

The graph for each of the following functions f is a parabola. (Compare to Problems 21 through 30.) For each function, algebraically simplify $[f(x + h) - f(x)]/h$, then substitute $h = 0$. Set the resulting expression equal to 0 and solve this equation. Compare the solution to the first coordinate of the vertex of f.

31. $f(x) = x^2$
32. $f(x) = -x^2$
33. $f(x) = x^2 - 3$
34. $f(x) = x^2 + 5$
35. $f(x) = x^2 + 2x$
36. $f(x) = x^2 - 4x$
37. $f(x) = 1 - x^2$
38. $f(x) = 4 + x^2$
39. $f(x) = (x + 1)^2 - 2$
40. $f(x) = (x - 2)^2 + 1$

3-2 PROBLEM SET

1. The height h of a projectile t seconds after it is fired at an initial velocity of 160 ft/sec is given by $h = -16t^2 + 160t$. Determine the maximum height of the projectile.

2. The height h of a projectile t seconds after it is fired at an initial velocity of 256 ft/sec is given by $h = -16t^2 + 256t$. Determine the maximum height of the projectile.

3. In statistics, the size n of a random sample to estimate the true proportion of a population (with 95 percent confidence) is given by $n = p(1 - p) (1.96/E)^2$ (E is constant). Show that the maximum value for the expression $p(1 - p)$ is $\frac{1}{4}$ and occurs when $p = \frac{1}{2}$.

4. Show that the rectangle with perimeter P enclosing a maximum area is a square.

5. The height h that water rises in a curved tube placed in a river is proportional to the square of the velocity of the water (Torricelli's Law). If water traveling at 2 meters/second causes water to rise in the tube 0.08 meters, how high will the water rise if the velocity of the water is 1 meter/sec? (Hint: write a formula relating the water height h and the water velocity v.)

6. For a fixed note, the tension of a guitar string is proportional to the square of the length of the string. If a string 0.4 meters long and at a tension of 4 kg produces a note of C, what will the tension have to be for a similar string of 0.8 meters to produce the same note?

7. The water depth d in a shallow channel is proportional to the square of the velocity of wave propagation v in the channel. If the velocity is 1 meter/sec when the channel is 0.1 meter deep, determine the depth of the channel when the velocity measures 2 meters/sec.

8. The propeller thrust P (in kg) of a mtorboat is proportional to the square of the number of revolutions per minute (R) of the propeller. If 50 rpm produces 100 kg of thrust, determine the thrust at 75 rpm.

9. In Problem Set 1-5, we suggested that no part of a parabola is "straight." To extend the concept of how a

parabola is curved we use the technique of Problem Set 1-5 to define *convex up*. A (continuous) function f is convex up if for every pair of values a and b in the domain of f,

$$f\left(\frac{a+b}{2}\right) < \frac{f(a)+f(b)}{2}.$$

If

$$f\left(\frac{a+b}{2}\right) > \frac{f(a)+f(b)}{2}$$

the function is convex down. Use this definition to prove that $f(x) = x^2$ is convex up.

10. See Problem 9. Prove that $f(x) = -x^2$ is convex down.

11. See Problem 9. Draw a figure to represent $f(x) = x^2$. Label two points of f, $(a, f(a))$ and $(b, f(b))$. Connect these two points with a line segment and label the midpoint of the line segment. Now interpret the definition of convex up in terms of this figure.

12. See Problem 9. Draw a figure to represent $f(x) = -x^2$. Label two points of f $(a, f(a))$ and $(b, f(b))$. Connect these two points with a line segment and label the midpoint of the line segment. Now interpret the definition of convex down in terms of this figure.

PROBLEMS FOR TECHNOLOGY

Graph each of the following. Use the trace function to locate the vertex. Note whether the graph of the parabola opens up or down.

13. $f(x) = x^2 + 8x - 3$
14. $f(x) = 3x^2 - 5x + 2$
15. $f(x) = -2x^2 + 3x - 4$
16. $f(x) = -5x^2 + 7x - 1$
17. $f(x) = 2(x - 3)^2 - 4$
18. $f(x) = 3(x + 2)^2 + 1$
19. $f(x) = -3(x + 1)^2 + 2$
20. $f(x) = -2(x + 5)^2 - 3$

21. See Problems 13–20. Describe a relation between the direction the parabola opens and the coefficient of x^2.

22. See Problems 17–20. Discuss how the vertex of a parabola can be predicted from a rule of pairing of the form $f(x) = a(x - h)^2 + k$.

EXPLORATION

The following problems anticipate topics in the next section. Graph each of the following. Use the trace function to approximate the x-intercepts.

23. $f(x) = x^2 - 5$
24. $f(x) = x^2 - 7$
25. $f(x) = x^2 + 3$
26. $f(x) = x^2 + 2$
27. $f(x) = 3x^2 - 2$
28. $f(x) = 2x^2 - 3$
29. $f(x) = x^2 - x - 5$
30. $f(x) = x^2 + 3x - 7$
31. $f(x) = x^2 - 7x$
32. $f(x) = 3x^2$

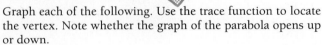

---

3   ROOTS AND APPROXIMATIONS

Round numbers are always false.
—Samuel Johnson

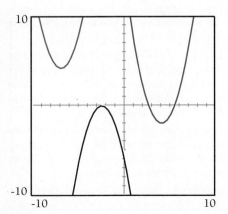

Three possibilities exist for the x-intercepts of a quadratic function. The parabola may intersect the x-axis two, one or no times. See Figure 3-16.

If the coefficients and x-intercepts of $f(x) = ax^2 + bx + c$ are rational numbers, we have reasonable hopes of determining the intercepts by factoring. The process is a result of the **zero factor theorem**.

Theorem 3-3 Zero Factor Theorem

> If r and s are real numbers such that $rs = 0$, then at least one of r or s is 0.

FIGURE 3-16 x-intercepts of three parabolas

Proof: Suppose that $rs = 0$ and that at least one of r and s is not 0. If $r \neq 0$, then $1/r$ is a real number. We have

$$\frac{1}{r}\, rs = \frac{1}{r} 0$$

$$s = 0$$

A similar argument would demand that $r = 0$ when $s \neq 0$. Hence at least one of r or s is 0.

EXAMPLE 3-5

Determining X-intercepts of a Quadratic Function.   Determine the x-intercepts for the following quadratic functions.

Illustration 1:

$$f(x) = x^2 - 6x + 9$$

Solution: See Figure 3-17. Set $f(x) = 0$

$$x^2 - 6x + 9 = 0$$
$$(x - 3)(x - 3) = 0 \qquad\qquad \text{(factor)}$$
$$x - 3 = 0 \quad \text{or} \quad x - 3 = 0 \qquad \text{(zero factor theorem)}$$
$$x = 3 \quad \text{or} \quad x = 3$$

Illustration 2:

$$g(x) = 3x^2 - 2x - 5$$

Solution: See Figure 3-18.

$$3x^2 - 2x - 5 = 0$$
$$(3x - 5)(x + 1) = 0$$
$$3x - 5 = 0 \quad \text{or} \quad x + 1 = 0$$
$$x = \tfrac{5}{3} \quad \text{or} \quad x = \text{-}1$$

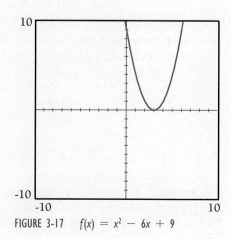

FIGURE 3-17 $f(x) = x^2 - 6x + 9$

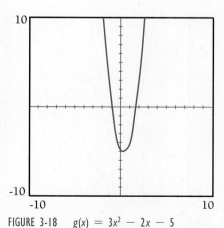

FIGURE 3-18 $g(x) = 3x^2 - 2x - 5$

Illustration 3:

$$h(x) = \tfrac{3}{2}x^2 - 24.$$

Solution: See Figure 3-19.

$$\tfrac{3}{2}(x^2 - 16) = 0 \qquad \text{(Difference of two squares)}$$

$$\tfrac{3}{2}(x - 4)(x + 4) = 0$$

$$x - 4 = 0 \quad \text{or} \quad x + 4 = 0 \qquad (\tfrac{3}{2} \neq 0)$$

$$x = 4 \quad \text{or} \quad x = \text{-}4$$

Illustration 4:

$$j(x) = (x + 1)^2 - 17$$

Solution: See Figure 3-20.

$$(x + 1)^2 - 17 = 0 \qquad \text{(Difference of two squares)}$$

$$[(x + 1) - \sqrt{17}][(x + 1) + \sqrt{17}] = 0$$

$$(x + 1) - \sqrt{17} = 0 \quad \text{or} \quad (x + 1) + \sqrt{17} = 0$$

$$x = \text{-}1 + \sqrt{17} \quad \text{or} \quad x = \text{-}1 - \sqrt{17} \qquad \text{(Irrational intercepts)}$$

What is a square root of 17? No rational number squared yields 17: $\sqrt{17}$ is irrational. However, many rational numbers are "close" to $\sqrt{17}$. Because rational numbers are more familiar than irrational numbers, we may choose one of the "close" rational numbers to stand in for $\sqrt{17}$. The process of selecting a "close" rational number is **approximation**.

Section 3.7 examines the question of precision: "how close is close enough?" For now, consider some approximation methods without regard to precision or accuracy.

One method of approximation is to summon the square root of 17 from our calculator and obtain the approximation 4.123105626. Although this method is fast and clean, it provides no insight into the nature of $\sqrt{17}$ or irrational numbers in general. Instead, we should wonder how the calculator approximated $\sqrt{17}$.

Substitution reveals that $x = \sqrt{17}$ is the solution of $x^2 - 17 = 0$. Rather than proceed in the usual manner, consider the following steps:

$$x^2 - 16 - 1 = 0$$

$$x^2 - 16 = 1 \qquad \text{(difference of 2 squares)}$$

$$(x + 4)(x - 4) = 1 \qquad \text{(because } x \neq \text{-}4, \text{ then)}$$

$$x - 4 = \frac{1}{x + 4}$$

$$x = 4 + \frac{1}{X + 4} \qquad \text{[Equation 3.1]}$$

This formula defines x in terms of itself: x also appears on the right side of the equation. *Recursion* is the name applied to representations of variables that reference themselves. We say that x is given recursively.

How will this be helpful? Notice the capitalization of X on the righthand

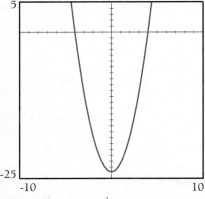

FIGURE 3-19 $h(x) = \left(\tfrac{3}{2}\right)x^2 - 24$

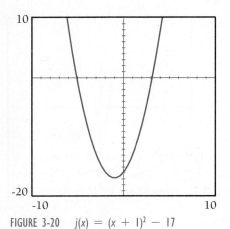

FIGURE 3-20 $j(x) = (x + 1)^2 - 17$

side of the equation. We capitalize here to distinguish this x from the lower case x on the left. Suppose that we substitute what the lower case x equals in place of X:

$$x = 4 + \cfrac{1}{4 + \cfrac{1}{X+4} + 4}$$

$$x = 4 + \cfrac{1}{8 + \cfrac{1}{X+4}}$$

[Equation 3.2]

Again, recursion. Now substitute [Equation 3.2] for X in [Equation 3.2] and simplify:

$$x = 4 + \cfrac{1}{8 + \cfrac{1}{4 + \cfrac{1}{8 + \cfrac{1}{X+4}} + 4}}$$

$$x = 4 + \cfrac{1}{8 + \cfrac{1}{8 + \cfrac{1}{8 + \cfrac{1}{X+4}}}}$$

The process can continue without resolution through infinitely many substitutions. The result is a continued fraction.

We begin to gain insight into the nature of irrational numbers. Whether represented by decimal or by fraction, some sort of infinite process stands between us and writing the irrational number. No wonder we settle for the symbol $\sqrt{17}$ rather than a decimal representation of the value of $\sqrt{17}$.

The continued fraction representation provides a method of approximating $\sqrt{17}$ with a rational number. The further "out" is the continued fraction, the less value is added by the successive denominators. Consider chopping off the infinite tail of the fraction at the position marked in the continued fraction.

$$x \cong 4 + \cfrac{1}{8 + \cfrac{1}{8 + \cfrac{1}{8 + 0}}}$$

$$x \cong 4 + \frac{65}{528}$$

$$x \cong 4.12310606$$

However, a quicker way is to use the $\boxed{\text{ANS}}$ on your graphing calculator. Be sure you are in calculation mode. Press 1 $\boxed{\text{ENTER}}$ to initialize ANS. Then enter the original recursion formula [Equation 3.1], replacing X with ANS:

$$4 + 1/(4 + \boxed{\text{ANS}})$$

Press ENTER several times. On each press, the new ANS replaces the original ANS in the formula. The value for ANS rapidly approaches 4.1231056.

Consider one last approximation method using the trace feature available with graphing technology. See Figure 3-21.

Although it is true that we could approximate the x-intercept here using the square root key, we might not always have such a simple key stroke to obtain a value. Using the trace feature, move the cursor along the graph of f until it is very close to the x-axis. Observe the y-coordinate as you adjust the cursor close to the x-axis. Your goal is to get y as close to 0 as you can. If y changes sign, you have crossed the axis. When y is as close to 0 as you can get, the x value is your intercept. For improved accuracy, zoom in on the intercept. Now consider more physical applications.

EXAMPLE 3-6

Illustration 1:

Applications of Quadratic Functions. Suppose that a projectile is fired upward at 320 ft/sec. From physics, it is known that gravitation simultaneously causes the projectile to fall toward the earth at a rate proportional to the square of the time it falls. These statements describe the height, h in feet, of the projectile as a function of the time, t in seconds, of flight (Figure 3-22):

$$h(t) = 320t - 16t^2$$

How long is the projectile in the air?

Solution: The projectile ceases to be airborne the instant it strikes the ground. We seek the time t when the height is $h = 0$.

$$0 = 320t - 16t^2$$
$$0 = 16t(20 - t)$$
$$16t = 0 \quad \text{or} \quad 20 - t = 0$$
$$t = 0 \quad \text{or} \quad t = 20 \qquad \text{(seconds)}$$

Interpret these values as the times at which the projectile is at ground level ($h = 0$): $t = 0$ is the time projectile is launched. And $t = 20$ is the instant it hits ground level. The time of the flight is 20 seconds.

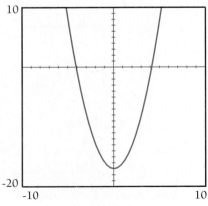

FIGURE 3-21 $f(x) = x^2 - 17$

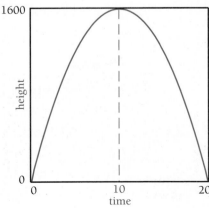

FIGURE 3-22 $h(t) = 320t - 16t^2$

Note: The function $h(t)$ is nonsensical for a value of $t = 30$ or $t = -5$ even though evaluation produces a value for h. Why? Because h is a model for a physical situation; it is not the situation itself. When the projectile strikes the ground, it is no longer a free falling object; therefore the formula no longer applies.

Illustration 2:   Consider a fence to enclose a rectangular area of 525 square meters (0.0525 hectares) in such a manner that the length of the rectangle is exactly 20 meters longer than the width. What should be the dimensions of the rectangle?

Solution: Let w be the width of the rectangle, then $w + 20$ is the length. Since

$$\text{area} = (\text{length})(\text{width})$$
$$525 = (w)(w + 20) \qquad \text{(Figure 3-23)}$$
$$0 = w^2 + 20w - 525$$
$$0 = (w - 15)(w + 35)$$
$$0 = w - 15 \quad \text{or} \quad 0 = w + 35$$
$$w = 15 \quad \text{or} \quad w = -35$$

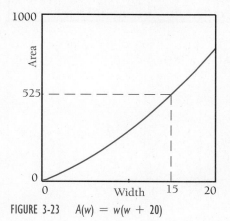

FIGURE 3-23 $A(w) = w(w + 20)$

Comment: Once more the equation is a model for reality. The equation yields -35 meters as a solution. A field may not have a width of -35 meters. Physical constraints force us to reject this answer. The only feasible solution is 15 meters for the width. Note that if 15 meters is the width, then the length is $w + 20$ or 35 meters.   ▬

The preceding example illustrates quadratic models. The second illustration is not as mundane as it seems. Areas of rectangles may represent more than geometrical figures. In Figure 3-24, distance = (rate)(time), so areas of rectangles can model distance. In Figure 3-25, work = (average force)(distance), so areas of rectangles can model work. In Figure 3-26, total cost = (cost / item)(# items), so areas of rectangles can model economics. Moreover, quadratic functions *model* the area of rectangles.

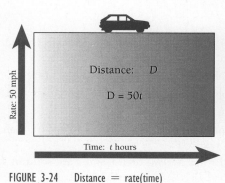

FIGURE 3-24   Distance = rate(time)

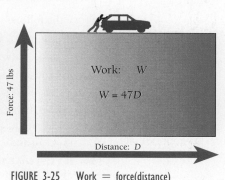

FIGURE 3-25   Work = force(distance)

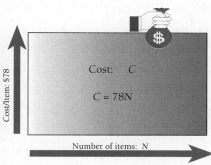

FIGURE 3-26   Cost = cost per item (no. of items)

S U M M A R Y

- To determine the x-intercepts of a quadratic function f, solve the equation

$$f(x) = 0$$

- If factoring cannot solve a quadratic equation, various methods exist for approximating the irrational roots. Among these approximation methods are fixed point iteration and continued fractions. Another method is to use the trace feature of a graphing calculator.
- Quadratic functions serve as models for many physical situations.

3-3   E X E R C I S E S

Solve the following.

1. $x^2 - 3x + 2 = 0$

2. $x^2 + 3x + 2 = 0$

3. $x^2 + x - 2 = 0$

4. $x^2 - x - 2 = 0$

5. $x^2 - 5x - 6 = 0$

6. $x^2 + 5x - 6 = 0$

7. $x^2 - x - 6 = 0$

8. $x^2 + x - 6 = 0$

9. $x^2 + 7x + 6 = 0$

10. $x^2 - 7x + 6 = 0$

Determine the x-intercepts for the following functions.

11. $f(x) = x^2 + 6x + 8$

12. $f(x) = x^2 - 6x + 8$

13. $f(x) = x^2 - 9x + 8$

14. $f(x) = x^2 + 9x + 8$

15. $f(x) = x^2 - 7x - 8$

16. $f(x) = x^2 + 7x - 8$

17. $f(x) = x^2 + 2x - 8$

18. $f(x) = x^2 - 2x - 8$

19. $f(x) = 2x^2 + 11x + 5$

20. $f(x) = 2x^2 - 11x + 5$

Graph the following. Label the x-intercepts.

21. $f(x) = 2x^2 + 7x + 5$

22. $f(x) = 2x^2 - 7x + 5$

23. $f(x) = 2x^2 - 9x - 5$

24. $f(x) = 2x^2 + 9x - 5$

25. $f(x) = 3x^2 + 19x + 6$

26. $f(x) = 3x^2 - 19x + 6$

27. $f(x) = 3x^2 - 7x - 6$

28. $f(x) = 3x^2 + 7x - 6$

29. $f(x) = 4x^2 + x - 5$

30. $f(x) = 4x^2 - x - 5$

Each of the following functions has an x-intercept on the indicated interval, use continued fractions to approximate the indicated x-intercept.

31. $f(x) = x^2 - 2$, $1 < x < 2$

32. $f(x) = x^2 - 3$, $1 < x < 3$

33. $f(x) = x^2 - 5$, $2 < x < 3$

34. $f(x) = x^2 - 6$, $2 < x < 3$

35. $f(x) = x^2 - 6$, $-3 < x < -2$

36. $f(x) = x^2 - 5$, $-3 < x < -2$

Each of the following functions has an x-intercept on the indicated interval. Use the trace feature of your graphing technology to approximate the indicated x-intercept.

37. $f(x) = x^2 + 4x - 1$, $0 < x < 1$

38. $f(x) = x^2 + 3x - 2$, $0 < x < 1$

39. $f(x) = x^2 + 4x - 1$, $-5 < x < -4$

40. $f(x) = x^2 + 3x - 2$, $-4 < x < -3$

Prove the following.

41. If $a(x - r)(x - s) = 0$, $a \neq 0$, then r and s are solutions of the equation.

42. If r and s are solutions of $x^2 + dx + e = 0$, then $rs = e$.

43. If r and s are solutions of $x^2 + dx + e = 0$, then $r + s = -d$.

44. Prove Theorem 3-3.

45. If r and s are solutions of $ax^2 + bx + c = 0$ ($a \neq 0$), then $rs = c/a$ and $r + s = -b/a$.

46. If $rs = c/a$ and $r + s = -b/a$, then r and s are solutions of $ax^2 + bx + c = 0$.

3-3   PROBLEM SET

1. Suppose that the wind pressure on a flat surface is proportional to the square of the velocity of the wind. If the wind exerts a 50 kg force on the object at 5 km/hr, express the pressure as a function of the wind velocity.

2. Suppose that the power produced by a windmill is proportional to the square of the diameter of the blades (where the wind velocity is constant). If a windmill with a 2 meter diameter blade generates 1000 Watts, express the power as a function of the blade diameter.

3. Light intensity is proportional to the reciprocal of the square of the distance from the light source to the illuminated object. If a light source measures 100 lumens at 5 cm, express the light intensity as a function of the distance to the light source.

4. Torricelli's Law states the height of water forced up into an L-shaped tube placed in a stream of water is proportional to the square of the velocity of the water. If the height of the water is $\frac{1}{64}$ of a foot when the velocity is 1 ft/sec, express the height as a function of the water velocity.

PROBLEMS FOR TECHNOLOGY

Use the trace function of your graphing technology to approximate the x-intercepts of the following functions.

5. $f(x) = x^2 - 17$

6. $f(x) = x^2 - 15$

7. $f(x) = x^2 - 8$

8. $f(x) = x^2 - 6$

9. $f(x) = x^2 - 3x + 1$

10. $f(x) = x^2 + 5x + 1$

11. $f(x) = 2x^2 - 5x - 17$

12. $f(x) = 3x^2 + 2x - 11$

13. Discuss any difficulties you encountered.

14. Compare and contrast the roots found in Problems 5–8 with the roots found in Problems 9–10.

 The continued fraction approximation method implemented recursively on your graphing calculator is known as *fixed-point iteration*. Refer to the continued fraction recursion formula in the example approximation of the positive root of $f(x) = x^2 - 17$. Use fixed point iteration to approximate the positive roots of the following.

15. $f(x) = x^2 - 7$

16. $f(x) = x^2 - 11$

17. $f(x) = x^2 - 33$

18. $f(x) = x^2 - 23$

19. Yet another method for approximating x-intercepts is Newton's method. Suppose $f(x) = ax^2 + bx + c$, $a \neq 0$. Also suppose that x_0 is "close" to an x-intercept of f; x_0 is the seed or starting approximation. Then Newton's method indicates an improved approximation x_1 where

$$x_1 = x_0 - \frac{f(x_0)}{2ax_0 + b}, \quad 2ax_0 + b \neq 0.$$

Consider the example from the text, $f(x) = x^2 - 17$. Use $x_0 = 4$ as a seed. Then,

$$2ax_0 + b = 2(1)4 + 0$$
$$= 8$$
$$f(x_0) = 4^2 - 17 = -1$$

So
$$x_1 = x_0 - \frac{f(x_0)}{2ax_0 + b}$$
$$= 4 - \frac{-1}{8}$$
$$= 4.125$$

4.125 is our first improved approximation of $\sqrt{17}$. Now if we wish to improve this approximation, let x_1 take on the role of x_0. Then our second approximation is

$$x_2 = x_1 - \frac{f(x_1)}{2ax_1 + b}.$$

In general, the nth approximation is obtained from its preceding approximation $(n - 1)$ by the formula
$$x_n = x_{n-1} - \frac{f(x_{n-1})}{2ax_{n-1} + b}, \quad 2ax_{n-1} + b \neq 0.$$ How would you implement this formula with your calculator's ANS key? (a) Apply Newton's method twice (x_2) to obtain an approximation for an x-intercept of $f(x) = x^2 - 5$. (b) Use continued fractions to approximate the same x-intercept. (c) Use fixed point iteration to approximate the same x-intercept. (d) Use a calculator to approximate $\sqrt{5}$. (e) Compare the approximation in parts a, b, c, and d. Which methods seem to come closest to the calculator approximation?

20. See the discussion of Newton's method in Problem 19. (a) Apply Newton's method twice (x_2) to obtain an approximation for an x-intercept of $f(x) = x^2 - 10$. (b) Use continued fractions to approximate the same x-intercept. (c) Use the trace feature to approximate the same x-intercept. (d) Use a calculator to approximate $\sqrt{10}$. (e) Compare the approximation in parts a, b, c, and d. Which methods seem to come closest to the calculator approximation?

EXPLORATION

The following problems anticipate topics in the next section. Graph each of the following sets of functions in the same window. Discuss the relationship among their x-intercepts.

21. $f(x) = x^2 - 4x + 3$, $g(x) = x + 2$,
$h(x) = f \circ g(x)$

22. $f(x) = x^2 + 2x - 8$, $g(x) = x - 1$,
$h(x) = f \circ g(x)$

23. $f(x) = 2x^2 - 8x + 3$, $g(x) = x + 2$,
$h(x) = f \circ g(x)$

24. $f(x) = 3x^2 - 6x + 1$, $g(x) = x + 1$,
$h(x) = f \circ g(x)$

25. $f(x) = ax^2 + bx + c$, $g(x) = x - \dfrac{b}{2a}$,

$h(x) = f \circ g(x)$, where $a = 2, b = 3$ and $c = -5$

26. Same as Problem 25 except that $a = 3, b = -4$ and $c = -2$.

---

4   THE QUADRATIC FORMULA

Each problem I solved became a rule which afterwards served to solve other problems.
—RENE DESCARTES

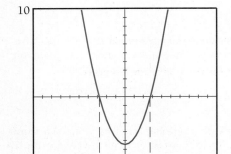

FIGURE 3-27 $f(x) = a(x^2 - d)$

FIGURE 3-28 $f(x) = ax^2 + bx + c$

Factoring is not always a suitable method for solving a quadratic equation. Irrational x-intercepts can frustrate factoring attempts. The exception is the quadratic form known as the difference of two squares (Figure 3-27):

$$f(x) = a(x^2 - d), \qquad a \neq 0, \quad d > 0$$
$$= a(x + \sqrt{d})(x - \sqrt{d})$$

Notice that there is no linear term, bx, in a difference of two squares. If we could eliminate the linear term from

$$f(x) = ax^2 + bx + c, \qquad a \neq 0$$

to form $f(x) = a(x^2 - d)$

then we could factor the quadratic into linear factors. Better yet, if the process for converting to a difference of two squares is mechanical, then it should be possible to obtain an algorithm (process) or formula that will apply to the general quadratic function. Our goal is to eliminate the linear term bx from (Figure 3-28)

$$f(x) = ax^2 + bx + c$$

The most common method to accomplish this is to complete the square. Let $f(x) = 0$ and solve for x by completing the square.

$$ax^2 + bx + c = 0$$

Because $a \neq 0$, we may divide by a:

$$x^2 + \frac{b}{a}x + \frac{c}{a} = 0$$

To complete the square, add and subtract $(b/2a)^2$ on the left side of the equation:

$$x^2 + \frac{b}{a}x + \left(\frac{b}{2a}\right)^2 - \left(\frac{b}{2a}\right)^2 + \frac{c}{a} = 0$$

The first three terms are a perfect square. Factor them:

$$\left(x + \frac{b}{2a}\right)^2 - \left(\frac{b}{2a}\right)^2 + \frac{c}{a} = 0$$

$$\left(x + \frac{b}{2a}\right)^2 - \frac{b^2}{4a^2} + \frac{c}{a} = 0$$

$$\left(x + \frac{b}{2a}\right)^2 - \frac{b^2}{4a^2} + \frac{4ac}{4a^2} = 0$$

$$\left(x + \frac{b}{2a}\right)^2 - \frac{b^2 - 4ac}{4a^2} = 0$$

Factor the resulting difference of squares:

$$\left(x + \frac{b}{2a} + \frac{\sqrt{b^2 - 4ac}}{2a}\right)\left(x + \frac{b}{2a} - \frac{\sqrt{b^2 - 4ac}}{2a}\right) = 0$$

Set each factor equal to 0 and solve for x:

$$x + \frac{b}{2a} + \frac{\sqrt{b^2 - 4ac}}{2a} = 0 \quad \text{or} \quad x + \frac{b}{2a} - \frac{\sqrt{b^2 - 4ac}}{2a} = 0$$

$$x = \frac{-b - \sqrt{b^2 - 4ac}}{2a} \quad \text{or} \quad x = \frac{-b + \sqrt{b^2 - 4ac}}{2a}$$

These results provide the theorem for the **quadratic formula**.

Theorem 3-4
Quadratic Formula

If $ax^2 + bx + c = 0$, $a \neq 0$, then $x = \dfrac{-b \pm \sqrt{b^2 - 4ac}}{2a}$

The transformations of Chapter 2 provide another method for deriving the quadratic formula. That derivation is left as an exercise.

EXAMPLE 3-7

Illustration 1:

Determine the x-intercept of each of the following quadratic functions. Remember that x-intercepts occur when $f(x) = 0$.

$$f(x) = x^2 - 3x + 2$$

Solution:

$a = 1$, $b = -3$, and $c = 2$, thus

$$x = \frac{-(-3) \pm \sqrt{(-3)^2 - 4(1)(2)}}{2(1)}$$

$$x = \frac{3 \pm 1}{2}$$

$$x = 2 \quad \text{or} \quad x = 1$$

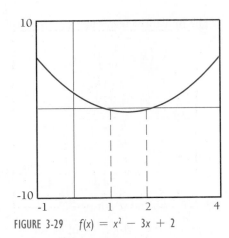

FIGURE 3-29 $f(x) = x^2 - 3x + 2$

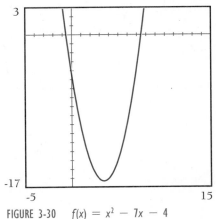

FIGURE 3-30 $f(x) = x^2 - 7x - 4$

$(2, 0)$ and $(1, 0)$ are the x-intercepts. See Figure 3-29.

Illustration 2: $f(x) = x^2 - 7x - 4$

Solution: See Figure 3-30.

$$x = \frac{7 \pm \sqrt{49 - 4(1)(-4)}}{2(1)}$$

$$x = \frac{7 \pm \sqrt{65}}{2}$$

$\left(\dfrac{7 + \sqrt{65}}{2}, 0\right)$ and $\left(\dfrac{7 - \sqrt{65}}{2}, 0\right)$ are the x-intercepts.

Note: Whereas the solution for Illustration 1 can be easily checked by substitution, irrational roots pose more difficult arithmetic. An easier check is as follows:

x_1 and x_2 are solutions of $ax^2 + bx + c = 0$, if $x_1 + x_2 = -b/a$ and $x_1 x_2 = c/a$.

Check:

$$\frac{7 + \sqrt{65}}{2} \times \frac{7 - \sqrt{65}}{2} = \frac{49 - 65}{4}$$

$$= -4 \qquad \text{(compare this to } \frac{c}{a}\text{)}$$

$$\frac{7 + \sqrt{65}}{2} + \frac{7 - \sqrt{65}}{2} = \frac{14}{2}$$

$$= 7 \qquad \text{(compare this to } -\frac{b}{a}\text{)}$$

Illustration 3: $f(x) = 5x^2 + x + 1$

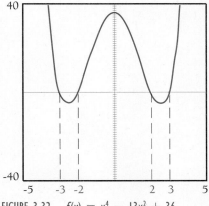

FIGURE 3-31　$f(x) = 5x^2 + x + 1$　　FIGURE 3-32　$f(x) = x^4 - 13x^2 + 36$

$$x = \frac{-1 \pm \sqrt{-19}}{10}$$

$\sqrt{-19}$ is not a real number: there are no x-intercepts. See Figure 3-31.

Illustration 4:

$$f(x) = x^4 - 13x^2 + 36$$

Solution: The function f is *not* a quadratic function, but the resemblance is close enough to inspire a similar attack. An expression is in *quadratic format* if it is in the form $ax^{2n} + bx^n + c$. The quadratic formula applies in solving for x^n. See Figure 3-32.

$$x^4 - 13x^2 + 36 = (x^2)^2 - 13(x^2) + 36$$

$$x^2 = \frac{13 \pm \sqrt{169 - 144}}{2}$$

$$x^2 = 9 \quad \text{or} \quad x^2 = 4$$

Now solve each of these quadratic equations.

$$x = 3 \quad \text{or} \quad x = -3 \quad \text{or} \quad x = -2 \text{ or } x = 2$$

Note: This fourth-degree function has 4 x-intercepts. Use graphing technology to verify the number of intercepts. Also note that f factors as

$$f(x) = (x - 3)(x + 3)(x + 2)(x - 2)$$ ▬

　　Observe from the preceding example the influence of the portion of the quadratic formula in the radicand. If the coefficients a, b, and c are rational numbers, then the radicand influences the nature of the solution. For that reason we call $b^2 - 4ac$ the **discriminant** of a quadratic equation. Sometimes we use D to stand for the discriminant: $D = b^2 - 4ac$. The following table summarizes the relationship of the discriminant to the roots of a quadratic equation.

Discriminant for $ax^2 + bx + c = 0$, where a, b and c are rational numbers, $a \neq 0$

Discriminant $D = b^2 - 4ac$	Nature of D	Nature and number of roots	Real roots	Graph
$D > 0$	Perfect square	Two rational roots	Two	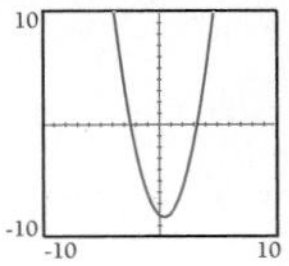
	Not a perfect square	Two irrational roots	Two	
$D = 0$		One rational root	One	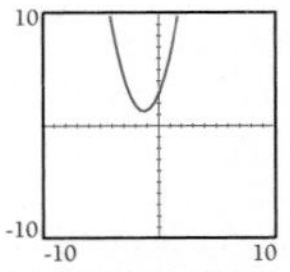
$D < 0$		Two imaginary roots	None	

Indicate the nature and number of roots for the following quadratic functions.

EXAMPLE 3-8

Illustration 1:

$$f(x) = x^2 + 3x + 7$$

Solution:

$$D = 3^2 - 4(1)(7)$$
$$= -19$$

Because D is negative, there are no real roots; the parabola does not intersect the x-axis. See Figure 3-33.

Illustration 2:

$$f(x) = x^2 + 3x - 7$$

Solution:

$$D = 3^2 - 4(1)(-7)$$
$$= 37$$

Because D is positive, there are two x-intercepts; see Figure 3-34. Because 37 is not a perfect square, the intercepts are irrational

$$\left(\frac{-3 \pm \sqrt{37}}{2}, 0\right)$$

Illustration 3:

$$f(x) = x^2 + 3x - 10$$

Solution:

$$D = 3^2 - 4(1)(-10)$$
$$= 49$$

D is a positive perfect square, there are two rational roots (2 and -5). See Figure 3-35. Note that this quadratic could have been factored by careful inspection.

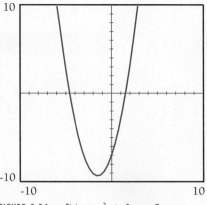

FIGURE 3-33 $f(x) = x^2 + 3x + 7$ FIGURE 3-34 $f(x) = x^2 + 3x - 7$

Illustration 4:

$$f(x) = x^2 - 6x + 9$$

Solution:

$$D = (-6)^2 - 4(1)(9)$$
$$= 0$$

Because D is 0 there is one rational root (3). See Figure 3-36.

Illustration 5:

$$f(x) = 4x^2 - 4\sqrt{2}x + 1$$

Solution: Although $D = 16$ and 16 is a perfect square, the rule does not apply because the coefficient $-4\sqrt{2}$ is not rational. By the quadratic formula the roots are $(\sqrt{2} \pm 1)/2$.

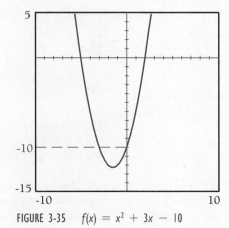

FIGURE 3-35 $f(x) = x^2 + 3x - 10$

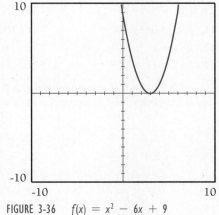

FIGURE 3-36 $f(x) = x^2 - 6x + 9$

SUMMARY

- To solve a quadratic equation $ax^2 + bx + c = 0$, use the quadratic formula

$$x = \frac{-b \pm \sqrt{b^2 - 4ac}}{2a}$$

- For an equation in quadratic form $ax^{2n} + bx^n + c = 0$, the quadratic formula solves for x^n:

$$x^n = \frac{-b \pm \sqrt{b^2 - 4ac}}{2a}$$

Use the two resulting equations to solve for x.

- The discriminant $D = b^2 - 4ac$ of the quadratic formula controls the number and nature of roots of a quadratic equation with real coefficients. If $D > 0$, then the equation has two real roots. If $D = 0$, then the equation has one real root. If $D < 0$, then the equation has no real roots; however, it does have two complex roots.

3-4   EXERCISES

Determine the x-intercepts of the following quadratic functions. Sketch the graph and label the intercepts.

1. $f(x) = x^2 + 4x + 3$
2. $f(x) = x^2 + 3x + 4$
3. $f(x) = x^2 - 3x - 4$
4. $f(x) = x^2 - 4x + 3$
5. $f(x) = x^2 + 9x - 2$
6. $f(x) = x^2 - 7x - 3$
7. $f(x) = 2x^2 - 9x + 4$
8. $f(x) = 2x^2 + 9x + 5$
9. $f(x) = 4x^2 - 20x + 20$
10. $f(x) = 9x^2 + 12x + 4$
11. $f(x) = 3x^2 + 5x + 7$
12. $f(x) = 2x^2 - 3x + 4$
13. $f(x) = 5x^2 + x - 1$
14. $f(x) = 4x^2 - x - 2$
15. $f(x) = 3x^2 + 2x + 3$
16. $f(x) = 2x^2 + 3x + 7$
*17. $f(x) = x^2 - \sqrt{2}x - 3$
*18. $f(x) = x^2 + \sqrt{7}x - 2$
*19. $f(x) = x^2 - \sqrt{5}$
*20. $f(x) = x^2 - \sqrt{7}$

For each of the following quadratic equations, (a) evaluate the discriminant, (b) indicate the nature and number of real roots.

21. $x^2 + 7x + 3 = 0$
22. $x^2 - 5x + 2 = 0$
23. $x^2 - 12x + 36 = 0$
24. $4x^2 - 12x + 9 = 0$
25. $x^2 - 25 = 0$
26. $x^2 - 24 = 0$
27. $x^2 + 3 = 0$
28. $x^2 + 9 = 0$
29. $x^2 - 7x - 18 = 0$
30. $x^2 - 3x - 28 = 0$

Solve for x.

31. $x^4 - 24x^2 - 25 = 0$
32. $x^4 - 14x^2 - 32 = 0$
33. $x^4 - 9x^2 + 20 = 0$
34. $x^4 - 8x^2 + 7 = 0$
35. $x^6 + 7x^3 - 8 = 0$
36. $x^6 - 7x^3 - 16 = 0$
*37. $x^8 - 17x^4 + 16 = 0$
*38. $x^8 - 15x^4 - 8 = 0$
39. $x^6 - 2x^3 - 3 = 0$
40. $x^6 - 3x^3 - 10 = 0$
41. $x - 5x^{1/2} + 6 = 0$
42. $x^{2/3} - x^{1/3} - 6 = 0$

Prove the following.

43. $\dfrac{-b + \sqrt{b^2 - 4ac}}{2a} \times \dfrac{-b - \sqrt{b^2 - 4ac}}{2a} = \dfrac{c}{a}$

44. $\dfrac{-b + \sqrt{b^2 - 4ac}}{2a} + \dfrac{-b + \sqrt{b^2 - 4ac}}{2a} = -\dfrac{b}{a}$

45. The sum of the roots of the quadratic equation $ax^2 + bx + c = 0$ is $-b/a$.

46. The product of the roots of the quadratic equation $ax^2 + bx + c = 0$ is c/a.

Use the results of Problems 43–46 to write quadratic functions with x-intercepts as follows.

47. -3, 5
48. 2, -7

49. $\frac{2}{3}$, 4

50. $-\frac{1}{2}$, -5

51. $3 + \sqrt{5}, 3 - \sqrt{5}$

52. $2 - \sqrt{7}, 2 + \sqrt{7}$

53. Wind pressure on a flat surface is proportional to the square of the velocity of the wind. Suppose that wind at 8 km/hr exerts 64 Newtons. Determine the wind velocity that would exert a force of 1000 Newtons.

54. Suppose the pressure on the blades of a windmill is $P = 100v^2$, where P is in kg/m^2 and v is the wind velocity in km/hr. Determine the wind velocity when the pressure is 2500 kg/m^2.

55. The power in watts produced by a windmill at a fixed wind velocity is given by $P = 25d^2$, where d is the diameter in meters of the windmill blade turning circle. If the power produced by the windmill is 2500 watts, determine the blade diameter.

56. The power produced by a windmill is proportional to the square of the diameter of the blades (in a wind of constant velocity). If a 2 meter blade generates 4000 watts, what size blades will produce 8000 watts?

57. Torricelli's Law for determining the height h in feet that water will rise in an L-shaped tube when placed in a stream of water moving at velocity V (ft/sec) is $V^2 = 64 h$. What velocity water forces the water $\frac{1}{4}$ ft up the tube?

58. See Problem 57. If water velocity forces water $\frac{1}{9}$ ft up the tube, determine the water's velocity.

59. The power P (in watts) in an electrical circuit is proportional to the square of the voltage E in the circuit. Determine the increase in power (and resulting increase in charges) if the power company increases the line voltage from 110 volts to 120 volts.

60. The power P (in watts) in an electrical circuit is proportional to the square of the amperage A carried in the circuit. Determine the decrease in power when the current drops from 10 amperes to 5 amperes.

* 61. Light intensity is proportional to the inverse of the square of the distance to the light source. If a light source produces 100 lumens at 5 cm, at what distance will the intensity be 5 lumens?

* 62. Suppose that for a given light source light intensity = 1000(lumens). At what distance (meters) will the intensity be 100 lumens?

3-4 PROBLEM SET

1. Solve $ax^2 + bx + c = 0$, $a \neq 0$, by transformation.

(a) Substitute $t - b/(2a)$ for x in the equation.

(b) Simplify the resulting equation (the t terms should add to 0).

(c) Combine the constant terms into a single term. Hint: get a common denominator and factor out -1.

(d) Divide through by a.

(e) Factor the resulting difference of two squares.

(f) Set each factor equal to 0 and solve.

(g) Since $x = t - b/(2a)$, then $t = x + b/(2a)$. Substitute $x + b/(2a)$ for t and then solve for x.

(h) Combine the two solutions for x using a common denominator and the $\pm$ notation.

2. Solve $x^2 - 3x - 10 = 0$ and $-10x^2 - 3x + 1 = 0$. Compare the roots.

3. Solve $6x^2 - 5x + 1 = 0$ and $x^2 - 5x + 6 = 0$. Compare the roots.

4. Solve $3x^2 - 2x - 5 = 0$ and $-5x^2 - 2x + 3 = 0$. Compare the roots.

5. Prove that if r is a solution of $cx^2 + bx + a = 0$, $a \neq 0$, $c \neq 0$, then $(c/a)r$ is a solution of $ax^2 + bx + c = 0$.

PROBLEMS FOR GRAPHING TECHNOLOGY

Use graphing technology to graph each of the following collection of functions on the same coordinate system.

6. $f(x) = x^2$, $g(x) = 2f(x) - 3$, $h(x) = f(2x - 3)$

7. $f(x) = x^2$, $g(x) = 3f(x) - 2$, $h(x) = f(3x - 2)$

8. $f(x) = x^2$, $g(x) = 2 - f(x)$, $h(x) = f(2 - x)$

9. $f(x) = x^2$, $g(x) = 3 - f(x)$, $h(x) = f(3 - x)$

10. $f(x) = x^2$, $g(x) = 1 - 3f(x)$, $h(x) = f(1 - 3x)$

11. $f(x) = x^2$, $g(x) = 1 - 2f(x)$, $h(x) = f(1 - 2x)$

12. $f(x) = x^2$, $g(x) = \sqrt{f(x)}$, $h(x) = f(\sqrt{x})$

13. $f(x) = x^2$, $g(x) = -\sqrt{f(x)}$, $h(x) = f(-\sqrt{x})$

14. $f(x) = x^2 - 10x + 3$, $g(x) = f(x + 5)$

15. $f(x) = x^2 + 8x - 1$, $g(x) = f(x - 4)$

16. Refer to Problems 6, 8, 10, 12 and 14. Discuss any patterns you detect.

17. Refer to Problems 7, 9, 11, 13 and 15. Discuss any patterns.

EXPLORATION

Each of the following problems anticipates concepts introduced in the next section. Graph each of the following.

18. (a) $y = 3x - 6$

   (b) Identify the x-intercept.

   (c) Identify the x-interval for which the graph is *above* the x-axis.

   (d) Identify the x-interval for which the graph is *below* the x-axis.

   (e) Is the function increasing or decreasing?

19. (a) $y = -5x + 10$

   (b) Identify the x-intercept.

   (c) Identify the x-interval for which the graph is *above* the x-axis.

   (d) Identify the x-interval for which the graph is *below* the x-axis.

   (e) Is the function increasing or decreasing?

20. Consider the inequality $3x - 6 > 0$. Refer to Problem 18. Because $y = 3x - 6$, the solution of $3x - 6 > 0$ consists of those x values for which $y > 0$. What interval represents the solution of $3x - 6 > 0$?

21. See Problems 19 and 20. What is the solution of $-5x + 10 < 0$?

22. Graph $y = 3x - 5$ and $y = 5 - 2x$ in the same window. Use the trace option to locate the intersection of the two lines. For what interval of x is the graph of $y = 5 - 2x$ *above* the graph of $y = 3x - 5$?

23. See Problem 22. For what interval of x is the graph of $y = 7 - 4x$ *below* the graph of $y = x - 8$?

Use graphing technology to graph each of the following quadratic functions. In each case, identify the x-intercepts. Also determine the interval(s) for which the graph is above the x-axis and the intervals for which the graph is below the x-axis.

24. $f(x) = x^2 - 3x - 10$ **25.** $f(x) = x^2 + 2x - 15$

26. $f(x) = 3x^2 + x - 5$ **27.** $f(x) = 2x^2 - 3x - 1$

28. $f(x) = x^2 + x + 3$ **29.** $f(x) = -x^2 + 3x - 7$

30. $f(x) = -2x^2 + x + 3$ **31.** $f(x) = -3x^2 - x + 5$

32. $f(x) = -2x^2 - x - 9$ **33.** $f(x) = -2x^2 + 2x - 7$

34. Discuss any patterns you observe in Problems 24–28.

35. Discuss any patterns you observe in Problems 29–33.

36. Use the results of Problem 26 to solve $3x^2 + x - 5 < 0$.

37. Use the results of Problem 27 to solve $2x^2 - 3x - 1 > 0$.

---

5   LINEAR AND QUADRATIC INEQUALITIES

The human mind has never invented a labor-saving machine equal to algebra.
—THE NATION, VOL. 33, P. 237

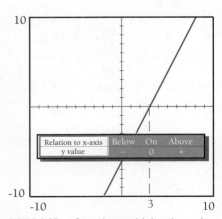

FIGURE 3-37 f on, above and below the x-axis

Suppose we do not need the detail provided by a complete graph of a linear function. We want to "abbreviate" the graphical information. Let us collapse some graphical information by indicating where the graph intersects the x-axis, those domain intervals for which the graph is above the x-axis and those intervals where the graph is below the x-axis.

The x-intercept of a function f is a domain value r for which $f(r) = 0$. For every other real number x, $f(x)$ is either positive or negative. In fact, $f(x) > 0$ for those x values where the graph of f is above the x-axis. Similarly, $f(x) < 0$ for those x values where the graph of f is below the x-axis. For example, every first-degree function has exactly one x-intercept. See Figure 3-37.

A **chart** indicates where a graph intersects the x-axis, where it is below the x-axis and where it is above the x-axis. Let 0 represent *on the x-axis*, a plus represent *above* the x-axis, and a minus represent *below* the x-axis; then the graph of Figure 3-37 has the following chart:

x-axis	- ∞	3	∞	Notes
$f(x) = 2x - 6$	−	0	+	

The first row of the chart indicates important values on the *x*-axis. The **infinity symbols** -∞ and ∞ indicate respectively there is no left or right domain boundary. The domain is all real numbers. Because $2x - 6 = 0$ implies $x = 3$ is the *x*-intercept, 3 is the only point labeled on the *x*-axis. The abbreviated *y* values appear of the second row. The 0 beneath the 3 reminds us that $f(3) = 0$ so that $(3, 0)$ is the *x*-intercept. The *left* negative sign indicates that the graph of *f* is below the *x*-axis for all values of *x* less than 3. Similarly, choose any $x > 3$ and *f* is above the *x*-axis as the + sign confirms.

To convince yourself that these choices are algebraically sound use selected test values. Select any number less than 3, say, 1, and evaluate $f(1)$. Because $f(1) = -4$, the choice of a negative sign to represent both below the *x*-axis is appropriate. Test values greater than 3 produce positive range components.

We assume that every elementary function is continuous at each point of its domain. We expect that a continuous graph cannot pass from below the *x*-axis to above the *x*-axis without crossing the *x*-axis. In numerical terms, a change in sign for *y* values from negative to positive occurs where *y* is 0. Hence, the *x*-intercept, where the functional value is 0, is the one point in the domain of a continuous graph where the functional values may change from positive to negative.

Refer to Theorem 3-1. Suppose that $f(x) = mx + b$ is a linear function where $m > 0$. Clearly, *f* is an increasing line. Visualizing the line as increasing from negative values to positive values allows us to construct a chart effortlessly:

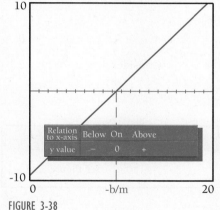

FIGURE 3-38

x-axis	- ∞	$\dfrac{-b}{m}$	∞	Notes
$f(x) = mx + b$	−	0	+	See Figure 3-38

Now suppose that $m < 0$. The graph of *f* is a decreasing line with a chart as follows:

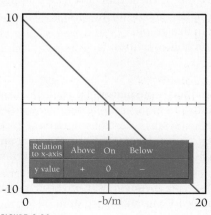

FIGURE 3-39

x-axis	- ∞	$\dfrac{-b}{m}$	∞	Notes
$f(x) = mx + b$	+	0	−	See Figure 3-39

EXAMPLE 3-9

Illustration 1:

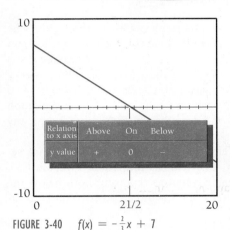

FIGURE 3-40 $f(x) = -\frac{2}{3}x + 7$

Illustration 2:

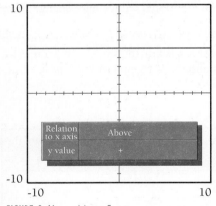

FIGURE 3-41 $g(x) = 5$

Forming Charts for Linear Functions. Construct a chart for each of the following linear functions.

$$f(x) = -\frac{2}{3}x + 7$$

Determine the x-intercept: $0 = -\frac{2}{3}x + 7$

$$\frac{2}{3}x = 7$$

$$x = \frac{21}{2}$$

Because the slope of the line is negative, the line is decreasing. Use graphing technology to verify the graph in Figure 3-40.

x-axis	$-\infty$	$\frac{21}{2}$	∞	Notes
$f(x) = -\frac{2}{3}x + 7$	+	0	−	See Figure 3-40

$$g(x) = 5$$

This constant function does not cross the x-axis. Consider its graph (Figure 3-41) to form its chart.

x-axis	$-\infty$	∞	Notes
$g(x) = 5$		+	See Figure 3-41

Note: A horizontal line is constantly above the x-axis: 5 does not change its sign.

Charts provide useful tabular information about functions. Information is sometimes more accessible in a table than in a graph.

Another use for charts is the solution of inequalities.

$$\text{Solve } -\frac{2}{3}x + 7 \geq 0$$

Although a method for solving a **linear inequality** is available from Chapter 0, that method does not generalize to more complicated inequalities. Indeed, the reason we repeat linear inequalities is to establish graphing technology as a powerful tool in the solution of general inequalities. Forming and interpreting charts to solve inequalities is a prelude to reading solutions directly from graphs. Even without graphing technology, charts formed with test values provide a useful method for solving inequalities. Conversely, without technology, the chart becomes valuable for visualizing a graph.

To solve $-\frac{2}{3}x + 7 \geq 0$, refer to the chart for $f(x) = -\frac{2}{3}x + 7$ as given in Example 3-9, Illustration 1. Interpret $-\frac{2}{3}x + 7 \geq 0$ in one of two ways: where (for what values of x) is $-\frac{2}{3}x + 7$ positive or 0; or where is $f(x) = -\frac{2}{3}x + 7$ above or on the x-axis.

From the chart or Figure 3-42, read the solution: $x \leq \frac{21}{2}$. The infinity symbol in a chart encourages the use of interval notation: $(-\infty, \frac{21}{2}]$.

EXAMPLE 3-10

Solving Linear Inequalities with Charts. Use charts to solve the following linear inequalities.

Illustration 1:

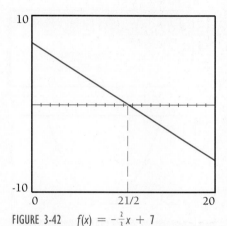

$$5x - 3 \geq 2x + 15$$

The chart method applies only to an inequality in which one side of the inequality is 0. First we rewrite $5x - 3 \geq 2x + 15$:

$$3x - 18 \geq 0$$

Form a chart for $3x - 18$. Use graphing technology for $y = 3x - 18$ or observe that the slope is positive.

x-axis	$-\infty$	6	∞	Notes
$3x - 18$	$-$	0	$+$	Positive Slope

We seek the x-interval for which $3x - 18$ is positive or 0. The solution appears on the chart as $[6, \infty)$.

FIGURE 3-42 $f(x) = -\frac{2}{3}x + 7$

Illustration 2:

Determine the domain of $g(x) = 1/\sqrt{5 - 7x}$.

Because the radicand must be nonnegative to produce real range values, then $5 - 7x \geq 0$. But a denominator may not be 0, therefore $5 - 7x > 0$ provides the domain.

x-axis	$-\infty$	$\frac{5}{7}$	∞	Notes
$5 - 7x$	$+$	0	$-$	Negative slope

From the chart, $Dg = (-\infty, \frac{5}{7})$.

Quadratic inequalities exhibit the same relation to quadratic equations that linear inequalities have with linear equations. For that reason, a chart allows the solution of quadratic inequalities by abbreviating graphical information about quadratic functions. The possibilities for entries in quadratic charts are more numerous than for linear charts.

Two intercepts x:	$-\infty$		x_1		x_2		∞	Note: $b^2 - 4ac > 0$
$ax^2 + bx + c$	+		0	−	0		+	$a > 0$. See Figure 3-43
$ax^2 + bx + c$	−		0	+	0		−	$a < 0$. See Figure 3-44

One intercept x:	$-\infty$		x_0		∞	Note: $b^2 - 4ac = 0$
$ax^2 + bx + c$	+		0	+		$a > 0$. See Figure 3-45
$ax^2 + bx + c$	−		0	−		$a < 0$. See Figure 3-46

No intercepts x:	$-\infty$		∞	Note: $D < 0$
$ax^2 + bx + c$		+		$a > 0$. See Figure 3-47
$ax^2 + bx + c$		−		$a < 0$. See Figure 3-48

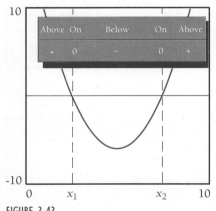

FIGURE 3-43

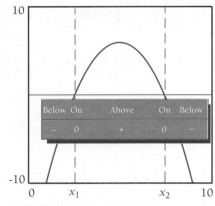

FIGURE 3-44

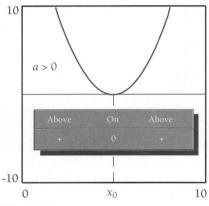

FIGURE 3-45

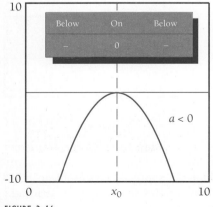

FIGURE 3-46

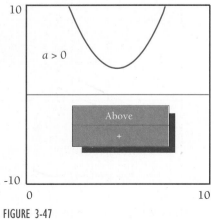

FIGURE 3-47

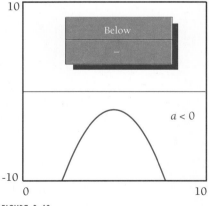

FIGURE 3-48

EXAMPLE 3-11

Illustration 1:

Solving Quadratic Inequalities. Solve the following inequalities.

$$x^2 - 5x + 6 \leq 0$$

Solution 1: Factor: $x^2 - 5x + 6 = (x - 3)(x - 2)$. Combine the charts of the two linear factors.

Two intercepts x:	$-\infty$	2		3	∞	Note:
$x - 3$	$-$		$-$	0	$+$	Increasing
$x - 2$	$-$	0	$+$		$+$	Increasing
$x^2 - 5x + 6$	$+$	0	$-$	0	$+$	$a > 0$. See Figure 3-49

Because the product of an even number of negative factors is positive and the product of an odd number of negative factors is negative, count the number of negative factors in each column to determine the signs in the last row. From the negative signs and zeros on the last row, read the x-axis solution for the inequality: [2, 3].

Solution 2: Because the graph of $f(x) = x^2 - 5x + 6$ is a parabola that opens up, use technology or imagine the chart directly.

Two intercepts x:	$-\infty$	2		3	∞	Note: $b^2 - 4ac > 0$
$x^2 - 5x + 6$	$+$	0	$-$	0	$+$	$a > 0$. See Figure 3-50

Again the solution is [2, 3].

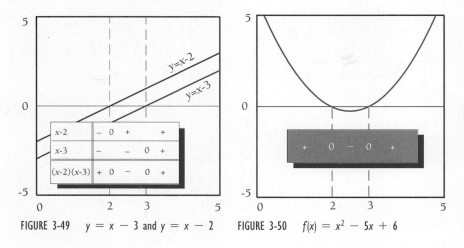

FIGURE 3-49 $y = x - 3$ and $y = x - 2$ FIGURE 3-50 $f(x) = x^2 - 5x + 6$

Illustration 2:

$$x^2 - 7x + 1 > 0$$

Solution: Determine the x-intercepts using the quadratic formula

$$x = \frac{7 \pm \sqrt{49 - 4}}{2} = \frac{7 \pm 3\sqrt{5}}{2}$$

Use graphing technology to form a chart based on the graph.

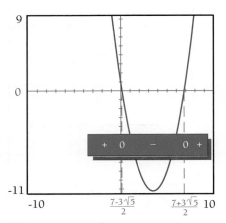

Two intercepts x: $-\infty$		$\dfrac{7 - 3\sqrt{5}}{2}$		$\dfrac{7 + 3\sqrt{5}}{2}$		∞	Note: $b^2 - 4ac > 0$
$x^2 - 7x + 1$	$+$	0	$-$	0	$+$		$a > 0$. See Figure 3-51

$$x \in \left(-\infty, \frac{7 - 3\sqrt{5}}{2}\right) \cup \left(\frac{7 + 3\sqrt{5}}{2}, \infty\right).$$

FIGURE 3-51 $y = x^2 - 7x + 1$

Note: Use graphing technology for the graph of $y = x^2 - 7x + 1$. Use the quadratic formula to determine the exact x-intercepts. Then read the solution for the inequality from the graph. Charts provide a convenient method for listing intercepts and intervals.

Illustration 3:

$$-x^2 + 5x - 2 \geq 0$$

Solution:

$$x = \frac{-5 \pm \sqrt{17}}{-2}$$

$$x = \frac{-5 - \sqrt{17}}{-2} \quad \text{or} \quad x = \frac{-5 + \sqrt{17}}{-2}$$

Charts are not always necessary. Directly from Figure 3-52, read the x-interval where the graph is above or on the x-axis:

$$x \in \left[\frac{-5 + \sqrt{17}}{-2}, \frac{-5 - \sqrt{17}}{-2}\right]$$

FIGURE 3-52 $y = -x^2 + 5x - 2$

Illustration 4:

$$x^2 - 6x + 9 > 0$$

Solution: See Figure 3-53.

$$x = \frac{6 \pm \sqrt{36 - 36}}{2}$$

$$= 3$$

$$x \in (-\infty, 3) \cup (3, \infty)$$

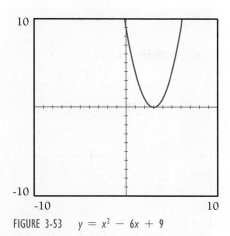

FIGURE 3-53 $y = x^2 - 6x + 9$

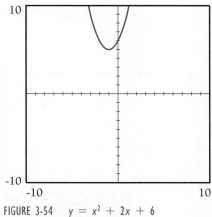

FIGURE 3-54 $y = x^2 + 2x + 6$

Illustration 5:

$$x^2 + 2x + 6 < 0$$

Solution:

$$x = \frac{-2 \pm \sqrt{-20}}{2}$$ (No real roots!)

See Figure 3-54; $x^2 + 2x + 6$ is never negative. No solution.

Illustration 6:

$$x^2 + 2x + 6 > 0$$

See Figure 3-54; $x^2 + 2x + 6$ is always positive:

$$x \in (-\infty, \infty).$$

SUMMARY

- A chart abbreviates graphical information about a function. The chart for a function f shows where f is below, on or above the x-axis. For a linear function of the form $f(x) = mx + b$ the possible charts are as follows.

x-axis	$-\infty$	$\dfrac{-b}{m}$	∞	Notes
$mx + b$	$-$	0	$+$	$m > 0$, See Figure 3-38

x-axis	$-\infty$		$\dfrac{-b}{m}$		∞	Notes
$mx + b$		+	0	–		$m > 0$, See Figure 3-39

x-axis	∞		∞	Note
b		Sign of b		$m = 0$

The following charts summarize quadratic cases:

Two intercepts x:	$-\infty$		x_1		x_2		∞	Note: $b^2 - 4ac > 0$
$ax^2 + bx + c$		+	0	–	0	+		$a > 0$. See Figure 3-43
$ax^2 + bx + c$		–	0	+	0	–		$a < 0$. See Figure 3-44

One intercept x:	$-\infty$		x_0		∞	Note: $b^2 - 4ac = 0$
$ax^2 + bx + c$		+	0	+		$a > 0$. See Figure 3-45
$ax^2 + bx + c$		–	0	–		$a < 0$. See Figure 3-46

No intercepts x:	$-\infty$		∞	Note: $b^2 - 4ac < 0$
$ax^2 + bx + c$		+		$a > 0$. See Figure 3-47
$ax^2 + bx + c$		–		$a < 0$. See Figure 3-48

3-5   EXERCISES

Construct the chart for the following linear functions.

1. $I(x) = x$

2. $j(x) = -x$

3. $k(x) = -2x + 8$

4. $w(x) = 5x + 3$

5. $p(x) = 0.7x - 4$

6. $r(x) = 5 - 3x$

7. $t(x) = 3 - \frac{1}{5}x$

8. $m(x) = 1.2x + 3$

9. $q(x) = \dfrac{x + 5}{2}$

10. $s(x) = \dfrac{3 - x}{4}$

Solve the following linear inequalities.

11. $3x - 5 < 0$

12. $2x + 5 \geqslant 0$

13. $3x - 5 \geqslant 0$

14. $2x + 5 < 0$

15. $5 - \frac{1}{2}x < 0$

16. $5 + \frac{1}{2}x \leqslant 0$

17. $\dfrac{2 + x}{3} > 0$

$\dfrac{-4}{5} > 0$

18. $\dfrac{-3x}{}$

Prove the following.

21. If $m > 0$ and $x > -b/m$, then $mx + b > 0$.

22. If $m < 0$ and $x < -b/m$, then $mx + b > 0$.

23. If $m > 0$ and $x < -b/m$, then $mx + b < 0$.

24. If $|x| - b < 0$, then $x \in (-b, b)$, where $b > 0$.

Solve each of the following quadratic inequalities.

25. $(x - 1)(x + 3) < 0$

26. $(x + 5)(x - 4) > 0$

27. $(3x + 1)(2 - x) > 0$

28. $(3 - x)(5 + 2x) < 0$

29. $x^2 - 7x + 12 \geqslant 0$

30. $x^2 - 7x + 12 \leqslant 0$

31. $x^2 - 9 \leqslant 0$

32. $x^2 - 25 \geqslant 0$

33. $x^2 - 8x + 16 < 0$

34. $x^2 - 10x + 25 > 0$

35. $x^2 + 7x + 6 > 0$

36. $x^2 + 7x + 6 < 0$

37. $x^2 - 4x + 3 \leqslant 0$

38. $x^2 - 4x + 3 \geqslant 0$

39. $2x^2 - 5x - 3 \geqslant 0$

40. $2x^2 - 5x - 3 \leqslant 0$

41. $x^2 + 6x + 1 < 0$

42. $x^2 + 6x + 1 > 0$

43. $3x^2 - x - 2 > 0$

44. $3x^2 - x - 2 < 0$

45. $5 - x - x^2 \leqslant 0$

46. $5 - x - x^2 \geqslant 0$

47. $4 + 5x - 2x^2 \geqslant 0$

48. $4 + 5x - 2x^2 \leqslant 0$

49. $x^2 + 3x + 5 < 0$

50. $x^2 + 3x + 5 > 0$

51. $x^2 - 5x + 7 > 0$

52. $x^2 - 5x + 7 < 0$

53. $x^2 - 4x + 4 \leqslant 0$

54. $x^2 - 4x + 4 \geqslant 0$

Prove the following.

55. If x is a real number and $a > 0$, then $ax^2 \geqslant 0$.

56. If x is a real number and $a < 0$, then $ax^2 \leqslant 0$.

57. If x and h are real numbers and $a < 0$, then $a(x - h)^2 \leqslant 0$.

58. If x and h are real numbers and $a > 0$, then $a(x - h)^2 \geqslant 0$.

59. If x and h and k are real numbers and $a > 0$, then $a(x - h)^2 + k \geqslant k$.

60. If x and h and k are real numbers and $a < 0$, then $a(x - h)^2 + k \leqslant k$.

61. The range of $f(x) = a(x - h)^2 + k$ is $[k, \infty)$ if $a > 0$.

62. The range of $f(x) = a(x - h)^2 + k$ is $(-\infty, k]$ if $a < 0$.

3-5   P R O B L E M   S E T

PROBLEMS FOR GRAPHING TECHNOLOGY

Graph each of the following. Use the graph to solve the indicated inequality.

1. Graph $y = 5x - 3$. Solve $5x - 3 < 0$.

2. Graph $y = 7x + 2$. Solve $7x + 2 > 0$.

3. Graph $y = 5 - 4x$. Solve $5 - 4x \geqslant 0$.

4. Graph $y = -3 - 4x$. Solve $-3 - 4x \leqslant 0$.

5. Graph $y = 2x - 5$ and $y = 7 - 3x$. Solve $2x - 5 < 7 - 3x$.

6. Graph $y = 3 - 2x$ and $y = x + 5$. Solve $3 - 2x > x + 5$.

Use graphing technology to solve the following.

7. $4 + 5x - 2x^2 \leqslant 0$

8. $x^2 + 3x + 5 < 0$

9. $x^2 + 3x + 5 > 0$

10. $x^2 - 5x + 7 > 0$

11. $x^2 - 5x + 7 < 0$

12. $x^2 - 4x + 4 \leqslant 0$

13. $x^2 - 4x + 4 \geqslant 0$

14. $4 + 5x - 2x^2 \geqslant 0$

EXPLORATION

The following problems anticipate concepts from the next section. Graph each of the pairs of functions on the same graph.

15. $f(x) = 3x - 5$, $g(x) = |3x - 5|$

16. $f(x) = 9 - x^2$, $g(x) = |9 - x^2|$

17. $f(x) = x^2 - 4$, $g(x) = |x^2 - 4|$

18. $f(x) = 4 - 2x$, $g(x) = |4 - 2x|$

19. $f(x) = x^2 - 7x + 12$, $g(x) = |f(x)|$

20. $f(x) = x^2 - 6x + 8$, $g(x) = |f(x)|$

21. From Problems 15–20, conjecture a relation between the graphs of $y = f(x)$ and $y = |f(x)|$.

22. Graph $f(x) = x^2 - 9$ and $g(x) = |f(x)| + 2$. Conjecture a relation between the graphs of $y = f(x)$ and $y = |f(x)| + b$.

23. Graph $f(x) = 3x - 2$ and $g(x) = 3|x| - 2$. Compare the results to the previous problems.

24. Graph $f(x) = 2 - 3x$ and $g(x) = 2 - 3|x|$. Compare the results to the previous problems.

6   ABSOLUTE VALUE FUNCTIONS AND INEQUALITIES

The moving power of mathematical invention is not reasoning but imagination.
—A. DeMorgan

Distance measures the separation between points. Time may be thought of as a temporal distance between two events in much the same manner that spatial distance separates objects. Absolute value represents the geometric concept of distance, see Figure 3-55.

DEFINITION 3-2
Geometry of Absolute Value

> The distance of a real number x from the origin is symbolized by $|x|$, read "absolute value of x."

Because $|x|$ represents distance, $|x|$ is never negative. We repeat the algebraic definition of absolute value for easy reference.

DEFINITION 3-3
Absolute Value: Algebraic Interpretation

> The **absolute value** of a real number x is given by
> $$|x| = \begin{cases} x & \text{if } x \geq 0 \\ -x & \text{if } x < 0 \end{cases}$$

Absolute Value—*Absolute* is from the Latin *absoluere* meaning "to free from." Normally we assume this means that the absolute value of a number is free from any sign. Weierstrauss was the first to use the term *absolute value* when, in 1841, he called the distance of $a + bi$ to the origin the absolute value of the complex number $a + bi$ and represented it by $|a + bi|$. Weierstrauss's symbol for absolute value has survived and includes the absolute value of real numbers. The definition for $|a + bi|$ is $\sqrt{(a + bi)(a - bi)}$. For any real number, $b = 0$, therefore $|a| = \sqrt{a^2}$.

The first definition of absolute value is visual (geometric). The second definition is algebraic in the truest sense: it provides an alternative representation for $|x|$. Convince yourself that the definitions are equivalent by examining examples of distances of both positive and negative numbers on a number line.

What is the distance between two points on a number line whose coordinates are a and b? Subtraction, $a - b$, provides a directed distance from b to a. See Figure 3-56. *Directed distance* means that $a - b$ is positive if a is to the right of b and negative if a is to the left of b. If the direction from b to a is of no concern, absolute value represents distance. The undirected distance from b to a is given by $|a - b|$. Because the distance from a to b is the same as the distance from b to a, we have the following theorem.

Theorem 3-5
Absolute Value as Distance

> The distance between two real numbers a and b is given by $|a - b|$; Note that $|a - b| = |b - a|$.

An *absolute value equation* is an equation in which an absolute value symbol encloses the variable. We may further classify an absolute value equation according to the use of the variable within the absolute value. For example,

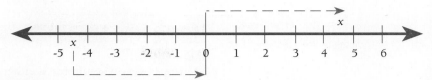

FIGURE 3-55   Distance of x from origin

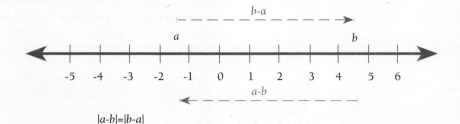

$|a-b|=|b-a|$

FIGURE 3-56

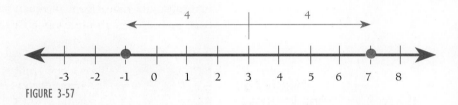

FIGURE 3-57

$|ax + b| = c$, where $ax + b$ is a real expression, is a linear absolute value equation because the degree of x is one.

Consider $|x - 3| = 4$. Geometrically the equation describes all numbers x whose distance from 3 is 4.

From Figure 3-57, -1 or 7 would satisfy the conditions of the equation. Check both of these in the equation. Similarly, $|x + 3| = 5$ is rewritten as $|x - (-3)| = 5$ for a geometric interpretation: all x whose distance from -3 is 5. Clearly, $|x - 2| = -3$ has no solution, because distance cannot be negative. But to solve an absolute value equation such as $|3x - 2| = |1 - x|$ requires a method more powerful than geometric visualization. Before examining such a method, let us summarize with some theorems.

Theorem 3-6
Solutions of Absolute Value Equations

If $|mx + b| = c$ and $m \neq 0$, then
1. If c is positive, the equation has two solutions.
2. If c is 0, the equation has one solution.
3. If c is negative, the equation has no solution.

Careful application of the algebraic definition of absolute value starts the solution of a wide variety of absolute value equations. Similarly, the proof of the following theorems begins with the definition of absolute value. These proofs are advanced exercises.

Theorem 3-7
Absolute Value Not Negative

If c is a real number, then $|c| \geqslant 0$. (In fact, $|c| \geqslant c$.)

Hint: Every real number is either positive, 0 or negative. Refer to the definition of absolute value and consider each of these cases separately.

The next theorem is the basis for the rule for signs in the multiplication of signed numbers. Usually, we determine the sign of a product then multiply the absolute values of the factors to determine the absolute value of the product. The hint for Theorem 3-7 is useful in the proof of Theorem 3-8.

Theorem 3-8
Product of Absolute Values

If a and b are real numbers, then $|a|\,|b| = |ab|$.

The proof of the next theorem combines the results of Theorems 3-7 and 3-8 with the fact that c^2 is never negative. Theorem 3-9 provides a way to eliminate absolute values from an equation by squaring both sides of the equation.

Theorem 3-9
Square of Absolute Value

If c is a real number, then $|c^2| = |c|^2 = c^2$.

Theorem 3-10 resembles Theorem 3-8.

Theorem 3-10
Quotient of Absolute Values

If a and b are real numbers, $b \neq 0$, then $|a|/|b| = |a/b|$.

To solve a linear absolute value equation, refer to the algebraic definition of absolute value. The next example illustrates this method.

Solving Absolute Value Equations. Solve each of the following.

EXAMPLE 3-12

Illustration 1:

$$|x + 3| = 5$$

Solution 1: Refer to Figure 3-58.

$$x = 2 \text{ or } x = -8$$

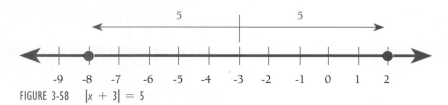

FIGURE 3-58 $|x + 3| = 5$

When a conditional sentence has more than one solution, it is sometimes more convient to list those solutions using set notation:

$$x \in \{2, -8\}$$

Solution 2: By definition

$$|x + 3| = \begin{cases} x + 3 & \text{if } x + 3 \geq 0 \\ -(x + 3) & \text{if } x + 3 < 0 \end{cases}$$

Therefore, $|x + 3| = 5$ may be written as

If $x + 3 \geq 0$	If $x + 3 < 0$
$x + 3 = 5$ $x = 2$	$-(x + 3) = 5$ $-x - 3 = 5$ $-x = 8$ $x = -8$

$$x = 2 \quad \text{or} \quad x = -8$$

Illustration 2: $|3x - 7| = 6$

Solution 1:

$$\tfrac{1}{3}|3x - 7| = \tfrac{1}{3}(6)$$

$$|x - \tfrac{7}{3}| = 2 \qquad\qquad\qquad \text{(see Figure 3-59)}$$

$$x \in \left\{\tfrac{1}{3}, \tfrac{13}{3}\right\}$$

Solution 2:

Rewrite $|3x - 7| = 6$

If $3x - 7 \geq 0$	If $3x - 7 < 0$
$3x - 7 = 6$ $3x = 13$ $x = \tfrac{13}{3}$	$-(3x - 7) = 6$ $-3x + 7 = 6$ $-3x = -1$ $x = \tfrac{1}{3}$

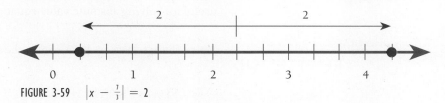

FIGURE 3-59 $|x - \tfrac{7}{3}| = 2$

Illustration 3:

$$|3 - 5x| = 0$$

Solution:

If $3x - 5x \geqslant 0$	If $3 - 5x < 0$
$3 - 5x = 0$	$-(3 - 5x) = 0$
$-5x = -3$	$3 - 5x = 0$
$x = \frac{3}{5}$	Duplicate of the first case

$x = \frac{3}{5}$, one solution. See Theorem 3-6.

Illustration 4:

$$|3x - 2| = -5$$

Solution: Although it is possible to rewrite this equation and solve the resulting equations, the solutions will not check in the original absolute value equation. See Theorem 3-6; absolute values cannot be negative. This equation has no solution. ▬

The next theorem, called the **triangle inequality**, has important applications in calculus. Many methods for proving the triangle inequality exist. Exercise Problems 117–122 will lead you through one method.

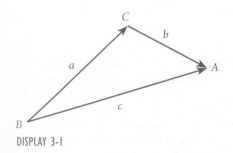

B

DISPLAY 3-1

Theorem 3-11
Triangle Inequality

> If a and b are real numbers, then $|a + b| \leqslant |a| + |b|$

Consider triangle ABC shown in Display 3-1. The distance from B to C is $|a|$. Similarly, the distance from C to A is $|b|$. The total distance along the path from B to C to A is $|a| + |b|$. But the shortest distance from B to A is $|c|$. Therefore, $|c| < |a| + |b|$. This inspires the name *triangle inequality*.

Notice that $|3 + 5| = |3| + |5|$. Also, $|-3 + (-5)| = |-3| + |-5|$.
Finally, $|(-3) + 5| < |-3| + |5|$. Under what conditions do you think the equality portion of the triangle inequality holds? Under what conditions will the inequality portion hold true?

Theorem 3-12
Absolute Value Equations

> If a and b are real numbers and $|a| = |b|$, then $a = b$ or $a = -b$.

One method to prove Theorem 3-12 is to apply Theorem 3-9 to remove the absolute values, then solve the resulting equation for a. Theorem 3-12 is useful for solving absolute value equations in the next example.

Solving Absolute Value Equations. Solve the following.

EXAMPLE 3 - 13

Illustration 1:

$$|3x - 2| = |1 - x|$$

Solution 1: Apply Theorem 3-12 to rewrite the problem:

$$3x - 2 = 1 - x \qquad \text{or} \qquad 3x - 2 = -1 + x$$
$$4x = 3 \qquad \text{or} \qquad 2x = 1$$
$$x = \tfrac{3}{4} \qquad \text{or} \qquad x = \tfrac{1}{2}$$

Solution 2: Apply Theorem 3-9 by squaring both sides:

$$|3x - 2|^2 = |1 - x|^2$$
$$(3x - 2)^2 = (1 - x)^2 \qquad \text{(Theorem 3-9)}$$
$$9x^2 - 12x + 4 = 1 - 2x + x^2$$
$$8x^2 - 10x + 3 = 0.$$

By the quadratic formula:

$$x = \frac{10 \pm \sqrt{100 - 96}}{16}$$

$$x = \tfrac{3}{4} \text{ or } x = \tfrac{1}{2}$$

Note: Squaring both sides of an equation is not always a safe procedure. For example, in the equation $x = 3$, the only solution is 3. Yet if we square both sides of the equation we get $x^2 = 9$, an equation with two solutions: 3, -3. The -3 solution will not check in the original equation $x = 3$. The process of squaring both sides of the equation introduced an extra value that is no solution in the original equation. Such extra values produced as a side effect of the solution process are **extraneous solutions**. Squaring both sides of an equation may introduce extraneous solutions. Most techniques for solving absolute value equations are susceptible to extraneous solutions. For this reason, you should always check proposed answers to absolute value equations. Any proposed value that does not check is extraneous and must be discarded. Any remaining values form the solution. If you check $\tfrac{3}{4}$ and $\tfrac{1}{2}$ you will discover both are solutions.

Illustration 2:

$$3x = |1 - x|$$

Solution 1: Rewrite as $3x - |1 - x| = 0$

If $1 - x \geq 0$, then $1 \geq x$	If $1 - x < 0$, then $1 < x$
$3x - (1 - x) = 0$	$3x - [-(1 - x)] = 0$
$3x - 1 + x = 0$	$3x + 1 - x = 0$
$4x = 1$	$2x = -1$
$x = \tfrac{1}{4}$	$x = -\tfrac{1}{2}$

$$x = \tfrac{1}{4} \quad \text{or} \quad x = -\tfrac{1}{2}$$

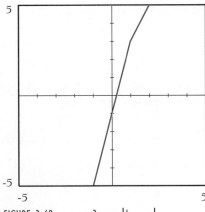

FIGURE 3-60 $y = 3x - |1 - x|$

See Figure 3-60. Note that $-\frac{1}{2}$ is no solution of $3x - |1 - x|$ because that solution was based on a case that required $1 < x$. Reject $-\frac{1}{2}$ as an extraneous solution. The solution consists of $x = \frac{1}{4}$.

Solution 2: Square both sides of the equation:

$$(3x)^2 = |1 - x|^2$$
$$9x^2 = 1 - 2x + x^2$$
$$8x^2 + 2x - 1 = 0$$
$$(4x - 1)(2x + 1) = 0$$
$$4x - 1 = 0 \quad \text{or} \quad 2x + 1 = 0$$
$$x = \tfrac{1}{4} \quad \text{or} \quad x = -\tfrac{1}{2}$$

As before, $-\frac{1}{2}$ is no solution of $3x - |1 - x|$. Again, $-\frac{1}{2}$ is an extraneous solution introduced by the solving process. Reject $-\frac{1}{2}$ as a solution. $x \in \{\frac{1}{4}\}$.

$$|x^2 - 7x + 2| = 10$$

Solution:

$$x^2 - 7x + 2 = 10 \quad \text{or} \quad -(x^2 - 7x + 2) = 10$$
$$x^2 - 7x - 8 = 0 \quad \text{or} \quad x^2 - 7x + 12 = 0$$
$$(x - 8)(x + 1) = 0 \quad \text{or} \quad (x - 3)(x - 4) = 0$$
$$x - 8 = 0 \quad \text{or} \quad x + 1 = 0 \quad \text{or} \quad x - 3 = 0 \quad \text{or} \quad x - 4 = 0$$
$$x = 8 \quad \text{or} \quad x = \text{-}1 \text{ or } x = 3 \quad \text{or} \quad x = 4.$$

Check each possible solution.   ▬

Illustration 3:

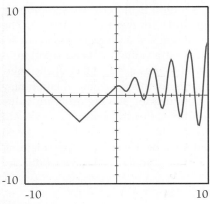

FIGURE 3-61 f

Consider the absolute value function $A(x) = |x|$. From the definition of absolute value, the rule of pairing can be rewritten

$$A(x) = \begin{cases} x & \text{if } x \geqslant 0 \\ \text{-}x & \text{if } x < 0 \end{cases}$$

Examine the graph of the function f shown in Figure 3-61.

Analytically, the absolute value function will affect the graph of f under composition as follows:

$$A \circ f(x) = |f(x)| = \begin{cases} f(x) & \text{if } f(x) \geqslant 0 \\ \text{-}f(x) & \text{if } f(x) < 0 \end{cases}$$

The graphic interpretation is clear: the portion of f above the x-axis remains there. The absolute value reflects the graph of f that lies below the x-axis to above the axis. Compare the graph of a function f in Figure 3-61 with the graph of $|f|$ in Figure 3-62.

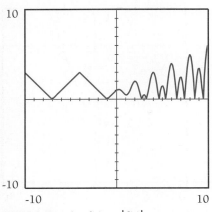

FIGURE 3-62 $A \circ f(x) = |f(x)|$

EXAMPLE 3-14

Graphs of Absolute Value Functions. Graph the following functions. Note the role of absolute value in each graph.

Illustration 1:

$$f(x) = |x^2 - 4|$$

Solution: Recognize $y = x^2 - 4$ as a parabola that opens up, its vertex at (0, -4). See Figure 3-63. Thus, $f(x) = |x^2 - 4|$ has the graph in Figure 3-64.

Illustration 2:

$$g(x) = |3x - 5|$$

Solution: Graph $y = 3x - 5$. See Figure 3-65. Then the graph of g is as in Figure 3-66.

Illustration 3:

$$h(x) = |5 - x| - 2$$

In this case, the function is not strictly the composite of the form $A \circ f$. It can be solved by at least two different methods.

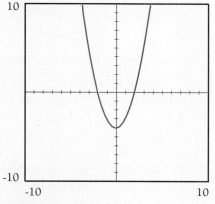

FIGURE 3-63 $f(x) = x^2 - 4$

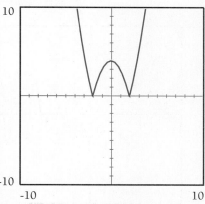

FIGURE 3-64 $f(x) = |x^2 - 4|$

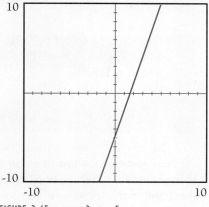

FIGURE 3-65 $y = 3x - 5$

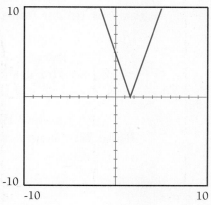

FIGURE 3-66 $g(x) = |3x - 5|$

Method 1: Use the definition of absolute value: $h(x) = |5 - x| - 2$.

$$h(x) = \begin{cases} (5 - x) - 2 & \text{if } 5 - x \geq 0 \\ -(5 - x) - 2 & \text{if } 5 - x < 0 \end{cases}$$

$$h(x) = \begin{cases} -x + 3 & \text{if } 5 \geq x \\ x - 7 & \text{if } 5 < x \end{cases}$$

Clearly, $x = 5$ is a critical value, where an acute change in the graph of h occurs (Figure 3-67). Note that $h(5) = -2$.

Method 2: Translate axes. Graph $y = |5 - x|$. See Figure 3-68. Shift the graph of $y = |5 - x|$ down 2 units to obtain the graph of $y = |5 - x| - 2$ (Figure 3-69).

Note: Set y equal to 0 to determine the x-intercepts of the graph:

$$|5 - x| - 2 = 0$$
$$|5 - x| = 2$$

$$\begin{array}{ccc} 5 - x = 2 & \text{or} & -(5 - x) = 2 \\ -x = -3 & \text{or} & -5 + x = 2 \\ x = 3 & \text{or} & x = 7 \end{array}$$

Set x equal to 0 to determine the y-intercept.

$$y = |5 - 0| - 2$$
$$y = 3$$

Absolute value inequalities have extensive applications in the development of the calculus. They are also useful in the control of error in computer approximations. At least two methods for the solution of absolute inequalities exist. The first method uses Theorems 3-13, 3-14 and 3-15.

Theorem 3-13
Absolute Value: Less than Positive

If a is positive, then $|x| < a$ and $-a < x < a$ are equivalent.

Theorem 3-14
Absolute Value: Less than 0

For $|x| < a$, if a is *not* positive, then there is no solution.

Theorem 3-15
Absolute Value: Greater than

For $a > 0$, $|x| > a$ is equivalent to $x > a$ or $x < -a$.

The proof of the last theorem follows immediately from the definition of absolute value. Careful use of the definition will lead to the proof of all three theorems. These proofs are left as optional exercises. For geometric insight, consider the graphs of $f(x) = |x| - a$ as shown in Figure 3-70.

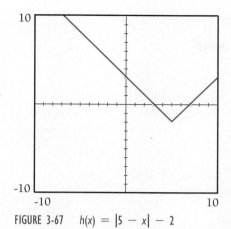

FIGURE 3-67 $h(x) = |5 - x| - 2$

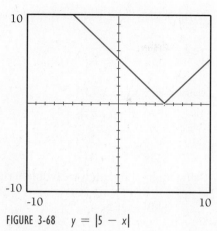

FIGURE 3-68 $y = |5 - x|$

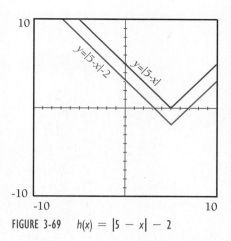

FIGURE 3-69 $h(x) = |5 - x| - 2$

Imagine an associated chart for each of these cases. The results of the theorems follow from the charts.

Function x-axis:	$-\infty$		$-a$		a		∞	Note
$|x| - a$		+	0	−	0	+		$a > 0$. See Figure 3-70

Function x-axis:	$-\infty$		0		∞	Note
$|x| - a$		+	0	+		$a = 0$. See Figure 3-71

Function x-axis:	$-\infty$			∞	Note
$|x| - a$			+		$a < 0$. See Figure 3-72

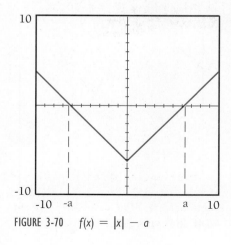

FIGURE 3-70 $f(x) = |x| - a$

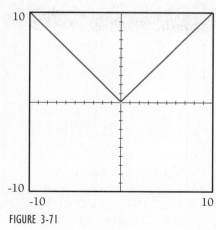

FIGURE 3-71

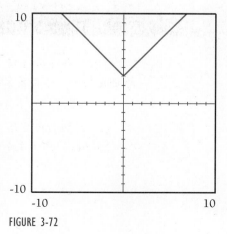

FIGURE 3-72

EXAMPLE 3 - 15

Solving Absolute Value Inequalities.   Solve each of the following.

Illustration 1:

$$|3x - 2| < 4$$

Solution: See Figure 3-73.

$$-4 < 3x - 2 < 4$$ (Theorem 3-13)

$$-2 < 3x < 6$$

$$-\tfrac{2}{3} < x < 2$$

Using interval notation we would write $x \in (-\tfrac{2}{3}, 2)$.

Illustration 2:

$$|3x - 2| \geqslant 4$$

Solution 1:

$$3x - 2 \geqslant 4 \quad \text{or} \quad -(3x - 2) \geqslant 4$$ (Theorem 3-15)

$$3x \geqslant 6 \quad \text{or} \quad -3x + 2 \geqslant 4$$

$$x \geqslant 2 \quad \text{or} \quad -3x \geqslant 2$$

$$x \geqslant 2 \quad \text{or} \quad x \leqslant -\tfrac{2}{3}$$

Solution 2: From the Figure 3-73 and Theorem 3-15, the solution is

$$x \in (-\infty, -\tfrac{2}{3}] \cup [2, \infty).$$

Illustration 3:

$$|3x - 2| < |1 - x|$$

Solution: The inequality is equivalent to $|3x - 2| - |1 - x| < 0$. From Example 3-13, Illustration 1, the roots for

$$|3x - 2| - |1 - x| = 0 \text{ are } \{\tfrac{1}{2}, \tfrac{3}{4}\}.$$

View the graph of $y = |3x - 2| - |1 - x|$ with graphing technology. See Figure 3-74. Otherwise use test values to complete the signs of the chart.

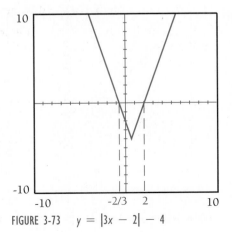

FIGURE 3-73 $y = |3x - 2| - 4$

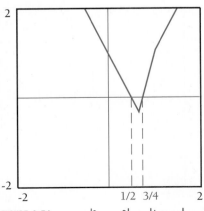

FIGURE 3-74 $y = |3x - 2| - |1 - x|$

Function x-axis:	$-\infty$		$\frac{1}{2}$		$\frac{3}{4}$		∞	Note
$\|3x - 2\| - \|1 - x\|$	?		0	?	0		?	Test values

Try any number smaller than $\frac{1}{2}$, say, 0:

$$|3(0) - 2| - |1 - (0)| = +1$$

Try a number between $\frac{1}{2}$ and $\frac{3}{4}$, say, $\frac{2}{3}$:

$$|3(\tfrac{2}{3}) - 2| - |1 - (\tfrac{2}{3})| = -\tfrac{1}{3}$$

Try a number larger than $\frac{3}{4}$, say, 100:

$$|3(100) - 2| - |1 - (100)| = +199$$

Thus the chart becomes as follows.

Function x-axis:	$-\infty$	0	$\frac{1}{2}$	$\frac{2}{3}$	$\frac{3}{4}$	100	∞	Test values
$\|3x - 2\| - \|1 - x\|$		+	0	−	0	+		Sign of results

Because we require $|3x - 2| - |1 - x|$ to be negative, the solution is $(\frac{1}{2}, \frac{3}{4})$.

Illustration 4:

$$3x \leqslant |1 - x|$$

Solution: Rewrite as $3x - |1 - x| \leqslant 0$. Determine the intercepts of $f(x) = 3x - |1 - x|$. From Example 3-13, Illustration 2, $x = \frac{1}{4}$. Use technology to produce the graph of $y = 3x - |1 - x|$ in Figure 3-75. Otherwise, test values (e.g., $x = -2$ and $x = 1$) provide the signs for the remainder of the chart.

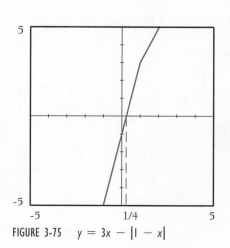

Function x-axis:	$-\infty$		$\frac{1}{4}$		∞	Test values or
$3x - \|1 - x\|$		−	0	+		Observe from graph

FIGURE 3-75 $y = 3x - |1 - x|$

Because we require $3x - |1 - x| \leqslant 0$ then the solution is $(-\infty, \frac{1}{4}]$.   ▬

S U M M A R Y

- Absolute value represents distance. The distance from a to b is $|b - a|$.
- Moreover, $|x|$ represents the distance of x from 0. Algebraically,

$$|x| = \begin{cases} x & \text{if } x \geq 0 \\ -x & \text{if } x < 0 \end{cases}$$

- The definition of the absolute value is a primary tool in the solution of absolute value equations.
- The graph of $A \circ f$, where A is the absolute value function, is the same as f, where f is above the x-axis. Otherwise, the graph of $A \circ f$ is a positive mirror image of the part of the graph of f below the x-axis.
- Charts and graphing technology provide an efficient method for solving absolute value inequalities.

3-6   E X E R C I S E S

Solve for x.

1. $|x + 3| = 2$

2. $|x - 3| = 2$

3. $|x - 5| = 4$

4. $|x + 5| = 4$

5. $|3x + 2| = 5$

6. $|3x - 2| = 5$

7. $|5x - 1| = 4$

8. $|5x + 1| = 4$

9. $|7 - 2x| = 3$

10. $|2x - 7| = 3$

11. $|4x - 3| = 5$

12. $|3 - 4x| = 5$

13. $\left|\frac{1}{2}x + \frac{1}{3}\right| = \frac{1}{5}$

14. $\left|\frac{1}{2}x - \frac{1}{3}\right| = \frac{1}{5}$

15. $\left|\frac{1}{3} - \frac{1}{2}x\right| = \frac{1}{5}$

16. $\left|-\frac{1}{2}x - \frac{1}{3}\right| = \frac{1}{5}$

17. $|5x - 3| = 0$

18. $|5x + 3| = 0$

19. $|3x + 2| = -1$

20. $|3x - 2| = -2$

Solve for x. Be sure to check all the answers.

21. $|2x - 1| = |x + 3|$

22. $|x - 3| = |2x + 1|$

23. $|5x + 2| = |3x - 1|$

24. $|5x - 2| = |3x + 1|$

25. $|3x + 7| + |x - 1| = 0$

26. $|3x + 7| - |x - 1| = 0$

27. $|2x - 3| - |5x + 2| = -1$

28. $|5x - 2| + |2x - 3| = 1$

29. $|x| + |x + 2| = 3$

30. $|x| - |x - 2| = -3$

31. $|x^2 - 4x + 6| = 3$

32. $|x^2 - 2x + 3| = 6$

33. $|x^2 + 5x - 4| = 1$

34. $|x^2 - 3x + 1| = 2$

Graph the following pairs of functions.

35. $f(x) = x + 5, f(x) = |x + 5|$

36. $g(x) = x - 2, g(x) = |x - 2|$

37. $r(x) = x^2 - 7x + 12, r(x) = |x^2 - 7x + 12|$

38. $t(x) = x^2 - 9x + 8, t(x) = |x^2 - 9x + 8|$

39. $q(x) = (x - 1)(x + 3)(x - 4),$
$q(x) = |(x - 1)(x + 3)\ (x - 4)|$

40. $p(x) = (x + 4)(x - 1)(x - 5),$
$p(x) = |(x + 4)(x - 1)\ (x - 5)|$

41. $t(x) = \dfrac{x + 4}{x - 5}, t(x) = \left|\dfrac{x + 4}{x - 5}\right|$

42. $z(x) = \dfrac{x - 1}{x + 3}, z(x) = \left|\dfrac{x - 1}{x + 3}\right|$

43. $m(x) = x^2 - 1, m(x) = |x^2 - 1| + 3$

44. $n(x) = x^2 - 4, n(x) = |x^2 - 4| - 2$

For each of the following functions f with graph as shown in Problems 45 through 54, sketch the graph of $|f|$.

45.
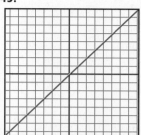

46.
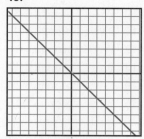

47.
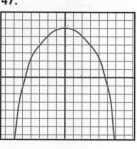

48.
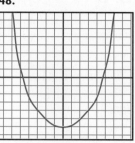

49.
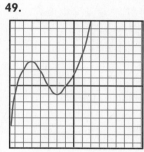

50.
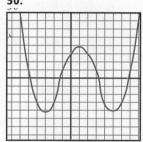

51.
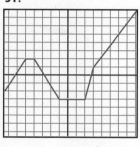

52.
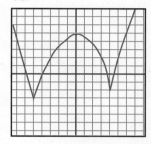

53.
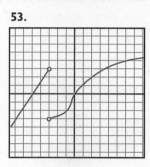

54.
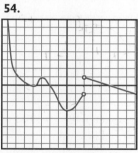

Solve each of the following. Express the answer in interval notation.

55. $|x - 2| < 3$

56. $|x - 2| \geqslant 3$

57. $|x + 5| > 4$

58. $|x + 5| < 4$

59. $|x - 3| \leqslant 2$

60. $|x - 3| > 2$

61. $|x - 5| \geqslant 0$

62. $|x + 5| \leqslant 0$

63. $|x - 1| < -3$

64. $|x - 1| \geqslant -3$

65. $|x + 1| > -3$

66. $|x + 1| < -3$

67. $|2x - 1| \leqslant 3$

68. $|2x + 1| \leqslant 3$

69. $|2x + 1| \geqslant 3$

70. $|2x - 1| > 3$

71. $|1 - 2x| > 3$

72. $|1 - 2x| < 3$

73. $|\frac{1}{2}x + 3| < \frac{1}{5}$

74. $|\frac{1}{2}x - 3| \leqslant \frac{1}{5}$

Construct a chart for each of the following functions.

75. $f(x) = |3x - 5|$

76. $f(x) = |5x - 3|$

77. $f(x) = |5 - 3x|$

78. $f(x) = |3 - 5x|$

79. $f(x) = |2x + 1| - |x|$

80. $f(x) = |x| - |2x + 1|$

81. $f(x) = |x - 3| - |2x + 1|$

82. $f(x) = |x + 3| - |2x - 7|$

***83.** $f(x) = 2x - |x + 3|$

***84.** $f(x) = |x + 1| - 2x$

Solve each of the following. Compare each to Problems 75–84.

85. $|3x - 5| > 0$

86. $|5x - 3| \geqslant 0$

87. $|5 - 3x| < 0$

88. $|3 - 5x| \leqslant 0$

89. $|2x + 1| - |x| \le 0$

90. $|x| - |2x + 1| < 0$

91. $|x - 3| \ge |2x + 1|$

92. $|x + 3| > |2x - 7|$

*93. $2x > |x + 3|$

*94. $|x + 1| \ge 2x$

Use graphing technology to solve the following inequalities. If necessary, estimate the intercepts using the trace function.

95. $|x^3 + 2x^2 - x| < 2$

96. $|4x - 5| > |2x - 1|$

97. $|3x + 2| \ge 2x$

98. $|x^2 - 4x + 3| > 5 - x$

99. $|x^2 - 4x + 3| \le |5 - x|$

100. $|3x + 2| < |2x|$

Write an absolute value inequality to describe each of the following.

101. All x whose distance from 3 is less than $\frac{1}{2}$.

102. All x whose distance from 2 is less than $\frac{1}{3}$.

103. All x whose distance from –5 is greater than $\frac{1}{5}$.

104. All x whose distance from –7 is greater than $\frac{1}{4}$.

105. All x whose distance from $-\frac{1}{2}$ is less than or equal 4.

106. All x whose distance from $\frac{1}{3}$ is greater than or equal 1.

107. For what values of a and b is $|a + b| = |a| + |b|$?

108. For what values of a and b is $|a + b| < |a| + |b|$?

109. Prove Theorem 3-7.

110. Prove Theorem 3-8.

111. Prove Theorem 3-9.

112. Prove Theorem 3-10.

113. Prove Theorem 3-13.

114. Prove Theorem 3-14.

115. Prove Theorem 3-15.

116. If a is real, prove $|a| \ge a$.

117. If a and b are real, prove $|a + b|^2 = (a + b)^2$.

118. If a and b are real, prove $|a + b|^2 = a^2 + 2ab + b^2$.

119. If a and b are real, prove $2|a| |b| \ge 2ab$.

120. If a and b are real, prove $(|a| + |b|)^2 = a^2 + 2|a| |b| + b^2$.

121. If a and b are real, prove $|a + b|^2 \le (|a| + |b|)^2$. See Problems 117–120.

122. Prove Theorem 3-11. See Problems 117–121.

123. If a and b are real, prove $|a - b| \ge |a| - |b|$. (Apply Theorem 3-11 to $|(a - b) + b|$.)

124. If a is real, then prove $|a|$ is not negative.

125. Prove or disprove: $|a|^3 = a^3$.

3-6   PROBLEM SET

1. Sketch the graph of $y = 3x - 2$. Sketch the graph of $y = 3|x| - 2$. Compare the graphs.

2. Sketch the graph of $y = x^2 - 3x + 2$. Sketch the graph of $y = |x|^2 - 3|x| + 2$. Compare the graphs.

3. Sketch the graph of $y = \sqrt{x}$. Sketch the graph of $y = \sqrt{|x|}$. Compare the graphs.

4. Sketch the graph of $y = x^3 - 8$. Sketch the graph of $y = |x|^3 - 8$. Compare the graphs.

5. Compare the graph of f with the graph of $f \circ A$, where $A(x) = |x|$.

6. If $A(x) = |x|$, compare the graphs of $y = A \circ f(x)$ and $y = f \circ A(x)$.

Consider the interval $(-3, 5)$. Half way between the endpoints is the point where $x = 1$: $[(-3 + 5)/2 = 1]$. The distance from the midpoint to either endpoint is

$$|\text{midpoint} - \text{endpoint}|.$$

Therefore the distance to the endpoint is 4. If we wish to use

absolute value to describe x where $-3 < x < 5$, then subtract the midpoint:

$$-3 - 1 < x - 1 < 5 - 1$$
$$-4 < x - 1 < 4$$

But this is equivalent to $|x - 1| < 4$. Describe each of the intervals in Problems 7–16 using both interval notation and an absolute value inequality.

7.

8.

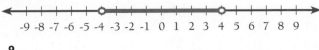

9.

10.

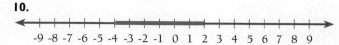

-9 -8 -7 -6 -5 -4 -3 -2 -1 0 1 2 3 4 5 6 7 8 9

11.

-9 -8 -7 -6 -5 -4 -3 -2 -1 0 1 2 3 4 5 6 7 8 9

12.

-9 -8 -7 -6 -5 -4 -3 -2 -1 0 1 2 3 4 5 6 7 8 9

13.

-9 -8 -7 -6 -5 -4 -3 -2 -1 0 1 2 3 4 5 6 7 8 9

14.

-9 -8 -7 -6 -5 -4 -3 -2 -1 0 1 2 3 4 5 6 7 8 9

15.

-9 -8 -7 -6 -5 -4 -3 -2 -1 0 1 2 3 4 5 6 7 8 9

16.

-9 -8 -7 -6 -5 -4 -3 -2 -1 0 1 2 3 4 5 6 7 8 9

PROBLEMS FOR GRAPHING TECHNOLOGY

Use graphing technology to compare the graphs of the following.

17. $f(x) = 5x - 3, g(x) = |f(x)|, h(x) = |f(x)| - 4$

18. $f(x) = 3 - 5x, g(x) = |f(x)|, h(x) = 4 - |f(x)|$

19. $f(x) = 4 - x^2, g(x) = |f(x)|, h(x) = 1 - |f(x)|$

20. $f(x) = x^2 - 9, g(x) = 1 + |f(x)|, h(x) = -|f(x)|$

21. $f(x) = x^2 - 7x + 3, g(x) = |f(x)|$

22. $f(x) = x^2 - 7x + 3, g(x) = |f(-x)|$

23. $f(x) = -x^2 + 8x + 2, g(x) = |f(x)|$

24. $f(x) = -x^2 + 8x + 2, g(x) = |f(x)|$

25. $f(x) = x^3 - 9x, g(x) = |f(x)|$

26. $f(x) = 9x - x^3, g(x) = |f(x)|$

27. Suppose that f is a function in which $f(x) > 0$ for all values of x. Describe the graph of $g(x) = |f(x)|$.

28. Suppose that f is a function in which $f(x) < 0$ for all values of x. Describe the graph of $g(x) = |f(x)|$.

29. Suppose that f is a function in which $f(x) > 0$ for $x > 0$ and $f(x) < 0$ for $x < 0$; $f(0) = 0$. Describe the graph of $g(x) = |f(x)|$.

30. See Problem 29. Describe the graph of $g(x) = |-f(x)|$.

Use graphing technology to graph the following sets of function in the same window. For each pair compare the x-intercepts. Also identify the intervals over which graph is above the x-axis (positive y values) and the intervals for which each graph is below the x-axis (negative y values).

31. $f(x) = 5x - 10, g(x) = |5x - 10|, h(x) = |5x - 10| - 5$

32. $f(x) = 8 - 3x, g(x) = |8 - 3x|, h(x) = |8 - 3x| - 7$

33. $f(x) = x^2 - 7, g(x) = |x^2 - 7|, h(x) = |x^2 - 7| - 2$

34. $f(x) = x^2 - 5x + 3, g(x) = |f(x)|, h(x) = g(x) + 1$

35. $f(x) = x^3 - 8, g(x) = |f(x)|$

36. $f(x) = |3x - 2| - |5x - 3|, g(x) = f(x) + 2$

37. Refer to the graph of h in Problem 31 to solve $|5x - 10| < 5$.

38. Refer to the graph of h in Problem 32 to solve $|8 - 3x| \geq 7$.

39. Solve $|x^2 - 7| \leq 2$.

40. Solve $|3x - 2| + 2 > |5x - 3|$.

7

TRUNCATION, ROUNDING, FLOORS AND CEILINGS

It would appear as though mathematics is the creation of human fallible minds rather than a fixed, eternally existing body of knowledge. The subject seems very much dependent on the creator.
—MORRIS KLINE

In a previous section, we suggested that an exact decimal representation of an irrational number is not possible. Instead, rational approximations provided insight into these irrational numbers. Some questions concerning the quality of these estimates remain. What constitutes a good **approximation**?

Consider a rational number, written in lowest terms, for which the prime factors of the denominator include only 2 or 5. Because the decimal numeration system is based on 10 ($10 = 2 \times 5$), such a rational number always has an exact finite decimal representation. For example, $\frac{1}{8} = 0.125$, $\frac{3}{20} = 0.15$.

However, if we write a rational number in lowest terms and the denominator contains any factor other than 2 or 5, the decimal representation is an infinite decimal that eventually settles into a repeating block of digits. For example

$$\frac{1}{3} = 0.3333\ldots,$$

$$\frac{1}{12} = 0.08333333\ldots,$$

$$\frac{1}{7} = 0.142857142857142857\ldots$$

To emphasize the repetitive nature of such rational numbers and to save the overuse of . . ., we use a bar to designate a block of digits as infinitely repeating:

$$\frac{1}{12} = 0.08\overline{3}, \qquad \frac{1}{7} = 0.\overline{142857}$$

Although it is possible for human beings to express these repeating infinite decimals by use of a bar, such representations are difficult in computers and calculators. As with irrational numbers, a finite decimal approximation often substitutes for the infinite repeating decimal. How are these approximations chosen?

Methods of approximation require the concepts of precision and accuracy to measure the "goodness" of an approximation. The precision of a decimal representation of a real number refers to the number of digits displayed in the representation, usually in relation to the decimal point. In computer science terminology, these representations are finite precision numbers. Precision information consists of the total number of digits, the number of digits after the decimal point and the number of digits before the decimal point. For example, 567.23 and 122.81 are of the same precision: five digits total, two after the decimal point, three before the decimal point.

Do not confuse precision with accuracy. Accuracy measures the error in an approximation. Consider approximations for π. Although 4.7893 is more precise than 3.1, 3.1 is a more accurate approximation of π because it is "closer" to π.

Suppose that a rational number (not a whole number) has an *exact* finite decimal representation given by a finite precision number. Examine the digits of this representation from left to right. The first nonzero digit encountered is the most significant digit. All the digits that follow the most significant digit are significant digits. An underline marks the significant digits in the following: 3.100250. The digit 3 is the most significant digit in this representation. For 0.001203, the digit 1 is the most significant digit.

Examine a rational approximation for π, the finite precision number 3.1415278. For approximations, care must be taken with significant digits. The digit 3 is the most significant digit. But the digits that follow 3 are significant only to the point where a digit fails to coincide with (or come within ± 1 of) the corresponding digit in the infinite decimal representation of π. Because $\pi = 3.1415926\ldots$, then 3, 1, 4, 1, 5 are all significant. The 5 is the least significant digit. But 2 was poorly chosen in the approximation. Therefore 2, 7 and 8 are not significant.

One measure of the accuracy of an approximation is the number of significant digits.

One method of approximation is **truncation**. The method applies easily to infinite repeating decimals. Normally, the number of digits to the right of the decimal point specifies the desired accuracy. Truncation consists of removing all digits beyond the specified number of digits.

Approximation by Truncation

EXAMPLE 3-16

Illustration 1:

Approximate $\frac{1}{7}$ with a finite precision decimal truncated to nine significant digits.

Solution: Recall that

$$\frac{1}{7} = \underline{0.142857142}857142857\ldots$$

$$\frac{1}{7} \cong 0.142857142$$

Examine Figure 3-76. Truncation on a positive number results in an approximation that is never larger than the actual number. Actually,

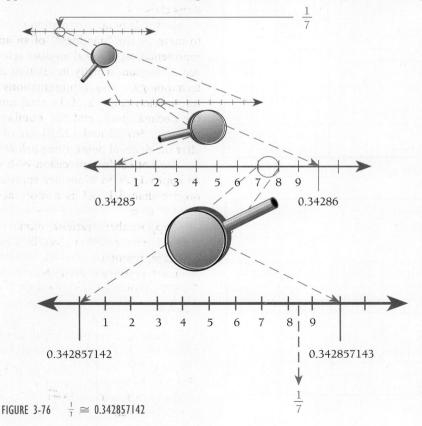

FIGURE 3-76 $\frac{1}{7} \cong 0.342857142$

0.142857143 is "closer" to $\frac{1}{7}$ than is 0.142857142 and satisfies the significance requirement for accuracy. In one case we have the choice of an approximation smaller than $\frac{1}{7}$, and in the other case, the choice of a larger number.

When the precision specified is zero decimal places (i.e., in the "units" position) the following functions define these methods.

DEFINITION 3-4
Floor or Greatest Integer Function

The floor (or round down or greatest integer) function is defined by

$$\lfloor x \rfloor = k,$$

where k is an integer and x is a real number such that $k \leq x < k + 1$ (Figure 3-77).

DEFINITION 3-5
Ceiling Function

The ceiling (or round up) function is defined by

$$\lceil x \rceil = k,$$

where k is an integer and x is a real number such that $k - 1 < x \leq k$ (Figure 3-78).

For example, $\lfloor 3.51 \rfloor = 3$; $\lceil 3.51 \rceil = 4$; $\lfloor -2.1 \rfloor = -3$; $\lceil -2.1 \rceil = -2$.

Consistently rounding up (or rounding down) to an integer is inappropriate in sequences of calculations because the accumulated error is always in one direction. A round off function R compensates for this.

$$R(x) = \lfloor x + 0.5 \rfloor$$

Thus, $R(3.51) = 4$; $R(-2.1) = -2$; and $R(2.3999) = 2$.

Unfortunately, even this method of rounding tends to round up too often in the case when the fractional part of the decimal is exactly 0.5. The resulting accumulated error may become significant in a computer program or in any calculations with large amounts of rounding. To overcome this bias toward rounding up, a modification of the rounding strategy called *rounding to an even digit* is advantageous. **Scientific rounding** is rounding to an even digit. Suppose that x is a decimal number to be rounded and that d represents the digit in x that immediately follows the decimal point (the "tenths" digit). Also suppose that u is the units digit of x, the digit just before the decimal point. Then

$$SR(x) = \begin{cases} \lfloor x \rfloor, \text{ if } d \in \{0, 1, 2, 3, 4\} \text{ or if } d = 5 \text{ is the least significant} \\ \qquad \text{digit and } u \text{ is even} \\ \lceil x \rceil \text{ if } d \in \{5, 6, 7, 8, 9\} \text{ (5 is not the least significant digit)} \\ \qquad \text{or if } d = 5 \text{ is the least significant digit and } u \text{ is odd} \end{cases}$$

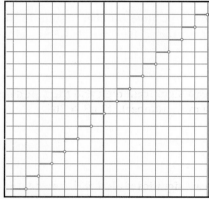

FIGURE 3-77 $y = \lfloor x \rfloor$

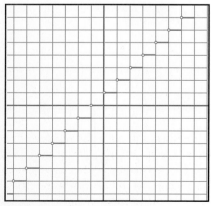

FIGURE 3-78 $y = \lceil x \rceil$

EXAMPLE 3-17

Scientific Rounding. Use scientific rounding to round the following to the nearest integer.

Illustration 1:

321.279

Solution: Because the tenths digit is 2, round down:

$$321.279 \cong 321$$

Illustration 2:

562.71119

Solution: Because the tenths digit is 7, round up:

$$562.71119 \cong 563$$

Illustration 3:

119.842

Solution: Because the tenths digit is 8, round up. Note that adding 1 to 9 in the units position causes a *carry* to the tens digit:

$$119.842 \cong 120$$

Illustration 4:

203.5

Solution: Because the tenths digit 5 is the least significant digit and the unit digit is odd, round up (rounding to the closest even number): $203.5 \cong 204$.

Illustration 5:

204.5

Solution: Because the tenths digit 5 is the least significant digit and the unit digit is even, round down (rounding to the nearest even number):

$$204.5 \cong 204$$

Illustration 6:

118.534

Solution: The tenths digit 5 is not the least significant digit, therefore we round up (odd and even are ignored):

$$118.534 \cong 119$$

Rounding to the nearest integer does not always provide the required precision. In representing money, for example, we may need a finite precision number to display two digits to the right of the decimal point. Specifying such precision is usually done one of two ways: "round to the nearest hundredth" or "round to two decimal places." We shall use the number of decimal places method because it poses few pronunciation problems.

DEFINITION 3-6
Rounding to n Decimal Places

The decimal representation of x rounded to n decimal places is given by $(0.1)^n SR(10^n x)$.

The formula is not difficult to apply. The factor 10^n serves the purpose of "shifting" the decimal point immediately behind the specified digit. Then the function SR applies. Finally, multiplication by 0.1^n "shifts" the decimal back to its original position. In practice, people simply imagine that the decimal is "moved" and restored. Computers are not known for their imagination.

Scientific Rounding. Round each of the following as indicated.

EXAMPLE 3-18

Illustration 1:

Round 3.1562715 to three decimal places.

Solution: Imagine the decimal point "behind" the 6 and apply the SR function:

$$3.15\ 6\ 2715 \cong 3.156$$

Illustration 2:

Round 3.1562715 to two decimal places.

Solution:

$$3.1562715 \cong 3.16$$

Illustration 3:

Round 3.1562715 to six decimal places.

Solution:

$$3.1562715 \cong 3.156272 \qquad \text{(round to even)}$$

Illustration 4:

Round 9.115185 to two decimal places.

Solution:

$$9.115185 \cong 9.12$$

(5 is not least significant: odd and even does not apply)

Illustration 5:

Round 9.115185 to five decimal places.

Solution:

$$9.1151\ 8\ 5 \cong 9.11518 \qquad \text{(round to even)}$$

We used fixed-point iteration to approximate x-intercepts in which successive approximations relied upon previous approximations. Such repetitive procedures are not unusual in computer-based approximations. How do we indicate to the computer when the desired accuracy is achieved? We make several assumptions about the approximations. In practice, conforming to these assumptions may be difficult.

1. At least two approximations will be calculated.

2. Except for the first approximation, each successive approximation is more accurate than the previous one.

3. The specified accuracy is given as n decimal places (0.1^n) and is within the precision capabilities of the computing environment.

The following procedure describes an algorithm for achieving the desired accuracy. However, indicating the process is easier than implementing it.

Initialize the first approximation;
repeat:
    store previous calculation
    make current approximation

until |(current approximation) − (previous approximation)|
    < $(0.5)(0.1^n)$

The "until" condition is the key to the required accuracy. For example, if $n = 3$, then $(0.5)(0.1^3) = 0.0005$. Therefore, if the absolute difference between two successive approximations is less then 0.0005, assume that the third decimal place has stabilized at the required precision. If the approximation method produced a significant digit, the required accuracy follows.

For example, to approximate a root of $x^2 + 3x - 1 = 0$ with a graphing calculator, rewrite the equation as follows:

$$x^2 + 3x = 1$$
$$x(x + 3) = 1$$
$$x = \frac{1}{(x + 3)}$$

Initialize the first approximation by pressing 1 then ENTER. This stores the first approximation 1 under ANS. Now enter the recursive formula 1/(ANS + 3). Press ENTER. The value 1 replaces ANS resulting in a display of 0.25. Each time you press enter, the previous value for ANS is substituted into the recursive formula and evaluated: this is an iteration. Press ENTER four more times and observe

0.3076923	(second iteration)
0.3023255	(third iteration)
0.3028169	(fourth iteration)
0.3027718	(fifth iteration)

Notice that the third decimal place in the fourth and fifth iteration appear stable at 2. Press ENTER twice more and observe:

0.3027759	(sixth iteration)

0.3027756   (seventh iteration)

The approximations now appear stable in the sixth decimal position. Try one more iteration to observe the seventh decimal position stabilize. To approximate the other root of $x^2 + 3x + 1 = 0$, initialize ANS at -1 and write the formula in the following form:

$$x(x + 3) = 1$$
$$x + 3 = 1/x$$
$$x = -3 + 1/x$$

3-7   E X E R C I S E S

Truncate, round up, round down, and round off 3.141592654 to the number of decimal places indicated.

1. 1   **2.** Nearest unit

3. Nearest ten   **4.** 2

5. 3   **6.** 4

7. 5   **8.** 6

9. 7   **10.** 8

Truncate, round up, round down, and round off -2.718281829 to the number of decimal places indicated.

11. 2   **12.** Nearest ten

13. Nearest unit   **14.** 1

15. 3   **16.** 4

17. 6   **18.** 5

19. 7   **20.** 8

Round each of the following to three decimal places (the nearest thousandth).

21. 2.7182818   **22.** 0.785398

23. 1.570796327   **24.** 5.4365636

25. 19.9997321   **26.** 19.9993721

27. 39.7435   **28.** 39.7345

29. 0.8955   **30.** 0.96235

31. 0.8955261   **32.** 0.96275

33. 0.8945   **34.** 0.9625

35. 0.8945103   **36.** 0.9625001

37. 0.89481   **38.** 0.9655

39. 0.89418   **40.** 0.96550102

 41. List two measures of the precision of a number.

42. Discuss the difference between precision and accuracy.

43. Emil must measure out 5.4 cc of sulphuric acid for his chemistry lab experiment. Discuss the relative importance of precision versus accuracy in Emil's experiment.

44. Kathy is writing a program for her computer science class. Her program must approximate the harmonic mean of three numbers to the nearest thousandth. Discuss the relative importance of precision and accuracy in Kathy's program.

PROBLEMS FOR TECHNOLOGY

Use fixed-point iteration on your graphing calculator for each of the following recursive formulas. In each case initialize the approximation at 1, then enter the formula. Note the number of times you press ENTER before the approximation stabilizes in the fifth decimal place.

45. $5 - 1/(2 + $ ANS $\times$ ANS $)$

46. $3 + 2/(1 + $ ANS $\times$ ANS $)$

47. $2 - 1/(3 + $ ANS $)$

48. $3 - 5/(4 - $ ANS $)$

49. $5(1 - 0.5 \wedge $ ANS $)$

50. $4(1 - 0.3 \wedge $ ANS $)$

Use fixed-point iteration to approximate the roots of the following equations. Discuss any difficulties you encounter.

51. $x^2 + 7x - 2 = 0$

52. $x^2 - 5x + 1 = 0$

53. $x^2 + 8x + 3 = 0$

54. $x^2 - 3x - 10 = 0$

CHAPTER SUMMARY

- A linear function f is a function given by $f(x) = mx + b$. The graph of f is a straight line. Unless otherwise stated, the domain of f is the set of real numbers. Thus, the y-intercept of the graph is $(0, b)$. The slope of the line is m. The graphs of two linear functions are parallel if they have the same slope. The graphs are perpendicular if the product of their slopes is -1.
- The graph of $A \circ f$, where $A(x) = |x|$ is the absolute value function, is the same as f, where f is above the x-axis. Otherwise, the graph of $A \circ f$ is a positive mirror image of the part of the graph of f below the x-axis.
- The graph of a quadratic function f given by $f(x) = ax^2 + bx + c$, $a \neq 0$, is a parabola. Completing the square converts a quadratic function in standard form to graphing form: $f(x) = a(x - h)^2 + k$, $a \neq 0$.
- The x-intercepts of a quadratic function f are the solutions of $f(x) = 0$.

Discriminant for $ax^2 + bx + c = 0$, where a, b and c are rational numbers; $a \neq 0$

Discriminant $D = b^2 - 4ac$	Nature of D	Nature and number of roots	Real roots	Graph
$D > 0$	Perfect square	Two rational roots	Two	
	Not a perfect square	Two irrational roots	Two	
$D = 0$		One rational root	One	
$D < 0$		Two imaginary roots, no real roots	None	

- A chart abbreviates graphical information about a function f. The chart shows where f is below, on and above the x-axis.
- For a linear function f given by $f(x) = mx + b$ the possible charts are as follows.

x-axis	$-\infty$	$\dfrac{-b}{m}$	∞	Notes
$mx + b$	$-$	0	$+$	$m > 0$

x-axis	$-\infty$		$\dfrac{-b}{m}$		∞	Notes
$mx + b$		$+$	0	$-$		$m < 0$

x-axis	∞		∞	Notes
b		Sign of b		$m = 0$

The following charts summarize quadratic cases:

Two intercepts x:	$-\infty$		x_1		x_2		∞	$b^2 - 4ac > 0$
$ax^2 + bx + c$		$+$	0	$-$	0	$+$		$a > 0$
$ax^2 + bx + c$		$-$	0	$+$	0	$-$		$a < 0$

One intercepts x:	$-\infty$		x_0		∞	$b^2 - 4ac = 0$
$ax^2 + bx + c$		$+$	0	$+$		$a > 0$
$ax^2 + bx + c$		$-$	0	$-$		$a < 0$

No intercepts x:	$-\infty$		∞	$b^2 - 4ac < 0$
$ax^2 + bx + c$		$+$		$a > 0$
$ax^2 + bx + c$		$-$		$a < 0$

KEY   WORDS   AND   CONCEPTS

Absolute value
Approximation
Chart
Completing the square
Constant function
Discriminant
Distance
Extraneous solution
Infinity symbol

Linear function
Linear inequality
Model
Parabola
Parallel
Parameter
Perpendicular
Quadratic formula
Quadratic function

Quadratic inequality
Scientific rounding
Slope
Triangle inequality
Truncation
x-intercept
Zero factor theorem

3   REVIEW   EXERCISES

SECTION 1

For each of the following linear functions, determine the slope and y-intercept. Determine whether the function is increasing, decreasing or neither. Graph the function.

1. $f(x) = -7$

2. $t(x) = -0.9x + 5$

3. $v(x) = 2 - \frac{3}{4}x$

4. $u(x) = \frac{5}{3}x - 4$

Write a rule of pairing for the linear function whose graph satisfies the following conditions.

5. Through $(0, 4)$ and perpendicular to $g(x) = -\frac{5}{3}x + 4$.

6. Through $(5, -4)$ and with slope of $\frac{2}{5}$.

7. Through $(3, -2)$ and $(7, 5)$.

8. Parallel to $j(x) = 5x - 4$ and with the same x-intercept as $k(x) = -4x + 2$

For each of the following pairs of linear functions, determine whether the functions' graphs are parallel, the same or intersect in a single point. If the lines intersect, indicate whether the lines are perpendicular.

9. $f(x) = 3x - 4$, $g(x) = 2 + 3x$

10. $f(x) = \frac{1}{2}x + 7$, $g(x) = -2x + 7$

11. $f(x) = 4 - x$, $g(x) = x + 5$

12. $f(x) = \frac{3}{5}x + 9$, $g(x) = -\frac{3}{5}x + 1$

Evaluate the difference quotient for the indicated points of the following functions.

13. $h(x) = \frac{1}{3}x$ for $(-5, h(-5))$ and $(2, h(2))$.

14. $t(x) = -4x + 3$ for $(4705, t(4705))$ and $(-0.034, t(-0.034))$.

15. $f(x) = kx - b$ for $(x, f(x))$ and $(x + h, f(x + h))$.

16. $f(x) = x^2 + b$ for $(x, f(x))$ and $(x + h, f(x + h))$.

SECTION 2

Graph the following quadratic functions. Annotate the vertex and intercepts.

17. $f(x) = -2x^2$

18. $f(x) = x^2 - 7x$

19. $f(x) = 3 - x^2$

20. $f(x) = (x - 5)^2 + 2$

21. $f(x) = (x + 4)^2 - 3$

22. $f(x) = -3(x - 1)^2 - 2$

23. $f(x) = x^2 + 6x - 5$

24. $f(x) = x^2 - 7x - 2$

For each of the following functions f, determine (a) $f(x + h)$, (b) $f(x + h) - f(x)$, and (c) $[f(x + h) - f(x)]/h$

25. $f(x) = 7x^2$

26. $f(x) = x^2 - 7$

27. $f(x) = 3x^2 - 2x$

28. $f(x) = 1 + 5x - 4x^2$

29. The height h of a projectile t seconds after it is fired at an initial velocity of 960 ft/sec is given by $h = -16t^2 + 960t$. Determine the maximum height of the projectile.

30. The height h of a projectile t seconds after it is fired from a height of 100 feet at an initial velocity of 1600 ft/sec is given by $h = -16t^2 + 1600t + 100$. Determine the maximum height of the projectile.

SECTION 3

Solve for x.

31. $x^2 - 3x - 10 = 0$ **32.** $x^2 - x = 12$

33. $x^2 - 5x = 6$ **34.** $x^2 - 17x + 70 = 0$

Determine the x-intercepts for the following functions.

35. $f(x) = x^2 + 6x - 16$ **36.** $f(x) = x^2 + 9x - 22$

37. $f(x) = x^2 - 12x + 35$ **38.** $f(x) = 2x^2 - 5x - 7$

Graph the following. Label the x-intercepts, if any.

39. $f(x) = 3x^2 + 7x - 5$ **40.** $f(x) = 2x^2 + 11x - 5$

41. $f(x) = 2x^2 - 19x + 6$ **42.** $f(x) = 4x^2 + x - 5$

Each of the following functions has an x-intercept on the indicated interval. Use continued fractions to approximate the indicated x-intercept.

43. $f(x) = x^2 - 7, 2 < x < 3$.

44. $f(x) = x^2 - 13, -4 < x < -3$.

Each of the following functions has an x-intercept on the indicated inverval. Use fixed point iteration to approximate the indicated x-intercept.

45. $f(x) = x^2 + x - 1, 0 < x < 1$.

46. $f(x) = x^2 - x - 3, -2 < x < -1$.

47. Suppose that the power produced by a windmill is proportional to the square of the diameter of the blades (where the wind velocity is constant). If a windmill with a 2.5 meter diameter blade generates 6250 watts, express the power as a function of the blade turning diameter.

48. Light intensity is proportional to the inverse of the square of the distance from the light source to the illuminated object. If a light source measures 2000 lumens at 10 cm, express the light intensity as a function of the distance to the light source.

SECTION 4

Determine the x-intercepts of the following quadratic functions. Graph and label the intercepts.

49. $f(x) = x^2 + 4x - 24$ **50.** $f(x) = x^2 - 4x - 21$

51. $f(x) = x^2 + 7x - 3$ **52.** $f(x) = 2x^2 - 5x + 1$

53. $f(x) = 5x^2 + x - 2$ **54.** $f(x) = x^2 - \sqrt{101}$

For each of the following quadratics, (a) evaluate the discriminant, (b) indicate the nature and number of real roots.

55. $x^2 + 7x - 3$ **56.** $9x^2 - 12x + 4$

57. $x^2 - 49$ **58.** $x^2 - 3x + 28$

Solve for x.

59. $x^4 - 48x^2 - 49 = 0$ **60.** $x - 8x^{1/2} + 7 = 0$

61. $x^6 + 26x^3 - 27 = 0$ **62.** $x^8 - 17x^4 + 16 = 0$

63. $x^6 - 2x^3 + 1 = 0$ **64.** $x^{2/3} - x^{1/3} - 12 = 0$

Write quadratic functions with x-intercepts as follows.

65. $-\frac{2}{5}, 10$ **66.** $4 + \sqrt{11}, 4 - \sqrt{11}$

67. Torricelli's law for determining the height h in feet that water will rise in an L-shaped tube when placed in a stream of water moving at velocity V (ft/sec) is $V^2 = 64 h$. What velocity of water forces the water $\frac{3}{8}$ ft up the tube?

68. The power P (in Watts) in an electrical circuit is proportional to the square of the voltage E in the circuit. Determine the increase in power (and resulting increase in charges) if the power company increases the line voltage from 115 Volts to 130 Volts.

Construct the chart for the following linear functions.

69. $I(x) = -x$ **70.** $w(x) = 7x + 35$

71. $p(x) = 0.2x - 10$ **72.** $s(x) = \dfrac{5 - x}{3}$

Use charts to solve the following linear inequalities.

73. $4x - 15 < 0$ **74.** $3x + 15 < 0$

75. $3 - \frac{2}{3}x < 0$ **76.** $\dfrac{2x - 7}{5} \geq 0$

SECTION 5

Solve each of the following quadratic inequalities by use of charts.

77. $(x - 4)(x + 3) < 0$ **78.** $(7 - x)(5 + 3x) < 0$

79. $x^2 + 12 \geq 7x$ **80.** $x^2 \geq 49$

81. $x^2 - 10x + 16 < 0$ **82.** $3x^2 - x - 4 < 0$

83. $4 - 2x - x^2 \leq 0$ **84.** $6 + 4x - 2x^2 \leq 0$

85. $x^2 + 3x + 7 < 0$ **86.** $x^2 - 6x + 10 < 0$

87. $x^2 - 8x + 16 \leq 0$ **88.** $x^2 - 6x + 9 \geq 0$

SECTION 6

Solve for x.

89. $|x + 9| = 4$ **90.** $|x - 5| = 7$

91. $|3 - 9x| = 11$ **92.** $|7 - 4x| = 5$

93. $\left|\frac{1}{2}x - \frac{1}{3}\right| = \frac{5}{12}$ **94.** $\left|\frac{1}{2}x + \frac{1}{3}\right| = \frac{1}{4}$

95. $|8x - 7| = 0$ **96.** $|3x - 5| = -7$

97. $|3x - 1| = |2x + 3|$ **98.** $|7x - 2| = |5x + 1|$

99. $|3x + 7| + |5x - 1| = 0$ **100.** $|3x - 2| + |2x - 3| = 1$
101. $|x| - |x + 2| = 5$ **102.** $|x^2 - 2x - 2| = 1$
103. $|x^2 - 5x + 3| = 1$ **104.** $|x^2 - 3x + 1| = 5$

Graph each of the following pairs of functions.

105. $f(x) = 2x + 7$ $g(x) = |f(x)|$
106. $r(x) = x^2 - 7x + 12$ $g(x) = |r(x)|$
107. $p(x) = (x + 3)(x - 4)(x - 6)$ $g(x) = |p(x)|$
108. $r(x) = (x - 1)(x + 5)$ $q(x) = |(x - 1)(x + 5)|$
109. $m(x) = x^2 - 9$ $n(x) = |x^2 - 9| - 1$
110. $f(x) = 2x - 3$ $h(x) = |f(x)| - 4$
111. $f(x) = 8 - x^2$ $h(x) = 1 - |f(x)|$
112. $f(x) = x^2 - 5x + 2$ $g(x) = |f(x)|$

Solve each of the following. Express the answer in interval notation.

113. $|x - 11| < 4$ **114.** $|x + 7| < 5$
115. $|x - 8| \leq 3$ **116.** $|x + 15| \leq 0$
117. $|x - 7| < -1$ **118.** $|x + 7| > -3$
119. $|3x - 13| \leq 5$ **120.** $|5x - 11| > 4$
121. $|2 - 3x| > 7$ **122.** $|\frac{1}{2}x - 10| \leq \frac{1}{5}$

Construct a chart for each of the following functions to indicate where the function is positive, where negative and where 0.

123. $f(x) = |5x - 12|$ **124.** $f(x) = |12 - 5x|$
125. $f(x) = |3x + 2| - |x|$ **126.** $f(x) = |x + 5| - x$

Solve each of the following.

127. $|5x - 12| > 0$ **128.** $|12 - 5x| \leq 0$
129. $|3x + 2| - |x| \leq 0$ **130.** $|x + 5| \geq x$

SECTION 7

Round each of the following to 3 decimal places (the nearest thousandth).

131. 5.7185   **132.** 5.65636
133. 1.99197321   **134.** 9.1047345
135. 0.8949999   **136.** 0.9695
137. 0.4945003   **138.** 0.96950102

Approximate the x-intercepts using fixed point iteration.

139. $f(x) = x^2 - x - 1$ **140.** $f(x) = x^2 + 2x - 5$

SIDELIGHT

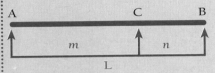

A   C   B

m n

L

FIGURE 3-79   Line segment AB, cut at C

The Golden Ratio

Geometry has two great treasures: one is the Theorem of Pythagoras; the other, the division of a line into extreme and mean ratio. The first we may compare to a measure of gold; the second we may name a precious jewel.
—JOHANNES KEPLER (1571–1630)

Imagine a line segment AB with length L (see Figure 3-79). Let C be a point on AB where the length from A to C is m and the length from C to B is n. If we locate C so that

$$\frac{L}{m} = \frac{m}{n}$$

then C is the golden cut and the ratio m/n is the golden ratio.

To the Greeks there was an aesthetically pleasing quality about this ratio. Many Greek buildings incorporated the ratio into the proportions of the structure. Many observers believe that many of the features of the Parthenon conform to the golden ratio (Display 3-2).

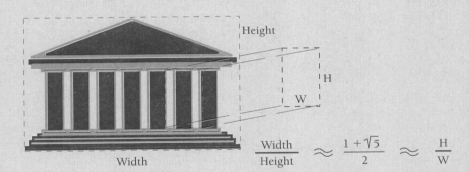

DISPLAY 3-2

$$\frac{\text{Width}}{\text{Height}} \approx \frac{1 + \sqrt{5}}{2} \approx \frac{\text{H}}{\text{W}}$$

The famous Greek sculptor Phidias made extensive use of the golden proportion. For that reason many have suggested that the ratio be represented by the Greek letter Φ (phi) in his honor.

What is the numerical value of Φ? Refer to Figure 3-79. Let $n = 1$ form the basis for the calculation. We need determine only the length of m. The total length L is $m + n$, so we have

$$\frac{m + 1}{m} = \frac{m}{1} \qquad\qquad (m \neq 0)$$

$$m + 1 = m^2$$

$$m^2 - m - 1 = 0$$

Because we are seeking the length of a line segment only the positive solution from the quadratic formula applies:

$$m = \frac{1 + \sqrt{5}}{2} \ (\cong 1.618033989)$$

Because $\Phi = \dfrac{m}{n}$, then $\Phi = \dfrac{1 + \sqrt{5}}{2}$.

In what manner is the golden ratio pleasing? Display 3-3 shows a house formed with proportions based on the "golden rectangle." Many designers find structures based on the ratio pleasing.

Even music is not immune to the aesthetics of the golden ratio. Experiments have shown that one of the more pleasing chords to the human ear is the Major sixth. The frequency ratio of this chord is 8:5 (approximately a golden ratio).

Some mathematicians do not believe that the golden ratio is so pervasive in nature and artifacts. Instead they attribute some of the "discoveries" of the golden ratio in nature to the subconscious desires of the observers. They point out, quite correctly, that many numerologist can find arithmetic patterns in most things. Collect enough measurements and perform enough arithmetic and eventually a result will be close to the pattern you seek.

Whether the golden ratio is a common pattern in nature or only a reflection of human imagination, the fact that many have sought the pattern indicates the inherent appeal of the golden ratio. The golden ratio Φ is an irrational number that has fascinated mathematicians with its beauty and its mysterious appearance in nature. In this respect Φ is not unlike the other more famous, but no less interesting, irrational numbers e and π.

33.2 ft

20 ft

DISPLAY 3-3

---

3 CHAPTER TEST

1. For the following linear function, determine the slope and y-intercept. Determine whether the graph is increasing, decreasing or neither. Sketch the graph of the function

$v(x) = \frac{1}{3}x + 2$

2. Write a rule of pairing for the linear function whose graph is through $(2, 5)$ and perpendicular to $f(x) = \frac{2}{3}x - 1$.

3. Sketch the graph of the following quadratic function. Label the coordinates of the vertex. Label the x-intercepts.

$f(x) = 3x^2 - 5x - 2$

4. Solve $2x^2 - 5x - 3 = 0$.

5. Use a chart to solve the linear inequality

$3x^2 - 5x - 2 < 0$.

6. Approximate the indicated x-intercept:

$f(x) = 2x^2 - 9, 2 < x < 3$.

7. Solve for x: $x^6 - 7x^3 - 8 = 0$.

Solve:

8. $|5 - 0.2x| < 0.1$

9. $|x + 3| = |2x - 5|$

10. $|x - 3| < |2x + 5|$

Polynomial Functions

This chapter generalizes constant, linear and quadratic functions to polynomial functions. Polynomial functions provide a unifying theme for some of our previous discussion. Polynomials also serve as useful tools in the development of other functions.

1 Synthetic Division and Radix Notation

Perhaps the best reason for regarding mathematics as an art is not so much that it affords an outlet for creative activity as that it provides spiritual values. It puts man in touch with the highest aspirations and loftiest goals. It offers intellectual delight and the exaltation of resolving the mysteries of the universe.
—Morris Kline

Have you observed a pattern that began with the extension of linear functions to quadratic functions? Note the addition of an x^2 term to a linear function produces a quadratic function. An obvious extension to these first- and second-degree functions is to use a third degree, x^3, term in the rule of pairing. Figure 4-1 illustrates the graphs of $y = x^n$ for $n = 1$ through 6. Note that all the even powered graphs are symmetric to the y-axis; the odd powered graphs are symmetric to the origin. Polynomials are a natural generalization from these patterns.

DEFINITION 4-1
Polynomial Function

A **polynomial function** P is a function given by

$$P(x) = a_n x^n + a_{n-1} x^{n-1} + \ldots + a_1 x^1 + a_0$$

where coefficients $a_0, a_1, \ldots, a_n, a_n \neq 0$, are real numbers, and n is a nonnegative integer called the *degree of P*. As a special case, $P(x) = 0$ is a polynomial without any degree.

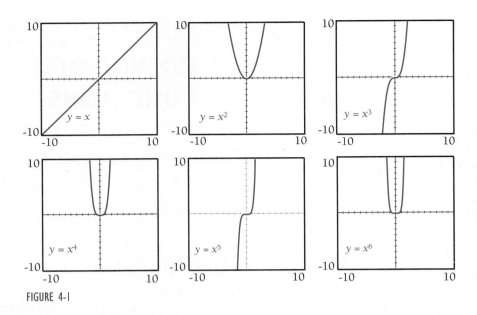

FIGURE 4-1

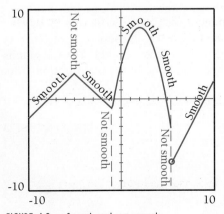

FIGURE 4-2   Smooth and not smooth

Because each polynomial consists of sums and products of x and constants, you can use the algebra of functions to construct a polynomial from the identity function $I(x) = x$ and constant functions. For example, the third-degree polynomial $P(x) = 2x^3 + x^2 - 5.7$ could be viewed as $P = 2I^3 + I^2 - 5.7$, where $I^3 = I(I)(I)$ and $I^2 = I(I)$ in the algebra of functions. Moreover, the algebra to construct a polynomial does not include division by a variable or fractional exponents on a variable; therefore the domain of a polynomial function is the set of real numbers.

Based on our experience with constant, linear and quadratic functions, we assume that the graphs of polynomial functions are continuous. From Figure 4-1, we also assume that the graphs of polynomial functions are smooth. By **smooth**, we mean a graph has no sharp points or sudden changes in direction. See Figure 4-2.

A function is well-behaved if its graph does not oscillate wildly between relatively "close" points of the graph. We expect the graphs of polynomials to be well-behaved. As a result, after plotting sufficient points of a polynomial, expect the graph to flow through the plotted points in a smooth, unbroken curve.

For a **polynomial** in standard form several characteristics are obvious. The first term of the polynomial is the term with the largest exponent. Each succeeding term is in descending order, sorted according to exponents. In standard form, the polynomial is simplified (according to GEMA from chapter 0). In this simplified form, if the polynomial has more than one term, then addition is the last indicated operation in an evaluation.

Many of the techniques of this chapter apply only to polynomials. Because recognition of a problem precedes selection of a tool to solve it, we must be adept at recognition.

EXAMPLE 4-1

Identification of Polynomial Functions. Identify each of the following as a polynomial or not. If a polynomial, simplify it to standard form and give the degree of the polynomial. Also indicate the number of terms for the simplified polynomial.

Illustration 1:

$$P(x) = (x - 3)(x + 5)$$

Solution:

$$P(x) = x^2 + 2x - 15$$

P is a second-degree polynomial with three terms.

Illustration 2:

$$P(x) = (x^2 - 5x^3) + 3(x - 2) - x^2$$

Solution:

$$P(x) = -5x^3 + 0x^2 + 3x - 6$$

P is a third-degree polynomial with four terms. We included and counted the $0x^2$ term only to indicate a relationship between the degree of a polynomial and the *expected* number of terms. For an nth degree polynomial, we expect $n + 1$ terms when the polynomial is written in standard form. If we do not count terms with a 0 coefficient the actual number of terms may be less than $n + 1$.

Illustration 3:

$$P(x) = (x^{10} - 3)^4(x^2 + 7)^2 + x^4$$

Solution: The arithmetic exceeds our space limitation and perhaps our ambition. Nonetheless, arithmetic should verify P as a polynomial of degree 44. We expect a term for each possible power of x (forty-four different ones) plus a constant term. Therefore the polynomial has a maximum of forty-five terms. Notice that the sum of two polynomials is a polynomial. More important, the product of two polynomials is a polynomial whose degree is the sum of the degrees of the polynomial factors.

Illustration 4:

$$P(x) = 5x - 3x^{2/3}$$

Solution: P is not a polynomial. The exponent of a variable in a polynomial must be a nonnegative integer. The exponent $\frac{2}{3}$ is not allowed.

Illustration 5:

$$P(x) = 3x - 5x^{-2}$$

Solution: P is not a polynomial. Polynomials do not have negative exponents.

Illustration 6:

$$P(x) = \frac{x + 3}{x - 5}$$

Solution: P is not a polynomial. Note that the divisor includes x (i.e., $x - 5$).

Illustration 7:

$$P(x) = 3x - \sqrt{x} + 7$$

Solution: P is not a polynomial. Note the presence of x inside the radical. The radical represents a fractional exponent.

Illustration 8:

$$f(x) = 17, \qquad g(x) = 0$$

Solution: The function f is a polynomial of degree 0, whereas g is a polynomial but has no degree.   ▬

From the preceding illustrations observe that addition (or subtraction) or multiplication of polynomials produces a polynomial. What of division? Polynomials resemble integers in structure and behavior. The sum or product of two integers is always an integer. Not so with division: $\frac{10}{2} = 5$, but $\frac{9}{4}$ is not an integer. The division of polynomials uses an algorithm similar to the division of counting numbers. The side effects in polynomial division have some interesting applications:

First let us review **division** with integers as a preview of division of polynomials. If P and $D \neq 0$ are counting numbers then the quotient Q and remainder R for P divided by D are whole numbers such that $P = DQ + R$ and $R < D$. For example, divide 31 by 4. The quotient is 7 and the remainder is 3. Therefore, we write $31 = 4(7) + 3$. The same process applies to **polynomial division**.

Theorem 4-1
Polynomial Division

> Suppose that $P(x)$ and $D(x) \neq 0$ are polynomials, then the **quotient** $Q(x)$ and **remainder** $R(x)$ for $P(x)$ divided by $D(x)$ are polynomials such that
>
> $$P(x) = D(x)\, Q(x) + R(x)$$
>
> and $R(x) = 0$, or the degree of $R(x)$ is less than the degree of $D(x)$.

Theorem 4-1 proclaims the existence of a quotient and remainder but provides no algorithm to them. Let us limit our discussion to linear divisors of the form $(x - r)$. For the divisor $(x - r)$, synthetic division provides a fast method for obtaining the quotient and remainder. For a first-degree divisor such as $(x - r)$, the degree of the remainder is smaller than first degree so that $R(x)$ is a constant. Hence, we write $R(x)$ as R.

Although synthetic division extends to higher degree divisors, the notation becomes clumsy. (See Problem Set 4-1.) Moreover most interesting applications only require division by a linear divisor.

Suppose that $P(x)$ is a polynomial:

$$P(x) = a_n x^n + a_{n-1}x^{n-1} + \ldots + a_1 x + a_0$$

Synthetic division is an abbreviated form of polynomial long division. To determine the quotient and remainder for $P(x)$ divided by $(x - r)$ proceed as follows:

1. Rewrite $P(x)$ in standard form. Be sure that all $n + 1$ terms are present. Use a 0 coefficient for each missing term.

2. Put the $n + 1$ coefficients of the polynomial in order according to the following. (Do not write the variable x. Position indicates degree.)

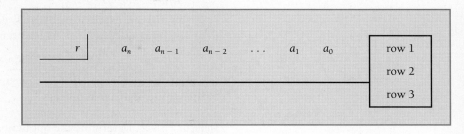

3. Bring a_n down into the first position below the line:

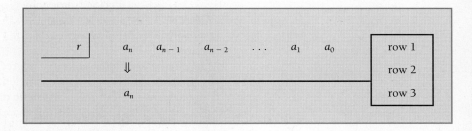

a_n is the first coefficient of the quotient. (This initializes the algorithm.)

4. Each succeeding coefficient of the quotient comes from repeating the following steps until reaching the a_0 column:

Multiply r times the last value entered in row 3.

Place the product in the next column on row 2.

Add the value from row 1 to the value just entered in row 2.

Place the sum in row 3.

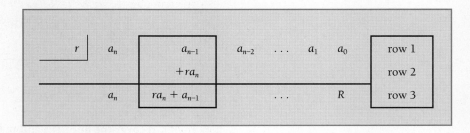

5. Because the divisor is first degree, the quotient is degree $n - 1$ and has one less term. To recover the quotient from the abbreviated format, begin with the first number in row 3 and use it as a coefficient for x^{n-1}.

Reduce the exponent by one (x^{n-2}) and use the next value from row 3 as its coefficient. Continue until reaching the constant term in the next to the last entry. The last entry in row 3 is not a coefficient of the quotient. The last entry is the remainder.

Now we can express the original polynomial in terms of the divisor quotient and remainder as given in Theorem 4-1.

$$P(x) = (x - r)(b_{n-1} x^{n-1} + b_{n-2}x^{n-2} + \ldots + b_1 + b_0) + R$$

where $b_{n-1} = a_n$, and $b_{n-2}, b_{n-3}, \ldots, b_0$ and R are the respective entries in row 3 of the synthetic division process.

EXAMPLE 4-2

Synthetic Division. Determine the quotient and remainder for

$$P(x) = 3x^4 - 20x^2 - 10x - 3$$

using the indicated divisor. Write the solution in the form

$$P(x) = (\text{divisor})(\text{quotient}) + \text{remainder}$$

Illustration 1:

Divide by $(x - 3)$. (Therefore, $r = 3$.)

Solution: The polynomial is fourth degree and must have five terms. The third-degree term is missing, so its coefficient is 0. For step 1, the polynomial is already in standard form. The synthetic division format becomes

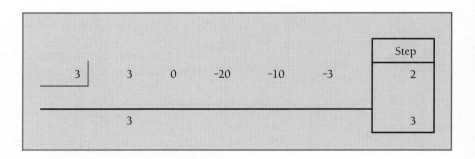

For step 5, recover the quotient and remainder from the last row and write in the required form:

$$P(x) = 3x^4 - 20x^2 - 10x - 3$$
$$= (x - 3)(3x^3 + 9x^2 + 7x + 11) + 30$$

Illustration 2:   Divide by $(x + 2) = [x - (-2)]$. (Therefore, $r = -2$.)

Solution:

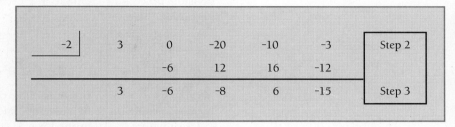

-2	3	0	-20	-10	-3	Step 2
		-6	12	16	-12	
	3	-6	-8	6	-15	Step 3

$$P(x) = 3x^4 - 20x^2 - 10x - 3$$
$$= (x + 2)(3x^3 - 6x^2 - 8x + 6) - 15$$

Note: To check these solutions, simplify the righthand side of the solution.

Illustration 3: $(8x^3 + 2) \div (2x + 1)$

Solution: Synthetic division does not apply directly. However, express the division as a rational expression and multiply the numerator and denominator by $\frac{1}{2}$ to obtain

$$\frac{8x^3 + 2}{2x + 1} \times \frac{\frac{1}{2}}{\frac{1}{2}} = \frac{4x^3 + 1}{x + \frac{1}{2}}$$

Now synthetic division applies:

$-\frac{1}{2}$	4	0	0	1	
		-2	1	$-\frac{1}{2}$	
	4	-2	1	$\frac{1}{2}$	Step 4

$$4x^3 + 1 = (x + \tfrac{1}{2})(4x^2 - 2x + 1) + \tfrac{1}{2}$$

This is not the original problem. To obtain the original problem, multiply both sides by 2:

$$2(4x^3 + 1) = 2(x + \tfrac{1}{2})(4x^2 - 2x + 1) + 2(\tfrac{1}{2})$$
$$8x^3 + 2 = (2x + 1)(4x^2 - 2x + 1) + 1$$

Note: The quotient remains unchanged. The essential difference between the original problem and the converted problem is the remainder. Multiply the remainder $\frac{1}{2}$ by 2 to obtain the remainder 1 for the original problem. The quotient is the same in both forms of the solution.

Consider $P(x) = 3x^4 - 20x^2 - 10x - 3$ once more. Let us evaluate the polynomial at 3. To determine $P(3)$ by direct substitution would be tedious, requiring exponentiation as well as multiplication and addition. The synthetic division process in Example 4-2 used only multiplication and addition. As promised, this side effect of division will be useful. Rewrite the polynomial as $P(x) = (x - 3)(3x^3 + 9x^2 + 7x + 11) + 30$ (from division by $(x - 3)$). Substituting 3 for x, obtain

$$P(3) = (3 - 3)[3(3)^3 + 9(3)^2 + 7(3) + 11] + 30$$

Because $(3 - 3) = 0$, the remaining calculation is not necessary. The right-hand side collapses to the remainder: $P(3) = 30$.

Evaluating $P(-2)$, in the form $P(x) = (x - 3)(3x^3 + 9x^2 + 7x + 11) + 30$ would not be helpful, because $(-2 - 3)$ as the first factor does not simplify the arithmetic. However, Illustration 2 of the previous example does offer some hope:

$$P(x) = (x + 2)(3x^3 - 6x^2 - 8x + 6) - 15$$
$$P(-2) = (-2 + 2)[3(-2)^3 - 6(-2)^2 - 8(-2) + 6] - 15$$
$$P(-2) = -15$$

Therefore, divide $P(x)$ by $(x - r)$ to obtain $P(x) = (x - r)Q(x) + R$, where $Q(x)$ is the quotient and R is the remainder. Evaluate for $x = r$ so that $P(r) = R$.

If $(x - r)$ divided into $P(x)$ has a remainder of 0, then $(x - r)$ is a factor of $P(x)$. Hence, $P(x) = (x - r)Q(x)$. Notice that $Q(x)$ is also a factor of $P(x)$. In general, a polynomial F is a factor of a polynomial P if for some polynomial Z with degree, $P = FZ$. These results are summarized in theorems in Section 4.3.

EXAMPLE 4-3

Determining Factors of Polynomials. Determine which of the following are factors of

$$P(x) = x^3 - 3x^2 - 13x + 15$$

Illustration 1:

$(x + 1)$

Solution:

-1	1	-3	-13	15	Note:
		-1	4	9	$r = -1$
	1	-4	-9	24	

$$P(x) = (x + 1)(x^2 - 4x - 9) + 24$$

Because the remainder is 24, not 0, then $(x + 1)$ is not a factor.

Illustration 2:

$(x + 3)$

Solution:

-3	1	-3	-13	15	Note:
		-3	18	-15	$r = -3$
	1	-6	5	0	

$$P(x) = (x + 3)(x^2 - 6x + 5) + 0$$

Because the remainder is 0, then $(x + 3)$ is a factor.

Note: The other linear factors of $P(x)$, if any, are the factors of $x^2 - 6x + 5$.

Illustration 3:

$$(x - 5)$$

Solution: From Illustration 2, the remaining factors of $P(x)$ are the factors of $x^2 - 6x + 5$. The resulting quotient used to determine the remaining factors is also known as the **depressed polynomial**.

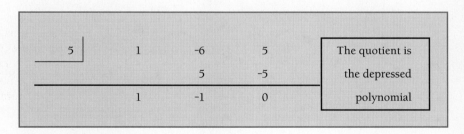

5	1	-6	5	The quotient is
		5	-5	the depressed
	1	-1	0	polynomial

$$x^2 - 6x + 5 = (x - 5)(x - 1)$$

Then
$$P(x) = (x + 3)(x^2 - 6x + 5)$$
$$= (x + 3)(x - 5)(x - 1)$$

The structure of integer arithmetic parallels arithmetic with polynomials. This is no accident. Our modern placeholder representation of an integer is a special kind of polynomial. The decimal system is a placeholder notation based upon the number 10. However, using a polynomial for base 10 notation for integers requires the use of special integer constants called **digits**, each of which has a value from 0 to 9. There are ten such digits in the decimal system: 0, 1, 2, 3, 4, 5, 6, 7, 8 and 9.

Notice that the decimal representation of three-hundred-forty-seven in polynomial form is given by

$$3(10^2) + 4(10^1) + 7 = 347$$

Base 10 is the usual base, so most people have no difficulty digesting the value of the base 10 representation of a number. But any integer greater than 1 could be a base for representing integers. Base 2 (binary), base 8 (octal) and base 16 (hexadecimal) are common for internal arithmetic in a computer. The base for the polynomial representation of a number is the representation's **radix**. The polynomial is the **radix** form of the number:

DEFINITION 4-2
Radix Notation

The radix form for a nonnegative integer is a polynomial evaluated at an integer $r > 1$, where the coefficients of the polynomial are nonnegative integer constants less than r. These coefficients are called *digits*.

The abbreviation of the radix form $ar^3 + br^2 + cr^1 + d$ is $abcd_r$, formed by "stringing" the digits and subscripting the radix. Where no base is subscripted, assume the radix to be 10. Represent negative integers as

$$-abcd_r = -1(abcd_r)$$

EXAMPLE 4-4

Radix Notation. Rewrite each of the following integers in radix notation. Express the value of the expression in base 10.

Illustration 1:

$$5312_8 \quad \text{(octal)}$$

Solution:

$$5312_8 = 5(8^3) + 3(8^2) + 1(8^1) + 2$$

Note: This is the polynomial $P(x) = 5x^3 + 3x^2 + 1x + 2$ evaluated at 8. Use synthetic division:

8	5	3	1	2	Note:
		40	344	2760	radix = 8
	5	43	345	2762	

Illustration 2:

$$5312_8 = 2762$$

Hexadecimal (base 16), requires sixteen digits. Customarily, the first ten digits are the same as base 10, then letters of the alphabet represent the remaining digits: 0, 1, 2, 3, 4, 5, 6, 7, 8, 9, A, B, C, D, E, F. Consider $2C3A_{16} = 2C3A_{\text{Hex}}$.

Solution: The decimal value of the digit C is 12; digit A has a value of 10.

$$2C3A_{16} = 2(16^3) + C(16^2) + 3(16^1) + A$$
$$= 2(16^3) + 12(16^2) + 3(16^1) + 10$$

16	2	12	3	10	Note:
		32	704	11312	radix = 16
	2	44	707	11322	

$$2C3A_{16} = 11322$$

Illustration 3: 100101_2 (binary)

Solution:

2	1	0	0	1	0	1	Note:
		2	4	8	18	36	radix = 2
	1	2	4	9	18	37	

$$100101_2 = 37$$

Illustration 4: 74_b for $b = 5, 10$ and 20

Solution: 74_5 is not valid: 7 is not a digit for base 5.

Solution:

$$74_{10} = 74$$ (obvious)

Solution:

$$74_{20} = 144$$ because

20		7	4	Note:
			140	radix = 20
		7	144	

In Example 4-4 the conversions to base 10 used multiplication and addition. To convert a base 10 representation to a different radix notation, this process is undone. The inverse process of multiply and add is subtract and divide:

Algorithm: Radix Conversion

To convert a base 10 representation of a number N to base b:

1. Divide N by b. Record the quotient and remainder.
2. Repeat
Divide the previous quotient by b
Record the new quotient and remainder until the new quotient is 0.

The remainders from last to first are the digits of N in radix b. If $b \geqslant 10$, substitute an appropriate letter to represent the digit.

EXAMPLE   4 - 5

Illustration:

Convert 54327 to the indicated radix.

Hexadecimal

Radix	Quotient	Remainder	Note
16	54327	Initialize	Divide by radix = 16
16	3395	7	$54327 = 16(3395) + 7$
16	212	3	$3395 = 16(212) + 3$
16	13	4	$212 = 16(13) + 4$
16	0	13	$13 = 16(0) + 13$

The usual choices for hexadecimal digits larger than 9 are A = 10, B = 11, C = 12, D = 13, E = 14, F = 15. Therefore use D as the leading digit and $54327 = D437_{16}$

S U M M A R Y

- A polynomial P is a function with rule of pairing expressible in the form $P(x) = a_n x^n + a_{n-1}x^{n-1} + \ldots + a_1 x^1 + a_0$, where $a_n \neq 0$, $n \in$ integer.
- If $n > 0$, then the degree of P is n. If $n = 0$, and $a_0 \neq 0$, then the degree of P is 0. If $n = 0$ and $a_0 = 0$, then the polynomial has no degree. The domain of a polynomial is all real numbers. The graph of P is smooth and continuous.

• Synthetic division provides a method for determining the quotient and remainder of an nth-degree polynomial and a first-degree polynomial. Synthetic division also provides a method of evaluating polynomials and converting bases in a placeholder numeral system.

4-1   EXERCISES

Simplify each of the following. Identify each as a polynomial or not. If a polynomial, convert to standard form and determine the degree and expected number of terms (counting zero coefficients).

1. $3(x - 5) + x^2$

2. $2(x + 7) - x^2$

3. $(x + 3)(x - 5)$

4. $(x - 3)(x + 8)$

5. $(x - 1)(x^2 + 6x - 2)$

6. $(x + 3)(x^2 - 5x + 1)$

7. $(x - 2)^3$

8. $(x + 3)^3$

9. $x^2 - \dfrac{2}{x} + 7$

10. $3x^3 - \dfrac{5}{x^2} + \sqrt{3}$

11. $3x - 5x^{-1} - \sqrt{2}$

12. $x^{-2} + x - 5$

13. $\sqrt{3}x - 1$

14. $5x - \sqrt{2}$

15. $\sqrt{x} + 3x - 1$

16. $5x^{-3} + \sqrt{x} + \sqrt{2}$

17. $3 - \sqrt[3]{x} - 7$

18. $\sqrt[4]{x} + 2 - 15x$

19. $(x - 1)^3 - (x + 1)^3$

20. $(x + 2)^3 - (x - 1)^3$

Determine the quotient and remainder for each of the following divisions. Express the answer in the form $P(x) = (x - r)Q(x) + R$.

21. $(x^2 + 2x + 3) \div (x + 1)$

22. $(x^2 - 2x + 3) \div (x - 1)$

23. $(x^2 - 5x + 7) \div (x - 2)$

24. $(3x^2 + 5x - 4) \div (x + 2)$

25. $(x^3 - 7x^2 + 2x - 3) \div (x + 5)$

26. $(x^3 + 4x^2 - 3x + 5) \div (x - 5)$

27. $(2x^3 + 3x - 5) \div (x - 3)$

28. $(4x^3 + 2x^2 - 7) \div (x + 3)$

29. $(5x^3 - 3x^2 + 6) \div (x + 4)$

30. $(x^3 - 2x + 3) \div (x - 4)$

31. $(2x^3 - 5x^2 + 3) \div (2x + 1)$

32. $(3x^3 - 7x^2 + 6) \div (3x + 1)$

33. $(x^3 - x^2 + 6) \div (3x - 2)$

34. $(x^3 - x^2 + 6) \div (2x - 3)$

Convert each of the following radix representations to base 10.

35. 45_8

36. 45_{16}

37. 45_{12}

38. 45_6

39. 21_5

40. 21_7

41. $A3_{16}$

42. $2C_{16}$

43. 10011_2

44. 1101_2

Convert the following base 10 integer to the indicated radix.

45. 53 $r = 8$

46. 53 $r = 16$

47. 53 $r = 14$

48. 53 $r = 5$

49. 201 $r = 16$

50. 201 $r = 8$

51. 23 $r = 2$

52. 15 $r = 2$

53. 32 $r = 2$

54. 32 $r = 8$

Determine which, if any, of $(x + 1)$, $(x - 1)$, $(x + 3)$ or $(x - 3)$ are factors of the following.

55. $3x - 9$

56. $5x + 5$

57. $x^3 - x^2 - 5x - 3$

58. $x^3 + x^2 - 5x + 3$

59. $x^3 + 3x^2 - x - 3$

60. $x^3 - 3x^2 - x + 3$

61. $x^5 - 3x^2 + 2$

62. $x^4 - 2x - 3$

63. $x^3 + 1$

64. $x^3 - 1$

65. $3x^3 + 2x^2 - 3x - 2$

66. $2x^3 - 5x^2 - 4x + 3$

67. Consider $P(b) = 5b + 3 = 53_b$. For what value b will tripling the base double the value of the expression? (Hint: $5(3b) + 3 = 2(5b + 3)$.)

68. Consider $34_b = 3b + 4$. For what value of b will $34_b = 2(34_5) - 1$?

69. Consider 15_b. For what radix b will reversing the digits to form 51_b quadruple the value of the expression?

70. For what value b will reversing the digits of 27_b (to form 72_b) double the value of the expression?

71. An open box is made from a 10 by 15 sheet of cardboard by cutting a square notch (x by x) at each corner, then folding up the sides. Express the volume V of the box as a function of the notch x. Graph the function. What values are acceptable for a volume? Examine the graph and determine the domain of the function.

72. The volume of a cylinder of radius r and height 10 is given by $V_c(x) = 10\pi r^2$. The volume of a sphere of radius r is given by $V_s(x) = \frac{4}{3}\pi r^3$. Suppose the cylin-

der is filled with water and a sphere of the same radius is submerged in the water. Model the volume of water V that remains in the cylinder as a function of r. Graph the function. If volume must be positive, what is the domain of function? Use the trace feature to estimate the largest volume of water and the value of r that produces it.

4-1   P R O B L E M   S E T

Synthetic division extends to division with higher degree divisors. The process is similar to division with a linear divisor. The process maintains the condition that the leading coefficient of the divisor be 1. A new step requires that additional terms be carried. Consider the following example: $(x^4 - 5x^3 + 6x^2 + 7x - 2) \div (x^2 + 2x - 1)$. Because x^2 in the divisor has a coefficient of 1, we may begin. Change the signs of the two remaining terms of the divisor and place in the usual format:

x	c						Carry row
-2	1	1	-5	6	7	-2	row 1
							row 2
							row 3

Be sure to save space above the top row to carry the product of the additional term in the divisor. Consider the usual starting process: bring down the 1 in the first column of the dividend. Then multiply 1 by each of the coefficients of the divisor. Note how -2(1) goes immediately to the second row of the next column. However, 1(1) is "carried" to the top of the third column. The -5 and -2 of the second row are immediately added for -7. But the 1 in the carry row of the third column waits until the next step.

x	c			1			Carry 1(1)
-2	1	1	-5	6	7	-2	
			-2				-2(1)
		1	-7				-5 + (-2)

The next step is a repetition of the previous step. Multiply each of -2 and 1 times the last number in the last row, -7. Place the product of -2 and -7 in the second row of the next column. Place the product of 1 and -7 in the carry row of column 4. Now add 14, 6 and the previously carried 1. Place the sum 21 in the last row of the third column. Delay adding the "carried" -7 until the next step.

x	c			1	-7		Carry 1(-7)
-2	1	1	-5	6	7	-2	
			-2	14			-2(-7)
		1	-7	21			1 + 6 + 14

Once more, multiply each of -2 and 1 times the last number in the third row, 21. Place the product of -2 and 21 in the second row. Place the product of 1 and 21 in the carry row of the last column. Now add -42, 7 and the carried -7 for the sum of -42 in column 4 of row 3. Because there are no columns beyond column 5, we add the numbers in that column to complete the division.

x	c				1	-7	21	Carry 1(21)
-2	1	1	-5	6	7	-2		
			-2	14	-42			-2(-42)
		1	-7	21	-42	19	-7 + 7 − 42 ; 21 − 2	

Because the divisor was second degree, the remainder must be first degree or less. The quotient is second degree. Then the last two columns of the process hold the coefficients of the remainder. This is the same number of columns as appear in the divisor of the synthetic division process. Now recover the quotient and remainder from the process and express the answer in a standard format.

$$x^4 - 5x^3 + 6x^2 + 7x - 2 =$$
$$(x^2 + 2x - 1)(x^2 - 7x + 21) + (-42x + 19)$$

Determine the quotient and remainder for each of the following.

1. $(x^4 - 3x^2 + x + 2) \div (x^2 - x + 1)$

2. $(x^4 - 3x^2 + x + 2) \div (x^2 + x - 1)$

3. $(x^3 + 1) \div (x^2 - x + 1)$

4. $(x^3 - 1) \div (x^2 + x + 1)$

5. $(x^4 - 16) \div (x^2 + 4)$

6. $(x^6 - 1) \div (x^2 + x + 1)$

7. $(x^6 - 1) \div (x^2 - x + 1)$

8. $(x^4 - 81) \div (x^2 + 9)$

9. Determine the quotient and remainder of $x^4 - 16$ divided by $x^2 - 4$. Compare this with the results of first dividing $x^4 - 16$ by $x + 2$ and then dividing the quotient by $x - 2$.

10. Determine the quotient and remainder of $x^4 - 81$ divided by $x^2 - 9$. Compare this with the results of first dividing $x^4 - 81$ by $x + 3$ and then dividing the quotient by $x - 3$.

PROBLEMS FOR GRAPHING TECHNOLOGY

Graph the following collections of functions. Use the same coordinate system for each collection. Note any similarities or differences among f, g and h in each case.

11. Discuss any patterns in each of the following collections.

(a) $f(x) = x^3$, $g(x) = f(x - 2)$, $h(x) = f(x) - 2$

(b) $f(x) = -x^4$, $g(x) = f(x + 4)$, $h(x) = f(x) + 4$

(c) $f(x) = x^3 - 4x^1$, $g(x) = f(x + 2)$, $h(x) = f(x) + 2$

(d) $f(x) = x^3$, $g(x) = f(x - 4)$, $h(x) = 2g(x) + 1$

(e) $f(x) = x^3 - x$, $g(x) = f(x + 3)$, $h(x) = 3 - \dfrac{g(x)}{4}$

12. Discuss any patterns in each of the following collections.

(a) $f(x) = x^4$, $g(x) = f(x - 3)$, $h(x) = f(x) - 3$

(b) $f(x) = -x^3$, $g(x) = f(x + 2)$, $h(x) = f(x) + 2$

(c) $f(x) = x^4 - 9x^2$, $g(x) = f(x - 3)$, $h(x) = f(x) - 3$

(d) $f(x) = x^3$, $g(x) = f(x + 3)$, $h(x) = \dfrac{g(x)}{3} - 2$

(e) $f(x) = x^3 + x$, $g(x) = (4 - x)$, $h(x) = 2 - g(x)$

13. Discuss any patterns in each of the following collections.

(a) $f(x) = x^3$, $g(x) = f(2x)$, $h(x) = 2f(x)$

(b) $f(x) = x^3$, $g(x) = \dfrac{x}{8}$, $h(x) = f(x) + g(x)$

(c) $f(x) = x^3$, $g(x) = \dfrac{x}{8}$, $h(x) = f[g(x)]$

(d) $f(x) = x^3$, $g(x) = \dfrac{x}{8}$, $h(x) = f(x)g(x)$

(e) $f(x) = x^3$, $g(x) = \dfrac{x}{8}$, $h(x) = \dfrac{f(x)}{g(x)}$

14. Discuss any patterns in each of the following collections.

(a) $f(x) = x^3$, $g(x) = f(3x)$, $h(x) = 3f(x)$

(b) $f(x) = x^3$, $g(x) = -\dfrac{x}{6}$, $h(x) = f(x) + g(x)$

(c) $f(x) = x^3$, $g(x) = -\dfrac{x}{6}$, $h(x) = f[g(x)]$

(d) $f(x) = x^3$, $g(x) = -\dfrac{x}{6}$, $h(x) = f(x)g(x)$

(e) $f(x) = x^3$, $g(x) = -\dfrac{x}{6}$, $h(x) = \dfrac{f(x)}{g(x)}$

EXPLORATION

The following problems anticipate concepts from the next section. Use graphing technology to produce the indicated graphs.

15. Graph $P(x) = 6x^3 - 11x^2 - 4x + 4$.

  (a) What is the degree of the polynomial?

  (b) How many x-intercepts does the polynomial have?

  (c) What is the smallest integer greater than or equal to all of the x-intercepts of the polynomial?

  (d) What is the largest integer less than or equal to all of the x-intercepts of the polynomial?

  (e) How many of the x-intercepts are positive?

  (f) How many of the x-intercepts are negative?

16. Compare the graph of P to the graph of $y = P(-x)$.

$$P(-x) = 6(-x)^3 - 11(-x)^2 - 4(-x) + 4.$$

How are the graphs similar? How different?

17. Compare the graph of P to the graph of $J(x) = P(x)/6$.

$$J(x) = x^3 - \frac{11}{6}x - \frac{2}{3}x + \frac{2}{3}$$

How are the graphs similar? How different?

18. (a) Verify that the roots of P are 2, $\frac{1}{2}$ and $-\frac{2}{3}$.

  (b) Which of the numerators of the roots divide the constant term 4?

  (c) Which of the denominators of the roots divide the coefficient of the cubic term 6?

<table>
<tr><td>2</td><td></td></tr>
</table>

CLUES FOR REAL ROOTS OF POLYNOMIAL

It is the man not the method that solves the problem.
—H. Maschke

Synthetic division provides a fast method for determining whether a given number is a root for a particular polynomial. With infinitely many numbers to choose from, randomly selecting a number to check is not a good strategy. In this section we explore methods for isolating possible roots of a given polynomial.

Recall that every first-degree function intersected the x-axis exactly once. Quadratic functions, on the other hand, may intersect the x-axis zero, one or two times. Because of our experience with linear and quadratic functions we expect that a polynomial of degree n may have as many as n roots.

For simplicity, we shall restrict our discussion to polynomials with integer coefficients at first. Later we expand the discussion to rational coefficients, real coefficients and optionally complex coefficients.

The first method for narrowing the choice of possible **polynomial roots** is Descartes's rule of signs. Descartes's rule requires some special terminology. If a polynomial is in standard form, for example $P(x) = 3x^5 - 4x^3 + 7x^2 + 2x - 5$, then a variation of sign occurs when two consecutive terms are of opposite signs. Do not consider zero coefficients.

Variation of signs
$$3x^5 - 4x^3 + 7x^2 + 2x - 5$$
⇑   ⇑⇑ ⇑   ⇑   ⇑

$P(x)$ has three variations of sign.

Theorem 4-2
Descartes's Rule of Signs

1. The number of positive roots of a real polynomial $P(x)$ either equals the number of variations of sign or is an even number less than the number of variations of sign.

2. The number of negative roots of a real polynomial $P(x)$ equals the number of variations of sign in $P(-x)$ or is less than that number by an even integer.

Note that $P(-x)$ transforms $P(x)$ by reversing the signs of all odd-degree terms while leaving the even-degree terms' signs unchanged. Graphically, the tranformation reflects the polynomial about the y-axis, converting the negative roots to positive roots so that the positive root principle of Descartes's rule applies. See Figure 4-3.

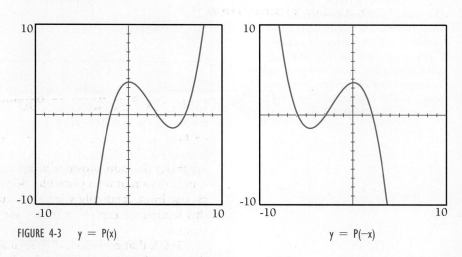

FIGURE 4-3 $y = P(x)$ $y = P(-x)$

EXAMPLE 4-6

Predicting the Number of Roots Using Descartes's Rule. Use Descartes's rule to predict the number of positive and negative roots of $P(x) = 3x^5 - 4x^3 + 7x^2 + 2x - 5$.

Solution: As previously noted there are three variation in signs. The number of positive real roots is either three or one. For negative roots, note that $P(-x) = -3x^5 + 4x^3 + 7x^2 - 2x - 5$ has two variations of signs. Then $P(x)$ has exactly two negative real roots or zero negative roots. Use graphing technology to verify that P has the predicted number of roots. See Figure 4-4. ■

Synthetic division not only checks possible roots, the process can offer additional clues to the alert user. For example, if a polynomial $P(x)$ is negative when evaluated at a and positive when evaluated at b, intuition about continuity suggest that $P(x)$ has a root between a and b. Because polynomials are well-behaved, a polynomial cannot change from a negative value to a positive value without passing through 0.

FIGURE 4-4

EXAMPLE 4-7

Finding Bounds on Roots. Determine an interval containing the roots of

$$P(x) = 2x^3 - 3x^2 + 2x - 3$$

From Descartes's rule of signs identify three variations of sign. $P(x)$ must have one or three positive roots. Because 0 is less than any positive root, 0 is a tentative endpoint for the interval sought. Try successive integers, 1,2,3, and so on until there is a change of sign in the remainder.

Clearly, $P(0) = -3$
Try 1:

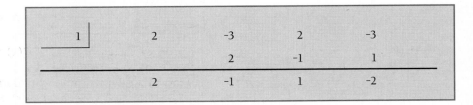

$P(1) = -2$

Because $P(0) = -3$ and $P(1) = -2$, the sign does not change. We have no indication of a root between 0 and 1.
    Try 2:

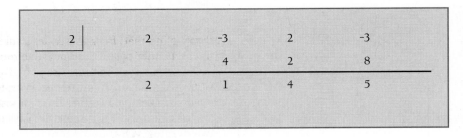

$P(2) = 5$

See Figure 4-5. $P(x)$ Changes sign from -2 to 5 on the interval from 1 to 2. There must be a root on $(1, 2)$. Substitute to verify that $P(1.5) = 0$. In fact, 1.5 is the only real root for $P(x)$. $P(x) = (x - 1.5)(x^2 + 2)$. Use graphing technology to verify that P has only one positive x-intercept.   ▬

Synthetic division can also provide upper and lower bounds for roots of a polynomial. These bounds can help focus our search for polynomial roots.

Calculus—Calculus is from the Latin meaning "a small stone used in reckoning" (from Greek *khalix*, "pebble"). Leibniz and Bernoulli debated both the name and the principal symbol for the integral calculus. Leibniz favored the name *calculus summatorius* and the elongated *s* as the symbol. Bernoulli preferred *calculus integralis* and the capital letter *I* for the sign. They compromised on elongated *s* and calculus integralis. In 1676 Leibniz introduced the term *calculus differentialis*. In 1684 he used *dx* for a differential.

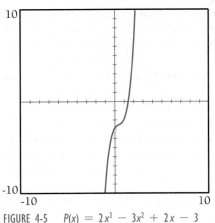

FIGURE 4-5 $P(x) = 2x^3 - 3x^2 + 2x - 3$

Theorem 4-3
Bounds for Polynomial Roots

Suppose $P(x) = a_n x^n + \ldots + a_1 x^1 + a_0, a_n \neq 0$ is a polynomial. If $P(x) = (x - t)Q(x) + R$, as determined by synthetic division, then

1. If $t > 0$ and neither R nor any coefficient of $Q(x)$ (row 3 of synthetic division) is negative, then every root r of $P(x)$ is less than t; in other words, t is an upper bound on the roots of $P(x)$.
2. If $t < 0$ and all the coefficients of $Q(x)$ and R alternate in sign (0 may be designated as +0 or -0 as appropriate), then every root r of $P(x)$ is greater than t; in other words, t is a lower bound on the roots of $P(x)$.

Proof:

1. Suppose the coefficients of $Q(x)$ and R are all nonnegative. Also, suppose that $t > 0$ and consider any $r > t$. Then $r - t > 0$. $Q(r) > 0$ because the sum and product of nonnegative numbers is nonnegative and $a_n > 0$. Now $P(r) = (r - t)Q(r) + R > 0$. Therefore, r cannot be a root of $P(x)$. No root of $P(x)$ is greater than t.
2. The second part of the theorem is similar to part 1, except that there are two cases to consider. In the first case, $Q(x)$ is of even degree. In the second case, $Q(x)$ is of odd degree. ▬

EXAMPLE 4-8

Determining Bounds for the Roots of a Polynomial. Determine upper and lower integer bounds for the roots of

$$P(x) = 2x^3 + x^2 - 14x - 7$$

Solution: Note that, from Descartes's rule of signs, there is exactly one positive root and either two or zero negative roots.

For the upper bounds, use graphing technology to graph the polynomial. If the window displays three intercepts, zoom in until the window just encloses the three intercepts, then choose and test integers close to Xmin and Xmax as bounds. If the graph displays two intercepts, zoom out until you discover a third. Why?

If only one intercept is displayed, then test integers close to Xmin and Xmax as boundaries. If these are boundaries, there is only one intercept and you may zoom in to establish closer bounds. If these are not bounds, then you must zoom out to discover whether there are more.

If no intercepts show, your window is misaligned. There *must* be at least one intercept. Either scroll the window (testing Xmin and Xmax could provide a clue for which way) or zoom out until you find at least one intercept. Then repeat the preceding steps.

Another strategy is to try the integers 1,2,3 and so on until the third row of the synthetic division process satisfies Theorem 4-3.

1	2	1	-14	-7
		2	3	-11
Failure	2	3	-11	-18

2	2	1	-14	-7
		4	10	-8
Failure	2	5	-4	-15

3	2	1	-14	-7
		6	21	21
Success	2	7	7	14

The third row is all positive, therefore 3 is an upper bound for the roots. But 3 is not the only upper bound, 4, 5, 6, and so on are also upper bounds. Try 2.9 and you will discover that 3 is not the smallest of all upper bounds; 3 is merely the smallest integral upper bound.

Similarly, to locate the lower bounds, try -1, -2, and so on until Theorem 4-3 is satisfied. This time the signs must alternate.

-1	2	1	-14	-7
		-2	1	13
Failure	2	-1	-13	6

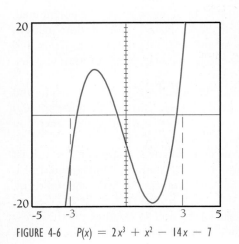

FIGURE 4-6 $P(x) = 2x^3 + x^2 - 14x - 7$

-2	2	1	-14	-7
		-4	6	16
Failure	2	-3	-8	9

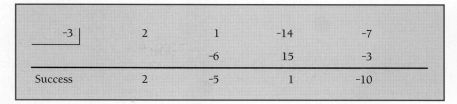

-3		2	1	-14	-7
			-6	15	-3
Success		2	-5	1	-10

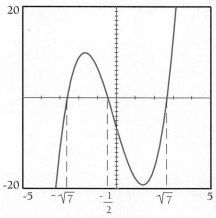

FIGURE 4-7 $P(x) = 2x^3 + x^2 - 14x - 7$

The signs in the third row alternate: -3 is a lower bound for the roots of $P(x)$. See Figure 4-6.

Summary: All the roots of $P(x)$ are in the interval (-3, 3). These bounds provide the width for a suitable window to display a graph of the function. The height of the window is not obvious. In quadratic functions determining the coordinates of the vertex by completing the square was easy. Unfortunately we have no simple algebraic procedure to determine similar points for general polynomials. Without calculus, predicting an appropriate window height for polynomials of degree greater than 2 is difficult.

Note: The roots of $P(x)$ are $-\frac{1}{2}$, $\sqrt{7}$ and $-\sqrt{7}$. See Figure 4-7.   ▬

S U M M A R Y

Suppose $P(x) = a_n x^n + \ldots + a_1 x + a_0$ is an nth degree polynomial.

* P can have at most n roots (x-intercepts).
* The graph P can change direction at most $n - 1$ times.
* The number of positive roots of P is the number of variations in sign of P or else is less than the number of variations in sign by an even number.
* The number of negative roots of P is the number of variations in sign of $P(-x)$ or else is less than the number of variations in sign by an even number.
* If b is a positive number and the third row of numbers in the synthetic division for $P(x)$ divided by $x - b$ consists of nonnegative numbers, then b is an upper bound on the roots of P.
* If b is a negative number and the third row of numbers in the synthetic division for $P(x)$ divided by $x - b$ alternates in sign, then b is a lower bound on the roots of P.

4-2 EXERCISES

For each of the following polynomials $P(x)$, determine (a) the number of variations of sign of $P(x)$, (b) the possible number of positive roots of $P(x)$, (c) the number of variations of sign of $P(-x)$, (d) the possible number of negative roots of $P(x)$.

1. $P(x) = x^3 + 6x - 5$
2. $P(x) = x^2 - 6x + 5$
3. $P(x) = x^3 + 7x^2 - 2x + 3$
4. $P(x) = x^3 - 7x^2 + 2x - 3$
5. $P(x) = 3x^4 - 1$
6. $P(x) = 3x^4 + x$
7. $P(x) = x^4 - x^2 + 5x + 2$
8. $P(x) = 2x^4 - x^3 - x - 1$
9. $P(x) = x^4 + x^2 + 2$
10. $P(x) = x^5 + x^3 + x$

Determine the smallest integer upper bound and the largest integer lower bound for the roots of the following polynomials.

11. $P(x) = x^3 - 8x + 3$
12. $P(x) = x^3 + 2x^2 - 6x - 4$
13. $P(x) = x^3 + 6x^2 + 10x + 4$
14. $P(x) = x^3 - 6x^2 + 10x - 3$
15. $P(x) = 4x^3 + 12x^2 - x - 3$
16. $P(x) = 9x^3 + 18x^2 - 4x - 8$
17. $P(x) = 3x^3 - 2x^2 + 3x - 2$
18. $P(x) = 3x^3 + 2x^2 + 6x + 4$
19. $P(x) = 9x^4 - 13x^2 + 4$
20. $P(x) = 2x^4 - 19x^2 + 9$

Prove the following.

21. Theorem 4-3, part 2, first case.
22. Theorem 4-3, part 2, second case.

4-2 PROBLEM SET

Consider the following theorem (offered without proof). Let $f(x) = a_n x^n + a_{n-1}x^{n-1} + \ldots + a_1 x^1 + a_0$, $a_n \neq 0$, be a polynomial of degree $n > 2$. If a_n, a_0 and the sum of the coefficients are odd, then $f(x)$ has no rational roots.

Example: $f(x) = 15x^4 - 3x^3 + 5x^2 + x + 845$

$a_n = 15$, and 15 is odd. $a_0 = 845$, and 845 is odd. $15 - 3 + 5 + 1 + 845 = 863$, and 863 is odd. f has no rational roots. We need not factor 845 and 15.

Example: $f(x) = 3x^3 - 5x^2 - 3x + 7$

$a_n = 3$ (odd). $a_0 = 7$ odd). $3 - 5 - 3 + 7 = 2$. 2 is even. The theorem does not apply. We must use other methods to determine whether f has any rational roots.

Use the preceding theorem to identify which of the following have no rational roots.

1. $f(x) = 5x^5 - 7x^3 + 6x^2 + 27$
2. $f(x) = 15x^4 + 7x^3 - 5x^2 + 31x + 105$
3. $f(x) = 13x^4 - 8x^3 + 16x^2 + 5x + 7$
4. $f(x) = 9x^9 - 10x^3 + 15x^2 + 24x + 39$

5. $f(x) = 2x^5 - 5x^3 + 2x^2 + 1$
6. $f(x) = 3x^4 - 7x^2 + 2x + 2$

PROBLEMS FOR TECHNOLOGY

Use graphing technology to determine the smallest width window with integral bounds that displays all of the x-intercepts of each of the following.

7. $P(x) = x^3 - 8x + 3$
8. $P(x) = x^3 + 2x^2 - 6x - 4$
9. $P(x) = x^3 + 6x^2 + 10x + 4$
10. $P(x) = x^3 - 6x^2 + 10x - 3$
11. $P(x) = 4x^3 + 12x^2 - x - 3$
12. $P(x) = 9x^3 + 18x^2 - 4x - 8$
13. $P(x) = 3x^3 - 2x^2 + 3x - 2$
14. $P(x) = 3x^3 + 2x^2 + 6x + 4$
15. $P(x) = 9x^4 - 13x^2 + 4$
16. $P(x) = 2x^4 - 19x^2 + 9$
17. $f(x) = 5x^5 - 7x^3 + 6x^2 + 27$
18. $f(x) = 15x^4 + 7x^3 - 5x^2 + 31x + 105$

EXPLORATION

The following problems anticipate topics from the next section. Use graphing technology to graph each of the following polynomials. In each case compare the x-intercepts to the factors of the polynomial.

19. $P(x) = x(x - 3)(x + 5)$

20. $P(x) = x(x + 2)(x - 7)$

21. $P(x) = (x - 5)(x + 7)(2x - 3)$

22. $P(x) = (3x - 2)(x + 6)(x - 4)$

23. $P(x) = (x - 2)(x + 3)^2$

24. $P(x) = (x - 5)^2(x + 4)$

25. $P(x) = (x - 3)^2(x + 5)^3$

26. $P(x) = (x + 2)^2(x - 6)^3$

27. Refer to Problems 19–22. Conjecture a relation between the factors of a polynomial and the roots of the polynomial.

28. Refer to Problems 23–26. Discuss the influence of squaring a factor on the graph of a polynomial. Discuss the effect of cubing a factor.

29. Graph $P(x) = x^3 - 6x^2 - 7x + 60$. Use the trace operation to determine the x-intercepts r_1, r_2, r_3 of the polynomial. Use this information to write $P(x)$ in factored form:

$$P(x) = (x - r_1)(x - r_2)(x - r_3)$$

30. See Problem 29. Write $P(x) = x^3 - 7x^2 + 8x + 16$ in factored form.

31. See Problem 29. Write $P(x) = x^3 - 3x^2 - 9x - 27$ in factored form.

32. See Problem 29. Write $P(x) = 2x^3 - 3x^2 - 32x - 15$ in factored form.

---

3   FACTORS OF POLYNOMIALS

. . . we cannot get more out of the mathematical mill than we put into it, though we may get it in a form infinitely more useful for our purpose.
—JOHN HOPKINSON

The component factors of a factored polynomial give us valuable information about the polynomial. From linear and quadratic equations we learned to expect an intimate relationship between the factors of a polynomial and its roots. The **root theorem** and **factor theorem** summarize the experience and consequences of the preceding sections.

Theorem 4-4
Root Theorem

If $P(x)$ is a polynomial of degree $n > 0$ and $x - r$ is a factor of $P(x)$, then r is a root of $P(x)$: $P(r) = 0$.

Proof: Because $x - r$ is a factor of $P(x)$, then $P(x) = (x - r)\ Q(x)$. Evaluate $P(r) = (r - r)Q(r) = 0$ to verify that r is a root. ■

Theorem 4-5
Factor Theorem

If $P(x)$ is a polynomial of degree $n > 0$ and r is a root of $P(x)$, then $x - r$ is a factor of $P(x)$: $P(x) = (x - r)Q(x)$, where $Q(x)$ is the polynomial quotient and the remainder is 0.

Proof: The proof is similar to Theorem 4-4 and is left as an exercise. ■

The preceding theorems suggest that whenever we find a linear factor of a polynomial, we also determine a root. Similarly, whenever a root of a polynomial becomes known, so does a factor. A more general result is known as the **Remainder Theorem**.

Theorem 4-6
Remainder Theorem

Suppose that $P(x)$ is a polynomial of degree $n > 0$. If $P(x) = (x - r)Q(x) + R$, where $Q(x)$ is the quotient and R is the remainder for $P(x)$ divided by $(x - r)$, then $P(r) = R$.

Proof: If $P(x) = (x - r)Q(x) + R$, then $P(r) = (r - r)Q(r) + R$; hence, $P(r) = R$.

Theorem 4-6 reinforces our experience from previous sections. Whenever we perform the synthetic division process the result includes not only a quotient and remainder but the polynomial evaluated at r.

Synthetic Division as Synthetic Substitution

EXAMPLE 4-9

(a) For each of the following determine the quotient and remainder.

(b) Evaluate the polynomial as indicated.

Illustration 1:

$$P(x) = x^3 - 3x^2 + 2x - 1$$

(a) Divide by $x - 2$.
(b) Evaluate $P(2)$.

Solution:

2	1	-3	2	-1
		2	-2	0
	1	-1	0	-1

(a) $P(x) = (x - 2)(x^2 - x + 0) - 1$
(b) $P(2) = -1$

Illustration 2:

$$P(x) = x^4 - 3x^3 + 2x^2 - 5$$

(a) Divide by $x + 3$.
(b) Evaluate $P(-3)$.

Solution:

-3	1	-3	2	0	-5
		-3	18	-60	180
	1	-6	20	-60	175

(a) $P(x) = (x + 3)(x^3 - 6x^2 + 20x - 60) + 175$
(b) $P(-3) = 175$

Illustration 3:

$$P(x) = x^3 - 2x^2 + 4$$

(a) Divide by $x + 4$.
(b) Evaluate $P(4)$.

Solution:

-4	1	-2	0	4
		-4	24	-96
	1	-6	24	-92

(a) $P(x) = (x + 4)(x^2 - 6x + 24) - 92$
(b) $P(-4) = -92$

However, we were to determine $P(4)$!

4	1	-2	0	4
		4	8	32
	1	2	8	36

(c) $P(4) = 36$

Illustration 4:

$$P(x) = x^3 - 7x^2 + 5x + 1$$

(a) Divide by $(x - 1)$.
(b) Evaluate $P(1)$.

Solution:

1	1	-7	5	1
		1	-6	-1
	1	-6	-1	0

(a) $P(x) = (x - 1)(x^2 - 6x - 1) + 0$; $(x - 1)$ is a factor of $P(x)$.

(b) $P(1) = 0$; 1 is a root of $P(x)$.

In Illustration 4 of Example 4-9, we determined a factor and, as a result, a root of a polynomial. If a polynomial is written as a product of linear factors, the associated roots are obvious. Even quadratic factors will quickly yield roots by the quadratic formula. All that remains is to develop a method for factoring polynomials. As in earlier algebra courses, it may be possible to factor a polynomial by trial and error if some strategy for formulating possible factors can be found.

Testing Possible Roots of a Polynomial

EXAMPLE 4-10

Illustration 1:   Determine whether 1, -1, 2, -2, 3 or -3 are roots of

$$P(x) = x^3 + x^2 - 9x - 9$$

Solution: Our method is trial and error:

1	1	1	-9	-9
Not		1	2	-7
root	1	2	-7	-16

-1	1	1	-9	-9
		-1	0	9
Success	1	0	-9	0

$$P(x) = (x + 1)(x^2 - 9)$$

Because $x^2 - 9 = (x + 3)(x - 3)$

then $P(x) = (x + 1)(x + 3)(x - 3)$

From the factors of $P(x)$ we read the roots as $\{-1, -3, 3\}$

Illustration 2: Write a polynomial with roots of 5, -3 and 17:

Solution: Because 5 is a root then $x - 5$ is a factor. Similarly, $(x + 3)$ and $(x - 17)$ are factors so that $P(x) = (x - 5)(x + 3)(x - 17)$ satisfies the specifications.

Note:
$$P_2(x) = 23(x - 5)(x + 3)(x - 17)$$

also meets the specifications, as does

$$P_3(x) = -5(x - 5)^2(x + 3)^4(x - 17)^{51}$$

What are the benefits of factoring a polynomial? Recall the charts developed in Chapter 2 for solving inequalities. The same charts are useful in sketching the graphs of polynomials. All that is necessary is to interpret a chart as follows.

Intercepts	$-\infty$		r_1		r_2		r_3		∞
$P(x)$	+	0		-	0		-	0	+
y related to the x-axis	Above	On		Below	On		Below	On	Above

EXAMPLE 4-11

Graphing Polynomials. Graph the following polynomials.

Illustration 1:
$$P(x) = (x + 3)(4 - x)(2x + 1)$$

Solution: The roots of $P(x)$ are $\{-3, 4, -\frac{1}{2}\}$

Chart

Intercepts	$-\infty$		-3		$-\frac{1}{2}$		4		∞	Note
$x + 3$	-		0	+		+		+		inc
$4 - x$	+			+		+		0	-	dec
$2x + 1$	-			-	0	+			+	inc
$P(x)$	+		0	-	0	+		0	-	
y to the x-axis	Above		On	Below	On	Above		On	Below	

This is graphed in Figure 4-8.

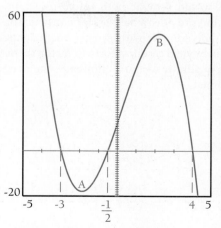

FIGURE 4-8 $P(x) = (x + 3)(4 - x)(2x + 1)$

Note: The *extreme points* of a function are the points where the graph changes direction. Extreme points and intercepts are among the most interesting points of a function. The extreme points of *P* are labeled *A* and *B*. Compare these extreme points to the vertex of a parabola. Accurately plotted extreme points would enhance the accuracy of any graph. Unfortunately, polynomials with degree larger than 2 require techniques of calculus to consistently locate extreme points. A good substitute is the graphing technology trace feature to estimate the coordinates of extreme points: $A \cong (-1.9, -18.2)$ and $B \cong (2.2, 50.5)$.

Illustration 2:

$$P(x) = (x - 1)^2(x^2 - 7)(x - 5)^3$$

Solution: A squared factor is never negative. On any interval, a cubed factor has the same sign as the same factor without the cube.

Intercepts	$-\infty$		$-\sqrt{7}$		1		$\sqrt{7}$		5		∞	Note
$(x - 1)^2$	+		+		0	+		+		+		squared
$x^2 - 7$	+	0	−			−	0	+		+		parabola
$x - 5$	−		−			−		−	0	+		cube
$P(x)$	−	0	+	0		+	0	−	0	+		
y to the *x*-axis	Below	On	Above	On		Above	On	Below	On	Above		

This is graphed in Figure 4-9.   ▬

All we lack is a method to acquire a suitable list of possible roots for a polynomial. Testing the possible roots would provide both actual roots and factors of the polynomial. The next section provides a method for constructing a list of all possible rational roots for polynomials with integer coefficients. Other kinds of polynomials may not easily yield to our efforts to factor them. Linear equations are easy to solve. The quadratic formula eliminates the need for trial and error in a quadratic polynomial.

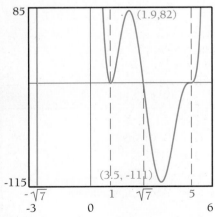

FIGURE 4-9
$P(x) = (x + 1)^2(x^2 - 7)(x - 5)^3$

SUMMARY

For the graphs of polynomials, the relation of factors to x-intercepts suggest the following:

- An nth degree polynomial intersects the x-axis at most n times. (The graph of a first-degree polynomial is a line that intersects the x-axis exactly once. The graph of a second-degree polynomial is a parabola that intersects the x-axis at most twice. Expect that a third-degree polynomial will intersect the x-axis no more than three times.)
- An nth degree polynomial changes direction at most $n - 1$ times. (The graph of a first-degree polynomial is a line that changes direction zero times. The graph of a second-degree polynomial is a parabola that changes direction one time. Expect the graph of a third-degree polynomial to change direction not more than twice.)
- Charts based on the factors of a polynomial show where the graph of the polynomial is above, below or on the x-axis. Graphing technology provides more detail for the graphs.

4-3 EXERCISES

Determine the roots for each of the following factored polynomials.

1. $P(x) = (x - 3)(x + 2)(x - 5)$
2. $P(x) = (x + 4)(x - 1)(x + 7)$
3. $P(x) = 5(x - 4)(x - 1)(2x + 3)$
4. $P(x) = 6(3x - 5)(x + 6)(x - 7)$
5. $P(x) = 3x(x - 4)(5x - 1)$
6. $P(x) = -2x(x + 3)(3 - 7x)$
7. $P(x) = -4x^2(5 - 3x)(x + 1)^2$
8. $P(x) = 5x^2(x - 1)^2(3x + 2)$
9. $P(x) = x^2(x^2 + 1)(x - 1)^2$
10. $P(x) = x^2(x^2 + 2)(x - 2)^2$

(a) Determine which, if any, of 1, -1, 2, -2, 3, -3 are roots of the following polynomials. (b) Consult the list of roots in part a to factor each polynomial.

11. $P(x) = x^3 - 2x^2 - 2x + 1$
12. $P(x) = x^3 - 2x^2 + 3x - 6$

13. $P(x) = x^3 + 2x^2 + 3x + 6$
14. $P(x) = 2x^3 - 3x^2 + 4x - 3$
15. $P(x) = x^3 - 7x + 6$
16. $P(x) = x^3 - 7x - 6$
17. $P(x) = x^4 - 1$
18. $P(x) = x^4 - 16$
19. $P(x) = x^3 + 8$
20. $P(x) = x^3 + 1$

Use synthetic division to evaluate each of the following polynomials for the value of r indicated.

21. $P(x) = x^3 + 2x^2 - 2x + 1, r = 2$
22. $P(x) = x^3 - x^2 + 3x - 1, r = 3$
23. $P(x) = 3x^3 - x^2 + 7x - 2, r = -3$
24. $P(x) = 2x^3 + x^2 - 5x + 1, r = -2$
25. $P(x) = x^4 + 3x + 1, r = 3$
26. $P(x) = x^4 - 3x^2 + 2, r = 2$
27. $P(x) = 2x^3 - 8x + 4, r = \frac{1}{2}$
28. $P(x) = 3x^3 - 9x^2 + 6, r = \frac{1}{3}$
29. $P(x) = x^3 - 5x^2 + 1, r = \sqrt{3}$
30. $P(x) = x^3 - 3x + 4, r = \sqrt{5}$

Use the chart method or technology to graph each of the following polynomials.

31. $P(x) = (x + 1)(x - 3)$

32. $P(x) = (x - 2)(x + 4)$

33. $P(x) = x^2(x - 1)$

34. $P(x) = x^3(x + 1)$

35. $P(x) = (x + 2)(2x - 1)x$

36. $P(x) = x(3x + 1)(x - 2)$

37. $P(x) = (x - 3)^2(x + 1)^3$

38. $P(x) = (x - 5)^3(x + 2)^2$

39. $P(x) = (x^2 - 5x + 1)(x + 1)$

40. $P(x) = (x^2 + 3x - 1)(x - 2)$

Write a third-degree polynomial that has as its roots the following.

41. 1, 2, 3

42. -1, -2, -3

43. 5, -2, 6

44. 3, 14, -5

45. 10, 0, -3

46. 0, -13, 25

47. $\pi, -e, \sqrt{13}$

48. $-\sqrt{19}, -\pi, \frac{2}{3}$

49. $-\frac{3}{4}, \frac{1}{2}, \frac{1}{2}$

50. $\frac{2}{3}, -\frac{1}{5}, -\frac{1}{5}$

 Graph the indicated line and polynomial on the same coordinate system.

51. $P(x) = (x - 1)(x - 2)(x + 1)$ with $f(x) = -1.89(x - 1)$. Note: the line f came from choosing a value close to the x-intercept 1. In this case we chose 1.1 for easy arithmetic. Partial substitution into $P(x)$ yields $f(x) = (x - 1)(1.1 - 2)(1.1 + 1)$. Therefore f is a line that approximates the polynomial near $x = 1$.

52. $P(x) = (x - 1)(x - 2)(x + 1)$ with $f(x) = 3.41(x - 2)$. Note: the line f came from choosing a value close to the x-intercept 2. In this case we chose 2.1 for easy arithmetic. Partial substitution into $P(x)$ yields $f(x) = (2.1 - 1)(x - 2)(2.1 + 1)$. Therefore f is a line that approximates the polynomial near $x = 2$.

53. If r is a root of a polynomial $P(x)$, prove r is a root of $cP(x)$ where c is any nonzero constant.

54. If r is a root of a polynomial $P(x)$ of degree 2 or greater, and t is a root of $Q(x)$ where $P(x) = (x - r)Q(x)$, prove t is a root of $P(x)$.

4-3 PROBLEM SET

Suppose that we have two distinct points of a function. Because two points determine a straight line, it would be easy to "fit" a linear function to these two points. In general, if we have n points of a function, we may "fit" an $(n - 1)$-degree polynomial to the points. If the n points are all on the x-axis, then the factor theorem provides an easy method to write a polynomial through the points. Multiply the resulting polynomial by a nonzero constant, and you will observe that the polynomial through these intercepts is not unique. Indeed, the simplest function containing any given number of points on the x-axis is given by $y = 0$.

If the n points are not all on the x-axis, we have more difficulty in writing a rule of pairing for a function to include the points. Several methods exist for producing smooth curves through the given points. One method to construct a curve through n given points is by use of a spline. (A spline is a flexible piece of wood, rubber or hard metal used in drawing curves. Splines are particularly useful in art and drafting for fitting a curve to some given points.) Another method is to construct a polynomial function that passes through the given points. Given the points $\{(x_1, y_1), (x_2, y_2), \ldots, (x_n, y_n)\}$ then a curve through these n points is given by the Lagrange interpolation polynomial:

$$P_n(x) = y_1 \frac{(x - x_2)(x - x_3) \ldots (x - x_n)}{(x_1 - x_2)(x_1 - x_3) \ldots (x_1 - x_n)} +$$

$$y_2 \frac{(x - x_1)(x - x_3) \ldots (x - x_n)}{(x_2 - x_1)(x_2 - x_3) \ldots (x_2 - x_n)} + \ldots +$$

$$y_n \frac{(x - x_1) \ldots (x - x_{n-1})}{(x_n - x_1)(x_n - x_2) \ldots (x_n - x_{n-1})}$$

For example, you would expect that a quadratic function would fit three noncollinear points such as $\{(1, 2), (3, 4), (-1, 5)\}$. Substitute to obtain the polynomial function:

$$P_3(x) = 2\frac{(x - 3)(x + 1)}{(1 - 3)(1 + 1)} + 4\frac{(x - 1)(x + 1)}{(3 - 1)(3 + 1)} +$$

$$5\frac{(x - 1)(x - 3)}{(-1 - 1)(-1 - 3)}$$

$$P_3(x) = -\tfrac{1}{2}(x-3)(x+1) + \tfrac{1}{2}(x-1)(x+1) + \tfrac{5}{8}(x-1)(x-3)$$

Notice that $P_3(x)$ is a second-degree polynomial.

1. (a) Write a polynomial function $P_2(x)$ for the points $\{(-1, 3), (4, 2)\}$.

(b) Verify that $P_2(-1) = 3$ and $P_2(4) = 2$.

2. (a) Write a polynomial function $P_2(x)$ for the points $\{-3, 2), (2, 5)\}$.

(b) Verify that $P_2(-3) = 2$ and $P_2(2) = 5$.

3. (a) Write a polynomial function $P_3(x)$ for the points $\{(1,2), (0, 0), (-2, 5)\}$.

(b) Verify that $P_3(-2) = 5$.

4. (a) Write a polynomial function $P_3(x)$ for the points $\{(-1, -1), (0, 5), (3, -2)\}$.

(b) Verify that $P_3(3) = -2$.

5. (a) Attempt to write a polynomial function $P_3(x)$ for $\{(1, 1), (2, 2), (3, 3)\}$

(b) Simplify the rule of pairing in part a.

(c) Graph P_3 with the three points. What do you notice?

6. (a) Attempt to write a polynomial function $P_3(x)$ for $\{(1, 1), (1, 2), (1, 3)\}$.

(b) Simplify the rule of pairing in part a. What problems do you encounter?

PROBLEMS FOR TECHNOLOGY

Graph each of the following collections of polynomials in the same window.

7. Describe the relationship between the functions f, g, h and P.

(a) $f(x) = x - 3$, $g(x) = x + 5$, $h(x) = x + 7$, $P(x) = f(x)g(x)h(x)$.

(b) $f(x) = x + 6$, $g(x) = x - 4$, $h(x) = x - 8$, $P(x) = f(x)g(x)h(x)$.

(c) $f(x) = 2x - 1$, $g(x) = 3x + 17$, $h(x) = x$, $P(x) = f(x)g(x)h(x)$.

(d) $f(x) = 3x - 7$, $g(x) = -x$, $h(x) = 4 - x$, $P(x) = f(x)g(x)h(x)$.

(e) $f(x) = x^2 - 5$, $g(x) = -x$, $h(x) = 3 - x$, $P(x) = f(x)g(x)h(x)$.

8. Describe the relationship between the functions f, g and P.

(a) $f(x) = x - 2$, $g(x) = 5 - x$, $P(x) = [f(x)]^2 g(x)$

(b) $f(x) = x - 2$, $g(x) = 5 - x$, $P(x) = [f(x)]^3 g(x)$

(c) $f(x) = 4 - x$, $g(x) = x$, $P(x) = [f(x)]^2 [g(x)]^3$

(d) $f(x) = 2$, $g(x) = (x - 1)$, $P(x) = f(x)[g(x)]^2$

(e) $f(x) = x^2 - 7$, $g(x) = 1 - x$, $h(x) = 5 - x$, $P(x) = f(x)g(x)h(x)$.

4   RATIONAL ROOT THEOREM

The solution of such questions as these [referring to the solution of cubic equations] depends on correct judgement, aided by the assistance of God.
—GANITA AIJA

Now that we have an efficient method for checking possible roots of a polynomial, we must devise a method for selecting likely candidates for the roots. One method you may have used for quadratic polynomials is trial and error. For example, in $3x^2 - 5x + 2$, clues for the factors—and therefore ultimately for the roots—come from the 3 and the 2. For higher degree polynomials, the rational root theorem provides similar candidates under special circumstances.

Theorem 4-7
Rational Root Theorem

Suppose that P is a polynomial,

$$P(x) = a_n x^n + \ldots + a_1 x + a_0,$$

where P is of degree n and $a_0, a_1 \ldots a_n$ are integers.
If P has any nonzero rational roots (reduced to lowest terms), the roots are of the form N/D, where

1. D divides a_n (D is a factor of a_n, $D \neq 0$)
2. N divides a_0 (N is a factor of a_0, $N \neq 0$).

Proof: Suppose that N/D is a rational root of P; $N \neq 0$ and $D \neq 0$ are integers with N/D reduced to lowest terms. Substituting N/D for x in the polynomial results in 0:

$$a_n \left(\frac{N}{D}\right)^n + a_{n-1}\left(\frac{N}{D}\right)^{n-1} + \ldots + a_1\left(\frac{N}{D}\right) + a_0 = 0$$

The least common denominator is D^n Multiply by D^n to obtain

$$a_n N^n + a_{n-1} N^{n-1} D + \ldots + a_1 N D^{n-1} + a_0 D^n = 0$$
$$a_n N^n + a_{n-1} N^{n-1} D + \ldots + a_1 N D^{n-1} = -a_0 D^n$$

Divide by N:

$$a_n N^{n-1} + a_{n-1} N^{n-2} D + \ldots + a_1 D^{n-1} = -a_0 \frac{D^n}{N}$$

The lefthand side of the equation is the sum and product of integers. The lefthand side is an integer. But this implies that N exactly divides $-a_0 D^n$. Because N/D was reduced to lowest terms, no positive integer except 1 divides both N and D, N and D have no other factors in common. As a consequence, N must divide a_0. The proof that D divides a_n is similar to the preceding and is left as an exercise.   ■

Notes on the Rational Root Theorem

1. The rational root theorem describes the characteristics of only the rational roots of a polynomial with integer coefficients; some polynomials may have no rational roots.
2. The coefficients of the polynomial must be integers, or the theorem does not apply.
3. If the roots of a polynomial are not rational, we have no general method of discovering their exact value.

Procedure for Using
the Rational Root Theorem

To determine the rational roots of a polynomial P with integer coefficients

$$P(x) = a_n x^n + \ldots + a_1 x + a_0$$

1. List all the positive factors of the leading coefficient a_n. These are the candidates for the denominators of the possible rational roots.
2. List all positive and negative integer factors of the constant term a_0. These are the candidates for the numerators of the rational roots.
3. List possible rational roots by forming all combinations of numerators and denominators from steps 1 and 2.
4. Test the rational root candidates one at a time using synthetic division. Eliminate from the list those that fail. Descartes's rule and other hints from Section 4.2 help to trim the list.
5. Whenever you discover a rational root r, list it as a root and $(x - r)$ as a factor. Do all subsequent tests on the quotient polynomial from the synthetic division process.
6. Repeat steps 1 through 5 for each quotient polynomial until you exhaust the rational root candidates or until the quotient polynomial is a quadratic. Because a root may come from a repeated factor, be sure to retry a discovered root in the quotient polynomial before moving to the next candidate.
7. If the quotient polynomial is a quadratic, apply the quadratic formula to determine the remaining roots. The quadratic formula also determines the remaining roots when they are not rational.
8. Because each root has a corresponding linear factor and the maximum number of linear factors for an nth-degree polynomial is n, the maximum number of roots is n.

EXAMPLE 4-12

Application of the Rational Root Theorem. Determine the rational roots of the following. Where possible determine all the real roots.

Illustration 1:

$$P(x) = 3x^3 - 5x^2 - 2x + 4$$

Solution: P exhibits two variation of signs. By Descartes's rule P will have either two positive roots or zero positive roots. Now apply the rational root theorem.

Factors of 4: 1, -1, 2, -2, 4, -4.   Factors of 3: 1, 3.

Possible rational roots: 1, -1, 2, -2, 4, -4, $\frac{1}{3}$, $-\frac{1}{3}$, $\frac{2}{3}$, $-\frac{2}{3}$, $\frac{4}{3}$, $-\frac{4}{3}$.

Test the possible rational roots:

1	3	-5	-2	4
		3	-2	-4
Success!	3	-2	-4	0

The first root is 1. The quotient polynomial is $3x^2 - 2x - 4$. Apply the quadratic formula:

$$\frac{2 \pm \sqrt{4 - 4(3)(-4)}}{2(3)} = \frac{2 \pm \sqrt{52}}{6} = \frac{1 \pm \sqrt{13}}{3}.$$

Answer:

$$x \in \left\{ 1, \frac{1 + \sqrt{13}}{3}, \frac{1 - \sqrt{13}}{3} \right\}.$$

Note that exactly two of the roots are positive. In factored form,

$$P(x) = 3(x - 1)\left[x - \frac{1 - \sqrt{13}}{3} \right]\left[x - \frac{1 + \sqrt{13}}{3} \right]$$

Illustration 2:

$$P(x) = 3x^4 + 5x^3 - x^2 - 5x - 2$$

Solution: The possible rational roots are $1, -1, 2, -2, \frac{1}{3}, -\frac{1}{3}, \frac{2}{3}, -\frac{2}{3}$.

$P(x)$ has one variation of sign. By Descartes's rule of signs the polynomial has one positive root. Try the possible positive roots.

2	3	5	-1	-5	-2
		6	22	42	74
Failure	3	11	21	37	72

Note the positive third row: 2 is an upper bound on the roots.

1	3	5	-1	-5	-2
		3	8	7	2
Success!	3	8	7	2	0

Now $P(x) = (x - 1)(3x^3 + 8x^2 + 7x + 2)$.

The rest of the analysis uses the quotient polynomial.

$$Q(x) = 3x^3 + 8x^2 + 7x + 2$$

The leading coefficient of $Q(x)$ is 3. Because $Q(x)$ does not vary signs there are no positive roots. Try negative roots.

-2	3	8	7	2
		-6	-4	-6
Failure	3	2	3	-4

-1	3	8	7	2
		-3	-5	-2
Success!	3	5	2	0

$$Q(x) = (x + 1)(3x^2 + 5x + 2)$$

So that $P(x) = (x - 1)(x + 1)(3x^2 + 5x + 2)$

The new quotient polynomial is a quadratic

$$Q(x) = 3x^2 + 5x + 2$$

The quadratic formula provides the roots of Q: $-1, -\frac{2}{3}$. It should not surprise you that -1 and $-\frac{2}{3}$ are among the possible rational roots previously listed. Therefore, $Q(x) = 3(x + 1)(x + \frac{2}{3})$.

The coefficient of 3 is necessary to restore the leading coefficient of both $Q(x)$ and the original polynomial. Now,

$$P(x) = 3(x + 1)(x + 1)(x - 1)(x + \tfrac{2}{3})$$

$$= 3(x + 1)^2(x - 1)(x + \tfrac{2}{3})$$

The roots of P are $-1, 1, -\frac{2}{3}$. Note that the root -1 comes from a repeated factor. Because the factor that produced -1 occurs twice, we say that -1 is a root of multiplicity 2.

Illustration 3:

$$P(x) = x^3 - \tfrac{1}{2}x^2 - 2x + 1$$

Solution: Because the polynomial has a noninteger coefficient, the rational root theorem does not apply. However, the roots of $P(x)$ are the same as the roots of $2P(x)$.

$$2P(x) = 2x^3 - x^2 - 4x + 2$$

The possible rational roots are 1, -1, 2, -2, $\frac{1}{2}$, $-\frac{1}{2}$. Try $\frac{1}{2}$.

$\frac{1}{2}$	2	-1	-4	2
		1	0	-2
Success!	2	0	-4	0

The quotient polynomial is $Q(x) = 2x^2 - 4$. From the quadratic formula the remaining roots are irrational: $x \in \{\frac{1}{2}, \sqrt{2}, -\sqrt{2}\}$. Now, $P(x) = (x - \frac{1}{2})(x - \sqrt{2})(x + \sqrt{2})$.

Illustration 4:

$$P(x) = 2x^3 + \tfrac{1}{2}x^2 - \tfrac{1}{2}x + 2$$

Solution: Because the polynomial has non-integer coefficients, the rational root theorem does not apply. However, the roots of $P(x)$ are the same as the roots of

$$4x^3 + x^2 - x + 4 \qquad\qquad \text{(multiply by 2)}$$

The possible rational roots are 1, -1, 2, -2, 4, -4, $\frac{1}{2}$, $-\frac{1}{2}$, $\frac{1}{4}$, $-\frac{1}{4}$. Try all these. None will work.

Answer: $P(x)$ has no rational roots. In Section 4-6 we show that every odd-degree polynomial with real coefficients has at least one x-intercept. Because P is third degree, $P(x)$ has at least one real root. This root(s) cannot be rational. In the next section we shall see how the zoom and trace features of graphing technology can help us estimate such irrational roots.

Illustration 5:

Prove that $\sqrt[3]{5}$ is not a rational number, that is, $\sqrt[3]{5}$ is irrational.

Proof: Use substitution to verify that $\sqrt[3]{5}$ is a root of $x^3 - 5$. But the possible rational roots of $x^3 - 5$ are 1, -1, 5, -5. Use substitution to verify that none of these possible rational roots is an actual root. Then $x^3 - 5$ has no rational roots. Because $\sqrt[3]{5}$ is a root, it must be irrational.   ▬

Now we use the factored polynomials to sketch some graphs.

Graphing Polynomials. Graph the following polynomials.

EXAMPLE 4-13

Illustration 1:

$$P(x) = x^3 - x$$

Solution: Notice that P is odd. Therefore P is symmetric to the origin. One nice feature of polynomials is that the exponents immediately reveal whether the polynomial is odd or even. We need plot points only on the positive x-axis to visualize the complete graph. First factor,

$$P(x) = x^3 - x$$
$$= x(x^2 - 1)$$
$$= x(x + 1)(x - 1)$$

Form a chart for the factors.

Intercepts	$-\infty$		-1		0		1		∞	Note
x	$-$		$-$	0	$+$			$+$		inc
$x + 1$	$-$	0	$+$		$+$			$+$		inc
$x - 1$	$-$		$-$		$-$		0	$+$		inc
$P(x)$	$-$	0	$+$	0	$-$		0	$+$		
y to the x-axis	Below	On	Above	On	Below		On	Above		

See Figure 4-10. Notice the symmetry to the origin. P intersects the x-axis three times and changes direction twice. Unfortunately, the location of the extreme points remains a mystery. Graphing technology can help us estimate them.

Our insight into the graph of P would improve if the relative "steepness" of the graph at each intercept were known. For first-degree polynomials, $f(x) = mx + b$, the graph is a straight line for which the slope m measures the steepness between any two points of the line. Recall the slope formula given as a difference quotient in function notation (Figure 4-11):

$$m = \frac{f(b) - f(a)}{b - a}$$

For polynomials of degree greater than 1, the problem becomes more difficult. The slope between two points is not constant. Moreover, because we want the slope at an intercept, we have only one point, not two, with which to work. See Figure 4-12. P intercepts the x-axis at $(-1, 0)$. Therefore, $P(-1) = 0$.

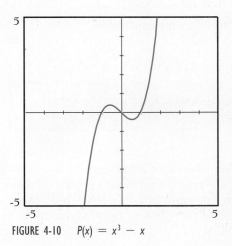

FIGURE 4-10 $P(x) = x^3 - x$

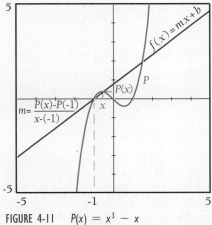

FIGURE 4-11 $P(x) = x^3 - x$

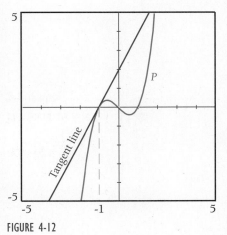

FIGURE 4-12

Now, form the difference quotient:

$$\frac{P(x) - P(-1)}{x - (-1)} = \frac{P(x) - 0}{x + 1}$$

$$= \frac{P(x)}{x + 1}$$

$$= \frac{x(x - 1)(x + 1)}{(x + 1)}$$

$$= x(x - 1)$$

The quotient of $P(x)/(x + 1)$ is $Q(x) = x(x - 1)$. Not only is Q the quotient, $Q(-1)$ provides the slope of the tangent line to $P(x) = x^3 - x$ at the point $(-1, 0)$. Because $Q(-1) = 2$, the slope of the tangent line to P at $(-1, 0)$ is 2. From the point slope formula for a line, the equation of the tangent line with slope 2 through $(-1, 0)$ is given by

$$y = 2[x - (-1)] + 0$$
$$y = 2(x + 1)$$

Use graphing technology to graph this line in the same window as P.

More generally, consider a polynomial P with a factor of $x - r$, $P(x) = (x - r)Q(x)$. Because $x - r$ is a factor, then r is a root and $P(r) = 0$. Therefore,

$$P(x) - P(r) = (x - r)Q(x)$$

$$\frac{P(x) - P(r)}{x - r} = Q(x)$$

However, the lefthand side of this equation is the diffence quotient. If x is "close to" the intercept r, then $Q(r)$ is approximately the slope of the graph of P at r. Because slope is a property of lines, we refer to the measure of the steepness of the polynomial at its intercept as the slope of the line tangent to the polynomial at its intercept. ▄

This line of reasoning leads us to the following conclusion.

Theorem 4-8	The slope of the tangent line to a polynomial P at the root r is given
Slopes of Tangents at Intercepts	by $m = Q(r)$, where $Q(x)$ is the quotient of $P(x)$ divided by $(x - r)$.

The techniques of algebra provide the slope of a line tangent to a polynomial at the polynomial's intercepts. The calculus provides a general method to determine slopes of tangent lines at any point of a function. Thus, calculus allows us to measure the steepness of the graph at most any point of a function. See Problem Set 4-4, problems 31-44.

Using Tangent Lines to Graph Polynomials. Graph the following polynomial.

$$P(x) = 3x^4 + 5x^3 - x^2 - 5x - 2$$

Solution: From Example 4-12, Illustration 2, factor $P(x)$ into $P(x) = 3(x + 1)^2(x - 1)(x + \frac{2}{3})$. The chart follows.

Intercepts	$-\infty$		-1		$\frac{-2}{3}$		1	∞	Note
$(x + 1)^2$		$+$	0	$+$		$+$		$+$	Squared
$x - 1$		$-$		$-$		$-$	0	$+$	inc
$x + \frac{2}{3}$		$-$		$-$	0	$+$		$+$	inc
$P(x)$		$+$	0	$+$	0	$-$	0	$+$	
y to the x-axis		Above	On	Above	On	Below	On	Above	

Theorem 4-8 provides the slope of the tangent line at each x-intercept.

Factor: $(x - r)$	x-intercept: $(r, 0)$	$Q(x) = P(x)/(x - r)$	$Q(r)$: slope of tangent line	Equation of tangent line
$x + 1$	$(-1, 0)$	$3(x + 1)(x - 1)(x + \frac{2}{3})$	0	$y = 0(x + 1)$
$x + \frac{2}{3}$	$(-\frac{2}{3}, 0)$	$3(x + 1)^2(x - 1)$	$-\frac{5}{9}$	$y = -\frac{5}{9}(x + \frac{2}{3})$
$x - 1$	$(1, 0)$	$3(x + 1)^2(x + \frac{2}{3})$	20	$y = 20(x - 1)$

Figure 4-13 shows the graph of each of the tangent lines. With the guidance of the chart and these tangent lines, graph P (Figure 4-14). Graphing technology should confirm the relation of P to the tangent lines.

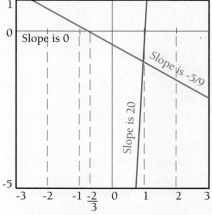

FIGURE 4-13 $P(x) = 3x^4 + 5x^3 - x^2 - 5x - 2$

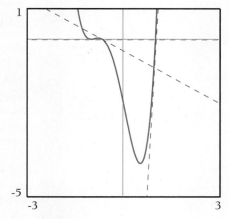

FIGURE 4-14 $P(x) = 3x^4 + 5x^3 - x^2 - 5x - 2$

Note: You may wonder of the importance of the factor 3 in the representation. What would happen if 3 were omitted? Suppose

$$P(x) = 3(x + 1)^2(x - 1)(x + \tfrac{2}{3}) \qquad \text{and}$$

$$P_2(x) = (x + 1)^2(x - 1)(x + \tfrac{2}{3})$$

Because $P_2(x) = \tfrac{1}{3} P(x)$, P_2 is simply a transformation of P. As you would expect the graph of P_2 is shown in Figure 4-15. Compare the graph of P with the graph of P_2. The graph of P_2 is the graph of P compressed along the y-axis to $\tfrac{1}{3}$ the height of P. The general shape and the x-intercepts remain intact.

Note the tangent lines drawn at $(1, 0)$ in Figures 4-14 and 4-15. To calculate the slope of the tangent line to P at $(1, 0)$, divide out the $x - 1$ factor in $P(x)$ to obtain $Q(x) = 3(x + 1)^2(x + \tfrac{2}{3})$. Verify that the slope of the tangent line to P at $(1, 0)$ is given by $Q(1) = 20$. Similarly for P_2, divide $x - 1$ into $P_2(x)$ to obtain $Q_2(x) = (x + 1)^2(x + \tfrac{2}{3})$. As you would expect, the tangent line to P_2 at $(1, 0)$ is given by $Q_2(1) = \tfrac{20}{3}$. The slope of the tangent line to P_2 is exactly $\tfrac{1}{3}$ the slope of the tangent line to P at $(1, 0)$. Use your graphing technology to graph $P(x)$ and $\tfrac{1}{3}P(x)$ on the same coordinate system to verify these results. From the slope-intercept form of a line we can write the equation for the line tangent to P at $(1, 0)$:

$$y = 20(x - 1) + 0$$

Graph $y = 20(x - 1)$ in the same window as P to verify that this line appears tangent to P at $(1, 0)$. ∎

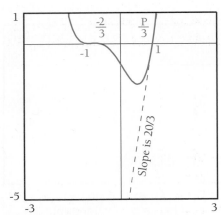

FIGURE 4-15
$P_2(x) = (x + 1)^2(x - 1)(x - \tfrac{2}{3})$

SUMMARY

- Suppose that $P(x) = a_n x^n + \ldots + a_0$ is a polynomial with rational coefficients. Let LCM be the least common multiple of the denominators of the coefficients of P. Then $P_2(x) = (\text{LCM})P(x)$ is a polynomial with integer coefficients that has the same roots as P. The graph of P_2 resembles the graph of P: the roots of P_2 are the roots of P. The graph of P_2 is a transformation of the graph of P stretched by a factor of LCM along the y-axis.
- Suppose that $P(x) = a_n x^n + \ldots + a_0$ is a polynomial with integer coefficients. (If the coefficients are rational numbers rather than integers, refer to the previous paragraph.) If P has any nonzero rational roots of the form n/d reduced to lowest terms, then n is a factor of a_0 and d is a factor of a_n.
- Suppose a polynomial is factored into linear factors: $P(x) = a(x - r_1)(x - r_2) \ldots (x - r_n)$. Then a chart for the factors of P helps identify where the graph is on, below and above the x-axis. The same chart allows for the solution of inequalities.

- Suppose that r is a root of a polynomial P. Then the slope of the tangent line to the polynomial at the point $(r, P(r))$ is $Q(r)$, where $Q(x)$ is the quotient of $P(x)$ divided by $(x - r)$. The slope of the tangent line measures the steepness of the polynomial at the point of tangency.
- Every polynomial of odd degree has at least one real root.

4-4   EXERCISES

For each of the following polynomials P: (a) list the possible rational roots, (b) determine all rational roots, (c) where possible determine all real roots, (d) factor the polynomials.

1. $P(x) = x^3 - 8x + 3$
2. $P(x) = x^3 + 2x^2 - 6x - 4$
3. $P(x) = x^3 + 6x^2 + 10x + 4$
4. $P(x) = x^3 - 6x^2 + 10x - 3$
5. $P(x) = 4x^3 + 12x^2 - x - 3$
6. $P(x) = 9x^3 + 18x^2 - 4x - 8$
7. $P(x) = 3x^3 - 2x^2 + 3x - 2$
8. $P(x) = 3x^3 + 2x^2 + 6x + 4$
9. $P(x) = 9x^4 - 13x^2 + 4$
10. $P(x) = 2x^4 - 19x^2 + 9$

Factor each of the following.

11. $P(x) = 4x^3 + 8x^2 - x - 2$
12. $P(x) = 9x^3 - 9x^2 - 4x + 4$
13. $P(x) = 6x^3 + 7x^2 - 16x - 12$
14. $P(x) = 6x^3 - 19x^2 + 19x - 6$
15. $P(x) = x^4 - 5x^3 - 3x^2 + 20x - 4$
16. $P(x) = x^4 + x^3 - 12x^2 - 9x + 27$
17. $P(x) = x^4 + 2x^2 - 7x - 10$
18. $P(x) = x^4 - 3x^3 + 3x^2 - 5x - 12$
19. $P(x) = 4x^4 - 13x^2 + 9$
20. $P(x) = 9x^4 - 40x^2 + 16$

Solve the following. Compare to Problems 11 through 20.

21. $4x^3 + 8x^2 - x - 2 = 0$
22. $9x^3 - 9x^2 - 4x + 4 = 0$
23. $6x^3 + 7x^2 - 16x - 12 = 0$
24. $6x^3 - 19x^2 + 19x - 6 = 0$

25. $x^4 - 5x^3 - 3x^2 + 20x - 4 = 0$
26. $x^4 + x^3 - 12x^2 - 9x + 27 = 0$
27. $x^4 + 2x^2 - 7x - 10 = 0$
28. $x^4 - 3x^3 + 3x^2 - 5x - 12 = 0$
29. $4x^4 - 13x^2 + 9 = 0$
30. $9x^4 - 40x^2 + 16 = 0$

Solve the following. Compare to Problems 11 through 20.

31. $4x^3 + 8x^2 - x - 2 > 0$
32. $9x^3 - 9x^2 - 4x + 4 \leq 0$
33. $6x^3 + 7x^2 - 16x - 12 \leq 0$
34. $6x^3 - 19x^2 + 19x - 6 > 0$
35. $x^4 - 5x^3 - 3x^2 + 20x - 4 \geq 0$
36. $x^4 + x^3 - 12x^2 - 9x + 27 < 0$
37. $x^4 + 2x^2 - 7x - 10 < 0$
38. $x^4 - 3x^3 + 3x^2 - 5x - 12 \geq 0$
39. $4x^4 - 13x^2 + 9 > 0$
40. $9x^4 - 40x^2 + 16 \leq 0$

Graph each of the following. Compare to Problems 11 through 20.

41. $P(x) = 4x^3 + 8x^2 - x - 2$
42. $P(x) = 9x^3 - 9x^2 - 4x + 4$
43. $P(x) = 6x^3 + 7x^2 - 16x - 12$
44. $P(x) = 6x^3 - 19x^2 + 19x - 6$
45. $P(x) = x^4 - 5x^3 - 3x^2 + 20x - 4$
46. $P(x) = x^4 + x^3 - 12x^2 - 9x + 27$
47. $P(x) = x^4 + 2x^2 - 7x - 10$
48. $P(x) = x^4 - 3x^3 + 3x^2 - 5x - 12$
49. $P(x) = 4x^4 - 13x^2 + 9$
50. $P(x) = 9x^4 - 40x^2 + 16$

For each of the following polynomials P determine the slope of the line tangent to P at the indicated intercept r. Write an equation for the tangent line to P at $(r, 0)$.

51. $P(x) = (x - 2)(x + 5)$, $r = 2$

52. $P(x) = (x + 3)(x - 4)$, $r = -3$

53. $P(x) = x(x - 3)(x + 6)$, $r = -6$

54. $P(x) = x(x + 4)(x - 3)$, $r = 0$

55. $P(x) = 4x^3 + 8x^2 - x - 2$, see Problem 41, r = leftmost intercept.

56. $P(x) = 9x^3 - 9x^2 - 4x + 4$, see Problem 42, r = leftmost intercept.

57. $P(x) = 6x^3 + 7x^2 - 16x - 12$, see Problem 43, r = rightmost intercept.

58. $P(x) = 6x^3 - 19x^2 + 19x - 6$, see Problem 44, r = rightmost intercept.

59. $P(x) = x^4 - 5x^3 - 3x^2 + 20x - 4$, see Problem 45, r = leftmost intercept.

60. $P(x) = x^4 + x^3 - 12x^2 - 9x + 27$, see Problem 46, r = leftmost intercept.

Prove the following.

61. $\sqrt{7}$ is irrational.

62. $\sqrt{5}$ is irrational.

63. $\sqrt{6}$ is irrational.

64. $\sqrt{10}$ is irrational.

65. Part 2 of the rational root theorem.

66. Expand the proof of the rational root theorem by discussing why it is essential that the rational root N/D be reduced to lowest terms.

67. Verify that if P is a polynomial and r is a root of P then the equation of the line tangent to P at $(r, 0)$ is $y = Q(r)(x - r)$ where Q is the quotient of $P(x)/(x - r)$.

4-4   P R O B L E M   S E T

Francois Vieta's (1540–1603) method provided the inspiration for the derivation of the quadratic formula in Problem set 3-5. A similar substitution method leads to the solution of general cubic equations. The following problem set develops this method based on the work of Giraolamo Cardano (1501–1576).

The following cubic equations are in the form $ax^3 + bx^2 + cx + d = 0$. Use the substitution $x = y - b/(3a)$ to eliminate the quadratic term from these equations. (The resulting equations are reduced cubic equations.)

1. $x^3 + 3x^2 - 4 = 0$

2. $x^3 + 6x^2 + 9x + 4 = 0$

3. $x^3 + 3x^2 - 6x + 20 = 0$

4. $x^3 + 3x^2 - 6x - 36 = 0$

5. $x^3 + 6x^2 + 9x + 12 = 0$

6. $x^3 + 6x^2 + 9x - 8 = 0$

In 1545 Gerolamo Cardan published a formula in *Ars Magna* (The Great Art) to solve a cubic equation of the form

$$x^3 - rx = s \quad \text{or}$$
$$x^3 = rx + s \qquad \text{(reduced cubic equation)}$$

Compare this to the following identity which you may verify by direct calculations: $(a + b)^3 - 3ab(a + b) = a^3 + b^3$.

If we choose a and b so that $3ab = r$ and $s = a^3 + b^3$, the solution for x is $x = a + b$. Solving this system of equations produces the following formulas:

Let $\quad t = \sqrt{\left(\dfrac{s}{2}\right)^2 - \left(\dfrac{r}{3}\right)^3}$

Now $\quad A = \sqrt[3]{\dfrac{s}{2} + t} \quad$ and $\quad B = \sqrt[3]{\dfrac{s}{2} - t}$

so that $x_1 = A + B$.

R. Bombelli (ca. 1526–1572) extended the formula to provide two additional solutions

$$x_2 = -\dfrac{A + B}{2} + i(A - B)\sqrt{\dfrac{3}{2}} \quad \text{and}$$

$$x_3 = -\dfrac{A + B}{2} - i(A - B)\sqrt{\dfrac{3}{2}}$$

and in the process invented complex numbers.

Use Cardan's formula to find a solution of each of the following.

7. $x^3 = 3x + 2$

8. $x^3 = 3x - 2$

9. $x^3 = 9x - 28$

10. $x^3 = 9x + 28$

*11. $x^3 = 3x - 10$

*12. $x^3 = 3x + 10$

For each of the following, (a) transform the equation to a reduced cubic equation, (b) solve the reduced cubic using Cardan's formula, (c) solve the original equation.

13. $x^3 + 3x^2 - 4 = 0$

14. $x^3 + 6x^2 + 9x + 4 = 0$

15. $x^3 + 3x^2 - 6x + 20 = 0$

16. $x^3 + 3x^2 - 6x - 36 = 0$

17. $x^3 + 6x^2 + 9x + 12 = 0$

18. $x^3 + 6x^2 + 9x - 8 = 0$

19. $x^3 - 3x^2 - 5x + 15 = 0$

20. $x^3 + 3x^2 + 4x + 12 = 0$

21. Show that if $ax^3 + bx^2 + cx + d = 0$, $a \neq 0$, then the substitution $x = y - b/(3a)$ eliminates the quadratic term.

22. Verify that $(m + n)^3 - 3mn(m + n) = m^3 + n^3$.

23. Refer to Problem 22. Let $a = -3mn$ and $b = m^3 - n^3$ and $x = m + n$. Compare these results to Cardan's reduced cubic equation.

*24. Refer to Problem 23. Solve the equations $a = -3mn$ and $b = m^3 - n^3$ for m and n. Use the result and the fact that $x = m + n$ to verify Cardan's formula.

EXPLORATION

The following problems anticipate concepts in the next section. Use the trace feature to approximate the x-intercepts of the following.

25. $P(x) = x^3 + 2$

26. $P(x) = x^3 - 3$

27. $P(x) = x^4 - 7x^2 - 1$

28. $P(x) = x^4 - 5x^2 + 2$

29. $P(x) = x^3 - 3x - 1$

30. $P(x) = x^3 - 4x^2 + 3$

TANGENTS TO A POLYNOMIAL

Synthetic division can be used to determine the slope of a tangent line to any point of a polynomial. The process is simple. Suppose that P is a polynomial and $(b, P(b))$ is a point of P. Divide $P(x)$ by $x - b$ to obtain a quotient of $Q(x)$ and a remainder of R. Set aside the remainder R. Now divide the quotient $Q(x)$ by $x - b$ to obtain a second quotient $Q_2(x)$ with a remainder T. This second remainder T represents the slope of the tangent line to P at the point $(b, P(b))$. For example, suppose we want the slope of the tangent line to $P(x) = x^3 - 4x + 1$ at $(1, P(1))$. Here, $b = 1$. Hence, we divide by $x - 1$:

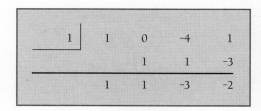

Set aside the remainder -2. Repeat the division of the resulting quotient.

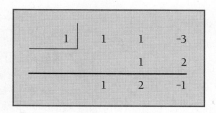

The remainder -1 is the slope.

From the first division, we know that $P(1) = -2$. As a result we know the slope of the tangent line to P through $(1, -2)$ is -1. Using the point-intercept formula the equation for the tangent line is given by

$$y - (-2) = -1(x - 1)$$

Hence, $\qquad y = -x - 1$

You may graph $P(x) = x^3 - 4x + 1$ and $y = -x - 1$ in the same window to confirm that the line is tangent to P at $(1, -2)$.

For each given polynomial P: (a) Determine the slope of the tangent line at the given point, (b) Determine an equation for the tangent line to the point, and (c) Graph the polynomial and tangent line in the same window.

31. $P(x) = x^2 - 5x + 1$ at $(2, P(2))$.

32. $P(x) = x^2 + 7x - 3$ at $(-3, P(-3))$.

33. $P(x) = x^2 - 4x + 1$ at $(-2, P(-2))$.

34. $P(x) = x^2 + 6x - 2$ at $(-3, P(-3))$.

35. $P(x) = x^2 - 4x + 1$ at $(2, P(2))$.

36. $P(x) = x^3 + 6x - 2$ at $(-2, P(-2))$.

37. $P(x) = x^3 - 4x^2 + 1$ at $(-2, P(-2))$.

38. $P(x) = x^3 + 2x^2 - 1$ at $(3, P(3))$.

39. $P(x) = x^3 - 2x^2 - 4x + 2$ at $(2, P(2))$.

40. $P(x) = x^3 + x^2 + 3x - 2$ at $(-3, P(-3))$.

41. For Problems 31, 33, 35, 37 and 39, which tangent lines have a positive slope and which have a negative slope? Describe whether the polynomial is increasing or decreasing close to the point of tangency.

42. For Problems 32, 34, 36, 38, and 40, which tangent lines have a positive slope and which have a negative slope? Describe whether the polynomial is increasing or decreasing close to the point of tangency.

43. Discuss the significance of the slope of the tangent line to the polynomial in Problem 35.

44. Discuss the significance of the slope of the tangent line to the polynomial in Problem 34.

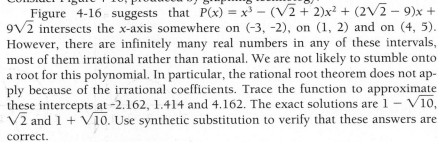

ITERATIVE APPROXIMATION OF IRRATIONAL ROOTS

5*

In Mathematicks he was greater Than Tycho Brahe, or Erra Pater: For he, by Geometrick scale, Could take the size of Pots of Ale; Resolve by Signs and Tangents streight, If Bread and Butter wanted weight; And wisely tell the hour o'the'day, The Clock doth strike, by Algebra.
—SAMUEL BUTLER

Several flaws are apparent in the methods of the previous section. Not only must we exhaust ourselves (or our computer) with trial and error in the hopes of discovering rational roots, but the method for listing possible rational roots limits us to polynomials with integer coefficients. Rewriting polynomials with rational coefficients allows us to include these in the process, but we have no method if any coefficients are irrational.

Further, suppose the polynomial has a degree of 3 or more. If the polynomial has no rational roots, we have little chance of guessing a solution. Consider Figure 4-16, produced by graphing technology.

Figure 4-16 suggests that $P(x) = x^3 - (\sqrt{2} + 2)x^2 + (2\sqrt{2} - 9)x + 9\sqrt{2}$ intersects the x-axis somewhere on $(-3, -2)$, on $(1, 2)$ and on $(4, 5)$. However, there are infinitely many real numbers in any of these intervals, most of them irrational rather than rational. We are not likely to stumble onto a root for this polynomial. In particular, the rational root theorem does not apply because of the irrational coefficients. Trace the function to approximate these intercepts at -2.162, 1.414 and 4.162. The exact solutions are $1 - \sqrt{10}$, $\sqrt{2}$ and $1 + \sqrt{10}$. Use synthetic substitution to verify that these answers are correct.

In this section we will examine some methods for producing successively better approximations for the irrational roots of a polynomial. A method that repeats a process to get an answer is an iterative method. Each application of the method is an iteration. Consider the following.

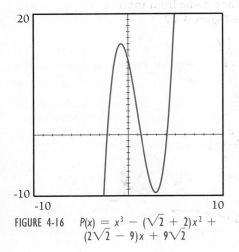

FIGURE 4-16 $P(x) = x^3 - (\sqrt{2} + 2)x^2 + (2\sqrt{2} - 9)x + 9\sqrt{2}$

Algorithm Outline

Repeat steps 1 and 2 until the approximate root c is within the desired tolerance.

1. Determine two real values a and b such that $P(a) > 0$ and $P(b) < 0$. A root of P must be between a and b. (It does not matter which of a or b is larger.)
2. Choose some value between a and b, say, c, to shorten the boundary on the approximation. Calculate $P(c)$. If $P(c) > 0$, then substitute c in place of a. If $P(c) < 0$, replace b with c.

Although the process sounds simple, major difficulties confront us.

Difficulties with the Algorithm

1. How do we "seed" the values for a and b? We must have starting values to initialize the variables in an iterative process.

2. By what method will we select c between a and b? This is the method used for each iteration. Some methods are less efficient than others.

3. When will c be "close enough" to the actual root? *Tolerance* describes how close is "close enough." We must have criteria for terminating an iterative process so that we know when to stop.

Our technique for seeding a and b will be guessing from a graph or by trial and error. Difficulty 2 has many answers. We will not attempt to measure the efficiency of our methods. As for difficulty 3, we shall use a simple method for deciding the accuracy of our approximation without claiming that the method is flawless.

The first method for approximating a root of a polynomial function is known as *linear interpolation*. Values for c come from the x-intercept of a line connecting the seed values a and b. See Figure 4-17.

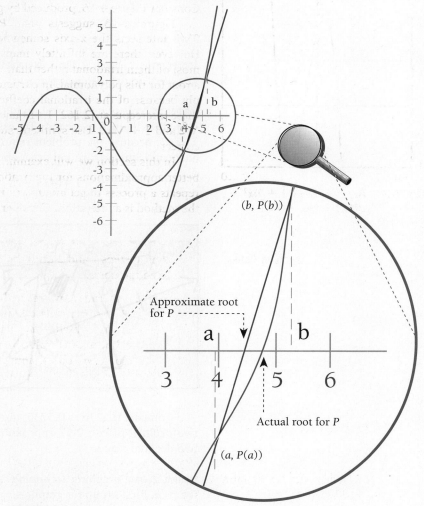

FIGURE 4-17

The equation of the line in Figure 4-17 is given by

$$y = m(x - b) + P(b)$$

$$y = \frac{P(b) - P(a)}{b - a} (x - b) + P(b)$$

To determine the x-intercept of the line, set $y = 0$ and solve for x:

$$0 = \frac{P(b) - P(a)}{b - a} (x - b) + P(b)$$

$$\frac{P(b) - P(a)}{b - a} (x - b) = -P(b)$$

$$x - b = -P(b)\frac{b - a}{P(b) - P(a)}$$

$$x = b + \frac{aP(b) - bP(b)}{P(b) - P(a)}$$

$$x = \frac{aP(b) - bP(a)}{P(b) - P(a)}$$

EXAMPLE 4-15

Approximation Using Linear Interpolation. Approximate a root of $P(x) = x^3 + \sqrt{2}x - 1$ to two decimal places.

Solution 1: Note that $P(0) = -1$ and $P(1) = \sqrt{2}$. There is a root between 0 and 1. Use these values as seeds. To ease the calculations, approximate $\sqrt{2}$ as 1.4142. Because the problem requires two-decimal precision, this should be adequate. Similarly, round other approximations during the process:

$$P(x) \cong x^3 + 1.4142x - 1$$

Negative (a)	Positive (b)	P(a)	P(b)	New = $\frac{aP(b) - bP(a)}{P(b) - P(a)}$	p(New)	New Replaces
0	1	-1	1.4142	0.414	-0.3436	a
0.414	1	-0.3436	1.4142	0.5286	-0.1048	a
0.5286	1	-0.1048	1.4142	0.5611	-0.0297	a
0.5611	1	-0.0297	1.4142	0.5908	0.0417	b
0.5611	0.5908	-0.0297	0.0417	0.5735	-0.00038	a
0.5735	0.5908	-0.00038	0.0417	0.5736	-0.00006	a

Compare 0.5736 to 0.5735, the approximations reach apparent stability at two decimal places: 0.57 approximates a root of $P(x) = x^3 + 1.4142x - 1$ to two decimal places.

Solution 2: Use graphing technology's zoom and trace features to approximate the root. First set up for graphics. See Section 1.2. Set the range using Xmin = 0, Xmax = 2, Ymin = -1 and Ymax = 1.

Now we are ready to graph P. Enter the formula for P as follows:

Casio: [GRAPH] [XθT] [x^y] [3] [+] [√] [2] [X] [XθT] [−] [1] [EXE]

Note how the multiplication key is set to ensure $\sqrt{2}$ rather than $\sqrt{2x}$.

Sharp: [+/−] [MENU] [1] [XθT] [a^b] [3] [>] [+] [√] [2] [>] [XθT] [−] [1] [+/−]

TI-81: [Y=] [X|T] [^] [3] [+] [√] [2] [x] [X|T] [−] [1] [GRAPH]

When the graph appears, use the trace key to establish approximate x-intercepts. You should see a sign change for y with x between 0.55 and 0.57. Press the zoom key. Choose BOX zoom and mark a box of approximately ($x = 0.51$, $y = -0.10$) to ($x = 0.61$, $y = 0.12$). After the zoom, choose the trace key once more. The sign of y changes for x approximately between 0.572 and 0.574. Because the decimal appears stable in the second digit, assume that the root is approximately 0.57. Compare the speed of this method to the previous method.   ▬

4-5   EXERCISES

Approximate the root of each of the following to one decimal place (nearest tenth). Examine the graph to initialize the process.

1. $y = x^3 - 3$

2. $y = x^3 - 4$

3. $y = x^4 - 7$

4. $y = x^4 - 5$

5. $y = x^3 + 2$

6. $y = x^5 + 3$

Each of the following has a root between 0 and 1. Approximate the root to one decimal place (nearest tenth):

7. $P(x) = x^3 - 2x + 0.1$

8. $P(x) = x^3 - 3x + 1$

9. $P(x) = x^3 + 3x - 1$

10. $P(x) = x^3 + 2x - 0.5$

4-5   PROBLEM SET

PROBLEMS FOR GRAPHING TECHNOLOGY

Use the graphing technology trace feature to estimate the x-intercepts of the following functions. Use the zoom feature to increase precision to three decimal places.

1. $f(x) = x^3 - 5x^2 + 7x - \sqrt{2}$

2. $f(x) = 2x^3 - \sqrt{5}x^2 + x - 1$

3. $f(x) = x^4 - 3x + 1$

4. $f(x) = x^4 + 7x - 2$

5. Discuss which step in graphing technology is used to *initialize* the values to begin the iterative approximation of roots.

6. Discuss which step in graphing technology is the process repeated for each *iteration* of approximating roots.

7. Describe the *criteria* that result in the termination of the graphing technology approximation of roots.

8. Compare and contrast the steps in linear interpolation with the steps used in graphing technology for approximating roots. Formulate the comparison in terms of the general algorithm steps.

EXPLORATION

Recursion is another iterative method. Consider the function $f(x) = x^4 - 7x + 2$. The graph of f indicates two positive roots between 0 and 2. To determine the roots recursively solve $x^4 - 7x + 2 = 0$ for x in terms of x.

First Method: From the graph there are two roots. Prime the [ANS] key with a number close to a root, say, 1: [1] [EXE].

Solve the equation recursively.

$$x^4 = 7x - 2$$
$$x = \sqrt[4]{7x - 2}$$

Enter the recursion equation in your calculator using the [ANS] key:

$$(7 \,[\text{ANS}] - 2)^{0.25}$$

Press [ENTER] until the answer stabilizes at about 1.806 . . .

Second Method:

$$x^4 - 7x = -2$$
$$x(x^3 - 7) = -2$$
$$x = \frac{-2}{x^3 - 7}$$
$$x = \frac{2}{7 - x^3}$$

Prime the [ANS] key again and enter $2 \div (7 - [\text{ANS}]^3)$. Press [ENTER] repeatedly until the answer stabilizes at about 0.286 . . .

Use either of these methods to approximate the roots of the following. You may encounter difficulties in some cases. Note the difficulties then reexamine the graphs. Conjecture a reason for your difficulties.

9. $f(x) = x^3 - 5x^2 + 7x - \sqrt{2}$
10. $f(x) = 2x^3 - \sqrt{5}\,x^2 + x - 1$
11. $f(x) = x^4 - 3x + 1$
12. $f(x) = x^4 + 7x - 2$

6   FUNDAMENTAL THEOREM OF ALGEBRA

Some proofs command assent. Others woo and charm the intellect. They evoke delight and an overpowering desire to say "Amen, Amen."
—LORD RAYLEIGH

Sometimes a single statement ties together diverse threads into a coherent knot. When the diverse threads can be woven into a pleasing tapestry, the results are more satisfying. If the warp of the fabric provides a unity or theme to the tapestry, we tend to call the results art. In mathematics, we call these unifying ideas fundamental theorems. This section explores the fundamental theorem of algebra.

For a parallel structure refer once more to integers. Recall that a positive integer *n* is *prime* if it has exactly two factors: itself and 1. There are infinitely many prime numbers, including 2, 3, 5, 7, 11, 13 and 17.

Except for 1, if a positive integer is not prime, it is composite. *Composite numbers* have three or more factors. The number 6 is a composite number. Factors of 6 include not only 1 and 6 but also 2 and 3. The number 1 is neither prime nor composite. Except for 2, all even numbers are composite. Therefore, except for 2, all prime numbers must be odd.

No one has discovered a formula to produce *all* prime numbers. At present, the only absolute test for a proposed prime number is division by possible integer factors. If we find no positive factors other than 1 and the number, the number is prime. Testing large numbers to determine whether they are prime can consume hours or days of time on high-speed computers. As a result, large prime numbers have extensive uses in public key encryption code systems.

The Fundamental Theorem of Arithmetic states that except for the order of factors, each composite number factors into unique prime factors. Whereas $12 = 3(4)$ and $12 = 2(6)$, if we insist on prime factors, $12 = (2^2)3$. No other prime factorization of 12 is possible. Excluding -1 any negative integer factors into -1 times a unique product of primes.

Applications of the fundamental theorem of arithmetic include public key encryption and reducing a fraction to its lowest terms. Both applications use prime factorization.

The structure of polynomial algebra parallels that of the integers. Recall the usefulness of factoring polynomials into linear factors. Unfortunately, some polynomials do not factor easily. Try as we might, $P(x) = x^2 - 2x + 2$ will not factor into linear factors with real coefficients. From the quadratic formula, the roots of $P(x)$ are $(2 \pm \sqrt{-4})/2$.

Every real number must be either positive or negative or 0. Let i represent $\sqrt{-1}$. Then $i^2 = -1$. Obviously, i cannot be 0. Every positive number squared is positive; therefore, i is not positive. Similarly, the square of every negative number is also positive; i is not negative. Because i is not negative, positive or 0, i is not a real number. Recall from Section 0.8 that i forms the basis for a set of numbers called *complex numbers*. Complex numbers provide the tools to make factoring complete.

A complex polynomial takes the same form as a real polynomial except that the coefficients are complex numbers and complex numbers provide the domain for the variable z. Suppose that

$$P(z) = a_n z^n + \ldots + a_1 z^1 + a_0, \qquad a_n \neq 0$$

is an nth degree complex polynomial where each coefficient is a complex number and z is a complex number. Then any complex number r such that $P(r) = 0$ is a **complex root** of P.

The following theorems provide results analogous to the fundamental theorem of arithmetic. The first of these is the fundamental theorem of algebra, whose name suggests its importance. Although the theorem may seem intuitively obvious based upon our experience with real polynomials, a rigorous proof of the fundamental theorem eluded mathematicians for years. The first acceptable proof was presented by Carl Friedrich Gauss (1777–1855).

Theorem 4-9
The Fundamental Theorem of Algebra

> Every complex polynomial of degree 1 or more has a complex root.

A corollary is a theorem for which the proof follows from a closely related theorem. The next theorem is a corollary to the fundamental theorem of algebra; in other words, the proof of the next theorem depends upon the fundamental theorem. Mathematicians believed the corollary to be correct long before Gauss proved the fundamental theorem. Once the Fundamental Theorem proof became available, the corollary followed immediately.

Corollary

> Counting duplicate roots, each nth-degree complex polynomial has n roots.

The corollary implies that every nth-degree complex polynomial has exactly n roots (counting multiple roots). Because each root of a polynomial has an associated linear factor for the polynomial, we deduce that an nth-degree complex polynomial factors into n factors (counting duplicate factors).

For example, consider $x^2 + 6x + 9 = (x + 3)^2$. Note that -3 is a root of $P(x) = x^2 + 6x + 9$. Because the factor $(x + 3)$ appears twice, we say that -3 is a root of multiplicity 2. For a polynomial with $(x - r)^n$ as a factor, r is a root of multiplicity n because $(x - r)$ is a factor n times.

EXAMPLE 4-16

Counting Complex Roots. For each of the following polynomials determine the number of complex roots. If the polynomial is factored, list the roots and multiplicity of each root.

Illustration 1:

$$P(z) = 3z^{17} - (1 + i)z^3 + 2$$

Solution: P has seventeen complex roots.

Illustration 2:

$$P(z) = 5(z - 42)^2(z + \tfrac{1}{2})^3(z - 1 + i)^3(z - 1 - i)^5$$

Solution: P is degree 13; 42 is a root of multiplicity 2; $-\frac{1}{2}$ is a root of multiplicity 3;

$1 - i$ is a root of multiplicity 3; $1 + i$ is a root of multiplicity 5.

Note: The degree of the polynomial equals the sum of the multiplicities of the roots: $2 + 3 + 3 + 5 = 13$.

The roots $1 + i$ and $1 - i$ in the last illustration are special, related complex numbers called *conjugates*. We repeat the definition from Section 0.7.

DEFINITION 4-3
Complex Conjugate

> The complex conjugate $\bar{z}$ of a complex number $z = a + bi$ is a complex number given by $\bar{z} = a - bi$, where a and b are real numbers and $i^2 = -1$. Read $\bar{z}$ as "z-bar."

Notice that a real number, such as 3, could be written as $3 + 0i$. Because any real number x could be written as $x + 0i$, you could argue that every real number is also a complex number. However, real numbers are conceptually different from complex numbers.

Conjugate pairs have some useful properties. For example, the sum or product of a conjugate pair is a real number: $z + \bar{z} = 2a$; $z\bar{z} = a^2 + b^2$. We leave the proof as an exercise.

Recall $1 + i$ and $1 - i$ are also the roots of $P(x) = x^2 - 2x + 2$ presented at the beginning of this section. Note particularly that P is a polynomial with real coefficients. This polynomial and the properties of conjugates lead us to the following theorem.

Theorem 4-10
Roots in Conjugate Pairs

> If a complex number $\bar{z}$ is a root of a complex polynomial P with real coefficients, then the conjugate $\bar{z}$ is also a root.

For any real number x in complex form $x + 0i$, the conjugate of $x + 0i$ is $x - 0i$: a real number is its own complex conjugate. Such a result does not violate Theorem 4-10, but neither is it an interesting or useful result. For that reason, we shall apply Theorem 4-10 to only those complex roots that are not real (then the conjugate is a distinct number).

EXAMPLE 4-17

Roots of Complex Polynomials. Determine the complex roots for each of the following polynomials.

Illustration 1:

$$P(x) = x^4 - 6x^3 + 15x^2 - 18x + 10$$

It is given that $2 + i$ is a root of P.

Solution: P has real coefficients, and $2 + i$ is a root. Therefore from Theorem 4-10, $2 - i$ is also a root. Use synthetic division to depress P to a second-degree polynomial.

$2 + i$	1	$-6 + 0i$	$15 + 0i$	$-18 + 0i$	10	
		$2 + i$	$-9 - 2i$	$14 + 2i$	-10	
$2 - i$	1	$-4 + i$	$6 - 2i$	$-4 + 2i$	0	Verify and depress
		$2 - i$	$-4 + 2i$	$4 - 2i$		
	1	-2	2	0		Verify and depress

The quotient is $x^2 - 2x + 2$. Apply the quadratic formula. The roots are $1 + i$ and $1 - i$. Note that the roots are in conjugate pairs. (Careful examination of the $\sqrt{\ }$ in the quadratic formula may give you further insight into the source of the conjugate pairs.)

Solution: The roots of P are $2 \pm i$ and $1 \pm i$.

$$P(x) = [x - (2 - i)][x - (2 + i)][x - (1 - i)][x - (1 + i)]$$

Illustration 2:

$$P(x) = x^3 - x^2 + 2x - 2$$

Solution: From the rational root theorem, the possible rational roots are 1, -1, 2, -2. Experimenting reveals that 1 is an actual root:

1	1	-1	2	-2
		1	0	2
	1	0	2	0

The quotient is the quadratic $x^2 + 2$. Solve $x^2 + 2 = 0$:

$$x = \pm \sqrt{2}\, i$$

Answer:

$1, \sqrt{2}\, i, -\sqrt{2}\, i;$
$P(x) = (x - 1)(x - \sqrt{2}\, i)(x + \sqrt{2}\, i)$

Note that the complex roots occur in conjugate pairs.

Illustration 3: $P(x) = x + i$

Solution: $x + i = 0$ implies $x = -i$. Because the polynomial is first degree it has only one root. Theorem 4-10 does not apply because the coefficients are not real.

Illustration 4: $P(x) = x^2 - 5x + 6$

Solution: From the quadratic formula the roots are 2 and 3. These roots are not complex numbers. These numbers are their own conjugates. Theorem 4-10 does not apply.   ▬

Illustration 2 of the previous example provides the clue that leads us to the following theorem.

Theorem 4-11
Corollary to Theorem 4-10

> If P is an odd-degree polynomial with real coefficients, then P has at least one real root.

Proof: Suppose that P is an odd-degree polynomial with real coefficients. Assume that P has only complex roots. By Theorem 4-10 if z is a complex root so also is $\bar{z}$. Pair each root with its conjugate. Because only real roots are their own conjugates, there is no root for which $z = \bar{z}$. We have an even number of roots, contradicting the statement that the degree of P is odd. To prove that only real roots are their own conjugates, first suppose there is a root $z = a + bi$ such that $z = \bar{z}$.

However, $\bar{z} = a - bi$. Hence, $a + bi = a - bi$. Therefore, $2bi = 0$; in which case $b = 0$. But b is the imaginary part of z. So that $z = a + 0i$. As a result, z is a real number.   ▬

Some graphing calculators have the ability to do arithmetic in either real or complex mode. Consult your user's manual. If your calculator has a complex mode, be sure to use the $x + yi$ form for the exercises of this section. The $r \angle \theta$ form will be discussed in Chapter 8. Chapter 8 examines more applications of the fundamental theorem of algebra.

SUMMARY

- An nth-degree complex polynomial has exactly n roots, counting the multiplicity of each root.
- For a complex polynomial with real coefficients, any complex roots must occur in conjugate pairs. As a result, every polynomial of odd degree with real coefficients must have at least one real root.

4-6 EXERCISES

Use synthetic division to determine whether the indicated value of z is a root of the given polynomial.

1. $P(z) = z^2 + 1, z = i$

2. $P(z) = z^2 + 1, z = -1$

3. $P(z) = z^2 - 2z + 2, z = 1 + i$

4. $P(z) = z^2 + 2z + 5, z = 1 + 2i$

5. $P(z) = z^2 - 2z + 2, z = 1 + 2i$

6. $P(z) = z^2 + 2z + 5, z = 1 + i$

7. $P(z) = z^2 - 4z + 5, z = 2 - i$

8. $P(z) = z^2 - 4z + 5, z = 2 + i$

9. $P(z) = z^3 + 3z^2 + 7z + 5, z = 1 - 2i$

10. $P(z) = z^3 - z^2 + 2, z = 1 - i$

For each of the following polynomials, determine all the roots beginning with the given root.

11. $P(z) = z^2 - 6z + 10, z = 3 - i$

12. $P(z) = z^2 + 4z + 5, z = -2 + i$

13. $P(z) = z^2 + 2z + 5, z = -1 + 2i$

14. $P(z) = z^2 - 2z + 10, z = 1 - 3i$

15. $P(z) = z^2 + 6z + 13, z = -3 - 2i$

16. $P(z) = z^2 + 4z + 13, z = 2 - 3i$

17. $P(z) = z^3 - 7z^2 + 16z - 10, z = 3 + i$

18. $P(z) = z^3 + 3z^2 + z - 5, z = -2 - i$

*19. $P(z) = z^2 + (i - 3)z - 3i, z = -i$

*20. $P(z) = z^2 + (2 - i)z - 2i, z = i$

Solve each of the following complex polynomial equations.

21. $3z + 2i = 1$

22. $3z - 5 + i = 0$

23. $(1 + i)z - 2 = i$

24. $(3 - i)z + i = 1$

25. $3z + iz - 2i + 1 = 0$

26. $5i - z + 3i - 2 = 1$

27. $z^2 + 2 = 0$

28. $3z^2 + 1 = 0$

29. $z^2 + z + 1 = 0$

30. $z^2 - z + 1 = 0$

31. $5z^2 - z + 2 = 0$

32. $2z^2 - 3z + 10 = 0$

33. $z^2 + iz - 3 = 0$

34. $z^2 - 3iz + 2 = 0$

35. $iz^2 + 3z - 2i = 0$

36. $iz^2 - z + 2i = 0$

37. $iz^2 + 2iz - 3i = 0$

38. $iz^2 - 5iz + i = 0$

39. $z^3 + 3z^2 + 7z + 5 = 0$

40. $z^3 - z^2 + 2 = 0$

Suppose that $z = a + bi$ and $\bar{z} = a - bi$ are complex numbers.

41. Prove $z + \bar{z} = 2a$.

42. Prove that the sum of z and $\bar{z}$ is a real number.

43. Prove that $z - \bar{z}$ is a complex number.

44. Prove that $z - \bar{z} = 2bi$.

45. Prove that $z\bar{z} = a^2 + b^2$.

46. Prove that the product of z and $\bar{z}$ is a real number.

47. If $z = \bar{z}$, prove that z is a real number.

48. If $z = -\bar{z}$, prove that $a = 0$.

49. Populations of rabbits fluctuate rapidly. Suppose ten rabbits escape to an isolated island and that their population $P(t)$ after t years is modeled by $P(t) = 6 - t(2 + t\{3 - t[4 + t]\})$. Graph P. Approximate the roots of P. What is the domain for the model? Does a model predict a time at which the population becomes extinct? At what time does the population exceed 100?

50. The volume of a sphere is given by $V(r) = \frac{4}{3}\pi r^3$, where r is the radius of the sphere. The surface area of a sphere is given by $S(r) = 4\pi r^2$. For what radius r is the surface area of a sphere numerically equal to the volume of the sphere? For what values of r is the volume larger than the area? For what values of r is the volume smaller than the area?

CHAPTER SUMMARY

- A polynomial function P is a function of the form

$$P(x) = a_n x^n + a_{n-1} x^{n-1} + \ldots + a_1 x^1 + a_0, \quad \text{where } a_n \neq 0, n \in \text{integer}$$

- If $n > 0$, the degree of P is n. The graph of P is smooth and continuous.
- Synthetic division provides a method for determining the quotient and remainder of an nth-degree polynomial and a first-degree polynomial.
- Suppose $P(x) = a_n x^n + \ldots + a_1 x + a_0$ is an nth-degree polynomial. P can have at most n roots.
- As far as graphs of polynomials are concerned, the relation of factors to x-intercepts suggests the following:
    An nth-degree polynomial intersects the x-axis at most n times.
    An nth-degree polynomial changes direction at most $n - 1$ times.
- If a polynomial is factored into linear factors: $P(x) = a_n(x - r_1)(x - r_2) \ldots (x - r_n)$, then a chart based on the factors of P helps identify where the graph is on, below and above the x-axis. The same chart allows for the solution of inequalities. Graphing technology provides more detail for the graphs.
- Suppose that $P(x) = a_n x^n + \ldots + a_0$ is a polynomial with integer coefficients. If P has any nonzero rational roots of the form n/d reduced to lowest terms, then n is a factor of a_0 and d is a factor of a_n (rational root theorem).
- Suppose that r is a root of a polynomial P. Then the slope of the tangent line to the polynomial at the point $(r, P(r))$ is $Q(r)$, where $Q(x)$ is the quotient of $P(x)$ divided by $(x - r)$. The slope of the tangent line measures the steepness of the polynomial at the point of tangency. The equation of the tangent line is given by $y = Q(r)(x - r)$. (Note that $P(r) = 0$.)
- Every polynomial of odd degree has at least one real root.
- An nth-degree complex polynomial has exactly n complex roots counting the multiplicity of each root.
- For a complex polynomial with real coefficients, any complex roots must occur in conjugate pairs. As a result, every polynomial of odd degree with real coefficients must have at least one real root.

KEY WORDS AND CONCEPTS

Complex roots
Depressed polynomial
Digits
Division
Domain
Factor theorem

Polynomial
Polynomial division
Polynomial function
Polynomial roots
Quotient
Radix

Radix notation
Remainder
Remainder theorem
Root theorem
Smooth
Synthetic division

4   R E V I E W   E X E R C I S E S

SECTION 1

Simplify each of the following. Identify each as a polynomial or not. If a polynomial, convert to standard form and determine the degree and number of terms (counting zero coefficients.)

1. $5(x^3 - 5) + 3x^2$

2. $(x - 3)(x^2 + 8)$

3. $(x^2 - 1)(x^2 + 5x)$

4. $7x^{30} - \dfrac{5}{x^2} + \sqrt{3}$

5. $\sqrt{3}x^{-1} - 1$

6. $4x^{-3}\sqrt{x} + \sqrt{2}$

7. $4x^3 - \sqrt[3]{x} - 7$

8. $(x - 5)^2 - (x - 1)^2$

Determine the quotient and remainder for each of the following divisions. Express the answer in the form $P(x) = (x - r)Q(x) + R$.

9. $(x^2 + 5x + 2) \div (x + 1)$

10. $(2x^2 + 7x - 3) \div (x + 2)$

11. $(x^3 - 5x^2 + 3x - 8) \div (x + 5)$

12. $(x^3 - 2x^2 + 3) \div (x - 1)$

Convert each of the following to base 10:

13. 35_8

14. $5B_{16}$

15. 10101_2

16. 1121_3

Convert the following to the indicated radix.

17. $53 \quad r = 7$

18. $24 \quad r = 2$

19. $27 \quad r = 3$

20. $33 \quad r = 16$

Determine which, if any, of $(x + 1)$, $(x - 1)$, $(x + 3)$ or $(x - 3)$ are factors of the following.

21. $5x - 10$

22. $x^4 - 2x - 3$

23. $x^3 + 27$

24. $3x^3 - 3$

SECTION 2

For each of the following polynomials $P(x)$, determine (a) the number of variations of sign of $P(x)$, (b) the possible number of positive roots of $P(x)$, (c) the number of variations of sign of $P(-x)$, (d) the possible number of negative roots of $P(x)$.

25. $P(x) = x^3 + 7x - 3$

26. $P(x) = 2x^4 - x^3 + x - 1$

27. $P(x) = x^4 - x^2 - 2$

28. $P(x) = x^4 + x^3 + x^2$

Determine the smallest integer upper bound and the largest integer lower bound for the roots of the following polynomials.

29. $P(x) = x^3 - 3x + 1$

30. $P(x) = 3x^3 + 4$

31. $P(x) = 9x^4 - 4$

32. $P(x) = 2x^4 - 3x^2 - 9$

SECTION 3

For each of the following factored polynomials, determine the roots.

33. $P(x) = (x - 4)(x + 23)(x - 17)$

34. $P(x) = 6(5x - 7)(x + 11)(x - 9)$

35. $P(x) = 7x(3x - 4)(5x - 3)$

36. $P(x) = x^3(x^2 - 7)(x - 3)^7$

(a) Determine which, if any, of 1, -1, 2, -2, 3, -3 are roots of the following polynomials. (b) Consult the list of roots in part a to factor each polynomial.

37. $P(x) = x^3 - 3x^2 - x + 3$

38. $P(x) = 2x^3 - 3x^2 - x + 4$

39. $P(x) = x^3 - 7x^2 + 6$

40. $P(x) = x^3 - 1$

Use synthetic division to evaluate each of the following polynomials for the value of r indicated.

41. $P(x) = x^3 + 2x^2 - 2x + 1, r = 4$

42. $P(x) = 3x^3 - x^2 + 7x - 2, r = -2$

43. $P(x) = x^3 - 5x^2 + 2, r = \sqrt{7}$

44. $P(x) = x^3 - 3x + 1, r = -\sqrt{5}$

Graph each of the following polynomials.

45. $P(x) = (x + 4)(x - 3)$

46. $P(x) = x^3(x + 3)$

47. $P(x) = (x + 3)(2x - 5)x$

48. $P(x) = (x^2 + 4x - 3)(x - 7)$

Write a third-degree polynomial that has as its roots the following:

49. $1, -2, 3$

50. $-3, 9, -5$

51. $0, 4, -3$

52. $-\sqrt{11}, -\pi, \dfrac{1}{2}$

53. $-\dfrac{3}{4}, -\dfrac{1}{2}, -\dfrac{1}{2}$

54. $\dfrac{2}{3}, -\dfrac{1}{5}, \dfrac{3}{5}$

SECTION 4

For each of the following polynomials P: (a) list the possible rational roots, (b) determine all rational roots, (c) where possible determine all real roots.

55. $P(x) = x^3 - 5x + 2$
56. $P(x) = x^3 - 2x^2 - 2x - 3$
57. $P(x) = 4x^3 + 12x^2 - x - 9$
58. $P(x) = 2x^4 - 10x^2 + 8$

Factor each of the following.

59. $P(x) = x^3 - 5x + 2$
60. $P(x) = x^3 - 2x^2 - 2x - 3$
61. $P(x) = 4x^3 + 12x^2 - x - 9$
62. $P(x) = 2x^4 - 10x^2 + 8$

Solve the following.

63. $4x^3 + 8x^2 - x - 5 = 0$
64. $x^3 - 19x^2 - 19x - 1 = 0$
65. $x^4 - 5x^3 - 3x^2 + 7x = 0$
66. $9x^4 - 24x^2 + 16 = 0$

Solve the following.

67. $4x^3 + 8x^2 - x - 5 > 0$
68. $x^3 - 19x^2 + 19x - 1 > 0$
69. $x^4 - 5x^3 - 3x^2 + 7x \geqslant 0$
70. $9x^4 - 24x^2 + 16 \leqslant 0$

Graph each of the following.

71. $P(x) = 4x^3 + 8x^2 - x - 5$
72. $P(x) = x^3 - 19x^2 + 19x - 1$
73. $P(x) = x^4 - 5x^3 - 3x^2 + 7x$
74. $P(x) = 9x^4 - 24x^2 + 16$

Determine the slope of the tangent line to the polynomial at the indicated root. Write an equation for the tangent line.

75. $P(x) = (x - 3)(x + 2), r = -2$
76. $P(x) = (x + 5)(x - 3), r = 3$
77. $P(x) = (x + 1)(x - 3)(x + 2), r = 3$
78. $P(x) = (x + 1)^2(x - 2), r = -1$

Prove the following.

79. $\sqrt{17}$ is irrational.
80. $\sqrt[3]{4}$ is irrational.

SECTION 5

Approximate the root of each of the following to one decimal place (nearest tenth).

81. $P(x) = x^4 - 3$
82. $P(x) = x^5 + 7$

Each of the following has a root between 0 and 1. Approximate the root to one decimal place (nearest tenth).

83. $P(x) = x^3 - 3x + 1$
84. $P(x) = x^3 + 5x^2 - 5$

SECTION 6

Use synthetic division to determine whether the indicated value of z is a root of the given polynomial.

85. $P(z) = z^4 + 1, z = -i$
86. $P(z) = z^2 - 4z + 5, z = 2 - 3i$
87. $P(z) = z^2 - 4z + 13, z = 2 + 3i$
88. $P(z) = z^3 - z^2 + 2, z = 2 + i$

For each of the following polynomials, determine all the roots beginning with the given root.

89. $P(z) = z^2 + 6z + 10, z = -3 + i$
90. $P(z) = z^2 - 2z + 26, z = 1 - 5i$
91. $P(z) = z^3 + 3z^2 + z - 5, z = -2 + i$
92. $P(z) = z^2 + (3 - i)z - 3i, z = i$

Solve each of the following complex polynomial equations.

93. $5z + 2i = 4$
94. $(2 - i)z + i = 3$
95. $5z + iz - 4i + 1 = 0$
96. $2z^2 - 1 = 0$
97. $z^2 + z + 2 = 0$
98. $2z^2 - 3z + 6 = 0$
99. $z^2 + iz - 3i = 0$
100. $iz^2 - z + 2 = 0$
101. $iz^2 + 2z + 3i = 0$
102. $z^3 - z^2 + 2z = 0$

Carl Friedrich Gauss

Mathematics is the queen of sciences, and the theory of numbers is the queen of mathematics.
—CARL FRIEDRICH GAUSS

Many people regard Carl Friedrich Gauss as the finest mathematician of the nineteenth century. Often called the *prince of mathematicians,* Gauss was born in 1777. His father, a laborer with little regard for formal education, often attempted to discourage his son from attending school. Were it not for his mother's intervention, Gauss may have ended up as a laborer also.

Gauss was a prodigy. At age 3, he detected an arithmetic error in his father's accounts. Not until age 10 did he enter his first arithmetic class. The child's ability astonished his teacher. Impressed, the teacher bought Carl an arithmetic book. Carl sped through the text leaving the teacher behind. The teacher conceded, "I can teach him nothing more."

By the time he was 14, Gauss had attracted the attention of Carl Wilhelm Ferdinand, Duke of Brunswick. The duke paid all the boy's bills when Gauss enrolled at Collegium Carolinum in February 1792. Proud of his "discovery," the duke continued support for Gauss throughout his life.

By the time he left Caroline College in 1795 Gauss had invented the method of "least squares" used extensively in fitting lines to data. Gauss was 18.

That same year he enrolled in the University of Göttingen. Just before his twentieth birthday Gauss decided to dedicate himself to mathematics.

Gauss also made many contributions to astronomy. He determined a precise astronomical formula for calculating the date of Easter. When Giuseppe Piazzi discovered Ceres in 1801, only Gauss was able to calculate the orbit of that asteroid. In 1807 Gauss became director of the University of Göttingen observatory. Gauss dedicated much of his later life to astronomy.

Gauss contributed to other areas of physics as well. He invented the electric telegraph. In his honor, the unit of measure of magnetic induction is the *gauss.*

The contributions of Gauss pervade much of modern mathematics. Among these are the first satisfactory proof of the fundamental theorem of algebra. Perhaps the best measure of greatness is praise from one's colleagues. Mathematicians were quick to recognize the genius of Gauss. When Laplace was asked who was the greatest mathematician in Germany, Laplace answered "Pfaff." "What about Gauss?" responded the puzzled questioner. Laplace replied, "Gauss is the greatest mathematician in the world."

Honor for Gauss was not limited to his contemporaries. His achievements span history. Ask a mathematician to name the three greatest mathematicians of all times and most will respond, "Archimedes, Newton, and Carl Friedrich Gauss."

4 CHAPTER TEST

1. Determine the quotient and remainder for $(x^3 + 2x^2 - x + 3) \div (x - 1)$. Write the solution in the form $P(x) =$ divisor(quotient) + remainder.

2. For $P(x) = 2x^3 + 5x^2 - x - 4$, determine (a) the number of variations of sign of $P(x)$, (b) the possible number of positive roots of $P(x)$, (c) the number of variations of sign of $P(-x)$, (d) the possible number of negative roots of $P(x)$, (e) determine an upper bound and a lower bound for the roots.

3. Use synthetic division to evaluate the following polynomial as indicated: $P(x) = x^3 + 2x^2 - x + 3$, determine $P(-1)$

In Problems 4 and 5, for each of the polynomials P (a) list the possible rational roots, (b) determine all rational roots, (c) where possible determine all real roots, (d) factor the polynomials.

4. $P(x) = x^3 - 8$

5. $P(x) = 2x^3 - 3x^2 - 4x + 6$

6. Solve $2x^3 - 3x^2 - 4x + 6 < 0$.

7. What is the slope of the line tangent to $P(x) = (x - 1)^2(x + 2)$, where $x = 1$? Write the equation of the tangent line.

8. Sketch the graph of $P(x) = (x - 1)^2(x + 2)$.

9. Prove $\sqrt[3]{2}$ is irrational.

*10. Determine all the roots $P(z) = z^3 - 3z^2 + 9z + 13$, given root $z = 2 - 3i$.

RATIONAL AND RADICAL FUNCTIONS

Previously, synthetic division proved helpful in factoring and graphing polynomial functions. However, division did not always result in a polynomial function: some had remainders. This chapter analyzes the quotients of polynomial functions.

The composition of functions provided an operation for forming the inverse of a function. The inverse of raising x to a power n (x^n) is taking the nth root of x, $\sqrt[n]{x}$ (radical notation). This chapter also investigates the behavior of radical functions.

We classify function formulas according to the operations on the independent variable x. Polynomial functions are algebraic combinations of constant functions with the identity function, $I(x) = x$. A formula for a polynomial uses a finite number of additions or multiplications.

Polynomials resembled integers in their structure. Arithmetic extends the integers to rational numbers by use of division: the indicated quotient of two integers (where the divisor is not 0) is a rational number. Similarly, we extend polynomials to rational functions. If we include taking a root, $\sqrt[n]{}$, among algebraic operations, then a function formed through the finite addition, multiplication, division or "rooting" of constant functions and the identity function is an **algebraic function**.

RATIONAL FUNCTIONS

Mathematicians assume the right to choose, within the limits of logical contradiction, what path they please in reaching their results.
—HENRY ADAMS

The simplest quotient of polynomials that is not a polynomial is the function $H(x) = 1/x$. The domain does not include 0. The graph does not intersect

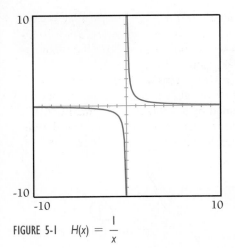

FIGURE 5-1 $H(x) = \dfrac{1}{x}$

Fraction—The Arabic word for fraction *al-kasr* derives from a stem meaning "to break." The Latin past participle of *frangere*, "to break," is *fractus*. Early English writers used "broken numbers." Chaucer used the word *fraction* in 1321. In 1556 Tartaglia offered modern fraction notation: "write the numerator above a little bar and the denominator below it."

the y-axis. Because $H(-x) = -H(x)$, H is an odd function, symmetric to the origin. Once the graph is established for positive values of x, reflect the graph through the origin to obtain the graph over the negative portion of the domain. Use graphing technology or plot points to convince yourself that Figure 5-1 represents the function H. The shape of the graph of H is a hyperbola.

The graph exhibits interesting behavior in the neighborhood of the x-axis and y-axis. Because $H \circ H(x) = x$, H is its own inverse. Any conclusions about the graph and the y-axis also describe the graph in relation to the x-axis.

Because 0 is not in the domain of the function, we cannot plot a point of the function for $x = 0$. Instead plot points "close" to 0 to discover the behavior of the graph "near" the y-axis. Use your graphing technology trace feature to verify that points of H "near" the y-axis include (0.5, 2), (0.4, 2.5), (0.3, $\frac{10}{3}$), (0.2, 5), (0.1, 10) and (0.01, 100).

Follow the path of H toward $x = 0$ along the righthand branch of the curve. We say that, as x approaches 0 through positive values, $H(x)$ increases without bounds. The idea of examining y values of a graph as x approaches a given value is known as a **limit**. Abbreviate the statement that $H(x)$ is increasing without bounds as x gets close to 0 from the positive direction as

$$\lim_{x \to 0^+} \frac{1}{x} = +\infty$$

read "the limit of $1/x$ as x approaches 0 from the right is positive infinity." Because of symmetry to the origin, the limit of $1/x$ as x approaches 0 from the left is negative infinity:

$$\lim_{x \to 0^-} \frac{1}{x} = -\infty$$

The idea of a limit is one of the foundations of calculus.

The y-axis serves as a graphing guide for H. As the x-values get closer to 0, the value of y increases (or decreases) without bounds; the graph gets closer to the y-axis, but never quite reaches it. The y-axis plays the role of a **vertical asymptote** for the graph of H. Because H is its own inverse, the x-axis serves as a **horizontal asymptote**.

As x gets larger, $1/x$ approaches 0, written

$$\lim_{x \to +\infty} \frac{1}{x} = 0$$

Also $$\lim_{x \to -\infty} \frac{1}{x} = 0$$

Asymptotes are excellent aids for graphing. Given an asymptote, we know the graph "follows the asymptote" towards infinity. We can anticipate the shape of the graph. This anticipation allows us to correct misrepresentations of asymptotes by graphing technology.

Functions in the same family usually share some characteristics. We expect functions formed by division of polynomials to exhibit some of the same attributes as the function H. The remainder of this chapter investigates the implied domain, vertical and horizontal asymptotes, x-intercepts and graphs of functions formed from quotients of polynomials.

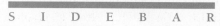
Recall the parallels between polynomials and integers. When dividing integers, two choices are apparent. One choice is to restrict the solution to integers. The last chapter treated division of polynomials in that manner. Synthetic division provided a polynomial quotient and a remainder.

The second division choice allows answers that are not integers. Therefore, we write 14/5 as the decimal 2.8 or in rational form, $\frac{14}{5}$. One rational number replaces the information contained by a quotient and remainder. We define rational numbers in terms of two integers, the numerator and the denominator. These comments motivate the definition of rational functions.

DEFINITION 5-1
Rational Functions

A **rational function** f is a function given by

$$f(x) = \frac{N(x)}{D(x)}, \qquad D(x) \neq 0$$

where the denominator $D(x)$ and numerator $N(x)$ are polynomials.

Because $D(x) = 1$ represents a polynomial, polynomial functions are a special case of rational functions in which the denominator is 1. The denominator demands attention in a rational function. What are the consequences of the domain restrictions inferred from the denominator of a rational function? As usual we frame our answers in both geometric and algebraic terms. Consider a rational function such as

$$f(x) = \frac{x + 3}{(x - 2)(x + 3)}$$

The denominator cannot be 0, so

$$D(x) \neq 0$$
$$(x - 2)(x + 3) \neq 0$$
$$x - 2 \neq 0 \quad \text{and} \quad x + 3 \neq 0$$
$$x \neq 2 \quad \text{and} \quad x \neq -3$$

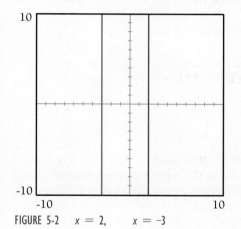

FIGURE 5-2 $x = 2$, $x = -3$

No values for f exist for $x = -3$ or $x = 2$, see Figure 5-2. The graph cannot intersect the lines $x = -3$ or $x = 2$. Because f is formed from polynomials, the graph of f should be smooth, well-behaved and continuous at every point of its domain. But -3 and 2 are not in the domain of f. "Breaks" must occur in the graph of f where $x = -3$ and $x = 2$.

More on f in a moment. First, let us consider the possible nature of a "break" in a graph. Imagine a graph of a function where the domain is all real numbers except possibly a, b and c. Figure 5-3 shows such a graph.

Examine the break in the graph where $x = a$. The small circle indicates a single point missing from an otherwise continuous region of the graph. Such a break is a removable discontinuity: plugging a point into the hole repairs the break in the graph.

Consider the break in the graph of Figure 5-3 at $x = b$. You should recognize this break as asymptotic behavior of the graph. The line given by $x = b$ is

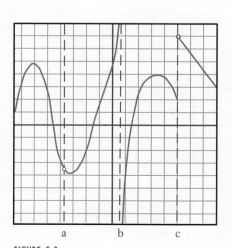

FIGURE 5-3

To properly define the concept of *continuity* requires ideas from calculus. Our use of *continuity* in this text is not intended to replace a rigorous calculus definition. Nor does the use of the term *discontinuous* reflect exact terminology from calculus. Our usage of *discontinuity* indicates "not continuous."

DISPLAY 5-1

A more colorful way of recalling a removable discontinuity is as "the little Dutch boy" discontinuity (Display 5-1), named after the story of the boy who plugged the ocean flow through a break in a dyke.

a vertical asymptote for f. Replacing a missing point cannot repair this break in the graph. The break is nonremovable.

For the last case, note the severe fracture in the graph of Figure 5-3 where $x = c$. The graph of f exhibits an abrupt change in y values for x values near c. The graph cannot be united into a continuous curve at $x = c$ without destroying the original shape of the graph. Replacing one point will not repair the break. The break is nonremovable.

Each of these breaks in a graph corresponds to information contained in the function formula. Now we return to our analysis of

$$f(x) = \frac{x + 3}{(x - 2)(x + 3)}$$

Graphing technology reveals the graph of f as shown in Figure 5-4.

Note that graphing technology is unable to display the removable discontinuity where $x = -3$. We annotated the circle to indicate the removable discontinuity. The break in the graph at $x = 2$ is due to a nonremovable asymptote. Depending on your technology and the window you choose, your graph may display an extra line or not reach the bounds of your window. You must correct this by hand.

Let us reduce the formula for f to lowest terms:

$$f(x) = \frac{1}{(x - 2)}, \qquad x \neq -3, \quad x \neq 2$$

Recall that

$$\frac{(x + 3)}{(x - 2)(x + 3)} = \frac{1}{(x - 2)}$$

only if $x \neq -3$ and $x \neq 2$. But $1/(x - 2)$ offers no hint that $x = -3$ is forbidden. Without the note about $x \neq -3$, we would have a different function. Call this different function g. Figure 5-5 illustrates the graph of $g(x) = 1/(x - 2)$, $x \neq 2$.

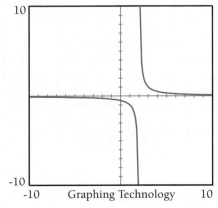
Graphing Technology

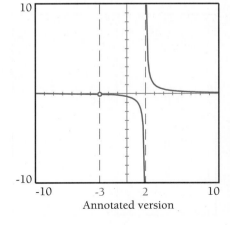
Annotated version

FIGURE 5-4 $f(x) = \dfrac{x + 3}{(x - 2)(x + 3)}$

DISPLAY 5-2   Chicken contemplating crossing an asymptote.

A more colorful description for an asymptotic discontinuity is the "why the chicken crossed the road" discontinuity.

DISPLAY 5-3

Finally, a colorful description for an abrupt *y* jump break is the "beam me up" discontinuity.

Notice that *g* has no break for $x = -3$. Reducing the rule of pairing to lowest terms removed the discontinuity at $x = -3$. No algebra can remove the restriction imposed by $x - 2$ in the denominator. The discontinuity at 2 is due to asymptotic behavior: the discontinuity is not removable. See the following table.

Domain analysis	Type of break	Note
$x \neq -3$	Removable	$(x + 3)$ reduced
$x \neq 2$	Asymptote	$(x - 2)$ did not reduce

Removable discontinuites and asymptotic discontinuities are the only types of break in the graph of a rational function, because we assume that an algebraic function is continuous, smooth and well-behaved at each point in its domain.

How does the remaining type of discontinuity usually occur? Consider the graph of a function whose rule of pairing is given piecewise.

$$p(x) = \begin{cases} x + 2 & \text{if } x > 1 \\ x^2 - 1 & \text{if } x \leq 1 \end{cases}$$

The graph of the function *p* comes in two pieces (Figure 5-6). The graph of *p* is a line to the right of $x = 1$. To the left and up to $x = 1$ the graph is a parabola. Note the "beam me up" discontinuity at $x = 1$. The definition of *p* is not a simple algebraic formula: *p* requires two formulas to describe one rule of pairing; hence the definition is formulated as a **piecewise function**. However, *p* is a piecewise continuous function: the graph segments of *p* each have a formula defined over a discrete subinterval of the domain of *p*. Then each segment is continuous over its defined interval. For larger segments of the function to be continuous, the segment endpoints must align. For example, in

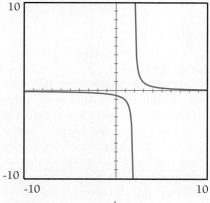

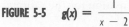

FIGURE 5-5 $g(x) = \dfrac{1}{x - 2}$

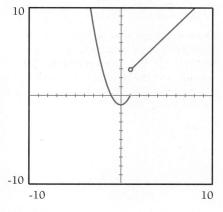

FIGURE 5-6   *p*

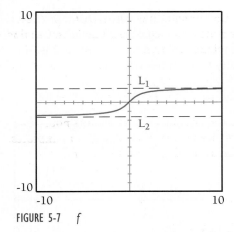

FIGURE 5-7 f

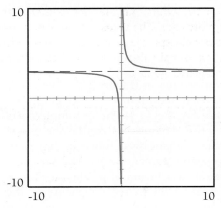

FIGURE 5-8 $t(x) = \dfrac{3x + 1}{x} = 3 + \dfrac{1}{x}$

the rule of pairing for p, replace $x + 2$ with $1 - x$ and the separate segments would connect at $x = 1$ into a continuous function.

Examination of the denominator of a rational function provides insight into vertical asymptotes. But rational functions may have asymptotes that are not vertical: one possibility is horizontal aysmptotes. Because we are discussing functions, the number of horizontal asymptotes is limited. No function has more than two horizontal asymptotes. Examine Figure 5-7.

Consider a function f and a horizontal asymptote L_1 associated with f on $(0, \infty)$. As x gets large, $f(x)$ gets close to L_1. Because f is a function, it cannot simultaneously *get close* to a second horizontal line.

A similar argument for the negative side of the graph establishes a maximum of two asymptotes: one for the negative side, one for the positive side. Sometimes the same line serves as an asymptote as x increases without bounds as well as when x decreases without bounds.

To identify nonvertical asymptotes requires another format for a rational function. The form is a compromise between the polynomial format and the rational format. The inspiration from arithmetic is "mixed numbers." Consider the rational number $\frac{13}{5}$. In the checked format for integer division: $13 = 2(5) + 3$. Divide by 5 to obtain a "mixed" number, half integer, half rational: $\frac{13}{5} = 2 + \frac{3}{5}$.

The fractional part of the expression should be reduced to lowest terms. Also, the absolute value of the numerator of the fractional part must be "smaller than the positive denominator." Because rational expressions are neither positive nor negative until they are evaluated, we have a different requirement for the fractional part of a rational function: the *degree* of the numerator must be smaller than the degree of the denominator.

Polynomial division converts rational functions into mixed number format. Other methods exist, but most involve tedious arithmetic. This chapter restricts problems to those with feasible arithmetic. Consider the function t displayed in Figure 5-8.

$$t(x) = \frac{3x + 1}{x}$$

$$= 3 + \frac{1}{x}$$

As x becomes larger, $1/x$ approaches 0, so that the graph of t gets closer to the line $y = 3$. As x approaches negative infinity, $1/x$ closes on 0 and the graph of t approaches the line $y = 3$. Then $y = 3$ is a horizontal asymptote for the function t. This is no surprise, because t is a translation of $H(x) = 1/x$. Compare Figure 5-8 to Figure 5-1.

More generally, suppose that $f(x) = mx + b + 1/x$. As x increases without bounds $1/x$ approaches 0 so that $f(x)$ gets closer to $y = mx + b$. If m is 0, then $y = b$ is a horizontal aysmptote for f. If m is not 0, then $y = mx + b$ is an oblique line. The line $y = mx + b$ is a **slant asymptote** for f. Thus, the "whole number" portion of a rational function in "mixed number" format provides asymptote information for the rational function.

EXAMPLE 5-1

Illustration 1:

Graphing Rational Functions. Determine the domain and *x*-intercepts of each of the following functions. Identify asymptotes for each of the functions. Use this information in the sketch of the graph of the function.

$$k(x) = \frac{x + 1}{(x - 2)(x + 4)}$$

Solution: Domain $x \neq 2$, $x \neq -4$. Because the rational expression does not reduce, no discontinuities are removable: $x = 2$ and $x = -4$ are vertical asymptotes. The degree of the numerator is 1 whereas the degree of the denominator is 2. In mixed format we have

$$k(x) = 0 + \frac{x + 1}{(x - 2)(x + 4)}$$

Now, $y = 0$ is a horizontal asymptote. When the numerator $(x + 1)$ is 0 the entire expression is 0. The *x*-intercept for the graph is -1.

Asymptotes are powerful graphing aids, helping us anticipate the behavior of the graph. One way to summarize this information is to construct a chart for $N(x)/D(x)$. The zeros of the numerator provide candidates for *x*-intercepts. Factors from the denominator require more care. The zeros for the denominator produce undefined values for the rational expression. We mark these zeros with an * and label the corresponding functional value with *U* or *A* to indicate undefined (a hole) or asymptote, respectively.

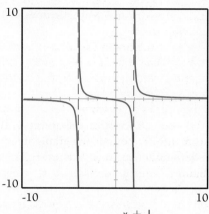

FIGURE 5-9 $k(x) = \dfrac{x + 1}{(x - 2)(x + 4)}$

x-axis	$-\infty$		-4		-1		2		$+\infty$	Note
$x + 1$	-		-		0	+			+	Numerator
$x - 2$	-		-			-	0*		+	Denominator
$x + 4$	-	0*	+			+			+	Denominator
$\dfrac{x + 1}{(x - 2)(x + 4)}$	-	A	+		0	-	A		+	
Graph	Below		Above		On	Below		Above		

Graphing technology (Figure 5-9) confirms the details of the chart.

Illustration 2:

$$f(x) = \frac{x^2 - 5x + 6}{x - 2}$$

Solution: Domain: $x \neq 2$. Use synthetic division to rewrite the formula in mixed format:

$$f(x) = (x - 3) + \frac{0}{x - 2}$$
$$= x - 3, \qquad x \neq 2$$

The fractional portion disappeared. Therefore, $x = 2$ identifies a removable discontinuity. Moreover, the graph coincides with $y = x - 3$ at all points except $x = 2$. The graph lies in a straight line and has no asymptotes. To convert $y = x - 3$ to the graph for f simply punch a hole in it at $x = 2$ (Figure 5-10).

Illustration 3:

$$j(x) = \frac{x^2 - 5x + 7}{x - 2}$$

Solution: Compare this to Illustration 2.

$$j(x) = \frac{x^2 - 5x + 6 + 1}{x - 2}$$ (or use synthetic division)

$$= \frac{x^2 - 5x + 6}{x - 2} + \frac{1}{x - 2}$$

$$= x - 3 + \frac{1}{x - 2}$$

Note the $1/(x - 2)$ term. Now, $x = 2$ represents a vertical asymptote and $y = x - 3$ is a slant asymptote. As x increases without bounds the graph of $f(x)$ approaches the line $y = x - 3$. The graph does not cross the x-axis: $x^2 - 5x + 7$ has no real roots (Figure 5-11).

Illustration 4:

$$k(x) = \frac{2x^3 + 8x^2 - 14x - 20}{x^2 + 2x - 8}$$

Solution: Rewrite k in both formats. Factor the numerator by using the rational root theorem:

$$2x^3 + 8x^2 - 14x - 20 = 2(x - 2)(x + 1)(x + 5)$$

Factor the denominator: $x^2 + 2x - 8 = (x - 2)(x + 4)$

Rewrite k:

$$k(x) = \frac{2(x - 2)(x + 1)(x + 5)}{(x - 2)(x + 4)}$$

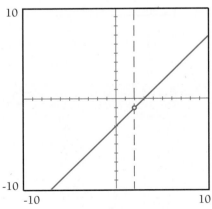

FIGURE 5-10 $f(x) = \dfrac{x^2 - 5x + 6}{x - 2}$

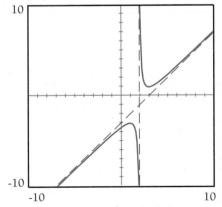

FIGURE 5-11 $j(x) = \dfrac{x^2 - 5x + 7}{x - 2}$

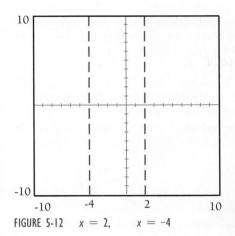

FIGURE 5-12 $x = 2$, $x = -4$

The domain of k is $x \ne 2$ and $x \ne -4$. The discontinuity at $x = 2$ is removable. The line given by $x = -4$ is a vertical asymptote (Figure 5-12).

From the numerator, the x-intercepts are -1 and -5. Because 2 is not in the domain of k, 2 is not an intercept. Punch a hole in the graph at $x = 2$. To rewrite in mixed format, use long division:

							$2x$	$+$	4
$x^2 + 2x - 8$			$2x^3$	$+$	$8x^2$	$-$	$14x$	$-$	20
			$-2x^3$	$-$	$4x^2$	$+$	$16x$		
					$4x^2$	$+$	$2x$	$-$	20
					$-4x^2$	$-$	$8x$	$+$	32
Remainder							$-6x$	$+$	12

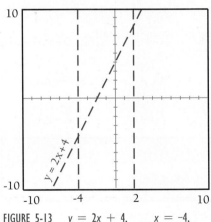

FIGURE 5-13 $y = 2x + 4$, $x = -4$,
$x = 2$

$y = 2x + 4$

Now, $k(x) = 2x + 4 + (-6)\dfrac{(x - 2)}{(x - 2)(x + 4)}$

$k(x) = 2x + 4 - \dfrac{6}{x + 4}$

As x increases or decreases without bounds, $-6/(x + 4)$ approaches 0. The graph of k approaches the line $y = 2x + 4$. Thus $y = 2x + 4$ gives a slant asymptote. See Figure 5-13. Now for the chart for k.

x-axis	$-\infty$	-5		-4		-1		2	∞	Note
$x - 2$	$-$		$-$		$-$		$-$	0	$+$	Numerator
$x + 1$	$-$		$-$		$-$	0	$+$		$+$	Numerator
$x + 5$	$-$	0	$+$		$+$		$+$		$+$	Numerator
$x + 4$	$-$		$-$	$0*$	$+$		$+$		$+$	Denominator
$x - 2$	$-$		$-$		$-$		$-$	$0*$	$+$	Denominator
$k(x)$	$-$	0	$+$	A	$-$	0	$+$	U	$+$	
Graph	Below	On	Above		Below	On	Above		Above	

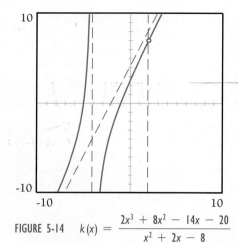

FIGURE 5-14 $k(x) = \dfrac{2x^3 + 8x^2 - 14x - 20}{x^2 + 2x - 8}$

Mark the zeros from the denominator with asterisks. The asterisks indicate that column as either undefined (*U*, a removable discontinuity) or as an asymptote (*A*). Technology completes the graph in Figure 5-14.

Note: We could have reduced the factored form of *k* to $(2x^2 + 12x + 10)/(x + 4)$ and then used synthetic division. However, the graph must conform to the domain restrictions from the original rule of pairing.

Either the chart or the graph in the last illustration provides a bonus. The solution of the inequality

$$\frac{2x^3 + 8x^2 - 14x - 20}{x^2 + 2x - 8} \geqslant 0$$

The lefthand side of the inequality is $k(x)$ from Illustration 4. Interpret this inequality as "what values of *x* make $k(x)$ positive or 0?" Read the answer from the last line of the chart for *k*.

$$x \in [-5, -4) \cup [-1, 2) \cup (2, \infty)$$

The interval endpoints at −4 and 2 are not part of the solution: *k* is undefined at these points rather than positive or 0.   ▬

EXAMPLE 5-2

Illustration 1:

Solving Rational Inequalities. Solve each of the following inequalities.

$$\frac{(x + 3)(x - 5)}{(x + 7)(2 - x)(x + 1)} \geqslant 0 \qquad \text{(Figure 5-15)}$$

Solution:

Factors	$x + 3$	$x - 5$	$x + 7$	$2 - x$	$x + 1$
Zeros	−3	5	−7	2	−1

x-axis	−∞	−7		−3		−1		2		5		∞
$x + 3$	−		−	0	+		+		+		+	
$x - 5$	−		−		−		−		−	0	+	
$x + 7$	−	0*	+		+		+		+		+	
$2 - x$	+		+		+		+	0*	−		−	
$x + 1$	−		−		−	0*	+		+		+	
$\dfrac{(x + 3)(x - 5)}{(x + 7)(2 - x)(x + 1)}$	+	A	−	0	+	A	−	A	+	0	−	
Graph	Above		Below	On	Above		Below		Above	On	Below	

We seek the values for *x* (top row) where the rational expression (bottom row) is positive or 0: $x \in (-\infty, -7) \cup [-3, -1) \cup (2, 5]$.

Illustration 2:

$$\frac{3(x-5)(x^2+1)(x-2)}{-2(x^2-6x+5)x^3} \le 0 \qquad \text{(Figure 5-16)}$$

Solution:

Factors	Real root	Notes
3	None	Constant
$x-5$	5	Increasing line
x^2+1	None	Parabola opens up
$x-2$	2	Increasing line
-2	None	Constant
x^2-6x+5	1, 5	(*Denominator) parabola opens up
x	0	(*Denominator) increasing line

Notice that the last factor x is a repeated factor of multiplicity 3.

x-axis	$-\infty$		0		1		2		5		∞
3	+		+		+				+		+
$x-5$	-		-		-				0		+
x^2+1	+		+		+				+		+
$x-2$	-		-		-		0		+		+
-2	-		-		-				-		-
x^2-6x+5	+		+		0*		-		0*		+
x^3	-		0*		+		+		+		+
$\dfrac{3(x-5)(x^2+1)(x-2)}{-2(x^2-6x+5)x^3}$	+		A		-		A	+	0		-	U	-
Graph	Above		Below		Above		On	Below			Below

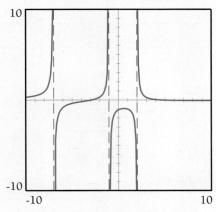

FIGURE 5-15 $y = \dfrac{(x+3)(x-5)}{(x+7)(2-x)(x+1)}$

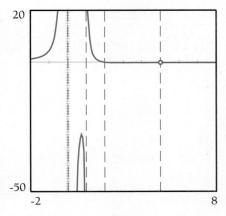

FIGURE 5-16 $x \in (0,1) \cup [2, 5) \cup (5, \infty)$

EXAMPLE 5-3

Illustration 1:

Applications. Suppose that Donald drives his car at 70 kph for 2 hours and 60 kph for 1 hour. What is his average speed for the three-hour trip?

Solution: Use the physics model given by distance = (average speed)(time). Therefore average speed = distance/time.

At 70 kilometers per hour, two hours of travel yields 140 kilometers distance 70(k/h)(2 h) = 140 k.

Each hour at 60 kph gives an additional 60 kilometers distance. The total distance is 200 kilometers. The average speed is (200 kilometers)/(3 hours): Average speed $\cong$ 66.6 kph.

Illustration 2:

Suppose that a trip from Chicago to Dallas is made at one speed and that the return trip from Dallas to Chicago is at a different speed. Determine a formula for the average speed of the round trip.

Solution: The distance from Chicago to Dallas is the same in both directions. In fact the actual distance is not important in our calculation. Call the distance D. Now let Rg be the rate going and Tg be the time going. Similarly, let Rr be the rate returning and Tr be the time returning. Finally let A represent the average rate for the trip. The following relations are apparent:

Total time of trip:	$T = Tg + Tr$
Total Distance of round trip is $2D$	
Distance of round trip	$AT = 2D$
Now,	$A = \dfrac{2D}{T}$
Recall that	$D = RgTg$
Also	$D = RrTr$
Therefore	$Tg = \dfrac{D}{Rg}$
	$Tr = \dfrac{D}{Rr}$
Because	$T = Tg + Tr$
then	$T = \dfrac{D}{Rg} + \dfrac{D}{Rr}$

Substitute this into the formula for the average rate:

$$A = \frac{2D}{\dfrac{D}{Rg} + \dfrac{D}{Rr}}$$

D divides out and we obtain

$$A = \frac{2}{\dfrac{1}{Rg} + \dfrac{1}{Rr}}$$

The actual distance D plays no role in the final formula. If $Rg = 500$ mph and $Rr = 600$ mph then average rate $\cong 545.45$ mph.

This average rate is the harmonic mean of Rg and Rr.

SUMMARY

Consider the rational function $f(x) = N(x)/D(x)$.

- The domain of f is the set of all real numbers such that $D(x) \neq 0$.

- Let $f(x) = \not{N}(x)/\not{D}(x)$, $\not{D}(x) \neq 0$, represent f reduced to lowest terms, then

  1. If $\not{D}(r) = 0$, $x = r$ is a vertical asymptote for f.
  2. If $D(r) = 0$ and $\not{D}(r) \neq 0$, $x = r$ is a removable discontinuity for f.

- If $D(x) = x - r$ and $f(x) = mx + b + R/(x - r)$, $m \neq 0$, then $y = mx + b$ is an asymptote if the constant $R \neq 0$.

- A chart for $f(x) = D(x)/N(x)$ abbreviates graphical information about f. A chart for $f(x)$ should distinguish between the zeros of $D(x)$ and the zeros of $N(x)$. Graphing technology provides the bottom line of a chart immediately, without analyzing the separate factors. However, removable discontinuities may not be obvious without the factors. Without technology, forming the chart from factors can aid you in sketching the graph of f. Once formed, the bottom line of a chart is invaluable in solving inequalities related to $f(x)$.

5-1   EXERCISES

For each of the following rational functions, (a) determine the domain and x-intercepts; (b) choose a window large enough to sketch the graph showing any vertical, horizontal or slant asymptotes as dashed lines; (c) sketch the graph. Indicate removable discontinuities with a small circle.

1. $f(x) = \dfrac{1}{x + 2}$

2. $f(x) = \dfrac{1}{x - 3}$

3. $f(x) = \dfrac{x + 5}{x - 3}$

4. $f(x) = \dfrac{x - 2}{x + 2}$

5. $f(x) = \dfrac{3x + 5}{2x - 1}$

6. $f(x) = \dfrac{2x - 1}{3x + 5}$

7. $f(x) = \dfrac{x - 2}{x^2 - 4}$

8. $f(x) = \dfrac{x + 3}{x^2 - 9}$

9. $f(x) = \dfrac{x^2 - 5x + 1}{x + 2}$

10. $f(x) = \dfrac{x^2 - 6x - 1}{x + 1}$

11. $f(x) = \dfrac{3(x - 2)}{(x + 4)(x - 5)}$

12. $f(x) = \dfrac{5(x + 4)}{(x - 3)(x + 2)}$

13. $f(x) = \dfrac{(x - 3)(x + 2)}{5(x + 4)}$

14. $f(x) = \dfrac{(x + 4)(x - 5)}{3(x - 2)}$

15. $f(x) = \dfrac{(1 - x)(x + 2)}{(x - 3)(2 - 3x)}$

16. $f(x) = \dfrac{(2 - 3x)(x - 3)}{(1 - x)(x + 2)}$

17. $f(x) = \dfrac{(x+5)(x-3)}{(x-2)(x+5)}$ **18.** $f(x) = \dfrac{(x-2)(x+5)}{(x+5)(x-3)}$

19. $f(x) = \dfrac{(x^2-3)(x)}{x^2+5x+2}$ **20.** $f(x) = \dfrac{x^2+5x+2}{(x^2-3)x}$

Solve the following inequalities.

21. $3(x-2)(x-5)(x+1) > 0$

22. $5(x-1)(x+2)(x-7) < 0$

23. $-2(3-x)(5+x)(x-1) < 0$

24. $-3(5-x)(x+3)(x+1) > 0$

25. $(2x+1)(3-x)(x^2-5) \geq 0$

26. $(x^2-7)(4-x)(3x-2) \leq 0$

27. $(x^2+3)(x^2-4x+1) \leq 0$

28. $(x^2-7x+2)(x^2+2) \geq 0$

29. $(x^2-16)(4x^2-9) > 0$

30. $(x^2-25)(9x^2-4) < 0$

31. $\dfrac{3(x-2)}{(x+4)(x-5)} < 0$ **32.** $\dfrac{5(x+4)}{(x-3)(x+2)} > 0$

33. $\dfrac{(x-3)(x+2)}{5(x+4)} \geq 0$ **34.** $\dfrac{(x+4)(x-5)}{3(x-2)} \leq 0$

35. $\dfrac{(1-x)(x+2)}{(x-3)(2-3x)} \leq 0$ **36.** $\dfrac{(2-3x)(x-3)}{(1-x)(x+2)} \geq 0$

37. $\dfrac{(x+5)(x-3)}{(x-2)(x+5)} > 0$ **38.** $\dfrac{(x-2)(x+5)}{(x+5)(x-3)} < 0$

39. $\dfrac{(x^2-3)(x^2)}{x^2+5x+2} < 0$ **40.** $\dfrac{x^2+5x+2}{(x^2-3)x^2} > 0$

41. $\dfrac{x^3+2x^2-6x-4}{x^3-8x+3} \leq 0$

42. $\dfrac{x^3-8x+3}{x^3+2x^2-6x-4} \geq 0$

43. $\dfrac{x^3-6x^2+10x-3}{x^3+6x^2+10x+4} \geq 0$

44. $\dfrac{x^3+6x^2+10x+4}{x^3-6x^2+10x-3} \leq 0$

45. $\dfrac{9x^3+18x^2-4x-8}{4x^3+12x^2-x-3} > 0$

46. $\dfrac{x^3+2x^2+6x+4}{3x^3-2x^2+3x-2} < 0$

47. $\dfrac{3x^3-2x^2+3x-2}{3x^3+2x^2+6x+4} \geq 0$

48. $\dfrac{4x^3+12x^2-x-3}{9x^3+18x^2-4x-8} \geq 0$

49. $\dfrac{9x^4-13x^2+4}{2x^4-19x^2+9} < 0$ **50.** $\dfrac{2x^4-19x^2+9}{9x^4-13x^2+4} < 0$

Solve the following. (Hint: to solve a rational inequality the righthand side should be 0. The lefthand side should be a single, factored rational expression.)

51. $\dfrac{x^2-3}{x-1} \geq 1$ **52.** $\dfrac{x^2-5}{x-2} < 1$

53. $\dfrac{4-2x}{x^2-x-2} < 1$ **54.** $\dfrac{6x+2}{x^2+2x-3} \geq 1$

Graph the following functions. Annotate the graphs, intercepts and asymptotes.

55. $f(x) = x^2 + \dfrac{1}{x}$ **56.** $f(x) = x^3 + \dfrac{1}{x}$

57. $f(x) = (x-2)^2 + \dfrac{1}{x-2}$ **58.** $f(x) = (x-1)^3 + \dfrac{1}{x-1}$

59. $f(x) = x^2 + \dfrac{1}{x-2}$ **60.** $f(x) = x^3 + \dfrac{1}{x-1}$

61. $f(x) = \dfrac{|x|}{x}$ **62.** $f(x) = \dfrac{x}{|x|}$

63. $f(x) = \begin{cases} -1 \text{ if } x \leq 0 \\ 1 \text{ if } x > 0 \end{cases}$ **64.** $f(x) = \begin{cases} x \text{ if } x \geq 0 \\ -x \text{ if } x < 0 \end{cases}$

65. $f(x) = \begin{cases} 2x \text{ if } x \geq 0 \\ -x \text{ if } x < 0 \end{cases}$ **66.** $f(x) = \begin{cases} x \text{ if } x > 0 \\ -x^2 \text{ if } x \leq 0 \end{cases}$

67. Show that the average rate formula

$$\dfrac{2}{\dfrac{1}{Rg}+\dfrac{1}{Rr}}$$

can be rewritten as

$$\dfrac{2(Rg)(Rr)}{Rg+Rr}$$

68. In a circuit with two resistors in parallel, the resistance equivalent is

$$R_{eq} = \dfrac{1}{\dfrac{1}{R_1}+\dfrac{1}{R_2}}$$

Show that

$$R_{eq} = \dfrac{R_1 R_2}{R_1+R_2}.$$

69. The intensity of light on a surface varies inversely as the square of the distance from light source to surface. If a point source of light provides 300 lumens of light from 2 meters, determine the constant of proportionality for this light source. Use a rational function to model the relationship of distance to intensity. Predict the light intensity at 20 meters. Examine the rule of pairing for the model and discuss the effect on light intensity as the distance increases without bounds.

70. The mutual force of attraction between two asteroids varies inversely as the square of the distance between the two objects. How would the distance between the two asteroids have to be changed to double the force of attraction between them? Discuss the attraction effects when the distance between the asteroids increases without bounds.

5-1   PROBLEM SET

A function is an *involution* if it is its own inverse. Therefore for an involution f, $f \circ f(x) = x$. Prove that each of the following functions is an involution.

1. $f(x) = -x + 4$

2. $f(x) = -x + 3$

3. $f(x) = -x - 5$

4. $f(x) = -x - 7$

5. $f(x) = \dfrac{1}{x}$

6. $f(x) = -\dfrac{1}{x}$

7. $f(x) = \dfrac{x + 3}{x - 1}$

8. $f(x) = \dfrac{x - 2}{x - 1}$

9. $f(x) = \dfrac{2x + 1}{x - 2}$

10. $f(x) = \dfrac{3x - 2}{x - 3}$

11. $f(x) = \dfrac{3x - 2}{2x - 3}$

12. $f(x) = \dfrac{5x - 3}{2x - 5}$

13. $f(x) = \dfrac{4x - 1}{3x - 4}$

14. $f(x) = \dfrac{2x - 5}{3x - 2}$

Graph the following involutions.

15. $f(x) = \dfrac{2x + 1}{x - 2}$

16. $f(x) = \dfrac{3x - 2}{x - 3}$

17. $f(x) = \dfrac{3x - 2}{2x - 3}$

18. $f(x) = \dfrac{5x - 3}{2x - 5}$

19. $f(x) = \dfrac{4x - 1}{3x - 4}$

20. $f(x) = \dfrac{2x - 5}{3x - 2}$

21. Form the inverse of $y = (ax + b)/(cx + d)$ by interchanging the roles of x and y and solving for y. Compare the results with the original function.

22. Prove that $f(x) = (ax + b)/(cx - a)$ is an involution. Are any restrictions necessary on a, b or c?

PROBLEMS FOR GRAPHING TECHNOLOGY

Use technology to graph the following. In each case, annotate vertical and skew asymptotes. Note whether asymptotic behavior is correctly displayed. You may have to switch between connected and dot mode to determine the correctness of asymptotic displays. Finally, determine whether the display accurately represents removable discontinuities.

23. $f(x) = \dfrac{1}{x - 3}$

24. $f(x) = \dfrac{5}{x + 2}$

25. $f(x) = \dfrac{x - 3}{x - 3}$

26. $f(x) = \dfrac{5x + 10}{x + 2}$

27. $f(x) = 2 + \dfrac{1}{x - 3}$

28. $f(x) = 1 - \dfrac{5}{x + 2}$

29. $f(x) = 2x\dfrac{x - 3}{x - 3}$

30. $f(x) = x\dfrac{5x + 10}{x + 2}$

31. $f(x) = 2x + \dfrac{1}{x - 3}$

32. $f(x) = x + \dfrac{5x + 10}{x + 2}$

EXPLORATION

The following problems anticipate concepts from the next section. Use graphing technology to graph the following in the standard window. Note any intervals for which no graph appears.

33. $f(x) = \sqrt{x}$

34. $f(x) = \sqrt{-x}$

35. $f(x) = \sqrt{5 - x}$

36. $f(x) = \sqrt{x - 4}$

37. $f(x) = \sqrt{x^2 - 9}$

38. $f(x) = \sqrt{25 - x^2}$

39. $f(x) = \sqrt{\dfrac{1}{x - 2}}$

40. $f(x) = \sqrt{\dfrac{1}{x + 3}}$

41. $f(x) = \sqrt{x^2}$

42. $f(x) = (\sqrt{x})^2$

43. $f(x) = \sqrt[3]{x + 1}$

44. $f(x) = \sqrt[3]{x - 2}$

45. $f(x) = \sqrt[4]{3 - x}$

46. $f(x) = \sqrt[5]{1 - x}$

47. Refer to the graphs in Problems 33–36. The shape of these graphs resembles a function we have previously studied. What is that shape? How are these graphs different from the earlier graphs?

48. Refer to Problems 37 and 38. Answer the same questions from Problem 47 for these functions.

49. Compare and contrast the graphs of Problems 41 and 42.

50. Which of the graphs in Problems 43–46 does not include the complete interval [-10, 10] in its domain? Discuss the differences in these graphs that contribute to the restricted domain.

2   RADICAL FUNCTIONS

In most sciences one generation tears down what another has built and what one has established another undoes. In Mathematics alone each generation builds a new story to the old structure.
—HERMANN HANKEL

This section examines radical functions. Recall that $\sqrt{2}$ is a root of $p(x) = x^2 - 2$. But from the rational root theorem, p has no rational roots. Therefore, $\sqrt{2}$ is not rational. See Figure 5-17.

The x-intercepts for $p(x) = x^2 - 2$ are not rational numbers. As division motivated the extension of integers to rational numbers, taking roots provokes the investigation of irrational numbers. Irrational numbers fill in the gaps of the real number line left vacant by rational numbers. Irrational numbers that come from taking "roots" of rational numbers inspire the radical functions covered in this section.

DEFINITION 5-2
Radical Functions

A **radical function** f is a function of the form

$$f(x) = \sqrt[n]{R(x)},$$

where $n > 1$, n is an integer called the *index*, and the radicand $R(x)$ is a rational function.

 1. If the index n is an even integer, the domain of f is restricted by $R(x) \geq 0$
 2. If the index n is an odd integer, then the domain of R is the domain of f.

As customary, the name *radical function* reflects the last indicated operation in the formula. Only *even* indexed radical functions require domain restrictions: even roots of a negative number are not real. Odd roots accept any real argument—positive, negative or 0. Restrictions on the domain of the radicand $R(x)$ may impose additional restrictions on the domain of the radical function f.

Consider the graphs of $f(x) = \sqrt{x}$, $g(x) = \sqrt[3]{x}$, $r(x) = \sqrt[4]{x}$ and $s(x) = \sqrt[5]{x}$ shown in Figure 5-18. Compare these to the graphs of $f^*(x) = x^2$, $g^*(x) = x^3$, $r^*(x) = x^4$ and $s^*(x) = x^5$ shown in Figure 5-19.

The domain of the even indexed radical functions is $[0, \infty)$. The

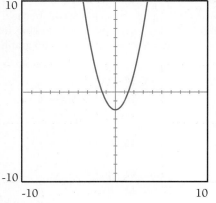

FIGURE 5-17 $p(x) = x^2 - 2$

Root—The Arab writers conceived a square number as growing out of a root, whereas Latin writers imagined that a side generated a geometric square. The Latin term was *latus*. Latin writers discovered the *latus* from the square. Their Arab counterparts "pulled out" (extracted) the root. The Arabic view of a root translated into the Latin equivalent for root, *radix*. From *radix* comes terms such as *radical* and *radish* (a tangy root). In 1621 Rudoph introduced the radical symbol $\sqrt{\ }$. Euler surmised that deforming the first letter of radix, *r*, inspired $\sqrt{\ }$.

S I D E B A R

In statistics, the spread of data is measured by taking the square root of average squared deviation from the mean. (See Display 5-4.)

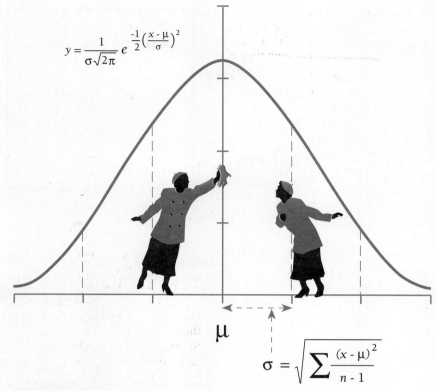

$$y = \frac{1}{\sigma\sqrt{2\pi}}\, e^{-\frac{1}{2}\left(\frac{x-\mu}{\sigma}\right)^2}$$

$$\sigma = \sqrt{\sum \frac{(x-\mu)^2}{n-1}}$$

DISPLAY 5-4   Normal bell-shaped curve

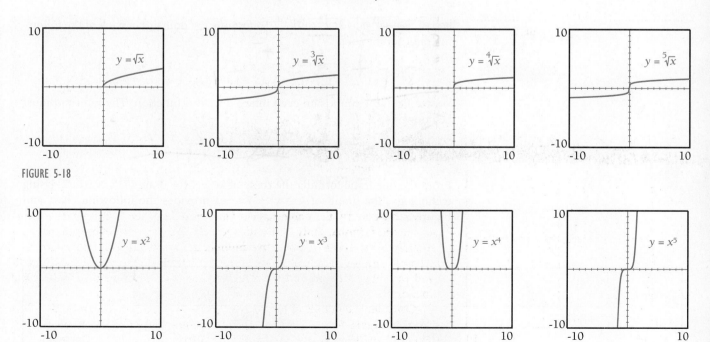

FIGURE 5-18

FIGURE 5-19

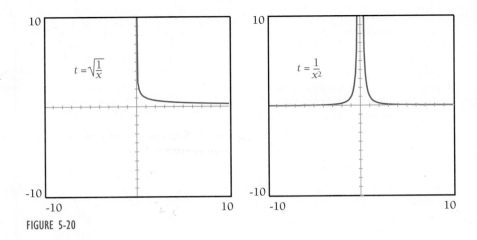

FIGURE 5-20

corresponding converse functions are not one to one. The odd powered polynomial functions are one to one. As a result, each corresponding inverse radical function is a mirror image across the line $y = x$.

Consider the graphs of $t(x) = \sqrt{1/x}$ and $t^*(x) = 1/x^2$ shown in Figure 5-20. Compare the asymptotes and restrictions on the domain.

This section analyzes the effects on a graph of composition of the radicand with a radical function. As usual, transformations stretch, contract and shift the graph about the plane. However, the first clue for anticipating the graph of a radical function is the domain.

EXAMPLE 5 - 4

Illustration 1:

Implied Domains of Radical Functions. Determine the domain for each of the following radical functions.

$$f(x) = \sqrt[4]{x^2 - 7x + 3}$$

Solution: The function f is an even indexed radical function. The radicand must be nonnegative.

$$R(x) \geqslant 0$$
$$x^2 - 7x + 3 \geqslant 0$$

From the quadratic formula the roots of $x^2 - 7x + 3$ are $(7 \pm \sqrt{37})/2$. Using technology, the graph of $y = x^2 - 7x + 3$ provides the following chart and Figure 5-21.

x-axis	$-\infty$	$\dfrac{7 - \sqrt{37}}{2}$		$\dfrac{7 + \sqrt{37}}{2}$		Note
$x^2 - 7x + 3$	$+$	0	$-$	0	$+$	Parabola
Graph	Above	On	Below	On	Above	Opens up

$$\text{Domain}_f = \left(-\infty, \frac{7 - \sqrt{37}}{2}\right] \cup \left[\frac{7 + \sqrt{37}}{2}, \infty\right)$$

$$f(x) = \sqrt{\frac{x - 2}{x + 5}}$$

Solution: Because the radical is even indexed, the radicand may assume only nonnegative values:

$$R(x) \geq 0$$

$$\frac{(x - 2)}{(x + 5)} \geq 0$$

Because

$$y = \frac{x - 2}{x + 5}$$

$$= 1 - \frac{7}{x + 5}$$

Therefore, $y = 1$ is a horizontal asymptote (Figure 5-22). Form a chart for $y = (x - 2)/(x + 5)$.

x-axis	$-\infty$		-5		2		∞	Note
$\dfrac{x - 2}{x + 5}$		$+$	A	$-$	0	$+$		Rational
Graph		Above		Below	On	Above		

$$\text{Domain}_f = (-\infty, -5) \cup [2, \infty).$$

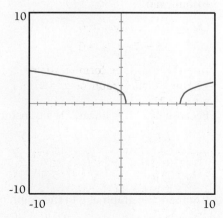

FIGURE 5-21 $f(x) = \sqrt[4]{x^2 - 7x + 3}$

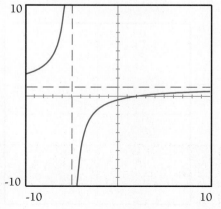

FIGURE 5-22 $f(x) = \dfrac{x - 2}{x + 5}$

Illustration 3:

$$g(x) = \sqrt[3]{\frac{x-2}{x+5}}$$

Solution: The index is odd. The only restriction is that the radicand produce a real number. And $(x-2)/(x+5)$ is real so long as the denominator is not 0.

$$\text{Domain}_g = \{x : x \neq -5\}.$$

Figure 5-22 establishes graph of $f(x) = (x-2)/(x+5)$. How does composition of f with a radical function g, $g(x) = \sqrt[n]{x}$, n is odd, affect the graph of g? Any root of 0 is 0 so that the x-intercepts of the radicand are x-intercepts of the radical function. Then, $x = 2$ is an x-intercept for both f and $g(x) = \sqrt[n]{f(x)}$.

Examine the vertical asymptote at $x = -5$. As x approaches -5 from the left the corresponding y value becomes infinitely large. A root of a very large number is very large. See Figure 5-22. As x approaches -5 from the right, y approaches $-\infty$. An odd root of a number with a large absolute value has a large absolute value. A vertical asymptote from the domain of a radicand indicates a vertical asymptote for the radical function.

What of horizontal asymptotes? Rewrite $g(x) = \sqrt[3]{(x-2)/(x+5)}$ so that the radicand is in mixed format: $g(x) = \sqrt[3]{1 - 7/(x+5)}$. As x increases without bounds $7/(x+5)$ approaches 0. Thus, $g(x)$ approaches $\sqrt[3]{1} = 1$. If radicand $R(x)$ has a horizontal asymptote of $y = c$, $c > 0$, then $f(x) = \sqrt[n]{R(x)}$ has a horizontal asymptote of $y = \sqrt[n]{c}$.

For complex radicands graphing technology fleshes out the details of a graph. Some technology requires the user to filter the x value inputs to avoid domain errors. In these programs, you must select a graphing window that does not exceed the domain of the function. Most technology automatically avoids attempts to graph outside the domain.   ▬

EXAMPLE 5-5

Graphing Radical Functions. Sketch the graph of each of the following radical functions.

Illustration 1:

$$f(x) = \sqrt{5x-3}$$

Solution:

$$5x - 3 \geqslant 0$$

$$x \geqslant \frac{3}{5}$$

Because $5x - 3$ is linear, f is a transformation of $g(x) = \sqrt{x}$ (Figure 5-23):

$$f(x) = \sqrt{5x-3}$$

$$= \sqrt{5}\sqrt{x - \frac{3}{5}}$$

Compare the graph of g to the graph of f given in Figure 5-24.

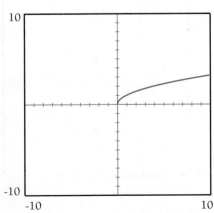

FIGURE 5-23 $g(x) = \sqrt{x}$

Illustration 2:

$$f(x) = \sqrt{\frac{x-2}{x+5}}$$

Solution: Graphing technology applied to the radicand provides the first line of the chart (also Figure 5-22 and Figure 5-25).

x-axis	$-\infty$		-5		2		∞	Note
$\dfrac{x-2}{x+5}$		+	A	−	0		+	Rational
$\sqrt{\dfrac{x-2}{x+5}}$		Above	A	U	On		Above	Domain

Analyzing the results of taking the square root of the first line of the chart produces the second line. Square roots of positive values are positive. The square root of 0 is 0. Square roots preserve vertical asymptotes. Square roots of negative numbers are undefined (represented by U).

Also from Example 5-4, $y = \sqrt{1}$ is a horizontal asymptote. Plot some points and obtain the graph in Figure 5-26.

Illustration 3:

$$f(x) = \sqrt[3]{\frac{5x-3}{x-1}}$$

Solution: The only restriction on the domain is that $x \neq 1$. Rewrite the radicand in mixed format to determine the horizontal asymptote.

$$f(x) = \sqrt[3]{\frac{5x-5+2}{x-1}} \qquad \text{(or use synthetic division)}$$

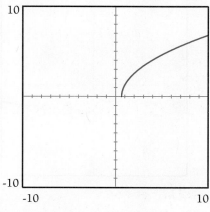

FIGURE 5-24 $f(x) = \sqrt{5x-3}$

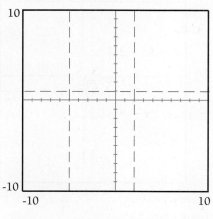

FIGURE 5-25 $x = -5, \qquad x = 2$

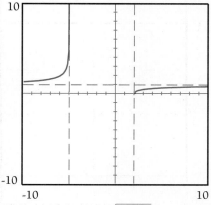

FIGURE 5-26 $f(x) = \sqrt{\dfrac{x-2}{x+5}}$

$$= \sqrt[3]{\frac{5(x-1)+2}{x-1}}$$

$$= \sqrt[3]{5 + \frac{2}{x-1}}$$

Then $y = \sqrt[3]{5}$ is a horizontal asymptote. Form a chart for the radicand to complete the analysis.

x-axis	$-\infty$		$\frac{3}{5}$		1		∞	Note
$\frac{5x-3}{x-1}$	+		0	−	A		+	Rational
$\sqrt[3]{\frac{5x-3}{x-1}}$	+		0	−	A		+	Domain

Mark the horizontal asymptote $y = \sqrt[3]{5}$ and the vertical asymptote $x = 1$. Label the x-intercept at $\frac{3}{5}$. See Figure 5-27. Use the chart to correct or clarify the application of your graphing technology. Figure 5-28 shows the resulting graph.   ▬

Consider $f(x) = \sqrt{x} - 5$. This transformation of x results in a graph with an x-intercept of 25 (outside the default window). See Figure 5-29.

If the formula for a radical function consists of two or more terms, determining an x-intercept is more complex. Radicals possess few manipulative algebraic properties. The following theorems summarize some useful properties associated with radicals.

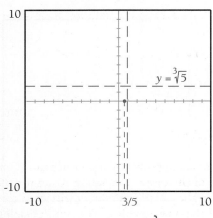

FIGURE 5-27 $x = 1,$ $x = \frac{3}{5},$ $y = \sqrt[3]{5}$

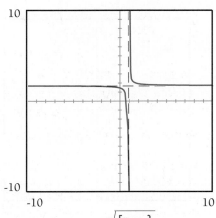

FIGURE 5-28 $f(x) = \sqrt[3]{\frac{5x-3}{x-1}}$

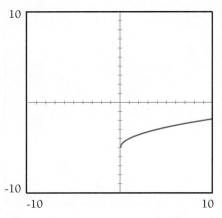

FIGURE 5-29 $f(x) = \sqrt{x} - 5$

Theorem 5-1
Squaring Property of Equality

If $x = a$, then $x^2 = a^2$.
(Also, $x^3 = a^3$, $x^4 = a^4$, etc.)

Proof: Multiply each side of the equation $x = a$ by a. Then $ax = a^2$. Now multiply each side of $x = a$ by x. Obtain $x^2 = ax$. Because a^2 and x^2 both equal ax, then $x^2 = a^2$. Extending the argument produces $x^3 = a^3$, etc.   ▄

Theorem 5-2
Square Root Property of Equality

If $x^2 = a^2$ then $x = a$ or $x = -a$.

Proof: Because $x^2 = a^2$ is a quadratic equation, solve it by factoring:

$$x^2 - a^2 = 0$$
$$(x - a)(x + a) = 0$$
$$x = a \text{ or } x = -a$$

The preceding theorems are not converse theorems. If $x = 3$, then $x^2 = 9$. The solutions to $x^2 = 9$ are $x = 3$ or $x = -3$. Two equations are equivalent if they have identical solutions. The equation $x = 3$ is not equivalent to $x^2 = 9$. Although all solutions of $x = 3$ are solutions of $x^2 = 9$, the converse is incorrect. Squaring each side of $x = 3$ results in an equation with an extra solution, $x = -3$. The following theorem, generalizes this result.   ▄

Theorem 5-3
Generalized Multiplication Property of Equality

Suppose that r_0 is a solution for $f(x) = g(x)$ and $m(x)$ is a nonzero polynomial. Then r_0 is a solution for $m(x)\,f(x) = m(x)\,g(x)$.

Proof: If $f(x) = g(x)$ then $f(x) - g(x) = 0$. Because r_0 is a solution then $f(r_0) - g(r_0) = 0$. Now consider

$$m(x)\,f(x) = m(x)\,g(x)$$

Rewrite as $m(x)[f(x) - g(x)] = 0$
Substitute r_0: $m(r_0)[f(r_0) - g(r_0)] = 0$
$$m(r_0)0 = 0 \qquad \text{(true)}$$

So that r_0 is a solution for $m(x)f(x) = m(x)g(x)$.   ▄

Note: Multiplication of an equation by a nonzero polynomial produces an equation whose solution includes the roots of the original equation. Examine the portion of the proof where $m(x)[f(x) - g(x)] = 0$. You should discover that the roots of $m(x)$ may become roots of the new equation. The multiplication by $m(x)$ is the source of **extraneous roots**. If $m(x)$ is a constant function then multiplication produces no extraneous roots.

For example, the only solution for the equation $x = 2$ is 2. Rewrite as $x - 2 = 0$. Multiply both sides of the equation by $x + 3$:

$$(x + 3)(x - 2) = (x + 3)0$$
$$x^2 + x - 6 = 0$$

Solve this quadratic to obtain the solution $\{-3, 2\}$. But -3 does not satisfy the original equation $x = 2$. Therefore, -3 is the extraneous root introduced by the multiplication of $x + 3$.

The moral is now clear: it is okay to multiply both sides of an equation by a polynomial—no solutions will be lost. But every solution of the resulting equation must be checked in the original equation. Those solutions that do not check are extraneous and must be discarded. Retain the answers that check as the solution to the original equation.

Why was $m(x)$ specified to be a polynomial? Why not a rational expression? Consider that multiplication by $1/x$ is equivalent to dividing by x. Note that the solutions to $x^2 - 3x = 2x - 6$ are $\{2, 3\}$.

Can we multiply both sides by $1/(x - 3)$?

$$x^2 - 3x = 2x - 6$$
$$x(x - 3) = 2(x - 3)$$
$$\frac{1}{x - 3}x(x - 3) = \frac{1}{x - 3}2(x - 3)$$
$$x = 2$$

The last equation is not equivalent to the first. The process of multiplying by $1/(x - 3)$ removed 3 as part of the solution. Because 3 is not in the domain of $1/(x - 3)$, it cannot be an answer for the resulting equation. Getting an extraneous root is not harmful so long as we check each proposed root. However, multiplying by a rational expression may cause an answer to be lost. Lost answers may be difficult to retrieve. To avoid this difficulty, check whether 3 is an answer for the original equation *before* multiplying by a rational function that would exclude it.

As the proof of Theorem 5-1 indicates, squaring each side of a conditional equation is equivalent to multiplying by a variable. Therefore, all answers of a "squared" equation must be checked in the original equation to determine whether they are extraneous.

Review Theorem 5-2. Instead of following the suggested proof, try taking the square root of both sides. You will lose the negative answers. Taking roots of both sides of a conditional equation should be avoided: answers might be lost.

EXAMPLE 5-6

Illustration 1:

Intercepts of Radical Functions. Determine any x-intercepts for the following functions.

$$f(x) = \sqrt{5x - 3} - 2$$

Solution: The function f is the same function as in Example 5-5, Illustration 1 shifted down 2 units. See Figure 5-30. $\text{Domain}_f = [\frac{3}{5}, \infty)$.

Figure 5-30 reveals that f intersects the x-axis. To determine the precise value of the intercept, set $f(x)$ equal to 0 and solve:

$$0 = \sqrt{5x - 3} - 2$$

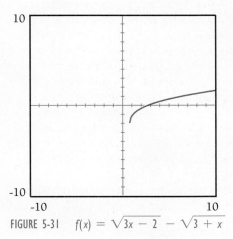

10

-10

-10   10

FIGURE 5-30 $f(x) = \sqrt{5x - 3} - 2$

Because squaring is the inverse operation of square root, square each side of the equation to "undo" the effect of the square root. However, the -2 is in the way. One strategy to eliminate radicals is to isolate them on one side of the equation, then remove them by applying the inverse operation.

$$2 = \sqrt{5x - 3}$$
$$4 = 5x - 3 \qquad \text{(square both sides)}$$

Solve the resulting linear equation.

$$7 = 5x$$
$$\frac{7}{5} = x$$

Squaring may have introduced extraneous roots. Check $\frac{7}{5}$ in the original equation.

$$0 = \sqrt{5\frac{7}{5} - 3} - 2$$
$$0 = \sqrt{4} - 2$$
$$0 = 0 \qquad \text{(true)}$$

So that $\frac{7}{5}$ is a correct solution. Therefore, $(\frac{7}{5}, 0)$ is the x-intercept of f.

Illustration 2:

$$f(x) = \sqrt{3x - 2} - \sqrt{3 + x}$$

10

-10

-10   10

FIGURE 5-31 $f(x) = \sqrt{3x - 2} - \sqrt{3 + x}$

Solution: Technology produced the graph of f as shown in Figure 5-31. From the first radical, $x \geq \frac{2}{3}$. From the second radical, $x \geq -3$. As a result, Domain$_f$ = $[\frac{2}{3}, \infty)$. Set $f(x) = 0$ to determine any x-intercepts.

$$0 = \sqrt{3x - 2} - \sqrt{3 + x}$$
Square both sides: $0^2 = (\sqrt{3x - 2} - \sqrt{3 + x})^2$
$$0 = 3x - 2 - 2\sqrt{(3x - 2)(3 + x)} + 3 + x$$

This is more complicated than the original equation. Isolate one of the square roots and then square to eliminate it:

$$\sqrt{3 + x} = \sqrt{3x - 2}$$
Now square both sides: $(\sqrt{3 + x})^2 = (\sqrt{3x - 2})^2$
$$3 + x = 3x - 2 \qquad \text{(linear equation)}$$
$$5 = 2x$$
$$\frac{5}{2} = x$$

Check: Checking for extraneous roots is mandatory.

$$\sqrt{3\frac{5}{2} - 2} - \sqrt{3 + \frac{5}{2}} = 0$$

$$\sqrt{\frac{11}{2}} - \sqrt{\frac{11}{2}} = 0 \qquad \text{(true)}$$

As a result, $\frac{5}{2}$ is the x-intercept.

Note: Figure 5-32 displays the graphs of the component parts of f. Use the trace feature of your technology to verify the point of intersection; $y = \sqrt{3x - 2}$ and $y = \sqrt{3 + x}$. Refer to Figure 5-31 or experiment with values to form a chart for f.

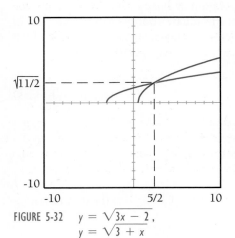

FIGURE 5-32 $y = \sqrt{3x - 2}$,
$y = \sqrt{3 + x}$

x-axis		$-\infty$	$\dfrac{2}{3}$		$\dfrac{5}{2}$		∞	Note
$\sqrt{3x - 2} - \sqrt{3 + x}$		U	0	$-$	0	$+$		

Interpret the chart several ways. The graph of f is below the x-axis on $[\frac{2}{3}, \frac{5}{2})$ and above the x-axis on $(\frac{5}{2}, \infty)$. The point of intersection of the separate components $y = \sqrt{3 + x}$ and $y = \sqrt{3x - 2}$ occurs where $x = \frac{5}{2}$. See Figure 5-32. Moreover, if $x > \frac{5}{2}$, then the graph of $y = \sqrt{3x - 2}$ is above" the graph of $y = \sqrt{3 + x}$. Similarly, the graph of $y = \sqrt{3x - 2}$ is "below" the graph of $y = \sqrt{3 + x}$ on $[\frac{2}{3}, \frac{5}{2})$.

Either the graph or chart provides solutions to related inequalities.

Illustration 3:

$$f(x) = \sqrt{3 + 2x} + \sqrt{3x}$$

Solution:

$$0 = \sqrt{3 + 2x} + \sqrt{3x}$$

Isolate the radicals, then square:

$$-\sqrt{3 + 2x} = \sqrt{3x}$$
$$3 + 2x = 3x$$
$$3 = x$$

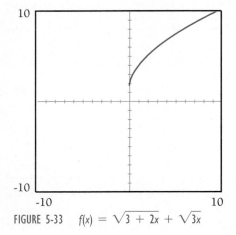

FIGURE 5-33 $f(x) = \sqrt{3 + 2x} + \sqrt{3x}$

Check:

$$0 = \sqrt{3 + 2(3)} + \sqrt{3(3)}$$
$$0 = 3 + 3 \qquad \text{(false)}$$

This method preserves all answers with some possible extraneous roots. Discard any extraneous roots; in this case, 3. The remaining roots form the solution. Because no roots remain, there is no solution. The graph of f does not intersect the x-axis. See Figure 5-33.

Illustration 4:

Consider $f(x) = \sqrt{x + 4} + 1$ and $g(x) = \sqrt{3x + 1}$. Determine the x-intercept of $f - g$.

Solution: The problem is equivalent to determining the points of intersection of f and g. See Figure 5-34. Use the trace feature of your technology to approxi-

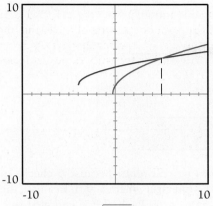

FIGURE 5-34 $y = \sqrt{x + 4} + 1$,
$y = \sqrt{3x + 1}$

mate the points of intersection. Confirm your solution with the following algebraic process:

At the point(s) of intersection the y values are equal so that $f(x) = g(x)$:

$$\sqrt{x + 4} + 1 = \sqrt{3x + 1}$$

Square both sides: $(\sqrt{x + 4} + 1)^2 = (\sqrt{3x + 1})^2$

$$x + 4 + 2\sqrt{x + 4} + 1 = 3x + 1$$

Squaring eliminated only the isolated radical $\sqrt{3x + 1}$. The lefthand side of the equation had two terms. Because $(a + b)^2 = a^2 + 2ab + b^2$, the lefthand side did not simplify, but became more complex. Raising to a power is effective only with isolated radicals. To eliminate $\sqrt{x + 4}$, isolate and square it.

$$2\sqrt{x + 4} = 2x - 4$$
$$\sqrt{x + 4} = x - 2$$
$$(\sqrt{x + 4})^2 = (x - 2)^2$$
$$x + 4 = x^2 - 4x + 4 \qquad \text{(a quadratic)}$$
$$0 = x^2 - 5x$$
$$0 = x(x - 5)$$
$$x = 0 \quad \text{or} \quad x = 5$$

Check:

$$\sqrt{0 + 4} + 1 = \sqrt{3(0) + 1}$$
$$3 = 1 \qquad \text{(discard 0 as extraneous; see graph)}$$
$$\sqrt{5 + 4} + 1 = \sqrt{3(5) + 1}$$
$$3 + 1 = 4 \qquad \text{(true)}$$
$$x = 5 \qquad \text{(the only solution)}$$

See Figure 5-34. The graphs of f and g intersect at $(5, 4)$. See Problem 85 in the Review Exercises.

Illustration 5:   Determine the point(s) of intersection of

$$f(x) = \sqrt[3]{x + 3} \text{ and } g(x) = \sqrt{x - 1}$$

Solution:

$$(x + 3)^{1/3} = (x - 1)^{1/2}$$

Eliminating the radicals means eliminating the denominators of the fractional exponents. Handling both fractional exponents simultaneously requires a common denominator; in this case, 6. Alternately, to eliminate the square root, square; to eliminate the cube root, cube. Squaring followed by cubing is equivalent to raising both sides to the sixth power. Why?

$$[(x + 3)^{1/3}]^6 = [(x - 1)^{1/2}]^6$$
$$(x + 3)^2 = (x - 1)^3$$
$$x^2 + 6x + 9 = x^3 - 3x^2 + 3x - 1$$
$$x^3 - 4x^2 - 3x - 10 = 0$$

From the rational root theorem, the possible rational roots are ±1, ±2, ±5, ±10. Of these, only 5 works:

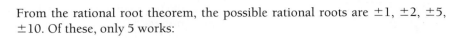

5	1	-4	-3	-10
		5	5	10
	1	1	2	0

The depressed polynomial $x^2 + x + 2$ has no real roots. Because 5 checks in the original equation, the graphs of f and g intersect where $x = 5$. Evaluate the functions at 5 to determine the second coordinate. Each of $f(5)$ and $g(5)$ is 2. The graphs intersect at $(5, 2)$.

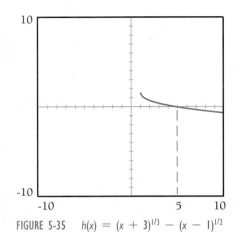

FIGURE 5-35 $h(x) = (x + 3)^{1/3} - (x - 1)^{1/2}$

Note: The solution to the inequality $\sqrt[3]{x + 3} < \sqrt{x - 1}$ is $(5, \infty)$. See Figure 5-35. The solution to $\sqrt[3]{x + 3} \geq \sqrt{x - 1}$ is $[1, 5]$.

SUMMARY

For a radical function $f(x) = \sqrt[n]{R(x)}$,

- To determine the domain of f:

    (a) Form a chart for the rational function radicand $R(x)$.
    (b) If n is odd, the domain of f is the domain of R.
    (c) If n is even, the domain of f is the solution to $R(x) \geq 0$.

- The vertical asymptotes for R are the vertical asymptotes of f.
- Rewrite the radicand in mixed format: if $f(x) = \sqrt[n]{c + [a/(mx + b)]}$ then $y = \sqrt[n]{c}$ is a horizontal asymptote for f.
- Use a chart to refine a graph produced by technology. Tracing a graph provides approximate solutions to radical equations.
- To algebraically solve an equation containing a variable in a radicand:
    Repeat until all radicals are eliminated:

    (a) Isolate a radical on one side of the equation.
    (b) Raise both sides of the equation to a power that will "undo" the radical (inverse operation).
    (c) Simplify both sides of the equation.
    Solve the resulting equation.
    Refer to the graph.

Check all proposed answers against the graph and in the original equation. Those that do not check are extraneous roots. Eliminate the extraneous roots. The remaining roots form the solution of the equation. If no root remains there is no solution.

5-2   EXERCISES

Determine the domain for each of the following radical functions.

1. $f(x) = \sqrt{5x - 2}$

2. $f(x) = \sqrt{2 - 3x}$

3. $f(x) = \sqrt{7 - 5x}$

4. $f(x) = \sqrt{3x + 8}$

5. $f(x) = \sqrt[4]{x^2 - 6x + 8}$

6. $f(x) = \sqrt[4]{x^2 - 5x - 14}$

7. $f(x) = \sqrt[3]{\dfrac{3(x - 5)}{2x + 1}}$

8. $f(x) = \sqrt[3]{\dfrac{2x + 1}{3(x - 5)}}$

9. $f(x) = \sqrt{\dfrac{x^2 + x}{x - 3}}$

10. $f(x) = \sqrt{\dfrac{2x - 3}{x^2 + x}}$

Graph each of the following radical functions. Annotate asymptotes using a dashed line.

11. $f(x) = \sqrt{x}$

12. $f(x) = \sqrt{x + 1}$

13. $f(x) = \sqrt{x - 1}$

14. $f(x) = \sqrt{4x}$

15. $f(x) = \sqrt{9x}$

16. $f(x) = \sqrt{\dfrac{1}{x}}$

17. $f(x) = \sqrt{\dfrac{1}{x + 2}}$

18. $f(x) = \sqrt{\dfrac{1}{x - 2}}$

19. $f(x) = \sqrt{\dfrac{25}{x}}$

20. $f(x) = \sqrt{\dfrac{9}{x}}$

21. $f(x) = \sqrt{\dfrac{x + 3}{x}}$

22. $f(x) = \sqrt{\dfrac{x}{x - 3}}$

23. $f(x) = \sqrt{\dfrac{x + 2}{x + 1}}$

24. $f(x) = \sqrt{\dfrac{x - 2}{x - 3}}$

25. $f(x) = \sqrt{\dfrac{4x + 4}{x + 1}}$

26. $f(x) = \sqrt{\dfrac{x - 1}{9x - 9}}$

27. $f(x) = \sqrt{\dfrac{3x + 2}{x - 2}}$

28. $f(x) = \sqrt{\dfrac{x - 2}{3x + 2}}$

29. $f(x) = \sqrt[3]{x + \dfrac{1}{x}}$

30. $f(x) = \sqrt[3]{x - \dfrac{1}{x}}$

Solve the following equations.

31. $\sqrt{3x - 5} = 2$

32. $\sqrt{2x + 3} = 5$

33. $\sqrt{5x - 2} = 3$

34. $\sqrt{3x + 2} = 5$

35. $\sqrt{x - 5} = \sqrt{3x + 2}$

36. $\sqrt{x + 5} = \sqrt{2x - 1}$

37. $\sqrt{2x + 3} = \sqrt{7 - x}$

38. $\sqrt{5 - 2x} = \sqrt{x + 4}$

39. $\sqrt{3x - 1} - \sqrt{x + 2} = 0$

40. $\sqrt{x - 5} - \sqrt{3x + 1} = 0$

41. $\sqrt{4x + 3} - \sqrt{2 - x} = 0$

42. $\sqrt{5 - 3x} - \sqrt{3 + x} = 0$

43. $\sqrt{5x + 2} + \sqrt{x - 1} = 0$

44. $\sqrt{3x - 1} + \sqrt{2x - 1} = 0$

45. $\sqrt{3x + 5} + \sqrt{2 - x} = 0$

46. $\sqrt{5 - 3x} + \sqrt{1 - x} = 0$

47. $x = \sqrt{3 - 2x}$

48. $\sqrt{x - 2} = x$

49. $x - \sqrt{3 + 2x} = 0$

50. $\sqrt{x + 6} - x = 0$

51. $2x = \sqrt{8x + 5}$

52. $\sqrt{3 - 4x} = 2x$

53. $2x = \sqrt{5 - 2x}$

54. $\sqrt{2 - 3x} = 3x$

55. $\sqrt{x - 2} + \sqrt{x + 5} = 7$

56. $\sqrt{2x - 1} - \sqrt{x - 1} = 1$

57. $\sqrt{x + 5} - \sqrt{x} = 1$

58. $\sqrt{x} + \sqrt{x - 5} = 1$

59. $\sqrt{x + 2} + 2 = \sqrt{3x + 4}$

60. $\sqrt{x + 7} - 2 = \sqrt{3x - 5}$

61. $\sqrt{x} = \sqrt[3]{x}$

62. $\sqrt{x} + \sqrt[3]{x} = 0$

63. $\sqrt{x + 1} = \sqrt[3]{x + 5}$

64. $\sqrt{x - 1} = \sqrt[3]{x + 3}$

65. $\sqrt[3]{x^2 + 4} = 5$

66. $\sqrt[3]{x^2 - 1} = 2$

67. $\sqrt{x^2} = x$

68. $\sqrt[3]{x^3} = x$

69. $\sqrt[3]{x^6} = x^2$

70. $\sqrt{x^4} = x^2$

Solve the following. (Hint: see exercises 31–40.)

71. $\sqrt{3x - 5} > 2$

72. $\sqrt{2x + 3} > 5$

73. $\sqrt{5x - 2} < 3$

74. $\sqrt{3x + 2} < 5$

75. $\sqrt{x - 5} > \sqrt{3x + 2}$

76. $\sqrt{x + 5} > \sqrt{2x - 1}$

77. $\sqrt{2x + 3} < \sqrt{7 - x}$

78. $\sqrt{5 - 2x} < \sqrt{x + 4}$

79. $\sqrt{3x - 1} - \sqrt{x + 2} > 0$

80. $\sqrt{x - 5} - \sqrt{3x + 1} > 0$

Graph the following functions. (Hint: see exercises 71–80)

81. $f(x) = \sqrt{3x - 5} - 2$

82. $f(x) = \sqrt{2x + 3} - 5$

83. $f(x) = \sqrt{5x - 2} - 3$

84. $f(x) = \sqrt{3x + 2} - 5$

85. $f(x) = \sqrt{x - 5} - \sqrt{3x + 2}$

86. $f(x) = \sqrt{x + 5} - \sqrt{2x - 1}$

87. $f(x) = \sqrt{2x + 3} - \sqrt{7 - x}$

88. $f(x) = \sqrt{5 - 2x} - \sqrt{x + 4}$

89. $f(x) = \sqrt{3x - 1} - \sqrt{x + 2}$

90. $f(x) = \sqrt{x - 5} - \sqrt{3x + 1}$

91. If $x = a$, prove $x^3 = a^3$.

92. If $x = a$, prove $x^4 = a^4$.

93. If $x^n = a^n$, prove $x^{n+1} = a^{n+1}$.

94. If $x^{n-1} = a^{n-1}$, prove $x^n = a^n$.

Challenge or defend the following statements for all real values of x.

95. $\sqrt[3]{x^3} = x$

96. $\sqrt{x^2} = x$

97. $\sqrt[2n]{x^{2n}} = x$, where n is a counting number.

98. $\sqrt[2n+1]{x^{2n+1}} = x$, where n is a counting number.

99. $\sqrt[n]{x^{2n}} = x^2$, where n is a counting number.

100. $\sqrt[n]{x^{3n}} = x^3$, where n is a counting number.

5-2   PROBLEM SET

Suppose that the difference quotient for a function f is defined by

$$\frac{\Delta f}{\Delta x} = \frac{f(x) - f(a)}{x - a} \qquad \text{(Definition A)}$$

Determine the difference quotient for each of the following functions.

1. $f(x) = \dfrac{1}{x}$

2. $f(x) = -\dfrac{1}{x}$

3. $f(x) = -\dfrac{1}{x^2}$

4. $f(x) = \dfrac{1}{x^2}$

5. $f(x) = \dfrac{1}{x + 2}$

6. $f(x) = \dfrac{1}{x - 3}$

*7. $f(x) = \sqrt{x}$

*8. $f(x) = -\sqrt{x}$
  *(Hint: rationalize the numerator)

*9. $f(x) = \dfrac{-1}{\sqrt{x}}$

*10. $f(x) = \dfrac{1}{\sqrt{x}}$

Suppose that the difference quotient for a function f is defined by

$$\frac{\Delta f}{\Delta x} = \frac{f(x + h) - f(x)}{h} \qquad \text{(Definition B)}$$

Determine the difference quotient for each of the following functions.

11. $f(x) = \dfrac{1}{x}$

12. $f(x) = -\dfrac{1}{x}$

13. $f(x) = -\dfrac{1}{x^2}$

14. $f(x) = \dfrac{1}{x^2}$

15. $f(x) = \dfrac{1}{x + 2}$

16. $f(x) = \dfrac{1}{x - 3}$

17. $f(x) = \sqrt{x}$

18. $f(x) = -\sqrt{x}$
  *(Hint: rationalize the numerator)

19. $f(x) = \dfrac{-1}{\sqrt{x}}$

20. $f(x) = \dfrac{1}{\sqrt{x}}$

21. Show that the A definition of difference quotient is equivalent to the B definition. Hint: let $h = x - a$.

22. Show that the B definition of difference quotient is equivalent to the A definition. Hint: let $a = x + h$.

PROBLEMS FOR GRAPHING TECHNOLOGY

Apply graphing technology to approximate solutions to the following. Use the trace feature to approximate the points of intersection. Identify the intervals where the graph of f is above the graph of g. Use this information to solve the indicated inequality.

23. $f(x) = x^2 - 4$ and $g(x) = \sqrt{x}$, solve $f(x) > g(x)$.

24. $f(x) = x^2 - 4$ and $g(x) = \sqrt{x + 2}$, solve $x^2 - 4 > \sqrt{x + 2}$.

Solve the following.

25. $x^2 - 4 \le \sqrt{x + 2}$

26. $x^2 - 4 \le \sqrt{x}$

27. $\sqrt{\dfrac{x}{x - 3}} < x$

28. $\sqrt{\dfrac{x}{x - 3}} > x^2$

29. $\dfrac{1}{x} \le \sqrt{x}$

30. $\sqrt[3]{\dfrac{x - 2}{x + 3}} > 1$

*31. $\sin x > \sqrt{x - 1}$

*32. $\sqrt{x + 2} > e^x$

CHAPTER SUMMARY

- Consider the rational function $f(x) = N(x)/D(x)$.
  The domain of f is the set of all real numbers such that $D(x) \neq 0$. Let $f(x) = \cancel{N}(x)/\cancel{D}(x)$, $\cancel{D}(x) \neq 0$, represent f reduced to lowest terms.

  1. If $\cancel{D}(r) = 0$, then $x = r$ is a vertical asymptote for f.
  2. If $D(r) = 0$ and $\cancel{D}(r) \neq 0$, $x = r$ is a removable discontinuity for f.
  3. If $D(x) = x - r$ and $f(x) = mx + b + R/(x - r)$, $m \neq 0$, then $y = mx + b$ is a slant asymptote if $R \neq 0$.

- For a radical function $f(x) = \sqrt[n]{R(x)}$,

  1. If n is odd, the domain of f is the domain of R.
  2. If n is even, the domain of f is the solution to $R(x) \geq 0$.

  Refer to the graph when solving a radical equation. Check all proposed answers against the graph and in the original equation. Eliminate those that do not check as extraneous roots.

KEY WORDS AND CONCEPTS

Algebraic function
Asymptote
Extraneous root
Horizontal asymptote

Limit
Piecewise function
Radical function

Rational function
Slant asymptote
Vertical asymptote

5   REVIEW EXERCISES

SECTION I

For each of the following rational functions, (a) determine the domain and x-intercepts, (b) choose a window large enough to sketch the graph and annotate any vertical, horizontal or slant asymptotes as dashed lines, (c) sketch the graph.

1. $f(x) = \dfrac{2}{x + 5}$

2. $f(x) = \dfrac{x - 3}{x + 7}$

3. $f(x) = \dfrac{5x + 12}{2x - 3}$

4. $f(x) = \dfrac{x + 7}{x^2 - 49}$

5. $f(x) = \dfrac{x^2 - 6x + 3}{x + 2}$

6. $f(x) = \dfrac{7(x + 5)}{(x - 4)(x + 2)}$

7. $f(x) = \dfrac{(x - 4)(x + 2)}{7(x + 5)}$

8. $f(x) = \dfrac{(5 - 3x)(x - 1)}{(1 - x)(x + 3)}$

9. $f(x) = \dfrac{2(x + 5)(x - 4)}{(4 - x)(x + 5)}$

10. $f(x) = \dfrac{x^2 + 7x + 2}{(x^2 - 5)x}$

Solve each of the following inequalities.

11. $3(x - 1)(x - 7)(x + 8) > 0$

12. $-3(15 - x)(x + 9)(x + 1) > 0$

13. $(5x + 1)(7 - x)(x^2 - 11) \geq 0$

14. $(x^2 - 7x + 2)(x^2 + 13) \geq 0$

15. $(x^2 - 25)(9x^2 - 16) > 0$

16. $\dfrac{6(x + 9)}{(x - 5)(x + 8)} > 0$

17. $\dfrac{(x - 6)(x + 3)}{9(x + 4)} \geq 0$

18. $\dfrac{(2 - 7x)(x - 3)}{(3 - x)(x + 5)} \geq 0$

19. $\dfrac{(x + 7)(x - 4)}{(x - 2)(x + 7)} > 0$

20. $\dfrac{x^2 + 5x + 1}{(x^2 - 5)x^2} > 0$

21. $\dfrac{x^3 + 2x^2 - 6x + 3}{x(x-1)} \leqslant 0$

22. $\dfrac{x^3 + 2x^2 + 6x + 5}{x(x+1)} < 0$

23. $\dfrac{x^3 + 2x^2 - 6x + 3}{x^2 + 6x + 5} \geqslant 0$ **24.** $\dfrac{x^4 - 10x^2 + 9}{x^2 - 9} < 0$

25. $\dfrac{x^2 - 3}{x + 1} \geqslant 2$ **26.** $\dfrac{4 - x}{x^2 - x} < 1$

Graph and annotate the following functions.

27. $f(x) = x^2 + \dfrac{5}{x}$ **28.** $f(x) = (x-2)^3 + \dfrac{1}{x-2}$

29. $f(x) = x^2 + \dfrac{1}{x}$ **30.** $f(x) = \dfrac{x+1}{|x+1|}$

31. $f(x) = \begin{cases} -x \text{ if } x \leqslant 0 \\ x^2 \text{ if } x > 0 \end{cases}$ **32.** $f(x) = \begin{cases} 2x \text{ if } x > 0 \\ -x^2 \text{ if } x \leqslant 0 \end{cases}$

33. Suppose that the value of data stored on a computer varies inversely as the time since its last access. For example, suppose that a customer on a mailing list is valued at \$10 if the customer has placed an order within the last year, but drops to \$5 if the last order was 2 years ago (\$10/2). Write an equation to model the value of a customer as a function of time. What is the value of a customer whose last order was 5 years ago? Discuss the value of the customer data as the time since the last order increases. Suppose it costs \$1.00/year to store the customer data. At what point would the accumulated cost of the data storage exceed the value of the customer data?

34. Suppose that the time it takes a rocket car to cross a flat plain varies inversely as the speed of the car. Suppose that the car crosses the plain in 3 minutes at 200 mph. Write a model to represent time as a function of speed. How much time would it take to make the trip at 300 mph? What speed would be required to make the trip in 15 minutes?

SECTION 2

Determine the domain for each of the following radical functions.

35. $f(x) = \sqrt{4x - 3}$ **36.** $f(x) = \sqrt{5x + 8}$

37. $f(x) = \sqrt[7]{x^2 - 6x + 9}$ **38.** $f(x) = \sqrt[3]{\dfrac{2x + 3}{3(x - 4)}}$

39. $f(x) = \sqrt{\dfrac{x^2 + 2x}{x - 2}}$ **40.** $f(x) = \sqrt{\dfrac{2x - 6}{x^2 - 9}}$

Graph each of the following radical functions.

41. $f(x) = \sqrt{4x}$ **42.** $f(x) = \sqrt{-9x}$

43. $f(x) = \sqrt{9 - x}$ **44.** $f(x) = \sqrt{\dfrac{1}{x - 1}}$

45. $f(x) = \sqrt{\dfrac{1}{x + 3}}$ **46.** $f(x) = \sqrt{\dfrac{9}{x - 1}}$

47. $f(x) = \sqrt{\dfrac{x + 5}{x}}$ **48.** $f(x) = \sqrt{\dfrac{x - 1}{x - 5}}$

49. $f(x) = \sqrt{\dfrac{9x + 9}{x + 1}}$ **50.** $f(x) = \sqrt[5]{x - \dfrac{2}{x}}$

Solve the following equations.

51. $\sqrt{4x - 7} = 2$ **52.** $\sqrt{5x + 2} = 4$

53. $\sqrt{x - 6} = \sqrt{2x + 3}$ **54.** $\sqrt{5 - x} = \sqrt{x + 7}$

55. $\sqrt{3x - 2} - \sqrt{2x} = 0$ **56.** $\sqrt{7 - 3x} - \sqrt{x + 19} = 0$

57. $\sqrt{5x + 1} + \sqrt{6x - 2} = 0$

58. $\sqrt{10 - 3x} + \sqrt{6 - 5x} = 0$

59. $x = \sqrt{5 - 4x}$ **60.** $\sqrt{2x + 8} - x = 0$

61. $2x + 1 = \sqrt{3x + 6}$ **62.** $\sqrt{1 - 3x} = 2x$

63. $\sqrt{x - 2} + \sqrt{2x + 4} = 6$ **64.** $\sqrt{x} + \sqrt{x - 8} = 2$

65. $\sqrt{x + 2} + 1 = \sqrt{3x - 5}$ **66.** $\sqrt{x} - \sqrt[3]{x} = 0$

67. $\sqrt{x + 1} = \sqrt{\dfrac{12x}{x + 5}}$ **68.** $\sqrt{\dfrac{3x^2 + 5}{x^2 - x}} = 2$

69. $\sqrt{x^2} = |x|$ **70.** $\sqrt[3]{x^3} = x$

71. $\sqrt{2x - 5} > 1$ **72.** $\sqrt{3x + 8} < 1$

73. $\sqrt{x - 5} > \sqrt{3x + 4}$

74. $\sqrt{x - 5} - \sqrt{3x + 4} > 0$

Sketch the graph for each of the following.

75. $f(x) = \sqrt{2x - 7} + 1$ **76.** $f(x) = \sqrt{2x - 7} - 5$

77. $f(x) = \sqrt{x - 5} - \sqrt{3x + 4}$ **78.** $f(x) = \sqrt{5 - x} - x$

79. $f(x) = \sqrt{3x + 4} - \sqrt{x - 5}$

80. $f(x) = \sqrt{x - 7} - \sqrt{x + 1}$

Determine the difference quotient for each of the following functions.

81. $f(x) = \dfrac{1}{x + 1}$ **82.** $f(x) = \dfrac{3}{x^2}$

83. $f(x) = \dfrac{5}{x + 2}$ **84.** $f(x) = \dfrac{1}{\sqrt{x - 1}}$

85. See Fig 5.34 and Example 5-6, Illustration 4. Give a graphical interpretation for the extraneous root 0. Hint: imagine that the missing "halves" of the parabolas are restored by squaring.

Radix Notation for Rational Numbers

Radix notation extends naturally to rational numbers. Consider the integer 65 expressed in octal as $101_8 = 1(8^2) + 0(8^1) + 1(8^0)$. When limited to polynomial form, only integers yield to the compact notation. But the pattern for the exponents establishes an extension beyond polynomial representation into rational expressions. Note the exponents: 2, 1, 0.

Suppose we extend the sequence of exponents to 2, 1, 0, −1, etc. $8^{-1} = \frac{1}{8}$, a rational number. Because negative integer exponents indicate the reciprocal of the base, rational numbers have a radix notation that uses both positive and negative exponents.

If $r > 1$ is an integer, representing the base of a numeration system and constants with integer values from 0 to $r − 1$ represent the digits of the numeration system, then the radix form of a rational number takes the form $a_n(r^n) + a_{n-1}(r^{n-1}) + \ldots + a_1(r^1) + a_0(r^0) + a_{-1}(r^{-1}) + \ldots + a_{-m}(r^{-m})$, where $a_n, a_{n-1}, \ldots, a_{-m}$ are legitimate digits for r.

Customarily, we abbreviate the radix notation by subscripting the radix (base) r and stringing the digits. The units position (r^0) marks the transition from positive exponents (integers) to negative exponents (fractions), so a special symbol in the string marks the end of the integral value of the representation. This symbol is the radix point, represented in the United States by a period. Appropriately, the period terminates the integer (polynomial) portion of the representation. For example,

$$123.45_8 = 1(8^2) + 2(8^1) + 3(8^0) + 4(8^{-1}) + 5(8^{-2}).$$

In the more familiar decimal system, consider

$$32.765 = 3(10^1) + 2(10^0) + 7(10^{-1}) + 6(10^{-2}) + 5(10^{-3}).$$

Finally, note

$$0.11_2 = 1(2^{-1}) + 1(2^{-2})$$
$$= \frac{1}{2} + \frac{1}{4}$$
$$= \frac{3}{4}$$
$$= 0.75_{10}.$$

Notice that base 3 notation displays the rational number $\frac{1}{3}$ as 0.1. The same rational number in base 10 (decimal) notation becomes 0.3333 . . . . In Chapter 12 we will explore some implications of both finite and infinite representations of rational numbers in radix form.

5   CHAPTER TEST

For each of the following rational functions, (a) determine the domain and x-intercepts, (b) label breaks in the graph as asymptotic or simple discontinuities, (c) choose a window large enough to sketch the graph showing any vertical, horizontal or slant asymptotes as dashed lines. Also indicate removable discontinuities with a small circle.

1. $f(x) = \dfrac{x - 2}{(x - 3)^2}$

2. $f(x) = \dfrac{5 - x}{x + 3}$

Solve each of the following inequalities:

3. $\dfrac{(x - 2)}{(x - 3)^2} < 0$

4. $\dfrac{(5 - x)}{(x + 3)} \geq 0$

5. $\dfrac{x^2 - 3x}{x - 3} > 2$

6. Determine the domain for the following radical function.

$$f(x) = \sqrt{2 - 3x}$$

Sketch the graph of each of the following radical functions. Use a dashed line to indicate any asymptotes.

7. $f(x) = \sqrt{2 - 3x}$

8. $f(x) = \sqrt{\dfrac{5 - x}{x + 3}}$

Solve the following.

9. $\sqrt{2x - 3} = -2$

10. $\sqrt{x - 1} - \sqrt{2x + 1} = 0$

11. $\sqrt{x + 1} + \sqrt{2x} = 7$

12. $\sqrt{2 - 3x} > 5$

6

EXPONENTIAL AND LOGARITHMIC FUNCTIONS

S I D E B A R

Pizza Giant is running a special: two pizzas for the price of one. But Melissa VanHooten is adamant about paying for everything she receives.

"I'll have a cheese pizza," said Melissa. "How much is it?"

"When you buy one, you get two," the clerk replied.

"If I get two, I'll pay for two," said Melissa.

"If you pay for two, you get four."

"If I get four, I most certainly will pay for four."

"If you pay for four, you get eight . . ."

Ten minutes later the clerk sighed, "Yes and if you pay for 512 pizzas . . ."

Previous chapters used the algebra of functions to generate polynomial, rational, and radical functions (algebraic functions) from the identity function ($I(x) = x$) and constant functions. The rule of pairing for each of these consisted of a finite number of additions, subtractions, multiplications, divisions or root operations. Are there functions without a formula expressible as a finite number of algebraic operations? Recall from Chapter 2 that some functions have no explicit formula at all. For example, if you fire x bullets blindly at a target, no formula predicts y, the number of bullseyes.

This chapter introduces functions that have no *finite* algebraic formula. Such functions are **transcendental functions**. Implicitly, any rule of pairing for a transcendental function does not consist of ordinary algebra. If there is a formula involving only algebraic operations, these operations must include some sort of infinite process. Compare this idea to the continued fraction approximation of irrational numbers in Chapter 2.

The transcendental functions introduced in this chapter are the exponential and logarithmic functions. Chapters 7 and 8 investigate other transcendental functions.

1 EXPONENTIAL FUNCTIONS

What logarithms are to mathematics that mathematics are to the other sciences.
—NOVALIS

Chapter 1 introduced average rate of change. That concept evolved into the slope of a linear function in Chapter 3. What controls how a quantity changes? Consider the growth of principal under compound interest.

Label	A	B	C	D	E
Amount	$1,000	$2,000	$4,000	$8,000	$16,000
Time	0 years	7 years	14 years	21 years	28 years

The measured time period is fixed, 7 years. The principal in this account doubles every 7 years. For that reason 7 years is the *doubling period*. Note the rate of change from B to C is

$$\frac{(\$4,000 - \$2,000)}{(14 \text{ years} - 7 \text{ years})} = \frac{\$2,000}{7 \text{ years}}$$

Whereas the average change from D to E is

$$\frac{(\$16,000 - \$8,000)}{(28 \text{ years} - 21 \text{ years})} = \frac{\$8,000}{7 \text{ years}}$$

The average rate of change is not the same on all 7 year periods. For each doubling period, the increase in the account equals the beginning amount for that period. This "current amount" is the controlling variable in calculating the rate of change.

The preceding savings problem is an example of compound growth problems. Other compound growth problems include growth of populations and chain reactions in fissionable material.

The formula for compound interest is given by

$$A = P\left(1 + \frac{r}{n}\right)^{nt}$$

where A is the current amount, P is the starting principal, r is the interest rate per year, n is the number of compounding periods during the year, and t is the number of years. For example, if you invest 1000 at 5% APR compounded monthly for 2 years, the amount in your account will be

$$A = 1000\left(1 + \frac{0.05}{12}\right)^{12(2)}$$

By calculator, $A \cong \$1104.94$

Now let $x = r/n$ so that $n = r/x$. As n increases without bounds, x approaches 0. Conversely, as x approaches 0, n increases without bounds. Substitute these into the formula to obtain $A = [P(1 + x)^{1/x}]^{rt}$

With your graphing calculator graph $y = (1 + x)^{1/x}$. Technically, the graph is not defined for $x = 0$. However, your calculator is unaware of this. Use the trace feature to let x approach 0 and thus approximate the apparent y-intercept. You should obtain a y-value of approximately 2.7182 . . . . Call this number e. As a result as n increases toward infinity the amount A in an account with beginning principal P and yearly interest rate r for t years becomes

$$A = Pe^{rt}$$

S I D E B A R

Growth of $1 at r APR is given by $A = (1 + r/n)^{nt}$ where t is years with n conversion periods per year. By calculator, if $r = 1$ and $t = 1$, then

Compound	*n* periods	*A* amount
yearly	1	2.00
semi annual	2	2.25
monthly	12	2.61
daily	365	2.71
minutely	525600	2.72
continuously	∞	$e \cong 2.72$

Compound growth's distinguishing characteristic is the connection of the rate of growth with the current size of the growing media. More succinctly, if A is a quantity whose rate of change with respect to time t varies directly as A, then A experiences compound growth: $(\Delta A)/(\Delta t) = kA$ (k is the constant of proportionality).

The letter e used to designate the constant $2.718281828459\ldots$ is in honor of Leonhard Euler (1707–1783), one of the first mathematicians to investigate compound growth.

Since there are now infinitely many compounding periods in the year, this formula represents continuously compounded interest. Note that A is a function of t, where t is in an exponent. Base e is an irrational number called the natural base. Your calculator has an approximation for e built into it.

Let us consider the $1,000 principal at 5 percent for two years again. This time, use continuous compounding to calculate the value.

$$A = 1000\, e^{0.05(2)}$$
$$A \cong \$1105.17.$$

Compared with the previous example, continuous compounding provides a better return than compounding monthly.

The continuously compounded interest model generalizes to all compound growth problems. Let A_0 represent the starting amount of a quantity (e.g., principal, population, fissionable material). Let k represent the rate of compound growth for a given time period (e.g., APR, half-life). The current amount $A(t)$ of the quantity is a function of the number of time periods t as given by

$$A(t) = A_0\, e^{kt}$$

The function A is a member of a family of functions known as *exponential functions*. These functions are characterized by the appearance of the independent variable in an exponent. We postpone a formal definition of general exponential functions.

Exponential Functions as a Model for Growth. Write an exponential function to model each of the following compound growth problems.

EXAMPLE 6-1

Illustration 1:

Suppose that population growth is proportional to the current population size. A gene-splicing researcher develops a bacteria culture that doubles in size every hour. If a lab technician counts 6 bacteria at the beginning of forming the culture, write a formula expressing the number of bacteria as a function of time.

Solution: Let t = time and y = number of bacteria. To model compound growth use the exponential function $y = A_0 e^{kt}$. Because the starting culture is 6 (when $t = 0$), we have

$$y = 6\, e^{kt} \qquad \text{[Equation 6.1]}$$

The population doubles each hour, therefore $y = 12$ when $t = 1$:

$$12 = 6\, e^{k(1)}$$
$$2 = e^k$$

Substitute 2 for e^k in Equation [6.1]: $y = 6(2^t)$.

Note: Evaluate the function for $t = 2$ and $t = 10$:

$$y = 6(2^2)$$
$$= 24 \qquad \text{(bacteria after 2 hours)}$$
$$y = 6(2^{10})$$
$$= 6144 \qquad \text{(bacteria after 10 hours)}$$

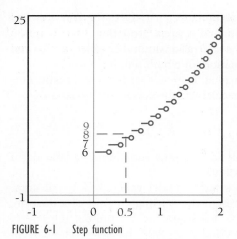

FIGURE 6-1   Step function

Interpreting the result for $t = 0.5$ is more difficult.

$$y = 6(2^{0.5})$$
$$\cong 8.485281374\ldots \qquad \text{(by calculator)}$$

What does 8.485 . . . bacteria mean? One interpretation is that the bacteria number 8 slightly before a half-hour elapses and increase to 9 shortly after a half-hour elapses. This type of interpretation problem occurs often. The exponential function is a continuous model for bacterial growth. Bacterial growth occurs a whole bacterium at a time. Bacterial growth is discrete (separate) like a step function. See Figure 6-1. Continuous functions have so many excellent attributes that we use them to approximate many discrete situations.

Illustration 2:   Scientists measure the decay of a radioactive material in terms of how long it takes for a given quantity of the material to decrease by half. This period is the half-life of the material. Suppose we have 100 grams of tritium, a radioactive isotope with a half-life of approximately 10 years. Write a formula modeling the amount of tritium as a function of elapsed time t (in years). Predict the amount of tritium after 75 years.

Solution: Let A = amount of tritium. Use the compound growth model

$$A = A_0\, e^{kt}$$

Because $A_0 = 100$, $A = 100\, e^{kt}$.

The half-life indicates that 100 grams will decay to 50 grams after 10 years:

$$50 \text{ grams} = 100 \text{ grams } e^{k(10 \text{ years})}$$

$$\frac{1}{2} = e^{10k \text{ years}}$$

$$\left(\frac{1}{2}\right)^{1/(10 \text{ years})} = (e^{10k \text{ years}})^{1/(10 \text{ years})}$$

$$\left(\frac{1}{2}\right)^{1/(10 \text{ years})} = e^{k}$$

$$A = 100 \text{ grams} \left(\frac{1}{2}\right)^{t/(10 \text{ years})}$$

To predict the amount of tritium after 75 years, substitute 75 years for t.

$$A = 100 \text{ grams} \left(\frac{1}{2}\right)^{(75 \text{ years})/(10 \text{ years})}$$

$$= 100 \text{ grams} \left(\frac{1}{2}\right)^{7.5}$$

$$\cong 0.552427172 \text{ grams} \qquad \text{(by calculator)}$$

Note: In the preceding examples, the formula began with base e. Algebraic manipulation converted the formula to a base appropriate for the specific type of growth. However, e, the natural base, is the common source of the conversions.

Illustration 3:

If $100.00 is invested at 6 % APR compounded continuously, then the value of the account after t years is

$$A = \$100\ e^{0.06t}$$
$$\text{If } t = 2 \text{ years, then } A = \$100\ e^{0.06(2)}$$
$$A = \$100\ e^{0.12}$$
$$\text{By calculator, } e^{0.12} \cong 1.1275$$
$$A = \$112.75 \qquad \text{(to the nearest cent)}$$
$$\text{If } t = 4 \text{ years, then } A = \$100\ e^{.24}$$
$$A = \$127.12$$

The interest rate r is the constant of proportionality for the compound growth model: $(\Delta A)/(\Delta t) = rA$. This model indicates the rate of growth of an account varies jointly as the interest rate, r, and the account amount, A.

Varies jointly—y varies jointly as x and z means that $y = kxz$, where k is the constant of variation.

An exponential function has a graphic representation. First we investigate the basic shape of the exponential curve, then we consider transformations. As always, graphing technology gives quick access to these graphs.

Graphs of Exponential Functions. Graph each of the following exponential functions.

EXAMPLE 6-2

Illustration 1:

$$f(x) = e^x$$

Solution: Use graphing technology or plot points. By calculator, confirm the following are points of f.

x	-3	-2	-1	0	1	2	3
$y = e^x$	0.05	0.14	0.37	1	2.72	7.39	20.09

Your technology should reveal a well-behaved graph of f as in Figure 6-2.

Note: Use the trace feature. What is the domain of $f(x) = e^x$? Your trace feature may not help you because the cursor jumps from one finite precision value to the next. Due to the discrete nature of our technology, we rely on intuition to conclude that f is continuous. Previously we considered exponents no more complicated than a rational number. Therefore, $e^{2/3}$ is a meaningful equivalent to $\sqrt[3]{e^2}$. Because e is positive, any root of e is defined so that the domain of f includes rational numbers. But what could be the meaning of $e^{\sqrt{2}}$? Consider a sequence of rational approximations of $\sqrt{2}$, say 1, 1.4, 1.41, 1.414, . . . where each approximation gets closer to $\sqrt{2}$. The approximations are

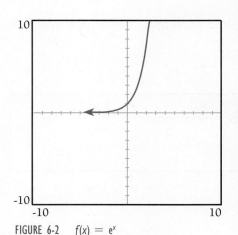

FIGURE 6-2 $f(x) = e^x$

rational numbers, so they are acceptable as exponents for e. Form a corresponding sequence of powers of e:

$$e^1 \cong 2.7182818284\ldots$$

$$e^{1.4} \cong 4.055199967\ldots$$

$$e^{1.41} \cong 4.095955404\ldots$$

$$e^{1.414} \cong 4.112372037\ldots$$

$$\ldots$$

These values appear to approach a number near 4. Define $e^{\sqrt{2}}$ as the limit of the approximation sequence: $\lim_{x\to\sqrt{2}}(e^x) = e^{\sqrt{2}}$. A simpler way to get this approximation is to enter $e^{\sqrt{2}}$ into your calculator:

$$e^{\sqrt{2}} \cong 4.113250378\ldots$$

Define other irrational exponents similarly and conclude that the domain of f is all real numbers.

Scroll to the left. Observe the y-values get close to 0 but never become 0. $\lim_{x\to-\infty}(e^x) = 0$. The x-axis is a horizontal asymptote for f. Trace back to $x = 0$. The y-intercept is $(0, 1)$.

From the graph, f increases over its domain. The range of f is $(0, \infty)$. Choose any positive number b and locate it on the y-axis. Because f is continuous, there is a real number k such that $e^k = b$. See Figure 6-3. This property of exponential functions validates our activities in Example 6-1. *Of great importance, f is one-to-one. Therefore, f has an inverse function.*

If $b > 0$ we can make use of $e^k = b$. Any function with rule of pairing of the form $y = A_0 e^{kx}$ may be rewritten as $y = A_0 b^x$. Because k is multiplied times x, experience with transformations indicates that the graph of $y = A_0 b^x$ is the same shape as $y = A_0 e^x$ stretched (or contracted) in the direction of the x-axis. A negative value for k reflects the graph about the y-axis.   ▬

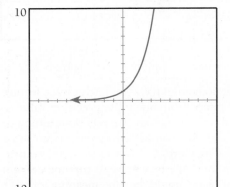

FIGURE 6-3 $y = b$, $x = k$, $y = e^x$

EXAMPLE 6-3

Illustration 1:

Graphs of General Exponential Equations. Graph the following.

$$f(x) = 2^x$$

Solution: Compare the graph of this function (Figure 6-4) to Example 6-2, Illustration 1.

x	-3	-2	-1	0	1	2	3
2^x	$\dfrac{1}{8}$	$\dfrac{1}{4}$	$\dfrac{1}{2}$	1	2	4	8

Figure 6-5 compares the graphs of $y = e^x$ and $y = 2^x$.

FIGURE 6-4 $y = 2^x$

Illustration 2:

$$f(x) = \left(\frac{1}{2}\right)^x$$

Solution:

$$\left(\frac{1}{2}\right)^x = 2^{-x}$$

Compare this function to the graph in Illustration 1.

x	-3	-2	-1	0	1	2	3
2^{-x}	8	4	2	1	$\frac{1}{2}$	$\frac{1}{4}$	$\frac{1}{8}$

Note: Here f is a decreasing function (Figure 6-6). In fact, f is the left-to-right mirror image of $y = 2^x$ from Illustration 1.

Illustration 3:

$$f(x) = 10^x$$

Solution: By technology, we have Figure 6-7.

Figure 6-8 compares all the graphs of this example to $y = e^x$.

Example 6-3 forms the basis for the following theorem.

Theorem 6-1
Increasing or Decreasing
Exponential Functions

If $f(x) = b^x$ represents an exponential function ($b = e^k$), then

1. f is increasing if $b > 1$,
2. f is decreasing if $0 < b < 1$.

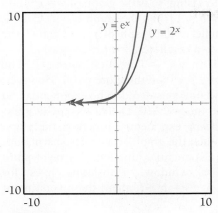

FIGURE 6-5 $y = e^x$, $y = 2^x$

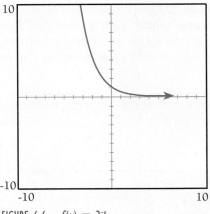

FIGURE 6-6 $f(x) = 2^{-x}$

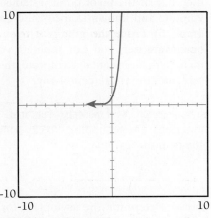

FIGURE 6-7 $f(x) = 10^x$

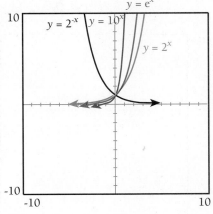

FIGURE 6-8 $y = e^x$, $y = 2^x$, $y = 2^{-x}$,
              $y = 10^x$

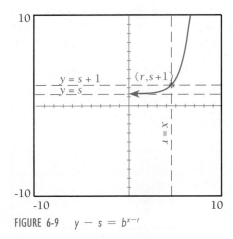

FIGURE 6-9 $y - s = b^{x-r}$

Suppose $b > 1$ and $b = e^k$. From the graph of $y = e^x$, if $b > 1$ then $k > 0$. But if $k > 0$, then the graph of $y = e^{kx}$ is a horizontal stretch of the graph of $y = e^x$. Because the graph of $y = e^x$ is increasing, so is the graph of $y = e^{kx}$, and as a result the graph of $f(x) = b^x$. If $0 < b < 1$, then the graph of $y = e^x$ indicates that $k < 0$. A negative value k reflects the graph of $y = e^x$ across the y-axis in the transformation to $y = e^{kx}$. As a result, $f(x) = b^x$ is decreasing.

Theorem 6-1 omits $b = 1$. If $b = 1$, then $e^k = 1$ implies that $k = 0$. The exponential function becomes $y = e^{0x}$ or $y = e^0$ and therefore $y = 1$. However, $y = 1$ is a familiar constant function and not considered to be exponential. So far, each exponential function contained the point $(0, 1)$. This is a **pivot point** in the shape of the graph. In an increasing exponential function, the graph rises slowly until it reaches this point, then the graph accelerates sharply upward. When sketching the graph of an exponential function, this pivot point provides a convenient center for a window. Translating axes for $y = b^x$ to (r, s) produces the equation $y - s = b^{x - r}$ where the pivotal point translates to $(r, s + 1)$. See Figure 6-9.

The complete transformation of $f(x) = e^x$ describes all elementary exponential functions and provides the following definition:

DEFINITION 6-1
General Exponential Functions

> An **exponential function** f is a function given by
>
> $$f(x) = A\, e^{kx+c} + d, \qquad k \neq 0$$
>
> ($e^k = b$ is a convenient base for the exponential function:
> $f(x) = Ab^{x+r} + d$.)

The graphs of all exponential functions have a similar shape. The effect of the constants in the formula is to stretch or contract the graph and shift the graph about the coordinate system. The shape of an exponential function is easy to memorize, so the best graphing strategy is to establish the horizontal asymptote, then use a pivot point to center a window. By plotting a point to either side of the pivot point, establish three points that, together with the asymptote and known shape of the graph, should be sufficient to visualize the graph. By anticipating the graph you are not as likely to be misled by technology. Moreover, you can annotate your graphs for clarity. Good annotations include the horizontal asymptote, the coordinates of a pivot point, the y-intercept, and the x-intercept if any.

EXAMPLE 6-4

Transformations on Exponential Functions. Graph each of the following exponential functions. Use technology to experiment with the graphs by changing the constants in the formula.

Illustration 1:

$$f(x) = 2(3^{x-5}) + 1$$

Solution: Determine the asymptote for an exponential function from the view of a transformation or by direct observation. Consider the $+1$ in the rule of pairing. The effect of $+1$ is to shift the entire graph up one unit. Because the horizontal asymptote of $y = e^x$ is $y = 0$, the asymptote of f will be $y = 1$. By direct observation, consider y values as x approaches negative infinity. A negative exponent for the base 3 produces a reciprocal. Thus, $2(3^{x-5})$ gets close to 0 as x decreases without bounds. Again, the horizontal asymptote is $y = 1$.

The easiest way to establish a pivot point is to work backwards. The x coordinate of an untransformed pivot point is 0. As in Chapter 1, ask what value for x produces a zero exponent. (A pivot point is not unique. Rather, it arises out of a given formula to produce an easily calculated exponent.)

$$x - 5 = 0$$
$$x = 5$$

To determine the y value of the pivot point, evaluate

$$f(5) = 2(3^{5-5}) + 1$$
$$f(5) = 3 \qquad\qquad \text{[pivot at (5, 3), see Figure 6-10]}$$

Now plot a point to either side of the pivot point, say $x = 4$ (Figure 6-11):

$$f(4) = 2(3^{4-5}) + 1$$
$$f(4) = 1 + \frac{2}{3}$$

and $x = 6$:

$$f(6) = 2(3^{6-5}) + 1$$
$$f(6) = 7$$

We annotate the asymptotes and points separate from the graph of f for two reasons: the size of figures in a textbook sometimes makes it difficult to distinguish a graph from its asymptote, and the annotations allow us to visualize the shape of f before technology fleshes out these points into the graph. When your own graphs have sufficient space, you should attempt to distinguish asymptotes by using "dashed" lines or a color different than that of the graph of the function. See Figure 6-12.

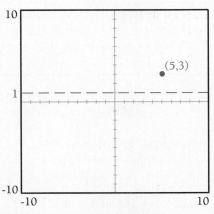

FIGURE 6-10   Annotate asymptote and pivot

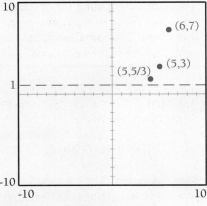

FIGURE 6-11   Asymptote and points

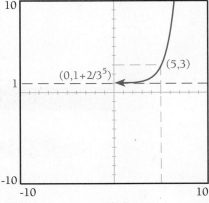

FIGURE 6-12 $f(x) = 2(3^{x-5}) + 1$

Illustration 2:

$$g(x) = -(5^{4-2x}) + 3$$

Solution:

Asymptote: $y = 3$

Pivotal point: $4 - 2x = 0$

$-2x = -4$

$x = 2$

Evaluate: $g(2) = -(5^{4-2(2)}) + 3$

$g(2) = 2$ (pivot, see Figure 6-13)

Plot points: $g(1) = -(5^{4-2(1)}) + 3$

$g(1) = -22$ (see Figure 6-14)

$g(3) = -(5^{4-2(3)}) + 3$ (see Figure 6-14)

$g(3) = 2.96$

Observe the effect of the negatives in the rule of pairing. The -2 coefficient of x reflected the graph from right to left. The -1 coefficient of the base 5 reflected the graph about the asymptote.

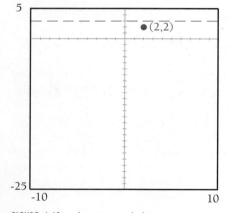

FIGURE 6-13   Asymptote and pivot

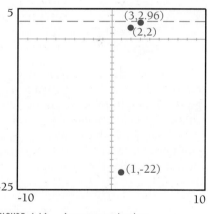

FIGURE 6-14   Asymptote and points

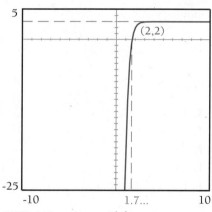

FIGURE 6-15 $g(x) = -(5^{4-2x}) + 3$

S U M M A R Y

- An exponential function models the *compound* growth of a quantity A when the change in A over time t is proportional to the current amount of the quantity: $A = A_0 \, e^{kt}$, where A_0 is the initial amount.
- For any exponential function $f(x) = Ab^{mx+c} + d$, $b > 0$, $b \neq 1$,

1. $\text{Domain}_f = (-\infty, \infty)$

2. $\text{Range}_f = \begin{cases} (d, \infty) & \text{if } A > 0 \\ (-\infty, d) & \text{if } A < 0 \end{cases}$

3. Asymptote for f is $y = d$

4. Pivot point for f is $\left(-\dfrac{c}{m}, A + d\right)$

5. Based on its graph, f is one to one. Therefore f has an inverse.

6-1 ⬤ EXERCISES

Graph the following functions indicating the domain, the range, and whether increasing or decreasing. Annotate asymptotes and pivot points.

1. $f(x) = 3^x$

2. $f(x) = \left(\dfrac{1}{4}\right)^x$

3. $f(x) = \left(\dfrac{1}{3}\right)^x$

4. $f(x) = 4^x$

5. $f(x) = 6^{-x}$

6. $f(x) = 6^x$

7. $f(x) = 5^x$

8. $f(x) = 5^{-x}$

9. $f(x) = e^{-x}$

10. $f(x) = -e^x$

11. $f(x) = 2^{x+1}$

12. $f(x) = 2^{x-1}$

13. $f(x) = 2^x + 1$

14. $f(x) = 2^x - 1$

15. $f(x) = 3^{x/2}$

16. $f(x) = 2^{x/3}$

17. $f(x) = (\sqrt{3})^x$

18. $f(x) = (\sqrt{2})^x$

19. $f(x) = 3^{x-1} + 2$

20. $f(x) = 3^{x+1} - 2$

21. $f(x) = 3^{x+2} - 1$

22. $f(x) = 3^{x-2} + 1$

23. $f(x) = -2(3^{x-1}) + 1$

24. $f(x) = 2(3^{x+1}) - 1$

25. $f(x) = 2(3^{1-x}) + 1$

26. $f(x) = -2(3^{x+1}) - 1$

27. $f(x) = 2(10^{2x-3}) - 4$

28. $f(x) = -3(5^{2-3x}) + 4$

29. $f(x) = -2(10^{3-2x}) - 4$

30. $f(x) = 3(5^{3x-2}) + 4$

31. Suppose the half life of a radioactive isotope of carbon, C^{14}, is approximately 5,700 years. As long as a plant is alive it absorbs C^{14}. When the plant dies absorption ceases. As a result, the age of an organic fossil may be estimated from the half-life of C^{14}. Suppose that when a plant died it contained 0.1 gram of C^{14}. Write a function to express the amount of C^{14} after t years.

32. Suppose that the half-life of Pb^{210} (lead) is 21 years. Write a function to model the amount of Pb^{210} after t years, if the initial amount is 2 grams.

33. The half life of Po^{218} (polonium) is 3.05 minutes. If A_0 is the initial amount of Po^{218}, express the amount of Po^{218} as a function of time t in minutes.

34. The half life of Th^{234} (thorium) is 24.1 days. Express the amount of Th^{234} as a function of time t in days if the initial amount of Th^{234} is A_0.

35. Suppose the population of a microbe colony doubles daily. If 20 microbes are currently in the colony, express the population size as a function of time. How many microbes are in the colony after 3 days? Suppose the microbes have been growing at the same rate for several days prior to the current colony of 20. How many were in the colony yesterday?

36. Suppose that a wildlife population of rabbits triples yearly when all predators are removed. If a wildlife refuge begins with 100 rabbits, express the number of rabbits as a function of time. How large will the population be in 5 years? If the predators have been missing for some time, how large was the population of rabbits 2 years ago?

37. Write a formula representing the value of a savings account with initial principal of $1000 compounded continuously at a rate of 5 percent per annum. What will be the value of the account in 3 years?

38. Write a formula modelling the value of a savings account after t years using a principal of $100 compounded continuously at a rate of 6 percent APR. What was the value of the account 2 years ago?

39. Rewrite the exponential function $f(x) = 9^x$ so that the base is 3.

40. Rewrite the exponential function $g(x) = 16^x$ so that the base is 4.

41. Rewrite $f(s) = 5^s$ so that the base is 25.

42. Rewrite $g(s) = 6^s$ so that the base is 36.

43. Prove Theorem 6-1.

44. Prove that $f(x) = (e^x + e^{-x})/2$ is an even function.

45. Prove that $g(x) = (e^x - e^{-x})/2$ is an odd function.

46. Graph $f(x) = (e^x + e^{-x})/2$. The name of this graph is a *catenary*. A catenary is the shape assumed by a flexible cable suspended from two fixed points. What shape have we previously studied that resembled a catenary?

47. Graph $g(x) = (e^x - e^{-x})/2$.

48. For $f(x) = e^x$ approximate $f(1), f(0.5)$ and $f(0.1)$.

49. For $g(x) = 1 + x + x^2/2 + x^3/6 + x^4/24$ approximate $g(1), g(0.5)$ and $g(0.1)$.

50. See Problems 48 and 49. Compare the values calculated. Also compare $f(0)$ with $g(0)$ and $f(-1)$ with $g(-1)$.

51. See Problems 44 and 45. Prove that $f(x) + g(x) = e^x$.

52. See Problem 51. Prove that $f(x) - g(x) = e^{-x}$.

53. See Problem 51. Prove that $[f(x)]^2 - [g(x)]^2 = 1$.

54. See Problem 51. Prove that $f(x)g(x) = (e^{2x} - e^{-2x})/4$.

6-1   PROBLEM SET

THE RULE OF 72

1. Use a calculator to approximate $e^{0.72}$.

2. Use a calculator to approximate $f(0.72)$, where $f(x) = e^x$.

3. The formula for continuous compounding of the amount A in a savings account with principal P after t years is given by $A = Pe^{rt}$, where r is the annual interest rate. Determine the amount in an account with $100 principal after 8 years if the interest rate is 0.09.

4. See Problem 3. Determine the amount in an account with $100 principal after 9 years if the interest rate is 0.08.

5. See Problem 3. Determine the amount in an account with $100 principal after 10 years if the interest rate is 0.072.

6. See Problem 4. Determine the amount in an account with $100 principal after 12 years if the interest rate is 0.06.

7. See Problem 5. Determine the amount in an account with principal P if the product of the interest rate and time is $rt = 0.72$.

8. See Problem 6. Show that if the interest rate for an account is r, the amount in an account approximately doubles in $0.72/r$ years.

9. The result in each of the preceding problems is known as the *rule of 72*. Suppose you are offered an interest rate of 0.04, use the rule of 72 to estimate the number of years it will take your account to double.

10. See Problem 9. Use the rule of 72 to estimate the doubling time for your account if your interest rate is 0.10.

11. Suppose you receive 0.10 interest. (a) How many doubling periods are there in 36 years? (b) If you deposit $10,000 for 36 years at 0.10 interest, what will be the value of your account?

12. Suppose you receive 0.12 interest. (a) How many doubling periods are there in 36 years? (b) If you deposit $10,000 for 36 years at 0.12 interest, what will be the value of your account?

13. See Problem 11. Suppose you receive 8 percent APR. (a) How many doubling periods are there in 36 years? (b) If you deposit $10,000 for 36 years at 8 percent APR, approximate the account value.

14. See Problem 12. Suppose you receive 6 percent APR. (a) How many doubling periods are there in 36 years? (b) If you deposit $10,000 for 36 years at 6 percent APR, approximate the account value.

15. See Problems 11 and 13. (a) Determine the value of $10,000 after 1 year at 0.10 interest compounded continuously. (b) Determine the value of $10,000 after 1 year at 0.08 interest compounded continuously. (c) See a and b; what is the difference in value of the two accounts after 1 year? (d) See Problems 11 and 13; what is the difference in the value of the two accounts after 36 years? (e) Discuss whether there is any significant difference between an account at 8 percent and an account at 10 percent over 36 years.

16. See Problems 12 and 14. (a) Determine the value of $10,000 after 1 year at 0.12 interest compounded continuously. (b) Determine the value of $10,000 after 1 year at 0.06 interest compounded continuously. (c) See a and b; what is the difference in value of the two accounts after 1 year? How did doubling the interest rate affect the value? (d) See Problems 12 and 14; what is the difference in the value of the two accounts after 36 years? (e) Attack or defend the following claim: if you double the interest from 0.06 to 0.12 then at most you will double the value of your account over 36 years.

17. List all the positive divisors of 69, 70, 71 and 72. Which of these numbers has the most divisors?

18. Calculate $e^{0.69}$, $e^{0.70}$ and $e^{0.72}$. Which of these values is "closer" to two? Discuss why we have a rule of 72 rather than a rule of 69. Hint: see Problem 17.

EXPLORATION

Recall that if f is a function then the slope of a secant line through the points $(x + h, f(x + h))$ and $(x, f(x))$ is given by

$$\text{Secant}(f, x, h) = \frac{f(x + h) - f(x)}{x + h - x}$$

$$= \frac{f(x + h) - f(x)}{h}$$

19. Sketch the graph of $f(x) = e^x$. Choose a point of the graph and label it $(x, f(x))$.

20. See Problem 19. Choose a second point of the graph to the right of x. Call it $(x + h, f(x + h))$. Draw a straight line through these two points. Write a formula to represent the slope of the line.

21. Suppose that $f(x) = e^x$. Write a formula for secant (f, x, h).

22. See problem 21. Simplify the secant formula by factoring out e^x. Hint: $e^{x+h} = e^x e^h$.

23. Use a calculator to estimate $(e^{0.1} - 1)/0.1$.

24. Use a calculator to estimate $(e^{0.01} - 1)/0.01$.

25. Use a calculator to estimate $(e^{0.001} - 1)/0.001$.

26. Use a calculator to estimate $(e^{0.0001} - 1)/0.0001$.

27. Use a calculator to estimate $(e^{0.00001} - 1)/0.00001$.

28. Use a calculator to estimate $(e^{0.000001} - 1)/0.000001$.

29. See Problems 21, 22 and 27. Use a calculator to estimate secant$(f, x, 0.00001)$.

30. See problems 22 and 28. Use a calculator to estimate secant$(f, x, 0.000001)$.

31. See Problem 29. If the slope of the secant line is used to approximate the rate of change of the exponential function at (x, e^x), conjecture a formula for the rate of change.

32. If the secant line in Problem 30 is approximately the tangent line to the exponential function at (x, e^x), conjecture a formula that would give the slope of the tangent line.

PROBLEMS FOR GRAPHING TECHNOLOGY

Graph each of the following sets of functions in the same window. In each case predict the shape of g and h compared to f before graphing.

33. Discuss any patterns in transformations you observe.
   (a) $f(x) = e^x$, $g(x) = e^{2x}$, $h(x) = e^{-2x}$
   (b) $f(x) = e^x$, $g(x) = e^{x+2}$, $h(x) = e^{x-2}$
   (c) $f(x) = e^x$, $g(x) = 2e^x$, $h(x) = -2e^x$
   (d) $f(x) = e^x$, $g(x) = e^x + 2$, $h(x) = e^x - 2$
   (e) $f(x) = e^x$, $g(x) = e^{3x-2}$, $h(x) = 3e^x - 2$
   (f) $f(x) = e^x$, $g(x) = e^{3x-2}$, $h(x) = 5e^{3x-2} - 7$

34. Discuss any patterns in transformations you observe.
   (a) $f(x) = e^x$, $g(x) = e^{-3x}$, $h(x) = e^{3x}$
   (b) $f(x) = e^x$, $g(x) = e^{x-3}$, $h(x) = e^{x+3}$
   (c) $f(x) = e^x$, $g(x) = -3e^x$, $h(x) = 3e^x$
   (d) $f(x) = e^x$, $g(x) = e^x - 3$, $h(x) = e^x + 3$
   (e) $f(x) = e^x$, $g(x) = e^{2x-3}$, $h(x) = 2e^x - 3$
   (f) $f(x) = e^x$, $g(x) = e^{5x-7}$, $h(x) = 3e^{5x-7} - 2$

35. (a) Graph $f(x) = e^{0.693x}$, $g(x) = 2^x$
   (b) Approximate $e^{0.693}$. Explain the results of part a.

36. (a) Graph $f(x) = e^{1.0986x}$, $g(x) = 3^x$
   (b) Approximate $e^{1.0986}$. Explain the results of part a.

37. Graph $y = e^{-x}$ and $y = 1/e^x$. Discuss the relation of these two graphs.

38. Graph $y = e^{x/2}$ and $y = (\sqrt{e})^x$. Discuss the relation of these two graphs.

39. The general equation for a catenary is given by $y = a(e^{x/a} + e^{-x/a})/2$. Sketch the graph of several catenarys using $a = 1$, $a = 2$, $a = 3$, $a = 4$, $a = -1$, $a = -2$. Conjecture the role of a in each of the graphs.

40. See Problem 39. Compare the graph of $y = (e^x + e^{-x})/2$ and the graph of $y = x^2 + 1$.

EXPLORATION

The following anticipate concepts from the next section. Compare the graphs of each of the following pairs of functions. Note any similarities.

41. $y = e^x$, $y = \ln x$

42. $y = e^{x+1}$, $y = -1 + \ln x$

43. $y = e^{2x}$, $y = 0.5 \ln x$

44. $y = e^{3x-2}$, $y = \dfrac{2 + \ln x}{3}$

45. $y = 5^x$, $y = \dfrac{\ln x}{\ln 5}$

46. $y = 7^x$, $y = \dfrac{\ln x}{\ln 7}$

47. $y = \ln(10x)$, $y = \ln x + \ln 10$

48. $y = \ln(x^2)$, $y = 2 \ln x$

49. $y = \ln(\sqrt{x})$, $y = \dfrac{\ln x}{2}$

50. $y = 2 \ln(3x - 1)$, $y = \dfrac{e^{x/2} + 1}{3}$

---

2   LOGARITHMIC FUNCTIONS

The miraculous powers of modern calculations are due to three inventions: the Arabic Notation, Decimal Fractions and Logarithms.
—F. CAJORI

From the previous section, every exponential function is one to one. As a result exponential functions have inverse functions. This section investigates the inverses of exponential functions.

Suppose that f is an exponential function defined as follows:

$$f: \quad y = b^x, \qquad b > 0, \quad b \neq 1$$

The domain of f is $(-\infty, \infty)$. The range of f is $(0, \infty)$ To form the inverse of f, f^{-1}, interchange the domain and range variables in the rule of pairing:

$$f^{-1}: \quad x = b^y, \qquad b > 0, \ b \neq 1$$

The domain of f^{-1} is $(0, \infty)$ and the range of f^{-1} is $(-\infty, \infty)$. We prefer the formula for f^{-1} solved for the dependent variable y. The following notational definition allows us to solve for y.

DEFINITION 6-2
Logarithmic Notation: $\log_b N$

$$b^p = N \text{ is equivalent to } \log_b N = p \qquad (b > 0, \quad b \neq 1)$$

Now $b^p = N$ is the exponential form of the expression, and $\log_b N = p$ is the *logarithmic* form of the same information. Read "$\log_b N = p$" as "logarithm base b of N equals p." Both forms of the expression convey the same information: the base is b, the exponent is p and the number produced is N.

$$\text{Base}^{\text{exponent}} = \text{number} \Leftrightarrow \log_{\text{base}} (\text{number}) = \text{exponent}$$

Algebra is the science of rewriting expressions, and logarithmic notation provides another, useful method for expressing information.

EXAMPLE 6-5

Rewriting Logarithmic and Exponential Expressions. Rewrite each of the following expressions as indicated.

Illustration 1:   Rewrite $3^2 = 9$ in logarithmic form.

Solution: The base is 3, the exponent is 2 and the number is 9: $\log_3 9 = 2$.

Illustration 2:

Rewrite $25^{1/2} = 5$ in logarithmic form.

Solution: The base is 25, the exponent is $\frac{1}{2}$ and the number is 5: $\log_{25} 5 = \frac{1}{2}$.

Illustration 3:

Rewrite $\log_4 16 = 2$ in exponential form.

Solution: The base is 4, the exponent is 2 and the number is 16: $4^2 = 16$.

Illustration 4:

Rewrite $\log_3 \left(\frac{1}{81}\right) = -4$ in an exponential form. $3^{-4} = \frac{1}{81}$.

Logarithm—From the Greek *logo* (ratio) and *arithmos* (number). John Napier coined the word in 1616 in his treatise on logarithms. Napier stated, "logarithmes doth cleave take away all the difficultie that heretofore hath beene in mathematicall calculations." Kepler introduced the contraction *log* in 1624.

An expression written in logarithmic form leaves the exponent "on line" rather than superscripted. Rewriting the formula for the inverse of an exponential function into logarithmic form allows us to solve for the dependent variable.

$$f^{-1}: \quad x = b^y, \qquad b > 0, \quad b \neq 1$$

The exponent y is the dependent variable. To solve for y requires that y be "on line" rather than superscripted. Logarithms provide the necessary conversion. The base is b, the exponent is y and the number is x.

$$f^{-1}: \quad \log_b x = y$$

The exponential function $f(x) = b^x$ has an inverse. The inverse, $f^{-1}(x) = \log_b x$, is a **logarithmic function**. Since f and f^{-1} are inverse functions the following theorems follow immediately from the composition of the two functions.

Theorem 6-2
Logarithm and Exponential are Inverses

$$\log_b(b^x) = x, \text{ where } x \text{ is any real number}$$

Theorem 6-3
Exponential and Logarithm are Inverses

$$b^{\log_b x} = x, \text{ where } x > 0$$

Evaluating Logarithmic Expressions. Evaluate each of the following.

EXAMPLE 6-6

Illustration 1:

$$\log_{13}(13^2)$$

Solution 1: By Theorem 6-2, $\log_{13}(13^2) = 2$.

Solution 2: Let $\qquad\qquad\qquad\qquad\qquad\qquad x = \log_{13}(13^2)$
$\qquad$ Rewrite in exponential form $\qquad 13^x = 13^2$

Because exponential functions are one to one, $x = 2$.

Note: Solution 2 indicates an important principle. If you have difficulty with a logarithmic expression, try rewriting the expression in exponential form.

Illustration 2: $\log_5(125^2)$

Solution I: The bases do not match. Theorem 6-2 does not apply.

Let $x = \log_5(125^2)$

Rewrite in exponential form. $5^x = 125^2$

The bases still do not match. $5^x = (5^3)^2$ (because $5^3 = 125$)

$5^x = 5^6$

$x = 6$

Solution 2:

$$\log_5(125^2) = \log_5[(5^3)^2]$$
$$= \log_5(5^6) \qquad \text{(now the bases match)}$$
$$= 6 \qquad \text{(by Theorem 6-2)}$$

Illustration 3: $7^{\log_7 6}$

Solution I: By Theorem 6-3, $7^{\log_7 6} = 6$

Solution 2:

Let $x = 7^{\log_7 6}$

Rewrite in logarithmic form. $\log_7 6 = \log_7 x$

Logarithmic functions are one- to- one, therefore $x = 6$.

Note: Solution 2 indicates an important tactic. If you have difficulty with an exponential expression, try converting to logarithmic form.

Illustration 4: $7^{\log_5 3}$

Solution: The bases do not match, Theorem 6-3 does not apply.

Let $x = 7^{\log_5 3}$

Rewrite in logarithmic form $\log_7 x = \log_5 3$.

Again the bases do not match.

We are unable to simplify $7^{\log_5 3}$

A calculator supplies an approximation: $7^{\log_5 3} \cong 3.774584542 \ldots$ ▬

The preceding example provides another interpretation of logarithms.

Suppose that $\log_b N = p$

then $b^p = N$

Replace p with $\log_b N$ and obtain $b^{\log_b N} = N$, as in Theorem 6-3. Note that p is the exponent for the base b that produces the number N. But $p = \log_b N$. Interpret $\log_b N$ as the exponent that when used with the base b produces the number N. More succinctly, a logarithm *is* an exponent.

What is the graph of a logarithmic function? Recall the graph of $E(x) = e^x$ (Figure 6-16).

The inverse function of E is given by $E^{-1}(x) = \log_e x$. Because of the usefulness of the natural base e, we have a special notation for $\log_e x$.

DEFINITION 6-3
Natural Logarithm and Common Logarithm

Abbreviate the *natural base* logarithm as $\log_e x = \ln x$.
(read "Lin of x" or "Ell En of x" or "natural log of x" or "log natural of x")
    Denote the *common base* logarithm by $\log_{10} x = \log x$
(read "log of x").

Because $\ln(x)$ is the inverse of $E(x)$, the graph of $\ln x$ is the reflection of $E(x)$ across the line $y = x$. It follows that the domain of $\ln x$ is $(0, \infty)$; the range is $(-\infty, \infty)$. See Figure 6-17.

Note that $\ln(x)$ has a vertical asymptote at $x = 0$. This vertical asymptote is a reflection of the horizontal asymptote $y = 0$ for e^x. The domain of $\ln x$ provides a reminder for the vertical asymptote: all x values are to the right of 0. Hence, $x = 0$ describes the vertical asymptote.

The pivotal point of $E(x)$ $(0, 1)$ became $(1, 0)$ on the graph of $\ln x$. Based on experience with exponential functions, you should expect that all logarithmic functions are transformations of the basic logarithmic shape expressed by the graph of $\ln x$. The following theorem confirms this suspicion.

Theorem 6-4
Change of Base Formula

$$\log_b N = \frac{\log_a N}{\log_a b}$$

where $a > 0, a \neq 1, b > 0, b \neq 1$ and $N > 0$.

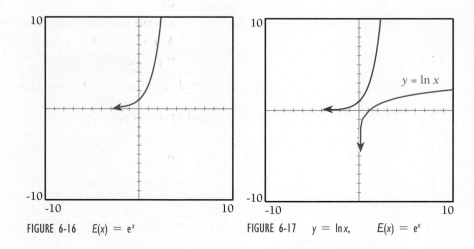

FIGURE 6-16 $E(x) = e^x$

FIGURE 6-17 $y = \ln x, \qquad E(x) = e^x$

Proof:

Let	$p = \log_b N$
Rewrite to obtain	$b^p = N$
Let	$q = \log_a b$
Then	$a^q = b$
Replace b with a^q,	$a^{qp} = N$
Rewrite as a logarithm:	$qp = \log_a N$
Replace q with $\log_a b$:	$p \log_a b = \log_a N$
Solve for p:	$p = \dfrac{\log_a N}{\log_a b}$
Because $p = \log_b N$,	$\log_b N = \dfrac{\log_a N}{\log_a b}$

(In solving for p, how did we know that $\log_a b \neq 0$?)

Two crucial aids are now available for sketching the graph: the vertical asymptote and the domain.

Graphing Logarithmic Functions. Graph the following logarithmic functions.

E X A M P L E   6 - 7

Illustration I:

$$f(x) = \log_4 x$$

Solution: The argument for a logarithmic function must be greater than 0. Therefore $x > 0$. Determining the domain simultaneously establishes the vertical asymptote ($x = 0$) and which "side" of the asymptote contains the graph. Next, establish a pivot point. Because $b^0 = 1$, $\log_b 1 = 0$.

$$f(1) = \log_4 1$$
$$f(1) = 0$$

Because the asymptote is such a powerful graphing aid, only one more point is needed to complete the familiar logarithmic shape. But logarithmic functions are not always easy to evaluate. Remember logarithms are exponents for a particular base. The only easy arguments will be powers of the base. The base is 4, so suitable domain components are 4, 16, 64, . . . $\frac{1}{4}$, $\frac{1}{16}$, $\frac{1}{64}$, ... For practice we evaluate several of these. See Figure 6-18. A window large enough to include 64 may obscure details of the graph.

$$f(4) = \log_4 4$$
$$f(4) = 1 \qquad \text{(because } 4^1 = 4\text{)}$$
$$f(16) = \log_4 16$$
$$f(16) = 2 \qquad \text{(because } 4^2 = 16\text{)}$$
$$f(64) = \log_4 64$$
$$f(64) = 3 \qquad \text{(because } 4^3 = 64\text{)}$$

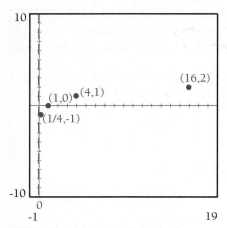

FIGURE 6-18   Asymptote and plotted points

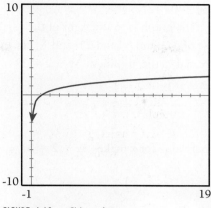

FIGURE 6-19 $f(x) = \log_4 x$ FIGURE 6-20 f and $y = \log_4 9$

$$f\left(\frac{1}{4}\right) = \log_4\left(\frac{1}{4}\right)$$

$$f\left(\frac{1}{4}\right) = -1 \qquad\qquad\qquad \left(\text{because } 4^{-1} = \frac{1}{4}\right)$$

$$f\left(\frac{1}{16}\right) = \log_4\left(\frac{1}{16}\right)$$

$$f\left(\frac{1}{16}\right) = -2 \qquad\qquad\qquad \left(\text{because } 4^{-2} = \frac{1}{16}\right)$$

Use the trace option to verify that $(1, 0)$, $(4, 1)$, $(16, 2)$, $(\frac{1}{4}, -1)$ on the graph of f is as given in Figure 6-19.

Note: To use *GraphWindows* to sketch the graph, store the formula under F1 by pressing [Alt] + [F1]. Enter [L][o][g][↓][4][↑][(][x][)][ENTER]. Press [F1] to display the graph. For a graphing calculator, use the change of base formula to sketch the graph $\log_4 x = [\ln x]/[\ln 4]$. Because $\ln 4$ is a constant, $\log_4 x = c \ln x$, where $c = 1/\ln 4$. Refer to the graph of $\log_4 x$ to confirm that the graph is the same as $\ln x$ compressed along the y-axis. The change of base formula assures us that any logarithmic function is a transformation of $\ln x$ and therefore has the same basic shape.

Examine the graph carefully. Although we do not know the exact value of $\log_4 9$ ($4^x = 9$, $x = ?$), Figure 6-20 indicates that $\log_4 9$ is between 1 and 2. Your trace feature should approximate it at about 1.6.

Illustration 2: $$f(x) = \log_5(3x - 2)$$

Solution: The argument of a logarithm must be positive.

$$3x - 2 > 0$$

$$x > \frac{2}{3}$$

Therefore, $x = \frac{2}{3}$ is a vertical asymptote.

The graph is to the right of the asymptote. Now locate some points. For a base of 5, 1 and 5 (and 25, and $\frac{1}{5}$, etc.) would be good arguments. What value or x makes the argument $3x - 2 = 1$?

$$3x - 2 = 1$$
$$3x = 3$$
$$x = 1$$

What value for x makes $3x - 2 = 5$?

$$3x - 2 = 5$$
$$3x = 7$$
$$x = \frac{7}{3}$$

The points to plot have domain components of 1 and $\frac{7}{3}$ (Figure 6-21).

$$f(1) = \log_5[3(1) - 2]$$
$$f(1) = \log_5 1$$
$$f(1) = 0.$$
$$f\left(\frac{7}{3}\right) = \log_5\left[3\left(\frac{7}{3}\right) - 2\right]$$
$$f\left(\frac{7}{3}\right) = \log_5 5$$
$$f\left(\frac{7}{3}\right) = 1$$

Annotate the asymptote, label $(1, 0)$ and $\left(\frac{7}{3}, 1\right)$ and graph f (Figure 6-22).

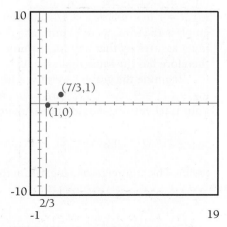

FIGURE 6-21   Annotated asymptote and points

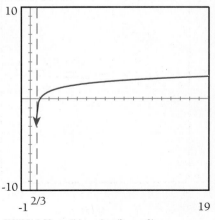

FIGURE 6-22 $f(x) = \log_5(3x - 2)$

Note: The effect of the constants 2 and 3 shift and stretch the graph. Figure 6-23 compares the graph of $y = \log_5 x$ to the graph of f.

Illustration 3:

$$f(x) = 2 \log(4 - x) + 1$$

Solution: The shape is logarithmic. The constants provide a transformation on the basic shape. First find the asymptote and domain.

$$4 - x > 0$$
$$4 > x$$

The asymptote is $x = 4$. The graph is to the left of the asymptote. Since no base is displayed, the assumed base is 10. Good choices for arguments for a base 10 logarithm include powers of 10, such as 1 and 10

$$4 - x = 1 \qquad \text{(argument of 1)}$$
$$\text{-}x = \text{-}3$$
$$x = 3$$
$$f(3) = 2 \log (4 - 3) + 1$$
$$f(3) = 2(0) + 1$$
$$f(3) = 1 \qquad \text{(plot (3, 1); Figure 6-24)}$$
$$4 - x = 10 \qquad \text{(argument of 10)}$$
$$\text{-}x = 6$$
$$x = \text{-}6$$
$$f(\text{-}6) = 2 \log [4 - (\text{-}6)] + 1 \qquad \text{(Figure 6-25)}$$
$$f(\text{-}6) = 2(1) + 1$$
$$f(\text{-}6) = 3 \qquad \text{(plot (-6, 3))}$$

Before calculators, approximating values for logarithmic functions was extremely tedious. Once values had been calculated, mathematicians saved their approximations in table form so that the calculations would not have to

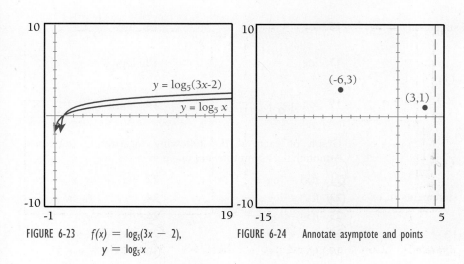

FIGURE 6-23 $f(x) = \log_5(3x - 2)$,
$y = \log_5 x$

FIGURE 6-24   Annotate asymptote and points

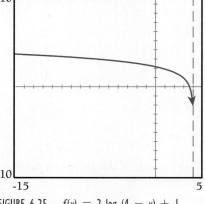

FIGURE 6-25 $f(x) = 2 \log (4 - x) + 1$

be repeated. One historical reason for the popularity of base 10 logarithms is that base 10, as the base for the decimal system, lends itself to table approximations. Table 1 of the Appendix is a table of base 10 logarithms. However, with the advent of calculators, the advantage of tables has given way. Indeed, because the natural base logarithm has more applications than the common base, some texts eliminate the special notation for base 10. In those texts, any logarithm without an explicit base is assumed to be the natural base. ▬

SUMMARY

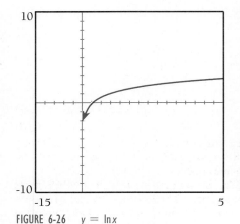

- $f(x) = b^x$ and $g(x) = \log_b x$ are inverse functions ($b > 0, b \neq 1$).
- Abbreviations: $\ln x = \log_e x$ and $\log x = \log_{10} x$.
- A logarithm of one base converts to a logarithm of another base by use of the transformation $\log_b x = \log_a x / \log_a b$. Because $\log_a b$ is a constant, multiplication by $1/\log_a b$ is a transformation that vertically stretches the shape of the graph.
- The shape of all logarithmic functions resembles $y = \ln x$ (Figure 6-26).
- Transformations on $\ln x$ yield other logarithmic functions. Graphs of all logarithmic function are the shape of $\ln x$ shifted, stretched or reflected.

FIGURE 6-26 $y = \ln x$

6-2   EXERCISES

Rewrite each of the following exponential statements in logarithmic form.

1. $2^4 = 16$

2. $4^2 = 16$

3. $5^3 = 125$

4. $3^5 = 243$

5. $3^{-2} = \dfrac{1}{9}$

6. $4^{-3} = \dfrac{1}{64}$

7. $36^{1/2} = 6$

8. $16^{1/4} = 2$

9. $27^{-2/3} = \dfrac{1}{9}$

10. $81^{-3/4} = \dfrac{1}{27}$

Rewrite each of the following logarithmic statements in exponential form:

11. $\log_3 9 = 2$

12. $\log_4 16 = 2$

13. $\log_3 243 = 5$

14. $\log_5 125 = 3$

15. $\log_5\left(\dfrac{1}{25}\right) = -2$

16. $\log_7\left(\dfrac{1}{49}\right) = -2$

17. $\log_{25} 5 = \dfrac{1}{2}$

18. $\log_{27} 3 = \dfrac{1}{3}$

19. $\log_{81}\left(\dfrac{1}{9}\right) = -\dfrac{1}{2}$

20. $\log_{27}\left(\dfrac{1}{3}\right) = -\dfrac{1}{3}$

Graph of each of the following logarithmic functions. Annotate the asymptote and one point.

21. $f(x) = \log_3 x$

22. $f(x) = \log_4 x$

23. $f(x) = \log_{1/3} x$

24. $f(x) = \log_{1/4} x$

25. $f(x) = \log(x + 3)$

26. $f(x) = \log(x - 3)$

27. $f(x) = \ln x$

28. $f(x) = \ln(-x)$

29. $f(x) = -\ln x$

30. $f(x) = \ln(x + 1)$

31. $f(x) = \ln x + 1$

32. $f(x) = \log_2(x - 2)$

33. $f(x) = \log_2 x - 2$

34. $f(x) = 3\log_2 x$

35. $f(x) = \log_2(3x)$

36. $f(x) = \log_2 x + 1$

37. $f(x) = \log_2(x + 1)$

38. $f(x) = \log_2\left(\dfrac{x}{3}\right)$

39. $f(x) = \dfrac{\log_2 x}{3}$

40. $f(x) = 3\log_2(x + 1)$

41. $f(x) = \log_3(2x) + 1$

42. $f(x) = \log_3(2x + 1)$

43. $f(x) = 2\log_3(x + 1)$

44. $f(x) = -\log_3(x + 2)$

45. $f(x) = \log_3(2 - x)$

46. $f(x) = 2\log_3(x - 1) + 5$

47. $f(x) = \log_3(2x + 1) - 4$

48. $f(x) = \log_3(3 - x) + 5$

49. $f(x) = 4\log_3(3x - 2) + 1$

50. $f(x) = -2\log_3(3 - 2x) + 1$

51. $f(x) = \log|x|$

52. $f(x) = \log|x + 1|$

53. $f(x) = \log|x - 1|$

54. $f(x) = \log|-x|$

55. $f(x) = |\log x|$

56. $f(x) = -|\log x|$

57. Graph $f(x) = \log_2(x^2)$, $x > 0$ and $g(x) = 2\log_2 x$.

58. Compare the graph of $f(x) = 3\log_2 x$ with $g(x) = \log_2(x^3)$.

59. Compare the graph of $f(x) = \log_2(8x)$ with $g(x) = 3 + \log_2 x$.

60. Compare the graph of $f(x) = \log_2 x - 3$ with $g(x) = \log_2(x/8)$.

61. Prove Theorem 6-2.

62. Prove Theorem 6-3.

6-2   P R O B L E M   S E T

EXPLORING AREA UNDER A CURVE

Recall the reciprocal function $H(x) = 1/x$. The shaded region in Figure 6-27a is bounded by H, the vertical lines given by $x = a$ and $x = b$ and the x-axis. One method to approximate this area is to divide the region into numerous rectangles and sum the areas of these rectangles. Figure 6-27b shows three rectangles whose sum approximates the area. The base of each rectangle is dx, and the corresponding height is the functional value, $H(x)$. The area of each rectangle is given by $H(x)dx$. Abbreviating sum with an elongated S we represent the area of the shaded region by

$\int_a^b H(x)dx$ (read as "Area under H from a to b"). Using area as a model for applications such as "work" or "total cost" indicates possible applications for the area under H.

In Figure 6-27a, the points $(a, 1/a)$, $(b, 1/b)$ on H are connected by a line segment. The line segment, together with $x = a$, $x = b$ and the x-axis, bounds a trapezoidal region that approximates the area under H from a to b. Intuitively, the closer a is to b the better the area of the trapezoid approximates the area under H.

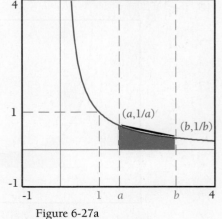

Figure 6-27a

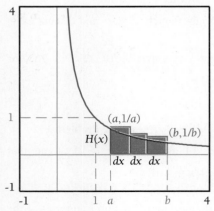

Figure 6-27b

FIGURE 6-27 H, $x = a$, $x = b$

1. Show that, $\int_a^b H(x)dx \cong (b^2 - a^2)/(2ab)$. (Hint: the area of a trapezoid is the average of the parallel sides times the distance between them.)

2. See Figure 6-28. Use a trapezoid to approximate $\int_{1/a}^a H(x)dx$.

3. See Figure 6-28. Use a trapezoid to approximate $\int_1^a H(x)dx$.

4. Show that $\int_b^a H(x)dx \cong -(b^2 - a^2)/(2ab)$.

5. Suppose $c > 0$. Approximate $\int_{ac}^{bc} H(x)dx$.

6. Suppose $c > 0$. Approximate $\int_{a/c}^{b/c} H(x)dx$.

7. Consider $y = cH(x)$, where $c > 0$. Approximate $\int_a^b cH(x)dx$.

8. Suppose $c > 0$. Approximate $c\int_a^b H(x)dx$.

For the remainder of the problem set, assume that the preceding are exact areas rather than approximations.

9. Show that $\int_a^b H(x)dx = -\int_b^a H(x)dx$.

10. Show that $\int_1^a H(x)dx = \int_{1/a}^1 H(x)dx$.

11. Show that $\int_a^b H(x)dx = \int_{ac}^{bc} H(x)dx$.

12. Show that $\int_a^b H(x) = \int_{a/c}^{b/c} H(x)$

13. If c is between a and b, show that
$$\int_a^b H(x)dx = \int_a^c H(x)dx + \int_c^b H(x)dx$$
Do not use trapezoids. Sketch a figure and discuss exact areas.

14. If $c > b$, argue that $\int_a^b H(x)dx = \int_a^c H(x)dx + \int_c^b H(x)dx$. See Problems 9 and 13.

15. If $0 < c < a$, argue that
$$\int_a^b H(x)dx = \int_a^c H(x)dx + \int_c^b H(x)dx. \text{ See Problem 14.}$$

16. See Problems 13, 14 and 15. If $c > 0$, argue that
$$\int_a^b H(x)dx - \int_a^c H(x)dx = \int_c^b H(x)dx.$$

Let us define a new function L in terms of the area under a curve. Suppose that $x > 0$. Define the corresponding y value for the new function L as

the area under H between 1 and x. The region defined has an area assuring a unique y value; hence, L is a function. See Figure 6-29.
$$L(x) = \int_1^x H(t)dt, \quad \text{where } t > 0 \text{ and } x > 0.$$

17. Argue that $L(b) - L(a) = \int_a^b H(t)\,dt$. See Problem 16.

18. Argue that $L(b/a) = \int_a^b H(t)\,dt$. See Problem 11 and definition of L.

19. Argue that $L(1/a) = -\int_1^a H(t)\,dt$. See Problem 18.

20. Argue that $-L(1/a) = L(a)$. See Problem 19.

21. Argue that $L(b/a) = L(b) - L(a)$. See Problem 17.

22. Argue that $L(ba) = L(b) + L(a)$. See Problems 20 and 21.

23. Argue that $L(a^2) = 2L(a)$. See Problem 22 and note $a^2 = aa$.

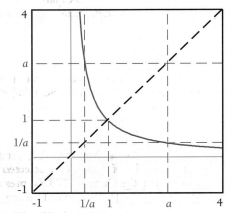

FIGURE 6-28 H, $x = \dfrac{1}{a}$, $x = a$

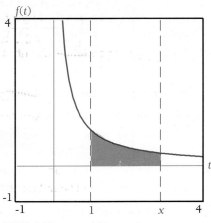

FIGURE 6-29 $y = H(t)$, $t = 1$, $t = x$

24. Argue that $L(a^{-2}) = -2L(a)$. See Problems 20 and 23.

25. Argue that $L(a^3) = 3L(a)$.

26. Argue that $L(a^{-3}) = -3L(a)$.

27. See Problems 23 and 25. Extend the pattern to $L(a^n)$, n is a counting number.

28. See Problems 24 and 26. Extend the pattern to $L(a^{-n})$, n is a counting number.

29. Examine Problems 20, 21, 22, 27 and 28. These results express the transformation properties of the function L. How do these differ from the transformation properties of linear functions (Chapter 2)?

30. Note that the rule of pairing for L is given by an area under a curve. Consider $L(3)$. So far we have only approximated this value by using a trapezoid $(L(3) \cong (3^2 - 1)/(2 \times 3) = \frac{4}{3})$. Because 3 is not "close" to 1, this is not a very good approximation. Discuss how to improve the approximation. (Hint: see Problem 13.)

AREA EXPERIMENTS

Refer to the discussion of area in the Problem Set. Your calculator can approximate an area under $H(x)$. Consider

$$\int_1^5 \frac{1}{x}\, dx.$$

Set the range to Xmin to -1, Xmax to 10, Ymin to -5, Ymax to 5. Be sure no other functions are selected and the graph screen is clear. For the Casio fx-7700G, press the following keys: [AC] [MODE] [SHIFT] [+] {All clear and Rectangular}.

Press [SHIFT] [G↔T] [(] [1] [÷] [XθT] [)] [SHIFT] [→] [1] [SHIFT] [→] [5] [EXE] {Enter H, set bounds, draw graph and calculate area} You will see Display 6-1.

For the TI-81, press [2nd] [PRGM] (DRAW). From the menu, select 7: Shade. The screen clears and the display reads [Shade (]

At the flashing cursor enter [0] [ALPHA] [.] [1] [÷] [X|T] [ALPHA] [.] [1] [ALPHA] [.] [1] [ALPHA] [.] [5] [)] [ENTER]. The display becomes [Shade (0, 1/x, 1, 1, 5)]

The entries represent, in order, the following:

0:	$y = 0$	the lower bound for shading.
$1/x$:	$y = 1/x$	the upper bound for shading.
1		the degree of shading (1 to 8)
1	$x = 1$	the lefthand bound for shading.
5	$x = 5$	the righthand bound for shading.

Press [ENTER] to produce the graph.

For the Sharp choose the Math mode. Choose [MATH]

C(Calc) 2. The symbol $\int_{[]}^{[]}$ appears. Enter 1 in the lower box, 5 in the upper box and $1/x$ in the middle to obtain $\int_1^5 1/x$. Choose C(Calc) 3 to obtain $\int_1^5 1/x\, dx$. Press [ENTER] for the approximation. No graph is shown. Use your graphing calculator to approximate the indicated areas.

31. $\int_1^2 \frac{1}{x}\, dx$

32. $\int_1^3 \frac{1}{x}\, dx$

33. $\int_1^{1/2} \frac{1}{x}\, dx$

34. $\int_1^{1/3} \frac{1}{x}\, dx$

35. $\int_1^{5/4} \frac{1}{x}\, dx$

36. $\int_1^{3/2} \frac{1}{x}\, dx$

37. $\int_1^{4/5} \frac{1}{x}\, dx$

38. $\int_1^{2/3} \frac{1}{x}\, dx$

39. See Problems 31, 33, 35, and 37. Use your calculator to approximate $\ln 2$, $\ln(\frac{1}{2})$, $\ln(\frac{5}{4})$, and $\ln(\frac{4}{5})$. Compare the results.

40. See Problems 32, 34, 36 and 38. Use your calculator to approximate $\ln 3$, $\ln(\frac{1}{3})$, $\ln(\frac{3}{2})$, and $\ln(\frac{2}{3})$. Compare the results.

41. Refer to Problems 31, 33, 35, 37 and 39. How are $\ln t$, $\ln(1/t)$ and $\int_1^t 1/x\, dx$ related?

42. Refer to Problems 32, 34, 36, 38 and 40. How are $\ln t$, $\ln(1/t)$ and $\int_1^{1/t} 1/x\, dx$ related?

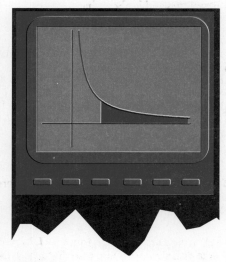

DISPLAY 6-1

EXPLORATION

The following problems anticipate concepts from the next section. Graph each of the following pairs of functions in the same window. In each case conjecture a relationship among the functions.

43. $fx = \ln(5x), g(x) = \ln 5 + \ln x$

44. $f(x) = \ln(x/5), g(x) = \ln x - \ln 5$

45. $f(x) = \ln(x/3), g(x) = \ln x - \ln 3$

46. $f(x) = \ln(7x), g(x) = \ln 7 + \ln x$

47. $f(x) = \ln(x^2), g(x) = 2\ln x$

48. $f(x) = \ln(x^3), g(x) = 3\ln x$

49. $f(x) = \ln(\sqrt[3]{x}), g(x) = \ln x/3$

50. $f(x) = \ln(\sqrt{x}), g(x) = \ln x/2$

51. $f(x) = \log(10x), g(x) = 1 + \log x$

52. $f(x) = \ln\left(\dfrac{1}{x}\right), g(x) = -\ln x$

---

3 LOGARITHMIC TRANSFORMATIONS AND APPLICATIONS

The whole of Mathematics consists in the organization of a series of aids to the imagination in the process of reasoning.
—A. N. WHITEHEAD

This section investigates applications of logarithmic functions. Most applications of logarithmic functions follow from the transformation properties of logarithms. Logarithmic transformations are quite different from the transformations based on linear functions in Chapter 2.

The following order of operations inspired the acronym GEMA: grouping symbols, exponents, multiplication (division), and addition (subtraction). An outstanding attribute of a logarithmic function is that it transforms exponentiation and multiplication one level down in the order of operations. In particular, logarithms transform exponentiation into multiplication. Logarithms transform multiplication to addition and division to subtraction. The following theorems express these concepts formally.

THEOREM 6-5
Multiplication Property of Log

$$\log_b(xy) = \log_b x + \log_b y, \qquad x > 0 \quad y > 0$$

(multiplication to addition)

THEOREM 6-6
Reciprocal Property of Log

$$\log_b\left(\frac{1}{x}\right) = -\log_b x, \qquad x > 0$$

(reciprocal to negation)

THEOREM 6-7
Division Property of Log

$$\log_b\left(\frac{x}{y}\right) = \log_b x - \log_b y, \qquad x > 0, \quad y > 0$$

(division to subtraction)

THEOREM 6-8
Exponentiation Property of Log

$$\log_b(x^p) = p\log_b x, \qquad x > 0 \qquad \text{(exponentiation to multiplication)}$$

Because logarithms are an alternate form for exponents, these properties are analogous to properties of exponents. The proof of Theorem 6-5 illustrates this similarity. The proofs of the remaining theorems are left as exercises.

Proof of Theorem 6-5:

$$\text{Let } p = \log_b x$$
$$\text{Then } b^p = x \qquad \text{(definition of logarithm)}$$

$$\text{Let } q = \log_b y$$
$$\text{Then } b^q = y \qquad \text{(definition of logarithm)}$$

$$\text{Multiply } x \text{ times } y \text{ to obtain } xy = b^p b^q \qquad \text{(property of}$$
$$\text{Then, } xy = b^{p+q} \qquad \text{exponents)}$$

$$\text{Rewrite in exponential form: } \log_b(xy) = p + q \qquad \text{(definition of logarithm)}$$

$$\text{Substitute for } p \text{ and } q \text{ to obtain } \log_b(xy) = \log_b x + \log_b y$$

The preceding theorems provide a method of rewriting logarithmic expressions. Rewriting logarithmic expressions takes two common forms: expansion and contraction. To expand a logarithmic expression means to rewrite the expression so that the argument of each logarithm is a prime expression containing no quotients. The exponent of the prime expression is one. The converse is to contract a logarithmic expression into a single logarithmic function. ∎

EXAMPLE 6-8

Expanding Logarithmic Expressions. Expand the following logarithmic expressions. Assume that all variables are positive. (Why?)

Illustration 1:

$$\log_b\left(\frac{2x}{y}\right)$$

Solution:

$$\log_b\left(\frac{2x}{y}\right) = \log_b 2 + \log_b x - \log_b y$$

Illustration 2:

$$\log_b\left(\frac{x^3}{y^2}\right)$$

Solution:

$$\log_b\left(\frac{x^3}{y^2}\right) = \log_b(x^3) - \log_b(y^2)$$
$$= 3\log_b x - 2\log_b y$$

Note: This problem could be done in a single step. Imagine that logarithm transforms the operations in its argument one level down in the order of operations: the divide becomes subtraction and the exponents become the coefficients (multipliers) of the logarithms.

Illustration 3:

$$\log_b\left(\frac{5x^4y^2}{z^5}\right)$$

Solution:

$$\log_b\left(\frac{5x^4y^2}{z^5}\right) = \log_b 5 + 4\log_b x + 2\log_b y - 5\log_b z$$

Illustration 4:

$$\log_b(x + y)$$

Solution: $\log_b (x + y)$ cannot be expanded.

Note: Logarithms do not transform additions down another level in the order of operations. GEMA lists no level below addition.

Illustration 5:

$$\log_b\left(\frac{\frac{2}{x} + 5}{7}\right)^4$$

Solution:

$$\log_b\left(\frac{\frac{2}{x} + 5}{7}\right)^4 = 4\log_b\left(\frac{2}{x} + 5\right) - 4\log_b 7$$

$$= 4\log_b\left(\frac{2 + 5x}{x}\right) - 4\log_b 7$$

$$= 4\log_b(2 + 5x) - 4\log_b x - 4\log_b 7$$

EXAMPLE 6-9

Contraction of Logarithmic Expressions. Contract each of the following logarithmic expressions.

Illustration 1:

$$\log_b x + \log_b y - \log_b z$$

Solution:

$$\log_b x + \log_b y - \log_b z = \log_b\left(\frac{xy}{z}\right)$$

Illustration 2:

$$3\log_b x - \left(\frac{1}{2}\right)\log_b y - 2\log_b z$$

Solution:

$$3\log_b x - \left(\frac{1}{2}\right)\log_b y - 2\log_b z = \log_b\left(\frac{x^3}{\sqrt{y}z^2}\right)$$

Illustration 3:

$$\log_b x + \log_a y$$

Solution:

$$\log_b x + \log_a y$$

This expression does not immediately contract because the bases do not match. From the change of base formula,

$$\log_b x = \frac{\log_a x}{\log_a b}$$

$$= \frac{1}{\log_a b} \log_a x$$

By Theorem 6-8, $= \log_a(x^{1/\log_a b})$

Now, $\log_b x + \log_a y = \log_a(x^{1/\log_a b}) + \log_a y$

$$= \log_a(yx^{1/\log_a b})$$

John Napier (1550–1617) developed logarithms as a device to ease the burden of arithmetic. Napier began by generating two sequences of numbers: the first formed by addition, the second formed by multiplication. The sequences were much like the following:

			↓	+	↓	=	↓	
Add 1	0	1	2	3	4	5	6	. . .
Multiply by 2	1	2	4	8	16	32	64	. . .
			↑	×	↑	=	↑	

Arrows mark the table to reflect that $4 \times 16 = 64$. The corresponding addition is $2 + 4 = 6$. In exponential form $4(16) = 2^2(2^4) = 2^{2+4} = 2^6 = 64$. From the Greek, *logarithm* translates literally as "ratio number." A ratio indicates comparison. Performing multiplication by comparison to addition justifies the name *logarithm*.

Modern digital computers usually store finite precision decimal repesentations of real numbers in a logarithmic inspired form. The next section examines such representations.

Now consider some applications of logarithms. Each example problem uses logarithmic transformations to obtain a solution.

Applications of Logarithms

EXAMPLE 6-10

Illustration 1:

Recall the bacteria growth model of Example 6-1, $y = 6(2^t)$. The time of growth is t hours, and y is the number of bacteria. At what time will there be 24 bacteria in the culture?

Solution:

Let $y = 24$ bacteria. Then

$$24 = 6(2^t)$$
$$4 = 2^t$$

Convert to logarithms to get the exponent "on line."

$$\log_2 4 = t$$
$$2^t = 4$$
$$2 = t$$

Illustration 2:

Suppose Larry invests \$1,000 at 5 percent compounded continuously and Sarah invests \$900 at 6 percent compounded continuously. How long before Sarah's account equals Larry's account?

Solution:

Model for Larry's account: $A = \$1{,}000\ e^{0.05t}$

Model for Sarah's account: $A = \$900\ e^{0.06t}$

Figure 6-30 displays the growth in both accounts. The point of intersection represents equal amounts in the accounts. Algebraically, set the two amounts equal: $\$1{,}000\ e^{0.05t} = \$900\ e^{0.06t}$ and solve for t.

$$\frac{\$1000}{\$900} = \frac{e^{0.06t}}{e^{0.05t}}$$

$$\frac{10}{9} = e^{0.01t}$$

$$\ln\left(\frac{10}{9}\right) = 0.01t$$

$$100\ln\left(\frac{10}{9}\right) = t$$

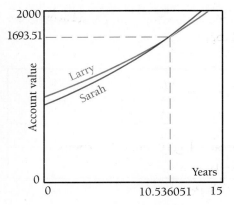

2000
1693.51

Account value

Larry

Sarah

Years

0

0   10.536051   15

FIGURE 6-30 $y = 1{,}000\ e^{.05t}$, $y = 900\ e^{.06t}$

Note 1: Use your calculator to verify the approximation $t \cong 10.53605156$ years (about 10 years, 6 months and 8 to 10 days). Try the trace feature of your technology to determine the point of intersection. Compare the precision of the answers. Convert the trace solution to years, months and days. Again compare the precision of the solutions.

Note 2: An interesting result is that Sarah's account eventually surpasses Larry's. Once her account is larger, the difference continues to grow. Why is this? Examine the interest rate. By definition of exponential functions, the interest rate is the constant of proportionality in the rule of pairing expressing the *rate of change*. If requested to determine the time during which Larry's account contains more money than Sarah's, model the problem with the inequality

$$1{,}000\ e^{0.05t} > 900\ e^{0.06t}.$$

The preceding illustrates solving exponential equations. In an exponential equation the variable is in an exponent.

Illustration 3:

Solve $5^{x-2} = 4^{x+1}$

Solution 1: Rewrite using logarithms. Either 4 or 5 is a reasonable base for the logarithm. Try base 4.

$$(5^{x-2}) = 4^{x+1}$$
$$\log_4(5^{x-2}) = x + 1$$
$$(x - 2) \log_4 5 = x + 1 \qquad \text{(Theorem 6-8)}$$
$$x \log_4 5 - 2 \log_4 5 = x + 1$$
$$x \log_4 5 - x = 2 \log_4 5 + 1$$
$$x (\log_4 5 - 1) = 2 \log_4 5 + 1 \qquad \text{(factor)}$$
$$x = \frac{2 \log_4 5 + 1}{\log_4 5 - 1}$$

Solution 2: Try base 5.

$$5^{x-2} = 4^{x+1}$$
$$x - 2 = \log_5(4^{x+1})$$
$$x - 2 = (x + 1) \log_5 4 \qquad \text{(Theorem 6-8)}$$
$$x - 2 = x \log_5 4 + 1 \log_5 4$$
$$x - x \log_5 4 = 2 + \log_5 4$$
$$x (1 - \log_5 4) = 2 + \log_5 4$$
$$x = \frac{2 + \log_5 4}{1 - \log_5 4}$$

Solution 3: Try the common base 10. Because logarithms are functions,

$$\log(5^{x-2}) = \log(4^{x+1})$$
$$(x - 2) \log 5 = (x + 1) \log 4$$
$$x \log 5 - 2 \log 5 = x \log 4 + 1 \log 4$$
$$x \log 5 - x \log 4 = 2 \log 5 + 1 \log 4$$
$$x [\log 5 - \log 4] = 2 \log 5 + \log 4$$
$$x = \frac{2 \log 5 + \log 4}{\log 5 - \log 4}$$

Note: The change of base formula confirms that these solutions provide the same answer in different forms. A calculator approximation is $x \cong 20.63770232$.

Solution 4: Use the trace feature of your calculator to determine the point of intersection of $y = 5^{x-2}$ and $y = 4^{x+1}$. Compare the precision of your answer to the approximation in the Note.

Illustration 4:

Suppose data collected from an experiment seems to have a pattern. The researcher suspects that a polynomial is the best model for the data. With a formula to model the experiment, the researcher can make predictions. If further

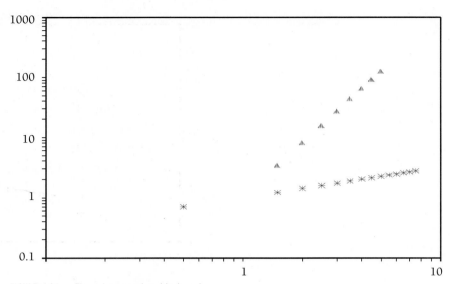

FIGURE 6-31   Three data sets, logarithmic scale

experiments conform to the predictions, the value of the model is confirmed. But what polynomial? First degree? Second degree? Third degree? With numerous collected data, the best choice may not be obvious.

One way to select a suitable polynomial model is to use logarithms.

Because $\log(x^n) = n \log x$

logarithms can transform a power function to a linear function. Special graph paper with logarithmic scales on the x and y axes gives the user a different view of the plotted data. See Figure 6-31. If the resulting points appear to be collinear, then a power function provides a model. Further, if the slope of the line is 2, the model is $y = x^2$. If the slope of the line is 3, then $y = x^3$ is the model. The model extends to many power functions. A slope of $\frac{1}{2}$ suggests that $y = \sqrt{x}$ is an appropriate power function. Figure 6-32 compares data sets graphed on a logarithmic scale with the corresponding power function model. ▬

Solving Logarithmic and Exponential Equations.   Solve the following equations:

EXAMPLE 6-11

Illustration 1:

$$\log(x + 2) + \log(x - 1) = 1$$

Solution: A variable is in the argument of a logarithm, so we label this equation a *logarithmic equation*. To gain access to the variable, rewrite the expression in exponential form. First, contract the lefthand side of the equation.

$$\log(x + 2) + \log(x - 1) = 1$$
$$\log[(x + 2)(x - 1)] = 1 \qquad \text{(Theorem 6-8)}$$
$$\log(x^2 + x - 2) = 1 \qquad \text{(note: base 10)}$$

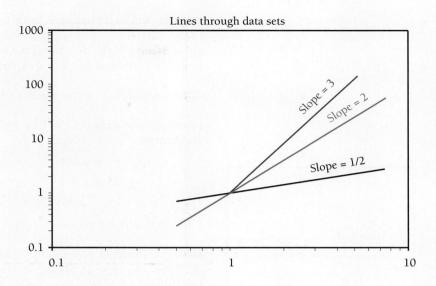

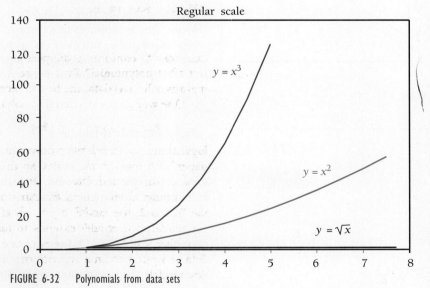

FIGURE 6-32   Polynomials from data sets

$$x^2 + x - 2 = 10^1 \qquad \text{(exponential form)}$$
$$x^2 + x - 12 = 0 \qquad \text{(quadratic equation)}$$
$$(x + 4)(x - 3) = 0$$
$$x = ^-4 \quad \text{or} \quad x = 3$$

Reject the solution of $^-4$. Substitution of $^-4$ in the original equation results in a negative argument for each logarithm. Hence, $^-4$ is not in the domain of the original equation: $x = 3$ provides the only solution.

Note: The contraction step in the solution introduced this extraneous root. Check $^-4$ in $\log[(x + 2)(x - 1)] = 1$. Each factor of the argument will be negative, but the product is positive. Contraction may introduce extraneous roots. Therefore, proposed solutions of logarithmic equations must be checked against the domain of the original equation. From the original equation, $x + 2 > 0$ and $x - 1 > 0$ exclude -4 as a solution.

Illustration 2:   Consider $3^{y+1} = 2^x$. Express y as a function of x.

Solution: Choose a convenient base and rewrite in logarithmic form. Base 3 looks convenient.

$$\log_3(2^x) = y + 1$$
$$x \log_3(2) = y + 1$$
$$y = -1 + x \log_3 2$$

Call this function f: $f(x) = [\log_3 2]x - 1$

Note: The function f is linear. From the change of base formula, $\log_3 2 = \ln 2)/\ln 3$. The slope of f is $\ln 2/\ln 3$, the y-intercept is $(0, -1)$. See Figure 6-33. Logarithms have numerous applications. The transformation properties of logarithms make them suitable for reducing unimaginably large (or ridiculously small) numbers to a comprehensible size. The Richter scale for measuring earthquake intensity is a logarithmic scale. The decibel system of measuring sound loudness is a logarithmic scale. Logarithmic scales also play a role in the measurement of chemical acidity and in atomic and galactic distances.

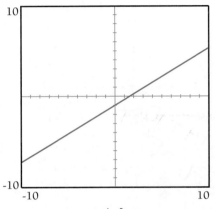

FIGURE 6-33 $f(x) = \dfrac{\ln 2}{\ln 3} x - 1$

E X A M P L E   6 - 12

More Applications of Logarithms

Illustration 1:   A measurement for sound intensity is the decibel (*deci* for 10 and *bel* for Alexander Graham Bell). The definition for a decibel is

$$\text{Db} = 10 \log\left(\frac{I}{I_0}\right)$$

where I is the sound intensity in Watts/cm^2 and the base intensity is $I_0 = 10^{-16}$ Watts/cm^2. If a normal speaking voice registers 10^{-14} Watts/cm^2, how many decibels is the voice?

Solution:

$$\text{Db} = 10 \log\left(\frac{10^{-14}}{10^{-16}}\right)$$

$$\text{Db} = 10 \log(10^2)$$
$$\text{Db} = 20$$

Illustration 2:   The pH (potency in Hydrogen) of a water solution is defined as the negative, base 10 logarithm of the molar concentration of hydronium ions. For pure water the pH is $+7$, because there are 10^{-7} moles of hydronium atoms per liter of

H_2O (water): pH = $-\log(10^{-7})$ pH = $-(-7)$. Determine the pH of a solution of nitric acid that contains 10^{-4} moles of hydronium ions per liter of H_2O.

Solution:

$$\text{pH} = -\log(10^{-4})$$
$$\text{pH} = 4$$

Note that in general, acidic solutions have a pH below 7 with the lower numbers being more acidic. Basic solutions have a pH above 7 with the higher values being more basic.

SUMMARY

- A logarithmic transformation transforms an algebraic operation one level down on the order of operations:

  1. $\log_b(xy) = \log_b x + \log_b y$

  2. $\log_b\left(\dfrac{x}{y}\right) = \log_b x - \log_b y$

  3. $\log_b(x^p) = p \log_b x$

- Logarithmic transformations are useful for solving logarithmic and exponential equations. Logarithms also play a role in the development of scales for measuring very small or very large quantities.

6-3 EXERCISES

Expand each of the following logarithmic expressions.

1. $\ln\left(\dfrac{15}{y}\right)$

2. $\log(3xy)$

3. $\log_3(27x^7)$

4. $\log_5(25y^{-3})$

5. $\ln(x^3y^2)$

6. $\log\left(\dfrac{x^4}{y^3}\right)$

7. $\ln\left(\dfrac{10}{\sqrt{x}}\right)$

8. $\log(\sqrt[3]{x^2})$

9. $\ln(x^2 - 7x + 12)$

10. $\log\left(\dfrac{2x-1}{x+5}\right)$

Contract each of the following logarithmic expressions.

11. $\log x + 3 \log y$

12. $2 \log x - \log y$

13. $\frac{2}{3} \log x - \log y$

14. $\ln x - 2 \ln y + 3 \ln w$

15. $5 \ln x - (\ln w + 2 \ln y)$

16. $3 \ln x - \frac{1}{2} \ln x + 2 \ln w$

17. $2 + 3 \log 2$

18. $1 - 2 \ln 3$

19. $\log x + \ln x$

20. $2 + 5 \log_3 x$

Solve for x.

21. $5^x = 7$

22. $7^x = 5$

23. $3^{x+2} = 10$

24. $4^{x-2} = 9$

27. $7^{x+3} = 4^{x-2}$

28. $5^{x+1} = 6^{x-7}$

29. $10^{x-3} = e^{x+1}$

30. $e^{2x} = 10^{1-x}$

31. $\log(x + 1) + \log(x + 4) = 1$

32. $\log(x - 1) + \log(x + 2) = 1$

33. $\log(x - 5) + \log(x + 4) = 1$

34. $\log(x + 3) + \log(x - 6) = 1$

35. $\log(x + 6) - \log(x - 3) = 1$

36. $\log(7x + 5) - \log(x - 1) = 1$

37. $\log(5) + \log(x - 2) + \log(x - 3) = 1$

38. $\log(2) + \log(x - 2) + \log(x + 2) = 1$

39. $2\log(5) + \log(x) + \log(x + 3) = 2$

40. $2\log(5) + \log(2x) + \log(x + 1) = 2$

41. $e^{2x} - 2e^x + 1 = 0$

42. $e^{2x} - 3e^x + 2 = 0$

43. $10^{2x} - 3(10^x) + 2 = 0$

44. $10^{2x} - 2(10^x) + 1 = 0$

45. $\dfrac{e^x - e^{-x}}{2} = 0$

46. $\dfrac{e^x + e^{-x}}{2} = 1$

47. $\dfrac{e^x - e^{-x}}{e^x + e^{-x}} = 1$

48. $\dfrac{e^x - e^{-x}}{e^x + e^{-x}} = \dfrac{1}{2}$

49. $\log(x + 1) + \log(x + 7) = 1$

50. $\log(x + 2) + \log(x + 3) = 1$

51. $\log(x + 1) - \log(x) = \log(4)$

52. $\log(3x - 2) - \log(x + 1) = \log 2$

53. $3\sqrt{\log(x)} = \log x$

54. $\sqrt{\log(x)} = 2\log x$

Given the formula $C = (1200/\ln 2)\,\ln(f_1/f_2)$, solve as indicated.

55. Solve for f_1.

56. Solve for f_2.

Given the formula $T = T_f + C\,e^{-kt}$, solve as indicated.

57. Solve for k.

58. Solve for t.

59. Solve for T_f.

60. Solve for C.

Given the formula $I = I_s(1 - e^{-Rt/L})$, solve as indicated.

61. Solve for R.

62. Solve for t.

63. Solve for L.

64. Solve for I_s.

65. Calculate the number of decibels for a whisper: $I = 10^{-16}$.

66. Calculate the number of decibels for heavy traffic: $I = 10^{-9}$.

67. Show that, if the sound intensity I doubles, then the number of decibels increases approximately 3 units.

68. Show that, if the sound intensity I increases by a factor of 5, then the number of decibels increases approximately 7 units.

69. The current amount of tritium is modeled by $A = 5$ grams $2^{-t/10}$ years. How many years will pass for the tritium to decay to 1 gram?

70. See Problem 69. How many years for the tritium to decay to 2 grams?

71. See Problem 69. After how many years will 10 grams of tritium be present? What does the answer mean?

72. See Problem 69. After how many years will 15 grams of tritium be present? What does the answer mean?

73. The colony of bacteria A after t hours is modeled by $A(t) = 81(2^{t+5})$. The colony of bacteria B after t hours is modeled by $B(t) = 81(3^{t-4})$. After how many hours will the two populations be equal? Which colony is the larger at the beginning? Which colony is growing faster?

74. One model for population growth in country x after t years is $x(t) = (\frac{1}{2})(1.5)^t$. The growth of population in country y after t years is modeled by $y(t) = 3(0.9)^t$. After how many years will the countries have equal populations? Which country started with more people?

75. If inflation grows at 3 percent per year, how many years will it take for prices to double?

76. If an automobile depreciates at 20 percent per year, in how many years will it be worth half its original value?

77. A convention in music is to divide an octave into 1200 cents. By formula, cents $= (1200/\ln 2)\,\ln(f_1/f_2)$, where f_1 and f_2 are frequencies of two notes played, $f_1 > f_2$. For a perfect fifth, $f_1/f_2 = 3/2$. Show that the cents for a perfect fifth is 701.955.

78. See Problem 77. Solve the equation for f_1/f_2. If the cents for a major third is 386.314, what is the ratio f_1/f_2?

79. See Problem 77. Solve the equation for f_2. If Mary sounds a tuning fork with a frequency of $f_1 = 400$ cycles per second (cps), what frequency will a lower frequency tuning fork make to produce a perfect fourth? (For a perfect fourth, cents $= 498.045$.)

80. See Problem 77. Solve the equation for f_1. If John plays a note of $f_2 = 800$ cps on his synthesizer, what is the frequency of a higher note to produce a major tone? (For a major tone, cents $= 203.91$.)

81. Prove $\log_a b = 1/\log_b a$, $a > 0$, $b > 0$, $a \neq 1$, $b \neq 1$.

82. Prove Theorem 6-6.

83. Prove Theorem 6-7.

84. Prove Theorem 6-8.

6-3   P R O B L E M   S E T

GENERALIZED EXPONENTS

Most people take for granted the exponent key $\boxed{x^y}$ or $\boxed{\wedge}$ on their calculator. Similarly, BASIC programmers use x^y for exponentiation. However, some are surprised that their calculator cannot do $(-1)^{3/5}$, when they know the answer is -1. The difficulty lies in the way calculators compute exponentials. The following problems assume that your calculator has no $\boxed{x^y}$ function key, but does have the standard $\boxed{e^x}$ and $\boxed{\ln}$ keys. For example, Pascal has no operator for exponentiation. Pascal does provide the functions EXP(x) = e^x, $x \in (-\infty, \infty)$ and LN(x), $x > 0$.

1. Suppose your Pascal program needs to approximate $5^{0.6}$. Use the EXP and LN functions to accomplish the approximation. Hint: let Ans = $5^{0.6}$. Take LN of both sides, expand, then solve for Ans.

2. Suppose your Pascal program needs to approximate $7^{1.3}$.

Use the EXP and LN functions to accomplish the approximation.

3. Suppose your Pascal program needs to approximate Ans = b^p. Rewrite the equation to express Ans in terms of EXP and LN.

4. Use the results in Problem 3 to define a general exponential expression b^p. The definition will place restrictions on the values for b and p. What are these restrictions?

5. Examine Problem 4. Some calculators use a piecewise function to allow for negative bases. Try to define such a piecewise function. Pay special attention to the exponent.

6. The Sharp EL-9300C can evaluate $\sqrt[5]{(-1)^3}$, but requires complex mode to work with $(-1)^{3/5}$. Discuss why some calculators are unable to evaluate $(-1)^{3/5}$. Hint: $\frac{3}{5} = 0.6$.

4   APPROXIMATIONS

God made the integers, all else is the work of man.
—KRONECKER

Mantissa—Henry Briggs (1561—1630) introduced *mantissa* in 1624. Of Etruscan origin, mantissa is from the Latin meaning "something of minor value." Briggs is also responsible for popularizing common base 10 logarithms.
Characteristic is from the Greek meaning "a distinctive trait."

Although $\log_2 8$ is easy to evaluate, $\log_2 8 = 3$, $\log_2 7$ is not obvious. The common logarithm lends itself to a decimal tabular representation because base 10 uses the same base as our decimal numeration system. Because electronic calculators are a recent innovation, easy-to-use table values popularized the common base for logarithmic approximations. Table A-1 of Appendix A contains approximate base 10 logarithms for two-decimal place arguments from 1.00 to 9.99. The logarithms are precise to 4 decimal places (see Section 3-7). We count in base 10, so Table A-1 provides approximate logarithms for any finite precision decimal with three significant digits. By the change of base formula (Theorem 6-4) any logarithm converts to base 10. Similarly, the change of base formula works with the built-in logarithmic function of a calculator to determine logarithms for bases not on the calculator.

The use of logarithmic tables is archaic. To equal the precision of a calculator would require a massive table. Although the discussion of tables is of historical interest, a modern reason to discuss tables is to acquire some concepts and terminology for understanding how computers and calculators store finite precision decimals.

In general, calculated logarithms in this section are approximations. In particular, all of the values in Table A-1 are approximations and thus calculations based upon these numbers are approximations.

....

EXAMPLE 6-13

Illustration 1:

Numerical Approximation of Logarithms. Evaluate the following logarithmic expressions.

$$\log 7.83$$

Solution 1: From Table A-1,

N	.00	.01	.02	.03	.04	¼	.01
				⇓			
7.7				⇓			
7.8	⇒	⇒	⇒	.8938			

$$\log 7.83 \cong 0.8938$$

Solution 2: With your calculator in real/math mode, press [LOG] 7.83 [ENTER].

Note: A **mantissa** is the logarithm of a number between 1 and the base (10 in this case). Table A-1 is a table of mantissas. The mantissa for 7.83 is 0.8938. Although these examples are easy on a calculator with a [LOG] key, the idea of mantissa can be visualized from tables.

Illustration 2:

$$\log 7830$$

Solution 1: Use your calculator. Compare your findings with the result to Illustration 1.

Solution 2: Table A-1 does not contain 7830. However, every real number can be written as a number between 1 and 10 times a power of 10.

$$7830 = 7.830(10^3)$$

Now, take advantage of logarithmic transformations.

$$\log 7830 = \log[7.830(10^3)]$$
$$\log 7830 = \log 7.830 + 3\log 10$$
$$\log 7830 \cong 0.8938 + 3(1)$$
$$\log 7830 \cong 3.8938$$

Note: The fractional part of log 7830 is 0.8938, the same mantissa as Illustration 1. The whole number portion of log 7830 is 3, called the **characteristic** of the logarithm. The characteristic of a logarithm indicates the position of the decimal point in the argument. Notice that the digits in these illustrations are the same. The mantissa indicates the digits of the argument.

Illustration 3: $\log 78.3$

Solution: The digits are as before, but the decimal occupies a different position:

$$\log 78.3 = \log[7.83(10^1)]$$
$$\log 78.3 = \log 7.83 + \log 10$$
$$\log 78.3 \cong 0.8938 + 1$$
$$\log 78.3 \cong 1.8938$$

Illustration 4: $\log 783000$

Solution: Because the digits are the same as the other illustrations, the mantissa is 0.8938. The decimal point is 5 digits to the right of the leading digit. The characteristic is 5. Use your calculator to verify that

$$\log 783000 \cong 5.8938$$

Illustration 5: $\log 0.000783$

Solution: The mantissa is 0.8939. The decimal is 4 places to the left of the first significant digit (see Section 3-7). The characteristic is –4.

$$\log 0.000783 = \log[7.83(10^{-4})]$$
$$\log 0.000783 = \log 7.83 - 4 \log 10$$
$$\log 0.000783 \cong 0.8938 - 4$$
By calculator, $\log 0.000783 \cong -3.1062382$

Note: The mantissa is positive. It is incorrect to write

$$\log 0.000783 \cong -4.8938.$$

Both characteristic and mantissa are negative in –3.1062382.

Illustration 6: $\log_2 7$

Solution: Convert to base 10 with the change of base formula. Use your calculator.

$$\log_2 7 = \frac{\log 7}{\log 2}$$
$$\log_2 7 \cong \frac{0.8451}{0.3010}$$
$$\log_2 7 \cong 2.8074$$

Note: By calculator $2^{2.8074} \cong 7$

Illustration 7:

$$\log_2 700$$

Solution:

$$\log_2 700 = \frac{\log 700}{\log(2)}$$

$$\log_2 700 \cong \frac{2.8451}{0.3010}$$

$$\log_2 700 \cong 9.4512$$

Note: Because $2^9 = 512$ and $2^{10} = 1024$, our answer is reasonable. By calculator, $2^{9.4512} \cong 700$.

Convert 700 to base 2 (see Section 4-2).

$$700_{ten} = 1010111100_{two}$$

Nine digits follow the leading digit in the base 2 representation: the characteristic is meaningful when the base of the logarithm is the same as the radix for the number representation.

Illustration 8:

$$\ln 13$$

Solution:

$$\ln 13 = \frac{\log 13}{\log e}$$

$$\ln 13 \cong \frac{1.1139}{0.4343}$$

Note that $\log e \cong 0.4343$

Now $\ln 13 \quad \cong 2.5649$

Note: A calculator should have an LN function key for direct evaluation. The natural base is so important that a calculator is more likely to have an LN key than a log key.

Illustration 9:

If $\log x = 3.7497$, approximate x.

Solution 1: Because logarithms are one-to-one functions, reading Table A-1 "backward" provides the solution. Because Table A-1 is a table of mantissas, separate the mantissa 0.74974 from the rest of the logarithm.

N	.00	.01	.02	.03	. . .	.08	.09
5.5	. . .		⇑				
5.6	⇐	⇐	.7497				
5.7	. . .						

$$\text{Log } 5.62 \cong 0.7497$$

Now use the characteristic to relocate the decimal.

$$\log 5620 \cong 3.7497$$
$$x \cong 5620$$

Solution 2: A logarithmic function is the inverse of an exponential function, so most calculators can approximate x using one of the following:

(a) Enter $10^{3.7497} \cong 5619.532$

(b) If the calculator has an [INV] key, enter 3.47497, then press [INV] [LOG] for the same approximation.

Illustration 10:   Approximate x where $\log x = 0.7497 - 2$

Solution: See Illustration 9. The mantissa provides the digits of x: 5.62. The characteristic relocates the decimal point two positions to the left.

$$x \cong 0.0562$$
$$\text{By calculator, } x \cong 10^{(0.7497-2)}$$
$$x \cong 0.0561953$$

Humans communicate in sentences, made up of words, which are made up of characters. The computer equivalent of a character of information is a byte of information. A byte consists of information stored in 8 bits. *Bit* is short for binary digit, which indicates an electronic switch that represents the digits 1 or 0 by being on or off, respectively. Graphing calculators often have in excess of 32,000 byes of memory. For personal computers, memory is usually measured in millions of bytes.

There are numerous schemes for representing finite precision decimals in the memory of a machine. Most of these methods split the numerical information into three fields. One field indicates whether the number is positive or negative. A second field locates the decimal point of the number. The last field stores the digits of the number.

For example, one method of storing a decimal uses 8 bytes (64 bits) of computer memory. The 64 bits are divided into three fields as shown in the following table:

Num bits	1	11	52
Purpose	sign	characteristic (exponent)	mantissa (digits)

The number of bits in the mantissa limits the precision of the decimals. The number of bits in the exponent limits how large or small a decimal can be represented. With this particular 8-byte representation, the range of positive finite precision decimals is $5 \times 10^{-324} \ldots 1.7 \times 10^{308}$ with 15 to 16 significant digits. Therefore, the structure of logarithmic functions still assists with complex computations even in high-speed computers.

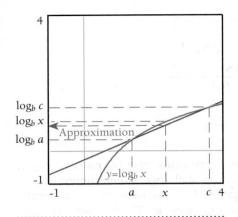

Suppose the precision of a number exceeds the precision of Table A-1. Linear interpolation has long been used to approximate missing values in tables for functions.

See Figure 6-34. The approximating linear function through $(a, \log_b a)$ and $(c, \log_b c)$ is

$$f(x) = \frac{\log_b c - \log_b a}{c - a}(x - a) + \log_b a.$$

The closer a is to c, the better f approximates $\log_b x$, where $a < x < c$.

FIGURE 6-34 $y = \log_b x,$ $x = a,$
$x = c$ and interpolating line

EXAMPLE 6-14

Approximating Logarithms. Use linear interpolation to approximate the following expressions.

Illustration 1:

$\log 2.647$

Solution: From Table A-1,

(a) $\log 2.64 \cong 0.4216$
(b) $\log 2.647 = ?$
(c) $\log 2.65 \cong 0.4232$

See Figure 6-35. The approximating linear function is

$$f(x) = \frac{0.4232 - .4216}{2.65 - 2.64}(x - 2.64) + 0.4216$$

$$f(x) = 0.16x - 0.0008$$

Now $\log 2.647$ is approximately $f(2.647)$.

$$\log 2.647 \cong 0.16(2.647) - 0.0008$$
$$\log 2.647 \cong 0.42272$$

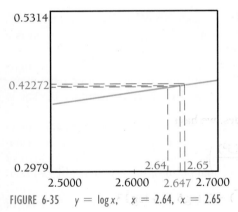

FIGURE 6-35 $y = \log x,$ $x = 2.64,$ $x = 2.65$

Note: By calculator $\log 2.647 \cong 0.422753941$, a difference of about 0.00003.

Illustration 2:

If $\log x = 1.7518$, determine x.

Solution:

(a) $\log 5.64 \cong 0.7513$ $(10^{0.7513} \cong 5.64)$
(b) $\log x = 0.7518$ $(10^{0.7518} = x)$
(c) $\log 5.65 \cong 0.7520$ $(10^{0.7520} \cong 5.65)$

Apply linear interpolation.

$$f(x) = \frac{0.7520 - 0.7513}{5.65 - 5.64}(x - 5.64) + 0.7513$$

$$f(x) = 0.07x + 0.3565$$

This time x is not known, only that $f(x) = 0.7518$.

$$0.7518 = 0.07x + 0.3565$$

Solve for x $$x = \frac{0.7518 - 0.3565}{0.07}$$

$$x \cong 5.647$$

By calculator $10^{0.7518} \cong 5.646768716$ —

Before calculators became popular, logarithms were commonly used to simplify complicated calculations. Although such logarithmic contortions are no longer necessary for arithmetic, logarithmic transformations can be invaluable in simplifying algebraic expressions in calculus. The following example indicates the transformation power of logarithms.

EXAMPLE 6-15

Applications of Logarithmic Transformations. Use logarithms to approximate the following.

Illustration 1:

$$\frac{\sqrt[3]{7\sqrt{3}}\,(5^3)}{13^4}$$

Solution: Let $A = \dfrac{\sqrt[3]{7\sqrt{3}}\,(5^3)}{13^4}$.

Applying the log function to both sides, we have

$$\log A = \log\left(\frac{\sqrt[3]{7\sqrt{3}}\,(5^3)}{13^4}\right)$$

$$\log A = \left(\frac{1}{3}\right)\log(7\sqrt{3}\,) + 3\log 5 - 4\log 13$$

$$\log A = \left(\frac{1}{3}\right)\log 7 + \left(\frac{1}{3}\right)\left(\frac{1}{2}\right)\log 3 + 3\log 5 - 4\log 13$$

$$\log A \cong \left(\frac{1}{3}\right)(.8451) + \left(\frac{1}{6}\right)(.4771) + 3(.69897) - 4(1.1139)$$

$$\log A \cong -1.99747$$

$$\cong -1.99747 + 2 - 2$$

$$\cong 0.00253 - 2$$

$$A \cong 0.01$$

By calculator, $A \cong 0.0100544$ —

SUMMARY

- Table A-1, Appendix A, provides approximate mantissas for the common logarithms of numbers between 1 and 10. To use the table, rewrite an argument as the product of a finite precision decimal between 1 and 10 times a power of 10 (scientific notation). Table A-1 provides the fractional portion (mantissa) of the logarithm. The exponent of the power of 10 provides the whole number portion (characteristic) of the logarithm.
- To approximate logarithms for bases other than 10 use the change of base formula:

$$\log_b N = \frac{\log N}{\log b}.$$

- For higher precision numbers, linear interpolation extends the usefulness of tables:

$$f(x) = \frac{\log b - \log a}{b - a}(x - a) + \log a$$

- Calculators provide quicker access to logarithmic values. Like Table A-1, however, calculators have limited precision.

6-4   EXERCISES

Approximate the following logarithms. Identify the mantissa and characteristic of each:

1. $\log 237$ **2.** $\log 0.0354$

3. $\log 0.00461$ **4.** $\log 873$

5. $\log 46100$ **6.** $\log 0.000873$

7. $\log 56.2$ **8.** $\log 47.9$

9. $\log 562000$ **10.** $\log 479000$

11. $\log_2 5$ **12.** $\log_3 7$

13. $\log_5 2$ **14.** $\log_7 3$

15. $\ln 10$ **16.** $\ln 2$

17. $\ln 4$ **18.** $\ln 100$

19. $\log_8 7$ **20.** $\log_7 8$

Approximate N to three significant figures.

21. $\log N = 2.9212$ **22.** $\log N = 0.8109 - 2$

23. $\log N = 0.7275 - 3$ **24.** $\log N = 3.4757$

25. $\log N = 0.4281 - 1$ **26.** $\log N = 0.9595 - 4$

27. $\log(10^2 N) = 0.4281$ **28.** $\log(10N) = 0.9595$

29. $\log\left(\dfrac{N}{10}\right) = 0.4281$ **30.** $\log\left(\dfrac{N}{10^2}\right) = 0.9595$

Write a linear interpolation function to approximate the following. Compare the interpolation approximation with a calculator approximation.

31. $\log 3.579$ **32.** $\log 7.863$

33. $\ln 9.694$ **34.** $\ln 2.392$

35. $\ln 985.3$ **36.** $\ln 873.4$

37. $\log_5 0.006854$ **38.** $\log_6 0.07932$

39. $\log_8 5637000$ **40.** $\log_7 87260$

Approximate N in each of the following.

41. $\log N = 3.9209$ **42.** $\log N = 1.6498$

43. $\ln N = 1.8946$ **44.** $\ln N = 1.2585$

45. $17 = 2^{N+1}$

46. $13^{N-2} = 3$

47. $3^N = 5^{N-2}$

48. $2^{N+1} = 7^{N-1}$

49. $e^{N+2} = 10$

50. $10^{N+1} = e^N$

55. $\log(\sqrt[3]{x^2})$

56. $\log(100y)$

57. $\log(x/1000)$

58. $\log(100/y)$

59. $\log(y^x)$

60. $\log(x^y)$

Suppose that $\log x = t$ and $\log y = s$. Express the following in terms of t and s.

51. $\log(xy)$

52. $\log(\dfrac{x}{y})$

53. $\log\left(\dfrac{y^4}{x^3}\right)$

54. $\log(x^2 y^3)$

*61. Let z be between x and y, $z = 0.3y + 0.7x$. Write a linear interpolation function to approximate $\log z$ in terms of t and s.

*62. Let z be between x and y, $z = 0.6y + 0.4x$. Write a linear interpolation function to approximate $\log z$ in terms of t and s.

6-4 PROBLEM SET

1. Complete the following table for log base 2.

N	$\text{Log}_2 N$	N_2
1	0.0000	1
2	1.0000	10
3		11
4		100
5		101
6		110
7		111
8		1000

Compare the characteristic of the logarithm to the representation of N in base 2.

2. Complete the following table for log base 3.

N	$\text{Log}_3 N$	N_3
1	0.0000	1
2		2
3	1.0000	10
4		11
5		12
6		20
7		21
8		22
9		100

Compare the characteristic of the logarithm to the representation of N in base 3.

C H A P T E R   S U M M A R Y

- For an exponential function $f(x) = Ab^{mx+c} + d, b > 0, b \neq 1$,

$$\text{Domain}_f = (-\infty, \infty); \text{Range}_f = \begin{cases} (d, \infty) & \text{if } A > 0 \\ (-\infty, d) & \text{if } A < 0 \end{cases}.$$

Asymptote for f is $y = d$. Pivot point for f is $(-c/m, A + d)$. f is one to one and has an inverse.

- $f(x) = b^x$ and $g(x) = \log_b x$ are inverse functions. Abbreviations: $\ln x = \log_e x$ and $\log x = \log_{10} x$. A logarithm of one base converts to a logarithm of another base by use of the transformation

$$\log_b x = \frac{\log_a x}{\log_a b}$$

- The graph of logarithmic functions with base $b > 1$ resembles $y = \ln x$. If $0 < b < 1$, the graph resembles $y = \ln(x)$ reflected about the x-axis.

- Logarithmic transformations transform algebraic operations down one level on the order of operations (GEMA):

  1. $\log_b(xy) = \log_b x + \log_b y$ (multiplication to addition)

  2. $\log_b\left(\dfrac{x}{y}\right) = \log_b x - \log_b y$ (division to subtraction)

  3. $\log_b(x^p) = p \log_b x$ (exponentiation to multiplication)

- Logarithmic transformations are useful for solving logarithmic and exponential equations. Logarithms play an important role in the development of measurement scales for quantities connected to very small or very large numbers. In calculus, the transformation properties of logarithms are useful in transforming complex algebraic expressions down one level of difficulty in the order of operations.

- Calculators provide quicker access to logarithmic values. Like Table A-1, calculators are limited in precision.

KEY   WORDS   AND   CONCEPTS

Characteristic	Logarithmic function	Pivot point
Exponential function	Mantissa	Transcendental function

6 REVIEW EXERCISES

SECTION 1

Graph each of the following functions indicating the domain, the range and whether increasing or decreasing. Annotate asymptotes.

1. $(x) = 7^x$
2. $f(x) = 9^x$
3. $f(x) = 3^{-2x}$
4. $f(x) = 12^{-x}$
5. $f(x) = e^{-x}$
6. $f(x) = 5^{x-1}$
7. $f(x) = 2^x - 5$
8. $f(x) = 8^{x/3}$
9. $f(x) = (\sqrt{3})^{x+1}$
10. $f(x) = 4^{x+1} - 2$
11. $f(x) = 3^{x/2} - 1$
12. $f(x) = 5(3^{x+1}) - 4$
13. $f(x) = 3(8^{1-x}) + 12$
14. $f(x) = 2(2^{3x-2}) + 1$
15. $f(x) = x + e^x$
16. $f(x) = x - e^x$

17. Suppose that the half life of Plutonium is 24,000 years. Write a function to model the amount of Plutonium after t years, if the initial amount is 5 grams.

18. Write a formula giving the value of a savings account with current principal of $1000 compounded continuously at a rate of 5 percent per annum after t years. If the account has drawn the same rate for the past several years, what was the value of the account 2 years ago?

19. Rewrite the exponential function $g(x) = 25^{x-3}$ so that the base is 5.

20. Rewrite $g(s) = 6^{2s-3}$ so that the base is 36.

21. Use the rule of 72 to estimate the amount in an account with $1000 principal after 18 years if the APR is 0.08 compounded continuously.

22. Use the rule of 72 to estimate the amount in an account with $1000 principal after 24 years if the interest rate is 0.06. Compare this to $1000 principal for 12 years with an interest rate of 0.12.

23. Use the rule of 72 to estimate the doubling time for your account if your interest rate is 0.05.

24. Suppose you receive 0.04 interest. (a) How many doubling periods are there in 36 years? (b) If you deposit $10,000 for 36 years at 0.04 interest, what will be the value of your account?

SECTION 2

Rewrite each of the following exponential statements in logarithmic form.

25. $2^{-4} = 0.0625$
26. $4^{-2} = 0.0625$
27. $25 = 125^{2/3}$
28. $4^{-3} = \dfrac{1}{64}$
29. $81^{1/2} = 3^2$
30. $81^{-3/4} = \dfrac{1}{3^3}$

Rewrite each of the following logarithmic statements in exponential form.

31. $\log_{27} 9 = \dfrac{2}{3}$
32. $\log_5 0.008 = -3$
33. $\log_{25}\left(\dfrac{1}{25}\right) = -1$
34. $\log_{125}\left(\dfrac{1}{5}\right) = -\dfrac{1}{3}$
35. $\log_{1/3} 9 = -2$
36. $\log_{4/9}\left(\dfrac{3}{2}\right) = -\dfrac{1}{2}$

Graph each of the following logarithmic functions.

37. $f(x) = \log_5 x$
38. $f(x) = \log_{1/4}(x - 3)$
39. $f(x) = \log(x + 7)$
40. $f(x) = \ln(5 - x)$
41. $f(x) = \ln(5x)$
42. $f(x) = \ln(x - 1)$
43. $f(x) = \ln x + \ln 5$
44. $f(x) = 2 \log_7 x$
45. $f(x) = \log_2(4x)$
46. $f(x) = \log_5\left(\dfrac{x}{3}\right)$
47. $f(x) = \dfrac{\log_5 x}{3}$
48. $f(x) = \log_4(2x - 1)$
49. $f(x) = 2 \log_4(x + 1)$
50. $f(x) = 4 \log_3(x - 1) + 5$
51. $f(x) = \log_{1/2}(2x + 1) - 4$
52. $f(x) = -2 \log_{1/3}(3 - 2x) + 1$
53. $f(x) = \log |x - 2|$
54. $f(x) = \log |2 - x|$
55. $f(x) = |\ln x|$
56. $f(x) = |3 - \ln x|$
57. Compare the graph of $f(x) = 4 \log_2(x)$ with $g(x) = \log_2(x^4)$.
58. Compare the graph of $f(x) = \log_2 x - 4$ with $g(x) = \log_2(x/16)$.

SECTION 3

Expand:

59. $\ln\left(\dfrac{x\sqrt{y}\sqrt{z}}{3w^2}\right)$
60. $\log[(x-3)^4 \sqrt[3]{z}]$

Contract:

61. $\log 17 - 3 \log z + 5 \log x - 0.5 \log y$
62. $3 + \ln y - 2 \ln x + (\ln z)/4$

Solve for x.

63. $49^x = 7$

64. $4^{x-2} = 64$

65. $5^{x-2} = 125$

66. $5^{x+1} = 7^{x-6}$

67. $10^{x-4} = e^{x+1}$

68. $e^{3x} = 10^{2-x}$

69. $\log(x + 1) + \log(4x + 4) = 2$

70. $\log(x - 3) + \log(10 - x) = 1$

71. $\log(x + 7) - \log(x - 2) = 1$

72. $\log(7x - 1) - \log(x + 2) = 2$

73. $\log(x - 2) - \log(x + 2) = 1$

74. $\log(2x) - \log(x + 1) = 0$

75. $e^{2x} - 2e^x - 3 = 0$

76. $10^{2x} - 4(10^x) + 4 = 0$

77. $\dfrac{e^x - e^{-x}}{2} = 1$

78. $\dfrac{e^x - e^{-x}}{e^x + e^{-x}} = \dfrac{1}{2}$

79. $\log_{12}(x + 1) + \log_{12}(7 - x) = 1$

80. $\log_7(3x + 4) + \log_7 x = 1$

81. Suppose that the starting salary for a beginning teacher in 1960 was $4,000 per year. Assuming continuously compounded 6 percent average annual inflation rate, what would be the equivalent salary in 1995? In what year does the salary become $30,000?

82. In a developing country, one model for the population growth (in thousands) of a city after t years is $x(t) = (\frac{1}{2})(1.7)^t$. The growth of population in a rural area after t years is modeled by $y(t) = 5(0.9)^t$. After how many years will the city and rural area be equal? Which started with more people?

SECTION 4

Approximate the following to four decimal places.

83. $\log 258$

84. $\log 963$

85. $\log 258000$

86. $\log 96.3$

87. $\log 5030$

88. $\log_3 0.007$

89. $\log_5(5^{1.38})$

90. $\ln(e^2)$

91. $\ln 40$

92. $\log_6 8$

Approximate N to 3 significant figures.

93. $\log N = 3.9212$

94. $\log N = 0.8109 - 3$

95. $\log N = 0.7275 - 1$

96. $\log N = 5.9595$

97. $\log(10^2 N) = 0.9876$

98. $\log(N/10^2) = 0.9876$

Write a linear interpolation formula to approximate the following. Confirm the approximation with a calculator.

99. $\log 3.572$

100. $\log 2.496$

101. $\log 3572000$

102. $\log 0.002496$

Approximate N in each of the following.

103. $\log N = 2.9209$

104. $\log N = 0.2858 - 3$

105. $15 = 3^{N+2}$

106. $2^{N+2} = 5^{N-1}$

107. $e^{2N-1} = 10$

108. $10^N = e^{N+2}$

Suppose $\log w = k$ and $\log u = m$, express the following in terms of k and m.

109. $\log(wu^2)$

110. $\log[(wu)^2]$

111. $\log(100w^3/u)$

112. $\log(\sqrt{w^2 x^5})$

Napier's Bones

SIDELIGHT

The sixteenth century was an age of adventure and discovery. The world was not flat. The frontier pushed from land across a vast sea. If you hoped to return from out of sight of land, you needed a method to navigate.

The map of the sky directed seafarers across mapless seas. Mathematics was the language that traced the paths marked by the stars. To follow the star beacons required a navigational tool called a *sextant*. A sextant measures the angular height of the north star above the horizon. That height corresponds to the latitude above the equator.

Time was important. The difference between local time and the time in Greenwich, England, provided the longitude. Ships track time with a time piece set to Greenwich time. It is inaccurate of course: how can a pendulum swing correctly under the influence of a swaying ship? An hourglass and a watchman sounding bells provided a partial solution. The watch was critical. If you fall asleep and fail to turn the hourglass, you lose track of Greenwich time. If you lose track of Greenwich time, you are lost at sea. Your ship has sailed off the edge of the world!

Instruments improved. Navigation relied on good arithmetic as well as precision instruments. To ensure good arithmetic, calculating devices became desirable. One such calculating device was Napier's logarithms. Napier created another, simpler method for multiplication known as *Napier's bones*.

Napier's bones required that sailors prepare special strips marked with digits for multiplication. The strip for each digit consists of the multiples of that digit through 9. Display 6-2 illustrates a strip prepared for the digit 6.

To multiply two numbers using Napier's bones, lay out the strips for the digits of one number, say the multiplier. The product of the numbers follows from the multiples (of the multiplicand) written on the strips. Consider 573 times 624. Lay out the strips for 5, 7 and 3.

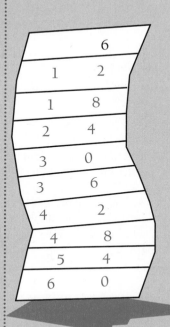

DISPLAY 6-2   Napier's bone for 6

Napier's bones for 5, 7 and 3

5		7		3		
1	0	1	4		6	Multiples of 2 (20)
1	5	2	1		9	
2	0	2	8	1	2	Multiples of 4
2	5	3	5	1	5	
3	0	4	2	1	8	Multiples of 6 (600)
3	5	4	9	2	1	
4	0	5	6	2	4	
4	5	6	3	2	7	

To multiply 573 by 4, choose the multiples of 4 row, then copy the digits across that row, summing those digits in adjacent columns of adjacent strips:

$$2 \quad 0 + 2 \quad 8 + 1 \quad 2: \quad 4 (573) = 2292$$

Similarly, for multiples of 2 and 6:

$$1 \quad 0+1 \quad 4+0 \quad 6: \qquad 2(573)=1146 \quad \text{and} \quad 20(573)=11460$$
$$3 \quad 0+4 \quad 2+1 \quad 8: \qquad 6(573)=3438 \quad \text{and} \quad 600(573)=343800$$

Now sum the multiples to get the final product:

343800	{600(573)}
11460	{20(573)}
2292	{4(573)}
357552	

$$624(573)=357552.$$

Napier's bones is no match for a hand-held calculator, but the method does exemplify a long-standing desire to ease the burden of calculation in everyday problems. The necessity for easy, accurate arithmetic has manifested itself through generations of invention. Necessity was the mother of Napier's bones. Certainly the calculator descended from the same family tree. Locked within the microchips of today are the same genes that Napier used to analyze our numerical system and create logarithms.

6 CHAPTER TEST

1. Sketch the graph, give the domain and range and indicate whether $f(x) = 2(\frac{1}{3})^{x+1} + 5$ is increasing, decreasing or neither.

2. Sketch the graph, give the domain and range and indicate whether $f(x) = 3 - \log_{1/2}(x - 2)$ is increasing, decreasing or neither.

3. (a) Contract $3 \log x - \frac{1}{2} \log y + \log 4 + \log 25$.
   (b) Expand $\log(\sqrt[3]{x}\sqrt{y}\,/z)$.

4. Approximate $\log_7 15$.

5. Solve $3^{x+2} = 5^{2x-1}$ and approximate the answer to two decimal places.

6. Approximate $\log\sqrt[5]{273}$. Use the approximation to determine $\sqrt[5]{273}$ to three decimal places.

7. Solve $\log(x + 3) + \log(x - 1) = 2$.

8. Suppose that a colony of bacteria doubles every hour. A culture starts with 7 bacteria. After how many hours will the number of bacteria exceed 3500?

9. Suppose that a computer depreciates by 30 percent of its current value each year. How many years will it take for a computer to depreciate to half its purchase price?

10. Prove $\log_b(x/y) = \log_b x - \log_b y$.

11. Use a calculator to approximate

$$(5^{1.63} - 1.5 \log_5 2.476)^{4.3}$$

TRIGONOMETRIC AND CIRCULAR FUNCTIONS

During a storm, a clap of thunder, like a stone dropped in water, generates waves of sound in the air. Unlike the waves in water, sound waves are not visible. We hear them. Imagine a ship in the storm. Sailors measure distance as a portion of the circumference of the earth: a nautical mile is $\frac{1}{21600}$ of the earth's circumference. But even short distances at sea are formidable during a storm. The ship's navigator confirms the ship's location by forming a triangle on navigation charts with the ship's position lying at one vertex of the triangle. Despite limited visibility from the storm, simple mathematics based on the direction and distance to known beacons pinpoints the ship's location. Now, the ship's crew is more confident. A quick radio message to shore can summon immediate help if needed.

Curiously, the sound of thunder, the nautical measurements, the triangulated location and the radio message are related, not simply by the vignette, but mathematically. That relationship is found in the study of trigonometric functions. This chapter and the next investigate the properties and use of trigonometric and circular functions.

Trigonometry is an ancient science, dating back to early Hindu references to the length of a bowstring related to the curve of a bow. For the Greeks, trigonometry served as a tool for the study of astronomy. Humans have always measured the passing of the seasons by studying the stars. Without astronomy, calendars cannot be accurate. Without accurate calendars to indicate planting and harvest times, agriculture becomes difficult. Without agriculture, we are nomadic food gatherers. Civilization itself becomes tenuous. Perhaps the future is written in the stars.

Literally, **trigonometry** means three (*tri*) angle (*gono*) measure (*metry*). Scale drawings and charts to accurately represent distances and angles between objects provided the original stimulus for trigonometry. Because the Greeks believed that the earth was a sphere, their measurement of angles was directly connected to arcs of a circle. Today, the ideas of angular measurement, rotation about a center and length of an arc of a circle are still intertwined.

Trigonometry was the title of an exposition by Bartolomeus Pitiscus in 1595. Meaning "three angle measure," trigonometry became synonymous with Pitiscus's subject.

MEASUREMENT: ANGLES, ARC LENGTH AND ROTATION

As long as algebra and geometry proceeded along separate paths, their advance was slow and their applications limited. But when these sciences joined company, they drew from each other fresh vitality and thenceforward marched on at a rapid pace toward perfection.
—LAGRANGE

An **angle** is the union of two half-lines and a common endpoint. The shared endpoint is the vertex of the angle. See Figure 7-1.

On the popular front, **degree** has been the dominant unit for measuring angles. *Degree* measure associates a circle with each measured angle. The center of the circle is at the vertex of the angle. Because all circles are similar in shape, the circumference is proportional to the radius; therefore the exact radius of the circle is unimportant. We may as well choose a radius of 1 for simplicity. See Figure 7-2. For historical reasons degree measurements are taken from 360 equally spaced marks about the circumference. The angular measurement between a successive pair of marks is 1 degree symbolized by 1°. A protractor is a semicircular device used for degree measurement. See Display 7-2.

vertex   side

FIGURE 7-1

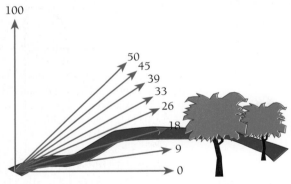

DISPLAY 7-1   Several inclines with superimposed gradient measure

Appropriate for measuring by protractor, **static angles** measure between 0° and 180°. Static angles measuring larger than 90° are **obtuse angles.** Static angles smaller than 90° are **acute angles.** A 90° angle is a **right angle.** A 30° angle is an acute angle that measures $\frac{1}{3}$ of a right angle.

A pair of perpendicular lines forms four right angles. See Figure 7-3. Divide one right angle into two equal angles and each is a 45° angle. Divide a right angle into three equal angles and each measures 30°. Remove the common side between two of the 30° angles and obtain a 60° angle.

For a measurement of angles finer than degrees, divide each degree into 60 uniform portions. A **minute** is $\frac{1}{60}$ of 1°. The symbol for 1 minute is 1′.

Since 1959, the official United States definition of a nautical mile is a unit of linear measure equal in length to 1 minute of the arc of the circumference of the earth. Because the circumference of the earth is approximately 40,000 kilometers, then a nautical mile is 40000 ÷ 360 ÷ 60 or approximately

S   I   D   E   B   A   R

Surveyors often measure inclines using a decimal-like scale called grads. A horizontal stretch of land measures 0 grads. A vertical cliff measures 100 grads. See Display 7-1. Inclines in between measure as a percent of 100 grads. The Babylonians gave us degree measure. Babylonians counted using base 60. Measures of 360°, 60′ per degree and 60″ per minute are reasonable scales in a base 60 system.

Static—adjective, from the Greek *statikos* meaning to cause to stand, to take a position. In terms of physics (mechanics), static implies that an object is at rest, motionless, in equilibrium. We use static to distinguish the fixed angles found in rigid figures like triangles from the measure of angular motion.

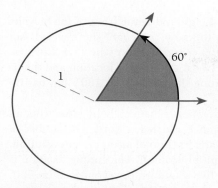

FIGURE 7-2   An angle and a unit circle

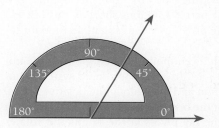

DISPLAY 7-2   A protractor measuring an angle

The Greek letter ω is pronounced "omega." The Greek letter θ is pronounced "theta." The Greek letter φ is pronounced "fie."

1.85185 kilometers (about 1.15 miles). In seagoing jargon, a ship doing 20 knots (20 nautical miles per hour) is traveling about 23 miles per hour and covers 1/360 of the distance around the earth every 3 hours.

Divide a minute into 60 uniform portions, or seconds. A **second** is $\frac{1}{60}$ of 1′. The symbol for 1 second is 1″. Thus an angle might measure 72° 23′ 13″, read "72 degrees 23 minutes 13 seconds." Many scientific calculators are capable of taking angular measure in the degree-minute-second format.

The connection of angles with circles implies rotation. We discuss measurements of rotation shortly. Now, we turn our attention to triangles.

A **triangle** consists of three noncollinear points and the line segments connecting them. The three points are the vertices of the triangle. The three connecting line segments are the sides of the triangle. See Figure 7-4.

The sides of a triangle are line segments. Begin with a pair of segments at a common vertex, then imagine the extension of the segments away from the vertex into half-lines. Each side lies within a half-line so that each pair of sides

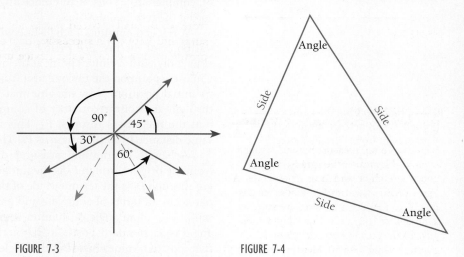

FIGURE 7-3   FIGURE 7-4

of the triangle defines an angle at each of the three vertices: hence the name *tri*angle. The angles of a triangle are static angles. Triangles are rigid figures. The size of an angle in a triangle cannot change without breaking or changing a side of the triangle. From plane geometry, you may recall that the sum of the angles of a triangle is 180°. Therefore, each angle is smaller than 180°. A right triangle is a triangle in which one of the three angles is a right angle. No plane triangle can have more than one right angle. Why?

In a right triangle, the sum of the acute angles is 90°. Two angles whose sum is 90° are **complementary**. *Complementary* comes from the same root word as complete. For example, a 30° angle and a 60° angle are complementary. Their sum is 90°. If a 30° angle and a 60° angle share a vertex and a side they complete each other to form one right angle. Refer to Figure 7-3. Greek letters such as ω, θ or φ label degree-measured angles to distinguish them from real numbers variables like x and y.

Now we connect the concepts of static angle, angle of rotation and measure of arc length. Imagine a bright mark on the tire of an exercise bicycle. In

DISPLAY 7-3

Dynamic—from the Greek *dynamikos,* meaning power or strength; having to do with motion; energetic, vigorous.

Ray—Originally from the Latin *radius* through the French *rai*, meaning a thin beam of light from a bright source.

Display 7-3, the mark begins parallel to the ground at the front of the bicycle. Suppose Jan uses the bicycle and wants to know "how far" she traveled. Jan cannot exercise and then measure the tire mark location when she finishes. Once the tire stops rotating, the angle between the starting location and the final location of the marker is static, limiting the measure to 180°. Jan has no clue to the "distance" pedaled.

To resolve Jan's problem, track the rotation dynamically by measuring an angle of rotation. To determine the distance pedaled, record the number of complete revolutions of the marker plus any fraction of a complete revolution. The starting point of the marker is the **initial** position. The ending location is the **terminal** position. In such a system, five rotations, or revolutions (at 360° each), translates as 1800°. Ten revolutions is 3600°. Moreover, 780° represents two complete revolutions plus 60° more. In terms of integer division, 2 is the quotient and 60°, the remainder in dividing 780° by 360°. See Figure 7-5. Static angle measurement could measure partial rotations short of, or in excess of, complete rotations. A **reference angle** provides a convenient method for locating the terminal position.

A **ray** is the union of a half-line with its endpoint. The endpoint is the initial point of the ray. See Figure 7-6. On a Cartesian coordinate system, consider a ray with its initial point at the origin and lying along the positive x-axis. Choose a point on the ray one unit from the origin and label it P_0. The ray is in its initial position. Now imagine that the ray pivots at the origin and rotates through an arbitrary number of complete rotations before coming to rest so that the ray passes through P_1. This is the ray's terminal position. P_1 is the same distance from the origin as P_0. The rotation transforms ray P_0 into ray P_1.

An **angle of rotation** measures the imagined rotation of a ray. The ray from the origin through P_0 is the initial side of the angle. The ray from the origin through P_1 is the terminal side of the angle. The measure of the angle is dynamically in terms of complete and partial revolutions about the origin. If the initial side of an angle of rotation coincides with the positive x-axis, then the angle is in standard position. If rotation is in a counterclockwise direction, then positive values measure the angle. If the rotation is in a clockwise direction, use negative values. See Figure 7-7.

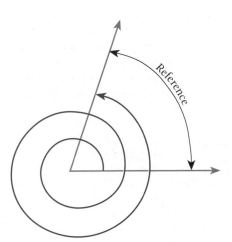

FIGURE 7-5

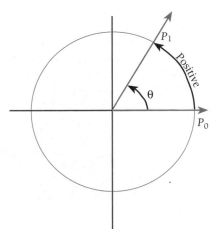

FIGURE 7-6

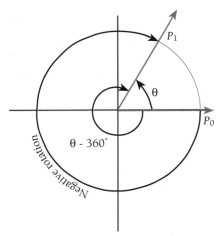

FIGURE 7-7

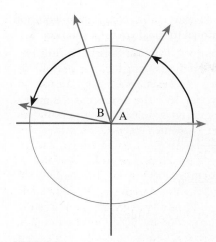

FIGURE 7-8

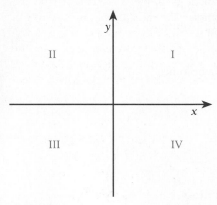

FIGURE 7-9

Two angles of rotation are **equal angles** if their measures are equal. This does not mean that angles with coincident terminal and initial sides must be equal. As Figure 7-7 suggests, many unequal angles of rotation in standard position have coincident terminal sides. In fact, if angle θ is in standard position and measures 30°, then an angle in standard position that measures 390° (360° + 30°) has a coincident terminal side. Similarly, all angles in standard position with measure expressible in the form 30° + k360° (k ∈ Integers) have coincident terminal sides. On the other hand, Figure 7-8 displays two equal angles that have different terminal (and initial) sides.

Let us return to Figure 7-7. Label the angle of rotation θ. If θ does not terminate on the x- or y-axis then the reference angle for θ is the acute static angle formed by the x-axis and the terminal side of θ. If θ terminates on the y-axis the reference angle is 90°. If θ terminates on the x-axis, the reference angle is 0°. Every angle of rotation in standard position can be identified by three pieces of information: the *number of complete revolutions*, the *quadrant in which the terminal side lies*, and the *reference angle*.

Consider positive angles in standard position (that is, with its initial side on the positive x-axis) terminating on an axis but measuring less than one complete revolution. If the angle terminates on the positive x-axis its measure is 0°. If the angle terminates on the positive y-axis, its measure is 90°. If the angle terminates on the negative x-axis, its measure is 180°. If the angle terminates on the negative y-axis, its measure is 270°. Label these values as quadrantal boundaries. Then an angle with terminal side between the positive x-axis and the positive y-axis, terminates in the first **quadrant** (symbolized by the Roman numeral I). Similarly, in a counterclockwise (positive) direction label the remaining quadrants II, III and IV. See Figure 7-9.

To connect these ideas to arc length, return to the exercise bike. Remove the bicycle from its stand, so that it rolls when pedaled. How far will the bicycle travel if the wheel turns 2500 revolutions? First experiment on a smaller scale: how far does the bike travel when the wheel rotates 150°? See Display 7-3.

Recall the relation of the circumference of a circle to its diameter.

Circumference = π Diameter.

Because the diameter is twice the radius, we have $C = 2\pi r$, where C is the circumference and r is the radius. Suppose the radius of the bicycle tire is 11.5 inches. Then the circumference is $2\pi(11.5)$ inches of approximately 72.2567 inches. Allow for tire wear and errors in measuring the radius by rounding the circumference to 72 inches. If the tire rolls while making a complete revolution, the bicycle travels about 72 inches (or 6 feet or more roughly 1.80 meters). See Display 7-4.

About 6 feet

DISPLAY 7-4

A degree measure is a proportional part of a complete revolution. Therefore the distance traveled is proportional to the degrees of rotation:

Distance = k degrees of rotation

$$D = k\,\theta$$

where k is the constant of proportionality. For 360° the distance is 1.8 meters. Hence, 180 cm = k 360°.

Solve for k to obtain $k = \dfrac{180 \text{ cm}}{360°}$

The formula becomes $D = \dfrac{1 \text{ cm}}{2°}\,\theta$

where D is in meters and θ is in degrees. For our example of 150°,

$$D = \frac{1 \text{ cm}}{2°}\,150°$$

$$D = 75 \text{ cm}$$

If we express the distance in inches rather than centimeters, the formula becomes $D = (1 \text{ in.}/5°)\,\theta$.

Here D measures an arc of the circle as if it were "straightened." See Display 7-4. Hence D measures the **arc length** of a sector of a circle. We do not limit ourselves to the circumference of the circle. As the wheel rolls over the street, it leaves an almost invisible layer of rubber to mark its path. The length of this path represents arc length for larger angles of rotation.

For 2500 revolutions we have

$$D = \frac{1 \text{ cm}}{2°}\;\frac{2500 \text{ revolutions}}{1}\;\frac{360°}{1 \text{ revolution}}$$

$$D = 450{,}000 \text{ cm (or 4.5 km)}$$

Using feet measurement, the formula becomes

$$D = \frac{6 \text{ ft}}{360°}\,\theta$$

$$D = \frac{1 \text{ ft}}{60°}\,\theta.$$

Then for 2500 revolutions,

$$D = 15{,}000 \text{ feet}\quad \text{(about 2.841 miles)}$$

Using both the metric and English systems is clumsy. For greater generality, consider the formula for arc length without specific units for the radius r:

$$D = \frac{2\pi r\theta}{360°}$$

How long did the trip take? That depends on how fast you pedal. These two concepts lead to the measures of angular and linear velocity.

Once more, imagine the mark on the turning wheel. *Velocity* measures a relative change in distance per unit of time. The **angular velocity** of a point

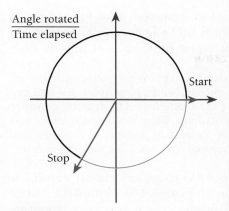

$$\frac{\text{Angle rotated}}{\text{Time elapsed}}$$

Start

Stop

FIGURE 7-10

moving along the circumference of a circle is the relative rate of rotation per unit of time: rotation/time = θ/t. See Figure 7-10. Suppose the wheel rotates 250 complete revolutions each minute. Any point on the wheel sweeps through 360° for each revolution. At 250 rev/min with 360°/rev, the point sweeps through 90,000°/min. The angular velocity of the wheel measures 90,000°/min. We could have chosen degrees/second or revolutions/hour. The choice of units depends upon the application. In any case, angular velocity measures the rate of rotation of a point on the circumference of a circle. Let ω represent angular velocity: ω = θ/t.

Although 250 rev/min (or 90,000°/min) conveys how fast you pedal, it does not measure how fast you go. Until we took the bicycle from the stand, there was only circular motion and angular velocity. The next goal is to translate this motion into linear motion and hence linear velocity.

Linear velocity measures the rate of change of distance per unit of time. Let us use arc length to obtain a suitable formula:

$$\text{Linear velocity} = \frac{\text{distance}}{\text{time}}$$

$$= \frac{d}{t}$$

Because the distance is the arc length, replace distance with $d = 2\pi r\theta/360°$.

$$\text{Linear velocity} = \frac{2\pi r\theta}{360°t}$$

$$= \frac{2\pi r}{360°}\frac{\theta}{t}$$

Alternately, because $\dfrac{\theta}{t}$ measures angular velocity, express the result as

$$\text{Linear velocity} = \frac{2\pi r\omega}{360°}, \qquad \text{where } \omega = \frac{\theta}{t}$$

Recall the measurements for the bicycle. The angular velocity is given by ω = 90,000°/minute and the circumference of the circle by $2\pi r = 6$ feet. Therefore, the linear velocity is

$$v = \frac{(6 \text{ feet})(90,000°/\text{minute})}{360°}$$

$$= 1500 \text{ ft/min} \quad (\text{about } 17 \text{ mph})$$

The formula for arc length could stand some simplification. The source of difficulty is the degree measure of rotation. If rotation, using degrees, can form a basis for determining arc length, then conversely, arc length can form a basis for measuring angles of rotation.

Examine the formula for linear velocity once more. Perhaps with a simpler measure of rotation, we could avoid dividing by 360. While we are wishing, why not get rid of the cumbersome 2π? Linear velocity is dependent on the remaining variable components of the formula. A larger wheel covers more

Radian—*Radian* first appeared in an exam given on June 5, 1873, by James Thomson at Queen's College, Belfast. Radian was also known as the *wheelwright's measure* because of the relation of the spoke of a wheel to the metal band wrapped around the wheel's circumference.

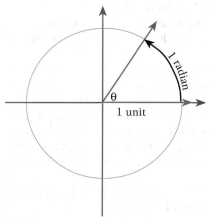

FIGURE 7-11

ground in one revolution, so *r* is essential. Distance covered also depends upon the amount of rotation, so we cannot wish away θ. Finally, time *t* is indispensable in any measure of velocity. To rid the formula of both the 360° and the 2π we introduce radian measure.

Because all circles are similar, the circumference is proportional to the radius. Hence, without loss of generality, choose a circle of radius 1, the unit circle, for reference. Rather than dividing the circle into 360 degrees, consider that the circumference is exactly 2π units long. But 2π is a real number and real numbers provide an ideal base for measuring distance. Let us use arc length on the unit circle to measure angles of rotation. This measure is called *radian measure*. See Figure 7-11.

Why choose a unit circle? Why not a circle of any radius? Actually it makes little difference. Examine the circle of radius 1 in Figure 7-11. Notice that the arc *x* subtends (spans across) central angle θ. Because *x* represents an arc length on the unit circle, *x* is the radian measure of the central angle θ. Suppose *x* is 1; then *x* equals the radius. Proportionally, *x* is $1/(2\pi)$ of a complete revolution. The measurement of θ is 360° $1/(2\pi)$ or approximately 57.29577951°. Because the ratio of the circumference of a circle to its radius is 2π, any arc of a circle equal in length to the radius of that circle subtends a central angle of one radian (approximately 57.29577951°). Similarly a 90° central angle always intercepts an arc equal in length to $\pi/(2)\ r$, where *r* is the radius of the circle.

Unfortunately, the radius *r* differs from circle to circle. An arc length uses the same units as the radius, so linear measurement complicates using an arc to measure an angle. Because the circumference of a circle is proportional to the radius, divide the radius *r* into an arc length and the like linear units divide out. The resulting angular measure, **radian** measure, is free of linear units (see Display 7-5). In a unit circle, the radius is 1 so that the division is not needed. For example, 90° coincides with π/2 rad. Many calculators use the abbreviation *rad* for radian measure. Where the context is clear, we often omit the *rad* label (see Figure 7-12).

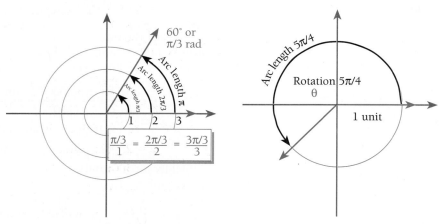

DISPLAY 7-5   Arc length, degree and radian measure

FIGURE 7-12

The arc length of a unit circle measures angles in a manner similar to degree measure. Radian measure is directly proportional to the degree measure of an angle of rotation: radian = k(degree). Because a complete rotation measures 360° in degrees and subtends a 2π rad arc,

$$2\pi \text{ rad} = k(360°)$$

$$k = \frac{2\pi \text{ rad}}{360°}$$

Therefore, $\qquad$ radian $= \dfrac{\pi \text{ rad}}{180°}$ degree

The inverse yields $\qquad$ degree $= \dfrac{180°}{\pi \text{ rad}}$ rad

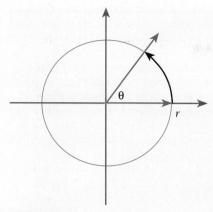

FIGURE 7-13

Now back to our arc length formula. Consider a circle of radius r. Let arc s of the circle subtend central angle θ. Then $s = (2\pi r/360°)\,\theta$.

Suppose the x represents the equivalent radian measure. Conversion of θ to x is given by $x = (2\pi \text{ rad}/360°)\theta$. Replace these in the formula for s to obtain $s = rx$.

Examine Figure 7-14. Note that each arc length is proportional to the radius of its circle: $s/r = x/1$. Solve this proportion for s to obtain the same results.

The arithmetic in this version of the formula is cleaner. For example, suppose a circle has a radius of 10 feet. What is the length of an arc of the circle that subtends a central angle measuring 90°? The new formula requires radian measure: $90° = \pi/2$. Now $s = 10$ feet$(\pi/2)$. The arc length is $s = 5\,\pi$ feet. By calculator, $s \cong 15.71$ feet.

Similarly, the formula for linear velocity simplifies when radian measure replaces degree measure. First measure the angular velocity in radians:

$$\omega = x/t \qquad\qquad\qquad\qquad \text{(radians/unit time)}$$

then $v = r\omega$, $\qquad$ where r is the radius

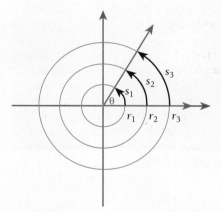

FIGURE 7-14

From the bicycle example, 2500 revolutions translate immediately into 5000π radians. The radius of the wheel is 11.5 inches. The linear distance traveled is given by

$$D = xr$$

$$D = 5000\pi \quad (11.5 \text{ in.})$$

$$D \cong 180641.5776 \text{ in.} \quad \text{(about 15053.4648 ft)}$$

Similarly, 250 revolutions per minute becomes an angular velocity of

$$\omega = 250(2\pi)/\text{min}$$

$$= 500\pi/\text{min}$$

From which the linear velocity quickly follows:

$$v = (11.5 \text{ inches})500\pi/\text{minute}$$

$$\cong 18064.15776 \text{ inches/minute}$$

$$\cong 17.10621 \text{ miles/hour}$$

SUMMARY

- Degree measure and radian measure are two ways to measure angles of rotation. To convert from degree measure to radian measure use $x = (\pi/180°)\theta$. To convert radian measure to degree measure use $\theta = (180°/\pi)x$.
- To determine the angular velocity of a point on the circumference of a circle rotating through x radians over time period t use $\omega = x/t$ (a similar formula applies to degree measure).
- The corresponding linear velocity is given by $v = r\omega$ (where r is the radius of the circle).
- To determine the velocity of a point on the circumference of a circle of radius r:

  1. Be sure the rotation is measured in radians x.
  2. Calculate the angular velocity, ω.
  3. Apply the formula $v = r\omega$.

- The following are equivalent radian and degree measures:

Degree:	0°	30°	45°	60°	90°	180°	270°	360°	3600°
Radian:	0	$\pi/6$	$\pi/4$	$\pi/3$	$\pi/2$	π	$3\pi/2$	2π	20π

7-1   EXERCISES

Convert each of the following radian measure to degree measure.

1. π **2.** $\dfrac{\pi}{2}$ **3.** $\dfrac{3\pi}{2}$

4. 2π **5.** 5π **6.** 6π

7. $-\dfrac{\pi}{4}$ **8.** $-\dfrac{\pi}{3}$ **9.** $\dfrac{5\pi}{3}$

10. $\dfrac{5\pi}{4}$ **11.** $\dfrac{\pi}{6}$ **12.** $\dfrac{\pi}{12}$

13. $\dfrac{7\pi}{12}$ **14.** $\dfrac{7\pi}{6}$ **15.** $\dfrac{11\pi}{6}$

16. $\dfrac{13\pi}{12}$ **17.** $.5$ **18.** 1.5

19. 1 **20.** 2

Convert each of the following degree measure to radian measure expressed as a multiple of π.

21. 75°   **22.** 100°   **23.** 30°

24. 150°   **25.** 95°   **26.** 135°

27. 180°   **28.** 45°   **29.** 120°

30. 60°   **31.** 250°   **32.** 210°

33. –30°   **34.** –60°   **35.** 720°

36. 1080°   **37.** 225°   **38.** 315°

39. 300°   **40.** 210°

41. Convert 20 rev/min to degrees/min.

42. Convert 30 rev/min to degrees/min.

43. Convert 720°/min to rev/min.

44. Convert 180°/min to rev/min.

45. Convert 20 rev/min to degrees/sec.

46. Convert 30 rev/min to degrees/sec.

47. Convert 720°/min to rev/hour.

48. Convert 180°/min to rev/hour.

49. Convert 20 rev/min to degress/hour.

50. Convert 30 rev/min to degrees/hour.

51. Convert 720°/min to radians/min.

52. Convert 180°/min to radians/min.

53. A point on the unit circle travels from 120° to 210° in 5 seconds. What is the angular velocity?

54. A point on a circle travels from 80° to 310° in 10 seconds. What is the angular velocity?

55. A point on a circle rotates from -35° to 145° in $\frac{1}{2}$ hour. What is the angular velocity?

56. A point on a circle rotates from 200° to -10° in 1.5 minutes. What is the angular velocity?

57. A point on the unit circle moves from π radian to 4π radians in 10 seconds. Determine the angular velocity in degrees/min.

58. A point on the unit circle moves from 0.5π radians to 3π radians in 5 minutes. Determine the angular velocity in degrees/second.

59. A 5.25″ floppy disk rotates at 300 revolutions per minute. Determine the angular velocity of the disk.

60. A 3.5″ floppy disk rotates at 360 revolutions per minute. Determine the angular velocity of the disk.

61. Determine the arc length for radius = 5 feet and central angle of 180°.

62. Determine the arc length for radius = 6 meters and central angle 135°.

63. Determine the arc length for radius = 10 meters and central angle 45°.

64. Determine the arc length for radius = 8 yards and central angle 90°.

65. Determine the arc length for radius = 10/π cm, central angle = π/3 rad.

66. Determine the arc length for radius = 50 mm, central angle = 2 rad.

67. The first track of a floppy disk is 2.25″ from the center of the disk. Suppose this track is divided into nine equal sectors. What is the length of one of these sectors?

68. The last track of a floppy disk is 1.5″ from the center of the disk. Suppose this track is divided into nine equal sectors. What is the length of one of these sectors?

*69. A video cassette tape has a take-up reel that turns at 60 revolutions per minute (rpm) when rewinding. The radius of the take-up spindle is 0.375″ in radius. The recording tape is 0.005″ thick. Suppose the tape

rewinds from the end for 2 minutes and 5 seconds. Approximate the length of the next complete wrap of tape.

*70. See Problem 69. Suppose that the total time to rewind is 3 minutes and 45 seconds. Approximate the circumference of the final wind of tape.

71. Determine the linear velocity in cm/min of a point on the circumference of a 50-cm radius wheel that is turning at 100 rpm.

72. Determine the linear velocity in meters/min of a point on the tip of a propeller 1 meter in radius turning at 50 rpm.

73. Suppose the minute hand of a clock is 5″ long. What is the linear velocity in inches/hour of the tip of the hand?

74. Suppose the sweep seconds hand of a pocket watch is 1″ long. What is the linear velocity in inches/min of the tip of the hand?

75. A 30″ diameter bicycle wheel rotates at between 336 and 337 revolutions per minute. Approximate the velocity of the bicycle to the nearest mile per hour.

76. A merry-go-round is 10 meters in diameter and makes 2 revolutions per minute. What is the speed in kilometers/hour of a horse on the outer circumference of the merry-go-round?

77. See Problem 75. Consider a point on a spoke of the bicycle wheel 12″ from the center of the wheel. Determine the linear velocity of the point.

78. See Problem 76. A horse on the interior of the merry-go-round is 2 meters from the horse on the outside circumference. Determine the velocity of the interior horse.

79. See Problem 69. Approximate the linear velocity of the tape at 2 minutes 5 seconds into the rewind.

80. See Problem 70. Approximate the tape velocity just before the rewind is complete (at 3 minutes and 45 seconds).

81. See Problems 75 and 77. Suppose that a disc of radius r rotates at fixed velocity l. Prove that any point in the interior of the disk has a linear velocity less than any point on the circumference of the disk.

82. See Problems 76 and 78. Prove that the velocity of a point on the circumference of a disk is twice the velocity of a point on the disk halfway between the center and the circumference.

83. See Problem 81. Discuss which of the points has a higher angular velocity.

84. See Problem 82. Discuss what is meant by the statement "the outside of a wheel turns faster than the inside of a wheel."

7-1   PROBLEM SET

Imagine a plane through the center of the earth. The intersection of the plane with the surface of the earth is a great circle. A nautical mile is the distance along a great circle that subtends a 1-minute angle at the center of the earth. The equator and lines of longitude perpendicular to it are examples of great circles.

1. Determine the length of a nautical mile if a great circle is 40,000 kilometers in circumference.

2. Determine the length of a nautical mile if a great circle is 25,000 miles in circumference.

3. How many nautical miles along the equator from 15°43′ east to 16°11′ east?

4. How many nautical miles along the equator from 35°4′ west to 34°57′ west?

5. The nautical measure of velocity is the knot, defined as 1 nautical mile per hour. If a ship averages 15 knots headway, how many nautical miles will it cover in 4 hours?

6. See Problem 5. If a ship averages 20 knots headway, how many nautical miles will it cover in 3 hours?

7. See Problem 5. Given the same information, how many kilometers will the ship cover in 4 hours?

8. See Problem 6. Given the same information, how many kilometers will the ship cover in 3 hours?

9. How long will it take a ship to circumnavigate the earth at an average speed of 36 knots?

10. How long will it take an airplane to circumnavigate the earth at an average speed of 360 knots?

PROBLEMS FOR TECHNOLOGY

Many calculators express degree measure in decimal format, as for example 72.6897°. Many people prefer that degrees be broken into minutes and seconds, where 60 minutes = 1° and 60 seconds = 1 minute. The abbreviation for minutes is a single quote and for seconds is a double quote: 60′ = 1° and 60″ = 1′. We read 65° 32′ 16″ as "65 degrees 32 minutes 16 seconds." Because minutes and seconds are a form of base 60 notation, to convert minutes and seconds to decimal measure use the nested form of a polynomial. Degrees° minutes′ seconds″ converts to a decimal as (Seconds ÷ 60 + Minutes) ÷ 60 + Degrees. Therefore, 65° 32′ 16″ = (16 ÷ 60 + 32) ÷ 60 + 65 = 65.53777°.

Converting from decimal degree to minutes and seconds reverses this process. The following steps extract the degree,

minute and second measure from the decimal representation 65.53777°.

(a) The whole number portion of the decimal represents degree measure. Copy it down and subtract it from the decimal:

Deg	Minute	Second
65°		

65.53777 − 65 = 0.53777

(b) Convert the decimal degree 0.53777° to minutes by multiplying by 60 min/deg: 60(0.53777°) = 32.2662′. The whole number portion, 32, of the resulting number represents the number of minutes. Copy 32 as the minute value and subtract 32 from the decimal for minutes.

Deg	Minute	Second
65°	32′	

32.2662′ − 32′ = 0.2662′

(c) Convert the remaining 0.2662′ to seconds by multiplying by 60 sec/min: 60(0.2662) = 15.972″. Round this to the nearest second to complete the conversion. 15.972″ ≅ 16″.

Deg	Minutes	Seconds
65°	32′	16″

Convert the following decimal degrees to degrees, minutes and seconds.

11. 35.87994°	12. 87.65234°
13. 125.77653°	14. 231.87653°
15. 309.88823°	16. 101.87654°
17. −13.90552°	18. −113.23885°
19. 513.87991°	20. 402.39029°

Convert the following degree, minute and second measures to decimal degrees.

21. 23° 42′ 37″	22. 15° 29′ 13″
23. 247° 47′ 59″	24. 310° 13′ 29″
25. 109° 9′ 43″	26. 206° 53′ 17″
27. 59° 17′ 45″	28. 114° 35′ 30″
29. 10° 30″	30. 5° 50′

31. Convert 114° 35′ 30″ to radian measure.

32. Convert 59° 17′ 45″ to radian measure.

Suppose that an angle in standard position has the following radian measure. Determine the quadrant in which the angle terminates.

33. 2

34. 3

35. -4

36. -2

37. 1.39

38. 2.43

39. 5.75

40. 6.92

2   CIRCULAR FUNCTIONS, TRIGONOMETRIC FUNCTIONS

Sine—*Jiba* is Hindi for the chord of an arc (bottom of a bowl or string on a bow). In 510, Aryubhata applied the name *jiba* to the relationship that evolved into the sine. Because Hindu writers often omitted vowels, early translators mistook *jiba* for *jaib*, which meant "fold of a garment." The Latin for "fold" or "hollow" is *sinus*. Through mistaken translation Regiomontanus (1436–1476) gave us the word *sine*.

Alexander is said to have asked Menaechmus to teach him geometry in a quick easy manner, but Menaechmus replied "O King, through the country there are roads for the royal and roads for the common citizen, but in geometry there is one road for all."

Examine the angle of rotation in Figure 7-15. Choose any point (x_0, y_0) on the terminal side of the angle. The slope of the line through the origin and this point is y_0/x_0. Similarly for (x_1, y_1) and (x_2, y_2). Then the slope

$$m = \frac{y_0}{x_0} = \frac{y_1}{x_1} = \frac{y_2}{x_2}$$

Dropping a line from any point (x_n, y_n) on the line perpendicular to the *x*-axis outlines a right triangle with vertices at $(0, 0)$, (x_n, y_n) and $(x_n, 0)$. The right angle is at $(x_n, 0)$. Finally, $m = y_n/x_n$.

In any right triangle the side opposite the right angle is the **hypotenuse**. In Figure 7-16, r_n is the hypotenuse. By the Pythagorean theorem or from the distance formula (see Chapter 1), we have $r_n = \sqrt{x_n^2 + y_n^2}$. What follows is the key concept in the development of trigonometry.

Hy-pot-e-nuse—From the Latin meaning "subtending" or "stretched under" (the right angle).

Consider a right triangle with an acute angle θ. The other acute angle ϕ must measure $90° - \theta$ ($\pi/2 - \theta$ in radians). Once any acute angle of a right triangle is known the other angle is apparent because they are complementary. What of the sides? Examine Figure 7-17.

There are two right triangles in Figure 7-17. The angle at vertex *A* of triangle *ABC* is equal to the angle at vertex *D* of triangle *DEF*. As a consequence, the angle at vertex *B* equals the angle at vertex *E*. From plane geometry, you

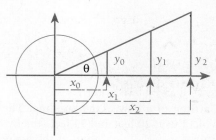

FIGURE 7-15

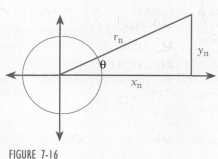

FIGURE 7-16

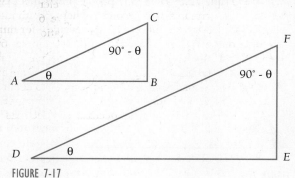

FIGURE 7-17

may recall that triangle *ABC* and triangle *DEF* are similar. *Similarity* implies that the sides of the triangles are in proportion. This is not unexpected. Examine the lines in Figures 7-15 and 7-16. For the fixed angle θ the slope

$$m = \frac{y_0}{x_0} = \frac{y_1}{x_1} = \frac{y_2}{x_2} = \ldots = \frac{y_n}{x_n}$$

The slope ratio is based on the similarity of the triangles formed.

To alter the shape of a right triangle, and thus the ratio of the sides of a right triangle, the acute angles must change. The right angle cannot change, or it would no longer be a right triangle. For a given acute angle in a right triangle, *selecting the angle size fixes the ratio of any pair of sides*. Alter the ratio of the sides—and thus the shape of the triangle—and the angle must change. This pairing of angles to ratios implies functions, because unique acute angles produce unique ratios of sides.

Applications of Similar Triangles

EXAMPLE 7-1

Illustration 1:

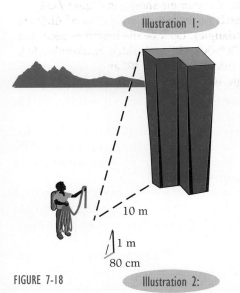

Janelle plans to rappel down a large rock. To be sure she takes enough line on her ascent, she needs to know the height of the rock. It is morning and the rock casts a 10-meter shadow. Because the rock face is approximately vertical, Janelle holds a meter stick perpendicular to the ground and measures its shadow. The meter stick shadow is 80 cm. See Figure 7-18. Janelle makes a working drawing of the rock and meter stick. Because the sun casts shadows at the same angle to the rock and the meter stick, the triangles formed are similar. Janelle forms a proportion.

$$\frac{80 \text{ cm}}{100 \text{ cm}} = \frac{10 \text{ m}}{X}$$

She solves for *X*:

$$X = \frac{1000 \text{ m}}{80}$$

$$X = 12.5 \text{ meters}$$

FIGURE 7-18

Illustration 2:

Ken is designing a garden area for the atrium of a new office building. The atrium base is a right triangle with sides of 15 feet, 20 feet and 25 feet (the hypotenuse). Ken plans to feature a sculptured waterfall and pond at the acute angle of the triangle opposite the 20-foot side. The waterfall and pond will be fabricated at a factory and shipped to the construction site. Although the outer portion of the pond is free-form the back must custom-fit the corner of the atrium. The factory must have a precision measurement of the acute angle where the pond will fit. Ken makes a scale drawing of the atrium (Figure 7-19): a right triangle with sides 15.0 inches, 20.0 inches, and 25.0 inches measured to the nearest tenth of an inch.

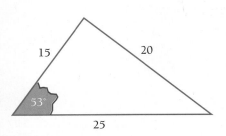

He carefully measures the angle in his scale drawing: 53° to the nearest degree. Because the triangles are similar, the angle in the atrium is also 53°.

FIGURE 7-19

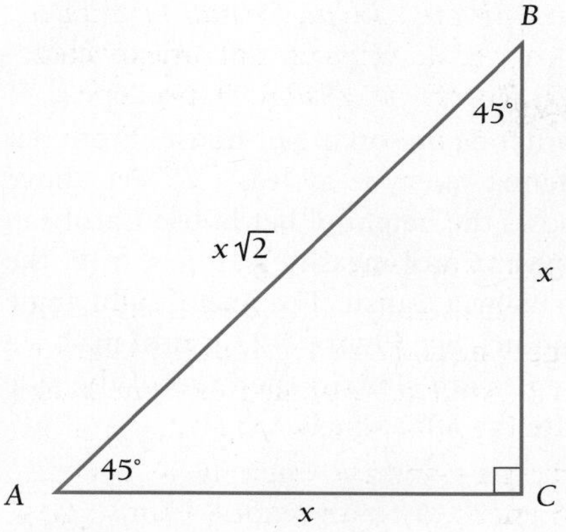

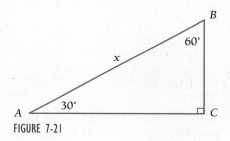

FIGURE 7-20

Two special right triangles deserve mention, the 45° right triangle and the 30°-60° right triangle. The 45° right triangle gets its name from the acute angles. If one acute angle is 45° then so is the other. See Figure 7-20.

From the rules of plane geometry, the sides of a triangle opposite equal angles must be equal in length. Examine Figure 7-20 again. If the length of side AC is x then the length of side BC is also x. The Pythagorean theorem implies the length of the hypotenuse must be

$$\sqrt{x^2 + x^2} = x\sqrt{2}$$

These are sides of a triangle, so x is positive. Now consider the ratio of any pair of sides: x/x or $x\sqrt{2}/x$ or $x/(x\sqrt{2})$. In each case the x divides out. In a 45° right triangle, the ratio of the sides opposite the acute angles is always 1. The ratio of a side to the hypotenuse is $1/\sqrt{2}$ or $\sqrt{2}/2$.

The 30°-60° triangle has different ratios. If one angle of a right triangle measures 30°, the other acute angle must be its complement at 60°. Consider the 30°-60° right triangle in Figure 7-21. To create a situation similar to the 45° right triangle, use symmetry to form the triangle shown in Figure 7-22.

Triangle ABQ has a 60° angle at each vertex. Since the angles of ABQ are equal, the sides are equal (equilateral triangle). Let x be the length of each side of triangle ABQ. Consider the original 30°-60° ABC embedded in triangle ABQ. Because of symmetry the length of the side opposite the 30° angle in ABC must be $x/2$. Side AC opposite the 60° angle is a mystery. Apply the Pythagorean theorem.

FIGURE 7-21

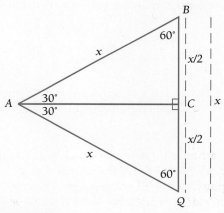

FIGURE 7-22

$$\left(\frac{x}{2}\right)^2 + (AC)^2 = x^2$$

$$(AC)^2 = x^2 - \frac{x^2}{4}$$

$$(AC)^2 = \frac{3x^2}{4}$$

$$AC = x\frac{\sqrt{3}}{2} \qquad (x > 0)$$

Examine various ratios of the sides of the triangle. In every case, x divides out. This is because x is simply the constant of proportionality for the relative "size" of various right triangles. A nice pattern develops.

1. $\dfrac{\text{Side opposite } 30°}{\text{hypotenuse}} = \dfrac{\sqrt{1}}{2} = \dfrac{1}{2}$

2. $\dfrac{\text{Side opposite } 45°}{\text{hypotenuse}} = \dfrac{\sqrt{2}}{2}$

3. $\dfrac{\text{Side opposite } 60°}{\text{hypotenuse}} = \dfrac{\sqrt{3}}{2}$

EXAMPLE 7-2

Illustration:

Application of Special Triangles

Carol is planning to install a television antenna on top of her house. From her neighbor's antenna she knows her antenna must be at least 37 feet above ground for clear reception. Once she knows the height of her house Carol can calculate the minimum length for the mast. Carol measures 15 feet from the base of the house and sights the roof top using a transit. The line of sight angle is 60° above horizontal (angle of inclination). See Figure 7-23. Carol makes a drawing. Because all 30°-60° right triangles are similar, the ratio of the side opposite the 60° angle to the side opposite the 30° angle is $\sqrt{3}/1$.

$$\text{Thus} \qquad \frac{X}{15} = \sqrt{3}$$

$$X = 15\sqrt{3}$$

Carol uses her calculator to approximate X.

$$X \cong 26 \text{ feet}$$

The transit is 5 feet above ground level, so the roof is 31 feet above ground. Carol will need at least 6 feet of mast.　▬

From the previous discussion, we know that a function exists between the angular measure of an acute angle of a right triangle and a selected ratio of sides. Consult Figure 7-24 as a reference for the following preliminary definitions.

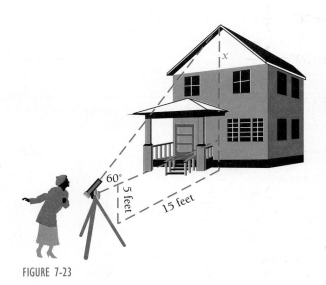

FIGURE 7-23

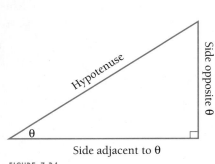

FIGURE 7-24

DEFINITION 7-1
Fundamental Trigonometric Functions

Let θ be the measure of an acute angle of a right triangle, $0° < \theta < 90°$, {$0 < \theta < \pi/2$ in radians}; then each of the following is a function:

FUNCTION NAME	RULE OF PAIRING
sine	$\sin(\theta) = \dfrac{\text{side opposite } \theta}{\text{hypotenuse}}$
cosine	$\cos(\theta) = \dfrac{\text{side adjacent } \theta}{\text{hypotenuse}}$
tangent	$\tan(\theta) = \dfrac{\text{side opposite } \theta}{\text{side adjacent } \theta}$

For a trigonometric function, the domain component θ is sometimes called the argument of the function. This terminology extends to composites of trigonometric functions. For example, in $\sin(5\theta + 3)$, $5\theta + 3$ is the argument of the sine function. Where the argument for a trigonometric function is a single variable or measurement, we often omit the parentheses. For example, $\tan \theta$ and $\cos 45°$ are acceptable abbreviations. However, $\sin \theta + 5 = \sin(\theta) + 5$ and is not the same as $\sin(\theta + 5)$.

Evaluating Trigonometric Functions at Special Values

EXAMPLE 7-3

Illustration:

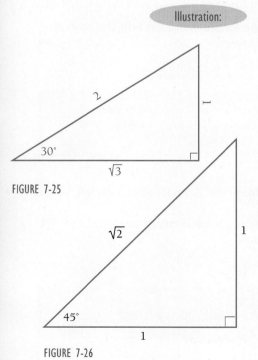

FIGURE 7-25

FIGURE 7-26

Evaluate each of the acute angle functions for the special angles (30°, 45° and 60°)

Solution 1:

$\theta = 30°$ (See Figure 7-25)

$$\sin 30° = \frac{1}{2}$$

$$\cos 30° = \frac{\sqrt{3}}{2}$$

$$\tan 30° = \frac{1}{\sqrt{3}} = \frac{\sqrt{3}}{3}$$

Solution 2:

$\theta = 45°$ (See Figure 7-26)

$$\sin 45° = \frac{1}{\sqrt{2}} = \frac{\sqrt{2}}{2}$$

$$\cos 45° = \frac{1}{\sqrt{2}} = \frac{\sqrt{2}}{2}$$

$$\tan 45° = \frac{1}{1} = 1$$

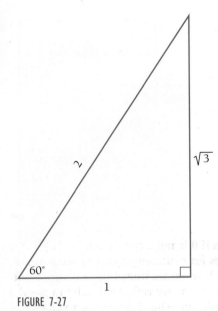

FIGURE 7-27

Solution 3:

$$\theta = 60°$$ (See Figure 7-27)

$$\sin 60° = \frac{\sqrt{3}}{2}$$

$$\cos 60° = \frac{1}{2}$$

$$\tan 60° = \frac{\sqrt{3}}{2} \div \frac{1}{2} = \sqrt{3}$$

Note, because $\pi/3 = 60°$, $\sin (\pi/3) = \sqrt{3}/2$, $\cos (\pi/3) = \frac{1}{2}$ and $\tan (\pi/3) = \sqrt{3}$. In general we have the following results:

θ radian	θ degree	$\sin \theta$	$\cos \theta$	$\tan \theta$
$\dfrac{\pi}{6}$	30°	$\dfrac{1}{2}$	$\dfrac{\sqrt{3}}{2}$	$\dfrac{\sqrt{3}}{3}$
$\dfrac{\pi}{4}$	45°	$\dfrac{\sqrt{2}}{2}$	$\dfrac{\sqrt{2}}{2}$	1
$\dfrac{\pi}{3}$	60°	$\dfrac{\sqrt{3}}{2}$	$\dfrac{1}{2}$	$\sqrt{3}$

Notice the similarity of the answers for $\sin \theta$ and $\cos \theta$ in the preceding example. The following theorem refines this discovery. ▬

Theorem 7-1
Complementary Relation

> If θ and ϕ are complemental angles, then $\sin \theta = \cos \phi$. Because θ and ϕ are complementary, restate as $\sin(90° - \theta) = \cos \theta$; $\sin(\pi/2 - \theta) = \cos \theta$, in radians.

The proof follows immediately from the definitions and is left as an exercise.

The name cosine is an abbreviation of complementary sine. Theorem 7-1 indicates the motivation behind the name.

Theorem 7-2
Tangent in Terms of Sine and Cosine

> $$\tan \theta = \frac{\sin \theta}{\cos \theta}, \qquad 0° < \theta < 90°; \quad 0 < \theta < \pi/2$$

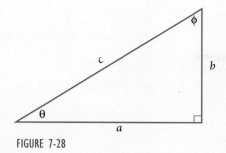

FIGURE 7-28

Proof: See Figure 7-28.

$$\frac{\sin \theta}{\cos \theta} = \frac{\dfrac{b}{c}}{\dfrac{a}{c}}$$

Then $$\frac{\sin \theta}{\cos \theta} = \frac{b}{a}$$

Because $$\frac{b}{a} = \tan \theta$$

$$\tan \theta = \frac{\sin \theta}{\cos \theta}$$

How do we evaluate sin, cos and tan if θ is not a special angle? Table A-5 of the Appendix lists approximate values for acute angle trigonometric functions. Because of Theorem 7-1, the table lists values from 1° to 45° in the lefthand column, labeled *degree*. The values in the far righthand column are the complements of the values in the left column. The right column values run from 45° to 89°. The table structure demonstrates the complementary relationship. Actual approximation of a trigonometric function is best done with a calculator.

EXAMPLE 7-4

Illustration 1:

Approximating Trigonometric Functions. Approximate each of the following.

sin 27°

Solution 1: See Table A-5. Locate 27° in the lefthand column. Locate value under column with heading of sin.

sin θ ≅ 0.4540

Solution 2: Be sure the calculator is in degree mode. For a scientific calculator, enter 27 then press [SIN]. To use a graphing calculator press [SIN] 27 [ENTER]

sin 27° ≅ 0.453990499

Illustration 2:

sin 63°

Solution 1: See Table A-5. Locate 63° in the righthand column. Because this is the complement column, look for the function value in the column labeled sin at the *bottom*.

sin 63° ≅ 0.8910

Solution 2: By calculator, sin 63° ≅ 0.891006524.

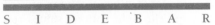 Illustration 3:   cos 27°

Solution: By calculator, cos 27° ≅ 0.891006524
Note: Compare this solution to Illustration 2. Can you explain why they are equal?

Illustration 4:   sin 0.21   (no degree symbol, assume radian measure)

Solution: sin 0.21 ≅ 0.208459899. Be sure your calculator is in radian mode.

Illustration 5:   tan 27°

Solution: Although Table A-5 contains values for tan, alternately by applying Theorem 7-2 we have:

$$\tan \theta = \frac{\sin \theta}{\cos \theta}$$

From Illustrations 1 and 3, get sin 27° and cos 27°.

$$\tan 27° \cong \frac{0.453990499}{0.891006524}$$

$$\tan 27° \cong 0.509525449$$

Note 1: Because 0.453990499/0.891006524 *demands* a calculator, most scientific calculators have a ⌜TAN⌝ key. Enter ⌜TAN⌝ 27 to duplicate the previous results.

Note 2: By calculator tan 63° ≅ 1.962610505.

$$\text{Also } \frac{1}{\tan 63°} \cong 0.509525449$$

$$\text{Then } \tan 27° \cong \frac{1}{\tan 63°}$$

$$\text{In general, } \tan \theta = \frac{1}{\tan(90° - \theta)}$$

S I D E B A R

Tangent is from the Latin *tangere* meaning to touch. A tangent to a circle is a line that intersects the circle in exactly one point.

  Secant is from the Latin *secare* meaning to cut. As a saw first touches a log its edge is tangent to the log's circumference. As the blade cuts into the log, the cutting edge becomes a secant line to the circumference. A secant line to a circle intersects the circle in exactly two points.

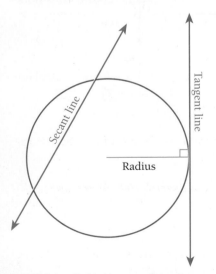

DISPLAY 7-6   Circle with tangent line and secant line

EXAMPLE 7-5

Illustration 1:

S I D E B A R

Because of the usefulness of the Pythagorean relation, the squares of sin θ and cos θ occur frequently. Writing numerous pairs of parentheses can become tedious, so we often use a special abbreviation:

$\sin^2\theta = (\sin\theta)^2$ and $\cos^2\theta = (\cos\theta)^2$.

In general, the same scheme abbreviates positive integer powers of trigonometric functions:

$$\text{TRIG}^n\theta = (\text{TRIG}\,\theta)^n.$$

Be careful with the notation TRIG θ^n. To avoid ambiguity, write the expression as TRIG(θ^n) or (TRIG θ^n). Many people are tempted to interpret TRIG θ^n as TRIG(θ^n). *GraphWindows* interprets the expression as (TRIG θ)^n. In general, graphing calculators will not handle the notation TRIG$^n\theta$.

Solving Right Triangles.   Solve as indicated.

See Figure 7-29. Consider a circle with radius of 1 containing an acute angle θ with vertex at the center of the circle. The angle intersects the circle at points p and q. At point p a tangent to the circle is drawn. Because θ is acute, the side of the angle through point q must intersect the tangent line at point x.

Determine the length of px. To solve this problem, recall from geometry that a tangent to a circle is perpendicular to a radius to the point of tangency The triangle formed is a right triangle. By definition, $\tan\theta = px/1$. Therefore, $px = \tan\theta$.

Note: Figure 7-29 justifies the name tangent for the tan θ function by relating tan θ to a tangent line to a circle. Examine Figure 7-30. A similar argument verifies that the sides of the right triangle are sin θ and cos θ as labeled. From the Pythagorean theorem,

$$(\sin\theta)^2 + (\cos\theta)^2 = 1$$

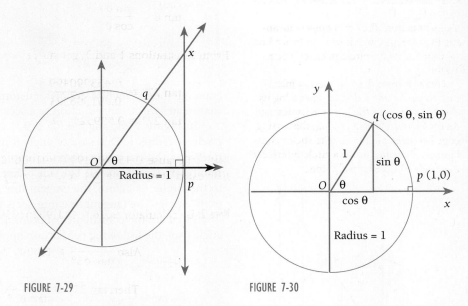

FIGURE 7-29   FIGURE 7-30

Illustration 2:

Refer to Figure 7-29. If $\theta = 27°$ determine the length of px. Also determine the coordinates of point q in Figure 7-20.

Solution: From Illustration 1, $px = \tan 27°$. From Example 7-4, Illustration 4,

$$\tan 27° \cong 0.509525449$$
$$\text{Then}\quad px \cong 0.509525449$$

From the note to Illustration 1, the lengths of the sides of the triangle are cos 27° and sin 27°. From Example 7-4, the coordinates are

$$(\cos 27°, \sin 27°) \cong (0.8910, 0.4540)$$

The note to Illustration 1 generalizes to the **Pythagorean relation**, between sin and cos. The proof is left as an exercise.

Theorem 7-3
The Pythagorean Relation

$$\sin^2 \theta + \cos^2 \theta = 1, \qquad 0° < \theta < 90°, \quad \{0 < \theta < \pi/2\}$$

Actually, there are six possible ratios of sides in a right triangle: sin θ, cos θ and tan θ are three fundamental relations; the other three are reciprocals of these original three. We discuss these in detail in following sections. First we define the reciprocal functions in terms of sine and cosine. See whether you can relate each of these new functions to ratios of sides in a right triangle.

DEFINITION 7-2
Reciprocal Trigonometric Functions

cosecant	$\csc \theta = \dfrac{1}{\sin \theta},$	$\sin \theta \neq 0$	
secant	$\sec \theta = \dfrac{1}{\cos \theta},$	$\cos \theta \neq 0$	
cotangent	$\cot \theta = \dfrac{\cos \theta}{\sin \theta},$	$\sin \theta \neq 0$ (sometimes denoted ctn θ)	

Naming these function follows custom. A geometric interpretation follows that relates the secant function to a secant line of a circle.

In Figure 7-31, angle θ, 0° < θ < 90°, has its vertex at the center of the circle of radius 1. One side of angle θ intersects the circle at *p*; the other side intersects the circle at *x*. From Example 7-5, Illustration 1, the coordinates of point *p* are (cos θ, sin θ). The side of angle θ defines a line *j* that is a secant line to the circle. As before, the tangent line at point *x* is perpendicular to one side of the angle. The tangent line intersects the secant line at point *q*.

What are the coordinates of point q? Because triangle *OPA* is similar to triangle *Oqx*, the following proportion applies.

$$\frac{\sin \theta}{\cos \theta} = \frac{qx}{1}$$

$$\tan \theta = qx \qquad \text{(as before)}$$

The coordinates of *q* are (1, tan θ).

$$\text{Also, } Oq = \frac{1}{\cos \theta}$$

$$Oq = \sec \theta \qquad \text{(by definition)}$$

The secant function describes the length of the segment *Oq* that lies within the secant line. Geometry justifies the name for the secant function.

As you might expect, the remaining two functions are complementary to the secant and the tangent; hence, the names cosecant and cotangent. The following theorem fleshes out the cofunction relationships.

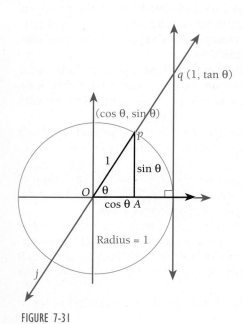

FIGURE 7-31

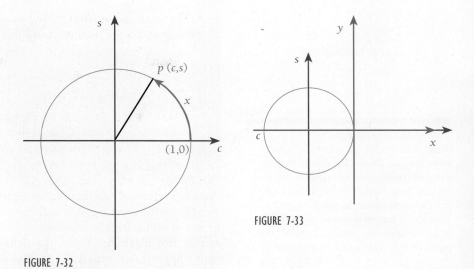

FIGURE 7-33

FIGURE 7-32

**Theorem 7-4
Complementary Relations**

Suppose $0° < \theta < 90°$, $\{0 < \theta < \pi/2\}$, then

$$\csc \theta = \sec(90° - \theta), \qquad \{\csc \theta = \sec(\pi/2 - \theta)\}$$
$$\cot \theta = \tan(90° - \theta), \qquad \{\cot \theta = \tan(\pi/2 - \theta)\}$$

The proof is left as an exercise.

Most scientific calculators do not have function keys for the reciprocal functions. It is easy to compose the fundamental functions with the $\boxed{^1/_x}$ key to evaluate the reciprocal functions. For example, consider $\csc 27°$. Enter 27 then press $\boxed{\text{SIN}}$ $\boxed{^1/_x}$ to obtain $\csc 27° \cong 2.202689264$.

Now we extend the concept of trigonometric functions to include angles of rotation. Our extension will be based on the unit circle using radian measure for the rotation. To emphasize this we call the resulting functions **circular functions.**

Imagine a circle of radius 1 with center at $(0, 0)$. Label the horizontal axis c and the vertical axis s. Then the equation for the circle is $c^2 + s^2 = 1$. See Figure 7-32.

For clarity, superimpose an xy axis over the cs axis so that the origin of the xy system is at $(1, 0)$ on the cs system. See Figure 7-33. Our goal is to define y as a function of x by referencing the cs coordinate system.

Consider a positive real number x. The segment from 0 to x on the x-axis has a length of $|x|$. On the unit circle, measure an arc of $|x|$ radians counterclockwise from $(1, 0)$. Such an arc is in standard position. Because of the unit circle, the arc length equals the radian measure. Label the end point of the arc p.

Let the coordinates of point p be (c, s). This discussion leads to the following definition.

DEFINITION 7-3
Circular Functions Sine and Cosine

Refer to Figure 7-33. Suppose that x is any real number. Then the sine and cosine are functions of x defined as follows:

sine: $\sin x = s$

cosine: $\cos x = c$

where (c, s) is the point on the circle $c^2 + s^2 = 1$ located $|x|$ radians around the circumference of the circle from $(1, 0)$. If $x > 0$, the arc from $(1, 0)$ to (c, s) is measured counterclockwise. If $x < 0$, the distance is measured clockwise.

S   I   D   E   B   A   R

Some texts prefer "wrapping x." Like a wheelwright, fix the left end of the segment of length x at the origin of the xy system, then "wrap" the segment counterclockwise about the circle from 0 if x is positive. If x were negative, we would wrap the segment clockwise. Name the point of the circle where x terminates p.

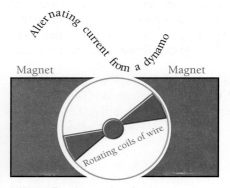

DISPLAY 7-7   Cross section of a dynamo

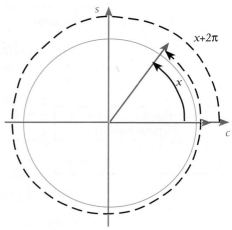

FIGURE 7-34

The definitions for sine and cosine produce range components by construction rather than by algebraic calculation: circular functions are **transcendental**, not algebraic. From the definition, several properties of sine and cosine become apparent. Later we show relationships between sine and cosine that indicate common properties. First consider sine.

Real numbers represent directed distances on a number line. We defined clockwise and counterclockwise rotation to correspond respectively to negative and positive directions. The absolute value of a real number measures arc length, which on a unit circle is equivalent to radian measure. Combine direction with arc length and the resulting domain of sine is all real numbers. Because (c, s) is on the unit circle, $-1 \leq s \leq 1$. The range of sine is $[-1, 1]$.

In 1831 Michael Farraday demonstrated that a conductor moving in a magnetic field produced electric current. Today, rotating complex coils of wire within a magnetic field is the primary method of generating the electrical needs of the world. See Display 7-7. As the coil turns within the magnetic field, the flow of current increases, peaks, decreases and then reverses the direction of flow. The pattern of generation is called *alternating current*. The circular rotation produces a current flow in a wavelike form. The power produced at any instant is directly related to the angle of rotation within the magnetic field. Sine is also connected to rotation, so sine might provide a suitable model for the generation of **wave forms**, such as alternating current. Next we consider the graph of sine to firm up this conjecture.

One interesting property of sine is its **periodic nature**. A circle of radius 1 has a circumference of 2π. Beginning at $(1, 0)$ on the unit circle, any arc of length x terminates at the same point as $x + 2\pi$. See Figure 7-34. Therefore, for any real value x.

$$\sin x = s$$
$$= \sin(x + 2\pi)$$

Sine is periodic with a fundamental period of 2π.

Few values of sine are obvious from the definition. The point where x terminates on the x-axis or y-axis is easy. The values of x that terminate on an axis are quadrantal values. See Figure 7-35. Because of periodicity, we need only examine values of x on $[0, 2\pi]$.

From Figure 7-35 the quadrantal values of sine and cosine are apparent:

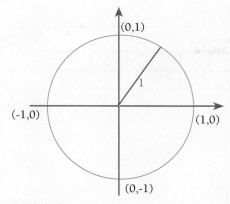

FIGURE 7-35

x	0	$\dfrac{\pi}{2}$	π	$\dfrac{3\pi}{2}$	2π
$\sin x$	0	1	0	-1	0
$\cos x$	1	0	-1	0	1

Note the expected repetition of values for $x = 0$ and $x = 2\pi$.

Consider a graphing window with x in $[0, 2\pi]$ and y in $[-1, 1]$. Because of periodicity, once the graph of sine is known in this window, the rest of the graph is repetitive (paste copies of the window in place as needed). Graphing technology can supply us with the graph of sine. However, this first time, we determine the shape within the window from plotting sample points directly from the unit circle.

Compare Figure 7-35 to Figure 7-32. Notice that the quadrantal values are critical, providing x-intercepts and local maximum and minimum values of the graph. Choose domain values within the window and form the corresponding unit circle for reference. The value s represents the directed length of a vertical line segment from the x-axis. Use s as the second coordinate with x in the window. Repeat the process for $x_1, x_2, x_3, \ldots, x_n$. Note that, when x exceeds π, the values for s are negative and the graph of sine is below the x-axis. See Figure 7-36.

A complete cycle of sine appears in the window in Figure 7-37. Memorize the characteristic sinusoidal shape of the graph. The shape is snakelike or more appropriately *sinuous*. Check your dictionary. Ironically, the name *sine*

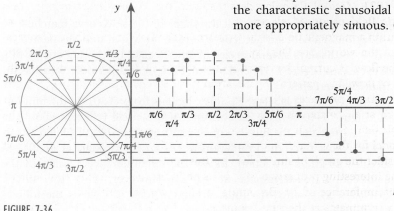

FIGURE 7-36

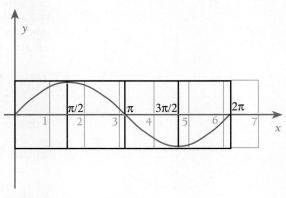

FIGURE 7-37

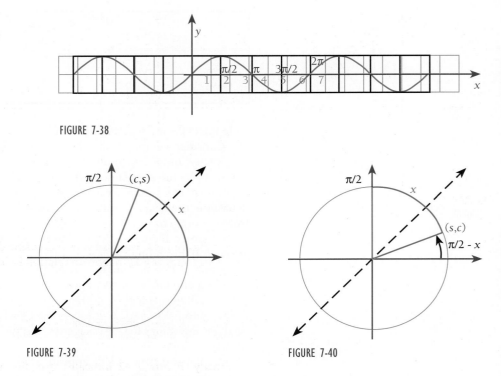

FIGURE 7-38

FIGURE 7-39   FIGURE 7-40

first appeared as a mistranslation, but the shape of the sine curve matches the
Latin description of a bent surface, curve or fold, for which *sine* was named.

Figure 7-38 displays several cycles of sine formed by pasting copies of a
complete cycle (see Figure 7-37) end to end. Because it is periodic, sine has infinitely many x-intercepts.

In Figure 7-39, (c, s) labels the point x radians from $(1, 0)$ on the unit circle, $x > 0$. Figure 7-40 reveals the converse of the circle reflected about the
line $I(x) = x$. Notice that the measurement of x now starts at $(0, 1)$ and terminates at (s, c). Now trace the evaluation of $\cos(\pi/2 - x)$: start at $(1, 0)$, go
to $\pi/2$. Because of the subtraction, follow the circumference x units in the
opposite direction to terminate at (s, c). However, s is the $\sin(x)$.
Hence $\cos(\pi/2 - x) = \sin x$. A similar argument for $x < 0$ yields the following relations.

Theorem 7-5
Complementary Relation

$$\cos\left(\frac{\pi}{2} - x\right) = \sin x$$

$$\sin\left(\frac{\pi}{2} - x\right) = \cos x \qquad \text{for all real } x$$

Because of this *complementary* relationship between sine and cosine, the
graph of cosine follows immediately. The graph of cosine is the graph of sine
reflected about the y-axis and translated $\pi/2$ units to the right.

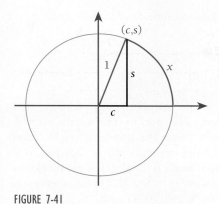

FIGURE 7-41

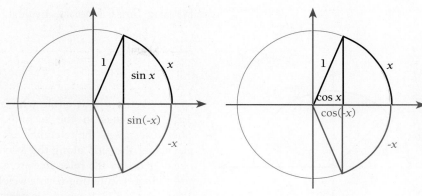

FIGURE 7-42

See Figure 7-41. Because sine and cosine depend upon the unit circle $c^2 + s^2 = 1$, the following Pythagorean relation holds.

Theorem 7-6
Pythagorean Relation

$$\cos^2 x + \sin^2 x = 1, \qquad \text{for all real } x$$

Finally, Figure 7-42 indicates that sine is an **odd** function because $\sin(-x) = -\sin x$, whereas cosine is an **even** function.

Theorem 7-7
Sine Odd
Cosine Even

Suppose x is a real number, then

$$\sin(-x) = -\sin x \qquad \text{(sine is odd)}$$
$$\cos(-x) = \cos x \qquad \text{(cosine is even)}$$

We used the names *sine* (and *cosine*) to represent both circular functions that use a real number argument x and trigonometric functions that use an angle measure argument θ. Because we applied the same name for both functions, you should suspect the two are related. All that remains is to specify the connection.

Consider Figure 7-43. To determine sin x and cos x, where x is not a quadrantal value, note that x terminates at $P(c, s)$. From P drop a line perpendicular to the c-axis. From P draw a segment to the center of the circle. The c-axis and the two segments form a right triangle. The lengths of the sides of the right triangle correspond to $|c|$ and $|s|$.

The absolute value of the shortest arc from P to the x-axis is the **reference arc** for x. Because the reference arc subtends and measures the acute central angle formed by the right triangle, denote the reference arc by θ. The hypotenuse of the triangle is 1. Once we determine sides $|c|$ and $|s|$ of the right triangle with static angle θ, evaluating cos x and sin x requires only providing the correct sign.

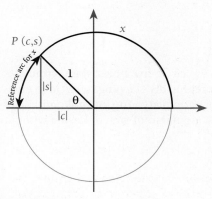

FIGURE 7-43

EXAMPLE 7-6

Illustration 1:

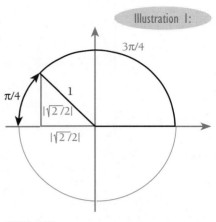

FIGURE 7-44

Illustration 2:

Illustration 3:

Evaluating Circular Functions. Evaluate the following functions as indicated.

$$\cos \frac{3\pi}{4}$$

Solution: Sketch a unit circle with a $3\pi/4$ arc in standard position. Note that $3\pi/4$ terminates in the second quadrant. The arc to the x-axis is $\pi/4$. Therefore $\pi/4$ is the reference arc for $3\pi/4$. Form the associated right triangle. Label the side along the x-axis c and the vertical side s. Clearly c is a negative. From the reference arc, the central angle θ for the triangle measures $\pi/4$ radian. Evaluating cosine we obtain $\cos(\pi/4) = \sqrt{2}/2$. The hypotenuse is 1, so $|c| = \sqrt{2}/2$. Because c is negative, $c = -\sqrt{2}/2$. As a result,

$$\cos \frac{3\pi}{4} = -\frac{\sqrt{2}}{2}$$

$$\sin \frac{5\pi}{6}$$

Solution: Arc $5\pi/6$ terminates in the second quadrant with a reference arc of $\pi/6$. Refer to side s. In the second quadrant s is positive. Because $\sin(\pi/6) = \frac{1}{2}$, we have

$$\sin \frac{5\pi}{6} = \frac{1}{2}$$

$$\cos \frac{43\pi}{4}$$

Solution: Because of periodicity, cosine values repeat every 2π. Rewrite the argument as a mixed fraction to determine the number of complete rotations around the circle.

$$\cos \frac{43\pi}{4} = \cos\left(10\pi + \frac{3\pi}{4}\right)$$

$$= \cos\left[5(2\pi) + \frac{3\pi}{4}\right]$$

See Figure 7-45. The argument "wraps" the circle five times, then terminates at the same position as $3\pi/4$. Rather than repeatedly "lapping" a circle to envision the terminal point, take advantage of periodicity. Any integral multiple of 2π may be added or subtracted from the argument of sine or cosine without affecting the value of the function:

$$\cos \frac{43\pi}{4} = \cos \frac{3\pi}{4}$$

From Illustration 1, $\cos \dfrac{43\pi}{4} = -\dfrac{\sqrt{2}}{2}$

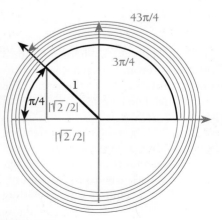

FIGURE 7-45

Illustration 4:

$$\sin \frac{199\pi}{6}$$

Solution 1:

$$\sin \frac{199\pi}{6} = \sin\left(33\pi + \frac{\pi}{6}\right) \qquad \text{(by division)}$$

$$= \sin\left(32\pi + \pi + \frac{\pi}{6}\right)$$

$$= \sin\left(\pi + \frac{\pi}{6}\right) \qquad \text{(by periodicity)}$$

Now $\pi + \pi/6$ terminates in the third quadrant. The reference arc is $\pi/6$. $\operatorname{Sin}(\pi/6) = \frac{1}{2}$. The value for s in quadrant III is negative. Therefore,

$$\sin(199\pi/6) = -\frac{1}{2}.$$

Note: Some scientific calculators cannot process arguments as large as $199\pi/6$ or $43\pi/4$. In these cases you should reduce the argument to a value between 0 and 2π yourself. On a scientific calculator this is easy:

1. Enter the argument (in radian mode).
2. Divide by 2π (number of revolutions).
3. Subtract the whole number portion.
4. Multiply by 2π (which produces remainder).
5. Press the appropriate function key.

Solution 2: By calculator (be sure you are in radian mode):

1. Enter: [1] [9] [9] [×] [π] [÷] [6] [=] (the argument).
2. Enter: [÷] [2] [÷] [π] [=] (divide by 2π).
3. Display: 16,58333333 (revolutions).
   Enter: [−] [1] [6] [=] (complete revolution).
4. Display 0.58333333 (partial revolution).
   Enter: [×] [2] [×] [π] [=] (back to radians).
5. Display 3.66519142 (execute function).
   Press [SIN].
   Display −0.4999999992.

Note the accumulated round-off error in the answer. Without the reduction of the argument, one calculator displayed E for error! Graphing calculators often handle large arguments better, but they are also subject to round-off errors.

Illustration 5:

$$\cos \frac{43\pi}{4} \qquad \text{(by graphing calculator)}$$

Solution: Press [COS] [(] [4] [3] [π] [÷] [4] [)] [ENTER].

Compare the answer to Illustration 1: $-\sqrt{2}/2 \cong -0.707106781$.

$$\cos \frac{43\pi}{4} = \cos \frac{3\pi}{4} \qquad \text{(adjust for periodicity)}$$

$$\cong -0.707106781 \qquad \text{(by calculator)}$$

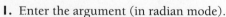

FIGURE 7-46

The previous example establishes a pattern for evaluating circular functions by using reference arcs. Be careful using a calculator for evaluating trigonometric functions or circular functions. Most calculators have at least two modes, one for degree measure and one for radian measure. If you are in degree measure and enter $\pi/4 \cong 0.785398163$, your calculator will assume 0.785398163 degrees rather than 0.785398163 radians. Because the domain of the circular functions $\sin x$ and $\cos x$ is the set of real numbers, you *must* be in radian mode to evaluate these functions. For example, $\cos 0.785398163° \cong 0.999906049$, but $\cos 0.785398163 \cong 0.707106781$.

Because we duplicated the names sine, cosine, and so on, in trigonometric and circular functions, the context determines whether arguments are in degrees or radians. From this point forward, we use **Greek letters** such as θ to indicate an argument in degree measure. Letters such as x will indicate a real variable argument in radian measure. For an angle of rotation measured in θ degrees and a real variable x, the relation of *trigonometric* sine to *circular* sine is given by $\sin(\pi\theta/180°) = \sin x$. The domain of a circular function makes it a better model for most applications requiring time (or some other real number measure) as an argument. Trigonometric functions are useful where triangles model the problem.

- If two right triangles have equal acute angles, the triangles are similar and the sides are in proportion. As a result, in a right triangle, the ratios of pairs of sides are uniquely connected to the measure of an acute angle. The second acute angle is the complement of the first. See Figure 7-47.

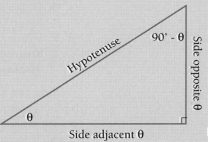

FIGURE 7-47

- Special right triangles have special ratios:

$$\frac{\text{Side opposite } 30°}{\text{hypotenuse}} = \frac{\sqrt{1}}{2} = \frac{1}{2}$$

$$\frac{\text{Side opposite } 45°}{\text{hypotenuse}} = \frac{\sqrt{2}}{2}$$

$$\frac{\text{Side opposite } 60°}{\text{hypotenuse}} = \frac{\sqrt{3}}{2}$$

• Let θ be the measure of an acute angle of a right triangle, $0° < \theta < 90°$, then each of the following is a function:

$$\text{sine} \qquad \sin \theta = \frac{\text{side opposite } \theta}{\text{hypotenuse}}$$

$$\text{cosine} \qquad \cos \theta = \frac{\text{side adjacent } \theta}{\text{hypotenuse}}$$

$$\text{tangent} \qquad \tan \theta = \frac{\text{side opposite } \theta}{\text{side adjacent } \theta}$$

Fundamental Function	Cofunction
$\sin \theta$	$\cos \theta$
$\cos \theta$	$\sin \theta$
$\tan \theta$	(not defined yet)

• To evaluate $\sin x$ or $\cos x$ where x has one of the following reference arcs:

REFERENCE ARC	SIN X	COS X	REFERENCE ANGLE
$\dfrac{\pi}{6}$	$\dfrac{1}{2}$	$\dfrac{\sqrt{3}}{2}$	30°
$\dfrac{\pi}{4}$	$\dfrac{\sqrt{2}}{2}$	$\dfrac{\sqrt{2}}{2}$	45°
$\dfrac{\pi}{3}$	$\dfrac{\sqrt{3}}{2}$	$\dfrac{1}{2}$	60°

I. Determine the quadrant where x terminates. Add or subtract integral multiples of 2π until the argument is between 0 and 2π.

QUADRANT	INTERVAL	SIGN OF SINE	SIGN OF COSINE
I	$\left(0, \dfrac{\pi}{2}\right)$	+	+
II	$\left(\dfrac{\pi}{2}, \pi\right)$	+	−
III	$\left(\pi, \dfrac{3\pi}{2}\right)$	−	−
IV	$\left(\dfrac{3\pi}{2}, 2\pi\right)$	−	+

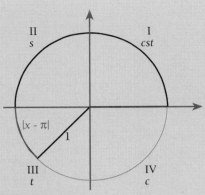

FIGURE 7-48

2. Let x represent the argument (arg) of a trigonometric function. If arg is not between 0 and $\pi/2$, subtract integer multiples of π from arg until $|x - n\pi| < \pi/2$. Now $|x - n\pi|$ is the reference arc. If the reference arc is in the reference arc table, use the exact value from the table and the sign from the sign table to write the value of the function. A calculator approximates the values.

7-2   E X E R C I S E S

Determine the size of the side labeled x in each of the pairs of similar triangles in Problems 1–10.

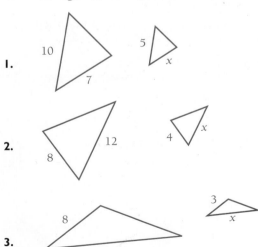

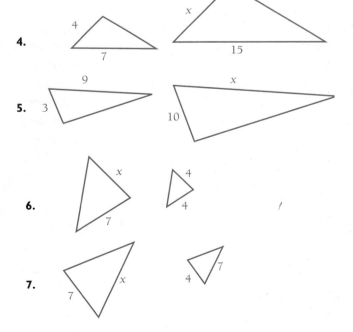

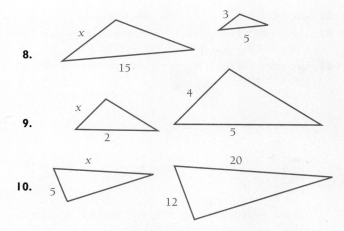

8.

9.

10.

45. $\cos 50°$

47. $\sin 50°$

49. $\tan 50°$

51. $\sin 37°$

53. $\cos 7°$

*55. $\sin 27° \, 30'$

*57. $\cos 51° \, 20' \, 10''$

*59. $\tan 73° \, 45''$

*61. $\sin 1° \, 1' \, 1''$

*63. $\dfrac{\cos 45°}{\sin 45°}$

46. $\cos 40°$

48. $\sin 40°$

50. $\tan 40°$

52. $\sin 53°$

54. $\cos 86°$

*56. $\sin 69° \, 20'$

*58. $\cos 16° \, 40' \, 30''$

*60. $\tan 21° \, 15''$

*62. $\cos 89° \, 59' \, 59''$

*64. $\dfrac{\cos 45°}{\sin 45°}$

Refer to the 30°-60° right triangle and 45° right triangle shown in Figure 7-49 to solve for the indicated side.

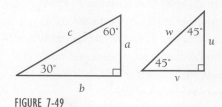

FIGURE 7-49

11. If $a = 5$, determine b.
12. If $a = 5$, determine c.
13. If $u = 7$, determine v.
14. If $v = 6$, determine u.
15. If $a = 6$, determine c.
16. If $a = 6$, determine b.
17. If $c = 10$, determine a.
18. If $c = 10$, determine b.
19. If $c = 8$, determine b.
20. If $c = 8$, determine b.
21. If $b = 4$, determine a.
22. If $b = 4$, determine c.
23. If $b = 6$, determine c.
24. If $b = 6$, determine a.
25. If $v = 5$, determine u.
26. If $u = 8$, determine v.
27. If $w = 10$, determine u.
28. If $w = 10$, determine v.
29. If $u = 20$, determine w.
30. If $v = 20$, determine w.

Give an exact value for the following trigonometric functions.

31. $\sin 30°$
32. $\cos 30°$
33. $\cos 60°$
34. $\sin 60°$
35. $\sin 45°$
36. $\cos 45°$
37. $\cos^2 30° + \sin^2 30°$
38. $\sin^2 60° + \cos^2 60°$
39. $2 \sin (30°) \cos (30°)$
40. $1 - \cos^2 45°$

Approximate the following.

41. $\sin 15°$
42. $\cos 15°$
43. $\cos 75°$
44. $\sin 75°$

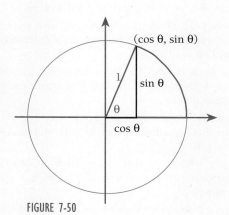

FIGURE 7-50

Refer to Figure 7-50 for the following.

65. Prove Theorem 7-5.

66. Prove that $\sin^2 \theta + \cos^2 \theta = 1$.

Evaluate each of the following circular functions. Give an exact value when feasible, otherwise use an approximation.

67. $\sin \pi$
68. $\cos \dfrac{\pi}{2}$
69. $\tan \dfrac{3\pi}{2}$
70. $\sin 2\pi$
71. $\cos 5\pi$
72. $\tan 6\pi$
73. $\sin \dfrac{\pi}{4}$
74. $\cos \dfrac{\pi}{3}$
75. $\tan \dfrac{5\pi}{3}$
76. $\sin \dfrac{5\pi}{4}$
77. $\cos \dfrac{\pi}{6}$
78. $\tan \dfrac{\pi}{12}$

79. $\sin \dfrac{7\pi}{12}$

80. $\cos \dfrac{7\pi}{6}$

81. $\tan \dfrac{11\pi}{6}$

82. $\sin \dfrac{13\pi}{12}$

83. $\cos 1.5$

84. $\tan 0.5$

85. $\sin \ 1.0$

86. $\cos \ 2.0$

87. Use graphs to show that $1 - x^2/2$ approximates $\cos x$ for $0 < x < \pi/6$.

88. Use graphs to show that $x - x^3/6$ approximates $\sin x$ for $0 < x < \pi/6$.

7-2   PROBLEM SET

1. See Figure 7-51. Determine the first quadrant point of intersection of $x^2 + y^2 = 1$ and $(x - 1)^2 + y^2 = 1$.

2. See Figure 7-51. Show that the triangle with vertices at $(0, 0)$, $(1, 0)$ and the first quadrant point of intersection (see Problem 1) is equilateral.

3. See Problems 1 and 2. Show that the arc in circle $x^2 + y^2 = 1$ from $(1, 0)$ to the first quadrant point of intersection measures $\pi/3$.

4. See Problem 3. Show that the arc in circle $x^2 + y^2 = 1$ from the first quadrant point of intersection to $(0, 1)$ measures $\pi/6$.

5. See Figure 7-52. Show that the arc from $(1, 0)$ to the first quadrant point of intersection of $x^2 + y^2 = 1$ and $y = x$ measures $\pi/4$.

6. See Figure 7-52. Solve the system $y = x$ and $x^2 + y^2 = 1$ to determine the coordinates of the first quadrant point of intersection.

7. Refer to Figures 7-52 and Figure 7-51. Also refer to Problem 1. Determine the first quadrant coordinates

of the endpoint of a $\pi/6$ arc starting at $(1, 0)$. (Hint: use symmetry about $y = x$.)

8. Refer to Problem 1, determine $\sin (\pi/3)$.

9. Refer to Problem 1, determine $\cos (\pi/3)$.

10. Refer to Problem 6, determine $\cos (\pi/4)$.

11. Refer to Problem 6, determine $\sin (\pi/4)$.

12. Refer to Problem 7, determine $\sin (\pi/6)$.

13. Refer to Problem 7, determine $\cos (\pi/6)$.

14. Compare Problems 8 and 13.

15. Compare Problems 9 and 12.

16. Compare Problems 10 and 11.

EXPLORATION

The following problems anticipate concepts from the next section. Graph each pair of functions in the same window. Indicate a relationship between the pair of functions.

17. $f(x) = \sin x$, $g(x) = \cos\left(\dfrac{\pi}{2} - x\right)$

18. $f(x) = \cos x$, $g(x) = \sin\left(\dfrac{\pi}{2} - x\right)$

19. $f(x) = \tan x$ $g(x) = \sin x/\cos x$

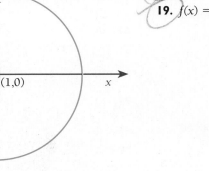

FIGURE 7-51   Two unit circles, one centered at $(0, 0)$ the other at $(1, 0)$.
$x^2 + y^2 = 1$ and $(x - 1)^2 + y^2 = 1$

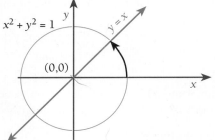

FIGURE 7-52   Unit circle $x^2 + y^2 = 1$ and $y = x$

20. $f(x) = \tan\left(\dfrac{\pi}{2} - x\right),$ $g(x) = 1/\tan x$

21. $f(x) = \sin^2 x + \cos^2 x,$ $g(x) = 1$

22. $f(x) = 1 - \sin^2 x,$ $g(x) = \cos^2 x$

23. $f(x) = 1 - \cos^2 x,$ $g(x) = \sin^2 x$

24. $f(x) = (\cos x)(\tan x),$ $g(x) = \sin x$

25. $f(x) = \sin x,$ $g(x) = x$

26. $f(x) = \cos x,$ $g(x) = -x$

27. Problems 17–24 demonstrate a relationship called an *identity*. Describe the identity relationship.

28. See Problem 27. Problems 25 and 26 do not demonstrate the identity relation. Describe how Problems 25 and 26 differ from 17–24.

3   ELEMENTARY IDENTITIES

Mathematics in general is fundamentally the science of self-evident things.
—FELIX KLEIN

All equations are not merely equal. Some are more equal than others. A conditional equation relates two expressions, at least one of which contains a variable. A conditional equation may be true or false when values are substituted for the variable(s). The following are conditional equations.

$$3x + 2 = 5x - 8$$ (true for $x = 5$, false otherwise)

$$x^2 + 1 = x + 3$$ (true only for $x = 2$ or $x = -1$)

An identity is also an equation that relates two expressions. However, an identity is true for the substitution of any value(s) from the common domain of the two expressions. To emphasize this more general equality, identities use $\equiv$ for equals. The following are identities.

$$3x + 7x \equiv 10x$$

$$\sin(90° - x) \equiv \cos x$$

Definitions provide one source for identities. For example, identities using sine and cosine are the building blocks of other circular functions.

DEFINITION 7-4
Circular Functions

tangent:	$\tan x \equiv \dfrac{\sin x}{\cos x},$	$\cos x \neq 0$
cotangent:	$\cot x \equiv \dfrac{\cos x}{\sin x},$	$\sin x \neq 0$
secant:	$\sec x \equiv \dfrac{1}{\cos x},$	$\cos x \neq 0$
cosecant:	$\csc x \equiv \dfrac{1}{\sin x},$	$\sin x \neq 0$

Values for these functions follow immediately from the definition. In the following table the values for tangent, cotangent, secant and cosecant come from the values of sine and cosine (*indicates values by calculator: (SIN), (COS), (TAN), (÷)).

x	$\sin x$	$\cos x$	$\tan x$	$\cot x$	$\sec x$	$\csc x$
$\dfrac{\pi}{6}$	$\dfrac{1}{2}$	$\dfrac{\sqrt{3}}{2}$	$\dfrac{1}{\sqrt{3}}$	$\sqrt{3}$	$\dfrac{2}{\sqrt{3}}$	2
$\dfrac{2\pi}{3}$	$\dfrac{\sqrt{3}}{2}$	$-\dfrac{1}{2}$	$-\sqrt{3}$	$-\dfrac{1}{\sqrt{3}}$	-2	$\dfrac{2}{\sqrt{3}}$
π	0	-1	0	Undefined	-1	Undefined
$\dfrac{5\pi}{4}$	$-\dfrac{\sqrt{2}}{2}$	$-\dfrac{\sqrt{2}}{2}$	1	1	$-\sqrt{2}$	$-\sqrt{2}$
$\dfrac{3\pi}{2}$	-1	0	Undefined	0	Undefined	-1
*4	-0.7568	-0.6535	1.1578	0.8637	-1.529	-1.3213

Established identities are useful in the simplification of expressions. Because trigonometric expressions can be quite complex, knowledge of identities is invaluable in simplifying trigonometric expressions.

One view of an identity is as an equation in which the lefthand expression produces exactly the same values as the righthand expression. From a function viewpoint, if $f(x) \equiv g(x)$ is an identity then f and g are the same functions with coincident graphs at all points of their common domain. Even though the rule of pairings for f and g may look different, the graphs are identical at common domain values. On the other hand, the conditional equation, $f(x) = g(x)$ indicates the points of intersection of the independent graphs of f and g. See Figure 7-53.

From an algebraic point of view, identities provide us with an alternative representation of an expression. You may be confused at this point. If you wonder how we recognize whether an equation is conditional or an identity, then you have pinpointed a major difficulty. Identities are not always immediately obvious.

Before rewriting an expression by use of an identity, it is important to verify that the substituted expression is indeed identical to the original expression. In the case of trigonometric functions, there are countless identities relating trigonometric expressions. Rather than attempt to memorize hundreds of such identities for future use, a better method is to memorize a few identities and develop a strategy to verify other possible identities as the need arises.

Verifying an identity is a formalized method of proving that two expressions represent the same function for common domain values. Verifying an identity is *not* like solving a conditional equation. One way to verify that two

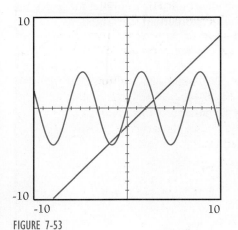

FIGURE 7-53

expressions are identical is to derive one expression from the other. Consider the following derivation in which $1 + \tan^2 \theta$ is transformed into $\sec^2 \theta$.

$$1 + \tan^2 \theta \equiv 1 + \tan^2 \theta \qquad \text{(reflexive property)}$$

$$\equiv 1 + \frac{\sin^2 \theta}{\cos^2 \theta} \qquad \text{(definition)}$$

$$\equiv \frac{\cos^2 \theta}{\cos^2 \theta} + \frac{\sin^2 \theta}{\cos^2 \theta} \qquad \text{(common denominator)}$$

$$\equiv \frac{\cos^2 \theta + \sin^2 \theta}{\cos^2 \theta}$$

$$\equiv \frac{1}{\cos^2 \theta} \qquad \text{(Pythagorean relation)}$$

$$\equiv \sec^2 \theta \qquad \text{(definition of secant)}$$

$$1 + \tan^2 \theta \equiv \sec^2 \theta$$

Strategy to Verify an Identity

> Problem: verify $f(x) \equiv g(x)$.
> Begin with $f(x) \,||\, g(x)$.
>
> Goal: reduce each side separately to a common expression, then use this scratch work to prove the identity by writing it as a derivation.
>
> **1.** Do not add, subtract, multiply or divide on both sides of the equation. Work with each side separately. Remember the transformations are from top to bottom, not left to right. Because of this it may be safer to work on one side of the identity only. Working with one side only more closely resembles the steps of a derivation. Choose the most complex side and simplify it. (Hints: to avoid addition or subtraction to both sides, add $h(x) - h(x)$ to one side of the expression only. To avoid multiplication or division on both sides, multiply one side only by $h(x)/h(x)$. Be sure to note the extended restriction on the domain: $h(x) \neq 0$.)
>
> **2.** It is more difficult to complicate an expression than to simplify it; therefore separately simplify each side of the proposed identity.
>
> **3.** Sometimes it helps to convert all trigonometric functions to sine and cosine, then simplify the expressions. Combine rational expressions into a single term. Sums of rational expressions require a least common denominator.
>
> **4.** Make use of the known identities
>
> $$\sin^2 \theta + \cos^2 \theta = 1 \quad \text{and} \quad \sin(90° - \theta) = \cos \theta$$
>
> to further reduce the number of trigonometric functions in each expression.

Strategy to Verify an Identity—Cont'd

5. Once each side of the identity is reduced to sine and cosine and simplified so that the two sides are the same expression, then "shake out" the steps to obtain a derivation of the identity. See Display 7-8. (Note: if you add or subtract or multiply on both sides of the equation, you will not be able to "shake out" the steps into a derivation. This is the reason for the restriction in step 1.)

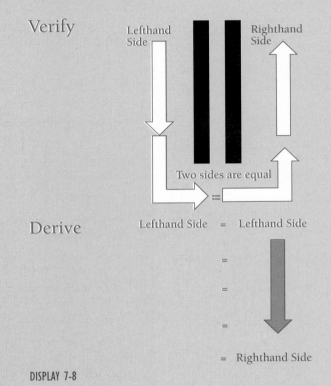

DISPLAY 7-8

6. Because the "shake out" to obtain a derivation for the identity is obvious when we carefully follow steps 1–5, we sometimes skip the "shake out." The steps documenting the reduction of each side of the equation to a common expression represent a verification of the identity. Verifying an identity requires less effort and thus is more common than deriving an identity. Some instructors insist that verifications be done by working on only one side of the proposed identity. Usually the more complicated side is chosen, and you must rewrite it to resemble the simpler side. Working on one side only helps avoid difficulties that might arise from ignoring the cautions in step 1 and more closely resembles a derivation. Ask your instructor whether you should work on one side only when verifying an identity.

Verifying an Identity. Verify that the following are identities.

E X A M P L E 7 - 7

Illustration 1:

$$1 + \cos\theta \equiv \frac{\sin^2\theta}{(1 - \cos\theta)}$$

Analysis: For the two sides to be equal, either the left side must become a fraction or the right side must eliminate its denominator. We will convert the sines to cosines on the right and attempt to reduce the right side to lowest terms to eliminate the denominator.

Solution:

$1 + \cos\theta$ $\qquad$ $\dfrac{\sin^2\theta}{1 - \cos\theta}$ $\qquad$ (common domain)

$\dfrac{1 - \cos^2\theta}{1 - \cos\theta}$ $\qquad$ ($\cos\theta \neq 1$)

$\dfrac{(1 - \cos\theta)(1 + \cos\theta)}{1 - \cos\theta}$

$1 + \cos\theta \equiv 1 + \cos\theta$ $\qquad$ (where $\cos\theta \neq 1$)

Note: Because $\sin^2\theta + \cos^2\theta \equiv 1$, it is easy to verify that $1 - \cos^2\theta \equiv \sin^2\theta$ is a valid identity. In general, if $f(x) + g(x) \equiv h(x)$ is a valid identity, then by substitution, so is $f(x) \equiv h(x) - g(x)$.

Note: In the proof of an identity the "direction" of equality is from top to bottom, rather than left to right. Any left side of the equation must be equal to all the other left sides. Similarly for the righthand sides, so that, when a match between the left and right sides is made, all forms of each side will be equal. Because of this requirement, the steps of the proof can be "shaken out" into a derivation of the righthand side from the left:

$$1 + \cos\theta \equiv 1 + \cos\theta$$
$$\equiv (1 + \cos\theta)\frac{1 - \cos\theta}{1 - \cos\theta} \qquad (\text{note: } \cos\theta \neq 1)$$
$$\equiv \frac{1 - \cos^2\theta}{1 - \cos\theta}$$
$$\equiv \frac{\sin^2\theta}{1 - \cos\theta}$$

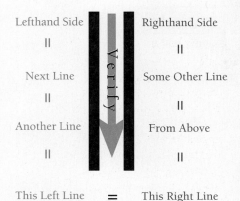

Lefthand Side $\quad$ Righthand Side

$\|$ $\qquad$ $\|$

Next Line $\qquad$ Some Other Line

$\|$ $\qquad$ $\|$

Another Line $\qquad$ From Above

$\|$ $\qquad$ $\|$

This Left Line $\quad = \quad$ This Right Line

Two sides are equal

DISPLAY 7-9 Identity with equalities from top to bottom

The righthand side of the derivation traces down the lefthand side of the verification then back up the right side.

Illustration 2:

$$\cot x + \tan x \equiv \csc(x)\sec(x)$$

Analysis: Because sine and cosine have the most familiar properties, convert all functions to sine and cosine.

Solution 1:

$$
\begin{array}{c|c}
\tan x + \cot x & \csc(x)\sec(x) \\
\dfrac{\sin x}{\cos x} + \dfrac{\cos x}{\sin x} & \dfrac{1}{\sin x}\,\dfrac{1}{\cos x} \\
\dfrac{\sin^2 x}{\sin x \cos x} + \dfrac{\cos^2 x}{\sin x \cos x} & \dfrac{1}{\sin x \cos x} \\
\dfrac{\sin^2 x + \cos^2 x}{\sin x \cos x} & \dfrac{1}{\sin x \cos x}
\end{array}
$$

$$\frac{1}{\sin x \cos x} \equiv \frac{1}{\sin x \cos x}$$

Note: A common strategy for proving an identity is to convert all the trigonometric functions to sine and cosine, then simplify each side into a single expression. Once each side simplifies to an expression of sines or cosines, the remainder of the proof relies on recognition or inspiration. In Illustration 2, recognition that $\sin^2 x + \cos^2 x = 1$ completes the proof. The derivation "shakes out" as follows:

$$\cot x + \tan x \equiv \cot x + \tan x$$
$$\equiv \frac{\cos x}{\sin x} + \frac{\sin x}{\cos x}$$
$$\equiv \frac{\sin^2 x}{\sin(x)\cos(x)} + \frac{\cos^2 x}{\sin x \cos x}$$
$$\equiv \frac{\sin^2 x + \cos^2 x}{\sin x \cos x}$$
$$\equiv \frac{1}{\sin x \cos x}$$
$$\equiv \frac{1}{\sin x}\,\frac{1}{\cos x}$$
$$\equiv \csc(x)\sec(x)$$

Solution 2: Use graphing technology to graph

$$f(x) = \cot x + \tan x \quad \text{and} \quad g(x) = \csc x \sec x$$

on the same coordinate system. Coincident graphs support, but do not confirm, that the proposed identity is correct. With two distinct graphs, we have clear evidence of a conditional equation, not an identity. See Figure 7-54.

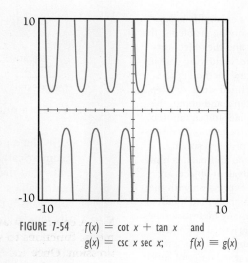

FIGURE 7-54 $f(x) = \cot x + \tan x$ and
$g(x) = \csc x \sec x;$ $f(x) \equiv g(x)$

Illustration 3:   Use identities to simplify $(\sec \theta \csc \theta - \cot \theta)(\sin \theta - \csc \theta)$.

Solution:

$$(\sec \theta \csc \theta - \cot \theta)(\sin \theta - \csc \theta)$$

$$\equiv \left(\frac{1}{\sin \theta \cos \theta} - \frac{\cos \theta}{\sin \theta} \right)\left(\sin\theta - \frac{1}{\sin \theta} \right)$$

$$\equiv \left(\frac{1}{\sin \theta \cos \theta} - \frac{\cos^2 \theta}{\sin \theta \cos \theta} \right)\left(\frac{\sin^2 \theta}{\sin \theta} - \frac{1}{\sin \theta} \right)$$

$$\equiv \left(\frac{1 - \cos^2 \theta}{\sin \theta \cos \theta} \right)\left(\frac{\sin^2 \theta - 1}{\sin \theta} \right)$$

$$\equiv \left(\frac{\sin^2 \theta}{\sin \theta \cos \theta} \right)\left(\frac{-\cos^2 \theta}{\sin \theta} \right)$$

$$\equiv -\cos \theta$$

SUMMARY

S I D E B A R

Computer languages exemplify at least two approaches to solving problems: the procedural method and the object-oriented approach. The identifying characteristic of a procedural method is a sequence of well-defined steps: do step 1, do step 2, . . ., do step n and obtain the answer. Solving a quadratic equation by using the quadratic formula is an example of a procedural problem solving.

The object-oriented approach is more subtle. Various equivalent objects are substituted into an expression with the goal of finding a match. A match indicates a solution. Proving identities relies more on an object-oriented approach than a procedural approach. To "prove" an identity discover a sequence of substitutions that transform the two sides of the equation into a match.

Definitions

tangent:	$\tan x \equiv \dfrac{\sin x}{\cos x},$	$\cos x \neq 0$
cotangent:	$\cot x \equiv \dfrac{\cos x}{\sin x},$	$\sin x \neq 0$
secant:	$\sec x \equiv \dfrac{1}{\cos x},$	$\cos x \neq 0$
cosecant:	$\csc x \equiv \dfrac{1}{\sin x},$	$\sin x \neq 0$

• Consider a proposed identity $f(x) \equiv g(x)$. Interpret the identity geometrically and algebraically. Geometrically, $f(x) \equiv g(x)$ indicates that the graphs of f and g are the same for all common domain values. Analysis of the graphs of trigonometric and circular functions is the subject of the next section. Algebraically, the identity $f(x) \equiv g(x)$ means that, subject to the common domain restriction, $f(x)$ can replace $g(x)$ in any expression containing $g(x)$. Similarly, $g(x)$ can substitute for $f(x)$. To prove $f(x) \equiv g(x)$, use substitutions of known identities into the expression $f(x) \equiv g(x)$ until the two sides match. The following substitutions are useful:

1. Transform all trigonometric functions to sine and cosine.
2. Simplify expressions as much as possible.
3. Strategies:
   (a) Multiply an expression by 1 (for example, $\sin \theta / \sin \theta$).
   (b) Add 0 to an expression (for example, $\cos \theta - \cos \theta$).
   (c) Recognize forms of well-known identities such as $\sin^2 \theta + \cos^2 \theta \equiv 1$ and $\sin(90° - \theta) \equiv \cos \theta$.
4. Be sure that each resulting side of the equation is equal to the side preceding it.

• To convert the verification of an identity $f(x) \equiv g(x)$ to the derivation of the identity, rewrite as follows:

1. Copy the lefthand side $f(x)$ and set it equal to itself: $f(x) \equiv f(x)$.
2. Under the new righthand side:

   Copy each step from the lefthand verification from top to bottom.

   Copy the righthand side from the bottom up to $g(x)$.

   Because each step of the verification requires equality from top to bottom as well as left to right, the resulting sequence of steps derives $g(x)$ from $f(x)$.

7-3 EXERCISES

Evaluate each of the following as indicated.

1. $\sin 30°$ **2.** $\cos 30°$ **3.** $\tan 30°$

4. $\sec 30°$ **5.** $\csc 30°$ **6.** $\cot 30°$

7. $\cot 60°$ **8.** $\tan 60°$ **9.** $\tan 60°$

10. $\cot 60°$ **11.** $\sec 45°$ **12.** $\csc 45°$

13. $\csc 45°$ **14.** $\sec 45°$ **15.** $\cot 45°$

16. $\tan 45°$ **17.** $\sec 60°$ **18.** $\csc 60°$

*19. $\cot 50°$ ***20.** $\cot 40°$ ***21.** $\sec 50°$

*22. $\sec 40°$ ***23.** $\csc 50°$ ***24.** $\csc 40°$

25. Prove $\tan \theta = \cot(90° - \theta)$.

26. Prove $\sec \theta = \csc(90° - \theta)$.

27. Prove that $\sec^2 \theta = 1 + \tan^2 \theta$.

28. Prove that $\csc^2 \theta = 1 + \cot^2 \theta$.

29. Prove that $\sin^2 \theta = 1 - \cos^2 \theta$.

30. Prove that $\cos^2 \theta = 1 - \sin^2 \theta$.

Prove the following identities.

31. $\sin \theta \equiv \cos \theta \tan \theta$

32. $\cos \theta \equiv \sin \theta \cot \theta$

33. $\cos \theta \sec \theta \equiv 1$

34. $\sin \theta \csc \theta \equiv 1$

35. $\tan \theta \cot \theta \equiv 1$

36. $\sec \theta \equiv \csc \theta \tan \theta$

37. $1 + \tan^2 \theta \equiv \sec^2 \theta$

38. $\csc^2 \theta \equiv 1 + \cot^2 \theta$

39. $\csc^2 \theta - 1 \equiv \cot^2 \theta$

40. $\tan^2 \theta + 1 \equiv \sec^2 \theta$

41. $\sin \theta \csc \theta - \sin^2 \theta \equiv \cos^2 \theta$

42. $\cos \theta \sec \theta - \cos^2 \theta \equiv \sin^2 \theta$

43. $\dfrac{\sin \theta}{1 + \cos \theta} \equiv \dfrac{1 - \cos \theta}{\sin \theta}$

44. $\dfrac{\cos \theta}{1 + \sin \theta} \equiv \dfrac{1 - \sin \theta}{\cos \theta}$

45. $\sin^2 \theta - \cos^2 \theta \equiv \sin^4 \theta - \cos^4 \theta$

46. $\cos^2 \theta - \sin^2 \theta \equiv \cos^4 \theta - \sin^4 \theta$

47. $\cos^2 \theta + \tan^2 \theta \cos^2 \theta \equiv 1$

48. $\sin^2 \theta \cot^2 \theta + \sin^2 \theta \equiv 1$

49. $\dfrac{\sec \theta}{\sin \theta} - \dfrac{\sin \theta}{\cos \theta} \equiv \cot \theta$

50. $\dfrac{\csc \theta}{\cos \theta} - \dfrac{\cos \theta}{\sin \theta} \equiv \tan \theta$

51. $\dfrac{\sin x + \cos x}{\cos x} \equiv 1 + \tan x$

52. $\dfrac{\sin x + \cos x}{\sin x} \equiv 1 + \cot x$

53. $\dfrac{\sec t - \cos t}{\tan t} \equiv \dfrac{\tan t}{\sec t}$

54. $\dfrac{\csc t - \sin t}{\cot t} \equiv \dfrac{\cot t}{\csc t}$

55. $\dfrac{\sec w - \csc w}{\sec w + \csc w} \equiv \dfrac{\sin w - \cos w}{\sin w + \cos w}$

56. $\dfrac{\sec w + \csc w}{\csc w - \sec w} \equiv \dfrac{\sin w + \cos w}{\cos w - \sin w}$

57. $\sec^4 t - \tan^4 t \equiv \sec^2 t + \tan^2 t$

58. $\csc^4 t - \cot^4 t \equiv \csc^2 t + \cot^2 t$

59. $\tan^2 x - \sin^2 x \equiv \tan^2 x \sin^2 x$

60. $\cot^2 x - \cos^2 x \equiv \cot^2 x \cos^2 x$

61. $\csc x - \sin x \equiv \cot x \cos x$

62. $\sec x - \cos x \equiv \tan x \sin x$

63. $(\csc x + \cot x)^2 \equiv \dfrac{1 + \cos x}{1 - \cos x}$

64. $(\sec x + \tan x)^2 \equiv \dfrac{1 + \sin x}{1 - \sin x}$

65. $\sin^2 x - \cos^2 x \equiv 1 - 2\cos^2 x$

66. $\cos^2 x - \sin^2 x \equiv 1 - 2\sin^2 x$

67. $\dfrac{1}{1 - \cos t} + \dfrac{1}{1 + \cos t} \equiv 2\csc^2 t$

68. $\dfrac{1}{1 + \sin t} + \dfrac{1}{1 - \sin t} \equiv 2\sec^2 t$

69. $\tan w + \cot w = \csc w \sec w$

70. $\dfrac{1}{\tan w + \cot w} = \sin w \cos w$

71. $\dfrac{(\sin x + \cos x)^2 - 1}{2 \cos x} \equiv \sin x$

72. $1 - (\sin x - \cos x)^2 \equiv 2 \sin x \cos x$

73. $\dfrac{\tan x}{\sec x + 1} \equiv \dfrac{\sec x - 1}{\tan x}$

74. $\dfrac{\csc x + 1}{\cot x} \equiv \dfrac{\cot x}{\csc x - 1}$

75. $\dfrac{\cos B}{1 + \sin B} + \dfrac{1 + \sin B}{\cos B} \equiv 2 \sec B$

76. $\dfrac{\cos B}{1 + \sin B} - \dfrac{1 - \sin B}{\cos B} \equiv 0$

77. The odd and even exercise Problems 31–50 are in pairs. Each even-numbered problem derives from the previous odd-numbered problem. Show that identity number 32 may be obtained from identity number 31 by substituting $90° - \theta$ for θ.

78. See Problem 77. Suppose $f(x) \equiv g(x)$ is a valid identity, then another valid identity follows by replacing each trigonometric function with its complemental function. Verify this process by showing how the identity in Problem 49 may be obtained from the identity in Problem 50.

79. Convert the verification of the identity in Problem 49 to a derivation.

80. Convert the verification of the identity in Problem 50 to a derivation.

Express each of the following in terms of sine and cosine. Simplify the resulting expressions.

81. $\dfrac{\tan x - \tan y}{1 + \tan x \tan y}$

82. $\dfrac{\cot x - \cot y}{1 + \cot x \cot y}$

83. $\dfrac{1}{\tan \theta + \cot \phi}$

84. $\dfrac{1 - \tan \phi}{\cot \theta - 1}$

85. $\dfrac{\sec x - \cos x}{\tan x}$

86. $\dfrac{\csc x - \sin x}{\cot x}$

87. $\dfrac{1 + \csc \phi}{\sec \phi} - \cot \phi$

88. $\dfrac{1 + \sec \theta}{\csc \theta} - \tan \theta$

89. $\dfrac{\tan \theta + \sin \theta}{1 + \sec \theta}$

90. $\dfrac{\cot \alpha + \cos \alpha}{1 + \csc \alpha}$

91. Prove that if $f(x) + g(x) \equiv h(x)$ is a valid identity, so is $f(x) \equiv h(x) - g(x)$.

92. Prove that if $f(x) - g(x) \equiv h(x)$ is a valid identity, so is $f(x) \equiv h(x) + g(x)$.

93. Find the error in the following argument. Assume $x = 1$. Then $x + 1 = 2$. Multiply by x, then $x^2 + x = 2x$. Subtract 2: $x^2 + x - 2 = 2x - 2$. Factor: $(x + 2)(x - 1) = 2(x - 1)$. Divide both sides by $x - 1$: $x + 2 = 2$. Solve for x: $x = 0$. We began with $x = 1$, so we have proven that $0 = 1$.

94. Consider the line given by $y = mx + b$, where $m \neq 0$. The line forms an angle θ with the x-axis. Prove that $\tan \theta = m$.

95. Consider the line given by $y = mx + b$, where $m \neq 0$ and $b \neq 0$. The line forms an angle θ with the x-axis. Suppose that a line is drawn from the origin perpendicular to the given line. Prove that the slope of the perpendicular is $-\cot \theta$.

96. Show that if a line passes through (r, s) and forms an angle θ with the x-axis, the equation for the line is given by $\cos \theta\,(y - s) = \sin \theta\,(x - r)$.

Verify the following identities where $0 < x < \dfrac{\pi}{2}$.

97. $\log|\tan x| \equiv \log|\sin x| - \log|\cos x|$

98. $\ln|\cot x| \equiv \ln|\cos x| + \ln|\csc x|$

99. $\ln|\sec x| \equiv -\ln|\cos x|$

100. $-\log|\csc x| \equiv \ln|\sin x|$

7-3   PROBLEM SET

In each of the following equations make the indicated substitution, then simplify.

1. $x^2 + y^2 = 25$.   Let $x = 5 \cos \theta$ and $y = 5 \sin \theta$.

2. $x^2 + y^2 = 36$.   Let $x = 6 \cos \theta$ and $y = 6 \sin \theta$.

3. $(x - 3)^2 + (y + 2)^2 = 49$.   Let $x = 3 + 7 \cos \theta$ and $y = -2 + 7 \sin \theta$.

4. $(x + 5)^2 + (y - 4)^2 = 9$.   Let $x = -5 + 3 \cos \theta$ and $y = 4 + 3 \sin \theta$.

5. $\dfrac{(x - 2)^2}{9} + \dfrac{(y + 3)^2}{16} = 1$.   Let $x = 2 + 3 \cos \theta$ and $y = -3 + 4 \sin \theta$.

6. $\dfrac{(x + 1)^2}{4} + \dfrac{(y - 3)^2}{9} = 1$.   Let $x = -1 + 2 \sin \theta$ and $y = 3 + 3 \sin \theta$.

7. $\dfrac{x^2}{9} - \dfrac{y^2}{25} = 1.$ Let $x = 3 \sec t$ and $y = 5 \tan t.$

8. $\dfrac{y^2}{16} - \dfrac{x^2}{9} = 1.$ Let $y = 4 \sec t$ and $y = 3 \tan t.$

9. $\dfrac{(x-2)^2}{25} - \dfrac{y^2}{9} = 1.$ Let $x = 2 + 5 \sec t$ and $y = 3 \tan t.$

10. $\dfrac{x^2}{36} - \dfrac{(y+4)^2}{9} = 1.$ Let $x = 6 \sec t$ and $y = -4 + 3 \tan t.$

For trigonometric functions, three useful forms of the Pythagorean theorem are

$$1 - \sin^2\theta = \cos^2\theta$$
$$\tan^2\theta + 1 = \sec^2\theta$$
$$\sec^2\theta - 1 = \tan^2\theta$$

In each of these, the lefthand sum or difference collapses to a single squared term on the right. These identities are ideal for eliminating square roots with radicands consisting of a variable squared in a sum or difference with a constant. For example, in the expression $\sqrt{x^2 - 9}$, the radicand resembles the lefthand side of the identity $\sec^2\theta - 1 = \tan^2\theta$. Therefore, if we let $x = 3 \sec\theta$, the expression simplifies quickly:

$$\sqrt{x^2 - 9} = \sqrt{(3 \sec\theta)^2 - 9}$$
$$= \sqrt{9 \sec^2\theta - 9}$$
$$= \sqrt{9(\sec^2\theta - 1)}$$
$$= 3\sqrt{\tan^2\theta}$$

Moreover, if $0 \leq \theta < \dfrac{\pi}{2}$, then $0 \leq \tan\theta$, so that

$$\sqrt{x^2 - 9} = 3 \tan\theta$$

For each of the following expressions, choose a substitution based on a trigonometric form of the Pythagorean theorem. Use the substitution to simplify the expression.

11. $\sqrt{16 - x^2}$

12. $\sqrt{x^2 - 16}$

13. $\sqrt{16 + x^2}$

14. $\sqrt{25 - x^2}$

15. $\dfrac{\sqrt{25 - x^2}}{x}$

16. $\dfrac{x}{\sqrt{16 - x^2}}$

17. $\dfrac{\sqrt{x^2 - 9}}{x}$

18. $\dfrac{x}{\sqrt{9 + x^2}}$

19. $\sqrt{(x-2)^2 + 9}$

20. $\sqrt{(x+1)^2 - 9}$

PROBLEMS FOR GRAPHING TECHNOLOGY

Use graphing technology to graph the following pairs of functions $g(x)$ and $h(x)$ on the same coordinate system. From the graphs determine whether $h(x) = g(x)$ is a conditional equation (two distinct graphs). If the graphs appear coincident, indicate an "apparent" identity.

21. $g(x) = \sin(2x),\ h(x) = 2 \sin x \cos x$

22. $g(x) = \cos(2x),\ h(x) = 2 \cos^2 x - 1$

23. $g(x) = \cos x,\ h(x) = 1 - \dfrac{x^2}{2} + \dfrac{x^4}{24}$

24. $g(x) = \sin x,\ h(x) = x - \dfrac{x^3}{6} + \dfrac{x^5}{120}$

25. $g(x) = (\sin x + \cos x)^2,\ h(x) = \sin^2 x + \cos^2 x$

26. $g(x) = \cos(-x),\ h(x) = -\cos x$

27. $g(x) = 1 - 2 \cos x,\ h(x) = 1 + 2 \sin x$

28. $g(x) = \sin(-x),\ h(x) = -\sin(x)$

29. $g(x) = \sin(x),\ h(x) = \cos\left(\dfrac{\pi}{2} - x\right)$

30. $g(x) = \tan^2 x,\ h(x) = \dfrac{1 - \cos x}{1 + \cos^2}$

31. $g(x) = \tan(2x),\ h(x) = \dfrac{2 \tan x}{1 - \tan^2 x}$

32. $g(x) = \tan\left(\dfrac{x}{2}\right),\ h(x) = \dfrac{\sin x}{1 + \cos x}$

EXPLORATION

The following problems anticipate concepts from the next section. Use graphing technology to produce the graphs of each of the following functions for the given values of k. In each, conjecture the effect of k upon the graph of the function.

33. $y = \sin(kx),\ k = 1, 2, 3$

34. $y = \sin(kx),\ k = -1, -2, -3$

35. $y = \cos(kx),\ k = -1, -2, -3$

36. $y = \cos(kx),\ k = 1, 2, 3$

37. $y = k \sin x,\ k = -1, -2, -3$

38. $y = k \sin x,\ k = 1, 2, 3$

39. $y = k \cos x,\ k = 1, 2, 3$

40. $y = k \cos x,\ k = -1, -2, -3$

41. $y = \sin(x + k),\ k = 1, 2, 3$

42. $y = \cos(x + k),\ k = -1, -2, -3$

43. $y = \cos x + k,\ k = 1, 2, 3$

44. $y = \sin x + k,\ k = -1, -2, -3$

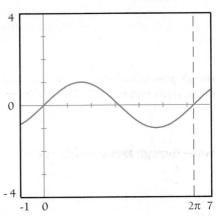

FIGURE 7-55

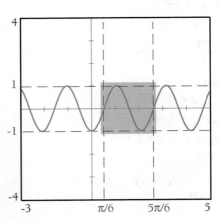

FIGURE 7-56

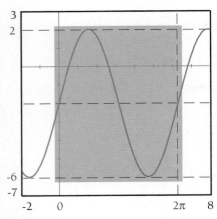

FIGURE 7-57

4   ANALYTICAL TRIGONOMETRY

The major part of every meaningful life is the solution of problems; a considerable part of the professional life of technicians, engineers, scientists, etc., is the solution of mathematical problems. It is the duty of all teachers and of teachers of mathematics in particular, to expose their students to problems more than to facts.
—PAUL R. HALMOS

What is the graph of cosine? How shall we transform graphs of sine and cosine? In what manner do they stretch? This section examines transformations of the graphs of sine and cosine. An appropriate window's view and fundamental properties of graphs such as symmetry and periodicity combine to produce a complete picture of the graph. First we develop graphing techniques for the sine function. The graph of cosine operates in the same manner. Technology supplies the basic sine curve in Figure 7-55.

Let $L(x) = mx + b$ be a transformation. Then L applies to $f(x) = \sin x$ in two different ways:

1. $L \circ f(x) = m \sin x + b$ affects the graph window of sine vertically.
2. $f \circ L(x) = \sin(mx + b)$ affects the graph window of sine horizontally.

Let us examine $f(x) = \sin(3x - \pi/2)$. For any argument, the range values of sine are between -1 and 1:

$$-1 \leq \sin(\text{argument}) \leq 1$$

No vertical adjustment in our basic window is necessary. The domain is another matter. Although we may trace a complete cycle starting at any domain component, the most familiar complete cycle occurs when the argument takes on values from 0 to 2π.

$$0 \leq \text{argument} \leq 2\pi$$

$$0 \leq 3x - \frac{\pi}{2} \leq 2\pi \qquad \text{(solve for } x)$$

$$\frac{\pi}{2} \leq 3x \leq \frac{5\pi}{2}$$

$$\frac{\pi}{6} \leq x \leq \frac{5\pi}{6}$$

As x assumes values from $\pi/6$ to $5\pi/6$, the argument $3x - \pi/2$ takes on values from 0 to 2π. Therefore, a window bounded by $x = \pi/6$ and $x = 5\pi/6$ will contain a copy of the familiar complete cycle of sine.

Figure 7-56 displays $f(x) = \sin(3x - \pi/2)$ in the window and extended through several cycles.

The next example is $g(x) = 4 \sin x - 2$. Again start from the familiar known shape of sine as given in Figure 7-55. For the domain window, assume the argument is between 0 and 2π. Then

$$0 \leq x \leq 2\pi$$

contains a complete cycle. For the range, transform from $-1 \leq \sin x \leq 1$:

$$-1 \leq \sin x \leq 1$$
$$-4 \leq 4 \sin x \leq 4$$
$$-6 \leq 4 \sin x - 2 \leq 2$$
$$-6 \leq g(x) \leq 2$$

An appropriate window for g has the bounds $y = -6$ and $y = 2$. Into this window copy the familiar complete cycle of sine from Figure 7-55. The results appear in Figure 7-57.

Graphs of Sine and Cosine.   Graph the following through two complete cycles.

$$f(x) = 5 \sin(2x - \pi) + 1$$

Solution:

Domain window: $0 \leq 2x - \pi \leq 2\pi$

$$\pi \leq 2x \leq 3\pi$$

$$\frac{\pi}{2} \leq x \leq \frac{3\pi}{2}$$

Range window: $-1 \leq \sin \arg \leq 1$

$$-5 \leq 5 \sin \arg \leq 5$$
$$-4 \leq 5 \sin \arg + 1 \leq 6$$
$$-4 \leq 5 \sin(2x - \pi) + 1 \leq 6$$
$$-4 \leq f(x) \leq 6$$

Fit one cycle of sine into the window. Then duplicate the window to obtain two cycles. The results appear in Figure 7-58.

$$f(x) = \frac{1}{2} \cos(5x + 2) - \pi$$

Solution: Figure 7-59 shows a familiar complete cycle of cosine in a window where

$$0 \leq x \leq 2\pi \quad \text{and} \quad -1 \leq y \leq 1$$

To graph a suitable cycle for f requires construction of an equivalent window.

Domain $0 \leq 5x + 2 \leq 2\pi$ (solve for x)

$$-2 \leq 5x \leq 2\pi - 2$$

$$-\frac{2}{5} \leq x \leq \frac{2\pi - 2}{5}$$

E X A M P L E   7 - 8

Illustration 1:

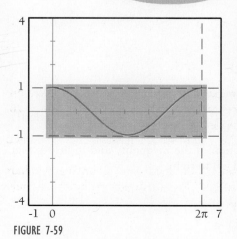

FIGURE 7-58

Illustration 2:

FIGURE 7-59

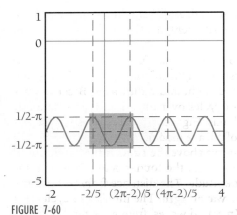

FIGURE 7-60

Now, for the range, let arg = 5x + 2.

$$-1 \le \cos \arg \le 1 \qquad \text{(build to } f(x)\text{)}$$

$$-\frac{1}{2} \le \left(\frac{1}{2}\right)\cos \arg \le \frac{1}{2}$$

$$-\frac{1}{2} - \pi \le \left(\frac{1}{2}\right)\cos \arg - \pi \le \frac{1}{2} - \pi$$

$$-\frac{1}{2} - \pi \le \left(\frac{1}{2}\right)\cos(5x + 2) - \pi \le \frac{1}{2} - \pi$$

$$-\frac{1}{2} - \pi \le f(x) \le \frac{1}{2} - \pi$$

Refer to Figure 7-59 to copy the fundamental shape into the defined window. Duplicate the window for a second cycle. See Figure 7-60.

Illustration 3:

$$g(x) = -2 \sin\left(\frac{1}{2}x - \frac{\pi}{12}\right)$$

Solution:

Domain
$$0 \le \frac{1}{2}x - \frac{\pi}{12} \le 2\pi$$

$$\frac{\pi}{12} \le \frac{1}{2}x \le \frac{25\pi}{12}$$

$$\frac{\pi}{6} \le x \le \frac{25\pi}{6}$$

Range
$$-1 \le \sin \arg \le 1$$

$$2 \ge -2 \sin \arg \ge -2 \qquad \text{(alert!)}$$

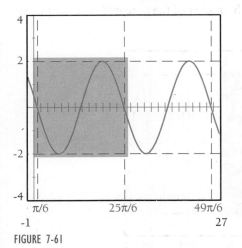

FIGURE 7-61

The multiplication by -2 in the determination of the range reverses the direction of the inequality and indicates that the window and its graph are reflected (mirror image) from top to bottom. After locating the window, invert the familiar sine shape from Figure 7-55. The results appear in Figure 7-61.

Illustration 4:

$$h(x) = 3 \sin(\pi - 4x)$$

Solution:

Domain
$$0 \le \pi - 4x \le 2\pi$$

$$-\pi \le -4x \le \pi$$

$$\frac{\pi}{4} \ge x \ge -\frac{\pi}{4} \qquad \text{(alert!)}$$

The division by -4 reversed the direction of the inequality and indicates that the graph window should be reflected left to right.

Range
$$-1 \le \sin \arg \le 1$$

$$-3 \le 3 \sin \arg \le 3$$

$$-3 \le 3 \sin(\pi - 4x) \le 3$$

$$-3 \le h(x) \le 3$$

"Wait a minute. Something's wrong here. We have alternating current, ac, from the power company. But if alternating current is a wave form for electricity, how can the voltage be fixed at 110 Volts?"

"Calm down, Larry. You are absolutely correct. The voltage is not fixed. It varies just like the sine curve."

"OK, Sheila, does that mean that 110 Volts is the maximum voltage? Like where the sine curve peaks? Maybe we have an amplitude of 110 volts?"

"That's not what my engineering professor says. A maximum voltage is not truly representative of the entire power curve."

"Well, if we are going to talk about representatives, then we must be looking for an average. Is 110 volts the mean of the voltages?"

"Look at the curve, Larry. Half the time the voltage is positive. The other half is an equivalent negative voltage. Average those and you get 0, not 110."

"OK. I give up. What does the 110 Volts represent?"

"Actually, 110 volts is the effective voltage. Forces between wires vary during the cycle as the square of the voltage . . ."

"Wait, You're losing me!"

"At any rate, the square of the voltage is a more useful measure, so we average the squares and take the square root. This number is the effective or root mean square, rms for short, voltage."

". . . And the rms is 110 Volts."

"You got it. In a well designed ac machine the maximum voltage is about $\sqrt{2}$ rms. This places the maximum voltage close to 156 Volts."

"Sheila, you've made the situation more shocking . . ."

After locating the graphing window, reflect the familiar sine shape from left to right to account for the negative coefficient of x (Figure 7-62). ■

One of the most useful applications of circular functions is in modeling wave forms. Wave forms provide a visual insight into many natural phenomena. Among these phenomena are sound, light and radiowaves. Because wave forms are usually measured in terms of cycles over time, it is appropriate that we use t (seconds) for our horizontal axis. Figure 7-63 illustrates special terminology to describe various aspects of a wave model.

The maximum height of the sine curve above the t-axis is the amplitude of the wave. In sound waves, amplitude models the loudness of the sound: the louder the sound, the higher is the amplitude. The distance from one crest to the next crest (a complete cycle) is the **wavelength**. The number of cycles in 1 second is the **frequency** of the wave. In sound waves, frequency represents the pitch of the sound: the higher the frequency (more cycles per second), the higher is the pitch.

For some wave forms, the speed of propagation of the wave provides another method for measuring wavelength. Suppose that sound travels 331.3 meters in 1 second. Suppose that the frequency of the note middle C on a piano

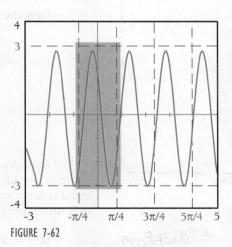

FIGURE 7-62

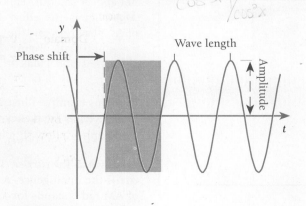

FIGURE 7-63

is 512 cycles per second. In terms of time the wavelength is $\frac{1}{512}$ of a second. (With 512 cycles in 1 second, one cycle takes $\frac{1}{512}$ second. The frequency of a sound is the reciprocal of its wavelength.) Multiply the speed of sound (in meters/second) times the wave length (in seconds) to determine the wave length in meters:

$$331.3 \left(\frac{\text{meters}}{\text{second}} \right) \frac{1}{512} \text{ second} \cong 0.647 \text{ meters}$$

Suppose that A represents the amplitude of a wave form and f represents the frequency, then the intensity y of the wave at time t is

$$y = A \sin(2\pi f t)$$

The model wave form of middle C with amplitude 100 is given by

$$y = 100 \sin[2\pi(512)t]$$

Figure 7-64 illustrates two similar sounds started at slightly different times. The graph of the second is the graph of the first translated a fraction of a second. In this case, the sounds are "out of phase." The slight delay in the propagation of the second sound is the **phase shift** of the wave. This is equivalent to the translation $y = A \sin(2\pi f t - B)$. Analyze the graph in terms of a graphing window. You will discover that the actual horizontal translation of axis is

$$\text{phase shift} = \frac{B}{2\pi f}$$

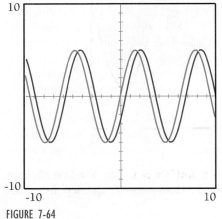

FIGURE 7-64

More Graphs of Sine and Cosine

EXAMPLE 7-9

Illustration 1:

Graph $f(t) = 2 \sin(2\pi t) \sin(100\pi t)$

Solution: First imagine the graph of $y = \sin(100\pi t)$. The graph is sinusoidal with a fundamental period of $2\pi/100\pi = \frac{1}{50}$. In 1 second $y = \sin(100\pi t)$ passes through 50 complete cycles. See Figure 7-65. Note that the curve does not appear smooth. This is a distortion produced by graphing technology.

What is the effect of $2 \sin(2\pi t)$ on $y = \sin(100\pi t)$? Here, $2 \sin(2\pi t)$ plays the role of the amplitude: $2 \sin(2\pi t)$ modifies the amplitude of $y = \sin(100\pi t)$. Figure 7-66 is the graph of $y = 2 \sin(2\pi t) \sin(100\pi t)$ with xMin = -1 and xMax = 1. Note the outline of $y = 2 \sin(2\pi t)$ superimposed over the graph of f.

Note: This example illustrates the ability of a high-frequency wave to carry patterns of lower frequency waves. In this case the high-frequency wave

$$y = \sin(100\pi t)$$

is known as the *carrier*. The image of $y = 2 \sin(2\pi t)$ superimposed on the carrier is the intelligence. A simple example of this process is AM radio. The AM of AM radio stands for amplitude modulation. The carrier of AM radio is the radio frequency. The intelligence is the lower frequency sound waves.

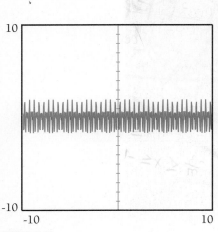

FIGURE 7-65

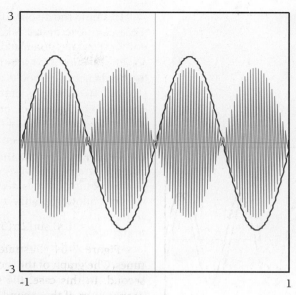

FIGURE 7-66

Illustration 2: Use a graphing calculator to graph $y = 6 \sin(10\pi x)$ in the window $x\text{Min} = -10$, $x\text{Max} = 10$, $y\text{Min} = -10$, $y\text{Max} = 10$. Guess the frequency from the graph.

Solution: Set the range, graphics mode and clear other graphs. In *GraphWindows*, set number of points to plot to 94. To graph the function with the Casio, press $\boxed{\text{GRAPH}}$ $\boxed{6}$ $\boxed{\text{SIN}}$ $\boxed{(}$ $\boxed{1}$ $\boxed{0}$ $\boxed{\text{SHIFT}}$ $\boxed{\text{EXP}}$ $\boxed{\text{X}\theta\text{T}}$ $\boxed{)}$ $\boxed{\text{EXE}}$. To graph on a Sharp EL9300, press $\boxed{\triangle}$ $\boxed{\text{MENU}}$ $\boxed{1}$ $\boxed{6}$ $\boxed{\text{SIN}}$ $\boxed{(}$ $\boxed{1}$ $\boxed{0}$ $\boxed{\text{2ndF}}$ $\boxed{\text{EXP}}$ $\boxed{\text{X}\theta\text{T}}$ $\boxed{)}$ $\boxed{\triangle}$. To graph with the TI-81, press $\boxed{\text{Y=}}$ $\boxed{6}$ $\boxed{\text{SIN}}$ $\boxed{(}$ $\boxed{1}$ $\boxed{0}$ $\boxed{\text{2nd}}$ $\boxed{\wedge}$ $\boxed{\text{X|T}}$ $\boxed{)}$ $\boxed{\text{GRAPH}}$. The graph will appear as seen in Display 7-10.

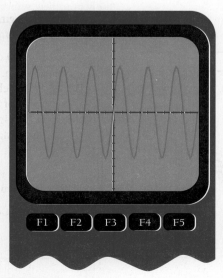

DISPLAY 7-10

DISPLAY 7-11

Examine the display. Between $x = 0$ and $x = 10$ there appears to be about three complete cycles. Therefore one cycle would appear between $x = 0$ and $x = \frac{10}{3}$. You might conjecture that the wavelength is approximately $\frac{10}{3}$. But according to our discussion the fundamental period should be $2\pi/10\pi$ or 0.2. If the fundamental period is 0.2 then we should see 100 complete cycles between -10 and 10. Which is correct, the calculator or our calculations? To resolve this question, zoom in on one cycle of the display: box in from $x = 0$ to $x = 3.5$ and $y = -7$ to $y = 7$. When you complete the zoom, you will suddenly see the missing cycles. See Display 7-11.

The problem is in the calculator display. Because calculators and computer programs produce graphs by connecting line segments between closely plotted points, there is *always* a chance that the graph is more active than the "closely" plotted points indicate. Recall a similar problem with graphing asymptotes in Chapter 5. In the case of periodic curves, the problem may be compounded by an interaction with the number of points plotted in a particular window and the frequency of the graph. In these cases, "false" cycles (false harmonics) appear. Display 7-12 shows how connecting points of a sine curve along equally spaced intervals can produce the false curve. You might also try graphing $y = 6 \sin(100\pi x)$ on $X\min = -10$ to $X\max = 0$, $Y\min = -10$ to $Y\max = 10$ to simulate the false harmonics on the curve. The resulting graph may look like a carrier wave with superimposed intelligence as in the first illustration. Zooming in on successively smaller intervals will dispel this notion.

One method for avoiding this problem is to always analyze the function yourself. Be aware of the period. Change the size of the window until the graph is a better representation. In *GraphWindows* you may adjust the number of points plotted as well as the window size.

Illustration 3:

Graph $y = 5 \cos(3x - \pi)$ over two cycles.

Solution: The analysis of the domain and range is like sine. Therefore we have $-5 \leq y \leq 5$ and one complete cycle for $\pi/3 \leq x \leq \pi$. The major difference is in the graph. Figure 7-67 shows the shape of cosine on $[0, 2\pi]$. Fit this into our new graphing window and we have Figure 7-68.

In the next section we examine the graphs of other circular functions.

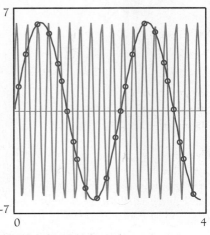

DISPLAY 7-12   False harmonics

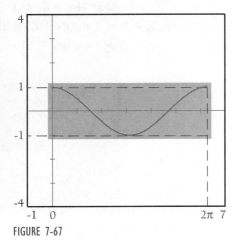

FIGURE 7-67

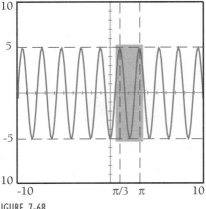

FIGURE 7-68

- For $f(x) = A\sin(Bx + C) + D$, the amplitude of the graph is $|A|$. Transformations on a basic graphing window indicate the fundamental period is $2\pi/|B|$. Similarly, the phase shift $-C/B$ translates the graph horizontally. The vertical shift is D units. Figure 7-69 illustrates one cycle of f.

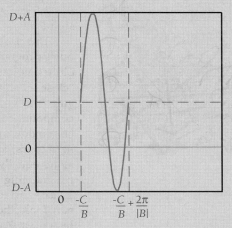

FIGURE 7-69

- To determine an appropriate window for sketching a cycle,

  1. For x-value bounds, set the argument between 0 and 2π and solve for x:

  $$0 \le Bx + C \le 2\pi$$

  $$-\frac{C}{B} \le x \le \frac{2\pi - C}{B} \qquad (if\ B > 0)$$

  $$-\frac{C}{B} \ge x \ge \frac{2\pi - C}{B} \qquad (if\ B < 0:\ reflect\ left\ to\ right)$$

  2. For y-value bounds, set the sine (or cosine) root between -1 and 1 and build up into the function

  $$-1 \le \sin(Bx + C) \le 1$$

  $$-A + D \le A\sin(Bx + C) + D \le A + D \qquad (if\ A > 0)$$

  $$-A + D \ge A\sin(Bx + C) + D \ge A + D \qquad (if\ A < 0:\ reflect\ top\ to\ bottom)$$

- Take care when using technology for graphs of sine or cosine. Because of the window selected or the interaction of the period of the graph with the number of points plotted, the graph may not accurately represent the function.

- Because sine and cosine are complementary functions, $\cos x = \sin(\pi/2 - x)$ and $\sin x = \cos(\pi/2 - x)$. Sine is an odd function, $\sin(\pi/2 - x) = -\sin(x - \pi/2)$, so the graph of cosine is the graph of sine translated $\pi/2$ to the right and reflected about the x-axis. Similarly, cosine is even, so $\cos(\pi/2 - x) = \cos(x - \pi/2)$.

7-4   E X E R C I S E S

Graph each of the following over two complete cycles. Indicate the amplitude, period, frequency and phase shift.

1. $f(x) = 5 \sin(2x)$

2. $f(x) = 3 \cos(4x)$

3. $f(x) = -3 \cos(2x)$

4. $f(x) -5 \sin(4x)$

5. $f(x) = \dfrac{1}{2} \sin(4x)$

6. $f(x) = \dfrac{1}{3} \cos(2x)$

7. $f(x) = \cos\left(x - \dfrac{\pi}{4}\right)$

8. $f(x) = \sin\left(x + \dfrac{\pi}{3}\right)$

9. $f(x) = \sin(x + \pi)$

10. $f(x) = \cos(x - \pi)$

11. $f(x) = \cos \dfrac{x}{3}$

12. $f(x) = \sin \dfrac{x}{2}$

13. $f(x) = -2 \sin x + 5$

14. $f(x) = 3 \cos x - 4$

15. $f(x) = 3 \cos\left(x - \dfrac{\pi}{6}\right) - 2$

16. $f(x) = -4 \sin\left(x + \dfrac{\pi}{3}\right) + 5$

17. $f(x) = \pi \sin(x - 1) + 3$

18. $f(x) = -\pi \cos(x + 2) - 4$

19. $f(x) = 2 \cos(\pi x + 3) - 1$

20. $f(x) = 3 \sin(\pi x - 2) + 1$

Write the rule of pairing to represent each of the following wave forms. Assume the amplitude in each case is 1.

21. Sound frequency of 1000 cycles per second (cps or Hertz). If sound travels at 331.3 m/sec, determine the wave length of the sound.

22. Sound frequency of 500 cps (Hertz). If sound travels at 331.3 m/sec, determine the wavelength of the sound.

23. Radio frequency of 3 megaHertz (3,000,000 cps). If radio travels at 300,000 km/sec, determine the wavelength of the radiowave.

24. Radio frequency of 1.5 megaHertz (1,500,000 cps). If radio travels at 300,000 km/sec, determine the wavelength of the radiowave.

25. Consider green light (frequency of 600,000 gigaHertz = 6×10^{14}). If light travels at 300,000 km/sec, determine the wavelength of the light wave.

26. Consider red light (frequency of 500,000 gigaHertz = 5×10^{14}). If light travels at 300,000 km/sec, determine the wavelength of the light wave.

27. See Problems 21 and 23. Use the radiowave as a carrier and the sound wave as intelligence and write a rule of pairing for the resulting wave.

28. See Problems 22 and 24. Use the radiowave as a carrier and the sound wave as intelligence and write a rule of pairing for the resulting wave.

29. See Problems 23 and 25. Use the light wave as a carrier and the radiowave as intelligence.

30. See Problems 24 and 26. Use the light wave as a carrier and the radiowave as intelligence.

7-4   P R O B L E M   S E T

Graph each of the following pairs of functions on the same coordinate system. Discuss any patterns observed between the graphs of parts a and b.

1. (a) $y = x$

(b) $y = x \sin x$

2. (a) $y = x^2$

(b) $y = x^2 \sin x$

3. (a) $y = -x^2$

(b) $y = -x^2 \sin x$

4. (a) $y = -x$

(b) $y = -x \sin x$

5. (a) $y = \dfrac{1}{x}$

(b) $y = \dfrac{1}{x} \sin x$

6. (a) $y = \dfrac{1}{x^2}$

(b) $y = \dfrac{1}{x^2} \sin x$

7. (a) $y = x$

(b) $y = x + \sin x$

8. (a) $y = -x$

(b) $y = -x + \sin x$

9. (a) $y = |x|$

(b) $y = |x| + \sin x$

10. (a) $y = x$

(b) $y = x - \sin x$

11. (a) $y = |x|$

(b) $y = |\sin x|$

12. (a) $y = |x|$

(b) $y = \sin |x|$

13. (a) $y = |x|$

(b) $y = \cos |x|$

14. (a) $y = |x|$

(b) $y = |\cos x|$

15. (a) $y = \sin x$

(b) $y = \sin x \sin(50x)$

16. (a) $y = 5 \sin x$

(b) $y = 5 \sin x \sin(50x)$

17. (a) $y = 5 \cos x$

(b) $y = 5 \cos x \cos(100x)$

18. (a) $y = \cos x$ (b) $y = \cos x \cos(100x)$

19. (a) $y = \sin(2x)$ (b) $y = 2 \sin x \cos(x)$

20. (a) $y = \sin(4x)$ (b) $y = 2 \sin(2x) \cos(2x)$

PROBLEMS FOR GRAPHING TECHNOLOGY

Set up your graphing window for $Xmin = -10$, $Xmax = 10$, $Ymin = -10$, $Ymax = 10$. By experimentation, determine the following.

21. Graph $y = \sin(n\pi x)$ for $n = 1, 3, 5, \ldots, 19$. For which of these values did the graph exhibit false harmonics instead of a true representation of the curve?

22. Graph $y = \sin(n\pi x)$ for $n = 2, 4, 6, \ldots, 20$. For which of these values did the graph exhibit false harmonics instead of a true representation of the curve?

23. Graph $y = 6 \sin(150x)$. Discuss the validity of the graph.

24. Reset your $Xmin$ value to 0.001. Graph $y = 6 \sin(1/x)$. Zoom in on the graph between $x = 0.001$ and 1. Discuss what you see.

25. See Problem 23. Graph $y = 6 \sin(75x)$. Discuss the shape of the resulting graph. Compare it to the graph from Problem 23.

26. See Problem 24. Perform the same experiment by $y = 7 \sin(1/x^2)$.

27. In the same window, graph $y = x$ and $y = x + \sin(4x)$. Discuss the results.

28. In the same window, graph $y = -x$ and $y = -x + \sin(4x)$. Discuss the results.

29. In the same window, graph $y = 5 \sin x$ and $y = 5 \sin x + \sin(4x)$. Discuss the results.

30. In the same window, graph $y = 5 \cos x$ and $y = 5 \cos(x) - \sin(4x)$.

EXPLORATION

The following problems anticipate concepts from the next section. Use graphing technology to sketch the graph of the following functions for the given values of k. In each case describe the effect of k on the graph of the function.

31. $y = \tan(kx)$, $k = 1, 2, 3$

32. $y = \sec(kx)$, $k = 1, 2, 3$

33. $y = \sec(kx)$, $k = -1, -2, -3$

34. $y = \tan(kx)$, $k = -1, -2, -3$

35. $y = k \sec x$, $k = 1, 2, 3$

36. $y = k \tan x$, $k = -1, -2, -3$

37. $y = k \tan x$, $k = 1, 2, 3$

38. $y = k \sec x$, $k = -1, -2, -3$

39. $y = \tan(x + k)$, $k = 1, 2, 3$

40. $y = \sec(x + k)$, $k = 1, 2, 3$

41. $y = \tan(x - k)$, $k = 1, 2, 3$

42. $y = \sec(x - k)$, $k = 1, 2, 3$

43. $y = \csc x + k$, $k = 1, 2, -3$

44. $y = \cot x + k$, $k = -1, -2, 3$

5

ANALYTICAL TRIGONOMETRY: OTHER FUNCTIONS

It is easier to square the circle than to get round a mathematician.
—A. DE MORGAN

Because tangent and secant are defined as a quotient with cosine as a denominator, these functions are undefined whenever $\cos x = 0$. The x-intercepts of cosine mark the asymptotes of secant and tangent. Similarly, cotangent and cosecant are undefined when $\sin x = 0$. Compare tangent, secant, cotangent and cosecant to the rational functions of Chapter 5. When using technology, hand annotate asymptotes for trigonometric functions as was done with rational functions.

EXAMPLE 7-10

Illustration 1:

Graphs of Other Circular Functions

Sketch the graph of $y = \sec x$.

Solution: By definition $\sec x = 1/\cos x$. Because $\cos x = 0$ when

$$x = \frac{\pi}{2} + k\,\pi$$

$$= (2k + 1)\,\frac{\pi}{2} \qquad\qquad (k \in \text{Integers})$$

then the vertical lines shown in Figure 7-70 are asymptotes for $\sec x$. Moreover, cosine has a period of 2π; therefore, so does secant. As a result, $\sec(x + 2\pi) = \sec x$. Observe the portions of the graph of cosine where

$$0 < \cos x \leqslant 1$$

$$0 < \frac{1}{\sec x} \leqslant 1$$

Because secant has the same sign as cosine, $0 < 1 \leqslant \sec x$. Similarly, where $-1 \leqslant \cos x < 0$,

$$-1 \leqslant \frac{1}{\sec x} < 0$$

These values of secant are negative. Multiplying by $\sec x$ reverses the direction of the inequality:

$$-\sec x \geqslant 1 > 0$$

Divide by -1: $\sec x \leqslant -1 < 0$

Imagine that the secant's reciprocal relationship to cosine produces a graph that resembles cosine "turned inside-out" about the horizontal lines $y = 1$, $y = -1$. See Figure 7-71. Reciprocating a number between 0 and 1 produces a y value larger than 1. Indeed, because $\cos x$ approaches 0 as x approaches $\pi/2$, $\sec x$ increases without bounds as x approaches $\pi/2$ from the left. Therefore, secant has vertical asymptotes wherever cosine is 0.

See Figure 7-71. Now plot some points of secant on $[0, 2\pi]$. See Figure 7-72. Use the asymptotes, horizontal bounds $y = 1$ and $y = -1$, and plotted points to sketch one cycle of secant, then duplicate several cycles. See Figure 7-73. To graph with technology enter secant as $y = 1/\cos x$. Why? Cosine is an even function and as a result so is secant. Hence secant must also be symmetric to the y-axis.

Illustration 2:

Consider that $\qquad \tan x = \dfrac{\sin x}{\cos x}$

Now $\qquad\qquad \tan(-x) = \dfrac{\sin(-x)}{\cos(-x)}$

$$= -\,\frac{\sin x}{\cos x}$$

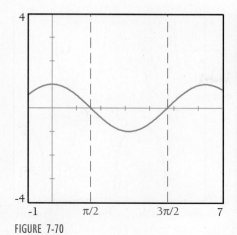

FIGURE 7-70

since sine is odd and cosine is even. Therefore, $\tan(-x) = -\tan x$. Tangent is an odd function, symmetric to the origin. Because cosine forms the denominator in the definition of tangent, the asymptotes of tangent are the same as the asymptotes of secant. Examine Figure 7-74. Note that $\tan(x + \pi) = \sin(x + \pi)/\cos(x + \pi)$

Because

$$\sin(x + \pi) = -\sin x$$

$$\cos(x + \pi) = -\cos x$$

Then

$$\tan(x + \pi) = \frac{-\sin x}{-\cos x}$$

$$\tan(x + \pi) = \frac{\sin x}{\cos x}$$

$$\tan(x + \pi) = \tan x$$

As a result, tangent has a period of π. See Figure 7-75. We need only one cycle of tangent. The remainder of the graph is obvious from periodicity. Most calculators have a tangent key. Select an initial window between the vertical asymptotes at $-\pi/2$ and $\pi/2$ and enter $y = \tan x$. Figure 7-76 displays the

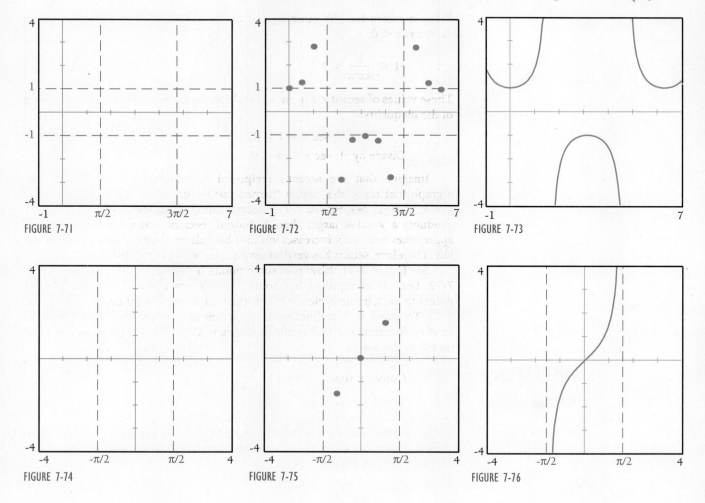

FIGURE 7-71

FIGURE 7-72

FIGURE 7-73

FIGURE 7-74

FIGURE 7-75

FIGURE 7-76

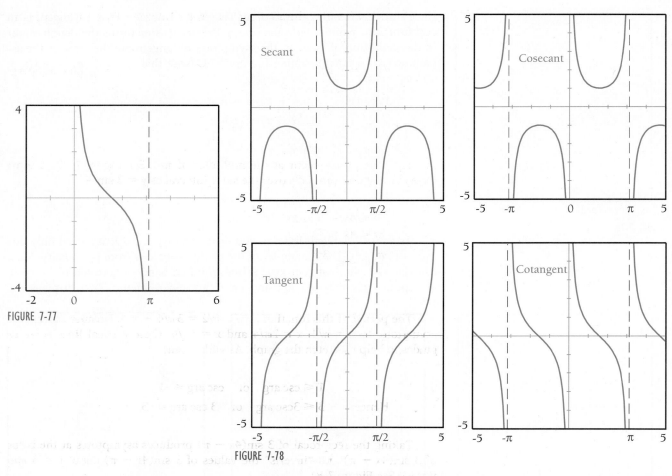

FIGURE 7-77

FIGURE 7-78

graph of one cycle of tangent. A similar analysis produces one cycle of cotangent in Figure 7-77. For technology enter $y = 1/\tan x$ or $y = \cos(x)/\sin(x)$.

The fundamental shape of the graph of cosecant follows in a like manner. Figure 7-78 summarizes the graphs of the quotient functions.

Notice the similarity of secant and cosecant. Secant and cosecant are complementary functions sharing much the same relationship that sine and cosine share. Tangent and cotangent have a dual relation: tangent and cotangent are both complementary functions and reciprocal functions.

Graphs of Transformations of Circular Functions. Graph the following through two cycles.

EXAMPLE 7-11

Illustration 1:

$$y = 3 \csc(4x - \pi)$$

Solution: Because cosecant is the reciprocal of sine, the period is the same as the period of $y = \sin(4x - \pi)$. Asymptotes correspond to the x-intercepts of $y = \sin(4x - \pi)$. See Figure 7-79.

Sine is 0 whenever the argument is $k\pi$, $k \in$ Integers. Thus the asymptotes occur for

$$4x - \pi = k\pi \qquad\qquad (k \in \text{Integers})$$
$$4x = (k + 1)\,\pi$$
$$x = (k + 1)\,\frac{\pi}{4}$$

The asymptotes occur at the multiples of $\pi/4$. See Figure 7-80. A complete cycle of cosecant falls over the same interval as $y = 3\sin(4x - \pi)$

$$0 \leqslant 4x - \pi \leqslant 2\pi$$
$$\pi \leqslant 4x \leqslant 3\pi$$
$$\frac{\pi}{4} \leqslant x \leqslant \frac{3\pi}{4}$$

The period of this function is $\pi/2$: $\pi/2 = 3\pi/4 - \pi/4$. Because cosecant is asymptotic to $x = \pi/4$, $x = 2\pi/4$ and $x = 3\pi/4$, these vertical lines serve as guides to help us sketch the graph. As with secant

$$1 \leqslant \csc\text{ arg} \quad\text{or}\quad \csc\text{ arg} \leqslant \text{-}1$$
$$\text{Hence,}\qquad 3 \leqslant 3\csc\text{ arg} \quad\text{or}\quad 3\csc\text{ arg} \leqslant \text{-}3$$

Taking the reciprocal of $3\sin(4x - \pi)$ produces asymptotes at the zeros of $3\sin(4x - \pi)$ and "inverts" the values of $3\sin(4x - \pi)$ about $y = 3$ and $y = \text{-}3$. See Figure 7-81.

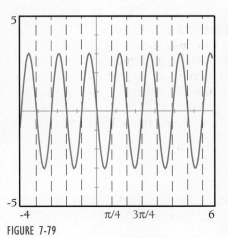

FIGURE 7-79

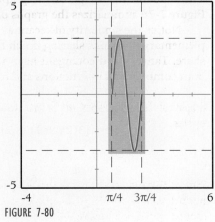

FIGURE 7-80

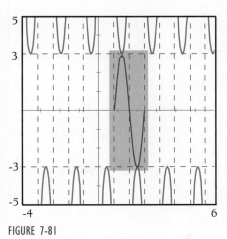

FIGURE 7-81

Illustration 2:

$$y = -2 \cot\left(2x - \frac{\pi}{3}\right)$$

Solution:

$$y = -2 \; \frac{\cos\left(2x - \dfrac{\pi}{3}\right)}{\sin\left(2x - \dfrac{\pi}{3}\right)}$$

Because $y = \cot x$ has a period of π then $y = -2\cot(2x - \pi/3)$ completes a cycle for

$$0 < 2x - \frac{\pi}{3} < \pi$$

$$\frac{\pi}{3} < 2x < \frac{4\pi}{3}$$

$$\frac{\pi}{6} < x < \frac{2\pi}{3}$$

Therefore the period is $\pi/2$. The shape of $y = \cot x$ on $(0, \pi)$ transforms into a window on $(\pi/6, 2\pi/3)$; $y = \pi/6$ and $y = 2\pi/3$ are vertical asymptotes. In general vertical asymptotes for this function occur where $\sin(2x - \pi/3) = 0$. See Figure 7-82. Hence we have asymptotes for

$$2x - \frac{\pi}{3} = k\pi$$

$$2x = (3k + 1)\frac{\pi}{3}$$

asymptotes: $x = (3k + 1)\dfrac{\pi}{6}$

Similarly, $y = -2\cot(2x - \pi/3)$ intercepts the x-axis where $\cos(2x - \pi/3) = 0$:

$$2x - \frac{\pi}{3} = (2k + 1)\frac{\pi}{2}$$

$$2x = (2k + 1)\frac{\pi}{2} + \frac{\pi}{3}$$

$$2x = 3(2k + 1)\frac{\pi}{6} + \frac{2\pi}{6}$$

$$2x = [3(2k + 1) + 2]\frac{\pi}{6}$$

$$2x = (6k + 5)\frac{\pi}{6}$$

$$x = (6k + 5)\frac{\pi}{12}$$

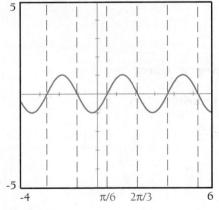

FIGURE 7-82

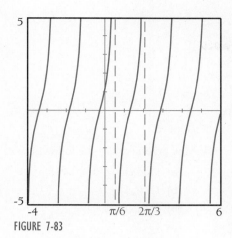

FIGURE 7-83

The x-intercepts for $y = -2 \cot(2x - \pi/3)$ are half-way between the asymptotes.

To account for the effect of the sign of -2 on the graph, reflect the graph shape from top to bottom; $|-2|$ stretches the graph vertically. We have no guidelines for this vertical effect, so we plot values of the function or have technology demonstrate the expected "stretch" in the graph. See Figure 7-83. For example,

$$\text{if } x = \frac{\pi}{3}, \text{ then} \qquad y = -2 \cot\left(\frac{2\pi}{3} - \frac{\pi}{3}\right)$$

$$y = -2 \cot\left(\frac{\pi}{3}\right)$$

$$y = -\frac{2}{\sqrt{3}} \qquad\qquad (\cong -1.1547)$$

$$\text{If } x = \frac{5\pi}{12}, \text{ then} \qquad y = -2 \cos\left[2\left(\frac{5\pi}{12}\right) - \frac{\pi}{3}\right]$$

$$y = -2 \cos\left(\frac{5\pi}{6} - \frac{\pi}{3}\right)$$

$$y = -2 \cos\left(\frac{\pi}{2}\right)$$

$$y = 0.$$

SUMMARY

- For graphs of secant, cosecant, tangent and cotangent see Figure 7-78.
- The following are complementary functions:

  secant and cosecant: $\sec x = \csc\left(\dfrac{\pi}{2} - x\right)$, $\csc x = \sec\left(\dfrac{\pi}{2} - x\right)$

  tangent and cotangent: $\tan x = \cot\left(\dfrac{\pi}{2} - x\right)$, $\cot x = \tan\left(\dfrac{\pi}{2} - x\right)$

- The following are reciprocal functions:

  sine and cosecant: $\sin x = \dfrac{1}{\csc x}$, $\csc x = \dfrac{1}{\sin x}$

  cosine and secant: $\cos x = \dfrac{1}{\sec x}$, $\sec x = \dfrac{1}{\cos x}$

  tangent and cotangent: $\tan x = \dfrac{1}{\cot x}$, $\cot x = \dfrac{1}{\tan x}$

7-5   E X E R C I S E S

Graph each of the following over two complete cycles.

1. $f(x) = \tan(2x)$

2. $f(x) = \sec x + 1$

3. $f(x) = \sec\left(x - \dfrac{\pi}{6}\right)$

4. $f(x) = \tan\left(x + \dfrac{\pi}{4}\right)$

5. $f(x) = \csc x - 3$

6. $f(x) = \cot\left(x - \dfrac{\pi}{2}\right)$

7. $f(x) = \cot(x - \pi) + 1$

8. $f(x) = \csc(x + \pi) - 1$

9. $f(x) = 3\tan\left(x - \dfrac{\pi}{3}\right) + 1$

10. $f(x) = -2\sec\left(x - \dfrac{\pi}{6}\right) - 5$

11. $f(x) = 3\sec x$

12. $f(x) = -5\tan x$

13. $f(x) = -4\csc x$

14. $f(x) = 3\cot x$

15. $f(x) = \sec(5x)$

16. $f(x) = \tan(-4x)$

17. $f(x) = \cot(\pi - 3x) + 1$

18. $f(x) = \csc(2 - 5x) - \pi$

19. $f(x) = \pi\tan(2x - \pi/3) - 2$

20. $f(x) = -2\sec(3x - \pi/2) + \pi$

Graph $L(x) = 2x - \pi/3$. Now graph each of the following.

21. (a) $f(x) = \sec x$ (b) $f \circ L(x)$

22. (a) $f(x) = \csc x$ (b) $L \circ f(x)$

23. (a) $f(x) = \sec x$ (b) $L \circ f(x)$

24. (a) $f(x) = \csc x$ (b) $f \circ L(x)$

25. (a) $f(x) = \cot x$ (b) $f \circ L(x)$

26. (a) $f(x) = \tan x$ (b) $L \circ f(x)$

27. (a) $f(x) = \cot x$ (b) $L \circ f(x)$

28. (a) $f(x) = \tan x$ (b) $f \circ L(x)$

*29. (a) $f(x) = \tan x$ (b) $L \circ f \circ L(x)$

*30. (a) $f(x) = \sec x$ (b) $L \circ f \circ L(x)$

7-5   P R O B L E M   S E T

Recall that the slope of the line through $(x, f(x))$ and $(x + h, f(x + h))$ is given by $m = [f(x + h) - f(x)]/h$. This quotient represents the average rate of change of the function over the interval $(x, x + h)$. If h is a "small" number then the quotient approximates the instantaneous rate of change of the function at x. Approximate each of the following as indicated.

1. Discuss any pattern you see in the answers to these problems.

(a) m, where $f(x) = \sin x$, $x = \pi/3$ and $h = 0.0001$, and $\cos(\pi/3)$

(b) m, where $f(x) = \sin x$, $x = \pi/6$ and $h = 0.0001$, and $\cos(\pi/6)$

(c) m, where $f(x) = \sin x$, $x = \pi/4$ and $h = 0.0001$, and $\cos(\pi/4)$

(d) m, where $f(x) = \sin x$, $x = 0$ and $h = 0.0001$, and $\cos 0$

2. Discuss any pattern you see in these problems.

(a) m, where $f(x) = \cos x$, $x = \pi/3$ and $h = 0.0001$, and $\sin(\pi/3)$

(b) m, where $f(x) = \cos x$, $x = \pi/4$ and $h = 0.0001$, and $\sin(\pi/4)$

(c) m, where $f(x) = \cos x$, $x = \pi/2$ and $h = 0.0001$, and $\sin(\pi/2)$

(d) m, where $f(x) = \cos x$, $x = \pi/3$ and $h = 0.0001$, and $\sin(\pi/3)$

PROBLEMS FOR GRAPHING TECHNOLOGY

If the identity $f(x) \equiv g(x)$ is valid then the graphs of $f(x)$ and $g(x)$ coincide for all x in their common domain. In Figure 7-84, the graph of $y = \sin x$ coincides with the graph $y = \cos(\pi/2 - x)$. Therefore, $\sin x \equiv \cos(\pi/2 - x)$ is an identity. However, in Figure 7-84, the graph of $y = 2\tan x$ and $y = 1$ are distinct graphs. Then $2\tan x = 1$ is a conditional equation and the points where $y = 1$ and $y = 2\tan x$ intersect form the solution of the equation.

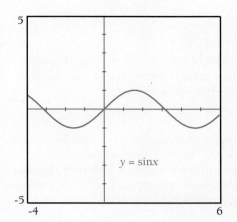
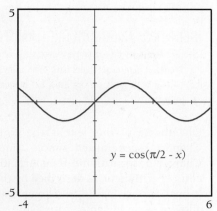
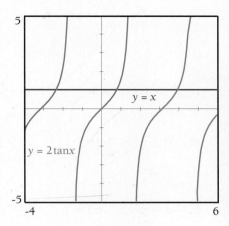

FIGURE 7-84

Identify each of the following as conditional equations *or* identities by examining their graphs. Be aware that graphs are not always as accurate as we might like. It is correct in these cases to say the equation "seems" to be an identity.

3. $\tan x = \dfrac{\sin x}{\cos x}$ **4.** $\cot x = \dfrac{\cos x}{\sin x}$

5. $\sin^2 x + \cos^2 x = 1$ **6.** $1 + \tan^2 x = \sec^2 x$

7. $\sin x = \cos x$

8. $\sin(2x) = 2\cos(x)\sin(x)$

9. $1 + \cot^2 x = \csc^2 x$

10. $\cos(2x) = \cos^2(x) - \sin^2(x)$

11. $\cos(2x) = 2\cos^2 x - 1$

12. $\cos(2x) = 1 - 2\sin^2 x$

13. $\sin^2 x = \dfrac{1 - \cos(2x)}{2}$

14. $\cos^2 x = \dfrac{1 + \cos(2x)}{2}$

*15. $\cos x = \dfrac{\sin(x + 0.0001) - \sin x}{0.0001}$

*16. $\sin x = \dfrac{\cos x - \cos(x - 0.0001)}{0.0001}$

*17. $\sin x = \csc x$ ***18.** $\cos x = \sec x$

*19. $\tan x = \cot x$ ***20.** $\sec x = \csc x$

**21. $e^x = \dfrac{e^{x+0.0001} - e^x}{0.0001}$

**22. $\dfrac{1}{x} = \dfrac{\ln(x + 0.0001) - \ln(x)}{0.0001}$

EXPLORATION

The following problems anticipate concepts from the next section. Graph each of the following pairs of functions f and g in the same window. From the graph, conjecture whether $f(x) = g(x)$ is an identity.

23. $f(x) = \sin(2x)$, $g(x) = 2\sin x \cos x$

24. $f(x) = \cos(2x)$, $g(x) = \cos^2 x - \sin^2 x$

25. $f(x) = \cos(2x)$, $g(x) = 2\cos^2 x - 1$

26. $f(x) = \cos(2x)$, $g(x) = 1 - 2\sin^2 x$

27. $f(x) = \cos(3x) + \cos x$, $g(x) = 2\cos(2x)\cos x$

28. $f(x) = \sin(3x) - \sin x$, $g(x) = 2\cos(2x)\sin x$

29. $f(x) = 2\sin^2 x$, $g(x) = 1 - \cos(2x)$

30. $f(x) = \tan \dfrac{x}{2}$, $g(x) = \dfrac{\sin x}{1 + \cos x}$

31. $f(x) = \tan \dfrac{x}{2}$, $g(x) = \dfrac{1 - \cos x}{\sin x}$

32. $f(x) = \tan^2 x$, $g(x) = 1 - \cos(2x)$

6   IDENTITIES: TRIGONOMETRIC TRANSFORMATIONS

Scientific discoveries do not progress necessarily on the lines of direct usefulness. Very many applications of the theories of pure mathematics have come many years, sometimes centuries, after the actual discoveries themselves. The weapons were at hand, but the men were not able to use them.
—A. J. FORSYTH

One theme of this text is the concept of a function as a transformation. Chapter 2 generalized simple translation of axes into transformations. In Chapter 6, logarithmic transformations proved useful in reducing the difficulty of arithmetic as described by the order of operations. Each of these transformations was useful in its own way. Transformations using linear functions allowed us to imagine shapes shifted, stretched or compressed. A logarithmic transformation converted exponentiation to multiplication and multiplication to addition. Now, it is reasonable to wonder whether circular functions possess useful transformation properties. Because rotations form the basis of the definition of circular functions, perhaps properties of circular functions would be useful in the rotation of graphic images. Chapter 9 pursues rotation as an application of circular functions. For now, we develop techniques of rewriting expressions involving circular functions.

Identities provide the primary method for rewriting expressions of circular function. Section 7-3 explored some of these identities. One such identity is the complemental relation of sine and cosine: $\cos(\pi/2 - x) \equiv \sin x$. We also discovered that sine is an odd function, $\sin(-x) \equiv -\sin x$; whereas cosine is an even function, $\cos(-x) \equiv \cos x$. These two identities follow immediately from the observed symmetry of sine and cosine.

Are there properties of circular functions similar to the transformation properties of logarithms? This section examines properties of circular functions with multiple arguments. The process for proving an identity remains the same. Where appropriate, derive the identiy. At other times, verification is sufficient. The first identity we investigate is a formula for the sum of two angles.

Chord—A line segment with endpoints on a circle.

Examine Figure 7-85. The arc labeled a and b is the same size as the arc labeled a and $-b$. Equal arcs have equal chords so that the distance from point

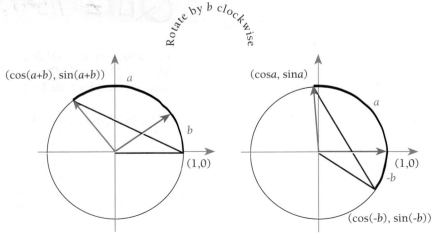

FIGURE 7-85

(cos a, sin a) to point (cos($-b$), sin($-b$), sin($-b$)) is the same as the distance from (cos($a + b$), sin($a + b$)) to (1, 0). Because cosine is even and sine is odd, replace (cos($-b$), sin($-b$)) with (cos b, $-$sin b) and apply the **distance formula** (from Chapter 1, Section 1):

$$[\cos(a + b) - 1]^2 + [\sin(a + b) - 0]^2 = [\cos a - \cos b]^2 + [\sin a + \sin b]^2$$

$$\cos^2(a + b) - 2\cos(a + b) + 1 + \sin^2(a + b) = \cos^2 a - 2\cos(a)\cos(b) + \cos^2 b + \sin^2 a + 2\sin(a)\sin(b) + \sin^2 b$$

By the Pythagorean relation,

$$\cos^2(a + b) + \sin^2(a + b) \equiv 1,$$

also

$$\cos^2 a + \sin^2 a \equiv 1$$

and

$$\cos^2 b + \sin^2 b \equiv 1$$

Therefore,

$$2 - 2\cos(a + b) = 2 - 2\cos(a)\cos(b) + 2\sin(a)\sin(b)$$
$$\cos(a + b) \equiv \cos(a)\cos(b) - \sin(a)\sin(b)$$

which is the sum formula.

For cos($a - b$),

$$\cos(a - b) \equiv \cos[a + (-b)]$$
$$\cos(a - b) \equiv \cos(a)\cos(-b) - \sin(a)\sin(-b)$$
$$\cos(a - b) \equiv \cos(a)\cos(b) + \sin(a)\sin(b)$$

These results form the next theorem.

Theorem 7-8
Sum Formula for Cosine

$$\cos(a \pm b) \equiv \cos(a)\cos(b) \mp \sin(a)\sin(b)$$

What about sin($a + b$)? Because sine and cosine are complementary functions, sin($a + b$) $\equiv$ cos[$\pi/2 - (a + b)$]

Apply the sum formula for cosine:

$$\sin(a + b) \equiv \cos\left[\frac{\pi}{2} - (a + b)\right]$$

$$\equiv \cos\left[\left(\frac{\pi}{2} - a\right) - b\right]$$

$$\equiv \cos\left(\frac{\pi}{2} - a\right)\cos(b) + \sin\left(\frac{\pi}{2} - a\right)\sin(b)$$

$$\sin(a + b) \equiv \sin(a)\cos(b) + \cos(a)\sin(b)$$

To determine a formula for $\sin(a - b)$, proceed as with $\cos(a - b)$. The derivation for $\sin(a - b)$ is left as an exercise.

Theorem 7-9 Sum Formula for Sine	$$\sin(a \pm b) \equiv \sin(a)\cos(b) \pm \cos(a)\sin(b)$$

Because $(2a) \equiv (a + a)$, a corollary to the two previous theorems follows.

Theorem 7-10 Double Angle Formulas	$$\cos(2a) \equiv \cos^2(a) - \sin^2(a)$$ $$\cos(2a) \equiv 2\cos^2(a) - 1$$ $$\cos(2a) \equiv 1 - 2\sin^2(a)$$ $$\sin(2a) \equiv 2\sin(a)\cos(a)$$

The proofs are left as exercises.

E X A M P L E   7 - 12

Application of Sum and Double Angle Formulas. Evaluate the following as indicated.

Illustration 1:

$$\sin\left(\frac{7\pi}{12}\right)$$

Solution:

$$\sin\left(\frac{7\pi}{12}\right) = \sin\left(\frac{\pi}{3} + \frac{\pi}{4}\right)$$

$$= \sin\left(\frac{\pi}{3}\right)\cos\left(\frac{\pi}{4}\right) + \cos\left(\frac{\pi}{3}\right)\sin\left(\frac{\pi}{4}\right)$$

$$= \frac{\sqrt{3}}{2}\frac{\sqrt{2}}{2} + \frac{1}{2}\frac{\sqrt{2}}{2}$$

Answer:

$$\sin\left(\frac{7\pi}{12}\right) = \frac{\sqrt{6} + \sqrt{2}}{4}$$

Note: By calculator, $\sin(7\pi/12) \cong 0.965925826$, whereas $(\sqrt{6} + \sqrt{2})/4 \cong 0.965925826$.

Illustration 2:   Suppose $\sin x = 0.2$, where $\pi/2 < x < \pi$, determine $\sin(2x)$.

Solution:

$$\sin(2x) = 2\sin(x)\cos(x)$$

Although $\sin x$ is known, we must calculate $\cos x$ to complete the solution. By the Pythagorean relation,

$$\sin^2 x + \cos^2 x = 1$$
$$(0.2)^2 + \cos^2 x = 1$$
$$\cos^2 x = 1 - 0.04$$
$$\cos x = \pm \sqrt{0.96}$$

Because x is in the second quadrant, choose $\cos x = -\sqrt{0.96}$.

$$\sin(2x) = 2 \sin(x) \cos(x)$$
$$= 2(0.2)(-\sqrt{0.96})$$

Answer:

$$\sin(2x) = -0.4\sqrt{0.96}$$

Note:

$$-0.4\sqrt{0.96} = -\frac{4}{10}\sqrt{\frac{96}{100}} = -\frac{2}{5}\sqrt{\frac{16(6)}{100}} = -\frac{4\sqrt{6}}{25}$$

Illustration 3: Suppose $\sin x = 2\sqrt{2}/5$, determine $\cos(2x)$.

Solution:

$$\cos(2x) = 1 - 2\sin^2 x$$
$$= 1 - 2\left(\frac{2\sqrt{2}}{5}\right)^2$$
$$= 1 - \frac{16}{25}$$

Answer:

$$\cos(2x) = \frac{9}{25}$$

Illustration 4: Suppose $\sin \theta = \frac{3}{5}$, $\cos \phi = -\frac{5}{13}$, θ in quadrant II and ϕ in quadrant III. Determine $\cos(\theta - \phi)$.

Solution:

$$\cos(\theta - \phi) = \cos(\theta) \cos(\phi) + \sin(\theta) \sin(\phi)$$

Although $\sin \theta$ and $\cos \phi$ are known, we need $\sin \phi$ and $\cos \theta$. Use the Pythagorean relation:

$$\sin^2 \theta + \cos^2 \theta = 1$$
$$\left(\frac{3}{5}\right)^2 + \cos^2 \theta = 1$$
$$\cos^2 \theta = \frac{16}{25}$$

$$\cos \theta = -\frac{4}{5} \qquad (\theta \text{ in quadrant II})$$

$$\sin^2 \phi + \cos^2 \phi = 1$$

$$\sin^2 \phi + \left(-\frac{5}{13}\right)^2 = 1$$

$$\sin^2 \phi = \frac{144}{169}$$

$$\sin \phi = -\frac{12}{13} \qquad (\phi \text{ in quadrant III})$$

Now,

$$\cos(\theta - \phi) = \cos(\theta)\cos(\phi) + \sin(\theta)\sin(\phi)$$

$$= \left(-\frac{4}{5}\right)\left(-\frac{5}{13}\right) + \left(\frac{3}{5}\right)\left(-\frac{12}{13}\right)$$

$$= \frac{20}{65} - \frac{36}{65}$$

Answer:

$$\cos(\theta - \phi) = -\frac{16}{65}$$

The derivation of a half-angle formula makes subtle use of the double angle formula. Begin with

$$\cos a \equiv \cos\left(\frac{2a}{2}\right)$$

Now apply the second formula from Theorem 7-10.

$$\cos a \equiv \cos\left[2\left(\frac{a}{2}\right)\right]$$

$$\cos a \equiv 2\cos^2\left(\frac{a}{2}\right) - 1$$

Solve for $\cos\left(\frac{a}{2}\right)$:

$$\cos^2\left(\frac{a}{2}\right) = \frac{1 + \cos a}{2}$$

$$\cos\left(\frac{a}{2}\right) = \pm\sqrt{\frac{1 + \cos a}{2}}$$

Which sign is correct, plus or minus? The sign depends upon the quadrant in which $a/2$ terminates. If $a/2$ terminates in the first or fourth quadrant, choose the positive sign. If $a/2$ terminates in quadrants II or III, choose the negative sign.

Derive a formula for $\sin(a/2)$ from the third formula in Theorem 7-10.

Theorem 7-11 Half-Angle Formulas	$$\cos \frac{a}{2} \equiv \pm \sqrt{\frac{1 + \cos a}{2}}$$ $$\sin \frac{a}{2} \equiv \pm \sqrt{\frac{1 - \cos a}{2}}$$ $$\tan \frac{a}{2} \equiv \frac{1 - \cos a}{\sin a} \equiv \frac{\sin a}{1 + \cos a}$$

E X A M P L E   7 - 13

Application of Half-Angle Formulas. Suppose that $\sin \theta = 0.6$ and $0 < \theta < \pi/2$, evaluate each of the following.

Illustration 1:

$$\cos \theta$$

Solution:

$$\cos^2 \theta + \sin^2 \theta = 1$$
$$\cos^2 \theta = 1 - \sin^2 \theta$$
$$\cos^2 \theta = 1 - 0.6^2$$
$$\cos^2 \theta = 0.64$$

Answer:

$$\cos \theta = 0.8 \qquad \left(\text{because } 0 < \theta < \frac{\pi}{2} \right)$$

Note: Use $\cos \theta = 0.8$ in the next three illustrations.

Illustration 2:

$$\cos \frac{\theta}{2}$$

Solution:

$$\cos \frac{\theta}{2} = \sqrt{\frac{1 + \cos \theta}{2}} \qquad \left(\text{because } 0 < \frac{\theta}{2} < \frac{\pi}{2} \right)$$

$$= \sqrt{\frac{1 + 0.8}{2}}$$

Answer:

$$\cos \frac{\theta}{2} = \sqrt{0.9}$$

Approximation: $\cong 0.948683298$

Illustration 3:

$$\sin \frac{\theta}{2}$$

Solution:

$$\sin \frac{\theta}{2} = \sqrt{\frac{1 - \cos \theta}{2}} \qquad \left(\text{because } 0 < \frac{\theta}{2} < \frac{\pi}{2} \right)$$

$$= \sqrt{\frac{1 - 0.8}{2}}$$

Answer:

$$\sin \frac{\theta}{2} = \sqrt{0.1}$$

Approximation $\cong 0.316227766$

Illustration 4:

$$\tan \frac{\theta}{2}$$

Solution 1:

By definition $\tan \dfrac{\theta}{2} = \dfrac{\sin \frac{\theta}{2}}{\cos \frac{\theta}{2}}$

From Illustrations 1 and 2 $\tan \dfrac{\theta}{2} = \dfrac{\sqrt{0.1}}{\sqrt{0.9}}$

$$\tan \frac{\theta}{2} = \frac{1}{\sqrt{9}}$$

Answer:

$$\tan \frac{\theta}{2} = \frac{1}{3}$$

Solution 2:

From Theorem 7-11, $\tan \dfrac{\theta}{2} = \dfrac{1 - \cos a}{\sin a}$

$$\tan \frac{\theta}{2} = \frac{1 - 0.8}{0.6}$$

Answer:

$$\tan \frac{\theta}{2} = \frac{1}{3}$$

Illustration 5: $\cot \dfrac{\pi}{12}$

Solution:

$$\frac{\pi}{12} = \frac{\frac{\pi}{6}}{2}$$

Use the half-angle formula for the tangent and the reciprocal relation of cotangent and tangent.

$$\tan \frac{\pi}{12} = \tan \left(\frac{\frac{\pi}{6}}{2} \right)$$

$$= \frac{1 - \cos \dfrac{\pi}{6}}{\sin \dfrac{\pi}{6}}$$

$$= \frac{1 - \dfrac{\sqrt{3}}{2}}{\dfrac{1}{2}}$$

$$= 2 - \sqrt{3}$$

Now $\cot \dfrac{\pi}{12} = \dfrac{1}{\tan \dfrac{\pi}{12}}$

Answer:

$$\cot \frac{\pi}{12} = \frac{1}{2 - \sqrt{3}}$$

Alternate: $= 2 + \sqrt{3}$
Approximation: $\cong 3.732050808$

Note: Direct approximation of cot ($\pi/12$) by calculator should match the preceding approximation to at least eight decimal places.

Illustration 6:   Derive a formula for $\cos(A + B) + \cos(A - B)$.

Solution:

$$\cos(A + B) + \cos(A - B) \equiv \cos(A + B) + \cos(A - B)$$
$$\equiv \cos A \cos B - \sin A \sin B +$$
$$\cos A \cos B + \sin A \sin B$$
$$\equiv 2 \cos A \cos B$$

Note:

As a result, $\cos A \cos B \equiv \dfrac{\cos(A + B) + \cos(A - B)}{2}$

Illustration 7:   Refer to Illustration 6 and substitute $u = A + B$ and $v = A - B$.

Solution: If $u = A + B$ and $v = A - B$, then $u + v = 2A$ so that $A = (u + v)/2$. Similarly, $(u - v) = 2B$ and $B = (u - v)/2$. Substitute into $\cos(A + B) + \cos(A - B) \equiv 2 \cos A \cos B$ to obtain $\cos u + \cos v \equiv 2 \cos [(u + v)/2] \cos [(u - v)/2]$

Illustration 8:   Evaluate $\cos 75° + \cos 15°$.

Solution: Refer to Illustration 7: $\cos u + \cos v \equiv 2 \cos [(u + v)/2] \cos (u - v)/2]$. Now,

$$\cos 75° + \cos 15° = 2 \cos \frac{75° + 15°}{2} \cos \frac{75° - 15°}{2}$$

$$= 2 \cos 45° \cos 30°$$

$$= 2 \frac{\sqrt{2}}{2} \frac{\sqrt{3}}{2}$$

$$= \frac{\sqrt{6}}{2}$$

Illustration 6 leads to the family of product formulas in the following theorem. Their derivation is left as an exercise.   ▬

Theorem 7-12
Product of Functions Formulas

$$\cos A \cos B \equiv [\cos(A + B) + \cos(A - B)]/2$$
$$\cos A \sin B \equiv [\sin(A + B) - \sin(A - B)]/2$$
$$\sin A \cos B \equiv [\sin(A + B) + \sin(A - B)]/2$$
$$\sin A \sin B \equiv [\cos(A - B) - \cos(A + B)]/2$$

Similarly, Illustration 7 suggests the following family of sum formulas.

Theorem 7-13
Sum of Functions Formulas

$$\cos u + \cos v \equiv 2 \cos \frac{u+v}{2} \cos \frac{u-v}{2}$$

$$\cos u - \cos v \equiv -2 \sin \frac{u+v}{2} \sin \frac{u+v}{2}$$

$$\sin u + \sin v \equiv 2 \sin \frac{u+v}{2} \cos \frac{u-v}{2}$$

$$\sin u - \sin v \equiv 2 \cos \frac{u+v}{2} \sin \frac{u-v}{2}$$

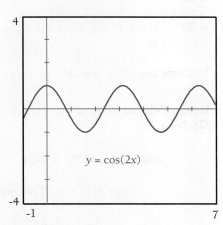

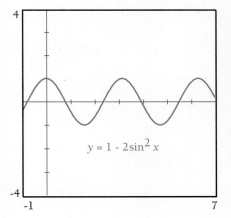

FIGURE 7-86 $\cos(2x) \equiv 1 - 2\sin^2 x$

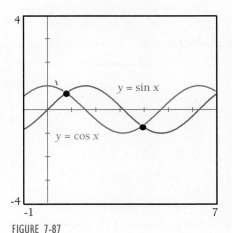

FIGURE 7-87

Consider the graph of $f(x) = \cos(2x)$. Analysis tells us the graph is a cosine curve with period π. The graph of $g(x) = 1 - 2\sin^2 x$ is more difficult to analyze: neither transformations with logarithms nor linear functions provide for the effect of squaring. However, our calculator rapidly plots points to produce the graph of g. Figure 7-86 provides a comparison of the graphs of f and g.

Note that the graphs are identical for every point in the common domain of f and g: $f(x) \equiv g(x)$. Because of the identity we could anticipate the graph of g from the graph of f. Using a similar method, what is the graph of $h(x) = \cos^2 x$?

Figure 7-87 displays $h(x) = \sin x$ and $j(x) = \cos x$ on the same coordinate system. Note where h and j intersect—for example, at $x = \pi/4$—these are points where $\sin x = \cos x$. Sine and cosine are not *identical*, although they sometimes coincide: $\sin x = \cos x$ is a *conditional equation*, not an identity.

SUMMARY

$$\cos(a \pm b) \equiv \cos(a)\cos(b) \mp \sin(a)\sin(b) \qquad \text{(sums of arguments)}$$
$$\sin(a \pm b) \equiv \sin(a)\cos(b) \pm \cos(a)\sin(b)$$
$$\cos(2a) \equiv \cos^2 a - \sin^2 a \qquad \text{(double angle formulas)}$$
$$\equiv 2\cos^2(a) - 1$$
$$\equiv 1 - 2\sin^2 a$$
$$\sin(2a) \equiv 2\sin a \cos a$$
$$\cos\frac{a}{2} \equiv \pm\sqrt{\frac{1 + \cos a}{2}} \qquad \text{(half-angle formulas)}$$
$$\sin\frac{a}{2} \equiv \pm\sqrt{\frac{1 - \cos a}{2}}$$
$$\tan\frac{a}{2} \equiv \frac{1 - \cos a}{\sin a} \equiv \frac{\sin a}{1 + \cos a}$$
$$\cos A \cos B \equiv [\cos(A + B) + \cos(A - B)]/2 \qquad \text{(product formulas)}$$
$$\cos A \sin B \equiv [\sin(A + B) - \sin(A - B)]/2$$
$$\sin A \cos B \equiv [\sin(A + B) + \sin(A - B)]/2$$
$$\sin A \sin B \equiv [\cos(A - B) - \cos(A + B)]/2$$
$$\cos u + \cos v \equiv 2\cos\frac{u + v}{2}\cos\frac{u - v}{2} \qquad \text{(sums of functions)}$$
$$\cos u - \cos v \equiv -2\sin\frac{u + v}{2}\sin\frac{u - v}{2}$$
$$\sin u + \sin v \equiv 2\sin\frac{u + v}{2}\cos\frac{u - v}{2}$$
$$\sin u - \sin v \equiv 2\cos\frac{u + v}{2}\sin\frac{u - v}{2}$$

7-6   EXERCISES

Verify the following identities.

1. $\sec\left(\dfrac{\pi}{2} - x\right) \equiv \csc x$

2. $\cot\left(\dfrac{\pi}{2} - x\right) \equiv \tan x$

3. $\sec x + \tan x \equiv \dfrac{1 + \sin x}{\cos x}$

4. $\dfrac{1 + \cos x}{\sin x} \equiv \csc x + \cot x$

5. $\sin(x)\cos(x) \equiv \dfrac{\tan x}{1 + \tan^2 x}$

6. $\sin(x)\cos(x) \equiv \dfrac{\cot x}{1 + \cot^2 x}$

7. $\cot^2 x + 1 \equiv \csc^2 x$

8. $\dfrac{1}{\sec^2 x} + \dfrac{1}{\csc^2 x} \equiv 1$

9. $[\sin x + \cos x]^2 + [\sin x - \cos x]^2 \equiv 2$

10. $[\sec x - \tan x][\sec x + \tan x] \equiv 1$

11. $\cot(A + B) \equiv \dfrac{\cot A \cot B - 1}{\cot A + \cot B}$

12. $\sec(A + B) \equiv \dfrac{\sec A \sec B}{1 - \tan A \tan B}$

13. $\cos(r + s)\cos(r - s) \equiv \cos^2 r - \sin^2 s$

14. $\sin(r + s) + \sin(r - s) \equiv 2 \sin r \cos s$

15. $\cos(A + B) + \cos(A - B) \equiv 2 \cos A \cos B$

16. $\sin(A + B) + \sin(A - B) \equiv \sin^2 A - \sin^2 B$

17. $\tan \dfrac{x}{2} \equiv \csc x - \cot x$

18. $\cos(2x) \equiv \cos^4 x - \sin^4 x$

19. $\sin(2x) \equiv \dfrac{2 \tan x}{1 + \tan^2 x}$

20. $\sec(2x) \equiv \dfrac{\tan x + \cot x}{\cot x - \tan x}$

21. $\tan(2x) \equiv \dfrac{\tan(5x) - \tan(3x)}{1 + \tan(5x)\tan(3x)}$

22. $\cos(2x) \equiv \dfrac{1 - \tan^2 x}{\sec^2 x}$

Evaluate the following without using a calculator.

23. $\sin\left(\dfrac{\pi}{4} + \dfrac{\pi}{6}\right)$ 24. $\cos\left(\dfrac{5\pi}{12}\right)$

25. $\cos\left(\dfrac{\pi}{12}\right)$ 26. $\sin\left(\dfrac{\pi}{4} - \dfrac{\pi}{6}\right)$

27. $\cos\left(\dfrac{7\pi}{12}\right)$ 28. $\sin\left(\dfrac{\pi}{4} + \dfrac{\pi}{3}\right)$

29. $\sin\left(-\dfrac{\pi}{6}\right)$ 30. $\cos\left(-\dfrac{\pi}{6}\right)$

31. $\sin\left(-\dfrac{3\pi}{4}\right)$ 32. $\cos\left(-\dfrac{3\pi}{4}\right)$

Simplify the following trigonometric expressions by rewriting the expression to contain one trigonometric function.

33. $\cos(7x)\cos(3x) - \sin(7x)\sin(3x)$

34. $\cos(5x)\sin(3x) - \sin(5x)\cos(3x)$

35. $1 - 2\cos^2(6x)$

36. $\dfrac{\tan x + \tan(2x)}{1 - \tan x \tan(2x)}$

37. $\cos(5x)\cos x + \sin(5x)\sin x$

38. $\sin(8x)\cos(2x) + \cos(8x)\sin(2x)$

Simplify the following trigonometric expressions by rewriting them into a product of sines or cosines.

39. $\sin(7x) + \sin(3x)$

40. $\cos(9x) + \cos x$

41. $\cos(5x) - \cos(9x)$

42. $\sin(4x) - \sin(6x)$

In problems 43–55 use the following information: $\sin x = 0.1$; $\cos r = 0.3$; $\cos y = 0.2$; $\sin s = 0.5$. Each of x, y, r and s terminate in quadrant I. Evaluate as indicated.

43. Determine $\cos x$. (Hint: use $\sin^2 x + \cos^2 x = 1$.)

44. Determine $\sin r$. (Hint: see Problem 43.)

45. Determine $\sin y$.   46. Determine $\cos s$.

47. Determine $\cos(x + y)$.   48. Determine $\sin(r + s)$.

49. Determine $\sin(x - y)$.   50. Determine $\cos(r - s)$.

51. Determine $\tan(x + y)$.   52. Determine $\tan(r - s)$.

53. Determine $\sin(2r)$.   54. Determine $\cos(2s)$.

55. Determine $\sin(2r + 2s)$.

Derive the following identities.

56. $\cos(2x) \equiv 2 \sin^2 x - 1$

57. $\sin(2x) \equiv 2 \sin(x)\cos(x)$

58. $\cos(2x) \equiv \cos^2 x - \sin^2 x$

59. $\cos(2x) \equiv 2 \cos^2 x - 1$

60. $\sin \dfrac{x}{2} \equiv \pm\sqrt{\dfrac{1 - \cos x}{2}}$

61. $\cos \dfrac{x}{2} \equiv \pm\sqrt{\dfrac{1 + \cos x}{2}}$

62. $\tan \dfrac{x}{2} \equiv \dfrac{1 - \cos x}{\sin x}$ 63. $\tan \dfrac{x}{2} \equiv \dfrac{\sin x}{1 + \cos x}$

64. $\cot \dfrac{x}{2} \equiv \dfrac{1 + \cos x}{\sin x}$ 65. $\cot \dfrac{x}{2} \equiv \dfrac{\sin x}{1 - \cos x}$

7-6   P R O B L E M   S E T

Derive formulas for the following.

1. $\cos(A + B) + \cos(A - B)$

2. $\cos(A - B) - \cos(A + B)$

3. $\sin(A + B) - \sin(A - B)$

4. $\sin(A + B) + \sin(A - B)$

5. $\cos(A)\cos(B)$. (Hint: see the results to Problem 1.)

6. $\sin(A)\sin(B)$. (Hint: see the results to Problem 2.)

7. $\cos(A)\sin(B)$. (Hint: see the results to Problem 3.)

8. $\sin(A)\cos(B)$. (Hint: see the results to Problem 4.)

Use the substitution $u = A + B$, $v = A - B$ (hence $A = (u + v)/2$, $B = (u - v)/2$) to derive formulas for the following.

9. $\cos\left(\dfrac{u + v}{2}\right)\cos\left(\dfrac{u - v}{2}\right)$ (see Problem 5)

10. $\sin\left(\dfrac{u + v}{2}\right)\sin\left(\dfrac{u - v}{2}\right)$ (see Problem 6)

11. $\cos\left(\dfrac{u + v}{2}\right)\sin\left(\dfrac{u - v}{2}\right)$ **12.** $\sin\left(\dfrac{u + v}{2}\right)\cos\left(\dfrac{u - v}{2}\right)$

13. $\cos v$ **14.** $\cos u$

15. $\sin u$ **16.** $\sin v$

Refer to the results of problems 1–16 to evaluate the following.

17. $\cos\dfrac{7\pi}{12} + \cos\dfrac{\pi}{12}$ **18.** $\cos\dfrac{7\pi}{12} - \cos\dfrac{\pi}{12}$

19. $\sin\dfrac{7\pi}{12} + \sin\dfrac{\pi}{12}$ **20.** $\sin\dfrac{7\pi}{12} - \sin\dfrac{\pi}{12}$

21. $\sin\dfrac{\pi}{4}\cos\dfrac{\pi}{12}$ **22.** $\cos\dfrac{\pi}{4}\sin\dfrac{\pi}{12}$

23. $\cos\dfrac{\pi}{4}\cos\dfrac{\pi}{12}$ **24.** $\sin\dfrac{\pi}{4}\sin\dfrac{\pi}{12}$

7   *APPROXIMATING CIRCULAR FUNCTIONS

What is exact about mathematics but exactness? And is not this a consequence of the inner sense of truth?
—GOETHE

S I D E B A R

The caterpillar stretched, each row of legs undulating. He puffed his reply to Alice out through his water pipe, "Precisely."

"Precisely?" asked Alice. "Precisely, what?"

"Your enquiry contains the essence of the required answer, more or less."

"More or less? What are you talking about? How could my question contain the answer more or less?"

"Exactly," replied the caterpillar. "Exactly correct, roughly speaking, give or take a little, for all practical purposes."

"This is maddening. . . . I don't believe you know the answer. Nor do I see how my question contains the answer, when I don't know the answer."

The caterpillar stretched, each row of legs undulating. He puffed his reply to Alice out through his water pipe, "Precisely."

Chapter 4 explored approximating irrational roots of polynomial functions. Polynomial approximations of circular functions are desirable because polynomials tend to behave themselves and are easy functions to evaluate. The more common approximating polynomials are the Taylor (Brook Taylor, 1685–1731) and Maclaurin (Colin Maclaurin, 1698–1746) polynomials. The derivation of these polynomials requires calculus.

Because the evaluation of a polynomial approximation requires only a finite number of additions and multiplications, calculators and computers sometimes use polynomials to approximate transcendental functions.

Let us examine a polynomial to approximate an exponential function. The fourth-degree Maclaurin polynomial to approximate e^{-x} is given by

$$e^{-x} \cong 1 - \frac{x}{1} + \frac{x^2}{2} - \frac{x^3}{6} + \frac{x^4}{24}$$

Substitute $-x$ for x to obtain a polynomial for e^x:

$$e^x \cong 1 + \frac{x}{1} + \frac{x^2}{2} + \frac{x^3}{6} + \frac{x^4}{24}$$

The seventh-degree Maclaurin polynomial to approximate sine is given by

$$\sin x \cong x - \frac{x^3}{6} + \frac{x^5}{120} - \frac{x^7}{5040}$$

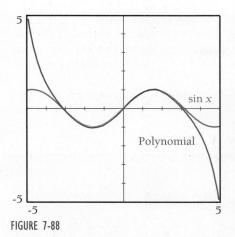

FIGURE 7-88

A seventh-degree polynomial intercepts the x-axis no more than seven times. On the other hand, the sine function has infinitely many x-intercepts. Clearly, the approximation is not perfect. Figure 7-88 illustrates the graph of sine and its seventh-degree polynomial approximation.

Note that the approximation is exact at $x = 0$. However, the farther x is from the origin, the poorer the approximation. When x is far enough from the origin, the polynomial moves toward infinity, while sine cycles about the x-axis. This explains why many excellent calculators generate an error message rather than provide an approximation of sine for values of x too far from the origin. Even an x as small as 8π is unavailable on some calculators.

The sixth-degree Maclaurin polynomial to approximate cosine is given by

$$\cos x \cong 1 - \frac{x^2}{2} + \frac{x^4}{24} - \frac{x^6}{720}$$

Not surprisingly, the Maclaurin polynomial for sine is odd, while the polynomial for cosine is even. If you are confused by the denominators of these approximations, be reassured. There is a simple pattern to the denominators. That pattern rests in the concept of factorial.

DEFINITION 7-5
Factorial

Suppose that n is a counting number, then n!, read "*n* **factorial**," is defined as the product of all counting numbers up to and including n:

$$n! = (1)(2)(3)(4) \ldots (n)$$

As a special case, define $0! = 1$.

EXAMPLE 7-14

Evaluating n!

Illustration:

Evaluate 0!, 1!, 3!, 5!, 10! and 100!

Solution:

By definition
$$0! = 1$$
$$1! = 1$$
$$3! = (1)(2)(3)$$
$$= 6$$
$$5! = (1)(2)(3)(4)(5)$$
$$= 120$$
$$10! = (1)(2)(3)(4)(5)(6)(7)(8)(9)(10)$$
$$= 3,628,800$$

No wonder we use an exclamation point for factorial! Because the calculations get tedious and because factorials are useful, most scientific calculators have a factorial key: the sequence 1 0 & should provide 10! quickly. You may have to use a menu to access factorial on a graphing calculator. In any case,

S I D E B A R
James Stirling's (1692–1770) original approximation of $n!$ involved logarithms. The formula given here known as Stirling's formula is actually by A. De Moivre (1667–1754).

100! is so large that most calculators cannot compute it. When you realize that 20! (approximately $2.432902008 \times 10^{18}$) is larger than the national debt, you begin to understand why. Stirling's formula approximates $n!$ where n is a very large number: $n! \cong n^n \sqrt{2\pi n} \, e^{-n}$. Using this formula we were able to approximate 20!. But our calculator is not able to do 100^{100}.

Now we can generalize the approximating polynomials for sine and cosine.

SUMMARY

The Maclaurin approximating polynomials for sine and cosine are

$$\sin x \cong \frac{x^1}{1!} - \frac{x^3}{3!} + \frac{x^5}{5!} - \ldots + (-1)^{n+1} \frac{x^{2n-1}}{(2n-1)!} \quad (2n-1 \text{ degree})$$

$$\cos x \cong \frac{x^0}{0!} - \frac{x^2}{2!} + \frac{x^4}{4!} - \frac{x^6}{6!} + \ldots + (-1)^{n+1} \frac{x^{2n}}{(2n)!} \quad (2n \text{ degree})$$

EXAMPLE 7-15

Approximating Circular Functions with Maclaurin Polynomials

Illustration 1: Use a fifth-degree Maclaurin polynomial to approximate sin 1.

Solution: The fifth-degree Maclaurin polynomial approximation is

$$\sin x \cong \frac{x}{1} - \frac{x^3}{3!} + \frac{x^5}{5!}$$

Therefore, $\sin 1 \cong \dfrac{1}{1} - \dfrac{1^3}{6} + \dfrac{1^5}{120}$

$$\sin 1 \cong 0.841666666$$

A calculator [SIN] key gives sin 1 $\cong 0.841470984$.

Illustration 2: Use a fourth-degree Maclaurin polynomial to approximate cos 2.

Solution: The fourth-degree Maclaurin polynomial approximation is

$$\cos x \cong 1 - \frac{x^2}{2!} + \frac{x^4}{4!}$$

Therefore, $\cos 2 \cong 1 - \dfrac{2^2}{2} + \dfrac{2^4}{24}$

$$\cos 2 \cong -0.33333333$$

Calculator [COS] key produces cos 2 $\cong -0.416146836$.

This last approximation is not as good as Illustration 1. The source of the excessive error is twofold. First, 2 is not as close to the origin as 1. Maclaurin polynomials give their best approximation "near" the origin. Because we will not move 2 closer to the origin, we rely upon improvement in the second source of error, the degree of the polynomial. The higher the degree, the better is the approximation. The next highest degree is degree 6. The sixth-degree term in that approximation is $-x^6/6!$. Evaluate at 2 and add to the previous approximation:

$$\cos 2 \cong -0.33333333 - 64/720$$
$$\cos 2 \cong -0.422222221$$

7-7   E X E R C I S E S

Write an approximating Maclaurin polynomial as indicated.

1. third-degree for $\sin x$

2. second-degree for $\cos x$

3. fourth-degree for $\cos x$

4. fifth-degree for $\sin x$

5. seventh-degree for $\sin x$

6. sixth-degree for $\cos x$

7. 10th-degree for $\cos x$

8. 11th-degree for $\sin x$

* 9. sixth-degree for e^x

*10. sixth-degree for e^{-x}

Evaluate each of the following.

11. $6!$

12. $7!$

13. $\dfrac{6!}{4!}$

14. $\dfrac{(7)(6)(5)(4)(3)(2)(1)}{(5)(4)(3)(2)(1)}$

15. $\dfrac{(6)(5)(4)(3)(2)(1)}{(4)(3)(2)(1)}$

16. $\dfrac{7!}{5!}$

17. $\dfrac{10!}{6!\,4!}$

18. $\dfrac{9!}{7!\,2!}$

19. $\dfrac{20!}{16!\,4!}$

20. $\dfrac{15!}{12!\,3!}$

Write a seventh-degree Maclaurin polynomial to approximate sine and use it to approximate each of the following:

21. $\sin 1$

22. $\sin 2$

23. $\sin -2$

24. $\sin -1$

25. $\sin 0$

Write a 6th degree polynomial to approximate cosine and use it to approximate each of the following.

26. $\cos 0$

27. $\cos -1$

28. $\cos 1$

29. $\cos 2$

30. $\cos -2$

31. Write a first-degree Maclaurin polynomial to approximate sine. What kind of polynomial is it? Graph this polynomial and sine on the same coordinate system. For how many points does it appear that the approximation is "perfect"?

32. Write a zero-degree Maclaurin polynomial to approximate cosine. What kind of polynomial is it? Graph this polynomial and cosine on the same coordinate system. For how many points does it appear that the approximation is "perfect"?

33. See Problem 31. Answer the same question for a third-degree Maclaurin polynomial.

34. See Problem 32. Answer the same question for a second-degree Maclaurin polynomial.

7-7   P R O B L E M   S E T

Consider another approximating polynomial for sine:

$$\sin z \cong z\left(1 - \frac{z^2}{1^2\pi^2}\right)\left(1 - \frac{z^2}{2^2\pi^2}\right)\left(1 - \frac{z^2}{3^2\pi^2}\right)\cdots$$

Use the first *three* factors of this expression to approximate the following.

1. $\sin 0$

2. $\sin \pi$

3. $\sin \pi/2$

4. $\sin 3\pi/2$

5. $\sin \pi/4$

6. $\sin \pi/6$

7. $\sin 2\pi/3$

8. $\sin \pi/3$

9. $\sin \pi/5$

10. $\sin \pi/8$

CHAPTER SUMMARY

- To convert from degree measure θ to radian measure x use $x = (\pi/180°)\theta$.
  To convert radian measure x to degree measure θ use $\theta = (180°/\pi)x$.
- The angular velocity of a point on the circumference of a circle rotating through x radians over time period t: $\omega = \dfrac{x}{t}$

  The linear velocity of a point on the circumference of a circle of radius r: $v = r\omega$
- Special right triangles have special ratios:

$$\frac{\text{Side opposite } 30°}{\text{hypotenuse}} = \frac{\sqrt{1}}{2} = \frac{1}{2}$$

$$\frac{\text{Side opposite } 45°}{\text{hypotenuse}} = \frac{\sqrt{2}}{2}$$

$$\frac{\text{Side opposite } 60°}{\text{hypotenuse}} = \frac{\sqrt{3}}{2}$$

- The following define acute angle fundamental trigonometric functions: let θ be the measure of an acute angle of a right triangle, $0° < \theta < 90°$, then each of the following is a function.

$$\text{sine: } \sin\theta = \frac{\text{side opposite } \theta}{\text{hypotenuse}}$$

$$\text{cosine: } \cos\theta = \frac{\text{side adjacent } \theta}{\text{hypotenuse}}$$

$$\text{tangent: } \tan\theta = \frac{\text{side opposite } \theta}{\text{side adjacent } \theta}$$

Fundamental function:	$\sin\theta$	$\tan\theta$	$\sec\theta$
Cofunction:	$\cos\theta$	$\cot\theta$	$\csc\theta$

REFERENCE ARC	SIN X	COS X	(REFERENCE ANGLE)
$\dfrac{\pi}{6}$	$\dfrac{1}{2}$	$\dfrac{\sqrt{3}}{2}$	30°
$\dfrac{\pi}{4}$	$\dfrac{\sqrt{2}}{2}$	$\dfrac{\sqrt{2}}{2}$	45°
$\dfrac{\pi}{3}$	$\dfrac{\sqrt{3}}{2}$	$\dfrac{1}{2}$	60°

- To evaluate sine and cosine, if x terminates in

QUADRANT	$\left(0, \dfrac{\pi}{2}\right)$	$\left(\dfrac{\pi}{2}, \pi\right)$,	$\left(\pi, \dfrac{3\pi}{2}\right)$	$\left(\dfrac{3\pi}{2}, 2\pi\right)$
sine	+	+	−	−
cosine	+	−	−	+

If the argument, arg, is not between 0 and $\pi/2$ then subtract integer multiples of π from arg until $|\text{arg} - n\pi| < \pi/2$. Now $|\text{arg} - n\pi|$ is the reference arc. If the reference arc is in the reference arc table, use the value from the reference arc table with the sign from the sign table to write the value of the function.

- Use the following definitions for tangent, cotangent, secant and cosecant:

$$\tan x = \frac{\sin x}{\cos x}, \qquad \cos x \neq 0$$

$$\cot x = \frac{\cos x}{\sin x}, \qquad \sin x \neq 0$$

$$\sec x = \frac{1}{\cos x}, \qquad \cos x \neq 0$$

$$\csc x = \frac{1}{\sin x}, \qquad \sin x \neq 0$$

- Consider a proposed identity $f(x) \equiv g(x)$. Interpret the identity geometrically and algebraically. Geometrically, $f(x) \equiv g(x)$ indicates that the graphs of f and g are the same for all common domain values. Algebraically, the identity $f(x) \equiv g(x)$ means that, subject to the common domain restriction, $f(x)$ can replace $g(x)$ in any expression containing $g(x)$. Similarly, $g(x)$ can substitute for $f(x)$. To prove $f(x) \equiv g(x)$, use substitutions of known identities into identity $f(x) \equiv g(x)$ until the two sides match. The following strategies are useful:

  1. Transform all trigonometric functions to sine and cosine.
  2. Simplify expressions as much as possible.
  3. Be sure that each resulting side of the equation is equal to the side preceding it.

- For $f(x) = A\sin(Bx + C) + D$, the amplitude of the graph is $|A|$. The fundamental period is $2\pi/|B|$. The phase shift $-C/B$ translates the graph horizontally. The graph is translated D units vertically.
- To determine an appropriate window for sketching a cycle,
  1. For x-value bounds, set the argument between 0 and 2π and solve for x:

$$0 \leq Bx + C \leq 2\pi$$
$$-\frac{C}{B} \leq x \leq \frac{2\pi - C}{B} \qquad \text{(if } B > 0\text{)}$$

$$-\frac{C}{B} \geq x \geq \frac{2\pi - C}{B} \qquad \text{(if } B < 0, \text{ reflect left to right)}$$

2. For y-value bounds, set the sine (or cosine) root between -1 and 1 and build up into the original function:

$$-1 \leqslant \sin(Bx + C) \leqslant 1$$

$$-A + D \leqslant A \sin(Bx + C) + D \leqslant A + D \qquad (\text{if } A > 0)$$

$$-A + D \geqslant A \sin(Bx + C) + D \geqslant A + D \qquad (\text{if } A < 0, \text{ reflect top to bottom})$$

- Because sine and cosine are complementary functions, $\sin x = \cos(\pi/2 - x)$ and $\cos x = \sin(\pi/2 - x)$, the graph of cosine is the graph of sine translated $\pi/2$ to the left.
- The following pairs of functions are complementary functions:

$$\sec x = \csc\left(\frac{\pi}{2} - x\right), \qquad \csc x = \sec\left(\frac{\pi}{2} - x\right)$$

$$\tan x = \cot\left(\frac{\pi}{2} - x\right), \qquad \cot x = \tan\left(\frac{\pi}{2} - x\right)$$

- The following pairs of functions are reciprocal functions:

$$\sin x = \frac{1}{\csc x}, \qquad \csc x = \frac{1}{\sin x}$$

$$\cos x = \frac{1}{\sec x}, \qquad \sec x = \frac{1}{\cos x}$$

$$\tan x = \frac{1}{\cot x}, \qquad \cot x = \frac{1}{\tan x}$$

$$\cos(a \pm b) \equiv \cos a \cos b \mp \sin a \sin b \qquad \text{(sum formulas)}$$

$$\sin(a \pm b) \equiv \sin a \cos b \pm \cos a \sin b$$

$$\cos(2a) \equiv \cos^2 a - \sin^2 a \qquad \text{(double angle formulas)}$$

$$\equiv 2 \cos^2 a - 1$$

$$\equiv 1 - 2 \sin^2 a$$

$$\sin(2a) \equiv 2 \sin a \cos a$$

$$\cos\frac{a}{2} \equiv \pm \sqrt{\frac{1 + \cos a}{2}} \qquad \text{(half-angle formulas)}$$

$$\sin\frac{a}{2} \equiv \pm \sqrt{\frac{1 - \cos a}{2}}$$

$$\tan\frac{a}{2} \equiv \frac{1 - \cos a}{\sin a} \equiv \frac{\sin a}{1 + \cos a}$$

$$\cos A \cos B \equiv [\cos(A + B) + \cos(A - B)]/2 \qquad \text{(function product formulas)}$$

$$\cos A \sin B \equiv [\sin(A + B) - \sin(A - B)]/2$$

$$\sin A \cos B \equiv [\sin(A + B) + \sin(A - B)]/2$$

$$\sin A \sin B \equiv [\cos(A - B) - \cos(A + B)]/2$$

$$\cos u + \cos v = 2 \cos \frac{u+v}{2} \cos \frac{u-v}{2}$$ (function sum formula)

$$\cos u - \cos v \equiv -2 \sin \frac{u+v}{2} \sin \frac{u-v}{2}$$

$$\sin u + \sin v \equiv 2 \sin \frac{u+v}{2} \cos \frac{u-v}{2}$$

$$\sin u - \sin v \equiv 2 \cos \frac{u+v}{2} \sin \frac{u-v}{2}$$

KEY WORDS AND CONCEPTS

Acute angle
Angle
Angle of rotation
Angular velocity
Arc length
Circular function
Complementary
Degree
Distance formula
Equal angles
Even
Factorial
Frequency

Greek letter
Hypotenuse
Initial
Linear velocity
Minute
Obtuse angle
Odd
Periodic
Phase shift
Pythagorean relation
Pythagorean theorem
Quadrant
Radian

Ray
Reference angle
Reference arc
Right angle
Second
Static angle
Terminal
Transcendental
Triangle
Trigonometry
Wave form
Wavelength

7 REVIEW EXERCISES

SECTION 1

Convert each of the following radian measures to degree measures.

1. -3π

2. $\frac{5\pi}{18}$

3. $\frac{\pi}{9}$

4. $\frac{-7\pi}{36}$

5. $\frac{25\pi}{18}$

6. 5

Convert each of the following degree measures to radian measures expressed as a multiple of π.

7. $15°$

8. $1500°$

9. $5°$

10. $-210°$

11. $-300°$

12. $760°$

13. $-72°$

14. $36°$

15. Convert 50 rev/min to degrees/min.

16. Convert 300 rev/min to degrees/sec.

17. Convert 1800°/min to rev/hour.

18. Convert 7200°/min to radians/min.

19. A point on a circle travels from 10° to 310° in 5 seconds. What is the angular velocity?

20. A point on a circle rotates from 20° to -10° in 0.5 minutes. What is the angular velocity?

21. A point on the unit circle moves from 0.5π radians to 4π radians in 1 minute. Determine the angular velocity in degrees/sec.

22. A compact disk (CD) rotates once per second. Determine the angular velocity of the disk.

23. Suppose the last track of a compact disk is 1.25″ from the center of the disk and the first track is 2.25″ from the center. How much longer is the first track than the last track?

24. See Problems 22 and 23. Determine the linear velocity of a point on the first track of the CD.

SECTION 2

Refer to the 30°-60° right triangle and 45° right triangle shown in Figure 7-89 to solve for the indicated side.

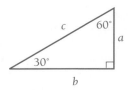

FIGURE 7-89

25. If $a = 5$, determine b.

26. If $v = 6$, determine u.

27. If $u = 8$, determine v.

28. If $b = 4$, determine c.

29. If $w = 10$, determine u.

30. If $b = 20$, determine c.

Evaluate the following trigonometric functions as indicated.

31. $\cos 30°$

32. $\sin 60°$

33. $\sin 45°$

34. $\sin^2 45° + \cos^2 45°$

35. $2 \sin 60° \cos 60°$

36. $1 - \cos^2 30°$

Approximate the following.

37. $\cos 43°$

38. $\sin 47°$

39. $\tan 5°$

40. $\sin 23°$

41. $\cos 21° \, 27' \, 10''$

42. $\tan 35° \, 15''$

43. $\sin 45°$

44. $\dfrac{\cos 45°}{\sin 45°}$

45. $\cos(2.5\pi)$

46. $\cos \dfrac{\pi}{5}$

47. $\sin \dfrac{5\pi}{12}$

48. $\sin \dfrac{\pi}{7}$

49. $\cos 0.9$

50. $\cos 1.3$

SECTION 3

Evaluate each of the following as indicated.

51. $\sin 60°$

52. $\sec 60°$

53. $\csc 60°$

54. $\tan 60°$

55. $\tan 45°$

56. $\cot 45°$

57. $\csc 45°$

58. $\cos 45°$

59. $\sec 50°$

60. $\csc 40°$

Prove the following identities.

61. $\sin \theta \equiv \cos \theta \tan \theta$

62. $\cos \theta \sec \theta - \cos \theta \equiv \sin^2 \theta$

63. $\dfrac{\cos \theta}{1 + \sin \theta} \equiv \sec \theta - \tan \theta$

64. $\sin^2 \theta \cot^2 \theta + \sin^2 \theta \equiv 1$

65. $\dfrac{\sin x + \cos x}{\sin x} \equiv 1 + \cot x$

66. $\dfrac{\sec w + \csc w}{\csc w - \sec w} \equiv \dfrac{\sin w + \cos w}{\cos w - \sin w}$

67. $\cot^2 x - \cos^2 x \equiv \cot^2 x \cos^2 x$

68. $\dfrac{1 + \sec x}{\csc x} \equiv \tan x + \sin x$

69. $\cos^2 x - \sin^2 x \equiv 1 - 2 \sin^2 x$

70. $\dfrac{1}{1 + \sin t} + \dfrac{1}{1 - \sin t} \equiv 2 \sec^2 t$

Express each of the following in terms of sine and cosine. Simplify the resulting expressions.

71. $\dfrac{\cot x - \cot y}{1 + \cot x \cot y}$

72. $\dfrac{\sec \phi - \tan \phi}{\cot \theta - \csc \phi}$

73. $\dfrac{\sec x + \cos x}{\tan x} - \csc x$

74. $\dfrac{\csc x - \sin x}{\cot x}$

75. $\dfrac{1 + \csc \phi}{\sec \phi} - \cot \phi$

76. $\dfrac{\cot \alpha + \cos \alpha}{1 + \csc \alpha}$

SECTION 4

Sketch the graph of each of the following over two complete cycles.

77. $f(x) = 3 \sin(5x)$

78. $f(x) = -3 \sin(-5x)$

79. $f(x) = \sin\left(\dfrac{1}{2}x\right)$

80. $f(x) = \sin\left(x - \dfrac{\pi}{3}\right)$

81. $f(x) = \sin(3x + \pi)$

82. $f(x) = -3\sin\dfrac{x}{2}$

83. $f(x) = -2\sin x + 4$

84. $f(x) = -5\sin\left(x + \dfrac{\pi}{3}\right) + 2$

85. $f(x) = \pi\sin(x - 2) - 3$

86. $f(x) = 2\sin(\pi x - 4) + 3$

87. Middle C on a piano is about 512 cps (Hertz). If sound travels at 331.3 m/sec, determine the wavelength of the sound.

88. Infrared light has a frequency of about 480,000 gigaHertz (about 4.8×10^{14}). If light travels at 300,000 km/sec, determine the wavelength of the light wave.

SECTION 5

Sketch the graph of each of the following over two complete cycles:

89. $f(x) = \tan(3x)$

90. $f(x) = \tan\left(x - \dfrac{\pi}{4}\right)$

91. $f(x) = \csc x + 2$

92. $f(x) = \csc(x + \pi)$

93. $f(x) = 2\tan\left(x + \dfrac{\pi}{3}\right) + 5$

94. $f(x) = -4\tan x$

95. $f(x) = \csc(-4x)$

96. $f(x) = \tan(4 + x)$

97. $f(x) = \cot(\pi - 4x) + 2$

98. $f(x) = -\sec(x - \pi/2) + \pi$

Sketch the graph of $L(x) = 3x - \pi$. Now sketch the graph of each of the following. Use graphing technology.

99. (a) $f(x) = \csc x$ **(b)** $L \circ f(x)$

100. (a) $f(x) = \tan x$ **(b)** $L \circ f(x)$

101. (a) $f(x) = \tan x$ **(b)** $f \circ L(x)$

SECTION 6

Verify the following identities.

102. $\sec\left(\dfrac{\pi}{2} - x\right) \equiv \csc x$

103. $\sin(x)\cos(x) \equiv \dfrac{\tan x}{1 + \tan^2 x}$

104. $\dfrac{1}{\sec^2 x} + \dfrac{1}{\csc^2 x} \equiv 1$

105. $[\sin x + \cos x]^2 + [\sin x - \cos x]^2 \equiv 2$

106. $[\sec x - \tan x][\sec x + \tan x] \equiv 1$

Evaluate the following without using a calculator.

107. $\cos\left(\dfrac{\pi}{4} + \dfrac{\pi}{6}\right)$

108. $\sin\left(\dfrac{\pi}{4} - \dfrac{\pi}{6}\right)$

109. $\sin\left(\dfrac{7\pi}{12}\right)$

110. $\cos\left(-\dfrac{\pi}{6}\right)$

111. $\cos\left(-\dfrac{3\pi}{4}\right)$

112. $\sin\left(-\dfrac{3\pi}{4}\right)$

113. If $\cos r = 0.4$ and r terminates in quadrant I, determine $\sin r$.

114. If $\sin s = 0.7$ and s terminates in quadrant I, determine $\cos s$.

115. See Problems 113 and 114. Determine $\sin(r + s)$.

116. See Problems 113 and 114. Determine $\cos(r - s)$.

Derive the following identities.

117. $\cos(2x) \equiv 1 - 2\sin^2 x$

118. $\cos(2x) \equiv 2\cos^2 x - 1$

119. $\sin\dfrac{x}{2} \equiv \pm\sqrt{\dfrac{1 - \cos x}{2}}$

120. $\tan\dfrac{x}{2} \equiv \dfrac{\sin x}{1 + \cos x}$

121. $\cot\dfrac{x}{2} \equiv \dfrac{\sin x}{1 - \cos x}$

Derive a formula for the following.

122. $\cos(A - B) - \cos(A + B)$

123. $\sin(A + B) + \sin(A - B)$

124. $\sin(A)\sin(B)$

125. $\sin(A)\cos(B)$

126. $\sin\left(\dfrac{u + v}{2}\right)\cos\left(\dfrac{u - v}{2}\right)$

Evaluate the following.

127. $\cos\dfrac{7\pi}{12} + \cos\dfrac{\pi}{12}$

128. $\sin\dfrac{7\pi}{12} - \sin\dfrac{\pi}{12}$

129. $\sin\dfrac{\pi}{4}\cos\dfrac{\pi}{12}$

130. $\sin\dfrac{\pi}{4}\sin\dfrac{\pi}{12}$

SECTION 7

131. Write a third-degree Maclaurin polynomial for $\sin x$.

132. Write a fourth-degree Maclaurin polynomial for $\cos x$.

Brook Taylor and Colin Maclaurin

Brook Taylor was one of the major contributors to mathematical thought in the eighteenth century. Of English descent, Taylor was born in 1685. Colin Maclaurin, a Scotsman, was born in 1698. Although these two men were not direct colleagues, their names have been linked by calculus textbooks for years.

In 1715, Taylor published an expansion theorem that allowed for approximation of transcendental functions (such as sine and cosine) by use of polynomials. Taylor applied his method to the solution of numerical equations in 1717. The approximating polynomials are called *Taylor polynomials*, and the general method is called the *Taylor series*. We shall study some simple series in Chapter 12.

Taylor's method was complex and very general. In 1742 (after Taylor's death in 1731), Colin Maclaurin was defending Newton's invention of calculus in a paper entitled "Treatise of Fluxions," when he used a special, simple case of Taylor's method. Although Maclaurin gave full credit to Taylor for the method and acknowledged his use as only a special case, Maclaurin's name became attached to the approximating polynomials.

Maclaurin had no need of credit that was not rightfully his. He was an accomplished mathematician who began as a prodigy. He entered the University of Glasgow at age 11 and by 15 he had finished his master's degree. One of the important topics of the age was on the true shape of the earth. In 1740, in his prize-winning treatise on the tides, Maclaurin proved that the natural shape for a fluid body (similar to the earth) under uniform angular rotation was an oblate spheroid. Today, we still describe the earth as having that shape.

Maclaurin also developed determinants as a method for solving systems of linear equations. Although it is thought that Maclaurin did this work in 1729, the results were not published until 1748, two years after his death in 1746.

Taylor's series ultimately brought both Taylor and Maclaurin fame, but the importance of Taylor's work was not fully recognized until after his death. In 1775, Euler applied Taylor's methods in his differential calculus. Later Lagrange used the series as a foundation for his theory of functions.

Taylor was creative in other areas, too. Since the Renaissance, artists had attempted to form a theory of perspective. But Brook Taylor was one of the first to build a true theory of perspective. Taylor's work in perspective has found application into our current age. Aerial photographs are often an aid in surveying land. Taylor's theory of perspective plays a role in this modern science.

If you prefer music to art, then Taylor's work on vibrating strings may interest you. Taylor derived the fundamental frequency of a stretched vibrating string by applying the methods of calculus. Moreover, he indicated that the shape of the string at any time during the vibration is given by

$$y = A \sin\left(\frac{\pi x}{L}\right)$$

where L is the length of the string. If you have ever plucked a taut rubber band to hear its twang then you have probably noticed the sinusoidal curve that Taylor described. Taylor's result for the fundamental frequency of the string is

$$v = \frac{1}{2L}\sqrt{\frac{gT}{m}}$$

where T is the tension on the string, m is the mass per unit length and g is the acceleration due to gravity. Next time you hear a violin played, or see a landscape painting think of Brook Taylor and, with him, Colin Maclaurin.

7   CHAPTER TEST

1. Determine the linear velocity of a point on the first track of a music CD, if the first track is 2.25″ from the center of the CD and the CD makes 1 revolution per second.

2. Suppose that the side opposite the 30° angle in a right triangle is 10 meters long. Determine the length of the other side and the hypotenuse.

Verify the following identities.

3. $\tan^2 x - \sin^2 x \equiv \tan^2 x \sin^2 x$

4. $\dfrac{1 + \sec x}{\csc x} \equiv \tan x + \sin x$

5. $1 + \cot^2 x \equiv \dfrac{\tan x}{\sin x \cos x}$

Change all functions to sine or cosine and simplify.

6. $\dfrac{1 + \csc \phi}{\sec \phi} - \cot \phi$

Graph the following.

7. $f(x) = -2 \sin x + 4$

8. $f(x) = -5 \sin\left(x + \dfrac{\pi}{3}\right) + 2$

9. $f(x) = \cot(\pi - 4x) + 2$

Evaluate the following.

10. $\sin\left(-\dfrac{3\pi}{4}\right)$

11. Suppose that $\sin x = \frac{3}{5}$ and $\cos y = \frac{5}{13}$. Both x and y are in the first quadrant. Determine $\sin(x + y)$.

12. Suppose that $\cos A = -\frac{5}{13}$ and $\sin A > 0$. Determine $\sin A$ and $\tan A$.

APPLICATIONS OF TRIGONOMETRIC FUNCTIONS

1 SOLVING RIGHT TRIANGLES

Let no one ignorant of geometry enter my door.
—PLATO

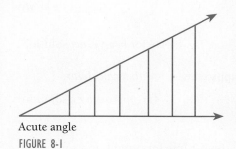

Acute angle
FIGURE 8-1

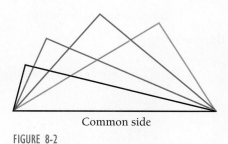

Common side
FIGURE 8-2

Six pieces of data describe any triangle: the length of each of the three sides and the measure of each of the three angles. For a right triangle, one angle is fixed at 90°. How much of the data must be measured before the remaining data is known? Suppose one other measurement is made.

If you know an acute angle, then the second acute angle is the complement. However, the lengths of the three sides are indeterminate: in a right triangle, one acute angle describes infinitely many similar right triangles. See Figure 8-1.

Suppose instead, you measure the length of a side of a right triangle. Unfortunately, no other information reveals itself. Infinitely many right triangles may share a common side. See Figure 8-2.

One measurement in a right triangle is *not* sufficient to uniquely define that triangle. Suppose we take two measurements of a right triangle. If both the measurements are acute angles the results are the same as in Figure 8-1. However, if at least one of the two measurements is a side, the triangle described is unique. The Pythagorean theorem and trigonometric functions provide the measure of the remaining data. To **solve a right triangle** means to determine the unmeasured data of the triangle from given measurements. To solve a right triangle, in addition to the right angle you must know two measurements, one of which must be a side.

Solving a Right Triangle. Solve as indicated.

EXAMPLE 8-1

Illustration 1:

In right triangle ABC, one acute angle is $A = 25°$, the side opposite angle A is $a = 10$. Determine the remaining measurements for the triangle.

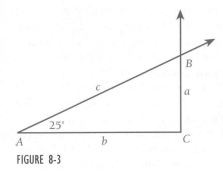

FIGURE 8-3

Solution: Draw a figure to represent the triangle. First sketch a right angle. It is not clear whether B or C should be the right angle. Let us use C. Construct a 25° angle on one side of angle C. Label the angle A. Extend one side of A until it intersects the other side of C. Label this point B. See Figure 8-3.

Triangle Data	A	B	C	c	b	a
Measure	25°	?	90°	?	?	10

Angle B is the complement of angle A: $B = 65°$. Two sides remain. Because ratios of sides of a right triangle define trigonometric functions, use a trigonometric function to determine one of the sides. Consider side b. Relate b to some known value through a trigonometric function. But which trigonometric function? The only known side is a, which is 10. Because b and a are opposite sides, either tangent or cotangent is appropriate. Tangent is on your calculator, so choose it.

$$\tan 25° = \frac{a}{b} \qquad \text{(definition of tangent)}$$

$$\tan 25° = \frac{10}{b} \qquad \text{(because } a = 10)$$

$$b = \frac{10}{\tan 25°} \qquad \text{(solve for } b, \text{ an } exact \text{ solution)}$$

Be sure you are in the degree mode to approximate b *with a calculator.*

$$b \cong 21.44506921 \qquad \text{(an } approximate \text{ solution)}$$

From the Pythagorean theorem,

$$c^2 = b^2 + a^2 \qquad \text{(substitute for } a \text{ and } b \text{ and solve for } c)$$

$$c \cong \sqrt{559.8909932}$$

$$c \cong 23.66201583 \qquad \text{(approximate solution for } c)$$

Note that c was also available from the sine ratio:

$$\sin 25° = \frac{10}{c} \qquad \text{(definition of sine)}$$

$$c = \frac{10}{\sin 25°}$$

$$c \cong 23.66201583 \qquad \text{(solve and approximate)}$$

The precision of the approximate answers for c and b is questionable. The specifications of the problem gave no hint of the precision of the original measurements. Was a measured to the nearest 10? Nearest unit? Nearest tenth? Certainly, if the measure had been given as 10.0000, we would expect this to indicate precision to four decimal places. But integer measures are usually taken as exact in textbooks. Such is not the case in the physical world where measurements are never exact. A similar difficulty arises with the 65° measurement of angle B. The problem with the answers for c and b arise from a simple maxim: *the precision of any answer cannot be greater than the precision of the measurements that led to the answer.* Most calculators maintain a precision of at least eight digits. Therefore, sin 25° is more precise in a calculator than for most measurements. A calculator cannot decide the precision of entered measurements and so displays as many decimals as it can. You are left the job of rounding to a final approximation. For example, suppose that the initial measures were accurate to one decimal place, then rounding c to 23.7 would approximate a reasonable solution for the triangle. Unless otherwise indicated, we assume that all measures are accurate enough to justify the inclusion of all decimals produced by a calculator. Although this is not a good idea for solving physical problems, it does provide a safer method for checking the process of a solution.

Illustration 2:

Consider right triangle ABC with right angle at C and $c = 26$, $a = 10$.

Solution: Sketch a right triangle. (Figure 8-4). Label the hypotenuse $c = 26$. Label $a = 10$. The Pythagorean theorem produces b:

$$c^2 = b^2 + a^2$$

$$676 = b^2 + 100 \qquad \text{(substitute for } a \text{ and } c\text{)}$$
$$b^2 = 576 \qquad \text{(solve for } b\text{)}$$
$$b = 24$$

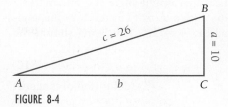

FIGURE 8-4

There are two solutions to this equation, 24 and –24. We reject –24 because the length of a side of a triangle may not be negative. Now choose an angle, say, angle A. All sides are known, so that any trigonometric function related to A applies. Because sine is familiar, use it.

$$\sin A = \frac{\text{opposite}}{\text{hypotenuse}} \qquad \text{(definition of sine)}$$

$$\sin A = \frac{a}{c} \qquad \text{(for this triangle)}$$

$$\sin A = \frac{10}{26} \qquad \text{(substitute for } a \text{ and } c\text{)}$$

$$\sin A \cong 0.384615384 \qquad \text{(calculator approximation)}$$

How do we determine A? Not only are the acute angles associated with unique ratios, but the ratios produce unique acute angles. In other words, each of the triangle trigonometric functions has an inverse function. Conveniently, scientific calculators have built in inverse trigonometric functions. Either a unique function key such as $\boxed{N}$ is available, or a sequence of keys such as 2nd SIN produces the inverse. Be sure the calculator is in the degree mode. By calculator, $A \cong 22.61986495°$. The next section discusses inverse circular functions.

Illustration 3:

A prefabricated roof truss spans 24 feet. Moreover, the king post at the middle of the truss is 5 feet. See Figure 8-5. Determine the inclination of the roof as measured by an angle θ.

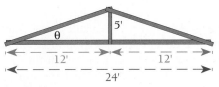

FIGURE 8-5

Solution: The inclination θ relates directly to the tangent ratio: $\frac{5}{12}$. The ratio $\frac{5}{12}$ also measures the slope of a cross section of the roof. The slope of an inclined line is the tangent of its angle of inclination.

$$\tan \theta = \frac{5}{12} \qquad \text{(definition of tangent)}$$

$$\theta = \tan^{-1} \frac{5}{12} \qquad \text{(apply the inverse)}$$

By calculator: $\theta \cong 22.61986495$

(or $\theta \cong 22° \, 37' \, 12''$)

Illustration 4:

An engineer plans to prefabricate a bridge and swing it across a river, spanning from point B to point C. The engineer measures a right angle at C and marks a point A 10 meters from C. See Figure 8-6. With a transit (theodolite) the engineer measures the angle A between b and c: $A = 57°$. How long should the bridge be?

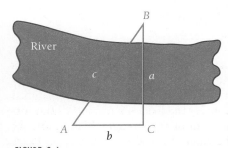

FIGURE 8-6

$$\tan 57° = \frac{a}{10 \text{ meters}} \qquad \text{(definition of tangent)}$$

$$a = (10 \text{ meters}) \tan 57° \qquad \text{(solve for } a)$$

$$a \cong (10 \text{ meters})1.539864964 \qquad \text{(approximate tangent)}$$

$$a \cong 15.39864964 \text{ meters.} \qquad \text{(by calculator)}$$

Illustration 5:

A cable for a tram is to stretch from the top of a cliff (point B) to a point A 100 meters from a point C directly below B at the base of the cliff. See Figure 8-7. If the angle of inclination at A is 30°, what is the minimal length of the cable? How high is the cliff?

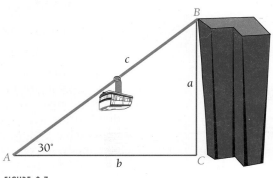

FIGURE 8-7

Solution:

$$\cos A = \frac{b}{c} \qquad \text{(definition of cosine)}$$

$$\cos 30° = \frac{100 \text{ meters}}{c} \qquad \text{(substitute } b = 100\text{)}$$

$$\frac{\sqrt{3}}{2} = \frac{100 \text{ meters}}{c} \qquad \left(\text{because } \cos 30° = \frac{\sqrt{3}}{2}\right)$$

$$c = \frac{200 \text{ meters}}{\sqrt{3}} \qquad \text{(solve for } c\text{)}$$

Cable: $c \cong 115.4700538$ meters   (by calculator)

Also, $\tan 30° = \dfrac{a}{b}$ (definition of tangent)

$$\frac{1}{\sqrt{3}} = \frac{a}{100 \text{ meters}} \qquad \text{(substitution)}$$

$$a = \frac{100 \text{ meters}}{\sqrt{3}} \qquad \text{(solve for } a\text{)}$$

Height: $a \cong 57.73502692$ meters   (by calculator)

Because of the precision of the original measurements, round the height to 58 meters. Rounding up the length of the cable is not sufficient. The distance measured for the cable is along a straight line. However, the cable must have some "sag" in it and will sag more under the load of a tram. Without further specifications on the weight of the tram or the strength of the cable we cannot provide a better estimate for the cable. Instead note that the cable is longer than 116 meters. This is why the problem requested a minimal, rather than exact, length.

 Incline a meter stick 40° to the ground. How long is the shadow at noon?

Solution: See Figure 8-8.

$$\cos 40° = \frac{\text{length}}{1 \text{ meter}} \qquad \text{(definition of cosine)}$$

length $= (1 \text{ meter}) \cos 40°$ (substitute and solve)

length $\cong 0.766044443$ meter   (by calculator)

Note: Although right triangles become irrelevant, two limits to the angle of inclination are of interest. What is the length of the shadow if the meter stick is inclined 0° (parallel) to the ground? What is the length of the shadow if the meter stick is inclined 90° (perpendicular) to the ground?

One more comment about precision and significant digits: the decimal places of the answer probably exceeds the accuracy of our meter stick. Almost paradoxically, we wonder "to how many decimal places is this meter stick

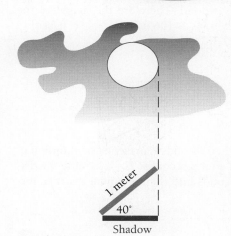

1 meter

40°

Shadow

FIGURE 8-8

1 meter long?" We leave the official answer at 0.766044443 meters even though 0.766 meters may be more realistic.

Illustration 7:   Jeff belongs to the Society for Anachronistic Pursuits (SAP). As part of the yearly festival, Jeff designs a working catapult. Jeff sets the launch angle at 62.5°. The launch velocity is 108.3 ft/sec. Disregard all influences except gravity to determine the projectile range.

Solution: Figure 8-9 represents the launch of the projectile. The length of a solid arrow models the velocity of the projectile at 62.5° angle. This velocity resolves into a vertical velocity v_v and horizontal velocity h_v, modeled by dashed arrows. The arrow models form a right triangle.

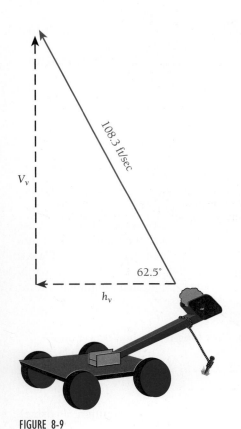

FIGURE 8-9

$$\cos 62.5° = \frac{h_v}{108.3 \text{ ft/sec}} \qquad \text{(definition of cosine)}$$

$$h_v \cong 50 \text{ ft/sec} \qquad \text{(Solve and approximate)}$$

$$\sin 62.5° = \frac{v_v}{108.3 \text{ ft/sec}} \qquad \text{(definition of sine)}$$

$$v_v \cong 96 \text{ ft/sec} \qquad \text{(Solve and approximate)}$$

Gravity does not affect the horizontal velocity. The projectile travels 50 feet horizontally each second after launch. If the projectile has been in flight for t seconds the horizontal distance from the catapult is given by $H_s \cong (50 \text{ ft/sec})t$.

To determine how far the projectile travels, we must know the duration of the flight. The flight ends when the projectile strikes the ground. At that instant, the vertical height is 0 ft.

The vertical velocity is subject to the influence of gravity. Gravity changes the vertical velocity by -16 ft/sec every second. Therefore the height of the projectile after t seconds is approximated by $V_s \cong (96 \text{ ft/sec})t - 16t^2 \text{ ft/sec}^2$

Suppose the vertical height is $V_s = 0$,

$$0 \cong (96 \text{ ft/sec})t - 16t^2 \text{ ft/sec}^2$$
$$0 \cong 16t \text{ ft/sec}(6 - t/\text{sec})$$
$$t \cong 0 \quad \text{or} \quad t \cong 6 \text{ sec}$$

Because 0 was the moment of launch, 6 seconds represents the duration of flight. The projectile strikes the ground after 6 seconds. To determine the distance from the catapult, substitute 6 seconds into the horizontal distance formula: $H_s = (50 \text{ ft/sec})(6 \text{ sec})$

$$H_s = 300 \text{ ft.}$$

- To solve a right triangle means to determine the measure of all sides and angles. To do this, you need the measure of at least two other parts of the triangle, one of which must be a side.
- Sketch a figure and label known values before solving a right triangle. In triangle *ABC*, label angles using capital letters and label the side opposite a given angle with the same letter in lower case. For example, label the side opposite angle *A* as *a*. Side *a* connects the remaining two vertices *B* and *C*. Similarly, side *b* is opposite angle *B*, and side *c* is opposite angle *C*.

8-1   E X E R C I S E S

Refer to triangle *ABC* in Figure 8-10. For the following given values of the triangle, solve for the remaining parts of the triangle.

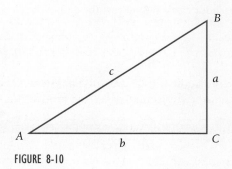

FIGURE 8-10

Triangle ABC in Figure 8-11 is not a right triangle. All the angles are acute. From *C* a line *h* is perpendicular to *c*. Segment *h* intersects *c* at point *p*. Point *p* divides *c* into segments of length *x* and *y*. Segment *h* divides angle *C* into angles θ and φ.

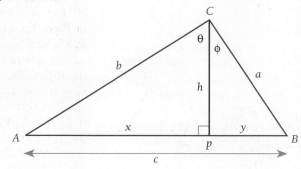

FIGURE 8-11

1. If $A = 50°$ and $b = 10$ **2.** If $B = 35°$ and $c = 12$
3. If $A = 20°$ and $a = 5$ **4.** If $B = 70°$ and $b = 10$
5. If $A = 40°$ and $c = 8$ **6.** If $B = 70°$ and $a = 9$
7. If $B = 30°$ and $b = 10$ **8.** If $A = 43°$ and $c = 12$
9. If $B = 82°$ and $a = 5$ **10.** If $A = 17°$ and $a = 9$
11. If $c = 15$ and $a = 7$ **12.** If $c = 10$ and $a = 3$
13. If $b = 15$ and $a = 7$ **14.** If $b = 10$ and $a = 3$
15. If $c = 10$ and $b = 2$ **16.** If $c = 12$ and $a = 10$
*17. If $c = 7$ and $a = 10$ ***18.** If $c = 9$ and $b = 10$
*19. If $c = 11$ and $b = 7$ ***20.** If $c = 6$ and $a = 9$

21. (a) If angle *A* is 40° and angle *B* is 70° determine the size of angle *C*. Determine the size of angles θ and φ.
(b) If $b = 10$, determine the length of *h* and *x*.
(c) Determine *y* and *a*.
(d) Determine *c*.

22. (a) If angle *A* is 50° and angle *B* is 80° determine the size of angle *C*. Determine the size of angles θ and φ.

(b) If $b = 8$, determine the length of h and x.

(c) Determine y and a.

(d) Determine c.

*23. See Figure 8-11. Suppose $A = 45°$ and $B = 60°$ and $a = 10$. Solve triangle ABC.

*24. See Figure 8-11. Suppose $A = 30°$ and $B = 45°$ and $a = 10$. Solve triangle ABC.

25. An observer spots a plane so that the angle of the line of sight is inclined 30° to the ground. If the plane is at 1000 meters altitude, how far is the plane from the observer?

26. The angle of inclination to a point "above" an observer is the angle formed by the line of sight to the point and a horizontal line from the observer in the direction of the point. If the point is "below" the observer, the same angle is an angle of depression. At 100 meters a surveyor measures the angle of elevation to the top of a building as 30°. What is the height of the building?

27. What is the length of a noontime shadow cast by a meter stick inclined 60° to the ground?

28. The pointer on a sundial casts a shadow 10 cm long at noon. If the pointer is inclined 60° to the face of the dial, how long is the pointer?

29. A meter stick held perpendicular to the ground casts a shadow 1.7321 meters long. How many degrees is the sun above the horizon?

30. A model railroader knows that the sharpest incline a model locomotive can traverse is a 1 cm rise in height for each 10 cm traveled horizontally. What is the angle formed by the track and horizontal at this incline?

31. A hiker treks a path 30° east of due north. After hiking for 10 km, how far east is the hiker from the starting point?

32. Given the same information as in Problem 31, how far north is the hiker from the starting point?

33. A road is inclined 5° to a level grade. How much altitude will a traveler gain after traveling 1 kilometer?

34. A new combat plane is able to climb at 75° to the horizon immediately upon takeoff. What will its horizontal distance from takeoff be when its altitude is 10000 kilometers?

35. A projectile is launched at 200ft/sec at 30° angle to the ground. How long and how far is the flight?

36. An arrow is shot at a 50° angle to the ground to strike a distant target. Estimate the distance to the target if the initial velocity of the arrow is 100 ft/sec.

37. Wild Woman Wilma proposes to jump ten Toyotas in her Volkswagon convertible. She plans to hit the UP ramp at a minimum of 66 ft/sec (45 miles/hour). See Figure 8-12. Both the UP ramp and DOWN ramp have bases of 22 feet and an incline of 15°. What should be the distance between the UP ramp and DOWN ramp? If each Toyota parks in a 6 foot width, can she clear ten?

38. See Problem 37. Bad Boy Bob is Wilma's boyfriend. He thinks the jump is close and that Wilma should increase her speed to 50 miles/hour. But if Wilma goes too fast, she may go past the end of the Down ramp. The resulting shock could destroy the VW's suspension and cause Wilma to lose control and have a crash. Will 50 miles/hour carry Wilma past the end of the Down ramp?

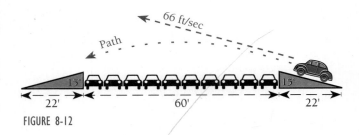

FIGURE 8-12

8-1   P R O B L E M   S E T

EXPLORATION

The following problems anticipate concepts from the next section. Use graphing technology for each of the following.

1. (a) Graph $y = \sin x$ and $y = 0.2$ in the same window. In how many points do these two functions intersect.?

(b) Use the trace feature to determine a point of intersection in the first quadrant.

(c) Use the trace feature to determine a point of intersection in the second quadrant.

(d) Use the $\boxed{\text{SIN}^{-1}}$ key to approximate $\sin^{-1} 0.2$.

(e) Use the $\boxed{\text{SIN}^{-1}}$ key to graph $y = \sin^{-1} x$

2. (a) Graph $y = \cos x$ and $y = 0.8$ in the same window. In how many points do these two functions intersect?

(b) Use the trace feature to determine a point of intersection in the first quadrant.

(c) Use the trace feature to determine a point of intersection in the second quadrant.

(d) Use the $\boxed{COS^{-1}}$ key to approximate $\cos^{-1} 0.8$.

(e) Use the $\boxed{COS^{-1}}$ key to graph $\cos^{-1} x$.

3. Graph $y = \tan x$ and $y = \tan^{-1} x$ (use the $\boxed{TAN^{-1}}$ key) in the same window. Compare the graphs. How are they similar? How are they different?

4. Consider the graph of $y = \sin x$. The function is not one to one. Choose an interval from the domain of sine such that if you erased all portions of the graph outside the interval, the portion remaining in the interval would be one to one and still include the complete range of sine.

5. See Problem 4. Choose a similar interval for cosine.

6. See Problem 4. Choose a similar interval for tangent.

INVERSE CIRCULAR FUNCTIONS

Euclid alone has looked on Beauty bare . . .
—EDNA ST. VINCENT MILLAY

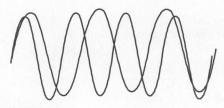
Observe the line f and sine curve g shown in Figure 8-13. The functions f and g are not identical. The line f intersects the graph of the sine curve g in infinitely many points. Because $g(x) = 5 \sin x + 1$ has a period of 2π, once the intersection of $f(x) = 3$ and $g(x) = 5 \sin x + 1$ is known for $x \in [0, 2\pi)$, the remainder of the solution follows. The intersection occurs where the function values are equal. We seek values of x where

$$f(x) = g(x)$$
$$3 = 5 \sin x + 2 \qquad \text{(substitute formulas for } f \text{ and } g\text{)}$$

The equation $f(x) = g(x)$ is not an identity. It is a conditional equation. What remains is to solve for x. The variable is "bound up" in the argument of the sine function. None of the usual methods for extracting x apply directly. However, it is often possible to solve for the circular function. In this case, compare

$$3 = 5y + 1 \text{ to}$$
$$3 = 5 \sin x + 1.$$

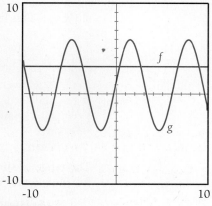

FIGURE 8-13

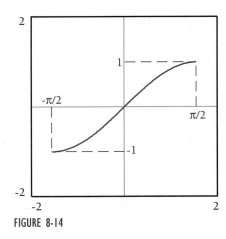

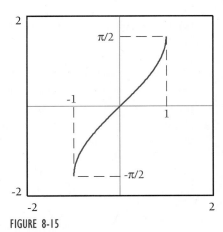

FIGURE 8-14   FIGURE 8-15

The first equation is a first-degree equation in y. The second equation is first-degree in sin x. Solve the first equation for y and obtain $y = \frac{2}{5}$. Solve the second equation for sin x and obtain $\sin x = \frac{2}{5}$.

Dealing with sin $x = \frac{2}{5}$ is more complex than $\frac{2}{5} = e^x$. The exponential function is one to one; the sine function is not. Sine has no inverse function. Therefore we cannot unbind x by applying a function to both sides of the equation to "undo" sine. Recall that the function f given by $f(x) = x^2$ is not one to one when the domain is all real numbers. However, restrict the domain of f to nonnegative numbers and the inverse f^{-1} is given by $f^{-1}(x) = \sqrt{x}$. A similar technique applies to trigonometric functions.

This section develops **inverses for trigonometric functions** with the restricted domains. Later, we apply these concepts to the solution of conditional trigonometric equations.

To obtain a one-to-one function from sine restrict the domain to $[-\pi/2, \pi/2]$. See Figure 8-14. The restricted function is one to one. The inverse function is a reflection about the line $y = x$. See Figure 8-15.

Because inverse sine is not the inverse of the original sine function, we refer to the inverse function with a name descriptive of its purpose. Consider that the domain of sine is drawn from an arc measure. The range of the inverse must be values that measure arcs: given a sine value t, find the arc x for which $\sin x = t$. This leads to the following definition.

DEFINITION 8-1
Arc Sine

> The inverse sine function is defined by $y = \sin^{-1}x$ if an only if $\sin y = x$, for $-\pi/2 \leq y \leq \pi/2$ and $-1 \leq x \leq 1$. . . . Inverse sine is also known as **arc sine**.

The range of $\sin^{-1}$ is given as $[-\pi/2, \pi/2]$. From the range of the original sine function comes the domain of $\sin^{-1}$, the interval $[-1, 1]$.

Note that $\sin^{-1} x$ is an odd function. Therefore $\sin^{-1}(-\frac{1}{2}) = -\sin^{-1}(\frac{1}{2})$. Now we must determine an arc for which the sine is $\frac{1}{2}$. Because $\sin(\pi/6) = \frac{1}{2}$, then

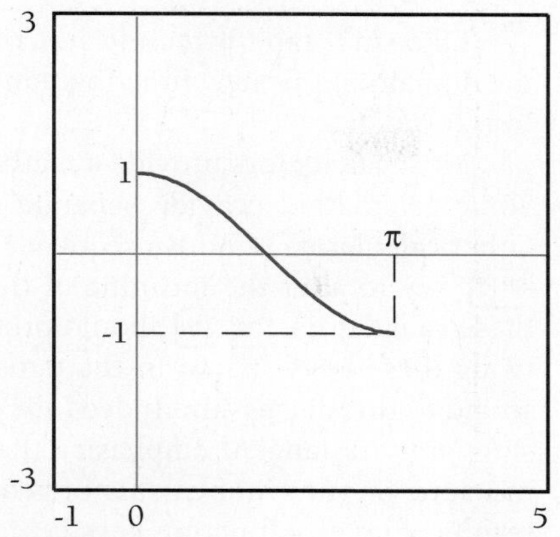

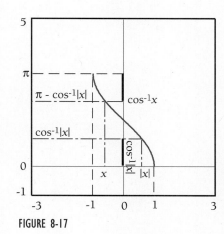

FIGURE 8-16

FIGURE 8-17

$\sin^{-1}\left(-\frac{1}{2}\right) = -\pi/6$. The same domain restriction on cosine does not produce a one-to-one function. See Figure 8-16.

Restrict the domain for cosine to $[0, \pi]$ to obtain a one-to-one function. Unfortunately, the resulting inverse (see Figure 8-17) is neither odd nor even. To determine the arc from a given cosine requires more effort than for inverse sine.

DEFINITION 8-2
Arc Cosine

> The inverse cosine function is defined by $y = \cos^{-1} x$ if and only if $\cos y = x$, where $0 \leq y \leq \pi$ and $-1 \leq x \leq 1$. Inverse cosine is also known as **arc cosine**.

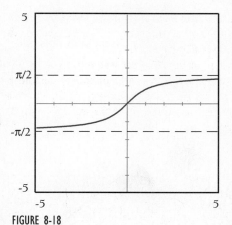

FIGURE 8-18

The range of $\cos^{-1}$ is $[0, \pi]$. Like $\sin^{-1}$, the domain of $\cos^{-1}$ is $[-1, 1]$. If $x > 0$ then $\cos^{-1} x \in [0, \pi/2]$. For $x < 0$, $\cos^{-1} x = \pi - \cos^{-1}|x|$. Consider $\cos^{-1}(-\frac{1}{2})$.

$$\cos^{-1}\left(-\frac{1}{2}\right) = \pi - \cos^{-1}\frac{1}{2}$$

$$\cos^{-1}\left(-\frac{1}{2}\right) = \pi - \frac{\pi}{3}$$

$$\cos^{-1}\left(-\frac{1}{2}\right) = \frac{2\pi}{3}$$

Finally, an analysis of the tangent function suggests that restricting the domain to $(-\pi/2, \pi/2)$ produces a one-to-one function. See Figure 8-18.

DEFINITION 8-3
Arc Tangent

> The inverse tangent function is defined by $y = \tan^{-1}x$ if and only if $\tan y = x$, where $-\pi/2 < y < \pi/2$. The inverse tangent is also known as **arc tangent**.

Like sin^{-1}, **tan^{-1}** is an odd function: tan^{-1}(-x) = tan^{-1} x. For the first time, the domain of this new function, tan^{-1}, is all real numbers. The range of tan^{-1} is (-π/2, π/2).

Most calculators provide a method for calculating these inverse functions. Some calculators provide separate function keys for [SIN^{-1}], [TAN^{-1}] and [COS^{-1}]. Other calculators require a sequence of keystrokes, using an [2nd] inverse key or [SHIFT] key to alter the meaning of the usual sine, cosine and tangent keys. In these calculators, [2nd] [TAN] should produce the same result as [TAN^{-1}]. Be cautious using these keys. When in the radian mode, calculator results should correspond to this discussion. Indeed the names of these functions, arc sine, arc cosine, and arc tangent emphasize that our discussion is geared toward radian measure. Be sure your calculator is in the radian mode before you interpret the results of inverse function keys.

Some programming languages abbreviate the inverse functions as ASIN, ACOS and ATAN. Normally these functions provide answers in radian measure. Direct degree measure results may not be available. Some textbooks use the abbreviation *arcsin* for sin^{-1}. The prefix *arc* emphasizes that the range values are in radian rather than degree measure. Similar comments apply for *arccos* in place of cos^{-1} and *arctan* for tan^{-1}.

EXAMPLE 8-2

Evaluating Inverse Circular Functions. Evaluate the following inverse functions.

Illustration 1:

$$\tan^{-1}(-1)$$

Solution

$$\tan^{-1}(-1) = -\tan^{-1} 1 \qquad \text{(because tan^{-1} is odd)}$$

$$\tan^{-1}(-1) = -\frac{\pi}{4} \qquad \left(\text{because } \tan\frac{\pi}{4} = 1\right)$$

Illustration 2:

$$\sin\left[\sin^{-1}\left(\frac{1}{2}\right)\right]$$

Solution:

$$\sin\left[\sin^{-1}\left(\frac{1}{2}\right)\right] = \sin\frac{\pi}{6} \qquad \left(\text{because } \sin^{-1}\left(\frac{1}{2}\right) = \frac{\pi}{6}\right)$$

Answer:

$$\sin\left[\sin^{-1}\left(\frac{1}{2}\right)\right] = \frac{1}{2} \qquad \text{(no surprise here)}$$

Illustration 3:

$$\cos^{-1}\left(\cos\frac{5\pi}{3}\right)$$

Solution:

$$\cos^{-1}\left(\cos\frac{5\pi}{3}\right) = \cos^{-1}\frac{1}{2} \qquad \left(\text{because } \cos\frac{5\pi}{3} = \frac{1}{2}\right)$$

Answer:

$$\cos^{-1}\left(\cos\frac{5\pi}{3}\right) = \frac{\pi}{3} \qquad \text{(surprise)}$$

Note: Although $\cos^{-1}$ represents an inverse function, it is not the inverse of the complete cosine function. Rather it is the inverse of the cosine with domain restricted to $[0, \pi]$. Because $5\pi/3$ is not in that domain, the composition of the two functions does not behave like the identity function.

Illustration 4:

$$\tan\left[\sin^{-1}\left(-\frac{\sqrt{3}}{2}\right)\right]$$

Solution:

$$\tan\left[\sin^{-1}\left(-\frac{\sqrt{3}}{2}\right)\right] = \tan\left[-\sin^{-1}\left(\frac{\sqrt{3}}{2}\right)\right] \qquad \text{(because } \sin^{-1} \text{ is odd)}$$

$$\tan\left[\sin^{-1}\left(-\frac{\sqrt{3}}{2}\right)\right] = \tan\left(-\frac{\pi}{3}\right) \qquad \left(\sin^{-1}\left(\frac{\sqrt{3}}{2}\right) = \frac{\pi}{3}\right)$$

Answer:

$$\tan\left[\sin^{-1}\left(-\frac{\sqrt{3}}{2}\right)\right] = -\sqrt{3} \qquad \left(\tan\left(-\frac{\pi}{3}\right) = -\sqrt{3}\right)$$

Illustration 5:

$$\cos(\sin^{-1}x)$$

Let $y = \sin^{-1} x$, then $\sin y = x$, $-\pi/2 \leq y \leq \pi/2$.

$$\cos(\sin^{-1} x) = \cos y$$

From the Pythagorean relation,

$$\sin^2 y + \cos^2 y = 1 \qquad \text{(see Figure 8-19)}$$
$$\cos y = \pm\sqrt{1 - \sin^2 y}$$

Because the range of $\sin^{-1}$ is restricted to $-\pi/2$ to $\pi/2$, then $-\pi/2 \leq y \leq \pi/2$. Cosine is positive on this interval so that $\cos y = \sqrt{1 - \sin^2 y}$.

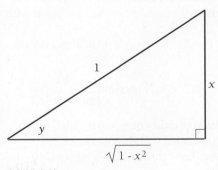

FIGURE 8-19

Answer: $\cos(\sin^{-1} x) = \sqrt{1 - x^2}$

Illustration 6:

$$\sin(\cos^{-1} A + \sin^{-1} B)$$

Solution: Let $x = \cos^{-1} A$. Then $\cos x = A$ and $0 \le x \le \pi$. In these quadrants, sine is positive. By the Pythagorean relation, $\sin x = \sqrt{1 - A^2}$ (Figure 8-20). Let $y = \sin^{-1} B$, then $\sin y = B$, $-\pi/2 \le y \le \pi/2$. Now $\cos y$ is positive in quadrant I or IV. Hence, $\cos y = \sqrt{1 - B^2}$ (Figure 8-21).

$$\sin(\cos^{-1} A + \sin^{-1} B) = \sin(x + y) \text{ (apply the sum formula)}$$
$$\sin(\cos^{-1} A + \sin^{-1} B) = \sin x \cos y + \cos x \sin y$$

Answer: $\sin(\cos^{-1} A + \sin^{-1} B) = \sqrt{1 - A^2}\,\sqrt{1 - B^2} + AB$

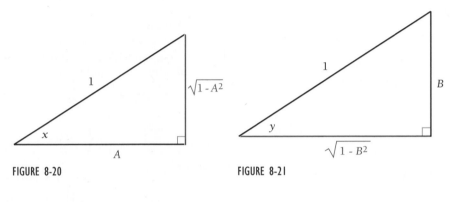

FIGURE 8-20   FIGURE 8-21

Illustration 7:

$$\sin^{-1} 2$$

Solution: No arc has a sine of 2: $-1 \le \sin x \le 1$. As a result, 2 is not in the domain of $\sin^{-1}$. Because $\sin^{-1} 2$ is undefined you should receive an error message if you evaluate $\sin^{-1} 2$ by calculator. ▬

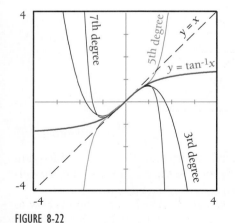

FIGURE 8-22

How do we approximate values for other inverse circular functions? Although many calculators have keys for $\tan^{-1}$, $\sin^{-1}$ and $\cos^{-1}$, few calculators have keys for the other inverse functions. Of the three listed, $\tan^{-1}$ is most likely to be represented on a calculator key or in a programming language. This is no accident. First, the approximating formulas for $\tan^{-1}$ are quite efficient and easy to implement. Of more importance, the five other inverse circular functions can be approximated in terms of $\tan^{-1}$. Therefore, we begin with $\tan^{-1}$.

For values of x close to the origin, the $2n - 1$-degree Maclaurin polynomial approximation for $\tan^{-1}$ is given by

$$\tan^{-1} x \cong x - \frac{x^3}{3} + \frac{x^5}{5} - \frac{x^7}{7} + \frac{x^9}{9} - \ldots + (-1)^{n+1} \frac{x^{2n-1}}{(2n-1)}, |x| < 1$$

Because $\tan^{-1}$ is odd so is the degree of the approximating polynomial. The formula is general, producing better approximations with higher degree.

Figure 8-22 displays the graphs of the first-, third-, fifth- and seventh- degree Maclaurin polynomials for approximating $\tan^{-1}$. Compare these to the graph of $\tan^{-1}$. The larger degree polynomials "fit" $\tan^{-1}$ more tightly. The approximations are "good" when x is between -1 and 1.

The domain of $\tan^{-1}$ is all real numbers. How, then, do we approximate larger values of x? If x is larger than 1, then $1/x$ is smaller than 1. Let $y = \tan^{-1} x$. Then $\tan y = x$. Because tangent and cotangent are complementary functions,

$$\tan\left(\frac{\pi}{2} - y\right) = \cot y$$

$$= \frac{1}{\tan y}$$

$$= \frac{1}{x}$$

Apply $\tan^{-1}$ to both sides of the equation:

$$\tan^{-1}\left[\tan\left(\frac{\pi}{2} - y\right)\right] = \tan^{-1}\frac{1}{x}$$

$$\frac{\pi}{2} - y = \tan^{-1}\frac{1}{x} \qquad \text{(property of inverse functions)}$$

$$\frac{\pi}{2} - \tan^{-1} x = \tan^{-1}\frac{1}{x} \qquad \text{(substitute for } y\text{)}$$

$$\tan^{-1} x = \frac{\pi}{2} - \tan^{-1}\frac{1}{x} \qquad \text{(solve for } \tan^{-1}x\text{)}$$

Because $1/x < 1$ whenever $x > 1$, use the polynomial approximation for $\tan^{-1} 1/x$ and the identity $\tan^{-1} x \equiv \pi/2 - \tan^{-1} 1/x$ whenever $x > 1$. In the case of $x = 1$, $\tan^{-1} 1 = \pi/4$ is a well-known value and needs no approximation.

In Chapter 3, we approximated roots of quadratic equations using continued fractions. Many transcendental functions have continued fraction representations. Among the better known is the continued fraction representation of $\tan^{-1}$.

$$\tan^{-1} x = \cfrac{x}{1 + \cfrac{(1x)^2}{3 + \cfrac{(2x)^2}{5 + \cfrac{(3x)^2}{7 + \cfrac{(4x)^2}{9 + \ldots}}}}}$$

As with irrational roots of quadratic equations, truncate the continued fraction to approximate the $\tan^{-1}$. The further "down" the fraction we truncate, the better is the approximation.

But a $\tan^{-1}$ key is probably on your calculator. The key question is how to use $\boxed{\text{TAN}^{-1}}$ for approximating the inverse circular functions that have no

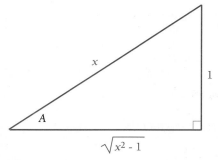

FIGURE 8-23

function key. Consider the approximation of $\csc^{-1} x$. The first step in the solution of many problems is an appeal to the definition. We begin in that manner.

$$\text{Let } A = \csc^{-1} x$$

Then $\csc A = x$ (definition of $A = \csc^{-1} x$)

We seek a connection to $\tan^{-1}$. Because circular functions are more familiar, first attempt a connection to tangent. If A is a small positive arc subtending an acute central angle, then the triangle in Figure 8-23 helps us visualize the relation.

Based on the definition of $\csc A = x/1$, label the side opposite angle A as 1 and the hypotenuse as x. By the Pythagorean relation, the remaining side of the triangle must be $\sqrt{x^2 - 1}$. Read critical information directly from this triangle:

$$\tan A = \frac{1}{\sqrt{x^2 - 1}}$$ (definition of tangent)

The same relation can be obtained through well-chosen identities. Start with the definition of cosecant: $\csc A = x/1$. Because cosecant is the reciprocal of sine: $\sin A = 1/x$. Then $\cos A$ follows from the Pythagorean relation:

$$\cos A = \sqrt{1 - \frac{1}{x^2}}$$

$$\cos A = \frac{\sqrt{x^2 - 1}}{x}$$ (simplify)

From sine and cosine obtain tangent:

$$\frac{\sin A}{\cos A} = \frac{\dfrac{1}{x}}{\dfrac{\sqrt{x^2 - 1}}{x}}$$

$$\tan A = \frac{1}{\sqrt{x^2 - 1}}$$ (definition of tangent)

Either way, we have an expression relating A and x through the tangent function. For our limited domain and range, $\tan^{-1}$ is one to one. Apply $\tan^{-1}$ to both sides of the equation:

$$\tan^{-1} (\tan A) = \tan^{-1} \left(\frac{1}{\sqrt{x^2 - 1}} \right)$$

On the left, $\tan^{-1}$ and $\tan$ "undo" each other:

$$A = \tan^{-1} \left(\frac{1}{\sqrt{x^2 - 1}} \right)$$

But A represented $\csc^{-1} x$. Therefore,

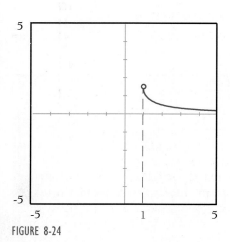

FIGURE 8-24

$$\csc^{-1} x = \tan^{-1} \left(\frac{1}{\sqrt{x^2 - 1}} \right), \qquad \text{for } x > 1. \quad \text{(See Figure 8-24)}.$$

EXAMPLE 8-3

Illustration 1:

Approximating Inverse Circular Functions. Approximate the following.

$$\tan^{-1} 0.1$$

Solution 1: By calculator, $\tan^{-1} 0.1 \cong 0.099668652$.

Solution 2: By using a seventh-degree Maclaurin polynomial

$$\tan^{-1} 0.1 \cong 0.1 - \frac{0.1^3}{3} + \frac{0.1^5}{5} - \frac{0.1^7}{7}$$

$$\tan^{-1} 0.1 \cong 0.099668652$$

Note: The accuracy of solution 2 relates to the closeness of x to 0. In this case, 0.1 is close to 0. If we approximate $\tan^{-1} 0.9$, the calculator key for $\tan^{-1}$ yields 0.732815101, whereas the seventh-degree Maclaurin polynomial gives a less accurate 0.693161871. In any case, the polynomial should not be used for values of x where $|x| \geq 1$.

Solution 3: Rather than attempt the dirty arithmetic of a continued fraction, we shall demonstrate a variation of the continued fraction. A "quick and dirty" approximation of $\tan^{-1} x$ is $x/(1 + 0.28x^2)$ (see Exercise 8-2, Problem 81):

$$\tan^{-1} 0.1 \cong \frac{0.1}{1 + 0.28(0.1)^2}$$

$$\tan^{-1} 0.1 \cong 0.099720781$$

Illustration 2:

$$\csc^{-1} 1.5$$

Solution: Use the conversion to $\tan^{-1}$ previously discussed:

$$\csc^{-1} 1.5 = \tan^{-1}\left(\frac{1}{\sqrt{1.5^2 - 1}}\right)$$

$$\csc^{-1} 1.5 = \tan^{-1} 0.894427191$$

By calculator, $\csc^{-1} 1.5 \cong 0.729727656$

SUMMARY

- $\sin^{-1} x = y$ means $\sin y = x$ and $-\pi/2 \leq y \leq \pi/2$.
- $\cos^{-1} x = y$ means $\cos y = x$ and $0 \leq y \leq \pi$.
- $\tan^{-1} x = y$ means $\tan y = x$ and $-\pi/2 < y < \pi/2$.
- $\tan^{-1}$ has several efficient approximation formulas. Approximate the other inverse circular functions by expressing them in terms of $\tan^{-1}$.

8-2   E X E R C I S E S

Determine the exact value of the following.

1. $\sin^{-1} 0.5$

2. $\cos^{-1} \dfrac{\sqrt{3}}{2}$

3. $\cos^{-1}(-0.5)$

4. $\sin^{-1}\left(-\dfrac{\sqrt{3}}{2}\right)$

5. $\tan^{-1}(-1)$

6. $\tan^{-1} \sqrt{3}$

7. $\sin^{-1} \dfrac{\sqrt{2}}{2}$

8. $\cos^{-1}(-0.5)$

9. $\cos^{-1} \dfrac{\sqrt{2}}{2}$

10. $\sin^{-1}\left(-\dfrac{\sqrt{2}}{2}\right)$

Approximate the following using a calculator.

11. $\sin^{-1} 0.3$

12. $\cos^{-1}(-0.3)$

13. $\cos^{-1}(-0.6)$

14. $\sin^{-1} 0.8$

15. $\sin^{-1}\left(\dfrac{5}{13}\right)$

16. $\cos^{-1}\left(\dfrac{4}{5}\right)$

17. $\cos^{-1}\left(\dfrac{12}{13}\right)$

18. $\sin^{-1}\left(-\dfrac{3}{5}\right)$

19. $\tan^{-1} 2$

20. $\tan^{-1}(-5)$

Refer to the definitions of $\sin^{-1}$, $\cos^{-1}$ and $\tan^{-1}$.

21. Formulate a definition for $\cot^{-1}$.

22. Formulate a definition for $\csc^{-1}$.

*23. Formulate a definition for $\sec^{-1}$. (Note: most calculus books restrict the range to $[0, \pi/2) \cup [\pi, 3\pi/2)$ while many precalculus texts use $[0, \pi/2) \cup (\pi/2, \pi]$.)

*24. Refer to Problem 23. One concern in calculus is whether a function is increasing or decreasing on a particular interval. Discuss how this might influence the choice of range restriction for $\sec^{-1}$ in calculus.

Refer to your definitions (Problems 21–23) to evaluate the following.

25. $\sec^{-1} \sqrt{2}$

26. $\csc^{-1} 2$

27. $\cot^{-1}(-1)$

28. $\csc^{-1}\left(\dfrac{-2\sqrt{3}}{3}\right)$

29. $\sec^{-1} 2$

30. $\csc^{-1} \sqrt{2}$

31. $\sec^{-1}(-2)$

32. $\csc^{-1}(-1)$

Evaluate as indicated.

33. $\sin(\sin^{-1} 0.4)$

34. $\cos[\cos^{-1}(-0.3)]$

35. $\tan(\tan^{-1} 5)$

36. $\sin(\sin^{-1} 0.4)$

37. $\cos^{-1}\left(\cos \dfrac{\pi}{3}\right)$

38. $\sin^{-1}\left(\sin \dfrac{\pi}{4}\right)$

39. $\sin^{-1}\left(\sin \dfrac{5\pi}{6}\right)$

40. $\cos^{-1}\left[\cos\left(-\dfrac{\pi}{4}\right)\right]$

41. $\cos^{-1}\left(\cos \dfrac{5\pi}{6}\right)$

42. $\sin^{-1}\left[\sin\left(-\dfrac{\pi}{4}\right)\right]$

43. $\sin^{-1}\left(\cos \dfrac{\pi}{6}\right)$

44. $\cos^{-1}\left(\sin \dfrac{\pi}{3}\right)$

45. $\cos^{-1}\left(\sin \dfrac{\pi}{6}\right)$

46. $\sin^{-1}\left(\cos \dfrac{\pi}{3}\right)$

47. $\cos(\sin^{-1} 0.5)$

48. $\sin\left(\cos^{-1} \dfrac{\sqrt{3}}{2}\right)$

49. $\sin\left[\cos^{-1}\left(-\dfrac{\sqrt{3}}{2}\right)\right]$

50. $\cos[\sin^{-1}(-0.5)]$

51. $\cos(\sin^{-1} x)$

52. $\sin(\cos^{-1} y)$

Express each of the following as a function of $\tan^{-1}$.

53. $\sin^{-1} x$

54. $\cos^{-1} x$

55. $\cot^{-1} x$

56. $\sec^{-1} x$

Use the formulas from Problems 53–56 and a calculator to approximate the following.

57. $\tan^{-1} 0.5$

58. $\tan^{-1} 0.3$

59. $\tan^{-1}(-0.3)$

60. $\tan^{-1}(-0.5)$

61. $\sin^{-1} 0.2$

62. $\sin^{-1}(-0.4)$

63. $\cos^{-1}(-0.2)$

64. $\cos^{-1} 0.4$

65. $\sec^{-1} 2$

66. $\csc^{-1}(-2)$

67. $\cot^{-1}(-0.5)$

68. $\cot^{-1}(-0.3)$

69. $\sin^{-1} 3$

*70. $\cos^{-1}(-1.5)$

*71. $\csc^{-1} 0.5$

*72. $\sec^{-1}(-0.3)$

Verify the following identities.

73. $2 \sin^{-1} x + 2 \cos^{-1} x = \pi$

74. $\cot^{-1} x = \tan^{-1} \dfrac{1}{x}$

75. $\tan^{-1} x + \tan^{-1}\left(\dfrac{1}{x}\right) = \dfrac{\pi}{2}$

76. $\cos^{-1} x = \tan^{-1} \sqrt{\dfrac{1 - x^2}{x}}$

For Problems 77–79, simplify each of the expressions. (Each expression is a "truncation" of the coefficient of x^2 from the continued fraction representation of $\tan^{-1}$.)

77. $\cfrac{1}{3 + \cfrac{(2x)^2}{5}}$

78. $\cfrac{1}{3 + \cfrac{(2x)^2}{5 + \cfrac{(3x)^2}{7}}}$

79. $\cfrac{1}{3 + \cfrac{(2x)^2}{5 + \cfrac{(3x)^2}{7 + \cfrac{(4x)^2}{9}}}}$

80. Evaluate each of Problems 77–79 for $x = 0$ and for $x = 1$. Round your answer to two decimal places.

81. Refer to Problems 77–79. Discuss a possible source for $x/(1 + 0.28x^2)$ as an approximation for $\tan^{-1}$.

82. Refer to Problems 77–79. Every truncation of the continued fraction representation of $\tan^{-1}$ can be simplified to a rational function of the form $x/(1 + cx^2)$, where c is dependent on both x and the truncation point of the continued fraction. Whatever the value of c, the rational function always provides the exact value for $\tan^{-1} 0$. Substituting 1 for x in the coefficient c of x^2 approximates the coefficient c at 0.28. This "adjusts" the graph of $y = x/(1 + cx^2)$ to fit $\tan^{-1}$ closely at 1. Discuss whether this is a good strategy. How does this affect the approximations for $\tan^{-1} x$, where $0 < x < 1$? Experi-

ment with various values for c on your graphing calculator.

Solve each of the following for y in terms of x.

83. $x = 2 \sin^{-1}(y + 1)$ **84.** $x = -5 \tan^{-1}(1 - y)$

85. $x = \cot(2 - 3y)$ **86.** $x = 3 \cos(y + \pi)$

87. From the earth, the sun appears as a disc whose diameter subtends an angle of approximately 0.5 degrees. If the mean distance from the earth to the sun is 93 million miles, approximate the actual diameter of the sun to the nearest 100,000.

88. The mean distance of the earth to the moon is 238,855 miles. The actual diameter of the moon is 2,160 miles (to the nearest mile). Determine the apparent diameter of the moon as viewed from the earth as an angle measured in degrees to the nearest tenth of a degree (or to the nearest minute).

89. See Problems 87 and 88. Discuss the relative sizes of the sun and moon in terms of both their actual and apparent diameters.

90. See Display 8-2. A large sign is placed 10 feet above eye level over the expressway. The sign is 4 feet high. As a driver approaches the sign at a distance d, the apparent height of the sign is measured by θ as shown in the diagram. Express θ as a function of d.

91. See Problem 90. Suppose we begin to time a car when it is 2,000 feet from the sign, traveling at 80 ft/sec (about 55 mph). Express the distance from the car to the sign as a function of time (t in seconds).

92. See Problems 90 and 91. Express θ as a function of t. Does there appear to be a time (or distance) at which $\theta = 0$? Give a physical interpretation of the event at which $\theta = 0$.

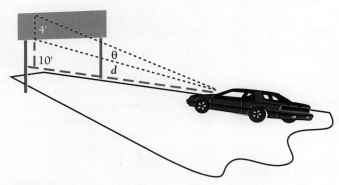

DISPLAY 8-2

8-2 PROBLEM SET

The following identities tie together circular functions, exponential functions and complex numbers:

$$e^{ix} \equiv \cos x + \sin x.$$

Because sine is odd and cosine is even then

$$e^{-ix} \equiv \cos x - \sin x.$$

Recall that $i^2 = -1$. Use these identities and assume that complex exponents obey the same rules as real exponents for the following:

1. Show that $e^{i\pi} + 1 = 0$ (Euler's identity).
2. Show that $e^{-i\pi} + 1 = 0$.
3. Show that $e^{ix}e^{-ix} = 1$.
4. Show that $e^{ix}e^{iy} = e^{i(x+y)}$.
5. Simplify the product $[\cos x + i \sin x][\cos x - i \sin x]$.
6. Simplify the product $[\cos x + i \sin x][\cos y - i \sin y]$.
7. Prove $\cos^2 x + \sin^2 x \equiv 1$ (see Problems 3 and 5).
8. Express $e^{i(x+y)}$ in terms of cosine and sine.
9. Show that $e^{-x} = \cos(ix) + i \sin(ix)$.
10. See Problems 4, 6 and 8. Rewrite the identity in Problem 4 in terms of sine and cosine.
11. Show that $e^x = \cos(ix) - i \sin(ix)$.
12. Use Problem 10 to prove that
$$\cos(x + y) = \cos x \cos y - \sin x \sin y.$$
13. Show that $\cos(ix) = \dfrac{e^x + e^{-x}}{2}$.
14. Show that $\cos x = \dfrac{e^{ix} + e^{-ix}}{2}$.
15. Show that $\sin x = \dfrac{e^{ix} - e^{-ix}}{2i}$.
16. Show that $-i \sin(ix) = \dfrac{e^x - e^{-x}}{2}$.
17. Define a function named *hyperbolic sine* as $\sinh x = i \sin(ix)$. Express $\sinh x$ in terms of e^x.
18. Define a function named *hyperbolic cosine* as $\cosh x = \cos(ix)$. Express $\cosh x$ in terms of e^x.
*19. See Problem 12.
Prove $\sin(x + y) = \sin x \cos y + \cos x \sin y$.
*20. See Problems 17 and 18. Prove $\cosh^2 x - \sinh^2 x = 1$.

PROBLEMS FOR GRAPHING TECHNOLOGY

Graph each pair of functions f and g in the same window.

21. Compare and contrast the graphs of $f(x) = \sin(\sin^{-1} x)$ and $g(x) = \sin^{-1}(\sin x)$.
22. Compare and contrast the graphs of $f(x) = \tan(\tan^{-1} x)$ and $g(x) = \tan^{-1}(\tan x)$.
23. For $f(x) = \tan^{-1} x$ discuss which function g appears to be "closer" to f in terms of intervals of x.

    (a) $g(x) = \dfrac{x}{1 + 0.28x^2}$

    (b) $g(x) = x - \dfrac{x^3}{3}$

24. For $f(x) = \tan^{-1} x$ discuss which function g appears to be "closer" to f in terms of intervals.

    (a) $g(x) = \dfrac{x}{1 + 0.33x^2}$

    (b) $g(x) = x - \dfrac{x^3}{3} + \dfrac{x^5}{5}$

25. (a) Use your calculator trace feature to estimate the location of the highest and lowest points of the graph $f(x) = e^{-x/4} \cos(4x)$ in the window xMin $= -1$, xMax $= 3$, yMin $= -2$, yMax $= 2$.

    (b) In the same window graph $y = 16 \sin(4x) + \cos(4x)$. Compare the x-intercepts of this graph to high and low points of f.

    (c) Determine the x-intercepts of $y = 16 \sin(4x) + \cos(4x)$ by solving the equation $16 \sin(4x) + \cos(4x) = 0$ on $[-1, 3]$.

    (d) Compare the estimated high and low points of f with the solution to part c. (One topic in calculus is how to produce the equation in part b from the equation in part a.)

EXPLORATION

The following problems anticipate concepts from the next section. Use graphing technology to graph each of the following pairs of functions, f and g, in the same window. Use the trace feature to approximate points of intersection of the two functions on the interval $[0, 2\pi)$. If the graphs of f and g appear to coincide label the functions as *identities*.

26. $f(x) = 3 \sin x$, $g(x) = 2$
27. $f(x) = 5 \cos x - 1$, $g(x) = 3$

28. $f(x) = 3 \cos x$, $g(x) = 2 \sin x$

29. $f(x) = 5 \sin(3x - \pi)$, $g(x) = 2$

30. $f(x) = \tan x$, $g(x) = 1.5$

31. $f(x) = 3 \sin^2 x - 1$, $g(x) = 2 \sin x$

32. In each of the preceding, determining the point(s) of intersection is equivalent to solving the equation $f(x) = g(x)$. Such equations are conditional equations rather than identities. Discuss the difference between conditional equations and identities.

33. You found the points of intersection only on the interval $[0, 2\pi)$. How many points of intersection would you expect on the entire domain of the functions?

34. Compare the solution(s) of Problem 28 to the solutions of 30. Are they connected? In what way?

35. Approximate the solution of $3 \cos(5x - 2) = 4$ on $[0, 2\pi)$.

36. Solve $5 \sin(3x + \pi) = 10$. Discuss any difficulties you encounter.

37. See Problem 35. Give the solution to $3 \cos(5x - 2) = 4$ on $(-\infty, \infty)$.

     CONDITIONAL EQUATIONS

Inverse trigonometric functions provide the tools to solve **conditional equations** involving circular functions. Recall the line given by $f(x) = 3$ and sine curve given by $g(x) = 5 \sin x + 1$ that opened the previous section. The graphs of these functions are shown in Figure 8-25.

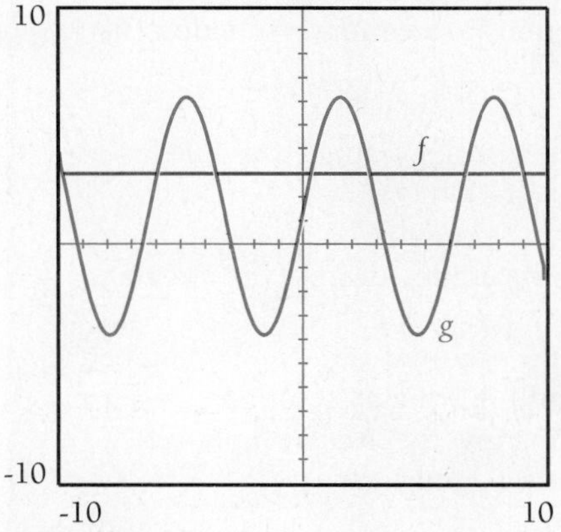

    The graph of f intersects the graph of g in infinitely many points. Because $g(x) = 5 \sin x + 1$ has a fundamental period of 2π, once the points of intersection between $f(x) = 3$ and $g(x) = 5 \sin x + 1$ are known for $x \in [0, 2\pi)$, the remainder of the solution will follow from periodicity. We must solve

$$f(x) = g(x)$$
$$3 = 5 \sin x + 1$$

To *solve* this conditional equation means to determine the values of the variable x for which the two functions are equal. From the last section, our strategy is to first solve the equation for the circular function.

$$5 \sin x + 1 = 3$$
$$5 \sin x = 2$$
$$\sin x = \frac{2}{5}$$

Now apply inverse sine to unbind x.

$$\sin^{-1}(\sin x) = \sin^{-1}\left(\frac{2}{5}\right)$$
$$x = \sin^{-1}\left(\frac{2}{5}\right)$$

By calculator, $x \cong 0.411516846$.

    However, this single solution is in the first quadrant, and there should be infinitely many solutions. Fortunately, the periodicity of sine supplies what

FIGURE 8-25

$\sin^{-1}$ did not. Because sine repeats every 2π, many of the solutions of the equation can be expressed as $x = \sin^{-1}(\frac{2}{5}) + 2k\pi$, where k is an integer.

Examine Figure 8-25. We have not described all of the answers. Another solution occurs in the interval $(\pi/2, \pi)$. The original solution is the reference arc for this solution. Therefore, $x = \pi - \sin^{-1}(\frac{2}{5})$ is a solution. Apply the periodicity of sine to this solution and obtain

$$x = \pi - \sin^{-1}\left(\frac{2}{5}\right) + 2k\pi$$

$$x = (2k + 1)\pi - \sin^{-1}\left(\frac{2}{5}\right) \qquad \text{(combining terms of } \pi\text{)}$$

Hence the complete solution of $3 = 5 \sin x + 1$ consists of $x = (2k + 1)\pi - \sin^{-1}(\frac{2}{5})$ or $x = \sin^{-1}(\frac{2}{5}) + 2k\pi$.

There is no single procedure for solving a conditional equation where the variable is bound in the argument of a circular function. However, the following form a reasonable strategy.

Strategy for Solving Conditional Circular Equations

1. Use identities to rewrite the equation. Inspiration may be required. Convert all circular functions to a common function (usually sine or cosine).
2. Identify the type of equation as if the circular function were a simple variable (some suggest substituting a single variable for the function).
3. Solve for the function (or substituted variable).
4. Use inverse functions to unbind the argument from the circular function.
5. Solve the resulting equation for the required variable. Remember, the original equation may have infinitely many solutions. Because of the limitations imposed by using inverse circular functions, these five steps lead only to an **initial solution**.
6. Extend the initial solution to determine all solutions for the argument between 0 and 2π. These form the **primary solutions**.
7. If necessary, write out the general solution. The general solution consists of all solutions of the equation. The general solution is usually inferred from the primary solutions and the periodicity of the circular functions.

EXAMPLE 8-4

Illustration 1:

Solving Trigonometric Equations. Solve as indicated.

Determine the solutions of $2 \sin(5x - \pi) + \sqrt{3} = 0$.

Solution: The equation is first-degree in sine. Solve for $\sin(5x - \pi)$:

$$2\sin(5x - \pi) = -\sqrt{3}$$

$$\sin(5x - \pi) = -\frac{\sqrt{3}}{2} \qquad \text{[Equation 8.1]}$$

Apply the inverse ($\sin^{-1}$) for the initial solution:

$$5x - \pi = \sin^{-1}\left(-\frac{\sqrt{3}}{2}\right) \qquad \text{(solve for } x\text{)}$$

$$5x - \pi = -\frac{\pi}{3} \qquad \text{[Equation 8.2]}$$

$$5x = \frac{2\pi}{3}$$

$$x = \frac{2\pi}{15}$$

For the primary solutions we must determine the arcs between 0 and 2π with a reference arc of $\left|-\pi/3\right|$ (see [Equation 8.2]):

$$\frac{\pi}{3}, \quad \frac{2\pi}{3}, \quad \frac{4\pi}{3} \quad \text{and} \quad \frac{5\pi}{3}$$

Refer to [Equation 8.1]. Which arcs produce a negative value for sine? From the graph in Figure 8-26, $4\pi/3$ and $5\pi/3$ have negative values of sine. Therefore, from [Equation 8.1]

$$5x - \pi = \frac{4\pi}{3} \quad \text{or} \quad 5x - \pi = \frac{5\pi}{3} \qquad \text{[Equation 8.3]}$$

$$5x = \frac{7\pi}{3} \quad \text{or} \quad 5x = \frac{8\pi}{3}$$

$$x = \frac{7\pi}{15} \quad \text{or} \quad x = \frac{8\pi}{15}$$

Finally for the general solution, recall that sine is periodic with a fundamental period of 2π. Therefore,

$$5x - \pi = \frac{4\pi}{3} + 2k\pi \quad \text{or} \quad 5x - \pi = \frac{5\pi}{3} + 2k\pi \quad \text{(see Equation 8.3)}$$

$$5x = \frac{7\pi}{3} + 2k\pi \quad \text{or} \quad 5x = \frac{8\pi}{3} + 2k\pi$$

$$x = \frac{7\pi}{15} + \frac{2k\pi}{5} \quad \text{or} \quad x = \frac{8\pi}{15} + \frac{2k\pi}{5}$$

$$x = \frac{(6k + 7)\pi}{15} \quad \text{or} \quad x = \frac{(3k + 4)2\pi}{15}$$

Notice that the initial solution is a special case of the general solution: let $k = -1$ in $x = (3k + 4)2\pi/15$.

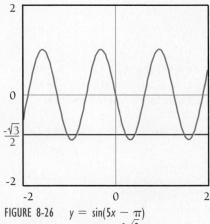

FIGURE 8-26 $y = \sin(5x - \pi)$

and $y = -\dfrac{\sqrt{3}}{2}$

Illustration 2:   Solve $\sin x = \cos x$.

Solution: From graphs of sine and cosine, $\cos x \neq 0$ when $\sin x = 0$. Therefore we may divide by $\cos x$:

$$\frac{\sin x}{\cos x} = 1$$

$$\tan x = 1$$

Initial solution: $x = \tan^{-1} 1$

$$x = \frac{\pi}{4} \qquad \text{(reference arc for primary solutions)}$$

The primary solutions lie in the first and third quadrants. The primary solutions are $x = \pi/4$ or $x = 5\pi/4$. Because the tangent has a period of π, the general solutions follow immediately:

$$x = \frac{\pi}{4} + k\pi, \quad \text{or} \quad x = \frac{5\pi}{4} + k\pi$$

$$x = (4k + 1)\frac{\pi}{4} \quad \text{or} \quad x = (4k + 5)\frac{\pi}{4}$$

$$x = (4k + 1)\frac{\pi}{4} \quad \text{or} \quad x = (4k + 4 + 1)\frac{\pi}{4}$$

$$x = (4k + 1)\frac{\pi}{4} \quad \text{or} \quad x = [4(k + 1) + 1]\frac{\pi}{4}$$

But these are two descriptions of the same values: $4(k + 1) + 1$ generates the same values as $(4k + 1)$. For example, if $k = 2$, then $4(k + 1) + 1 = 13$, which is the same as $(4k + 1)$ when $k = 3$. Therefore combine the description of the general solutions:

Answer:

$$x = (4k + 1)\frac{\pi}{4}$$

Illustration 3: $2\cos^2 x - \sin x = 1$ \hfill (quadratic in cosine)

Solution: Substitute $1 - \sin^2 x$ for $\cos^2 x$ \hfill (obtain single circular function)

$2(1 - \sin^2 x) - \sin x = 1$ \hfill (Quadratic in sine)

$2 - 2\sin^2 x - \sin x = 1$

$2\sin^2 x + \sin x - 1 = 0$

$(2\sin x - 1)(\sin x + 1) = 0$ \hfill (factor)

$2\sin x - 1 = 0 \quad \text{or} \quad \sin x + 1 = 0$ \hfill (set each factor = 0)

$\sin x = \dfrac{1}{2} \quad \text{or} \quad \sin x = -1$ \hfill (and solve)

Initial: $x = \dfrac{\pi}{6} \quad \text{or} \quad x = -\dfrac{\pi}{2}$

Primary: $x = \dfrac{\pi}{6}$ or $x = \dfrac{5\pi}{6}$ or $x = \dfrac{3\pi}{2}$

General: $x = \dfrac{\pi}{6} + 2k\pi,$ $x = \dfrac{5\pi}{6} + 2k\pi,$ $x = \dfrac{3\pi}{2} + 2k\pi$

$x = \dfrac{12k + 1}{6}\,\pi,$ $x = \dfrac{12k + 5}{6}\,\pi,$ $x = \dfrac{4k + 3}{2}\,\pi$

Illustration 4: $3\cos^2 x - 5\cos x - 1 = 0$ (quadratic in cosine)

Solution: See Figure 8-27. The illustration is in quadratic form, so use quadratic formula:

$$\cos x = \frac{5 \pm \sqrt{25 - 4(3)(\text{-}1)}}{2(3)} \qquad \text{(solve for cosine)}$$

$$= \frac{5 \pm \sqrt{37}}{6}$$

Initial: $x = \cos^{-1}\!\left(\dfrac{5 + \sqrt{37}}{6}\right)$ or $x = \cos^{-1}\!\left(\dfrac{5 - \sqrt{37}}{6}\right)$

Approximation: $x \cong \cos^{-1} 1.847$ or $x \cong \cos^{-1}(\text{-}0.18046)$

7

-5

-10   10

FIGURE 8-27 $y = 3\cos^2 x - 5\cos x - 1$

Note: 1.847 is not in the domain of $\cos^{-1}$. Discard the first initial solution. This leaves $x \cong \cos^{-1}(\text{-}0.18046)$

$x \cong 1.7523$ radians   (approximation of initial solution)

However, $y = 3\cos^2 x - 5\cos x - 1$ is an even function, symmetrical to the y-axis, so that $\text{-}\cos[(5 - \sqrt{37}/6)] \cong \text{-}1.7523$ is also a solution.

The general solution is $x = 2k\pi \pm \cos^{-1}[(5 - \sqrt{37})/6]$.

Illustration 5: $\sin(2x) = \cos x$

Solution: The arguments of the two circular functions are different. The first goal is to match arguments. Use the double angle formula for sine:

$\sin(2x) = \cos x$

$2\sin x \cos x = \cos x$ (double angle formula)

$2\sin x \cos x - \cos x = 0$ (get equal to 0)

$\cos x\,[2\sin x - 1] = 0$ (factor a common factor)

$\cos x = 0$ or $2\sin x - 1 = 0$ (set each factor = 0)

$\sin x = 0.5$ (and solve)

Initial: $x = \dfrac{\pi}{2}$ or $x = \dfrac{\pi}{6}$

Primary: $x = \dfrac{\pi}{2},$ $x = \dfrac{3\pi}{2}$ or $x = \dfrac{\pi}{6},$ $x = \dfrac{5\pi}{6}$

General: $x = (2k + 1)\dfrac{\pi}{2}$ or $x = \dfrac{\pi}{6} + 2k\pi$, $x = \dfrac{5\pi}{6} + 2k\pi$

Therefore, $x = \dfrac{2k + 1}{2}\pi$, or $x = \dfrac{12k + 1}{6}\pi$, $x = \dfrac{12k + 5}{6}\pi$.

Illustration 6:

$\tan(2x) = 1$

Solution: Apply $\tan^{-1}$ to obtain initial solutions for $2x$:

$$\tan^{-1}[\tan(2x)] = \tan^{-1} 1$$

Initial: $2x = \dfrac{\pi}{4}$ $\left(\text{thus, } x = \dfrac{\pi}{8}\right)$

Primary: $2x = \dfrac{\pi}{4}$ or $2x = \dfrac{5\pi}{4}$

Because tangent has a period of π, the solutions for $2x$ repeat every π:

$$2x = \dfrac{\pi}{4} + k\pi \quad \text{or} \quad 2x = \dfrac{5\pi}{4} + k\pi$$

$$2x = (4k + 1)\dfrac{\pi}{4} \quad \text{or} \quad 2x = (4k + 5)\dfrac{\pi}{4}$$

As in Illustration 2, $(4k + 1)\dfrac{\pi}{4}$ and $(4k + 5)\dfrac{\pi}{4}$ produce duplicate values:

Use the first: $2x = (4k + 1)\dfrac{\pi}{4}$ (divide by 2)

General: $x = (4k + 1)\dfrac{\pi}{8}$

Illustration 7:

$x = 2 \sin x$

Solution 1: This equation does not resemble the type of conditional equations previously discussed. The variable x appears not only in the argument of the trigonometric function, but also as a linear term. Applying $\sin^{-1}$ to both sides of the equation may free x from the argument of sine, but it will bind x on the lefthand side of the equation into the argument for $\sin^{-1}$. We need another method for solving this equation.

Recall the graphical representation of conditional equations. Figure 8-28 shows the graphs of $y = x$ and $y = 2 \sin x$ on the same coordinate system. The points of intersection of these two graphs represent the solution to $x = 2 \sin x$. From the graph a solution occurs at $x = 0$. Check 0 in $x = 2 \sin x$ to verify the solution. The graph suggests two more solutions, one negative and one positive. Consider the positive solution. Use the trace feature of graphing technology to estimate x. First rewrite the equation into a more convenient form:

$$x - 2 \sin x = 0$$

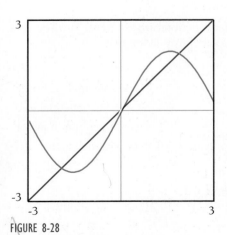

FIGURE 8-28

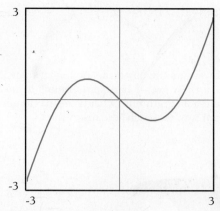

FIGURE 8-29 $y = x - 2\sin x$

See Figure 8-29 for the graph of $y = x - 2\sin x$. Compare this to Figure 8-28. The x-intercepts of $y = x - 2\sin x$ correspond to the solutions of the original equation. Zoom in on the positive solution. Apply the trace feature to extimate the x-intercept. Repeat the zoom to achieve greater accuracy. Even one zoom should produce an approximation between 1.88 and 1.90.

Solution 2: Use the [ANS] key of your technology to apply fixed point iteration. First write the equation in the form $x = 2\sin x$.

Based on the graph, initialize Ans close to a solution, say 1.5. Enter 1.5. Enter the righthand side of the formula replacing x with Ans: [2] [SIN] [ANS]. Press [EXE] or [ENTER] several times until the answer stabilizes at about $1.895\ldots$. To obtain the negative solution initialize Ans with a value like -2.2.

From the symmetry of the graph to the origin, the negative solution is approximately -1.895. This process provides only estimates. If you check 1.895 in the original equation, the two sides of the equation will not be identical: However, they will be close.   ▬

S U M M A R Y

See the strategy for solving conditional circular equations.

8-3   E X E R C I S E S

For each of the following equations, (a) determine an initial solution, (b) determine the primary solutions, (c) write a formula for the general solution.

1. $\sin x + 1 = 0$
2. $\cos x - 1 = 0$
3. $2\cos x - 1 = 0$
4. $2\sin x + 1 = 0$
5. $2\sin x - \sqrt{3} = 0$
6. $2\cos x + \sqrt{3} = 0$
7. $\tan x - 1 = 0$
8. $\cot x + 1 = 0$
9. $\sin x = \cos x$
10. $\sin x + \cos x = 0$
11. $\cos^2 x - 1 = 0$
12. $\sin^2 x = 1$
13. $\cos x \sin x = 0$
14. $\cot x \tan x = 0$
15. $\cos^2 x = \cos x$
16. $\sin^2 x - \sin x = 0$
17. $4\sin^2 x - 3 = 0$
18. $4\cos^2 x = 3$
19. $2\cos^2 x - \cos x - 1 = 0$
20. $2\sin^2 x + \sin x - 1 = 0$
21. $\sin(2x) + 1 = 0$
22. $\cos(2x) - 1 = 0$
23. $2\cos(3x) - 1 = 0$
24. $2\sin(3x) + 1 = 0$
25. $2\sin(2x) - \sqrt{3} = 0$
26. $2\cos(2x) + \sqrt{3} = 0$
27. $\tan \dfrac{x}{2} - 1 = 0$
28. $\cot \dfrac{x}{2} + 1 = 0$

29. $\sin \dfrac{x}{3} = \cos \dfrac{x}{3}$
30. $\sin \dfrac{x}{3} + \cos \dfrac{x}{3} = 0$
31. $\cos^2(3x) - 1 = 0$
32. $\sin^2(3x) = 1$
33. $\cot x \tan x = 0$
34. $\cos x \sin x = 0$
35. $\cos^2\left(x + \dfrac{\pi}{2}\right) = \cos\left(x + \dfrac{\pi}{2}\right)$
36. $\sin^2\left(x - \dfrac{\pi}{3}\right) - \sin\left(x - \dfrac{\pi}{3}\right) = 0$
37. $4\cos^2(2x) = 3$
38. $4\sin^2(2x) - 3 = 0$
39. $\cos(2x) - \cos x = 0$
40. $\cos(2x) + \sin x = 0$
41. $\sin x - 2 = 0$
42. $\sec x = 0.5$
43. $\sin^2 x + \sin x - 2 = 0$
44. $\cos^2 x - 3\cos x + 2 = 0$
45. $2\cos^2 x - 3\cos x - 2 = 0$
46. $2\sin^2 x + 5\sin x + 2 = 0$
47. $2\sin^2 x + \cos^2 x = 0$
48. $\sin^2 x - 3\cos^2 x = 0$
49. $\sin^2(2x) = 2\sin x \cos x$
50. $\sin x = \tan \dfrac{x}{2}$

8-3   P R O B L E M   S E T

1. When Bart goes surfing the first thing he does is check out the waves. Today Bart observes that the wave period averages 15 seconds from peak to peak. The height of the waves from crest to valley averages 4 feet. Use the sine function to write a formula to model the wave action. Bart likes to "catch" a wave on the downside, no higher than 1 foot to the crest of the wave. If Bart starts timing at the moment a wave crests beneath him, what is the maximum average time he can wait to "catch" the next wave?

15 seconds

2. One measure of the apparent height of an object is the height h of the projection of the object on the back surface of the eye. Another measure of the apparent height is the angle θ of light the object subtends as the light enters the lens of the eyes. One way to estimate the distance of an object of a known size is from its apparent height. Express the distance d of a 6 ft person from the eye as a function of the apparent height θ of the person. The diameter of a penny is 0.75 in. What is the maximum distance from the eye you can hold a penny to "block out" its vision of a 6 ft person standing 100 ft away?

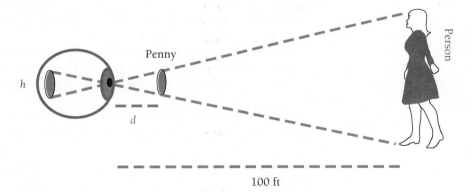

100 ft

3. A racketball court is 30′ wide and 60′ long. Burt serves from midcourt left. His goal is to hit the front wall and bounce to the back corner. The angle θ at which the ball strikes is the angle at which it rebounds. Determine how far from the server's wall the ball should strike the front wall and the total distance the ball travels from server to corner. Also what is angle ϕ?

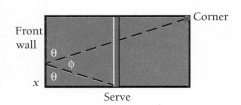

4. The intensity of an electron beam in a video display tube is given by $y = \cos^4 \theta$, where θ is the angle of deflection of the beam from the center of the screen. Suppose that the intensity is measured at 0.675 (67.5% of the intensity at dead center), what is the angle of deflection?

PROBLEMS FOR TECHNOLOGY

Approximate the primary solutions of the following.

5. $x + \cos x = 0$

6. $x - \cos x = 0$

7. $1 - 2x = \sin x$

8. $3 - x = \sin x$

9. $x^2 = \sin x$

10. $x^2 = \cos x$

11. $\tan x = 3 \sin x$

12. $2 \cos x = \cot x$

13. $\csc x = 2 \cos x$

14. $\sec x = 3 \sin x$

15. $\tan^{-1} x = \sin x$

16. $\tan^{-1} x = \cos x$

17. $e^{-x} = \sin x$

18. $e^{-x} = \cos x$

19. $\ln(x) = \cos x$

20. $\ln(x) = \sin x$

21. $|x| = \cos x$

22. $|x| + \sin x = 0$

23. $|\sin x| = |\cos x|$

24. $|\sin x - \cos x| = 1$

Use fixed-point iteration to approximate a solution for x in each of the following. (Compare these to Problems 5–10)

25. $x + \cos x = 0$

26. $x - \cos x = 0$

27. $1 - 2x = \sin x$

28. $3 - x = \sin x$

29. $x^2 = \sin x$

30. $x^2 = \cos x$

---

4   LAW OF SINES, LAW OF COSINES, GENERAL TRIANGLES

Plato said that God geometrizes continually.
—PLUTARCH

One of the first applications of trigonometry was measuring the universe. Before precision laser measurements squeezed the accuracy of the distance to the moon to within the nearest foot, astronomers used observation and trigonometry to estimate the distance. A team of astronomers would set up two observation posts a known distance apart. Each post would then simultaneously observe selected features of the moon. See Figure 8-30. The difference in the observers' line of sight against the fixed star background defines an angle known as the *parallax* of the moon.

Solving the triangle formed by the lines of sight provided the distance to the moon. Unfortunately, the triangle formed was not a right triangle. This section extends the solution of right triangles to the solution of general triangles.

Consider triangles that are not right triangles. To extend trigonometry to general triangles begin with the height of the triangle. In each view of triangle

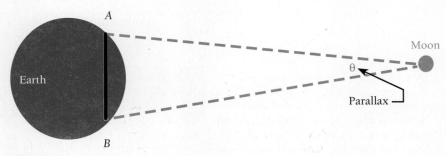

FIGURE 8-30

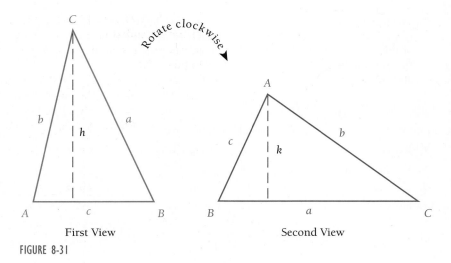

First View   Second View

FIGURE 8-31

ABC in Figure 8-31, label the bottom side as the base. Measure the height of each triangle with a line perpendicular to the base from the vertex of the opposite angle. In the first view, height *h* forms two right triangles. Solve the right triangle with sides *b* and *h* in the first view of triangle *ABC*. Then, $\sin A = h/b$:

$$h = b \sin A$$

For the triangle with sides *a* and *h* in the first view,

$$h = a \sin B$$

So that $a \sin B = b \sin A$

$$\frac{\sin B}{b} = \frac{\sin A}{a} \qquad \text{(divide by } ab\text{)}$$

From the second view of triangle *ABC*, a similar process uses height *k*.

$$k = b \sin C$$

Also, $k = c \sin B$

Then, $b \sin C = c \sin B$

$$\frac{\sin C}{c} = \frac{\sin B}{b} \qquad \text{(divide by } bc\text{)}$$

Combining these we obtain the **Law of Sines.**

Theorem 8-1
Law of Sines

For a general triangle *ABC*, $\dfrac{\sin A}{a} = \dfrac{\sin B}{b} = \dfrac{\sin C}{c}$

Unfortunately, the argument does not quite prove the Law of Sines. The difficulty occurs when one of the angles is obtuse. Suppose that angle *A* is

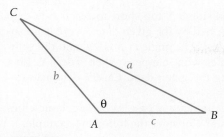

FIGURE 8-32

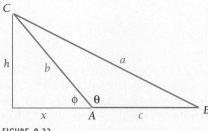

FIGURE 8-33

larger than 90°, see Figure 8-32. Extend side AB. Drop a perpendicular line h to the extended side AB (h is the height of ABC where side c is the base). Moreover h completes a right triangle with the extension of AB. See Figure 8-33.

Note that θ in Figure 8-33 is angle A of triangle ABC. The other angle at A is $180° - \theta = \phi$, but ϕ is acute. Also, ϕ and θ are supplementary angles: their sum is 180°. From Figure 8-33, $\sin \phi = h/b$. Divide by a and obtain $(\sin \phi)/a = h/(ab)$. Also $\sin B = h/a$. Divide both sides by b:

$$\frac{\sin B}{b} = \frac{h}{ab}$$

Thus, $$\frac{\sin \phi}{a} = \frac{\sin B}{b}$$

$$\text{Now } \sin \theta = \sin(180° - \phi)$$
$$= \sin 180° \cos \phi - \cos 180° \sin \phi$$
$$= \sin \phi$$

Because $A = \theta$, then $(\sin B)/b = (\sin A)/a$.

Equality of the remaining ratio, $(\sin C)/c$ follows similarly by extending side CA. If $A = 90°$, then triangle ABC is a right triangle and $\sin B = b/a$. But then $(\sin B)/b = 1/a$. Whenever $A = 90°$, then $\sin A = 1$, as expected.

As with right triangles, at least three measurements of a triangle are needed to solve the general triangle. In a right triangle, the right angle counts as one of the measurements.

The Law of Sines works best when solving triangles in which two angles and one side are known. The third angle follows from the sum of the angles being 180°. The Law of Sines provides the remaining sides through the solution of simple proportions.

Various combinations of measures pose different challenges for solving a triangle. Three measurements might include three sides, or two sides and an angle or two angles and a side. To facilitate the solution of general triangles, we classify various cases of givens. These cases are directly connected to geometry congruency theorems for triangles.

Figure 8-34 shows three possible solutions to a triangle given two sides, a and b, and an angle, B, not between them. The solution with these givens is not unique, so side-side-angle (SSA) is known as the **ambiguous case**.

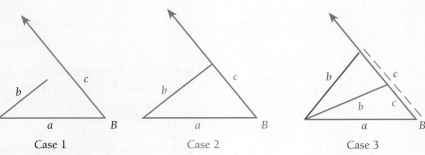

Case 1   Case 2   Case 3

FIGURE 8-34

Case 1 in Figure 8-34 occurs when side b is too short to reach side c. In Case 1 the triangle cannot exist as described by the given data. There is no solution. In Case 2, side b just reaches side c. The angle at A is 90° and the solution is unique. Finally, Case 3 illustrates what happens when side b is long enough to form two triangles satisfying the given data. Case 3 demonstrates why SSA is the ambiguous case.

Figure 8-34 is the geometry of the ambiguous case. What algebraic clues indicate which case represents given data? Consider the following example.

Solving General Triangles. Given triangle ABC. Solve as indicated.

EXAMPLE 8-5

Illustration 1:

Suppose $A = 30°$, $B = 50°$ and $c = 4$ cm.

Solution: With two angles given, the third angle $C = 180° - (A + B)$: $C = 100°$. Now sketch a triangle with appropriately sized angles labeled A, B and C. See Figure 8-35. Label the side opposite angle C as $c = 4$ cm. Two angles with any side is not the ambiguous case. A unique solution exists (see note). By the Law of Sines,

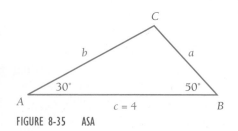

FIGURE 8-35   ASA

$$\frac{\sin 30°}{a} = \frac{\sin 50°}{b} = \frac{\sin 100°}{4 \text{ cm}}$$

$$\frac{\sin 30°}{a} = \frac{\sin 100°}{4 \text{ cm}} \qquad \text{(choose first and last ratios)}$$

$$a = \frac{4 \text{ cm} \sin 30°}{\sin 100°} \qquad \text{(solve for } a\text{)}$$

By calculator, $\quad a \cong \dfrac{4 \text{ cm} (0.5)}{0.984807753}$

$$a \cong 2.030853224 \text{ cm}$$

Similarly, $\dfrac{\sin 50°}{b} = \dfrac{\sin 100°}{4 \text{ cm}} \qquad$ (choose second and last ratios)

$$b = \frac{4 \text{ cm} \sin 50°}{\sin 100°} \qquad \text{(solve for } b\text{)}$$

$$b \cong \frac{4 \text{ cm}(0.766044443)}{0.984807753}$$

$$b \cong 3.111447653$$

Note: Although approximations are often difficult to check, we should examine the solutions for reasonableness. If the sketch in Figure 8-35 is close to scale, it should model the relative size of the sides. C is the largest angle and should be opposite the largest side; $c = 4$ cm is the largest side. Similarly, expect a to be larger than b because angle A is larger than B. Notice that C is twice the size of B. But side c is not exactly twice the size of side b. Do not expect side measurements to be directly proportional to angle measurements.

Illustration 2:   Suppose $A = 30°$, $a = 3$ m and $c = 8$ m.

Solution: We are given two sides and an angle not between them (SSA). This is the ambiguous case. Sketching a figure is difficult, because it is not clear whether such a triangle can exist. The only possible attack is to determine angle C. C and A will provide B. We will have three angles and can complete the solution as in Illustration 1.

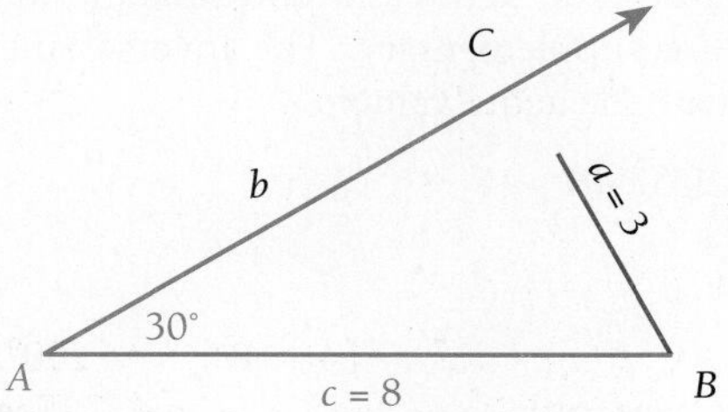

$$\frac{\sin A}{a} = \frac{\sin C}{c}$$

$$\frac{\sin 30°}{3 \text{ m}} = \frac{\sin C}{8 \text{ } m}$$

$$\sin C = \frac{8 \text{ m} \sin 30°}{3 \text{ m}}$$

$$\sin C = \frac{8(0.5)}{3}$$

$$\sin C = 1.33333333$$

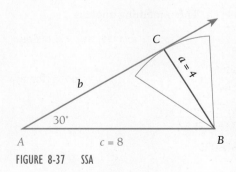

FIGURE 8-36   SSA

This is impossible! The range of sine is $[-1, 1]$. For triangles, we are further limited to $[0, 1]$. Careful construction of a figure reveals the interpretation. See Figure 8-36. Side a is not long enough to reach side b. The triangle does not exist.

Illustration 3:   Suppose $A = 30°$, $a = 4$ cm and $c = 8$ cm.

Solution: Again, we have the ambiguous case. See Illustration 2.

$$\frac{\sin 30°}{4 \text{ cm}} = \frac{\sin C}{8 \text{ cm}}$$

$$\sin C = \frac{8 \text{ cm} \sin 30°}{4 \text{ cm}}$$

$$\sin C = 1$$

From the extended definition, the only possible value for C is $90°$. Immediately, $B = 60°$. ABC is a $30°$-$60°$ right triangle, see Figure 8-37. The Pythagorean theorem applies.

FIGURE 8-37   SSA

$$(8 \text{ cm})^2 = (4 \text{ cm})^2 + b^2$$

$$b^2 = 48 \text{ cm}^2$$

$$b = 4 \sqrt{3} \text{ cm}$$

Illustration 4:   Suppose $A = 30°$, $a = 5'$ and $c = 8'$.

Solution: Once more, we have the ambiguous case. See Illustration 3.

$$\frac{\sin 30°}{5'} = \frac{\sin C}{8'}$$

$$\sin C = \frac{8' \sin 30°}{5'}$$

$$\sin C = 0.8$$

Although 0.8 is within the range of sine, there are two possible values for C. (See Figure 30.) The definition indicates both an acute and obtuse angle for which the sine is 0.8. The two angles are supplementary. The inverse sine function of your calculator will approximate the acute value.

$$C_1 = 53.13010236° \quad \text{or} \quad C_1 = 53°7'48''$$

The obtuse angle is the supplement of C_1.

$$C_2 = 180° - C_1$$
$$C_2 = 126.8698976° \quad \text{or} \quad C_2 = 126°52'12''$$

See Figure 8-38. Notice two choices now exist. For the first choice (Figure 8-39), use C_1 and the remaining angle is

$$B_1 = 180° - (A + C_1)$$

$$B_1 = 96°52'12''$$

From the Law of Sines,

$$\frac{\sin 96°52'12''}{b_1} = \frac{\sin 30°}{5'}$$

$$b_1 = \frac{5' (0.9928201093)}{0.5}$$

$$b_1 = 9.928201093$$

For the second choice (Figure 8-40), use C_2. The remaining angle is

$$B_2 = 180° - (A + C_2)$$
$$B_2 = 23°7'38''$$

From the Law of Sines:

$$\frac{\sin 23°7'48''}{b_2} = \frac{\sin 30°}{5'}$$

$$b_2 = \frac{5'(0.3928186802)}{0.5}$$

$$b_2 = 3.928186802$$

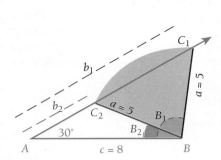

FIGURE 8-38   SSA

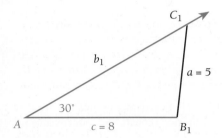

FIGURE 8-39

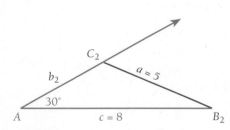

FIGURE 8-40

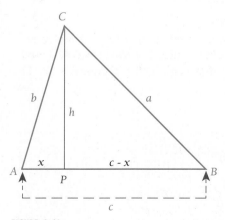

FIGURE 8-41

The preceding discussion generalized the application of the sine function to all plane triangles. In Illustration 3 of Example 8-5, the **Pythagorean theorem** applied only because the triangle was a right triangle. Can we expand the Pythagorean theorem to include general triangles? Consider the acute triangle in Figure 8-41.

The height h dropped from angle C to P on side c forms two right triangles. Label side AP of triangle APC as x. As a result the length of PB is $c - x$. The Pythagorean theorem applies to each of the right triangles.

$$x^2 + h^2 = b^2$$
$$h^2 = b^2 - x^2$$

and
$$h^2 + (c - x)^2 = a^2$$
$$h^2 + c^2 - 2cx + x^2 = a^2$$

Substitute $b^2 - x^2$ for h^2 in the last equation:

$$b^2 - x^2 + c^2 - 2cx + x^2 = a^2$$
$$b^2 + c^2 - 2cx = a^2$$

Finally, note that $\cos A = x/b$. $x = b \cos A$. Substitute for x:

$$b^2 + c^2 - 2bc \cos A = a^2$$

This formula, known as the **Law of Cosines**, relates the three sides of a general triangle in a manner reminiscent of the Pythagorean theorem. A generalization of a theorem should include the theorem as a special case. If angle $A = 90°$, then $\cos 90° = 0$, and the $-2bc \cos A$ term disappears. As a result, with a as the hypotenuse, the Law of Cosines includes the Pythagorean theorem as a special case.

To complete the derivation of Law of Cosines, consider a triangle with an obtuse angle. In Figure 8-42, A is an obtuse angle. Drop a perpendicular line from B to some point P on an extended side AC. Two right triangles result: PAB and PCB. The Pythagorean theorem applies to both:

$$x^2 + h^2 = c^2$$

and
$$h^2 + (x + b)^2 = a^2$$
$$h^2 + x^2 + 2bx + b^2 = a^2 \qquad \text{(expand } (x + b)^2)$$

Substitute c^2 for $x^2 + h^2$ in the second equation:

$$c^2 + b^2 + 2bx = a^2$$

From triangle PAB, $\cos(180° - A) = x/c$.

Substitute
$$x = c \cos(180° - A)$$
$$c^2 + b^2 + 2bc \cos(180° - A) = a^2$$

Because A is acute, then A is the reference angle for $180° - A$ which terminates in the second quadrant. The cosine is negative in the second quadrant, so we have $\cos(180° - A) = -\cos A$. Hence, $c^2 + b^2 - 2bc \cos A = a^2$.

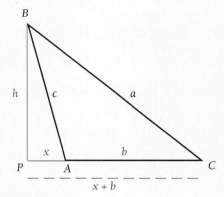

B

h c a

x b

P A C

$x + b$

FIGURE 8-42

Theorem 8-2
Generalized Pythagorean Theorem:
Law of Cosines

For any plane triangle ABC:
$$a^2 = b^2 + c^2 - 2bc \cos A$$
$$b^2 = a^2 + c^2 - 2ac \cos B$$
$$c^2 = a^2 + b^2 - 2ab \cos C$$

The previous discussion establishes the first identity. The proof of the remaining formulas is similar and is left as an exercise.

As a summary of the Law of Sines and Law of Cosines, consider the possible combinations of three measurements given to solve a triangle.

Abbreviation	Measurements	Solution	Start with
AAA	three angles	not unique	Not applicable
AAS	two angles, one side	Unique solution	Law of Sines
SAS	two sides, angle between	Unique solution	Law of Cosines
SSA	two sides, angle not between	Ambiguous case	Either
SSS	3 sides	Unique solution	Law of Cosines

EXAMPLE 8-6

Solving General Triangles. Solve the following as indicated.

Illustration 1:

For triangle ABC, $A = 30°$, $b = 10$ and $c = 5 + 5\sqrt{3}$, determine B, C and a.

Solution: The Law of Cosines relates a to A, b and c (SAS):

$$a^2 = b^2 + c^2 - 2bc \cos A$$
$$a^2 = 10^2 + (5 + 5\sqrt{3})^2 - 2(10)(5 + 5\sqrt{3}) \cos 30°$$
$$a^2 = 100 + 100 + 50\sqrt{3} - 150 - 50\sqrt{3}$$
$$a^2 = 50$$
$$a = 5\sqrt{2} \qquad \text{(-}5\sqrt{2} \text{ is not acceptable. Why?)}$$

The Law of Sines relates B to a, A and b:

$$\frac{\sin B}{b} = \frac{\sin A}{a}$$

$$\frac{\sin B}{10} = \frac{\dfrac{1}{2}}{5\sqrt{2}}$$

$$\sin B = \frac{1}{2} \frac{10}{5\sqrt{2}}$$

$$\sin B = \frac{\sqrt{2}}{2}$$

By memory or calculator, $B = 45°$. Because the sum of the angles of the triangle is 180°, angle $C = 180° - (A + B)$. $C = 105°$.

Illustration 2:

Two persons 100 meters apart observe a hot air balloon in the vertical plane between them. Mary measures an angle of inclination to the balloon of 70°. Jean measures an angle of inclination to the balloon of 60°. Determine the distance from each person to the balloon. Also determine the altitude of the balloon. Round your answers to the nearest meter.

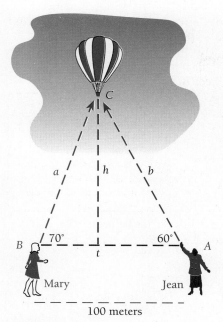

FIGURE 8-43

Solution: Sketch a diagram to model the situation, see Figure 8-43, (ASA). Immediately,

$$C = 180° - (A + B)$$
$$C = 50°$$

Apply the Law of Sines:

$$\frac{\sin 60°}{a} = \frac{\sin 50°}{100 \text{ m}}$$

$$a = 100 \text{ m} \frac{\sin 60°}{\sin 50°}$$

$$a = 113.0515875 \text{ m}$$

To find b, apply the Law of Cosines:

$$b^2 = 100^2 + 113.0515875^2 - 2(100)(113.0515875) \cos 70°$$
$$b^2 = 15047.4774$$
$$b = 122.6681597$$

The altitude h comes from solving right triangle BtC, where

$$\sin 70° = \frac{h}{a}$$

$$h = 106.2337425 \text{ meters}$$

The distance to Jean is approximately 123 meters. The distance to Mary is approximately 113 meters. The altitude of the balloon is approximately 106 meters.

Illustration 3:   Suppose $A = 30°$, $a = 5'$ and $c = 8'$.

Solution: This is the same (SSA) triangle as Example 8-5, Illustration 4. This time we apply the Law of Cosines first. Because angle A is known, use

$$a^2 = b^2 + c^2 - 2bc \cos A$$
$$5^2 = b^2 + 8^2 - 2b(8) \cos 30°$$
$$25 = b^2 + 64 - 16b \frac{\sqrt{3}}{2}$$

The equation is a quadratic in b. Rewrite in standard form and apply the quadratic formula.

$$b^2 - 8\sqrt{3}\, b + 39 = 0$$

$$b = \frac{8\sqrt{3} \pm \sqrt{(8\sqrt{3})^2 - 4(1)(39)}}{2}$$

$$= 4\sqrt{3} \pm 3$$

$$b_1 \cong 9.92820323' \quad \text{or} \quad b_2 \cong 3.92820323' \text{ (as before)}$$

With three sides known, apply the Law of Cosines to get angle B. First use b_1 to get angle B_1.

$$b^2 = a^2 + c^2 - 2ac \cos B$$
$$(4\sqrt{3} + 3)^2 = 5^2 + 8^2 - 2(5)(8) \cos B_1$$
$$80 \cos B_1 = 32 - 24\sqrt{3}$$
$$\cos B_1 = \frac{4 - 3\sqrt{3}}{10}$$
$$B_1 = \cos^{-1}\left(\frac{4 - 3\sqrt{3}}{10}\right)$$
$$B_1 \cong 96° \, 52' \, 12''$$

Because $C_1 = 180° - (A + B_1)$, $C_1 \cong 53° \, 52' \, 12''$ (as before).
For the second solution, use side b_2 to get angle B_2:

$$(4\sqrt{3} - 3)^2 = 5^2 + 8^2 - 2(5)(8) \cos B_2$$
$$B_2 = \cos^{-1}\left(\frac{4 + 3\sqrt{3}}{10}\right)$$
$$B_2 \cong 23° \, 7' \, 48''$$

Therefore, $C_2 \cong 126° \, 52' \, 12''$ (as before)

The intensity of an electron beam on the phosphors of a video display tube is directly proportional to the fourth power of the cosine of the deflection angle of the beam. See Figure 8-44. Suppose that the intensity of a particular tube is 100 units at the center of the screen. Write a function expressing the intensity as a function of the deflection angle. Predict the intensity at a point of the display that subtends an angle 10° from the center of the screen.

Solution: Let θ be the deflection angle and I the intensity, then

$$I = k \cos^4 \theta$$

models the fourth power cosine proportion. For this specific tube, $I = 100$ when $\theta = 0°$. Hence, $100 = k \cos^4 0°$.

$$100 = k(1)$$
$$k = 100$$

The specific model becomes

$$I = 100 \cos^4 \theta$$

To predict I when $\theta = 10°$, substitute into the function:

$$I = 100 \cos^4 10°$$
$$I = 100(.984807753)^4$$
$$I = 94.0601866$$

Notice that 10° from center the intensity of the electron beam is approximately 94 percent of the intensity at the center of the display tube.   ▬

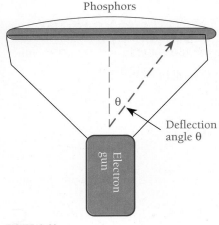

Illustration 4:

Phosphors

Electron gun

Deflection angle θ

FIGURE 8-44

SUMMARY

- To solve a general triangle, sketch a triangle *ABC* illustrating all given information (Figure 8-45). Given at least three pieces of information concerning the triangle (except three angles only), the remaining parts of the triangle follow by applying one of the following:

Law of Sines: $\dfrac{\sin A}{a} \equiv \dfrac{\sin B}{b} \equiv \dfrac{\sin C}{c}$

Law of Cosines: $a^2 \equiv b^2 + c^2 - 2bc \cos A$

$b^2 \equiv a^2 + c^2 - 2ac \cos B$

$c^2 \equiv a^2 + b^2 - 2ab \cos C$

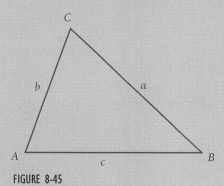

FIGURE 8-45

- The ambiguous case for a triangle occurs when two sides and an angle not between the sides form the given information. Depending on the information one of the triangles in Figure 8-46 occurs.

1. The triangle may not exist, leading to no solution.
2. The triangle may be a single right triangle.
3. The triangle may have two separate solutions.

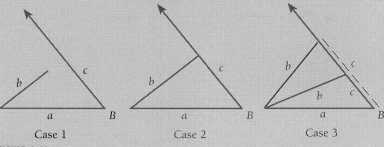

Case 1   Case 2   Case 3

FIGURE 8-46

8-4   E X E R C I S E S

Sketch a figure to represent the triangle(s) described by each of the following. Use the Law of Sines to determine the indicated value. (Take special care with the ambiguous case.) A capital letter indicates an angle and a lower case letter indicates the side opposite the corresponding angle.

1. $a = 10, B = 60°, A = 50°$. Determine b.

2. $a = 8, B = 40°, A = 60°$. Determine b.

3. $a = 8, b = 16, A = 30°$. Determine B.

4. $a = 6, b = 8.484, A = 45°$. Determine B.

5. $a = 5, b = 4, B = 40°$. Determine A.

6. $a = 10, b = 7, B = 20°$. Determine A.

7. $b = 7, B = 30°, A = 50°$. Determine a.

8. $b = 9, A = 45°, B = 55°$. Determine a.

9. $a = 10, b = 4, B = 30°$. Determine A.

10. $a = 9, b = 3, B = 40°$. Determine A.

Sketch a figure to represent the triangle described by each of the following. Use the Law of Cosines to determine the indicated value for each triangle.

11. $b = 4, c = 5, A = 60°$. Determine a.

12. $b = 3, c = 6, A = 60°$. Determine a.

13. $a = 7, c = 3, B = 50°$. Determine b.

14. $b = 4, a = 6, c = 40°$. Determine c.

15. $a = 5, b = 6, c = 7$. Determine B.

16. $a = 5, b = 6, c = 7$. Determine C.

17. $a = 4, b = 3, c = 6$. Determine A.

18. $a = 4, b = 3, c = 6$. Determine B.

19. $a = 5, b = 4, C = 40°$. Determine c.

20. $a = 4, b = 7, C = 50°$. Determine c.

Sketch a figure to represent the triangle(s) described by each of the following. Solve the triangle for the three remaining parts of the triangle.

21. $a = 5, b = 3, c = 7$

22. $a = 4, b = 5, c = 6$

23. $A = 25°, b = 10, c = 11$

24. $A = 50°, B = 80°, c = 10$

25. $B = 40°, C = 70°, a = 10$

26. $a = 5, b = 4, B = 40°$

27. $a = 10, b = 7, B = 70°$

28. $B = 55°, a = 8, c = 10$

29. $a = 4, b = 10, c = 8$

30. $b = 3, c = 6, A = 30°$

Given triangle ABC. Prove each of the following.

31. Area of triangle $ABC = \dfrac{1}{2} bc \sin A$

32. Area of triangle $ABC = \dfrac{1}{2} ac \sin B$

33. Area of triangle $ABC = \dfrac{1}{2} ab \sin C$

34. Use the results of Problem 31–33 to prove the Law of Sines.

For each of the following, determine the area of triangle ABC.

35. $b = 5, c = 3, A = 30°$

36. $a = 7, b = 4, A = 50°, B = 70°$

37. $a = 7, b = 5, c = 8$

38. $a = 5, b = 10, c = 6$

39. $c = 10, A = 80°, B = 70°$

40. $a = 8, b = 9, B = 45°$

Solve as indicated.

41. Two straight roads cross at a 50° angle. If a house is 5 km from the crossroads and a bus stop is 10 km from the crossroads along the other road, what is the shortest possible distance between the house and bus stop? Is there more than one model?

42. Suppose you construct a pyramid out of 4 triangles, where each triangle has two sides each 2 meters long and one side 1 meter long. See Figure 8-47. Determine the height of the finished pyramid.

43. See Problem 42. Determine the angle between two opposing faces of the pyramid at the vertex of the pyramid.

44. A plane travels 100 km due north from an airport while a second plane flies 80 km north 40° east (40° clockwise from due north). What is the distance between the two planes?

45. See Figure 8-30. Suppose the distance between observer post A and B is 1,000 miles. If the observed parallax for the moon is 0.00417 radians, approximate the distance of the moon.

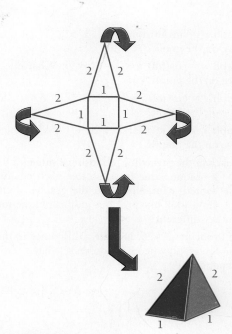

FIGURE 8-47

46. See Problem 45. In the special case that A and B are at opposite ends of the earth's diameter, the parallax is referred to as the *geocentric parallax of the moon*. Suppose the earth's diameter is 8,000 miles and both line of sights

to the moon are 240,000 miles. Determine the geocentric parallax of the moon.

Consider general triangle ABC.

47. Prove $b^2 = a^2 + c^2 - 2ac \cos B$.

48. Prove $c^2 = a^2 + b^2 - 2ab \cos C$.

49. Show that the intensity of an electron beam deflected $21.47°$ from center of a video display tube is approximately 75 percent of the intensity at center.

50. Show that the intensity of an electron beam deflected $32.75°$ from center of a video display tube is approximately 50 percent of the intensity at center.

51. See Problem 49. Discuss, from a trigonometric point of view, why it is easier to design a "deep" video display tube with a uniform bright display, than it is to design a "shallow" display tube with equivalent brightness.

52. See Problem 49. One way to compensate for the loss of brightness away from the center of the screen is to increase the intensity (and thus power consumption) of the electron beam as a function of the deflection angle from the center of the screen. Using this method, a manufacturer creates two video display tube models with the same uniform brightness and the same size screens. One of the tubes is $28''$ deep; the other is $22''$ deep. Which tube is most likely to consume more power? Give a trigonometric argument to justify your answer.

8-4   P R O B L E M   S E T

I. Consider triangle ABC. Let $s = \dfrac{a + b + c}{2}$.

(a) Show that $s - a = \dfrac{b + c - a}{2}$

(b) Show that $s - b = \dfrac{a + c - b}{2}$.

(c) Show that $s - c = \dfrac{a + b - c}{2}$.

(d) Use the Law of Cosines to verify that $\frac{1}{2}bc(1 + \cos A) = s(s - a)$

(e) Use the Law of Cosines to verify that $\frac{1}{2}bc(1 - \cos A) = (s - b)(s - c)$.

(f) Show that $\frac{1}{4}b^2c^2(1 - \cos^2 A) = s(s - a)(s - b)(s - c)$.

(g) Show that $\frac{1}{4}b^2c^2 \sin^2 A = s(s - a)(s - b)(s - c)$.

(h) Show that $\frac{1}{2}bc \sin A = \sqrt{s(s - a)(s - b)(s - c)}$.

(i) Show that the area of triangle ABC is given by area $= \sqrt{s(s - a)(s - b)(s - c)}$.

This result is known as Hero's formula (Hero, first century A.D., is also known as Heron). Use Hero's formula to determine the area of each of the following triangles.

2. $a = 9, b = 12, c = 15$

3. $a = 5, b = 12, c = 13$

4. $a = 7, b = 8, c = 9$

5. $a = 10, b = 4, c = 6$

6. $a = 10, b = 4, c = 4$

7. Discuss the results of Problems 5 and 6. What does the answer to these problems indicate about the triangles?

EXPLORATION

The following problems anticipate concepts from the next section. Use graphing technology to graph each of the following: First set the graphing mode to polar coordinates. Consult your user's manual for the method. Set the range for t to $t\text{Min} = 0$, $t\text{Max} = 7$. Describe the shape of each graph. Which "appear" to be functions?

8. $r = 2 \cos \theta$ **9.** $r = 3 \sin \theta$ **10.** $r = \tan \theta$

11. $r = 5 \cos \theta$ **12.** $r = 4 \sin \theta$ **13.** $r = -3 \cos \theta$

14. Graph $r = k \cos \theta$, for $k = 1, 2, -3$. What effect does k have on the graph?

15. Graph $r = k \sin \theta$, for $k = 1, 2, 3$. What effect does k have on the graph?

16. $r = \cos(2\theta)$ **17.** $r = \sin(3\theta)$

18. $r = \sin(4\theta)$ **19.** $r = \cos(5\theta)$

20. $r = \sin(6\theta)$ **21.** $r = \cos(6\theta)$

22. Graph $r = \sin(k\theta)$ for $k = 1, 3, 5, 7$. What effect does k have on the graph?

23. Graph $r = \cos(k\theta)$ for $k = 2, 4, 6$. What effects does k have on the graph?

24. Graph $r = \cos(k\theta)$ for $k = 1, 3, 5, 7$. What effect does k have on the graph?

25. Graph $r = \sin(k\theta)$ for $k = 2, 4, 6$. What effect does k have on the graph?

26. In each of the preceding, choosing a value for θ produced a unique value for r, so that each equation represented a function. Did the graphs of these functions look like previous graphs of functions? How are they different?

27. See Problems 22–25. Is there a difference in the effect of k on the graph when k is odd rather than even? Describe the pattern.

---

5   ## POLAR COORDINATES

Geometry has been throughout, of supreme importance in the history of knowledge.
—BERTRAND RUSSELL

The Cartesian co-ordinate system is not the only method for locating points in a plane. Another common method is polar coordinates. In **polar coordinates**, designate a reference point as an origin called the *pole*. From the pole, select a direction marked by a ray. Usually the ray corresponds to the positive x-axis of the Cartesian coordinate system. Refer to Figure 8-48. The polar address of any point in the plane is an ordered pair of numbers (r, θ), where r (for radius) measures the distance from the point to the origin and θ is the angle (in degrees or radians) formed by rotating the radius from the reference ray to the point. Counterclockwise rotations produce positive values for θ, whereas clockwise rotations produce negative values of θ.

EXAMPLE 8-7

Illustration 1:   **Plotting Points in Polar Coordinates.** Plot the following points in polar coordinates.

$$A = \left(3, \frac{\pi}{4}\right), \qquad B = (2, \pi), \qquad C = \left(4, \frac{5\pi}{4}\right), \qquad D = \left(2.5, -\frac{3\pi}{4}\right)$$

Solution: See Figure 8-49.

Illustration 2: $$E = (3, 135°), \qquad F = (3, -225°), \qquad G = (3, 495°)$$

Solution: See Figure 8-50.

Note: All three addresses represent the same point in polar coordinates. We cannot determine a unique address for a point from its graph.

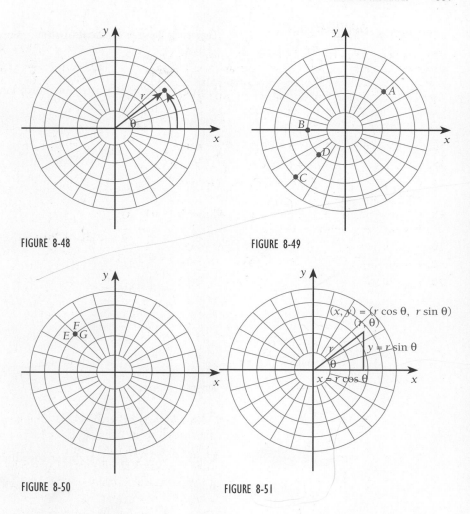

FIGURE 8-48

FIGURE 8-49

FIGURE 8-50

FIGURE 8-51

Because we can imagine either a Cartesian coordinate system or polar coordinate system for the window view of a graph, some relationship exists between the two systems. Superimpose the two coordinate systems so that they share an origin and the reference ray coincides with the positive x-axis. See Figure 8-51. The point shown corresponds to the address (x, y) under the Cartesian coordinate system and (r, θ) with polar coordinates. The triangle formed is a right triangle, so the trigonometric functions of Chapter 7 apply.

Summary of Cartesian and
Polar Relationships

$$\sin \theta = \frac{y}{r} \qquad \cos \theta = \frac{x}{r}$$

$$y = r \sin \theta \qquad x = r \cos \theta$$

$$\frac{y}{x} = \tan \theta \qquad r^2 = x^2 + y^2$$

EXAMPLE 8-8

Conversion Between Coordinate Systems. Convert from one coordinate system to the other.

Illustration 1:

Convert $(4, \pi/6)$ to Cartesian coordinates.

Solution:

$$r = 4 \quad \text{and} \quad \theta = \frac{\pi}{6} \qquad \text{(polar coordinates)}$$

$$x = 4\cos\frac{\pi}{6} \qquad \text{(formula for } x \text{ coordinate)}$$

$$= 4\frac{\sqrt{3}}{2} \qquad \text{(evaluate cosine)}$$

$$= 2\sqrt{3}$$

$$y = 4\sin\frac{\pi}{6} \qquad \text{(formula for } y \text{ coordinate)}$$

$$= 4\left(\frac{1}{2}\right) \qquad \text{(evaluate sine)}$$

$$= 2$$

Answer:

$$\left(4, \frac{\pi}{6}\right) \Leftrightarrow (2\sqrt{3}, 2)$$

Illustration 2:

Convert $(2\sqrt{2}, 2\sqrt{2})$ to polar coordinates.

Solution:

$$x = 2\sqrt{2} \quad \text{and} \quad y = 2\sqrt{2} \qquad \text{(Cartesian coordinates)}$$

$$r^2 = x^2 + y^2 \qquad \text{(formula for } r)$$

$$= (2\sqrt{2})^2 + (2\sqrt{2})^2 \qquad \text{(substitute } x \text{ and } y)$$

$$= 8 + 8$$

$$= 16$$

$$r = 4 \qquad \text{(square root of } r)$$

$$\sin\theta = \frac{2\sqrt{2}}{4} = \frac{\sqrt{2}}{2} \qquad \text{(formula with } \theta)$$

$$\text{so that} \quad \theta = \frac{\pi}{4} \qquad \text{(degree measure } \theta = 45°)$$

Answer:

$$(2\sqrt{2}, 2\sqrt{2}) \Leftrightarrow \left(4, \frac{\pi}{4}\right)$$

Note that other solutions for θ include $9\pi/4$ or $17\pi/4$.

One reason for exploring polar coordinates is that many graphs have simpler equations with polar coordinate representation. For example, $r = 3$ in polar coordinates represents a circle centered at the origin with radius of 3. Similarly, $\theta = \pi/6$ represents a line through the origin with slope of $1/\sqrt{3}$. Consider a few simple examples.

EXAMPLE 8-9

Conversion of Formulas to Polar Coordinates. Convert each of the following to polar coordinates.

Illustration 1:

A line

$$y = 3x$$

Solution: Substitute $y = r \sin \theta$ and $x = r \cos \theta$ (conversion formulas):

$$r \sin \theta = 3 r \cos \theta \qquad \text{(simplify)}$$

$$\frac{r \sin \theta}{r \cos \theta} = 3$$

Answer: $\tan \theta = 3$ or $\theta = \tan^{-1} 3$

Illustration 2:

A circle

$$x^2 + y^2 = 16$$

Solution: Substitute $y = r \sin \theta$ and $x = r \cos \theta$ (conversion formulas).

$$(r \sin \theta)^2 + (r \cos \theta)^2 = 16$$
$$r^2(\sin^2 \theta + \cos^2 \theta) = 16 \qquad \text{(Pythagorean relation)}$$
$$r^2(1) = 16$$

Answer: $r = 4$

Illustration 3:

A parabola

$$y = x^2$$

Solution:

As before, $r \sin \theta = r^2 \cos^2 \theta$ (conversion formulas)
$$\sin \theta = r \cos^2 \theta \qquad \text{(divide by cosine)}$$

Answer: $r = \tan \theta \sec \theta$

....

EXAMPLE 8-10

Conversion of Formulas to Cartesian Coordinates. Convert each of the following to Cartesian coordinates.

Illustration 1:

$$r^2 = \theta$$

Solution:

$$r^2 = x^2 + y^2$$

so that $$x^2 + y^2 = \theta$$

$$\tan(x^2 + y^2) = \tan\theta$$

Because $$\tan\theta = \frac{y}{x}$$

$$\tan(x^2 + y^2) = \frac{y}{x}$$

Illustration 2:

A line

$$\theta = \frac{\pi}{4}$$

Solution: We know $\tan\theta = y/x$, so we take the tangent of both sides.

$$\tan\theta = \tan\frac{\pi}{4}$$

$$y/x = 1$$

$$y = x$$

Illustration 3:

A circle

$$r = \sin\theta$$

Solution: Because $y/r = \sin\theta$, then

$$r = \frac{y}{r}$$

$$r^2 = y$$

But $$r^2 = x^2 + y^2$$

Therefore, $$x^2 + y^2 = y$$

Note: Complete the square and rewrite the equation in the form $x^2 + (y - \frac{1}{2})^2 = \frac{1}{4}$ to recognize a circle of radius $\frac{1}{2}$ centered at $(0, 1/2)$.   ▬

Suppose we have a formula that specifies r as a function of θ, $r = f(\theta)$. By carefully plotting points we can determine the graph of f directly on polar coordinates.

Consider the graph of $r = 3 + 2\cos\theta$. Because cosine is periodic, the graph repeats every 360°. Unlike a function in rectangular coordinates, we do not examine the graph from left to right to imagine the shape of its curve;

DISPLAY 8-3

rather polar coordinates trace a curve much like a weather radar screen. Imagine the radar beam sweeping through a complete rotation (counterclockwise rather than clockwise) and as it sweeps through angle θ it illuminates points of the curve r units from the origin. See Display 8-3. Form a table to analyze the values of r as θ moves from 0 to 360°.

θ	$\cos \theta$	$2 \cos \theta$	$r = 3 + 2 \cos \theta$
0°	1	2	5
60°	$\frac{1}{2}$	1	4
90°	0	0	3
120°	$-\frac{1}{2}$	-1	2
180°	-1	-2	1
240°	$-\frac{1}{2}$	-1	2
270°	0	0	3
300°	$\frac{1}{2}$	1	4
360°	1	2	5

Figure 8-52 plots these points. Summarize these plotted values as "trends" in the graph.

θ	$\cos \theta$	$2 \cos \theta$	$r = 3 + 2 \cos \theta$
0° < θ < 90°	Decreases from 1 to 0	Decreases from 2 to 0	Decreases from 5 to 3
90° < θ < 180°	Decreases from 0 to -1	Decreases from 0 to -2	Decreases from 3 to 1
180° < θ < 270°	Increases from -1 to 0	Increases from -2 to 0	Increases from 1 to 3
270° < θ < 360°	Increases from 0 to 1	Increases from 0 to 2	Increases from 3 to 5

Use the trends and plotted points to sketch a graph. See Figure 8-53. Graphs with such heartlike shapes are called *cardioids*.

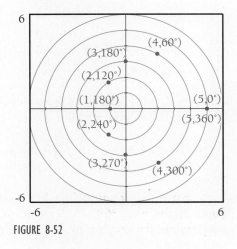

FIGURE 8-52

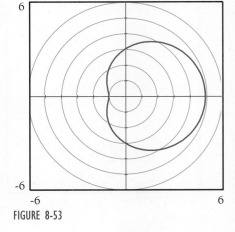

FIGURE 8-53

EXAMPLE 8-11

Graphing in Polar Coordinates.   Graph each of the following polar functions.

Illustration 1:

$$r = 4$$

Solution: The variable θ plays no role in the value of r. We seek all points 4 units from the origin. This is the description of a circle, see Figure 8-54.

Illustration 2:

$$\theta = \frac{\pi}{3}$$

Solution: This time θ is fixed: all points that form an angle of $\pi/3$ lie along a straight line, see Figure 8-55.

Illustration 3:

$$r = \theta$$

Solution: The values of r and θ are equal. There is no trigonometric function in the rule of pairing, so periodicity does not aid in visualizing the graph. As θ gets larger so does r. Plot some points for positive values of θ. See Figure 8-56. The resulting curve is the *spiral of Archimedes*. However there was no restriction on θ. We must plot negative values of θ as well. Negative values of θ produce negative values for r. See Figure 8-57. Interpret a negative r as indicating a point π radians in the opposite direction of θ. The resulting $(\theta, -r)$ plots as $(\theta + \pi, r)$, which is symmetric to (θ, r) about the origin.

Illustration 4:

$$r = 3 \sin(2\theta)$$

Solution: Even though $\sin(2\theta)$ completes a cycle in π units, we examine θ from 0 to 2π. This is because we are plotting an angle against a radius. The angles are different in 0 to 2π, so the resulting points may be different. See Figures 8-58 through 8-60.

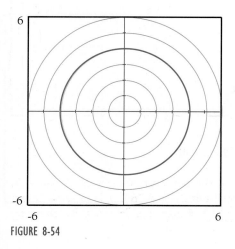

FIGURE 8-54

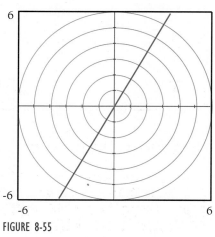

FIGURE 8-55

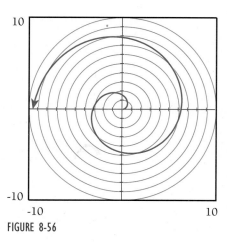

FIGURE 8-56

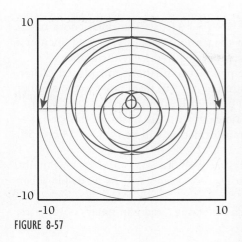

FIGURE 8-57

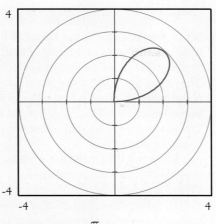

FIGURE 8-58   0 to $\frac{\pi}{2}$

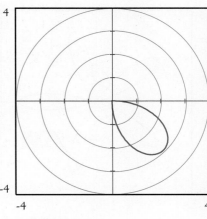

FIGURE 8-59 $\frac{\pi}{2}$ to π

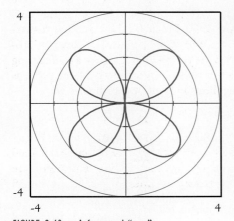

FIGURE 8-60   A four-petal "rose"

θ	2θ	$\sin(2\theta)$	$3\sin(2\theta)$
0 to $\frac{\pi}{4}$	0 to $\frac{\pi}{2}$	0 to 1	0 to 3
$\frac{\pi}{4}$ to $\frac{\pi}{2}$	$\frac{\pi}{2}$ to π	1 to 0	3 to 0
$\frac{\pi}{2}$ to $\frac{3\pi}{4}$	π to $\frac{3\pi}{2}$	0 to -1	0 to -3
$\frac{3\pi}{4}$ to π	$\frac{3\pi}{2}$ to 2π	-1 to 0	-3 to 0
π to $\frac{5\pi}{4}$	2π to $\frac{5\pi}{2}$	0 to 1	0 to 3
$\frac{5\pi}{4}$ to $\frac{3\pi}{2}$	$\frac{5\pi}{2}$ to 3π	1 to 0	3 to 0
$\frac{3\pi}{2}$ to $\frac{7\pi}{4}$	3π to $\frac{7\pi}{2}$	0 to -1	0 to -3
$\frac{7\pi}{4}$ to 2π	$\frac{7\pi}{2}$ to 4π	-1 to 0	-3 to 0

Illustration 5: Use graphing technology to graph $r = 4\cos(5\theta)$.

Solution: A Casio plots polar graphs directly. Set your range with Xmax = 5, Xmin = -5, Ymin = -5 and θmin = 0 with θmax = 6.2832 (about 2π). Select polar graphing by pressing [MODE] [SHIFT] [−]. Enter the formula by pressing [GRAPH] [4] [COS] [(] [5] [XθT] [)] [EXE].

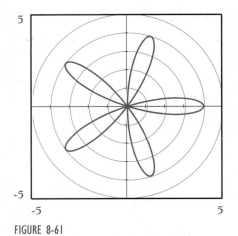

5

-5

-5   5

FIGURE 8-61

For the Sharp calculator, use the SET-UP E 2 to select polar coordinates. Set the range to the same values as for the Casio. Select A MENU 1 to enter a function of θ stored under R1. Finally, press A to display the graph.

The TI-81 uses a variation of the parametric methods introduced earlier. The key to graphing is the transformation of $r = f(\theta)$ from polar to rectangular coordinates as given by $(x, y) = (f(\theta) \cos \theta, f(\theta) \sin \theta)$. First use the MODE to select Param over Function. Set your range as with the Casio. In this case, $r = f(\theta)$ is $r = 4 \cos(5\theta)$. Therefore the transformation will be $(x, y) = (4 \cos(5\theta) \cos \theta, 4 \cos(5\theta) \sin \theta)$.

Press Y=. Enter these formulas respectively for $X_{1t} =$ and $Y_{1t} =$. Finally, press GRAPH. Whatever calculator you use, the graph will be as displayed in Figure 8-61.

GraphWindows can display scratch functions *f* and *g* in polar coordinates. Consult your user's manual.

8-5   E X E R C I S E S

Graph each of the following in a polar coordinate system.

1. $(3, 90°)$

2. $(2, 0)$

3. $(2, 180°)$

4. $\left(3, \frac{3\pi}{2}\right)$

5. $(1, 45°)$

6. $\left(2, -\frac{\pi}{4}\right)$

7. $\left(\frac{1}{2}, 360°\right)$

8. $\left(1.5, \frac{\pi}{3}\right)$

9. $(2.5, -90°)$

10. $\left(\frac{1}{3}, \frac{5\pi}{6}\right)$

Convert each of the following to Cartesian coordinates.

11. $(5, 0°)$

12. $(4, 0)$ $(4, 0)$

13. $(2, 90°)$

14. $\left(3, \frac{\pi}{2}\right)$

15. $(4, 45°)$

16. $\left(6, \frac{\pi}{4}\right)$

17. $(6, 60°)$

18. $\left(8, \frac{\pi}{6}\right)$

19. $(10, 30°)$

20. $\left(4, \frac{\pi}{3}\right)$

Convert each of the following to polar coordinates with $0° \leq \theta \leq 90°$.

21. $(1, \sqrt{3})$

22. $(\sqrt{3}, 1)$

23. $(\sqrt{2}, \sqrt{2})$

24. $(4, 0)$

25. $(0, 5)$

26. $(3\sqrt{2}, 3\sqrt{2})$

27. $(3\sqrt{3}, 3)$

28. $(2, 2\sqrt{3})$

29. $(2, 0)$

30. $(0, 7)$

Convert to Cartesian coordinates.

31. $(5, 225°)$

32. $\left(5, \frac{3\pi}{4}\right)$

33. $(4, 270°)$

34. $(4, \pi)$

35. $(6, 300°)$

36. $\left(6, \frac{2\pi}{3}\right)$

37. $(5, -210°)$

38. $\left(5, -\frac{\pi}{6}\right)$

39. $(5, 150°)$

40. $\left(5, \frac{\pi}{6}\right)$

Convert to polar coordinates. Use radian measure.

41. $(5, 5)$

42. $(3, 3)$

43. $(-5, 5)$

44. $(3, -3)$

45. $(-5, -5)$

46. $(-3, 3)$

47. $(3, -3\sqrt{3})$

48. $(-4\sqrt{3}, 4)$

49. $(-3, -3\sqrt{3})$

50. $(4, 4\sqrt{3})$

Graph of each of the following in polar coordinates.

51. $r = 5$

52. $r = \dfrac{1}{2}$

53. $\theta = 60°$

54. $\theta = \dfrac{3\pi}{4}$

55. $r = \dfrac{1}{3}$

56. $r = 6$

57. $\theta = -150°$

58. $\theta = -\dfrac{\pi}{4}$

59. $r = \sqrt{3}$

60. $r = \sqrt{2}$

Convert the following equations in Cartesian form to polar form.

61. $y = 2x$

62. $y = -x$

63. $y = -4x$

64. $y = \sqrt{3}x$

65. $y = \dfrac{1}{2}x$

66. $y = -5x$

67. $x^2 + y^2 = 4$

68. $x^2 + y^2 = 25$

*69. $x^2 - y^2 = 25$ (Hint: compare to $\cos(2\theta)$.)

*70. $2y = \dfrac{1}{x}$ (Hint: compare to $\sin(2\theta)$.)

Graph each of the following.

71. $r = 3$

72. $r = 2$

73. $r = 2\theta$

74. $r = \dfrac{\theta}{2}$

75. $r = 3\sin\theta$

76. $r = 3\cos\theta$

77. $r = 2\sin(2\theta)$

78. $r = 2\cos\theta$

79. $r = 4\sin(3\theta)$

80. $r = 4\cos(3\theta)$

81. $r = 5\sin(4\theta)$

82. $r = 5\cos(4\theta)$

83. $r = -3\sin(5\theta)$

84. $r = -3\cos(5\theta)$

85. $r = 5\cos(6\theta)$

86. $r = 5\sin(6\theta)$

87. $r = 1 + \sin\theta$

88. $r = 1 - \cos\theta$

89. $r = 2 - \sin\theta$

90. $r = 2 + \cos\theta$

91. $r = 2 + 2\cos\theta$

92. $r = 2 - 2\sin\theta$

93. $r = \tan\theta$

94. $r = \cot\theta$

95. $r = \csc\theta$

96. $r = \sec\theta$

97. $r = 2^\theta$

98. $r = e^\theta$

99. $r = 1 + \tan^2(\theta/2)$

100. $r = \sec^2(\theta/2)$

8-5   PROBLEM SET

For the moment, assume that the polar coordinates $(-r, \theta)$, where $r > 0$ is meaningful. Convert each of the following to rectangular coordinates. Graph the resulting points on the same coordinate system.

1. (a) $(-5, 30°)$ **(b)** $(5, 210°)$

2. (a) $\left(-4, \dfrac{5\pi}{6}\right)$ **(b)** $\left(4, \dfrac{11\pi}{6}\right)$

3. (a) $(-2, 225°)$ **(b)** $(2, 45°)$

4. (a) $\left(-2, -\dfrac{\pi}{4}\right)$ **(b)** $\left(2, \dfrac{3\pi}{4}\right)$

5. (a) $(-2, 135°)$ **(b)** $(2, 315°)$

6. (a) $\left(2, -\dfrac{5\pi}{4}\right)$ **(b)** $\left(2, \dfrac{3\pi}{4}\right)$

7. (a) $(-2, -135°)$ **(b)** $(2, 45°)$

8. (a) $(10, 0)$ **(b)** $(-10, \pi)$

9. (a) $(10, 90°)$ **(b)** $(-10, -90°)$

10. (a) $\left(-10, -\dfrac{\pi}{2}\right)$ **(b)** $\left(-10, \dfrac{3\pi}{2}\right)$

11. Prove $-|r|\sin\theta = |r|\sin(\theta + 180°)$.

12. Prove $-|r|\cos\theta = |r|\cos(\theta + \pi)$.

13. Examine the positive x-axis. Locate any real number r on the positive x-axis. Now locate $-r$ on x-axis. Through how many degrees would the positive x-axis have to rotate for r to coincide with $-r$?

14. Locate 4 on the positive y-axis and -4 on the negative y-axis. Through how many degrees would the y-axis have to rotate for these two numbers to exchange positions?

15. Make up a definition for plotting polar coordinates of the form $(-|r|, \theta)$. Compare your definition to the results of Problems 1–10.

16. Examine the definition you created in Problem 15. Is the definition consistent with our interpretation of "negative" on the real number line? Explain.

PROBLEMS FOR GRAPHING TECHNOLOGY

Use graphing technology to explore the following graphs. Be sure that you set your graphing device to polar mode. Also the range for t or θ should be set to tMin = 0, tMax = 6.5. Set your graphing window to xMin = -5, xMax = 5, yMin = -5, yMax = 5. Consult your manual for details.

17. Graph $r = 3 \cos(k\theta)$, for $k = 2, 4, 6, 8$. How does k affect the graph?

18. Graph $r = 3 \sin(k\theta)$, for $k = 2, 4, 6, 8$. How does k affect the graph?

19. Graph $r = 3 \cos(k\theta)$, for $k = 1, 3, 5, 7$. How does k affect the graph?

20. Graph $r = 3 \sin(k\theta)$, for $k = 1, 3, 5, 7$. How does k affect the graph?

21. See Problems 17 and 19. Formulate a rule for the effect of k on the graph of $r = 3 \cos(k\theta)$. Test your rule on $k = 9, 10$. Test your rule for $k = 2.5$.

22. See Problems 18 and 20. Formulate a rule for the effect of k on the graph of $r = 3 \sin(k\theta)$. Test your rule on $k = 9, 10$. Test your rule for $k = 2.5$

23. Graph $r = k \cos(3\theta)$, for $k = 2, 3, 4, 5$. How does k affect the graph?

24. Graph $r = k \sin(3\theta)$, for $k = 2, 3, 4, 5$. How does k affect the graph?

25. Predict the shape of the graphs of $r = 4 \cos(5\theta)$ and $r = 5 \cos(4\theta)$. Verify your prediction with your graphing technology.

26. Predict the shape of the graphs of $r = 4 \sin(5\theta)$ and $r = 5 \sin(4\theta)$. Verify your prediction with your graphing technology.

27. Graph $r = k\theta$, for $k = 0.25, 0.5, 1, 1.5$. Describe the shape of the graphs.

28. Graph $r = k$, for $k = 1, -2, -3, 4$. Describe the shape of the graphs.

6   TWO-DIMENSIONAL VECTORS: DIRECTION COSINES

Time is said to have only one dimension, and space to have three dimensions. . . . The mathematical quaternion partakes of both these elements. . . .
> And how the One of Time, of Space the Three,
> Might in the Chain of Symbols girdled be.

—W. R. HAMILTON

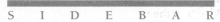

To a physicist, a **vector** is a directed distance. Two important concepts are attributed to vectors. The first of these is magnitude. The **magnitude of a vector** can represent many measurable quantities, such as force, velocity or mass. The second of these attributes is direction. The magnitude acts in a particular direction. In this manner we distinguish forces acting in opposite directions, but signed numbers could also do that. Vectors are more versatile. Vectors can describe forces acting at a 30° angle. A convenient metaphor for vectors is as a directed line segment. The direction is often indicated by placing a small arrowhead on one end of the segment. See Figure 8-62.

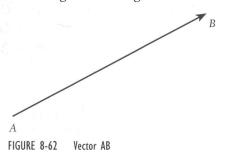

FIGURE 8-62   Vector AB

This section introduces vectors in a plane. An ordered pair of points summarizes information about a directed line segment, so we designate one point as the initial point and the other point as the terminal point. This produces the following definition.

DEFINITION 8-4
Definition of a Vector

The two-dimensional vector **AP** is the directed line segment from **initial point** $A(x_0, y_0)$ to **terminal point** $P(x_1, y_1)$.

Before defining the direction of a vector, we introduce the coordinates of a vector.

DEFINITION 8-5
Coordinates of a Vector

The **coordinates of a vector AP** are given by $\langle x_1 - x_0, y_1 - y_0 \rangle$, where the initial point is $A(x_0, y_0)$ and the terminal point is $P(x_1, y_1)$. Moreover, two vectors are equal if they have equal coordinates.

EXAMPLE 8-11

Illustration:

Coordinates of a Vector

Consider the following points, $A = (1, 4)$, $B = (3, 5)$, $C = (-2, 4)$, $D = (5, 1)$, $E = (6, 5)$, $O = (0, 0)$, plotted in Figure 8-63. Determine the coordinates of vectors **AB**, **AC**, **CB**, **OD**, **AE** and **BD**. Which of the vectors are equal?

Solution:

$$\mathbf{AB} = \langle 3 - 1, 5 - 4 \rangle$$
$$= \langle 2, 1 \rangle$$
$$\mathbf{AC} = \langle -2 - 1, 4 - 4 \rangle$$
$$= \langle -3, 0 \rangle$$
$$\mathbf{CB} = \langle 3 - (-2), 5 - 4 \rangle$$
$$= \langle 5, 1 \rangle$$
$$\mathbf{OD} = \langle 5 - 0, 1 - 0 \rangle$$
$$= \langle 5, 1 \rangle$$
$$\mathbf{AE} = \langle 6 - 1, 5 - 4 \rangle$$
$$= \langle 5, 1 \rangle$$
$$\mathbf{BD} = \langle 5 - 3, 1 - 5 \rangle$$
$$= \langle 2, -4 \rangle$$

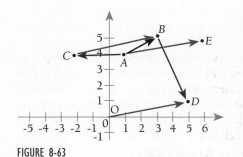

FIGURE 8-63

Note: The initial point of vector **OD** is at the origin. Vector **OD** is said to be in **standard position**. Vectors **CB**, **OD** and **AE** are all equal. Suppose that **UV** represents a vector with coordinates $\langle x_0, y_0 \rangle$ then **UV** is equal to a vector with initial point at the origin and terminal point (x_0, y_0). That is, the two vectors have the same direction and the same magnitude. Any discussion of vectors in standard position also applies to equal vectors not in standard position. By discussing vectors with initial points at the origin, we concentrate on the terminal points, which also serve as the **coordinates of the vector**. In fact, we might simply refer to a vector as $\vec{V}$ instead of $\vec{OV}$ when we know it is in standard position. Also, $\vec{0}$ represents the **zero vector**: $\vec{0} = \langle 0, 0 \rangle$. ▬

If two vectors are equal they are of the same length or magnitude. The next theorem indicates how to calculate the magnitude.

Theorem 8-3
Magnitude of a Vector

> Suppose the vector $\vec{V}$ has coordinates $\langle a, b \rangle$. The magnitude of $\vec{V}$, written $\|\vec{V}\|$, $= \sqrt{a^2 + B^2}$.

In the preceding example, the magnitudes of **CB** and **AE** are equal to

$$\|OD\| = \sqrt{5^2 + 1^2}$$
$$= \sqrt{26}$$

Discussion of vectors in standard position will not cause loss of generality, so we are ready to explore the direction of a vector. The easiest method for defining a vector's direction is with an angle of rotation between 0° and 360°. Unfortunately, this method does not extend nicely to three dimensions, and in three dimensions vectors prove most valuable. Otherwise, polar coordinates from the last section or complex numbers from the next section would have proved adequate for nineteenth century physicists. Instead we define the direction of a vector in terms of both the x-axis and y-axis to take advantage of the known properties of inverse cosine. The method requires spinning out a few diverse threads to be knitted into a final product. We begin with unit vectors.

When James Maxwell identified the components of quaternions as scale and direction, he was reiterating concepts from physics that demanded measurements include the magnitude of a quantity (mass, speed, etc.) and the direction in which the quantity acts. As usual, some unit defined the magnitude. This leads to the following definition.

DEFINITION 8-6
Unit Vector

> A **unit vector** $\vec{V}$ is a vector such that $\|\vec{V}\| = 1$.

Consider vector $\vec{V} = \langle \frac{3}{5}, -\frac{4}{5} \rangle$. Because $\|\vec{V}\| = \sqrt{\frac{9}{25} + \frac{16}{25}}$, then $\|\vec{V}\| = 1$. $\vec{V}$ is a unit vector. See Figure 8-64.

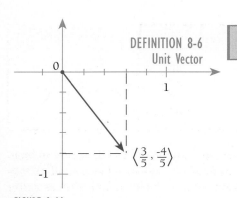

$\left\langle \frac{3}{5}, \frac{-4}{5} \right\rangle$

FIGURE 8-64

DEFINITION 8-7
Scalar Multiplication

> The **scalar** product $\alpha\vec{V}$ of a real number α and a vector $\vec{V} = \langle r, s \rangle$ is a vector given by $\alpha\vec{V} = \langle \alpha r, \alpha s \rangle$.

Recall unit vector $\vec{V} = \langle \frac{3}{5}, -\frac{4}{5} \rangle$. Now $5\vec{V} = \langle 3, -4 \rangle$. Moreover, the magnitude of $5\vec{V}$ is given by $\|5\vec{V}\| = \sqrt{3^2 + 4^2} = 5$. Multiplying a unit vector by 5 produces a vector five times longer. It follows that multiplying a unit vector by 0.5 will produce a vector half the original length. The results are general and follow from the definitions of scalar multiplication and magnitude.

Theorem 8-4
Scalar multiplication and Magnitude

> If $\vec{V}$ is a vector and α is a real number, then $\|\alpha\vec{V}\| = |\alpha| \, \|\vec{V}\|$

As a result of Theorem 8-4, any vector other than $\vec{0}$ can be transformed into a unit vector. Simply multiply the vector by the reciprocal of its magnitude.

Theorem 8-5
Creating Unit Vectors

> Suppose $\vec{V}$ is a vector other than the **0** vector, then the vector $\dfrac{1}{\|\vec{V}\|}\vec{V}$ is a unit vector. (Abbreviate $\dfrac{1}{\|\vec{V}\|}\vec{V}$ as $\dfrac{\vec{V}}{\|\vec{V}\|}$.)

Intuitively, $\vec{V}$, $\vec{V}/\|\vec{V}\|$ and $\alpha\vec{V}$, $\alpha > 0$, all have the same direction but not necessarily the same magnitude. Confirmation requires refining the concept of direction. We use static angles to measure direction and so limit our measurements to $[0°, 180°]$. The direction of a nonzero vector in standard position is given in terms of two static angles, the angle θ formed by the vector and the positive x-axis and the angle ϕ formed by the vector and the positive y-axis. See Figure 8-65 on p. 552. The following table indicates values for θ and ϕ for each of the vectors in Figure 8-65.

Vector	θ	ϕ	Quadrant
$\vec{A}$	60°	30°	I
$\vec{B}$	150°	60°	II
$\vec{C}$	135°	135°	III
$\vec{D}$	30°	120°	IV

At first glance, this seems a clumsy way to represent the direction of a vector. In fact, it is clumsy. But we are not finished. Suppose that each of vectors $\vec{A}, \vec{B}, \vec{C}$, and $\vec{D}$ is a unit vector. Let us extend the table to include the cosines of θ and ϕ.

Vector	θ	$\cos \theta$	ϕ	$\cos \phi$	Quadrant	(x, y)
$\vec{A}$	60°	$\dfrac{1}{2}$	30°	$\dfrac{\sqrt{3}}{2}$	I	$\left\langle \dfrac{1}{2}, \dfrac{\sqrt{3}}{2} \right\rangle$
$\vec{B}$	150°	$-\dfrac{\sqrt{3}}{2}$	60°	$\dfrac{1}{2}$	II	$\left\langle -\dfrac{\sqrt{3}}{2}, \dfrac{1}{2} \right\rangle$
$\vec{C}$	135°	$-\dfrac{\sqrt{2}}{2}$	135°	$-\dfrac{\sqrt{2}}{2}$	III	$\left\langle -\dfrac{\sqrt{2}}{2}, -\dfrac{\sqrt{2}}{2} \right\rangle$
$\vec{D}$	30°	$\dfrac{\sqrt{3}}{2}$	120°	$-\dfrac{1}{2}$	IV	$\left\langle \dfrac{\sqrt{3}}{2}, -\dfrac{1}{2} \right\rangle$

Experimentation should convince you that the coordinates of the last column are the coordinates of the vectors $\vec{A}$, $\vec{B}$, $\vec{C}$ and $\vec{D}$. The coordinates are the same as the cosines.

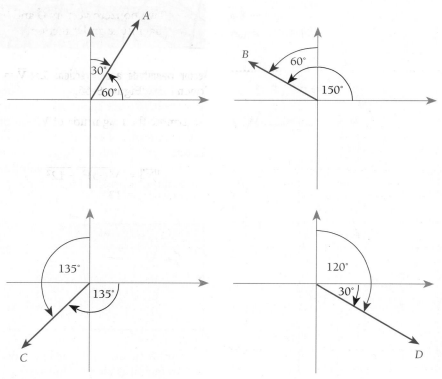

FIGURE 8-65

DEFINITION 8-8
Direction Cosines

The **direction cosines** of a nonzero vector $\vec{V} = \langle a, b \rangle$ are the coordinates of the unit vector $\dfrac{\vec{V}}{\|\vec{V}\|} = \langle \dfrac{a}{\|\vec{V}\|}, \dfrac{b}{\|\vec{V}\|} \rangle$

Moreover, the angle θ that $\vec{V}$ forms with the positive x-axis is $\cos^{-1} \dfrac{a}{\|\vec{V}\|}$ and the angle ϕ that $\vec{V}$ forms with the positive y-axis is $\cos^{-1} \dfrac{b}{\|\vec{V}\|}$.

Two vectors have the same direction if they have the same direction cosines. This is only another way to say that each is a positive scalar multiple of the same unit vector. The following theorem summarizes the discussion by tying the physicist's concepts of direction and magnitude with the mathematical concepts of scalar multiples, unit vectors, and direction cosines of vectors.

Theorem 8-6
Equality of Vectors

Two nonzero vectors $\vec{U}$ and $\vec{V}$ are equal if they are equal in both magnitude and direction.

EXAMPLE 8-12

Vector Magnitude and Direction. Let $\vec{V} = \langle -5, 12 \rangle$ be the vector in each of the following. See Figure 8-66.

Illustration 1:

Determine the magnitude of $\vec{V}$.

Solution:

$$\|\vec{V}\| = \sqrt{(-5)^2 + 12^2}$$
$$\|\vec{V}\| = 13.$$

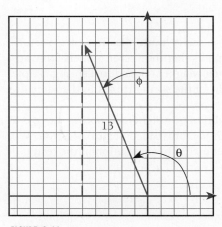

FIGURE 8-66

Illustration 2:   Determine a unit vector in the same direction as $\vec{V}$.

Solution: Because the magnitude of $\vec{V}$ is 13,

$$\frac{\vec{V}}{\|\vec{V}\|} = \langle \frac{-5}{13}, \frac{12}{13} \rangle$$

Illustration 3:   Determine the direction cosines of vector $\vec{V}$.

Solution: From the coordinates in Illustration 2, the direction cosines are $\cos \theta = -\frac{5}{13}$; $\cos \phi = \frac{12}{13}$.

Illustration 4:   Determine the angle vector $\vec{V}$ makes with the positive x-axis.

Solution: From Illustration 3, $\theta = \cos^{-1}(-\frac{5}{13})$. By calculator, $\theta \cong 112.6198649°$ (or $\theta \cong 1.965587446$ radians).

Illustration 5:   Determine the angle vector $\vec{V}$ makes with the positive y-axis.

Solution: Again, from Illustration 3, $\phi = \cos^{-1}(\frac{12}{13})$. By calculator, $\phi \cong 22.61986495°$.

Illustration 6:   Determine the coordinates of a vector in the same direction as $\vec{V}$, but with a magnitude of 10.

Solution: Refer to the unit vector in Illustration 2. Multiply it by the scalar 10:

$$10\frac{\vec{V}}{\|\vec{V}\|} = \langle 10 \times \frac{-5}{13}, 10 \times \frac{12}{13} \rangle$$

$$= \langle \frac{-50}{13}, \frac{120}{13} \rangle$$

Note: Multiply by –10 and get a vector 10 units long but in exactly the opposite direction from $\vec{V}$ (180° "away" from $\vec{V}$).   ▬

What is a physical interpretation of scalar multiplication of a vector? Among many interpretations, one example is that the vector models force applied to a spring. Let the vector represent a weight attached to a spring. Triple the weight on the spring and the "stretch" of the spring triples in the same direction.

Because it is possible to add several different weights to a spring, there is a corresponding model for the addition for vectors, called **vector addition**. We give an algebraic definition and offer a geometric interpretation of the results.

DEFINITION 8-9
Vector Addition

The vector sum $\vec{U} + \vec{V}$ of vectors $\vec{U} = \langle a, b \rangle$ and $\vec{V} = \langle r, s \rangle$ is a vector given by $\vec{U} + \vec{V} = \langle a + r, b + s \rangle$.

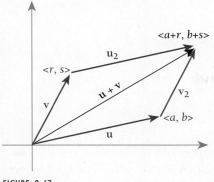

FIGURE 8-67

For a geometrical interpretation, consider Figure 8-67.

Note that $\vec{U}_2 = \vec{U}$ and $\vec{V}_2 = \vec{V}$. Then

$$\vec{U} + \vec{V} = \vec{U}_2 + \vec{V}$$
$$= \vec{U} + \vec{V}_2$$

Also, the four vectors outline a parallelogram, Locate the vertex of the parallelogram where $\vec{U}_2$ and $\vec{V}_2$ come together. The coordinates of this point are $(a + r, b + s)$. Use this as a terminal point for a vector from the origin. This new vector forms the diagonal of the parallelogram and equals $\vec{U} + \vec{V}$ as defined. The interpretation of the sum of two vectors as the vector diagonal from their common initial point is called the **parallelogram law**. In physics, the sum of the two vectors is also known as the **resultant**.

Applications of Vectors

EXAMPLE 8-13

Illustration 1:

An airplane flies due north at 400 kph while a wind blows toward the east at 30 kph. Each of these is a directed magnitude and thus may be represented by vectors. From physics, the actual course (direction) and (ground) speed of the airplane is the resultant of the two vectors. See Figure 8-68. Determine the direction of the plane and the ground speed.

Solution: Let $\vec{A}$ represent the plane's velocity (directed distance over time) and $\vec{B}$ represent the wind's velocity. Then the speed (undirected distance over time) is the magnitude of $\vec{A} + \vec{B}$ and the direction follows from the direction cosines of $\vec{A} + \vec{B}$.

$$\vec{A} = \langle 0 \text{ kph}, 400 \text{ kph} \rangle$$
$$\vec{B} = \langle 30 \text{ kph}, 0 \text{ kph} \rangle$$
$$\vec{A} + \vec{B} = \langle 30 \text{ kph}, 400 \text{ kph} \rangle$$
$$\|\vec{A} + \vec{B}\| = \sqrt{(30 \text{ kph})^2 + (400 \text{ kph})^2}$$
$$= \sqrt{160{,}900} \text{ kph}$$
$$\cong 401.12342 \text{ kph} \qquad \text{(by calculator)}$$

The direction cosines are $\langle 30 \text{ kph}/(\sqrt{160900} \text{ kph}), 400 \text{ kph}/(\sqrt{160900} \text{ kph}) \rangle$ or approximately $\langle 0.074789948, 0.99719931 \rangle$. Note how the kph "divided out."

$$\theta \cong 85.71084667° \text{ (north of due east) or}$$
$$\phi \cong 4.289153329° \text{ (east of due north)}$$

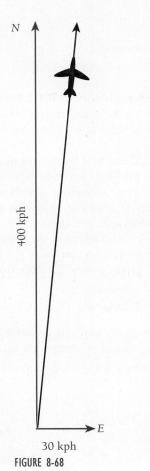

FIGURE 8-68

Illustration 2:

If $\vec{U} = \langle 5, -2 \rangle$ and $\vec{V} = \langle -4, 7 \rangle$, determine $3\vec{U} + 2\vec{V}$.

Solution:

$$3\vec{U} + 2\vec{V} = 3\langle 5, -2 \rangle + 2\langle -4, 7 \rangle$$
$$= \langle 15, -6 \rangle + \langle -8, 14 \rangle$$
$$= \langle 7, 8 \rangle$$

Note: $\|3\vec{U} + 2\vec{V}\| = \sqrt{49 + 64}$ (approximately 10.63014581)   ▬

Theorem 8-7
Properties of a Vector Space

Suppose that $\vec{U}$, $\vec{V}$ and $\vec{W}$ are (two-dimensional) vectors and that α and β are scalars (real numbers), then

$$\vec{U} + \vec{V} = \vec{V} + \vec{U}$$
$$(\vec{U} + \vec{V}) + \vec{W} = \vec{U} + (\vec{V} + \vec{W})$$
$$\beta(\vec{U} + \vec{V}) = \beta\vec{U} + \beta\vec{V}$$
$$\alpha(\beta\vec{U}) = (\alpha\beta)\vec{U}$$
$$(\alpha + \beta)\vec{U} = \alpha\vec{U} + \beta\vec{U}$$
$$\vec{U} + \vec{0} = \vec{U} \qquad (\vec{0} \text{ is the zero vector: } \vec{0} = \langle 0, 0 \rangle)$$
$$1\vec{U} = \vec{U}$$
$$\vec{U} + (-1\vec{U}) = \vec{0}$$

Collections of vectors and scalars satisfying Theorem 8-7 form a **vector space**. The proofs of the properties follow from the definitions of scalar multiplication and vector addition. Some individual parts of the proof are left as advanced exercises.

What is the angle between two vectors? If two vectors are not in standard position, define the angle between them as the angle between two vectors in standard position equal, respectively, to the original vectors. Therefore, without loss of generality, consider two vectors $\vec{U}$ and $\vec{V}$ in standard position. Label the static angle formed by the two vectors as θ. See Figure 8-69.

Connect the terminal points of $\vec{U}$ and $\vec{V}$ to form a vector, say, from $\vec{U}$ to $\vec{V}$. Call this vector $\vec{W}$. From Figure 8-69, $\vec{U} + \vec{W} = \vec{V}$ so that $\vec{W} = \vec{V} - \vec{U}$. Note the general triangle formed by the vectors. To determine θ, apply the Law of Cosines:

$$\|\vec{V} - \vec{U}\|^2 = \|\vec{U}\|^2 + \|\vec{V}\|^2 - 2\|\vec{U}\|\|\vec{V}\|\cos\theta.$$

Before continuing, let us consider notation to cut down on our work. The lefthand side of the equation offers some impressive computation. Suppose $\vec{V} = \langle r, s \rangle$ and $\vec{U} = \langle a, b \rangle$ then $\|\vec{U}\|^2 = a^2 + b^2$ and $\|\vec{V}\|^2 = r^2 + s^2$. Moreover, $\vec{V} - \vec{U} = \langle r - a, s - b \rangle$. Then,

$$\|\vec{V} - \vec{U}\|^2 = (r - a)^2 + (s - b)^2.$$
$$\|\vec{V} - \vec{U}\|^2 = r^2 - 2ra + a^2 + s^2 - 2sb + b^2$$
$$= r^2 + s^2 + a^2 + b^2 - 2(ar + bs)$$
$$= \|\vec{V}\|^2 + \|\vec{U}\|^2 - 2(ar + bs)$$

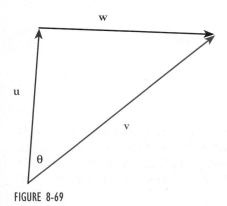

FIGURE 8-69

The following definition provides notation to abbreviate $(ar + bs)$.

> The **dot product** of vectors $\vec{U} = \langle a, b \rangle$ and $\vec{V} = \langle r, s \rangle$ is the real number given by $\vec{U} \cdot \vec{V} = ar + bs$.

There are several advantages to this notation. First, it allows us to complete the arithmetic we began in a more familiar fashion:

$$\|\vec{V} - \vec{U}\|^2 = \|\vec{V}\|^2 - 2\vec{V} \cdot \vec{U} + \|\vec{U}\|^2$$

Second, it allows for a compact rewriting of the formula for the magnitude of a vector. If $\vec{U} = \langle a, b \rangle$ then $\vec{U} \cdot \vec{U} = aa + bb$ or $\vec{U} \cdot \vec{U} = a^2 + b^2$. As a result, $\vec{U} \cdot \vec{U} = \|\vec{U}\|^2$.

Now back to the Law of Cosines:

$$\|\vec{V} - \vec{U}\|^2 = \|\vec{U}\|^2 + \|\vec{V}\|^2 - 2\|\vec{U}\| \cdot \|\vec{V}\| \cos \theta$$

$$\|\vec{V}\|^2 - 2\vec{U} \cdot \vec{V} + \|\vec{U}\|^2 = \|\vec{V}\|^2 + \|\vec{U}\|^2 - 2\|\vec{V}\| \cdot \|\vec{U}\| \cos \theta$$

$$-2\vec{U} \cdot \vec{V} = -2\|\vec{V}\| \cdot \|\vec{U}\| \cos \theta$$

$$\cos \theta = \frac{\vec{U} \cdot \vec{V}}{\|\vec{U}\| \cdot \|\vec{V}\|}$$

Thus, the angle between the two vectors is given by

$$\theta = \cos^{-1} \frac{\vec{U} \cdot \vec{V}}{\|\vec{U}\| \cdot \|\vec{V}\|}$$

Angles between Vectors

Illustration 1:

Determine the angle between $\vec{U} = \langle -3, 4 \rangle$ and $\vec{V} = \langle 5, -12 \rangle$.

Solution:

$$\|\vec{U}\| = 5$$
$$\|\vec{V}\| = 13$$
$$\vec{U} \cdot \vec{V} = -3(5) + 4(-12)$$
$$= -63$$

Now
$$\theta = \cos^{-1}\left(\frac{-63}{5(13)}\right)$$
$$\theta \cong 165.7499673°$$

Illustration 2:

Suppose that vectors $\vec{U}$ and $\vec{V}$ form an angle θ. From the terminal point of $\vec{U}$ drop a line perpendicular to a line containing $\vec{V}$. Label the intersection of the lines P. The projection of vector $\vec{U}$ on vector $\vec{V}$ is the vector with an initial point the same as $\vec{V}$ and a terminal point at P. Determine the magnitude of the projection of $\vec{U}$ on $\vec{V}$.

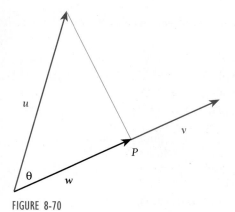

FIGURE 8-70

Solution: Let $\vec{W}$ be the projection of $\vec{U}$ on $\vec{V}$. From Figure 8-70,

$$\|\vec{W}\| = \|\vec{U}\|\cos\theta$$

But $\cos\theta = \dfrac{\vec{U} \cdot \vec{V}}{\|\vec{U}\| \cdot \|\vec{V}\|}$ (derivation of dot product)

$$\|\vec{W}\| = \|\vec{U}\|\dfrac{\vec{U} \cdot \vec{V}}{\|\vec{U}\| \cdot \|\vec{V}\|}$$ (substitute and simplify)

$$\|\vec{W}\| = \dfrac{\vec{U} \cdot \vec{V}}{\|\vec{V}\|}$$

You will explore properties of the dot product in the problem set.   ▬

SUMMARY

- The vector from $Q(x_0, y_0)$ to $P(x_1, y_1)$ has coordinates $QP = \langle x_1 - x_0, y_1 - y_0 \rangle$
- If $\vec{U} = \langle a, b \rangle$ and $\vec{V} = \langle r, s \rangle$ are two-dimensional vectors and α is a real number, then

$$\vec{U} + \vec{V} = \langle a + r, b + s \rangle$$
$$\alpha\vec{U} = \langle \alpha a, \alpha b \rangle$$
$$\vec{U} \cdot \vec{V} = ar + bs$$
$$\|\vec{U}\| = \sqrt{\vec{U} \cdot \vec{U}}$$

- The static angle between $\vec{U}$ and $\vec{V}$ is given by $\theta = \cos^{-1}\left(\dfrac{\vec{U} \cdot \vec{V}}{\|\vec{U}\| \cdot \|\vec{V}\|}\right)$.

- The magnitude of the projection of $\vec{U}$ on $\vec{V}$ is given by $\dfrac{\vec{U} \cdot \vec{V}}{\|\vec{V}\|}$.

- The collection of two-dimensional vectors using vector addition with real numbers for scalar multiplication form a vector space.

8-6   EXERCISES

Determine the magnitude and direction cosines for each of the following vectors. Also determine the angle the vector makes with the positive x-axis and the angle the vector makes with the positive y-axis.

1. $\langle 3, -4 \rangle$

2. $\langle -3, -4 \rangle$

3. $\langle 1, 0 \rangle$

4. $\langle 0, 1 \rangle$

5. $\langle -12, 5 \rangle$

6. $\langle -5, -12 \rangle$

7. $\langle 1, 1 \rangle$

8. $\langle -1, -1 \rangle$

9. $\langle \dfrac{1}{2}, \dfrac{\sqrt{3}}{2} \rangle$

10. $\langle \dfrac{\sqrt{2}}{2}, \dfrac{\sqrt{2}}{2} \rangle$

11. $\langle \dfrac{\sqrt{2}}{2}, -\dfrac{\sqrt{2}}{2} \rangle$

12. $\langle -\dfrac{1}{2}, -\dfrac{\sqrt{3}}{2} \rangle$

13. $\langle -6, 6 \rangle$

14. $\langle -5, -5 \rangle$

15. $\langle 0, -1 \rangle$

16. $\langle -1, 0 \rangle$

17. $\langle 0, 0 \rangle$

18. $\langle 4\sqrt{3}, -4 \rangle$

19. $\langle -5\sqrt{2}, 5\sqrt{2} \rangle$

20. $\langle 10, 10 \rangle$

Suppose that $\vec{U} = \langle 5, 2 \rangle$ and $\vec{V} = \langle -3, 2 \rangle$ determine the following.

21. $\vec{U} + \vec{V}$

22. $\vec{U} - \vec{V}$

23. $\vec{V} - \vec{U}$

24. $2\vec{U} + \vec{V}$

25. $\vec{U} + 3\vec{V}$

26. $5\vec{U} - 2\vec{V}$

27. $3\vec{U} - 5\vec{V}$

28. $6\vec{U} + \vec{V}$

29. $2\vec{U} + 4\vec{V}$

30. $3\vec{V} - 4\vec{U}$

Suppose that $\vec{i} = \langle 1, 0 \rangle$ and $\vec{j} = \langle 0, 1 \rangle$, determine the following.

31. $\vec{i} + \vec{j}$

32. $\vec{i} - \vec{j}$

33. $3\vec{i} + 4\vec{j}$

34. $5\vec{j} - 12\vec{i}$

35. $4\vec{j} + 3\vec{i}$

36. $-12\vec{i} + 5\vec{j}$

37. $\|\vec{i}\| + \|\vec{j}\|$

38. $\|\vec{i}\| - \|\vec{j}\|$

39. $x\vec{i} + y\vec{j}$

40. $r\vec{i} + s\vec{j}$

Determine the angle between the indicated vectors.

41. $\langle 1, 0 \rangle$ and $\langle 0, 1 \rangle$

42. $\langle -1, 0 \rangle$ and $\langle 0, 1 \rangle$

43. $\langle 1, 0 \rangle$ and $\langle 5, 5 \rangle$

44. $\langle 0, 10 \rangle$ and $\langle -5, 5 \rangle$

45. $\langle 3, -3\sqrt{3} \rangle$ and $\langle -4, -4 \rangle$

46. $\langle -2\sqrt{3}, 2 \rangle$ and $\langle -5, -5\sqrt{3} \rangle$

47. $\langle 2, -1 \rangle$ and $\langle 1, -2 \rangle$

48. $\langle 5, -3 \rangle$ and $\langle 3, -5 \rangle$

49. $\langle 6, 4 \rangle$ and $\langle 2, -3 \rangle$

50. $\langle 9, -6 \rangle$ and $\langle 2, 3 \rangle$

Determine the magnitude of each of the following projections.

51. The projection of $\langle 1, 0 \rangle$ on $\langle 0, 1 \rangle$.

52. The projection of $\langle -1, 0 \rangle$ on $\langle 0, 1 \rangle$.

53. The projection of $\langle 1, 0 \rangle$ on $\langle 5, 5 \rangle$.

54. The projection of $\langle 0, 10 \rangle$ on $\langle -5, 5 \rangle$.

55. The projection of $\langle 3, -3\sqrt{3} \rangle$ on $\langle -4, -4 \rangle$.

56. The projection of $\langle -2\sqrt{3}, 2 \rangle$ on $\langle -5, -5\sqrt{3} \rangle$.

57. The projection of $\langle 2, -1 \rangle$ on $\langle 1, -2 \rangle$.

58. The projection of $\langle 5, -3 \rangle$ on $\langle 3, -5 \rangle$.

59. The projection of $\langle 6, 4 \rangle$ on $\langle 2, -3 \rangle$.

60. The projection of $\langle 9, -6 \rangle$ on $\langle 2, 3 \rangle$.

61. Prove that, if $\vec{U} \cdot \vec{V} = 0$, then $\vec{U}$ is perpendicular to $\vec{V}$.

62. Prove that, if $\vec{U} = -\vec{V}$, then the angle between $\vec{U}$ and $\vec{V}$ is $180°$.

63. Suppose $\vec{V} = \langle a, b \rangle$, where neither a nor b is 0. Determine a vector $\vec{U}$ so that $\vec{U}$ and $\vec{V}$ are perpendicular.

64. Suppose $\vec{V} = \langle a, b \rangle$, where neither a nor b is 0. Determine a vector $\vec{U}$ so that $\vec{U} \cdot \vec{V} = 0$.

65. Prove $\vec{U} \cdot \vec{V} = \vec{V} \cdot \vec{U}$.

66. Prove $\|\vec{U}\| = \sqrt{\vec{U} \cdot \vec{U}}$.

67. Suppose the magnitudes of $\vec{U}$ and $\vec{V}$ are known, as well as the angle θ between $\vec{U}$ and $\vec{V}$. Prove that $\vec{U} \cdot \vec{V} = \|\vec{U}\| \, \|\vec{V}\| \cos \theta$.

68. Prove $\|\vec{U} + \vec{V}\|^2 = \|\vec{U}\|^2 + 2\vec{U} \cdot \vec{V} + \|\vec{V}\|^2$.

69. If the angle between two vectors is $90°$ the two vectors are orthogonal (*ortho* = right, *gonal* = angle). Prove that two vectors are orthogonal if their dot product is 0 (the zero vector will be orthogonal to every other vector).

70. Prove that, if the two vectors are orthogonal (see Problem 69), then their dot product is 0.

8-6 PROBLEM SET

1. Suppose a force of 10 lbs acts upward on an object while a force of 5 lbs pulls to the right. Determine a vector to represent a single force equivalent to these two.

2. Suppose a force pulls straight down on an object with 10 lbs of force while a spring pulls the object to the right with 6 lbs of force. Determine a vector to represent a single force equivalent to these two.

3. Suppose that water travels downstream at 5 miles per hour. A swimmer swims at an angle 30° from a right angle to the current toward the direction of the current. If the swimmer's velocity relative to the water is 3 miles per hour, represent these velocities as vector quantities. What is the swimmer's velocity relative to land?

4. A boat crosses a river by angling 45° upstream and across the current. If the boats speed relative to the water is 10 knots and the water speed is 8 knots, represent these as vector quantities. What is the resultant of the vectors? If the resultant represents the boat's velocity relative to land, will the boat strike land upstream or downstream from its launch point?

5. Use vector methods to prove that if the diagonals of a parallelogram are perpendicular, the parallelogram's sides are all equal (a rhombus).

6. Use vector methods to prove that the base angles of an equilateral triangle are equal.

Suppose $\vec{U}$, $\vec{V}$ and $\vec{W}$ are two-dimensional vectors and a and b are real numbers (scalars). Prove the following.

7. $\vec{U} + \vec{V} = \vec{V} + \vec{U}$

8. $(\vec{U} + \vec{V}) + \vec{W} = \vec{U} + (\vec{V} + \vec{W})$

9. $b(\vec{U} + \vec{V}) = b\vec{U} + b\vec{V}$

10. $a(b\vec{U}) = (ab)\vec{U}$

11. $(a + b)\vec{U} = a\vec{U} + b\vec{U}$

12. $\vec{U} + \vec{0} = \vec{U}$ $(\vec{0} = \langle 0, 0 \rangle)$

13. $1\vec{U} = \vec{U}$

14. $\vec{U} + (-1\vec{U}) = \vec{0}$

15. $\vec{U} \cdot \vec{V} = \vec{V} \cdot \vec{U}$

16. $(\vec{U} + \vec{V}) \cdot \vec{W} = \vec{U} \cdot \vec{W} + \vec{V} \cdot \vec{W}$

17. $a(\vec{U} \cdot \vec{V}) = (a\vec{U}) \cdot \vec{V}$

18. $a(\vec{U} \cdot \vec{V}) = \vec{U} \cdot (a\vec{V})$

19. $\vec{0} \cdot \vec{U} = 0$

20. $\vec{V} \cdot \vec{V} = \|\vec{V}\|^2$

PROBLEMS FOR TECHNOLOGY

Most graphing calculators can do vector arithmetic. To operate with vectors enter the matrix mode: (Vectors are a special case of matrices. Chapter 11 develops matrices). Once in the matrix mode you have access to vectors $\vec{A}$ and $\vec{B}$. Dimension each of $\vec{A}$ and $\vec{B}$ to 1 row and 2 columns so they behave like vectors. You may have to select EDIT mode to enter the values for the coordinates of the vector. Under $\vec{A}$ enter 5 for the first coordinate (1, 1) and –3 for the second coordinate (1, 2). Store $\langle 2, 3 \rangle$ under $\vec{B}$.

To determine the scalar product $4\vec{A}$ on a Casio, enter 4 [F1]. Immediately the vector changes to $\vec{C}$ and displays [20, -12]. Press [MODE] [0] to return to the main matrix menu. Now select the add key [F3] to determine $\vec{A} + \vec{B}$. Immediately, the calculator displays [3 0]. The dot product is clumsy under matrix mode. For vector $\vec{B}$ press the B^t key [F2].

Store the results from $\vec{C}$ in $\vec{B}$ by pressing [F2]. Return to the main matrix menu with [MODE] [0]. Then choose the multiply key [F5]. The dot product appears as a single entry in $\vec{C}$. See your Casio manual for more details.

For the Sharp, choose MENU A 3 to select the matrix mode. For scalar multiplication enter [4] [2ndF] [A] [ENTER]. To add the two vectors press [MAT] [A] [+] [MAT] [B] [ENTER]. For the dot product first select vector $\vec{A}$ with [MAT] [A], then press [MAT] [E] [5] followed by [MAT] [B] [ENTER]. See your Sharp manual.

With the TI-81 the details are a little different. Choose matrix mode by pressing [MATRX]. The screen clears and you have the following choices:

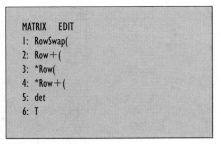

Use the arrow key and ENTER to select Edit. Choose vector $\vec{A}$. Alter the size of $\vec{A}$ to 1×2. Enter 5 into the 1, 1 position and –3 into the 1, 2 position. Press [2nd][CLEAR]. When the screen clears, press [2nd] [1] [ENTER]. $\vec{A}$ will be displayed on the screen. Now press [4] [2nd] [1]. This will multiply $\vec{A}$ by 4. Choose [MATRX] and select Edit $\vec{B}$ to store $\langle 2, 3 \rangle$. Press [2nd][CLEAR] to return to the main screen. Press [2nd] [1] [+] [2nd] [2] [ENTER] and the sum of $\vec{A} + \vec{B}$ appears on the screen. Then press [2nd] [1] [×] [2nd] [2] [MATRX] [6] [ENTER] and the dot product is displayed.

 Redo Exercise 8-6, Problems 21–30 using your calculator. Use the experience to answer the following.

21. Write a step-by-step algorithm to teach someone to do vector addition on your calculator.

22. Write a step-by-step algorithm to teach someone to do a dot product on your calculator.

7   COMPLEX NUMBERS AND DEMOIVRE'S THEOREM

It is no paradox to say that in our most theoretical moods we may be nearest to our most practical applications.
—A. N. WHITEHEAD

Although it takes two real numbers to locate a point in the real plane, these numbers do not have to be in rectangular coordinate format. In fact, because a complex number is formed from two real numbers, a single complex number can locate a point in the **complex plane**. Consider the complex number $z = a + bi$ graphed in Figure 8-71.

Locate z by using (a, b) as rectangular coordinates, or in polar coordinates use $(|z|, \theta)$, where θ is the angle the segment from the origin to z makes with the positive x-axis and $|z|$ is the length of the segment. Conversions between forms follows from Section 8-5.

Theorem 8-8
Conversion to Polar Form I

> If $z = a + bi$ in rectangular form, then the polar coordinates of z are given by $(|z|, \theta)$, where $|z| = \sqrt{a^2 + b^2}$; if $|z| \neq 0$ then $\sin \theta = b/|z|$, $\cos \theta = a/|z|$.

A little effort can improve the conversion to polar form by removing some of the guesswork. This conversion is best done with a calculator.

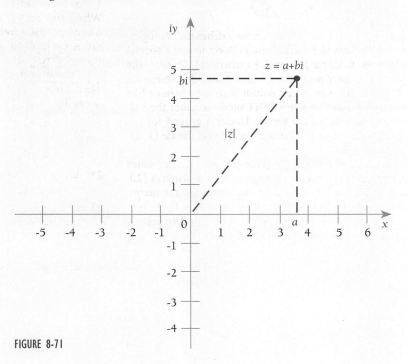

FIGURE 8-71

Theorem 8-9
Conversion to Polar Form II

If $z = a + bi$ in rectangular form and the polar form is $(|z|, \theta)$, where $|z| = \sqrt{a^2 + b^2}$, then if $|z| \neq 0$ and $b \neq 0$ then $\theta = (|b|/b) \cos^{-1}(a/|z|)$

if $b = 0$, then $\begin{cases} \theta = 0°, & \text{if } a \geq 0 \\ \theta = 180°, & \text{if } a < 0 \end{cases}$

if $\theta < 0°$, then add $360°$ to θ.

Note that $|b|/b$ provides the correct sign for $\cos^{-1}$. That sign is the same sign as b. For a complex number $a + bi$, b is the *imaginary part* of the complex number, whereas a is the *real part* of $a + bi$. If the complex number is in polar form $(|z|, \theta)$, then $|z|$ is the absolute value or *magnitude* or *modulus* of the complex number. The angle θ is the *argument*.

Theorem 8-10
Conversion to Rectangular Form

If z is located by $(|z|, \theta)$ in polar coordinates where the Cartesian coordinates are (a, b), then $a = |z| \cos \theta$, $b = |z| \sin \theta$.

A convenient way to write $z = a + bi$ is

$$z = |z| \cos \theta + |z| i \sin \theta$$
$$= |z| [\cos \theta + i \sin \theta]$$

DEFINITION 8-11
Polar Form of a Complex Number

The **polar form of a complex number** z is given by $z = |z|[\cos \theta + i \sin \theta]$ abbreviated $z = |z| \text{ cis } \theta$, where $|z|$ is the magnitude of z and θ is the argument of z.

EXAMPLE 8-15

Illustration 1:

Conversion of Rectangular to Polar Forms. Convert each of the following complex numbers to polar form.

$$z = \sqrt{2} - \sqrt{2}\, i$$

Solution:

$$|z| = \sqrt{(\sqrt{2})^2 + (\sqrt{2})^2} \qquad \text{(distance formula)}$$
$$= 2$$

The sign of the imaginary part is negative, the value of the real part is $\sqrt{2}$. Hence,

$$\theta = -\cos^{-1} \frac{\sqrt{2}}{2}$$
$$\theta = -45°$$

Because $\theta < 0°$, add 360°.

$$\theta = 315°$$

Answer:

$$z = 2 \text{ cis } 315°$$

Note that $\theta = -45°$ and $\theta = 675°$ are also acceptable. Every complex number has infinitely many polar representations. Generally, we prefer $0° \leq \theta < 360°$. This is the reason for adding 360°.

Illustration 2:

$$z = -\sqrt{3} + i$$

Solution:

$$|z| = \sqrt{3 + 1}$$

$$|z| = 2$$

$$\theta = + \cos^{-1}\left(-\frac{\sqrt{3}}{2}\right)$$

$$\theta = 150°$$

Answer:

$$z = \text{cis } 150°$$

EXAMPLE 8-16

Conversion of Polar to Rectangular Forms. Convert each of the following complex numbers to rectangular form.

Illustration 1:

$$z = 3 \text{ cis } 225°$$

Solution:

$$z = 3[\cos 225° + i \sin 225°]$$

$$= 3\left[-\frac{\sqrt{2}}{2} + i\left(-\frac{\sqrt{2}}{2}\right)\right]$$

$$= \frac{-3\sqrt{2}}{2} - \frac{3\sqrt{2}}{2}i$$

Illustration 2:

$$z = 6 \text{ cis } 60°$$

Solution:

$$z = 6[\cos 60° + i \sin 60°]$$

$$= 6\left[\frac{1}{2} + i\frac{\sqrt{3}}{2}\right]$$

$$= 3 + 3\sqrt{3}i$$

Although complex numbers are easy to add in rectangular form, they are more difficult to multiply and divide. In polar form, multiplication and division are simple.

Theorem 8-11
Product and Quotients of Complex Numbers

The product of complex numbers $z = |z| \text{ cis } \theta$ and $w = |w| \text{ cis } \phi$ is a complex number given by $zw = |z||w| \text{ cis}(\theta + \phi)$

The quotient is given by $z/w = \dfrac{|z|}{|w|} \text{ cis}(\theta - \phi), w \neq 0 + 0i.$

As proof (of the first part of the theorem), suppose $z = |z| \text{ cis } \theta$ and $w = |w| \text{ cis } \phi$. Then,

$zw = |z| \text{ cis } \theta \, |w| \text{ cis } \phi$ (substitution)

$= |z||w| (\cos \theta + i \sin \theta)(\cos \phi + i \sin \phi)$ (definition of cis)

$= |z||w| (\cos \theta \cos \phi + i \cos \theta \sin \phi + i \sin \theta \cos \phi + i^2 \sin \theta \sin \phi)$ (simplify)

$= |z||w| [(\cos \theta \cos \phi - \sin \theta \sin \phi) + i(\cos \theta \sin\phi + \sin \theta \cos \phi)]$ (factor i)

$= |z||w| [\cos(\theta + \phi) + i \sin(\theta + \phi)]$ (sum formulas)

$= |z||w| \text{ cis}(\theta + \phi)$ (definition of cis)

The proof of the quotient is left as an exercise.

Because positive integer exponents indicate repeated multiplication, the theorem extends to powers of a complex number. The following theorem is by Abraham DeMoivre (1667–1754). Although **DeMoivre's theorem** appears obvious, the actual proof uses a process known as *finite mathematical induction*. We discuss induction in Chapter 12. Therefore, the proof of DeMoivre's theorem awaits Chapter 12.

Theorem 8-12
DeMoivre's Theorem

If $z = |z| \text{ cis } \theta$ and n is a counting number then $z^n = |z|^n \text{ cis}(n\theta)$.

EXAMPLE 8-17

Illustration 1:

Products and Quotients of Complex Numbers. Suppose $z = 3 \text{ cis } 50°$ and $w = 4 \text{ cis } 70°$. Determine the following.

zw

Solution:

$zw = [3 \text{ cis } 50°][4 \text{ cis } 70°]$ (add the arguments)

$= (3)(4) \text{ cis}(50° + 70°)$ (multiply the magnitudes)

$= 12 \text{ cis } 120°$

Illustration 2:

$$\frac{w}{z}$$

Solution:

$$\frac{w}{z} = \frac{4 \text{ cis } 70°}{3 \text{ cis } 50°} \qquad \text{(subtract the arguments)}$$

$$= \frac{4}{3} \text{ cis}(70° - 50°) \qquad \text{(divide the magnitudes)}$$

$$= \frac{4}{3} \text{ cis } 20°$$

Illustration 3:

$$z^2 = [3 \text{ cis } 50°]^2 \qquad \text{(multiply exponent times argument)}$$

Solution:

$$z^2 = 3^2 \text{ cis}[2(50°)] \qquad \text{(raise magnitude to exponent)}$$
$$= 9 \text{ cis } 100°$$

Illustration 4:

$$\sqrt{w}$$

Solution:

Let	$q = \sqrt{w}$	
then	$q^2 = w$	(definition of square root)
Now	$q^2 = 4 \text{ cis } 70°$	(substitution for w)
Suppose that	$q = r \text{ cis } \theta$	(cis notation for q)
	$q^2 = r^2 \text{ cis}(2\theta)$	(square both sides)
	$r^2 \text{ cis}(2\theta) = 4 \text{ cis } 70°$	(because $q^2 = w$)
So that	$r^2 = 4 \quad$ and $\quad 2\theta = 70°$	(magnitudes are equal)
	$r = 2 \quad$ and $\quad \theta = 35°$	(arguments are equal, solve)
	$\sqrt{w} = 2 \text{ cis } 35°$	

Note: Illustration 4 leads us to believe that DeMoivre's theorem may apply even when n is not a counting number.

$$\sqrt{w} = w^{1/2}$$
$$= [4 \text{ cis } 70°]^{1/2} \qquad \text{(multiply exponent times argument)}$$
$$= 4^{1/2} \text{ cis}(\tfrac{1}{2} 70°) \qquad \text{(raise magnitude to exponent)}$$
$$= 2 \text{ cis } 35° \quad \text{(as before)}$$

In addition, because

$$4 \text{ cis } 70° = 4 \text{ cis } 430° \qquad \text{(add 360°)}$$
$$\text{then } w^{1/2} = 4^{1/2} \text{ cis}(\tfrac{1}{2} 430°)$$
$$= 2 \text{ cis}(215°)$$

Girolam Cardano (Jerome Cardan) (1501–1576) usually receives credit for the formula for solving reduced cubic equations. Actually, Cardan's formula is based on the work of Nicolo of Brescia (ca. 1499–1557) often called *Tartaglia* (the stammerer).

Tartaglia developed the formula for public competition with a rival mathematician named Fior. Cardan, who was not known for his high principles, managed to persuade Tartaglia to give him the formula by swearing to keep it secret. Within a few years, however, Cardan had the formula published in his Ars magna. A bitter dispute concerning who was the true author of the method followed.

A consequence of Cardan's formula, and not the quadratic formula, led to the development of complex numbers. Although Cardan claimed the formula was inapplicable in those cases where a negative number appeared as the radicand of a square root, Rafael Bombelli (ca. 1556–1573) decided that such expressions could be manipulated in the same manner as "numbers."

Bombelli's insight came before modern notation for radicals. Bombelli represented $\sqrt{17}$ by Rq17, where Rq represented square root (Rc represented cube root). As a result, Bombelli's representation of $\sqrt{-17}$ was dimRq17.

As a result, there are two complex square roots of 4 cis 70°: 2 cis 35° and 2 cis 215°.   ∎

Multiplication and division of complex numbers in polar is fast and easy. However, addition or subtraction of complex numbers in polar form would be tedious at best. Addition and subtraction of complex numbers is best accomplished in Cartesian form.

Recall the fundamental theorem of algebra from Section 4-6. If we count multiple roots, then an *n*th-degree polynomial should have exactly *n* complex roots. A polynomial as simple as $T(z) = z^4 - 1$, factors into $T(z) = (z^2 - 1)(z^2 + 1)$. Apply the quadratic formula to get $\{-1, 1, -i, i\}$. But how would you get the five roots of $P(z) = z^5 - 32$? You may guess that 2 is a root, but what of the other four?

The roots of P satisfy the equation $z^5 = 32$. Express 32 as a complex number in polar form. Then,

$$z^5 = 32 \text{ cis} 0°.$$

Of course the representation of 32 is not unique: 32 cis 360°, 32 cis 720°, 32 cis 1080°, 32 cis 1440°, 32 cis 1800°, all work. By DeMoivre's theorem, we have the following:

For $z^5 = 32$ cis 0°

             $z = 2$ cis 0°   (or 2 as expected)

But $z^5 = 32$ cis 360°   implies

$$(z^5)^{1/5} = 32^{1/5} \cos \frac{360°}{5}$$

             $z = 2$ cis 72°

Also, if $z^5 = 32$ cis 720°

then $z = 2$ cis $\dfrac{720°}{5}$

              $= 2$ cis 144°

Similarly, for $z^5 = 32$ cis 1080°, $z = 2$ cis 216°; for $z^5 = 32$ cis 1440°, $z = 2$ cis 288°; for $z^5 = 32$ cis 1800°, $z = 2$ cis 360°. But this last root is a repetition of $z = 2$ cis 0°. Because there are exactly five roots for this polynomial, we need list only five representations of 32. All solutions after that are repetitions. Also, note that the roots themselves are exactly 360°/5 or 72° apart. See Figure 8-72.

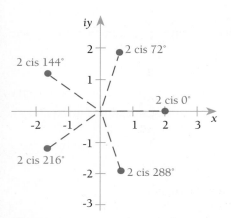

FIGURE 8-72

SUMMARY

- The polar form of a complex number z is given by $z = |z|$ cis θ, where θ is the angle made by the segment from the origin to z and the positive x-axis and $|z|$ is the length of the segment. The angle θ is the argument and $|z|$ is the magnitude of the complex number.

  1. $z = a + bi$

      $= |z|$ cis θ (polar form)

      where $|z| = \sqrt{a^2 + b^2}$ and $\theta = \dfrac{|b|}{b} \cos^{-1} \dfrac{a}{|z|}$, when $b \neq 0$ and

      $|z| \neq 0$. If $b = 0$, then $\theta = 0°$ when $a \geqslant 0$ and $\theta = 180°$ when $a < 0$.

  2. Convert $z = |z|$ cis θ to rectangular form, $z = a + bi$, by rewriting as $z = |z|[\cos \theta + i \sin \theta]$ and evaluating $\sin \theta$ and $\cos \theta$.

- From Section 0-7, if $z = a + bi$ then the conjugate of z is $\bar{z} = a - bi$. In polar form if $z = |z|$ cis θ, then $\bar{z} = |z|$ cis$(-\theta)$.

  1. $[|z|$ cis $\theta][|w|$ cis $\phi] = |z|\,|w|$ cis$(\theta + \phi)$.

  2. $\dfrac{|z| \text{ cis } \theta}{|w| \text{ cis } \phi} = \dfrac{|z|}{|w|}$ cis$(\theta - \phi)$.

  3. $[|z|$ cis $\theta]^n = |z|^n$ cis$(n\theta)$, where n is a counting number.

8-7   EXERCISES

Convert each of the following complex numbers to polar form, $|z|$ cis θ.

1. $\dfrac{\sqrt{3}}{2} - \dfrac{1}{2}i$ **2.** $3\sqrt{2} + 3\sqrt{2}i$

3. $-\sqrt{2} + \sqrt{2}i$ **4.** $1 - \sqrt{3}i$

5. $-3 + 4i$ **6.** $-3 - 4i$

7. $-1 - \sqrt{3}i$ **8.** $-\sqrt{2} + \sqrt{2}i$

9. $2i$ **10.** $-2i$

Convert each of the following complex numbers to rectangular form, $a + bi$.

11. 5 cis 30°   **12.** 6 cis 45°

13. 3 cis 135°   **14.** 2 cis 150°

15. 2 cis 240°   **16.** 3 cis 300°

17. 4 cis 70°   **18.** 4 cis 80°

19. -5 cis 30°   **20.** -6 cis 45°

If $z = 9$ cis 40°, $w = 8$ cis 60° and $q =$ cis 70°, determine each of the following.

21. zw **22.** zq

23. $\dfrac{z}{w}$ **24.** $\dfrac{z}{q}$

25. zwq **26.** $\dfrac{zw}{q}$

27. z^3 **28.** w^4

29. $z^{1/2}$ **30.** $\dfrac{3}{w}$

Prove the following.

31. If $z = r$ cis θ, then $\bar{z} = r$ cis$(-\theta)$.

32. The conjugate of $\bar{z}$ is z.

33. If $z = r$ cis θ, then $-z = r$ cis$(\theta + 180)$.

34. If $z = r \text{ cis } \theta$, then $-z = r \text{ cis}(\theta - 180)$.

35. If $z = r \text{ cis } \theta$, then $z^2 = r^2 \text{ cis}(2\theta)$.

36. If $z = r \text{ cis } \theta$, then $z^3 = r^3 \text{ cis}(3\theta)$.

37. If $z = r \text{ cis } \theta$, then $z^4 = r^4 \text{ cis}(4\theta)$.

38. If $z = r \text{ cis } \theta$, then $z^5 = r^5 \text{ cis}(5\theta)$.

39. If $z = r \text{ cis } \theta$, then $z^{1/3} = \sqrt[3]{r} \text{ cis } \dfrac{\theta}{3}$.

40. If $z = r \text{ cis } \theta$, then $z^{1/4} = \sqrt[4]{r} \text{ cis } \dfrac{\theta}{4}$.

41. If $z = r \text{ cis } \theta$, conjecture what $z^{p/q}$ equals.

42. If $z = r \text{ cis } \theta$, conjecture what $\sqrt[q]{z^p}$ equals.

43. If $z = r \text{ cis } \theta$, conjecture what z^{π} equals.

44. If $z = r \text{ cis } \theta$, conjecture what $z^{\sqrt{2}}$ equals.

45. Show that $|z| = \sqrt{z\bar{z}}$ using polar coordinates.

46. If $|z| = 1$, show that $\dfrac{z}{\bar{z}} = z^2$ using polar coordinates.

8-7   P R O B L E M   S E T

Determine the roots of the following. Use the following steps: (1) convert the complex number to polar form; $r \text{ cis } \theta$; (2) apply Demoivre's theorem to $r \text{ cis } \theta$, then list the remaining roots by adding $360° K/n$ [$K = 1 \ldots n - 1$; $r \text{ cis}(\theta + 360°K/n)$]; (3) convert back to rectangular form.

1. Determine the five fifth roots of 1.

2. Determine the six sixth roots of 1.

3. Determine the six sixth roots of -1.

4. Determine the five fifth roots of -1.

5. Determine the three third roots of i.

6. Determine the three third roots of $-i$.

7. Solve $z^3 = 1 + 1i$.

8. Solve $z^3 = -\sqrt{3} + 1i$.

9. Determine the three third roots of $2 - 2\sqrt{3}i$.

10. Determine the three third roots of $2\sqrt{2} + 2\sqrt{2}i$.

C H A P T E R   S U M M A R Y

- $\sin^{-1} x = y$ means $\sin y = x$ and $-\dfrac{\pi}{2} \leq y \leq \dfrac{\pi}{2}$.

- $\cos^{-1} x = y$ means $\cos y = x$ and $0 \leq y \leq \pi$.

- $\tan^{-1} x = y$ means $\tan y = x$ and $-\dfrac{\pi}{2} < y < \dfrac{\pi}{2}$.

- $\tan^{-1}$ has several efficient approximation formulas. Approximate the other five inverse circular functions by expressing them in terms of $\tan^{-1}$.

- Given three pieces of information concerning the triangle (except three angles only), the remaining parts of the triangle can be determined by using the following:

Law of Sines: $\dfrac{\sin A}{a} \equiv \dfrac{\sin B}{b} \equiv \dfrac{\sin C}{c}$

Law of Cosines: $a^2 \equiv b^2 + c^2 - 2bc \cos A$

$b^2 \equiv a^2 + c^2 - 2ac \cos B$

$c^2 \equiv a^2 + b^2 - 2ab \cos C$

- The ambiguous case for a triangle occurs when two sides and an angle not between the sides form the given information.
- The relationship of rectangular coordinates (x, y) to polar coordinates (r, θ) for a given point is given by the following:

$$\sin \theta = \frac{y}{r} \qquad \cos \theta = \frac{x}{r}$$

$$y = r \sin \theta \qquad x = r \cos \theta$$

$$\frac{y}{x} = \tan \theta \qquad r^2 = x^2 + y^2$$

- The vector from $Q(x_0, y_0)$ to $P(x_1, y_1)$ has coordinates $\mathbf{QP} = \langle x_1 - x_0, y_1 - y_0 \rangle$.
- If $\vec{U} = \langle a, b \rangle$ and $\vec{V} = \langle r, s \rangle$ are two-dimensional vectors and α is a real number, then

$$\vec{U} + \vec{V} = \langle a + r, b + s \rangle$$
$$\alpha \vec{U} = \langle \alpha a, \alpha b \rangle$$
$$\vec{U} \cdot \vec{V} = ar + bs$$
$$|\vec{U}| = \sqrt{\vec{U} \cdot \vec{U}}$$

The static angle between $\vec{U}$ and $\vec{V}$ is given by

$$\theta = \cos^{-1}\left(\frac{\vec{U} \cdot \vec{V}}{\|\vec{U}\| \, \|\vec{V}\|}\right)$$

- A complex number z expressed in polar form is given by $z = |z| \operatorname{cis} \theta$, where θ is the angle made by the segment from the origin to z and the positive x-axis and $|z|$ is the length of the segment. In rectangular form $z = a + bi = |z| \operatorname{cis} \theta$, where $|z| = \sqrt{a^2 + b^2}$ and

$$\theta = \frac{|b|}{b} \cos^{-1} \frac{a}{|z|},$$

when $b \neq 0$ and $|z| \neq 0$. If $b = 0$, then $\theta = 0°$ when $a \geq 0$ and $\theta = 180°$ when $a < 0$. Convert $z = |z| \operatorname{cis} \theta$ to $z = a + bi$ by rewriting it as $z = |z| [\cos \theta + i \sin \theta]$ and evaluating $\sin \theta$ and $\cos \theta$.

1. $[|z| \operatorname{cis} \theta][|w| \operatorname{cis} \phi] = |z| \, |w| \operatorname{cis}(\theta + \phi)$

2. $\dfrac{|z| \operatorname{cis} \theta}{|w| \operatorname{cis} \phi} = \dfrac{|z|}{|w|} \operatorname{cis}(\theta - \phi)$

3. $[|z| \operatorname{cis} \theta]^n = |z|^n \operatorname{cis}(n\theta)$

KEY WORDS AND CONCEPTS

Ambiguous case	Arc Tangent	Coordinates of a vector
Arc cosine	Complex plane	$\cos^{-1}$
Arc sine	Conditional equation	DeMoivre's theorem

KEY WORDS AND CONCEPTS, CONT'D.

Direction cosines
Dot product
General solution
Initial point
Initial solution
Inverse trigonometric function
Law of Cosines
Law of Sines
Magnitude of a vector
Parallelogram law

Polar coordinates
Polar form of a complex number
Primary solution
Pythagorean theorem
Restricted function
Resultant
Scalar
Scalar multiplication
$\sin^{-1}$

Solve a right triangle
Standard position
$\tan^{-1}$
Terminal point
Unit vector
Vector
Vector addition
Vector space
Zero vector

8   REVIEW EXERCISES

SECTION I

Solve for triangle *ABC* in Figure 8-73.

1. **(a)** If $B = 40°$ and $AB = 7$, solve triangle *ABC*.

   (b) If $B = 80°$ and $AC = 1$, solve triangle *ABC*.

2. If $B = 70°$ and $BC = 5$, solve triangle *ABC*.

3. If $A = 54°$ and $AB = 10$, solve triangle *ABC*.

4. If $A = 23°$ and $BC = 9$, solve triangle *ABC*.

5. If $AB = 7$ and $BC = 3$, solve triangle *ABC*.

6. If $AC = 7$ and $BC = 3$, solve triangle *ABC*.

7. If $AB = 5$ and $BC = 2$, solve triangle *ABC*.

8. If $AB = 5$ and $AC = 2$, solve triangle *ABC*.

9. If $AB = 60$ and $BC = 20$, solve triangle *ABC*.

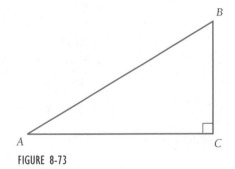

FIGURE 8-73

   Triangle *ABC* in Figure 8-74 is not a right triangle. All the angles are acute. From *C* a line *h* is dropped perpendicular to *AB*; *h* intersects *AB* at point *p*; *p* divides *AB* into segments of length *x* and *y*; *h* divides angle *C* into angles *a* and *b*.

10. If angle *A* is 60° and angle *B* is 70° determine the size of angle *C*. Determine the size of angles *a* and *b*.

11. If angle *A* is 60°, angle *B* is 70° and $AC = 10$, determine the length of *h* and *x*.

12. If angle *A* is 60°, angle *B* is 70° and $AC = 10$, determine *y* and *BC*.

13. If angle *A* is 60°, angle *B* is 70° and $AC = 10$, determine *AB*.

14. At 50 meters a surveyor measures the angle of elevation to a building as 60°. What is the height of the building to the nearest tenth of a meter?

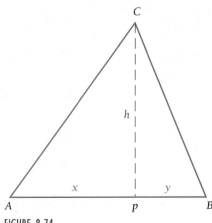

FIGURE 8-74

15. The pointer on a sundial casts a shadow 12 cm long at noon. If the pointer is inclined 45° to the face of the dial, how long is the pointer to the nearest tenth of a centimeter?

16. A skier knows that the sharpest incline she cares to ski drops an average of 6 meters for each 10 meters forward. Determine the average angle of descent to the nearest degree.

17. A hiker treks a path 45° east of due north. If he hikes for 6 km, to the nearest tenth of a kilometer how far east of his starting point is he?

SECTION 2

Determine the exact value of the following.

18. $\cos^{-1} 0.5$

19. $\csc^{-1}\left(-\dfrac{2}{\sqrt{3}}\right)$

20. $\cot^{-1}(-1)$

21. $\sin^{-1}(-0.5)$

22. $\cos^{-1}\dfrac{\sqrt{2}}{2}$

23. $\sin^{-1}\left(-\dfrac{\sqrt{2}}{2}\right)$

Approximate the following using a calculator or table values:

24. $\cos^{-1} 0.7$

25. $\sin^{-1} 0.8$

26. $\sin^{-1}\left(-\dfrac{12}{13}\right)$

27. $\cos^{-1}(-5)$

28. Define $\cos^{-1}$.

29. Define $\csc^{-1}$.

Evaluate the following:

30. $\cot^{-1}(-1)$

31. $\sec^{-1}(-1)$

32. $\sin(\sin^{-1} 0.2)$

33. $\sin(\sin^{-1} 0.7)$

34. $\cos^{-1}\left(-\cos\dfrac{\pi}{3}\right)$

35. $\cos^{-1}\left(\cos\dfrac{\pi}{4}\right)$

36. $\sin^{-1}\left(\cos\dfrac{5\pi}{6}\right)$

37. $\cos^{-1}\left(-\sin\dfrac{\pi}{3}\right)$

38. $\cos^{-1}\left(\sin\dfrac{\pi}{6}\right)$

39. $\cos(-\sin^{-1} 0.5)$

40. $\cos(\sin^{-1}\sqrt{1 - r^2})$

41. $\sin(\cos^{-1} z)$

Express each of the following as a function of $\tan^{-1}$.

42. $\sin^{-1} x$

43. $\sec^{-1}\sqrt{x^2 - 1}$

Use a calculator to approximate the following.

44. $\tan^{-1} 0.9$

45. $\tan^{-1} 1.5$.

46. $\sec^{-1}(-9)$

47. $\sin^{-1}(-0.205)$

48. $\cot^{-1}(-0.4)$

49. $\cos^{-1}(1.5)$

50. $\sin^{-1} 4.32$

51. $\sec^{-1}(-0.33)$

Verify the following identities.

52. $\tan^{-1} x = \cot^{-1}\dfrac{1}{x}$

53. $\cos^{-1} x = \cot^{-1}\dfrac{x}{\sqrt{1 - x^2}}$

Solve each of the following for y in terms of x.

54. $x = 3\sin^{-1}(2y - 1)$

55. $x - 5\cos(\pi y) = 0$

56. Suppose a viewer stands 20 miles from a tall building. At that distance the buildings apparent height is measured by a 1° angle. Estimate the actual height of the building measured to the nearest hundred feet. Assume that the relative elevation of the observer and building are equal and do not influence the approximation.

57. A large picture is placed with its center at eye level in a gallery. See Figure 8-75. The picture is 2 feet high. As a viewer approaches the picture at a distance d, the apparent height of the picture is measured by θ as shown in the diagram. Express θ as a function of d. Estimate θ at 25 feet.

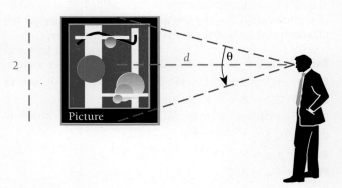

FIGURE 8-75

SECTION 3

Determine the primary solutions for each of the following equations. Write a formula for the general solution.

58. $\cos x + 1 = 0$

59. $2 + \csc x = 0$

60. $2 \sin x + \sqrt{3} = 0$

61. $\tan x - 1 = 0$

62. $\sin x + \cos x = 0$

63. $\tan x - \cot x = 0$

64. $\cos(2x) - 1 = 0$

65. $\sec x \sin x = 0$

66. $\sin^2 x = \cos x + 1$

67. $\sin^2 x + \sin x = 0$

68. $4 \cos^2 x - 3 = 0$

69. $\csc^2 x + \csc x - 2 = 0$

70. $\sec(2x) - 1 = 0$

71. $2 \cos(3x) + 1 = 0$

72. $2 \cos(2x) - \sqrt{3} = 0$

73. $\tan \dfrac{x}{2} + 1 = 0$

74. $\sin \dfrac{x}{3} = \cos x$

75. $\sin^2(3x) = 1$

76. $\cos x \sin x = 0$

77. $\cos^2\left(x + \dfrac{\pi}{3}\right) - \cos\left(x + \dfrac{\pi}{3}\right) = 0$

78. $\cos x - 2 = 0$

79. $\sin^2 x - 3 \sin x + 2 = 0$

80. $2 \sin^2 x + 3 \cos x = 0$

81. $\cos^2 x - 3 \sin^2 x = 0$

82. $\cos^2(2x) = 2 \sin x \cos x$

83. $1 - \cos x = \tan \dfrac{x}{2}$

SECTION 4

Sketch a figure to represent the triangle(s) described by each of the following. Use the Law of Sines or Law of Cosines to determine the indicated value.

84. $a = 10, B = 50°, A = 60°$. Determine b.

85. $a = 4, b = 5.656, A = 45°$. Determine B.

86. $a = 7, b = 8, c = 5$. Determine C.

87. $a = 10, b = 6, B = 15°$. Determine A.

88. $a = 6, b = 3, B = 40°$. Determine A.

89. $b = 4, c = 8, A = 60°$. Determine a.

90. $b = 4, a = 8, C = 40°$. Determine c.

91. $a = 4, b = 8, c = 6$. Determine B.

92. $b = 10, A = 65°, B = 55°$. Determine a.

93. $a = 5, b = 4, C = 50°$. Determine c.

Sketch a figure to represent the triangle(s) described by each of the following. Then solve the triangle for the three remaining parts of the triangle.

94. $a = 4, b = 3, c = 6$.

95. $A = 50°, B = 60°, c = 10$.

96. $B = 50°, C = 70°, a = 10$.

97. $B = 65°, a = 8, c = 10$.

98. $a = 4, b = 12, c = 8$.

99. $b = 4, c = 8, A = 30°$.

Solve as indicated.

100. Suppose that you must construct a pyramid out of four triangles, where each triangle has two sides each 3 meters long and one side 1 meter long. Determine the height of the finished pyramid.

101. A plane travels 500 km due north from an airport while a second plane flies 20 km north 50° east (50° clockwise from due north). What is the distance between the two planes?

SECTION 5

Graph each of the following in a polar coordinate system.

102. $(5, 270°)$

103. $(2, -90°)$

104. $\left(-3, \dfrac{3\pi}{2}\right)$

105. $\left(\dfrac{5}{3}, \dfrac{5\pi}{6}\right)$

Convert each of the following to Cartesian coordinates.

106. $(7, 0°)$

107. $(5, -45°)$

108. $\left(8, -\dfrac{\pi}{2}\right)$

109. $\left(-8, \dfrac{\pi}{6}\right)$

110. $(-10, 30°)$

111. $\left(4, \dfrac{\pi}{3}\right)$

Convert each of the following to polar coordinates with $0° \leqslant \theta \leqslant 90°$.

112. $(4, 0)$

113. $(1, \sqrt{3})$

114. $(2, 2\sqrt{3})$

115. $(0, 5)$

116. $(0, 7)$

117. $(2, 0)$

Convert to polar coordinates. Use radian measure.

118. $(4, 4)$

119. $(6, 6)$

120. $(-7, 7)$

121. $(-5, 5)$

122. $(5, -5\sqrt{3})$

123. $(2, 2\sqrt{3})$

Convert to Cartesian coordinates.

124. $(6, 240°)$

125. $(-6, 60°)$

126. $\left(5, -\dfrac{\pi}{6}\right)$

127. $\left(5, \dfrac{\pi}{6}\right)$

Graph each of the following in polar coordinates.

128. $r = 4$

129. $\theta = -\dfrac{3\pi}{4}$

130. $r = \dfrac{5}{2}$

131. $\theta = \dfrac{\pi}{3}$

132. $r = 3\sin\theta$

133. $r = 2 - \cos\theta$

Convert the following equations in Cartesian form to polar form.

134. $y = 5x$

135. $y = 3x$

136. $y = -2x$

137. $y = -5x$

138. $x^2 + y^2 = 16$

139. $9x^2 + 16y^2 = 144$

SECTION 6

Determine the magnitude and direction cosines for each of the following vectors. Also determine the angle the vector makes with the positive x-axis and the angle the vector makes with the positive y-axis.

140. $\langle 4, -3 \rangle$

141. $\langle 5, -12 \rangle$

142. $\langle 1, 2 \rangle$

143. $\langle \dfrac{\sqrt{2}}{2}, \dfrac{\sqrt{2}}{2} \rangle$

144. $\langle \sqrt{2}, \sqrt{2} \rangle$

145. $\langle 4\sqrt{3}, -4 \rangle$

146. $\langle -3\sqrt{2}, 3\sqrt{2} \rangle$

147. $\langle 1, 1 \rangle$

Suppose that $\vec{U} = \langle 6, 4 \rangle$ and $\vec{V} = \langle -5, 3 \rangle$ determine the following.

148. $\vec{U} + \vec{V}$

149. $3\vec{U} + \vec{V}$

150. $\vec{U} - 2\vec{V}$

151. $5\vec{U} + \vec{V}$

152. $5\vec{U} + 3\vec{V}$

153. $2\vec{V} - 3\vec{U}$

Suppose that $\vec{i} = \langle 1, 0 \rangle$ and $\vec{j} = \langle 0, 1 \rangle$, determine the following.

154. $\vec{i} + 3\vec{j}$

155. $5\vec{i} - 12\vec{j}$

156. $5\vec{j} - 3\vec{i}$

157. $2\vec{i} + 7\vec{j}$

158. $\|2\vec{i}\| - \|\vec{j}\|$

159. $t\vec{i} + u\vec{j}$

Determine the angle between the indicated vectors.

160. $\langle -5, 0 \rangle$ and $\langle 0, 4 \rangle$

161. $\langle 0, 2 \rangle$ and $\langle -3, 3 \rangle$

162. $\langle -\sqrt{3}, 1 \rangle$ and $\langle -4, -4\sqrt{3} \rangle$

163. $\langle 7, -2 \rangle$ and $\langle 2, -7 \rangle$

164. Suppose a force pulls straight down on an object with 20 lb. of force while a spring pulls the object to the right with 12 lb. of force. Determine a vector to represent a single force equivalent to these two.

165. A boat crosses a river by angling 45° upstream and across the current. If the boat's speed relative to the water is 5 knots and the water speed is 3 knots, represent these as vector quantities. What is the resultant of the vectors. If the resultant represents the boat's velocity relative to land, will the boat strike land upstream or downstream from its launch point?

SECTION 7

Convert each of the following complex numbers from rectangular form to polar form.

166. $\dfrac{\sqrt{3}}{2} + \dfrac{1}{2}i$

167. $3 - 3\sqrt{3}i$

168. $-3 - 4i$

169. $-5\sqrt{2} + 5\sqrt{2}i$

Convert each of the following polar form complex numbers to rectangular form.

170. $7\sqrt{3}\operatorname{cis} 30°$

171. $\sqrt{2}\operatorname{cis} 45°$

172. $-5\operatorname{cis} 135°$

173. $10\operatorname{cis} 240°$

174. $4\operatorname{cis} 80°$

175. $-5\operatorname{cis} 30°$

If $z = 7\operatorname{cis} 20°$, $w = 5\operatorname{cis} 50°$ and $q = \operatorname{cis} 30°$, determine each of the following.

176. zw

177. $\dfrac{z}{q}$

178. zwq

179. w^4

180. z

181. $\dfrac{10}{w}$

Use Demoivre's theorem to determine the roots of the following. Convert the numbers back to rectangular form.

182. Determine the four fourth roots of 1.

183. Determine the four fourth roots of $-i$.

184. Solve $z^5 = -\sqrt{3} + 1i$.

185. Solve $z^3 = 2\sqrt{2} + 2\sqrt{2}i$.

Music of the Spheres

The Greeks did not believe the world was flat. As early as 230 BC, the Greek mathematician and astronomer Eratosthenes attempted to approximate the circumference of the earth. Because the Greeks saw perfection in the circle, it was natural to them that not only was the earth a sphere, but that the mechanism for the motion of heavenly bodies consisted of wheels within wheels, spheres within spheres.

Because of their fascination with astronomy, the Greeks developed a form of trigonometry for triangles drawn on a sphere rather than on a plane. The sides of such triangles are great circles of the sphere. Examples of great circles on a globe include lines of longitude and the equator. As any sailor will tell you, the shortest path between two positions on the earth is along a great circle through the two points. In that respect, these geodesic lines behave analogously to straight lines in a plane.

Peculiar consequences are associated with spherical triangles. Examine a globe. Note that lines of longitude intersect the equator at a right angle. In particular, the prime meridian through Greenwich and the meridian 90° west of Greenwich not only intersect the equator at 90°, they intersect each other at 90°!

The sum of the angles of this spherical triangle is 270°. In a spherical triangle, the sum of the angles is always greater than 180°.

In a plane the sum of the angles of a triangle is exactly 180°. This result is a consequence of the parallel postulate. Briefly, the parallel postulate claims that through a point outside a "line" exactly one line could be drawn parallel to the given line. It took hundreds of years, but mathematicians came to realize that the parallel postulate was a characteristic of planes. Examine a globe. Choose any geodesic (great circle "line") on the sphere. From a point on the sphere outside the geodesic, *no* geodesic can be drawn parallel to the given geodesic. All geodesics intersect.

Because of the books he wrote summarizing and formalizing the methods of geometry, Euclid (about 300 BC) is often called the *father of geometry*. The familiar geometry of the plane is a form of Euclidean geometry. The Russian mathematician Nicolai Ivanovitch Lobaschevsky (1793–1856) and the Hungarian mathematician Janos Bolyai (1802–1860) usually receive credit for developing non-Euclidean geometry. Some evidence indicates that the genius Karl Gauss developed similar concepts. Whatever the source, the concept was revolutionary. As with any revolutionary idea the impact produced a profound reevaluation of the way we look at the universe.

Georg F. B. Riemann (1826–1866) and Hermann Minkowski (1869–1909) refined and developed the concepts of non-Euclidean geometry. The work of these two mathematicians later influenced Albert Einstein (1879–1955) as he developed the theory of relativity.

At different times civilization has debated the nature of the world. Is the world flat? Is it round? In practice both views are useful. For most practical matters, flat is a good answer. We are most influenced by our immediate vicinity. When you design a house or lay out your garden, you treat the world as if it were flat.

The world may be globally round, but locally, near us, it is flat. Only when we travel great distances does the curvature of the earth affect our decisions. A similar debate evolved in astrophysics. Isaac Newton "takes" the universe to be flat. In Newton's universe, Euclid describes the nature of lines and triangles. But Albert Einstein takes a larger view of the universe. In Einstein's universe, Riemann's or Minkowski's geometry may offer a better model. In the "local" sense, Newton may satisfy our needs. Flatness is sufficient. But cosmically, will Einstein's universe serve us better?

How do we tell the shape of the universe? Try the following metaphor. When you are outside, sketch a large triangle in the sand. Within the precision of your tools, carefully total the measurements of the angles of your triangle: the sum should be exactly 180°: locally your universe is flat. Now imagine that you could construct a triangle that covered thousands of nautical miles. At that scale, the sum of the angles of the triangle should be sufficiently larger than 180° to convince us that the world is round.

Now, use a cosmic pencil to form a triangle using the diameter of the earth as one side and the moon at the vertex of the opposite angle. Newton applies here. This is the local universe. Expect the angles to sum to exactly 180°. If only we might construct a large enough triangle. . . .

8   CHAPTER TEST

Determine the exact value.

1. $\sec 135°$

2. $\cot 30°$

3. $\sin 150°$

4. $\cos 120°$

5. If $B = 60°$, $C = 90°$ and $b = 20$, solve triangle ABC.

Verify the following identities.

6. $\sin^2 \theta \cot^2 \theta - \sin^2 \equiv 0$

7. $\sec^2 \theta + \sec \theta \tan \theta \equiv \dfrac{1}{1 - \sin \theta}$

Solve triangle ABC as indicated.

8. $b = 10, B = 50°, A = 10$. Determine a.

9. $a = 5, b = 3, c = 6$. Determine B.

10. A sailboat sails due south from a buoy at 12 knots. After 30 minutes, the skipper sites a light house 30° west of north. The charts indicate the distance from the buoy to the lighthouse is 10 nautical miles. How far is the sailboat from the lighthouse?

11. Show that $\cos 120° = 1 - \cos^2 60°$.

12. (a) Convert $2 + 3i$ to polar form.

(b) Convert $7 \text{ cis } 45°$ to rectangular form.

13. Graph in polar coordinates $r = 4 \sin(2\theta)$.

If $z = 9 \text{ cis } 60°$ and $w = 64 \text{ cis } 160°$, determine the following.

14. $z\overline{w}$

15. $\dfrac{w}{z}$

16. $\sqrt[4]{w}$

If $\vec{U} = \langle 1, 2 \rangle$ and $\vec{V} = \langle 1, -2 \rangle$ determine the following.

17. $\|\vec{V}\|$

18. $\vec{U} \cdot \vec{V}$

19. $3\vec{U} + \vec{V}$

20. The angle between $\vec{U}$ and $\vec{V}$.

Analytical Geometry

This chapter develops a classic topic of analysis, the conic sections. Throughout the text we have investigated graphs of various shapes. The most familiar were the line, circle, and parabola of Chapter 1. Since the Greeks, geometry has unified the familiar shapes of circles, ellipses, parabolas and hyperbolas as plane sections of a cone. But pure geometry requires powerful visualization skills. In 1637, Rene Descarte wrote *Discours de la methode pour bien conduire sa raisonet chercher la verite dans les sciences* (*A Discourse of Correct Methods for Reasoning and Seeking Truth in the Sciences*) in which the appendix *La geometrie* introduces the concepts of **analytical geometry.** This discourse lent the power of algebra to the analysis of geometry. Modern analysis was born.

1 Conic Sections

Mathematics is the science which draws necessary conclusions
—Benjamin Peirce

Sometimes a summary offers the opportunity to construct a unifying theme, to gather separate observations into a framework. If a theory results, the generalizations aid our insight and understanding. Conic sections provide a unifying theme for the graphs and equations of Chapter 1.

Consider a right circular cone. See Display 9-1 (page 578). To visualize the generation of a cone, imagine that two lines A and L intersect to form an acute angle θ. Now suppose that A becomes an axis for rotation so that L is spun rapidly about A without altering angle θ. The surface generated by the rotation of L is a cone. A is the axis of the cone. The line L at any particular moment of the rotation is an edge of the cone. From Display 9-1 the cone has two **nappes**, one above the intersection of L and the axis and one below.

Conic sections are figures formed by the intersection of a right circular cone with a plane. Display 9-2 (page 578) illustrates the graphs of the conic sections as they are formed by various intersecting planes.

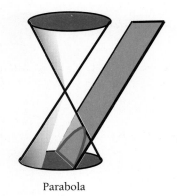

Parabola

DISPLAY 9-2a

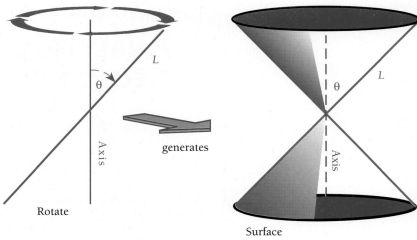

Rotate

generates

Surface

DISPLAY 9-1   Cones with conic sections

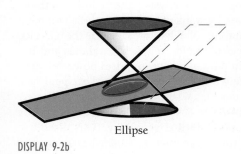

Ellipse

DISPLAY 9-2b

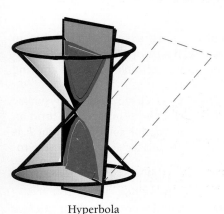

Hyperbola

DISPLAY 9-2c

The cone provides a unifying connection among these various geometric shapes. Is there an algebraic equivalent? Does some algebraic pattern unite the shapes with a corresponding algebraic pattern? The answer, of course, is yes. We would have been disappointed if a geometric pattern did not have an algebraic equivalent. The derivation of these formulas requires the distance formula. See Chapter 1 to review the distance formula.

This section provides an overview of the topics of this chapter. **General quadratic equations** in two variables represent graphs formed by intersecting a plane and a right circular cone. In particular, the general quadratic equation in x and y is given by

$$ax^2 + bxy + cy^2 + dx + ey + f = 0 \text{ (Not all of a, b and c are zero.)}$$

Note that if a, b and c are all 0, the equation is no longer quadratic, and we have the linear equation

$$dx + ey + f = 0$$

Similarly, select $b = 0$ and $c = 0$, then the equation for a parabola appears. Try selecting values to produce an equation representing a circle. In the next sections we shall also see what values produce equations representing ellipses and hyperbolas. The constants a, b, c, d, e and f are the **parameters** that determine the type of equation produced by the general quadratic equation.

Suppose we have enough information to specify a particular conic section, say, a circle. For example, suppose we know the center and the radius of the circle. From Chapter 1, we know one form of the equation of a circle using this information. That form is $(x - h)^2 + (y - k)^2 = r^2$. From the specification for the circle, we determine the parameters h, k and r. In general, with a known format for the equation and with sufficient data specified, we are able to determine the parameters of the format from the specifications.

EXAMPLE 9-1

Illustration 1:

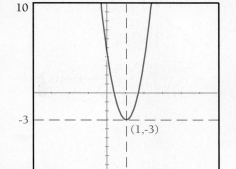

FIGURE 9-1 $y = 7(x - 1)^2 - 3$

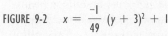

FIGURE 9-2 $x = \dfrac{-1}{49}(y + 3)^2 + 1$

Illustration 2:

Determining Parameters in a Standard Form

Determine an equation of a parabola with vertex at $(1, -3)$ and passing through the point $(0, 4)$. There are two possible solutions. The parabola may open vertically *or* horizontally.

Solution 1: From Chapter 1, recall that the general form for the equation of a parabola is

$$y = ax^2 + bx + c \quad \text{or} \quad y = a(x - h)^2 + k$$

Each requires three parameters: (a, b, c) or (a, h, k). Because the vertex provides two parameters in the second format, we shall use it. Substitute for the vertex $y = a(x - 1)^2 - 3$. (See Figure 9-1.) The only remaining parameter to determine is a. Substitute the information for the second point $(0, 4)$ for (x, y): $4 = a(0 - 1)^2 - 3$. Solve for a.

$$4 = a - 3$$
$$7 = a$$

Thus an equation for the parabola satisfying the specifications is

$$y = 7(x - 1)^2 - 3$$

Solution 2: The parabola may be horizontally oriented. Because of our experiences with converses in Chapter 2, we expect the general form to be given by

$$x = a(y - k)^2 + h$$

Substitute the vertex and obtain

$$x = a(y + 3)^2 + 1$$

Because the point $(0, 4)$ satisfies the equation, substitute and solve for a:

$$0 = a(4 + 3)^2 + 1$$
$$0 = 49a + 1$$
$$a = \frac{-1}{49}$$

The final equation is $x = \frac{-1}{49}(y + 3)^2 + 1$ (Figure 9-2).

Determine an equation for a line parallel to $y = \frac{1}{2}x + 3$.

Solution: The general form of a line, $y = mx + b$, requires two parameters, m and b. Because parallel lines of this form have the same slope, $m = \frac{1}{2}$. Now,

$$y = \frac{1}{2}x + b$$

No available information defines the remaining parameter b. The result suggests an infinite "family" of lines, all parallel to the given line. See Figure 9-3. With more information we can obtain b to produce a unique line.

Consider the family of lines from Illustration 2. Which of these passes through the point (-2, 5)?

Solution: We now have additional information. Substitute (-2, 5) for (x, y) in

$$y = \frac{1}{2}x + b$$

$$5 = \frac{1}{2}(-2) + b \qquad \text{(Solve for b)}$$

$$6 = b$$

$$\text{Thus, } y = \frac{1}{2}x + 6$$

The reflecting properties of conic sections have interesting applications. Figure 9-4 displays the reflection properties of a parabola, an ellipse and an hyperbola.

Recall the definition of the *focus* of a parabola from Chapter 1. Notice that rays of light parallel to the axis of a parabola reflect to the focus of the parabola. Conversely, light originating from the focus of a parabola would reflect in parallel beams. The reflecting property of parabolas is useful in the design of solar reflectors and automobile headlights. See Figure 9-5.

In the automobile headlight the reflector is a segment of a parabola rotated about its axis to generate the reflecting surface: a paraboloid. All light from the bulb striking the reflecting surface exits in parallel beams. This forms the bright spot of your headlight beam. The light missing the reflector forms a lighter halo about the center spot.

Other uses for the reflecting properties of a parabola include parabolic microphones to pick up sounds at a distance, satellite antennas to amplify satellite signals, solar collectors to intensify the heat from the sun and telescopes to collect and magnify light from distant stars. Isaac Newton (1642–1727) suggested using the reflecting properties of a parabola in the design of telescopes. As a result, telescopes of this design are said to have a Newtonian focus (Figure 9-6).

Just as a circle has a center and a parabola has a focus, an ellipse has two foci. These foci are important in the reflection properties of an ellipse: light (or sound, etc.) originating from one focus of an ellipse will reflect off the elliptical boundary toward the other focus. One of the more interesting uses of the reflection properties is the removal of kidney stones by a process known as extracorporeal shockwave lithotripsy.

Lithotripsy is a treatment that crushes kidney stones by use of shockwaves applied from outside the body. To use lithotripsy, technicians lower the patient into a tub of water. The tub contains a cup formed by rotating half an ellipse along its major axis: a halved ellipsoid. At the focus of the ellipsoid, internal to the cup, is an electrode. Doctors use a fluoroscope to position the pa-

Illustration 3:

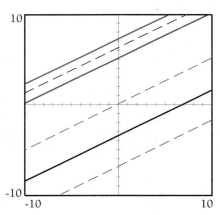

FIGURE 9-3   Family of lines: $y = \frac{1}{2}x + b$

Parabola comes from the Greek *parabole* meaning "juxtaposition" or "parallelism." A parabola is the intersection of a cone and a plane parallel to an "edge" of the conical surface. The plane intersects the axis of the cone at the same angle as a surface edge of the cone (Display 9-2a). Early writers measured the angle with the base of the cone. This angle is the complement of angle formed with the axis. As a result, the one angle increases as the other decreases. Because of this difference the definitions may at first appear to be backwards.

Ellipse comes from the Greek *elleipsis* meaning "falling short." A plane intersecting a cone so that the angle of the plane with the base of the cone falls short of the angle required for the plane to form a parabola forms an ellipse (Display 9-2b). Apolonius and Archimedes created the term *ellipse* about 200 B.C.

Hyperbola comes from the Greek *hyperbole* meaning "throwing beyond." Suppose a plane intersects a cone to form a parabola. Increasing the angle of the plane to the base of the cone produces an intersection that is a hyperbola (Display 9-2c).

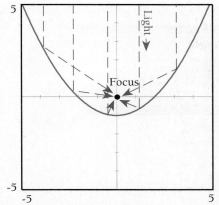
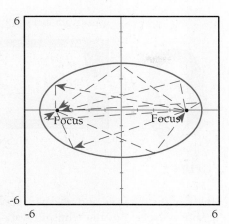
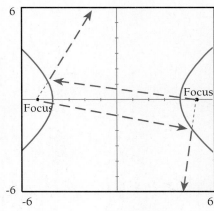

FIGURE 9-4   Reflection properties

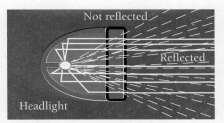

FIGURE 9-5   Automobile headlight

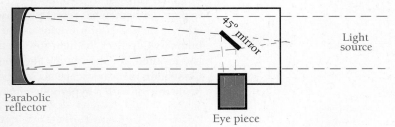

FIGURE 9-6   Newtonian focus

tient in the tub so that a kidney stone is at the focus of the ellipsoid external to the cup. A high energy charge through the electrode vaporizes the water at the focus, generating a powerful shockwave. With the water and the human body as a transmission medium, the cup concentrates the shockwave on the kidney stone at the external focus. Some kidney stones require several thousand shocks over a period of half an hour to shatter stones into sand-size fragments, which the kidney excretes. Despite the resulting soreness and bruising, lithotripsy reduces the recovery time from a month for conventional surgery to just a few days.

A large room designed so that the walls and ceiling outline an ellipsoid creates a "whispering chamber." See Display 9-3. Have a person stand at one

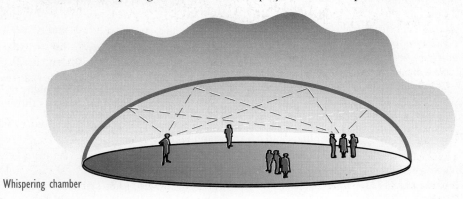

DISPLAY 9-3   Whispering chamber

Cassegrain focus

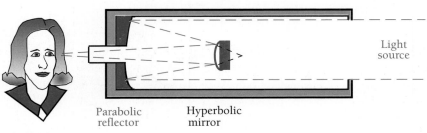

Light
source

Parabolic
reflector

Hyperbolic
mirror

DISPLAY 9-4

focus of the ellipsoid and whisper. Those near that person may not understand a soft whisper. But the ellipsoid walls collect the sound from one focus and concentrates it at the other focus. A person standing at the second focus should hear the amplified whisper.

Like ellipses, hyperbolas also have two foci. The reflecting properties of hyperbolas somewhat resemble those of an ellipse. If we aim light (or sound, etc.) at the focus of a branch of the hyperbola from outside the branch, the branch will reflect the light toward the other focus. This property of hyperbolas resulted in shortening the focal length of reflecting telescopes. The Cassegrain focus telescope accomplishes the shorter focus by replacing the 45° plane mirror with a hyperbolic mirror (Display 9-4).   ▬

9-1   E X E R C I S E S

Determine the indicated parameter(s) for the given equation.

1. $xy = k$, determine k if the graph includes $(1, 3)$.

2. $xy = -k$, determine k if the graph includes $(2, 4)$.

3. $y = mx + b$, determine m and b if the graph includes $(0, 5)$ and $(3, 2)$.

4. $y = mx + b$, determine m and b if the graph includes $(0, -3)$ and $(2, 4)$

5. $ax + by = 15$, determine a and b if the graph includes $(3, 0)$ and $(0, 5)$.

6. $ax + by = 12$, determine a and b if the graph includes $(4, 0)$ and $(0, -3)$.

7. $y = ax^2 + bx + 2$, determine a and b if the graph includes $(1, 0)$ and $(2, 4)$.

8. $y = ax^2 + bx + 1$, determine a and b if the graph includes $(0, 1)$ and $(1, 3)$.

9. $ax^2 + by^2 = 1$, determine a and b if the graph includes $(0, 3)$ and $(5, 0)$.

10. $ax^2 + by^2 = 1$, determine a and b if the graph includes $(-3, 0)$ and $(0, 2)$.

11. $ax^2 - by^2 = 1$, determine a and b if the graph includes $(1, 0)$ and $(0, -2)$.

12. $ay^2 - bx^2 = 1$, determine a and b if the graph includes $(1, 0)$ and $(0, -2)$.

Determine which of the following quadratics represent graphs that pass through the point $(1, 2)$.

13. $x^2 + y^2 = 5$

14. $x^2 - y^2 = -3$

15. $x^2 + 5x - y - 4 = 0$

16. $y^2 + y - 3x = 2$

17. $x^2 - 3xy + y^2 + 4x = 5$

18. $2y^2 - 3y + x^2 - x = 2$

19. $(x - 3)^2 + y^2 = 8$

20. $(x - 1)^2 + (y - 2)^2 = 16$

9-1 PROBLEM SET

PROBLEMS FOR GRAPHING TECHNOLOGY

Many graphs in this chapter are not functions. As such they are not readily adaptable to graphing calculators or programs like *GraphWindows*. There is a method around this limitation. The method uses parametric descriptions of the curves. Recall that a parameter is a descriptive measure. We want to describe curves that are collections of points of the form (x, y). The variable t is often uses a descriptive measure for *both* x and y. For example, consider the parabola given by $y = x^2$. Represent this by $(x, y) = (T, T^2)$. This equation indirectly describes the relationship of x and y: $y = T^2$ and $x = T$ so that $y = x^2$. Similarly, describe $x = y^2$ by $(x, y) = (T^2, T)$. In each of these cases, t is a parameter. The graph of these points do not constitute y as a function of x and cannot be graphed in the usual graphing mode. To graph $x = y^2$ using a CASIO fx-7700G use the following key strokes:

[AC] [MODE] [SHIFT] [x]	(all clear, select parametric)
[RANGE]	(set range)

Be sure to set Tmin = -10, Tmax = 10 on the second screen.

[GRAPH]	(go to graph mode)
Graph $(x, y) = ($	(screen display)
[XθT] [SHIFT] [√] [SHIFT] [→]	(enter formula)
Graph $(x, y) = (T^2,$	(screen display)
[XθT] [)] [EXE]	(finish formula and graph)

For the SHARP calculator, use [SET-UP] to select COORD [E]. Select *XYT* [3] to use parametric mode. Set [RANGE] values to tMin = -10 and tMax = 10. Press [∿] [MENU]. Enter formulas $X1T = T^2$ and $Y1T = T$. Press [∿] to display the graph.

The method for the TI-81 varies slightly. Use [MODE] to change the default to Param. Use [RANGE] to set Tmin = -10 and Tmax = 10. Set the range for x and y to the same values. Press [y=]. The screen displays three pairs of equations: $X_{1t} =$, $Y_{1t} =$, $X_{2t} =$, $Y_{2t} =$, $X_{3t} =$, $Y_{3t} =$. Enter T^2 for X_{1t} and T for Y_{1t}. Then press [GRAPH].

The resulting graph is a parabola. Since the parameter T, rather than x, determines which points are plotted, you may need to zoom or scroll the graph window to display a parametric curve.

Use a graphing calculator in parametric mode to graph each of the following. If a conic section, identify by name.

1. $(x, y) = (T, T^2)$
2. $(x, y) = (T^2, T)$
3. $(x, y) = (-T^2, T)$
4. $(x, y) = (T, -T^2)$
5. $(x, y) = (T^2, 4T)$
6. $(x, y) = (T^2, -4T)$
7. $(x, y) = (5 \text{ COS } T, 5 \text{ SIN } T)$
8. $(x, y) = (4 \text{ SIN } T, 4 \text{ COS } T)$
9. $(x, y) = (5 \text{ COS } T, 3 \text{ SIN } T)$
10. $(x, y) = (3 \text{ COS } T, 5 \text{ SIN } T)$

2 THE PARABOLA

One cannot escape the feeling that these mathematical formulas have an independent existence and an intelligence of their own, that they are wiser than we are, wiser even than their discoverers, that we get more out of them than was originally put into them.
—HEINRICH HERTZ

We first introduced parabolas in Chapter 1. From a geometric standpoint, a parabola is the intersection of the cone with a plane *parallel* to an edge of cone. See Display 9-5 on p. 584. Hence we have the name *parabola*. In fact the plane intersects the axis of the cone at precisely the same angle θ as formed by the axis and an edge of the cone. If the parallel plane intersects the cone at the vertex, the parabola degenerates to a straight line coincident with an edge of the cone.

Parabola

DISPLAY 9-5   Section of cone to produce parabola

This section includes parabolas that do not represent functions. As before, we begin with a definition based on the distance formula, illustrated graphically in Figure 9-7.

DEFINITION 9-1
Parabola

A **parabola** is the set of points in a plane equally distant from a line called the **directrix** and a point not on the line called the **focus**. The midpoint (h, k) of the line segment from the focus perpendicular to the directrix is the **vertex** of the parabola.

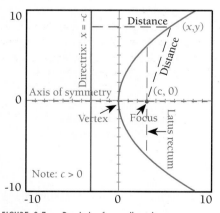

FIGURE 9-7   Parabola, focus, directrix

Consider a focus $P = (c, 0)$ and a directrix $x = -c$. Note that $(0, 0)$ will be the vertex of the parabola. The vertex must be a point of the parabola. Why? (See Figure 9-7). Let (x, y) be a point of the parabola. Calculate the distances from (x, y) to the focus and from (x, y) to the directrix. Set these two distances equal, square both sides and obtain the equation for a parabola:

$$\sqrt{(x - c)^2 + (y - 0)^2} = \sqrt{[x - (-c)]^2 + (y - y)^2}$$
$$x^2 + y^2 - 2cx + c^2 = x^2 + 2cx + c^2 \quad \text{(square both sides)}$$
$$-4cx = -y^2$$
$$4cx = y^2$$

Substitute $a = 1/(4c)$ and obtain $x = ay^2$. Compare to the equation for Figure 9-7. The parabola "changes direction" at the vertex.

Notice that if $a > 0$ and $y \neq 0$ then $ay^2 > 0$ so that $x \geq 0$. Except for the vertex, the graph is "to the right of" the y-axis. A parabola oriented in this direction "opens right." Now suppose that $a < 0$. Then $ay^2 < 0$ when $x \neq 0$. Except for the vertex, the graph is to the left of the y-axis. In this orientation the parabola opens left. We have the following theorem.

Theorem 9-1
Orientation of Parabolas

For parabolas of the form $y = ax^2$ or $x = ay^2$, the parabola opens in a positive direction if a is positive. The parabola opens in a negative direction if a is negative. Moreover the parabola is symmetric to the axis indicated by the first degree term: $y = ax^2$ opens along the y-axis; $x = ay^2$ opens along the x-axis. The vertex of the parabola is at the origin $(0, 0)$.

As in Chapter 1, translate the parabola off the origin by substituting $y - k$ for y and $x - h$ for x. Then we have the following.

Theorem 9-2
Graphing Form of a Parabola

The graph of $y - k = a(x - h)^2$ is a parabola with vertex at (h, k).
The graph of $x - h = a(y - k)^2$ is a parabola with vertex at (h, k).

Figure 9-7 displays additional terminology associated with parabolas. The line through the focus and vertex of the parabola is the **axis of symmetry**. For $y - k = a(x - h)^2$ the axis of symmetry is given by $x = h$. For $x - h = a(y - k)^2$ the axis of symmetry is given by $y = k$.

Examine $y^2 = 4cx$ in Figure 9-7. Construct a line at the focus perpendicular to the axis of symmetry. The line segment formed with endpoints on the parabola is the **latus rectum** (literally, perpendicular width). The coordinates of the endpoints are $(c, 2c)$ and $(c, -2c)$. Thus, the length of the latus rectum is $4|c|$.

Analyzing Graphs of Parabolas

EXAMPLE 9-2

Illustration 1:

Analyze the graph of $y^2 = 12x$.

Solution: The equation is in the form $y^2 = 4cx$. The graph is a parabola that opens to the right. The vertex is at $(0, 0)$. Because $4c = 12$, then $c = 3$, so that the focus is at $(3, 0)$ and the directrix is $x = -3$. The length of the latus rectum is 12. The latus rectum measures the width of the graph as in Figure 9-8.

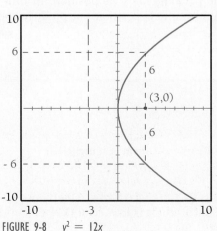

FIGURE 9-8 $y^2 = 12x$

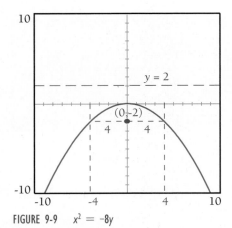

FIGURE 9-9 $x^2 = -8y$

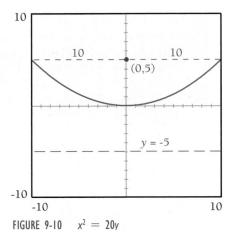

FIGURE 9-10 $x^2 = 20y$

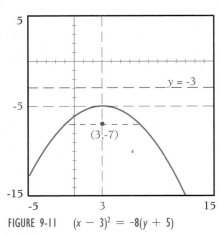

FIGURE 9-11 $(x - 3)^2 = -8(y + 5)$

Illustration 2:   Analyze $x^2 = -8y$

Solution: The parabola opens down: $c = -2$. The focus is at $(0, -2)$ and the directrix is $y = 2$. The length of the latus rectum is 8. The graph of this is in Figure 9-9.

Illustration 3:   Write an equation for a parabola with focus at $(5, 0)$ and vertex at the origin.

Solution: With this orientation of vertex and focus the parabola opens up. See Figure 9-10. The equation is of the form $x^2 = 4cy$. Because $c = 5$, the equation is $x^2 = 20y$.

Illustration 4:   Analyze $(x - 3)^2 = -8(y + 5)$.

Solution: This problem is identical to Illustration 2 with the vertex translated to $(3, -5)$. Expect an identical graph with vertex relocated at $(3, -5)$. See Figure 9-11. Check the length of the latus rectum. It is unchanged. Therefore, the translation did not alter the "width" of the parabola, only its position. In other words, the translation "shifts" the graph to a new position but produces no "stretch" in the shape.   ▬

S U M M A R Y

- The graph of an equation expressible in the form (Figure 9-12)

$$(y - k)^2 = 4c(x - h)$$

is a parabola with vertex located at (h, k). The axis of symmetry of the graph is $y = k$. The focus of the parabola is located at $(h + c,$

k). The directrix of the parabola is $x = h - c$. The length of the latus rectum is $4|c|$. If $c > 0$ then the parabola opens right. If $c < 0$ then the parabola opens left.

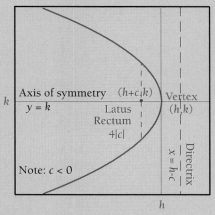

FIGURE 9-12

- The graph of an equation expressible in the form

$$(x - h)^2 = 4c(y - k)$$

is a parabola with vertex located at (h, k). See Figure 9-13. The axis of symmetry of the graph is $x = h$. The focus of the parabola is located at $(h, k + c)$. The directrix of the parabola is $y = k - c$. The length of the latus rectum is $4|c|$. If $c > 0$ then the parabola opens up. If $c < 0$ then the parabola opens down.

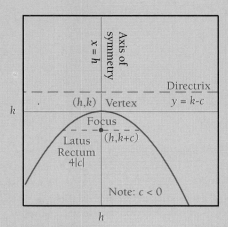

FIGURE 9-13

9-2   E X E R C I S E S

Graph each of the following. For each graph give the coordinates of the vertex. Also, annotate the axis of symmetry, the directrix and the focus.

1. $y^2 = x$ 2. $y^2 = -x$

3. $x^2 = -y$ 4. $x^2 = y$

5. $y^2 = 4x$ 6 $y^2 = -4x$

7. $x^2 = -4y$ 8. $x^2 = 4y$

9. $y^2 = -16x$ 10. $x^2 = 16y$

11. $(y - 1)^2 = (x + 2)$ 12. $(y + 2)^2 = -(x - 1)$

13. $(x - 1)^2 = -(y + 2)$ 14. $(x + 2)^2 = (y - 1)$

15. $(y + 5)^2 = 4x$ 16. $(x - 3)^2 = -4y$

17. $x^2 = -4y + 8$ 18. $y^2 = 4x - 8$

19. $(y - 3)^2 = -16(x + 2)$ 20. $(x - 4)^2 = 16(y - 3)$

Find an equation(s) for the parabola described.

21. Vertex (0, 0), focus (3, 0).

22. Vertex (0, 0), focus (0, 3).

23. Vertex (0, 0), focus (0, -3).

24. Vertex (0, 0), focus (-3, 0).

25. Vertex (0, 0), directrix $y = 2$.

26. Vertex (0, 0), directrix $x = 2$.

27. Vertex (0, 0), directrix $x = -2$.

28. Vertex (0, 0), directrix $y = -2$.

29. Focus (4, 0), directrix $x = -4$.

30. Focus (0, -4), directrix $y = 4$.

31. Latus rectum from (2, -4) to (2, 4), vertex (0, 0).

32. Latus rectum from (-4, -2) to (4, -2), vertex (0, 0).

33. Focus (10, 2), vertex (8, 2).

34. Focus (3, 5), vertex at (3, 3).

35. Focus (1, 3), directrix $x = 3$.

36. Focus (1, 3), directrix $y = 5$.

37. Focus (1, 3), directrix $y = -1$.

38. Focus (1, 3), directrix $x = -1$.

39. Latus rectum from (2, -4), to (2, 4).

40. Latus rectum from (-4, -2) to (4, -2).

41. Give an equation to represent the parabolic cross section of the solar reflector shown in Figure 9-14.

42. Given the same information as in Problem 41 except that the focal length is 100 mm, determine the representative equation.

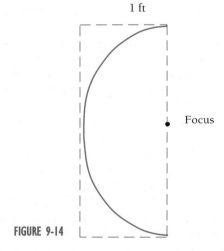

1 ft

Focus

FIGURE 9-14

9-2   P R O B L E M   S E T

1. Suppose a satellite antenna dish has a parabolic cross section and is 3 meters in diameter. Determine the focus if the dish is 0.5 meters deep.

2. A "directional" microphone used on the sidelines to pick up the sound of a football game makes use of a parabolic reflector to concentrate sound waves on the microphone. If the parabolic reflector is 2 feet in diameter and 6 inches deep, determine the location of the focus.

3. A secant line to a parabola is a linear function that intersects the parabola twice. If (r, r^2) and (s, s^2) are points of $y = x^2$, determine the slope of the line through (r, r^2) and (s, s^2).

4. See Problem 3. If $(r, r^2 + c)$ and $(s, s^2 + c)$ are points of $y = x^2 + c$, determine the slope of the line through these two points.

5. See Problem 3. The slope quotient is undefined where $r = s$. Remove this discontinuity (see Chapter 5) by reducing the quotient to lowest terms.

6. See Problems 4 and 5. Remove the discontinuity of the slope quotient in Problem 4 by reducing it to the lowest terms.

7. See Problem 5. A tangent line to $y = x^2$ is a linear function that intersects $y = x^2$ exactly once. One way to accomplish this is to set $r = s$ so that (r, r^2) and (s, s^2) become the same point. Use this method to determine the slope of the tangent line to $y = x^2$ at the point (r, r^2).

8. See Problem 7. Determine the slope of the tangent line to $y = x^2 + c$ at $(r, r^2 + c)$.

9. See Problem 7. Write an equation for the tangent line to $y = x^2$ at (r, r^2).

10. See Problem 8. Write an equation for the tangent line to $y = x^2 + c$ at $(r, r^2 + c)$.

11. See Problems 1–10. Write an equation for the tangent line to $y = x^2$ at $(3, 9)$.

12. See Problems 1–11. Write an equation for the tangent line to $y = x^2 + 2$ at $(3, 11)$.

13. See Problem 11. Write an equation for the tangent line to $y = x^2 + 5$ at $(3, 14)$.

14. See Problem 3. Graph $y = x^2 + c$ for $c = 1$ and $c = 3$. What is the slope of the tangent line at $x = 3$ for each of these parabolas?

15. See Problems 11 and 13. Discuss the relation of the two answers.

16. See Problems 13 and 14. Discuss the role of c in the slope of tangent lines to $y = x^2 + c$.

17. See Problems 1–10. Determine the slope of the tangent line to $y = 3x^2 + c$ at $(r, 3r^2 + c)$.

18. See Problems 1–11. Determine the slope of the tangent line to $y = ax^2$ at (r, ar^2).

19. Write an equation for the tangent line to $y = 3x^2 + 7$ at $x = r$.

20. Write an equation for the tangent line to $y = ax^2$ at $x = r$.

PROBLEMS FOR GRAPHING TECHNOLOGY

Most forms for parabolas separate the variables, x on one side of the equation, y on the other. This makes converting to parametric form easier. Consider $y = a(x - h)^2 + k$ or $y = ax^2 + bx + c$. For either y is a function of x. Graphing technology handles them in function mode or parametrically:

$$(x, y) = ((T, a(T - h)^2 + k) \text{ or } (x, y) = (T, aT^2 + bT + c)$$

However, when y is the quadratic term, y is *not* a function of x. Technology requires a parametric expression of the formula. In this, $x = a(y - k)^2 + h$ becomes $(x, y) = (a(T - k)^2 + h, T)$ whereas $x = ay^2 + by + c$ becomes $(x, y) = (aT^2 + bT + c, T)$. Thus, to graph $y = 2x^2 + 3x - 1$ in parametric mode, enter

$$(x, y) = (T, 2T^2 + 3T - 1)$$

To graph $x = -3y^2 + 5y - 2$ requires parametric mode. Enter

$$(x, y) = (-3T^2 + 5T - 2, T)$$

Graph each of the following using graphing technology. In each case, indicate the direction the parabola opens and whether it is a function or not.

21. $y^2 = x$
22. $y^2 = -x$
23. $x^2 = -y$
24. $x^2 = y$
25. $y^2 = 4x$
26. $y^2 = -4x$
27. $x^2 = -4y$
28. $x^2 = 4y$
29. $y^2 = -16x$
30. $x^2 = 16y$
31. $(y - 1)^2 = (x + 2)$
32. $(y + 2)^2 = -(x - 1)$
33. $(x - 1)^2 = -(y + 2)$
34. $(x + 2)^2 = (y - 1)$
35. $(y + 5)^2 = 4x$
36. $(x - 3)^2 = -4y$
37. $x^2 = -4y + 8$
38. $y^2 = 4x - 8$
39. $(y - 3)^2 = -16(x + 2)$
40. $(x - 4)^2 = 16(y - 3)$

3 ELLIPSES

It is true that a mathematician who is not somewhat of a poet, will never be a perfect mathematician.
—WEIERSTRASS

Recall that θ represents the angle between the axis of the cone and an edge of the cone. Imagine a plane intersecting the axis of the cone at some static angle α, where $\theta < \alpha \leq 90°$. The resulting conic section is an ellipse. See Figure 9-15 on p. 590. If $\alpha = 90°$ then the figure is circle. As with parabolas, if the

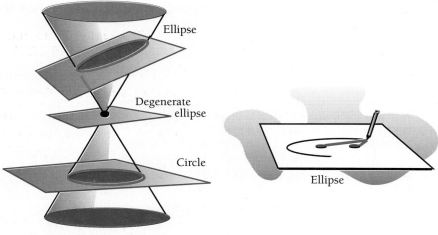

FIGURE 9-15   Ellipse, circle, degenerate ellipse

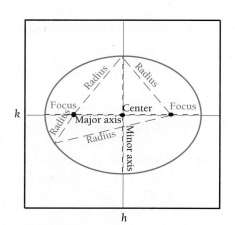

FIGURE 9-16   The ellipse

intersection of the plane and cone occurs at the vertex of the cone then the resulting point is called a **degenerate ellipse**.

One method for drawing an ellipse is to attach a loose string between two thumb tacks inserted into cardboard. If the pencil point keeps the string taut while tracing a curve on the cardboard, the resulting shape is elliptical.

The stretched string keeps constant the sum of the distances from the tack points to the pencil. This idea translates into the following definition, which is demonstrated graphically in Figure 9-16.

DEFINITION 9-2
Ellipse

An **ellipse** is the set of points in the plane whose summed distances from each of two fixed points is constant. Each of the fixed points is a focus (foci is the plural form). The distance from one focus to any point of the ellipse is a radius. The midpoint between the two foci is the center of the ellipse. The line segment with endpoints on the ellipse and passing through the foci is the major axis of the ellipse. The line segment with endpoints on the ellipse and perpendicular to the major axis at the center is the minor axis of the ellipse.

Intuitively, ellipses are symmetric to each of the major axis and minor axis as well as to the center. We derive a formula for an ellipse by choosing the origin as a convenient center for the ellipse. Later we can translate this to a new location. Refer to Figure 9-17.

Suppose the major axis is along the x axis. Let $(c, 0)$ and $(-c, 0)$ represent the foci of the ellipse. The center is at $(0, 0)$. Label the points where the ellipse intersects the x- and y-axes: $(a, 0)$ $(-a, 0)$, $(0, b)$, and $(0, -b)$. As a result of the labeling, $(a, 0)$ is a point of the graph and the distance from $(a, 0)$ to $(c, 0)$ plus the distance from $(a, 0)$ to $(-c, 0)$ is the constant described in the definition. Thus $(a - c) + [a - (-c)] = 2a$, where $2a$ is the constant sum.

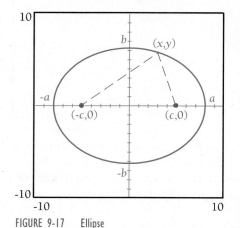

FIGURE 9-17   Ellipse

Let (x, y) be a point of the ellipse. Apply the definition as follows:

distance $[(-c, 0)$ to $(x, y)]$ + distance $[(c, 0)$ to $(x, y)] = 2a$

$$\sqrt{(x + c)^2 + (y - 0)^2} + \sqrt{(x - c)^2 + (y - 0)^2} = 2a$$

$$\sqrt{(x + c)^2 + (y - 0)^2} = 2a - \sqrt{(x - c)^2 + (y - 0)^2}$$

Square both sides to remove the radicals (see Section 5-2):

$$(x + c)^2 + y^2 = 4a^2 - 4a\sqrt{(x - c)^2 + (y - 0)^2} + (x - c)^2 + y^2$$

$$x^2 + 2cx + c^2 + y^2 = 4a^2 - 4a\sqrt{(x - c)^2 + (y - 0)^2} + x^2 - 2cx + c^2 + y^2$$

$$4cx - 4a^2 = -4a\sqrt{(x - c)^2 + (y - 0)^2}$$

$$\frac{a^2 - cx}{a} = \sqrt{(x - c)^2 + (y - 0)^2} \qquad \text{(square both sides)}$$

$$\frac{a^4 - 2cxa^2 + c^2x^2}{a^2} = x^2 - 2cx + c^2 + y^2$$

$$a^4 - 2cxa^2 + c^2x^2 = a^2x^2 - 2cxa^2 + a^2c^2 + a^2y^2$$

$$a^2x^2 - c^2x^2 + a^2y^2 = a^4 - a^2c^2$$

$$(a^2 - c^2)x^2 + a^2y^2 = a^2(a^2 - c^2)$$

The distance from $(-c, 0)$ to $(0, b)$ equals the distance from $(c, 0)$ to $(0, b)$. Because the total of these two distances is known to be $2a$, the distance from $(c, 0)$ to $(0, b)$ is a. But $(0, b)$ to $(c, 0)$ is the hypotenuse for the right triangle from $(b, 0)$ to $(c, 0)$ to the origin. So $b^2 + c^2 = a^2$. Then $a^2 - c^2 = b^2$. Substitute for $a^2 - c^2$ and obtain $b^2x^2 + a^2y^2 = a^2b^2$.

Since neither a nor b can be 0, divide by a^2b^2:

$$\frac{x^2}{a^2} + \frac{y^2}{b^2} = 1$$

Theorem 9-3
Graphing Form of Ellipse

The graph of an equation of the form $x^2/a^2 + y^2/b^2 = 1$ is an ellipse. The center of the ellipse is $(0, 0)$. The intercepts of the ellipse are $(a, 0)$, $(-a, 0)$, $(0, b)$ and $(0, -b)$. (These are the vertices of the ellipse.)

If $a > b$, then the major axis is on the x-axis and the foci are $(c, 0)$, $(-c, 0)$, where $c^2 = a^2 - b^2$.

If $a < b$, then the major axis is on the y-axis and the foci are $(0, c)$, $(0, -c)$, where $c^2 = b^2 - a^2$.

If $a = b$, then the major axis and minor axis are equal in length and the graph is a circle of radius $|a|$.

The vertices on the minor axis of an ellipse are sometimes called **covertices**. We use the term *covertices* only for clarity. As with parabolas, the vertices of an ellipse mark points at which the ellipse changes direction. The change in direction is less dramatic at covertices than at the major axis vertices.

Translating the ellipse off center gives a general formula for an ellipse.

Theorem 9-4
Translated Ellipses

The graph of an equation of the form
$$\frac{(x-h)^2}{a^2} + \frac{(y-k)^2}{b^2} = 1$$
is an ellipse. The center of the ellipse is (h, k). The graph passes through the points $(h + a, k)$, $(h - a, k)$, $(h, k + b)$, $(h, k - b)$. These are the vertices of the ellipse.

If $a > b$, then the major axis is through $(h + a, k)$ and $(h - a, k)$ and the foci are $(h + c, k)$ $(h - c, k)$, where $c^2 = a^2 - b^2$.

If $a < b$, then the major axis is through $(h, k, + b)$ and $(h, k - b)$ and the foci are $(h, k + c)$ $(h, k - c)$ where $c^2 = b^2 - a^2$.

If $a = b$, then the graph is a circle of radius $|a|$ centered at (h, k).

EXAMPLE 9-3

Equations for Ellipses

Illustration 1:

Write an equation for an ellipse centered at the origin with x-intercepts at $(5, 0)$ $(-5, 0)$ and y-intercepts at $(0, 4)$ $(0, -4)$.

Solution: Clearly $a = 5$ and $b = 4$, therefore

$$\frac{x^2}{25} + \frac{y^2}{16} = 1$$

The foci comes from $c = \sqrt{25 - 16}$. Hence the foci are on the x-axis at $(3, 0)$ and $(-3, 0)$.

Illustration 2:

Write an equation for an ellipse centered at the origin with x-intercepts of $(13, 0)$ and $(-13, 0)$ and with foci at $(5, 0)$ and $(-5, 0)$.

Solution: Because $a = 13$ and $c = 5$, then $b = \sqrt{169 - 25}$. Therefore $b = 12$. Then the equation is

$$\frac{x^2}{169} + \frac{y^2}{144} = 1$$

Illustration 3:

Graph $\dfrac{(x - 3)^2}{25} + \dfrac{(y + 2)^2}{9} = 1$

Solution: The graph is an ellipse centered at $(3, -2)$. Use this center to select a graphing window. As the center translates from the origin to $(3, -2)$ so the x-intercepts translate from $(5, 0)$ $(-5, 0)$ to be 5 units to either side of the new center. Similarly, the y-intercepts translate from $(0, 3)$ $(0, -3)$ to be 3 units above the below the new center. Use these vertices to create a box about the new center. This is called a **graphing box**. The box has vertical sides 5 units left and right of $(3, -2)$ and horizontal sides 3 units above and below $(3, -2)$. See Figure 9-18. Now the graphing box serves as a guide for sketching the ellipse in Figure 9-19. Note that the foci lie along a horizontal major axis: $c = \sqrt{25 - 9}$; $c = 4$. Thus the foci are 4 units either side of the center. Because the center is at $(3, -2)$, this places the foci at $(-1, -2)$ and $(7, -2)$.

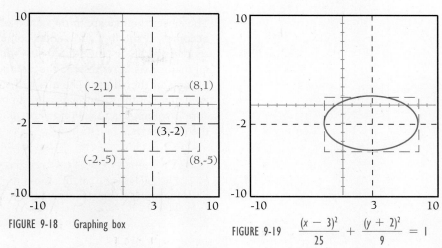

FIGURE 9-18   Graphing box

FIGURE 9-19 $\dfrac{(x-3)^2}{25} + \dfrac{(y+2)^2}{9} = 1$

Illustration 4:   Graph $4(x-2)^2 + 9y^2 - 18y = 27$.

Solution: This equation is not in the correct format. The presence of the $-18y$ term suggests completing the square to combine all y values into a single factor:

$$4(x-2)^2 + 9(y^2 - 2y) = 27$$
$$4(x-2)^2 + 9(y^2 - 2y + 1) - 9 = 27$$
$$4(x-2)^2 + 9(y-1)^2 = 36$$
$$\frac{(x-2)^2}{9} + \frac{(y-1)^2}{4} = 1$$

Identify the graph as an ellipse with center at $(2, 1)$. The graph is illustrated in Figure 9-20. The foci are c units either side of the center where $c = \sqrt{5}$. ▬

A **circle** is a special case of an ellipse. In some sense, a circle is "rounder" than a general ellipse. Also, some ellipses are more elongated than others. One measure of "out of roundness" for an ellipse is **eccentricity**. See Figure 9-21.

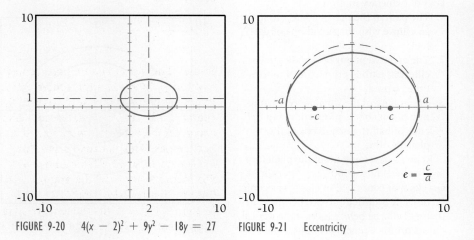

FIGURE 9-20 $4(x-2)^2 + 9y^2 - 18y = 27$

FIGURE 9-21   Eccentricity

The closer the foci are to the center of the ellipse, the more circular the ellipse becomes. If the foci and center coincide, we have a perfect circle. Suppose c represents the distance of a focus from the center of an ellipse. For a horizontal major axis, a represents the distance from the center to one end of the major axis of the ellipse. The a is half the length of the major axis. Because half the length of the minor axis is given by $b = \sqrt{a^2 - c^2}$, b is implied in any measure involving a and c. For a vertical major axis, b represents half the length of the major axis.

DEFINITION 9-3
Eccentricity

The eccentricity of an ellipse is given by $e = c/m$, where c is the distance from the center to a focus and m is half the length of the major axis.

CAUTION! Do not confuse the e used for eccentricity of conic sections with the natural base e used for logarithmic and exponential functions. Usually context indicates which e we mean.

If $c = 0$, then the focus is at the center and we have a circle. Moreover, the focus must lie on the major axis and thus $c < m$ so that $e < 1$. The closer e is to 0 the more circular is the ellipse. The closer e is to 1 the more elongated is the ellipse. For the ellipse in Illustration 1 of Example 9-3, the eccentricity is $\frac{3}{5}$.

The idea of eccentricity is useful in describing the paths of planets about the sun. When Kepler proposed that the paths of planets were ellipses, he challenged several important attributes that philosophers associated with the universe. One such idea is that the path of planets should be circular, because circles were perfect. The other was that the speed of a planet should remain constant. Neither of these ideas held up under Kepler's vision. The eccentricity of a planet's orbit measured its variation from a circle. The earth's eccentricity is about 0.017. The earth's path is close to a circle, varying its distance to the sun between 1 and 2 million miles from an average of 93 million miles. On the other hand, Pluto has an extreme elliptical path. The eccentricity of Pluto is about 0.25. Although Pluto is about 39.5 times further from the earth than the sun, its large eccentricity brought it closer to the sun than Neptune in 1989.

Figure 9-22 shows a family of ellipses all with eccentricity of 0.5. Figure 9-23 shows several ellipses all with the same major axis but with differing eccentricities.

S I D E B A R

Johann Kepler (1571–1630) was a mathematician and astronomer. Kepler believed fervently in a master mathematical plan for the universe. Giving up perfect circles and constant velocity as descriptions of the paths of planets was not easy for him. Fortunately, Kepler was able to replace these descriptions of planetary motion with some laws that he found as aesthetically pleasing. These are Kepler's three laws of planetary motion:

1. The path of the planet about the sun is an ellipse with the sun at one of the foci.

2. The area swept out by a planet in its elliptical path is proportional to the time elapsed.

3. The square of the time of one revolution of a planet is proportional to the cube of half the major axis of the planet's orbit.

Kepler's development of the planetary laws came almost 2000 years after the Greeks first conceived of ellipses as conic sections. It is difficult to predict the time that will ellapse before today's theoretical quests produce practical applications.

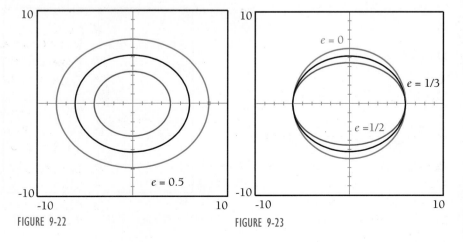

FIGURE 9-22   FIGURE 9-23

S U M M A R Y

• The graph of an equation of the form

$$\frac{(x-h)^2}{a^2} + \frac{(y-k)^2}{b^2} = 1$$

is an ellipse. The ellipse is centered at (h, k). A box to graph the ellipse is bounded by $x = h + a, x = h - a, y = k + b, y = k - b$. If $a^2 > b^2$ then the major axis is horizontal. If $a^2 < b^2$, then the major axis is vertical. If $a^2 = b^2$, then the graph is a circle. The foci of the ellipse are on the major axis c units from the center of the ellipse: $c = \sqrt{|a^2 - b^2|}$.

• The eccentricity of an ellipse is given by $e = c/m$, where m is half the length of the major axis and c is the distance of a focus from the center. For a horizontal major axis, $m = a$, for a vertical major axis, $m = b$. If $e = 0$, then ellipse is a circle. The closer e is to 1, the more elongated is the ellipse.

9-3   E X E R C I S E S

Determine the vertices and foci of the following ellipses. Sketch the graph.

1. $\dfrac{x^2}{4} + \dfrac{y^2}{9} = 1$

2. $\dfrac{x^2}{9} + \dfrac{y^2}{4} = 1$

3. $\dfrac{x^2}{16} + \dfrac{y^2}{25} = 1$

4. $\dfrac{x^2}{25} + \dfrac{y^2}{16} = 1$

5. $4x^2 + 9y^2 = 36$

6. $9x^2 + 4y^2 = 36$

7. $9y^2 + x^2 = 9$

8. $y^2 + 9x^2 = 9$

9. $y^2 + 9x^2 = 1$

10. $9y^2 + x^2 = 1$

11. $\dfrac{(x-2)^2}{4} + \dfrac{y^2}{9} = 1$

12. $\dfrac{x^2}{9} + \dfrac{(y+3)^2}{4} = 1$

13. $\dfrac{(x-2)^2}{16} + \dfrac{(y+4)^2}{25} = 1$

14. $\dfrac{(x-1)^2}{25} + \dfrac{(y+3)^2}{16} = 1$

15. $4(x-3)^2 + 9y^2 = 36$

16. $9x^2 + 4(y+5)^2 = 36$

17. $9y^2 + x^2 + 2x = 8$

18. $y^2 - 4y + 9y^2 = 5$

19. $y^2 - 6y + 9x^2 + 8 = 0$

20. $9y^2 + x^2 + 10x + 24 = 0$

Write an equation to represent the ellipse(s) described by each of the following:

21. Vertices at (5, 0), (-5, 0), (0, 3) and (0, -3).

22. Vertices at (0, 4), (0, -4), (3, 0) and (-3, 0).

23. Foci at (4, 0) and (-4, 0), vertices at (5, 0) and (-5, 0).

24. Foci at (3, 0) and (-3, 0), vertices at (0, 4), (0, -4).

25. Foci at (0, 3) and (0, -3), vertices at (4, 0) and (-4, 0).

26. Foci at (0, 4) and (0, -4), vertices at (0, 5) and (0, -5).

27. Center at (0, 0), major axis length 10, minor axis length 6 on the y-axis.

28. Center at (0, 0), major axis length 10, focus at (-4, 0).

29. Center at (0, 0), minor axis length 6, focus at (4, 0).

30. Center at (0, 0), major axis length 26, minor axis length 24 on the x-axis.

31. Center at (0, 0), major axis on x-axis of length 10, eccentricity $= \frac{3}{5}$.

32. Center at (0, 0), major axis on y-axis of length 10, eccentricity $= \frac{2}{5}$.

33. Center at (2, 3), vertices at (5, 3), (-1, 3), (2, 7) and (2, -1).

34. Center at (3, -2), vertices at (7, -2), (-1, -2), (3, 3) and (3, -7).

35. Like a parabola, each focus of an ellipse has an associated directrix. For

$$\frac{x^2}{a^2} + \frac{y^2}{b^2} = 1$$

the directrix associated with focus $(c, 0)$ is given by $x = a^2/c$, where $c = \sqrt{a^2 - b^2}$. Show that $x = a/e$, where e is the eccentricity of the ellipse.

36. See Problem 35. Determine the directrix for

$$\frac{x^2}{16} + \frac{y^2}{9} = 1$$

37. Suppose a football team builds the "domed" stadium in Figure 9-24 in which each cross section of the dome is a semi-ellipse. If the width of the dome is 300 yards and the highest point of the dome is 60 yards, how high is the ceiling 20 yards from an outside wall?

38. Suppose that 1 astronomical unit is half the length of the major axis of the ellipse that describes the path of the earth about the sun. One astronomical unit is about 93 million miles. The eccentricity of the earth's orbit is approximately 0.017. According to Kepler the sun is at one of the foci of the earth's orbit. How far is the sun from the center of the earth's orbit?

39. Suppose that the path of the moon about the earth is an ellipse in which the length of the semi-major axis (half the major axis) is 240,000 miles. If the eccentricity of the moon's orbit is 0.055 and Kepler's law applies, how far from the center of the moon's orbit is (the center of) the earth?

40. See Problem 38. Write an equation representing the path of earth as an ellipse centered "near" the sun.

41. See Problem 39. Write an equation representing the path of the moon as an ellipse centered "near" the earth.

42. See Figure 9-25. The apogee of a satellite with an elliptical orbit about the earth is the farthest distance from the center of the earth to the satellite. For example the apogee of the moon is 253,000 miles. The perigee of a satellite is the closest approach to the center of the earth. For example, the moon's perigee is 221,000 miles. Assuming that the earth's center is at one focus of the moon's elliptical orbit, how far is the other focus from the center of the earth?

43. See Problem 42. The roots for apogee and perigee are from Greek words *gaia* meaning earth. The prefix *apo* means "away from." The prefix *peri* means "around." Show that the apogee plus perigee equals the major diameter of the moon's orbit.

44. See Problem 42. How far from the center of the earth is the center of the moon's elliptical orbit?

45. See Problem 43. Argue that for an elliptical orbit, the eccentricity of the orbit is given by

eccentricity = (apogee − perigee)/(apogee + perigee)

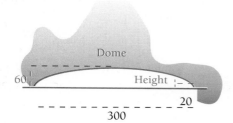

FIGURE 9-24

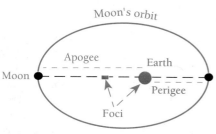

FIGURE 9-25

9-3

PROBLEM SET

The definitions of ellipse and parabola can be combined using the idea of eccentricity. See Exercise 9-3, Problem 35. Let $(c, 0)$ represent a focus of our figure. Let $x = a^2/c$ be the directrix and $e = c/a$ be the eccentricity of the figure. We exclude circles from our discussion, since a circle has eccentricity of $e = 0$, and as a result $c = 0$. But this causes the directrix to become nonexistent. Now consider the following alternative definition of a conic section: A conic section is the set of points in the plane whose distance from the focus equals the eccentricity times the distance to the directrix.

Recall the application of eccentricity to conic sections. If

$e = 1$, the conic section is a parabola. If $0 < e < 1$ then the conic section is an ellipse. Use the eccentricity definition of a conic section to write an equation to represent the figures in Problems 1–4.

1. A parabola with focus $(1, 0)$ and directrix $x = 4$.

2. The figure for focus $(1, 0)$, directrix $x = 4$ and eccentricity $e = 1$.

3. The figure for focus $(1, 0)$, directrix $x = 4$ and eccentricity $e = \frac{1}{2}$.

4. An ellipse with focus $(1, 0)$, directrix $x = 9$ and an eccentricity of $\frac{1}{3}$.

 The next two problems develop the idea of a tangent to an ellipse.

5. Consider the ellipse given by $x^2/a^2 + y^2/b^2 = 1$.

(a) Express y as a function of x by solving for y and selecting the "top half" of the ellipse. Call this function f, where $y = f(x)$.

(b) Show that $(r, (b/a)\sqrt{a^2 - r^2})$, where $0 < r \ |a|$ is a point of f.

(c) Suppose that t is in the domain of f. Evaluate $f(t)$.

(d) A secant line to f is a function that contains two points of f. Calculate the quotient

$$\frac{f(r) - f(t)}{r - t}$$

representing the slope of a secant line through $(r, f(r))$, $(t, f(t))$.

(e) Rationalize the numerator of the difference quotient by multiplying the numerator and denominator by $\sqrt{a^2 - r^2} + \sqrt{a^2 - t^2}$.

(f) The difference quotient is discontinuous where $r = t$. Remove this discontinuity by reducing the quotient to lowest terms.

(g) The slope of the tangent line at $(r, f(r))$ is obtained by allowing $r = t$ in the slope of the secant line. Substitute r for t in the difference quotient and simplify.

(h) Write an equation representing the tangent line to $x^2/9 + y^2/16 = 1$ at $(2, \frac{4}{3}\sqrt{5})$.

6. See Problem 5.

(a) Another way to get the equation of the tangent line to $x^2/a^2 + y^2/b^2 = 1$ through the point (r, s) is to use the formula $rx/a^2 + sy/b^2 = 1$. Use this formula to write an equation for the tangent line to $x^2/9 + y^2/16 = 1$ at $(2, \frac{4}{3}\sqrt{5})$. Compare this to the answer in Problem 5h.

(b) Show that $\dfrac{br}{as} = \dfrac{r}{\sqrt{a^2 - r^2}}$.

(c) Determine the slope of the line $\dfrac{rx}{a^2} + \dfrac{sy}{b^2} = 1$.

(d) Show that the slope of the line in problem 5g equals the slope of the tangent line to $x^2/a^2 + y^2/b^2 = 1$ through the point (r, s) as given in problem 6.

PROBLEMS FOR GRAPHING TECHNOLOGY

Consider how closely $x^2/a^2 + y^2/b^2 = 1$ resembles the Pythagorean relation $\cos^2 T + \sin^2 T = 1$. In fact, let $x^2/a^2 = \cos^2 T$, then $x = a \cos T$. Similarly, $y^2/b^2 = \sin^2 T$ implies $y = b \sin T$. This inspires the parametric representation of the ellipse $x^2/a^2 + y^2/b^2 = 1$:

$$(x, y) = (a \cos T, b \sin T)$$

Use the parametric mode of graphing technology to graph the following. Experiment with $t\text{Min} = 0$ and $t\text{Max} = 7$. Does this produce suitable graphs? Explain why the range of T need not be larger.

7. $\dfrac{x^2}{4} + \dfrac{y^2}{9} = 1$ **8.** $\dfrac{x^2}{9} + \dfrac{y^2}{4} = 1$

9. $\dfrac{x^2}{16} + \dfrac{y^2}{25} = 1$ **10.** $\dfrac{x^2}{25} + \dfrac{y^2}{16} = 1$

11. $4x^2 + 9y^2 = 36$ **12.** $9x^2 + 4y^2 = 36$

13. $9y^2 + x^2 = 9$ **14.** $y^2 + 9x^2 = 9$

15. $y^2 + 9x^2 = 1$ **16.** $9y^2 + x^2 = 1$

17. Consider

$$\frac{(x - h)^2}{a^2} + \frac{(y - k)^2}{b^2} = 1.$$

Show that if $\cos^2 T = (x - h)/a^2$, then $x = h \pm a \cos T$.

18. See Problem 17. Show that if $\sin^2 T = (y - k)^2/b^2$, then $y = k \pm b \sin T$.

19. See Problems 17 and 18. Graph $(x - 2)^2/9 + (y + 5)^2/16 = 1$ by using the parametric form $(X, Y) = (2 + 3 \cos T), -5 + 4 \sin T)$.

20. See Problems 17, 18 and 19. Graph $(x + 1)^2/25 + (y - 3)^2/16 = 1$.

21. Graph $\dfrac{(x - 1)^2}{25} + \dfrac{(y + 3)^2}{16} = 1.$

22. Graph $\dfrac{(x - 2)^2}{9} + \dfrac{(y + 4)^2}{25} = 1.$

4   HYPERBOLAS

Recall that θ measures the angle between the axis of a cone and an edge. Imagine a plane slicing the cone so that the angle α formed by the plane and the cone's axis is between 0° and θ: 0° ≤ α < θ. If the plane is parallel to the axis of the cone, the angle between the plane and the axis is considered to be 0°. See Display 9-6. In the case where the plane passes through the vertex of the cone, the hyperbola degenerates into a pair of intersecting lines.

Note particularly that the plane intersects *both* nappes of the cone. As a result we expect hyperbolas to have two distinct branches to their graphs. See Figure 9-26. As before, a definition of a hyperbola paves the way to the derivation of an equation.

DEFINITION 9-4
Hyperbola

A hyperbola is the set of points in the plane, the difference of whose distances to two fixed points is a positive constant. The two fixed points are the foci of the hyperbola. The midpoint of the segment connecting the foci is the center of the hyperbola.

The major difference between the definitions of hyperbolas and ellipses is that the ellipse adds distances whereas the hyperbola subtracts them. Because of the similarities, the equation for a hyperbola resembles the equation for an ellipse. The following theorem provides specifics.

Theorem 9-5
Graphing Form of a Hyperbola

The graph of an equation expressible in the form

$$\frac{x^2}{a^2} - \frac{y^2}{b^2} = 1$$

is a hyperbola centered at the origin. The x-intercepts of the hyperbola are $(a, 0)$ and $(-a, 0)$. The hyperbola has no y-intercepts. See Figure 9-27.

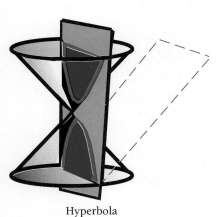

Hyperbola

DISPLAY 9-6

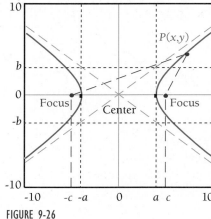

FIGURE 9-26

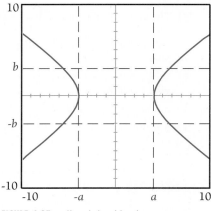

FIGURE 9-27   Hyperbola with x-intercepts

The equation of the hyperbola, like the ellipse, is second degree in both variables. See Figure 9-26. Draw a line through the foci of an hyperbola. The line segment formed with endpoints on the hyperbola is the **transverse axis** of the hyperbola. From Theorem 9-5 the length of the transverse axis is $2a$. Construct a line perpendicular to the transverse axis at the center of the hyperbola. From the center, mark points $|b|$ units in each direction. The resulting line segment is the **conjugate axis**. The length of the conjugate axis is $2b$. The derivation of the formula follows from the definition and is similar to the derivation of the formula for an ellipse.

Theorem 9-6
Graphing Form of a Hyperbola

> The graph of an equation of the form
> $$\frac{y^2}{b^2} - \frac{x^2}{a^2} = 1$$
> is a hyperbola centered at the origin. The y-intercepts of the hyperbola are $(0, b)$ and $(0, -b)$. The hyperbola has no x-intercepts. See Figure 9-28.

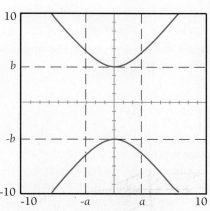

FIGURE 9-28   Hyperbola with y-intercepts

Theorem 9-6, is illustrated in Figure 9-28. The transverse axis joins $(0, b)$ and $(0, -b)$ and is therefore $2b$ units long. Constructing a conjugate axis at the center produces a line segment from $(a, 0)$ to $(-a, 0)$ that is $2a$ units long.

The derivations of Theorems 9-5 and 9-6 are left as exercises. These theorems provide the standard form for a hyperbola. As expected the standard form for a hyperbola resembles the standard form of an ellipse. The distinguishing characteristic is that the terms of an ellipse in standard form are added, but the terms of a hyperbola in standard form are subtracted.

Hyperbolas, like other conic sections, have vertices where the graph changes direction. Test a standard form for a hyperbola for symmetry. In each case, the hyperbola would be symmetric to the x-axis, y-axis and origin. With the exception of intercepts, each point of a hyperbola has a mirror image in the other quadrants. This duplication suggests that the graph of a hyperbola changes direction at the intercepts. Therefore the intercepts are the vertices of the hyperbola. The vertices are the endpoints of the transverse axis. A hyperbola does not pass through the endpoints of the conjugate axis.

EXAMPLE 9-4

Illustration 1:

Graphing Hyperbolas. Sketch the graph of each of the following.

$$\frac{x^2}{16} - \frac{y^2}{9} = 1$$

Solution: The x-intercepts are $(4, 0)$ and $(-4, 0)$. There are no y-intercepts. Compare this equation to the ellipse given by

$$\frac{x^2}{16} + \frac{y^2}{9} = 1$$

The x-intercepts are $(4, 0)$ and $(-4, 0)$ just like the hyperbola. See Figure 9-29. Unlike the hyperbola, the y-intercepts are $(0, 3)$ and $(0, -3)$. The intercepts of this related ellipse bounds both the transverse axis and the conjugate axis.

Construct dashed lines through the opposite vertices of the box that bounds the transverse and conjugate axis. These lines are **asymptotes** for the hyperbola. Asymptotes serve as graphing guides because the hyperbola "gets closer" to the asymptotes as the points get further from the center. To argue the feasibility of this claim, solve the preceding equation for y to obtain

$$y = \pm \sqrt{\frac{9}{16}(x^2 - 16)}$$

$$= \pm \frac{3}{4}\sqrt{x^2 - 16}$$

$$= \pm \frac{3}{4}\sqrt{x^2\left(1 - \frac{16}{x^2}\right)}$$

$$= \pm \frac{3}{4}x\sqrt{1 - \frac{16}{x^2}}$$

As in Chapter 5, as x gets larger, $16/x^2$ approaches 0 so that the equation approaches the skew asymptotes $y = \frac{3}{4}x$ and $y = -\frac{3}{4}x$. See Figure 9-30.

These asymptotes for the hyperbola are the lines through the opposite vertices of the graphing box associated with the hyperbola.

Illustration 2:

$$\frac{y^2}{9} - \frac{x^2}{16} = 1$$

Solution: As in Illustration 1, we obtain the box $x = 4$, $x = -4$, $y = 3$ and $y = -3$. Extend diagonals through the box for the asymptotes. In this case, the y-intercepts are $(0, 3)$ and $(0, -3)$. There are no x-intercepts: the hyperbola is vertically oriented. The conjugate axis of this hyperbola is the transverse axis of the hyperbola of Illustration 1 and vice versa. We say that two such hyperbolas are *conjugate hyperbolas.*

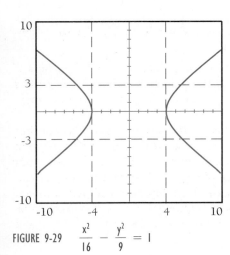

FIGURE 9-29 $\dfrac{x^2}{16} - \dfrac{y^2}{9} = 1$

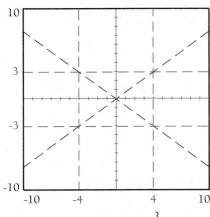

FIGURE 9-30   Asymptotes: $y = \pm\dfrac{3}{4}x$

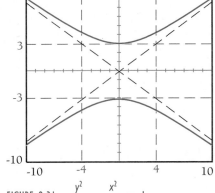

FIGURE 9-31 $\dfrac{y^2}{9} - \dfrac{x^2}{16} = 1$

$$10x^2 + 20 = 5y^2$$

Solution: Rewrite the equation in an appropriate format. Until the equation is rewritten we are not certain whether to expect an ellipse, a hyperbola or possibly some other figure.

$$10x^2 + 20 = 5y^2$$
$$10x^2 - 5y^2 = -20$$
$$\frac{y^2}{4} - \frac{x^2}{2} = 1$$

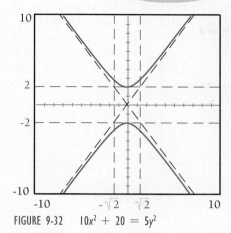

FIGURE 9-32 $10x^2 + 20 = 5y^2$

Identify the graph as a hyperbola. Construct the box formed by $x = \sqrt{2}$, $x = -\sqrt{2}$, $y = 2$ and $y = -2$ (Figure 9-32). Extend the box diagonals to use as asymptotes. The y-axis contains the transverse axis. The y-intercepts are $(0, 2)$ and $(0, -2)$. The x-axis contains the conjugate axis. We sketch the graph using the y-intercepts and the asymptotes as guides.

$$\frac{x^2}{16} - \frac{y^2}{9} = 0$$

Solution: Compare this equation to Illustration 1. The x-intercepts and y-intercepts are each 0. In effect, the graph is no longer an hyperbola. Factor the left-hand side of the equation, then solve for y:

$$\left(\frac{x}{4} + \frac{y}{3}\right)\left(\frac{x}{4} - \frac{y}{3}\right) = 0$$

$$\frac{x}{4} + \frac{y}{3} = 0 \quad \text{or} \quad \frac{x}{4} - \frac{y}{3} = 0$$

$$y = -\frac{3}{4}x \quad \text{or} \quad y = \frac{3}{4}x$$

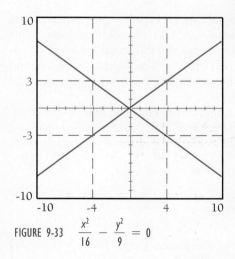

FIGURE 9-33 $\dfrac{x^2}{16} - \dfrac{y^2}{9} = 0$

Thus the graph of $x^2/16 - y^2/9 = 0$ consists of two lines through the origin, with slopes of $\frac{3}{4}$ and $-\frac{3}{4}$, as in Figure 9-33.

These two lines are the asymptotes for the hyperbola in Illustration 1. Because of the resemblance of $x^2/16 - y^2/9 = 0$ to Illustration 1, the graph is called a **degenerate hyperbola.** We have a visualization of asymptotes: imagine them from forcing the hyperbola to degenerate.

As with ellipses, translate axes to obtain windows for hyperbolas not centered at the origin. Determining values that make the squared terms 0 help center the window for the graph. Acceptable windows for hyperbolas are not as obvious as those for ellipses. The graphing box bounded by the vertices where the ellipse changes direction is an ideal window for an ellipse; however, the same graphing box produces only asymptotes for a similar hyperbola. Compare the ellipse $x^2/4 + y^2/9 = 1$ with the hyperbolas $x^2/4 - y^2/9 = 1$ and $y^2/9 - x^2/4 = 1$ shown in Figure 9-34.

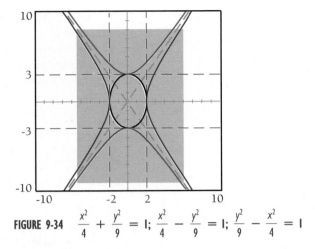

FIGURE 9-34 $\dfrac{x^2}{4} + \dfrac{y^2}{9} = 1; \ \dfrac{x^2}{4} - \dfrac{y^2}{9} = 1; \ \dfrac{y^2}{9} - \dfrac{x^2}{4} = 1$

Obviously, a window for a hyperbola must be larger than for a similar ellipse. Choosing points on the asymptotes several units out from the graphing box marks a suitable window. The shaded box in Figure 9-34 mark one such possible window.

Illustration 5:

$$\frac{x^2}{16} - \frac{y^2}{36} = 1$$

Solution: This hyperbola has the same x-intercepts as the hyperbola of Illustration 1; however, this hyperbola is wider. Like ellipses, hyperbolas come in wide and narrow sizes. As with ellipses, measure this variation as the eccentricity of a hyperbola. Let a represent half the length of the transverse axis and c the distance from the center of the hyperbola to a focus. The eccentricity of the hyperbola is the ratio of c to a: $e = c/a$.

Recall that b represents half the length of the converse axis. In the problem set, problem 7, you show that $c^2 = a^2 + b^2$. For this illustration, because $a^2 = 16$ and $b^2 = 36$, then $c^2 = 52$. The eccentricity of this hyperbola is given by

$$e = \frac{\sqrt{52}}{4}$$

$$e \cong 1.802775638$$

The eccentricity of the hyperbola of Illustration 1 is $\frac{5}{3}$. The eccentricity of the hyperbola of Illustration 3 is $\frac{\sqrt{6}}{2}$. The hyperbola of Illustration 3 is narrower than the hyperbola of Illustration 1.   ■

Theorem 9-7
Hyperbolas Not Centered at the Origin

The graph of
$$\frac{(x - h)^2}{a^2} - \frac{(y - k)^2}{b^2} = \pm 1$$
is a hyperbola centered at (h, k).

The results of the theorem follow from translating axes and combining the formats of Theorems 9-5 and 9-6.

Sketch the graph of the following.

EXAMPLE 9-5

Illustration 1:

$$\frac{(x - 2)^2}{16} - \frac{(y + 4)^2}{9} = 1$$

Solution: Locate the center of the hyperbola at (2, ⁻4). Construct a new window using auxiliary axes, x' and y', through (2, ⁻4). The conjugate axis lies in $x = 2$. The transverse axis lies in $y = ⁻4$. On the auxiliary axes outline a box using vertical lines 4 units on either side of (2, ⁻4) and horizontal lines 3 units above and below (2, ⁻4). Extend diagonals through the box to construct asymptotes for the hyperbola. Because the x term of the equation is positive and the y term is negative, the graph is horizontally oriented: the graph intersects x' but not y'. Use the intersection of the box with the x'-axis and the asymptotes to sketch the graph in Figure 9-35.

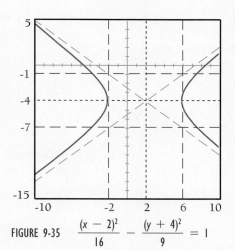

FIGURE 9-35 $\dfrac{(x - 2)^2}{16} - \dfrac{(y + 4)^2}{9} = 1$

$$25x^2 - 50x - 9y^2 + 36y + 214 = 0$$

Solution: Rewrite in graphing form by completing the square in x and y.

$$25x^2 - 50x - 9y^2 + 36y = ⁻214$$
$$25(x^2 - 2x) - 9(y^2 - 4y) = ⁻214$$
$$25(x^2 - 2x + 1 - 1) - 9(y^2 - 4y + 4 - 4) = ⁻214$$
$$25[(x - 1)^2 - 1] - 9[(y - 2)^2 - 4] = ⁻214$$
$$25(x - 1)^2 - 25 - 9(y - 2)^2 + 36 = ⁻214$$
$$25(x - 1)^2 - 9(y - 2)^2 = ⁻225$$
$$\frac{(y - 2)^2}{25} - \frac{(x - 1)^2}{9} = 1$$

The results are shown in Figure 9-36. In the next section, we establish that an equation of the form $xy = k$, $(k \neq 0)$ also represents a hyperbola. Solve for y to obtain

$$y = \frac{k}{x}, \qquad (x \neq 0)$$

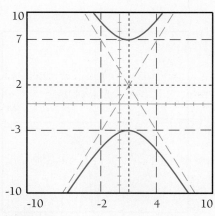

FIGURE 9-36
$25x^2 - 50x - 9y^2 + 36y + 214 = 0$

This is a rational function like those of Chapter 5, Section 1. The graph is familiar. Equations of this form express inverse variation. The constant k is the constant of variation. Express this concept with the phrase "y varies inversely with (as) x." For example, the pressure exerted by a fixed amount of gas varies inversely with the volume of the gas: the higher is the volume, the lower is the pressure.

....

EXAMPLE 9-6

Inverse Variation. Graph each of the following.

Illustration 1:

$$xy = 1$$

Note that rewriting this equation in two different forms

$$y = \frac{1}{x}$$

$$x = \frac{1}{y}$$

indicates that neither $x = 0$ nor $y = 0$ satisfies this equation. In fact, the lines $x = 0$ and $y = 0$ are asymptotes for the graph (see Chapter 5). Sketch the graph in Figure 9-37.

Illustration 2:

$$xy = -4$$

Solution: Here, $y = -4/x$. The asymptotes and graph (Figure 9-38) follow from Chapter 5.

Illustration 3:

The volume of a gas y (in liters at a fixed temperature) is inversely proportional to the pressure x (grams/square centimeter) exerted on the gas. If the gas occupies 2 liters at 10 g/cm², (a) determine the constant of proportional-

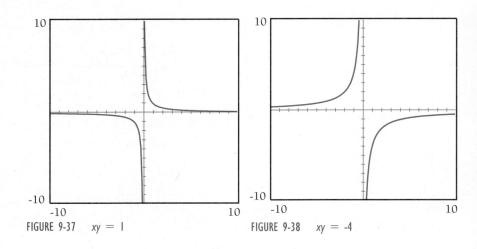

FIGURE 9-37 $xy = 1$ FIGURE 9-38 $xy = -4$

ity, (b) write a formula relating x and y and (c) determine the volume y of gas at 100 g/cm².

Solution: The phrase "y varies inversely to x" implies there is a k such that

$$y = \frac{k}{x} \quad \text{or} \quad xy = k$$

Substitute 10 for x and 2 for y:

$$(2\ l)(10\ \text{g/cm}^2) = k$$

(a) $20\ lg/cm^2 = k$

(b) Thus, $y = \dfrac{20\ lg}{cm^2\ x}$

(c) Now suppose that

$$x = 100\ \text{g/cm}^2$$

$$y = \frac{20\ lg}{cm^2\ 100\ g/cm^2}$$

$$= 0.2\ l\ \text{(liters)}$$

SUMMARY

- To sketch the graph of equations in the form

$$\frac{(x-h)^2}{a^2} - \frac{(y-k)^2}{b^2} = \pm 1$$

do the following:

1. Locate the center of the hyperbola at (h, k). Identify the transverse and conjugate axes from $x = h$ and $y = k$ (see step 5).
2. Construct auxiliary axes, x' and y', with their origin at (h, k).
3. Sketch a box as a graph aid about (h, k): the vertical sides should be $|a|$ units to either side of (h, k); the horizontal sides should be $|b|$ units above and below (h, k).
4. Extend the diagonals of the box to form the hyperbola's asymptotes:

$$y = \pm \frac{b}{a}(x - h) + k$$

5. (a) If

$$\frac{(x-h)^2}{a^2} - \frac{(y-k)^2}{b^2} = 1,$$

then the hyperbola has two horizontal branches, as in Figure 9-39. The transverse axis is in $y = k$. The foci are at $(h + c, k)$, $(h - c, k)$, where $c^2 = a^2 + b^2$. The eccentricity is given by $e = c/a$.

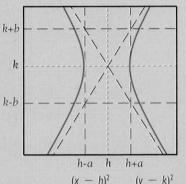

FIGURE 9-39 $\dfrac{(x - h)^2}{a^2} - \dfrac{(y - k)^2}{b^2} = 1$

(b) If
$$\frac{(y - k)^2}{b^2} - \frac{(x - h)^2}{a^2} = 1,$$

then the hyperbola has two vertical branches, as in Figure 9-40. The transverse axis is in $x = h$. The foci are at $(h, k + c)$, $(h, k - c)$ where $c^2 = a^2 + b^2$. The eccentricity is given by $e = c/b$.

(c) Equations expressible in the form $xy = k$ $(k \neq 0)$ have hyperbolas as graphs with the x-axis and y-axis as aymptotes; $(k, 1)$ and $(1, k)$ are points of the graph.

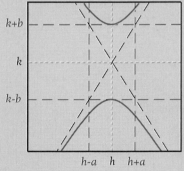

FIGURE 9-40 $\dfrac{(y - k)^2}{b^2} - \dfrac{(x - h)^2}{a^2} = 1$

• Equations of the form $x^2/a^2 - y^2/b^2 = 0$ or $xy = 0$ represent degenerate hyperbolas. The degenerate hyperbola's graph consists of two lines. To determine these lines, factor the lefthand side of these equations, then set each factor equal to 0. These degenerate hyperbolas correspond to the asymptotes of the hyperbolas formed by replacing 0 with 1 in the equation.

Graph the following. For Problems 1-20 annotate the graph with asymptotes, transverse and conjugate axes and foci. Indicate the eccentricity for each hyperbola in Problems 1–10.

1. $\dfrac{x^2}{4} - \dfrac{y^2}{9} = 1$

2. $\dfrac{y^2}{4} - \dfrac{x^2}{9} = 1$

3. $16y^2 - 25x^2 = 400$

4. $16x^2 - 25y^2 = 400$

5. $x^2 - y^2 = 100$

6. $x^2 - y^2 = 81$

7. $(x - 3)^2 - (y + 2)^2 = 25$

8. $(x + 1)^2 - (y - 2)^2 = -16$

9. $\dfrac{(x + 1)^2}{9} - \dfrac{(y - 3)^2}{16} = -1$

10. $\dfrac{(x - 3)^2}{25} - \dfrac{(y + 1)^2}{9} = 1$

11. $\dfrac{(x - 3)^2}{4} - \dfrac{(y + 0.5)^2}{16} = 1$

12. $\dfrac{(x - 1.3)^2}{81} - \dfrac{(y + 1)^2}{9} = 1$

13. $16(y + 2)^2 - 9(x + 1)^2 = 144$

14. $4(x - 1)^2 - 25(y - 2)^2 = 100$

15. $x^2 + 6x - 4y^2 = 27$

16. $y^2 + 6y - 4x^2 = 27$

17. $y^2 + 4y - 9x^2 - 18x = 14$

18. $x^2 + 8x - 4y^2 - 8y = 4$

19. $3x^2 - 5y^2 = -15$

20. $7y^2 - 2x^2 = -14$

21. $xy = 4$

22. $xy = 9$

23. $xy = -9$

24. $-2xy = 0$

25. $y = \dfrac{3}{x}$

26. $y = -\dfrac{2}{x}$

27. $3xy = 0$

28. $-3xy = -12$

*29. $(x - 1)(y + 2) = 1$

*30. $(x + 1)(y - 2) = -1$

31. $x^2 - y^2 = 0$

32. $\dfrac{x}{4} - \dfrac{y}{9} = 1$

33. $x^2 - y^2 = 9$

34. $\dfrac{x^2}{4} - \dfrac{y^2}{9} = 0$

35. $x - y^2 = 9$

36. $\dfrac{x^2}{4} - \dfrac{y}{9} = 1$

37. $x^2 - y^2 = -9$

38. $\dfrac{x^2}{4} + \dfrac{y^2}{9} = 1$

39. $x - y = 9$

40. $\dfrac{x^2}{4} - \dfrac{y^2}{9} = -1$

Write an equation for the hyperbola with its center at the origin that satisfies the given conditions.

41. Foci at (0, -5) (0, 5) and vertices at (0, -3), (0, 3).

42. Foci at (0, -5) (0, 5) and vertices at (0, -4), (0, 4).

43. Foci at (-3, 0) (3, 0) and vertices at (-1, 0), (1, 0).

44. Foci at (-4, 0) (4, 0) and vertices at (-1, 0), (1, 0).

45. Transverse axis from (-2, 0) to (2, 0), conjugate axis from (0, -3) to (0, 3).

46. Transverse axis from (0, -4) to (0, 4), conjugate axis from (-2, 0) to (2, 0).

47. Vertices at (-1, 0), (1, 0) and passing through $(2, 3\sqrt{3})$.

48. Vertices of (-2, 0) and (2, 0), asymptotes of $y = \pm\frac{2}{3}x$.

49. Vertices of (-3, 0) and (3, 0), asymptotes of $y = \pm\frac{4}{3}x$.

50. The x-intercepts of ± 4, asymptotes of $y = \pm\frac{4}{5}x$.

51. The frequency y (in Hertz) at which a string vibrates is inversely proportional to the length x (in cm) of the string. If the string vibrates at 200 Hz when it is 10 cm long, determine the length which will cause it to vibrate at 750 Hz.

52. The amperage flowing in a circuit is inversely proportional to the resistance in the circuit. Let amps be y and resistance be x. If $y = 15$ amps when $x = 75$ ohms, determine y when x is 50 ohms.

53. The f-stop number (y) on a camera lens is inversely proportional to the diameter x (in mm) of the lens aperture. If the f-stop of a (50 mm focal length) lens is 2 when the diameter aperture is 25 mm, determine the diameter when the f-stop is 1.4.

54. The time of rotation of a planet (y in hours) in inversely proportional to the surface velocity x (in km/hr) of the planet. If $y = 1$ hour when $x = 1750$ km/hr, determine x when $y = 12$ hours.

55. Discuss the nature of the graph of $(x - 3)y = 1$.

56. Discuss the nature of the graph of $x(y + 5) = 1$.

57. Discuss the nature of the graph of $y = 1/(x - 3)$.

58. Discuss the nature of the graph of $y = (1/x) - 5$.

59. Discuss the nature of the graph of $(x - h)(y - k) = 1$.

60. Discuss the nature of the graph of $y = \dfrac{1}{(x - h)} + k$.

61. Note that $(x + 4)/(x + 3) = [1/(x + 3)] + 1$. Sketch the graph of $y = (x + 4)/(x + 3)$.

62. See Problem 61. Sketch the graph of $y = (x - 5)/(x - 4)$.

63. Prove Theorem 9-5.

64. Prove Theorem 9-6. Hint: see Problem 63.

65. Prove Theorem 9-7. Hint: see Problem 63.

66. By factoring, show that $x^2/a^2 - y^2/b^2 = 0$ implies two

linear equations (which are the asymptotes for the hyperbola $x^2/a^2 - y^2/b^2 = 1$).

67. By factoring, show that $y^2/b^2 - x^2/a^2 = 0$ implies two linear equations (which are the asymptotes for the hyperbola $y^2/b^2 - x^2/a^2 = 1$).

9-4   PROBLEM SET

1. Refer to Figure 9-26.

(a) Show that $2a$ is the constant in the definition of hyperbola.

(b) Use the definition of hyperbola and the distance formula to show that $x^2 + c^2 + y^2 = a^2 + 2ax + c^2x^2/a^2$.

(c) Show that $\dfrac{x^2}{a^2} - \dfrac{y^2}{c^2 - a^2} = 1$.

2. See Figure 9-26. Examine the triangle formed from vertices at $P(x, y)$, $C_1(-c, 0)$ and $C_2(c, 0)$. Notice that $PC_1 < PC_2 + C_1C_2$.

(a) Show that $a < c$.

(b) Show that $c^2 - a^2 > 0$.

(c) Since $c^2 - a^2 > 0$ then replace $c^2 - a^2$ with b^2 to complete the derivation of $x^2/a^2 - y^2/b^2 = 1$.

3. See Problems 1 and 2. Show that $c^2 = a^2 + b^2$.

4. See Problem 3. Determine the coordinates of the foci and eccentricity of $x^2/144 - y^2/25 = 1$.

5. (a) Suppose two listening stations are positioned at precise locations to measure the difference in time it takes a signal from a third location to arrive at the stations. Draw a diagram to represent the stations and the signal source. Position the listening stations on the x-axis at equal distances from the origin.

(b) Use an equation to describe all possible locations for the source of the signal. Suggest a method for pinpointing the signal.

6. Refer to Problem Set 9-3. Define a secant line for the hyperbola $x^2/a^2 - y^2/b^2 = 1$.

7. See Problem 6. Define a tangent line for the same hyperbola.

PROBLEMS FOR GRAPHING TECHNOLOGY

Recall the Pythagorean relation $1 + \tan^2(T) = \sec^2(T)$. When written in the form $\sec^2(T) - \tan^2(T) = 1$ there is a strong resemblance to the equation for a hyperbola: $x^2/a^2 - y^2/b^2 = 1$. This inspires the following parametric representation of a hyperbola:

$(X, Y) = (a\ \text{SEC}\ T, b\ \text{TAN}\ T)$

To graph $x^2/9 - y^2/16 = 1$ using a graphing technology use the parametric representation $(X, Y) = (3 \div (\cos T), 4\ \text{TAN}\ T)$. Set $t\text{Min} = 0$ and $t\text{Max} = 7$. Adjust the window according to center of the hyperbola and the size of the transverse and conjugate axes.

Use graphing technology in parametric mode to graph following.

8. $\dfrac{x^2}{4} - \dfrac{y^2}{9} = 1$ **9.** $\dfrac{y^2}{4} - \dfrac{x^2}{9} = 1$

10. $16y^2 - 25x^2 = 400$ **11.** $16x^2 - 25y^2 = 400$

12. $x^2 - y^2 = 100$ **13.** $x^2 - y^2 = 81$

14. Show that if $\dfrac{(x + 3)^2}{16} = \sec^2(T)$ then $x = -3 \pm 4 \sec T$.

15. Show that if $\dfrac{(y - 4)^2}{9} = \tan^2(T)$ then $y = 4 \pm 3 \tan(T)$.

16. Use the results of Problems 14 and 15 to graph
$$\dfrac{(x + 3)^2}{16} - \dfrac{(y + 4)^2}{9} = 1.$$

17. See Problem 16. Graph $\dfrac{(y + 3)^2}{9} - \dfrac{(x - 2)^2}{25} = 1$.

18. Graph $\dfrac{(x - 3)^2}{4} - \dfrac{(y + 0.5)^2}{16} = 1$.

19. Graph $\dfrac{(x - 1.3)^2}{81} - \dfrac{(y + 1)^2}{9} = 1$.

20. Graph $16(y + 2)^2 - 9(x + 1) = 144$.

21. Graph $4(x - 1)^2 - 25(y - 2)^2 = 100$.

22. Calculate the eccentricity and graph $x^2/4 - y^2/k^2 = 1$, $k = 1$, for $k = 1, 2, 3$ and 4. Discuss the relation of k, the eccentricity and the graph.

23. Calculate the eccentricity and graph $y^2/9 - x^2/k^2 = 1$, for $k = 1, 2, 3$ and 4. Discuss the relation of k, the eccentricity and the graph.

5 ROTATION OF GRAPHS AND GENERAL CONIC SECTIONS

Everything that the greatest minds of all times have accomplished toward the comprehension of forms by means of concepts it gathered into one great science, mathematics.
—J. F. HERBERT

Let us recap transformations. Translations provide methods for relocating a graph. More general transformations based on composition with a linear function can also stretch or shrink a graph. Logarithmic transformations reduce the complexity of arithmetic operations by one level of difficulty under the order of operations (GEMA). Do the circular functions of Chapters 7 and 8 provide any useful types of transformations?

Consider two hyperbolas with rules of pairing as follow:

$$xy = 1 \quad \text{and} \quad x^2 - y^2 = 2$$

The graphs of these hyperbolas are shown in Figure 9-41. Although the graphs have the same shape, no transformation we have studied converts one graph into the other: linear motion is not sufficient. Use the origin as a center and rotate the points of $x^2 - y^2 = 2$ by $\pi/4$ radians. This form of transformation is **rotation** about the origin.

To determine the algebraic equivalent of rotation, consider the rotation of a coordinate system through angle θ as shown in Figure 9-42. The point P has

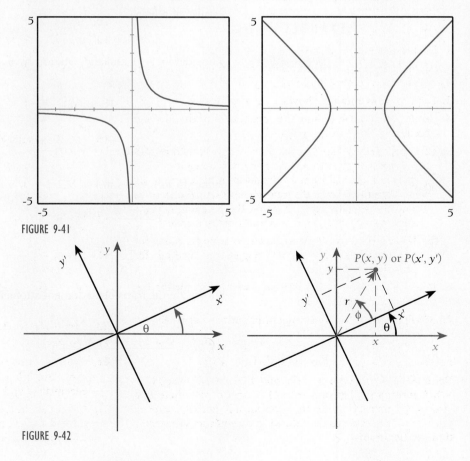

FIGURE 9-41

FIGURE 9-42

coordinates (x, y) in the original system and coordinates (x', y') in the rotated coordinate system.

Let r be the distance of P from the origin and ϕ the angle formed by the x'-axis and vector to P, $\langle x, y \rangle$. Then, in the $x'y'$ system $x' = r \cos \phi$ and $y' = r \sin \phi$. In the xy system P is given by $(x, y) = (r \cos(\phi + \theta), r \sin(\phi + \theta))$

As a result,

$$x = r \cos(\phi + \theta)$$
$$x = r (\cos \phi \cos \theta - \sin \phi \sin \theta) \qquad \text{(sum the formula)}$$
$$x = r \cos \phi \cos \theta - r \sin \phi \sin \theta \qquad \text{(distribute)}$$
$$x = x' \cos \theta - y' \sin \theta \quad \text{(substitute } x' \text{ for } r \cos \phi \text{ and } y' \text{ for } r \sin \phi)$$

Similarly,

$$y = r \sin(\phi + \theta)$$
$$y = r(\cos \phi \sin \theta + \sin \phi \cos \theta) \qquad \text{(sum formula)}$$
$$y = r \cos \phi \sin \theta + r \sin \phi \cos \theta$$
$$y = x' \sin \theta + y' \cos \theta \quad \text{(substitute } x' \text{ for } r \cos \phi \text{ and } y' \text{ for } r \sin \phi)$$

In summary, the transformation formulas are $x = x' \cos \theta - y' \sin \theta$ and $y = x' \sin \theta + y' \cos \theta$.

EXAMPLE 9-7

Rotation of a Graph

Illustration 1:

Consider $xy = 1$ and the rotation by $\theta = \pi/4$. The transformation equations are

$$x = x' \cos \frac{\pi}{4} - y' \sin \frac{\pi}{4}$$

$$y = x' \sin \frac{\pi}{4} + y' \cos \frac{\pi}{4}$$

Since $\sin \pi/4 = \cos \pi/4 = \sqrt{2}/2 = 1/\sqrt{2}$, the equations become

$$x = \frac{x' - y'}{\sqrt{2}}$$

$$y = \frac{x' + y'}{\sqrt{2}}$$

Substitute the transformation equations into $xy = 1$.

$$xy = 1$$
$$\frac{(x' - y')(x' + y')}{\sqrt{2}\,\sqrt{2}} = 1$$
$$\frac{x'^2 - y'^2}{2} = 1$$
$$x'^2 - y'^2 = 2$$

Illustration 2:

Now consider $x^2 - y^2 = 2$ and transform with the substitution formed when $\theta = -\pi/4$, $\cos(-\pi/4) = 1/\sqrt{2}$ and $\sin(-\pi/4) = -1/\sqrt{2}$. The transformation equations are

$$x = \frac{x' + y'}{\sqrt{2}}$$

$$y = \frac{y' - x'}{\sqrt{2}}$$

Substitute into

$$x^2 - y^2 = 2:$$

$$\left(\frac{x' + y'}{\sqrt{2}}\right)^2 - \left(\frac{y' - x'}{\sqrt{2}}\right)^2 = 2$$

$$\frac{x'^2 + 2x'y' + y'^2}{2} - \frac{y'^2 - 2x'y' + x'^2}{2} = 2$$

$$\frac{x'^2 + 2x'y' + y'^2 - y'^2 + 2x'y' - x'^2}{2} = 2$$

$$\frac{4x'y'}{2} = 2$$

$$x'y' = 1$$

One definition of an *affine transformation* defines its properties: an affine transformation of points in a plane preserves collinearity. By this we mean that if three points are in a straight line before the transformation, they will lie in an equivalent straight line after the transformation. This geometric description of an affine transformation provides visual insight.

For analytical purposes an algebraic definition of an *affine transformation* is more useful. An affine transformation is a transformation given by equations of the form

$x = ax' + by' + j$
$y = cx' + dy' + k$ (where $ad - bc \neq 0$)

Affine transformations are a generalization of properties discussed in geometry. First came the concept of congruent figures. One triangle is congruent to another if a rigid motion of one produces the other. Then came the concept of similarity: we were allowed to uniformly shrink or stretch a triangle to obtain a similar triangle. Now affine transformations (rubber surface geometry) allows not only rigid motions and rotations but also nonuniform expansion and contraction of a figure, as in Display 9-7.

Because θ represents a fixed amount of rotation, $\sin\theta$ and $\cos\theta$ are constant. The argument θ plays the role of a parameter in these equations. Hence, the transformation

$x = x' \cos\theta - y' \sin\theta$
$y = x' \sin\theta + y' \cos\theta$

is linear (first degree) in (x, y) and (x', y'). The transformation is an affine transformation.

From Section 9-1, recall that the general quadratic equation in two variables

$$Ax^2 + Bxy + Cy^2 + Dx + Ey + F = 0$$

represents a conic section.

When $B = 0$, algebraic manipulation allowed us to rewrite the equation in one of the standard forms for a line, parabola, hyperbola, ellipse or degenerate conic. Suppose $B \neq 0$ and the xy term is present. Our experience with the hyperbola given by $xy = k$ suggests that a general quadratic with an xy term represents a rotated conic section. The presence of an xy term complicates the identification of a conic section. Suppose we could rotate the graph to one of the usual orientations. Such rotation would eliminate the xy term and allow us to identify the graph from the equation.

To determine a suitable rotation, work backward. Assume $B \neq 0$. Substitute the transformation formulas

$x = x' \cos\theta - y' \sin\theta$ and $y = x' \sin\theta + y' \cos\theta$

DISPLAY 9-7 An affine transformation that includes stretching, shrinking and rotation.

into each term of the general quadratic:

$$Ax^2 = A(x' \cos\theta - y' \sin\theta)^2$$
$$= Ax'^2 \cos^2\theta - \mathbf{2Ax'y' \cos\theta \sin\theta} + Ay'^2 \sin^2\theta$$
$$Bxy = B(x' \cos\theta - y' \sin\theta)(x' \sin\theta + y' \cos\theta)$$
$$= Bx'^2 \cos\theta \sin\theta + \mathbf{Bx'y' \cos^2\theta} - \mathbf{Bx'y' \sin^2\theta} - By'^2 \sin\theta \cos\theta$$
$$= \mathbf{Bx'y'(\cos^2\theta - \sin^2\theta)} + B(x'^2 - y'^2) \sin\theta \cos\theta$$
$$Cy^2 = C(x' \sin\theta + y' \cos\theta)^2$$
$$= Cx'^2 \sin^2 0 + \mathbf{2Cx'y' \sin\theta \cos\theta} + Cy'^2 \cos^2\theta$$
$$Dx = Dx' \cos\theta - Dy' \sin\theta$$
$$Ey = Ex' \sin\theta + Ey' \cos\theta$$

In each of these expansions, bold type highlights the $x'y'$ terms. We require that the sum of these terms have no $x'y'$ terms. Then the combined coefficients of $x'y'$ must equal 0:

$$-2A \cos\theta \sin\theta + B(\cos^2\theta - \sin^2\theta) + 2C \sin\theta \cos\theta = 0$$

$$B(\cos^2\theta - \sin^2\theta) = (A - C) 2 \cos\theta \sin\theta$$

$$B \cos(2\theta) = (A - C) \sin(2\theta)$$

Because $B \neq 0$,
$$\frac{\cos(2\theta)}{\sin(2\theta)} = \frac{A - C}{B}$$

By definition of cotangent,
$$\cot(2\theta) = \frac{A - C}{B}$$

Then,
$$2\theta = \cot^{-1}\left(\frac{A - C}{B}\right)$$

$$\theta = 0.5\cot^{-1}\left(\frac{A - C}{B}\right)$$

(Note: $0 \leq \theta < \pi$.)   ▬

....

EXAMPLE 9-8

Illustration 1:

Rotating to a Standard Form. Graph the following.

$$xy = 1$$

Solution:

$$A = 0, \quad B = 1, \quad C = 0 \quad \text{(determine the rotation angle)}$$

$$\cot(2\theta) = \frac{0 - 0}{1}$$

$$\cot(2\theta) = 0$$

$$2\theta = \frac{\pi}{2}$$

$$\theta = \frac{\pi}{4}$$

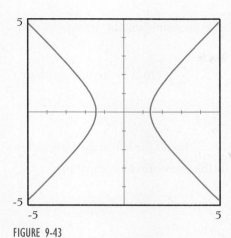

FIGURE 9-43

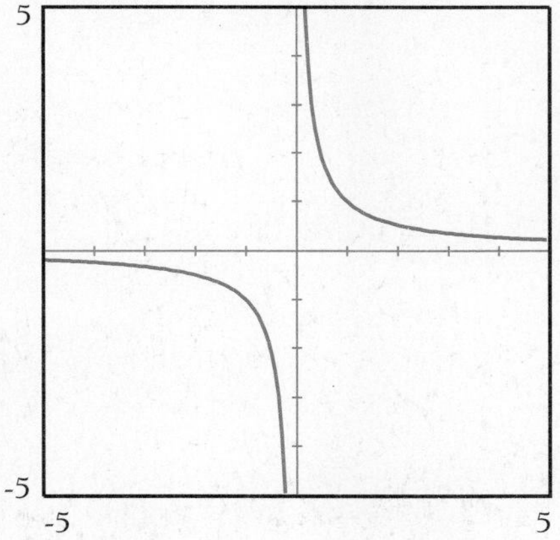

FIGURE 9-44

Then $\cos \pi/4 = \sqrt{2}/2$ and $\sin \pi/4 = \sqrt{2}/2$. The transformation equations are

$$x = \frac{\sqrt{2}}{2}x' - \frac{\sqrt{2}}{2}y'$$

$$y = \frac{\sqrt{2}}{2}x' + \frac{\sqrt{2}}{2}y'$$

Substitute into $xy = 1$

$$\left(\frac{\sqrt{2}}{2}x' - \frac{\sqrt{2}}{2}y'\right)\left(\frac{\sqrt{2}}{2}x' + \frac{\sqrt{2}}{2}y'\right) = 1$$

$$\frac{x^2}{2} - \frac{y^2}{2} = 1$$

From the previous section, this equation represents a hyperbola with transverse axis along the x-axis and x-intercepts at $\pm\sqrt{2}$. The conjugate axis along the y-axis spans $\pm\sqrt{2}$. See Figure 9-43.

The graph of $xy = 1$ is the same shape as the graph in Figure 9-43 but viewed from a new $x'y'$ axis system rotated $\pi/4$ radians (45°) from the original. This produces the graph shown in Figure 9-44. As a check, calculate the distance from $(0, 0)$ to the point $(1, 1)$ on the graph of $xy = 1$. You should obtain $\sqrt{2}$ to reinforce our view that the graph of $xy = 1$ is the graph of $x^2/2 - y^2/2 = 1$ rotated $\pi/4$ radians.

Illustration 2:

$$2x^2 - 4xy + 5y^2 = 9$$

Solution: To sketch the graph, first determine its shape. We seek a rotation of the graph that will eliminate the xy term from the equation. Then we may identify and sketch the resulting graph. Constructing the resulting graph on a rotated axis superimposed over the original axis produces the required graph.

$$A = 2, \quad B = -4, \quad C = 5 \quad \text{(calculate the angle of rotation)}$$

Then $\cot(2\theta) = \dfrac{2 - 5}{-4}$

$$\cot(2\theta) = \frac{3}{4}$$

$$2\theta = \cot^{-1}\left(\frac{3}{4}\right)$$

$$\theta = 0.5\,\cot^{-1}(0.75)$$

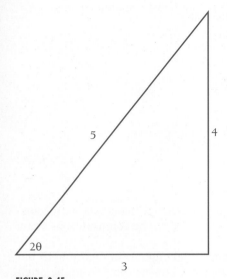

FIGURE 9-45

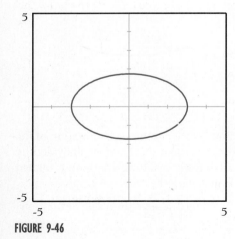

FIGURE 9-46

We do not need an exact value of θ to complete the transformation. Because $\cot(2\theta) = \frac{3}{4}$ then

$$0 < 2\theta < \frac{\pi}{2} \qquad\qquad (2\theta \text{ is in the first quadrant})$$

$$0 < \theta < \frac{\pi}{4}$$

From Figure 9-45 conclude that $\cos(2\theta) = \frac{3}{5}$. Recall the half angle formulas to calculate $\sin(\theta)$ and $\cos(\theta)$ as required by the transformation equations.

$$\cos\theta = \cos\left(\frac{1}{2}\,2\theta\right)$$

$$= \sqrt{\frac{1 + \cos(2\theta)}{2}}$$

$$= \sqrt{\frac{1 + \dfrac{3}{5}}{2}}$$

$$= \frac{2}{\sqrt{5}}$$

$$\sin\theta = \sin\left(\frac{1}{2}\,2\theta\right)$$

$$= \sqrt{\frac{1 - \cos(2\theta)}{2}}$$

$$= \sqrt{\frac{1 - \dfrac{3}{5}}{2}}$$

$$= \frac{1}{\sqrt{5}}$$

The transformation equations are

$$x = \frac{2x'}{\sqrt{5}} - \frac{y'}{\sqrt{5}}$$

$$y = \frac{x'}{\sqrt{5}} + \frac{2y'}{\sqrt{5}}$$

Substitute into the original equation and simplify:

$$2\left(\frac{2x'}{\sqrt{5}} - \frac{y'}{\sqrt{5}}\right)^2 - 4\left(\frac{2x'}{\sqrt{5}} - \frac{y'}{\sqrt{5}}\right)\left(\frac{x'}{\sqrt{5}} + \frac{2y'}{\sqrt{5}}\right) + 5\left(\frac{x'}{\sqrt{5}} + \frac{2y'}{\sqrt{5}}\right)^2 = 9$$

$$\frac{2(4x'^2 - 4x'y' + y'^2)}{5} - \frac{4(2x'^2 + 4x'y' - x'y' - 2y'^2)}{5} +$$

$$\frac{5(x'^2 + 4x'y' + 4y'^2)}{5} = 9$$

$$8x'^2 - 8x'y' + 2y'^2 - 8x'^2 - 16x'y' + 4x'y' + 8y'^2 + 5x'^2 + 20x'y' + 20y'^2$$
$$= 45$$

$$5x'^2 + 30y'^2 = 45$$

Notice that the $x'y'$ terms added to 0 as desired. Now

$$\frac{x'^2}{9} + \frac{y'^2}{\frac{3}{2}} = 1$$

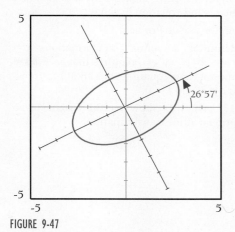

FIGURE 9-47

Recognize this equation as the equation of an ellipse with x'-intercepts of ±3 and y'-intercepts of $\pm\frac{\sqrt{6}}{2}$? The graph is shown in Figure 9-46 (on page 614). Rotate the graph $x'y'$ axes through θ radians in the xy system. Because $\cos\theta = 2/\sqrt{5}$ then $\theta = \cos^{-1}(2/\sqrt{5})$. By calculator,

$$\theta \cong 0.464 \text{ radians} \quad \text{or}$$
$$\theta \cong 26.57°$$

Figure 9-47 shows the graph of $\frac{x'^2}{9} + \frac{y^2}{3} = 1$ rotated by θ to obtain the required graph. ▬

SUMMARY

- Consider the graph of $f(x,y) = c$. The transformation equations

$$x = x' \cos\theta - y' \sin\theta$$
$$y = x' \sin\theta + y' \cos\theta$$

provide a transformation to the equation $f(x',y') = c'$ in terms of the $x'y'$ coordinate system. The $x'y'$ system's x'-axis forms an angle of θ with the x-axis of the original xy system.

- To simplify the general quadratic equation $Ax^2 + Bxy + Cy^2 + Dx + Ey + F = 0$, where $B \neq 0$, by eliminating the xy term, rotate the graph axes by angle θ, where

$$\cot(2\theta) = \frac{A - C}{B}$$

- The resulting equation graphed on the rotated $x'y'$-axes coincides with graph of the original equation on the original axis system.

9-5   EXERCISES

Find a new representation for each of the following equations by rotating through the given angle. Sketch the graph of the original equation and the graph of the rotated equation in the same window.

1. $x + y = 5, \theta = \dfrac{\pi}{4}$.

2. $x - y = 5, \theta = \dfrac{\pi}{4}$.

3. $xy = -1, \theta = \dfrac{\pi}{4}$.

4. $xy = 5, \theta = \dfrac{\pi}{4}$.

Determine θ for a rotation to eliminate the xy term in each of the following.

5. $2x^2 + xy + y^2 + 1 = 0$

6. $2x^2 - xy + 2y^2 - 3x = 0$

7. $3x^2 - xy + 2y^2 - 2 = 0$

8. $3x^2 + xy + 3y^2 + y = 0$

9. $3x^2 + \sqrt{3}xy + 2y^2 = 0$

10. $2x^2 - \sqrt{3}xy + 3y^2 = 0$

11. Show that $x^2 + y^2 = 9$ does not change under rotation for any angle.

12. Show that $x^2 + y^2 = 16$ does not change under rotation for any angle.

For the rotation equations $x = x' \cos \theta - y' \sin \theta$ and $y = x' \sin \theta + y' \cos \theta$.

13. Solve the system for x'. (Hint: multiply the first equation by $\cos \theta$ and the second equation by $\sin \theta$.)

14. Solve the system for y' (see Problem 13).

15. The formal definition of affine transformations $x = ax' + by' + c, y = dx' + ey' + f$ requires that $ae - bd \neq 0$. Show that the rotation equations are an affine transformation with $c = 0$ and $f = 0$.

16. See Problems 13, 14 and 15. Show that the system solved for x' and y' form an affine transformation.

9-5   PROBLEM SET

Consider the equation $Ax^2 + Bxy + Cy^2 + Dx + Ey + F = 0$. Use the rotation equations $x = x' \cos \theta - y' \sin \theta$, $y = x' \sin \theta + y' \cos \theta$. To transform the equation to $A'x'^2 + B'x'y' + C'y'^2 + D'x' + E'y' + F' = 0$.

1. Show that $A' = A \cos^2 \theta + B \sin \theta \cos \theta + C \sin^2 \theta$.

2. Show that
$B' = -2A \sin \theta \cos \theta + B(\cos^2 \theta - \sin^2 \theta) + 2C \sin \theta \cos \theta$.

3. Show that $C' = A \sin^2 \theta - B \sin \theta \cos \theta + C \cos^2 \theta$.

4. See Problem 2. Calculate B'^2.

5. See Problems 1 and 3. Calculate $A'C'$.

6. See Problems 4 and 5. Calculate $B'^2 - 4A'C'$.

7. See Problem 6. $B^2 - 4AC$ is invariant (does not change value) under rotation. We can use this value to help identify a conic section as follows:

If $B^2 - 4AC > 0$ then the quadratic represents an hyperbola or degenerate hyperbola.
If $B^2 - 4AC < 0$ then the quadratic represents an ellipse or degenerate ellipse.
If $B^2 - 4AC = 0$ then the quadratic represents a parabola. Identify the shape of the graph of $2x^2 - 5xy - 3y^2 + x = 0$.

See Problem 7. Identify the shape of the graph of each of the following.

8. $3x^2 - 6xy + 3y^2 - 2 = 0$.

9. $x^2 + 6xy - 3y^2 + 2x - 5 = 0$.

10. $2x^2 - 3xy + 5y^2 - 6y + 4 = 0$.

11. $5x^2 + 5xy + 5y^2 - x + 2y - 3 = 0$.

12. $8x^2 - 8xy + 2y^2 - 3 = 0$.

13. $3x^2 + 7xy - 5y^2 + 4x = 0$.

14. $3x^2 + 7xy + 5y^2 - 6y + 2 = 0$.

 Recall that the equation $Ax^2 + Bxy + Cy^2 + Dx + Ey + F = 0$ represents a conic section with appropriate limitations on the coefficients of the variables. How might we graph these equations using graphing technology?

First suppose that $C = 0$. Then if B and E are not both 0, the equation is linear in y and we can solve for y to obtain a function to graph.

$$Bxy + Ey = -(Ax^2 + Dx + F)$$

$$y(Bx + E) = -(Ax^2 + Dx + F)$$

$$y = \frac{-Ax^2 - Dx - F}{Bx + E}$$

Otherwise, suppose that $C \neq 0$. Then the equation is quadratic in y. Apply the quadratic formula. You should obtain two functions represented by

$$y = \frac{-(Bx + E) \pm \sqrt{(Bx + E)^2 - 4C(Ax^2 + Dx + F)}}{2C}.$$

Combine the graphs of both these functions in the same window to obtain the graph of the conic section.

15. Show that the discriminant of the conic section is equivalent to $(B^2 - 4AC)x^2 + (2BE - 4CD)x + (E^2 - 4CF)$.

16. Under what circumstances will the radicand in the quadratic formula form of the conic section be linear in x?

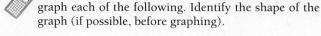 Use graphing technology in the rectangular mode to graph each of the following. Identify the shape of the graph (if possible, before graphing).

17. $2x^2 + xy + y^2 + 1 = 0.$

18. $3x^2 - xy + 2y^2 - 2 = 0.$

19. $2x^2 - xy + 2y^2 - 3x = 0.$

20. $3x^2 + xy + 3y^2 + y = 0.$

21. $3x^2 + \sqrt{3}xy + 2y^2 = 0.$

22. $2x^2 - \sqrt{3}xy + 3y^2 = 0.$

23. $x^2 - 2xy + y^2 + x + 2y + 1 = 0.$

24. $x^2 + xy - 2y^2 + 3x - 1 = 0.$

25. $x^2 + xy + 5x - 2y = 0.$

26. $3x^2 - xy + y^2 - 2x - 5 = 0.$

EXPLORATION

27. Convert $r = 1/(1 + \cos \theta)$ to rectangular coordinates. Hint: replace $\cos \theta$ with x/r then simplify. After simplifying replace r with $\sqrt{x^2 + y^2}$. Simplify again. See Section 8-5. Identify the shape of the graph from the rectangular form.

28. Convert $r = 1/(1 - \sin \theta)$ to rectangular coordinates. Hint: replace $\sin\theta$ with y/r then simplify. After simplifying replace r with $\sqrt{x^2 + y^2}$. Simplify again. See Section 8-5. Identify the shape of the graph from the rectangular form.

29. See Problem 27. Convert $r = 2/(1 + 2 \cos \theta)$ to rectangular coordinates. Identify the shape of the graph from the rectangular form.

30. See Problem 28. Convert $r = 0.5/(1 - 0.5 \sin \theta)$ to rectangular coordinates. Identify the shape of the graph from the rectangular form.

31. See Problem 27. Convert $r = 0.25/(1 + 0.25 \cos \theta)$ to rectangular coordinates. Identify the shape of the graph from the rectangular form.

32. See Problem 28. Convert $r = 3/(1 - 3 \sin \theta)$ to rectangular coordinates. Identify the shape of the graph from the rectangular form.

33. See Problem 27. Convert $r = e/(1 + e \cos \theta)$ to rectangular coordinates. Identify the shape of the graph from the rectangular form when $e = 1$. When $0 < e < 1$. When $e > 1$.

34. See Problem 28. Convert $r = e/(1 - e \sin \theta)$ to rectangular coordinates. Identify the shape of the graph from the rectangular form when $e = 1$, when $0 < e < 1$, when $e > 1$.

35. See Problems 27 or 28. Convert $r = de/(1 - e \sin \theta)$, $d \neq 0$, to rectangular coordinates. Identify the shape of the graph from the rectangular form when $e = 1$, when $0 < e < 1$, when $e > 1$. (Note: e is the eccentricity of the conic section.)

36. See Problems 27 or 28. Convert $r = de/(1 + e \cos \theta)$, $d \neq 0$, to rectangular coordinates. Identify the shape of the graph from the rectangular form when $e = 1$, when $0 < e < 1$, when $e > 1$. (Note: e is the eccentricty of the conic section.)

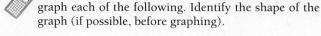 Use graphing technology in the polar mode to graph the following.

37. $r = \sin \theta$

38. $r = \cos \theta$

39. $r = 3$

40. $r = 2$

41. $r = 2\theta$

42. $r = \dfrac{\theta}{2}$

43. $r = \tan \theta$

44. $r = \text{ctn } \theta$

45. $r = \cos(2\theta)$

46. $r = \sin(2\theta)$

47. $r = \sin(3\theta)$

48. $r = \cos(3\theta)$

49. $r = \cos(5\theta)$

50. $r = \sin(5\theta)$

51. $r = \sec \theta$

52. $r = \csc \theta$

53. $r = 3 + 2 \cos \theta$

54. $r = 3 - 2 \cos \theta$

55. $r = 3 - 2 \sin \theta$

56. $r = 3 + 2 \sin \theta$

57. $r = \dfrac{4}{1 - 2 \sin \theta}$

58. $r = \dfrac{1}{3 - \cos \theta}$

59. $r = \dfrac{2}{4 + \cos \theta}$

60. $r = \dfrac{2}{1 + 1 \sin \theta}$

61. $r = \dfrac{5}{3 - \cos \theta}$

62. $r = \dfrac{4}{1 - 2 \sin \theta}$

63. $r = \dfrac{5}{1 + \cos \theta}$

64. $r = \dfrac{4}{1 + \cos \theta}$

65. $r = \dfrac{3}{4 - \sin \theta}$

66. $r = \dfrac{5}{1 - 3 \sin \theta}$

67. Discuss the shape of the curves in Problems 57, 59, 61, 63 and 65.

68. Discuss the shape of the curves in Problems 58, 60, 62, 64 and 66.

69. See Problem 59. Convert $r = 2/(4 + \cos\theta)$ to the form $r = de/(1 \pm e\cos\theta)$ by multiplying the numerator and denominator by $\frac{1}{4}$. Identify the value of e and d.

70. See Problem 58. Convert $r = 1/(3 - \cos\theta)$ to the form $r = de/(1 \pm e\cos\theta)$ by multiplying the numerator and denominator by $\frac{1}{3}$. Identify the value of e and d.

71. See Problems 57–66. What is the orientation (up/down or left/right) of equations of the form $r = de/(1 \pm e\sin\theta)$?

72. See Problems 57–66. What is the orientation (up/down or left/right) of equations of the form $r = de/(1 \pm e\cos\theta)$?

73. See Problems 57–66. What is the shape of graphs where $0 < e < 1$?

74. See Problems 57–66. What is the shape of graphs where $e > 1$?

 75. See Problems 57–66. Discuss the influence of e on the shape of the graphs.

76. See Problems 57–66. Discuss the influence of d on the shape of the graphs. Hint: try graphing $r = d/(1 \pm 2\cos\theta)$, for $d = 1$, $d = 2$, $d = 3$, $d = 4$, etc.

CHAPTER SUMMARY

• Conic sections are formed by the intersection of a plane and a right circular cone of two nappes.

PARABOLAS

The graph of an equation of the form $(y - k)^2 = 4c(x - h)$ is a parabola with vertex located at (h, k). The axis of symmetry of the graph is $y = k$. The focus of the parabola is located at $(h + c, k)$. The directrix of the parabola is $x = h - c$. The width of the latus rectum is $4|c|$. If $c > 0$, then the parabola opens right. If $c < 0$, then the parabola opens left.

The graph of an equation of the form $(x - h)^2 = 4c(y - k)$ is a parabola with vertex located at (h, k). The axis of symmetry of the graph is $x = h$. The focus of the parabola is located at $(h, k + c)$. The directrix of the parabola is $y = k - c$. The width of the latus rectum is $4|c|$. If $c > 0$, then the parabola opens up. If $c < 0$, then the parabola opens down.

ELLIPSES

The graph of an equation of the form

$$\frac{(x - h)^2}{a^2} + \frac{(y - k)^2}{b^2} = 1$$

is an ellipse. The ellipse is centered at (h, k). A window to graph the ellipse is bounded by $x = h + a$, $x = h - a$, $y = k + b$, $y = k - b$. If $a^2 > b^2$, then the major axis is horizontal. If $a^2 < b^2$, then the major axis is vertical. If $a^2 = b^2$, then the graph is a circle. The foci of the ellipse are on the major axis c units from the center of the ellipse $c = \sqrt{|a^2 - b^2|}$.

The eccentricity of an ellipse is given by $e = c/m$, where m is half the length of the major axis and c is the distance of a focus from the center. If $e = 0$, then ellipse is a circle. The closer e is to 1, the more elongated is the ellipse.

HYPERBOLAS

The graph of equation of the form

$$\frac{(x - h)^2}{a^2} - \frac{(y - k)^2}{b^2} = \pm 1$$

is a hyperbola. To graph it, perform these steps.

1. Locate the center of the hyperbola at (h, k). Identify the transverse and conjugate axes from $x = h$ and $y = k$ (see step 5).
2. Construct auxiliary axes, x' and y', with their origin at (h, k).
3. Sketch a box as a graph aid about (h, k): the vertical sides should be $|a|$ units to either side of (h, k); the horizontal sides should be $|b|$ units above and below (h, k).
4. Extend the diagonals of the box to form the hyperbola's asymptotes:

$$y = \pm \frac{b}{a} (x - h) + k$$

5. (a) If

$$\frac{(x - h)^2}{a^2} - \frac{(y - k)^2}{b^2} = 1,$$

then the hyperbola has two horizontal branches. The transverse axis is $y = k$. The foci are located at $(h + c, k)$ and $(h - c, k)$, where $c^2 = a^2 + b^2$. The eccentricity of the hyperbola is $e = c/a$.
   (b) If

$$\frac{(y - k)^2}{b^2} - \frac{(x - h)^2}{a^2} = 1,$$

then the hyperbola has two vertical branches. The transverse axis is $x = h$. The foci are located at $(h, k + c)$ and $(h, k - c)$, where $c^2 = a^2 + b^2$. The eccentricity of the hyperbola is $e = c/b$.
   Equations of the form $xy = k$ $(k \neq 0)$ have hyperbolas as graphs with the x-axis and y-axis as asymptotes; $(k, 1)$ and $(1, k)$ are points of the graph. Equations expressible in the form $x^2/a^2 - y^2/b^2 = 0$ or $xy = 0$ represent degenerate hyperbolas.

- Consider the graph of $f(x,y) = c$. The transformation equations

$$x = x' \cos \theta - y' \sin \theta$$
$$y = x' \sin \theta + y' \cos \theta$$

provide a transformation to the equation $f(x', y') = c'$ in terms of the $x'y'$ coordinate system. The $x'y'$ system's x'-axis forms an angle of θ with the x-axis of the original xy system.
   To simplify the general quadratic equation

$$Ax^2 + Bxy + Cy^2 + Dx + Ey + F = 0, \qquad \text{where } B \neq 0$$

by eliminating the xy term, rotate the graph by angle θ where

$$\cot(2\theta) = \frac{A - C}{B}$$

KEY WORDS AND CONCEPTS

Analytical geometry
Asymptotes
Axis of symmetry
Circle
Conic section
Conjugate axis
Covertices
Degenerate ellipse

Degenerate hyperbola
Directrix
Eccentricity
Ellipse
Focus
General quadratic equation
Graphing box
Hyperbola

Latus rectum
Nappe
Parabola
Parameter
Polar coordinates
Rotation
Transverse axis
Vertex

9   REVIEW EXERCISES

SECTION 1

Use the given information to determine the parameters for the indicated equation.

1. $xy = -k$, determine k if the graph includes $(3, 5)$.

2. $y = mx + b$, determine m and b if the graph includes $(0, -2)$, and $(1, 4)$.

3. $ax + by = 12$, determine a and b if the graph includes $(5, 0)$ and $(0, -2)$.

4. $y = ax^2 + bx + 2$, determine a and b if the graph includes $(0, 1)$ and $(1, 3)$.

5. $ax^2 + by^2 = 1$, determine a and b if the graph includes $(-5, 0)$ and $(0, 3)$.

6. $ay^2 - bx^2 = 1$, determine a and b if the graph incudes $(2, 0)$ and $(0, -3)$.

Find an equation(s) for the parabola described.

19. Vertex $(0, 0)$, focus $(0, 2)$.

20. Vertex $(0, 0)$, focus $(-5, 0)$.

21. Vertex $(0, 0)$, directrix $x = 4$.

22. Vertex $(0, 0)$, directrix $y = -2$.

23. Focus $(0, -1)$, directrix $y = 1$.

24. Latus rectum from $(-3, -1)$ to $(3, -1)$, vertex $(0, 0)$.

25. Focus $(1, 2)$, vertex at $(1, 0)$.

26. Focus $(0, 3)$, directrix $y = 3$.

27. Focus $(1, 3)$, directrix $x = -1$.

28. Latus rectum from $(-4, 5)$ to $(4, 5)$.

SECTION 3

Determine the vertices and foci of the following ellipses. Sketch the graph.

SECTION 2

Annotate and sketch the graphs of the following.

7. $y^2 = 2x$

8. $x^2 = 2y$

9. $y^2 = 6x$

10. $y^2 = 4x$

11. $y^2 = -16x$

12. $x^2 = 8y$

13. $(y - 4)^2 = (x + 7)$

14. $(x + 1)^2 = (y - 3)$

15. $(y + 6)^2 = 2x$

16. $y^2 = 6x - 10$

17. $(y - 1)^2 = -16(x + 5)$

18. $(x - 3)^2 = 8(y - 5)$

29. $\dfrac{x^2}{9} + \dfrac{y^2}{4} = 1$

30. $\dfrac{x^2}{36} + \dfrac{y^2}{16} = 1$

31. $4x^2 + 25y^2 = 100$

32. $4y^2 + x^2 = 16$

33. $y^2 + 9x^2 = 1$

34. $x^2 + 9(y + 1)^2 = 36$

35. $\dfrac{(x - 3)^2}{16} + \dfrac{(y + 2)^2}{25} = 1$

36. $\dfrac{x^2}{25} + \dfrac{(y + 2)^2}{9} = 1$

37. $9y^2 + x^2 + 4x = 5$

38. $4y^2 + x^2 + 10x + 9 = 0$

Write an equation to represent the ellipse(s) described by

39. Vertices at $(0, 5)$, $(0, -5)$, $(3, 0)$ and $(-3, 0)$.

40. Foci at $(3, 0)$ and $(-3, 0)$, vertices at $(0, 7)$, $(0, -7)$.

41. Center at $(0, 0)$, major axis length 10, focus at $(-3, 0)$.

42. Center at $(0, 0)$, major axis length 26, minor axis length 10, minor axis on y axis.

43. Center at $(0, 0)$, major axis on y-axis of length 5, eccentricity $= \frac{2}{5}$.

44. Center at $(4, -2)$, vertices at $(7, -2)$, $(1, -2)$, $(4, 3)$ and $(4, -7)$.

SECTION 4

Sketch the graph of each of the following.

45. $\dfrac{x^2}{16} - \dfrac{y^2}{9} = 1$

46. $16x^2 - 4y^2 = 64$

47. $x^2 - y^2 = 100$

48. $(x + 1)^2 - (y - 2)^2 = -16$

49. $\dfrac{(x + 5)^2}{9} - \dfrac{(y - 2)^2}{16} = -1$

50. $\dfrac{(x - 2.5)^2}{81} - \dfrac{(y + 2)^2}{9} = 1$

51. $32(y + 3)^2 - 18(x - 1)^2 = 288$

52. $y^2 + 6y - x^2 = 27$

53. $y^2 + 4y - 9x^2 - 18x = 41$ **54.** $5y^2 - 2x^2 = -20$

55. $xy = 5$ **56.** $-3xy = 0$

57. $y = \dfrac{5}{x}$ **58.** $4xy = -12$

59. $(x - 3)(y + 2) = 1$ **60.** $\dfrac{x}{9} - \dfrac{y}{4} = 1$

61. $x^2 - y^2 = 1$ **62.** $\dfrac{x^2}{9} - \dfrac{y}{4} = 1$

63. $x^2 - y^2 = -2$ **64.** $\dfrac{x^2}{9} - \dfrac{y^2}{4} = -1$

Write an equation for the hyperbola with center at the origin that satisfies the given conditions.

65. Foci at $(0, -5)$, $(0, 5)$ and vertices at $(0, -4)$, $(0, 4)$.

66. Foci at $(-4, 0)$ $(4, 0)$ and vertices at $(-1, 0)$, $(1, 0)$.

67. Vertices at $(-1, 0)$, $(1, 0)$ and passing through $(2, 3\sqrt{3})$

68. x-intercepts of ± 4, asymptotes of $y = \pm \frac{4}{5}x$.

SECTION 5

Find a new representation for each of the following equations by rotating through the given angle. Sketch the graph of the original equation and the graph of the rotated equation.

69. $x + y = 1$, $\theta = \dfrac{\pi}{4}$.

70. $xy = 4$, $\theta = \dfrac{\pi}{4}$.

Determine θ for a rotation to eliminate the xy term in each of the following.

71. $2x^2 + xy + y^2 + 5 = 0$

72. $3x^2 + xy + 3y^2 + y - x = 0$

73. $3x^2 + \sqrt{3}\,xy + 2y^2 + x = 0$

74. $2x^2 - \sqrt{3}\,xy + 3y^2 + 1 + x = 0$

75. Show that $x^2 + y^2 = 1$ does not change under rotation for any angle.

76. Show that $4x^2 + 4y^2 = 5$ does not change under rotation for any angle.

SIDELIGHT   The Bernoullis

Perhaps the most illustrious family in the history of mathematics is the Bernoulli family. The Bernoullis were a Swiss family who first gained prominence in the sciences and mathematics with the brothers Jakob Bernoulli (1654–1705) and Johann Bernoulli (1667–1748). These two brothers were fortunate enough to be born in an age of new thought and exploration. Their own insight led them to the application of the new calculus to a wide variety of problems. Each brother made many contributions to calculus and mathematical physics but they were only the beginning.

Johann Bernoulli had three sons. Nicolaus Bernoulli was born in 1695 and, though he showed great promise, died at a young age in 1726. Daniel Bernoulli was born in 1700 and lived until 1782. Daniel was the most famous of Johann's sons. He established much of his reputation in the study of probability. Finally, the youngest son was Johann Bernoulli II (1710–1790). In his early years he studied law, but later became a professor of mathematics.

Johann II in his turn had three sons: Johann III, Daniel II and Jakob II. In turn, Daniel II had a son Cristoph whose own son was Johann Gustav. Each of these men were reputable mathematicians.

Slightly off the main family tree, Johann and Jakob had a nephew named Nicolaus who was a well-known mathematician. Nicolaus had a son, Nicolaus II who was also a mathematician.

Few families can claim to have produced so many able practitioners of the same science. As a result the Bernoulli name is attached to some of the more interesting mathematical discoveries of the seventeenth and eighteenth centuries. It is generally believed that Jakob Bernoulli introduced the use of polar coordinates in 1691.

9   CHAPTER TEST

1. Determine values of a and b for $y = ax^2 + 1x + b$ so that the graph includes $(1, 5)$ and $(0, 2)$.

Graph Problems 2–6 in rectangular coordinates.

2. $(x - 3)^2 = 4 - (y + 1)^2$

3. $9(x - 1)^2 + \dfrac{(y + 5)^2}{16} = 9$

4. $\dfrac{(x - 2)^2}{25} = \dfrac{(y + 3)^2}{9} - 1$

5. $xy = 0$

6. $(x - 2)(y + 3) = 1$

7. Write an equation for a hyperbola with x-intercepts at $(4, 0)$ and $(-4, 0)$ and asymptotes of $y = \pm \frac{3}{4} x$.

8. Write an equation for an ellipse with center at $(0, 0)$, major axis on the x axis length 5, and minor axes of 4.

9. Annotate the graph of $y = 0.25x^2$ to show the focus, directrix, latus rectum and axis of symmetry. Discuss how the latus rectum affects the graph.

10. Determine the equation for the graph of $x - 3y = 2$, if the graph is rotated by $\theta = \pi/4$.

Systems of Equations

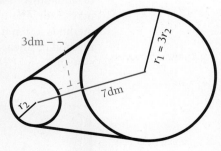

10

A mechanical engineer designs a pulley system in which the distance between the center of the pulleys is fixed at 7 dm and the clearance between the pulleys must be 3 dm. See Display 10-1. The relationship between the radii r_1 and r_2 of the pulleys is

$$r_1 + r_2 + 3 = 7$$

so that $\quad r_1 + r_2 = 4$

Further, the power and speed requirements of the pulley system demand that the radius of the first pulley be three times the radius of the second pulley:

$$r_1 = 3r_2$$

or $\quad r_1 - 3r_2 = 0$

Therefore the size of the pulleys is modeled by a *pair* of linear equations in two variables:

$$\begin{cases} r_1 + r_2 = 4 \\ r_1 - 3r_2 = 0 \end{cases}$$

Methods for determining values of r_1 and r_2 to satisfy these design parameters form the motivation for the topics of this chapter. The type of problem is a **system of linear equations.**

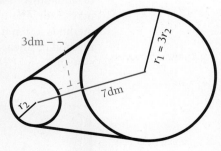

Two pulley system

DISPLAY 10-1

623

LINEAR SYSTEMS

When the boy begins to understand that the visible point is preceded by an invisible point, that the shortest distance between two points is conceived as a straight line before it is ever drawn with the pencil on the paper, he experiences a feeling of pride of satisfaction. And justly so, for the fountain of all thought has been opened to him, the difference between the ideal and the real, potentia et actua, has become clear to him; henceforth the philosopher can reveal him nothing new, as a geometrician he has discovered the basis of all thought.

—GOETHE

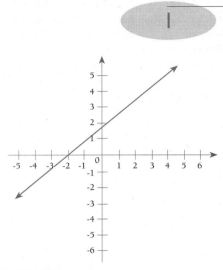

FIGURE 10-1   Dependent

Recall from Chapter 1, that an equation of the form

$$ax + by = c$$

represents a linear equation in two variables. The graph of the equation in the xy plane is a straight line. Consider two lines simultaneously:

$$\begin{cases} ax + by = c \\ dx + ey = f \end{cases}$$

This pair of equations represents a system of two linear equations in two variables. To solve the system means to determine a point (r, s) such that (r, s) satisfies both of the equations. If (r, s) is a solution, then $ar + bs = c$ and $dr + es = f$. Geometrically, (r, s) is a point that lies on both lines. If the two lines are distinct and not parallel, then (r, s) is the unique point of intersection of the two lines.

If the two equations represent the *same* line, then every point of each line is a common point of intersection. There are infinitely many points of intersection. We say that the two such equations are **dependent**. Geometrically, the slopes and y-intercepts of the two lines are respectively equal. See Figure 10-1.

If the two equations represent distinct parallel lines, then they share no common point of intersection. There is no solution to the system and the system is **inconsistent**. Geometrically, the slopes of the two lines are equal, but the y-intercepts are different. See Figure 10-2.

The most interesting system is lines that intersect in a unique point. These two lines have different slopes (or one line is vertical and the other is not). Such a system is **independent** and **consistent**. See Figure 10-3.

Graphing equations often leads to approximate solutions for systems of equations. The graphing method has several flaws. First, exact answers may elude one no matter how accurate the graph. More important, graphing methods do not extend well to systems of three variables. The algebraic method of linear combination avoids these flaws. Consider the simplest linear system.

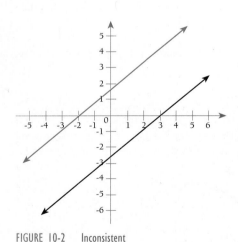

FIGURE 10-2   Inconsistent

Theorem 10-1
Simplest System

The solution of the system $\begin{cases} x = r \\ y = s \end{cases}$ is (r, s).

Note that $x = r$ is a vertical line, whereas $y = s$ is a horizontal line. For example, the solution to $x = 4$ and $y = -2$ is $(4, -2)$. For more complicated systems our goal is to convert these systems to a simpler system. Linear combinations provide a tool to accomplish this goal.

Form a combination of two equations by multiplying each equation by a nonzero constant, then adding the equations. If the two equations represent lines, the resulting equation is called a **linear combination**. For example, begin with the equations $x + 2y = 4$ and $3x - y = 5$. To eliminate y multiply the first equation by 1 and the second equation by 2, then "add" the respective sides of the equations:

$$1(x + 2y) = 1(4) \rightarrow 1x + 2y = 4$$
$$2(3x - y) = 2(5) \rightarrow \underline{6x - 2y = 10}$$
$$7x \qquad\quad = 14$$

Divide by 7 to obtain $\qquad x = 2$

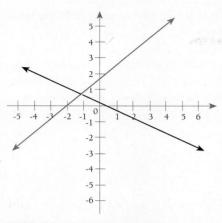

FIGURE 10-3   Independent and consistent

If we had multiplied the first equation by 3 and the second equation by -1, the combination would have eliminated the variable x. As a result, using linear combinations to eliminate a variable is called the **method of elimination**. The resulting linear combination, $y = 1$, completes our goal of the simplest system. The solution of $x = 2$ and $y = 1$ is $(2, 1)$. Check this point in the original equations to verify that it is also the solution to the original system. See Figure 10-4. Use the trace feature of your technology to support the algebraic solution.

Geometrically, the linear combination of two lines produces a third line that passes through the common point of intersection. The next theorem summarizes these observations.

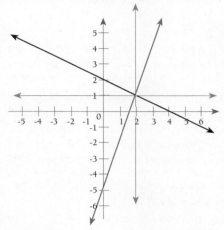

FIGURE 10-4 $x + 2y = 4$,
$3x - y = 5$ $\qquad x = 2$, $\quad y = 1$

Theorem 10-2
Linear Combinations

Suppose that the system

$$\begin{cases} ax + by = c \\ dx + ey = f \end{cases}$$

is consistent and independent with a solution (r, s). Then if k and m are nonzero constants, the line

$$k(ax + by) + m(dx + ey) = kc + mf$$

represents a line that passes through the common point of intersection (r, s).

This new line is called a *linear combination* of the two original equations.

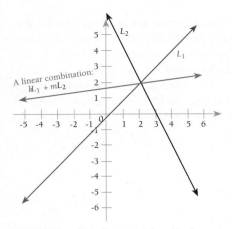

A linear combination:
$l_{L_1} + mL_2$

FIGURE 10-5   General system and combination

For the proof, substitution of (r, s) into the combination shows that (r, s) satisfies the new equation. See Figure 10-5. When two systems have the same solution, the systems are **equivalent**. Hence,

$$\begin{cases} x + 2y = 4 \\ 3x - y = 5 \end{cases} \quad \text{and} \quad \begin{cases} x = 2 \\ y = 1 \end{cases} \text{ are equivalent systems.}$$

Solving Linear Systems of Two Equations in Two Variables.   Solve each of the following.

EXAMPLE 10-1

Illustration 1:

$$\begin{cases} 5x - 2y = 12 \\ 3x + 4y = 2 \end{cases}$$

Solution: Use the method of elimination. Decide which variable to eliminate, say y. For a combination to eliminate y, the coefficients of y must be exact opposites. Currently the coefficients are -2 and 4. We need a common multiple for these two. Choose 4. Multiply the first equation by 2 and the second equation by 1. The coefficients of y become exact opposites. Combine the resulting equations to eliminate y:

$$\begin{aligned} L_1: \quad 5x - 2y = 12 &\to 2 \times L_1 \to 10x - 4y = 24 \\ L_2: \quad 3x + 4y = 2 &\to 1 \times L_2 \to \underline{3x + 4y = 2} \\ & \quad 2L_1 + 1L_2 \to 13x = 26 \\ & \quad\quad\quad\quad\quad x = 2 \end{aligned}$$

To eliminate x multiply by 3 and -5, respectively:

$$\begin{aligned} L_1: \quad 5x - 2y = 12 &\to 3 \times L_1 \to 15x - 6y = 36 \\ L_2: \quad 3x + 4y = 2 &\to -5 \times L_2 \to \underline{-15x - 20y = -10} \\ & 3L_1 + 5L_2 \to -26y = 26 \\ & \quad\quad\quad\quad\quad y = -1 \end{aligned}$$

Therefore, $\begin{cases} 5x - 2y = 12 \\ 3x + 4y = 2 \end{cases}$ is equivalent to $\begin{cases} x = 2 \\ y = -1 \end{cases}$.

The solution is (2, -1). Check the solution by direct substitution into the original system:

$$L_1: 5(2) - 2(-1) = 12 \quad \text{is true.}$$
$$L_2: 3(2) + 4(-1) = 2 \quad \text{is true.}$$

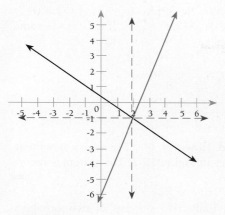

FIGURE 10-6

Also use technology to check the graph. The result should look like Figure 10-6.

Note: A variation on the method of elimination is substitution. **Substitution** eliminates a variable by replacing it with an equivalent expression. For example, once we established that $y = -1$, replacing y with -1 eliminates y from L_1: $5x - 2(-1) = 12$, so that $x = 2$.

Illustration 2:

$$\begin{cases} r_1 + r_2 = 4 \\ r_1 - 3r_2 = 0 \end{cases}$$

Solution: This is the introductory pulley problem. Because $r_1 = 3r_2$, substitute into the first equation to eliminate the r_1 term:

$$r_1 + r_2 = 4$$
$$(3r_2) + r_2 = 4 \qquad \text{(substitute } 3r_2 \text{ for } r_1)$$
$$4r_2 = 4$$
$$r_2 = 1$$

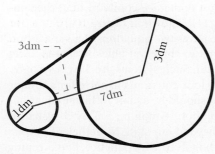

DISPLAY 10-2   Pulleys

Because $r_2 = 1$, substitute into $r_1 = 3r_2$ to determine r_1:

$$r_1 = 3(1)$$
$$r_1 = 3$$

The solution is $(r_1, r_2) = (3, 1)$. The radii of the pulleys are 3 dm and 1 dm, respectively.

Illustration 3:

$$\begin{cases} 2x + y = 3 \\ 4x + 2y = 6 \end{cases}$$

Solution: To eliminate y multiply the first equation by 2 and the second by -1:

$$L_1: 2x + y = 3 \rightarrow 2L_1 \rightarrow 4x + 2y = 6$$
$$L_2: 4x + 2y = 6 \rightarrow -1L_2 \rightarrow \underline{-4x - 2y = -6}$$
$$2L_1 + 1L_2 \rightarrow \qquad\qquad 0 = 0$$

All variables were eliminated! In fact $L_2 = 2L_1$. These equations represent the same line. The equations are dependent. (Note: $0 = 0$ is true.) There are infinitely many points in common. The solution is $\{(x, y): 2x + y = 3\}$. Check the graphs. They are the same line.

Illustration 4:

$$\begin{cases} 2x + \ y = 4 \\ 4x + 2y = 6 \end{cases}$$

Solution: As before eliminate y:

$$L_1: 2x + \ y = 4 \rightarrow \ 2L_1 \rightarrow \ 4x + 2y = \ 8$$
$$L_2: 4x + 2y = 6 \rightarrow -1L_2 \rightarrow \underline{-4x - 2y = -6}$$
$$2L_1 + 1L_2 \rightarrow \qquad\qquad 0 = \ 2$$

Once more all variables were eliminated. However, $0 = 2$ is a false statement. Graphing technology shows parallel lines in Figure 10-7. The system is inconsistent. There is no point of intersection.   ▬

The method of elimination is not limited to systems of two variables. Consider a linear equation in three variables: $ax + by + cz + d = 0 \ (abc \neq 0)$.

Graphing a three-variable equation requires a three-dimensional coordinate system. In a three-dimensional coordinate system the graph of a linear equation is a plane. See Figure 10-8.

Consider two such planes simultaneously, like lines they might coincide, be parallel or intersect. See Figure 10-9. If the intersection of two planes is a line, the intersection contains infinitely many points.

For a linear system in three variables to have a single point solution requires three equations and therefore three planes. If no two of the three planes are parallel and if the three planes do not contain a common line, then the planes intersect in exactly one point (r, s, t) as shown in Figure 10-10.

A linear system with two variables requires two equations. This is a 2 by 2 system. If we extend the linear equations to three variables, we need three equations for a unique point of intersection. Such systems are 3 by 3 systems. Linear equations in four variables require four equations to solve for a unique point of intersection. In general, a system should be square, n linear equations in n variables, to anticipate a unique solution. Graphing a 3 by 3 system is difficult. The graph of a 4 by 4 system is beyond our skill. Nonetheless, consistent and independent n by n systems can be solved with patience, bookkeeping skills and the method of elimination.

The strategy in the method of elimination for a 3 by 3 system is to reduce the system to a 2 by 2 and finally to a 1 by 1. A 1 by 1 system has one variable and provides one of the equations needed for a simpler equivalent system. Repeat the process to obtain two more 1 by 1 systems. Substitution sometimes speeds the process.

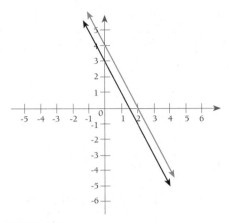

FIGURE 10-7

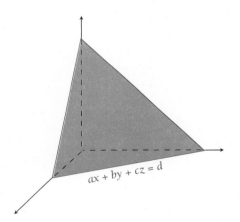

FIGURE 10-8 $ax + by + cz = d$

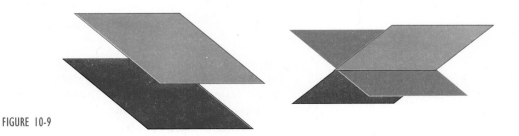

FIGURE 10-9

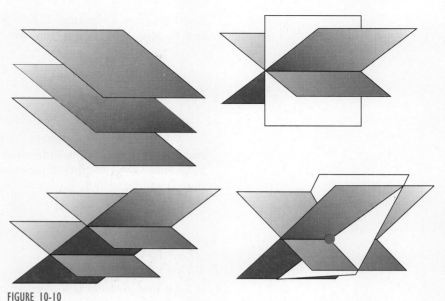

FIGURE 10-10

EXAMPLE 10-2

Illustration 1:

Solve the following systems.

$$\begin{cases} x + y + z = 2 \\ 2x + y - z = 9 \\ x - 2y + z = -1 \end{cases}$$

Solution: Choose to eliminate z. This leaves two variables. Two variables require two equations for a 2 by 2 system. Eliminate z from the first two equations (P_1 and P_2) for one equation. Eliminate z from the last two equations (P_2 and P_3) for the second equation:

$$\begin{array}{lrcl} P_1: & x + y + z = 2 \to 1P_1 \to & x + y + z = & 2 \\ P_2: & 2x + y - z = 9 \to 1P_2 \to & \underline{2x + y - z = \ 9} \\ & P_1 + P_2 \to & 3x + 2y \quad\ = 11 \\ P_2: & 2x + y - z = 9 \to 1P_2 \to & 2x + y - z = \ 9 \\ P_3: & x - 2y + z = -1 \to 1P_3 \to & \underline{x - 2y + z = -1} \\ & P_2 + P_3 \to & 3x - y \quad\ = 8 \end{array}$$

We have the 2 by 2 system

$$\begin{cases} 3x + 2y = 11 \\ 3x - y = \ 8 \end{cases}.$$

Solve this system by eliminating y.

From the second equation, if $3x - y = 8$ then $y = 3x - 8$. Replacing y in the first equation with $3x - 8$ eliminates y:

$$3x + 2y = 11$$
$$3x + 2(3x - 8) = 11$$
$$9x - 16 = 11$$
$$x = 3.$$

We have a value for x after eliminating z and y. To recover the values for y and z, work in the reverse order used to eliminate them. Use the 2 by 2 system. Substitute 3 for x in either of the equations (say the first, L_1) to obtain an equation containing only y:

$$L_1: 3(3) + 2y = 11$$
$$9 + 2y = 11$$
$$2y = 2$$
$$y = 1$$

All that remains is to recover z. Return to the original 3 by 3 system. Choose one of the equations, say P_1, and substitute 3 for x and 1 for y:

$$P_1: (3) + (1) + z = 2$$
$$4 + z = 2$$
$$z = -2$$

This completes the solution:

$$\begin{cases} x + y + z = 2 \\ 2x + y - z = 9 \\ x - 2y + z = -1 \end{cases} \quad \text{is equivalent to} \quad \begin{cases} x = 3 \\ y = 1. \\ z = -2 \end{cases}$$

The solution is $(3, 1, -2)$.

Illustration 2:

$$\begin{cases} 2x + y + z + w = 6 \\ x + \qquad\qquad w = 4 \\ -2z - w = 3 \\ x + 3y + \qquad w = 7 \end{cases}$$

Solution: Eliminate w. This leaves three variables. We need three equations to form a system:

$$E_1: 2x + y + z + w = 6 \rightarrow 1E_1 \rightarrow 2x + y + z + w = 6$$
$$E_2 \quad x + \qquad\qquad w = 4 \rightarrow -1E_2 \rightarrow \underline{-x - \qquad\qquad w = -4}$$
$$E_1 - E_2 \rightarrow x + y + z \qquad = 2$$

$$E_1: 2x + y + z + w = 6 \rightarrow 1E_1 \rightarrow 2x + y + z + w = 6$$
$$E_3: \qquad\qquad -2z - w = 3 \rightarrow 1E_3 \rightarrow \underline{\qquad\qquad -2z - w = 3}$$
$$E_1 + E_3 \rightarrow 2x + y - z \qquad = 9$$

$$E_1: 2x + y + z + w = 6 \rightarrow 1E_1 \rightarrow 2x + y + z + w = 6$$
$$E_4: \quad x + 3y + \qquad w = 7 \rightarrow -1E_4 \rightarrow \underline{-x - 3y - \qquad w = -7}$$
$$E_1 - E_4 \rightarrow x - 2y + z \qquad = -1$$

We have reduced the 4 by 4 system to the 3 by 3 system:

$$x + \ y + z = \ 2$$
$$2x + \ y - z = \ 9$$
$$x - 2y + z = \text{-}1$$

This is the same 3 by 3 system as in Illustration 1. Repeat the steps of Illustration 1 and obtain $x = 3, y = 1$ and $z = \text{-}2$. Finally, recover w from the 4 by 4 system. Choose any of the original equations, say E_2, and substitute 3 for x, 1 for y and -2 for z:

$$E_2\text{: } (3) + w = 4$$
$$w = 1$$

The solution is (3, 1, -2, 1). Check this point in all four of the original equations to verify that it is a solution. ■

One problem solving strategy is to break complex problems into simpler components. In fact, algebra means to disassemble and reunite. This strategy was the motivation for the factor theorem of Chapter 4. Factoring led to reducing rational functions to lowest terms. But reduction to lowest terms is not always the best method for expressing a rational function. For comparison, note that $\frac{3}{4}$ is reduced to lowest terms but the equation $\frac{3}{4} = \frac{1}{2} + \frac{1}{4}$ expresses $\frac{3}{4}$ as a sum of simpler fractions. Simplify $1/(x + 3) + 1/(x - 2)$ to verify the sum is $(2x + 1)/(x^2 + x - 6)$. Could we have started with the sum and broken it down into the original terms? The Calculus sometimes requires breaking a rational expression into a sum of simpler component rational expressions using a process known as *partial fraction decomposition*. One application of solving linear systems is partial fraction decomposition. Partial fraction decomposition depends upon the following theorem.

Theorem 10-3
Partial Fractions Theorem

If $f(x) = n(x)/d(x)$, where the degree of $n(x)$ is less than the degree of $d(x)$ and $f(x)$ is reduced to lowest terms, then $f(x)$ can be decomposed into a sum of rational functions:

$$f(x) = f_1(x) + f_2(x) + \ldots + f_r(x)$$

where each of $f_1, f_2, \ldots, f_r$ is a rational function in one of the following forms:

$$f_i(x) = \begin{cases} \dfrac{A}{(px + q)^n} \\ \dfrac{Bx + C}{(ax^2 + bx + c)^m} \end{cases}, \text{ where } i = 1, 2, \ldots, r$$

and $(px + q)^n$ is a factor of $d(x)$ or $(ax^2 + bx + c)^m$ is a factor of $d(x)$, $a \neq 0$ and $ax^2 + bx + c$ is prime.

For example, consider $f(x) = (3x - 7)/(x^2 - 5x + 6)$. The degree of the numerator is less than the degree of the denominator, and after factoring, $f(x)$ is seen to be reduced to lowest terms:

$$f(x) = \frac{(3x - 7)}{(x - 3)(x - 2)}.$$

By the theorem,

$$f(x) = \frac{A_1}{x - 3} + \frac{A_2}{x - 2}.$$

Adding these two fractions yields

$$f(x) = \frac{A_1(x - 2) + A_2(x - 3)}{(x - 3)(x - 2)} = \frac{(A_1 + A_2)x + (-2A_1 - 3A_2)}{(x - 3)(x - 2)}.$$

The new numerator must equal the original numerator:

$$3x - 7 = (A_1 + A_2)x + (-2A_1 - 3A_2).$$

These are equal only if the corresponding coefficients are equal:

$$A_1 + A_2 = 3$$
$$-2A_1 - 3A_2 = -7$$

Solve this system of equations for A_1 and A_2 ($A_1 = 2$ and $A_2 = 1$) to obtain

$$f(x) = \frac{2}{x - 3} + \frac{1}{x - 2}.$$

If a factor in the denominator has a counting number exponent, then every possible source of that factor from the sum must be considered.

For example the decomposition of

$$g(x) = \frac{2x^4 + 2x^3 - 7x^2 + 0x - 2}{(x - 2)(x + 1)^2(x^2 - x + 1)}$$

is

$$g(x) = \frac{A_1}{x - 2} + \frac{A_2}{x + 1} + \frac{A_3}{(x + 1)^2} + \frac{A_4x + A_5}{x^2 - x + 1}.$$

Note particularly, the repetition of the $(x + 1)$ factor. A denominator with a factor of $(x + r)^5$ would have five terms, with denominators of $(x + r)$, $(x + r)^2$, $(x + r)^3$, $(x + r)^4$ and $(x + r)^5$, respectively.

Return to $g(x)$. Combine the four terms in the decomposition form of $g(x)$ into a single rational expression by finding a common denominator and adding. Set the coefficients of the new numerator equal to the corresponding coefficients of the original expression. This produces five linear equations in five variables. Solving this system yields $A_1 = 1$, $A_2 = 0$, $A_3 = 2$, $A_4 = 1$ and $A_5 = 0$. As a result,

$$g(x) = \frac{1}{x - 2} + \frac{2}{(x + 1)^2} + \frac{x}{x^2 - x + 1}$$

SUMMARY

- We are able to solve a linear equation having only one variable. These 1 by 1 systems are the simplest.
- To solve a 2 by 2 system of linear equations, apply a strategy known as the *method of elimination,* which uses linear combinations or substitution to reduce the system to a 1 by 1 system. Working backward, recover the eliminated variable by substitution. This provides a solution for the original system.
- To solve a 3 by 3 system of linear equations, apply the method of elimination to reduce the system to a 2 by 2 system. Repeat the method on the 2 by 2 system. Once more, use substitution to recover the eliminated variables in the reverse order of elimination.
- To solve an *n* by *n* system of linear equations, perform these steps.

  1. If $n = 1$ or $n = 2$ or $n = 3$, use the previously described method.

  2. If the current version of the system is *n* by *n* ($n > 3$), then reduce the system to an $n - 1$ by $n - 1$ system using the method of elimination. Repeat this process until the current system is 1 by 1. Solve for the variable in the 1 by 1 system. Recover the eliminated variables (suppose the current solved system is *m* by *m*), substituting the *m* solved variables into the next $m + 1$ by $m + 1$ system. This leaves one variable in that system. Solve for it. Keep repeating the process until $m + 1$ equals the original *n*.

- If all variables are eliminated in one of these methods and we obtain an equation of the form $0 = c$ (where *c* is some constant), then the original system was either dependent or inconsistent.
- One application of solving systems of linear equations is partial fraction decomposition.

10-1 EXERCISES

Solve the following systems.

1. $\begin{cases} x + y = 5 \\ x - y = 3 \end{cases}$

2. $\begin{cases} x + y = 7 \\ x - y = 3 \end{cases}$

5. $\begin{cases} x - y = 4 \\ x - y = 5 \end{cases}$

6. $\begin{cases} x + y = 3 \\ x + y = 5 \end{cases}$

3. $\begin{cases} 2x + y = 4 \\ x + y = 1 \end{cases}$

4. $\begin{cases} 2x + y = 3 \\ x + y = 4 \end{cases}$

7. $\begin{cases} 2x + 3y = 1 \\ 3x + 2y = 4 \end{cases}$

8. $\begin{cases} 3x + y = 5 \\ 6x + 2y = 10 \end{cases}$

9. $\begin{cases} x + 2y = 4 \\ 3x + 6y = 12 \end{cases}$

10. $\begin{cases} 2x + 3y = 7 \\ 3x + 2y = 3 \end{cases}$

11. $\begin{cases} y = 2x + 1 \\ y = 4x - 3 \end{cases}$

12. $\begin{cases} y = 4 - x \\ y = 2 + x \end{cases}$

13. $\begin{cases} y = 5 \\ x = 4 \end{cases}$

14. $\begin{cases} x = 2 \\ y = 3 \end{cases}$

15. $\begin{cases} x = -1.5 \\ y = 3.8 \end{cases}$

16. $\begin{cases} x = 4.1 \\ y = -7.6 \end{cases}$

17. $\begin{cases} x + y = 3 \\ x - y = 4 \end{cases}$

18. $\begin{cases} x + y = 5 \\ x - y = 2 \end{cases}$

19. $\begin{cases} 2x + 3y = 5 \\ x + y = 2 \end{cases}$

20. $\begin{cases} 2x + 3y = 2 \\ x + y = 5 \end{cases}$

21. $\begin{cases} 3x - y = 5 \\ y = 2x + 2 \end{cases}$

22. $\begin{cases} 3x - y = 2 \\ y = 2x + 5 \end{cases}$

23. $\begin{cases} 0.3x + 0.2y = 0.5 \\ 0.5x - 0.4y = 1.2 \end{cases}$

24. $\begin{cases} 0.3x + 0.2y = 1.2 \\ 0.5x - 0.4y = 0.5 \end{cases}$

25. $\begin{cases} \frac{1}{3}x - \frac{1}{4}y = \frac{1}{2} \\ \frac{2}{5}x - \frac{1}{3}y = \frac{1}{5} \end{cases}$

26. $\begin{cases} \frac{1}{3}x - \frac{1}{4}y = \frac{1}{3} \\ \frac{2}{5}x - \frac{1}{3}y = \frac{1}{4} \end{cases}$

27. $\begin{cases} y = \frac{2}{3}x + 3 \\ y = -\frac{1}{2}x - 2 \end{cases}$

28. $\begin{cases} y = \frac{2}{3}x + 2 \\ y = -\frac{1}{2}x - 3 \end{cases}$

29. $\begin{cases} 3r - 2s = 5 \\ 5r + 3s = 1 \end{cases}$

30. $\begin{cases} 3r - 2s = 1 \\ 5r + 3s = 5 \end{cases}$

31. Determine the radii of two pulleys with design parameters given by $\begin{cases} r_1 + r_2 = 10 \\ r_1 - 4r_2 = 0 \end{cases}$

32. Determine the radii of two pulleys with design parameters given by $\begin{cases} r_1 + r_2 = 12 \\ r_1 - 2r_2 = 0 \end{cases}$

33. The resistance R and temperature T of a wire are related by the formula $R = aT + b$. If the resistance is 50 ohms at 10° and the resistance is 60 ohms at 20°, determine the values of a and b.

34. See Problem 33. If the resistance of a wire is 60 ohms at 50° and the resistance is 65 ohms at 100°, determine a and b for the wire.

35. Determine how much 20 percent alcohol solution and 50 percent alcohol solution must be mixed to produce 10 liters of 40 percent alcohol solution.

36. Determine how much D_5 (5%) glucose and pure glucose (100%) must be mixed to produce 500 cc of D_{10} (10%) glucose.

37. An employee can split her time between two tasks at her company. Task A pays $5.00/hour. Task B pays $4.00/hour. If the employee earns $198 in a 40 hour week, how did she split her time between the tasks?

38. The cost of producing item A is $5.00 each. The cost of producing item B is $10.00 each. Item A sells for $7.00 each. Item B sells for $12.00 each. If it costs $1000.00 per week to produce the two items and the gross income for the weeks production is $1300.00, determine the total number of item A produced and the total number of item B produced.

39. By gathering data an economist determines that the demand D for a product is related to the price P of the product by the following formula: $D = 1000 - 5P$. Also by experimenting, the economist discovers that the supply S for the same product is a function of the price and is given by $S = 200 + 3P$. The economist knows that market equilibrium occurs when supply is equal to demand. Determine the price at which market equilibrium will occur.

40. If a projectile is fired toward the earth at 136 ft/sec, its velocity after t seconds is given by $V = 136 + 32t$. A built-in sensor in the projectile measures its velocity and tracks the elapsed time. When the velocity is 100 times the elapsed time, a parachute opens to slow the projectile. How many seconds until the parachute deploys and what is the velocity at that time?

Solve the following.

41. $\begin{cases} x + y + z = 3 \\ x - y + z = 1 \\ 3x - y - z = 1 \end{cases}$

42. $\begin{cases} x + y + z = 4 \\ x - y + z = 0 \\ 3x - y - z = 0 \end{cases}$

43. $\begin{cases} x + y + z = 2 \\ x - y + z = 4 \\ 3x - y - z = 2 \end{cases}$

44. $\begin{cases} x + y + z = 0 \\ x - y + z = 2 \\ 3x - y - z = -4 \end{cases}$

45. $\begin{cases} 2x - y + z = 4 \\ x + y - 2z = 4 \\ x + y + z = 4 \end{cases}$

46. $\begin{cases} 2x - y + z = 0 \\ x + y - 2z = 5 \\ x + y + z = -1 \end{cases}$

47. $\begin{cases} 2x - y + z = 2 \\ x + y - 2z = -5 \\ x + y + z = 4 \end{cases}$

48. $\begin{cases} 2x - y + z = -1 \\ x + y - 2z = -4 \\ x + y + z = 2 \end{cases}$

49. $\begin{cases} x - y + 2z = -2 \\ 2x + y - z = 0 \\ x + y = 2 \end{cases}$

50. $\begin{cases} x - y + 2z = -9 \\ 2x + y - z = 7 \\ x + y = 3 \end{cases}$

51. $\begin{cases} x - y + 2z + w = 5 \\ 2x + y - z - w = 2 \\ x + y = 2 \\ z + w = 1 \end{cases}$

52. $\begin{cases} x - y + 2z + w = 6 \\ 2x + y - z - w = 0 \\ x + y = 3 \\ z + w = 4 \end{cases}$

53. Solve $\begin{cases} ax + by = c \\ dx + ey = f \end{cases}$

Hint: multiply the first equation by e and the second equation by $-b$, then combine the equations. This will allow you to solve for x. Repeat the process by multiplying

the first equation by d and the second equation by $-a$, then combine to eliminate x.

54. Solve $\begin{cases} dx + ey = f \\ ax + by = c \end{cases}$ See the Hint for Problem 53.

10-1   P R O B L E M   S E T

PARTIAL FRACTIONS

Use partial fractions to decompose each of the following rational functions.

1. $f(x) = \dfrac{3x - 1}{(x - 3)(x + 1)}$

2. $f(x) = \dfrac{x + 5}{(x - 1)(x + 2)}$

3. $f(x) = \dfrac{9}{x^2 - x - 20}$

4. $f(x) \dfrac{2}{x^2 + 4x + 3}$

5. $f(x) = \dfrac{2x + 1}{(x + 1)^2}$

6. $f(x) = \dfrac{3x - 1}{(x - 1)^2}$

7. $f(x) = \dfrac{x^2 + x + 2}{(x - 1)(x + 1)^2}$

8. $f(x) = \dfrac{x^2 + 5x + 1}{(x + 2)^2(x - 3)}$

9. $f(x) = \dfrac{3x^2 - x + 2}{x^3 + x}$

10. $f(x) = \dfrac{x^2}{(x + 1)(x^2 + x + 1)}$

PROBLEMS FOR GRAPHING TECHNOLOGY

The trace feature of your graphing technology can approximate the solutions to many nonlinear 2 by 2 systems of equations.

For example, to solve $\begin{cases} y = x^2 - 5 \\ y = 3\sin x \end{cases}$, graph $y = x^2 - 5$

and $y = 3 \sin x$. Zoom or scroll to display any points of intersection. There are two for this system. Activate the trace

function and move the cursor to the observed points. The approximate points of intersection are (1.45, -2.98) and (2.55, 1.67). To improve accuracy, zoom in. The next section discusses algebraic methods for selected nonlinear systems.

Use graphing technology to approximate the solution of the following.

11. $\begin{cases} y = 2x - 5 \\ y = 3 - 5x \end{cases}$

12. $\begin{cases} y = 4x + 7 \\ y = 5 - 7x \end{cases}$

13. $\begin{cases} y = x^2 \\ y = 3x + 1 \end{cases}$

14. $\begin{cases} y = 5 - x^2 \\ y = 3x - 1 \end{cases}$

15. $\begin{cases} y = 1 - x^2 \\ y = 2x^2 - 3 \end{cases}$

16. $\begin{cases} y = 5 - x^2 \\ y = 3x^2 \end{cases}$

17. $\begin{cases} y = 5 - x^2 \\ y = x^3 \end{cases}$

18. $\begin{cases} y = x^2 - 2 \\ y = x^3 - 2x^2 \end{cases}$

19. $\begin{cases} y = x + 3 \\ y = e^x \end{cases}$

20. $\begin{cases} y = e^{-x} \\ y = x^2 \end{cases}$

21. $\begin{cases} y = 1 - x^2 \\ y = \ln x \end{cases}$

22. $\begin{cases} y = \ln x \\ y = \dfrac{1}{x} \end{cases}$

23. $\begin{cases} y = \dfrac{1}{x} \\ y = 2^x \end{cases}$

24. $\begin{cases} y = \sin x \\ y = \ln(x) \end{cases}$

25. $\begin{cases} y = 1 - x \\ y = \cos x \end{cases}$

26. $\begin{cases} y = \sec x \\ y = \log(x) \end{cases}$

27. $\begin{cases} y = \sin^{-1} x \\ y = e^x \end{cases}$

28. $\begin{cases} y = \cos^{-1} x \\ y = \cos x \end{cases}$

29. $\begin{cases} y = \sqrt{x} \\ y = \sin x \end{cases}$

30. $\begin{cases} y = \sqrt{x} \\ y = \ln x \end{cases}$

31. Discuss any difficulty you encountered in approximating solutions.

32. See Problems 11–30. Discuss any difficulty you had choosing a window.

2   QUADRATIC SYSTEMS

In my own case, I got along fine without knowing the distributive law until my sophomore year in college; meanwhile I had drawn lots of graphs.
—DONALD E. KNUTH

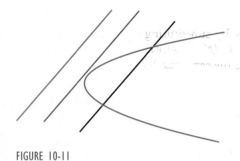

FIGURE 10-11

This section investigates systems of two equations in two variables more complex than linear systems. A 2 by 2 **quadratic system** is a system of equations consisting of two quadratic equations or a quadratic and a linear equation. We shall explore strategies for solving 2 by 2 quadratic systems, but we will by no means exhaust the possibilities.

Consider first a quadratic system consisting of a line and a parabola. Geometrically, there are three possibilities. The line and parabola can intersect twice, once, or not at all. See Figure 10-11. Therefore the maximum number of solutions is 2. Algebraically, the system resembles

$$\begin{cases} y = mx + b \\ y = ax^2 + bx + c \end{cases}$$

or

$$\begin{cases} y = mx + b \\ x = ay^2 + by + c \end{cases}.$$

Because a linear equation can always be solved for a variable, substitution is a good strategy to eliminate that variable from the system.

Another quadratic system consists of two parabolas with systems resembling $\begin{cases} y = dx^2 + ex + f \\ y = ax^2 + bx + c \end{cases}$ or $\begin{cases} x = dy^2 + ey + f \\ y = ax^2 + bx + c \end{cases}$

In the first system, the parabolas are oriented in the same direction. If they are distinct parabolas, they can intersect at most twice. See Figure 10-12. In the second system, the parabolas are oriented at right angles. This system can have up to four points of intersection. See Figure 10-13. In the preceding systems, either the methods of substitution or combination work.

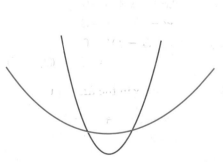

FIGURE 10-12

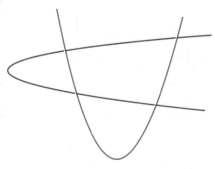

FIGURE 10-13

EXAMPLE 10-2

Illustration 1:

Solving a Linear-Quadratic System. Solve the following systems. Technology can help predict the number of solutions. Remember, tracing provides approximate solutions.

$$\begin{cases} y = x + 1 \\ y = x^2 + 4x + 3 \end{cases}$$

Solution: See Figure 10-14. Eliminate y by substituting $x + 1$ from the linear equation for y in the quadratic equation. (Note: subtracting the first equation from the second equation will produce the same results.)

$$x + 1 = x^2 + 4x + 3$$
$$0 = x^2 + 3x + 2$$
$$(x + 1)(x + 2) = 0$$
$$x = -1 \quad \text{or} \quad x = -2$$

This yields the x-coordinates of the points of intersection: $(-1, \quad)$, $(-2, \quad)$.

The corresponding y values follow by substituting each x value into one of the original equations. Because the arithmetic is simpler in the linear equation, choose it:

$$\text{if } x = -1, \quad \text{then } y = (-1) + 1$$
$$y = 0$$
$$\text{if } x = -2, \quad \text{then } y = (-2) + 1$$
$$y = -1$$

As a result the points of intersection are $(-1, 0)$ $(-2, -1)$.

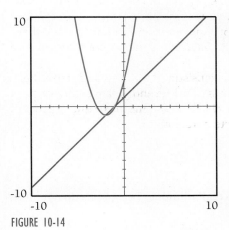

FIGURE 10-14

Illustration 2:

$$\begin{cases} y = -x^2 + 4x - 1 \\ y = 2x^2 - 2x + 2 \end{cases} \text{ban}$$

Solution: Substitute:

$$-x^2 + 4x - 1 = 2x^2 - 2x + 2$$
$$3x^2 - 6x + 3 = 0$$
$$x^2 - 2x + 1 = 0$$
$$(x - 1)^2 = 0$$
$$x = 1 \quad (1, \quad)$$

Substitute 1 for x in the first equation:

$$y = -(1)^2 + 4(1) - 1$$
$$y = 2$$

The solution is $(1, 2)$. See Figure 10-15.

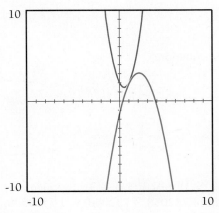

FIGURE 10-15

Illustration 3:

$$\begin{cases} y = x - 2 \\ y = x^2 \end{cases}$$

Solution: Substitute:

$$x = x^2 - 2$$
$$x^2 - x + 2 = 0$$

From the quadratic formula:

$$x = \frac{1 \pm \sqrt{1^2 - 4(1)(2)}}{2(1)}$$

$$x = \frac{1 \pm \sqrt{-7}}{2}$$

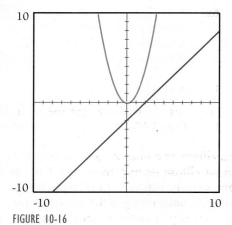

FIGURE 10-16

Because the radicand is -7, there are no real solutions. The graphs do not intersect. See Figure 10-16.

Illustration 4:

$$\begin{cases} y = x^2 + x \\ y = 2x^2 + x - 3 \end{cases}$$

Solution: Substitute:

$$x^2 + x = 2x^2 + x - 3$$
$$x^2 - 3 = 0$$

From the quadratic formula (or factoring), we determine

$$x = \pm \sqrt{3}$$

Substitute these values for x into the first equation:

$$y = (\sqrt{3})^2 + \sqrt{3}$$
$$= 3 + \sqrt{3}.$$
$$y = (-\sqrt{3})^2 + (-\sqrt{3})$$
$$= 3 - \sqrt{3}$$

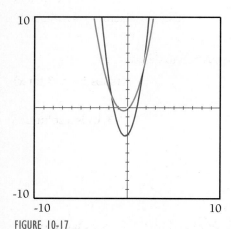

FIGURE 10-17

The points of intersection are $(\sqrt{3}, 3 + \sqrt{3})$, $(-\sqrt{3}, 3 - \sqrt{3})$. By calculator approximation or use of trace, we obtain (1.732058, 4.732058), (-1.7320508, 1.2679492). See Figure 10-17.

Illustration 5:

$$\begin{cases} y = x^2 - 2x + 1 \\ 2x = y^2 - 3y + 2 \end{cases}$$

Solution: Substitute $x^2 - 2x + 1$ from the first equation for y in the second equation:

$$2x = (x^2 - 2x + 1)^2 - 3(x^2 - 2x + 1) + 2$$
$$2x = x^4 - 4x^3 + 6x^2 - 4x + 1 - 3x^2 + 6x - 3 + 2$$
$$x^4 - 4x^3 + 3x^2 = 0$$
$$x^2(x^2 - 4x + 3) = 0$$
$$x^2(x - 3)(x - 1) = 0$$
$$x = 0, \quad x = 3, \quad x = 1$$

Substitute each of these back into the first equation to determine the corresponding values of y: $(0, 1)$, $(3, 4)$, $(1, 0)$. See Figure 10-18. ■

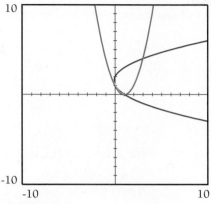

FIGURE 10-18

Systems consisting of a line and an ellipse or a line and a hyperbola also yield to substitution. A parabola with an ellipse or hyperbola may have as many as four points of intersection, as we will see in Figure 10-21. Substituting might result in a fourth-degree polynomial. With luck, we can avoid the fourth degree. Otherwise we must rely upon the rational root theorem or tracing the graph using technology. Similarly, two ellipses, two hyperbolas or an ellipse with a hyperbola could have up to four points of intersection. The next few illustrations by no means exhaust this theme.

Solving a Quadratic-Quadratic System. Solve the following systems.

EXAMPLE 10-3

Illustration 1:

$$\begin{cases} x^2 + y^2 = 9 \\ x = y^2 - 3 \end{cases}$$

Solution: Because $x = y^2 - 3$, then $x + 3 = y^2$. Substitute in the first equation.

$$\text{Then } x^2 + (x + 3) = 9$$
$$x^2 + x - 6 = 0 \qquad \text{(set = 0 and factor)}$$
$$(x + 3)(x - 2) = 0 \qquad \text{(set each factor = 0 and solve)}$$
$$x = -3 \quad \text{or} \quad x = 2$$

Substitute these values into the second equation to determine y.

$$-3 = y^2 - 3 \qquad \text{(substitute -3 for } x)$$
$$y^2 = 0$$
$$y = 0 \qquad ((-3, 0) \text{ is a solution)}$$

Substitute 2 for x:

$$2 = y^2 - 3$$
$$y^2 = 5$$
$$y = \pm\sqrt{5}$$

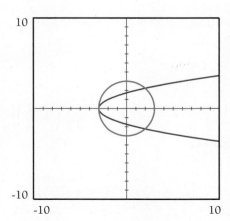

FIGURE 10-19

The points of intersection are $(-3, 0)$, $(2, \sqrt{5})$ and $(2, -\sqrt{5})$. See Figure 10-19.

Illustration 2:

$$\begin{cases} 2x^2 + y^2 = 18 \\ x^2 - y^2 = 9 \end{cases}$$

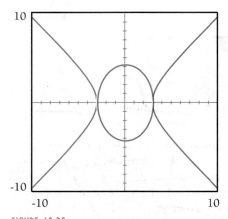

FIGURE 10-20

Solution: Because this system resembles a linear system, try a combination of the two equations.

$$\begin{aligned} 2x^2 + y^2 &= 18 \\ \underline{x^2 - y^2 = 9} \\ 3x^2 &= 27 \\ x^2 &= 9 \\ x &= \pm 3 \end{aligned}$$

Determining y is more of a challenge. Substitute 3 for x in the second equation:

$$\begin{aligned} (3)^2 - y^2 &= 9 \\ y^2 &= 0 \\ y &= 0 \end{aligned}$$

Similarly, $x = -3$ produces $y = 0$. See Figure 10-20. Because the substitution of x produced a quadratic in y, there could have been two y values for each of these x values. This would have produced four points of intersection. The points of intersection are (3, 0), (-3, 0)

Illustration 3:

$$\begin{cases} x^2 + 9y^2 = 5 \\ 3y = 3 - x^2 \end{cases}$$

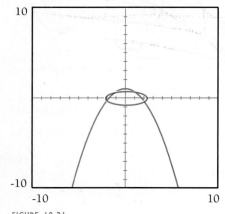

FIGURE 10-21

Solution: Because $3y = 3 - x^2$, then $x^2 = 3 - 3y$. Substitute for x^2 in the first equation.

$$(3 - 3y) + 9y^2 = 5 \qquad \text{(set = 0 and solve)}$$
$$9y^2 - 3y - 2 = 0$$
$$y = \frac{2}{3} \quad \text{or} \quad y = \frac{-1}{3} \qquad \text{(by quadratic formula)}$$

Substitute $\frac{2}{3}$ for y in the second equation:

$$\begin{aligned} 3\left(\frac{2}{3}\right) &= 3 - x^2 \\ x^2 &= 1 \\ x &= \pm 1 \end{aligned} \qquad \left\{\left(1, \frac{2}{3}\right), \left(-1, \frac{2}{3}\right)\right\}$$

Substitute $\frac{-1}{3}$ for y in the second equation:

$$\begin{aligned} 3\left(\frac{-1}{3}\right) &= 3 - x^2 \\ x^2 &= 4 \\ x &= \pm 2 \end{aligned} \qquad \left\{\left(2, \frac{-1}{3}\right), \left(-2, \frac{-1}{3}\right)\right\}$$

The points of intersection are $\left(-2, \frac{-1}{3}\right)$, $\left(2, \frac{-1}{3}\right)$, $\left(-1, \frac{2}{3}\right)$, $\left(1, \frac{2}{3}\right)$. See Figure 10-21.

SUMMARY

Use technology to predict the number of solutions in a quadratic system.

- To solve a system consisting of a linear equation and a quadratic equation, substitute from the linear into the quadratic. This eliminates one variable. The resulting equation has no more than two solutions. Solve the resulting equation. Substitute the solution into the linear equation to determine the other variable of each pair.
- To solve a quadratic system in which one of the equations represents a parabola, use the first-degree variable for the equation for the parabola to substitute into the second equation. This eliminates one variable. The resulting equation in one variable may be as complex as fourth degree. Solve for the variable in this equation. Substitute the solution(s) to this equation into the equation for the parabola to determine the other variable of each pair. Pair corresponding values of the variables to identify the points of intersection. For distinct graphs, there will be as few as none but no more than four points of intersection.
- For a quadratic system in which both equations are second degree in both variables, substitute or combine to eliminate one of the variables. Hope for the best and continue as in other cases.

10-2   EXERCISES

Solve the following systems.

1. $\begin{cases} y = x \\ y = x^2 + 2x \end{cases}$

2. $\begin{cases} y = x \\ y = x^2 - 2x \end{cases}$

3. $\begin{cases} y = -2x \\ y = x^2 - 3 \end{cases}$

4. $\begin{cases} y = -3x \\ y = x^2 - 4 \end{cases}$

5. $\begin{cases} y = x + 2 \\ y = x^2 \end{cases}$

6. $\begin{cases} y = x + 6 \\ y = x^2 \end{cases}$

7. $\begin{cases} y = 2x - 1 \\ y = -x^2 + 2 \end{cases}$

8. $\begin{cases} y = 2 - 3x \\ y = -x^2 + x + 7 \end{cases}$

9. $\begin{cases} y = \frac{1}{2}x + 1 \\ y = x^2 + \frac{1}{2} \end{cases}$

10. $\begin{cases} y = \frac{1}{3}x - 1 \\ y = x^2 - \frac{5}{3} \end{cases}$

11. $\begin{cases} y = x^2 \\ y = 2x^2 + x \end{cases}$

12. $\begin{cases} y = x^2 \\ y = 2 - x^2 \end{cases}$

13. $\begin{cases} y = x^2 \\ y = 3 - 2x^2 \end{cases}$

14. $\begin{cases} y = x^2 \\ y = 3x^2 + 8x \end{cases}$

15. $\begin{cases} y = x^2 + x \\ y = 2x^2 + 3x - 3 \end{cases}$

16. $\begin{cases} y = x^2 - x \\ y = 2x^2 + 2x - 4 \end{cases}$

17. $\begin{cases} y = x + 1 \\ y = x^2 + 3x - 1 \end{cases}$

18. $\begin{cases} y = x - 1 \\ y = x^2 - 2x + 1 \end{cases}$

31. $\begin{cases} y = 3^x \\ y = 3^{2x} - 6 \end{cases}$ (Hint: let $u = 3^x$)

19. $\begin{cases} y = x - 2 \\ y = x^2 + 3x + 2 \end{cases}$

20. $\begin{cases} y = x^2 + 1 \\ y = 2x^2 + x - 3 \end{cases}$

32. $\begin{cases} y = 5^{2x} - 10 \\ y = 3(5^x) \end{cases}$ (Hint: let $u = 5^x$)

21. $\begin{cases} y = x - 2 \\ y = x^2 - x - 1 \end{cases}$

22. $\begin{cases} y = x + 3 \\ y = x^2 - 3x - 1 \end{cases}$

33. $\begin{cases} \dfrac{1}{x^2} - \dfrac{1}{y^2} = 5 \\ \dfrac{1}{x^2} + \dfrac{1}{y^2} = 13 \end{cases}$ (Hint: let $u = 1/x$ and $v = 1/y$)

23. $\begin{cases} y = x \\ y = x^2 - x + 5 \end{cases}$

24. $\begin{cases} y = -x \\ y = 2x^2 + x + 3 \end{cases}$

25. $\begin{cases} x^2 - y^2 = 4 \\ 3x^2 + y^2 = 12 \end{cases}$

26. $\begin{cases} x^2 + 12x + y^2 = 8 \\ x^2 - y^2 = 6 \end{cases}$

34. $\begin{cases} \dfrac{1}{x^2} - \dfrac{2}{y^2} = 2 \\ \dfrac{3}{x^2} + \dfrac{1}{y^2} = 13 \end{cases}$ (Hint: let $u = 1/x$ and $v = 1/y$)

27. $\begin{cases} y = x^2 - 2 \\ 4x^2 + 9y^2 = 36 \end{cases}$

28. $\begin{cases} xy = 1 \\ y = x^2 \end{cases}$

29. $\begin{cases} xy = 1 \\ x^2 + y^2 = 2 \end{cases}$

30. $\begin{cases} xy = 2 \\ x^2 - y^2 = 1 \end{cases}$

10-2   P R O B L E M   S E T

PROBLEMS FOR GRAPHING TECHNOLOGY

Graph each system of polar functions on the same coordinates.

1. $\begin{cases} r = \sin \theta \\ r = \cos \theta \end{cases}$

(a) How many points of intersection are there? Graph $r = \cos \theta$ and then trace it to approximate the points of intersection.

(b) Regraph $r = \sin \theta$ and then trace it to approximate the points of intersection.

(c) Compare your answers to part a and b. Discuss any difficulty you encounter.

(d) One point of $r = \sin \theta$ is $(0, 0)$. A point of $r = \cos \theta$ is $(0, \pi/2)$. Discuss how the graphs of these two points may mislead you when you examine the graphs of these functions.

2. $\begin{cases} r = 1 - \sin \theta \\ r = 1 + \cos \theta \end{cases}$

(a) How many points of intersection are there? Graph $r = 1 + \cos \theta$ and then trace it to approximate the points of intersection.

(b) Regraph $r = 1 - \sin \theta$ and then trace it to approximate the points of intersection.

(c) Compare your answers to parts a and b. Discuss any difficulty you encounter.

(d) One point of $r = 1 - \sin \theta$ is $(0, \pi/2)$. A point of $r = 1 + \cos \theta$ is $(0, \pi)$. Discuss how the graphs of these two points may mislead you when you examine the graphs of these functions.

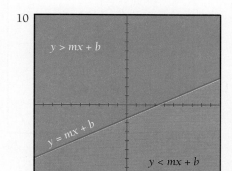

FIGURE 10-22

3 SYSTEMS OF INEQUALITIES

Mathematical language is not only the simplest and most easily understood of any, but the shortest also.
—H. L. BROUGHMAN

The line is one of the simplest graphs we have studied. Equations of lines are given by a linear equation in two variables:

$$ax + by + c = 0 \qquad (ab \neq 0)$$

In the xy plane, the rule of pairing for a line is an equation. What if the rule of pairing were a linear inequality?

Consider the collection of points (x, y) described by the following:

$$ax + by + c < 0, \qquad ax + by + c = 0, \qquad ax + by + c > 0$$

By the trichotomy property these descriptions exhaust all possible points in a plane. Every point in the plane is described by exactly one of these sentences. We already know that the points satisfying $ax + by + c = 0$ lie on a straight line. See Figure 10-22.

Note the line in Figure 10-22 divides the plane into three distinct regions: those points *on* the line, the half plane *below* the line, and the half plane *above* the line. These three regions correspond respectively to the equation and the two inequalities. We are already familiar with the line as graph of the linear equation. But more subtle is to decide which **half-plane** goes with which inequality. The simplest decision method is substituting trial points. Choose any point (r, s) from either side of the boundary line and substitute it into the given inequality. If (r, s) satisfies the given inequality, we chose the correct half-plane, otherwise the other half-plane is correct.

Similarly, graphs of quadratic inequalities are regions of the plane having conic sections as boundaries. Each of these conic sections partitions the plane into three regions: points *on* the boundary curve, points *inside* the curve and points *outside* the curve. Although *inside* and *outside* are more intuitive for regions bounded by an ellipse, the name of the region does not matter: substituting a trial point determines the correct region.

In general, if f is a function and the graph of $y = f(x)$ is a curve, then $y > f(x)$ represents the region of the plane *above* the curve. The region *below* the curve is the graph of $y < f(x)$. The graphs of $y \geq f(x)$ and $y \leq f(x)$ include the boundary curve. Usually graphing technology can shade the graphs of inequalities in two variables when the boundary curve is a function. Consult your manual. In any case, technology can usually provide the boundary curve.

Represent each of the following using a graph.

E X A M P L E   1 0 - 4

Illustration 1:

$$5x - 3y < 15$$

Solution: The boundary of the graph is the line given by

$$5x - 3y = 15$$

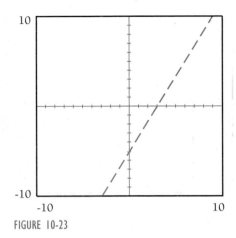

FIGURE 10-23

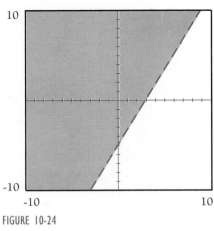

FIGURE 10-24

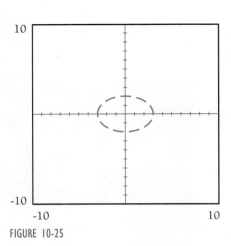

FIGURE 10-25

Because the expression is strictly "less than," the boundary is not part of the graph. We draw the boundary line dashed or dotted (as in Figure 10-23) to indicate that it is not part of the solution. Most technology lacks the detail of distinguishing included and excluded boundaries. As a result you should annotate the boundaries.

Which region makes up the graph? Select a convenient trial point for easy arithmetic. Try $(0, 0)$.

$$5(0) - 3(0) < 15 \qquad \text{is true.}$$

Then $(0, 0)$ is part of the solution. The graph is the half-plane that includes $(0, 0)$. Shade in that half plane to represent the solution, as in Figure 10-24.

Illustration 2:

$$4x^2 + 9y^2 > 36$$

Solution: The boundary for this inequality is the ellipse

$$\frac{x^2}{9} + \frac{y^2}{4} = 1$$

The boundary is not part of the graph. Use a dashed or dotted curve for the ellipse, as in Figure 10-25.

With the boundary in place, a test value indicates the region of the plane described by the inequality. Once more $(0, 0)$ is an excellent candidate for testing: $4(0)^2 + 9(0)^2 > 36$ is false. Therefore, $(0, 0)$ is not part of the solution. This eliminates the boundary and the interior of the ellipse from the solution. All that remains is the exterior of the ellipse. Shade the exterior as a graph to represent the inequality. See Figure 10-26.

Illustration 3:

$$4x^2 + 9y^2 \le 36$$

Solution: Compare this inequality with Illustration 2. The boundary is the same. The equal portion of "less than or equal" implies that the boundary is part of the graph. Draw the boundary as a solid curve. See Figure 10-27.

Because $(0, 0)$ satisfies the inequality, the graph is the boundary with the interior of the ellipse. See Figure 10-28.

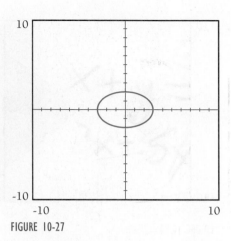

FIGURE 10-26

FIGURE 10-27

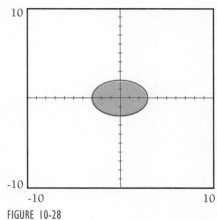

FIGURE 10-28

Illustration 4:

$$y \le x^2 - 3$$

Solution: The boundary is a parabola and also a function. Include the boundary as a solid curve. See Figure 10-29.

Try $(0, 0)$ once more. $0 \le 0^2 - 3$ is false. Conclude that the graph includes the region *below* the parabola. Because the boundary was a function, we expected the graph to be below the curve based upon the $\le$ See Figure 10-30.

Illustration 5:

$$4x^2 - 9y^2 > 36$$

Solution: The boundary is the dashed hyperbola shown in Figure 10-31. Some might argue that a hyperbola divides the plane into more than three parts. As a practical matter, we limit ourselves to three: (1) the hyperbola itself, (2) the region *between* the branches of the hyperbola and (3) the two part region *outside* the branches of the hyperbola. In this case, $(1, 1)$ is between the branches.

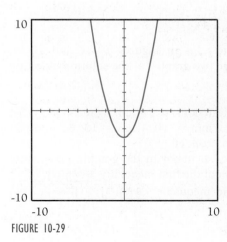

FIGURE 10-29

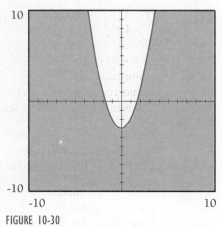

FIGURE 10-30

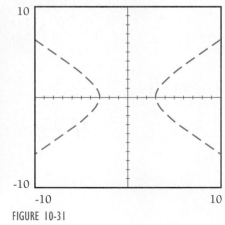

FIGURE 10-31

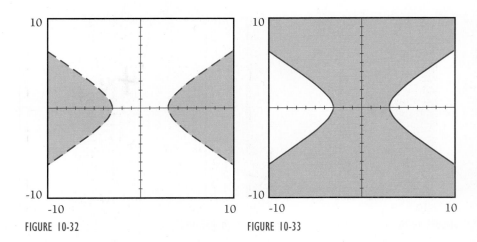

FIGURE 10-32   FIGURE 10-33

Because $(1)^2 - 9(1)^2 > 36$ is false, the correct region is outside the branches. See Figure 10-32.

Illustration 6:

$$4x^2 - 9y^2 \leqslant 36$$

Solution: Compare this to Illustration 5. The graph is in Figure 10-33. ▬

Suppose we simultaneously consider two or more inequalities in two variables. A **system of inequalities** in two variable consists of two or more inequalities. The graph of solution of a system of inequalities consists of those points (or regions, if any) that satisfy all of the individual inequalities of the system. This is equivalent to determining the region(s) (if any) where all of the graphs of the individual inequalities of the system overlap. Systems of inequalities require careful graphing techniques.

Graph the following systems of inequalities.

EXAMPLE 10-5

Illustration 1:

$$\begin{cases} y \leqslant 3 - x \\ y \geqslant x^2 - 1 \end{cases}$$

Solution 1: Sketch each graph separately, as in Figures 10-34 and 10-35. Determine the overlapping region of the two graphs for the solution in Figure 10-36.

Solution 2: Because both boundaries are functions, use graphing technology. Set the range to Xmin = -10, Xmax = 10, Ymin = -10, Ymax = 10. Be sure you are in standard modes: Norm, Function, Rect, etc.

For the Casio fx-7700G, place the calculator in the graphing inequality mode: [AC] [MODE] [SHIFT] [÷]. Enter the graph of the first inequality: [GRAPH] [F2] [3] [–] [XθT] [EXE]. Enter the graph of the second inequality: [GRAPH] [F3] [XθT] [SHIFT] [√] [–] [1] [EXE].

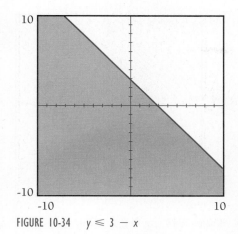

FIGURE 10-34 $y \leq 3 - x$

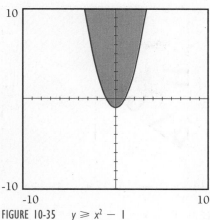

FIGURE 10-35 $y \geq x^2 - 1$

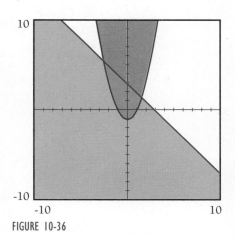

FIGURE 10-36

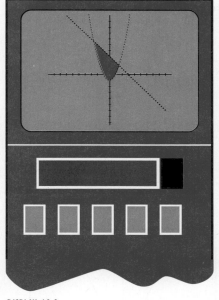

DISPLAY 10-3

For the Sharp EL9200/9300, store $3 - x$ under Y1 and $x^2 - 1$ under Y2. Continue to press [ENTER] through Y3 and Y4. The next screen will request FILL BELOW and FILL ABOVE. Select Y1 under FILL BELOW and Y2 under FILL ABOVE. Press [▲▼] then Y1 and Y2 are drawn and the region between them will be filled.

The TI-81 does shading with [2nd] [PRGM] [7] (DRAW Shade). Graph with [3] [−] [X|T] [ALPHA] [.] [X|T] [x²] [−] [1] [ENTER]. More responsibility lies with the user of the TI-81. The first entered function is the "lower" bound of shading. The second entered function is the "upper" bound of shading. Shading is done between the graphs, but not based upon the use of > or <. For example, for Shade enter [X|T] [x²] [−] [1] [ALPHA] [.] [3] [−] [X|T] [ENTER]. You get the same boundaries, but a different shaded region.

For *GraphWindows* prefix the formula with $y >$ or $y <$ as needed. Store $y < x - 2$ under F1. The graph of F1 will be the region below the line $y = x - 2$. Each of F1 through F4 is graphed in different color. Overlapping regions blend these colors. Press F1 to graph. Press F1 again and the inequality will be ungraphed without affecting the remaining graphs.

Your graphing technology may not differ from these examples, but you should refer to your user's manual. The overlapping region of the two inequalities appears in Display 10-3.

Illustration 2:

$$x > 0$$
$$y \geq 0$$
$$9x^2 - 4y^2 \geq 36$$
$$4x^2 + 36y^2 \leq 144$$

Solution: Graph each of the inequalities of the system separately, as we have in Figures 10-37 through 10-40 (See page 648).

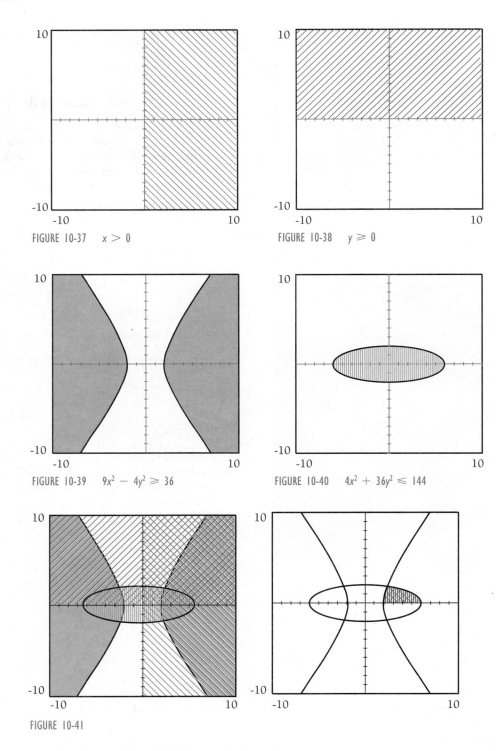

FIGURE 10-37 $x > 0$

FIGURE 10-38 $y \geqslant 0$

FIGURE 10-39 $9x^2 - 4y^2 \geqslant 36$

FIGURE 10-40 $4x^2 + 36y^2 \leqslant 144$

FIGURE 10-41

Combine these into one graph, as in Figure 10-41. The solution for the system is the region common to all the inequalities

SUMMARY

- To graph a single inequality, do the following.

  1. Imagine the inequality replaced with an equal sign. The result-ing equation represents the boundary of the graph. If the origi-nal inequality was less than or equal to, or greater than or equal to, then the boundary is part of the solution and should be drawn as a solid curve. Otherwise, draw the boundary as a dashed or dotted curve.

  2. The boundary divides the plane into two or more regions. Choose a point from each region and test it in the inequality. If the point satisfies the inequality, that region is part of the solu-tion. Shade that region. If the point does not satisfy the inequal-ity, that region is not part of the solution: do not shade it. If the remaining regions cannot be conjectured from the first point tried, use trial and error on each region separately.

- To graph a system of inequalities, do the following.

  1. Graph each inequality of the system separately. (See the preced-ing.)

  2. The final solution is the common overlap of all the separate so-lutions. Use careful shading to draw a combined view of all the separate inequalities into one graph where only the common solution is heavily shaded.

- If the boundary curve is a function represented by $y = f(x)$, then the inequality $y < f(x)$ is the region below the boundary. For $y > f(x)$, the graph is the region above the boundary. Because of this, graphing technology can shade regions to represent graphs of in-equalities when the boundary curves are functions.

10-3   E X E R C I S E S

Graph the following.

1. $y \leqslant 2x$

2. $y \geqslant 3x - 1$

3. $y < 3 - 2x$

4. $y > 2 - 3x$

5. $y \geqslant x^2$

6. $y \leqslant 1 - x^2$

7. $y > 5 - x^2$

8. $y < x^2 - 2$

9. $y \leqslant (x - 3)^2$

10. $y \geqslant (3 - x)^2$

11. $x^2 + y^2 < 9$

12. $x^2 + y^2 > 16$

13. $x^2 - y^2 \leqslant 16$

14. $x^2 - y^2 \geqslant 9$

15. $y^2 - x^2 > 9$

16. $y^2 - x^2 < 16$

17. $y^2 + 16x^2 \geqslant 16$

18. $x^2 + 9y^2 \leqslant 9$

19. $9x^2 - y^2 < 9$

20. $x^2 - 16y^2 > 16$

Graph the following systems.

21. $\begin{cases} y \le 2x \\ y > 5 - x^2 \end{cases}$

22. $\begin{cases} y \ge 3x - 1 \\ y \le 1 - x^2 \end{cases}$

23. $\begin{cases} y < 3 - 2x \\ y \ge x^2 \end{cases}$

24. $\begin{cases} y > 2 - 3x \\ y < x^2 - 2 \end{cases}$

25. $\begin{cases} y \le (x - 3)^2 \\ x^2 + y^2 < 9 \end{cases}$

26. $\begin{cases} y \ge (3 - x)^2 \\ x^2 + y^2 > 16 \end{cases}$

27. $\begin{cases} y^2 - x^2 > 9 \\ y > 5 - x^2 \end{cases}$

28. $\begin{cases} x^2 + y^2 > 16 \\ y > 5 - x^2 \end{cases}$

29. $\begin{cases} x^2 + 9y^2 \le 9 \\ y > 5 - x^2 \end{cases}$

30. $\begin{cases} x^2 + y^2 > 16 \\ x^2 - 16y^2 > 16 \end{cases}$

31. $\begin{cases} x^2 + y^2 \le 25 \\ y > 5 - x^2 \\ y < x \end{cases}$

32. $\begin{cases} x^2 + y^2 \le 16 \\ y > 4 - x^2 \\ y > x \end{cases}$

33. $\begin{cases} x > 0 \\ x + y \le 9 \\ y < 3 \\ x < 2 \end{cases}$

34. $\begin{cases} x + 3y \le 9 \\ y < 2 \\ x > 1 \\ x < 5 \end{cases}$

35. $\begin{cases} x^2 + y^2 \le 36 \\ x^2 + y^2 > 9 \\ y > x \end{cases}$

36. $\begin{cases} x^2 - y^2 \le 36 \\ x^2 - y^2 > 9 \\ y + x > 0 \end{cases}$

10-3   P R O B L E M   S E T

Consider a linear equation of three variables (x, y, z), where the equation is solved for z as a formula of x and y: $z = f(x, y)$. The graph of f is a plane in three space. Suppose we restrict the values for x and y to a subset of the xy plane described by a system of linear inequalities. A system of linear inequalities defines a region known as a *convex set* (or convex region). A set is convex if, for any two points in the set, the line segment connecting them is also in the set.

Figure 10-42 shows a convex set with linear boundaries compared with a region that is not convex. Like polygons in geometry, the points of intersection of the boundary segments of the convex set are called *vertices*.

Examine Display 10-4. Display 10-4 shows a linear equation of the form $z = f(x, y)$, where (x, y) lies in a convex region. Intuition (and the display) leads us to conclude that the largest value for z occurs at a vertex of the convex set. The smallest value for z also occurs at a vertex. The smallest and largest values for z over the convex set are known as optimal values for $z = f(x, y)$. If we wish to optimize values of $z = f(x, y)$ over a convex set with linear boundaries, we need only evaluate f at each vertex. Whichever vertex value (x, y) produces the largest value for z would be one optimum value.

Whichever vertex value (x, y) produces the smallest value for z provides the other optimal value. Although this trial substitution method is not very efficient, it does work. In general, a process of determining optimal values in problems such as this is called *linear programming*. The equation $z = f(x, y)$ to optimize is called the *objective* function. The set

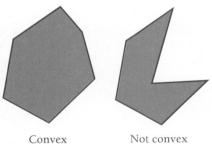

Convex   Not convex

FIGURE 10-42

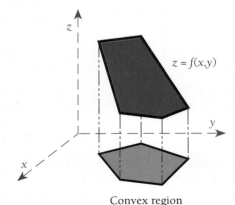

Convex region

DISPLAY 10-4

of linear inequalities that describe the convex set are the constraints. The vertices of the convex set provide the feasible solutions to the linear programming problem.

1. (a) Graph
$$\begin{cases} x \geq 0 \\ y \geq 0 \\ x + 5y \leq 10 \\ 4x - 5y \leq 20 \end{cases}$$

(b) Use this graph. Determine the point of intersection of $x = 0$ and $x + 5y = 10$. Determine the point of intersection of $x + 5y = 10$ and $4x - 5y = 20$. Determine the point of intersection of $4x - 5y = 20$ and $y = 0$. Determine the point of intersection of $x = 0$ and $y = 0$. Label these points on the graph as vertices. These are the feasible solutions.

(c) Why did we not worry about $4x - 5y = 20$ intersecting $x = 0$ or $x + 5y = 10$ intersecting $y = 0$?

(d) Consider the objective function $C = 10x - 10y$. Evaluate the objective function at each labeled vertex. What is the largest value for C? What is the smallest value for C?

2. See the system of inequalities for Problem 1. Consider the objective function $C = 20y - 5x$. Suppose we want the optimal solutions for C.

(a) What are the feasible solutions?

(b) What is the maximum value for C? At which vertex?

(c) What is the minimum value for C? At which vertex?

3. Maximize $C = 15x + 3y$ with constraints
$$\begin{cases} x \geq 0 \\ y \geq 0 \\ 2x + y \leq 16 \\ 5x + 4y \leq 50 \end{cases}$$

4. A data storage company rents short-term storage for data for its temporary archives. Storage of data on media x costs $7 per gigabyte. Storage on media y costs $4 per gigabyte. The company can never afford more than $560 invested in short-term storage. Therefore one constraint on their expenses is $7x + 4y \leq 560$. The cost of transferring data is different for the two storage media. It costs $7 per gigabyte to transfer data to and from media x but only $3 per gigabyte to transfer data to and from media y. The maximum amount the company can afford for data transfer expenses is $490. A second constraint on their storage is $7x + 3y \leq 490$. Security, insurance and overhead for extra space influence the profit for data storage on the two media. The profit for storing data on media x is 2 per giga-

byte whereas the profit for storage on media y is $1 per gigabyte. Therefore profit is given by $P = 2x + y$. Determine the number of units of x and y to maximize the companies profit.

PROBLEMS FOR GRAPHING TECHNOLOGY

Use graphing technology to graph the following inequalities. Annotate each graph by naming the boundary and indicating whether it is included.

5. $y \leq 2x$

6. $y \geq 3x - 1$

7. $y < 3 - 2x$

8. $y > 2 - 3x$

9. $y \geq x^2$

10. $y \leq 1 - x^2$

11. $y > 5 - x^2$

12. $y < x^2 - 2$

13. $y \leq (x - 3)^2$

14. $y \geq (3 - x)^2$

15. $\begin{cases} y \leq 2x \\ y > 5 - x^2 \end{cases}$

16. $\begin{cases} y \geq 3x - 1 \\ y \leq 1 - x^2 \end{cases}$

17. $\begin{cases} y < 3 - 2x \\ y \geq x^2 \end{cases}$

18. $\begin{cases} y > 2 - 3x \\ y < x^2 - 2 \end{cases}$

19. $\begin{cases} y < 3^{-2x} \\ y \geq \sin x \end{cases}$

20. $\begin{cases} y < \ln x \\ y \geq \cos x \end{cases}$

21. $\begin{cases} y < x \\ y \geq e^x \\ y < 9 - x^2 \end{cases}$

22. $\begin{cases} y < -x \\ y \geq x^2 - 4 \\ y > 2 \cos x \end{cases}$

23. $\begin{cases} y < 5 - 2x \\ y \geq x^2 - 5 \\ y < \ln x \end{cases}$

24. $\begin{cases} y < 3x \\ y \geq x^2 \\ y < e^{-x} \end{cases}$

25. In the same window graph,

(a) Graph $y < \sqrt{16 - x^2}$ and $y > -\sqrt{16 - x^2}$.

(b) Graph $y > \sqrt{x^2 - 16}$ and $y < -\sqrt{x^2 - 16}$.

(c) Describe a method for graphing the interior of a conic section that is not a function.

(d) What is the shape of the boundary for this combined inequality? How could this method be used for graphing the exterior of a conic section?

CHAPTER SUMMARY

- To solve an n by n system of linear equations, reduce the system to an $n - 1$ by $n - 1$ system using the method of elimination. Repeat this process until the current system is 1 by 1. Solve for the variable in the 1 by 1 system. Recover the eliminated variables. If the current solved system is m by m, then substitute the m solved variables into the next $m + 1$ by $m + 1$ system. This will leave one variable in that system. Solve for it. Repeat this substitution until $m + 1$ equals the original n.
- If a combination eliminates all variables and results in an equation of the form $0 = c$ (where c is some constant), then the original system was either dependent or inconsistent.
- To solve a quadratic system in which one of the equations represents a parabola, use the first-degree variable from the equation for the parabola to substitute into the second equation. This eliminates one variable. The resulting equation may be as complex as fourth degree. Solve for this variable. Substitute the solutions into the equation for the parabola. In that manner determine the pairs of values for each point (For distinct graphs, there will be as few as none but no more than four points of intersection.
- For a quadratic system in which both equations are second degree in both variables, substitute or combine if possible to eliminate one of the variables.
- To graph a single inequality, follow these steps.

  1. Imagine the inequality symbol replaced with an equal sign. The resulting equation represents the boundary of the graph. If the original inequality was less than or equal to, or greater than or equal to, then the boundary is part of the solution and should be drawn as a solid curve. Otherwise draw the boundary as a dashed or dotted curve.
  2. The boundary divides the plane into two or more regions. Choose a point from each region and test it in the inequality. If the point satisfies the inequality, that region is part of the solution. Shade that region. If the point does not satisfy the inequality, that region is not part of the solution. If the remaining regions cannot be conjectured from the first point tried, use trial and error on each region separately.

- To graph a system of inequalities, follow these steps.

  1. Graph each inequality of the system separately. (See the preceding)
  2. The final solution is the common overlap of all the separate solutions. Use careful shading to draw a combined view of all the separate inequalities into one graph, where only the common solution is heavily shaded.

KEY WORDS AND CONCEPTS

Consistent	Inconsistent	Quadratic system
Dependent	Independent	Substitution
Equivalent	Linear combination	System of linear equations
Half-plane	Method of elimination	Systems of inequalities

10   R E V I E W   E X E R C I S E S

SECTION 1

Solve the following systems.

1. $\begin{cases} 2x + y = 8 \\ x - 3y = 11 \end{cases}$ **2.** $\begin{cases} 3x + y = 9 \\ x + y = 1 \end{cases}$ **3.** $\begin{cases} 2x - 2y = 4 \\ x - y = 5 \end{cases}$

4. $\begin{cases} 3x + 2y = 5 \\ 6x + 4y = 10 \end{cases}$ **5.** $\begin{cases} x - 2y = 4 \\ 3x + 6y = 12 \end{cases}$ **6.** $\begin{cases} y = 5 - x \\ y = 1 + x \end{cases}$

7. $\begin{cases} y = 7 \\ x = -4 \end{cases}$ **8.** $\begin{cases} x = \pi \\ y = -e \end{cases}$ **9.** $\begin{cases} x + y = 7 \\ x - y = 11 \end{cases}$

10. $\begin{cases} 2x + 3y = 3 \\ x + y = 4 \end{cases}$ **11.** $\begin{cases} 3x - y = 4 \\ y = 2x - 4 \end{cases}$

12. $\begin{cases} 0.5x + 0.2y = 1.3 \\ 0.5x - 0.4y = 0.8 \end{cases}$ **13.** $\begin{cases} \dfrac{1}{5}x - \dfrac{1}{4}y = \dfrac{1}{10} \\ \dfrac{3}{5}x - \dfrac{1}{3}y = \dfrac{1}{2} \end{cases}$

14. $\begin{cases} y = \dfrac{4}{3}x + 2 \\ y = -\dfrac{3}{5}x - 3 \end{cases}$ **15.** $\begin{cases} 5r - 2s = 5 \\ 4r + 3s = 1 \end{cases}$ **16.** $\begin{cases} 3r - 5s = 1 \\ 5r + 3s = 5 \end{cases}$

17. Determine the radii of two pulleys with design parameters given by $\begin{cases} 3r_1 + r_2 = 19 \\ 2r_1 - r_2 = 1 \end{cases}$

18. The resistance R and temperature T of a wire are related by the formula $R = aT + b$. If the resistance is 35 ohms at 15° and the resistance is 40 ohms at 55°, determine the values of a and b.

Solve the following.

19. $\begin{cases} x + y + z = 4 \\ x - y + z = 2 \\ 3x - y - z = 16 \end{cases}$ **20.** $\begin{cases} x + y + z = 3 \\ x - y + z = 9 \\ 3x - y - z = 5 \end{cases}$

21. $\begin{cases} 2x - y + z = 1 \\ x + y - 2z = -18 \\ x + y + z = 6 \end{cases}$ **22.** $\begin{cases} 2x - y + z = 6 \\ x + y - 2z = -11 \\ x + y + z = 4 \end{cases}$

23. $\begin{cases} x - y + 2z = 7 \\ 2x + y - z = 22 \\ x + y = 11 \end{cases}$ **24.** $\begin{cases} x - y + 2z = 13 \\ 2x + y - z = 11 \\ x + y = 7 \end{cases}$

25. $\begin{cases} x - y + 2z + w = 9 \\ 2x + y - z - w = -4 \\ x + y = 0 \\ z + w = 5 \end{cases}$ **26.** $\begin{cases} x - y + 2z + w = 4 \\ 2x + y - z - w = -1 \\ x + y = 1 \\ z + w = 1 \end{cases}$

Use partial fractions to decompose the following.

*27. $f(x) = \dfrac{x - 2}{(x - 5)(x + 1)}$ ***28.** $f(x) = \dfrac{x + 5}{(x^2 - 4x + 3)}$

SECTION 2

Solve the following systems.

29. $\begin{cases} y = x \\ y = x^2 - 5x \end{cases}$ **30.** $\begin{cases} y = -2x \\ y = x^2 - 8 \end{cases}$ **31.** $\begin{cases} y = 3x - 2 \\ y = x^2 \end{cases}$

32. $\begin{cases} y = 1 - 2x \\ y = -x^2 - x + 7 \end{cases}$ **34.** $\begin{cases} y = x^2 \\ y = 3 - x^2 \end{cases}$ **33.** $\begin{cases} y = x^2 + \dfrac{1}{3} \\ y = \dfrac{1}{2}x + 1 \end{cases}$

35. $\begin{cases} y = 2x^2 \\ y = 3 - x^2 \end{cases}$ **36.** $\begin{cases} y = x^2 - x \\ y = x^2 + 2x - 9 \end{cases}$

37. $\begin{cases} y = x + 5 \\ y = x^2 + 5x - 1 \end{cases}$ **38.** $\begin{cases} y = x^2 + x + 1 \\ y = 2x^2 + x - 3 \end{cases}$

39. $\begin{cases} y = x + 3 \\ y = x^2 - x \end{cases}$ **40.** $\begin{cases} y = -4x \\ y = x^2 + x + 4 \end{cases}$

41. $\begin{cases} x^2 - y^2 = 5 \\ x^2 + y^2 = 13 \end{cases}$ **42.** $\begin{cases} xy = 8 \\ y = x^2 \end{cases}$

43. $\begin{cases} xy = 3 \\ x^2 + y^2 = 10 \end{cases}$ **44.** $\begin{cases} xy = 8 \\ x^2 - y^2 = 12 \end{cases}$

45. $\begin{cases} y = 3^{2x} - 12 \\ y = 4(3^x) \end{cases}$ **46.** $\begin{cases} \dfrac{1}{x^2} - \dfrac{2}{y^2} = 1 \\ \dfrac{2}{x^2} + \dfrac{1}{y^2} = 22 \end{cases}$

(Hint: let $u = 3^x$)   (Hint: let $u = 1/x$ and $v = 1/y$)

SECTION 3

Graph the following.

47. $y \le 3x - 5$ **48.** $y > 7 - 5x$

49. $y \ge x^2 - 6$ **50.** $y \le 1 - 2x + x^2$

51. $y > 5 + 4x - x^2$ **52.** $y \ge (4 - x)^2$

53. $x^2 + y^2 < 16$ **54.** $y^2 - x^2 < 36$

55. $4x^2 + 9y^2 \le 36$ **56.** $x^2 - 16y^2 > 64$

Graph the following systems.

57. $\begin{cases} y \le 3x \\ y > 4 - x^2 \end{cases}$ **58.** $\begin{cases} y > 5 - 2x \\ y < x^2 - 1 \end{cases}$ **59.** $\begin{cases} x^2 + y^2 > 9 \\ y > 3 - x^2 \end{cases}$

60. $\begin{cases} y > 0 \\ x + y > 1 \\ y < 3 \\ x < 5 \end{cases}$ **61.** $\begin{cases} x^2 - y^2 \le 16 \\ x^2 + y^2 > 9 \\ y + x > 0 \end{cases}$ **62.** $\begin{cases} x^2 + y^2 \le 25 \\ y > 5 - x^2 \\ y > x \end{cases}$

63. $\begin{cases} y > (x - 5)^2 \\ x^2 + (y - 3)^2 < 16 \end{cases}$ **64.** $\begin{cases} x^2 + 9y^2 \le 81 \\ y > 9 - \dfrac{x^2}{9} \end{cases}$

Arrays, Determinants and Matrices

The inspiration for arrays was Hamilton's quaternions. Arrays provided a tool by which physicists could represent natural phenomona. James Maxwell provided the first impetus to the evolution of quaternions to vectors, but the deciding break of vectors with quaternions came from a pamphlet by the mathematical physicist Josiah Gibbs, entitled *Elements of Vector Analysis*. At approximately the same time a telephone engineer named Oliver Heaviside developed a similar approach for the study of electromagnetic theory. Both these men developed vectors as having the direction and length associated with quaternions but separate from the quaternion idea. Vectors were a further abstraction eliminating the distracting elements of quaternions. At last physicists and engineers had a powerful tool for attacking the physical problems of the day. Vectors and vector analysis not only provided a tool for solving problems, they provided a new way of looking at problems.

Abstracting information by leaving out distracting details was not a new process. Consider the simple 2 by 2 linear system of equations:

$$\begin{cases} ax + by = c \\ dx + ey = f \end{cases}$$

Mathematicians had already abbreviated this problem by eliminating the details of the variable names. They wrote down only the coefficients of the variables in the shape of a rectangle.

$$\begin{vmatrix} a & b \\ d & e \end{vmatrix}$$

They then defined an arithmetic operation on these numbers that would help determine the solution of the system quickly. Such notation was called a *determinant*. Gottfried Liebniz (1646–1716) is usually given credit for originating the idea of determinants. (Leibniz also invented calculus independent of Isaac Newton.) Many mathematicians contributed to the use of determinants in problem solving. Among them were Cramer, Vandermonde, Laplace, Cauchy, Jacobi and Gauss. However, one of the most ardent developers of determinants was James Sylvester (1814–1897). Sylvester was a close friend of another mathematician whose name is associated with determinants, Arthur Cayley (1821–1895). Arthur Cayley is usually listed as one of the three most prolific writers of mathematics. (Euler is number one; second place is somewhat a toss up between Cayley and Cauchy.) Arthur Cayley is usually given credit for the invention of matrices.

Some people believe, incorrectly, that Cayley invented matrices from the inspiration of vectors. Cayley denies this. Instead he claims his inspiration was affine transformations of the form

$$\begin{cases} x' = ax + by \\ y' = cx + dy \end{cases}$$

and determinants. From the affine transformation and the determinant Cayley introduced the notation $\begin{bmatrix} a & b \\ c & d \end{bmatrix}$.

Although arrays provided a powerful impetus to physics both as a new tool and a new concept, determinants and arrays were new notational devices. They did not immediately spur a new way of looking at nature, instead they provided a compact notation for handling ideas already in existence (systems of equations, transformations, etc.)

In the next chapter we introduce matrices and determinants. However we shall not follow history. Determinants will be introduced as a consequence of matrices rather than a predecessor.

10 CHAPTER TEST

Solve the following.

1. $\begin{cases} 3x - 2y = 4 \\ 2x + 3y = 1 \end{cases}$

2. $\begin{cases} 2x - y + z = 3 \\ 3x + y - 2z = 1 \\ x + 2y + z = 3 \end{cases}$

3. $\begin{cases} y = 3x - 2 \\ y = x^2 + x \end{cases}$

4. $\begin{cases} xy = 4 \\ x^2 + y^2 = 9 \end{cases}$

Graph the following.

5. $(x - 3)^2 + (y + 2)^2 < 9$

6. $\begin{cases} y \leqslant 3 - x \\ y^2 - x^2 > 16 \end{cases}$

7. An employee can split her time between two tasks at her company. Task A pays \$9.00/hour. Task B pays \$5.00/hour. If the employee earns \$300 in a 40 hour week, how did she split her time between the tasks?

Matrices and Determinants

11

Many books use tables to present data. Tabular form provides a convenient and compact method for storing and displaying information. Tables helped inspire an entire category of computer program, the electronic spreadsheet. Tables are a special case of a more general form for storing data known as *matrices*. The English mathematician Arthur Cayley (1821–1895) introduced matrices as an efficient notational device. Tabular formats as computational devices were established before Cayley extracted matrices as a separate idea. In a paper published in 1855, Cayley referred to *affine transformations* and *determinants* as the inspiration for matrices. This chapter introduces matrices and determinants as powerful notational devices.

1

*Matrices: Address Routines

Many arts there are which beautify the mind of man; of all other none do more garnish and beautify it than those arts which are called mathematical.
—H. Billingsley

A **matrix** is a rectangular array of numbers. Matrices organize information into rows and columns of **elements**. Monthly calendars organize appointments by weeks, the rows, and days of the week, the columns. When February consists of exactly four weeks of seven days each, it fits the shape of a matrix (Display 11-1, page 658).

Name this array of numbers **Feb**. Identifying Week 3, Thur locates a particular day of the month, in this case 19. Week 3, Thur is the address of 19, which is an element of the matrix **Feb**. The **dimension** (or **order**) **of a matrix** measures the size of the matrix in terms of the number of rows and the number of columns. Ignoring the labels naming the days and weeks, matrix **Feb** has four rows and seven columns. We say that matrix **Feb** is of dimension 4×7 (read "4 by 7").

DISPLAY 11-1   February

	Sun	Mon	Tues	Wed	Thur	Fri	Sat
Week 1	1	2	3	4	5	6	7
Week 2	8	9	10	11	12	13	14
Week 3	15	16	17	18	19	20	21
Week 4	22	23	24	25	26	27	28

Giving the dimension of a matrix is reminiscent of describing systems of linear equations. A system of two linear equations in two variables is a 2 by 2 system. The resemblance is not accidental. One use of matrices is to abbreviate the representation of systems of linear equations. Examine the 3 by 3 system

$$\begin{bmatrix} 3x + 2y - z = 4 \\ 5x - 7y + 6z = 9 \\ 8x + 9y - 4z = 2 \end{bmatrix}$$

Let us abbreviate the lefthand side of the equations in this system into a matrix

$$C = \begin{bmatrix} 3 & 2 & -1 \\ 5 & -7 & 6 \\ 8 & 9 & -4 \end{bmatrix}$$

The first column represents the coefficients of x. The second column contains the coefficients of y; and the third column, z. Row 1 presents coefficients of the first equation. Row 2 shows the coefficients of the second equation. Row 3 gives the coefficients of the third equation. Matrix C is of order 3×3. Abbreviate this as $C_{3\times3}$. Because the number of rows equals the number of columns, this is a **square matrix**.

Because of the source of its elements, C is called the *coefficient* matrix of the system. As with **Feb**, the rows and columns of C could have been labeled. But the purpose of a matrix is to abbreviate. Therefore, we omit labels and address the elements by row and column number. The element in row 2, column 3 of C is 6. Matrices are commonly named using capital letters. Matching lower-case letters indicate individual elements of the matrix. Because C contains nine elements, we distinguish the elements using a pair of subscripts. For example, $c_{3,2}$ indicates the element of C in the third row, second column. Thus, $c_{3,2} = 9$. The first subscript is the row and the second subscript is the column. Hence, $c_{3,2}$ is different than $c_{2,3}$. In fact, $c_{2,3} = 6$.

Consider a matrix K of order 1×3, $K = [4\ 9\ 2]$. Matrices with only one row are called *row matrices*. We could have written the components one above another, instead of left to right. Such matrices with only one column are known as *column matrices*. We rewrite row matrix K as a column matrix:

$$K^T = \begin{bmatrix} 4 \\ 9 \\ 2 \end{bmatrix}$$

The superscript T indicates that the row matrix has been transposed into a column. Read the notation as "K transpose." The elements of $\mathbf{K}^T$ are the constants from the righthand side of the equations in the given 3 by 3 system of linear equations. Matrix $\mathbf{K}^T$ is the constant matrix for the system of equations.

Consider the following 3×5 matrix $\mathbf{B}$:

$$\mathbf{B} = \begin{bmatrix} 1 & 2 & 3 & 4 & 5 \\ 0 & 9 & 1 & 4 & 0 \\ 1 & 2 & 9 & 8 & -8 \end{bmatrix}$$

Clearly there are three rows and five columns. The element in the third row, fourth column is $b_{3,4} = 8$. To indicate that $\mathbf{B}$ is an 3×5 matrix, we write $\mathbf{B}_{3\times5}$. Alternately, we notate a matrix using bracket notation with one representative element in lower case, $[b_{ij}]_{3\times5}$, where b_{ij} represents the element in the ith row and jth column.

DEFINITION 11-1
Equality of Matrices

> Two matrices of the same dimension are equal if their corresponding elements are equal.

For example, if we want $\begin{bmatrix} 1 & 2 & x \\ 3 & 7 & 5 \end{bmatrix} = \begin{bmatrix} 1 & 2 & 4 \\ y & 7 & 5 \end{bmatrix}$ then $x = 4$ and $y = 3$. They are then **equal matrices.**

DEFINITION 11-2
The Zero Matrix

> The **zero matrix** of dimension $m \times n$ is a matrix in which every element is 0. Denote the zero matrix by $0_{m\times n} = [0_{ij}]$.

For example the 2×3 zero matrix is $0_{2\times3} = \begin{bmatrix} 0 & 0 & 0 \\ 0 & 0 & 0 \end{bmatrix}$.

DEFINITION 11-3
Transpose of a Matrix

> The **transpose of a matrix B** is a matrix $\mathbf{B}^T$ formed by interchanging the rows and columns of $\mathbf{B}$ (i.e., if $\mathbf{B}_{m\times n} = [b_{ij}]_{m\times n}$, then $\mathbf{B}^T = [b_{ji}]_{n\times m}$).

For brevity, we omit the subscript indicating the dimension of a matrix when the dimension is obvious. Consider the 2×3 matrix

$$\mathbf{C} = \begin{bmatrix} 1 & 2 & 3 \\ 4 & 5 & 6 \end{bmatrix}$$

Then $\mathbf{C}^T$ is the 3×2 matrix

$$\mathbf{C}^T = \begin{bmatrix} 1 & 4 \\ 2 & 5 \\ 3 & 6 \end{bmatrix}$$

In many respects, the transpose of a matrix resembles the converse of a relation. Each row of a matrix is a column in its transpose. Similarly, each column in a matrix is a row in its transpose. For example, 6 occupies position $c_{2,3}$ in C, but 6 is in row 3 column 2 of C^T.

The transpose of row matrix $V = [1\ 2\ 3\ 4]$ is column matrix

$$V^T = \begin{bmatrix} 1 \\ 2 \\ 3 \\ 4 \end{bmatrix}$$

The coefficient matrix of a system of three equations in three variables is square. The following definition applies to more than square matrices, but because of systems of equations, square matrices are our major focus.

DEFINITION 11-4
Main Diagonal of a Matrix

The **main diagonal of matrix** $A_{m \times n} = [a_{ij}]$ consists of those elements of A for which their row number equals their column number.

For matrix

$$A = \begin{bmatrix} 1 & 2 & 3 \\ 4 & 5 & 6 \\ 7 & 8 & 9 \end{bmatrix}$$

the main diagonal consists of components given by $a_{1,1} = 1$, $a_{2,2} = 5$ and $a_{3,3} = 9$. Imagine a line from the upper lefthand corner of the matrix to the lower righthand corner. This line, like the diagonal of a rectangle, slashes through the components of the main diagonal. We might imagine $[1\ 5\ 9]$ as a matrix formed from the main diagonal of the matrix.

SUMMARY

- A matrix is a rectangular array of elements organized into rows and columns. A boldface capital letter often names a matrix. Subscripts indicate the number of rows and columns.
- The dimension (or order) of a matrix is given by the number of rows and columns. An alternate matrix notation describes the matrix using brackets and a general element: $A_{m \times n} = [a_{ij}]_{m \times n}$ represents a matrix of dimension $m \times n$ where a_{ij} represents a single element of the array found in the ith row and jth column.
- The $m \times n$ zero matrix, $0_{m \times n}$, is a matrix in which all elements are 0.
- Two matrices are equal if their corresponding elements are equal.
- The transpose of a matrix B, denoted by B^T, is formed by interchanging the rows and columns of B.
- The main diagonal of a matrix consists of those elements for which the row and column numbers are equal.

For the matrix $A = \begin{bmatrix} 1 & 2 & 3 & 4 & 5 \\ 9 & 8 & 7 & 6 & 5 \end{bmatrix}$, identify each of the following.

1. $a_{2,4}$ **2.** $a_{1,3}$

3. $a_{1,2}$ **4.** $a_{2,5}$

5. column 4   **6.** row 1

7. row 2   **8.** column 2

9. A^T **10.** $(A^T)^T$

For the matrix $B = \begin{bmatrix} 1 & 2 & 3 \\ 4 & 5 & 6 \\ 7 & 8 & 9 \end{bmatrix}$ identify each of the following.

11. $b_{3,2}$ **12.** $b_{2,3}$

13. column 3   **14.** row 3

15. row 1   **16.** column 1

17. $(B^T)^T$ **18.** B^T

19. the main diagonal of B

20. the main diagonal of B^T

For each of the following matrices A, determine A^T.

21. $\begin{bmatrix} 1 & 0 \\ 0 & 1 \end{bmatrix}$ **22.** $\begin{bmatrix} 1 & 3 \\ 3 & 4 \end{bmatrix}$

23. $\begin{bmatrix} 5 & 2 \\ 3 & 4 \end{bmatrix}$ **24.** $\begin{bmatrix} -2 & 1 \\ 4 & 3 \end{bmatrix}$

25. $\begin{bmatrix} 1 & 3 & 6 \\ -3 & 7 & 8 \end{bmatrix}$ **26.** $\begin{bmatrix} 6 & -4 & 3 \\ 4 & 1 & 5 \end{bmatrix}$

27. $\begin{bmatrix} 1 & 5 \\ 4 & -2 \\ 1 & 3 \end{bmatrix}$ **28.** $\begin{bmatrix} 2 & 0 \\ 3 & -1 \\ 4 & 2 \end{bmatrix}$

29. $\begin{bmatrix} 1 & 2 & 1 \\ -1 & 4 & 5 \\ 3 & 0 & 4 \end{bmatrix}$ **30.** $\begin{bmatrix} 3 & 4 & 0 \\ -2 & 1 & 1 \\ 0 & 0 & 3 \end{bmatrix}$

31. Under what circumstances will $A = A^T$?

32. Show that if A is an $m \times n$ matrix and $A = A^T$, then $m = n$.

33. Form a 3×3 matrix in which $a_{ij} = i + j$.

34. Form a 3×3 matrix in which $a_{ij} = i - j$.

35. Suppose that a bank pays 5 percent simple interest on a deposit. Form a 4×4 matrix in which a_{ij} is the simple interest paid on i dollars after j years.

36. Suppose that a loan is to be repaid at 10 percent simple interest. Form a 4×4 matrix in which a_{ij} is the amount of interest charged on j dollars after i years.

37. Form a 4×4 matrix in which a_{ij} is the total cost of buying i tons of steel at j dollars per ton.

38. Form a 4×4 matrix in which a_{ij} is the average price per item of i items bought for a total price of j dollars.

39. If $\begin{bmatrix} 1 & x & 6 \\ -3 & 7 & z \end{bmatrix} = \begin{bmatrix} w-4 & 3 & 6 \\ -3 & y+2 & 8 \end{bmatrix}$. Determine values for x and y.

40. See Problem 39. Determine the values of z and w.

Determine each of the following.

41. If $a_{5,7} = 3$ in matrix A, what element of A^T will be 3?

42. If $b_{2,6} = 5$ in matrix B, what element of B^T will be 5?

43. Suppose that $a_{3,3}$ is an element of square matrix A. Where will this element be in A^T?

44. Suppose that $b_{5,5}$ is an element of square matrix B. Where will this element be in B^T?

45. Prove that if a_{ii} is on the main diagonal of A, then a_{ii} occupies the same position in A^T.

46. Prove that the main diagonals of B and B^T are the same.

47. Suppose A is a 3×3 matrix defined as follows

$$\begin{cases} a_{ij} = 0, & \text{if } i \neq j \\ a_{ij} = 1, & \text{if } i = j \end{cases}$$

write A in standard matrix format (bracket notation).

48. Suppose B is a 4×4 matrix defined as follows $\begin{cases} a_{ij} = 0, & \text{if } i \neq j \\ a_{ij} = 1, & \text{if } i = j \end{cases}$, write B in standard matrix format (bracket notation).

49. Suppose that $A = \begin{bmatrix} 1 & 2 \\ 3 & 4 \end{bmatrix}$, $B = \begin{bmatrix} 4 & 1 \\ 5 & 2 \end{bmatrix}$. Now suppose that C is defined as follows: $[c_{ij}] = [a_{ij} + b_{ij}]$. Write C in standard format.

50. Suppose that $A = \begin{bmatrix} 1 & 2 \\ 3 & 4 \end{bmatrix}$. Suppose that C is defined as follows: $[c_{ij}] = [5a_{ij}]$. Write C in standard format.

51. A matrix is symmetric if it is equal to its transpose. Does this mean the matrix must be square? Why?

52. See Problem 51. Is the 3×3 zero matrix symmetric?

53. See Problem 51. Is the matrix in Problem 27 symmetric?

54. See Problem 51. Is the matrix in Problem 28 symmetric?

55. Give an example of a symmetric matrix in which $a_{1,1} = 5$.

56. Give an example of a symmetric matrix in which $a_{1,1} = -1$.

57. Give an example of a symmetric matrix in which $a_{2,1} = 3$.

58. Give an example of a symmetric matrix in which $a_{1,2} = 4$.

59. Suppose **A** is a 3×3 matrix defined as follows $[a_{ij}] = [2i + j]$. Write **A** in standard matrix format.

60. Suppose **B** is a 3×3 matrix defined as follows: $[a_{ij}] = [(-1)^{i+j}]$. Write **B** in standard matrix format.

11-1   P R O B L E M   S E T

PROBLEMS FOR TECHNOLOGY

The next sections develop an algebra for matrices and applications. Much of the drudgery of matrix manipulation can be done by calculator. The first task is to store a matrix in the calculator. Check your manual. To enter matrix mode in the Casio, press [AC] [MODE] [0]. Select matrix A by pressing [F1]. To change the dimension of the matrix, press [F6]. Select [F1] for the DIM process. Enter new row and column sizes at the prompt. Complete the setting with [EXE]. The screen displays a 3×3 matrix A. Navigate through the array and edit the elements to enter matrix

$$A = \begin{bmatrix} 1 & 2 & 3 \\ 4 & 5 & 6 \\ 7 & 8 & 9 \end{bmatrix}$$

At the bottom of the screen are matrix functions assigned to the function keys. For example, [F2] represents A^T. Press [F2] and observe the effect. The transpose is stored under matrix **C**. The original matrix **A** is still in the memory.

On the Sharp EL9300/9200, enter the matrix mode by pressing [▦] [MENU] [3]. To dimension a matrix press [MATH] [C]. Select a matrix to dimension and press [ENTER]. Enter the new dimension. When you press [ENTER] you will be given the opportunity to enter the elements. Press [QUIT] to finish. [MATH] [B] allows you to edit matrix entries. To take the transpose of A enter [MATH] [E] [5] [2ndF] [MAT] [A] [ENTER].

The [MATRX] key is the gateway to matrix operations on the TI-81. Press [MATRX] then select EDIT from the menu. At the new menu select 1 (Matrix A). Change the dimension of A to 3×3. As with the Casio, enter 1, 2, 3, 4, 5, 6, 7, 8, 9 as elements of matrix A. Press [2nd] [CLEAR] to return to the main screen. Now press [2nd] [1] [ENTER]. This will display matrix A on the screen. If the entire matrix does not fit on the screen, use the arrow keys to scroll left or right. Now press [2nd] [1]

[MATRX] [6] [ENTER]. This displays the transpose of **A** on the screen.

Use a calculator with matrix capabilities to accomplish each of the following.

1. Enter the matrix $A = \begin{bmatrix} 1 & 3 & 2 \\ 5 & 1 & 1 \\ 0 & 9 & 1 \end{bmatrix}$.

2. Determine A^T.

3. Determine $(A^T)^T$.

4. Compare A^T and $(A^T)^T$.

5. Compare the contents of **A** and $(A^T)^T$. Discuss when they will be equal.

6. Discuss when **A** and A^T will be equal.

7. Edit the original matrix **A** as follows: change $a_{1,2}$ to 5; change $a_{2,3}$ to 9; change $a_{3,1}$ to 2.

8. Determine the transpose of **A**.

9. Compare the entries in A^T and **A**.

10. See Problems 8 and 9. If $A^T = A$, then **A** is a symmetric matrix. See Exercise Problem 31. Is **A** a symmetric matrix?

11. Redimension matrix **A** to a 4×4 matrix. What happens to the original entries in **A** when you redimension?

12. Edit matrix **A**, using the numbers from 1–16 as entries in order from upper lefthand corner to lower righthand corner.

13. Determine A^T.

14. Determine $(A^T)^T$. Compare the result to **A**.

2 ADDITION OF MATRICES

Mathematics . . . the ideal and norm of all careful thinking.
—G. STANLEY HALL

The arithmetic of matrices is analogous to the arithmetic of vectors. The simplest of these arithmetic operations is matrix addition. Although we do not offer a physical interpretation of matrix addition similar to the parallelogram law for vectors, examples will indicate uses for **matrix addition**. As with vectors, the addition of two matrices is defined in terms of adding corresponding components.

DEFINITION 11-5
Matrix Addition

> The sum $A + B$ of two $m \times n$ matrices, $\mathbf{A}$ and $\mathbf{B}$, is an $m \times n$ matrix whose elements are the sum of the corresponding elements of $\mathbf{A}$ and $\mathbf{B}$. In particular, if $\mathbf{C} = \mathbf{A} + \mathbf{B}$, then $c_{ij} = a_{ij} + b_{ij}$.

Two matrices must be of the same dimension before they can be added. The sum of $\mathbf{A}_{2\times3}$ and $\mathbf{B}_{4\times2}$ is not defined. However, if $\mathbf{A} = \begin{bmatrix} 2 & \text{-}1 & 5 \\ 0 & 3 & 6 \end{bmatrix}$ and $\mathbf{C} = \begin{bmatrix} 4 & 1 & \text{-}5 \\ 2 & 7 & 10 \end{bmatrix}$, then $\mathbf{A} + \mathbf{C} = \begin{bmatrix} 6 & 0 & 0 \\ 2 & 10 & 16 \end{bmatrix}$.

Theorem 11-1
Matrix Addition is Commutative

> Suppose $\mathbf{A}$ and $\mathbf{B}$ are matrices of the same dimension, then $\mathbf{A} + \mathbf{B} = \mathbf{B} + \mathbf{A}$.

Note $\mathbf{C} + \mathbf{A} = \begin{bmatrix} 6 & 0 & 0 \\ 2 & 10 & 16 \end{bmatrix}$. The proof follows from the definition of addition and the fact that real numbers are commutative under addition.

Theorem 11-2
Matrix Addition is Associative

> If $\mathbf{A}$, $\mathbf{B}$ and $\mathbf{C}$ are matrices of the same order, then
> $$\mathbf{A} + (\mathbf{B} + \mathbf{C}) = (\mathbf{A} + \mathbf{B}) + \mathbf{C}$$

Once more the theorem follows from the definition of matrix addition. The details of the proof are left as an exercise. The associative property indicates that grouping symbols are not essential in a matrix addition. Therefore, we may write $\mathbf{A} + \mathbf{B} + \mathbf{C}$ without ambiguity.

The last section defined the $m \times n$ zero matrix, $\mathbf{0}_{m\times n}$. All entries of the zero matrix are 0. Because 0 is the additive identity for real numbers, each zero matrix becomes an additive identity for matrices of the same order. The theorem for **additive identity** formalizes this discussion. When used without specifying the dimension in a sum, assume **0** is the same dimension as the other matrices.

For any $m \times n$ matrix **A**, **A** + **0** = **A**.

EXAMPLE 11-1

Determine the indicated sums.

Illustration 1:

$$\begin{bmatrix} 3 & 1 & 6 \\ 5 & 2 & 1 \end{bmatrix} + \begin{bmatrix} 1 & 0 & 3 \\ -1 & 2 & 4 \end{bmatrix}$$

Solution:

$$\begin{bmatrix} 3 & 1 & 6 \\ 5 & 2 & 1 \end{bmatrix} + \begin{bmatrix} 1 & 0 & 3 \\ -1 & 2 & 4 \end{bmatrix} = \begin{bmatrix} 3+1 & 1+0 & 6+3 \\ 5+(-1) & 2+2 & 1+4 \end{bmatrix}$$

$$= \begin{bmatrix} 4 & 1 & 9 \\ 4 & 4 & 5 \end{bmatrix}$$

Illustration 2:

If $\mathbf{A} = \begin{bmatrix} 5 & 2 \\ 3 & 4 \end{bmatrix}$, determine $\mathbf{A} + \mathbf{A}^T$.

Solution 1:

$$\mathbf{A} + \mathbf{A}^T = \begin{bmatrix} 5 & 2 \\ 3 & 4 \end{bmatrix} + \begin{bmatrix} 5 & 3 \\ 2 & 4 \end{bmatrix}$$

$$= \begin{bmatrix} 10 & 5 \\ 5 & 8 \end{bmatrix}$$

A matrix is symmetric if it equals its transpose. Because $\begin{bmatrix} 10 & 5 \\ 5 & 8 \end{bmatrix}^T = \begin{bmatrix} 10 & 5 \\ 5 & 8 \end{bmatrix}$, the matrix $\mathbf{A} + \mathbf{A}^T$ is symmetric.

Solution 2: Use a calculator with matrix features. Refer to your user's manual.

For the TI-81, Select [MATRX] . Choose Edit from the Menu. Select 1 (matrix A) and dimension it to 2×2. Enter the values. Press [2nd][CLEAR]. At the main screen, enter [2nd][1][+][2nd][1][MATRX][6][ENTER].The answer appears.

For the Sharp EL9300/9200, press [▦][MENU][3]. Then [MENU][C][0][1]. Set the dimension to 2×2. Enter the values. Press [QUIT]. Pressing [2ndF][MAT][A][+][MATH][E][5][2ndF][MAT][A][ENTER] displays the solution.

For the Casio enter [AC][MODE][0]. Followed by [F1][F6][F1][2][EXE][2][EXE]. Enter values for the matrix and finish with [EXE].

The next keys complete the operation [F2][F2][PRE][F3].

Illustration 3:   Suppose $A = \begin{bmatrix} 1 & 3 & 4 \\ 2 & -1 & 5 \end{bmatrix}$. Determine $A + A + A$.

Solution:

$$A + A + A = \begin{bmatrix} 1 & 3 & 4 \\ 2 & -1 & 5 \end{bmatrix} + \begin{bmatrix} 1 & 3 & 4 \\ 2 & -1 & 5 \end{bmatrix} + \begin{bmatrix} 1 & 3 & 4 \\ 2 & -1 & 5 \end{bmatrix}$$

$$= \begin{bmatrix} 3 & 9 & 12 \\ 6 & -3 & 15 \end{bmatrix}$$

Note: $A + A^T$ cannot be determined for A in this illustration. Why not?

Illustration 4:   The pay schedule for a company is a matrix P in which p_{ij} represents the salary for an employee with i years of experience and rank j in the company. $P = [\text{CurrentSalary}_{ij}]$. The company decides to give a raise and forms a raise matrix R in which the entry in the ij position is the raise to be given an employee with i years of experience and rank j in the company. $R = [\text{AmountofRaise}_{ij}]$. Then the new salary schedule N is

$$N = P + R = [\text{CurrentSalary}_{ij} + \text{AmountofRaise}_{ij}] \quad \blacksquare$$

Recall that $4x$ is shorthand for $x + x + x + x$. Illustration 3 of the previous example suggests the definition for **scalar multiplication**.

DEFINITION 11-6
Scalar Multiplication

> The scalar product of a real number c and a matrix $A_{m \times n}$ is a matrix in which each element is the corresponding element of A multiplied by c; that is, $cA = [ca_{ij}]$.

For example

$$5 \begin{bmatrix} 2 & -3 \\ 10 & 0 \end{bmatrix} = \begin{bmatrix} 5(2) & 5(-3) \\ 5(10) & 5(0) \end{bmatrix} = \begin{bmatrix} 10 & -15 \\ 50 & 0 \end{bmatrix}$$

The definition of scalar multiplication inspires a number of theorems. The proofs of these theorems, although not difficult, are notationally tedious. Examples suggest methods for proving the theorems for 2×2 matrices. The proofs for 2×2 cases are left as exercises. Proofs for higher dimension matrices are analogous to the 2×2 proofs.

Theorem 11-4
Scalar Multiplication is Associative

> If b and c are scalars and A is a matrix, then $b(cA) = (bc)A$.

Try multiplying a matrix by 3 and the resulting matrix by 2. These results are the same as multiplying the original matrix by 6.

Theorem 11-5
Scalar Distribution I

If c is a scalar and **A** and **B** are matrices of the same order, then $c(\mathbf{A} + \mathbf{B}) = c\mathbf{A} + c\mathbf{B}$

For example, add then multiply:

$$5\left(\begin{bmatrix} 1 & 2 \\ 3 & 4 \end{bmatrix} + \begin{bmatrix} 0 & 2 \\ -3 & 5 \end{bmatrix}\right) = 5\begin{bmatrix} 1 & 4 \\ 0 & 9 \end{bmatrix} = \begin{bmatrix} 5 & 20 \\ 0 & 45 \end{bmatrix}$$

The results are the same for multiplying then adding:

$$5\begin{bmatrix} 1 & 2 \\ 3 & 4 \end{bmatrix} + 5\begin{bmatrix} 0 & 2 \\ -3 & 5 \end{bmatrix} = \begin{bmatrix} 5 & 10 \\ 15 & 20 \end{bmatrix} + \begin{bmatrix} 0 & 10 \\ -15 & 25 \end{bmatrix} = \begin{bmatrix} 5 & 20 \\ 0 & 45 \end{bmatrix}$$

Theorem 11-6
Scalar Distribution II

If c and b are scalars and **A** is a matrix, then $(c + b)\mathbf{A} = c\mathbf{A} + b\mathbf{A}$.

Theorem 11-7
Additive Inverse

The **additive inverse** $-\mathbf{A}$ of a matrix **A** is a matrix given by $-\mathbf{A} = -1\mathbf{A}$. As a result $\mathbf{A} + (-\mathbf{A}) = \mathbf{0}$.

If $\mathbf{A} = \begin{bmatrix} 1 & -2 \\ 3 & 5 \end{bmatrix}$, then $-1\mathbf{A} = \begin{bmatrix} -1 & 2 \\ -3 & 5 \end{bmatrix}$. $\mathbf{A} + (-1\mathbf{A}) = \mathbf{0}$.

Based upon Theorem 11-7, we can now define **subtraction of matrices**. If **A** and **B** are matrices of the same order, then $\mathbf{A} - \mathbf{B} = \mathbf{A} + (-\mathbf{B})$.

Theorem 11-8
0 Scalar Multiplication

For any matrix A, $0\mathbf{A} = \mathbf{0}$.

Perform the indicated operations.

EXAMPLE 11-2

Illustration 1:

$$3\begin{bmatrix} 1 & 3 & 4 \\ 2 & -1 & 5 \end{bmatrix}$$

Solution:

$$3\begin{bmatrix} 1 & 3 & 4 \\ 2 & -1 & 5 \end{bmatrix} = \begin{bmatrix} 3(1) & 3(3) & 3(4) \\ 3(2) & 3(-1) & 3(5) \end{bmatrix}$$

$$= \begin{bmatrix} 3 & 9 & 12 \\ 6 & -3 & 15 \end{bmatrix}$$

Illustration 2:

$$-1\begin{bmatrix} 1 & 5 \\ 2 & -1 \end{bmatrix}$$

Solution:

$$-1\begin{bmatrix} 1 & 5 \\ 2 & -1 \end{bmatrix} = \begin{bmatrix} -1 & -5 \\ -2 & 1 \end{bmatrix}$$

Illustration 3:

$$0\begin{bmatrix} 5 & 1 & 6 \\ 4 & 2 & 1 \end{bmatrix}$$

Solution:

$$0\begin{bmatrix} 5 & 1 & 6 \\ 4 & 2 & 1 \end{bmatrix} = \begin{bmatrix} 0 & 0 & 0 \\ 0 & 0 & 0 \end{bmatrix}$$

Illustration 4:

$$\begin{bmatrix} 8 & 1 & 10 \\ 4 & -2 & 11 \end{bmatrix} - \begin{bmatrix} 5 & 1 & 6 \\ 4 & 2 & 1 \end{bmatrix}$$

Solution:

$$\begin{bmatrix} 8 & 1 & 10 \\ 4 & -2 & 11 \end{bmatrix} - \begin{bmatrix} 5 & 1 & 6 \\ 4 & 2 & 1 \end{bmatrix} = \begin{bmatrix} 8 & 1 & 10 \\ 4 & -2 & 11 \end{bmatrix} + (-1)\begin{bmatrix} 5 & 1 & 6 \\ 4 & 2 & 1 \end{bmatrix}$$

$$= \begin{bmatrix} 8 & 1 & 10 \\ 4 & -2 & 11 \end{bmatrix} + \begin{bmatrix} -5 & -1 & -6 \\ -4 & -2 & -1 \end{bmatrix}$$

$$= \begin{bmatrix} 3 & 0 & 4 \\ 0 & -4 & 10 \end{bmatrix}$$

Sometimes information in matrix form needs to be extracted as a scalar. For example, suppose that matrix H = [$8.00 $7.50 $9.00] represents the hourly pay rate of three employees. Let matrix J store the hours each will work on a job being bid. Estimate J = [40 35 37]. To complete the bid requires an estimate of the *total* labor cost. For an individual Gross Pay = (Hourly Rate)(Hours Worked). Thus, the first worker works 40 hours at $8.00 an hour for Gross Pay of $320.00. Similarly, the second worker's pay is 35($7.50) = $262.50. Finally, pay for the last worker is 37($9.00) = $333.00. Total the three salaries and obtain a labor estimate of $915.50. The steps in this calculation were to multiply the corresponding elements of the row matrices and total the results into a single number or scalar. This process generalizes to the following definition.

DEFINITION 11-7
The Dot Product

The dot product of two row matrices A and B of the same dimension is the sum of the products of corresponding elements. Therefore, the dot product produces a real number scalar. The notation for the dot product is $A \cdot B$.

For example $[1\ 2\ 3] \cdot [2\ -1\ 5] = 1(2) + 2(-1) + 3(5) = 15$. To extend the definition of two column matrices $\mathbf{A}$ and $\mathbf{B}$ of the same dimension, use $\mathbf{A} \cdot \mathbf{B} = \mathbf{A}^T \cdot \mathbf{B}^T$. Because $\mathbf{A}^T$ and $\mathbf{B}^T$ are row matrices the results are similar. If $\mathbf{A}$ is a row matrix and $\mathbf{B}$ is a column matrix, then define $\mathbf{A} \cdot \mathbf{B} = \mathbf{A} \cdot \mathbf{B}^T$. Because *corresponding* elements can usually be identified without forming a transpose, we write the answer directly:

$$[1\ \ 0\ -5\ \ 4] \cdot \begin{bmatrix} 2 \\ 3 \\ -1 \\ 6 \end{bmatrix} = 1(2) + 0(3) + (-5)(-1) + 4(6) = 31$$

SUMMARY

- $\mathbf{A}_{m \times n} + \mathbf{B}_{m \times n}$ is an $m \times n$ matrix formed by adding entries in corresponding positions.
- $c\mathbf{A}_{m \times n}$ is an $m \times n$ matrix formed by multiplying each entry of matrix $\mathbf{A}$ by the scalar c.
- $\mathbf{0}_{m \times n}$ is the $m \times n$ zero matrix in which every entry is 0.
- $\mathbf{A}_{m \times n} - \mathbf{B}_{m \times n} = \mathbf{A}_{m \times n} + (-\mathbf{B}_{m \times n})$
- The dot product of two matrices, $\mathbf{A}_{1 \times n} \cdot \mathbf{B}_{1 \times n}$, is a scalar formed by summing the products of the corresponding elements of the two matrices.

11-2 EXERCISES

Perform the indicated operations.

1. $\begin{bmatrix} 1 & 2 \\ 3 & 1 \end{bmatrix} + \begin{bmatrix} 5 & -1 \\ 2 & 2 \end{bmatrix}$

2. $\begin{bmatrix} 1 & 3 \\ -1 & 4 \end{bmatrix} + \begin{bmatrix} 4 & 3 \\ 1 & 1 \end{bmatrix}$

3. $\begin{bmatrix} 1 & 4 \\ 1 & 3 \end{bmatrix} + \begin{bmatrix} 1 & 4 \\ 1 & 3 \end{bmatrix}$

4. $\begin{bmatrix} 5 & 1 \\ -1 & 2 \end{bmatrix} + \begin{bmatrix} 5 & 1 \\ -1 & 2 \end{bmatrix}$

5. $2\begin{bmatrix} 1 & 4 \\ 1 & 3 \end{bmatrix}$

6. $2\begin{bmatrix} 5 & 1 \\ -1 & 2 \end{bmatrix}$

7. $\begin{bmatrix} 1 & 3 & 2 \\ -1 & 4 & 5 \end{bmatrix} + \begin{bmatrix} 2 & 1 \\ 4 & 3 \\ 1 & 1 \end{bmatrix}^T$

8. $\begin{bmatrix} 1 & 3 & 2 \\ -1 & 4 & 5 \end{bmatrix}^T + \begin{bmatrix} 2 & 1 \\ 4 & 3 \\ 1 & 1 \end{bmatrix}$

9. $\left(3\begin{bmatrix} 1 & 1 & 0 \\ 0 & 1 & 0 \\ -1 & 1 & 1 \end{bmatrix} + \begin{bmatrix} 1 & 0 & 0 \\ 0 & 1 & 0 \\ 0 & 0 & 1 \end{bmatrix}\right)^T$

10. $\begin{bmatrix} 5 & 1 \\ -1 & 3 \end{bmatrix} + \begin{bmatrix} 2 & -3 \\ 5 & 7 \end{bmatrix}$

11. $\begin{bmatrix} 1 & 3 \\ -1 & 4 \end{bmatrix} - \begin{bmatrix} -1 & 4 \\ -2 & 3 \end{bmatrix}$

12. $2\begin{bmatrix} 1 & 0 & -1 \\ 2 & 1 & 1 \\ 1 & 1 & 0 \end{bmatrix} + \begin{bmatrix} 1 & 1 & 1 \\ 1 & 2 & 1 \\ 3 & 1 & 1 \end{bmatrix}^T$

13. $3\begin{bmatrix} 2 & 1 \\ 5 & 1 \end{bmatrix} - 2\begin{bmatrix} 3 & 1 \\ -1 & 0 \end{bmatrix}$

14. $3\begin{bmatrix} 1 & -1 \\ 0 & 1 \end{bmatrix} - \dfrac{1}{2}\begin{bmatrix} 4 & 6 \\ 8 & -4 \end{bmatrix}$

15. $[3\ 5] \cdot [7\ -2]$

16. $[4\ -3] \cdot [0\ -2]$

17. $[1\ 4\ -2] \cdot [5\ -4\ 1]$

18. $[-1\ 1\ 4] \cdot [3\ 2\ 5]$

19. $[1\ 2] \cdot \begin{bmatrix} 3 \\ -2 \end{bmatrix}$

20. $[5\ -2] \cdot \begin{bmatrix} 2 \\ 5 \end{bmatrix}$

Solve for x in each of the following.

21. $[3\ x] \cdot [-2\ 5] = 0$

22. $[3\ -4] \cdot [x\ 2] = 0$

23. Attempt to define the dot product of two matrices of different dimensions. Discuss the difficulties encountered.

24. Attempt to define the sum of two matrices of different dimensions. Discuss the difficulties encountered.

25. Let $A = [x_1, x_2, x_3]$, where x_i represents the number of months that car i ($i = 1, 2, 3$) is in use. Let $E = [c_1, c_2, c_3]$, where c_i is the operating cost each month for car i ($i = 1, 2, 3$). Represent the total operating cost for all three cars as a matrix operation.

26. Let $M = [p_a, p_b, p_c]$, where the elements of the matrix represent the price per kilogram of coffee type a, b and c, respectively. Let $M = [A, B, C]$ represent the number of kilograms of coffees A, B and C in a particular blend. Represent the total cost of the blend using a matrix operation.

27. Suppose matrix P contains annual salaries for engineers, where a row corresponds to the years of experience and a column to the number of graduate courses taken. For example, the salary of an engineer with ten years of experience who has taken seven graduate courses is the entry in row 10 column 7. If salaries are increased 3.5 percent for inflation, express the new salary schedule as a scalar multiple of the old.

28. Using a computer, Teri records student grades in a column matrix. Suppose matrix R contains the student test scores for the mid semester exam, matrix H contains the homework averages and matrix F contains the final exams. If the midsemester grade counts 30 percent, the homework counts 25 percent and the final counts 45 percent of each students grade, express the final average matrix A as a sum of scalar multiples of R, H and F.

Verify the following theorems for matrices of order 2×2.

29. Theorem 11-1

30. Theorem 11-2

31. Theorem 11-3

32. Theorem 11-4

33. Theorem 11-5

34. Theorem 11-6

35. Theorem 11-7

36. Theorem 11-8

11-2 P R O B L E M S E T

PROBLEMS FOR TECHNOLOGY

If a matrix A is displayed on the Casio fx7700-G, you may enter a scalar by pressing digits. The scalar appears at the lowest left corner of the display. Press F1 to perform scalar multiplication and store the results in C. To do addition, use the PRE key to return to the opening matrix screen.

If matrix A is stored in the Sharp EL9300/9200 in matrix mode, pressing ▦ 7 MATH E 1 A ENTER displays the answer to 7A.

On the TI-81, if you have CLEARed to the initial screen, entering a number followed by the matrix (2nd A or 2nd B) and ENTER displays the scalar multiple. Check your user's manual for details.

Use a calculator with matrix capabilities to determine the following. Make a note of the tasks that were easy on your calculator and those that were difficult.

1. $\begin{bmatrix} 3 & 5 \\ 2 & 1 \end{bmatrix} + \begin{bmatrix} 3 & -5 \\ 2 & 1 \end{bmatrix}$

2. $\begin{bmatrix} 5 & 3 \\ -7 & 4 \end{bmatrix} + \begin{bmatrix} 7 & 6 \\ 1 & 9 \end{bmatrix}$

3. $\begin{bmatrix} 9 & 4 \\ 2 & 3 \end{bmatrix} + \begin{bmatrix} 9 & 4 \\ 2 & 3 \end{bmatrix}$

4. $\begin{bmatrix} 6 & 1 \\ -3 & 2 \end{bmatrix} + \begin{bmatrix} 6 & 1 \\ -3 & 2 \end{bmatrix}$

5. $5\begin{bmatrix} 3 & 4 \\ -2 & 8 \end{bmatrix}$

6. $4\begin{bmatrix} 6 & 2 \\ -3 & 1 \end{bmatrix}$

7. $\begin{bmatrix} 2 & 3 & 7 \\ -3 & 5 & 1 \end{bmatrix} + \begin{bmatrix} 1 & 3 \\ 6 & 2 \\ 8 & 5 \end{bmatrix}^T$

8. $\begin{bmatrix} 2 & 7 & 6 \\ -3 & 0 & 5 \end{bmatrix}^T + \begin{bmatrix} 1 & 9 \\ 3 & 8 \\ 0 & 4 \end{bmatrix}$

9. $\left(5\begin{bmatrix} 1 & 0 & 1 \\ 0 & 1 & 0 \\ -1 & 1 & 1 \end{bmatrix} + \begin{bmatrix} 0 & 1 & 0 \\ 1 & 1 & 0 \\ 1 & 0 & 0 \end{bmatrix} \right)^T$

10. $\begin{bmatrix} 2 & 3 \\ -5 & 1 \end{bmatrix} + \begin{bmatrix} 2 & -3 \\ 5 & 4 \end{bmatrix}$

11. $\begin{bmatrix} 6 & 3 \\ -1 & 5 \end{bmatrix} - \begin{bmatrix} -6 & 4 \\ -5 & 3 \end{bmatrix}$

12. $3\begin{bmatrix} 0 & 0 & -1 \\ 5 & 1 & 0 \\ 2 & 1 & 0 \end{bmatrix} + \begin{bmatrix} 2 & 1 & 0 \\ 1 & 5 & 1 \\ 2 & 0 & 1 \end{bmatrix}^T$

13. Based on the notes you made, discuss whether your calculator is "good" or "bad" at matrix operations.

3   MULTIPLICATION OF MATRICES

As far as I am concerned matrices are mysteries and linear transformations light the way.
—P.R. HALMOS

Compared to addition, multiplication of matrices is somewhat unusual. The motivation for the definition of matrix multiplication came from applying successive affine transformations. We duplicate that process for 2×2 matrices, then generalize the results. Consider the transformation given by

$$\begin{cases} x' = ax + by \\ y' = cx + dy \end{cases} \quad \{A\}$$

Suppose transformation **A** is applied to a function f. The transformation could result in a shift of the graph of f followed by a stretch in some direction. The exact change it makes in the graph is not important for the moment. Because the position of the variables in this transformations is standard, abbreviate the transformation by writing the coefficients of the transformation in

matrix format: $\mathbf{A} = \begin{bmatrix} a & b \\ c & d \end{bmatrix}$.

Suppose that we wish to apply a second transformation to the results of the first transformation. The second transformation **B** is given by

$\begin{cases} x'' = ex' + fy' \\ y'' = gx' + hy' \end{cases}$, which is abbreviated as $\mathbf{B} = \begin{bmatrix} e & f \\ g & h \end{bmatrix}$.

Transformation **B** may represent a rotation of the graph. Once more, the exact purpose is not important at this time. Substitute from the equations for transformation **A** into the equations for transformation **B** to obtain a transformation representing the effects of applying **A**, then **B**.

$$x'' = e(ax + by) + f(cx + dy)$$
$$y'' = g(ax + by) + h(cx + dy)$$

Remove grouping symbols.

$$x'' = aex + bey + cfx + dfy$$
$$y'' = agx + bgy + chx + dhy$$

Group together x terms and y terms to obtain transformation **C**:

$$\begin{cases} x'' = (ae + cf)x + (be + df)y \\ y'' = (ag + ch)x + (bg + dh)y \end{cases}$$

Transformation **C** provides equations for the direct transformation from (x, y) to (x'', y''). Transformation **C** eliminates the need to go through two separate transformations **A** and **B**. The new transformation accomplishes at one time the results of the two separate transformations. In some respects, the new transformation is a composition of the two previous transformations.

Abbreviate the new transformation in matrix form:

$$\mathbf{C} = \begin{bmatrix} ae+cf & be+df \\ ag+ch & bg+dh \end{bmatrix}$$

Although it is obvious that the new matrix is of the same dimension as the original two matrices, the pattern for the entries may not be clear. We want the new matrix to represent the *matrix product* of the original two matrices. We write out the product we plan to define and look for a pattern:

$$\mathbf{BA} = \mathbf{C}$$

$$\begin{bmatrix} e & f \\ g & h \end{bmatrix} \begin{bmatrix} a & b \\ c & d \end{bmatrix} = \begin{bmatrix} ae+cf & be+df \\ ag+ch & bg+dh \end{bmatrix}$$

Why $\mathbf{BA}$ and not $\mathbf{AB}$? First we apply $\mathbf{A}$ to a function f: $\mathbf{A}(f)$. Then apply $\mathbf{B}$ to the result: $\mathbf{B}[\mathbf{A}(f)]$. Therefore, $\mathbf{BA}$ represents the transformation we seek.

The definition of **matrix multiplication** must describe the calculation of each element of the product. Consider an element from the product $\mathbf{C}$, say $c_{1,1} = ae + cf$. This is familiar. Recall the dot product from the previous section. Look at row 1 of matrix $\mathbf{B}$ as a row matrix: $[e\,f]$. Look at column 1 of matrix $\mathbf{A}$ as a column matrix $[a\,c]^T$. The dot product of row 1 with column 1 is $ae + cf$. Similarly, select row 2 of matrix $\mathbf{B}[g\,h]$ and calculate the dot product with column 1 of matrix $\mathbf{A}[a\,c]^T$ to obtain $ag + ch$. Note that $ag + ch$ is the entry in the second row, first column of the product.

$$\begin{bmatrix} e & f \\ g & h \end{bmatrix} \begin{bmatrix} a & b \\ c & d \end{bmatrix} = \begin{bmatrix} ae + cf & be + df \\ ag + ch & bg + dh \end{bmatrix}$$

For a last sample, choose row 1 from matrix $\mathbf{B}$ and column 2 from the matrix $\mathbf{A}$. Dot the row and column to obtain $be + df$, which is the element in the first row, second column of the product $\mathbf{C}$. We have sufficient inspiration to formulate a definition.

DEFINITION 11-8
Matrix Multiplication

> The matrix product of two matrices $\mathbf{B}_{m \times n}\, \mathbf{A}_{n \times r}$ is a matrix $\mathbf{C}$ of dimension $m \times r$, where the element in the ith row and jth column of $\mathbf{C}$, c_{ij}, is given by the dot product of row i from $\mathbf{B}$ with column j from $\mathbf{A}$.

Of special importance is that the number of columns in $\mathbf{B}$ must equal the number of rows in $\mathbf{A}$: they are **conformable matrices**. In other words, if the dimensions of $\mathbf{B}$ and $\mathbf{A}$ are $(m \times n)$ and $(r \times s)$, respectively, then $\mathbf{BA}$ exists only when $n = r$:

$$\mathbf{B}_{m \times n} \mathbf{A}_{n \times s}$$
⇑ ⇑ must be equal

Products of Matrices. Determine the indicated product.

EXAMPLE 11-3

Illustration 1:

$$\begin{bmatrix} 1 & 3 \\ -1 & 4 \end{bmatrix} \begin{bmatrix} 1 & 5 \\ -2 & 0 \end{bmatrix}$$

Solution: The first matrix has two columns whereas the second matrix has two rows so that the product exists. Let **A** and **B** be the two matrices, where the product is $\mathbf{A}_{2\times 2}\mathbf{B}_{2\times 2}$. The innermost two subscripts are equal.

To construct the product, note that the product matrix will be of dimension 2×2 (the outer two subscripts). Calculate each element of the product matrix from a dot product of a row and column. For example, the element in the 2, 1 position will come from row 2 in matrix **A** dotted with column 1 in matrix **B**:

$$\begin{bmatrix} 1 & 3 \\ -1 & 4 \end{bmatrix} \begin{bmatrix} 1 & 5 \\ -2 & 0 \end{bmatrix}$$

$$[-1 \ \ 4] \cdot \begin{bmatrix} 1 \\ -2 \end{bmatrix} = 1(-1) + 4(-2) = -9$$

For the complete product matrix we have

$$\begin{bmatrix} [1\ 3]\cdot\begin{bmatrix}1\\-2\end{bmatrix} & [1\ 3]\cdot\begin{bmatrix}5\\0\end{bmatrix} \\ [-1\ 4]\cdot\begin{bmatrix}1\\-2\end{bmatrix} & [-1\ 4]\cdot\begin{bmatrix}5\\0\end{bmatrix} \end{bmatrix} = \begin{bmatrix} 1(1)+3(-2) & 1(5)+3(0) \\ -1(1)+4(-2) & -1(5)+4(0) \end{bmatrix}$$

$$= \begin{bmatrix} -5 & 5 \\ -9 & -5 \end{bmatrix}$$

Illustration 2: If $\mathbf{A}_{2\times 3} = \begin{bmatrix} 1 & 3 & 2 \\ -1 & 4 & 5 \end{bmatrix}$ and $\mathbf{B}_{3\times 2} = \begin{bmatrix} 2 & 1 \\ 4 & 3 \\ 1 & 1 \end{bmatrix}$, determine **AB**.

Solution: **A** has three columns and **B** has three rows (innermost subscripts). The dimension of the product will be 2×2 (outermost subscripts). As a sample entry consider row 1 column 2 of the product. This will be the dot product of row 1 from **A** with column 2 from **B**:

$$[1 \ \ 3 \ \ 2] \cdot \begin{bmatrix} 1 \\ 3 \\ 1 \end{bmatrix} = 1(1) + 3(3) + 2(1) = 12.$$

Calculate every entry:

$$\begin{bmatrix} [1\ 3\ 2]\cdot\begin{bmatrix}2\\4\\1\end{bmatrix} & [1\ 3\ 2]\cdot\begin{bmatrix}1\\3\\1\end{bmatrix} \\ [-1\ 4\ 5]\cdot\begin{bmatrix}2\\4\\1\end{bmatrix} & [-1\ 4\ 5]\cdot\begin{bmatrix}1\\3\\1\end{bmatrix} \end{bmatrix} = \begin{bmatrix} 1(2)+3(4)+2(1) & 1(1)+3(3)+2(1) \\ -1(2)+4(4)+5(1) & -1(1)+4(3)+5(1) \end{bmatrix}$$

$$= \begin{bmatrix} 16 & 12 \\ 19 & 16 \end{bmatrix}$$

Illustration 3:

See matrices **A** and **B** from Illustration 2. Determine $\mathbf{B}_{3\times2}\mathbf{A}_{2\times3}$. The matrices are conformable because the innermost subscripts are equal. The product will be a 3×3 matrix (outermost subscripts):

$$\begin{bmatrix}2&1\\4&3\\1&1\end{bmatrix}\begin{bmatrix}1&3&2\\-1&4&5\end{bmatrix}=\begin{bmatrix}[2\;1]\cdot\begin{bmatrix}1\\-1\end{bmatrix}&[2\;1]\cdot\begin{bmatrix}3\\4\end{bmatrix}&[2\;1]\cdot\begin{bmatrix}2\\5\end{bmatrix}\\[4\;3]\cdot\begin{bmatrix}1\\-1\end{bmatrix}&[4\;3]\cdot\begin{bmatrix}3\\4\end{bmatrix}&[4\;3]\cdot\begin{bmatrix}2\\5\end{bmatrix}\\[1\;1]\cdot\begin{bmatrix}1\\-1\end{bmatrix}&[1\;1]\cdot\begin{bmatrix}3\\4\end{bmatrix}&[1\;1]\cdot\begin{bmatrix}2\\5\end{bmatrix}\end{bmatrix}$$

$$=\begin{bmatrix}2(1)+1(-1)&2(3)+1(4)&2(2)+1(5)\\4(1)+3(-1)&4(3)+3(4)&4(2)+3(5)\\1(1)+1(-1)&1(3)+1(4)&1(2)+1(5)\end{bmatrix}$$

$$=\begin{bmatrix}1&10&9\\1&24&23\\0&7&7\end{bmatrix}$$

Note that the elements in the third row second column is formed from the dot product of row 3 from **B** and column 2 from **A**:

$$[1\;1]\cdot\begin{bmatrix}3\\4\end{bmatrix}=7$$

Illustration 4:

$$\mathbf{A}=\begin{bmatrix}3\\4\end{bmatrix},\quad\mathbf{B}=\begin{bmatrix}1&5\\-2&0\end{bmatrix}$$

Solution: We cannot determine $\mathbf{A}_{2\times1}\mathbf{B}_{2\times2}$. The inner subscripts do not match, **A** is not conformable to **B**. However, we can calculate $\mathbf{B}_{2\times2}\mathbf{A}_{2\times1}$. The product will be of dimension 2×1.

$$\begin{bmatrix}1&5\\-2&0\end{bmatrix}\begin{bmatrix}3\\4\end{bmatrix}=\begin{bmatrix}1(3)+5(4)\\-2(3)+0(4)\end{bmatrix}$$

$$=\begin{bmatrix}23\\-6\end{bmatrix}$$

From the illustrations, matrix multiplication is not generally commutative: $\mathbf{AB}\neq\mathbf{BA}$. However some properties of matrix multiplication are similar to real number multiplication. The following theorems elaborate on this point.

Theorem 11-9
Matrix Multiplication is Associative

$$\mathbf{A}_{n\times m}(\mathbf{B}_{m\times p}\,\mathbf{C}_{p\times t})=(\mathbf{A}_{n\times m}\mathbf{B}_{m\times p})\mathbf{C}_{p\times t}$$

Suppose $A = \begin{bmatrix} 1 & 2 \\ 3 & 4 \end{bmatrix}$, $B = \begin{bmatrix} 2 & -1 \\ 3 & 4 \end{bmatrix}$ and $C = \begin{bmatrix} 2 & 0 \\ 5 & -1 \end{bmatrix}$, then

$$AB = \begin{bmatrix} 8 & 7 \\ 18 & 13 \end{bmatrix} \text{ and } BC = \begin{bmatrix} -1 & 1 \\ 26 & -4 \end{bmatrix}.$$

Finally $A(BC) = \begin{bmatrix} 51 & -7 \\ 101 & -13 \end{bmatrix}$ and $(AB)C = \begin{bmatrix} 51 & -7 \\ 101 & -13 \end{bmatrix}$ as indicated in the theorem.

Theorem 11-10
Matrix Multiplication is Distributive

$$A_{n \times m}(B_{m \times p} + C_{m \times p}) = (A_{n \times m}B_{m \times p}) + (A_{n \times m}C_{m \times p})$$

Using the same A, B, AB and C as before, $B + C = \begin{bmatrix} 4 & -1 \\ 8 & 3 \end{bmatrix}$ and $AC = \begin{bmatrix} 12 & -2 \\ 26 & -4 \end{bmatrix}$. Thus, $A(B + C) = \begin{bmatrix} 20 & 5 \\ 44 & 9 \end{bmatrix}$ and $AB + AC = \begin{bmatrix} 20 & 5 \\ 44 & 9 \end{bmatrix}$ as indicated in the theorem.

DEFINITION 11-9
Multiplicative Identity

The $n \times n$ dimension identity matrix I is given by

$$I_{n \times n} = [t_{ij}], \text{ where } t_{ij} = \begin{cases} 0 & \text{if } i \neq j \\ 1 & \text{if } i = j \end{cases}$$

From the definition, an identity matrix is a square matrix in which all of the elements on the main diagonal are 1. All elements off the main diagonal are 0. Consider $I_{3 \times 3}$:

$$I_{3 \times 3} = \begin{bmatrix} 1 & 0 & 0 \\ 0 & 1 & 0 \\ 0 & 0 & 1 \end{bmatrix}$$

If I is conformable to a matrix A, then $IA = A$. The verification that I acts as an identity for matrix multiplication is left as an exercise.

Theorem 11-11
Matrix Identity for Multiplication

If A is a square matrix and I is an identity matrix of the same dimension as A, then $AI = A$ and $IA = A$. Thus, I is the **matrix identity for multiplication**.

Compare the role of the matrix identity I to identities we have encountered before. In particular, recall the multiplicative identity, 1, and the compositional identity given by $I(x) = x$.

EXAMPLE II-4

Illustration I:

More on Matrix Multiplication. Determine the indicated product.

$$\begin{bmatrix} 1 & 1 & 2 \\ -1 & 3 & 5 \end{bmatrix} \begin{bmatrix} 1 & 1 & 5 & 1 \\ 2 & 1 & 0 & -1 \\ 3 & 4 & 1 & 3 \end{bmatrix}$$

Solution: The product exists because the first matrix is 2×3 and the second 3×4. The product will be 2×4. Let C represent the product. Consider the calculation of $c_{2,3}$. Entry $c_{2,3}$ is the dot product of the row 2 of the first matrix and column 3 of the second matrix: $[-1 \ 3 \ 5] \cdot [5 \ 0 \ 1]^T = -1(5) + 3(0) + 5(1) + = 0$.

The number of elements compounds the problem of arithmetic with the difficulty in bookkeeping. For a systematic method to matrix multiplication start with row 1 of the first matrix and dot it, in sequence, with each column of the second matrix. This generates, in sequence, the elements of the first row of the product matrix. Then take the second row and dot it with the columns from the second matrix to generate the second row of the product matrix. Continue this pattern until all rows of the first matrix are exhausted:

$$\begin{bmatrix} 1 & 1 & 2 \\ -1 & 3 & 5 \end{bmatrix} \begin{bmatrix} 1 & 1 & 5 & 1 \\ 2 & 1 & 0 & -1 \\ 3 & 4 & 1 & 3 \end{bmatrix} = \begin{bmatrix} 9 & 10 & 7 & 6 \\ 20 & 22 & 0 & 11 \end{bmatrix}$$

Illustration 2:

$$\begin{bmatrix} 1 & 3 & 4 \\ 1 & 0 & 1 \end{bmatrix} \begin{bmatrix} 2 & 1 & 0 \\ 3 & 5 & 5 \end{bmatrix}$$

Solution: The product does not exist: 2×3 and 2×3. Try to dot the first row with the first column and observe the difficulty.

Illustration 3:

$$[1 \ 2 \ 1 \ 3] \begin{bmatrix} 3 \\ -1 \\ 2 \\ 1 \end{bmatrix}$$

Solution: The product exists (1×4 to 4×1). The product is a 1×1 matrix. The dot product of the two matrices is the only element in the product.

$$c_{1,1} = [1, 2, 1, 3] \cdot [3, -1, 2, 1]^T = 6$$

$$[1 \ 2 \ 1 \ 3] \begin{bmatrix} 3 \\ -1 \\ 2 \\ 1 \end{bmatrix} = [6]$$

Illustration 4:

$$\begin{bmatrix} 3 \\ -1 \\ 2 \\ 1 \end{bmatrix} [1\ 2\ 1\ 3]$$

Solution: The product exists (4×1 to 1×4). The product is a 4×4 matrix:

$$\begin{bmatrix} 3 \\ -1 \\ 2 \\ 1 \end{bmatrix} [1\ 2\ 1\ 3] = \begin{bmatrix} 3 & 6 & 3 & 9 \\ -1 & -2 & -1 & -3 \\ 2 & 4 & 2 & 6 \\ 1 & 2 & 1 & 3 \end{bmatrix}$$

Illustration 5:

Multiplication by the Identity Matrix

$$\begin{bmatrix} 1 & 0 & 0 \\ 0 & 1 & 0 \\ 0 & 0 & 1 \end{bmatrix}_{3\times3} \begin{bmatrix} 1 & 2 & 3 \\ 4 & 5 & 6 \\ 7 & 8 & 9 \end{bmatrix}_{3\times3} = \begin{bmatrix} 1 & 2 & 3 \\ 4 & 5 & 6 \\ 7 & 8 & 9 \end{bmatrix}_{3\times3}$$

Illustration 6:

Suppose $A = \begin{bmatrix} 3 & 5 \\ 2 & 7 \end{bmatrix}$, $X = \begin{bmatrix} x \\ y \end{bmatrix}$ and $C = \begin{bmatrix} 2 \\ 1 \end{bmatrix}$, rewrite $AX = C$ without using matrix notation.

Solution:

$$AX = C$$

$$\begin{bmatrix} 3 & 5 \\ 2 & 7 \end{bmatrix}\begin{bmatrix} x \\ y \end{bmatrix} = \begin{bmatrix} 2 \\ 1 \end{bmatrix}$$

This is equivalent to

$$\begin{bmatrix} 3x + 5y \\ 2x + 7y \end{bmatrix} = \begin{bmatrix} 2 \\ 1 \end{bmatrix}$$

Two matrices are equal only if their corresponding elements are equal. Thus,
$$\begin{cases} 3x + 5y = 2 \\ 2x + 7y = 1 \end{cases}$$

The last equations form a system of two linear equations in two variables. This system is equivalent to the original $AX = C$, a **matrix equation**. If we had a matrix A^{-1} that acted as a multiplicative inverse for A so that $A^{-1}A = I$, we could multiply through the equation by A^{-1}:

$$A^{-1}AX = A^{-1}C$$
$$IX = A^{-1}C$$
$$X = A^{-1}C$$

The righthand side would yield a 2×1 matrix with constant entries, say r and s. Then $\begin{bmatrix} x \\ y \end{bmatrix} = \begin{bmatrix} r \\ s \end{bmatrix}$, so that $x = r$ and $y = s$. The steps solved a linear system by using matrix multiplication. A later section details inverse matrices.

EXAMPLE 11-5

Applications of Matrix Multiplication

Illustration 1:

A Multiplicative Inverse

$$\text{Determine} \begin{bmatrix} \frac{1}{3} & -\frac{2}{3} & 1 \\ 0 & 1 & -1 \\ -\frac{1}{3} & -\frac{1}{3} & 1 \end{bmatrix} \begin{bmatrix} 2 & 1 & -1 \\ 1 & 2 & 1 \\ 1 & 1 & 1 \end{bmatrix}$$

$$\begin{bmatrix} \frac{1}{3} & -\frac{2}{3} & 1 \\ 0 & 1 & -1 \\ -\frac{1}{3} & -\frac{1}{3} & 1 \end{bmatrix} \begin{bmatrix} 2 & 1 & -1 \\ 1 & 2 & 1 \\ 1 & 1 & 1 \end{bmatrix} = \begin{bmatrix} 1 & 0 & 0 \\ 0 & 1 & 0 \\ 0 & 0 & 1 \end{bmatrix}$$

Illustration 2:

Solution of a System of Linear Equations

$$\text{Solve} \begin{cases} 2x + y - z = 1 \\ x + 2y + z = 8 \\ x + y + z = 6 \end{cases}$$

Solution: Rewrite the problem in matrix form: $\mathbf{AX} = \mathbf{C}$:

$$\begin{bmatrix} 2 & 1 & -1 \\ 1 & 2 & 1 \\ 1 & 1 & 1 \end{bmatrix} \begin{bmatrix} x \\ y \\ z \end{bmatrix} = \begin{bmatrix} 1 \\ 8 \\ 6 \end{bmatrix}$$

Matrix $\mathbf{A}$ is the coefficient matrix. Illustration 1 provides an inverse matrix for $\mathbf{A}$. Multiply both sides of the equation by $\mathbf{A}^{-1}$.

$$\begin{bmatrix} \frac{1}{3} & -\frac{2}{3} & 1 \\ 0 & 1 & -1 \\ -\frac{1}{3} & -\frac{1}{3} & 1 \end{bmatrix} \begin{bmatrix} 2 & 1 & -1 \\ 1 & 2 & 1 \\ 1 & 1 & 1 \end{bmatrix} \begin{bmatrix} x \\ y \\ z \end{bmatrix} = \begin{bmatrix} \frac{1}{3} & -\frac{2}{3} & 1 \\ 0 & 1 & -1 \\ -\frac{1}{3} & -\frac{1}{3} & 1 \end{bmatrix} \begin{bmatrix} 1 \\ 8 \\ 6 \end{bmatrix}$$

$$\begin{bmatrix} 1 & 0 & 0 \\ 0 & 1 & 0 \\ 0 & 0 & 1 \end{bmatrix} \begin{bmatrix} x \\ y \\ z \end{bmatrix} = \begin{bmatrix} \frac{1}{3} & -\frac{2}{3} & 1 \\ 0 & 1 & -1 \\ -\frac{1}{3} & -\frac{1}{3} & 1 \end{bmatrix} \begin{bmatrix} 1 \\ 8 \\ 6 \end{bmatrix}$$

$$\begin{bmatrix} x \\ y \\ z \end{bmatrix} = \begin{bmatrix} 1 \\ 2 \\ 3 \end{bmatrix}$$

The solution to the system is the point $(1, 2, 3)$.

Illustration 3:

Rotations. Consider the 2×2 identity $\mathbf{I} = \begin{bmatrix} 1 & 0 \\ 0 & 1 \end{bmatrix}$. From $\mathbf{I}$ form a new matrix $\mathbf{R}$

by interchanging rows 1 and 2. $\mathbf{R} = \begin{bmatrix} 0 & 1 \\ 1 & 0 \end{bmatrix}$. Now let matrix

$$\mathbf{T} = \begin{bmatrix} \text{FL} & \text{FR} \\ \text{BL} & \text{BR} \end{bmatrix}$$

where T represents the location of tires on a car (FR = Front Right, BL = Back Left, etc.). Here is one scheme to "rotate" the tires: use $\mathbf{RT}$.

$$\mathbf{RT} = \begin{bmatrix} 0 & 1 \\ 1 & 0 \end{bmatrix} \begin{bmatrix} \text{FL} & \text{FR} \\ \text{BL} & \text{BR} \end{bmatrix}$$

$$= \begin{bmatrix} \text{BL} & \text{BR} \\ \text{FL} & \text{FR} \end{bmatrix}$$

Interchanging the first and last row of $\mathbf{I}$ formed matrix $\mathbf{R}$. Multiplying $\mathbf{RT}$ transformed $\mathbf{T}$ by reversing the first and last rows.   ▬

SUMMARY

- The product of two matrices, $\mathbf{A}_{n \times m} \mathbf{B}_{m \times r} = \mathbf{C}_{n \times r}$, where the element in the ij position of $\mathbf{C}$ is the dot product of row i of $\mathbf{A}$ and column j of $\mathbf{B}$. $\mathbf{I}_{n \times n} = [t_{ij}]$, where $t_{ij} = 1$ when $i = j$ (main diagonal) and $t_{ij} = 0$ otherwise. $\mathbf{I}$ is the identity matrix of dimension $n \times n$.
- A system of n linear equations in n variables can be represented by the matrix equation $\mathbf{AX} = \mathbf{C}$, where $\mathbf{A}$ is the matrix of coefficients of the variables, $\mathbf{X}$ is the matrix of variables and $\mathbf{C}$ is the matrix of constants.
- If $\mathbf{A}$ is a square matrix and there exists a matrix $\mathbf{A}^{-1}$ such that $\mathbf{A}^{-1}\mathbf{A} = \mathbf{I}$, then $\mathbf{A}^{-1}$ is the (lefthand) multiplicative inverse of $\mathbf{A}$.

11-3   EXERCISES

Determine the indicated products.

1. $[1 \ 3 \ 2] \begin{bmatrix} 2 \\ 1 \\ -1 \end{bmatrix}$

2. $[1 \ 5 \ -1] \begin{bmatrix} 3 \\ 1 \\ 4 \end{bmatrix}$

3. $\begin{bmatrix} 2 \\ 1 \\ -1 \end{bmatrix} [1 \ 3 \ 2]$

4. $\begin{bmatrix} 3 \\ 1 \\ 4 \end{bmatrix} [1 \ 5 \ -1]$

5. $\begin{bmatrix} 2 & -1 \\ 3 & 2 \end{bmatrix} \begin{bmatrix} 5 & 2 \\ -1 & 4 \end{bmatrix}$ **6.** $\begin{bmatrix} 4 & 3 \\ -2 & 2 \end{bmatrix} \begin{bmatrix} 1 & 7 \\ 2 & 3 \end{bmatrix}$

7. $\begin{bmatrix} 5 & 2 \\ -1 & 4 \end{bmatrix} \begin{bmatrix} 2 & -1 \\ 3 & 2 \end{bmatrix}$ **8.** $\begin{bmatrix} 1 & 7 \\ 2 & 3 \end{bmatrix} \begin{bmatrix} 4 & 3 \\ -2 & 2 \end{bmatrix}$

9. $\begin{bmatrix} 1 & 3 & 1 \\ 2 & 1 & -1 \end{bmatrix} \begin{bmatrix} 1 & 2 & 1 \\ -1 & 0 & 1 \\ 1 & 1 & 1 \end{bmatrix}$

10. $\begin{bmatrix} 1 & 2 & 1 \\ -1 & 0 & 1 \\ 1 & 1 & 1 \end{bmatrix} \begin{bmatrix} 1 & 3 & 1 \\ 2 & 1 & -1 \end{bmatrix}$

Verify Problems 11–16 for 2 × 2 matrices.

11. Theorem 11-9

12. Theorem 11-10

13. Theorem 11-11

14. Note that $\mathbf{IA} = \mathbf{A}$. Show that $\mathbf{AI} = \mathbf{A}$. (The lefthand identity is also the righthand identity.)

15. $(\mathbf{AB})^T = \mathbf{B}^T\mathbf{A}^T$

16. $(s\mathbf{A})\mathbf{B} = s(\mathbf{AB})$

17. Prove: if $\mathbf{AB} = \mathbf{I}$ and $\mathbf{CA} = \mathbf{I}$, then $\mathbf{C} = \mathbf{B}$. (Hint: use associativity of matrix multiplication.)

18. Suppose $\mathbf{A}^{-1}\mathbf{A} = \mathbf{I}$ and $\mathbf{B}$ is a matrix such that $\mathbf{AB} = \mathbf{I}$. Show that $\mathbf{A}^{-1} = \mathbf{B}$. (A righthand inverse is the same as the lefthand inverse.)

Suppose that **A** is the hours/seatcover matrix:

	Cutting	Fabrication	Installation
Deluxe	1	5	2
Regular	3	3	2

B is the number of seatcovers/week matrix:

	Deluxe	Regular
Number	[100	200]

and **C** is the cost of labor/hour matrix:

	Cost
Cutting	4.00
Fabrication	5.00
Installation	6.00

Use these matrices or their transposes to determine each of the following.

19. The matrix for hours/week for each operation. (Hint: hours/week = (seatcover/week)(hours/seatcover).)

20. The matrix for the cost/seatcover. See Problem 19.

21. The matrix for cost/week.

22. See Problem 21. Use the transpose of the matrices to calculate the cost/week matrix.

11-3   P R O B L E M   S E T

PROBLEMS FOR TECHNOLOGY

Use a calculator with matrix capabilities to determine each of the following.

1. $[4 \ -2 \ 5] \begin{bmatrix} 7 \\ 2 \\ -3 \end{bmatrix}$ **2.** $[5 \ 1 \ -4] \begin{bmatrix} 4 \\ -5 \\ 2 \end{bmatrix}$

3. $\begin{bmatrix} 3 \\ 2 \\ -1 \end{bmatrix} [3 \ 2 \ -1]$ **4.** $\begin{bmatrix} 3 \\ 5 \\ 11 \end{bmatrix} [4 \ 2 \ -2]$

5. $\begin{bmatrix} 3 & -2 \\ 0 & 5 \end{bmatrix} \begin{bmatrix} 0 & 1 \\ -1 & 0 \end{bmatrix}$ **6.** $\begin{bmatrix} 0 & -1 \\ -1 & 0 \end{bmatrix} \begin{bmatrix} 4 & 1 \\ 9 & 5 \end{bmatrix}$

7. $\begin{bmatrix} 11 & 3 \\ -1 & 5 \end{bmatrix} \begin{bmatrix} 7 & -6 \\ 3 & 5 \end{bmatrix}$ **8.** $\begin{bmatrix} 5 & 7 \\ 8 & -7 \end{bmatrix} \begin{bmatrix} 6 & 5 \\ -7 & 3 \end{bmatrix}$

9. $\begin{bmatrix} 13 & 8 & 5 \\ 9 & 0 & -2 \end{bmatrix} \begin{bmatrix} 9 & 3 & -7 \\ -8 & 19 & 7 \\ 8 & 7 & 0 \end{bmatrix}$

10. $\begin{bmatrix} 8 & 5 & 9 \\ -7 & 7 & 0 \\ 3 & 6 & 9 \end{bmatrix} \begin{bmatrix} 6 & -3 & 9 \\ 1 & 3 & -6 \end{bmatrix}$

11. Redo Exercises 11-3 Problems 1–10 on your graphing calculator. Compare the advantages and disadvantages of using a calculator.

12. See Problem 19 in Exercises 11-3. What if two more columns were appended to the hours/seatcover matrix: a material column and a trim column. How well can your calculator handle the new matrix?

<table>
<tr><td>4</td></tr>
</table>

*ELEMENTARY ROW OPERATIONS

Almost everything, which the mathematics of our century has brought forth in the way of original scientific ideas, attaches to the name of GAUSS.
—L. KRONECKER

Chapter 10 presented the method of elimination for solving a system of linear equations. The method of elimination relied on a few simple rules for manipulating the system of equations. Following these rules produced equivalent systems of equations. Equivalent systems have the same solutions. Thus, using these rules preserves the solution. The following summary reformulates these rules.

Summary of Operations on a System of Linear Equations

> The following transformations preserve the solution of a system of linear equations:
>
> **1.** Any two equations in a system may be interchanged.
> **2.** Any equation may be multiplied by a nonzero scalar.
> **3.** Any equation may be replaced by itself plus a scalar multiple of another equation (linear combination).

Rule 3 is the primary engine to reduce the system to a simpler system. Rule 3 allows for the elimination of a variable through the careful choice of a scalar multiplier. Consider the linear system $\begin{cases} x + y - z = 0 \\ 2x + y + z = 7. \\ x - 2y + z = 0 \end{cases}$

Linear systems provide much of the inspiration for the development of matrices. Form a coefficient matrix A, variable matrix X and a constant matrix C, then rewrite the given system in matrix form:

$$\begin{matrix} \mathbf{AX} & = & \mathbf{C} \end{matrix}$$
$$\begin{bmatrix} 1 & 1 & -1 \\ 2 & 1 & 1 \\ 1 & -2 & 1 \end{bmatrix} \begin{bmatrix} x \\ y \\ z \end{bmatrix} = \begin{bmatrix} 0 \\ 7 \\ 0 \end{bmatrix}$$

We can duplicate the solution of the system by working with the rows of these matrices in the same manner we worked with individual equations of these matrices. Reducing the coefficient matrix to the identity matrix is equivalent to reducing the system to three equations, each with one variable and a coefficient of 1. To emulate working both sides of the equation simultaneously, combine the coefficient matrix A and constant matrix C to form an **augmented matrix** symbolized as $A|C$ given by

$$\begin{bmatrix} 1 & 1 & -1 & \bigm| & 0 \\ 2 & 1 & 1 & \bigm| & 7 \\ 1 & -2 & 1 & \bigm| & 0 \end{bmatrix}$$

To remind us of the equal signs, a vertical line separates the coefficients and constants in the augmented matrix. In one sense, the system is more abstract: we no longer track the equal signs nor specific variables. The first column represents the coefficients of x, the second column coefficients of y and the third column coefficients of z. Row 1 is the first equation, row 2 is the second equation, row 3 is the third equation. As we previously manipulated equations, we now manipulate rows. The goal is to rewrite the matrix into the form

$$\begin{bmatrix} 1 & 0 & 0 & | & r \\ 0 & 1 & 0 & | & s \\ 0 & 0 & 1 & | & t \end{bmatrix}$$

The solution for the system is $1x = r$, $1y = s$ and $1z = t$. Note that on the main diagonal each element is 1. Actually the first 3 rows and 3 columns resemble $I_{3\times3}$. Our primary goal is to get zeros *below* the main diagonal and, if possible, ones along the main diagonal. Once this is accomplished the values above the main diagonal will be simple.

One method for accomplishing this task is known as the **Gauss-Jordan elimination method.** This method uses elementary row operations on the augmented matrix. **Elementary row operations** are analogous to the manipulation of a system of linear equations.

Elementary Row Operations

Given an augmented matrix representing a system of linear equations. The following row operations on that matrix produce a matrix representing an equivalent system of equations (the corresponding notation is shown in parenthesis, read → as "replaces,"):

E_1: Interchange any two rows of the matrix

$(R_i \leftrightarrow R_j)$.

E_2: Replace any row with a nonzero multiple of itself

$(aR_i \rightarrow R_i, \quad a \neq 0)$.

E_3: Replace any row with itself plus a scalar multiple of another row

$(R_i + aR_j \rightarrow R_i)$. (Read as $R_i + aR_j$ replaces R_i.)

To use row operations requires tedious arithmetic, careful bookkeeping and a clear vision of the ultimate goal. Let us try to reduce the augmented matrix for our original system. Remember the goal: ones along the diagonal and zeros below it. A good strategy to accomplish this is to work by columns from left to right. When working on column r, get the coefficient on the main diagonal (position r, r) to be 1. Accomplish this by replacing *row r* with a scalar multiple of itself. Next get all the elements in the column below the main diagonal to be 0. Accomplish this by replacing each row containing a nonzero element in column r below the diagonal with itself plus a scalar multiple of row r.

$$\begin{bmatrix} 1 & 1 & -1 & | & 0 \\ 2 & 1 & 1 & | & 7 \\ 1 & -2 & 1 & | & 0 \end{bmatrix}$$

Start with column 1. Because the element in the 1, 1 position is already 1, our first task is to get a 0 in the 2, 1 position. Accomplish this by replacing row 2 with itself plus (-2)row 1:

$$(-2)R_1 = [-2 \ -2 \ 2 \ 0] \qquad R_2 = [2 \ 1 \ 1 \ 7]$$

$$R_2 + (-2)R_1 = [0 \ -1 \ 3 \ 7]$$

$$R_2 + (-2)R_1 \rightarrow R_2 \quad \begin{bmatrix} 1 & 1 & -1 & 0 \\ 0 & -1 & 3 & 7 \\ 1 & -2 & 1 & 0 \end{bmatrix}$$

$$R_3 + (-1)R_1 \rightarrow R_3 \quad \begin{bmatrix} 1 & 1 & -1 & 0 \\ 0 & -1 & 3 & 7 \\ 0 & -3 & 2 & 0 \end{bmatrix}$$

Now the first column has a one in the main diagonal and zeros below the main diagonal. Move to column 2. Replace row 2 with -1 times itself to get a one on the main diagonal.

$$(-1)R_2 \rightarrow R_2 \quad \begin{bmatrix} 1 & 1 & -1 & 0 \\ 0 & 1 & -3 & -7 \\ 0 & -3 & 2 & 0 \end{bmatrix}$$

$$R_3 + 3R_2 \rightarrow R_3 \quad \begin{bmatrix} 1 & 1 & -1 & 0 \\ 0 & 1 & -3 & -7 \\ 0 & 0 & -7 & -21 \end{bmatrix}$$

For column 3:

$$(-\tfrac{1}{7})R_3 \rightarrow R_3 \quad \begin{bmatrix} 1 & 1 & -1 & 0 \\ 0 & 1 & -3 & -7 \\ 0 & 0 & 1 & 3 \end{bmatrix}$$

Applying the same strategy to get zeros above the main diagonal, provides a solution to the original linear system. Start with column 2.

$$R_1 + (-1)R_2 \rightarrow R_1 \quad \begin{bmatrix} 1 & 0 & 2 & 7 \\ 0 & 1 & -3 & -7 \\ 0 & 0 & 1 & 3 \end{bmatrix}$$

$$R_2 + 3R_3 \rightarrow R_2 \quad \begin{bmatrix} 1 & 0 & 2 & 7 \\ 0 & 1 & 0 & 2 \\ 0 & 0 & 1 & 3 \end{bmatrix}$$

$$R_1 + (-2)R_3 \rightarrow R_1 \quad \begin{bmatrix} 1 & 0 & 0 & 1 \\ 0 & 1 & 0 & 2 \\ 0 & 0 & 1 & 3 \end{bmatrix}$$

Rewrite in matrix form:

$$\begin{bmatrix} 1 & 0 & 0 \\ 0 & 1 & 0 \\ 0 & 0 & 1 \end{bmatrix} \begin{bmatrix} x \\ y \\ z \end{bmatrix} = \begin{bmatrix} 1 \\ 2 \\ 3 \end{bmatrix}$$

$$\begin{bmatrix} x \\ y \\ z \end{bmatrix} = \begin{bmatrix} 1 \\ 2 \\ 3 \end{bmatrix}.$$

This solves the system. The point of intersection is (1, 2, 3). ▬

Actually, we need not reduce the coefficient matrix to the identity in order to solve the system. It is usually sufficient to get the matrix in row echelon form.

DEFINITION 11-10
Row Echelon Form

A matrix **A** is in **row echelon form** if it meets the following requirements:

1. In every row, the first nonzero number is 1.
2. Rows are arranged in order according to leading zeros: rows with fewer leading zeros come *before* rows with more leading zeros.

There are two consequences of condition 2: (a) any rows of all zeros are the last rows, and (b) the first nonzero number in a row is above a column of zeros.

Row echelon form is not as strict a requirement as obtaining the identity matrix. But row echelon form is the goal of the Gauss-Jordan method. The advantage of the method is that the system solved need not be square. Consider the system $\begin{cases} x + y - 2z = 3 \\ x - 2y + z = 2 \end{cases}$.

Here is a system of two linear equations in three variables. The system is not square. Nonetheless we create an augmented matrix and strive for row echelon form.

$$\left[\begin{array}{ccc|c} 1 & 1 & -2 & 3 \\ 1 & -2 & 1 & 2 \end{array}\right]$$

The strategy for the Gauss-Jordan elimination method is straightforward. Begin with row 1. If necessary, interchange rows so that row 1 has a nonzero value in the furthest left position (this should be column 1). For our matrix, row 1 already has a nonzero entry in column 1. Next, replace each row below row 1 with itself plus a scalar multiple of row 1. Choose the scalar so that each entry below the first nonzero entry of row 1 becomes 0.

Once the first column is in the correct form, move to row 2 and the columns below the first nonzero entry. This time we would like the first nonzero entry of row 2 in column 2 (or as far left as possible). As before use row operations to 0 the entries below the first nonzero entry of row 2. Repeat this process for all remaining rows. The following row operations chronicle these steps.

$$\left[\begin{array}{ccc|c} 1 & 1 & -2 & 3 \\ 1 & -2 & 1 & 2 \end{array}\right]$$

$$R_2 + (-1)R_1 \to R_2 \quad \left[\begin{array}{ccc|c} 1 & 1 & -2 & 3 \\ 0 & -3 & 3 & -1 \end{array}\right]$$

$$-\tfrac{1}{3}R_2 \to R_2 \quad \left[\begin{array}{ccc|c} 1 & 1 & -2 & 3 \\ 0 & 1 & -1 & \tfrac{1}{3} \end{array}\right]$$

Although the matrix is in row echelon form, do one more replacement to get 0 in the second entry of row 1:

$$R_1 - R_2 \rightarrow R_1 \quad \begin{bmatrix} 1 & 0 & -1 & \Big| & \frac{8}{3} \\ 0 & 1 & -1 & \Big| & \frac{1}{3} \end{bmatrix}$$

The matrix is still in row echelon form. In addition, the main diagonal of the matrix marks a region resembling the 2×2 identity. Rewriting the augmented matrix in equation format, we have

$$\begin{cases} x & -z = \frac{8}{3} \\ y - z = \frac{1}{3} \end{cases}$$

Solve the first equation for x in terms of z: $x = z + \frac{8}{3}$.

Solve the second equation for y in terms of z: $y = z + \frac{1}{3}$.

Not only will these solutions check in the original system, they make geometric sense. The original system represented the intersection of two planes. The intersection of two planes is a line. Think of z as a parameter. Hence, $(x, y, z) = (z + \frac{8}{3}, z + \frac{1}{3}, z)$. These first-degree parametric equations represent a line in three-dimensional space.

Suppose that we interchange rows 1 and 2 of $A = \begin{bmatrix} a & b \\ c & d \end{bmatrix}$ to obtain

$A^E = \begin{bmatrix} c & d \\ a & b \end{bmatrix}$. The superscripted E indicates that the new matrix was obtained from an elementary row operation on A. Notice that the interchange can be accomplished by performing the same elementary row operation of $I = \begin{bmatrix} 1 & 0 \\ 0 & 1 \end{bmatrix}$ to obtain $I^E = \begin{bmatrix} 0 & 1 \\ 1 & 0 \end{bmatrix}$, then multiplying A by I^E:

$$I^E A = A^E$$

Similar experiments with other elementary row operations leads to the following theorem.

Theorem 11-12
Elementary Row Operations and the Identity

Let A be an $n \times n$ matrix that can be converted to matrix A^E by a sequence of elementary row operations E. Let I^E be the matrix obtained from the $n \times n$ identity by performing the same sequence of elementary row operations on I. Then $I^E A = A^E$.

EXAMPLE 11-6

Illustration 1:

Elementary Row Operations and Multiplication. Perform the indicated elementary row operations.

Suppose $A = \begin{bmatrix} 3 & 1 & 5 \\ 2 & 1 & -1 \end{bmatrix}$. Perform elementary row operation $E: R_1 + R_2 \rightarrow R_1$.

Solution 1:

$$R_1 + R_2 \rightarrow R_1$$
$$[3 \ 1 \ 5] + [2 \ 1 \ -1] \rightarrow R_1$$
$$[5 \ 2 \ 4] \rightarrow R_1$$
$$A^E = \begin{bmatrix} 5 & 2 & 4 \\ 2 & 1 & -1 \end{bmatrix}$$

Solution 2: Perform row operation E on the 2×2 identity.

$$I = \begin{bmatrix} 1 & 0 \\ 0 & 1 \end{bmatrix}$$
$$R_1 + R_2 \rightarrow R_1$$
$$[1 \ 0] + [0 \ 1] \rightarrow R_1$$
$$[1 \ 1] \rightarrow R_1$$
$$I^E = \begin{bmatrix} 1 & 1 \\ 0 & 1 \end{bmatrix}$$

According to Theorem 11-12,

$$A^E = I^E A$$
$$= \begin{bmatrix} 1 & 1 \\ 0 & 1 \end{bmatrix} \begin{bmatrix} 3 & 1 & 5 \\ 2 & 1 & -1 \end{bmatrix}$$
$$= \begin{bmatrix} 5 & 2 & 4 \\ 2 & 1 & -1 \end{bmatrix}$$

Illustration 2:

Recall that matrix $A = \begin{bmatrix} 1 & 1 & -1 \\ 2 & 1 & 1 \\ 1 & -2 & 1 \end{bmatrix}$, the coefficient matrix from the system of linear equations used to motivate elementary row operations at the beginning of this section. Let us augment A with the 3×3 identity matrix, then repeat the same sequence of eight elementary row operations. Thus, we perform the row operations simultaneously on A and I.

$$A|I = \begin{bmatrix} 1 & 1 & -1 & | & 1 & 0 & 0 \\ 2 & 1 & 1 & | & 0 & 1 & 0 \\ 1 & -2 & 1 & | & 0 & 0 & 1 \end{bmatrix}$$

$$R_2 + (-2)R_1 \rightarrow R_2 \quad \begin{bmatrix} 1 & 1 & -1 & | & 1 & 0 & 0 \\ 0 & -1 & 3 & | & -2 & 1 & 0 \\ 1 & -2 & 1 & | & 0 & 0 & 1 \end{bmatrix}$$

$$R_3 + (-1)R_1 \rightarrow R_3 \quad \begin{bmatrix} 1 & 1 & -1 & | & 1 & 0 & 0 \\ 0 & -1 & 3 & | & -2 & 1 & 0 \\ 0 & -3 & 2 & | & -1 & 0 & 1 \end{bmatrix}$$

$$(-1)R_2 \rightarrow R_2 \quad \begin{bmatrix} 1 & 1 & -1 & | & 1 & 0 & 0 \\ 0 & 1 & -3 & | & 2 & -1 & 0 \\ 0 & -3 & 2 & | & -1 & 0 & 1 \end{bmatrix}$$

$$R_3 + 3R_2 \rightarrow R_3 \quad \begin{bmatrix} 1 & 1 & -1 & | & 1 & 0 & 0 \\ 0 & 1 & -3 & | & 2 & -1 & 0 \\ 0 & 0 & -7 & | & 5 & -3 & 1 \end{bmatrix}$$

$$(-\tfrac{1}{7})R_3 \rightarrow R_3 \quad \begin{bmatrix} 1 & 1 & -1 & | & 1 & 0 & 0 \\ 0 & 1 & -3 & | & 2 & -1 & 0 \\ 0 & 0 & 1 & | & \frac{-5}{7} & \frac{3}{7} & \frac{-1}{7} \end{bmatrix}$$

$$R_1 + (-1)R_2 \rightarrow R_1 \quad \begin{bmatrix} 1 & 0 & 2 & | & -1 & 1 & 0 \\ 0 & 1 & -3 & | & 2 & -1 & 0 \\ 0 & 0 & 1 & | & \frac{-5}{7} & \frac{3}{7} & \frac{-1}{7} \end{bmatrix}$$

$$R_2 + 3R_2 \rightarrow R_2 \quad \begin{bmatrix} 1 & 0 & 2 & | & -1 & 1 & 0 \\ 0 & 1 & 0 & | & \frac{-1}{7} & \frac{2}{7} & \frac{-3}{7} \\ 0 & 0 & 1 & | & \frac{-5}{7} & \frac{3}{7} & \frac{-1}{7} \end{bmatrix}$$

$$R_1 + (-2)R_3 \rightarrow R_1 \quad \begin{bmatrix} 1 & 0 & 0 & | & \frac{3}{7} & \frac{1}{7} & \frac{2}{7} \\ 0 & 1 & 0 & | & \frac{-1}{7} & \frac{2}{7} & \frac{-3}{7} \\ 0 & 0 & 1 & | & \frac{-5}{7} & \frac{3}{7} & \frac{-1}{7} \end{bmatrix}$$

$$\text{Now } \mathbf{I}^E = \begin{bmatrix} \frac{3}{7} & \frac{1}{7} & \frac{2}{7} \\ -\frac{1}{7} & \frac{2}{7} & -\frac{3}{7} \\ -\frac{5}{7} & \frac{3}{7} & -\frac{1}{7} \end{bmatrix}$$

But these elementary row operations converted **A** to the identity: $\mathbf{A}^E = \mathbf{I}$. By Theorem 11-12, $\mathbf{I}^E\mathbf{A} = \mathbf{A}^E$. Because $\mathbf{A}^E = \mathbf{I}$, then $\mathbf{I}^E\mathbf{A} = \mathbf{I}$. This implies that $\mathbf{I}^E = \mathbf{A}^{-1}$. To check this note that

$$\begin{bmatrix} \frac{3}{7} & \frac{1}{7} & \frac{2}{7} \\ -\frac{1}{7} & \frac{2}{7} & -\frac{3}{7} \\ -\frac{5}{7} & \frac{3}{7} & -\frac{1}{7} \end{bmatrix} \begin{bmatrix} 1 & 1 & -1 \\ 2 & 1 & 1 \\ 1 & -2 & 1 \end{bmatrix} = \begin{bmatrix} 1 & 0 & 0 \\ 0 & 1 & 0 \\ 0 & 0 & 1 \end{bmatrix}$$

Therefore, if a square matrix **A** has an inverse, we have a method to generate the inverse: augment **A** with the $n \times n$ identity to form **A|I**. Use elementary row operations to reduce **A** to the identity. These operations conver **A|I** to a *reduced* row echelon form. The corresponding transformation of **I** is the multiplicative inverse for **A**.

Although we have a method, there is no guarantee that it will work. Some matrices *do not* have an inverse. Predicting which matrix has an inverse is not always easy. In these cases we begin the process of transforming **A|I** only to discover, after much effort, that we cannot reach our destination. ■

SUMMARY

- Elementary row operations are the matrix equivalent of the operations used in solving a system of linear equations. These operations are

    E_1: Interchange any two rows of the matrix ($R_i \leftrightarrow R_j$).

    E_2: Replace any row with a nonzero scalar multiple of itself ($aR_i \rightarrow R_i$, $a \neq 0$).

    E_3: Replace any row of the matrix with itself plus a scalar multiple of another row ($R_i + aR_j \rightarrow R_i$).

- Convert matrices to row echelon form by application of elementary row operations. The strategy of Gauss-Jordan elimination begins by labeling row number 1 as the *top* row.

    1. First exchange rows until the top row has a nonzero element as far to the left as possible. Make this element 1 by multiplying the top row by the multiplicative inverse of the first nonzero element. Move any all-zero rows to the last row. These all-zero rows require no further attention.

    2. Replace every row below the top row with itself plus a scalar multiple of the top row. Choose the scalar multiple so that the sum replaces the first nonzero value of the row with 0 (select the additive opposite of the first nonzero as the scalar.)

    Label the row immediately below the top row as the *new* top row. Repeat steps 1 and 2 until you have worked your way to the last non-zero row. On the last row do at most step 2.

- Suppose a matrix $A_{n \times m}$ can be converted to a matrix A^E by the sequence of elementary row operations E. If the matrix I^E is formed by applying the sequence of elementary row operations to $I_{n \times n}$, then $I^E A = A^E$.

- Suppose that $A_{n \times n}$ can be reduced to the identity $I_{n \times n}$ by the sequence of elementary row operations E. Then the same sequence of elementary row operations applied to $I_{n \times n}$ produces the multiplicative inverse of A, if it exists; that is, $I^E = A^{-1}$, where $A^{-1}A = I$.

11-4   E X E R C I S E S

Solve the following systems by rewriting in augmented matrix form and then reducing to row echelon form.

1. $\begin{cases} 3x - 2y = 4 \\ 2x + y = 5 \end{cases}$

2. $\begin{cases} 2x + y = 5 \\ 3x + 2y = -3 \end{cases}$

3. $\begin{cases} 3x - 2y = 9 \\ 2x + y = -1 \end{cases}$

4. $\begin{cases} 2x + y = -2 \\ 3x - 2y = -10 \end{cases}$

5. $\begin{cases} x + y + z = 2 \\ x - y + z = 0 \\ 2x + y - z = 6 \end{cases}$

6. $\begin{cases} x - y + z = 3 \\ x + y + z = 1 \\ 2x + y - z = 0 \end{cases}$

7. $\begin{cases} x + y + z = 5 \\ x - y + z = 3 \\ 2x + y - z = 6 \end{cases}$

8. $\begin{cases} x - y + z = 0 \\ x + y + z = 4 \\ 2x + y - z = 3 \end{cases}$

9. $\begin{cases} w + x + y + z = 2 \\ w + x + y - z = 0 \\ w + x - y + z = 0 \\ w - x + y + z = 0 \end{cases}$

10. $\begin{cases} w - x + y + z = 4 \\ w + x - y + z = 0 \\ w + x + y - z = 0 \\ w + x + y + z = 2 \end{cases}$

11. $\begin{cases} x + y + z = 5 \\ x - y + z = 7 \end{cases}$

12. $\begin{cases} x + y + z = 3 \\ x - y + z = -1 \end{cases}$

*13. $\begin{cases} 2x + y - z = 4 \\ 3x - y + z = 6 \end{cases}$

*14. $\begin{cases} x - 2y + z = 5 \\ x - 2y - z = 7 \end{cases}$

*15. $\begin{cases} x - y + z = 0 \\ x - \quad z + w = 7 \\ y + z - w = 0 \end{cases}$

*16. $\begin{cases} x + y + z = 2 \\ x - y + z = 4 \\ 3x + y + 3z = 8 \end{cases}$

A system of homogeneous equations is a linear system in which the constant terms are all 0. These systems always have the origin as a solution $(0, 0)$, $(0, 0, 0)$, $(0, 0, 0, 0)$, etc., depending on the number of variables. However, homogeneous systems may have more than one solution. In these cases, give the solutions parametrically. Solve the following homogeneous systems:

17. $\begin{cases} 3x - 2y = 0 \\ 2x + y = 0 \end{cases}$

18. $\begin{cases} 5x - 3y = 0 \\ 4x + 5y = 0 \end{cases}$

19. $\begin{cases} 3x - 2y = 0 \\ 6x - 4y = 0 \end{cases}$

20. $\begin{cases} 5x - 3y = 0 \\ -10x + 6y = 0 \end{cases}$

21. $\begin{cases} x + y + z = 0 \\ x - y + z = 0 \\ 2x + y - z = 0 \end{cases}$

22. $\begin{cases} 3x + 2y + z = 0 \\ x - y + z = 0 \\ 2x + y - z = 0 \end{cases}$

23. $\begin{cases} 2x - y + z = 0 \\ x + y + z = 0 \\ 4x + y + 3z = 0 \end{cases}$

24. $\begin{cases} 2x - y + z = 0 \\ x + y + z = 0 \\ 3x - 3y + z = 0 \end{cases}$

Use elementary row operations on each of the following matrices of the form $[\mathbf{A} \mid \mathbf{I}]$ to reduce these matrices to the modified row echelon form $[\mathbf{I} \mid \mathbf{I}^E]$, where E represents the sequence of row operations performed. Verify that $\mathbf{I}^E\mathbf{A} = \mathbf{I}$ by direct multiplication.

25. $\begin{bmatrix} 3 & -2 & | & 1 & 0 \\ 2 & 1 & | & 0 & 1 \end{bmatrix}$

26. $\begin{bmatrix} 2 & 1 & | & 1 & 0 \\ 3 & -2 & | & 0 & 1 \end{bmatrix}$

27. $\begin{bmatrix} 1 & 1 & 1 & | & 1 & 0 & 0 \\ 1 & -1 & 1 & | & 0 & 1 & 0 \\ 2 & 1 & -1 & | & 0 & 0 & 1 \end{bmatrix}$

28. $\begin{bmatrix} 1 & -1 & 1 & | & 1 & 0 & 0 \\ 1 & 1 & 1 & | & 0 & 1 & 0 \\ 2 & 1 & -1 & | & 0 & 0 & 1 \end{bmatrix}$

29. $\begin{bmatrix} 1 & 1 & 1 & 1 & | & 1 & 0 & 0 & 0 \\ 1 & 1 & 1 & -1 & | & 0 & 1 & 0 & 0 \\ 1 & 1 & -1 & 1 & | & 0 & 0 & 1 & 0 \\ 1 & -1 & 1 & 1 & | & 0 & 0 & 0 & 1 \end{bmatrix}$

30. $\begin{bmatrix} 1 & -1 & 1 & 1 & | & 1 & 0 & 0 & 0 \\ 1 & 1 & -1 & 1 & | & 0 & 1 & 0 & 0 \\ 1 & 1 & 1 & -1 & | & 0 & 0 & 1 & 0 \\ 1 & 1 & 1 & 1 & | & 0 & 0 & 0 & 1 \end{bmatrix}$

31. Use $\mathbf{I}^E$ from Problem 25. Determine $\mathbf{I}^E[4 \ 5]^T$. Compare this to the answer for Problem 1.

32. Use $\mathbf{I}^E$ from Problem 26. Determine $\mathbf{I}^E[5 \ -3]^T$. Compare this to the answer for Problem 2.

33. Use $\mathbf{I}^E$ from Problem 27. Determine $\mathbf{I}^E[2 \ 0 \ 6]^T$. Compare this to the answer for Problem 5.

34. Use $\mathbf{I}^E$ from Problem 28. Determine $\mathbf{I}^E[3 \ 1 \ 0]^T$. Compare this to the answer for Problem 6.

35. Use $\mathbf{I}^E$ from Problem 29. Determine $\mathbf{I}^E[2 \ 0 \ 0 \ 0]^T$. Compare this to the answer for Problem 9.

36. Use $\mathbf{I}^E$ from Problem 30. Determine $\mathbf{I}^E[4 \ 0 \ 0 \ 2]^T$. Compare this to the answer for Problem 10.

37. Use $\mathbf{I}^E$ from Problem 31 to solve Problem 3.

38. Use $\mathbf{I}^E$ from Problem 32 to solve Problem 4.

39. Use $\mathbf{I}^E$ from Problem 33 to solve Problem 7.

40. Use $\mathbf{I}^E$ from Problem 34 to solve Problem 8.

11-4   P R O B L E M   S E T

PROBLEMS FOR TECHNOLOGY

Sharp calculators and TI calculators can perform elementary row operations on a matrix. The Casio fx-770G does not directly have this ability, but the row operations can be simulated by performing them on the identity matrix and multiplying the resulting matrix times the given matrix. The process with the Casio can be quite tedious. However, all these calculators can readily produce the multiplicative inverse of a matrix where that inverse exists. Be sure you are in matrix mode.

On the Casio, select the matrix then press the inverse key [F4]. On the Sharp press [2ndF] [MAT] [A] [2ndF] [x^{-1}] [ENTER]. On the TI-81, select matrix A with [2nd] [1], then press [x^{-1}] [ENTER].

Use a calculator with matrix arithmetic capabilities to reduce the following matrices to row echelon form.

1. $\begin{bmatrix} 3 & 1 & 2 & | & 7 \\ 1 & -1 & 5 & | & 4 \\ 2 & 1 & -1 & | & 1 \end{bmatrix}$

2. $\begin{bmatrix} 1 & -1 & 5 & | & -3 \\ 3 & 1 & 2 & | & 2 \\ 2 & 1 & -1 & | & 5 \end{bmatrix}$

3. $\begin{bmatrix} 3 & 1 & 2 & | & 1 & 0 & 0 \\ 1 & -1 & 5 & | & 0 & 1 & 0 \\ 2 & 1 & -1 & | & 0 & 0 & 1 \end{bmatrix}$

4. $\begin{bmatrix} 1 & -1 & 5 & | & 1 & 0 & 0 \\ 3 & 1 & 2 & | & 0 & 1 & 0 \\ 2 & 1 & -1 & | & 0 & 0 & 1 \end{bmatrix}$

5. Determine the inverse of $\begin{bmatrix} 3 & 1 & 2 \\ 1 & -1 & 5 \\ 2 & 1 & -1 \end{bmatrix}$. Compare to Problems 1 and 3.

6. Determine the inverse of $\begin{bmatrix} 1 & -1 & 5 \\ 3 & 1 & 2 \\ 2 & 1 & -1 \end{bmatrix}$. Compare to Problems 2 and 4.

7. Multiply the inverse from Problem 5 times $[7\ 4\ 1]^T$. Compare to Problem 1.

8. Multiply the inverse from Problem 7 times $[-3\ 2\ 5]^T$. Compare to Problem 2.

9. Solve $\begin{cases} 3x + y + 2z = 7 \\ x - y + 5z = 4. \\ 2x + y - \ z = 1 \end{cases}$

Compare your results to Problems 1, 3, 5 and 7.

10. Solve $\begin{cases} x - y + 5z = -3 \\ 3x + y + 2z = \ 2. \\ 2x + y - \ z = \ 5 \end{cases}$

Compare your results to Problems 2, 4, 6 and 8.

5   *DETERMINANTS AND INVERSES

What science can there be more noble, more excellent, more useful for men, more admirably high and demonstrative, than this of the mathematics?
—BENJAMIN FRANKLIN

Cayley developed matrices for notational convenience and conciseness. An important application of matrices is the solution of a system of linear equations. Consider a system of n linear equations in n variables. Because of the difficulties in tracking variables named x, y, z, etc., we use subscripts to distinguish variables and their coefficients: c_5 respresents the coefficient of the fifth variable x_5 in the system. The n equations in n variables are given by

$$\begin{cases} a_1x_1 + a_2x_2 + a_3x_3 + \ldots + a_nx_n = k_1 \\ b_1x_1 + b_2x_2 + b_3x_3 + \ldots + b_nx_n = k_2 \\ \qquad\qquad\qquad \vdots \\ m_1x_1 + m_2x_2 + m_3x_3 + \ldots + m_nx_n = k_n \end{cases}$$

We abbreviate this using matrices to obtain the following:

$$\begin{bmatrix} a_1 & a_2 & a_3 & \dots & a_n \\ b_1 & b_2 & b_3 & \dots & b_n \\ & & \vdots & & \\ m_1 & m_2 & m_3 & \dots & m_n \end{bmatrix} \begin{bmatrix} x_1 \\ x_2 \\ x_3 \\ \vdots \\ x_n \end{bmatrix} = \begin{bmatrix} k_1 \\ k_2 \\ k_3 \\ \vdots \\ k_n \end{bmatrix}$$

or

$$AX = C$$

where **A** is the matrix of coefficients, **X** is the variable matrix and **C** is the constant matrix.

To solve a linear equation in one variable such as $\frac{3}{5}x = 4$, we needed a multiplicative inverse ($\frac{5}{3}$) to multiply on each side of the equation. A multiplicative inverse for matrix **A** is desirable for the same reason. We use A^{-1} to represent an inverse matrix. In fact, if $A^{-1}A = I$, then $AA^{-1} = I$. This brings us to the following definition.

DEFINITION 11-11
Inverse Matrix

Suppose that **A** is an $n \times n$ matrix and there exists an $n \times n$ matrix A^{-1} such that $A^{-1}A = I$ then A^{-1} is called the multiplicative **inverse matrix** of **A**.

If we can determine A^{-1} for a given matrix **A**, then we can solve $AX = C$.

$$A^{-1}(AX) = A^{-1}C$$
$$(A^{-1}A)X = A^{-1}C$$
$$IX = A^{-1}C$$
$$X = A^{-1}C$$

The last section presented one method for producing an inverse matrix. We give that method the dignity of restating it as a theorem.

Theorem 11-13
Identity by Row Echelon Reduction

Suppose that $A_{n \times n}$ has an inverse A^{-1}. Reducing $[A \mid I]$ to the row echelon form, $[I \mid I^E]$, creates the inverse of **A**: $A^{-1} = I^E$.

Example 11-6, Illustration 2 of the last section gives an example for Theorem 11-13. Unfortunately, tedious arithmetic saturates this method for producing an inverse. The Casio fx-7700G, Sharp EL9300/9200 and TI-81/85 can calculate inverse matrices. The problem set of the last section indicates how to use a calculator to get an inverse. Often the inverse is an approximation because the entries are fractions with infinite decimal representations.

Because of tedious arithmetic, it is to our advantage to be able to predict whether a given matrix has an inverse. Curiously, the determinant of the matrix is often used for this purpose. The reason this is curious is because determinants came *before* matrices. We define a *determinant* as a function that takes an $n \times n$ matrix as an argument and returns a real number as its range component. The notation for a determinant is det(**A**), where **A** is a matrix. Alternately, replace the square brackets that enclose the components of a matrix with vertical bars to indicate the determinant. For example, det($[a_{ij}]$) = $|a_{ij}|$. The remainder of this section presents the actual process for obtaining the determinant from the matrix.

The simplest square matrix is the 1×1 matrix, $[a]$. The determinant for this matrix is the entry itself, a. The effort begins with a 2×2 matrix.

DEFINITION 11-12
2 × 2 Determinant

Let $A_{2\times2} = \begin{bmatrix} a & b \\ c & d \end{bmatrix}$. The determinant of A is det(A) = $ad - bc$.

Now the determinant of $\begin{bmatrix} 1 & -3 \\ 2 & 7 \end{bmatrix}$ is given by

$$\begin{vmatrix} 1 & -3 \\ 2 & 7 \end{vmatrix} = 1(7) - 2(-3) = 13$$

Multiplying down the main diagonal provides 1(7), then we subtract the product up the other diagonal 2(-3). This simple process produces a unique real number for each given 2×2 matrix. For higher dimension square matrices the calculations get rapidly out of hand. One way to attack determinants of higher dimension matrices is to reduce them to a lower dimension matrix. Reduce these in turn to the next lower dimension with the ultimate goal of obtaining a 2×2 matrix. Begin this process with the concept of a cofactor.

DEFINITION 11-13
Cofactors

Suppose that $A_{n\times n} = \begin{bmatrix} a_{11} & a_{12} & a_{13} \ldots a_{1n} \\ a_{21} & a_{22} & a_{23} \ldots a_{3n} \\ & & \cdot \\ & & \cdot \\ & & \cdot \\ a_{n1} & a_{n2} & a_{n3} \ldots a_{nn} \end{bmatrix}$, where $n > 1$ and

a_{ij} is the entry in the ith row and jth column of **A**. Then the **cofactor** of a_{ij} is $c_{ij} = (-1)^{i+j} |A^C|$, where A^C is the $(n-1) \times (n-1)$ submatrix formed by deleting the ith row of **A** and the jth column of **A**.

EXAMPLE 11-7

Determining Cofactors. Determine the indicated cofactors.

Illustration 1: If $A = \begin{bmatrix} 1 & -3 \\ 2 & 7 \end{bmatrix}$, determine the cofactors of **A**.

Solution: Begin with c_{11}. Deleting row 1 and column 1 of **A** leaves only 7:

$$c_{11} = (-1)^{1+1}|7| = 1(7) = 7$$

For c_{12}, delete row 1 and column 2 of **A**:

$$c_{12} = (-1)^{1+2}|2| = -2$$

Similarly, $c_{21} = (-1)^{2+1}|-3| = (-1)(-3)$

Caution! Remember that $|-3|$ is the determinant of $[-3]$, not the absolute value of -3.

$$c_{21} = 3$$

Finally, $c_{22} = (-1)^{2+2}|1|$

$$= 1$$

Illustration 2:   Determine cofactors of **A**, where $\mathbf{A} = \begin{bmatrix} 1 & 1 & -1 \\ 2 & 1 & 1 \\ 1 & -2 & 1 \end{bmatrix}$.

Solution: Start with c_{11}. First delete row 1 and column 1 of **A**:

$$\mathbf{A} = \begin{bmatrix} 1 & 1 & -1 \\ 2 & 1 & 1 \\ 1 & -2 & 1 \end{bmatrix}$$

$$c_{11} = (-1)^{1+1} \begin{vmatrix} 1 & 1 \\ -2 & 1 \end{vmatrix} = 1[\, 1(1) - (-2)(1) \,] = 3$$

For c_{12} delete row 1 and column 2 of **A**:

$$\mathbf{A} = \begin{bmatrix} 1 & 1 & -1 \\ 2 & 1 & 1 \\ 1 & -2 & 1 \end{bmatrix}$$

Then $c_{12} = (-1)^{1+2} \begin{vmatrix} 2 & 1 \\ 1 & 1 \end{vmatrix} = -1[\, 2(1) - 1(1) \,] = -1$

Similarly for c_{13}, $c_{13} = (-1)^{1+3} \begin{vmatrix} 2 & 1 \\ 1 & -2 \end{vmatrix} = 1[\, 2(-2) - 1(1) \,] = -5$.

For c_{22}, delete row 2 and column 2 from **A**.

$$\mathbf{A} = \begin{bmatrix} 1 & 1 & -1 \\ 2 & 1 & 1 \\ 1 & -2 & 1 \end{bmatrix}$$

$$c_{22} = (-1)^{2+2} \begin{vmatrix} 1 & -1 \\ 1 & 1 \end{vmatrix} = 1[\, 1(1) - 1(-1) \,] = 2$$

Continue the process for the remaining cofactors:

$$c_{21} = (-1)^{2+1} \begin{vmatrix} 1 & -1 \\ -2 & 1 \end{vmatrix} = 1$$

$$c_{23} = (-1)^{2+3} \begin{vmatrix} 1 & 1 \\ 1 & -2 \end{vmatrix} = 3$$

$$c_{31} = (-1)^{3+1} \begin{vmatrix} 1 & -1 \\ 1 & 1 \end{vmatrix} = 2$$

$$c_{32} = (-1)^{3+2} \begin{vmatrix} 1 & -1 \\ 2 & 1 \end{vmatrix} = -3$$

$$c_{33} = (-1)^{3+3} \begin{vmatrix} 1 & 1 \\ 2 & 1 \end{vmatrix} = -1.$$

A convenient method for listing the cofactors of **A** is the **cofactor matrix** of **A**, represented by **C(A)**. Each entry in the cofactor matrix is the cofactor of the corresponding entry in **A**. Thus in **C(A)** the entry in the c_{32} position is -3.

$$\mathbf{C(A)} = \begin{bmatrix} 3 & -1 & -5 \\ 1 & 2 & 3 \\ 2 & -3 & -1 \end{bmatrix}$$

For Illustration 1, the cofactor matrix is $\begin{bmatrix} 7 & -2 \\ 3 & 1 \end{bmatrix}$.

DEFINITION 11-14
The Determinant

> The **determinant** of a square matrix **A** (with dimension larger than 1×1) is the dot product of row 1 of **A** and row 1 of **C(A)**, where **C(A)** is the cofactor matrix of **A**.
>
> Actually, the dot product of any two corresponding rows (or columns) of **A** and **C(A)** produces the determinant of **A**.
>
> $$\det(\mathbf{A}) = [a_{i1} \ a_{i2} \ a_{i3} \ldots a_{in}] \cdot [c_{i1} \ c_{i2} \ c_{i3} \ldots c_{in}]$$

The following theorems indicate the effect of elementary row operations on the value of a determinant.

Theorem 11-14
Determinants and Interchanging Rows

> Interchanging two rows of a matrix reverses the sign of its determinant.

For example, $\begin{vmatrix} 1 & 2 \\ 3 & 4 \end{vmatrix} = -2$, whereas $\begin{vmatrix} 3 & 4 \\ 1 & 2 \end{vmatrix} = 2.$

Theorem II-I5
Scalar Multiples of Rows and Determinants

> Multiplying any row of $\mathbf{A}$ by $\mathbf{k} \neq 0$, multiplies the determinant by $\mathbf{k}$.

Consider again the matrix $\begin{bmatrix} 1 & 2 \\ 3 & 4 \end{bmatrix}$ with determinant -2. Multiply row 2 by -3

and evaluate the determinant: $\begin{vmatrix} 1 & 2 \\ -9 & -12 \end{vmatrix} = 6$. The determinant changed by a factor of -3.

Theorem II-I6
Replacing a Row with Itself Plus a Scalar Multiple of Another Row

> Replacing a row of a matrix with itself plus a scalar multiple of another row does not change the determinant of the matrix.

In elementary row operation notation, replacing a row r_k with itself plus a scalar multiple of another row r_j is expressed as $r_k + cr_j \rightarrow r_k$. Replace row 1 of $\begin{bmatrix} 1 & 2 \\ 3 & 4 \end{bmatrix}$ with itself plus 5 times row 2 to obtain $\begin{bmatrix} 16 & 22 \\ 3 & 4 \end{bmatrix}$. The determinant is still -2.

E X A M P L E   II - 8

Illustration I:

Calculating Determinants

Recall the 2×2 matrix and cofactor matrix from Example 11-7.

$$\mathbf{A} = \begin{bmatrix} 1 & -3 \\ 2 & 7 \end{bmatrix}, \qquad \mathbf{C(A)} = \begin{bmatrix} 7 & -2 \\ 3 & 1 \end{bmatrix}$$

Applying the definition of determinant for a 2×2 matrix establishes

$$\det(\mathbf{A}) = 13$$

We wish to confirm that the general definition for determinant is consistent with the 2×2 definition. Consider the dot products of the following corresponding pairs of rows and columns:

	A	C(A)	Dot product
Row 1	[1 -3]	[7 -2]	$1(7) + (-3)(-2) = 13$
Row 2	[2 7]	[3 1]	$2(3) + 7(1) = 13$
Column 1	[1 2]	[7 3]	$1(7) + 2(3) = 13$
Column 2	[-3 7]	[-2 1]	$-3(-2) + 7(1) = 13$

Illustration 2: Calculate the determinant of $A = \begin{bmatrix} 1 & 1 & -1 \\ 2 & 1 & 1 \\ 1 & -2 & 1 \end{bmatrix}$.

Solution: From Example 11-7, the cofactor matrix of A is given by

$$C(A) = \begin{bmatrix} 3 & -1 & -5 \\ 1 & 2 & 3 \\ 2 & -3 & -1 \end{bmatrix}$$

Choose any corresponding rows from A and $C(A)$ and calculate the dot product. The result is the determinant of A. For example, choose row 1:

$$\det(A) = [1 \ 1 \ -1] \cdot [3 \ -1 \ -5]$$
$$= 1(3) + 1(-1) + (-1)(-5)$$
$$= 7$$

Illustration 3: Calculate the determinant of $B = \begin{bmatrix} 2 & 3 & -1 \\ 3 & 1 & 4 \\ 0 & 0 & 1 \end{bmatrix}$.

Solution: We only need a corresponding pair of rows from each of B and $C(B)$ to calculate the determinant. Without $C(B)$, we make our decision based on the rows of B. Examine row 3 carefully. Notice the two zero entries. For row 3, the determinant will be $\det(B) = [0 \ 0 \ 1] \cdot [c_{31} \ c_{32} \ c_{33}]$, where c_{31}, c_{32} and c_{33} represent the cofactors of 0, 0 and 1, respectively. However, the dot product is $\det(B) = 0(c_{31}) + 0(c_{32}) + 1(c_{33}) = c_{33}$.

We need no cofactors except cofactor c_{33}. The savings in arithmetic justifies the choice of the third row. To calculate cofactor c_{33} delete the third row and third column of B. Then

$$c_{33} = (-1)^{3+3} \begin{vmatrix} 2 & 3 \\ 3 & 1 \end{vmatrix} = 1[\,2(1) - 3(3)\,] = -7$$

$$\det(B) = -7$$

Illustration 4: Calculate the determinant of $A = \begin{bmatrix} 1 & 2 & 3 \\ 1 & 3 & 1 \\ 2 & 3 & 1 \end{bmatrix}$.

Solution 1: Choose a row, say row 1, and immediately write the dot product of row 1 from A with row 1 of the cofactor matrix. This is not as difficult as it sounds. Remember, each cofactor is simply the appropriate sign times the determinant of a submatrix formed by removing a row and column of the original matrix. The determinant of the submatrix is called a **minor** and the method we use is called *expanding by minors*.

$$\begin{vmatrix} 1 & 2 & 3 \\ 1 & 3 & 1 \\ 2 & 3 & 1 \end{vmatrix} = 1(-1)^{1+1} \begin{vmatrix} 3 & 1 \\ 3 & 1 \end{vmatrix} + 2(-1)^{1+2} \begin{vmatrix} 1 & 1 \\ 2 & 1 \end{vmatrix} + 3(-1)^{1+3} \begin{vmatrix} 1 & 3 \\ 2 & 3 \end{vmatrix}$$

Each of the three terms of the expansion has three factors: the first factor is an entry from the chosen row; the second factor is $(-1)^{i+j}$, where i and j are the row and column the entry came from (if $i + j$ is even, this factor is 1, otherwise it is -1); and the third factor is a determinant, which comes from the submatrix formed by deleting the row and column from which the first factor was chosen. Since the minors are 2×2, we can easily finish the arithmetic.

$$\begin{vmatrix} 1 & 2 & 3 \\ 1 & 3 & 1 \\ 2 & 3 & 1 \end{vmatrix} = 1(1)[3(1) - 3(1)] + 2(-1)[1(1) - 2(1)] + 3(1)[1(3) - 2(3)]$$

$$= 1(0) + -2(-1) + 3(-3)$$

$$= -7$$

Solution 2: Unfortunately, there are few good choices of a row in this matrix. From the previous illustration, the desirability of zeros in a row should be obvious. Refer back to Theorems 11-14, 11-15 and 11-16. Elementary row operations provide a method for manipulating a matrix so that the row choices become more desirable. We manipulate the matrix in determinant form to remind ourselves that Theorems 11-14 and 11-15 alter the value of the determinant. Choose row 1. Because the entry in the 1, 1 position is 1, concentrate on converting the entries in the 2, 1 and 3, 1 positions to zeros. Like Gauss-Jordan elimination, begin by zeroing the column beneath the 1, 1 position:

$$\begin{vmatrix} 1 & 2 & 3 \\ 1 & 3 & 1 \\ 2 & 3 & 1 \end{vmatrix}$$

$$R_2 - R_1 \rightarrow R_2 \quad \begin{vmatrix} 1 & 2 & 3 \\ 0 & 1 & -2 \\ 2 & 3 & 1 \end{vmatrix}$$

$$R_3 - 2R_1 \rightarrow R_3 \quad \begin{vmatrix} 1 & 2 & 3 \\ 0 & 1 & -2 \\ 0 & -1 & -5 \end{vmatrix}$$

If we expanded about column 1, we would be finished. Instead we work toward row echelon form to demonstrate more of the elementary row operations. Next zero out the last entry in column 2:

$$R_3 + R_2 \rightarrow R_3 \quad \begin{vmatrix} 1 & 2 & 3 \\ 0 & 1 & -2 \\ 0 & 0 & -7 \end{vmatrix}$$

Expand along the first column.

$$\begin{vmatrix} 1 & 2 & 3 \\ 0 & 1 & -2 \\ 0 & 0 & -7 \end{vmatrix} = 1(-1)^{1+1} \begin{vmatrix} 1 & -2 \\ 0 & -7 \end{vmatrix} = 1[1(-7) - 0(-2)] = -7$$

Solution 3: Use a calculator with matrix arithmetic capabilities. Be sure you are in matrix mode. One the Casio fx-7700G, store the matrix in the usual manner, then press the determinant key [F3]. On the Sharp 9300, store the matrix under **A**, press MATH E 6 2ndF MAT A ENTER . On the TI-81, store the matrix under **A**, then clear to the main screen. Enter 2nd 1 MATRX to select **A** and display the matrix menu. Select determinant from the menu and press ENTER . ■

The next theorems make a connection between determinants and **inverse matrices**.

Theorem 11-17
Existence of an Inverse

$$\mathbf{A}_{n \times n} \text{ has an inverse } \mathbf{A}^{-1} \text{ if and only if } \det(\mathbf{A}) \neq 0.$$

All the matrices of Example 11-8 have inverses, because none of their determinants is 0. However, $\begin{bmatrix} 1 & 2 \\ 2 & 4 \end{bmatrix}$ has no inverse because $\begin{vmatrix} 1 & 2 \\ 2 & 4 \end{vmatrix} = 0$.

Theorem 11-17 is called an *existence theorem*. An existence theorem indicates that there is an answer. Theorem 11-17 offers no clue to finding $\mathbf{A}^{-1}$. Although we know an elementary row process for determining an inverse, Theorem 11-18 offers a connection of inverses to determinants and cofactors. Theorem 11-18 is an *algorithmic* theorem.

Theorem 11-18
Inverses, Determinants and Cofactors

$$\text{Suppose that } \det(\mathbf{A}_{n \times m}) \neq 0, \text{ then } \mathbf{A}^{-1} = \frac{1}{\det(\mathbf{A})} [C(\mathbf{A})]^T.$$

The calculation of an inverse matrix requires a nonzero determinant and the transpose of the cofactor matrix, as the following examples illustrate.

EXAMPLE 11-9

Illustration 1:

Using Determinants and Cofactors for an Inverse. Determine the inverse for the following.

$$\mathbf{A} = \begin{bmatrix} 1 & -3 \\ 2 & 7 \end{bmatrix}.$$

Solution: Example 11-8, Illustration 1 provides

$$\begin{vmatrix} 1 & -3 \\ 2 & 7 \end{vmatrix} = 13 \quad \text{and} \quad C(\mathbf{A}) = \begin{bmatrix} 7 & -2 \\ 3 & 1 \end{bmatrix}$$

Thus, $\mathbf{A}^{-1} = \frac{1}{13} \begin{bmatrix} 7 & -2 \\ 3 & 1 \end{bmatrix}^T$

$$= \frac{1}{13} \begin{bmatrix} 7 & 3 \\ -2 & 1 \end{bmatrix}$$

$$= \begin{bmatrix} \frac{7}{13} & \frac{3}{13} \\ \frac{-2}{13} & \frac{1}{13} \end{bmatrix}$$

Verify the inverse by direct calculation:

$$\mathbf{A}^{-1}\mathbf{A} = \begin{bmatrix} \frac{7}{13} & \frac{3}{13} \\ \frac{-2}{13} & \frac{1}{13} \end{bmatrix} \begin{bmatrix} 1 & -3 \\ 2 & 7 \end{bmatrix}$$

$$= \begin{bmatrix} \frac{7}{13}1 + \frac{3}{13}2 & \frac{7}{13}(-3) + \frac{3}{13}7 \\ \frac{-2}{13}1 + \frac{1}{13}2 & \frac{-2}{13}(-3) + \frac{1}{13}7 \end{bmatrix}$$

$$= \begin{bmatrix} 1 & 0 \\ 0 & 1 \end{bmatrix}$$

Illustration 2:

$$\mathbf{A} = \begin{bmatrix} 1 & 1 & -1 \\ 2 & 1 & 1 \\ 1 & -2 & 1 \end{bmatrix}$$

Solution: Example 11-8 provides both the determinant of $\mathbf{A}$, $\det(\mathbf{A}) = 7$, and the cofactor matrix, $C(\mathbf{A}) = \begin{bmatrix} 3 & -1 & -5 \\ 1 & 2 & 3 \\ 2 & -3 & -1 \end{bmatrix}$. Then, by Theorem 11-18,

$$\mathbf{A}^{-1} = \frac{1}{7} \begin{bmatrix} 3 & -1 & -5 \\ 1 & 2 & 1 \\ 2 & -3 & -1 \end{bmatrix}^T$$

$$= \frac{1}{7} \begin{bmatrix} 3 & 1 & 2 \\ -1 & 2 & -3 \\ -5 & 3 & -1 \end{bmatrix}$$

$$= \begin{bmatrix} \frac{3}{7} & \frac{1}{7} & \frac{2}{7} \\ \frac{-1}{7} & \frac{2}{7} & \frac{-3}{7} \\ \frac{-5}{7} & \frac{3}{7} & \frac{-1}{7} \end{bmatrix}$$

Note: Compare this to $\mathbf{A}^{-1}$ as found in Example 11-6, Illustration 2.

Illustration 3:

Use matrices to solve the system:

$$\begin{cases} x + y - z = 0 \\ 2x + y + z = 14 \\ x - 2y + z = -7 \end{cases}$$

First write the equation in matrix format:

$$\mathbf{AX} = \mathbf{C}$$

$$\begin{bmatrix} 1 & 1 & -1 \\ 2 & 1 & 1 \\ 1 & -2 & 1 \end{bmatrix} \begin{bmatrix} x \\ y \\ z \end{bmatrix} = \begin{bmatrix} 0 \\ 14 \\ -7 \end{bmatrix}$$

Then

$$\mathbf{X} = \mathbf{A}^{-1}\mathbf{C}$$

So that

$$\begin{bmatrix} x \\ y \\ z \end{bmatrix} = \begin{bmatrix} \frac{3}{7} & \frac{1}{7} & \frac{2}{7} \\ \frac{-1}{7} & \frac{2}{7} & \frac{-3}{7} \\ \frac{-5}{7} & \frac{3}{7} & \frac{-1}{7} \end{bmatrix} \begin{bmatrix} 0 \\ 14 \\ -7 \end{bmatrix}$$

$$\begin{bmatrix} x \\ y \\ z \end{bmatrix} = \begin{bmatrix} 0 \\ 7 \\ 7 \end{bmatrix}$$

Although this may seem a tedious process, remember that your calculator can do most of the work for you. Store the coefficient matrix under **A** and have the calculator compute $\mathbf{A}^{-1}$. Store the constant matrix. Multiply the inverse matrix times the constant matrix and you will have solved the system.

If your calculator calculates an inverse matrix using finite precision decimals, you may get round off errors in rational solutions. Another method for solving a linear system uses determinants that separate the numerator and denominator of the solution. It is **Cramer's rule**. Cramer's rule predates the invention of matrices by about 100 years.

S I D E B A R

Cramer's rule is named for Gabriel Cramer (1704–1752). However Cramer's method was actually created by Colin Maclaurin in about 1729. Cramer used determinants in the treatment of conics in a treatise about 1750. See Chapter 9. Cayley did not introduce matrix notation for more than 100 years after Cramer's publication.

Theorem 11-19
Cramer's Rule

The solution for variable x_i in the matrix equation $\mathbf{AX} = \mathbf{C}$, where $\det(\mathbf{A}) \neq 0$, $\mathbf{X} = [x_1\, x_2\, x_3 \ldots x_n]^T$ and $\mathbf{C} = [c_1\, c_2\, c_3 \ldots c_n]^T$, is given by

$$x_i = \frac{\det(\mathbf{A}^{Ci})}{\det(\mathbf{A})}$$

where $\mathbf{A}^{Ci}$ is formed by replacing column i of **A** with column matrix **C**

Solving Equations with Cramer's Rule

EXAMPLE 11-10

Illustration:

Use Cramer's rule to solve $\begin{cases} 3x - 2y + z = 10 \\ 2x + y - 3z = -9 \\ x + y + 2z = 5 \end{cases}$

Solution: Rewrite in matrix form $\begin{bmatrix} 3 & -2 & 1 \\ 2 & 1 & -3 \\ 1 & 1 & 2 \end{bmatrix} \begin{bmatrix} x \\ y \\ z \end{bmatrix} = \begin{bmatrix} 10 \\ -9 \\ 5 \end{bmatrix}$.

Minor along row 3 to get the determinate of the coefficient matrix:

$$\begin{vmatrix} 3 & -2 & 1 \\ 2 & 1 & -3 \\ 1 & 1 & 2 \end{vmatrix} = 1(-1)^{3+1}\begin{vmatrix} -2 & 1 \\ 1 & -3 \end{vmatrix} + 1(-1)^{3+2}\begin{vmatrix} 3 & 1 \\ 2 & -3 \end{vmatrix} + 2(-1)^{3+3}\begin{vmatrix} 3 & -2 \\ 2 & 1 \end{vmatrix}$$

$$= -1[-2(-3) - 1(1)] + (-1)[3(-3)-2(1)] + 2[3(1) - 2(-2)] = 30$$

To solve for x, form A^{C_1} by replacing column 1 with $\begin{bmatrix} 10 \\ -9 \\ 5 \end{bmatrix}$.

$$A^{C_1} = \begin{bmatrix} 10 & -2 & 1 \\ -9 & 1 & -3 \\ 5 & 1 & 2 \end{bmatrix}$$

$$\begin{vmatrix} 10 & -2 & 1 \\ -9 & 1 & -3 \\ 5 & 1 & 2 \end{vmatrix} = 5[-2(-3) - 1(1)] - 1[10(-3) - (-9)1] + 2[10(1) - (-9)(-2)]$$
$$= 30$$

Now $x = \dfrac{\det(A^{C_1})}{\det(A)} = \dfrac{30}{30} = 1$

Similarly, $\det(A^{C_2}) = \begin{vmatrix} 3 & 10 & 1 \\ 2 & -9 & -3 \\ 1 & 5 & 2 \end{vmatrix} = -60$

Then $y = \dfrac{-60}{30} = -2$

Finally, $\det(A^{C_3}) = \begin{vmatrix} 3 & -2 & 10 \\ 2 & 1 & -9 \\ 1 & 1 & 5 \end{vmatrix} = 90$

Then, $z = \dfrac{90}{30} = 3$

$$\begin{bmatrix} x \\ y \\ z \end{bmatrix} = \begin{bmatrix} 1 \\ -2 \\ 3 \end{bmatrix}$$

Now for a completely different application of determinants. Consider two points P_1 and P_2 with different x coordinates. There is a unique linear function containing these two points. Chapter 1 discussed a method for writing the equation for that line. However, there may be numerous higher degree polynomials that contain these same points.

What is the smallest degree polynomial that will contain $n + 1$ given points, where all points have different x coordinates? An answer to that question is not always apparent. Indeed we may need to modify the question to obtain the security of an algorithm. Based on our experience with two points with linear functions and three points with quadratic functions, we propose the following theorem.

Through $n + 1$ given points of a function, there is a polynomial P of degree $\leq n$ such that the graph of P contains all $n + 1$ points.

Theorem 11-20 guarantees that through five arbitrary points of a function, there is a polynomial of degree 4 or less whose graph contains the five points. Certainly higher degree polynomials exist that contain the same five points. The actual degree of the polynomial that contains the five points varies with the nature of the five points. For example, if all five points are collinear, then a first-degree polynomial (as well as a fifth-degree polynomial) passes through the five points. However, no quadratic polynomial could contain three much less five collinear points.

Theorem 11-20 is an existence theorem. The theorem offers no clue on how to produce the described polynomial. One method from Chapter 4 is the Lagrange approximating polynomial. Determinants provide another method for producing a polynomial through $n + 1$ points of a function. The technique is based upon the following property of determinants: *If any two rows (or columns) of a matrix are equal, the determinant of the matrix is 0.*

Given $n + 1$ points of a function, form an $(n + 2) \times (n + 2)$ matrix. The first row of the matrix will be $[y \;\; 1 \;\; x \;\; x^2 \;\; x^3 \ldots x^n]$.

The next $n + 1$ rows of the matrix come from substituting the coordinates of the $n + 1$ given points for their respective values of x and y in row 1. This generates an $(n + 2) \times (n + 2)$ matrix. Call this matrix **A**. Because we want row 1 to equal row 2 or row 1 to equal row 3 or row 1 to equal row 4, etc., we set the determinant of **A** equal to 0. By setting the determinant $= 0$, we form a determinant equation with y and powers of x. Solve this equation for y for the required polynomial.

EXAMPLE   11-11

Writing Polynomials Using Determinants. Determine a polynomial as indicated through the given points.

Illustration 1:

Determine a first-degree polynomial through $(0, -1)$ and $(2, 3)$.

Solution:

$$\text{Set } \begin{vmatrix} y & 1 & x \\ -1 & 1 & 0 \\ 3 & 1 & 2 \end{vmatrix} = 0$$

$$\begin{vmatrix} y & 1 & x \\ -1 & 1 & 0 \\ 3 & 1 & 2 \end{vmatrix} = y(-1)^{1+1} \begin{vmatrix} 1 & 0 \\ 1 & 2 \end{vmatrix} + 1(-1)^{1+2} \begin{vmatrix} -1 & 0 \\ 3 & 2 \end{vmatrix} + x(-1)^{1+3} \begin{vmatrix} -1 & 1 \\ 3 & 1 \end{vmatrix}$$

$$0 = 2y + 2 - 4x$$

$$y = 2x - 1$$

Note: A calculator is of little use in this problem until after expanding the determinant using minors.

Illustration 2:   Determine a polynomial containing $(0, -1)$, $(1, 3)$ and $(2, 9)$.

Solution:

Set $\begin{vmatrix} y & 1 & x & x^2 \\ -1 & 1 & 0 & 0 \\ 3 & 1 & 1 & 1 \\ 9 & 1 & 2 & 4 \end{vmatrix} = 0$. Minor the determinant about row 1.

$$y\begin{vmatrix} 1 & 0 & 0 \\ 1 & 1 & 1 \\ 1 & 2 & 4 \end{vmatrix} - 1\begin{vmatrix} -1 & 0 & 0 \\ 3 & 1 & 1 \\ 9 & 2 & 4 \end{vmatrix} + x\begin{vmatrix} -1 & 1 & 0 \\ 3 & 1 & 1 \\ 9 & 1 & 4 \end{vmatrix} - x^2\begin{vmatrix} -1 & 1 & 0 \\ 3 & 1 & 1 \\ 9 & 1 & 2 \end{vmatrix} = 0$$

$$2y + 2 - 6x - 2x^2 = 0$$

$$y = x^2 + 3x - 1$$

S U M M A R Y

- $\begin{vmatrix} a & b \\ c & d \end{vmatrix} = ad - bc$

- If A is a square matrix of order $n \times n$, $n > 1$, then $C(A)$ is the cofactor matrix of A. Each component of $C(A)$ is $c_{ij} = (-1)^{i+j} \det(A^C)$, where A^C is of order $(n-1) \times (n-1)$ formed by deleting the ith row and jth column of A.

- If A is an $n \times n$ matrix, $\det(A) = A_j \cdot C(A)_j$, where A_j is the jth row of A and $C(A)_j$ is the jth row of the cofactor matrix of A.

- If $\det(A) \neq 0$, $A^{-1} = \dfrac{1}{\det(A)} C(A)^T$.

- If $\det(A) \neq 0$, and $AX = C$, then $X = A^{-1}C$.

- If $\det(A) \neq 0$, and $AX = C$, then
  $$x_i = \frac{\det(A^{C_i})}{\det(A)}$$
  where x_i is the ith variable in matrix X and A^{C_i} is obtained from A by replacing the ith column of A with column matrix C.

- Suppose (x_0, y_0), (x_1, y_1), (x_2, y_2), . . .(x_n, y_n) are $n + 1$ points of a function. Then a polynomial containing these points is given by

$$\begin{vmatrix} y & 1 & x & x^2 \ldots x^n \\ y_0 & 1 & x_0 & x_0^2 \ldots x_0^n \\ y_1 & 1 & x_1 & x_1^2 \ldots x_1^n \\ y_2 & 1 & x_2 & x_2^2 \ldots x_2^n \\ & & \cdots & \\ y_n & 1 & x_n & x_n^2 \ldots x_n^{\,n} \end{vmatrix} = 0$$

11-5 EXERCISES

For each of the following matrices: (1) calculate the determinant, (b) determine the inverse when it exists.

1. $\begin{bmatrix} 3 & 4 \\ 2 & -1 \end{bmatrix}$

2. $\begin{bmatrix} 2 & -1 \\ 3 & 4 \end{bmatrix}$

3. $\begin{bmatrix} 4 & -1 \\ 2 & 3 \end{bmatrix}$

4. $\begin{bmatrix} 2 & 3 \\ 4 & -1 \end{bmatrix}$

5. $\begin{bmatrix} 1 & 2 & 1 \\ 1 & -1 & 3 \\ 2 & 2 & -1 \end{bmatrix}$

6. $\begin{bmatrix} 1 & -1 & -1 \\ 3 & 1 & 2 \\ 1 & 1 & 1 \end{bmatrix}$

7. $\begin{bmatrix} 1 & 1 & 3 \\ 2 & 1 & 2 \\ 1 & -1 & -2 \end{bmatrix}$

8. $\begin{bmatrix} 3 & 1 & 2 \\ 1 & -2 & 3 \\ 1 & 1 & 1 \end{bmatrix}$

9. $\begin{bmatrix} 1 & 1 & 1 & 2 \\ 1 & 1 & 0 & 3 \\ 1 & 0 & -1 & 1 \\ 1 & 2 & 2 & 0 \end{bmatrix}$

10. $\begin{bmatrix} 1 & 2 & 1 & 1 \\ 1 & 0 & 0 & 2 \\ 1 & 3 & 1 & 0 \\ 0 & 1 & 1 & 2 \end{bmatrix}$

Solve the following systems. (Hint compare these problems to the matrices in Problems 1–10.)

11. $\begin{cases} 3x + 4y = 10 \\ 2x - y = 3 \end{cases}$

12. $\begin{cases} 2x - y = 5 \\ 3x + 4y = 2 \end{cases}$

13. $\begin{cases} 4x - y = 7 \\ 2x + 3y = -1 \end{cases}$

14. $\begin{cases} 2x + 3y = 0 \\ 4x - y = 3 \end{cases}$

15. $\begin{cases} x + 2y + z = 4 \\ x - y + 3z = 3 \\ 2x + 2y - z = 3 \end{cases}$

16. $\begin{cases} x - y - z = -2 \\ 3x + y + 2z = 8 \\ x + y + z = 4 \end{cases}$

17. $\begin{cases} x + y + 3z = 4 \\ 2x + y + 2z = 5 \\ x - y - 2z = 1 \end{cases}$

18. $\begin{cases} 3x + y + 2z = 6 \\ x - 2y + 3z = 9 \\ x + y + z = 2 \end{cases}$

19. $\begin{cases} x + y + z + 2w = 6 \\ x + y + 3w = 7 \\ x - z + w = 2 \\ x + 2y + 2z = 5 \end{cases}$

20. $\begin{cases} x + 2y + z + w = 1 \\ x + 2w = 3 \\ x + 3y + z = 0 \\ y + z + 2w = 1 \end{cases}$

Solve the following.

21. A buyer purchases batches of three items for a store: wallets, purses and key chains. Suppose that the total number of items purchased is 60. Wallets cost $5 each, purses cost $6 each and keychains cost $1 each. The total invoice is $190. If the number of keychains is equal to the total number of wallets and purses combined, how many of each were purchased?

22. Hot dogs, hamburgers and sandwiches are to be made for a large picnic. It costs $0.50 to make a hot dog, $0.60 to make a hamburger and $0.40 to make a sandwich. The total budget allotted for these items is $51.00. To feed all the people at the picnic will require a total of 100 items. If as many people prefer hamburgers as prefer hotdogs and sandwiches total, how many of each should be made?

Verify each of the following for 2 × 2 matrices.

23. Theorem 11-13

24. Theorem 11-14

25. Theorem 11-15

26. Theorem 11-16

27. Theorem 11-17

28. Theorem 11-18

29. Theorem 11-19

30. Theorem 11-20.

31. If one column of a matrix is all zeros, the determinant is 0.

32. If two columns of a matrix are identical, then the determinant is 0.

33. If two rows of a matrix are identical, then the determinant is 0.

Verify each of the following for 3 × 3 matrices.

34. If one row of a matrix is all zeros, the determinant is 0.

35. If two rows of a matrix are identical, then the determinant is 0.

36. Theorem 11-13

37. Theorem 11-14

38. Theorem 11-15

39. Theorem 11-16

40. Theorem 11-17

41. Theorem 11-18

42. Theorem 11-19

43. Theorem 11-20.

44. If one row of a matrix is a scalar multiple of another then the determinant is 0.

45. If A is of order 2×2, then $\det(A) = \det(A^T)$.

46. If A is of order 3×3, then $\det(A) = \det(A^T)$.

47. For a 2×2 matrix A, let $B = C(A)$. Determine $C(B)$. Discuss the results.

48. Attack or defend: for a 2×2 matrix A, $C[C(A)] = A$.

Determine a formula for a nth polynomial through the given points.

49. $(1, 2)$ $(2, 0)$

50. $(1, -1)$ $(3, 2)$

51. $(0, -1)$ $(1, 5)$ $(2, 13)$

52. $(0, 3)$ $(-1, 4)$ $(1, 6)$

53. $(1, 0)$ $(2, 3)$ $(3, 8)$

54. $(1, 1)$ $(2, 10)$ $(-1, 1)$

***55.** Suppose that A and B are three-dimensional vectors, $A = a\vec{i} + b\vec{j} + c\vec{k}$ and $B = d\vec{i} + e\vec{j} + f\vec{k}$ (i.e., $A = \langle a\ b\ c \rangle$ and $B = \langle d\ e\ f \rangle$). Then $A \times B$ read "A cross B" is defined as $A \times B = \begin{vmatrix} \vec{i} & \vec{j} & \vec{k} \\ a & b & c \\ d & e & f \end{vmatrix}$. The cross product of two vectors A and B, produces a vector perpendicular to both A and B. Show that $A \times B = -(B \times A)$.

***56.** Determine a vector perpendicular to $\langle 1, -1, 1 \rangle$ and $\langle 2, 3, 0 \rangle$.

***57.** Determine a vector perpendicular to $\langle 2, 0, -1 \rangle$ and $\langle 3, 1, 1 \rangle$.

***58.** Show that $A \times 0 = 0$.

***59.** Show that if $A = kB$, then $A \times B = 0$.

***60.** Show that $A \cdot (A \times B) = 0$.

61. Solve $\begin{vmatrix} 2-x & 0 \\ 0 & 3+x \end{vmatrix} = 0$ for x.

62. Solve $\begin{vmatrix} 3-x & 0 \\ 0 & 4-x \end{vmatrix} = 0$ for x.

63. If $A = \begin{bmatrix} 5 & 0 \\ 0 & -1 \end{bmatrix}$, solve $|A - xI| = 0$ for x.

64. If $A = \begin{bmatrix} 5 & 0 \\ 0 & -1 \end{bmatrix}$, solve $|A - xI| = 0$ for x.

65. Suppose that A is the matrix in Problem 1. Solve $|A - xI| = 0$ for x.

66. Suppose that A is the matrix in Problem 2. Solve $|A - xI| = 0$ for x.

11-5   P R O B L E M   S E T

PROBLEMS FOR TECHNOLOGY

Use a calculator with matrix arithmetic capabilities to solve the following.

1. $\begin{cases} 3x + 4y = 10 \\ 2x - y = 3 \end{cases}$

2. $\begin{cases} 2x - y = 5 \\ 3x + 4y = 2 \end{cases}$

3. $\begin{cases} 4x - y = 7 \\ 2x + 3y = -1 \end{cases}$

4. $\begin{cases} 2x + 3y = 0 \\ 4x - y = 3 \end{cases}$

5. $\begin{cases} x + 2y + z = 4 \\ x - y + 3z = 3 \\ 2x + 2y - z = 3 \end{cases}$

6. $\begin{cases} x - y - z = -2 \\ 3x + y + 2z = 8 \\ x + y + z = 4 \end{cases}$

7. $\begin{cases} x + y + 3z = 4 \\ 2x + y + 2z = 5 \\ x - y - 2z = 1 \end{cases}$

8. $\begin{cases} 3x + y + 2z = 6 \\ x - 2y + 3z = 9 \\ x + y + z = 2 \end{cases}$

9. $\begin{cases} x + y + z + 2w = 6 \\ x + y + 3w = 7 \\ x - z + w = 2 \\ x + 2y + 2z = 5 \end{cases}$

10. $\begin{cases} x + 2y + z + w = 1 \\ x + 2w = 3 \\ x + 3y + z = 0 \\ y + z + 2w = 1 \end{cases}$

CHAPTER SUMMARY

- A matrix is a rectangular array of numbers called *entries* organized into rows and columns. Represent matrices with bold face capital letters or by using a typical element and bracket notation: $\mathbf{A}_{m\times n} = [a_{ij}]$

- If $\mathbf{B}$ is a matrix, the transpose of $\mathbf{B}$ represented by $\mathbf{B}^T$ is formed by interchanging the rows and columns of $\mathbf{B}$.

- $\mathbf{A}_{m\times n} + \mathbf{B}_{m\times n}$ is an $m \times n$ matrix formed by adding entries in corresponding positions.

  $c\mathbf{A}_{m\times n}$ is an $m \times n$ matrix formed by multiplying each entry of matrix $\mathbf{A}$ by the scalar c.

  $\mathbf{A}_{m\times n} - \mathbf{B}_{m\times n} = \mathbf{A}_{m\times n} + (-\mathbf{B}_{m\times n})$

  $\mathbf{A}_{n\times m}\mathbf{B}_{m\times r} = \mathbf{C}_{n\times r}$, where the element in the ij position of $\mathbf{C}$ is the dot product of row i of $\mathbf{A}$ and column j of $\mathbf{B}$.

  $\mathbf{I}_{n\times n} = [t_{ij}]$, where $t_{ij} = 1$ when $i = j$ (main diagonal) and $t_{ij} = 0$ otherwise. $\mathbf{I}$ is the identity matrix.

  If $\mathbf{A}$ is a square matrix and there exists a matrix $\mathbf{A}^{-1}$ such that $\mathbf{A}^{-1}\mathbf{A} = \mathbf{I}$, then $\mathbf{A}^{-1}$ is the multiplicative inverse of $\mathbf{A}$.

- A system of linear equations can be represented by $\mathbf{AX} = \mathbf{C}$, where $\mathbf{A}$ is the matrix of coefficients of the variables, $\mathbf{X}$ is the matrix of variables and $\mathbf{C}$ is the matrix of constants.

- Elementary row operations are the matrix equivalent of the operations used in solving a system of linear equations. These operations are
  E_1: Interchange any two rows of the matrix ($R_i \leftrightarrow R_j$).
  E_2: Multiply any row of the matrix by a nonzero scalar ($aR_1 \rightarrow R_i, a \neq 0$).
  E_3: Replace any row of the matrix with itself plus a scalar multiple of another row ($R_i + aR_j \rightarrow R_i$).

- The Gauss-Jordan elimination method uses elementary row operations to convert a matrix to row echelon form. The row echelon form is useful for the solution of linear systems.

- Suppose that $\mathbf{A}_{n\times n}$ can be reduced to the identity $\mathbf{I}_{n\times n}$ by a sequence of elementary row operations E. Then the same sequence of elementary row operations applied to $\mathbf{I}_{n\times n}$ will produce the multiplicative inverse of $\mathbf{A}$. That is,

  $\mathbf{I}^E = \mathbf{A}^{-1}$,   where $\mathbf{A}^{-1}\mathbf{A} = \mathbf{I}$

- If $\mathbf{A} = \begin{bmatrix} a & b \\ c & d \end{bmatrix}$, then $\det(\mathbf{A}) = \begin{vmatrix} a & b \\ c & d \end{vmatrix} = ad - bc$. If $\mathbf{A}$ is a square matrix of larger dimension than 1×1, then $C(\mathbf{A})$ is the cofactor matrix of $\mathbf{A}$. Each component of $C(\mathbf{A})$ is $c_{ij} = (-1)^{i+j} \det(\mathbf{A}^C)$, where $\mathbf{A}^C$ is the matrix formed by deleting the ith row and jth column of $\mathbf{A}$. As a result each element of the cofactor matrix is the minor of the corresponding component in $\mathbf{A}$. Moreover, $\det(\mathbf{A})$ is the dot product of any row of $\mathbf{A}$ with the corresponding row of $C(\mathbf{A})$. Finally, if $\det(\mathbf{A}) \neq 0$,

  $$\mathbf{A}^{-1} = \frac{1}{\det(\mathbf{A})} C(\mathbf{A})^T$$

- If $\det(\mathbf{A}) \neq 0$, and $\mathbf{AX} = \mathbf{C}$, then $x_i = \det(\mathbf{A}^{C_i})/\det(\mathbf{A})$, where x_i is the ith variable in matrix $\mathbf{X}$ and $\mathbf{A}^{C_i}$ is obtained from $\mathbf{A}$ by replacing the ith column of $\mathbf{A}$ with $\mathbf{C}$ (Cramer's rule).

KEY WORDS AND CONCEPTS

Additive inverse
Additive identity
Augmented matrix
Cofactor
Cofactor matrix
Conformable matrices
Cramer's rule
Determinant
Dimension of a matrix
Element

Elementary Row Operations
Equal matrices
Gauss-Jordan elimination method
Identity by row echelon reduction
Inverse matrix
Main diagonal of a matrix
Matrix
Matrix addition
Matrix equation
Matrix identity for multiplication

Matrix multiplication
Minors
Multiplicative identity
Order of a matrix
Row echelon form
Scalar multiplication
Square matrix
Subtraction of matrices
Transpose of a matrix
Zero matrix

11 REVIEW EXERCISES

SECTION 11-1

For $\mathbf{A} = \begin{bmatrix} 1 & 2 & 3 & 4 \\ 5 & 9 & 8 & 7 \\ 6 & 5 & 3 & -2 \end{bmatrix}$, $\mathbf{B} = \begin{bmatrix} 2 & 5 & 1 & 3 \\ 4 & 0 & 5 & 6 \\ -1 & 7 & 8 & 9 \\ 0 & 2 & 5 & 1 \end{bmatrix}$ identify each of the following.

1. $a_{2,3}$

2. $a_{3,2}$

3. column 3

4. row 2

5. $\mathbf{A}^T$

6. $(\mathbf{A}^T)^T$

7. $b_{3,2}$

8. column 3

9. row 1

10. $\mathbf{B}^T$

11. the main diagonal of $\mathbf{B}$

12. the main diagonal of $\mathbf{B}^T$

For each of the following matrices $\mathbf{A}$, determine $\mathbf{A}^T$.

13. $\begin{bmatrix} 2 & 1 \\ 0 & 3 \end{bmatrix}$

14. $\begin{bmatrix} -5 & 1 \\ 8 & 2 \end{bmatrix}$

15. $\begin{bmatrix} 4 & 3 & 8 \\ -1 & 5 & 2 \end{bmatrix}$

16. $\begin{bmatrix} 6 & 2 & 9 \\ -2 & 8 & 2 \\ 7 & 5 & 3 \end{bmatrix}$

17. Form a 3×3 matrix in which $a_{ij} = i \cdot j$.

18. Form a 3×3 matrix in which $a_{ij} = 2i - j$.

19. Suppose that a loan is to be repaid at 10 percent simple interest. Form a 3×5 matrix in which a_{ij} is the amount of interest charged on j dollars after i years.

20. Form a 5×4 matrix in which a_{ij} is the price per item of i items bought for a total price of j dollars.

21. If $\begin{bmatrix} 3x & 5 & t-1 \\ -3 & z & 8 \end{bmatrix} = \begin{bmatrix} x-4 & q & 6 \\ -3 & y+2 & y \end{bmatrix}$, determine values for x and y.

22. If $7_{6,2}$ is an element of matrix $\mathbf{A}$, what element of $\mathbf{A}^T$ will be a seven?

23. Suppose that $10_{5,3}$ is an element of square matrix $\mathbf{A}$. If $\mathbf{A}$ is symmetric, what other location also has a 10 as an element?

24. Suppose $\mathbf{A}$ is a 5×3 matrix defined as follows
$\begin{cases} a_{ij} = 0 & \text{if } i \neq j \\ a_{ij} = 1 & \text{if } i = j \end{cases}$, write $\mathbf{A}$ in standard matrix format (bracket notation).

25. Suppose $\mathbf{B}$ is a 2×3 matrix defined as follows
$\begin{cases} a_{ij} = 0 & \text{if } i \neq j \\ a_{ij} = 1 & \text{if } i = j \end{cases}$, write $\mathbf{B}$ in standard matrix format (bracket notation).

26. Suppose that $\mathbf{A} = \begin{bmatrix} 5 & 2 \\ 9 & 4 \end{bmatrix}$ and $\mathbf{B} = \begin{bmatrix} 2 & 1 \\ 5 & -1 \end{bmatrix}$. Now suppose that $\mathbf{C}$ is defined as follows: $[c_{ij}] = [a_{ij} - b_{ij}]$. Write $\mathbf{C}$ in standard format.

27. Suppose that $\mathbf{A} = \begin{bmatrix} 1 & 2 \\ 3 & 4 \end{bmatrix}$. Suppose that $\mathbf{C}$ is defined as follows: $[c_{ij}] = [7a_{ij}]$. Write $\mathbf{C}$ in standard format.

28. Suppose $\mathbf{A}$ is a 2×3 matrix defined as follows: $[a_{ij}] = [i - j]$. Write $\mathbf{A}$ in standard matrix format.

SECTION 11-2

Perform the indicated operations.

29. $\begin{bmatrix} 2 & 5 \\ 3 & 4 \end{bmatrix} + \begin{bmatrix} 3 & -1 \\ 7 & 2 \end{bmatrix}$ **30.** $\begin{bmatrix} 4 & 3 \\ -1 & 2 \end{bmatrix} + \begin{bmatrix} 4 & 3 \\ -1 & 2 \end{bmatrix}$

31. $5\begin{bmatrix} 1 & 4 \\ 1 & 3 \end{bmatrix}$ **32.** $3\begin{bmatrix} 4 & 3 \\ -1 & 2 \end{bmatrix}$

33. $\begin{bmatrix} 2 & 3 & 5 \\ -1 & 0 & 2 \end{bmatrix} + \begin{bmatrix} 2 & 3 \\ 5 & -1 \\ 0 & 2 \end{bmatrix}^T$

34. $\begin{bmatrix} 1 & 0 & 4 \\ -3 & 5 & 5 \end{bmatrix}^T + \begin{bmatrix} 1 & 0 \\ 4 & -3 \\ 5 & 5 \end{bmatrix}$

35. $\left(5\begin{bmatrix} 1 & 1 & 0 \\ 0 & 1 & 0 \\ -1 & 1 & 1 \end{bmatrix} - 2\begin{bmatrix} 1 & 0 & 0 \\ 0 & 1 & 0 \\ 0 & 0 & 1 \end{bmatrix} \right)^T$

36. $\begin{bmatrix} 1 & 0 & -1 \\ 2 & 1 & 1 \\ 1 & 1 & 0 \end{bmatrix} + 2\begin{bmatrix} 1 & 1 & 0 \\ 1 & 0 & 1 \\ 3 & -1 & 1 \end{bmatrix}^T$

SECTION 11-3

Determine the indicated products.

37. $[2 \ -1 \ 3]\begin{bmatrix} 5 \\ 7 \\ -1 \end{bmatrix}$ **38.** $\begin{bmatrix} 2 \\ 0 \\ 3 \end{bmatrix}[3 \ 5 \ -2]$

39. $\begin{bmatrix} 2 & 3 & 0 \\ -1 & 2 & 1 \\ 1 & 5 & 1 \end{bmatrix}\begin{bmatrix} 4 & 3 & 5 \\ 2 & 3 & -1 \end{bmatrix}$ **40.** $\begin{bmatrix} 4 & -3 \\ 3 & 7 \end{bmatrix}\begin{bmatrix} 0 & 2 \\ -3 & 4 \end{bmatrix}$

41. $\begin{bmatrix} 2 & 8 \\ 5 & -1 \end{bmatrix}\begin{bmatrix} 4 & 3 \\ -2 & 2 \end{bmatrix}$ **42.** $\begin{bmatrix} 1 & 3 & 1 \\ 2 & 1 & -1 \end{bmatrix}\begin{bmatrix} 6 & 0 & 2 \\ -3 & 0 & 2 \\ 5 & 0 & 1 \end{bmatrix}$

Verify each of the following for 2×2 matrices.

43. If $AB = I$ and $CA = I$, then $B = C$.

44. $(AB)^T = B^TA^T$

SECTION 11-4

Solve the following systems by rewriting them in augmented matrix form and then reducing to row echelon form:

45. $\begin{cases} 5x - 2y = 17 \\ 3x + y = 8 \end{cases}$ **46.** $\begin{cases} 3x + 2y = 0 \\ 4x - 3y = 17 \end{cases}$

47. $\begin{cases} x + y + z = 2 \\ x - y + z = 6 \\ 2x + y - z = 3 \end{cases}$ **48.** $\begin{cases} x - y + z = 2 \\ x + y + z = 4 \\ 2x + y - z = 13 \end{cases}$

49. $\begin{cases} w - x + y + z = 3 \\ w + x - y + z = 1 \\ w + x + y - z = 5 \\ w + x + y + z = 3 \end{cases}$

50. $\begin{cases} x + y - z = 2 \\ x + y + z = 3 \end{cases}$

51. $\begin{cases} x + y - z = 4 \\ x + z + w = 3 \\ y + z + w = 0 \end{cases}$ **52.** $\begin{cases} x + y + z = 5 \\ x + y - z = 4 \\ 3x + 3y + z = 4 \end{cases}$

Solve the following homogeneous systems.

53. $\begin{cases} 5x - 3y = 0 \\ 3x + 2y = 0 \end{cases}$ **54.** $\begin{cases} 2x - 3y = 0 \\ -4x + 6y = 0 \end{cases}$

55. $\begin{cases} x + y + z = 0 \\ x - y + z = 0 \\ -x + y + 2z = 0 \end{cases}$ **56.** $\begin{cases} 2x + y - z = 0 \\ x + y + z = 0 \\ 3x + y - 3z = 0 \end{cases}$

Use elementary row operations on each of the following matrices of the form $[A \mid I]$ to reduce these matrices to the modified row echelon form $[I \mid I^E]$. Verify that $I^E A = I$ by direct multiplication.

57. $\left[\begin{array}{cc|cc} 5 & -2 & 1 & 0 \\ 3 & 1 & 0 & 1 \end{array} \right]$

58. $\left[\begin{array}{cc|cc} 4 & 3 & 1 & 0 \\ 3 & -2 & 0 & 1 \end{array} \right]$

59. $\left[\begin{array}{ccc|ccc} 1 & -1 & 2 & 1 & 0 & 0 \\ 1 & 0 & 1 & 0 & 1 & 0 \\ 2 & 1 & -1 & 0 & 0 & 1 \end{array} \right]$

60. $\left[\begin{array}{cccc|cccc} 1 & 1 & -1 & 1 & 1 & 0 & 0 & 0 \\ 1 & -1 & 1 & 1 & 0 & 1 & 0 & 0 \\ 1 & 1 & 1 & -1 & 0 & 0 & 1 & 0 \\ 0 & 0 & 0 & 1 & 0 & 0 & 0 & 1 \end{array} \right]$

SECTION 11-5

Calculate the determinant and inverse (if it exists) for each of the following matrices.

61. $\begin{bmatrix} 5 & -2 \\ 2 & 3 \end{bmatrix}$ **62.** $\begin{bmatrix} 2 & 3 \\ 4 & -1 \end{bmatrix}$

63. $\begin{bmatrix} 1 & 0 & 1 \\ 1 & -1 & 0 \\ 2 & 2 & -1 \end{bmatrix}$ **64.** $\begin{bmatrix} 1 & 0 & 1 & 0 \\ 1 & 2 & 0 & 2 \\ 1 & 3 & 1 & 0 \\ 0 & 1 & 1 & 2 \end{bmatrix}$

Solve the following systems. (Hint: compare these problems
to the matrices in problems 61–64)

65. $\begin{cases} 5x - 2y = 10 \\ 2x + 3y = 3 \end{cases}$

66. $\begin{cases} 2x + 3y = 0 \\ 4x - y = 3 \end{cases}$

67. $\begin{cases} x + \quad z = 4 \\ x - y \quad = 3 \\ 2x + 2y - z = 3 \end{cases}$

68. $\begin{cases} x + \quad z \quad = 1 \\ x + 2y + \quad 2w = 3 \\ x + 3y + z \quad = 0 \\ z + 2w = 1 \end{cases}$

SIDELIGHT   More on Rotations

Note that the determinant of $\mathbf{R} = \begin{bmatrix} \sin\theta\ \cos\theta \\ -\cos\theta\ \sin\theta \end{bmatrix}$ is given by

$$\det(\mathbf{R}) = \sin^2\theta + \cos^2\theta$$
$$= 1$$

We began this chapter by indicating that transformations inspired Arthur Cayley to invent matrix notation. We close this chapter by indicating how matrices can compactly represent one of our previous transformations.

Think of $\mathbf{R}$ as a transformation. Suppose that (x, y) is a point in the plane. Represent this point as a column vector in matrix form:

$$\mathbf{X} = \begin{bmatrix} x \\ y \end{bmatrix}$$

Multiplying $\mathbf{R}$ times $\mathbf{X}$ transforms $\mathbf{X}$ to new coordinates. Label the resulting transformation of $\mathbf{X}$ as $\mathbf{X}'$, where $\mathbf{X}' = \begin{bmatrix} x' \\ y' \end{bmatrix}$. Now $\mathbf{X}' = \mathbf{RX}$.

$$\mathbf{X}' = \begin{bmatrix} \sin\theta\ \cos\theta \\ -\cos\theta\ \sin\theta \end{bmatrix}\begin{bmatrix} x \\ y \end{bmatrix}$$
$$= \begin{bmatrix} x\sin\theta + y\cos\theta \\ -x\cos\theta + y\sin\theta \end{bmatrix}$$
$$\begin{bmatrix} x' \\ y' \end{bmatrix} = \begin{bmatrix} x\sin\theta + y\cos\theta \\ -x\cos\theta + y\sin\theta \end{bmatrix}$$

As a result, $x' = x\sin\theta + y\cos\theta$ and $y' = y\sin\theta - x\cos\theta$. These are rotation transformations from Chapter 9. Matrices provide a convenient way to represent transformations including rotations.

CHAPTER TEST

1. If $\begin{bmatrix} 2 & x+y & 1 \\ -3 & 3 & z \end{bmatrix} = \begin{bmatrix} w\text{-}3 & 5 & 1 \\ -3 & x-y & 3 \end{bmatrix}$, determine values for x and y.

2. Suppose A is a 3×3 matrix defined as follows: $[a_{ij}] = [ij - 1]$. Write A in standard matrix format.

Perform the indicated matrix arithmetic.

3. $5\begin{bmatrix} 1 & -1 \\ 0 & 1 \end{bmatrix} - \dfrac{1}{3}\begin{bmatrix} 3 & 6 \\ 9 & -3 \end{bmatrix}$

4. $\begin{bmatrix} 1 & 3 & 2 \\ -1 & 4 & 5 \end{bmatrix}^T + \begin{bmatrix} 0 & 2 \\ 5 & 3 \\ 1 & -1 \end{bmatrix}$

5. $\begin{bmatrix} 2 & -1 \\ 3 & 2 \end{bmatrix}\begin{bmatrix} 3 & 4 \\ -1 & 3 \end{bmatrix}$

6. $[2 \ -1 \ 3]\begin{bmatrix} 1 \\ -2 \\ 1 \end{bmatrix}$

7. Reduce the following to row echelon form:
$$\begin{bmatrix} 4 & -2 & | & 1 & 0 \\ -2 & 1 & | & 0 & 1 \end{bmatrix}$$

8. Write the following in matrix form and solve.
$$\begin{cases} 4x - 2y = 9 \\ -2x + y = -1 \end{cases}.$$

9. (a) If $A = \begin{bmatrix} 1 & 2 & 1 \\ 1 & -1 & 3 \\ 2 & 1 & 4 \end{bmatrix}$, determine det(A).

   (b) What is the inverse of A?

10. Solve $\begin{vmatrix} 5-x & 1 \\ 0 & 4+x \end{vmatrix} = 0$ for x.

Sequences, Series and Induction

1   Sequences, Series and Recursion

Great fleas have little fleas
upon their backs to bite'em,
And little fleas have lesser fleas,
and so on ad infinitum.
And the great fleas themselves, in turn,
have greater fleas to go on;
While these again have greater still,
and greater still, and so on.
—A. DeMorgan

The common use of the word *sequence* often implies an "ordering" of events. Consider the sequence of steps in assembling a bicycle. Observe the sequence of activities in baking bread. Note the sequence of events that lead to an historic occurrence. As usual we require more precision to our word usage, so that we define *sequence* as follows.

DEFINITION 12-1
Sequence

> A **sequence** is a function whose domain is the set of counting numbers. The range components of a sequence are the terms of the sequence.

To emphasize that the domain of a sequence is the set of counting numbers, sequences have special terminology to distinguish them from other functions. In a sequence the range component paired with the number 1 is the **first term of the sequence**. In a similar manner, the sequence has second, third and fourth terms that are the range components paired with 2, 3 and 4, respectively. If n is a counting number and f is a sequence, then $f(n)$ is the nth term of the sequence. To notationally distinguish sequences from functions with other

Table 12-1
Sequences

Domain component	Function notation	Subscript notation	Term name
1	$f(1)$	f_1	First term
2	$f(2)$	f_2	Second term
3	$f(3)$	f_3	Third term
. . .	. . .	. . .	. . .
n	$f(n)$	f_n	nth term
. . .	. . .		
. . .	. . .		

domains, use the special subscript notation f_n for $f(n)$. Both the use of the subscript notation and the domain variable n remind us that the function f is a sequence (domain counting numbers) rather than our usual real function. To summarize, see Table 12-1. Like other functions, the name of the sequence, b, and the rule of pairing, $b_n = b(n)$ are often used interchangeably.

EXAMPLE 12-1

Illustration 1:

Evaluating Sequences. Indicate the first, fifth, and 10th term of each of the following sequences.

$$c_n = \frac{n}{n + 1}$$

Solution:

$$\text{First term:} \quad c_1 = \frac{1}{1 + 1}$$

$$= \frac{1}{2}$$

$$\text{Fifth term:} \quad c_5 = \frac{5}{5 + 1}$$

$$= \frac{5}{6}$$

$$\text{Tenth term:} \quad c_{10} = \frac{10}{10 + 1}$$

$$= \frac{10}{11}$$

Illustration 2:

The sequence b is a sequence whose first sixteen terms are 10, 9, 8, 7, 6, 5, 4, 3, 2, 1, 0, -1, -2, -3, -4, -5, ...

Solution:

$$b_1 = 10$$
$$b_5 = 6$$
$$b_{10} = 1$$

Illustration 3:

$$a_n = n^2 - 1$$

Solution:

$$a_1 = 0$$
$$a_5 = 5^2 - 1 = 24$$
$$a_{10} = 99$$

Note: Although we can calculate $a_{0.5} = (0.5)^2 - 1 = -0.75$, this is not allowed. Remember, *the domain of a sequence is limited to counting numbers*. The number 0.5 is not in the domain of a sequence.

Illustration 4:

$$b_n = (-1)^{n+1}$$

Solution:

$$b_1 = 1$$
$$b_5 = 1$$
$$b_{10} = -1$$

Note: The first six terms of this sequence are 1, -1, 1, -1, 1, -1, ... Although we were not asked, it should be obvious that $b_{1000} = -1$ and $b_{5031} = 1$.

Illustration 5:

$$a_n = \begin{cases} 3 & \text{if } n = 1 \\ 2a_{n-1} & \text{if } n > 1 \end{cases}$$

Solution 1/Recursive Definitions: A sequence is defined recursively if the rule of pairing evaluates a function term by referring to an earlier term of the sequence. The definition for a_n is recursive. The **recursion** is in the second formula for a_n. In particular, $2a_{n-1}$ makes use of the $n - 1$ term to evaluate the nth term. The $n - 1$ term is the term immediately before the nth term. Therefore to evaluate a particular term of this sequence, we refer to the previous term. To evaluate a_5, we must first know a_4. To evaluate a_4 we must have the term a_3. The process continues to the first term, which is known to be 3.

$$a_1 = 3$$
$$a_2 = 2a_1 \quad = 2(3) \quad = 6$$
$$a_3 = 2a_2 \quad = 2(6) \quad = 12$$
$$a_4 = 2a_3 \quad = 2(12) \quad = 24$$
$$a_5 = 2a_4 \quad = 2(24) \quad = 48$$
$$a_6 = 2a_5 \quad = 2(48) \quad = 96$$
$$a_7 = 2a_6 \quad = 2(96) \quad = 192$$
$$a_8 = 2a_7 \quad = 2(192) = 384$$
$$a_9 = 2a_8 \quad = 2(384) = 768$$
$$a_{10} = 2a_9 \quad = 2(768) = 1536$$
$$. . .$$
$$a_n = 2a_{n-1} = 3(2)^{n-1}$$

Solution 2/Graphing Calculator: Use the [ANS] key of your graphing calculator to generate the terms of the sequence. Enter [3] to initialize the sequence. Then enter [2] [ANS] [ENTER] for the next term of the sequence. Each successive press of [ENTER] displays the next term of the sequence. ▄

The algebra of functions from Chapter 1 extends to sequences to build new sequences from old.

Theorem 12-1
Algebra of Sequences

> Suppose that a_n and b_n represent two sequences, then $a_n + b_n$, $a_n b_n$, a_n/b_n ($b_n \neq 0$), and ca_n (where c is a real number) represents sequences as defined by the algebra of functions.

E X A M P L E   12-2

Combining Sequences. For $a_n = 2n - 1$ and $b_n = n^2$, form the indicated sequence. (Note: the first three terms of a_n are $1, 3, 5, \ldots$. The first three terms of b_n are $1, 4, 9, \ldots$.)

Illustration 1:

Sum of Two Sequences

$$a_n + b_n$$

Solution:

$$a_n + b_n = 2n - 1 + n^2$$

(Note: the first three terms of $a_n + b_n$ are $2, 7, 14, \ldots$.)

Illustration 2:

Product of Two Sequences

$$a_n b_n$$

Solution:

$$a_n b_n = n^2(2n - 1)$$

(Note: the first three terms of $a_n b_n$ are $1, 12, 45, \ldots$.)

Illustration 3:

Quotient of Two Sequences

$$\frac{a_n}{b_n}$$

Solution:

$$\frac{a_n}{b_n} = \frac{2n - 1}{n^2}$$

(Note: the first three terms of a_n/b_n are $1, 3/4, 5/9, \ldots$.)

Illustration 4:

Constant Times a Sequence

$$5a_n$$

Solution:

$$5a_n = 5(2n - 1)$$

(Note: the first three terms of $5a_n$ are 5, 15, 25, ...)   ▬

Suppose we know the first r terms of a sequence but do not have a formula. Producing a rule of pairing for a sequence to include these first r terms is often difficult. Moreover, a sequence to include these terms is not unique. As we shall see, infinitely many different sequences include any given first r terms. As a result, r terms of a sequence cannot uniquely define a sequence.

Theorem 12-2
r Terms Do Not Define a Unique Sequence

> Suppose that r terms of a sequence are given and the sequence a_n includes these r terms, then so does the sequence $c_n = a_n + (n - 1)$ $(n - 2)$ $(n - 3)$ . . . $(n - r)b_n$, where b_n is any sequence.

Theorem 12-2 is partially algorithmic in that it indicates how to construct a new sequence to include the same given terms, once *at least one rule of pairing is known*. For example, if we are given the first three terms of a sequence 2, 4, 6, then $a_n = 2n$ is a sequence that includes these three terms. However, from a_n we can construct many other sequences that begin with the same three terms:

$$c_n = 2n + (n - 1)(n - 2)(n - 3) \qquad \{2, 4, 6, 14, . . .\}$$
$$d_n = 2n + (n - 1)(n - 2)(n - 3)n \qquad \{2, 4, 6, 32, . . .\}$$
$$e_n = 2n + (n - 1)(n - 2)(n - 3)(5n + 7) \qquad \{2, 4, 6, 170, . . .\}$$
$$f_n = 2n + (n - 1)(n - 2)(n - 3)n^2 \qquad \{2, 4, 6, 104, . . .\}$$

We could continue this pattern indefinitely. Verify the listed terms for each of these sequences ($n = 1, 2, 3$ and 4). In each case, the first three terms are the same. Now compare $a_4 = 8$ with each of the fourth terms from c_n, d_n, e_n and f_n. What is the purpose of $(n - 1)(n - 2)(n - 3)$ in these formulas?

Because the domain of sequences must be counting numbers but the terms need not be, we do not define the general composition of two sequences. However, in the case where the range of a sequence contains counting numbers, such compositions may exist by the definition in Chapter 1.

There is at least one more method to form a new sequence from a given sequence. The method is based upon summing the terms of a sequence.

DEFINITION 12-2
Summation Notation

> The sum of the first r terms of a sequence a_n is symbolized by $\sum_{k=1}^{r} a_k$.
> In other words, $\sum_{k=1}^{r} a_k = a_1 + a_2 + a_3 + . . . + a_r$.

In the **summation notation**, the variable k is called the *index* of the summation. The index serves as a counter for the numbers of summed terms. Here $k = 1$ indicates the series begins with the first term a_1.

DEFINITION 12-3
Series

> The sequence of **partial sums** of a sequence a_n is a sequence S given by
>
> $$S_n = \sum_{k=1}^{n} a_k$$
>
> S_n is the nth partial sum. Another name for a sequence of partial sums S is **series**.

Sequences of Partial Sums

EXAMPLE 12-3

Illustration 1:

Suppose that $a_n = 2n$. Determine the first six terms of the series formed from a_n:

$$S_n = \sum_{k=1}^{n} 2k$$

Solution:

$$S_1 = \sum_{k=1}^{1} 2k = 2(1) \qquad\qquad = 2$$

$$S_2 = \sum_{k=1}^{2} 2k = 2(1) + 2(2) \qquad\qquad = 6$$

$$S_3 = \sum_{k=1}^{3} 2k = 2(1) + 2(2) + 2(3) \qquad\qquad = 12$$

$$S_4 = \sum_{k=1}^{4} 2k = 2 + 4 + 6 + 2(4) \qquad\qquad = 20$$

$$S_5 = \sum_{k=1}^{5} 2k = 2 + 4 + 6 + 8 + 2(5) \qquad\quad = 30$$

$$S_6 = \sum_{k=1}^{6} 2k = 2 + 4 + 6 + 8 + 10 + 2(6) = 42$$

Note: $S_6 = S_5 + 2(6)$. In general, $S_{n+1} = S_n + a_{n+1}$, where S_n represents the series formed from a_n. Alternately, $S_n = S_{n-1} + a_n$ where $n > 1$.

Illustration 2:

Determine the first, third, fourth and sixth terms of the series given by

$$S_n = \sum_{k=1}^{n} k^2$$

Solution:

$$S_1 = \sum_{k=1}^{1} k^2 = 1^2 \qquad\qquad = 1$$

$$S_3 = \sum_{k=1}^{3} k^2 = 1^2 + 2^2 + 3^2 \qquad\qquad = 14$$

$$S_4 = \sum_{k=1}^{4} k^2 = 1^2 + 2^2 + 3^2 + 4^2 \qquad\qquad = 30$$

$$S_6 = \sum_{k=1}^{6} k^2 = 1^2 + 2^2 + 3^2 + 4^2 + 5^2 + 6^2 = 91$$

SUMMARY

- A sequence is a function whose domain is the set of counting numbers. The special notation a_n is equivalent to $a(n)$ and represents the range component paired with n. The range component a_n is called the nth term of the sequence. Use n (also i, j and k) rather than x, y or z for the domain component to emphasize that the domain consists of counting numbers rather than real numbers.
- The sum, product or quotient of two sequences as defined by the algebra of functions is a sequence. Also, if a_n is a sequence, then ca_n, where c is a constant, represents the sequence formed by multiplying each term of a_n by c.
- Suppose that a_n represents a sequence, then the sum of the first r terms of the sequence is denoted by

$$S_r = \sum_{k=1}^{r} a_k$$

If a_n represents a sequence then

$$S_n = \sum_{k=1}^{n} a_k$$

represents the sequence of partial sums of a_n.

12-1   EXERCISES

Determine the first five terms for each of the sequences described by the following.

1. $a_n = 2n + 1$

2. $b_n = n^2$

3. $c_n = n^3$

4. $d_n = 3n - 1$

5. $b_n = \dfrac{n + 1}{n + 2}$

6. $a_n = \dfrac{n - 1}{3n}$

7. $d_n = \begin{cases} 5 & \text{if } n = 1 \\ 2d_{n-1} & \text{if } n > 1 \end{cases}$

8. $c_n = \begin{cases} 3 & \text{if } n = 1 \\ -1c_{n-1} & \text{if } n > 1 \end{cases}$

9. $a_n = (-1)^n 2n$

10. $b_n = (-1)^{n+1}(2n - 1)$

11. $c_n = \begin{cases} 1 & \text{if } n = 1 \\ 3 & \text{if } n = 2 \\ c_{n-2} + c_{n-1} & \text{if } n > 2 \end{cases}$

12. $d_n = \begin{cases} 2 & \text{if } n = 1 \\ -1 & \text{if } n = 2 \\ d_{n-2} - d_{n-1} & \text{if } n > 2 \end{cases}$

13. $d_n = 2n$

14. $a_n = 3n$

15. $b_n = 5n - 1$

16. $c_n = 4n + 1$

17. $a_n = \begin{cases} 1 & \text{if } n = 1 \\ na_{n-1} & \text{if } n > 1 \end{cases}$

18. $b_n = \begin{cases} 1 & \text{if } n = 1 \\ n + b_{n-1} & \text{if } n > 1 \end{cases}$

19. $a_n = n!$

20. $b_n = \displaystyle\sum_{k=1}^{n} k$

For each of the following pairs of sequences, a_n and b_n, determine (a) a rule of pairing and the values of the first three terms of $a_n + b_n$ (b) a rule of pairing and the values of the first three terms of $a_n b_n$.

21. $a_n = 2n + 1, b_n = n^2$

22. $a_n = 1 - 2n, b_n = n^2$

23. $a_n = (n + 1)^2, b_n = -1 - 2n$

24. $a_n = (n - 1)^2, b_n = 2n - 1$

25. $a_n = (-1)^n, b_n = 3n$

26. $a_n = (-1)^{n+1}, b_n = 2n$

27. $a_n = 2^n, b_n = \sqrt{n}$

28. $a_n = 3^n, b^n = \dfrac{1}{n}$

29. $a_n = \dfrac{1}{n + 1}, b_n = 3^{n-1}$

30. $a_n = \sqrt[3]{n}, b_n = 3n - 1$

Evaluate each of the following summations.

31. $\displaystyle\sum_{k=1}^{5} k$

32. $\displaystyle\sum_{k=1}^{4} (k + 1)$

33. $\displaystyle\sum_{k=1}^{4} 3k$

34. $\displaystyle\sum_{k=1}^{5} (2k - 1)$

35. $\displaystyle\sum_{k=1}^{3} \dfrac{1}{k}$

36. $\displaystyle\sum_{k=1}^{6} \dfrac{1}{k^2}$

37. $\displaystyle\sum_{k=1}^{6} k^3$

38. $\displaystyle\sum_{k=1}^{3} (k + 1)^2$

39. $\displaystyle\sum_{k=1}^{5} (-1)^k$

40. $\displaystyle\sum_{k=1}^{7} (-1)^{k+1}$

Determine the third and fifth terms for each of the following series.

41. $S_n = \displaystyle\sum_{k=1}^{n} k$

42. $S_n = \displaystyle\sum_{k=1}^{n} (-k)$

43. $S_n = \displaystyle\sum_{k=1}^{n} (2k)$

44. $S_n = \displaystyle\sum_{k=1}^{n} (3k + 1)$

45. $S_n = \displaystyle\sum_{k=1}^{n} (k^3 + 1)$

46. $S_n = \displaystyle\sum_{k=1}^{n} \dfrac{1}{k^2}$

47. $S_n = \displaystyle\sum_{k=1}^{n} \dfrac{1}{k + 1}$

48. $S_n = \displaystyle\sum_{k=1}^{n} (1 + k)^2$

49. $S_n = \displaystyle\sum_{k=1}^{n} (-1)^k(k + 1)$

50. $S_n = \displaystyle\sum_{k=1}^{n} \dfrac{(-1)^{k+1}}{k}$

For each of the following, determine rules of pairing for two sequences with the given first four terms.

51. $2, 4, 6, 8, \ldots$

52. $1, 3, 5, 7, \ldots$

53. $-1, 1, -1, 1, \ldots$

54. $1, -1, 1, -1, \ldots$

55. $1, -4, 9, -16, \ldots$

56. $-1, 8, -27, 64, \ldots$

57. $\dfrac{1}{2}, \dfrac{2}{3}, \dfrac{3}{4}, \dfrac{4}{5}, \ldots$

58. $\dfrac{1}{2}, \dfrac{3}{4}, \dfrac{5}{6}, \dfrac{7}{8}, \ldots$

59. $2, 5, 8, 11, \ldots$

60. $10, 6, 2, -2, \ldots$

Verify each of the following for $n = 3$.

61. $\displaystyle\sum_{k=1}^{n} (ca_k) = c \sum_{k=1}^{n} a_k$

62. $\displaystyle\sum_{k=1}^{n} (a_k + b_k) = \sum_{k=1}^{n} a_k + \sum_{k=1}^{n} b_k$

63. $\displaystyle\sum_{k=1}^{n} (a_k b_k) \neq \sum_{k=1}^{n} a_k \sum_{k=1}^{n} b_k$

64. $\displaystyle\sum_{k=1}^{n} c = nc$

Model the following using summation notation.

65. Ray mixes paint into a 1 gallon bucket from a series of cans of colored paint. The first can contains $\frac{1}{2}$ gallon of paint. The second can contains $\frac{1}{4}$ gallon of paint. The third can contains $\frac{1}{8}$ gallon of paint. Each successive can holds half the contents of the preceeding can. If Ray pours the first five such cans into the bucket, represent the total in the bucket using summation notation. Evaluate this summation. Is the bucket full? If not full, how much room is left in the bucket? Can you predict the number of such cans it will take to overflow the bucket?

66. Kay needs to shim up a $\frac{1}{2}$ ft gap between the floor and a structural timber. She uses precut shims. The first shim is $\frac{1}{3}$ ft. The next shim is $\frac{1}{9}$ ft. The next shim is $\frac{1}{27}$ ft. Each

successive shim is one-third the previous shim. If Kay uses the first four such shims, represent the total gap filled by the shims using summation notation. Evaluate this summation. Is the gap filled? If not filled, how much space is left? Can you predict the number of such shims it will take to fill the space?

67. Suppose the terms of a_n are counting numbers. Describe or define what it would mean for a_n to be an *increasing sequence* of counting numbers.

68. See Problem 67. Suppose a_n is an increasing sequence of counting numbers and b_n is any sequence, then the composition of b with a exists: $b(a_n)$. The resulting sequence is called a *subsequence* of b. Suppose that a_n is the sequence of odd numbers 1, 3, 5, 7, ... List the first six terms for any sequence b, then list the first three terms of the subsequence of b formed by composing b and a. Describe the process of producing the subsequence.

12-1   PROBLEM SET

The note to Example 12-3, Illustration 1 indicated a recursive definition to series: $S_n = S_{n-1} + a_n$, where $S_n = \sum_{k=1}^{n} a_k$. Rewriting the recursive form produces $a_n = S_n - S_{n-1}$. This allows us to determine a formula for a_n, $n > 1$, when given a formula for S_n. For example, $S_n = n^2$ then $S_{n-1} = (n-1)^2$. Thus, $a_n = n^2 - (n-1)^2 = 2n - 1$. Because $a_1 = S_1$, check the validity of the formula for a_1 by substitution: $S_1 = 1^2$ and $a_1 = 2(1) - 1 = 1$.

1. What is the purpose of requiring $n > 1$ in the discussion?

2. In the discussion, if $n = 1$, what does S_{n-1} represent?

For each of the following, $S_n = \sum_{k=1}^{n} a_k$.

3. If $S_n = n(n + 1)$, determine a_n.

4. If $S_n = 2n(n + 1)$, determine a_n.

5. If $S_n = \dfrac{n}{n + 1}$, determine a_n.

6. If $S_n = \dfrac{n(n + 1)(n + 2)}{3}$, determine a_n.

7. If $S_n = \dfrac{n(3n + 1)}{2}$, determine a_n.

8. If $S_n = \dfrac{n(n + 1)(2n + 1)}{6}$, determine a_n.

9. If $S_n = \dfrac{n^2(n + 1)^2}{4}$, determine a_n.

10. If $S_n = 2^{n+1} - 2$, determine a_n.

11. If $S_n = 2^{n+1} - 3$, determine a_n.

12. If $S_n = 2^{n+1} - 4$, determine a_n.

13. If $S_n = 2^{n+1} - 10$, determine a_n.

14. See Problems 10–13. Check a_1 and S_1 in each case. Discuss any discrepancies. Refer back to Problems 1 and 2 in your discussion.

PROBLEMS FOR GRAPHING TECHNOLOGY

Use the [ANS] key of your graphing calculator to generate the first five terms of the following sequences. If the sequence is not given recursively, rewrite it in a recursive form. If a sequence causes special problems, discuss those problems.

15. $a_1 = 5$, $a_n = a_{n-1} + 4$, for $n > 1$.
16. $a_1 = -3$, $a_n = a_{n-1} + 5$, for $n > 1$.
17. $a_1 = 15$, $a_n = a_{n-1} - 5$, for $n > 1$.
18. $a_1 = 10$, $a_n = a_{n-1} - 4$, for $n > 1$.
19. $a_1 = 2$, $a_n = 3a_{n-1}$, for $n > 1$.
20. $a_1 = -3$, $a_n = 2a_{n-1}$, for $n > 1$.
21. $a_1 = 3$, $a_n = -2a_{n-1}$, for $n > 1$.
22. $a_1 = -2$, $a_n = -3a_{n-1}$, for $n > 1$.
23. $a_n = 3 + 5n$.
24. $a_n = 2 - 3n$.

25. $a_n = 3 - 2n$.

26. $a_n = 1 + n/2$.

27. $a_n = 3^n$.

28. $a_n = 2^n$.

29. $a_n = (0.5)^n$.

30. $a_n = (0.1)^n$.

EXPLORATION

The following problems anticipate concepts from the next section. Each of the following problems pairs a sequence a with a similar real function f. Graph f then use the trace feature to determine $f(1), f(2), f(3)$ and $f(4)$. Compare these values with a_1, a_2, a_3 and a_4.

31. (a) $a_n = 2 + 3n, f(x) = 2 + 3x$

(b) $a_n = 3 + 2n, f(x) = 3 + 2x$

(c) $a_n = 1 - 3n, f(x) = 1 - 3x$

(d) What is the shape of graph of f?

(e) In the next section, sequences such as these are called *arithmetic sequences*. Make up a definition for *arithmetic sequences*.

(f) How would the graph of a differ from the graph of f?

32. (a) $a_n = 2(3)^n$, $f(x) = 2(3)^x$

(b) $a_n = 3(2)^n$, $f(x) = 3(2)^x$

*(c) $a_n = (-3)^n$, $f(x) = -(3)^x$

(d) What is the shape of graph of f?

(e) In the next section, sequences such as these are called *geometric sequences*. Make up a definition for *geometic sequences*.

(f) How would the graph of a differ from the graph of f?

(g) The function f in part c does not fit the pattern of the other functions. What would happen if it did? Discuss why.

---

2   # ARITHMETIC AND GEOMETRIC SEQUENCES

Mathematics is a science continually expanding; and its growth, unlike some political and industrial events, is attended by universal acclamation.
—H. S. WHITE

Some sequences hold special interest because of their attractive properties and applications. Among these are the arithmetic and geometric sequences and their associated series. We generate succeeding terms of an arithmetic sequence by adding a constant to the previous term. In geometric sequences we multiply rather than add.

DEFINITION 12-4
Arithmetic Sequence

> An **arithmetic sequence** a is a sequence expressible in the form
>
> $$a_n = a_1 + (n - 1)d \qquad (n > 1)$$
>
> where a_1 represents the first term of the sequence and d is the common difference between successive terms.

The term **common difference** is descriptive. In an arithmetic sequence the difference between any two successive terms of the sequence is a constant, represented in the definition by d. In other words, if $n > 1$, then

$$a_n - a_{n-1} = d$$

As a result

$$a_n = a_{n-1} + d$$

The second equation formulates a recursive definition for an arithmetic sequence. From the recursive definition, except for the first term, each new term of an arithmetic sequence comes from adding d to the previous term. Addition generates the terms of an arithmetic sequence.

EXAMPLE   12-4

Arithmetic Sequences. Determine whether each of the following are successive terms of an arithmetic sequence. If the sequence could be arithmetic, write a formula for the arithmetic sequence including these terms.

Illustration 1:

$$3, 7, 11, 15, 19, 23, \ldots$$

Solution: Let a_n represent the terms of the sequence. The first term is $a_1 = 3$. Calculate common differences among successive terms:

$$a_2 - a_1 = 4; \qquad a_3 - a_2 = 4; \qquad a_4 - a_3 = 4;$$
$$a_5 - a_4 = 4; \qquad a_6 - a_5 = 4$$

In each case the difference is the same. The pattern of the first six terms is arithmetic. Because the common difference is given by $d = 4$, then

$$a_n = a_1 + (n - 1)d$$
$$a_n = 3 + (n - 1)4$$
$$a_n = 4n - 1$$

(Note: $a_{10} = 4(10) - 1 = 39$.)

Illustration 2:

$$8, 6, 4, 2, 0, -2, -4, \ldots$$

Solution:

$$a_1 = 8.$$

The difference between any two successive terms is $d = -2$. The first six terms are arithmetic in pattern.

$$a_n = 8 + (n - 1)(-2)$$
$$a_n = 10 - 2n$$

Illustration 3:

$$3, 6, 10, 14, 17, 21, \ldots$$

Solution:

$$a_1 = 3$$

The difference between successive terms is not constant:

$$a_2 - a_1 = 6 - 3 = 3, \qquad \text{but}$$
$$a_3 - a_2 = 10 - 6 = 4$$

The pattern is not arithmetic. We attempt no rule of pairing. Note: determinants and the method for generating polynomials from Chapter 11 could produce a fifth-degree formula in n through the given six points of this sequence

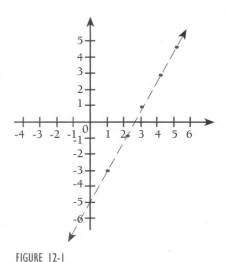

FIGURE 12-1

(use $(1, 3)$, $(2, 6)$, $(3, 10)$, $(4, 14)$, $(5, 17)$ and $(6, 21)$). However, for arithmetic sequences the news is better. The formula for an arithmetic sequence is first degree (in n). In general, first-degree sequence formulas represent arithmetic sequences and vice versa. The common difference is analogous to the slope of a line. The common difference represents the increase or decrease from one term to the next. Plotting the discrete points of $a_n = -5 + 2n$ produces separate points that lie in a straight line. See Figure 12-1. Compare to $y = -5 + 2x$ and the linear nature becomes more obvious. ▬

Now consider the series associated with a given arithmetic sequence. As you would suspect, such a series is an arithmetic series.

DEFINITION 12-5
Arithmetic Series

An **arithmetic series** S in a series expressible as $S_n = \sum_{k=1}^{n} a_k$, where a_k is an arithmetic sequence.

To determine a formula for S_n we expand the summation notation of Definition 12-5 as follows:

(substitute for each a_n)
(reverse the order of the terms)
(add)

$$S_n = a_1 + a_2 + \ldots + a_n$$
$$S_n = a_1 + (a_1 + d) + \ldots + [a_1 + (n - 1)d]$$
$$\underline{S_n = [a_1 + (n - 1)d] + [a_1 + (n - 2)d] + \ldots + a_1}$$
$$2S_n = 2a_1 + (n - 1)d + 2a_1 + (n - 1)d + \ldots + 2a_1 + (n - 1)d$$
$$2S_n = n[2a_1 + (n - 1)d]$$
$$S_n = \frac{2a_1 + (n - 1)d}{2}$$

$$S_n = n\,\frac{a_1 + a_n}{2}$$

These steps outline the proof of the following theorem.

Theorem 12-3
Formula for the *n*th Term of an
Arithmetic Series

If a_n is an arithmetic sequence, then $S_n = \sum_{k=1}^{n} a_n = n\,\dfrac{a_1 + a_n}{2}$

In other words, the sum of the first n terms of an arithmetic sequence is the number of terms times the "average" (arithmetic mean) of the first and nth term.

EXAMPLE 12-5

Illustration 1:

Arithmetic Series. Determine the indicated sum.

Determine the sum of the counting numbers from 1 to 1000.

Alternate statement of problem: if $a_k = 1 + (k - 1)1$, determine S_{1000}.

Alternate statement of problem: determine $\displaystyle\sum_{k=1}^{1000} k$.

Solution:

$$a_1 = 1, \qquad a_{1000} = 1000, \qquad n = 1000$$

$$S_{1000} = 1000 \, \frac{1 + 1000}{2}$$

$$= 500{,}500$$

Illustration 2:

Determine the sum of the first ten terms for the arithmetic sequence

$$a_n = 5 + (n - 1)(\text{-}2)$$

Even without the clue, we recognize this as an arithmetic sequence because the formula is first degree. The number of terms is $n = 10$. First, determine the first and tenth terms:

$$a_1 = 5 + (1 - 1)(\text{-}2) \ = 5$$
$$a_{10} = 5 + (10 - 1)(\text{-}2) = \text{-}13$$
$$\text{Now, } S_{10} = 10 \, \frac{5 + (\text{-}13)}{2} \qquad = \text{-}40$$

Arithmetic sequences generate successive terms using addition of the common difference. From the order of operations, we know that addition has the lowest priority in mixed arithmetic. Let us take the rule of pairing for an arithmetic sequence and jump all operations up one level according to GEMA. Multiplication becomes exponentiation: instead of multiplying d by $n - 1$, use $n - 1$ as an exponent. Also addition becomes multiplication: rather than add the first term a_1 to the remainder of the formula, multiply it. The new rule of pairing is given by

$$a_n = a_1 d^{n-1}$$

In the original formula d was the common difference, because addition of d generated successive terms. In the new formula multiplication generates successive terms. As a check, the difference in successive terms produced d in an arithmetic sequence. Similarly, division checks that successive terms of the new formula come from multiplication. For that reason, replace d with r and call r the common ratio. This leads to the following definition.

DEFINITION 12-6
Geometric Sequence

A **geometric sequence** a is a sequence expressible in the form

$$a_n = a_1 r^{n-1}$$

where $a_1 \neq 0$ is the first term, r ($r \neq 0$, $r \neq 1$) represents the **common ratio**.

SIDEBAR
John Napier's invention of logarithms was in part inspired by the comparison of an arithmetic and geometric sequence. In fact the literal translation (as evolved from the Greek through the Latin) of logarithm is "ratio (logos) number." Napier's goal in inventing logarithms was to make arithmetic easier. Hence, we use logarithms to determine products by doing addition.

Because a_1 and r are constant and n is the variable, the formula for a geometric sequence is exponential in n.

The name *common ratio* is descriptive. In a geometric sequence the ratio of two consecutive terms is the constant r.

$$\frac{a_n}{a_{n-1}} = r \qquad n > 1$$

As a result, the recursive form for the rule of pairing of a geometric sequence is

$$a_n = a_{n-1} r \qquad (\text{or } a_{n+1} = a_n r)$$

Geometric Sequences. Determine which of the following represent the first few terms of geometric sequences. If geometric, write a formula for the sequence.

EXAMPLE 12-6

Illustration 1:

$$2, -6, 18, -54, \ldots$$

Solution:

$$a_1 = 2 \qquad \text{(first term of a sequence)}$$
$$\frac{a_2}{a_1} = -3 \qquad \text{(check for a common ratio)}$$
$$\frac{a_3}{a_2} = -3$$
$$\frac{a_4}{a_3} = -3$$

The terms represent a geometric sequence with common ratio of $r = -3$. The rule of pairing for the sequence is $a_n = 2(-3)^{n-1}$.

Illustration 2:

$$5, \frac{5}{2}, \frac{5}{4}, \frac{5}{8}, \frac{5}{16}, \ldots$$

Solution:

$$a_1 = 5$$
$$\frac{\frac{5}{4}}{\frac{5}{2}} = \frac{1}{2}$$

In fact, the ratio of successive terms is consistently $\frac{1}{2}$. The formula for the corresponding geometric sequence is $a_n = 5(\frac{1}{2})^{n-1}$.

Illustration 3:

$$8, 4, 1, -1, -\frac{1}{4}, \ldots$$

Solution:

$$a_1 = 8$$

There is no common ratio:

$$\frac{a_2}{a_1} = \frac{4}{8} = \frac{1}{2}$$

but $\dfrac{a_3}{a_2} = \dfrac{1}{4} = \dfrac{1}{4}$

Consider the geometric series S associated with a geometric sequence a_n:

$$S_n = \sum_{k=1}^{n} a_n$$

$$= \sum_{k=1}^{n} a_1 r^{k-1}$$

The following theorem gives a formula for the nth term of a geometric series. The Problem Set will coach you through the steps of its derivation.

Theorem 12-4
Partial Sum for a Geometric Series

The sum of the first n terms of a geometric sequence $a_n = a_1 r^{n-1}$ is given by

$$S_n = \sum_{k=1}^{n} a_1 r^{k-1} = a_1 \frac{1 - r^n}{1 - r}$$

Geometric Series. Determine the indicated sum.

EXAMPLE 12-7

Illustration 1:

Determine the sum of the first six terms of $a_n = 3\left(-\frac{1}{2}\right)^{n-1}$.

Alternate statement of problem: evaluate $S_6 = \sum_{k=1}^{6} 3\left(-\frac{1}{2}\right)^{k-1}$.

Solution:

$$a_1 = 3, \qquad r = -\frac{1}{2}, \qquad n = 6 \qquad \text{(by Theorem 12-4)}$$

$$S_6 = 3 \frac{1 - \left(-\frac{1}{2}\right)^6}{1 - \left(-\frac{1}{2}\right)}$$

$$S_6 = \frac{63}{32}$$

Illustration 2:

Gail wants a new car, but it costs $17,367.59. On the first day of February, Gail decides she will save a penny in her coin bank. Gail does not think she can save fast enough, so on February 2, she doubles the previous day's deposit and puts two cents in the bank. The following day, she doubles her deposit and puts in four cents. The next day she deposits eight cents. She decides to continue doubling her deposits through the twenty-eight days of February. How much will be in her coin bank at the end of the month? How close will she be to buying her new car?

Solution: Because Gail doubles her deposit each time, the sequence of deposits is geometric with a first term of 1 (cent to start) and a common ratio of 2 (for doubles). Gail makes deposits for twenty-eight days, so we need the sum of the first twenty-eight terms of the sequence. Apply Theorem 12-4.

$$S_{28} = 1\frac{1 - 2^{28}}{1 - 2}$$

By calculator, $S_{28} = 268435455$ (cents) or $2,684,354.55. The amount is adequate to buy the car.

SUMMARY

- An arithmetic sequence has a rule of pairing expressible in one of the following forms:

$$a_n = a_1 + (n - 1)d$$

$$\text{or } a_n = \begin{cases} a_1 & \text{if } n = 1 \\ a_{n-1} + d & \text{if } n > 1 \end{cases} \quad \text{(recursive formula)}$$

where a_1 is the first term and d is the common difference, $d \neq 0$. The sum of the first n terms of an arithmetic sequence a_n is given by

$$S_n = n\frac{a_1 + a_n}{2}$$

- A geometric sequence has a rule of pairing expressible as one of the following:

$$a_n = a_1 r^{n-1}$$

$$\text{or } a_n = \begin{cases} a_1 & \text{if } n = 1 \\ a_{n-1}r & \text{if } n > 1 \end{cases} \quad \text{(recursive formula)}$$

where $a_1 \neq 0$ is the first $a_1 \neq 0$ and r is the common ratio, $r \neq 0$, $r \neq 1$. The sum of the first n terms of a geometric sequence $a_n = a_1 r^{n-1}$ is given by

$$S_n = a_1\frac{1 - r^n}{1 - r}$$

12-2   EXERCISES

Determine a rule of pairing for the arithmetic sequence determined by each of the following.

1. $a_1 = 2, d = 3$ **2.** $a_1 = 3, d = 2$

3. $a_1 = -1, d = \dfrac{1}{2}$ **4.** $a_1 = \dfrac{1}{2}, d = -1$

5. $a_1 = 0, d = -3$ **6.** $a_2 = 6, d = 4$

7. $a_2 = 9, d = 2$ **8.** $a_2 = 0, d = 3$

9. $3, 6, 9, 12, 15, \ldots$ **10.** $7, 4, 1, -2, \ldots$

Determine a rule of pairing for the geometric sequence determined by each of the following.

11. $a_1 = 2, r = 3$ **12.** $a_1 = 3, r = 2$

13. $a_1 = 5, r = -2$ **14.** $a_1 = 3, r = -5$

15. $a_2 = 1, r = \dfrac{1}{2}$ **16.** $a_1 = -1, r = \dfrac{1}{3}$

17. $4, 2, 1, \dfrac{1}{2}, \ldots$ **18.** $2, -4, 8, 32, \ldots$

19. $-3, 6, -12, 24, \ldots$ **20.** $6, 2, \dfrac{2}{3}, \dfrac{2}{9}, \ldots$

For each of the following sequences a_n, (a) determine whether a_n is arithmetic, geometric or neither; and (b) if a_n is arithmetic or geometric, write a rule of pairing for a_n.

21. $3, 8, 13, 18, \ldots$ **22.** $3, 6, 12, 24, \ldots$

23. $6, 2, 0, -2, -4, \ldots$ **24.** $5, 2, -1, -4, \ldots$

25. $0, 2, 4, 6, 8, \ldots$ **26.** $0, 3, 6, 9, \ldots$

27. $1, -2, 4, -8, \ldots$ **28.** $2, 4, 6, 12, \ldots$

29. $1, \dfrac{1}{10}, \dfrac{1}{100}, \dfrac{1}{1000}, \ldots$

30. $3, \dfrac{3}{10}, \dfrac{3}{100}, \dfrac{3}{1000}, \ldots$

Determine the indicated sums.

31. $\sum_{k=1}^{10} [3 + (k-1)2]$ **32.** $\sum_{k=1}^{20} [2 + (k-1)3]$

33. $\sum_{k=1}^{20} [5 + (k-1)(-1)]$ **34.** $\sum_{k=1}^{10} [3 + (k-1)(-2)]$

35. $\sum_{k=1}^{100} [1 + (k-1)3]$ **36.** $\sum_{k=1}^{100} [0 + (k-1)2]$

37. $\sum_{k=1}^{5} 3\left(\dfrac{1}{2}\right)^{k-1}$ **38.** $\sum_{k=1}^{6} 2\left(\dfrac{1}{3}\right)^{k-1}$

39. $\sum_{k=1}^{7} 2(-2)^{k-1}$ **40.** $\sum_{k=1}^{4} 3(-3)^{k-1}$

41. $\sum_{k=1}^{20} (3k)$ **42.** $\sum_{k=1}^{4} 3^k$

43. $\sum_{k=1}^{5} 2^k$ **44.** $\sum_{k=1}^{30} (2k)$

45. $\sum_{k=1}^{4} (5k - 1)$ **46.** $\sum_{k=1}^{5} 4^{k-1}$

47. $\sum_{k=1}^{4} 5^{k-1}$ **48.** $\sum_{k=1}^{5} (4k - 1)$

49. $\sum_{k=1}^{10} 5(0.1)^k$ **50.** $\sum_{k=1}^{8} 37(0.01)^k$

51. If a retirement fund pays a basic allowance of $6000/year and raises the basic allowance by $500 each year, is this an arithmetic or geometric sequence? What will the yearly allowance be in ten years? What will be the total paid over the ten year period?

52. If $100 is invested the first year and the amount of money invested in each succeeding year is double that of the preceding year, is this an arithmetic or geometric sequence? How much money will be invested in year 5 of the plan? What is the total investment over the five year period?

53. Suppose that $1000 is set aside by a gambler for high risks in one year. In each succeeding year the gambler sets aside $100 less than the previous year. Each year the gambler loses all the high-risk money. Over a six year period, what was the total amount lost?

54. If a monk decides to walk 100 kilometers by walking half the remaining distance each day, how far will he be from his goal after five days?

55. Let P be the principal invested for a period at interest rate r. Then the value of the account at the end of the period is $A = P(1 + r)$. Show that if the principal and interest are left in the account through another period, the value of the account will be given by $A = P(1 + r) + P(1 + r)r = P(1 + r)^2$.

56. See Problem 55. Show that if the accumulated principal and interest are left for another period (the third), the value of the account is given by
$$A = P(1 + r)^2 + P(1 + r)^2 r = P(1 + r)^3$$

57. See Problem 55. Show that the formula for the accumulated value of the account is a geometric sequence with $(1 + r)$ as the common ratio. Write the formula.

58. The formula from Problem 57 gives the value of an account based upon compound interest. The time period for each term of the sequence is the conversion period. If the given interest rate is quoted on an annual basis (APR) and the number of conversion periods during the year is n, then the effective rate for each conversion period is r/n. Let t be the number of years for the account, show that the value of the account is given by

$$A = P\left(1 + \frac{r}{n}\right)^{nt}$$

59. Refer to Problem 58. If Ted invests $1000 at 6 percent APR compounded quarterly (four times per year), what is the value of the account after 1.5 years?

60. Refer to Problem 58. If Julie invests $2000 at 4 percent APR compounded semiannually (two times per year), what is the value of the account after 2 years?

61. Refer to Problem 58. If Lara invests $2500 at 5 percent APR compounded monthly, what is the value of the account after 1 year?

62. Refer to Problem 58. If Layne invests $1000 at 3.65 percent APR compounded daily, what is the value of the account after 1 month (30 days)?

63. Refer to Problem 58. If Scott invests $1000 at 6 percent APR compounded quarterly, use logarithms to estimate how long until the value of account exceeds $1195.61.

64. Refer to Problem 58. If Trey invests $1000 at 5 percent APR compounded monthly, use logarithms to estimate how long until the value of the account exceeds $1195.61?

65. Show that if $a_n = a_1 + (n - 1)d$, then $a_n - a_{n-1} = d$.

66. Show that if $a_n = a_1 r^{n-1}$, $a_1 \neq 0$, $r \neq 0$, then $\dfrac{a_n}{a_{n-1}} = r$.

67. Show that, if a_n and b_n represent arithmetic sequences, then $a_n + b_n$ is an arithmetic sequence (i.e., show there is a common difference).

68. Show that, if a_n and b_n are geometric sequences, then $a_n b_n$ is a geometric sequence (see Problem 67).

69. Discuss why the common ratio of a geometric sequence should not be allowed to be 0 in the definition of geometric sequence.

70. Discuss why the common ratio of a geometric sequence should not be allowed to be 1 in the definition of geometric sequence.

12-2   P R O B L E M   S E T

Consider the series $S_n = \sum\limits_{k=1}^{n} (a_k - a_{k+1})$. Expand the series and combine like terms:

$$\sum_{k=1}^{n} (a_k - a_{k+1}) = (a_1 - a_2) + (a_2 - a_3) + (a_3 - a_4) + \ldots$$
$$+ (a_{n-1} - a_n) + (a_n - a_{n+1})$$
$$= a_1 - a_{n+1}$$

All of the middle terms add to 0. Series of this form are known as *telescoping series*. With telescoping series, we obtain a formula for the sum immediately. For example,

$$\sum_{k=1}^{n} \left(\frac{1}{k} - \frac{1}{k + 1}\right) = 1 - \frac{1}{n + 1} = \frac{n}{n + 1}.$$

Determine a formula for each of the following.

1. $\sum\limits_{k=1}^{n} (3^k - 3^{k+1})$

2. $\sum\limits_{k=1}^{n} (2^k - 2^{k+1})$

3. $\sum\limits_{k=1}^{n} (4^k - 4^{k+1})$

4. $\sum\limits_{k=1}^{n} (5^k - 5^{k+1})$

5. $\sum\limits_{k=1}^{n} (r^k - r^{k+1})$

6. $\sum\limits_{k=1}^{n} (ar^k - ar^{k+1})$

7. $\sum\limits_{k=1}^{n} [k^2 - (k + 1)^2]$

8. $\sum\limits_{k=1}^{n} [k^3 - (k + 1)^3]$

9. $\sum\limits_{k=1}^{n} [k - (k + 1)]$

10. $\sum\limits_{k=1}^{n} [k^4 - (k + 1)^4]$

11. This problem derives the formula for the partial sum of a geometric series. Let $S_n = a + ar + ar^2 + \ldots + ar^{n-1}$, where a is the first term and r ($r \neq 0$, $r \neq 1$) is the common ratio.

  (a) Multiply both sides of $S_n = a + ar + ar^2 + \ldots + ar^{n-1}$ by r. The left side is rS_n. Be sure to simplify the righthand side of the equation.

  (b) Refer to part (a). In the result what is term $n - 1$?

  (c) Subtract the results in part a from the original equation for S_n. Hint: the lefthand side is $S_n - rS_n$. The righthand side should telescope.

  (d) Factor S_n from the two terms on the left. If you divide both sides by $(1 - r)$ you should have the formula.

  (e) Refer to part (d). How do we know we can divide by $(1 - r)$?

12. Suppose that $S_n = a\,\dfrac{1 - r^n}{1 - r}$.

   (a) Determine a formula for S_{n+1}.

   (b) Determine a formula for $S_{n+1} - S_n$.

   (c) See part (b). Simplify the formula for $S_{n+1} - S_n$ by factoring out r^n and reducing to lowest terms.

   (d) See part (b). Let $a_n = S_{n+1} - S_n$. What type of sequence is a_n?

PROBLEMS FOR TECHNOLOGY

Use a graphing calculator to trace the values of each of the following real functions. Approximate $f(1)$, $f(2)$, $f(3)$, $f(4)$ and $f(5)$. Compare these values to the first five terms of the corresponding sequence a_n. Finally graph the sequence.

13. $f(x) = 1 + x$, $a_n = 1 + n$

14. $f(x) = 2x$, $a_n = 2n$

15. $f(x) = 3 + 4x$, $a_n = 3 + 4n$

16. $f(x) = 2x - 1$, $a_n = 2n - 1$

17. $f(x) = 1 - x$, $a_n = 1 - n$

18. $f(x) = 2^x$, $a_n = 2^n$

19. $f(x) = 2\left(\dfrac{1}{2}\right)^{x-1}$, $a_n = 2\left(\dfrac{1}{2}\right)^{n-1}$

20. $f(x) = 3(2)^{x-1}$, $a_n = 3(2)^{n-1}$

*21. $f(x) = (-1)^{x-1}$, $a_n = (-1)^{n-1}$

*22. $f(x) = (-2)^x$, $a_n = (-2)^n$

23. Compare and contrast the real functions to their sequence counterparts in terms of the domain of the functions.

24. See Problems 21 and 22. Compare real functions and their sequence counterparts in terms of the nature of their formula.

3

*INFINITE DECIMALS AND INFINITE BINARY NUMBERS

Leibniz believed he saw the image of creation in his binary arithmetic in which he employed only two characters, unity and zero. Since God may be represented by unity, and nothing by zero, he imagined the Supreme Being might have drawn all things from nothing, just as in the binary arithmetic all numbers are expressed by unity with zero.

—LAPLACE

Chapter 1 indicated that many numbers, for example, $\sqrt{17}$, have infinitely many digits in their decimal representation. In fact, if we include infinitely repeating zeros, every real number has an infinite (digit) decimal representation. Write 2.3 as 2.3000000 . . ., etc., or $\sqrt{3} = 1.7321\ldots$. Every real number falls into one of two categories: those in which the **infinite repeating decimal** representation eventually settles into a repeating block of digits and those in which there is no fixed repeating block of digits. Usually those real numbers that settle into repeating zeros are called **terminating** or finite **decimals** and the trailing zeros are not displayed.

    Consider 0.4747474747474747 . . . . To avoid redundancy, we abbreviate the repeating block by use of a bar: 0.474747474747 . . . $= 0.\overline{47}$ ($0.4\overline{747}$ also represents the decimal, but is not as brief).

    Actually, $0.\overline{47} = \frac{47}{100} + \frac{47}{10000} + \frac{47}{1000000} + \ldots$. Therefore, $0.\overline{47}$ represents a geometric series in which the first term is $\frac{47}{100}$ and the common ratio is $\frac{1}{100}$. Because there are infinitely many terms in this sum, we need a notation to

Some texts are more formal with the topics of Theorem 12-5. These texts first discuss the convergence of geometric series. Once established that the series converge, the formula $a_1/(1-r)$ follows as the limit of the series. Finally, comes the summation notation using the infinity sign to represent the limit of the series. This text bundles them together. The idea of adding all the terms of an infinite series to obtain a sum exceeds the claims of algebra. Indeed, one of the goals of calculus is to allow the human mind to approach the concept of infinity. However, this is not a calculus text, so we offer the notation with the hope that intuition fills the cracks formed by lack of rigor.

represent them. Replace the upper bound on the summation symbol with the infinity symbol, ∞, to indicate this sum. As a result

$$0.\overline{47} = \sum_{k=1}^{\infty} \frac{47}{100}\left(\frac{1}{100}\right)^{k-1}$$

Recall that

$$\sum_{k=1}^{n} a_1 r^{k-1} = a_1 \frac{1-r^n}{1-r}$$

$$= \frac{a_1}{1-r} - \frac{a_1 r^n}{1-r}$$

As n gets larger, (i.e., as n gets close to infinity, ∞) comparison to the exponential function $y = b^x$ indicates two possibilities:

1. If $|r| > 1$, then $|r|^n$ becomes infinitely large (compare to $b > 1$).
2. If $|r| < 1$, then $|r|^n$ approaches 0 (compare to $0 < b < 1$).

Remember, a geometric sequence does not allow $r = 1$ or $r = 0$. These comments outline the proof of the following theorem.

Theorem 12-5

Sum of an Infinite Geometric Series

The **sum of an infinite geometric series** S is given by

$$S_\infty = \sum_{k=1}^{\infty} a_1 r^{k-1} = \frac{a_1}{1-r}, \text{ where } |r| < 1.$$

The series in Theorem 12-5 converges to $a_1/(1-r)$ whenever $-1 < r < 1$: it is a **convergent series**. If $|r| > 1$ or $r = -1$, then the **series** is **divergent**. Convergence of a geometric series is not defined for $r > 1$. For infinite repeating decimals the common ratio is positive and always 0.1 or less. Hence, the series representation of an infinite repeating decimal is convergent. In an infinite repeating decimal, the first term and the common ratio are rational numbers. Because the formula for the series is simple arithmetic, the infinite repeating decimal converges to a rational number. The formula resembles a common fraction, so the infinite repeating decimal is equal to a common fraction.

EXAMPLE 12-8

Illustration 1:

Infinite Repeating Decimals. Rewrite each of the following infinite repeating decimals as a ratio of two integers (a common fraction).

$$0.\overline{47}$$

Solution: As previously noted, $0.\overline{47}$ represents an infinite geometric series in which the first term is $\frac{47}{100}$ and the common ratio is $\frac{1}{100}$. Because the common ratio is between -1 and 1, Theorem 12-5 applies.

$$0.\overline{47} = \frac{\frac{47}{100}}{1 - \frac{1}{100}}$$

$$M = \frac{47}{99}$$

Your calculator or long division suggests that $\frac{47}{99} = 0.\overline{47}$.

Illustration 2: $0.\overline{6}$

Solution:

$$0.\overline{6} = \frac{6}{10} + \frac{6}{100} + \frac{6}{1000} + \frac{6}{10000} + \ldots$$

Note: The common ratio for an infinite repeating decimal is always $(0.1)^d$, where d is the number of digits in the repeating block. Here, $d = 1$, so that the common ratio is $(0.1)^1$. In Illustration 1, $d = 2$ and the common ratio was given by $(0.1)^2 = 0.01$ or $\frac{1}{100}$.

$$a_1 = 0.6, \qquad r = 0.1$$

$$0.\overline{6} = \frac{0.6}{1 - 0.1}$$

$$= \frac{2}{3}$$

Illustration 3: $5.14231231231231231231231231\ldots = 5.14\overline{231}$

Solution:

$$\begin{aligned} 5.14\overline{231} &= 5.14 + 0.00\overline{231} \\ &= 5.14 + \sum_{k=1}^{\infty} 0.00231\,(0.001)^{k-1} \end{aligned}$$

$a_1 = 0.00231$ and $r = 0.001$ (because $d = 3$ (digits))

$$\sum_{k=1}^{\infty} 0.00231\,(0.001)^{k-1} = \frac{0.00231}{1 - 0.001} = \frac{231}{99900}$$

Now,

$$\begin{aligned} 5.14\overline{231} &= 5.14 + \frac{231}{99900} \\ &= \frac{514}{100} + \frac{231}{99900} \\ &= \frac{513717}{99900} \end{aligned}$$

Illustration 4: 0.576

Solution:

$$0.576 = 0.576\overline{0}$$

Although we may carry out the computations with a common ratio of 0.1, the process is not necessary: $0.576 = \frac{576}{1000}$. The repeating zero may be ignored. For this reason many people refer to 0.576 as a *terminating decimal*. The repeating block for a terminating decimal is $\overline{0}$.

The next theorem summarizes the results of the previous example.

Theorem 12-6
Repeating Decimals Represent
Rational Numbers

> An infinite repeating decimal (or terminating decimal) represents a rational number.

The converse of the theorem is also true.

Theorem 12-7
Rational Numbers Produce
Repeating Decimals

> The decimal representation of a rational number is an infinite repeating decimal (or a finite decimal if the repeating digits are $\overline{0}$).

To predict whether a given rational number produces a repeating infinite decimal representation or a finite decimal refer to the following theorem.

Theorem 12-8
Finite Versus Infinite
Repeating Decimals

> Suppose that p/q represents a rational number reduced to lowest terms, $q \neq 0$. If the denominator q contains only powers of 2 or 5 as factors, then the decimal representation of p/q is a finite decimal. If q contains any other prime factors, (such as 3 or 7 or 11) then the decimal representation of p/q is an infinite repeating decimal.

The preceding theorems imply that irrational numbers have infinite decimal representations without a repeating block of digits. The next example compares the decimal representation of rational and irrational numbers.

EXAMPLE 12-9

Identifying Decimals as Rational or Irrational Numbers. Identify each of the following as a rational or irrational real number.

Illustration 1:

$$10.576933333333\ldots$$

Solution:

$$10.5769333333\ldots = 10.5769\overline{3} \qquad \text{(assumed)}$$

Because the decimal eventually settles into a repeating block of one digit the number is rational.

Illustration 2:

$$3.57694218681172934199 3242\ldots$$

Solution: As far as we can see there is no repeating block of digits. The number is apparently irrational.

Illustration 3:

$$5.6$$

Solution:

$$5.6 = 5.6\overline{0}$$

The decimal terminates in zeros. It is a finite decimal and thus represents a rational number.

Illustration 4:

$$2.67896789678967896789\ldots$$

Solution:

$$2.67896789678967896789\ldots = 2.\overline{6789} \qquad \text{(assumed)}$$

The decimal settles into a block of four repeating digits, hence the number is rational.

Illustration 5:

$$5.01001000100001000001\ldots$$

Solution: Even though there is a pattern to the digits (you can probably predict what the next digits will be), the decimal does not settle into a repeating *block* of digits. Remember, the block size must contain a fixed number of digits. Apparently the number represented is irrational. ■

Infinite repeating decimals are not the only example of infinite geometric series.

EXAMPLE 12-10

Infinite Geometric Series. For each of the following geometric series, determine whether the series converges. If the series is convergent, determine the value to which the series converges.

Illustration 1:

$$S_n = \sum_{k=1}^{n} 3\left(-\frac{1}{2}\right)^{k-1}$$

Solution: The first term is 3. The common ratio is $-\frac{1}{2}$. Because the common ratio is between -1 and 1, the series converges by Theorem 12-5. Again, by Theorem 12-5, the series converges to

$$S_\infty = \sum_{k=1}^{\infty} 3\left(-\frac{1}{2}\right)^{k-1} = \frac{3}{1-\left(-\frac{1}{2}\right)} = 2$$

Illustration 2:

Consider the base 2 fraction given by $0.111111111\ldots_2 = 0.\overline{1}_2$

Solution: Expand the representation. $0.1111111\ldots_2 = 1/2 + 1/2^2 + 1/2^3 + \ldots$. The first term is $1/2$ and the common ratio is $1/2$. Because the common ratio is between -1 and 1 the series converges.

$$0.11111\ldots_2 = 0.\overline{1}_2$$

$$= \sum_{k=1}^{\infty} \frac{1}{2}\left(\frac{1}{2}\right)^{k-1} = \frac{\frac{1}{2}}{1-\frac{1}{2}} = 1$$

Note: This result is sometimes confusing. You will try a similar problem in the exercises $(0.\overline{9})$.

Illustration 3:

$$0.001001001001001001001\ldots_2 = 0.\overline{001}_2$$

Solution: The rule for repeating digits is the same as for base 10. The block size is 3, and the base is 2, therefore the common ratio is $1/2^3$. Similarly, the first terms is $1/2^3$.

$$0.\overline{001}_2 = \frac{1}{2^3} + \frac{1}{2^6} + \frac{1}{2^9} + \frac{1}{2^{12}} + \ldots \qquad \text{(base 2 radix)}$$

$$= \sum_{k=1}^{\infty} \frac{1}{8}\left(\frac{1}{8}\right)^{k-1} = \frac{\frac{1}{8}}{1 - \frac{1}{8}}$$

$$= \frac{1}{7} \qquad \text{(base 10 fraction)}$$

Note: It might not surprise you that $1/7$ has an infinite digit representation with base 2 radix. After all it also had an infinite digit representation in base 10. Base 2 radix is common for internal arithmetic in computers. Recall that, for base 10 representations, any fraction reduced to lowest terms will have a finite (terminating) decimal representation if the denominator contains no prime factors other than 2 or 5. A similar condition applies to base 2 fractions reduced to lowest terms. A fraction expressed in base 2 radix will have a finite (terminating) radix notation if the denominator contains no prime factors other than 2. Another popular mode of representing numbers in computers is hexadecimal (base 16 notation). Consider the next illustration.

Illustration 4:

$$0.199999\ldots_{16} = 0.1\overline{9}_{16}$$

Solution:

$$0.199999\ldots_{16} = \frac{1}{16} + \frac{9}{16^2} + \frac{9}{16^3} + \frac{9}{16^4} + \ldots$$

$$= \frac{1}{16} + \sum_{k=1}^{\infty} \frac{9}{16^2}\left(\frac{1}{16}\right)^{k-1}$$

$$= \frac{1}{16} + \frac{\frac{9}{16^2}}{1 - \frac{1}{16}}$$

$$= 0.1 \qquad \text{(base 10)}$$

Note:

$$0.1 = \frac{1}{10}$$

The denominator contains a factor of 5. This means that the hexadecimal representation must be infinite. In a similar manner for binary, $0.1_{10} = 0.1100_2$. Because computers are finite precision machines, such radix representations are rounded. As a result, arithmetic with perfectly legitimate finite decimals suffers rounding errors when converted to binary (or hexadecimal).

Illustration 5:

$$S_n = \sum_{k=1}^{n} \frac{1}{3} \left(\frac{3}{2} \right)^{k-1}$$

Solution: The common ratio of this geometric series is *not* between −1 and 1. The series diverges. There is no value for S_∞.

S U M M A R Y

- For the geometric series, $S_n = \sum_{k=1}^{n} a_1(r)^{k-1}$. The series converges if $-1 < r < 1$, in which case we write $S_\infty = \sum_{k=1}^{\infty} a_1(r)^{k-1} = a_1/(1-r)$. The series diverges if $r > 1$ or $r < -1$.
- An infinite repeating decimal that eventually repeats in a block of d digits is an infinite geometric series with common ratio given by $r = (0.1)^d$. (The same rule applies to radix notation.)
  Nonrepeating infinite decimals represent irrational numbers.
- Gratuitous notes about some peculiarities of the infinite:
  There are as many rational numbers as there are counting numbers (rational numbers are countable).
  There are more irrational numbers than there are counting (or rational) numbers (irrational numbers are not countable).
- More gratuitous notes about why $0.\overline{9} = 1$ and other mysteries:
  Real numbers are dense: between any two real numbers there is another real number.
  There is a rational number arbitrarily close to any irrational number. Thus we may approximate an irrational number with a rational number to any degree of precision.
  If $0.\overline{9} \neq 1$, then there must be a number between $0.\overline{9}$ and 1. Try to find such a number. If there is no number between $0.\overline{9}$ and 1, the two numbers must be equal.

12-3   E X E R C I S E S

Assume that any pattern of repeating digits is displayed and identify each of the following as representing a rational or irrational number.

1. $3.\overline{69}$

2. $2.\overline{74}$

3. $5.6161616161616161\ldots$

4. $3.283283283283283\ldots$

5. $4.63836394351\ldots$

6. $7.911991199119\ldots$

7. $3.211211211211\ldots$

8. $13.58215832697146\ldots$

9. $0.9988227766554477\ldots$

10. $0.554233427742994255426642\ldots$

Convert each of the following rational numbers to infinite repeating decimals. Use long division or your calculator.

11. $\dfrac{1}{3}$

12. $\dfrac{1}{2}$

13. $\dfrac{1}{5}$

14. $\dfrac{1}{9}$

15. $\dfrac{1}{6}$

16. $\dfrac{1}{7}$

17. $\dfrac{3}{11}$

18. $\dfrac{2}{13}$

19. $\dfrac{5}{7}$

20. $\dfrac{5}{6}$

Convert each of the following infinite repeating decimals to the form p/q, where p and q are integers.

21. $0.\overline{3}$

22. $0.\overline{5}$

23. $0.\overline{2}$

24. $0.\overline{1}$

25. 0.7

26. $0.\overline{61}$

27. $0.\overline{27}$

28. 0.5

29. $1.\overline{9}$

30. $2.\overline{9}$

Convert each of the following finite or infinite repeating radix representation to a base 10 common fraction.

31. 0.13

32. 0.2_7

33. $0.\overline{01}_2$

34. $0.\overline{01}_4$

35. $0.\overline{01}_8$

36. $0.\overline{01}_{16}$

37. $0.\overline{3}_4$

38. $0.\overline{7}_8$

39. $0.\overline{F}_{16}$(F represents 15_{10})

40. $0.\overline{B}_{12}$ ($B = 11_{10}$)

Determine whether each of the following infinite geometric series converges or diverges. If the series converges, indicate its sum:

41. $\displaystyle\sum_{k=1}^{\infty} 3\left(\frac{1}{2}\right)^{k-1}$

42. $\displaystyle\sum_{k=1}^{\infty} 2\left(\frac{1}{4}\right)^{k-1}$

43. $\displaystyle\sum_{k=1}^{\infty} \frac{1}{4}\,(2)^{k-1}$

44. $\displaystyle\sum_{k=1}^{\infty} \frac{1}{2}\left(-\frac{1}{3}\right)^{k-1}$

45. $\displaystyle\sum_{k=1}^{\infty} 5\left(\frac{1}{4}\right)^{k-1}$

46. $\displaystyle\sum_{k=1}^{\infty} \frac{1}{2}\,(-3)^{k-1}$

47. $\displaystyle\sum_{k=1}^{\infty} \left[-2\left(\frac{1}{3}\right)^{k-1}\right]$

48. $\displaystyle\sum_{k=1}^{\infty} \left[-4\left(\frac{1}{5}\right)^{k-1}\right]$

49. $\displaystyle\sum_{k=1}^{\infty} (0.2)^{k-1}$

50. $\displaystyle\sum_{k=1}^{\infty} (0.5)^{k}$

51. $\displaystyle\sum_{kf=1}^{\infty} \left(-\frac{1}{6}\right)^{k}$

52. $\displaystyle\sum_{k=1}^{\infty} \left(-\frac{1}{5}\right)^{k-1}$

53. $\displaystyle\sum_{k=1}^{\infty} (-0.1)^{k}$

54. $\displaystyle\sum_{k=1}^{\infty} 3(-0.3)^{k-1}$

 These next problems lead to some insight into the rational number $0.\overline{9}$.

55. Guess the value of $10(0.\overline{3})$.

56. Guess the value of $100(0.\overline{3})$.

57. Refer to Problem 55. Guess the value of $10(0.\overline{3}) - 0.\overline{3}$.

58. Refer to Problem 56. Guess the value of $100(0.\overline{3}) - 0.\overline{3}$.

59.. Refer to Problems 55 and 57. Let $x = 0.\overline{3}$. Determine the value of $10x - x$. Solve this equation for x. Express your answer as a common fraction.

60. Refer to Problems 56 and 58. Let $x = 0.\overline{3}$. Determine the value of $100x - x$. Solve this equation for x. Express your answer to a common fraction.

61. Guess the value of $3(0.\overline{3})$. Convert your answer to a common fraction.

62. Guess the value of $2(0.\overline{3})$. Convert this answer to a common fraction.

63. See Problems 55–62. Convert $0.\overline{9}$ to a fraction. Does the answer surprise you? Explain.

64. See Problems 55–62. Guess the sum of the $0.\overline{3}$ from Problem 60 and $2(0.\overline{3})$ from Problem 62. Now add the common fraction form of these two numbers. Does the answer surprise you? Explain.

65. See the note on the denseness of real numbers in the chapter summary. The usual way to determine a rational number between two rational numbers is to add them up and divide by 2. Which of 1 and $0.\overline{9}$ would you guess is smaller. Try to add them up and divide by 2 to determine a number *between* the two. Does the answer surprise you? Explain.

66. See Problems 55–65. Argue that 1 and $0.\overline{9}$ *must* be two different ways to represent the same number.

12-3   P R O B L E M   S E T

Because summation notation abbreviates a sum, the properties of addition apply to summation notation. As a result we have the following:

$$\sum_{k=1}^{n} (a_k + b_k) = \sum_{k=1}^{n} a_k + \sum_{k=1}^{n} b_k \qquad [1]$$

$$\sum_{k=1}^{n} ca_k = c \sum_{k=1}^{n} a_k \qquad \text{(distributive property) } [2]$$

$$\sum_{k=1}^{n} (a_k - b_k) = \sum_{k=1}^{n} a_k - \sum_{k=1}^{n} b_k \qquad [3]$$

$$\sum_{k=1}^{n} c = cn \qquad [4]$$

1. Verify property 1 for $n = 5$.

2. Verify property 2 for $n = 5$.

3. Verify property 3 for $n = 5$.

4. Verify property 4 for $n = 5$.

5. Verify that $\sum_{k=1}^{n+1} a_k = a_{n+1} + \sum_{k=1}^{n} a_k$ for $n = 5$.

6. Suppose that $S_n = \sum_{k=1}^{n} a_1 r^{k-1}$, show that $rS_n = \sum_{k=1}^{n} a_1 r^k$.

7. See Problem 6. Show that

$$S_n - rS_n = \sum_{k=1}^{n} a_1 r^{k-1} - \sum_{k=1}^{n} ar^k.$$

8. See Problem 7. Show that $S_n(1 - r) = a_1 \sum_{k=1}^{n} (r^{k-1} - r^k)$.

9. See Problem 8. Refer to Problem Set 12-2 on telescoping series. Show that $S_n(1 - r) = a_1(1 - r^n)$.

10. See Problem 9. If $r \neq 1$, show that $S_n = a_1 \dfrac{1 - r^n}{1 - r}$.

---

4   *MATHEMATICAL INDUCTION (FORMAL RECURSION)

The great science (mathematics) occupies itself at least just as much with the power of imagination as with the power of logical conclusion.
—F. J. HERBERT

Suppose someone claims a statement is true. How can you test that claim? The method of testing may vary from one type of statement to the next. But if we limit ourselves to mathematical statements about counting numbers, then one method for verifying that the statement is true uses the principle of finite mathematical induction. For example, consider the following claim about the sum of the terms of an arithmetic sequence:

$$St_n: \quad \sum_{k=1}^{n} a_k = n \, \frac{a_1 + a_n}{2}$$

For easy reference, name the statement St_n, where n represents a counting number. St_1 represents the statement with n replaced by 1. Then St_1 is

$$\text{St}_1: \quad \sum_{k=1}^{1} a_k = 1 \frac{a_1 + a_1}{2}$$

$$= \frac{2a_1}{2}$$

$$= a_1$$

Because the sum of the first one term is a_1, then St_1 is true. Try St_2.

$$\text{St}_2: \quad \sum_{k=1}^{2} a_k = 2 \frac{a_1 + a_2}{2}$$

$$= a_1 + a_2$$

Because the sum of the first two terms is $a_1 + a_2$, then St_2 is true. It is not so obvious that St_k is true for any counting number k. In any case, we do not want to attempt the verification of the statement for values of k larger than 2. Instead we rely on the following theorem for **finite mathematical induction.**

Theorem 12-9
Finite Mathematical Induction

> Suppose that a statement St_n, where n represents a counting number, is given. If (1) the statement is true for $n = 1$ (St_1 is true), and (2) whenever the statement is assumed true for $n = k$, then the statement is true for $n = k + 1$ ($\text{St}_k \Rightarrow St_{k+1}$), then the statement is true for every counting number n.

Intuitively, the theorem is true, operating like a recursive algorithm. Suppose we verify a given statement for $n = 1$ so that St_1 is known to be true. We will have initialized our algorithm. Now suppose that we establish that St_{k+1} follows from St_k being true. Because St_1 is true from part 1, then St_2 must be true because of part 2. Once we know that St_2 is true, then part 2 establishes that St_3 must also be true. Similarly, with St_3 noted as true, part 2 indicates that St_4 is also true. Continue the process until you establish the statement true for any given counting number n.

In many respects, the principle of finite mathematical induction resembles the looping structures in computer programs. The resemblance to recursion structures is even stronger. Part 1 of the theorem establishes an initial or starting value for n. Although, in general, n need not be 1, we begin with 1 to establish a statement for all counting numbers. Suppose we were particular and wanted to know whether the statement is true for $n = 1567$. Direct verification may be too time consuming to be feasible. Instead we formulate a repetitive process that can work its way to 1567 from 1: the process is to link St_k *to* St_{k+1}. This is often done by showing that ordinary algebraic manipulation can convert St_k into St_{k+1}. Implementing the process to convert St_k into St_{k+1} on a computer would involve the following steps:

Begin with $k = 1$

Convert St_1 into St_2

Convert St_2 into St_3

Convert St_3 into St_4

. . .

Convert St_{1566} into St_{1567}

Because we have a mechanical process (a transformation, usually algebraic) to convert St_k into St_{k+1}, we abbreviate the steps indicated as follows:

Begin with $k = 1$

Repeat

    Convert St_k into St_{k+1} then increase k by 1

until $k = 1566$

We can express this to the computer even more generally,

Get value of n from user   (establish the criteria)

Begin with $k = 1$

Repeat

    Convert St_k into St_{k+1} and increase k by 1

until $k = n$ (stop when the criteria is met)

The key elements are in place. Beginning with 1 initializes the algorithm. If we do not have the initial value St_1, we have no hope of producing St_2 or working our way to the desired value of n. Of equal importance is to establish a process that can convert St_k into St_{k+1}. After establishing the two parts of the process, the computer can mechanically work its way to the given value of n. Because n may be any counting number, it follows that the statement is true for all counting numbers. This is precisely the **principle of finite mathematical induction**. If you wonder about the *finite* portion, remember that the statement is established for any counting number n. These counting numbers may be large, but they are all finite.

EXAMPLE 12-11

Illustration 1:

Finite Mathematical Induction. Prove each of the following statements is true for any counting number n.

$$St_n: \quad 1 + 2 + 3 + \ldots + n = \frac{n(n + 1)}{2}$$

Proof: Use finite mathematical induction:

1. Verify the statement for $n = 1$:

$$St_1: \quad 1 = \frac{1(1 + 1)}{2} \qquad \text{(true)}$$

2. Assume that the statement is true for some number k and show that the statement for $k + 1$ follows ($\text{St}_k \Rightarrow St_{k+1}$).

Note: We need to obtain St_{k+1} from St_k. It will be helpful to know the appearance of St_{k+1}. In fact, comparing the needed St_{k+1} to St_k can provide guidance in the required manipulation for the process. To determine the appearance of St_{k+1} substitute $k + 1$ for n in the original statement. We must produce

$$\text{St}_{k+1}: \quad 1 + 2 + 3 + \ldots + k + \boxed{(k + 1)} = \frac{(k + 1)(k + 1 + 1)}{2}$$

Now assume that St_k is true:

$$\text{St}_k: \quad 1 + 2 + 3 + \ldots + k = \frac{k(k + 1)}{2} \qquad \text{(add } (k + 1) \text{ to both sides)}$$

$$1 + 2 + 3 + \ldots + k + (k + 1) = (k + 1) + \frac{k(k + 1)}{2} \qquad \text{(simplify the right side)}$$

$$1 + 2 + 3 + \ldots + k + (k + 1) = \frac{2(k + 1)}{2} + \frac{k(k + 1)}{2}$$

$$= \frac{2(k + 1) + k(k + 1)}{2} \qquad \text{(common factor)}$$

$$= \frac{(k + 1)(2 + k)}{2}$$

$$\text{St}_{k+1}: \quad 1 + 2 + 3 + \ldots + k + (k + 1) = \frac{(k + 1)(k + 1 + 1)}{2}$$

Therefore, St_{k+1} follows from St_k by adding $(k + 1)$ to both sides St_k. These steps satisfy Theorem 12-9. We have proved the statement true for all counting numbers n by using finite mathematical induction.

Illustration 2:

$$\text{St}_n: \quad 1^2 + 2^2 + 3^2 + \ldots + n^2 = \frac{n(n + 1)(2n + 1)}{6}$$

Proof:

1. $\text{St}_1:$ $\quad 1^2 = \frac{1(1 + 1)[2(1) + 1]}{6}$ $\qquad$ (true)

Note: Observe the statement for $n = k + 1$ to establish our goal:

$$\text{St}_{k+1}: \quad 1^2 + 2^2 + 3^2 + \ldots k^2 + \boxed{(k + 1)^2} = \frac{(k + 1)(k + 1 + 1)[2(k + 1) + 1]}{6}$$

$$= \frac{(k + 1)(k + 2)(2k + 3)}{6}$$

2. $St_k \Rightarrow St_{k+1}$.   Assume St_k is true.

$$St_k: \quad 1^2 + 2^2 + 3^2 + \ldots + k^2 = \frac{k(k+1)(2k+1)}{6} \quad \text{(add } (k+1)^2 \text{ to both sides)}$$

$$1^2 + 2^2 + 3^2 + \ldots + k^2 + (k+1)^2 = (k+1)^2 + \frac{k(k+1)(2k+1)}{6}$$

$$= \frac{6(k+1)^2}{6} + \frac{k(k+1)(2k+1)}{6}$$

$$= \frac{6(k+1)^2 + k(k+1)(2k+1)}{6}$$

$$= \frac{(k+1)[6(k+1) + k(2k+1)]}{6}$$

$$= \frac{(k+1)[6k+6+2k^2+k]}{6}$$

$$= \frac{(k+1)[2k^2+7k+6]}{6}$$

$$St_{k+1}: \quad 1^2 + 2^2 + 3^2 + \ldots + k^2 + (k+1)^2 = \frac{(k+1)(k+2)(2k+3)}{6}$$

By the principle of finite mathematical induction, the statement is true for any counting number n.

Note: We may evaluate the series $\sum_{k=1}^{5} k^2$ by using this formula:

$$1^2 + 2^2 + 3^2 + 4^2 + 5^2 = \frac{5(5+1)[2(5)+1]}{6}$$

$$= 55$$

SUMMARY

To prove a statement about counting numbers true for all counting numbers:

1. Show the statement to be true for $n = 1$ (St_1 is true).
2. (a) Note what the statement looks like for $n = k+1$ (St_{k+1}). This establishes a road map for future manipulations.
   (b) Assume that the statement is true for $n = k$ (St_k). Use algebraic manipulation to obtain St_{k+1} from St_k.

These two steps satisfy Theorem 12-9 and establish the statement to be true for any counting number by the principle of finite mathematical induction.

12-4   E X E R C I S E S

Prove each of the following statements true for all counting numbers.

1. $1 + 3 + 5 + \ldots + (2n - 1) = n^2$

2. $2 + 4 + 6 + \ldots + (2n) = n(n + 1)$

3. $\dfrac{1}{2} + \dfrac{1}{4} + \dfrac{1}{8} + \ldots + \left(\dfrac{1}{2}\right)^n = 1 - \left(\dfrac{1}{2}\right)^n$

4. $\dfrac{1}{3} + \dfrac{1}{9} + \dfrac{1}{27} + \ldots + \left(\dfrac{1}{3}\right)^n = \dfrac{1 - \left(\frac{1}{3}\right)^n}{2}$

5. If $1 < x$, then $x^{n-1} < x^n$.

6. If $0 < x < 1$, then $x^{n-1} > x^n$.

7. $\dfrac{1}{1(2)} + \dfrac{1}{2(3)} + \dfrac{1}{3(4)} + \ldots + \dfrac{1}{n(n + 1)} = \dfrac{n}{n + 1}$

8. $1(2) + 2(3) + 3(4) + \ldots + n(n + 1) =$

$\dfrac{n(n + 1)(n + 2)}{3}$

9. $1^3 + 2^3 + 3^3 + \ldots + n^3 = \left[\dfrac{n(n + 1)}{2}\right]^2$

10. $2 + 5 + 8 + \ldots + (3n - 1) = \dfrac{n(3n + 1)}{2}$

11. Suppose n is a counting number and r cis θ represents a complex number. Prove $(r \text{ cis } \theta)^n = r^n \text{ cis}(n\theta)$ (DeMoivre's theorem).

12. Suppose n is a counting number and r cis θ represents a complex number. Prove $\sqrt[n]{r \text{ cis } \theta} = \sqrt[n]{r} \text{ cis}(\theta/n)$.

Refer to the results of Problems 1–10 to evaluate each of the following series.

13. $\displaystyle\sum_{k=1}^{10} (2k - 1)$

14. $\displaystyle\sum_{k=1}^{10} (2k)$

15. $\displaystyle\sum_{k=1}^{5} \left(\dfrac{1}{2}\right)^k$

16. $\displaystyle\sum_{k=1}^{4} \left(\dfrac{1}{3}\right)^k$

17. $\displaystyle\sum_{k=1}^{8} (2k)$

18. $\displaystyle\sum_{k=1}^{8} (2k - 1)$

19. $\displaystyle\sum_{k=1}^{10} \dfrac{1}{k(k + 1)}$

20. $\displaystyle\sum_{k=1}^{5} [k(k + 1)]$

21. $\displaystyle\sum_{k=1}^{5} k^3$

22. $\displaystyle\sum_{k=1}^{6} (3k - 1)$

23. Discuss why a statement must be verified for $n = 1$ (St_1 is true) before mathematical induction can be valid.

24. Discuss how St_k implies St_{k+1} extends the validity of a statement from $n = 1$ to all counting numbers in mathematical induction.

25. Todd is great at Super StarShip Invaders. Eric is not as good. The enemy force has infinitely many ships each numbered with a counting number. Todd takes his turn first. He hits the first invader, then the second and third. He feels sorry for Eric and gives him the first ship. Then he hits the fourth, fifth, sixth and seventh ship. So he gives Eric the second and third. Todd blasts the eighth, ninth, tenth, eleventh, etc. through the fifteenth. He gives Eric ships 4 through 7. Todd continues with his amazing skill, but he continues the same pattern of donation to Eric but always keeping one extra for himself. All at once the players notice that the infinitely many ships have been cleared from the sky.

Who has the most ships? Who has ship 1? Who has ship 2,000,001? Will finite mathematical induction help you with this problem? Why?

26. The Secret Police believes it has a foolproof way to store secret documents. It plans to build infinitely many keyed boxes and number them with the counting numbers. Then it will place a secret document in box 1, lock it and lock the key for box 1 in box 2. Then it will place the key for box 2 in box 3, and lock box 3. The key for box 3 will go into box 4 which is locked in turn. The key for box 4 will go in box 5, etc. This will continue until all the boxes are locked.

How will the Secret Police unlock the boxes to retrieve the secret? Are there any boxes not locked? Are there any keys not in a locked box? Does finite mathematical induction play any role in this scheme?

27. As a sequence can be defined recursively, so can other concepts. Recall the definition of polynomial function from Chapter 4. Then consider the following recursive definition of a polynomial function:

A polynomial function is either

1. The identity function $I(x) = x$ or a constant function $k(x) = c$ or

2. $f + g$ or fg where f and g are polynomial functions.

Use this recursive definition to justify each of the following claims:

(a) $f(x) = 3x$ is a polynomial function.

(b) $f(x) = x^2$ is a polynomial function.

(c) $f(x) = x^2 + 3x$ is a polynomial function.

(d) $f(x) = 5x^2 - 3x + 2$ is a polynomial function.

(e) Refer to the non-recursive definition in Chapter 4. Discuss its connection to the recursive definition.

5 BINOMIAL THEOREM

There is no science which teaches the harmonies of nature more clearly than mathematics, . . .
—PAUL CARUS

Chapter 0 indicated that $(a + b)^n$ had no obvious simplification shortcut. Exponents do not distribute across addition; the levels of operations are not adjacent in the order of operations (GEMA). However, there is a pattern for the simplification of $(a + b)^n$. That pattern is identified in the binomial theorem. First, consider the tedious direct computation method for expanding a binomial to a power:

$$(a + b)^0 = 1$$
$$(a + b)^1 = a + b$$
$$(a + b)^2 = a^2 + 2ab + b^2$$
$$(a + b)^3 = a^3 + 3a^2b + 3ab^2 + b^3$$
$$(a + b)^4 = a^4 + 4a^3b + 6a^2b^2 + 4ab^3 + b^4$$
$$(a + b)^5 = a^5 + 5a^4b + 10a^3b^2 + 10a^2b^3 + 5ab^4 + b^5$$

and so on

Note that the number of terms on the righthand side is always one more than the exponent on the binomial. Also note that the sum of the exponents of each term is the same as the exponent on the binomial. The number of terms and pattern for the exponents is easy. The coefficient of each term is not so obvious. To fully develop the pattern for the expansion of a binomial requires additional terminology. First we repeat the definition of factorial notation.

DEFINITION 12-7
Factorial Notation

Suppose that n is a counting number, then $n! = 1(2)(3) \ldots (n)$. Furthermore, $0! = 1$ (for notational convenience).

Factorial notation provides one method for describing the coefficients in the binomial expansion. For example, consider the a^3b^2 term in the expansion of $(a + b)^5$. The coefficient of the term is 10. Evaluate

$$\frac{5!}{3!2!} = \frac{(1)(2)(3)(4)(5)}{(1)(2)(3)(1)(2)} = 10$$

This indicates that the coefficient of the $a^r b^{n-r}$ term in the expansion of $(a + b)^n$ is given by

$$\frac{n!}{r!(n - r)!}$$

In each term the sum of the exponents is n. Once the exponent of a is identified as r, the exponent for b must be $n - r$. In fact, given any two of the exponents from $(a + b)^n$ and $a^r b^{n-r}$, the third exponent follows. Based on this dependency, the next definition abbreviates the coefficient of a term in the binomial expression; that is, the **binomial coefficient**.

DEFINITION 12-8
Binomial Coefficient

If n and r are nonnegative integers with $n \geq r$, then

$$\binom{n}{r} = \frac{n!}{(n - r)!\, r!}$$

Evaluating Factorials and Binomial Coefficients. Evaluate the following.

EXAMPLE 12-12

Illustration 1:

$$5!$$

Solution:

$$5! = 1(2)(3)(4)(5)$$
$$5! = 120$$

Note: Most calculators, graphing calculators and *GraphWindows* can evaluate factorial notation.

Illustration 2:

$$\binom{5}{3}$$

Solution:

$$\binom{5}{3} = \frac{5!}{(5 - 3)!\, 3!}$$

$$= \frac{5!}{2!\, 3!}$$

$$= \frac{1(2)(3)(4)(5)}{1(2)\, 1(2)(3)}$$

$$= 10$$

Illustration 3:

$$\binom{150}{3}$$

Solution:

$$\binom{150}{3} = \frac{150!}{147!\,3!}$$

Most calculators cannot handle a number as large as 150!. Fortunately, humans have more insight than calculators. We divide out the factors common to the numerator and denominator as the factorials are expanded.

$$\binom{150}{3} = \frac{1(2)(3)(4)\ldots(146)(147)(148)(149)(150)}{1(2)(3)(4)\ldots(147)\,1(2)(3)}$$

$$= \frac{(148)(149)(150)}{2(3)} \qquad \text{(now a calculator can help)}$$

$$= 551{,}300$$

Note: In general, we may divide out common factors to reduce $\binom{n}{r}$.

$$\binom{n}{r} = \frac{n(n-1)(n-2)\ldots(n-r+1)(n-r)!}{r!\,(n-r)!}$$

$$= \frac{n(n-1)(n-2)\ldots(n-r+1)}{r!}$$

$$\binom{n}{n-r} = \frac{n(n-1)(n-2)\ldots(r+1)\,r!}{(n-r)!\,r!} = \frac{n(n-1)(n-2)\ldots(r+1)}{(n-r)!}$$

Finally, $\binom{n}{r} = \binom{n}{n-r}$

Summarizing the previous discussion and definitions results in the **binomial theorem.**

Theorem 12-10
Binomial Theorem

$$(a+b)^n = \binom{n}{0}a^nb^0 + \binom{n-1}{1}a^{n-1}b^1 + \ldots + \binom{n}{r}a^{n-r}b^r + \ldots + \binom{n}{n}a^0b^n$$

⇑ first term   ⇑ second term   ⇑ $r+1$ term   ⇑ $n+1$ term

where n is a counting number.

For any counting number $r+1$, $0 < r+1 \leq n+1$, the $r+1$ term in the expansion is given by $\binom{n}{r}a^{n-r}b^r$.

....

EXAMPLE 12-13

Binomial Theorem. Determine the following.

Illustration 1:

Determine the eighth term a^3b^7 in the expansion of $(a + b)^{10}$.

Solution: Here, $n = 10$, $r + 1 = 8$, so that $r = 7$. Therefore, the eighth term is

$$\binom{10}{7} a^{10-7}b^7 = \frac{10!}{3!\,7!}\, a^3b^7$$

$$= 120a^3b^7$$

Illustration 2:

Expand (determine all the terms of) $(a + b)^5$.

Solution: There will be six terms.

First term: $r = 0.$ $\dfrac{5!}{5!\,0!}\, a^5b^0 = 1a^5$

Second term: $r = 1.$ $\dfrac{5!}{4!\,1!}\, a^4b^1 = 5a^4b$

Third term: $r = 2.$ $\dfrac{5!}{3!\,2!}\, a^3b^2 = 10a^3b^2$

Fourth term: $r = 3.$ $\dfrac{5!}{2!\,3!}\, a^2b^3 = 10a^2b^3$

Fifth term: $r = 4.$ $\dfrac{5!}{1!\,4!}\, a^1b^4 = 5a^1b^4$

Sixth term: $r = 5.$ $\dfrac{5!}{0!\,5!}\, a^0b^5 = 1b^5$

$$(a + b)^5 = a^5 + 5a^4b + 10a^3b^2 + 10a^2b^3 + 5ab^4 + b^5$$

Note: The coefficients are symmetrical about the "center" of the expansion. Also notice the relation of the factorials in the denominator of the coefficient of each term to the exponents on the variables of that term. In particular, *if the term is a^mb^n* **then the coefficient** is $(m + n)!/(m!\,n!)$.

Illustration 3:

Expand $(2x - 3y^2)^4$.

Solution:

$$(2x - 3y^2)^4 = [(2x) + (-3y^2)]^4$$

Let $a = 2x$, $b = -3y^2$ and $n = 4$. There will be five terms.

Term 1: $r = 0.$ $\dfrac{4!}{4!\,0!}\, (2x)^4(-3y^2)^0 = 16x^4$

Term 2: $r = 1.$ $\dfrac{4!}{3!\,1!}\, (2x)^3(-3y^2)^1 = -96x^3y^2$

Term 3: $r = 2.$ $\dfrac{4!}{2!2!}\ (2x)^2(-3y^2)^2 = 216x^2y^4$

Term 4: $r = 3.$ $\dfrac{4!}{1!\ 3!}\ (2x)^1(-3y^2)^3 = -216xy^6$

Term 5: $r = 4.$ $\dfrac{4!}{0!\ 4!}\ (2x)^0(-3y^2)^4 = 81y^8$

$$(2x - 3y^2)^4 = 16x^4 - 96x^3y^2 + 216x^2y^4 - 216xy^6 + 81y^8$$

Note: The coefficients $\begin{pmatrix}4\\0\end{pmatrix}, \begin{pmatrix}4\\1\end{pmatrix}, \begin{pmatrix}4\\2\end{pmatrix}, \begin{pmatrix}4\\3\end{pmatrix}, \begin{pmatrix}4\\4\end{pmatrix}$ were symmetrical about the center of the expression until the arithmetic was done for $(2x)^{n-r}$ and $(-3y^2)^r$. Note how the signs of the terms alternated as a result of the $(-3y^2)$. ▬

SUMMARY

- If n is a counting number then $n! = 1(2)(3) \ldots (n)$. $0! = 1$. If n and r are counting numbers with $r \leqslant n$, then

$$\begin{pmatrix} n \\ r \end{pmatrix} = \frac{n!}{(n - r)!\ r!}.$$

- $(a + b)^n = \dfrac{n!}{n!0!}\ a^n b^0 + \ldots + \dfrac{n!}{(n - r)!r!}\ a^{n-r}b^r + \ldots + \dfrac{n!}{0!n!}\ a^0 b^n$

$$\text{Term } r + 1 \Uparrow$$

where the expansion has $n + 1$ terms.

12-5   EXERCISES

Determine the indicated term in the expansion of each of the following.

1. $(a + b)^7$, fifth term $(r = 4)$
2. $(a + b)^6$, fifth term $(r = 4)$
3. $(a + b)^7$, fourth term $(r = 3)$
4. $(a + b)^6$, third term $(r = 2)$
5. $(3x^2 + y)^6$, third term
6. $(x^2 + 3y)^7$, fifth term
7. $(x - 2y^3)^5$, fourth term
8. $(y - 2x^3)^8$, sixth term

9. $\left(\dfrac{1}{3} + \dfrac{2}{3}\right)^4$, third term

10. $\left(\dfrac{1}{4} + \dfrac{3}{4}\right)^5$, third term

Expand the following powers of binomials.

11. $(a + b)^6$
12. $(a + b)^7$
13. $(x - y)^7$
14. $(x - y)^6$
15. $(x + 2)^5$
16. $(x + 3)^4$
17. $(x - 2y)^4$
18. $(y - 2x)^5$
19. $(2x^3 + 3y^2)^5$
20. $(3x^2 + 2y^3)^4$

21. $(\sqrt{x} + 3)^4$

22. $(\sqrt[3]{x} + 2)^3$

23. $(x^{-1} + y)^3$

24. $(3x + y^{-2})^4$

25. Prove $\binom{n}{r} = \binom{n}{n-r}$.

26. Prove $\binom{a+b}{a} = \binom{a+b}{b}$

27. Prove $\dfrac{(n-1)!}{(k-1)!\,(n-k)!} = \dfrac{k\,(n-1)!}{k!\,(n-k)!}$

28. Prove $\dfrac{(n-1)!}{k!\,(n-1-k)!} = \dfrac{(n-k)(n-1)!}{k!\,(n-k)!}$

29. Prove $k(n-1)! + (n-k)(n-1)! = n!$

30. Prove $\dfrac{(n-1)!}{(k-1)!\,(n-k)!} + \dfrac{(n-1)!}{k!\,(n-1-k)!} =$

$\dfrac{n!}{k!\,(n-k)!}$

31. How many terms does $(x + y + z)^2$ have?

32. See Problem 31. How many terms does $(x + y + z)^3$ have?

33. See Problem 31. What is the degree of each term?

34. See Problem 32. What is the degree of each term?

35. Suppose that $(x + y + z)^n$ is expanded and that the coefficient of the term of the form $x^r y^s z^t$ (where r, s and t are nonnegative integers and $r + s + t = n$) is $n!/(r!\ s!\ t!)$ Determine the coefficient of $x^2 y$ in the expansion of $(x + y + z)^3$.

36. See Problem 35. Determine the coefficient of xyz in the expansion of $(x + y + z)^3$.

37. Determine the coefficient of $x^2 y z^2$ in the expansion of $(x + y + z)^5$.

38. Determine the coefficient of xyz^3 in the expansion of $(x + y + z)^5$.

*39. Prove that the $r + 1$ term of $(a - b)^n$ is

$$\frac{(-1)^r\, n!}{(n-r)!\ r!}\, a^{n-r} b^r.$$

*40. Use mathematical induction to prove the binomial theorem.

---

6   PERMUTATIONS AND COMBINATIONS

Hamlet: "And therefore as a stranger give it welcome.
    There are more things in heaven and earth, Horatio,
    Than are dreamt of in your philosophy."
—WILLIAM SHAKESPEARE

Chapter 0 discussed addition as quick counting and multiplication as quick addition. Exponents abbreviated multiplication. To close Chapter 12, we formalize this interpretation of arithmetic operations. This section examines counting principles, permutations and combinations. These discrete mathematics topics are particularly useful in probability and statistics, where counting numerous possibilities and choices are useful skills.

Theorem 12-11
The Addition Counting Principle

> If A represents an event with n outcomes and B represents an event with m outcomes different than A, the number of outcomes from A or B is given by $m + n$.

Note the key word *or* in Theorem 12-11. Suppose Marge is to choose a car from among twenty five automobiles at Larry's cars or thirty automobiles at Tom's cars, then Marge has a choice among $25 + 30$ or fifty five cars. It is im-

Giovanni's Pizza offers a pizza with your choice of any two of seven different toppings. As part of his advertising campaign Giovanni wonders how many different kinds of two-topping pizzas he serves. His nephew Guido suggests that they have a big party, make every possible kind of pizza and count them as they are eaten. Giovanni indicates he needs a faster, and cheaper, way to count.

portant that none of the choices overlap. Suppose Larry and Tom share adjacent lots. At the boundary of their lots they have 10 cars on consignment that either Tom or Larry can sell. Larry's twenty five cars include these ten. Tom's thirty cars also include these ten. These ten cars are common choices of the two lots. The **addition counting principle** counts the overlapping cars twice, indicating more choices than Marge actually has. We reduce the count by ten to compensate for an overlap of ten common choices.

The importance of nonoverlapping events in Theorem 12-11 generates special terminology to describe such choices. **Mutually exclusive events** are events that have no common outcomes. Here *mutually exclusive* means that a selection from *A* precludes a choice *B* and vice versa.

Consider the extension of counting principles to the **multiplication counting principle**.

Theorem 12-12
The Multiplication Counting Principle

> Suppose *A* represents an event with *n* outcomes and *B* represents an event with *m* outcomes. Then the number of outcomes in the event *A* **and** *B* is *nm*.

In set terminology, two sets of outcomes are mutually exclusive if the intersection of the two sets is empty. In Theorem 12-11, *A* and *B* represent sets of outcomes.

The key word in Theorem 12-12 is *and*. Susan has been on a limited budget but has been able to accumulate a wardrobe with variety by purchasing items that matched what she already owned. Early one morning, the electricity goes out as Susan is about to select a skirt and blouse. She has ten blouses and six skirts. But she is not worried about the dark. She knows all her blouses match all her skirts. From how many different outfits can Susan choose?

Susan is choosing a skirt *and* blouse, so Theorem 12-12 applies. Susan has 10(6) or sixty different outfits from which to choose.

Suppose that Susan's skirts did not match all blouses and Susan found a flashlight. Then Susan's selection of a particular blouse narrows her search to a matching skirt. We must then adjust the count of available skirts to reflect the influence of the blouse on Susan's choice.

Because it is important that one event does not affect another event in the multiplication counting principle, we have a special name to describe such events: they are **independent events**. When a sequence of events is not independent, we adjust the count on subsequent events to reflect the influence of previous events.

Counting Principles. Determine the number of outcomes in the following.

E X A M P L E   12 - 14

Illustration 1:

Suppose the phone company limits the first digit of a home phone number to digits other than 0, 1, 8 or 9. How many first digits may form a home telephone number?

Solution: The choices are limited to 2, 3, 4, 5, 6 or 7. Note the keyword *or*. Add (or in this case, counting is feasible) the number of separate choices and obtain 6.

Illustration 2:   See Illustration 1. How many different home phone numbers will be available for a given area code?

Solution: After the first digit, assume that any digit from 0 to 9 is available. Therefore the number of outcomes available for each digit is as follows:

Phone digit	1	2	3	-	4	5	6	7
# choices	6	10	10		10	10	10	10

Because forming a phone numbers requires a selection of digit 1 *and* a selection of digit 2 *and* a selection of digit 3 *and*, etc., through digit 7 and these events are independent, apply Theorem 12-12. Multiply the number of selections. The number of different home phone numbers is 6 million.

Illustration 3:   Karen has not been as thoughtful as Susan in her clothing selections. Karen has fifteen blouses: seven are blue, five are green, three are brown. Karen has ten skirts: four are blue, three are green and three are brown. Only the same color blouse and skirt go together. From how many different outfits can Karen choose?

Solution: Karen can choose to dress in blue *or* in green *or* in brown. Count the blue outfits, green outfits and brown outfits, then add these together by Theorem 12-11. For a blue outfit, Karen must choose a blue skirt *and* a blue blouse. Therefore the number of blue outfits follows from Theorem 12-12: 7(4) = 28. Similarly, the number of green outfits is given by 5(3) = 15. The number of brown outfits is 3(3) or 9. The total number of outfits is 28 + 15 + 9 or 52.

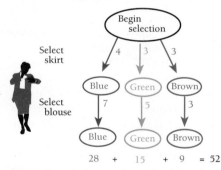

Select skirt

Select blouse

DISPLAY 12-1

Tree Diagram: One way to organize complicated counting problems with several layers of *and* and *or* is a **tree diagram**. Imagine the *and* events as pathways along which several choices must be made. The *or* choices are branches into two or more pathways. See Display 12-1. To determine the final number of choices, multiply down each *and* path, then total the products across the *or* branches.

Illustration 4:   Jeremy plays a game that requires him to scramble the letters in the word *games*, and choose three letters while blindfolded. Each successive letter is drawn without replacing the previous letter. Jeremy wins if he draws more vowels than consonants. How many different ways are there to draw three letters from the five? Of these draws, how many will produce a win for Jeremy?

Solution: Jeremy's first draw is from five letters. Because the first letter is not replaced, the second draw is from four letters. The final draw must be from among three letters. As a result the number of different ways to draw three letters from five (without replacement) is 5(4)(3) or 60.

S I D E B A R

Guido tells his uncle that the number of choices for a two-topping pizza is 7^2 or 49, since there are seven choices for the first topping and seven choices for the second topping. After muttering to himself for several minutes, Giovanni tells him that putting the same topping twice does not qualify as a two-topping pizza. Once you make a choice from the first seven choices, there remain only six choices if the pizza is a two-topping pizza. Because the selections are not independent, Guido must adjust his counting for the second choice.

Guido shouts, "Yes, Yes! Then the number of two-topping combination pizzas is 7(6) or 42."

Guido's uncle smiles and says, "What would you like for lunch, a sausage and onion pizza or an onion and sausage pizza?" Then he wanders off muttering to himself.

To organize the problem, construct a tree diagram. See Display 12-2. The choices in each case indicate the choice of a vowel or consonant. The "weights" along the choice paths indicate how many ways that choice can be made. Each weight is assigned by keeping track of how many vowels and consonants have been drawn along the path to that point.

Games (3 consonants, 2 vowels)

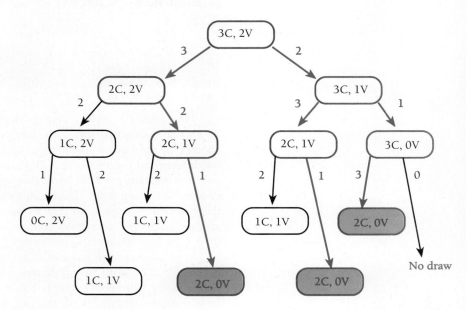

$$3(2)(1) + 3(2)(2) + 3(2)(2) + 3(2)(1) + 2(3)(2) + 2(3)(1) + 2(1)(3) + 0 = 60$$

$$3(2)(1) + 2(3)(1) + 2(1)(3) = 18$$

DISPLAY 12-2

You should expect that making three successive choices with two outcomes at each choice would cause the path to branch into eight distinct paths. However, notice that the count on the last branch of the path on the far right is 0: once you have drawn two vowels, it is impossible to draw another. At the bottom of each path is the count of the number of ways to follow that path. Note that the total of these paths is 60 as predicted. The three shaded boxes terminate the paths that produce a win for Jeremy. The Product of the weights on each of these paths is six. As a result, Jeremy can win in eighteen ways out of the possible sixty outcomes. The game does not seem to be fair for Jeremy.

In Illustration 4 of the previous example, Jeremy was fortunate not to have to spell a word with the letters he drew. This would require that the letters be placed into a particular order. Placing the same letters in a different order would not be considered the same word. Many situations require that selections be made and that the order of selections be part of the count. Consider drawing names out of a hat to win first second and third prize in a contest. Having your name drawn among the three does not guarantee you the best prize. In this situation, the order of choice makes a difference. Take the same three winners, draw them in a different order and you have a different prize distribution. In how many different ways could we distribute prizes among the winning three by reordering the draw of their names? In mathematics we refer to such reordering of choices as a **permutation. In counting permutations** we are counting the number of ways to rearrange outcomes.

Theorem 12-13
Counting Permutations

> The number of different ways to arrange n outcomes is $n!$.

The number of ways to arrange three people in line for a photograph is $3!$. To choose the first person in the line you have three choices. This leaves two choices for the second position and one choice for the last position. The count for different arrangements is $3(2)(1)$, but this is precisely $3!$. The formula remains consistent even for arranging one person in line: you have only one choice. The count is $1!$. If your philosophy allows it you might imagine that there are no persons to arrange in the photograph. There is only one way to arrange no people. This gives you one selection for the arrangement. Because $0! = 1$ the formula remains consistent.

Now consider the number of ways to rearrange the letters of the word *bookkeeper*. The word has ten letters. Thus the number of permutations of the letters in the word *bookkeeper* is $10!$. As your calculator will tell you, $10!$ is a large number (3628800), so that we will not list all of the permutations of *bookkeeper*. However, here are a few:

> *bookkeeper, obokkeeper, oobkkeeper, ookbkeeper, . . .*
>
> *bookkeeper, bookkeeper, bokokeeper, bokkoeeper, bokkeoeper, . . .*
>
> *bookkeeper, kbookeeper, . . . !*

You may think you have detected an error. It appears that *bookkeeper* appears three different times in the list. It only appears that way, because we are unable to distinguish duplicate letters. Numbering the letters may help.

$B_1O_1O_2K_1K_2E_1E_2P_1E_3R_1$ is a different permutation than

$B_1O_2O_1K_1K_2E_1E_2P_1E_3R_1$ and $B_1O_1O_2K_1K_2E_3E_2P_1E_1R_1$.

In one of these we swapped the position of the Os. In the other we rearranged the Es. The changes count as permutations, but they are not distinguishable without the subscripts. Because the two Os can be rearranged $2!$ ways we have overcounted distinguishable permutations by a factor of $2!$. Similarly for the Ks. There are three Es, which are overcounted by $3!$. To compensate for this overcounting, divide out the number of ways duplicate items can be rearranged. Thus the number of distinguishable permutations of the letters of bookkeeper is given by $10!/(1!\, 2!\, 2!\, 3!\, 1!\, 1!) = 151200$.

The denominator reflects the distinguishable letters. The first distinguishable letter is B. There is one B, thus the denominator contains 1! to reflect the number of ways that the Bs can be rearranged. Similarly, E is distinguishable from the other letters but appears 3 times; thus, we have a 3! in the denominator.

Suppose we wish to arrange eleven colored blocks in a line. Five of the blocks are red, four of the blocks are white and two of the blocks are blue. The blocks are otherwise indistinguishable. Then the blocks may be arranged in $11!/(5!\ 4!\ 2!)$ or 6930 distinguishable orders. Note that $5 + 4 + 2$ is the same as the total objects being rearranged.

Theorem 12-14
Counting Distinguishable Permutations

> The number of **distinguishable permutations of** M items is $M!/(m_1!\ m_2!\ \ldots\ m_n!)$, where item 1 occurs m_1 times, item 2 appears $m_2, \ldots$ and item n appears m_n times. Also, $M = \sum_{k=1}^{n} m_k$.

S I D E B A R

Guido runs after his uncle.

"Wait. I've got it. When we make a pizza with onion and sausage, it really doesn't matter whether we put on the sausage first or put on the onion first. The results are indistinguishable. We over counted by the number of ways two ingredients can be rearranged."

"We've over counted," says Giovanni. "What do you mean by we? And what do you plan to do about the overcount?"

"Divide it out. Let's see, you can rearrange two ingredients 2! ways. I divide that into 7(6) and get 21 different pizzas. Of course. Since the order of selection is not important, this is not a permutation of seven objects taken two at a time."

"Exactly. That is why we call it a *combination* pizza, not a permutation pizza."

In a permutation the order of the items is important. A different order of the same distinguishable items is a different count.

Suppose we wished to select and order some, but not all, items from a collection. Consider selecting three officers from a club with ten members. The first selected will be the president. The second selected will be the vice president and the last selected will be the treasurer. The order of selection is important. How many different slates of officers are there. The multiplication principle provides a correct answer, $10(9)(8)$ or 720. Because order is important, interpret this as a permutation. Now count the number of ways to rearrange all ten members of the club, 10!, where the first three members listed will form the slate of officers. After the first three members it makes no difference about the arrangement of the remaining seven members: these are not distinguishable among the slate of officers. The number of slates of officers is given by

$$\frac{10!}{(10 - 3)!} = 720.$$

Theorem 12-15
Permutation of n Objects Taken r at a Time

> The **permutation of** n **items taken** r **at a time is given by**
> $$_nP_r = \frac{n!}{(n - r)!} \quad \text{(clearly } n \geq r\text{)}.$$

Suppose we select r objects from a collection of n objects, but the order of selection is of no importance. All that is important is whether an object is selected or not selected. Selecting some objects without regard to order from a larger collection of objects is referred to as a *combination*.

Theorem 12-16
Combinations of n Objects Taken r at
a Time

The **combination of n objects taken r at a time**, symbolized by $_nC_r$ is given by

$$_nC_r = \frac{n!}{(n - r)! \, r!} \qquad (n \geq r)$$

Note how closely the formula resembles a permutation. Because order of selection is not important in a combination, rearrangements within the group selected or within the group not selected should not influence the count. These rearrangements are not distinguishable. With r selected from n, we have $n - r$ not selected. From Theorem 12-14 the number of distinguishable arrangements is

$$\frac{n!}{(n - r)! \, r!}$$

If the formula in Theorem 12-16 looks familiar, it is. The following theorem reminds us.

Theorem 12-17

Suppose that $_nC_r$ represents the combination of n items taken r at a time, then

$$_nC_r = \binom{n}{r}$$

where $\binom{n}{r}$ is the binomial coefficient.

···

EXAMPLE 12-15

Combinations and Permutations. Determine the following.

Illustration 1:

A contest has four prizes: first, second, third and fourth. The prizes are not equal in value. In how many possible ways could the four prizes be awarded among 100 entrants?

Solution: Because the order of award makes a difference, this is a permutation of 100 items taken 4 at a time.

$$_{100}P_4 = \frac{100!}{(100 - 4)!}$$
$$= 94,109,400$$

Illustration 2:

A club has 100 members and must select a committee of 4. How many different committees of four can be formed?

Solution: In a committee, order is not important. This is a combination of 100 items taken 4 at a time.

$$_{100}C_4 = \frac{100!}{(100 - 4)! \, 4!}$$
$$= 3{,}921{,}225$$

Note: Compare this to the previous illustration. You should conclude that

$$_nC_r = \frac{_nP_r}{r!}$$

Expect the answer for a combination to never be larger than a corresponding permutation: counting rearrangements of the same objects usually increases the count.

Illustration 3:   Determine $(a + b)^3$ without using the binomial theorem.

Solution: Use the distributive property, but avoid exponents.

$$(a + b)^3 = (a + b)(a + b)(a + b)$$
$$= (a + b)[aa + ab + ba + bb]$$
$$= aaa + aab + aba + abb + baa + bab + bba + bbb$$

Gather like terms. To combine these into a single term, count identical terms. Terms are alike if they contain exactly the same number of each variable factor without regard to the order. Consider terms like aab. Like terms include aba and baa. Are any other distinguishable ordering of these three factors possible? Because there are two factors, with a appearing twice and b appearing once, then by Theorem 12-14, there are exactly $3!/(2!1!)$ or three arrangements of aab. Multiplication produced all three of these arrangements. Therefore, by count, the coefficient of a^2b is 3. In a similar manner, count the coefficients for the other like terms and obtain $(a + b)^3 = a^3 + 3a^2b + 3ab^2 + b^3$.

Illustration 4:   See Illustration 3. Recall that $(a + b)^n = (a + b)(a + b)(a + b) \ldots (a + b)$. To simplify this expression we would also multiply then gather like terms. As in Illustration 3, each term contains a total of n factors consisting of repetitions of a and b. Gather all of the terms in which b appears r times. This term and others like it, each contain a $(n - r)$ times. To count the number of terms like $a^{n-r}b^r$, count the number of distinguishable permutations of the factors. Again by Theorem 12-14, we have $n!/[(n - r)! \, r!]$. The coefficient of $a^{n-r}b^r$ is

$$\frac{n!}{(n - r)! \, r!}$$

Counting theory supports the conclusion of the binomial theorem.   ▬

SUMMARY

- If A represents an event with n outcomes and B represents an event with m outcomes exclusive of A, the number of outcomes from A or B is $m + n$. Suppose A represents an event with n outcomes and B represents an event with m outcomes. If A and B are independent events then the number of outcomes in the event A and B is nm.
- An ordering of items is a permutation. Selecting in order r items from a collection of n items is a permutation of n items taken r at a time. The number of permutations of n items taken r at a time is given by

$$_nP_r = \frac{n!}{(n-r)!}$$

If the number selected is n ($r = n$), then the number of permutations is $n!$. Suppose we count permutations of a collection in which item A_1 is repeated a_1 times, item A_2 is repeated a_2 times, Item A_3 is repeated a_3 times, etc., A_r is repeated a_r times. Then the number of distinguishable permutations of the n items is

$$\frac{\left(\sum_{k=1}^{r} a_k\right)!}{a_1!\, a_2!\, a_3! \ldots a_r!}$$

- In statistics, a subset of a given set is a sample of that set. Order of selection is not important in a sample. A combination of n items selected r at a time is a size r sample of the n items. A combination of n items taken r at a time is given by

$$_nC_r = \binom{n}{r} = \frac{n!}{(n-r)!\, r!}$$

12-6 ▸ E X E R C I S E S

Determine the following.

1. $_5P_3$ 2. $_6C_4$ 3. $_5C_3$ 4. $_6P_4$
5. $_5C_2$ 6. $_6P_2$ 7. $_5P_2$ 8. $_6C_2$
9. $_7P_3$ 10. $_7C_4$

11. In how many ways can a committee of four be selected from a group of twelve people?

12. In how many ways can two different prizes be awarded among ten entries?

13. See Problem 11. In how many ways can we select eight people out of twelve to avoid being on the committee of four?

14. See Problem 12. Suppose all the prizes are indistinguishable and of equal value. How will this change the number of permutations?

15. How many ways can you select a group of eight people from a club with twenty people in it?

16. Fifteen televisions are delivered to a retailer. The retailer is overstocked and must select seven sets to put on sale at cost. How many ways are there to select the seven sets?

17. See Problem 15. Suppose twelve of the club members are male and eight are female. How many ways can you select five females from the club?

18. See Problem 16. Ten of the televisions are portables, the other five a consoles. How many ways can you select three consoles from the fifteen sets?

19. See Problem 17. How many ways can you select three males from the club?

20. See Problem 18. How many ways can you select four portables from the fifteen sets?

21. See Problem 19. How many ways can you select the group of eight people so that three are male *and* five are female?

22. See Problem 20. How many ways can you select four portables *and* three consoles from the fifteen sets?

23. How many five-digit zip codes can there be?

24. See Problem 23. How many nine-digit zip codes can there be?

25. An automobile can be custom ordered with your choice of five colors, three choices of radio, two choices of engines and four choices of interior. How many different versions of the car is the manufacturer offering?

26. If area codes may not begin with a 0 or 1, how many different area codes are available?

27. When Megan registers next spring two of her courses are required. She will be able to choose the remaining three courses as electives from among the following: one course from among five humanities courses, one course from among three science courses and one course from among four social science courses. How many possible schedules can Megan plan for in the spring?

28. See Problem 27. Juan's schedule offers more flexibility than Megan's. For his five courses Juan may choose any two courses from among five humanities courses, any course from among three science courses and any two courses from among the four social science courses. How many possible schedules can Juan plan for in the spring?

29. With eleven horses in a race, how many possible ways can the horses *win*, *place* and *show*? Is this a permutation, combination or neither?

30. See Problem 29. Any horse in the *win*, *place* or *show* categories finishes *in the money*. How many different ways may the eleven horses finish *in the money*? Is this a permutation, combination or neither?

31. Prove $_nP_r = r!(_nC_r)$.

32. Prove $_nC_r = {_nC_{n-r}}$.

33. Suppose a hat contains ten marbles. Six marbles are red the rest are blue. How many distinguishable permutations are there of these marbles?

34. A *word* contains nine letters. Six of the letters are a and three of the letters are b. Determine the number of distinguishable permutations of these letters.

35. Consider $(a + b)^{10}$. Compare the expansion of this binomial to Problem 33. What do these two problems have in common?

36. Compare the expansion of $(x + y)^9$ to Problem 34. What do these two problems have in common?

37. Review the meaning of summation notation: $\sum_{k=1}^{r} a_k$. Suppose we wanted a notation to express the product of the first r terms of a sequence rather than a sum. Use Π instead of Σ to create a definition for the product of the first r terms of a series.

38. See Problem 37. Use your definition to rewrite the definition of $n!$.

39. See Problem 37 and 38. Use your definition to rewrite the definition of distinguishable permutations.

40. See Problems 37, 38 and 39. Use your definition to express $1(3)(5)(7)(9)(11)(13)$.

- A sequence is a function whose domain is the set of counting numbers. Sequences use the special notation a_n, equivalent to $a(n)$, to represent the range component paired with n. The range component a_n is the nth term of the sequence. We use n (also i, j and k) rather than x, y or z for the domain component to emphasize that the domain consists of counting numbers rather than real numbers.
- The sum, product or quotient of two sequences as defined by the algebra of functions is a sequence. Also, if $a_n = a(n)$ represents a sequence, then ca_n, where c is a constant, represents the sequence in which each term of a_n is multiplied by c.
- Suppose that a_n represents a sequence, then the sum of the first r terms of the sequence is denoted by

$$S_r = \sum_{k=1}^{r} a_k$$

If a_n represents a sequence then

$$S_n = \sum_{k=1}^{n} a_k$$

is the sequence of partial sums of a_n.
- An arithmetic sequence has a rule of pairing expressible in one of the following forms:

$$a_n = a_1 + (n-1)d$$

$$\text{or} \quad a_n = \begin{cases} a_1 & \text{if } n = 1 \\ a_{n-1} + d & \text{if } n > d \end{cases} \quad \text{(recursive formula)}$$

where a_1 is the first term and d is the common difference, $d \neq 0$. The sum of the first n terms of an arithmetic sequence a_n is given by

$$S_n = n\,\frac{a_1 + a_n}{2}$$

- A geometric sequence has a rule of pairing expressible as one of the following:

$$a_n = a_1 r^{n-1}$$

$$\text{or} \quad a_n = \begin{cases} a_1 & \text{if } n = 1 \\ a_{n-1} r & \text{if } n > d \end{cases} \quad \text{(recursive formula)}$$

where $a_1 \neq 0$ is the first term and r is the common ratio, $r \neq 0$, $r \neq 1$.

The sum of the first n terms of a geometric sequence $a_n = a_1 r^{n-1}$ is given by

$$S_n = a_1\,\frac{1 - r^n}{1 - r}.$$

- For the geometric series

$$S_n = \sum_{k=1}^{n} a_1(r)^{k-1}$$

The series converges if $-1 < r < 1$, in which case we write

$$S_\infty = \sum_{k=1}^{\infty} a_1(r)^{k-1} = \frac{a_1}{1-r}$$

The series diverges if $r > 1$ or $r < -1$.

- An infinite repeating decimal that eventually repeats in a block of d digits is an infinite geometric series with common ratio given by $r = (0.1)^d$. (The same rule applies to other radix notation.) Nonrepeating infinite decimals represent irrational numbers.
- To prove a statement about counting numbers true for all counting numbers by the principle of finite mathematical induction, accomplish the following steps:

  1. Show the statement to be true for $n = 1$ (St_1 is true).
  2. (a) Note what the statement looks like for $n = k + 1$ (St_{k+1}). This establishes a road map for future manipulations.
  (b) Assume that the statement is true for $n = k$ (St_k). Use algebraic manipulation to obtain St_{k+1} from St_k.

These two steps satisfy Theorem 12-9 and establish the statement to be true for any counting number by the principle of finite mathematical induction.
- If n is a counting number then $n! = 1(2)(3) \ldots (n)$. $0! = 1$.

If n and r are counting numbers with $r \le n$, then

$$\binom{n}{r} = \frac{n!}{(n-r)!\, r!}$$

- $(a+b)^n = \dfrac{n!}{0!\, n!} a^n b^0 + \ldots + \dfrac{n!}{r!\,(n-r)!} a^{n-r} b^r + \ldots + \dfrac{n!}{n!\, 0!} a^0 b^n$

$$\text{Term } r+1 \Uparrow$$
where the expansion has $n + 1$ term (binomial theorem).
- If event A has n outcomes and event B has m outcomes and A and B are mutually exlusive, then the event A or B has $n + m$ outcomes (addition principal). If event A has n outcomes and event B has m outcomes and events A and B are independent, then event A **and** B has mn outcomes (multiplication principal).
- An ordering of objects is a permutation. Selecting in order r objects from a collection of n objects is a permutation of n objects taken r at a time. The number of permutations of n objects taken r at a time is given by $_nP_r = n!/(n-r)!$. If the number selected is n ($r = n$), then the number of permutations is $n!$. For permutations of a collection in which object A_1 is repeated a_1 times, item A_2 is repeated a_2 times, Item A_3 is repeated a_3 times, etc., A_r is repeated a_r times, the number of distinguishable permutations of the n items is

$$\frac{\left(\sum_{k=1}^{r} a_k\right)!}{a_1!\, a_2!\, a_3! \ldots a_r!}$$

- In statistics, a subset of a given set is a sample of that set. Order of selection is not important in a sample. A combination of n items selected r at a time is a size r sample of the n items. A combination of n items taken r at a time is given by

$$_nC_r = \binom{n}{r} = \frac{n!}{(n-r)!\, r!}$$

KEY WORDS AND CONCEPTS

Addition Counting Principle
And
Arithmetic sequence
Arithmetic series
Binomial coefficient
Binomial theorem
Combinations of *n* objects taken
  r at a time
Common difference
Common ratio
Convergence
Convergent Series
Counting Permutations

Distinguishable Permutations
Divergent Series
Factorial Notation
Finite mathematical induction
First term of a sequence
Geometric Sequence
Independent choices
Infinite Repeating Decimals
Multiplication counting principle
Mutually exclusive choices
Or
Partial sum

Permutation of *n* objects taken *r*
  at a time
Permutation
Principle of finite mathematical
  induction
Recursion
Sequence
Series
Sum of an infinite geometric series
Summation notation
Terminating decimal
Tree diagram

12 REVIEW EXERCISES

SECTION 12-1

Determine the first five terms for each of the sequences described by the following.

1. $a_n = 3n - 2$

2. $d_n = 4n + 10$

3. $b_n = \dfrac{2n + 1}{2n}$

4. $c_n = \begin{cases} 5 & \text{if } n = 1 \\ -2c_{n-1} & \text{if } n > 1 \end{cases}$

5. $d_n = \begin{cases} 3 & \text{if } n = 1 \\ 7 & \text{if } n = 2 \\ d_{n-2} - d_{n-1} & \text{if } n > 2 \end{cases}$

6. $a_n = (-1)^n 3^n$

7. $d_n = 5n$

8. $c_n = 5 - 4n$

9. $a_n = \begin{cases} 3 & \text{if } n = 1 \\ na_{n-1} & \text{if } n > 1 \end{cases}$

10. $b_n = \sum\limits_{k=1}^{n} (2k)$

For each of the following pairs of sequences, a_n and b_n, determine (*a*) a rule of pairing and the values of the first three terms of $a_n + b_n$ and (*b*) a rule of pairing and the values of the first three terms of $a_n b_n$.

11. $a_n = 3n + 2, b_n = n^2$

12. $a_n = (n + 1)^2, b_n = 4n - 1$

13. $a_n = (-1)^n, b_n = n + 1$

14. $a_n = (-1)^{n+1}, b_n = 5n$

15. $a_n = 2^{n+1}, b_n = \sqrt{n}$

16. $a_n = \sqrt[n-1]{n}, b_n = n + 1$

Evaluate each of the following summations.

17. $\sum\limits_{k=1}^{5} k^2$

18. $\sum\limits_{k=1}^{5} (3k - 1)$

19. $\sum\limits_{k=1}^{3} \dfrac{1}{2k - 1}$

20. $\sum\limits_{k=1}^{3} (k - 2)^2$

21. $\sum\limits_{k=1}^{5} (-2)^k$

22. $\sum\limits_{k=1}^{7} (-1)^{k+1}$

Determine the third and fifth terms for each of the following series.

23. $S_n = \sum\limits_{k=1}^{n} (2k)$

24. $S_n = \sum\limits_{k=1}^{n} (2k - 1)$

25. $S_n = \sum\limits_{k=1}^{n} (k^2 + 1)$

26. $S_n = \sum\limits_{k=1}^{n} \dfrac{(-1)^{k+1}}{k + 1}$

For each of the following, determine rules of pairing for *two* sequences with the given first four terms.

27. $4, 8, 12, 16, \ldots$

28. $3, -3, 3, -3, \ldots$

29. $2, -4, 8, -16, \ldots$

30. $1, -8, 27, -64, \ldots$

31. $\frac{2}{3}, \frac{3}{4}, \frac{4}{5}, \ldots$

32. $12, 7, 2, -3, \ldots$

Verify each of the following for $n = 4$.

33. $\sum\limits_{k=1}^{n} (ca_k) = c \sum\limits_{k=1}^{n} a_k$

34. $\sum\limits_{k=1}^{n} c = nc$

SECTION 12-2

Determine a rule of pairing for the arithmetic sequence determined by each of the following.

35. $a_1 = 5, d = -3$

36. $a_1 = 1, d = -\frac{1}{2}$

37. $a_1 = 0, d = 5$
3

38. $a_2 = 10, d =$

39. $3, 7, 11, 15, 19, \ldots$

40. $7, 3, -1, -5, \ldots$

Determine a rule of pairing for the geometric sequence determined by each of the following.

41. $a_1 = 1, r = 5$

42. $a_1 = 25, r = -5$

43. $a_2 = 8, r = -\frac{1}{2}$

44. $-2, 4, -8, 32, \ldots$

45. $3, -6, 12, -24, \ldots$

46. $18, 6, 2, \frac{2}{3}, \frac{2}{9}, \ldots$

For each of the following sequences a_n, (a) determine whether a_n is arithmetic, geometric or neither, and (b) if a_n is arithmetic or geometric, write a rule of pairing for a_n.

47. $2, 7, 12, 17, \ldots$

48. $6, 3, 0, -3, \ldots$

49. $0, -2, 4, -6, 8, \ldots$

50. $1, 2, 4, 6, 12, \ldots$

51. $10, 1, \frac{1}{10}, \frac{1}{100}, \ldots$

52. $30, -3, \frac{3}{10}, -\frac{3}{100}, \frac{3}{1000}, \ldots$

Determine the indicated sums.

53. $\sum_{k=1}^{10} [5 + (k-1)3]$

54. $\sum_{k=1}^{10} [2 + (k-1)(-3)]$

55. $\sum_{k=1}^{100} [2 + (k-1)5]$

56. $\sum_{k=1}^{6} 6\left(\frac{1}{3}\right)^{k-1}$

57. $\sum_{k=1}^{7} 10(-2)^{k-1}$

58. $\sum_{k=1}^{4} 1(-3)^{k-1}$

59. $\sum_{k=1}^{20} (5k)$

60. $\sum_{k=1}^{30} (2k+3)$

61. $\sum_{k=1}^{4} (3k-2)$

62. $\sum_{k=1}^{5} (7k-5)$

63. $\sum_{k=1}^{10} 4(0.1)^k$

64. $\sum_{k=1}^{8} 29(0.01)^k$

65. If \$1000 is invested the first year and the amount of money invested in each succeeding year is triple the preceding year, is this an arithmetic or geometric sequence? How much money will be invested in year 5 of the plan? What is the total investment over the five-year period?

66. Show that if a_n and b_n are geometric then a_n/b_n is geometric ($|r_a| \neq |r_b|$).

SECTION 12-3

Assume that any pattern of repeating digits is displayed and identify each of the following as representing a rational or irrational number.

67. $4.8\overline{9}$

68. $3.283283283283283 \ldots$

69. $5.63236374351 \ldots$ **70.** $3.58210058320697146 \ldots$

71. $0.996662277666554477 \ldots$

72. $0.542334277429425426642 \ldots$

Convert each of the following rational numbers to infinite repeating decimals. Use long division or you calculator.

73. $\frac{2}{3}$

74. $\frac{3}{9}$

75. $\frac{4}{6}$

76. $\frac{5}{13}$

77. $\frac{6}{7}$

78. $\frac{3}{6}$

Convert each of the following infinite repeating decimals to the form p/q, where p and q are integers.

79. $2.\overline{3}$

80. $2.\overline{5}$

81. 5.7

82. $1.\overline{21}$

83. $3.\overline{17}$

84. $9.\overline{9}$

Convert each of the following finite or infinite repeating radix representation to a base 10 common fraction.

85. 0.15

86. 0.3_7

87. $0.\overline{11}_2$

88. $0.\overline{02}_{16}$

89. $0.\overline{1}_4$

90. $0.\overline{A}_{12}$ ($A = 10_{10}$)

Determine whether each of the following infinite geometric series converges or diverges. If the series converges, indicate its sum.

91. $\sum_{k=1}^{\infty} 5\left(\frac{1}{2}\right)^{k-1}$

92. $\sum_{k=1}^{\infty} 2\left(\frac{3}{4}\right)^{k-1}$

93. $\sum_{k=1}^{\infty} \frac{1}{4} (3)^{k-1}$

94. $\sum_{k=1}^{\infty} \frac{1}{2} (-2)^{k-1}$

95. $\sum_{k=1}^{\infty} \left[-2\left(\frac{2}{3}\right)^{k-1} \right]$

96. $\sum_{k=1}^{\infty} (0.9)^k$

97. $\sum_{k=1}^{\infty} 3(-0.7)^{k-1}$

98. $\sum_{k=1}^{\infty} \left(-\frac{3}{5}\right)^k$

SECTION 12-4

Prove each of the following statements true for all counting numbers.

99. $2 + 4 + 6 + \ldots + (2n) = n(n+1)$

100. $\frac{1}{5} + \frac{1}{25} + \frac{1}{125} + \ldots + \left(\frac{1}{5}\right)^n = \frac{1 - \left(\frac{1}{5}\right)^n}{2}$

101. If $0 < x < 1$, then $x^n > x^{n+1}$.

102. $1(2) + 2(3) + 3(4) + \ldots + n(n+1) = \frac{n(n+1)(n+2)}{3}$

Refer to the results of Problems 99–102 to evaluate each of the following series.

103. $\sum_{k=1}^{10} (2k)$

104. $\sum_{k=1}^{4} \left(\frac{1}{5}\right)^k$

105. $\sum_{k=1}^{5} [k(k+1)]$

106. $\sum_{k=1}^{8} (6k)$ (see Problem number 103)

SECTION 12-5

Determine the indicated term in the expansion of each of the following.

107. $(a + b)^9$, fifth term $(r = 4)$

108. $(a + b)^8$, third term $(r = 2)$

109. $(3x^2 + y)^8$, third term

110. $(y - 2x^3)^9$, sixth term

111. $\left(\dfrac{1}{3} + \dfrac{2}{3}\right)^5$, third term

112. $\left(\dfrac{1}{4} + \dfrac{3}{4}\right)^6$, third term

Expand the following powers of binomials.

113. $(a + b)^7$ **114.** $(x - y)^7$ **115.** $(x + 3)^5$

116. $(x + 5)^4$ **117.** $(x - 3y)^4$ **118.** $(5x^2 + y^3)^4$

SECTION 12-6

Determine the following.

119. $_8P_3$ **120.** $_6C_5$ **121.** $_9C_3$ **122.** $_6P_2$

123. $_7C_2$ **124.** $_6P_6$ **125.** $_6P_2$ **126.** $_{11}C_2$

127. $_9P_3$ **128.** $_7C_1$

129. In how many ways can a committee of five be selected from a group of thirteen people?

130. In how many ways can three different prizes be awarded among eleven entries?

131. How many ways can you select a group of seven people from a club with twenty-one people in it?

132. Seventeen televisions are delivered to a retailer. The retailer is overstocked and must select eight sets to put on sale at cost. How many ways are there to select the eight sets?

133. See Problem 131. How many ways can you select the group of seven people so that three are male *and* four are female from 10 males and 11 females?

134. See Problem 132. How many ways can you select five table tops *and* three consoles from the seventeen sets if 10 of the 17 are table tops?

135. An automobile can be custom ordered with your choice of eight colors, four choices of radio, three choices of engines and five choices of interior. How many different versions of the car is the manufacturer offering?

SIDELIGHT   Pascal's Triangle and Dangers of Induction

One of the earliest instances of a correct use of mathematical induction was by Blaise Pascal (1623–1662) in his treatise on arithmetic triangles. These arithmetic triangles are usually called Pascal's triangle. Pascal's triangles give a method for generating the coefficients in a binomial expansion. To form the triangle begin with $(a + b)^0$, which is known to be 1:

Binomial	Pascal's Triangle
$(a + b)^0$	1
$(a + b)^1$	1 1
$(a + b)^2$	1 2 1
$(a + b)^3$	1 3 3 1
$(a + b)^4$	1 4 6 4 1
$(a + b)^5$	1 5 10 10 5 1
$(a + b)^6$	1 6 15 20 15 6 1
$(a + b)^7$	1 7 21 35 35 21 7 1

Pascal's triangle generates the coefficients of each successive power by simple addition rather than complex multiplication and division, as required by the factorial notation of the binomial theorem. The coefficient of each new row comes from adding the two coefficients immediately above and to the left and right of the new coefficient. For example the first 35 in the coefficients for the power 7 came from adding 15 and 20 in the previous row. For the power of 8, the first coefficient will be 1. Then add $1 + 7$ for the next coefficient. The third coefficient is $7 + 21$. Calculate the remaining coefficients similarly.

Although Pascal's triangle is simpler than the binomial theorem, the disadvantages are obvious. How would you calculate the coefficients of $(a + b)^{30}$ using Pascal's triangle. Imagine 31 rows of coefficients to see the disadvantages.

Like Pascal's triangle, mathematical induction has been with us for more than three centuries. Mathematical induction has its advantages and disadvantages, too. Mathematical induction is a mechanical method of obtaining a proof. Although this may be great for verification of a statement, it gives us no insight into the source of the statement. For that reason, mathematical induction is sterile. Induction does not generate statements. We must have the statement in hand before using induction to verify it. Creation of the statement comes from elsewhere.

Mathematical induction is also replete with hazards. An imprecise statement, or ambiguity in terminology can lead to ridiculous errors. The following peculiar attempt at induction exemplifies some of the hazards.

Prove: All elves are equal in height.

Proof: Use mathematical induction.

St_1:   Clearly, if the number of elves is 1, then all 1 of these elves are equal in height.

Now assume the statement is true for any collection of k elves and show that the statement for $k + 1$ elves must follow:

$$St_k \Rightarrow St_{k+1}$$

Assume that any collection of k elves are the same height. Now suppose we are given a group of $k + 1$ elves. Select one of the elves and have him go help Santa. Then the remaining k elves are the same height because of St_k. When the missing elf returns, choose a different elf to send to Santa. Then the remaining group of k elves are the same height because of St_k. By the transitive property of equality, the first elf to see Santa and the last elf to see Santa must be the same height. Then the group of $k + 1$ elves are all the same height.

12   CHAPTER TEST

1. List the first three terms of the series S_n formed as a sequence of partial sums from the sequence $a_k = k^2$.

2. If $a_1 = 2$ and $a_k = a_{k-1} + 3$, when $k > 1$, write a nonrecursive formula for a_n.

3. Evaluate

   (a) $\displaystyle\sum_{k=1}^{10} (3k - 1)$ **(b)** $\displaystyle\sum_{k=1}^{10} 2\left(\frac{1}{2}\right)^{k-1}$

4. Evaluate $\displaystyle\sum_{k=1}^{\infty} 5\left(\frac{1}{9}\right)^{k-1}$

5. Expand $(4x - \frac{1}{2}y)^5$

6. Rewrite $3.5\overline{67}$ as a ratio of two integers.

7. Prove by induction: if $0 < r < 1$, then $r^{k+1} < r^k$.

8. (a) Determine the fourth term of $(2x - 3y)^8$.

   (b) Determine the tenth term of $(3x - 5y)^{11}$.

9. Determine rules of pairing for two sequences whose first four terms are

   (a) $1, 3, 5, 7, \ldots$

   (b) $8, -4, 2, -1, \ldots$

10. (a) Of twenty parts to be used on an assembly line, three are nonstandard. An inspector plans to select a sample of five parts for testing. In how many ways can the inspector choose a sample that has no defective parts?

   (b) The word *ATOYOTA* is a palindrome: it reads the same from right to left as it does from left to right. In how many distinguishable ways may the letters of *ATOYOTA* be rearranged? If we do not worry about the meaning of these words, how many will form palindromes?

APPENDIX A
TABLES

A-1   TABLE OF COMMON LOGARITHMS

x	0.00	0.01	0.02	0.03	0.04	0.05	0.06	0.07	0.08	0.09
1.00	0.00000	0.00432	0.00860	0.01284	0.01703	0.02119	0.02531	0.02938	0.03342	0.03743
1.10	0.04139	0.04532	0.04922	0.05308	0.05690	0.06070	0.06446	0.06819	0.07188	0.07555
1.20	0.07918	0.08279	0.08636	0.08991	0.09342	0.09691	0.10037	0.10380	0.10721	0.11059
1.30	0.11394	0.11727	0.12057	0.12385	0.12710	0.13033	0.13354	0.13672	0.13988	0.14301
1.40	0.14613	0.14922	0.15229	0.15534	0.15836	0.16137	0.16435	0.16732	0.17026	0.17319
1.50	0.17609	0.17898	0.18184	0.18469	0.18752	0.19033	0.19312	0.19590	0.19866	0.20140
1.60	0.20412	0.20683	0.20952	0.21219	0.21484	0.21748	0.22011	0.22272	0.22531	0.22789
1.70	0.23045	0.23300	0.23553	0.23805	0.24055	0.24304	0.24551	0.24797	0.25042	0.25285
1.80	0.25527	0.25768	0.26007	0.26245	0.26482	0.26717	0.26951	0.27184	0.27416	0.27646
1.90	0.27875	0.28103	0.28330	0.28556	0.28780	0.29003	0.29226	0.29447	0.29667	0.29885
2.00	0.30103	0.30320	0.30535	0.30750	0.30963	0.31175	0.31387	0.31597	0.31806	0.32015
2.10	0.32222	0.32428	0.32634	0.32838	0.33041	0.33244	0.33445	0.33646	0.33846	0.34044
2.20	0.34242	0.34439	0.34635	0.34830	0.35025	0.35218	0.35411	0.35603	0.35793	0.35984
2.30	0.36173	0.36361	0.36549	0.36736	0.36922	0.37107	0.37291	0.37475	0.37658	0.37840
2.40	0.38021	0.38202	0.38382	0.38561	0.38739	0.38917	0.39094	0.39270	0.39445	0.39620
2.50	0.39794	0.39967	0.40140	0.40312	0.40483	0.40654	0.40824	0.40993	0.41162	0.41330
2.60	0.41497	0.41664	0.41830	0.41996	0.42160	0.42325	0.42488	0.42651	0.42813	0.42975
2.70	0.43136	0.43297	0.43457	0.43616	0.43775	0.43933	0.44091	0.44248	0.44404	0.44560
2.80	0.44716	0.44871	0.45025	0.45179	0.45332	0.45484	0.45637	0.45788	0.45939	0.46090
2.90	0.46240	0.46389	0.46538	0.46687	0.46835	0.46982	0.47129	0.47276	0.47422	0.47567
3.00	0.47712	0.47857	0.48001	0.48144	0.48287	0.48430	0.48572	0.48714	0.48855	0.48996
3.10	0.49136	0.49276	0.49415	0.49554	0.49693	0.49831	0.49969	0.50106	0.50243	0.50379
3.20	0.50515	0.50651	0.50786	0.50920	0.51055	0.51188	0.51322	0.51455	0.51587	0.51720
3.30	0.51851	0.51983	0.52114	0.52244	0.52375	0.52504	0.52634	0.52763	0.52892	0.53020
3.40	0.53148	0.53275	0.53403	0.53529	0.53656	0.53782	0.53908	0.54033	0.54158	0.54283
3.50	0.54407	0.54531	0.54654	0.54777	0.54900	0.55023	0.55145	0.55267	0.55388	0.55509
3.60	0.55630	0.55751	0.55871	0.55991	0.56110	0.56229	0.56348	0.56467	0.56585	0.56703
3.70	0.56820	0.56937	0.57054	0.57171	0.57287	0.57403	0.57519	0.57634	0.57749	0.57864
3.80	0.57978	0.58092	0.58206	0.58320	0.58433	0.58546	0.58659	0.58771	0.58883	0.58995
3.90	0.59106	0.59218	0.59329	0.59439	0.59550	0.59660	0.59770	0.59879	0.59988	0.60097
4.00	0.60206	0.60314	0.60423	0.60531	0.60638	0.60746	0.60853	0.60959	0.61066	0.61172
4.10	0.61278	0.61384	0.61490	0.61595	0.61700	0.61805	0.61909	0.62014	0.62118	0.62221
4.20	0.62325	0.62428	0.62531	0.62634	0.62737	0.62839	0.62941	0.63043	0.63144	0.63246
4.30	0.63347	0.63448	0.63548	0.63649	0.63749	0.63849	0.63949	0.64048	0.64147	0.64246
4.40	0.64345	0.64444	0.64542	0.64640	0.64738	0.64836	0.64933	0.65031	0.65128	0.65225
4.50	0.65321	0.65418	0.65514	0.65610	0.65706	0.65801	0.65896	0.65992	0.66087	0.66181
4.60	0.66276	0.66370	0.66464	0.66558	0.66652	0.66745	0.66839	0.66932	0.67025	0.67117
4.70	0.67210	0.67302	0.67394	0.67486	0.67578	0.67669	0.67761	0.67852	0.67943	0.68034
4.80	0.68124	0.68215	0.68305	0.68395	0.68485	0.68574	0.68664	0.68753	0.68842	0.68931
4.90	0.69020	0.69108	0.69197	0.69285	0.69373	0.69461	0.69548	0.69636	0.69723	0.69810
5.00	0.69897	0.69984	0.70070	0.70157	0.70243	0.70329	0.70415	0.70501	0.70586	0.70672
5.10	0.70757	0.70842	0.70927	0.71012	0.71096	0.71181	0.71265	0.71349	0.71433	0.71517
5.20	0.71600	0.71684	0.71767	0.71850	0.71933	0.72016	0.72099	0.72181	0.72263	0.72346
5.30	0.72428	0.72509	0.72591	0.72673	0.72754	0.72835	0.72916	0.72997	0.73078	0.73159
5.40	0.73239	0.73320	0.73400	0.73480	0.73560	0.73640	0.73719	0.73799	0.73878	0.73957
5.50	0.74036	0.74115	0.74194	0.74273	0.74351	0.74429	0.74507	0.74586	0.74663	0.74741

x	0.00	0.01	0.02	0.03	0.04	0.05	0.06	0.07	0.08	0.09
5.60	0.74819	0.74896	0.74974	0.75051	0.75128	0.75205	0.75282	0.75358	0.75435	0.75511
5.70	0.75587	0.75664	0.75740	0.75815	0.75891	0.75967	0.76042	0.76118	0.76193	0.76268
5.80	0.76343	0.76418	0.76492	0.76567	0.76641	0.76716	0.76790	0.76864	0.76938	0.77012
5.90	0.77085	0.77159	0.77232	0.77305	0.77379	0.77452	0.77525	0.77597	0.77670	0.77743
6.00	0.77815	0.77887	0.77960	0.78032	0.78104	0.78176	0.78247	0.78319	0.78390	0.78462
6.10	0.78533	0.78604	0.78675	0.78746	0.78817	0.78888	0.78958	0.79029	0.79099	0.79169
6.20	0.79239	0.79309	0.79379	0.79449	0.79518	0.79588	0.79657	0.79727	0.79796	0.79865
6.30	0.79934	0.80003	0.80072	0.80140	0.80209	0.80277	0.80346	0.80414	0.80482	0.80550
6.40	0.80618	0.80686	0.80754	0.80821	0.80889	0.80956	0.81023	0.81090	0.81158	0.81224
6.50	0.81291	0.81358	0.81425	0.81491	0.81558	0.81624	0.81690	0.81757	0.81823	0.81889
6.60	0.81954	0.82020	0.82086	0.82151	0.82217	0.82282	0.82347	0.82413	0.82478	0.82543
6.70	0.82607	0.82672	0.82737	0.82802	0.82866	0.82930	0.82995	0.83059	0.83123	0.83187
6.80	0.83251	0.83315	0.83378	0.83442	0.83506	0.83569	0.83632	0.83696	0.83759	0.83822
6.90	0.83885	0.83948	0.84011	0.84073	0.84136	0.84198	0.84261	0.84323	0.84386	0.84448
7.00	0.84510	0.84572	0.84634	0.84696	0.84757	0.84819	0.84880	0.84942	0.85003	0.85065
7.10	0.85126	0.85187	0.85248	0.85309	0.85370	0.85431	0.85491	0.85552	0.85612	0.85673
7.20	0.85733	0.85794	0.85854	0.85914	0.85974	0.86034	0.86094	0.86153	0.86213	0.86273
7.30	0.86332	0.86392	0.86451	0.86510	0.86570	0.86629	0.86688	0.86747	0.86806	0.86864
7.40	0.86923	0.86982	0.87040	0.87099	0.87157	0.87216	0.87274	0.87332	0.87390	0.87448
7.50	0.87506	0.87564	0.87622	0.87679	0.87737	0.87795	0.87852	0.87910	0.87967	0.88024
7.60	0.88081	0.88138	0.88195	0.88252	0.88309	0.88366	0.88423	0.88480	0.88536	0.88593
7.70	0.88649	0.88705	0.88762	0.88818	0.88874	0.88930	0.88986	0.89042	0.89098	0.89154
7.80	0.89209	0.89265	0.89321	0.89376	0.89432	0.89487	0.89542	0.89597	0.89653	0.89708
7.90	0.89763	0.89818	0.89873	0.89927	0.89982	0.90037	0.90091	0.90146	0.90200	0.90255
8.00	0.90309	0.90363	0.90417	0.90472	0.90526	0.90580	0.90634	0.90687	0.90741	0.90795
8.10	0.90849	0.90902	0.90956	0.91009	0.91062	0.91116	0.91169	0.91222	0.91275	0.91328
8.20	0.91381	0.91434	0.91487	0.91540	0.91593	0.91645	0.91698	0.91751	0.91803	0.91855
8.30	0.91908	0.91960	0.92012	0.92065	0.92117	0.92169	0.92221	0.92273	0.92324	0.92376
8.40	0.92428	0.92480	0.92531	0.92583	0.92634	0.92686	0.92737	0.92788	0.92840	0.92891
8.50	0.92942	0.92993	0.93044	0.93095	0.93146	0.93197	0.93247	0.93298	0.93349	0.93399
8.60	0.93450	0.93500	0.93551	0.93601	0.93651	0.93702	0.93752	0.93802	0.93852	0.93902
8.70	0.93952	0.94002	0.94052	0.94101	0.94151	0.94201	0.94250	0.94300	0.94349	0.94399
8.80	0.94448	0.94498	0.94547	0.94596	0.94645	0.94694	0.94743	0.94792	0.94841	0.94890
8.90	0.94939	0.94988	0.95036	0.95085	0.95134	0.95182	0.95231	0.95279	0.95328	0.95376
9.00	0.95424	0.95472	0.95521	0.95569	0.95617	0.95665	0.95713	0.95761	0.95809	0.95856
9.10	0.95904	0.95952	0.95999	0.96047	0.96095	0.96142	0.96190	0.96237	0.96284	0.96332
9.20	0.96379	0.96426	0.96473	0.96520	0.96567	0.96614	0.96661	0.96708	0.96755	0.96802
9.30	0.96848	0.96895	0.96942	0.96988	0.97035	0.97081	0.97128	0.97174	0.97220	0.97267
9.40	0.97313	0.97359	0.97405	0.97451	0.97497	0.97543	0.97589	0.97635	0.97681	0.97727
9.50	0.97772	0.97818	0.97864	0.97909	0.97955	0.98000	0.98046	0.98091	0.98137	0.98182
9.60	0.98227	0.98272	0.98318	0.98363	0.98408	0.98453	0.98498	0.98543	0.98588	0.98632
9.70	0.98677	0.98722	0.98767	0.98811	0.98856	0.98900	0.98945	0.98989	0.99034	0.99078
9.80	0.99123	0.99167	0.99211	0.99255	0.99300	0.99344	0.99388	0.99432	0.99476	0.99520
9.90	0.99564	0.99607	0.99651	0.99695	0.99739	0.99782	0.99826	0.99870	0.99913	0.99957

A-3   BASE E

x	e^x	e^{-x}	Ln x	1/x
0.00	1.00000	1.00000	Undefined	Undefined
0.10	1.10517	0.90484	-2.30259	10.00000
0.20	1.22140	0.81873	-1.60944	5.00000
0.30	1.34986	0.74082	-1.20397	3.33333
0.40	1.49182	0.67032	-0.91629	2.50000
0.50	1.64872	0.60653	-0.69315	2.00000
0.60	1.82212	0.54881	-0.51083	1.66667
0.70	2.01375	0.49659	-0.35667	1.42857
0.80	2.22554	0.44933	-0.22314	1.25000
0.90	2.45960	0.40657	-0.10536	1.11111
1.00	2.71828	0.36788	0.00000	1.00000
1.10	3.00417	0.33287	0.09531	0.90909
1.20	3.32012	0.30119	0.18232	0.83333
1.30	3.66930	0.27253	0.26236	0.76923
1.40	4.05520	0.24660	0.33647	0.71429
1.50	4.48169	0.22313	0.40547	0.66667
1.60	4.95303	0.20190	0.47000	0.62500
1.70	5.47395	0.18268	0.53063	0.58824
1.80	6.04965	0.16530	0.58779	0.55556
1.90	6.68589	0.14957	0.64185	0.52632
2.00	7.38906	0.13534	0.69315	0.50000
2.10	8.16617	0.12246	0.74194	0.47619
2.20	9.02501	0.11080	0.78846	0.45455
2.30	9.97418	0.10026	0.83291	0.43478
2.40	11.02318	0.09072	0.87547	0.41667
2.50	12.18249	0.08208	0.91629	0.40000
2.60	13.46374	0.07427	0.95551	0.38462
2.70	14.87973	0.06721	0.99325	0.37037
2.80	16.44465	0.06081	1.02962	0.35714
2.90	18.17415	0.05502	1.06471	0.34483
3.00	20.08554	0.04979	1.09861	0.33333
3.10	22.19795	0.04505	1.13140	0.32258
3.20	24.53253	0.04076	1.16315	0.31250
3.30	27.11264	0.03688	1.19392	0.30303
3.40	29.96410	0.03337	1.22378	0.29412
3.50	33.11545	0.03020	1.25276	0.28571
3.60	36.59823	0.02732	1.28093	0.27778
3.70	40.44730	0.02472	1.30833	0.27027
3.80	44.70118	0.02237	1.33500	0.26316
3.90	49.40245	0.02024	1.36098	0.25641
4.00	54.59815	0.01832	1.38629	0.25000
4.10	60.34029	0.01657	1.41099	0.24390
4.20	66.68633	0.01500	1.43508	0.23810
4.30	73.69979	0.01357	1.45862	0.23256
4.40	81.45087	0.01228	1.48160	0.22727
4.50	90.01713	0.01111	1.50408	0.22222
4.60	99.48432	0.01005	1.52606	0.21739
4.70	109.94717	0.00910	1.54756	0.21277

x	e^x	e^{-x}	Ln x	1/x
4.80	121.51042	0.00823	1.56862	0.20833
4.90	134.28978	0.00745	1.58924	0.20408
5.00	148.41316	0.00674	1.60944	0.20000
5.10	164.02191	0.00610	1.62924	0.19608
5.20	181.27224	0.00552	1.64866	0.19231
5.30	200.33681	0.00499	1.66771	0.18868
5.40	221.40642	0.00452	1.68640	0.18519
5.50	244.69193	0.00409	1.70475	0.18182
5.60	270.42641	0.00370	1.72277	0.17857
5.70	298.86740	0.00335	1.74047	0.17544
5.80	330.29956	0.00303	1.75786	0.17241
5.90	365.03747	0.00274	1.77495	0.16949
6.00	403.42879	0.00248	1.79176	0.16667
6.10	445.85777	0.00224	1.80829	0.16393
6.20	492.74904	0.00203	1.82455	0.16129
6.30	544.57191	0.00184	1.84055	0.15873
6.40	601.84504	0.00166	1.85630	0.15625
6.50	665.14163	0.00150	1.87180	0.15385
6.60	735.09519	0.00136	1.88707	0.15152
6.70	812.40583	0.00123	1.90211	0.14925
6.80	897.84729	0.00111	1.91692	0.14706
6.90	992.27472	0.00101	1.93152	0.14493
7.00	1096.63316	0.00091	1.94591	0.14286
7.10	1211.96707	0.00083	1.96009	0.14085
7.20	1339.43076	0.00075	1.97408	0.13889
7.30	1480.29993	0.00068	1.98787	0.13699
7.40	1635.98443	0.00061	2.00148	0.13514
7.50	1808.04241	0.00055	2.01490	0.13333
7.60	1998.19590	0.00050	2.02815	0.13158
7.70	2208.34799	0.00045	2.04122	0.12987
7.80	2440.60198	0.00041	2.05412	0.12821
7.90	2697.28233	0.00037	2.06686	0.12658
8.00	2980.95799	0.00034	2.07944	0.12500
8.10	3294.46808	0.00030	2.09186	0.12346
8.20	3640.95031	0.00027	2.10413	0.12195
8.30	4023.87239	0.00025	2.11626	0.12048
8.40	4447.06675	0.00022	2.12823	0.11905
8.50	4914.76884	0.00020	2.14007	0.11765
8.60	5431.65959	0.00018	2.15176	0.11628
8.70	6002.91222	0.00017	2.16332	0.11494
8.80	6634.24401	0.00015	2.17475	0.11364
8.90	7331.97354	0.00014	2.18605	0.11236
9.00	8103.08393	0.00012	2.19722	0.11111
9.10	8955.29270	0.00011	2.20827	0.10989
9.20	9897.12906	0.00010	2.21920	0.10870
9.30	10938.01921	0.00009	2.23001	0.10753
9.40	12088.38073	0.00008	2.24071	0.10638
9.50	13359.72683	0.00007	2.25129	0.10526
9.60	14764.78157	0.00007	2.26176	0.10417
9.70	16317.60720	0.00006	2.27213	0.10309
9.80	18033.74493	0.00006	2.28238	0.10204
9.90	19930.37044	0.00005	2.29253	0.10101
10.00	22026.46579	0.00005	2.30259	0.10000

A-4   BINOMIAL COEFFICIENTS AND FACTORIALS

Factorial

X	X!
0	1
1	1
2	2
3	6
4	24
5	120
6	720
7	5040
8	40320
9	362880
10	3628800
11	39916800
12	479001600
13	6227020800
14	87178291200
15	1307674368000
16	20922789888000
17	355687428100000
18	6402373705700000
19	121645100410000000
20	2432902008200000000

Binomial Coefficient $\begin{pmatrix} n \\ r \end{pmatrix}$

$\frac{r}{n}$	0	1	2	3	4	5	6	7	8	9
2	1	2	1							
3	1	3	3							
4	1	4	6	4						
5	1	5	10	10						
6	1	6	15	20	15					
7	1	7	21	35	35					
8	1	8	28	56	70	56				
9	1	9	36	84	126	126				
10	1	10	45	120	210	252	210			
11	1	11	55	165	330	462	462			
12	1	12	66	220	495	792	924	792		
13	1	13	78	286	715	1287	1716	1716		
14	1	14	91	364	1001	2002	3003	3432	3003	
15	1	15	105	455	1365	3003	5005	6435	6435	
16	1	16	120	560	1820	4368	8008	11440	12870	11440
17	1	17	136	680	2380	6188	12376	19448	24310	24310

A-5   TRIGONOMETRIC TABLES

Degree	Radians	sin	cos	tan	cot	CoRad	CoDeg
				Trigonometric tables		Complement	
0°00′	0.00000	0.00000	1.00000	0.00000		1.57080	90°00′
10′	0.00291	0.00291	1.00000	0.00291	343.77371	1.56789	50′
20′	0.00582	0.00582	0.99998	0.00582	171.88540	1.56498	40′
30′	0.00873	0.00873	0.99996	0.00873	114.58865	1.56207	30′
40′	0.01164	0.01164	0.99993	0.01164	85.93979	1.55916	20′
50′	0.01454	0.01454	0.99989	0.01455	68.75009	1.55625	10′
1°00′	0.01745	0.01745	0.99985	0.01746	57.28996	1.55334	89°00′
10′	0.02036	0.02036	0.99979	0.02036	49.10388	1.55043	50′
20′	0.02327	0.02327	0.99973	0.02328	42.96408	1.54753	40′
30′	0.02618	0.02618	0.99966	0.02619	38.18846	1.54462	30′
40′	0.02909	0.02908	0.99958	0.02910	34.36777	1.54171	20′
50′	0.03200	0.03199	0.99949	0.03201	31.24158	1.53880	10′
2°00′	0.03491	0.03490	0.99939	0.03492	28.63625	1.53589	88°00′
10′	0.03782	0.03781	0.99929	0.03783	26.43160	1.53298	50′
20′	0.04072	0.04071	0.99917	0.04075	24.54176	1.53007	40′
30′	0.04363	0.04362	0.99905	0.04366	22.90377	1.52716	30′
40′	0.04654	0.04653	0.99892	0.04658	21.47040	1.52425	20′
50′	0.04945	0.04943	0.99878	0.04949	20.20555	1.52135	10′
3°00′	0.05236	0.05234	0.99863	0.05241	19.08114	1.51844	87°00′
10′	0.05527	0.05524	0.99847	0.05533	18.07498	1.51553	50′
20′	0.05818	0.05814	0.99831	0.05824	17.16934	1.51262	40′
30′	0.06109	0.06105	0.99813	0.06116	16.34986	1.50971	30′
40′	0.06400	0.06395	0.99795	0.06408	15.60478	1.50680	20′
50′	0.06690	0.06685	0.99776	0.06700	14.92442	1.50389	10′
4°00′	0.06981	0.06976	0.99756	0.06993	14.30067	1.50098	86°00′
10′	0.07272	0.07266	0.99736	0.07285	13.72674	1.49807	50′
20′	0.07563	0.07556	0.99714	0.07578	13.19688	1.49517	40′
30′	0.07854	0.07846	0.99692	0.07870	12.70620	1.49226	30′
40′	0.08145	0.08136	0.99668	0.08163	12.25051	1.48935	20′
50′	0.08436	0.08426	0.99644	0.08456	11.82617	1.48644	10′

CoDeg	CoRad	cos	sin	cot	tan	Radians	Degrees
Complement				Trigonometric tables			

Degree	Trigonometric tables					Complement	
	Radians	sin	cos	tan	cot	CoRad	CoDeg
5°00′	0.08727	0.08716	0.99619	0.08749	11.43005	1.48353	85°00′
10′	0.09018	0.09005	0.99594	0.09042	11.05943	1.48062	50′
20′	0.09308	0.09295	0.99567	0.09335	10.71191	1.47771	40′
30′	0.09599	0.09585	0.99540	0.09629	10.38540	1.47480	30′
40′	0.09890	0.09874	0.99511	0.09923	10.07803	1.47189	20′
50′	0.10181	0.10164	0.99482	0.10216	9.78817	1.46899	10′
6°00′	0.10472	0.10453	0.99452	0.10510	9.51436	1.46608	84°00′
10′	0.10763	0.10742	0.99421	0.10805	9.25530	1.46317	50′
20′	0.11054	0.11031	0.99390	0.11099	9.00983	1.46026	40′
30′	0.11345	0.11320	0.99357	0.11394	8.77689	1.45735	30′
40′	0.11636	0.11609	0.99324	0.11688	8.55555	1.45444	20′
50′	0.11926	0.11898	0.99290	0.11983	8.34496	1.45153	10′
7°00′	0.12217	0.12187	0.99255	0.12278	8.14435	1.44862	83°00′
10′	0.12508	0.12476	0.99219	0.12574	7.95302	1.44571	50′
20′	0.12799	0.12764	0.99182	0.12869	7.77035	1.44281	40′
30′	0.13090	0.13053	0.99144	0.13165	7.59575	1.43990	30′
40′	0.13381	0.13341	0.99106	0.13461	7.42871	1.43699	20′
50′	0.13672	0.13629	0.99067	0.13758	7.26873	1.43408	10′
8°00′	0.13963	0.13917	0.99027	0.14054	7.11537	1.43117	82°00′
10′	0.14254	0.14205	0.98986	0.14351	6.96823	1.42826	50′
20′	0.14544	0.14493	0.98944	0.14648	6.82694	1.42535	40′
30′	0.14835	0.14781	0.98902	0.14945	6.69116	1.42244	30′
40′	0.15126	0.15069	0.98858	0.15243	6.56055	1.41953	20′
50′	0.15417	0.15356	0.98814	0.15540	6.43484	1.41663	10′
9°00′	0.15708	0.15643	0.98769	0.15838	6.31375	1.41372	81°00′
10′	0.15999	0.15931	0.98723	0.16137	6.19703	1.41081	50′
20′	0.16290	0.16218	0.98676	0.16435	6.08444	1.40790	40′
30′	0.16581	0.16505	0.98629	0.16734	5.97576	1.40499	30′
40′	0.16872	0.16792	0.98580	0.17033	5.87080	1.40208	20′
50′	0.17162	0.17078	0.98531	0.17333	5.76937	1.39917	10′
10°00′	0.17453	0.17365	0.98481	0.17633	5.67128	1.39626	80°00′
10′	0.17744	0.17651	0.98430	0.17933	5.57638	1.39335	50′
20′	0.18035	0.17937	0.98378	0.18233	5.48451	1.39045	40′
30′	0.18326	0.18224	0.98325	0.18534	5.39552	1.38754	30′
40′	0.18617	0.18509	0.98272	0.18835	5.30928	1.38463	20′
50′	0.18908	0.18795	0.98218	0.19136	5.22566	1.38172	10′

Complement		Trigonometric tables					
CoDeg	CoRad	cos	sin	cot	tan	Radians	Degrees

Degree	Trigonometric tables					Complement	
Degree	Radians	sin	cos	tan	cot	CoRad	CoDeg
11°00′	0.19199	0.19081	0.98163	0.19438	5.14455	1.37881	79°00′
10′	0.19490	0.19366	0.98107	0.19740	5.06584	1.37590	50′
20′	0.19780	0.19652	0.98050	0.20042	4.98940	1.37299	40′
30′	0.20071	0.19937	0.97992	0.20345	4.91516	1.37008	30′
40′	0.20362	0.20222	0.97934	0.20648	4.84300	1.36717	20′
50′	0.20653	0.20507	0.97875	0.20952	4.77286	1.36427	10′
12°00′	0.20944	0.20791	0.97815	0.21256	4.70463	1.36136	78°00′
10′	0.21235	0.21076	0.97754	0.21560	4.63825	1.35845	50′
20′	0.21526	0.21360	0.97692	0.21864	4.57363	1.35554	40′
30′	0.21817	0.21644	0.97630	0.22169	4.51071	1.35263	30′
40′	0.22108	0.21928	0.97566	0.22475	4.44942	1.34972	20′
50′	0.22398	0.22212	0.97502	0.22781	4.38969	1.34681	10′
13°00′	0.22689	0.22495	0.97437	0.23087	4.33148	1.34390	77°00′
10′	0.22980	0.22778	0.97371	0.23393	4.27471	1.34099	50′
20′	0.23271	0.23062	0.97304	0.23700	4.21933	1.33809	40′
30′	0.23562	0.23345	0.97237	0.24008	4.16530	1.33518	30′
40′	0.23853	0.23627	0.97169	0.24316	4.11256	1.33227	20′
50′	0.24144	0.23910	0.97100	0.24624	4.06107	1.32936	10′
14°00′	0.24435	0.24192	0.97030	0.24933	4.01078	1.32645	76°00′
10′	0.24725	0.24474	0.96959	0.25242	3.96165	1.32354	50′
20′	0.25016	0.24756	0.96887	0.25552	3.91364	1.32063	40′
30′	0.25307	0.25038	0.96815	0.25862	3.86671	1.31772	30′
40′	0.25598	0.25320	0.96742	0.26172	3.82083	1.31481	20′
50′	0.25889	0.25601	0.96667	0.26483	3.77595	1.31191	10′
15°00′	0.26180	0.25882	0.96593	0.26795	3.73205	1.30900	75°00′
10′	0.26471	0.26163	0.96517	0.27107	3.68909	1.30609	50′
20′	0.26762	0.26443	0.96440	0.27419	3.64705	1.30318	40′
30′	0.27053	0.26724	0.96363	0.27732	3.60588	1.30027	30′
40′	0.27343	0.27004	0.96285	0.28046	3.56557	1.29736	20′
50′	0.27634	0.27284	0.96206	0.28360	3.52609	1.29445	10′
16°00′	0.27925	0.27564	0.96126	0.28675	3.48741	1.29154	74°00′
10′	0.28216	0.27843	0.96046	0.28990	3.44951	1.28863	50′
20′	0.28507	0.28123	0.95964	0.29305	3.41236	1.28573	40′
30′	0.28798	0.28402	0.95882	0.29621	3.37594	1.28282	30′
40′	0.29089	0.28680	0.95799	0.29938	3.34023	1.27991	20′
50′	0.29380	0.28959	0.95715	0.30255	3.30521	1.27700	10′

Complement		Trigonometric tables					
CoDeg	CoRad	cos	sin	cot	tan	Radians	Degrees

Degree	Radians	sin	cos	tan	cot	CoRad	CoDeg
		Trigonometric tables				Complement	
17°00′	0.29671	0.29237	0.95630	0.30573	3.27085	1.27409	73°00′
10′	0.29961	0.29515	0.95545	0.30891	3.23714	1.27118	50′
20′	0.30252	0.29793	0.95459	0.31210	3.20406	1.26827	40′
30′	0.30543	0.30071	0.95372	0.31530	3.17159	1.26536	30′
40′	0.30834	0.30348	0.95284	0.31850	3.13972	1.26245	20′
50′	0.31125	0.30625	0.95195	0.32171	3.10842	1.25955	10′
18°00′	0.31416	0.30902	0.95106	0.32492	3.07768	1.25664	72°00
10′	0.31707	0.31178	0.95015	0.32814	3.04749	1.25373	50′
20′	0.31998	0.31454	0.94924	0.33136	3.01783	1.25082	40′
30′	0.32289	0.31730	0.94832	0.33460	2.98868	1.24791	30′
40′	0.32579	0.32006	0.94740	0.33783	2.96004	1.24500	20′
50′	0.32870	0.32282	0.94646	0.34108	2.93189	1.24209	10′
19°00′	0.33161	0.32557	0.94552	0.34433	2.90421	1.23918	71°00′
10′	0.33452	0.32832	0.94457	0.34758	2.87700	1.23627	50′
20′	0.33743	0.33106	0.94361	0.35085	2.85023	1.23337	40′
30′	0.34034	0.33381	0.94264	0.35412	2.82391	1.23046	30′
40′	0.34325	0.33655	0.94167	0.35740	2.79802	1.22755	20′
50′	0.34616	0.33929	0.94068	0.36068	2.77254	1.22464	10′
20°00′	0.34907	0.34202	0.93969	0.36397	2.74748	1.22173	70°00′
10′	0.35197	0.34475	0.93869	0.36727	2.72281	1.21882	50′
20′	0.35488	0.34748	0.93769	0.37057	2.69853	1.21591	40′
30′	0.35779	0.35021	0.93667	0.37388	2.67462	1.21300	30′
40′	0.36070	0.35293	0.93565	0.37720	2.65109	1.21009	20′
50′	0.36361	0.35565	0.93462	0.38053	2.62791	1.20719	10′
21°00′	0.36652	0.35837	0.93358	0.38386	2.60509	1.20428	69°00′
10′	0.36943	0.36108	0.93253	0.38721	2.58261	1.20137	50′
20′	0.37234	0.36379	0.93148	0.39055	2.56046	1.19846	40′
30′	0.37525	0.36650	0.93042	0.39391	2.53865	1.19555	30′
40′	0.37815	0.36921	0.92935	0.39727	2.51715	1.19264	20′
50′	0.38106	0.37191	0.92827	0.40065	2.49597	1.18973	10′
22°00′	0.38397	0.37461	0.92718	0.40403	2.47509	1.18682	68°00′
10′	0.38688	0.37730	0.92609	0.40741	2.45451	1.18392	50′
20′	0.38979	0.37999	0.92499	0.41081	2.43422	1.18101	40′
30′	0.39270	0.38268	0.92388	0.41421	2.41421	1.17810	30′
40′	0.39561	0.38537	0.92276	0.41763	2.39449	1.17519	20′
50′	0.39852	0.38805	0.92164	0.42105	2.37504	1.17228	10′

CoDeg	CoRad	cos	sin	cot	tan	Radians	Degrees
Complement		Trigonometric tables					

Degree	Radians	sin	cos	tan	cot	CoRad	CoDeg
		Trigonometric tables				Complement	
23°00′	0.40143	0.39073	0.92050	0.42447	2.35585	1.16937	67°00′
10′	0.40433	0.39341	0.91936	0.42791	2.33693	1.16646	50′
20′	0.40724	0.39608	0.91822	0.43136	2.31826	1.16355	40′
30′	0.41015	0.39875	0.91706	0.43481	2.29984	1.16064	30′
40′	0.41306	0.40141	0.91590	0.43828	2.28167	1.15774	20′
50′	0.41597	0.40408	0.91472	0.44175	2.26374	1.15483	10′
24°00′	0.41888	0.40674	0.91355	0.44523	2.24604	1.15192	66°00
10′	0.42179	0.40939	0.91236	0.44872	2.22857	1.14901	50′
20′	0.42470	0.41204	0.91116	0.45222	2.21132	1.14610	40′
30′	0.42761	0.41469	0.90996	0.45573	2.19430	1.14319	30′
40′	0.43051	0.41734	0.90875	0.45924	2.17749	1.14028	20′
50′	0.43342	0.41998	0.90753	0.46277	2.16090	1.13737	10′
25°00′	0.43633	0.42262	0.90631	0.46631	2.14451	1.13446	65°00′
10′	0.43924	0.42525	0.90507	0.46985	2.12832	1.13156	50′
20′	0.44215	0.42788	0.90383	0.47341	2.11233	1.12865	40′
30′	0.44506	0.43051	0.90259	0.47698	2.09654	1.12574	30′
40′	0.44797	0.43313	0.90133	0.48055	2.08094	1.12283	20′
50′	0.45088	0.43575	0.90007	0.48414	2.06553	1.11992	10′
26°00′	0.45379	0.43837	0.89879	0.48773	2.05030	1.11701	64°00′
10′	0.45669	0.44098	0.89752	0.49134	2.03526	1.11410	50′
20′	0.45960	0.44359	0.89623	0.49495	2.02039	1.11119	40′
30′	0.46251	0.44620	0.89493	0.49858	2.00569	1.10828	30′
40′	0.46542	0.44880	0.89363	0.50222	1.99116	1.10538	20′
50′	0.46833	0.45140	0.89232	0.50587	1.97681	1.10247	10′
27°00′	0.47124	0.45399	0.89101	0.50953	1.96261	1.09956	63°00′
10′	0.47415	0.45658	0.88968	0.51319	1.94858	1.09665	50′
20′	0.47706	0.45917	0.88835	0.51688	1.93470	1.09374	40′
30′	0.47997	0.46175	0.88701	0.52057	1.92098	1.09083	30′
40′	0.48287	0.46433	0.88566	0.52427	1.90741	1.08792	20′
50′	0.48578	0.46690	0.88431	0.52798	1.89400	1.08501	10′
28°00′	0.48869	0.46947	0.88295	0.53171	1.88073	1.08210	62°00′
10′	0.49160	0.47204	0.88158	0.53545	1.86760	1.07920	50′
20′	0.49451	0.47460	0.88020	0.53920	1.85462	1.07629	40′
30′	0.49742	0.47716	0.87882	0.54296	1.84177	1.07338	30′
40′	0.50033	0.47971	0.87743	0.54673	1.82906	1.07047	20′
50′	0.50324	0.48226	0.87603	0.55051	1.81649	1.06756	10′

CoDeg	CoRad	cos	sin	cot	tan	Radians	Degrees
Complement		Trigonometric tables					

	Trigonometric tables					Complement	
Degree	Radians	sin	cos	tan	cot	CoRad	CoDeg
29°00′	0.50615	0.48481	0.87462	0.55431	1.80405	1.06465	61°00′
10′	0.50905	0.48735	0.87321	0.55812	1.79174	1.06174	50′
20′	0.51196	0.48989	0.87178	0.56194	1.77955	1.05883	40′
30′	0.51487	0.49242	0.87036	0.56577	1.76749	1.05592	30′
40′	0.51778	0.49495	0.86892	0.56962	1.75556	1.05302	20′
50′	0.52069	0.49748	0.86748	0.57348	1.74375	1.05011	10′
30°00′	0.52360	0.50000	0.86603	0.57735	1.73205	1.04720	60°00′
10′	0.52651	0.50252	0.86457	0.58124	1.72047	1.04429	50′
20′	0.52942	0.50503	0.86310	0.58513	1.70901	1.04138	40′
30′	0.53233	0.50754	0.86163	0.58905	1.69766	1.03847	30′
40′	0.53523	0.51004	0.86015	0.59297	1.68643	1.03556	20′
50′	0.53814	0.51254	0.85866	0.59691	1.67530	1.03265	10′
31°00′	0.54105	0.51504	0.85717	0.60086	1.66428	1.02974	59°00′
10′	0.54396	0.51753	0.85567	0.60483	1.65337	1.02684	50′
20′	0.54687	0.52002	0.85416	0.60881	1.64256	1.02393	40′
30′	0.54978	0.52250	0.85264	0.61280	1.63185	1.02102	30′
40′	0.55269	0.52498	0.85112	0.61681	1.62125	1.01811	20′
50′	0.55560	0.52745	0.84959	0.62083	1.61074	1.01520	10′
32°00′	0.55851	0.52992	0.84805	0.62487	1.60033	1.01229	58°00′
10′	0.56141	0.53238	0.84650	0.62892	1.59002	1.00938	50′
20′	0.56432	0.53484	0.84495	0.63299	1.57981	1.00647	40′
30′	0.56723	0.53730	0.84339	0.63707	1.56969	1.00356	30′
40′	0.57014	0.53975	0.84182	0.64117	1.55966	1.00066	20′
50′	0.57305	0.54220	0.84025	0.64528	1.54972	0.99775	10′
33°00′	0.57596	0.54464	0.83867	0.64941	1.53986	0.99484	57°00
10′	0.57887	0.54708	0.83708	0.65355	1.53010	0.99193	50′
20′	0.58178	0.54951	0.83549	0.65771	1.52043	0.98902	40′
30′	0.58469	0.55194	0.83389	0.66189	1.51084	0.98611	30′
40′	0.58759	0.55436	0.83228	0.66608	1.50133	0.98320	20′
50′	0.59050	0.55678	0.83066	0.67028	1.49190	0.98029	10′
34°00′	0.59341	0.55919	0.82904	0.67451	1.48256	0.97738	56°00′
10′	0.59632	0.56160	0.82741	0.67875	1.47330	0.97448	50′
20′	0.59923	0.56401	0.82577	0.68301	1.46411	0.97157	40′
30′	0.60214	0.56641	0.82413	0.68728	1.45501	0.96866	30′
40′	0.60505	0.56880	0.82248	0.69157	1.44598	0.96575	20′
50′	0.60796	0.57119	0.82082	0.69588	1.43703	0.96284	10′

Complement		Trigonometric tables					
CoDeg	CoRad	cos	sin	cot	tan	Radians	Degrees

	Trigonometric tables					Complement	
Degree	Radians	sin	cos	tan	cot	CoRad	CoDeg
35°00′	0.61087	0.57358	0.81915	0.70021	1.42815	0.95993	55°00′
10′	0.61377	0.57596	0.81748	0.70455	1.41934	0.95702	50′
20′	0.61668	0.57833	0.81580	0.70891	1.41061	0.95411	40′
30′	0.61959	0.58070	0.81412	0.71329	1.40195	0.95120	30′
40′	0.62250	0.58307	0.81242	0.71769	1.39336	0.94830	20′
50′	0.62541	0.58543	0.81072	0.72211	1.38484	0.94539	10′
36°00′	0.62832	0.58779	0.80902	0.72654	1.37638	0.94248	54°00′
10′	0.63123	0.59014	0.80730	0.73100	1.36800	0.93957	50′
20′	0.63414	0.59248	0.80558	0.73547	1.35968	0.93666	40′
30′	0.63705	0.59482	0.80386	0.73996	1.35142	0.93375	30′
40′	0.63995	0.59716	0.80212	0.74447	1.34323	0.93084	20′
50′	0.64286	0.59949	0.80038	0.74900	1.33511	0.92793	10′
37°00′	0.64577	0.60182	0.79864	0.75355	1.32704	0.92502	53°00′
10′	0.64868	0.60414	0.79688	0.75812	1.31904	0.92212	50′
20′	0.65159	0.60645	0.79512	0.76272	1.31110	0.91921	40′
30′	0.65450	0.60876	0.79335	0.76733	1.30323	0.91630	30′
40′	0.65741	0.61107	0.79158	0.77196	1.29541	0.91339	20′
50′	0.66032	0.61337	0.78980	0.77661	1.28764	0.91048	10′
38°00′	0.66323	0.61566	0.78801	0.78129	1.27994	0.90757	52°00′
10′	0.66613	0.61795	0.78622	0.78598	1.27230	0.90466	50′
20′	0.66904	0.62024	0.78442	0.79070	1.26471	0.90175	40′
30′	0.67195	0.62251	0.78261	0.79544	1.25717	0.89884	30′
40′	0.67486	0.62479	0.78079	0.80020	1.24969	0.89594	20′
50′	0.67777	0.62706	0.77897	0.80498	1.24227	0.89303	10′
39°00′	0.68068	0.62932	0.77715	0.80978	1.23490	0.89012	51°00′
10′	0.68359	0.63158	0.77531	0.81461	1.22758	0.88721	50′
20′	0.68650	0.63383	0.77347	0.81946	1.22031	0.88430	40′
30′	0.68941	0.63608	0.77162	0.82434	1.21310	0.88139	30′
40′	0.69231	0.63832	0.76977	0.82923	1.20593	0.87848	20′
50′	0.69522	0.64056	0.76791	0.83415	1.19882	0.87557	10′
40°00′	0.69813	0.64279	0.76604	0.83910	1.19175	0.87266	50°00
10′	0.70104	0.64501	0.76417	0.84407	1.18474	0.86976	50′
20′	0.70395	0.64723	0.76229	0.84906	1.17777	0.86685	40′
30′	0.70686	0.64945	0.76041	0.85408	1.17085	0.86394	30′
40′	0.70977	0.65166	0.75851	0.85912	1.16398	0.86103	20′
50′	0.71268	0.65386	0.75661	0.86419	1.15715	0.85812	10′

Complement		Trigonometric tables					
CoDeg	CoRad	cos	sin	cot	tan	Radians	Degrees

		Trigonometric tables				Complement	
Degree	Radians	sin	cos	tan	cot	CoRad	CoDeg
41°00′	0.71558	0.65606	0.75471	0.86929	1.15037	0.85521	49°00′
10′	0.71849	0.65825	0.75280	0.87441	1.14363	0.85230	50′
20′	0.72140	0.66044	0.75088	0.87955	1.13694	0.84939	40′
30′	0.72431	0.66262	0.74896	0.88473	1.13029	0.84648	30′
40′	0.72722	0.66480	0.74703	0.88992	1.12369	0.84358	20′
50′	0.73013	0.66697	0.74509	0.89515	1.11713	0.84067	10′
42°00′	0.73304	0.66913	0.74314	0.90040	1.11061	0.83776	48°00′
10′	0.73595	0.67129	0.74120	0.90569	1.10414	0.83485	50′
20′	0.73886	0.67344	0.73924	0.91099	1.09770	0.83194	40′
30′	0.74176	0.67559	0.73728	0.91633	1.09131	0.82903	30′
40′	0.74467	0.67773	0.73531	0.92170	1.08496	0.82612	20′
50′	0.74758	0.67987	0.73333	0.92709	1.07864	0.82321	10′
43°00′	0.75049	0.68200	0.73135	0.93252	1.07237	0.82030	47°00′
10′	0.75340	0.68412	0.72937	0.93797	1.06613	0.81740	50′
20′	0.75631	0.68624	0.72737	0.94345	1.05994	0.81449	40′
30′	0.75922	0.68835	0.72537	0.94896	1.05378	0.81158	30′
40′	0.76213	0.69046	0.72337	0.95451	1.04766	0.80867	20′
50′	0.76504	0.69256	0.72136	0.96008	1.04158	0.80576	10′
44°00′	0.76794	0.69466	0.71934	0.96569	1.03553	0.80285	46°00′
10′	0.77085	0.69675	0.71732	0.97133	1.02952	0.79994	50′
20′	0.77376	0.69883	0.71529	0.97700	1.02355	0.79703	40′
30′	0.77667	0.70091	0.71325	0.98270	1.01761	0.79412	30′
40′	0.77958	0.70298	0.71121	0.98843	1.01170	0.79122	20′
50′	0.78249	0.70505	0.70916	0.99420	1.00583	0.78831	10′
45°00′	0.78540	0.70711	0.70711	1.00000	1.00000	0.78540	45°00′

Complement		Trigonometric tables					
CoDeg	CoRad	cos	sin	cot	tan	Radians	Degrees

ANSWERS TO SELECTED ODD EXERCISES & PROBLEMS

CHAPTER 0

SECTION 0-1

Exercises 0-1

1. commutative property of multiplication
3. symmetric property of equality
5. commutative property of addition
7. multiplication property of equality
9. associative property of multiplication
11. additive identity property
13. multiplicative inverse
15. commutative property of addition
17. associative property of addition
19. reflexive property of equality

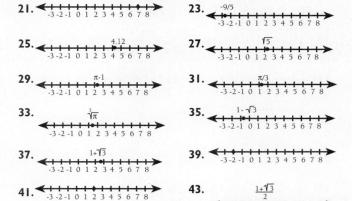

45. not a real number

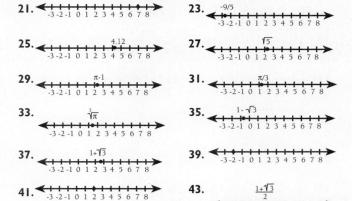

51. $x = 3$ 53. $x = -6$ 55. $x = 2, x = -2$
57. $x = -1, x = 4$

Problem Set 0-1

1. Answers vary but include ($5, $5). 3. -$5 5. $0
7. $8 11. S represents a positive cash flow.
15. The cash flow of ($500, $300) is positive.
19. The cash flow of ($0, $15) is negative. The cash flow of ($8, $0) is positive.
23. ($20, $0) 25. ($6, $2)

SECTION 0-2

Exercises 0-2

1. -5 3. 0 5. -98 7. 7 9. -32 11. $\frac{3}{10}$ 13. $\frac{40}{9}$
15. 5 17. $\frac{25}{12}$ 19. $\frac{5}{2}$ 21. 3.872 23. -3.3 25. 6.1
27. 6.29 29. 5.21 31. 105.744 33. 14.626 35. 2.951
37. 16.290 39. -1.877 41. 25 43. $\frac{4}{13}$ 45. 44.814
47. 2.065 49. not a real number 51. 3.6502E6
53. 3.84E-4 55. -4.567E8 57. 13400 59. -0.0000134
61. 1,010,000,000,000 63. 2.1875 inches
65. 2.275 inches 67. 2.577 seconds 69. $15,840,000/mile

SECTION 0-3

Exercises 0-3

1. x^3y^2 3. x^4y^5 5. $x^{5/6}$ 7. $\dfrac{x^7}{y^{10}}$ 9. $\dfrac{1}{y^{7/12}}$ 11. $x^{10}y^3$

13. $\dfrac{y^7}{x^3}$ **15.** $\dfrac{(y+x)}{x}$ **17.** 5   **19.** -3   **21.** xy^2z^3

23. $\dfrac{y\sqrt{xy}}{z^2}$ **25.** $\dfrac{y\sqrt{xyz}}{z^3}$ **27.** $\dfrac{\sqrt{3}}{3}$ **29.** $3\sqrt{x}$

31. $12\sqrt{x}-2\sqrt{y}$ **33.** $9\sqrt{x}+6\sqrt[3]{x^2}$ **35.** $\dfrac{-2\sqrt{x}}{x}$

37. $\dfrac{3-5\sqrt{x}}{x}$ **39.** $5\sqrt{3}+5\sqrt{2}$ **41.** $\sqrt{x}+\sqrt{2}$ **43.** 75

45. 10   **47.** $3x^3y^4$ **49.** $\dfrac{9x^3y^2\sqrt{z}}{z^2}$ **51.** False   **53.** True

55. False   **57.** True   **59.** False   **61.** False   **63.** True

65. False   **67.** False   **69.** False   **71.** $\dfrac{3}{5(\sqrt{7}+2)}$

73. $\dfrac{3}{7-2\sqrt{10}}$ **75.** $\dfrac{1}{\sqrt{x}+\sqrt{3}}$ **77.** $\dfrac{1}{\sqrt{x+h}+\sqrt{x}}$

79. $\dfrac{1}{\sqrt[3]{x^2}+\sqrt[3]{2x}+\sqrt[3]{4}}$ **81.**

83.   **85.**

87.   **89.**

91. $\sqrt[3]{x^2}$ **93.** $3\sqrt{x}$ **95.** $\sqrt{3x}$ **97.** $\sqrt[3]{x^2y}$

99. $\dfrac{y\sqrt[3]{x^2}+\sqrt[3]{y^2}}{y}$ **101.** $x^{1/2}$ **103.** $x^{3/4}y^{1/4}$ **105.** $x^{1/6}$

107. $x^{1/2}$ **109.** $(1+x^{1/2})^{1/2}$

SECTION 0-4

Exercises 0-4

1. $10x+1$ **3.** x^2+5x-6 **5.** $x^2-2x-15$
7. $x^2-2x-15$ **9.** $x^3-2x^2-13x+6$ **11.** $4x^2-49$
13. $9x^2-24x+16$ **15.** x^3-27 **17.** $4x^2-20x+25$
19. $10x^2+21x-10$ **21.** $5(x-3)$ **23.** $(x+4)(x-4)$
25. prime over the integers, prime over the real numbers
27. prime over the integers: $(x+\sqrt{7})(x-\sqrt{7})$
29. $(x+8)(x-2)$ **31.** $3(x+3)(x-3)$
33. $3(x+1)(x-5)$ **35.** $(x-3)(x^2+3x+9)$
37. $(x+11)(x-1)$
39. prime over the integers: $(x-5+2\sqrt{2})(x-5-2\sqrt{2})$
41. $(x+5)(x-3)$ **43.** $(x-2)(x-3)$ **45.** $(x-3)(x+2)$
47. $(x-4)(x-3)$ **49.** $(x-4)(x+3)$ **51.** $(x+6)(x-2)$
53. $(x-12)(x+1)$
55. prime over the integers: $(x-3+\sqrt{21})(x-3-\sqrt{21})$
57. $2(x+3)(x+4)$ **59.** $(2x+3)(x-8)$
61. $2(x+4)(x-3)$

63. prime over the integers: $(x-4+\sqrt{15})(x-4-\sqrt{15})$
65. prime over the integers: $(x+2+\sqrt{11})(x+2-\sqrt{11})$
67. prime over the integers:

$$\left(x+\frac{5}{2}+\frac{\sqrt{29}}{2}\right)\left(x+\frac{5}{2}-\frac{\sqrt{29}}{2}\right)$$

69. prime over the integers, prime over the real numbers
71. $(x-1)(x^2+x+1)$ **73.** $(3x+2)(9x^2-6x+4)$
75. $(x^2+1)(x^4-x^2+1)$
77. $(x-2)(x+2)(x^2+2x+4)(x^2-2x+4)$
79. $(a+2y)(3x-5b)$ **81.** $(x+7)(2x-3)$ **83.** $x(x-1)$
85. $x(x-4)(x-1)$ **87.** $(x+2)(x-2)(x+1)(x-1)$
89. $(y+x-3)(y-x+3)$
91. (a) $(x-3)(x-4)$ **(b)** $x=3$ or $x=4$
93. (a) $(x+1)(x+2)(x-2)$ **(b)** $x=-1$ or $x=-2$ or $x=2$
95. (a) $5x^2(x+1)(x-1)$ **(b)** $x=0$ or $x=1$ or $x=-1$
97. (a) $(2x-5)(2x-1)$ **(b)** $x=\frac{5}{2}$ or $x=\frac{1}{2}$
99. (a) $(x-1)(x-2)(x+3)(x-3)$
    (b) $x=1$ or $x=2$ or $x=-3$ or $x=3$
101. $x^{-1/2}(1+x)$ **103.** $\sqrt[3]{x}(\sqrt[3]{x}+1)$
105. $(x^{1/3}-2)(x^{1/3}+1)$ **107.** $x^{-1/3}(1+x)$
109. $x^{-1/2}y^{-1/2}(x+y)$

SECTION 0-5

Exercises 0-5

1. $\dfrac{4}{9}$ **3.** $\dfrac{(x-3)}{2}, x\neq -3$ **5.** $\dfrac{(x-3)}{(x+4)}, x\neq -4$

7. $\dfrac{3(x+1)}{(x-3)}, x\neq -3, x\neq -2$

9. $\dfrac{5(x-2)}{(x-4)(x-5)}, x\neq -4, x\neq 5$ **11.** $\dfrac{(3x+1)}{x^2}$

13. $\dfrac{(5-x)}{[(x+1)(x-1)]}$ **15.** $\dfrac{1}{(x-3)}, x\neq -3$

17. $1, x\neq 4, x\neq -2$ **19.** $\dfrac{2(x^2-x+6)}{3(x+2)^2}, x\neq 4$

21. $\dfrac{(x-3)}{3}, x\neq 0, x\neq -3$ **23.** $\dfrac{(x-5)}{5}, x\neq -5, x\neq 0$

25. $\dfrac{(x^2-2)}{2}, x\neq 0$ **27.** $\dfrac{2x}{15}, x\neq 0, x\neq -3, x\neq -5$

29. $\dfrac{-5(x+5)}{(x-5)}, x\neq 0, x\neq -5$ **31.** $x+3, x\neq 3, x\neq -3$

33. $2x+h, h\neq 0$ **35.** $x^2+3x+9, x\neq 3$
37. $3x^2-3xh+h^2, h\neq 0$ **39.** $x^3+3x^2+9x+27, x\neq 3$

SECTION 0-6

Exercises 0-6

1. 5.7   **3.** 9   **5.** 17   **7.** $\pi - 3 \cong 0.1416$ **9.** an identity
11. false   **13.** false   **15.** true   **17.** true   **19.** true

21. (number line, -3 to 8)   **23.** (number line, -7 to 4)

25. (number line, -3 to 8)   **27.** (number line, -3 to 8)

29. (number line, -3 to 8)   **31.** (number line, -3 to 8)

33. (number line, -3 to 8)   **35.** (number line, -3 to 8)

37. (number line, -3 to 8)   **39.** ϕ. No graph.

41. (number line, -3 to 8)   **43.** (number line, -3 to 8)

45. (number line, -3 to 8, π)   **47.** (number line, -3 to 8)

49. (number line, -7 to 4)

51. $(1, \infty)$ **53.** $(-2, \infty)$ **55.** $[-\frac{1}{2}, \infty)$ **57.** $[3, \infty)$

59. $\left(-\infty, \dfrac{\pi}{2}\right)$ **61.** $[-\pi, \pi]$ **63.** $\left[-\dfrac{\pi}{2}, \dfrac{3\pi}{2}\right]$

65. $\left[-\dfrac{3\pi}{4}, \dfrac{5\pi}{4}\right]$ **67.** $\left[-\dfrac{\pi}{4}, \dfrac{11\pi}{20}\right]$ **69.** $[2 - 4\pi, 4\pi + 2]$

81. $(-4, 4)$ **83.** $(-2, 4)$ **85.** $[3, 7]$ **87.** $[-11, -1]$

89. $(-7, 13)$ **91.** $|x| < 5$ **93.** $|x - 3| \leq 1$ **95.** $|x - 5| < 9$

97. $|x - 5| \leq 5$ **99.** $\left|x - \dfrac{7}{2}\right| < \dfrac{9}{2}$

SECTION 0-7

Exercises 0-7

1. -1   **3.** 1   **5.** -1   **7.** 1   **9.** -i   **11.** -1   **13.** i   **15.** -i

17. 1   **19.** -i   **21.** $1 + 3i$ **23.** 13   **25.** $\dfrac{(5 + 12i)}{13}$

27. $\dfrac{(5 - 12i)}{13}$ **29.** $9 - 8i$ **31.** $2i$ **33.** $3i\sqrt{2}$

35. $\sqrt{3} - 5i$ **37.** $-5 - 3i\sqrt{2}$ **39.** $\dfrac{3}{2} + \dfrac{i\sqrt{7}}{2}$

41. $z + \bar{z} = 6, z\bar{z} = 25$ **43.** $z + \bar{z} = 4, z\bar{z} = 13$
45. $z + \bar{z} = 0, z\bar{z} = 25$ **47.** $z + \bar{z} = -10, z\bar{z} = 25$
49. $z + \bar{z} = -3, z\bar{z} = \frac{7}{2}$ **77.** $x = 4, y = \frac{1}{2}$
79. $x = 4, y = 1$

CHAPTER 0 REVIEW EXERCISES

1. associative property of addition   **3.** distributive property
5. commutative property of addition   **11.** -6   **13.** 28

15. 12   **17.** 7   **19.** $\dfrac{5}{2}$ **21.** 10.1   **23.** 4.01   **25.** $x^9 y^9$

27. $\dfrac{5}{4}$ **29.** $\dfrac{xy^2\sqrt{yz}}{z^2}$ **31.** $6\sqrt{x} - 3\sqrt{y}$

33. $\sqrt{x} + \sqrt{7}, x \neq 7$
35. $\dfrac{7x^3 y^2\sqrt{z}}{z^3}$ **37.** $\dfrac{1}{\sqrt{x} + \sqrt{5}}, x \neq 5$

39. (number line, -7 to 4, $\frac{5 - \sqrt{7}}{2}$)

41. $\sqrt[4]{x^3}$ **43.** $x^{2/3}$ **45.** $x^{2/5}$ **47.** $9x + 2$ **49.** $x^2 - x - 56$
51. $5(x - 5)$ **53.** $(x + 1)(x - 11)$ **55.** $(x + 15)(x + 1)$
57. $(x - 1)(x - 6)$ **59.** $(x + 4)(x - 3)$
61. $2(x - 4)(x - 3)$
63. prime over the integers: $(x + 2 + \sqrt{6})(x + 2 - \sqrt{6})$
65. $(2x + 3)(4x^2 - 6x + 9)$ **67.** $2(y - z)(5x + 2a)$
69. $(x - 2)(x - 5)$ **71.** $(x + 2)(x - 2)(x + 1)(x - 1)$
73. (a) $(x - 6)(x - 2)$ (b) $x = 6$ or $x = 2$
75. (a) $(x + 1)(x - 2)(x + 3)(x - 3)$
  (b) $x = -1$ or $x = 2$ or $x = -3$ or $x = 3$
77. $\dfrac{\sqrt{x}(1 + x)}{x}$ **79.** $\dfrac{x - 2}{3}, x \neq -2$

81. $\dfrac{3(x + 1)}{x + 2}, x \neq -2, x \neq 2, x \neq -3$ **83.** $\dfrac{3 + x}{x^2}$

85. $\dfrac{2}{x - 3}, x \neq 3, x \neq -3$

87. $\dfrac{2(x^2 - 6)}{3(x - 2)(x + 2)}, x \neq 3, x \neq 2, x \neq -2$
89. $-x - 5, x \neq 5, x \neq 0$

91. $\dfrac{-14x + 78}{3}, x \neq 0, x \neq 3, x \neq -5$ **93.** $x + 5, x \neq 5$

95. $x^2 + x + 1, x \neq 1$ **97.** false   **99.** false, if x is negative.
111. $(1, \infty)$ **113.** $\left(-\infty, \dfrac{\pi}{3}\right)$ **115.** $\left[\dfrac{-\pi}{3}, \pi\right]$ **117.** $(-2, 2)$

119. $[14, 28]$ **121.** $|x| < 15$ **123.** $|x - 4| < 10$
125. -1   **127.** -i   **129.** i   **131.** -i   **133.** 1   **135.** $1 + 3i$

137. $\dfrac{21 + 20i}{29}$ **139.** $\dfrac{34 - 27i}{5}$ **141.** $7i$

143. $\sqrt{13} - 5i\sqrt{5}$ **145.** $-\dfrac{3}{2} + \dfrac{\sqrt{3}}{2}i$

147. $z + \bar{z} = 10$, $z\bar{z} = 169$
149. $z + \bar{z} = 8$, $z\bar{z} = 16$ **151.** $z + \bar{z} = -5$, $\bar{z} = 8$

CHAPTER 0 TEST

1. (a) -20   **(b)** $2x^2 - 2x - 12$ **3.** $\dfrac{3x\sqrt[3]{3x^2y^2z}}{yz^2}$

5. (a) $x(x - 1)$ **(b)** $(2x + 3)(2x - 5)$ **7.** $(-7, 15)$

9. $\dfrac{7 - 26i}{25}$

CHAPTER 1

SECTION 1-1

Exercises 1-1

1-19. See the figure below.   **21. (a)** 13 **(b)** $\left(4, \frac{1}{2}\right)$ **(c)** $-\frac{5}{12}$

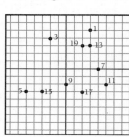

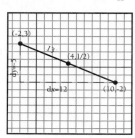

23. (a) 5 **(b)** $\left(1, \frac{11}{2}\right)$ **(c)** $-\frac{3}{4}$ **25. (a)** $\sqrt{2}$ **(b)** $\left(\frac{1}{2}, -\frac{1}{2}\right)$ **(c)** -1

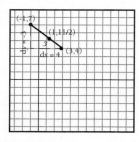

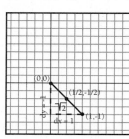

27. (a) $\sqrt{29}$**(b)** $\left(\frac{1}{2}, -1\right)$**(c)** $-\frac{2}{5}$ **29. (a)** $3\sqrt{a^2 + b^2}$

(b) $\left(\dfrac{5a}{2}, \dfrac{3b}{2}\right)$ **(c)** $-\dfrac{b}{a}$

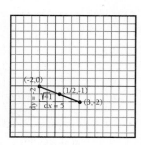

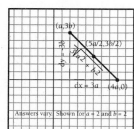

31. $d(P_1P_2) = 5\sqrt{2}$,
$d(P_2P_3) = 5$,
$d(P_1P_3) = 5$;
triangle is isosceles.

33. $d(P_1P_2) = 6$,
$d(P_2P_3) = 6\sqrt{3}$,
$d(P_1P_3) = 6$;
triangle is isosceles.

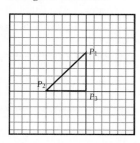

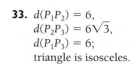

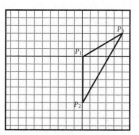

35. $d(P_1P_2) = 2$,
$d(P_2P_3) = 2$,
$d(P_1P_3) = 2\sqrt{2 - \sqrt{2}}$;
triangle is isosceles.

37. $d(P_1P_2) = 2\sqrt{2}$,
$d(P_2P_3) = \sqrt{17}$,
$d(P_1P_3) = \sqrt{13}$;
triangle is scalene.

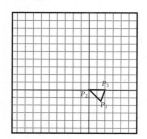

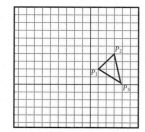

39. $d(P_1P_2) = \sqrt{10}$,
$d(P_1P_2) = \sqrt{1.25}$,
$d(P_1P_2) = \sqrt{16.25}$;
triangle is scalene.

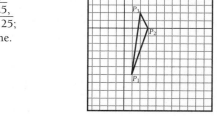

41. $\dfrac{\Delta y}{\Delta x}(P_1P_2) = 1$, $\dfrac{\Delta y}{\Delta x}(P_2P_3) = 1$; P_1, P_2, and P_3 are collinear.

43. $\dfrac{\Delta y}{\Delta x}(P_1P_2) = 3$, $\dfrac{\Delta y}{\Delta x}(P_2P_3) = 3$; P_1, P_2, and P_3 are collinear.

45. $\dfrac{\Delta y}{\Delta x}(P_1P_2) = \dfrac{-3}{2}$, $\dfrac{\Delta y}{\Delta x}(P_2P_3) = \dfrac{-4}{5}$; P_1, P_2, and P_3 are not collinear.

47. $\dfrac{\Delta y}{\Delta x}(P_1P_2) = 4$, $\dfrac{\Delta y}{\Delta x}(P_2P_3) = \dfrac{10.8}{9.7}$; P_1, P_2, and P_3 are not collinear.

49. $\dfrac{\Delta y}{\Delta x}(P_1P_2) = b + a$, $\dfrac{\Delta y}{\Delta x}(P_2P_3) = b + a$; P_1, P_2, and P_3 are collinear.

51. $d = 500$ ft

53. $d(P_1P_2) = 4\sqrt{2}, d(P_2P_3) = 8, d(P_1P_3) = 4\sqrt{2}$; Since $(4\sqrt{2})^2 + (4\sqrt{2})^2 = 8^2$, the triangle is a right triangle.

55. $d(P_1P_2) = 2\sqrt{2}, d(P_2P_3) = 2\sqrt{2}, d(P_1P_3) = 4, d(P_2P_4) = 4, d(P_3P_4) = 2\sqrt{2}$; Since $(2\sqrt{2})^2 + (2\sqrt{2})^2 = 4^2$, the quadrilateral is a rectangle. Alternately, since the diagonals are equal, the quadrilateral is a rectangle.

57. $M_1(P_1P_2) = \left(\frac{3}{2}, 2\right), M_2(P_1P_3) = \left(-\frac{1}{2}, \frac{5}{2}\right), d(M_1M_2) = \frac{\sqrt{17}}{2}$;

Thus, $d(P_2P_3) = \sqrt{17}, M_1M_2 = \frac{P_2P_3}{2}$

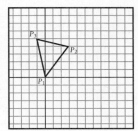

59. $x = 13$ **61.** $\dfrac{31 \text{ ft}}{1 \text{ sec}}$ **63.** $\dfrac{1}{20}$ **65.** $\dfrac{5.1 \text{ m}}{1 \text{ sec}^2}$

67. 10 ohms   **69.** $\dfrac{2000 \text{ flops}}{1 \text{ sec}}$

71. $d = \sqrt{(x_2 - x_1)^2 + (y_2 - y_1)^2}$ **75.** $d = \sqrt{(\Delta x)^2 + (\Delta y)^2}$

Problem Set 1-1

3. $(a + b, h)$

SECTION 1-2

Exercises 1-2

1. $\cong 105.744$ **3.** $\cong 14.626$ **5.** $\cong 2.951$ **7.** $\cong 16.290$
9. $\cong -1.878$

Problem Set 1-2

1. As a increases the line becomes steeper.

3. The graphs of $y = ax$ and $y = -ax$ are mirror images about the y axis.

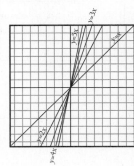

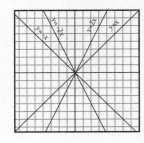

5. As $|a|$ increases the steepness of the line increases. As $|d|$ increases the steepness of the line decreases.

7. The coefficient a controls the steepness of the graph. The constant b controls where the graph crosses the y axis.

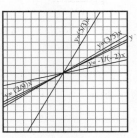

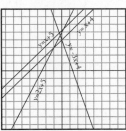

43.

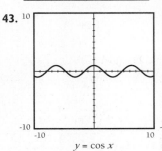

$y = \cos x$

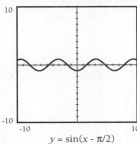

$y = \sin(x - \pi/2)$

45.

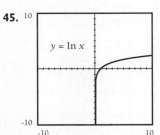

$y = \ln x$

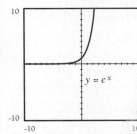

$y = e^x$

47.

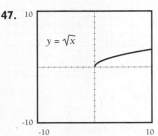

$y = \sqrt{x}$

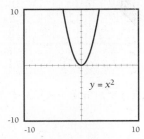

$y = x^2$

49.

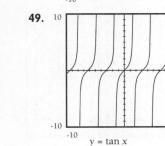

$y = \tan x$

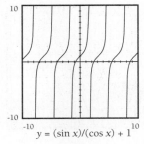

$y = (\sin x)/(\cos x) + 1$

SECTION 1-3

Exercises 1-3

1. $x = 3$ **3.** $x = -4$ **5.** $y = 1$ **7.** $y = -3$

9. $y = -x - 1$ **11.** $y = 9x - 29$ **13.** $y = -\frac{1}{7}x + \frac{26}{7}$

15. $y = -\frac{16}{9}x + \frac{5}{9}$ **17.** $y = mx$

19. $y = -(a + b)x + a^2 + ab + b^2$ **21.** $y = 2x + 3$

23. $y = 3x - 2$ **25.** $y = \frac{1}{2}x + \frac{7}{2}$ **27.** $y = -\frac{3}{5}x - \frac{14}{5}$

29. $y = 2$ **31.** **(a)** $y = -\frac{3}{2}x + 4$ **(b)** $m = -\frac{3}{2}$

   (c) x-intercept $= \frac{8}{3}$ **(d)** y-intercept $= 4$

31.

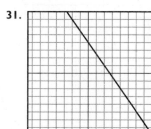

33. **(a)** $y = 3x + 2$
   (b) $m = 3$
   (c) x-intercept $= -\frac{2}{3}$
   (d) y-intercept $= 2$

35. **(a)** $y = -2x + \frac{5}{3}$
   (b) $m = -2$
   (c) x-intercept $= \frac{5}{6}$
   (d) y-intercept $= \frac{5}{3}$

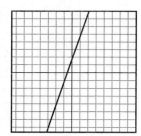

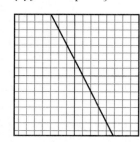

37. **(a)** $y = \frac{5}{2}x - \frac{3}{2}$
   (b) $m = \frac{5}{2}$
   (c) x-intercept $= \frac{3}{5}$
   (d) y-intercept $= -\frac{3}{2}$

39. **(a)** $y = \frac{3}{4}x - \frac{3}{8}$
   (b) $m = \frac{3}{4}$
   (c) x-intercept $= \frac{1}{2}$
   (d) y-intercept $= -\frac{3}{8}$

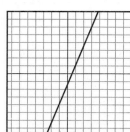

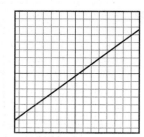

41. $y = -3x + 4$

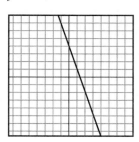

43. $y = -\frac{3}{2}x + 3$

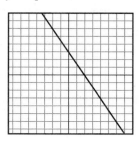

45. $y = \frac{1}{3}x - 3$

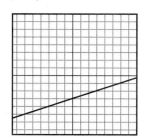

47. $y = -\frac{3}{5}x + 7$

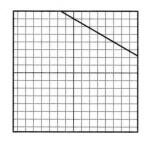

49. $y = 5x - 2$

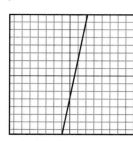

51. $x = 5$

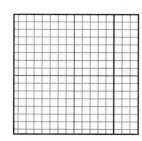

53. $y = -3$

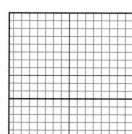

55. $y = -2x + \frac{5}{3}$

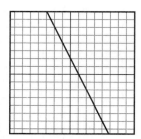

57. $y = \frac{5}{2}x - \frac{3}{2}$ **59.** $y = \frac{3}{4}x - \frac{3}{8}$

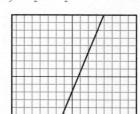 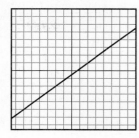

61. x-intercept : $(-2, 0)$, y-intercept : $(0, -1)$; $y = -\frac{1}{2}x - 1$
63. x-intercept : $(2, 0)$, y-intercept : none; $x = 2$
65. x-intercept : $(3, 0)$, y-intercept : $(0, -4)$; $y = \frac{4}{3}x - 4$
67. x-intercept: none, y-intercept : $(0, 7)$; $y = 7$
69. x-intercept : $(2, 0)$, y-intercept : $(0,1)$; $y = -\frac{1}{2}x + 1$

79.

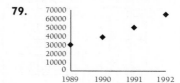

81. $\dfrac{\Delta y}{\Delta x} = \$11{,}000/\text{yr}$

Problem Set 1-3
1. (b) Graph is a semicircle of radius a centered at the origin.
3. (b) The pair of graphs form a circle of radius a centered at the origin.

SECTION 1-4

Exercises 1-4
1. $(x - 5)^2 + (y - 2)^2 = 9$ **3.** $(x + 3)^2 + (y - 1)^2 \cong 2.534$
5. $(x - 5)^2 + (y - 0)^2 = 25$ **7.** $x^2 + y^2 = 25$
9. $(x + 2)^2 + (y - 5)^2 = 225$
11. center $(0, 2)$, radius 3

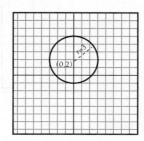

13. center $(-3, 5)$, radius 6   **15.** center $(-5, 0)$, radius $\sqrt{6}$

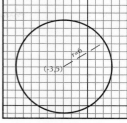 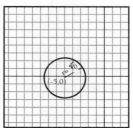

17. center $(-2, 3)$, radius 4   **19.** center $\left(\frac{3}{2}, -\frac{5}{2}\right)$, radius $\dfrac{\sqrt{38}}{2}$

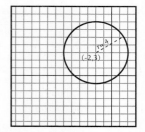 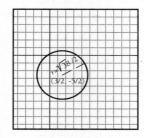

21. circle   **23.** line   **25.** neither   **27.** no graph
29. neither

31.   **33.**

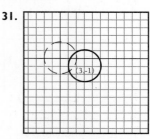 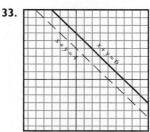

35.   **37.**

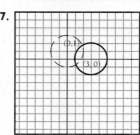

39.

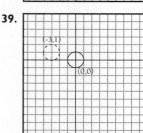

41. resulting equations are "simpler"
43. translates graph "down" 13 units
45. translates center from (0, 0) to (1, 3)
47. translates graph 3 units left and 1 unit up
49. translates graph down 2 units

Problem Set 1-4

1. (b) As $|a|$ increases the "width" of the graph becomes narrower.
3. (b) If a is positive the graph "opens" up. If a is negative the graph "opens" down.
5. (b) The constant h controls the horizontal translation of the graph.
7. (b) (h, k) is a point at which the graph "changes direction." The coefficient a controls the direction the graph opens and the graph "width."

SECTION 1-5

Exercises 1-5

1. Parabola opens up; the vertex is a minimum at (1, 3).

3. Parabola opens down; the vertex is a maximum at (-3, 8).

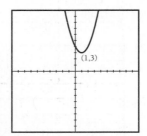
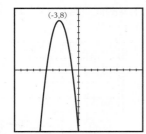

5. Parabola opens down; the vertex is a maximum at (-5, 0).

7. Parabola opens up; the vertex is a minimum at (0, 2).

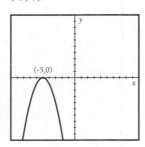
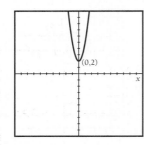

9. Parabola opens up; the vertex is a minimum at (2, 1.2).

11. Parabola opens up; the vertex is a minimum at (-3, -4).

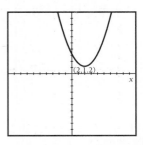
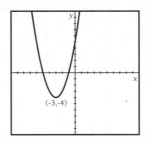

13. Parabola opens up; the vertex is a minimum at (4, -16).

15. Parabola opens down; the vertex is a maximum at $\left(\frac{3}{2}, \frac{7}{2}\right)$.

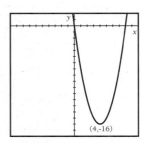
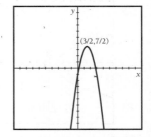

17. Parabola opens up; the vertex is a minimum at $\left(-\frac{5}{4}, \frac{7}{8}\right)$.

19. Parabola opens down; the vertex is a maximum at (0, 7).

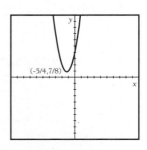
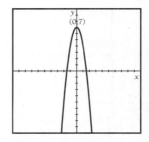

21. Parabola opens right; the vertex at (-1, -3) minimizes x.

23. Parabola opens left; the vertex at (4, 2) maximizes x.

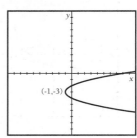
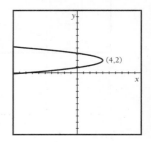

25. Parabola opens right; the vertex at $(-3, 0)$ minimizes x.

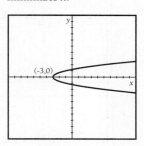

27. Parabola opens right; the vertex at $(0, 2)$ minimizes x.

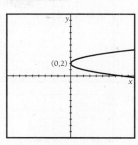

41.

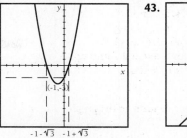

43.

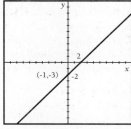

29. Parabola opens left; the vertex at $(-1, -\frac{1}{2})$ maximizes x.

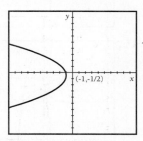

31. Parabola opens right; the vertex at $(-1, 2)$ minimizes x.

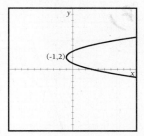

45.

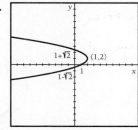

47.

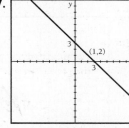

33. Parabola opens right; the vertex at $(-25, -5)$ minimizes x.

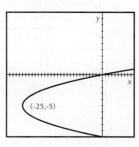

35. Parabola opens left; the vertex at $(19, -4)$ maximizes x.

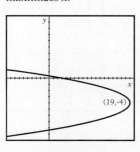

49.

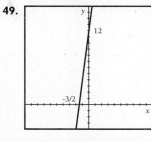

51.

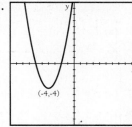

53.

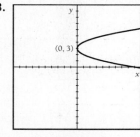

55.

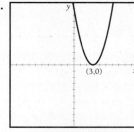

37. Parabola opens right; the vertex at $\left(\frac{11}{12}, \frac{5}{6}\right)$ minimizes x.

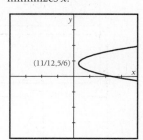

39. Parabola opens left; the vertex at $(6, 0)$ maximizes x.

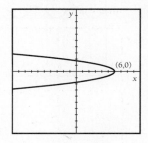

57.

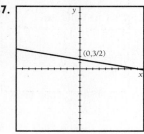

59.

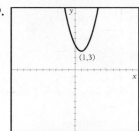

73. −2

75.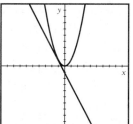

CHAPTER 1 REVIEW EXERCISES

1–9.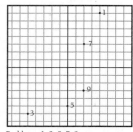

Problems: 1, 3, 5, 7, 9

11. **(a)** $d = \sqrt{340} \cong 18.349$ **(b)** Midpoint is $(4, 2)$.

(c) $\dfrac{\Delta y}{\Delta x} = \dfrac{-6}{7}$

13. **(a)** $d = 5$ **(b)** Midpoint is $\left(2, -\dfrac{3}{2}\right)$. **(c)** $\dfrac{\Delta y}{\Delta x} = \dfrac{-3}{4}$

15. **(a)** $d = \sqrt{16a^2 + 9b^2}$ **(b)** Midpoint is $\left(3a, \dfrac{3b}{2}\right)$.

(d) $\dfrac{\Delta y}{\Delta x} = \dfrac{-3b}{4a}$

17. $d(P_1P_2) = 7\sqrt{2}$, $d(P_2P_3) = 7$, $d(P_1P_3) = 7$; triangle is isoceles.

19. $d(P_1P_2) = 2$, $d(P_2P_3) = 1$, $d(P_1P_3) = \sqrt{5 - 2\sqrt{2}}$; triangle is scalene.

21. $d(P_1P_2) = 5$, $d(P_2P_3) = \sqrt{7.25}$, $d(P_1P_3) = \sqrt{18.25}$; triangle is scalene.

23. $\dfrac{\Delta y}{\Delta x}(P_1P_2) = 1$, $\dfrac{\Delta y}{\Delta x}(P_2P_3) = 1$: points are collinear.

25. $\dfrac{\Delta y}{\Delta x}(P_1P_2) = -2$, $\dfrac{\Delta y}{\Delta x}(P_2P_3) = -\dfrac{6}{5}$: points are not collinear.

27. $\dfrac{\Delta y}{\Delta x}(P_1P_2) = \dfrac{a^2 + b^2}{a} - b$, $\dfrac{\Delta y}{\Delta x}(P_1P_2) = -a - b$: points are not collinear.

29. $d(P_1P_2) = 5$, $d(P_2P_3) = 10$, $d(P_1P_3) = 5\sqrt{5}$.
Since $5^2 + 10^2 = (5\sqrt{5})^2$ the triangle is a right triangle.

31. $x = \dfrac{22}{3}$ **33.** $\dfrac{\Delta y}{\Delta x} = \dfrac{\$20}{1\ \text{kg}}$ **35.** $\dfrac{\Delta y}{\Delta x} = 20$ ohms

37.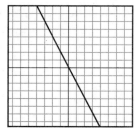

39.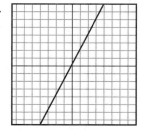

41.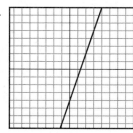

43.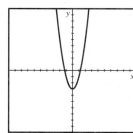

45.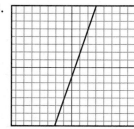

47.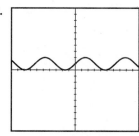

49.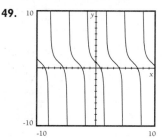

51.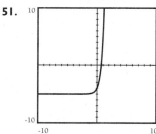

53. (a)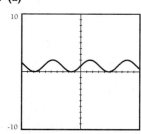

53. (b)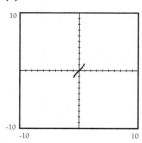

55. (a)

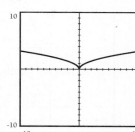

55. (b)

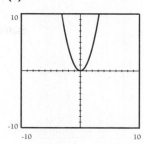

57. $y = 7$ **59.** $x = 5$ **61.** $y = -x$ **63.** $y = -\frac{19}{5}x + \frac{111}{5}$

65. $y = -\frac{10}{27}x - \frac{4}{27}$ **67.** $y = -2x + 5$ **69.** $y = \frac{1}{2}x + 6$

71. (a) $y = -\frac{4}{3}x + 4$ **(b)** $m = -\frac{4}{3}$ **(c)** x-intercept $= 3$

(d) y-intercept $= 4$ **73. (a)** $y = \frac{2}{3}x - \frac{5}{3}$ **(b)** $m = \frac{2}{3}$

(c) x-intercept $= \frac{5}{2}$ **(d)** y-intercept $= -\frac{5}{3}$

75. (a) $y = -\frac{3}{4}x + \frac{3}{8}$ **(b)** $m = -\frac{3}{4}$ **(c)** x-intercept $= \frac{1}{2}$

(d) y-intercept $= \frac{3}{8}$

77.

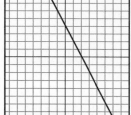

79.

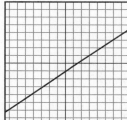

81.

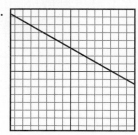

83.

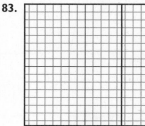

85.

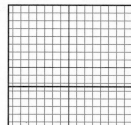

87.

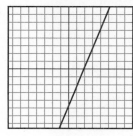

89. $(x - 3)^2 + (y - 4)^2 = 25$ **91.** $x^2 + y^2 = 169$

93. center $(0, 3)$, radius $= 5$ **95.** center $(0, -4)$, radius $= 5$

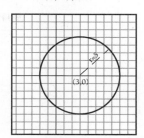

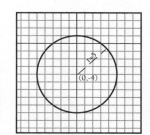

97.

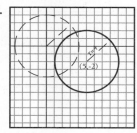

99.

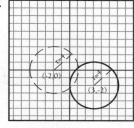

101. Parabola opens up; vertex at $(2,5)$ is a minimum.

103. Parabola opens down; vertex at $(-3, 0)$ is a maximum.

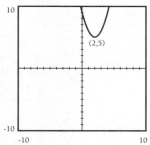

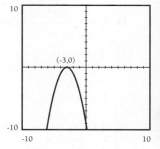

105. Parabola opens up; vertex at $(1, 1.2)$ is a minimum.

107. Parabola opens up; vertex at $(3, -9)$ is a minimum.

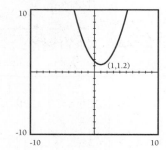

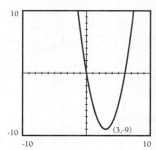

109. Parabola opens up; vertex at $\left(-\frac{5}{4}, -\frac{17}{8}\right)$ is a minimum.

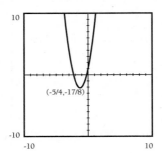

113. Parabola opens right; vertex at (-2, 0) minimizes x.

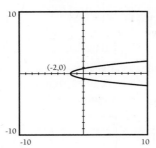

117. Parabola opens right; vertex at (-81, -9) minimizes x.

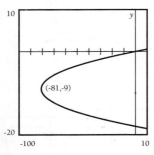

121.

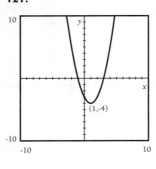

111. Parabola opens right; vertex at (-3, -2) minimizes x.

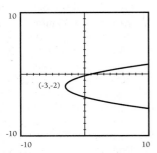

115. Parabola opens left; vertex at $\left(2, \frac{1}{2}\right)$ maximizes x.

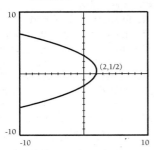

119. Parabola opens right; vertex at $\left(-\frac{1}{8}, \frac{5}{4}\right)$ minimizes x.

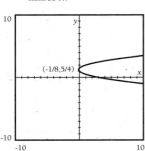

123.

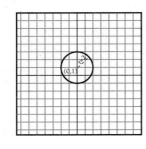

125.

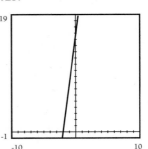

127.

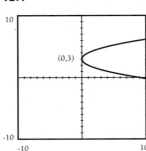

129.

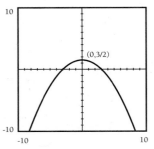

CHAPTER 1 TEST

1. i. $d = 162$, Midpoint is (3, -4). $\frac{\Delta y}{\Delta x} = -1$

1. ii. $\frac{\Delta y}{\Delta x}(P_1P_3) = \frac{4}{3}$, $\frac{\Delta y}{\Delta x}(P_2P_4) = -1$; Opposite sides are not parallel: not a rectangle.

3. (a) $y = 3x - 1$ **(b)** $y = -3x + 7$

5.

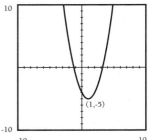

7.

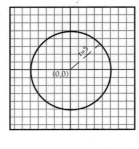

9. The equation has no graph.

CHAPTER 2

SECTION 2-1

Exercises 2-1

1. function **3.** not a function **5.** function

7. not a function   **9.** function   **11.** $f(x) = -\frac{3}{2}x + \frac{5}{2}$

13. $f(x) = \frac{1}{3}x^2 + \frac{2}{3}$ **15.** $f(x) = \frac{10}{x}$ **17.** $f(x) = \frac{x}{(3x + 1)}$

19. $f(x) = \frac{5}{(x^2 - x)}$ **21.** $(-\infty, \infty)$ **23.** $(-\infty, \infty)$ **25.** $[0, \infty)$

27. $(-\infty, \infty)$ **29.** $(-\infty, \frac{2}{3}]$ **31.** $(-\infty, 2) \cup (2, \infty)$

33. $(-\infty, -3) \cup (-3, 3) \cup (3, \infty)$ **35.** $(-7, \infty)$

37. $[0, \infty)$ **39.** $(0, \infty)$ **41.** $f(4) = \frac{65}{4}$ **43.** $f(2) = \frac{9}{2}$

45. $f(0)$ is undefined.   **47.** $f(-2) = \frac{7}{2}$ **49.** $f(-4) = \frac{63}{4}$

51. $g(-3) = -24$ **53.** $g(-1) = 0$ **55.** $g(0.5) = -0.375$
57. $g(a) = a^3 - a$
59. $g(a + b) = a^3 + 3a^2b + 3ab^2 + b^3 - a - b$ **61.** $f(5) = 16$
63. $f(5) = 3$ **65.** $f(5) = \frac{5}{7}$ **67.** $f(5) = 4$

69. $f(5) = \frac{9\sqrt{5}}{5}$ **71.** $f(5) = \sqrt[4]{72}$ **73.** $f(w) = 5w - w^2$

75. $f(z + 3) = -z^2 - z + 6$
77. $f(x + t) = 5x + 5t - x^2 - 2xt - t^2$ **79.** $f(3t) = 15t - 9t^2$
81. $f(-x) = -5x - x^2$ **83.** $f(x^2) = 5x^2 - x^4$
85. $f(x + h) = 5h - x^2 - 2xh - h^2$

87. $\dfrac{f(x + h) - f(x)}{h} = 5 - 2x - h$

89. $\dfrac{f(x) - f(x_0)}{(x - x_0)} = 5 - x - x_0$ **91.** $f(-x) = -5x$

93. $f(-x) = x^4 - 3x^2 + 5$ **95.** $f(-x) = -x^{15}$

97. $\dfrac{f(x + h) - f(x)}{h} = 2x + h$

99. $\dfrac{f(x + h) - f(x)}{h} = 2 - 6x - 3h$

101. $\dfrac{f(x + h) - f(x)}{h} = 5$

103. $\dfrac{f(x + h) - f(x)}{h} = 3x^2 + 3xh + h^2$

105. $\dfrac{f(x + h) - f(x)}{h} = \dfrac{\sqrt{x + h} - \sqrt{x}}{h}$

107. $f(z) = \frac{5}{2}z - 3$

109. $f(w) = \frac{2}{5}w + \frac{6}{5}$ **111.** $f(y) = -\frac{2}{3}y + 3$

113. $f(t) = -16t^2 + 120t$

115. $f(d) = \dfrac{15 + \sqrt{225 - d}}{4}$ or $f(d) = \dfrac{15 - \sqrt{225 - d}}{4}$

117. $c = f(n) = \$2.25n$

Problem Set 2-1
1. **(a)** $y = 5$ feet   **(b)** $y = f(x) = x$ feet
   (c) $x = (2 \text{ feet/sec})(t \text{ sec}) = 2t$ feet   **(e)** $y = f(t) = 2t$
   (g) $t \geqslant 0$ **(h)** $y \geqslant 0$
3. **(a)** $w = (1000 \text{ gm/m})(3.5 \text{ m}) = 3500$ gm
   (b) $f(x) = 10 - x, 0 < x \leqslant 10$ **(c)** $w = 1000(10 - x)$ gm
   (d) $w = 100t, 0 \leqslant t \leqslant 100$
5. Domain: $(-\infty, \infty)$, Range: $(-\infty, \infty)$
7. Domain: $(-\infty, \infty)$, Range: $(-\infty, 5]$
9. Domain: $(-\infty, \infty)$, Range: $(-\infty, \infty)$
11. Domain: $(-\infty, \infty)$, Range: $[0, \infty)$
13. Domain $(-\infty, \infty)$, Range: $[-1, 1]$
15. Domain: $(0, \infty)$, Range: $(-\infty, \infty)$
17. Domain: $[0, \infty)$, Range: $[0, \infty)$

SECTION 2-2

Exercises 2-2
1. function   **3.** function   **5.** not a function
7. not a function   **9.** not a function   **11.** function
13. function   **15.** not a function   **17.** function
19. function   **21.** Domain: $[-5, 5]$, Range: $[-2, 2]$
23. Domain: $[1, 7)$, Range: $\{-5, 2)$
25. Domain: $(-\infty, 7)$, Range: $(-\infty, 3]$
27. Domain: $(-3, \infty)$, Range: $(-3, \infty)$
29. Domain: $(-4, 6]$, Range: $[-4, 4]$
31. Domain: $(-\infty, \infty)$

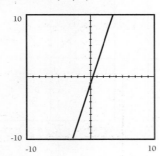

33. Domain: $(-\infty, \infty)$ **35.** Domain: $(-\infty, \infty)$

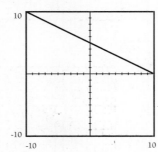   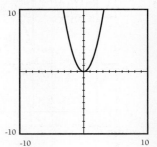

37. Domain: $(-\infty, \infty)$

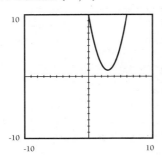

39. Domain: $(-\infty, \infty)$

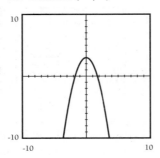

77.

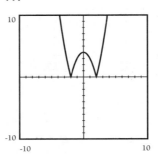

79.

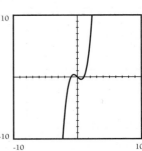

41. Domain: $(-\infty, \infty)$

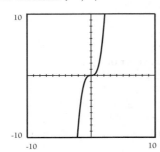

43. Domain: $[0, \infty)$

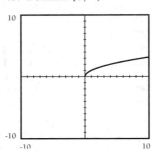

45. Domain: $(-\infty, 0) \cup (0, \infty)$

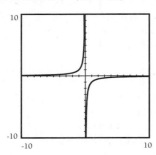

47. Domain: $(-\infty, \infty)$

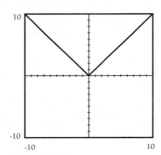

49. Domain: $[-3, 3]$

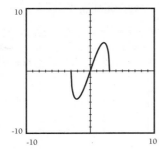

Problem Set 2-2

1. (a)

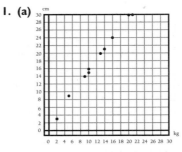

(b) mean weights = 12 kg, mean stretch = 18 cm
(c) Data given is not from a function.
(d) Good fit: *weight* = 1.5 *stretch*.
(e) theoretical domain: $[0, \infty)$, predict 37.5 kg for 25 cm
3. (g) *a*, *b* and *c* are parabolas. *d* and *e* resemble parabolas. Graphs *a*, *b*, *c*, and *d* "open up," whereas *e* "opens down."
5. Same straight line, except that *g* does not include (5,8). Thus the two functions are identical everywhere except where $x = 5$.

SECTION 2-3

Exercises 2-3
1. symmetric about the origin
3. symmetric about the *y*-axis
5. symmetric about the origin
7. symmetric about the *y*-axis
9. symmetric about the *x*-axis, *y*-axis and origin
11. symmetric about the *x*-axis, *y*-axis and origin
13. symmetric about the *x*-axis
15. not symmetric about either axis or the origin
17. not symmetric about either axis or the origin
19. symmetric about the origin

21. (a)
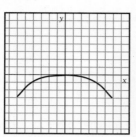

21. (b)
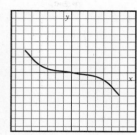

23. (a)
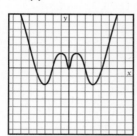

23. (b)
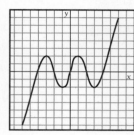

25. (a)
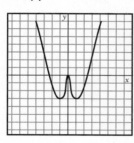

25. (b)
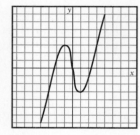

27. (a)
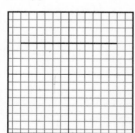

27. (b)
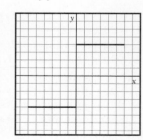

29. (a)
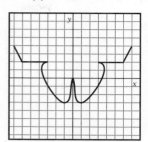

29. (b)
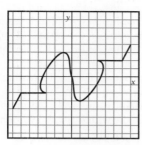

31. symmetric about the origin
33. not symmetric about either axis or the origin
35. symmetric about the y-axis
37. symmetric about the origin
39. symmetric about the y-axis
41. symmetric about the origin
43. not symmetric about either axis or the origin
45. symmetric about the x-axis, y-axis and origin
47. symmetric about the x-axis
49. symmetric about the x-axis, y-axis and origin
51. even, symmetric about y-axis
53. odd, symmetric about origin
55. even, symmetric about y-axis
57. odd, symmetric about origin
59. neither odd nor even   **61.** even, symmetric about y-axis
63. neither odd nor even   **65.** even, symmetric about y-axis
67. even, symmetric about y-axis
69. even, symmetric about y-axis
75. Signum function is odd.

Problem Set 2-3

1. (b) A relation R is symmetric to $y = x$ if, for every point $(x, y) \in R$, (y, x) also belongs to R.
3. (a) A graph is symmetric about vertical line $x = a$ if, for every point $(x - a, y)$ that belongs to the graph, $(a - x, y)$ also belongs to the graph.
5. (b) A graph is symmetric to the point (a, b) if for every point $(x - a, y - b)$ that belongs to the graph, $(a - x, b - y)$ also belongs to the graph.
9. (a) decreasing for x on $(-\infty, 0]$ and increasing for x on $[0, \infty)$
(b) decreasing for x on $(-\infty, 2]$ and increasing for x on $[2, \infty)$
(c) decreasing on intervals $(-\infty, -3]$ and $[0, 3]$, and increasing on intervals $[-3, 0]$ and $[3, \infty)$
(d) increasing for x on $(-\infty, \infty)$
(e) decreasing for x on $(0, \infty)$

SECTION 2-4

Exercises 2-4

1. increasing on $[-3, 5]$ **3.** decreasing on $[-4, 3)$
5. decreasing on $(-\infty, -1]$, increasing on $[-1, \infty)$
7. decreasing on $[-5, -2]$ and $[0, 6)$ **9.** increasing on $(-\infty, \infty)$

11. increasing   **13.** decreasing   **15.** decreasing
17. increasing   **19.** increasing
21. increasing on $(-\infty, 0]$, decreasing on $[0, \infty)$
23. increasing on $(-\infty, 3)$ and $[3, \infty)$
25. increasing on $(-\infty, -3]$, constant on $(-3, 4]$ and $(4, \infty)$
27. increasing on $(-\infty, \infty)$
29. constant on $(-\infty, 3]$, increasing on $(3, \infty)$
31. decreasing on $(-\infty, 2]$, constant on $(2, \infty)$
33. periodic with $k = 4$ **35.** periodic with $k = 3$
37. periodic with $k = 7$ **39.** periodic with $k = 7$
41. not periodic

Problem Set 2-4

5. h is the difference of f and g. $h(x)$ represents the difference in y values for each given value of x in $f(x)$ and $g(x)$.
7. h is the difference of f and g.   **9.** h is the sum of f and g.
11. h is the quotient of f and g.   **13.** h is the product of f and g.

SECTION 2-5

Exercises 2-5

1. $(f + g)(x) = 2x + 3$, $D_{f+g} = (-\infty, \infty)$; $(f - g)(x) = 2x - 3$, $D_{f-g} = (-\infty, \infty)$; $(fg)(x) = 6x$, $D_{fg} = (-\infty, \infty)$; $\left(\dfrac{f}{g}\right)(x) = \dfrac{2x}{3}$, $D_{f/g} = (-\infty, \infty)$

3. $(f + g)(x) = 2x + 3$, $D_{f+g} = (-\infty, \infty)$; $(f - g)(x) = 3 - 2x$, $D_{f-g} = (-\infty, \infty)$; $(fg)(x) = 6x$, $D_{fg} = (-\infty, \infty)$; $\left(\dfrac{f}{g}\right)(x) = \dfrac{3}{(2x)}$, $D_{f/g} = (-\infty, 0) \cup (0, \infty)$

5. $(f + g)(x) = \dfrac{(1 + x\sqrt{x})}{x}$, $D_{f+g} = (0, \infty)$; $(f - g)(x) = \dfrac{(1 - x\sqrt{x})}{x}$, $D_{f-g} = (0, \infty)$; $(fg)(x) = \dfrac{\sqrt{x}}{x}$, $D_{fg} = (0, \infty)$; $\left(\dfrac{f}{g}\right)(x) = \dfrac{\sqrt{x}}{x^2}$, $D_{f/g} = (0, \infty)$

7. $(f + g)(x) = 2x + 1$, $D_{f+g} = (-\infty, \infty)$; $(f - g)(x) = 5$, $D_{f-g} = (-\infty, \infty)$; $(fg)(x) = x^2 + x - 6$, $D_{fg} = (-\infty, \infty)$; $\left(\dfrac{f}{g}\right)(x) = \dfrac{x + 3}{x - 2}$, $D_{f/g} = (-\infty, 2) \cup (2, \infty)$

9. $(f + g)(x) = 3x + 6$, $D_{f+g} = (-\infty, \infty)$; $(f - g)(x) = x - 4$, $D_{f-g} = (-\infty, \infty)$; $(fg)(x) = 2x^2 + 11x + 5$, $D_{fg} = (-\infty, \infty)$; $\left(\dfrac{f}{g}\right)(x) = \dfrac{2x + 1}{x + 5}$, $D_{f/g} = (-\infty, -5) \cup (-5, \infty)$

11. $(f + g)(x) = x^2 + x^4$, $D_{f+g} = (-\infty, \infty)$; $(f - g)(x) = x^2 - x^4$, $D_{f-g} = (-\infty, \infty)$; $(fg)(x) = x^6$, $D_{fg} = (-\infty, \infty)$; $\left(\dfrac{f}{g}\right)(x) = \dfrac{1}{x^2}$, $D_{f/g} = (-\infty, 0) \cup (0, \infty)$

13. $(f + g)(x) = x^3 + x^5$, $D_{f+g} = (-\infty, \infty)$; $(f - g)(x) = x^3 - x^5$, $D_{f-g} = (-\infty, \infty)$; $(fg)(x) = x^8$, $D_{fg} = (-\infty, \infty)$; $\left(\dfrac{f}{g}\right)(x) = \dfrac{1}{x^2}$, $D_{f/g} = (-\infty, 0) \cup (0, \infty)$

15. $(f + g)(x) = x^2 + 3x - 2$, $D_{f+g} = (-\infty, \infty)$; $(f - g)(x) = x^2 - 3x + 2$, $D_{f-g} = (-\infty, \infty)$; $(fg)(x) = 3x^3 - 2x^2$, $D_{fg} = (-\infty, \infty)$; $\left(\dfrac{f}{g}\right)(x) = \dfrac{x^2}{3x - 2}$, $D_{f/g} = (-\infty, \frac{2}{3}) \cup (\frac{2}{3}, \infty)$

17. $(f + g)(x) = 8$, $D_{f+g} = (-\infty, \infty)$; $(f - g)(x) = 2x^2 - 10$, $D_{f-g} = (-\infty, \infty)$; $(fg)(x) = x^4 + 10x^2 - 9$, $D_{fg} = (-\infty, \infty)$; $\left(\dfrac{f}{g}\right)(x) = \dfrac{x^2 - 1}{9 - x^2}$, $D_{f/g} = (-\infty, -3) \cup (-3, 3) \cup (3, \infty)$

19. $(f + g)(x) = \sqrt{x} + \sqrt{2x + 3}$, $D_{f+g} = [0, \infty)$; $(f - g)(x) = \sqrt{x} - \sqrt{2x + 3}$, $D_{f-g} = [0, \infty)$; $(fg)(x) = \sqrt{2x^2 + 3x}$, $D_{fg} = [0, \infty)$; $\left(\dfrac{f}{g}\right)(x) = \dfrac{\sqrt{2x^2 + 3x}}{2x + 3}$, $D_{f/g} = [0, \infty)$

21. Answers vary but include: **(a)** $g(x) = 2x$ and $h(x) = 4$ **(b)** $m(x) = 2$ and $n(x) = x + 2$
23. Answers vary.   **25.** Answers vary.   **27.** Answers vary.
29. Answers vary.   **31.** Answers vary.   **33.** Answers vary.
35. Answers vary.   **37.** Answers vary.   **39.** Answers vary.

Problem Set 2-5

21. $h(x) = f(g(x))$ **23.** $h(x) = f(g(x))$ **25.** $h(x) = f(g(x))$
27. $h(x) = f(g(x))$ **29.** $h(x) = f(g(x))$

SECTION 2-6

Exercises 2-6

1. $f \circ g(2) = 3$, $g \circ f(2) = 6$ **3.** $f \circ g(2) = 6$, $g \circ f(2) = 3$
5. $f \circ g(2) = \dfrac{\sqrt{2}}{2}$, $g \circ f(2) = \dfrac{\sqrt{2}}{2}$
7. $f \circ g(2) = 3$, $g \circ f(2) = 3$ **9.** $f \circ g(2) = 15$, $g \circ f(2) = 10$
11. $f \circ g(2) = 256$, $g \circ f(2) = 256$
13. $f \circ g(2) = 32768$, $g \circ f(2) = 32768$
15. $f \circ g(2) = 16$, $g \circ f(2) = 10$ **17.** $f \circ g(2) = 24$, $g \circ f(2) = 0$
19. $f \circ g(2) = \sqrt[4]{7}$, $g \circ f(2) = \sqrt{2\sqrt{2} + 3}$
21. $f \circ g(2) = 2$, $g \circ f(2) = 2$ **23.** $f \circ g(2) = 2$, $g \circ f(2) = 2$
25. $f \circ g(2) = 2$, $g \circ f(2) = 2$
27. $f \circ g(2)$ is undefined, $g \circ f(2) = 1 - \sqrt{2}$
29. $f \circ g(2)$ is undefined, $g \circ f(2) = -\frac{3}{2}$
31. $f \circ g(x) = 3$, $D_{f \circ g} = (0, \infty)$
33. $f \circ g(x) = \dfrac{\sqrt{x}}{x}$, $D_{f \circ g} = (0, \infty)$
35. $f \circ g(x) = 2x + 11$, $D_{f \circ g} = (-\infty, \infty)$
37. $f \circ g(x) = x^8$, $D_{f \circ g} = (-\infty, \infty)$
39. $f \circ g(x) = 80 - 18x^2 + x^4$, $D_{f \circ g} = (-\infty, \infty)$
41. $f \circ g(x) = x$, $D_{f \circ g} = (-\infty, 0) \cup (0, \infty)$

43. $f \circ g(x) = \dfrac{\sqrt{x+1}}{5} - 1$, $D_{f \circ g} = [-1, \infty)$

45. $f \circ g(x) = x$, $D_{f \circ g} = [1, \infty)$

47. $f \circ g(x) = \sqrt{1-x}$, $D_{f \circ g} = (-\infty, 1]$

49. $f \circ g(x) = \dfrac{1}{x-2}$, $D_{f \circ g} = (-\infty, 2) \cup (2, \infty)$

51. Domain: $[0, \dfrac{2\pi}{3}]$,
Range: $[-1, 1]$

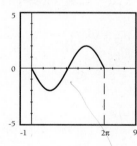

53. Domain: $[3, 2\pi + 3]$,
Range: $[-1, 1]$

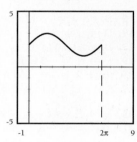

55. Domain: $[0, 2\pi]$,
Range: $[-2, 2]$

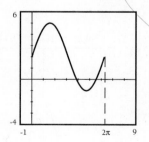

57. Domain: $[0, 2\pi]$,
Range: $[1, 3]$

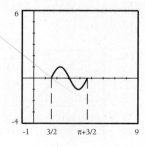

59. Domain: $[0, 2\pi]$,
Range: $[-1, 5]$

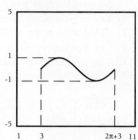

61. Domain: $[\dfrac{3}{2}, \pi + \dfrac{3}{2}]$,
Range: $[-1, 1]$

63. Domain: $[3, 2\pi + 3]$,
Range: $[-2, 2]$

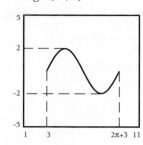

65. Domain: $[-1, 2\pi - 1]$,
Range: $[-5, -1]$

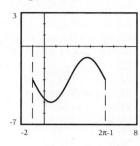

67. Domain: $[\dfrac{1}{2} - \pi, \dfrac{1}{2}]$,
Range: $[-3, 3]$

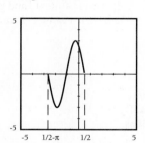

69. Domain: $[\dfrac{4}{3} - \dfrac{2\pi}{3}, \dfrac{4}{3}]$,
Range: $[-3, 1]$

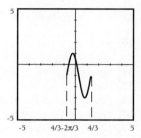

71. True **73.** True **75.** True **77.** True **79.** False

SECTION 2-7

Exercises 2-7

1. **(a)** function **(b)** not one-to-one
3. **(a)** function **(b)** not one-to-one
5. **(a)** function **(b)** is one-to-one

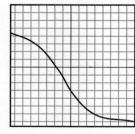

7. **(a)** function **(b)** is one-to-one

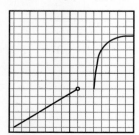

9. (a) function   **(b)** not one-to-one

11. f is one-to-one. $f^{-1}(x) = \dfrac{(x + 3)}{5}$

13. h is one-to-one. $h^{-1}(x) = \dfrac{(6 - 3x)}{2}$

15. j is one-to-one. $j^{-1}(x) = \dfrac{(x + 3.7)}{1.4}$

17. w is not one-to-one.

19. q is one-to-one. $q^{-1}(u) = 1 - \sqrt{1 - u}$

21. f is one-to-one. $f^{-1}(x) = \dfrac{1}{x}$

23. g is one-to-one. $g^{-1}(x) = \dfrac{(1 - 2x)}{(x - 1)}$

25. k is one-to-one. $k^{-1}(x) = x^2, x \geqslant 0$

27. p is one-to-one. $p^{-1}(r) = r^5$

29. z is one-to-one. $z^{-1}(x) = \dfrac{(x^2 - 3)}{2}, \ x \geqslant 0$

31. f and g are inverse functions.

33. f and g are not inverse functions.

35. f and g are not inverse functions.

37. f and g are not inverse functions.

39. f and g are inverse functions.

41.

43.

45.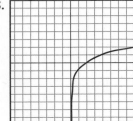

47.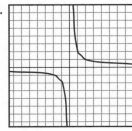

49.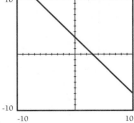

Problem Set 2-7

3. (a) $f^{-1}(x) = \dfrac{(x + 4)}{2}, g^{-1}(x) = \dfrac{(x - 3)}{5}$

(b) $f \circ g(x) = 10x + 2, g \circ f(x) = 10x - 17$

(c) $f^{-1} \circ g^{-1}(x) = \dfrac{(x + 17)}{10}, g^{-1} \circ f^{-1}(x) = \dfrac{(x - 2)}{10}$

(d) $(f \circ g)^{-1}(x) = \dfrac{(x - 2)}{10}, (g \circ f)^{-1}(x) = \dfrac{(x + 17)}{10}$

(e) $(f \circ g)^{-1}(x) = g^{-1} \circ f^{-1}(x), (g \circ f)^{-1}(x) = f^{-1} \circ g^{-1}(x)$

5. The functions are inverses.

7. The functions are inverses.

9. The functions are not inverses.

11. The functions are inverses.

13. The functions are not inverses.

CHAPTER 2 REVIEW EXERCISES

1. $y = f(x) = 3x - 5$ **3.** $y = f(x) = 2x^3$

5. $y = f(x) = x^2 + 8x + 1$ **7.** $f(x) = -\dfrac{5}{3}x + \dfrac{8}{3}$

9. $f(x) = -\dfrac{2}{x}$ **11.** $f(x) = \dfrac{3}{(3x^2 + x)}$ **13.** $D_f = (-\infty, \infty)$

15. $D_t = [-2, \infty)$ **17.** $D_t = (-\infty, 2]$

19. $D_g = (-\infty, -2) \cup (-2, \infty)$ **21.** $D_p = (6, \infty)$

23. $D_f = [-1, 0) \cup (0, \infty)$ **25.** $f(4) = 28$ **27.** $f(0) = 0$

29. $f(-4) = 36$ **31.** $f(-3) = -54$ **33.** $f(0.5) = 0.25$

35. $f(a + b) = 2a^3 + 6a^2b + 6ab^2 + 2b^3$ **37.** $f(3) = 0$

39. $f(3) = \frac{1}{2}$ **41.** $f(3) = \frac{3}{2}$ **43.** $f(w) = 4w - w^2$

45. $f(x + t) = 4x + 4t - x^2 - 2xt - t^2$

47. $f(-x) = -4x - x^2$ **49.** $f(x + h) - f(x) = 4h - 2xh - h^2$

51. $\dfrac{f(x) - f(x_0)}{(x - x_0)} = 4 - x - x_0$ **53.** $f(-x) = 2x^2 - 5$

55. $f(-x) = -x^5 + x$ **57.** $f(z) = \frac{7}{3}z - \frac{4}{3}$

59. $f(y) = -\frac{9}{5}x + \frac{8}{5}$

61.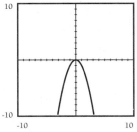

63.

65.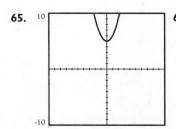

67.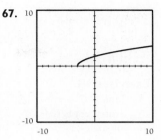

69.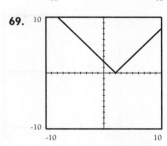

71. symmetric about the origin
73. symmetric about the origin
75. symmetric about the y-axis
77. symmetric about the y-axis
79. symmetric about the x-axis
81. f is even and symmetric about the y-axis.
83. t is even and symmetric about the y-axis.
85. q is odd and symmetric about the origin.
87. L is neither odd nor even.
89. T is even and symmetric about the y-axis.
91. increasing   **93.** decreasing
95. decreasing on $(-\infty, 1]$, increasing on $(1, \infty)$
97. increasing on $(-\infty, -3]$, constant on $(-3, 4]$, decreasing on $(4, \infty)$
99. constant on $(-\infty, 1]$, increasing on $(1, \infty)$
101. $(f + g)(x) = 3x + 2$, $D_{f+g} = (-\infty, \infty)$; $(f - g)(x) = 3x - 2$, $D_{f-g} = (-\infty, \infty)$; $(fg)(x) = 6x$, $D_{fg} = (-\infty, \infty)$; $\left(\dfrac{f}{g}\right)(x) = \dfrac{3x}{2}$, $D_{f/g} = (-\infty, \infty)$

103. $(f + g)(x) = \dfrac{1}{x} + \sqrt{x - 1}$, $D_{f+g} = [1, \infty)$; $(f - g)(x) = \dfrac{1}{x} - \sqrt{x - 1}$, $D_{f-g} = [1, \infty)$; $(fg)(x) = \dfrac{\sqrt{x - 1}}{x}$, $D_{fg} = [1, \infty)$; $\left(\dfrac{f}{g}\right)(x) = \dfrac{\sqrt{x - 1}}{x(x - 1)}$, $D_{f/g} = (1, \infty)$

105. $(f + g)(x) = 4x - 4$, $D_{f+g} = (-\infty, \infty)$; $(f - g)(x) = 2x + 6$, $D_{f-g} = (-\infty, \infty)$; $(fg)(x) = 3x^2 - 14x - 5$, $D_{fg} = (-\infty, \infty)$; $\left(\dfrac{f}{g}\right)(x) = \dfrac{3x + 1}{x - 5}$, $D_{f/g} = (-\infty, 5) \cup (5, \infty)$

107. $(f + g)(x) = x^3 + x$, $D_{f+g} = (-\infty, \infty)$; $(f - g)(x) = x^3 - x$, $D_{f-g} = (-\infty, \infty)$; $(fg)(x) = x^4$, $D_{fg} = (-\infty, \infty)$; $\left(\dfrac{f}{g}\right)(x) = x^2$, $D_{f/g} = (-\infty, 0) \cup (0, \infty)$

109. $(f + g)(x) = x^2 + 2x - 3$, $D_{f+g} = (-\infty, \infty)$; $(f - g)(x) = x^2 - 2x + 3$, $D_{f-g} = (-\infty, \infty)$; $(fg)(x) = 2x^3 - 3x^2$, $D_{fg} = (-\infty, \infty)$; $\left(\dfrac{f}{g}\right)(x) = \dfrac{x^2}{2x - 3}$, $D_{f/g} = (-\infty, \frac{3}{2}) \cup (\frac{3}{2}, \infty)$

111. $(f + g)(x) = \dfrac{x^2 - 1 + 3\sqrt{3x + 1}}{3}$, $D_{f+g} = [-\frac{1}{3}, \infty)$;

$(f - g)(x) = \dfrac{x^2 - 1 - 3\sqrt{3x + 1}}{3}$, $D_{f-g} = [-\frac{1}{3}, \infty)$;

$(fg)(x) = \dfrac{(x^2 - 1)\sqrt{3x + 1}}{3}$, $D_{fg} = [-\frac{1}{3}, \infty)$; $\left(\dfrac{f}{g}\right)(x) = \dfrac{(x^2 - 1)\sqrt{3x + 1}}{3(3x + 1)}$, $D_{f/g} = (-\frac{1}{3}, \infty)$

113. Answers vary, but include $g(x) = 3x$, $h(x) = 9$, $m(x) = 3$ and $n(x) = x + 3$.
115. Answers vary.   **117.** Answers vary.
119. Answers vary.   **121.** Answers vary.
123. $f \circ g(3) = 5$, $g \circ f(3) = 15$
125. $f \circ g(\sqrt{3}) = \dfrac{\sqrt{3}}{3}$, $g \circ f(\frac{1}{3}) = \dfrac{\sqrt{3}}{3}$

127. $f \circ g(3) = 7$, $g \circ f(3) = 4$
129. $f \circ g(3) = 1$, $g \circ f(3) = 25$
131. $f \circ g(3) = -3$, $g \circ f(3) = -27$
133. $f \circ g(3) = 3$, $g \circ f(3) = 3$
135. $f \circ g(3) = 3$, $g \circ f(3) = 3$
137. $f \circ g(3)$ is undefined, $g \circ f(3) = -1$
139. $f \circ g(x) = 5$, $D_{f \circ g} = (-\infty, \infty)$
141. $f \circ g(x) = 3x + 13$, $D_{f \circ g} = (-\infty, \infty)$
143. $f \circ g(x) = 2x^2 + x^4$, $D_{f \circ g} = (-\infty, \infty)$
145. $f \circ g(x) = \dfrac{\sqrt{5x - 1}}{5} - 1$, $D_{f \circ g} = (-\frac{1}{5}, \infty)$

147. $f \circ g(x) = \sqrt{x}$, $D_{f \circ g} = [0, \infty)$
149. Domain: $\left[0, \dfrac{2\pi}{5}\right]$, Range: $[-1, 1]$

151. Domain: $[0, 2\pi]$, Range: $[-4, 4]$

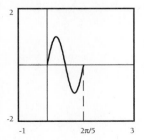

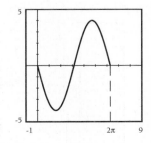

153. Domain: $[0, 2\pi]$,
Range: $[-3, 7]$

155. Domain: $[3, 3 + 2\pi]$,
Range: $[-5, 5]$

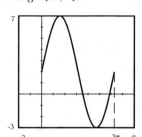

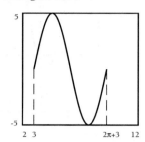

157. Domain: $\left[\dfrac{5}{3}, \dfrac{2\pi + 5}{3}\right]$,
Range: $[-4, 4]$

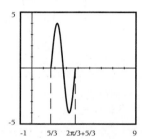

159. $f^{-1}(x) = \dfrac{(x + 30)}{50}$ **161.** $j^{-1}(x) = \dfrac{(15x + 9)}{10}$

163. $q^{-1}(u) = 2 - \sqrt{4 - u}$ **165.** $g^{-1}(x) = \dfrac{x}{x - 1}$

167. $p^{-1}(r) = r^2$ **169.** f and g are inverse functions.
171. f and g are inverse functions.
173. f and g are inverse functions.

CHAPTER 2 TEST

1. (a) a one-to-one function   **(b)** not a function
(c) a function, but not one-to-one
(d) a one-to-one function
5. $r = f(s) = (-\frac{3}{5})s^2 + \frac{2}{5}$

7. (a) $f \circ g(x) = x$ **(b)** $D_{f \circ g} = [-1, \infty)$ **(c)** $g \circ f(x) = |x|$
(d) $D_{g \circ f} = (-\infty, \infty)$

9. $f^{-1}(x) = \sqrt[3]{x + 1} = g(x)$: f and g are inverse functions.

CHAPTER 3

SECTION 3-1

Exercises 3-1

1. slope = 0,
y-intercept is $(0, -3)$,
constant

3. slope = 3,
y-intercept is $(0, 0)$,
increasing

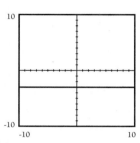

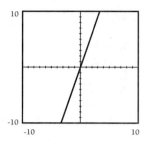

5. slope = -3,
y-intercept is $(0, 4)$,
decreasing

7. slope = 0.5,
y-intercept is $(0, -2)$,
increasing

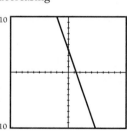

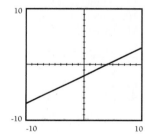

9. slope3 $= -\frac{1}{2}$, y-intercept is $(0, 4)$, decreasing

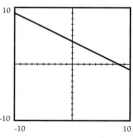

11. $y = -4x + 5$ **13.** $y = \frac{3}{2}x - 2$ **15.** $y = \frac{1}{2}x + 4$

17. $y = -6x + 11$ **19.** $y = 2x - \frac{10}{3}$ **21.** not perpendicular

23. perpendicular   **25.** perpendicular
27. not perpendicular   **29.** not perpendicular   **31.** 2
33. 1   **35.** 7   **37.** m **39.** $x + a$
41. (a) $k = 50$ **(b)** $y = 50x$ **(c)** $y = 3750$ mg

43. (a) $k = \frac{3}{5}$ **(b)** $y = \frac{3}{5}x$ **(c)** $y = 75$ cm

45. $k = 2.54$ **(b)** $y = 2.54x$ **(c)** $y = 101.6$ **(d)** $x = 15.748$

Problem Set 3-1

1. decreasing, y-intercept $= (0, 3)$, x-intercept $= \left(\frac{3}{2}, 0\right)$

3. increasing, y-intercept $= (0, -2)$, x-intercept $= \left(\frac{2}{3}, 0\right)$

5. increasing, y-intercept $= (0, -1)$, x-intercept $= \left(\frac{1}{\pi}, 0\right) \cong$ $(0.3, 0)$

7. increasing, y-intercept $= \left(0, -\frac{7}{3}\right)$, x-intercept $= \left(\frac{7}{2}, 0\right)$

9. increasing, y-intercept $= (0, 4)$, x-intercept $= \left(\frac{40}{37}, 0\right) \cong$ $(1.08, 0)$

11. a through e: decreasing on $(-\infty, 0]$, increasing on $[0, \infty)$

13. **(a)** opens up, increases on $[2, \infty)$, decreases on $(-\infty, 2]$, vertex at $(2, -4)$

(b) opens down, decreases on $[2, \infty)$, increases on $(-\infty, 2]$, vertex at $(2, 4)$

(c) opens up, increases on $\left[\frac{5}{6}, \infty\right)$, decreases on $\left(-\infty, \frac{5}{6}\right]$, vertex at $\left(\frac{5}{6}, -\frac{49}{12}\right)$

(d) opens up, increases on $\left[\frac{3}{4}, \infty\right)$, decreases on $\left(-\infty, \frac{3}{4}\right]$, vertex at $\left(\frac{3}{4}, -\frac{25}{8}\right)$

(e) opens down, decreases on $[2.5, \infty)$, increases on $(-\infty, 2.5]$, vertex at $(2.5, 8.25)$

(f) opens down, decreases on $[-1.5, \infty)$, increases on $(-\infty, -1.5]$, vertex at $(-1.5, -3.25)$

SECTION 3-2

Exercises 3-2

1. vertex $(0, 0)$

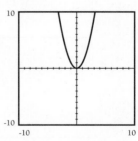

3. vertex $(0, -3)$

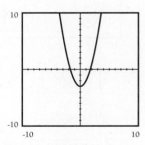

5. vertex $(-1, -1)$

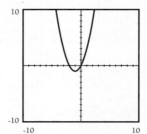

7. vertex $(0, 1)$

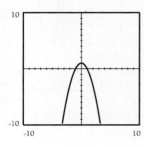

9. vertex $(-1, -2)$

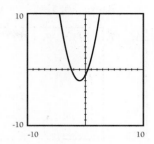

11. vertex $(4, 3)$

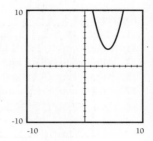

13. vertex $(-1, 3)$

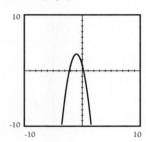

15. vertex $(-2, -5)$

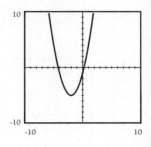

17. vertex $(3, -14)$

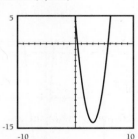

19. vertex $\left(-\frac{1}{2}, \frac{19}{4}\right)$

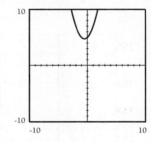

21. $2x + h, h \neq 0$ 23. $2x + h, h \neq 0$
25. $2x + h + 2, h \neq 0$ 27. $-2x - h, h \neq 0$
29. $2x + h + 2, h \neq 0$ 31. $x = 0$ 33. $x = 0$ 35. $x = -1$
37. $x = 0$ 39. $x = -1$

Problem Set 3-2

1. Maximum height of 400 ft occurs at 5 seconds.
5. $h = 0.02v^2$, $h = 0.02$ m 7. $d = 0.1v^2$, $d = 0.4$ m
13. opens up, vertex $(-4, -19)$
15. opens down, vertex $\left(\frac{3}{4}, \frac{-23}{8}\right)$
17. opens up, vertex $(3, -4)$
19. opens down, vertex $(-1, 2)$
23. x-intercepts $\cong (2.2, 0)$ and $(-2.2, 0)$
25. no x-intercepts 27. x-intercepts $\cong (0.8, 0)$ and $(-0.8, 0)$
29. x-intercepts $\cong (2.8, 0)$ and $(-1.8, 0)$
31. x-intercepts are $(0, 0)$ and $(7, 0)$

SECTION 3-3

Exercises 3-3

1. $\{1, 2\}$ **3.** $\{-2, 1\}$ **5.** $\{-1, 6\}$ **7.** $\{-2, 3\}$ **9.** $\{-6, -1\}$
11. x-intercepts: $(-2, 0)$ and $(-4, 0)$
13. x-intercepts: $(8, 0)$ and $(1, 0)$
15. x-intercepts: $(8, 0)$ and $(-1, 0)$
17. x-intercepts: $(-4, 0)$ and $(2, 0)$
19. x-intercepts: $\left(-\frac{1}{2}, 0\right)$ and $(-5, 0)$

21. x-intercepts:
$\left(-\frac{5}{2}, 0\right)$ and $(-1, 0)$

23. x-intercepts:
$\left(-\frac{1}{2}, 0\right)$ and $(5, 0)$

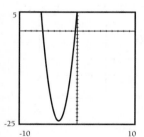

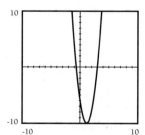

25. x-intercepts:
$(-6, 0)$ and $\left(-\frac{1}{3}, 0\right)$

27. x-intercepts:
$\left(-\frac{2}{3}, 0\right)$ and $(3, 0)$

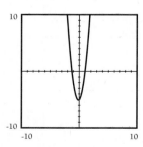

29. x-intercepts:
$\left(-\frac{5}{4}, 0\right)$ and $(1, 0)$

31. $x \cong 1.414214$ **33.** $x \cong 2.236068$ **35.** $x \cong -2.44949$
37. $x \cong 0.236$ **39.** $x \cong -4.236$

Problem Set 3-3

1. $p = 2v^2$ **3.** $I = 4d^2$

5. x-intercepts $\cong (-2.84, 0)$ and $(2.84, 0)$
7. x-intercepts $\cong (0.38, 0)$ and $(2.63, 0)$
11. x-intercepts $\cong (-2, 0)$ and $(4.32, 0)$
15. $x \cong \pm 2.64575$ **17.** $x \cong 5.74456$
19. (a) $x \cong 2.23611$ **(b)** $x \cong 2.236068$ **(c)** $x \cong 2.2105$
(d) $x \cong 2.236068$

SECTION 3-4

Exercises 3-4

1. x-intercepts:
$(-1, 0)$ and $(-3, 0)$

3. x-intercepts:
$(4, 0)$ and $(-1, 0)$

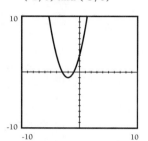

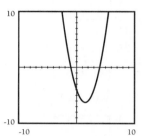

5. x-intercepts:
$\left(\dfrac{-9 \pm \sqrt{89}}{2}, 0\right) \cong (-9.217, 0)$
and $(0.217, 0)$

7. x-intercepts:
$\left(\frac{1}{2}, 0\right)$ and $(4, 0)$

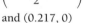

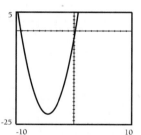

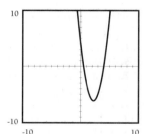

9. x-intercepts:
$\left(\dfrac{20 \pm \sqrt{80}}{8}, 0\right) \cong (1.382, 0)$
and $(3.618, 0)$

11. No x-intercepts

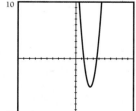

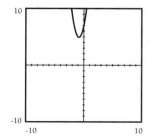

13. x-intercepts:

$$\left(\frac{-1 \pm \sqrt{21}}{10}, 0\right) \cong (-0.558, 0)$$

and $(0.358, 0)$

15. No x-intercepts

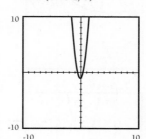

17. x-intercepts:

$$\left(\frac{\sqrt{2} \pm \sqrt{14}}{2}, 0\right) \cong (-1.164, 0)$$

and $(2.578, 0)$

19. x-intercepts:

$(\pm \sqrt[4]{5}, 0) \cong (-1.495, 0)$

and $(1.495, 0)$

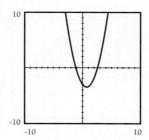
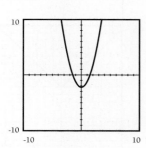

21. two irrational roots   **23.** one rational root
25. two rational roots
27. no real roots, but two complex roots
29. two rational roots
31. Real roots: $\{-5, 5\}$. Complex roots: $\{-5, 5, i, -i\}$
33. $\{-2, 2, \sqrt{5}, -\sqrt{5}\}$
35. Real roots $\{-2, 1\}$. Complex roots: $\{-2, 1, 1 \pm i\sqrt{3},$

$$\frac{-1 \pm i\sqrt{3}}{2}\}$$

37. Real roots $\{\pm 2, \pm 1\}$. Complex roots: $\{\pm 2, \pm 1, \pm 2i, \pm i\}$

39. Real roots $\{-1, \sqrt[3]{3}\}$. Complex roots: $\{-1, \sqrt[3]{3}, \frac{1 \pm i\sqrt{3}}{2},$

$$\frac{-1 \pm i\sqrt{3}}{2}\}$$

41. $\{4, 9\}$ **47.** $f(x) = x^2 - 2x - 15$
49. $f(x) = 3x^2 - 14x + 8$ **51.** $f(x) = x^2 - 6x + 4$
53. $P = 1v^2$, $v \cong 31.63$ k/hr   **55.** $P = 25d^2$, $d = 10$ m
57. $V^2 = 64h$, $V = 4$ ft/sec
59. $2300 k$ watts, where k is the constant of proportionality.
61. $I = \dfrac{2500}{d^2}$, $d = 10\sqrt{5} \cong 22.36$ cm

Problem Set 3-4
3. $\{\frac{1}{3}, \frac{1}{2}\}$, $\{2, 3\}$. The roots are reciprocals.
19. (b) $(2,0)$ **(c)** $(-\infty, 2)$ **(d)** $(2, \infty)$ **(e)** decreasing
21. $x > 2$ **23.** $(3, \infty)$
25. x-intercepts: $(3, 0)$ and $(-5, 0)$. Above: $(-\infty, -5) \cup (3, \infty)$.
Below: $(-5, 3)$
27. x-intercepts: $\left(\dfrac{3 \pm \sqrt{17}}{4}, 0\right)$. Above: $\left(-\infty, \dfrac{3 - \sqrt{17}}{4}\right) \cup$

$\left(\dfrac{3 + \sqrt{17}}{4}, \infty\right)$. Below: $\left(\dfrac{3 - \sqrt{17}}{4}, \dfrac{3 + \sqrt{17}}{4}\right)$

29. No x-intercepts. Below:$(-\infty, \infty)$

31. x-intercepts: $\left(\dfrac{-1 \pm \sqrt{61}}{6}, 0\right)$. Below: $\left(-\infty, \dfrac{-1 - \sqrt{61}}{6}\right) \cup$

$\left(\dfrac{-1 + \sqrt{61}}{6}, \infty\right)$. Above: $\left(\dfrac{-1 - \sqrt{61}}{6}, \dfrac{-1 + \sqrt{61}}{6}\right)$

33. No x-intercepts. Below: $(-\infty, \infty)$

37. $\left(-\infty, \dfrac{3 - \sqrt{17}}{4}\right) \cup \left(\dfrac{3 + \sqrt{17}}{4}, \infty\right)$

SECTION 3-5

Exercises 3-5

1.

x	$-\infty$		0		∞
$I(x)$		$-$	0	$+$	

3.

x	$-\infty$		4		∞
$k(x)$		$+$	0	$-$	

5.

x	$-\infty$		40/7		∞
$p(x)$		$-$	0	$+$	

7.

x	$-\infty$		15		∞
$t(x)$		$+$	0	$-$	

9.

x	$-\infty$		-5		∞
$q(x)$		$-$	0	$+$	

11. $(-\infty, \frac{5}{3})$ **13.** $[\frac{5}{3}, \infty)$ **15.** $(10, \infty)$ **17.** $(-2, \infty)$

19. $(-\infty, -\frac{1}{4}]$ **25.** $(-3, 1)$ **27.** $(-\frac{1}{3}, 2)$

29. $(-\infty, 3] \cup [4, \infty)$

31. $[-3, 3]$ **33.** $\varnothing$ **35.** $(-\infty, -6) \cup (-1, \infty)$ **37.** $[1, 3]$

39. $(-\infty, -\frac{1}{2}] \cup [3, \infty)$ **41.** $(-3 - 2\sqrt{2}, -3 + 2\sqrt{2})$

43. $(-\infty, -\frac{2}{3}) \cup (1, \infty)$

45. $\left(-\infty, \dfrac{-1 - \sqrt{21}}{2}\right] \cup \left[\dfrac{-1 + \sqrt{21}}{2}, \infty\right)$

47. $\left[\dfrac{5 - \sqrt{57}}{4}, \dfrac{5 + \sqrt{57}}{4}\right]$ **49.** $\varnothing$ **51.** $(-\infty, \infty)$

53. $[2, 2] = \{2\}$

Problem Set 3-5

1. $(-\infty, \frac{3}{5})$ **3.** $(-\infty, \frac{5}{4}]$ **5.** $(-\infty, \frac{12}{5})$

7. $\left(-\infty, \dfrac{5 - \sqrt{57}}{4}\right) \cup \left(\dfrac{5 + \sqrt{57}}{4}, \infty\right) \cong (-\infty, -0.64) \cup (3.14, \infty)$

9. $(-\infty, \infty)$ **11.** $\varnothing$ **13.** $(-\infty, \infty)$

SECTION 3-6

Exercises 3-6

1. $\{-1, -5\}$ **3.** $\{1, 9\}$ **5.** $\{-\frac{7}{3}, 1\}$ **7.** $\{1, -\frac{3}{5}\}$ **9.** $\{2, 5\}$

11. $\{-\frac{1}{2}, 2\}$ **13.** $\{-\frac{4}{15}, -\frac{16}{15}\}$ **15.** $\{\frac{4}{15}, \frac{16}{15}\}$ **17.** $\{\frac{3}{5}\}$

19. $\varnothing$ **21.** $\{4, -\frac{2}{3}\}$ **23.** $\{-\frac{3}{2}, -\frac{1}{8}\}$ **25.** $\varnothing$ **27.** $\{-2, \frac{2}{7}\}$

29. $\{\frac{1}{2}, -\frac{5}{2}\}$ **31.** $\{1, 3\}$ **33.** $\left\{\dfrac{-5 \pm 3\sqrt{5}}{2}, \dfrac{-5 \pm \sqrt{37}}{2}\right\}$

35.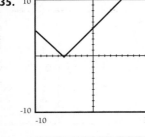

37.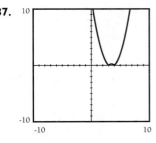

39.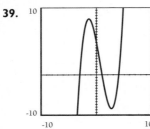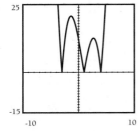

41.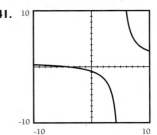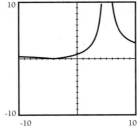

43.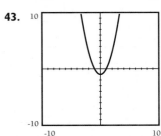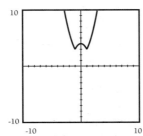

45.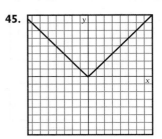

47.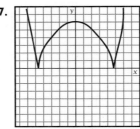

49.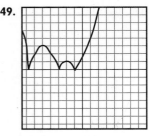

51.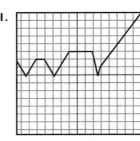

53.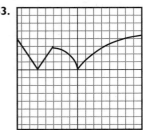

55. (-1, 5)   **57.** $(-\infty, -9) \cup (-1, \infty)$ **59.** [1, 5]   **61.** $(-\infty, \infty)$
63. $\varnothing$ **65.** $(-\infty, \infty)$ **67.** [-1, 2]   **69.** $(-\infty, -2] \cup [1, \infty)$
71. $(-\infty, -1) \cup (2, \infty)$ **73.** $\left(-\frac{32}{5}, -\frac{28}{5}\right)$

75.

x	$-\infty$		$\frac{5}{3}$		∞
$f(x)$		$+$	0	$+$	

77.

x	$-\infty$		$\frac{5}{3}$		∞
$f(x)$		$+$	0	$+$	

79.

x	$-\infty$		-1		$-\frac{1}{3}$		∞
$f(x)$		$+$	0	$-$	0	$+$	

81.

x	$-\infty$		-4		$\frac{2}{3}$		∞
$f(x)$		$-$	0	$+$	0	$-$	

83.

x	$-\infty$		3		∞
$f(x)$		$-$	0	$+$	

85. $\left(-\infty, \frac{5}{3}\right) \cup \left(\frac{5}{3}, \infty\right)$ **87.** $\varnothing$ **89.** $\left[-1, -\frac{1}{3}\right]$ **91.** $\left[-4, \frac{2}{3}\right]$
93. $(3, \infty)$ **95.** $\cong (-2.66, -2) \cup (-1, 1)$ **97.** $(-\infty, \infty)$
99. $\left[\dfrac{3 - \sqrt{17}}{2}, \dfrac{3 + \sqrt{17}}{2}\right]$
101. $|3 - x| < \frac{1}{2}$ **103.** $|x + 5| > \frac{1}{5}$ **105.** $\left|x + \frac{1}{2}\right| \leq 4$

Problem Set 3-6
7. $|x| < 3$ **9.** $|x - 1| \leq 3$ **11.** $|x - 3| \leq 2$
13. $|x + 2| < 2$ **15.** $\left|x + \frac{3}{2}\right| \leq \frac{7}{2}$ **37.** (1, 3)
39. $[-3, -\sqrt{5}] \cup [\sqrt{5}, 3]$

SECTION 3-7

Exercises 3-7
1. trunc = 3.1, up = 3.2, down = 3.1, off = 3.1
3. trunc = 0, up = 10, down = 0, off = 0
5. trunc = 3.141, up = 3.142, down = 3.141, off = 3.142
7. trunc = 3.14159, up = 3.14160, down = 3.14159, off = 3.14159
9. trunc = 3.1415926, up = 3.1415927, down = 3.1415926, off = 3.1415927
11. trunc = -2.71, up = -2.71, down = -2.72, off = -2.72

13. trunc = -2, up = -2, down = -3, off = -3
15. trunc = -2.718, up = -2.718, down = -2.719, off = -2.718
17. trunc = -2.718281, up = -2.718281, down = -2.718282, off = -2.718282
19. trunc = -2.7182818, up = -2.7182818, down = -2.7182819, off = -2.7182818
21. 2.718   **23.** 1.571   **25.** 20.000   **27.** 39.744
29. 0.896   **31.** 0.896   **33.** 0.894   **35.** 0.895   **37.** 0.895
39. 0.894   **45.** $\cong 4.962442$ **47.** $\cong 1.791288$
49. $\cong 4.823404$ **51.** $x \cong 0.274917$ or $x \cong -7.274917$
53. $x \cong -0.3944487$ or $x \cong -7.6055513$

CHAPTER 3 REVIEW EXERCISES
1. slope is 0, y-intercept is (0, -7), constant
3. slope is $-\frac{3}{4}$, y-intercept is (0, 2), decreasing

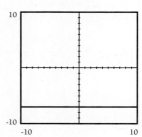
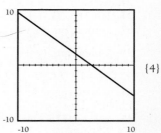 {4}

5. $y = \frac{3}{5}x + 4$ **7.** $y = \frac{7}{4}x - \frac{29}{4}$

9. parallel   **11.** perpendicular   **13.** $\frac{1}{3}$ **15.** k
17. vertex (0, 0), y-intercept (0, 0), x-intercept (0, 0)
19. vertex (0, 3), y-intercept (0, 3), x-intercepts: $(\sqrt{3}, 0)$, $(-\sqrt{3}, 0)$

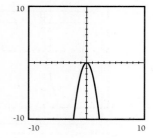
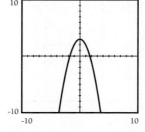

21. vertex $(-4, -3)$,
y-intercept $(0, 13)$,
x-intercepts:
$(-4, +\sqrt{3}, 0)$,
$(-4 - \sqrt{3}, 0)$

23. vertex $(-3, -14)$,
y-intercept $(0, -5)$,
x-intercepts:
$\left(\dfrac{-6 \pm \sqrt{3}}{2}, 0\right)$

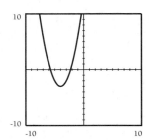

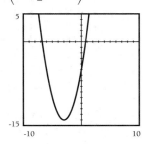

53. x-intercepts:
$\left(\dfrac{-1 \pm \sqrt{41}}{10}, 0\right) \cong (-0.74, 0)$
and $(0.54, 0)$

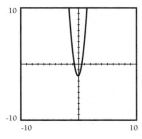

55. (a) 61   (b) two irrational roots
57. (a) 196   (b) two rational roots
59. Real roots: $\{-7, 7\}$. Complex roots: $\{-7, 7, -i, i\}$
61. Real roots: $\{-3, 1\}$. Complex roots: $\left\{-3, 1, \dfrac{3 \pm 3i\sqrt{3}}{2}, \dfrac{-1 \pm i\sqrt{3}}{2}\right\}$

25. $14x + 7h, h \neq 0$
27. $6x + 3h - 2, h \neq 0$ **29.** 14,400 ft at $t = 30$ sec
31. $\{-2, 5\}$ **33.** $\{-1, 6\}$ **35.** x-intercepts: $(2, 0), (-8, 0)$
37. x-intercepts: $(5, 0), (7, 0)$
39. x-intercepts: $\left(\dfrac{-7 \pm \sqrt{109}}{6}, 0\right) \cong (-2.91, 0)$ and $(0.57, 0)$

41. x-intercepts: $\left(\dfrac{19 \pm \sqrt{313}}{4}, 0\right) \cong (0.33, 0)$ and $(9.17, 0)$

43. $x \cong 2.64575$ **45.** $x \cong 0.61803$
47. $P = 1000d^2$

49. x-intercepts:
$\left(\dfrac{-4 \pm \sqrt{112}}{2}, 0\right) \cong (-7.29, 0)$
and $(3.29, 0)$

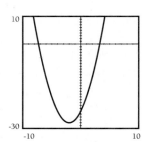

51. x-intercepts:
$\left(\dfrac{7 \pm \sqrt{61}}{2}, 0\right) \cong (-7.41, 0)$
and $(0.41, 0)$

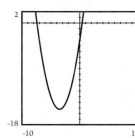

63. Real roots: $\{1\}$. Complex roots: $\left\{1, \dfrac{-1 \pm i\sqrt{3}}{2}\right\}$

65. $f(x) = 5x^2 - 48x - 20$ **67.** $V^2 = 64h, V = 2\sqrt{6}$ ft/ses

69.

x	$-\infty$		0		∞
$I(x)$		+	0	−	

71.

x	$-\infty$		50		∞
$p(x)$		−	0	+	

73.

x	$-\infty$		$\frac{15}{4}$		∞
$f(x)$		−	0	+	

$x \in \left(-\infty, \frac{15}{4}\right)$

75.

x	$-\infty$		$\frac{9}{2}$		∞
$3 - \left(\frac{2}{3}\right)x$		+	0	−	

$x \in \left(\frac{9}{2}, \infty\right)$

77.

x	$-\infty$		-3		4		∞
$(x - 4)(x + 3)$		+	0	−	0	+	

$x \in (-3, 4)$

79.

x	$-\infty$		3		4		∞
$(x-4)(x-3)$		+	0	−	0	+	

$x \in (-\infty, 3] \cup [4, \infty)$

81.

x	$-\infty$		2		8		∞
$(x-8)(x-2)$		+	0	−	0	+	

$x \in (2, 8)$

83.

x	$-\infty$		2		8		∞
$4 - 2x - x^2$		−	0	+	0	−	

$x \in (-\infty, -1 - \sqrt{5}] \cup [-1 + \sqrt{5}, \infty)$

85.

x	$-\infty$		∞
$x^2 + 3x + 7$		+	

$x \in \varnothing$

87.

x	$-\infty$		4		∞
$x^2 - 8x + 16$		+	0	+	

{4}

89. $\{-5, -13\}$ **91.** $\{-\frac{8}{9}, \frac{14}{9}\}$ **93.** $\{\frac{3}{2}, -\frac{1}{6}\}$ **95.** $\{\frac{7}{8}\}$

97. $\{4, -\frac{2}{5}\}$ **99.** $\varnothing$ **101.** $\varnothing$ **103.** $\left\{1, 4, \frac{5 \pm \sqrt{17}}{2}\right\}$

105.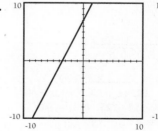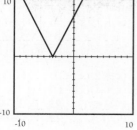

107.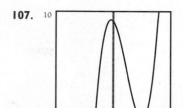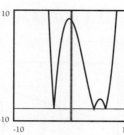

109.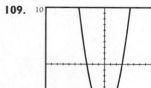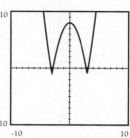

111.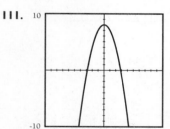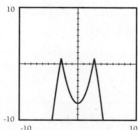

113. $(7, 15)$ **115.** $[5, 11]$ **117.** $\varnothing$ **119.** $\left[\frac{8}{3}, 6\right]$
121. $(-\infty, -\frac{5}{3}) \cup (3, \infty)$

123.

x	$-\infty$		$\frac{12}{5}$		∞
$|5x - 12|$		+	0	+	

125.

x	$-\infty$		-1		$-\frac{1}{2}$		∞
$f(x)$		+	0	−	0	+	

127. $x \in (-\infty, \frac{12}{5}) \cup (\frac{12}{5}, \infty)$ **129.** $x \in [-1, -\frac{1}{2}]$ **131.** 5.718

133. 1.992 **135.** 0.895 **137.** 0.495
139. $x \cong -0.61803$ or $x \cong 1.61803$

CHAPTER 3 CHAPTER TEST

1. slope is $\frac{1}{3}$, y-intercept $(0, 2)$, increasing

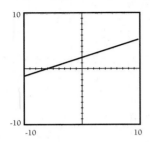

3. x-intercepts: $(-\frac{1}{3}, 0)$ and $(2, 0)$. vertex: $(\frac{5}{6}, -\frac{49}{12})$

5.

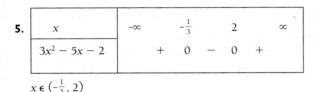

x	$-\infty$		$-\frac{1}{3}$		2		∞
$3x^2 - 5x - 2$		+	0	−	0	+	

$x \in (-\frac{1}{3}, 2)$

7. Real roots: $\{\sqrt[3]{9}, -\sqrt[3]{4}\}$.   **9.** $\{\frac{2}{3}, 8\}$

CHAPTER 4

SECTION 4-1

Exercises 4-1

1. $x^2 + 3x - 15$, polynomial, degree 2, expect 3 terms
3. $x^2 - 2x - 15$, polynomial, degree 2, expect 3 terms
5. $x^3 + 5x^2 - 8x + 2$, polynomial, degree 3, expect 4 terms
7. $x^3 - 6x^2 + 12x - 8$, polynomial, degree 3, expect 4 terms
9. not a polynomial   **11.** not a polynomial
13. $\sqrt{3}x - 1$, polynomial, degree 1, expect 2 terms
15. not a polynomial   **17.** not a polynomial
19. $-6x^2 + 0x - 2$, polynomial, degree 2, expect 3 terms
21. $P(x) = (x + 1)(x + 1) + 2$
23. $P(x) = (x - 2)(x - 3) + 1$
25. $P(x) = (x + 5)(x^2 - 12x + 62) - 313$
27. $P(x) = (x - 3)(2x^2 + 6x + 21) + 58$
29. $P(x) = (x + 4)(5x^2 - 23x + 92) - 362$
31. $P(x) = (2x + 1)(x^2 - 3x + \frac{3}{2}) + \frac{3}{2}$

33. $P(x) = (3x - 2)(\frac{1}{3}x^2 - \frac{1}{9}x - \frac{2}{27}) + \frac{158}{27}$

35. $48_8 = 37$ **37.** $45_{12} = 53$ **39.** $21_5 = 11$
41. $A3_{16} = 163$ **43.** $10011_2 = 19$ **45.** $53 = 65_8$
47. $53 = 3B_{14}$ **49.** $201 = C9_{16}$ **51.** $23 = 10111_2$
53. $32 = 100000_2$ **55.** $(x - 3)$ **57.** $(x + 1), (x - 3)$
59. $(x + 1), (x - 1), (x + 3)$ **61.** $(x - 1)$ **63.** $(x + 1)$

65. $(x + 1), (x - 1)$
67. No answer. $\frac{3}{5}$ is not an acceptable base.   **69.** 19
71. $V(x) = (15 - 2x)(10 - 2x)x, V(x) > 0, 0 < x < 5$

Problem Set 4-1

1. $x^4 - 3x^2 + x + 2 = (x^2 - x + 1)(x^2 + x - 3) + (-3x + 5)$
3. $x^3 + 1 = (x^2 - x + 1)(x + 1)$
5. $x^4 - 16 = (x^2 + 4)(x^2 - 4)$
7. $x^6 - 1 = (x^2 - x + 1)(x^4 + x^3 - x - 1)$
9. Quotients are the same.
15. (a) 3rd degree   (b) 3 x-intercepts   (c) 2   (d) -1   (e) 2
(f) 1

SECTION 4-2

Exercises 4-2

1. (a) 1   (b) 3 or 1   (c) 0   (d) 0
3. (a) 2   (b) 2 or 0   (c) 1   (d) 1
5. (a) 1   (b) 1   (c) 1   (d) 1
7. (a) 2   (b) 2 or 0   (c) 2   (d) 2 or 0
9. (a) 0   (b) 0   (c) 0   (d) 0
11. (a) Upperbound: 3   (b) Lower bound: -3
13. (a) No positive roots; upper bound: 0
(b) Lower bound: -4
15. (a) Upper bound: 1   (b) Lower bound: -3
17. (a) Upper bound: 1
(b) No negative roots. Lower bound: 0
19. (a) Upper bound: 1   (b) Lower bound: -1

Problem Set 4-2

1. no rational roots   **3.** no rational roots
5. Theorem does not apply.   **7.** [-3, 3]   **9.** [-4, 0]
11. [-3, 1]   **13.** [0, 1]   **15.** [-1, 1]   **17.** (-2, -1)

SECTION 4-3

Exercises 4-3

1. $\{3, -2, 5\}$ **3.** $\{4, 1, -\frac{3}{2}\}$ **5.** $\{0, 4, \frac{1}{3}\}$ **7.** $\{0, -1, \frac{5}{3}\}$

9. real: $\{0, 1\}$; complex: $\{0, 1, \pm i\}$

11. $(x + 1)(x^2 - 3x + 1) =$
$(x + 1)\left(x - \dfrac{3 + \sqrt{5}}{2}\right)\left(x - \dfrac{3 - \sqrt{5}}{2}\right)$

13. $(x + 2)(x^2 + 3)$ **15.** $(x + 3)(x + 1)(x - 2)$
17. $(x + 1)(x - 1)(x^2 + 1)$ **19.** $(x + 2)(x^2 - 2x + 4)$
21. $P(2) = 13$ **23.** $P(-3) = -113$ **25.** $P(3) = 91$
27. $P(\frac{1}{2}) = \frac{1}{4}$ **29.** $P(\sqrt{3}) = 3\sqrt{3} - 14$

31. x-intercepts: $(-1, 0), (3, 0)$ **33.** x-intercepts: $(0, 0), (1, 0)$

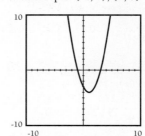
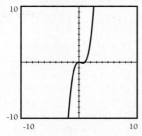

35. x-intercepts:
$(-2, 0), (0, 0), (\frac{1}{2}, 0)$

37. x-intercepts:
$(3, 0), (-1, 0)$

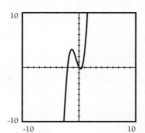
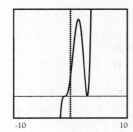

39. x-intercepts: $(-1, 0), \left(\dfrac{5 + \sqrt{21}}{2}, 0\right), \left(\dfrac{5 - \sqrt{21}}{2}, 0\right)$

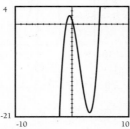

41. $P(x) = C(x - 1)(x - 2)(x - 3), C \neq 0$
43. $P(x) = C(x - 5)(x + 2)(x - 6), C \neq 0$
45. $P(x) = C(x)(x - 10)(x + 3), C \neq 0$
47. $P(x) = C(x - \pi)(x + e)(x - \sqrt{13}), C \neq 0$
49. $P(x) = C(x - \frac{3}{4})(x - \frac{1}{2})^2, C \neq 0$

Problem Set 4-3

1. (a) $P(x) = \dfrac{(-x + 14)}{5}$

3. (a) $P(x) = \dfrac{(3x^2 + x)}{2}$

5. (a) $P(x) = x$

SECTION 4-4

Exercises 4-4

1. (d) $P(x) = (x + 3)(x^2 - 3x + 1) =$
$(x + 3)\left(x - \dfrac{3 + \sqrt{5}}{2}\right)\left(x - \dfrac{3 - \sqrt{5}}{2}\right)$

3. (d) $P(x) = (x + 2)(x^2 + 4x + 2) = (x + 2)[x - (-2 + \sqrt{2})]$
$[x - (-2 - \sqrt{2})]$
5. (d) $P(x) = (x + 3)(2x + 1)(2x - 1)$
7. (d) $P(x) = (3x - 2)(x^2 + 1)$
9. (d) $P(x) = (x - 1)(x + 1)(3x + 2)(3x - 2)$
11. (d) $P(x) = (2x + 1)(2x - 1)(x + 2)$
13. (d) $P(x) = (x + 2)(3x + 2)(2x - 3)$
15. (d) $P((P_1P(x) = (x - 2)(x + 2)(x^2 - 5x + 1) = (x - 2)$
$(x + 2)\left(x - \dfrac{5 + \sqrt{21}}{2}\right)\left(x - \dfrac{5 - \sqrt{21}}{2}\right)$

17. (d) $P(x) = (x + 1)(x - 2)(x^2 + x + 5)$
19. (d) $P(x) = (x - 1)(x + 1)(2x + 3)(2x - 3)$
21. $\{-\frac{1}{2}, \frac{1}{2}, -2\}$ **23.** $\{-2, -\frac{2}{3}, \frac{3}{2}\}$ **25.** $\left\{2, -2, \dfrac{5 \pm \sqrt{21}}{2}\right\}$

27. $\{-1, 2\}$ **29.** $\{-\frac{3}{2}, \frac{3}{2}, -1, 1\}$ **31.** $(-2, -\frac{1}{2}) \cup (\frac{1}{2}, \infty)$

33. $(-\infty, -2] \cup [-\frac{2}{3}, \frac{3}{2}]$

35. $(-\infty, -2] \cup \left[\dfrac{5 - \sqrt{21}}{2}, 2\right] \cup \left[\dfrac{5 + \sqrt{21}}{2}, \infty\right)$

37. $(-1, 2)$ **39.** $(-\infty, -\frac{3}{2}) \cup (-1, 1) \cup (\frac{3}{2}, \infty)$

41.
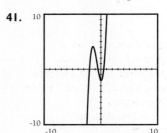

43.
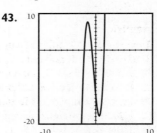

45.
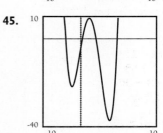

47.
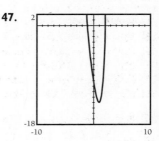

49.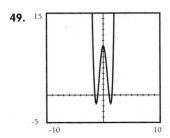

51. $y = 7x - 14$ **53.** $y = 54x + 324$ **55.** $y = 15x + 30$

57. $y = \dfrac{91}{2}x - \dfrac{273}{4}$ **59.** $y = -60x - 120$

Problem Set 4-4

1. $y^3 - 3y - 2 = 0$ **3.** $y^3 - 9y + 28 = 0$
5. $y^3 - 3y + 10 = 0$ **7.** $x = 2$ **9.** $x = -4$
11. $x = \sqrt{-5 + 2\sqrt{6}} + \sqrt{-5 - 2\sqrt{6}}$ **13. (c)** $\{1, -2\}$
15. $\{-5\}$ **17.** $x = \sqrt{-5 + 2\sqrt{6}} + \sqrt{-5 - 2\sqrt{6}} - 2$
19. $\{3, \pm\sqrt{5}\}$ (Note: Cardan's formula will become unwieldy, producing expressions formed from complex numbers. Factor or use the rational root theorem. As a result, we see that expressions of complex numbers may represent real numbers.)
25. $\cong (-1.26, 0)$ **27.** $\cong (2.67, 0), (-2.67, 0)$
29. $\cong (1.88, 0), (-0.34, 0), (-1.53, 0)$
31. (b) $y = -x - 3$ **33. (b)** $y = -8x - 3$ **35. (b)** $y = -3$
37. (b) $y = 28x + 33$ **39. (b)** $y = -6$

SECTION 4-5

Exercises 4-5

1. $\cong 1.4$ **3.** $\cong \pm 1.6$ **5.** $\cong -1.2$ **7.** $\cong 0.0$ ($\cong 0.05$)
9. $\cong 0.3$

Problem Set 4-5

1. $\cong 0.242$ **3.** $\cong 1.307$ **9.** $\cong 0.24176$
11. $\cong 1.30749, \cong 0.33767$

SECTION 4-6

Exercises 4-6

1. yes   **3.** yes   **5.** no   **7.** yes   **9.** no   **11.** $\{3 \pm i\}$
13. $\{-1 \pm 2i\}$ **15.** $\{-3 \pm 2i\}$ **17.** $\{3 \pm i, 1\}$ **19.** $\{i, 3\}$
21. $\left\{\dfrac{1}{3} - \dfrac{2i}{3}\right\}$ **23.** $\left\{\dfrac{3}{2} - \dfrac{i}{2}\right\}$ **25.** $\left\{-\dfrac{1}{10} + \dfrac{7i}{10}\right\}$
27. $\{\pm i\sqrt{2}\}$ **29.** $\left\{-\dfrac{1}{2} \pm \dfrac{i\sqrt{3}}{2}\right\}$ **31.** $\left\{\dfrac{1}{10} \pm \dfrac{i\sqrt{39}}{10}\right\}$
33. $\left\{\pm\dfrac{\sqrt{11}}{2} - i\right\}$
35. $\{2i, i\}$ **37.** $\{-3, 1\}$ **39.** $\{-1, -1 \pm 2i\}$

49. The population exceeds 100 when $t > 2.6$ years. No extinction in domain.

CHAPTER 4 REVIEW EXERCISES

1. 4 term polynomial of degree 3
3. 5 term polynomial of degree 4   **5.** not a polynomial
7. not a polynomial   **9.** $P(x) = (x + 1)(x + 4) - 2$
11. $P(x) = (x + 5)(x^2 - 10x + 53) - 273$ **13.** $35_8 = 29$
15. $10101_2 = 21$ **17.** $53 = 104_7$ **19.** $27 = 1000_3$
21. None are factors.   **23.** $(x + 3)$
25. (a) 1   **(b)** 1   **(c)** 0   **(d)** 0   **27. (a)** 1   **(b)** 1   **(c)** 1   **(d)** 1
29. upper bound: 2; lower bound: -2
31. upper bound: 1; lower bound: -1
33. $\{4, -23, 17\}$ **35.** $\{0, \frac{4}{3}, \frac{3}{5}\}$
37. (a) $\{1, -1, 3\}$ **(b)** $P(x) = (x - 1)(x + 1)(x - 3)$
39. (a) $\{1\}$
(b) $P(x) = (x - 1)(x^2 - 6x - 6) =$
$(x - 1)(x - 3 - \sqrt{15})(x - 3 + \sqrt{15})$
41. $P(4) = 89$ **43.** $P(\sqrt{7}) = 7\sqrt{7} - 33$

45. 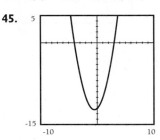   **47.**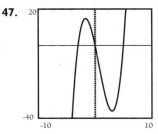

49. $P(x) = (x - 1)(x + 2)(x - 3)$
51. $P(x) = x(x - 4)(x + 3)$ **53.** $P(x) = (4x + 3)(2x + 1)^2$
55. $P(x) = (x - 2)(x^2 + 2x - 1) =$
$(x - 2)(x + 1 + \sqrt{2})(x + 1 - \sqrt{2})$
57. $P(x) = (x + 1)(4x^2 + 8x - 9) =$
$(x + 1)\left(x - \dfrac{-2 + \sqrt{13}}{2}\right)\left(x - \dfrac{-2 - \sqrt{13}}{2}\right)$
59. See Answer 55.   **61.** See Answer 57.
63. $\left\{-1, \dfrac{-1 \pm \sqrt{6}}{2}\right\}$ **65.** $\{0, 1, 2 \pm \sqrt{11}\}$
67. $\left(\dfrac{-1 - \sqrt{6}}{2}, -1\right) \cup \left(\dfrac{-1 + \sqrt{6}}{2}, \infty\right)$
69. $(-\infty, 2 - \sqrt{11}) \cup [0,1] \cup [2 + \sqrt{11}, \infty)$

71. 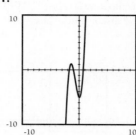   **73.**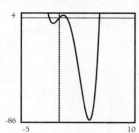

75. $y = -5x - 10$ **77.** $y = 20x - 60$ **81.** $\cong 1.3$
83. $\cong 0.3$ **85.** no   **87.** yes   **89.** $\{-3 - i, -3 + i\}$
91. $\{-2 + i, -2 - i, 1\}$ **93.** $\left\{\dfrac{4}{5} - \dfrac{2i}{5}\right\}$
95. $\left\{-\dfrac{1}{26} + \dfrac{21i}{26}\right\}$ **97.** $\left\{-\dfrac{1}{2} \pm \dfrac{i\sqrt{7}}{2}\right\}$
99. $\left\{-\dfrac{i}{2} \pm \dfrac{\sqrt{12i - 1}}{2}\right\}$ **101.** $\{3i, -i\}$

CHAPTER 4 TEST

1. $P(x) = (x - 1)(x^2 + 3x + 2) + 5$ **3.** $P(-1) = 5$
5. $P(x) = (2x - 3)(x^2 - 2) = (2x - 3)(x + \sqrt{2})(x - \sqrt{2})$
7. $y = 0$

CHAPTER 5

SECTION 5-1

Exercises 5-1

1. Domain: $(-\infty, -2) \cup (-2, \infty)$;
no x-intercepts;
asymptotes: $x = -2, y = 0$

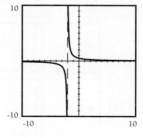

3. Domain: $(-\infty, 3) \cup (3, \infty)$;
x-intercept: $(-5, 0)$;
asymptotes: $x = 3, y = 1$

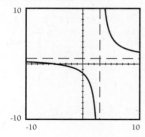

5. Domain: $\left(-\infty, \dfrac{1}{2}\right) \cup \left(\dfrac{1}{2}, \infty\right)$;
x-intercept: $\left(-\dfrac{5}{3}, 0\right)$;
asymptotes: $x = \dfrac{1}{2}, y = \dfrac{3}{2}$

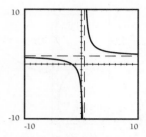

7. Domain: $(-\infty, -2) \cup (-2, 2) \cup (2, \infty)$;
no x-intercept;
removable discontinuity at $x = 2$;
asymptotes: $x = -2$, $y = 0$

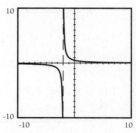

9. Domain: $(-\infty, -2) \cup (-2, \infty)$;
x-intercepts: $\left(\dfrac{5 \pm \sqrt{21}}{2}, 0\right)$;
asymptotes: $x = -2$, $y = x - 7$

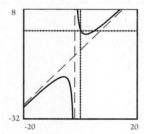

11. Domain: $(-\infty, -4) \cup (-4, 5) \cup (5, \infty)$;
x-intercept: $(2, 0)$;
asymptotes: $x = -4$, $x = 5, y = 0$

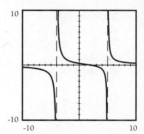

13. Domain: $(-\infty, -4) \cup (-4, \infty)$;
x-intercepts: $(3, 0)$ and $(-2, 0)$;
asymptotes: $x = -4$,
$$y = \dfrac{x}{5} - 1$$

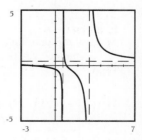

15. Domain: $\left(-\infty, \dfrac{2}{3}\right) \cup \left(\dfrac{2}{3}, 3\right) \ (3, \infty)$;
x-intercepts: $(-2, 0)$ and $(1, 0)$;
asymptotes: $x = 3$, $x = \dfrac{2}{3}, y = \dfrac{1}{3}$

17. Domain: $(-\infty, -5) \cup$ $(-5, 2) \cup (2, \infty)$; x-intercept: $(3, 0)$; removable discontinuity at $x = -5$; asymptotes: $x = 2$, $y = 1$

19. Domain: $\left\{x : x \ne \dfrac{-5 \pm \sqrt{17}}{2}\right\}$; x-intercepts: $(-\sqrt{3}, 0)$, $(0, 0)$ and $(\sqrt{3}, 0)$; asymptotes: $x = \dfrac{-5 - \sqrt{17}}{2}$, $x = \dfrac{-5 + \sqrt{17}}{2}$, $y = x - 5$

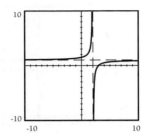

59. x-intercepts: $(1, 0)$, $\left(\dfrac{1 \pm \sqrt{5}}{2}, 0\right)$; asymptote: $x = 2$

61. no x-intercepts; discontinuity at $x = 0$

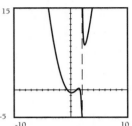

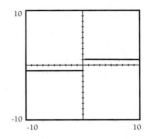

21. $(-1, 2) \cup (5, \infty)$ **23.** $(-\infty, -5) \cup (1, 3)$
25. $[-\sqrt{5}, -\frac{1}{2}] \cup [\sqrt{5}, 3]$ **27.** $[2 - \sqrt{3}, 2 + \sqrt{3}]$

29. $(-\infty, -4) \cup (-\frac{3}{2}, \frac{3}{2}) \cup (4, \infty)$ **31.** $(-\infty, -4) \cup (2, 5)$

33. $(-4, -2] \cup [3, \infty)$ **35.** $[-2, \frac{2}{3}) \cup [1, 3)$

37. $(-\infty, -5) \cup (-5, 2) \cup (3, \infty)$

39. $\left(\dfrac{-5 - \sqrt{17}}{2}, -\sqrt{3}\right) \cup \left(\dfrac{-5 + \sqrt{17}}{2}, 0\right) \cup (0, \sqrt{3})$

41. $[-2 - \sqrt{2}, -3) \cup \left[-2 + \sqrt{2}, \dfrac{3 - \sqrt{5}}{2}\right) \cup \left[2, \dfrac{3 + \sqrt{5}}{2}\right)$

43. $(-\infty, -2 - \sqrt{2}) \cup (-2, -2 + \sqrt{2}) \cup$ $\left[\dfrac{3 - \sqrt{5}}{2}, \dfrac{3 + \sqrt{5}}{2}\right] \cup [3, \infty)$

63. no x-intercept; discontinuity at $x = 0$

65. x-intercept: $(0, 0)$; no asymptotes

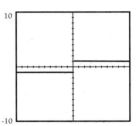

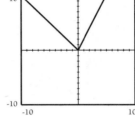

45. $[-\infty, -3) \cup [-2, -\frac{2}{3}) \cup (-\frac{1}{2}, \frac{1}{2}) \cup (\frac{2}{3}, \infty)$

47. $(-\infty, -\frac{2}{3}) \cup [\frac{2}{3}, \infty)$

49. $(-3, -1) \cup \left(-\dfrac{\sqrt{2}}{2}, -\dfrac{2}{3}\right) \cup \left(\dfrac{2}{3}, \dfrac{\sqrt{2}}{2}\right) \cup (1, 3)$

51. $[-1, 1) \cup [2, \infty)$ **53.** $(-\infty, -3) \cup (-1, 2) \cup (2, \infty)$

55. x-intercept: $(-1, 0)$; asymptote: $x = 0$

57. x-intercept: $(1, 0)$; asymptote: $x = 2$

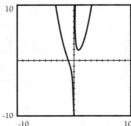

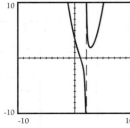

Problem Set 5-1

15. asymptotes: $x = 2$; $y = 2$

17. asymptote: $x = \frac{3}{2}$; $y = \frac{3}{2}$

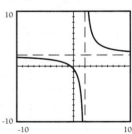

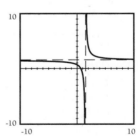

19. asymptotes: $x = \frac{4}{3}$; $y = \frac{4}{3}$

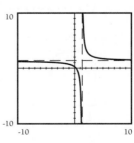

21. $y = \dfrac{b - xd}{xc - a}$ **23.** asymptotes: $x = 3$, $y = 0$

25. removable discontinuity at $x = 3$

27. asymptotes: $x = 3, y = 2$
29. removable discontinuity at $x = 3$
31. asymptotes: $x = 3, y = 2x$

SECTION 5-2

Exercises 5-2

1. Domain: $[\frac{2}{5}, \infty)$ **3.** Domain: $(-\infty, \frac{7}{5}]$

5. Domain: $(-\infty, 2] \cup [4, \infty)$ **7.** $(-\infty, -\frac{1}{2}) \cup (-\frac{1}{2}, \infty)$

9. $(3, \infty)$

11.

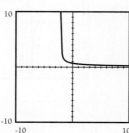

13.

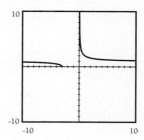

15.

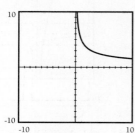

17. asymptotes: $x = -2$, $y = 0$

19. asymptotes: $x = 0$, $y = 0$

21. asymptotes: $x = 0$, $y = 1$

23. asymptotes: $x = -1$, $y = 1$

25.

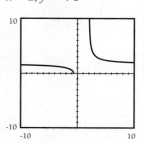

27. asymptotes: $x = 2, y = \sqrt{3}$

29. asymptote: $x = 0$

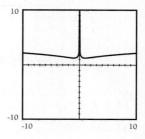

31. {3}   **33.** $\{\frac{11}{5}\}$

35. $\varnothing$. Solution must be in domain.   **37.** $\{\frac{4}{3}\}$ **39.** $\{\frac{3}{2}\}$

41. $\{-\frac{1}{5}\}$ **43.** $\varnothing$ **45.** $\varnothing$

47. {1}. Discard -3.   **49.** {3}. Discard -1.

51. $\{\frac{5}{2}\}$. Discard $-\frac{1}{2}$.   **53.** $\left\{\dfrac{-1 + \sqrt{21}}{4}\right\}$. Discard $\dfrac{-1 - \sqrt{21}}{4}$.

55. {11}   **57.** {4}   **59.** {7}. Discard -1.   **61.** {0, 1}

63. {3}   **65.** {11, -11}   **67.** $[0, \infty)$ **69.** $(-\infty, \infty)$

71. $(3, \infty)$ **73.** $[\frac{2}{5}, \frac{11}{5})$ **75.** $\varnothing$ **77.** $(-\frac{3}{2}, \frac{4}{3})$

79. $(\frac{3}{2}, \infty)$

81.

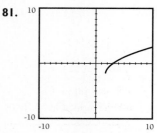

83.

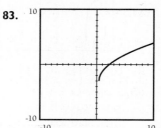

85.

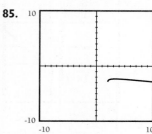

87.

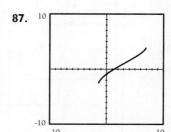

89.

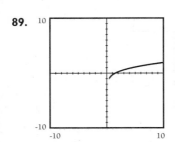

95. true   **97.** false   **99.** true

Problem Set 5-2

1. $\dfrac{-1}{ax}$ **3.** $\dfrac{x + a}{x^2 a^2}$ **5.** $\dfrac{-1}{(x+ 2)(a + 2)}$ **7.** $\dfrac{1}{\sqrt{x} + \sqrt{a}}$

9. $\dfrac{1}{a\sqrt{x} + x\sqrt{a}}$ **11.** $\dfrac{-1}{x(x + h)}$ **13.** $\dfrac{2x + h}{x^2(x + h)^2}$

15. $\dfrac{-1}{(x + h + 2)(x + 2)}$ **17.** $\dfrac{1}{\sqrt{x} + \sqrt{x + h}}$

19. $\dfrac{1}{\sqrt{x}(x + h) + x\sqrt{x + h}}$ **23.** $\cong (2.4, \infty)$ **25.** $\cong [-2, 2.4]$

27. $\cong (3.3, \infty)$ **29.** $\cong [0.95, \infty)$ **31.** $\cong (1.04, 1.86)$

CHAPTER 5 REVIEW EXERCISES

1. Domain: $(-\infty, -5) \cup (-5, \infty)$;
no x-intercepts;
asymptotes: $x = -5, y = 0$

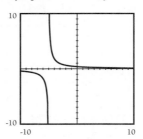

3. Domain: $(-\infty, \frac{3}{2}) \cup (\frac{3}{2}, \infty)$;
x-intercept: $(-\frac{12}{5}, 0)$;
asymptotes: $x = \frac{3}{2}$,
$y = \frac{5}{2}$

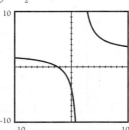

5. Domain: $(-\infty, -2) \cup (-2, \infty)$;
x-intercepts: $(3 \pm \sqrt{6}, 0)$;
asymptotes: $x = -2$,
$y = x - 8$

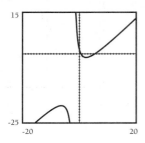

7. Domain: $(-\infty, -5) \cup (-5, \infty)$;
x-intercepts: $(-2, 0)$, $(4, 0)$;
asymptotes: $x = -5$,
$y = \dfrac{x}{7} - 1$

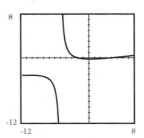

9. Domain: $(-\infty, -5) \cup (-5, 4) \cup (4, \infty)$;
no x-intercepts;
removeable discontinuities at $x = -5$,
and $x = 4$;

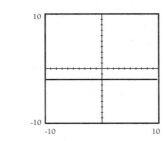

11. $(-8, 1) \cup (7, \infty)$ **13.** $[-\sqrt{11}, -\frac{1}{5}] \cup [\sqrt{11}, 7]$

15. $(-\infty, -5) \cup (-\frac{4}{3}, \frac{4}{3}) \cup (5, \infty)$ **17.** $(-4, -3] \cup [6, \infty)$

19. $(-\infty, -7) \cup (-7, 2) \cup (4, \infty)$

21. $\left(-\infty, \dfrac{-3 - \sqrt{21}}{2}\right) \cup \left(0, \dfrac{-3 + \sqrt{21}}{2}\right]$

23. $\left(-5, \dfrac{-3 - \sqrt{21}}{2}\right] \cup \left(-1, \dfrac{-3 + \sqrt{21}}{2}\right] \cup [1, \infty)$

25. $[1 - \sqrt{6}, -1] \cup [1 + \sqrt{6}, \infty)$

27.

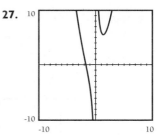

29.

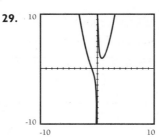

31.

33. 10 years   **35.** $\left[\frac{3}{4}, \infty\right)$ **37.** $(-\infty, \infty)$

39. $[-2, 0] \cup (2, \infty)$

41.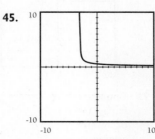

43.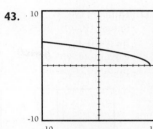

45.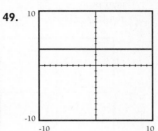

47.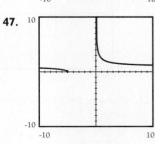

49.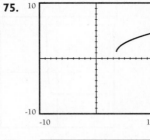

51. $\{\frac{11}{4}\}$ **53.** $\varnothing$ **55.** $\{2\}$ **57.** $\varnothing$ **59.** $\{1\}$; Discard -5.

61. $\{1\}$; Discard $-\frac{5}{4}$.   **63.** $\{6\}$; Discard 198.

65. $\{7\}$; Discard 2.   **67.** $\{1, 5\}$ **69.** $(-\infty, \infty)$

71. $(3, \infty)$ **73.** $\varnothing$

75.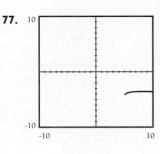

77.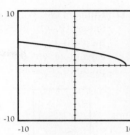

79.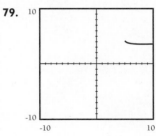

81. $\dfrac{-1}{(x + 1)(a + 1)}$ or $\dfrac{-1}{(x + h + 1)(x + h)}$

83. $\dfrac{-5}{(x + 2)(a + 2)}$ or $\dfrac{-5}{(x + h + 2)(x + 2)}$

CHAPTER 5 TEST

1. Domain: $(-\infty, 3) \cup (3, \infty)$;
x-intercept: $(2, 0)$;
asymptotes: $x = 3, y = 0$

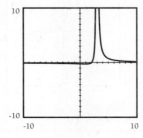

3. $(-\infty, 2)$ **5.** $(2, 3) \cup (3, \infty)$

7.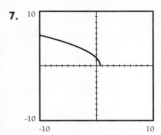

9. $\varnothing$ **11.** $\{8\}$; Discard 288.

CHAPTER 6

SECTION 6-1

Exercises 6-1

1. Domain: $(-\infty, \infty)$,
Range: $(0, \infty)$;
increasing;
asymptote: $y = 0$;
pivot: $(0, 1)$

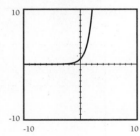

3. Domain: $(-\infty, \infty)$,
Range: $(0, \infty)$;
decreasing;
asymptote: $y = 0$;
pivot: $(0, 1)$

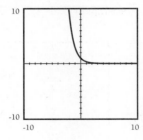

5. Domain: $(-\infty, \infty)$,
Range: $(0, \infty)$;
decreasing;
asymptote: $y = 0$;
pivot: $(0, 1)$

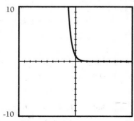

7. Domain: $(-\infty, \infty)$,
Range: $(0, \infty)$;
increasing;
asymptote: $y = 0$;
pivot: $(0, 1)$

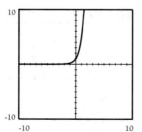

17. Domain: $(-\infty, \infty)$,
Range: $(0, \infty)$;
increasing;
asymptote: $y = 0$;
pivot: $(0, 1)$

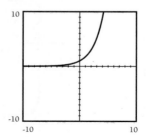

19. Domain: $(-\infty, \infty)$,
Range: $(2, \infty)$;
increasing;
asymptote: $y = 2$;
pivot: $(1, 3)$

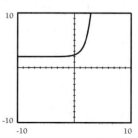

9. Domain: $(-\infty, \infty)$,
Range: $(0, \infty)$;
decreasing;
asymptote: $y = 0$;
pivot: $(0, 1)$

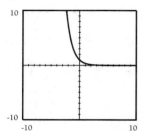

11. Domain: $(-\infty, \infty)$,
Range: $(0, \infty)$;
increasing;
asymptote: $y = 0$;
pivot: $(-1, 1)$

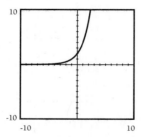

21. Domain: $(-\infty, \infty)$,
Range: $(-1, \infty)$;
increasing;
asymptote: $y = -1$;
pivot: $(-2, 0)$

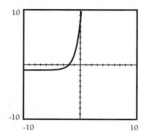

23. Domain: $(-\infty, \infty)$,
Range: $(-\infty, 1)$;
decreasing;
asymptote: $y = 1$;
pivot: $(1, -1)$

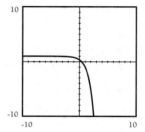

13. Domain: $(-\infty, \infty)$,
Range: $(1, \infty)$;
increasing;
asymptote: $y = 1$;
pivot: $(0, 2)$

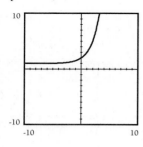

15. Domain: $(-\infty, \infty)$,
Range: $(0, \infty)$;
increasing;
asymptote: $y = 0$;
pivot: $(0, 1)$

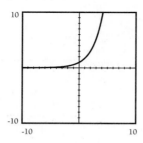

25. Domain: $(-\infty, \infty)$,
Range: $(1, \infty)$;
decreasing;
asymptote: $y = 1$;
pivot: $(1, 3)$

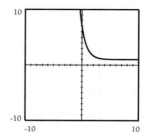

27. Domain: $(-\infty, \infty)$,
Range: $(-4, \infty)$;
increasing;
asymptote: $y = -4$;
pivot: $\left(\frac{3}{2}, -2\right)$

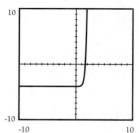

29. Domain: $(-\infty, \infty)$, Range: $(-\infty, -4)$;
increasing;
asymptote: $y = -4$;
pivot: $\left(\frac{3}{2}, -6\right)$

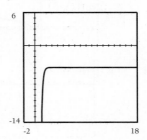

31. $A(t) = 0.1\left(\frac{1}{2}\right)^{t/5700}$ **33.** $A(t) = A_0\left(\frac{1}{2}\right)^{t/3.05}$

35. $P(t) = 20(2)^t$; $P(3) = 160$; $P(-1) = 20$, $t > 1$
37. $V(t) = \$1000e^{0.05t}$; $\$1161.83$
39. $f(x) = 3^{2x}$ **41.** $f(s) = 25^{s/2}$
49. $g(1) \cong 2.7083$; $g(0.05) \cong 1.6484$; $g(0.1) \cong 1.1052$

Problem Set 6-1

1. $\cong 2.0544$ **3.** $\$205.44$ **5.** $\$205.44$ **7.** $\cong 2.05443311P$
9. 18 years   **11. (a)** 5   **(b)** $\$365,982.34$
13. (a) 4   **(b)** $\cong \$160,000$
15. (a) $\$11,051.70$ **(b)** $\$10,832.87$ **(c)** $\$218.83$
   (d) $\$187,839.61$
17. 69: $\{1, 3, 23, 69\}$; 70: $\{1, 2, 5, 7, 10, 14, 35, 70\}$; 71: $\{1, 71\}$;
   But the most divisors is 72: $\{1, 2, 3, 4, 6, 8, 9, 12, 18, 24, 36, 72\}$.
21. Secant $(f, x, h) = \dfrac{e^{x+h} - e^x}{h}$, $h \neq 0$.   **23.** $\cong 1.051709$

25. $\cong 1.0005001$ **27.** $\cong 1.000005$ **29.** $\cong 1.000005e^x$
31. Average rate of change at (x, e^x) is e^x.
35. (b) approximately same graph
37. approximately same graph   **41.** inverse functions
43. inverse functions   **45.** inverse functions
47. same graph   **49.** same graph

SECTION 6-2

Exercises 6-2

1. $\log_2 16 = 4$ **3.** $\log_5 125 = 3$ **5.** $\log_3\left(\frac{1}{9}\right) = -2$

7. $\log_{36} 6 = \frac{1}{2}$ **9.** $\log_{27}\left(\frac{1}{9}\right) = -\frac{2}{3}$ **11.** $3^2 = 9$

13. $3^5 = 243$ **15.** $5^{-2} = \frac{1}{25}$ **17.** $25^{1/2} = 5$

19. $81^{-1/2} = \frac{1}{9}$

21. asymptote: $x = 0$;
pivot: $(1, 0)$

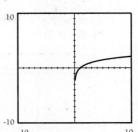

23. asymptote: $x = 0$;
pivot: $(1, 0)$

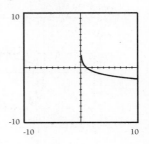

25. asymptote: $x = -3$;
pivot: $(-2, 0)$

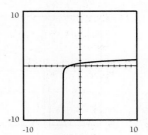

27. asymptote: $x = 0$;
pivot: $(1, 0)$

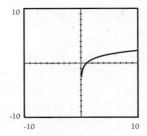

29. asymptote: $x = 0$;
pivot: $(1, 0)$

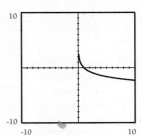

31. asymptote: $x = 0$;
pivot: $(1, 1)$

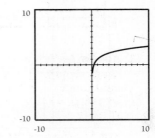

33. asymptote: $x = 0$;
pivot: $(1, -2)$

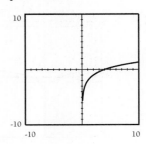

35. asymptote: $x = 0$;
pivot: $(1, \log_2 3)$

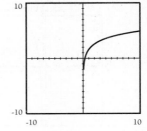

37. asymptote: $x = -1$;
pivot: $(0, 0)$

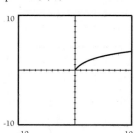

39. asymptote: $x = 0$;
pivot: $(1, 0)$

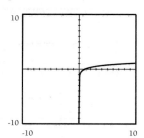

41. asymptote: $x = 0$;
pivot: $(\frac{1}{2}, 1)$

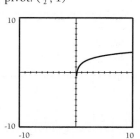

43. asymptote: $x = -1$;
pivot: $(0, 0)$

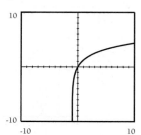

45. asymptote: $x = 2$;
pivot: $(1, 0)$

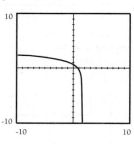

47. asymptote: $x = -\frac{1}{2}$;
pivot: $(0, -4)$

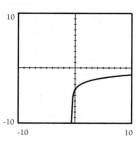

49. asymptote: $x = \frac{2}{3}$;
pivot: $(\frac{5}{3}, 5)$

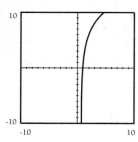

51. asymptote: $x = 0$

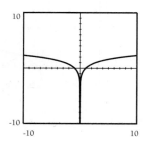

53. asymptote: $x = 1$

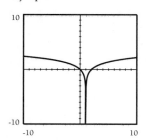

55. asymptote: $x = 0$

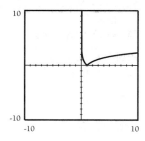

57. Same graph
59. same graph

Problem Set 6-2

3. $\dfrac{a^2 - 1}{2a}$ **5.** $\dfrac{b^2 - a^2}{2ab}$ **7.** $\cong \dfrac{b^2 - a^2}{2ab} c$

27. $L(a^n) = nL(a)$ **31.** $\cong 0.693147$ **33.** $\cong -0.693147$
35. $\cong 0.223143$ **37.** $\cong -0.223143$

41. $\ln\left(\dfrac{1}{t}\right) = -\ln(t)$; $\displaystyle\int_1^t \dfrac{1}{x}\,dx = \ln t$ **43.** same graph

45. same   **47.** same   **49.** same   **51.** same

SECTION 6-3

Exercise 6-3

1. $\ln 15 - \ln y$ **3.** $3 + 7 \log_3 x$ **5.** $3 \ln x + 2 \ln y$
7. $\ln 10 - \frac{1}{2} \ln x$ **9.** $\ln (x - 3) + \ln (x - 4)$ **11.** $\log(xy^3)$

13. $\log\left(\dfrac{\sqrt[3]{x^2}}{y}\right)$ **15.** $\ln\left(\dfrac{x^5}{w\,y^2}\right)$ **17.** $\log 800$

19. $\ln\left(x^{\frac{\ln 10\,e}{\ln 10}}\right)$ **21.** $x = \log_5 7$ **23.** $x = \log_3 10 - 2$

25. $x = 5$ **27.** $x = \dfrac{-\ln 5488}{\ln\left(\dfrac{7}{4}\right)}$ **29.** $x = \dfrac{1 + \ln 1000}{\ln 10 - 1}$

31. $x = 1$ **33.** $x = 6$ **35.** $x = 4$ **37.** $x = 4$ **39.** $x = 1$
41. $x = 0$ **43.** $x = 0$ or $x = \log 2$ **45.** $x = 0$ **47.** $\varnothing$
49. $x = -4 + \sqrt{19}$ **51.** $x = \frac{1}{3}$ **53.** $x = 1$ or $x = 10^9$

55. $f_1 = 2^{C/1200} f_2$ **57.** $k = \dfrac{1}{t} \ln\left(\dfrac{C}{T - T_f}\right)$

59. $T_f = T - Ce^{-kt}$ **61.** $R = \dfrac{L}{t} \ln\left(\dfrac{I_s}{I_s - I}\right)$

63. $L = \dfrac{-Rt}{\ln\left(\dfrac{I_s - I}{I_s}\right)}$ **65.** 0

69. $10 \log_2 5 \cong 23.219$ years   **71.** 10 years

73. $\cong 19.3856$ hours; colony A starts larger. Colony B grows faster.

75. $\cong 23.1049$ years   **79.** $f_2 = f_1 e^{-\text{cents}(\ln 2)/1200}$; 300 cps

Problem Set 6-3

1. $\cong 2.6265$ **3.** ANS := EXP(p *LN(b))

SECTION 6-4

Exercise 6-4

1. $\cong 2.374748$; mantissa $= 0.374748$; characteristic $= 2$

3. $\cong -2.336299$; mantissa $= 0.663701$; characteristic $= -3$

5. $\cong 4.663701$; mantissa $= 0.663701$; characteristic $= 4$

7. $\cong 1.749736$; mantissa $= 0.749736$; characteristic $= 1$

9. $\cong 5.749736$; mantissa $= 0.749736$; characteristic $= 5$

11. $\cong 2.321928$; mantissa $= 0.321928$; characteristic $= 2$

13. $\cong 0.430677$; mantissa $= 0.430677$; characteristic $= 0$

15. $\cong 2.302585$; mantissa $= 0.302585$; characteristic $= 2$

17. $\cong 1.386294$; mantissa $= 0.386294$; characteristic $= 1$

19. $\cong 0.935785$; mantissa $= 0.935785$; characteristic $= 0$

21. $\cong 834.065$ **23.** $\cong 0.00534$ **25.** $\cong 0.268$ **27.** $\cong 0.0268$

29. $\cong 26.798$ **31.** $\cong 0.553762$ **33.** $\cong 2.271507$

35. 6.892946 **37.** $\cong -3.096064$ **39.** $\cong 7.475499$

41. $\cong 8334.892445$ **43.** $\cong 6.649888$ **45.** $\cong 3.087463$

47. $\cong 6.301320$ **49.** $\cong 0.302585$ **51.** $t + s$ **53.** $4s - 3t$

55. $\dfrac{2t}{3}$ **57.** $t - 3$ **59.** $10^t s$ **61.** $f(z) = 0.3s + 0.7t$

CHAPTER 6 REVIEW EXERCISES

1. Domain: $(-\infty, \infty)$, Range: $(0, \infty)$; increasing; asymptote: $y = 0$

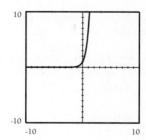

3. Domain: $(-\infty, \infty)$, Range: $(0, \infty)$; decreasing; asymptote: $y = 0$

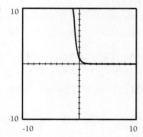

5. Domain: $(-\infty, \infty)$, Range: $(0, \infty)$; decreasing; asymptote: $y = 0$

7. Domain: $(-\infty, \infty)$, Range: $(-5, \infty)$; increasing; asymptote: $y = -5$

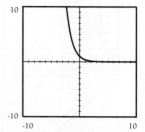

9. Domain: $(-\infty, \infty)$, Range: $(0, \infty)$; increasing; asymptote: $y = 0$

11. Domain: $(-\infty, \infty)$, Range: $(-1, \infty)$; increasing; asymptote: $y = -1$

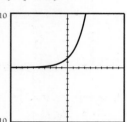
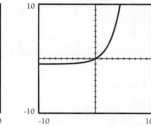

13. Domain: $(-\infty, \infty)$, Range: $(12, \infty)$; increasing; asymptote: $y = 12$

15. Domain: $(-\infty, \infty)$, Range: $(-\infty, \infty)$; increasing; asymptote: $y = x$

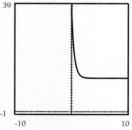
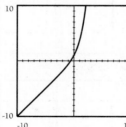

17. $A(t) = 5\left(\frac{1}{2}\right)^{t/24000}$ **19.** $g(x) = 5^{2(x-3)}$ **21.** $\cong \$4000$

23. $\cong 14.4$ years   **25.** $\log_2(0.0625) = -4$

27. $\log_{125}(25) = \frac{2}{3}$ **29.** $\log_{81}(3^2) = \frac{1}{2}$ **31.** $27^{2/3} = 9$

33. $25^{-1} = \frac{1}{25}$ **35.** $\left(\frac{1}{3}\right)^{-2} = 9$

37.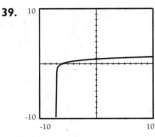

39.

41.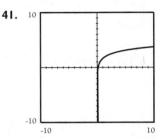

43.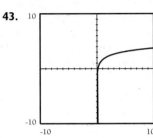

45.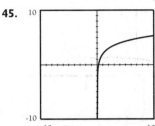

47.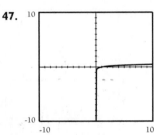

49.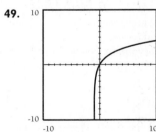

51.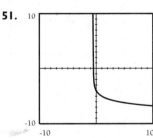

53.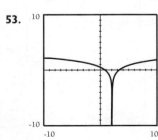

55.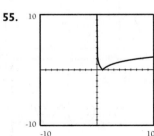

57. same

59. $\ln x + \frac{1}{2} \ln y + \frac{1}{4} \ln z - \ln 3 - 2 \ln w$

61. $\log\left(\dfrac{17x^5}{z^3\sqrt{y}}\right)$ **63.** $x = \frac{1}{2}$ **65.** $x = 5$

67. $\dfrac{1 + \ln(10000)}{\ln 10 - 1}$ **69.** $x = 4$ **71.** $x = 3$ **73.** $\varnothing$

75. $x = \ln 3$ **77.** $x = \ln(1 + \sqrt{2})$
79. $x = 5$ or $x = 1$ **81.** $\cong$ \$32,664.68; $\cong$ 33.5817 years
83. $\cong$ 2.4116   **85.** $\cong$ 5.4116   **87.** $\cong$ 3.7016
89. 1.3800   **91.** $\cong$ 3.6889   **93.** $\cong$ 8340.652
95. $\cong$ 0.534   **97.** $\cong$ 0.097   **99.** $\cong$ 0.552911
101. $\cong$ 6.552911   **103.** $\cong$ 833.4892445   **105.** $\cong$ 0.464974
107. $\cong$ 1.651293   **109.** $k + 2m$ **111.** $3k - m + 2$

CHAPTER 6 TEST

1. Domain: $(-\infty, \infty)$; Range: $(5, \infty)$; decreasing

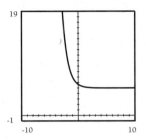

3. (a) $\log\left(\dfrac{100x^3}{\sqrt{y}}\right)$ **(b)** $\frac{1}{3}(\log x + \frac{1}{2} \log y) - \log z$

5. $x \cong 1.8$ **7.** $x = -1 + 2\sqrt{26}$ **9.** $\cong$ 2.31049 years
11. $\cong$ 60,384.53718

CHAPTER 7

SECTION 7-1

Exercises 7-1

1. 180°   **3.** 270°   **5.** 900°   **7.** -45°   **9.** 300°   **11.** 30°
13. 105°   **15.** 330°   **17.** $\cong$ 28.648°   **19.** $\cong$ 57.296°
21. $\dfrac{5\pi}{12}$ **23.** $\dfrac{\pi}{6}$ **25.** $\dfrac{19\pi}{36}$ **27.** π **29.** $\dfrac{2\pi}{3}$
31. $\dfrac{25\pi}{18}$ **33.** $-\dfrac{\pi}{6}$ **35.** 4π **37.** $\dfrac{5\pi}{4}$ **39.** $\dfrac{5\pi}{3}$
41. 7200°/min   **43.** 2 rev/min   **45.** 120°/sec
47. 120 rev/min   **49.** 432000°/min   **51.** 4π rad/min
53. 18°/sec   **55.** 360°/hr   **57.** 3240°/min   **59.** 1800°/sec
61. 5π feet   **63.** $\dfrac{5\pi}{2}$ meters   **65.** $\dfrac{10}{3}$ cm   **67.** $\dfrac{\pi}{2}$ inches
69. 2π inches   **71.** 10000π c/m   **73.** 10π in/hr
75. $\cong$30 m/hr   **77.** $\cong$24 m/hr   **79.** 6.28 in/sec
83. Angular velocity is the same for both.

Problem Set 7-1

1. $\cong$1.85 km   **3.** 28   **5.** 60   **7.** $\cong$111.11 km

9. 600 hours   **11.** 35° 52′ 48″   **13.** 125° 46′ 31″
15. 309° 53′ 18″   **17.** -13° 54′ 20″   **19.** 513° 52′ 48″
21. 23.71028°   **23.** 247.79972°   **25.** 109.16194°
27. 59.29583°   **29.** 10.00833°   **31.** 2.00000 rad   **33.** II
35. II   **37.** I   **39.** IV

SECTION 7-2

Exercises 7-2

1. 3.5   **3.** $\frac{45}{8}$ **5.** 30   **7.** $\frac{49}{4}$ **9.** $\frac{8}{5}$ **11.** $5\sqrt{3}$ **13.** 7

15. 12   **17.** 5   **19.** $4\sqrt{3}$ **21.** $\frac{4\sqrt{3}}{3}$ **23.** $4\sqrt{3}$ **25.** 5

27. $5\sqrt{2}$ **29.** $20\sqrt{2}$ **31.** $\frac{1}{2}$ **33.** $\frac{1}{2}$ **35.** $\frac{\sqrt{2}}{2}$ **37.** 1

39. $\frac{\sqrt{3}}{2}$ **41.** 0.25882   **43.** 0.25882   **45.** 0.64279

47. 0.76604   **49.** 1.19175   **51.** 0.60182   **53.** 0.99255
55. 0.46175   **57.** 0.62475   **59.** 3.43084   **61.** 0.01775
63. 1   **67.** 0   **69.** undefined   **71.** -1   **73.** $\frac{\sqrt{2}}{2}$

75. $-\sqrt{3}$ **77.** $\frac{\sqrt{3}}{2}$ **79.** $\frac{\sqrt{2}+\sqrt{6}}{4} \cong 0.96593$

81. $-\frac{\sqrt{3}}{3}$ **83.** $\cong 0.07074$ **85.** $\cong 0.84147$

Problem Set 7-2

1. $\left(\frac{1}{2}, \frac{\sqrt{3}}{2}\right)$ **3.** Central angle is 60°.

5. Central angle is 45°.   **7.** $\left(\frac{\sqrt{3}}{2}, \frac{1}{2}\right)$ **9.** $\frac{1}{2}$ **11.** $\frac{\sqrt{2}}{2}$

13. $\frac{\sqrt{3}}{2}$ **15.** They are equal.

17. (and 19, 21, 23, 25): Graphs are identical.
27. The functions f and g form an identity $f = g$, if their graphs are the same.

SECTION 7-3

Exercises 7-3

1. $\frac{1}{2}$ **3.** $\frac{\sqrt{3}}{3}$ **5.** 2   **7.** $\frac{\sqrt{3}}{3}$ **9.** $\sqrt{3}$ **11.** $\sqrt{2}$

13. $\sqrt{2}$ **15.** 1   **17.** 2   **19.** $\cong 0.83910$ **21.** $\cong 1.55572$

23. $\cong 1.30541$
81. $(\sin x \cos y - \cos x \sin y)/(\cos x \cos y + \sin x \sin y)$
83. $(\cos t\, \theta \sin \theta)/(\sin \theta \sin \theta + \cos \theta \cos \theta)$
85. $\sin x$ **87.** $\cos \phi$ **89.** $\sin \theta$
93. Cannot divide by $x - 1$ because $x = 1$.

Problem Set 7-3

1. $\cos^2 \theta + \sin^2 \theta = 1$ **3.** $\cos^2 \theta + \sin^2 \theta = 1$
5. $\cos^2 \theta + \sin^2 \theta = 1$ **7.** $\sec^2 t - \tan^2 t = 1$
9. $\sec^2 t - \tan^2 t = 1$
11. Let $x = 4 \cos \theta$; $4 \sin \theta$
13. Let $x = 4 \tan \theta$; $4 \sec \theta$
15. Let $x = 5 \cos \theta$; $\tan \theta$
17. Let $x = 3 \sec \theta$; $\sec \theta$
19. Let $x = 2 + 3 \tan \theta$; $\sin \theta$
21. identity   **23.** conditional (but close)   **25.** conditional
27. conditional   **29.** identity   **31.** identity
33. k affects the period.   **35.** k affects the period.
37. k affects range.   **39.** k affects range.
41. k translates horizontally.   **43.** k translates vertically.

SECTION 7-4

Exercises 7-4

1. amplitude = 5,
  period = π;
  frequency = $\frac{1}{\pi}$,
  phase shift = 0

3. amplitude = 3,
  period = π;
  frequency = $\frac{1}{\pi}$,
  phase shift = 0

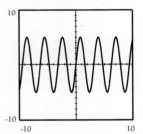

5. amplitude = $\frac{1}{2}$,
  period = $\frac{\pi}{2}$,
  frequency = $\frac{2}{\pi}$,
  phase shift = 0

7. amplitude = 1,
  period = 2π;
  frequency = $\frac{1}{2\pi}$,
  phase shift = $\frac{\pi}{4}$

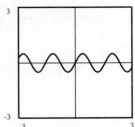
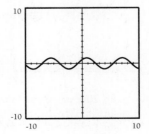

9. amplitude = 1,
period = 2π,
frequency = $\dfrac{1}{2\pi}$,
phase shift = $-\pi$

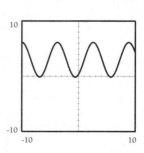

11. amplitude = 1,
period = 6π,
frequency = $\dfrac{1}{6\pi}$,
phase shift = 0

13. amplitude = 2,
period = 2π,
frequency = $\dfrac{1}{2\pi}$,
phase shift = 0

15. amplitude = 3,
period = 2π,
frequency = $\dfrac{1}{2\pi}$,
phase shift = $\dfrac{\pi}{6}$

17. amplitude = π,
period = 2π,
frequency = $\dfrac{1}{2\pi}$,
phase shift = 1

19. amplitude = 2,
period = 2,
frequency = $\dfrac{1}{2}$,
phase shift = $-\dfrac{3}{\pi}$

21. $y = \sin(2000\pi t)$, 0.3313 meters
23. $y = \sin(6\pi 10^6 t)$, 100 meters
25. $y = \sin(12\pi 10^{16} t)$, 0.0005 mm
27. $y = \sin(2000\pi t) \sin(6000000\pi t)$
29. $y = \sin(6\pi 10^6 t) \sin(12\pi 10^{14} t)$

Problem Set 7-4

1. Graph b "bounded" by graph a.
3. Graph b "bounded" by graph a.
5. Graph b "bounded" by graph a.
7. Graph b "centered along" graph a.
9. Graph b "centered along" graph a.
11. Both symmetric to y-axis.
13. Both symmetric to y-axis.
15. Graph b "bounded" by graph a.
17. Graph b "bounded" by graph a.
19. Graphs are identical.
21. Answers may vary according to technology used.
23. Answers may vary according to technology used.
25. Answers may vary according to technology used.
27. Answers may vary according to technology used.
29. Answers may vary according to technology used.
31. k affects period.   **33.** k affects period.
35. k affects range.   **37.** k affects "steepness."
39. k translates horizontally.   **41.** k translates horizontally.
43. k translates vertically.

SECTION 7-5

Exercises 7-5

Use technology to see graphs.

1.
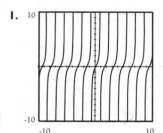

3.
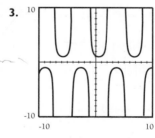

5.
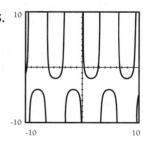

7.
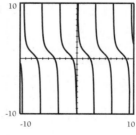

9.
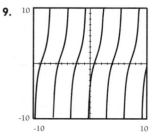

11.
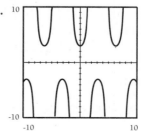

13.

15.

17.

19.

21. (b)

23. (b)

25. (b)

27. (b)

29. (b)

Problem Set 7-5

1. (a) $m \cong \cos\left(\dfrac{\pi}{3}\right)$ **(b)** $m \cong \cos\left(\dfrac{\pi}{6}\right)$ **(c)** $m \cong \cos\left(\dfrac{\pi}{4}\right)$

(d) $m \cong \cos 0$

3. identity   **5.** identity   **7.** conditional   **9.** identity
11. identity   **13.** identity   **15.** seems like an identity
17. conditional   **19.** conditional   **21.** seems like an identity
23. identity   **25.** identity   **27.** identity   **29.** identity
31. identity

SECTION 7-6

Exercises 7-6

23. $\dfrac{\sqrt{6} + \sqrt{4}}{4}$ **25.** $\dfrac{\sqrt{6} + \sqrt{4}}{4}$ **27.** $\dfrac{\sqrt{2} - \sqrt{6}}{4}$

29. $-\dfrac{1}{2}$ **31.** $-\dfrac{\sqrt{2}}{2}$ **33.** $\cos(10x)$ **35.** $-\cos(12x)$

37. $\cos(4x)$ **39.** $2 \sin (5x) \cos(2x)$
41. $-2 \sin(7x) \sin(-2x) = 2 \sin(7x) \sin(2x)$
43. $\sqrt{0.99} \cong 0.99499$ **45.** $\sqrt{0.96} \cong 0.97980$
47. $0.2\sqrt{0.99} - 0.1\sqrt{0.96} \cong 0.10102$
49. $0.02 - \sqrt{0.9504} \cong -0.95489$
51. $\dfrac{0.02 + \sqrt{0.9504}}{0.2\sqrt{0.99} - 0.1\sqrt{0.96}} \cong 9.848597$

   53. $0.6\sqrt{0.91} \cong 0.57236$
55. $0.3\sqrt{0.91} - 0.41\sqrt{3} \cong -0.42400$

Problem Set 7-6

1. $2 \cos A \cos B$ **3.** $2 \cos A \sin B$

5. $\dfrac{\cos(A + B) + \cos(A - B)}{2}$

7. $\dfrac{\sin(B + A) + \sin(B - A)}{2}$

9. $\dfrac{\cos u + \cos v}{2}$ **11.** $\dfrac{\sin u - \sin v}{2}$

13. $\cos\left(\dfrac{u + v}{2}\right)\cos\left(\dfrac{u - v}{2}\right) + \sin\left(\dfrac{u + v}{2}\right) \sin\left(\dfrac{u - v}{2}\right)$

15. $\sin\left(\dfrac{u + v}{2}\right)\cos\left(\dfrac{u - v}{2}\right) + \cos\left(\dfrac{u + v}{2}\right) \sin\left(\dfrac{u + v}{2}\right)$

17. $\dfrac{\sqrt{2}}{2}$ **19.** $\dfrac{\sqrt{6}}{2}$ **21.** $\dfrac{\sqrt{3} + 1}{4}$ **23.** $\dfrac{1 + \sqrt{3}}{4}$

SECTION 7-7

Exercise 7-7

1. $x - \dfrac{x^3}{3!}$ **3.** $1 - \dfrac{x^2}{2!} + \dfrac{x^4}{4!}$ **5.** $x - \dfrac{x^3}{3!} + \dfrac{x^5}{5!} - \dfrac{x^7}{7!}$

7. $1 - \dfrac{x^2}{2!} + \dfrac{x^4}{4!} - \dfrac{x^6}{6!} + \dfrac{x^8}{8!} - \dfrac{x^{10}}{10!}$

9. $1 + \dfrac{x^1}{1!} + \dfrac{x^2}{2!} + \dfrac{x^3}{3!} + \dfrac{x^4}{4!} + \dfrac{x^5}{5!} + \dfrac{x^6}{6!}$

11. 720   **13.** 30   **15.** 30   **17.** 210   **19.** 4845
21. $\cong 0.84147$ **23.** $\cong -0.90794$ **25.** $= 0$
27. $\cong 0.540278$ **29.** $\cong 0.42222$ **31.** $y = x$; 1 point

33. May exceed the resolution of your technology.

Problem Set 7-7

1. 0   **3.** ≅ 1.10447   **5.** ≅ 0.724806   **7.** ≅ 1.034270
9. ≅ 0.597154

CHAPTER 7 REVIEW EXERCISES

1. −540°   **3.** 20°   **5.** 250°   **7.** $\dfrac{\pi}{12}$ **9.** $\dfrac{\pi}{36}$

11. $-\dfrac{5\pi}{3}$ **13.** $-\dfrac{2\pi}{5}$ **15.** 18000°/min   **17.** 300 rev/hr

19. 60°/sec   **21.** 10.5°/sec   **23.** $2\pi''$ **25.** $5\sqrt{3}$ **27.** 8
29. $5\sqrt{2}$ **31.** $\dfrac{\sqrt{3}}{2}$ **33.** $\dfrac{\sqrt{2}}{2}$ **35.** $\dfrac{\sqrt{3}}{2}$

37. ≅ 0.731354   **39.** ≅ 0.087489   **41.** ≅ 0.930719
43. ≅ 0.707107   **45.** 0   **47.** ≅0.965926   **49.** ≅ 0.621610
51. $\dfrac{\sqrt{3}}{2}$ **53.** $\dfrac{2\sqrt{3}}{3}$ **55.** 1   **57.** $\sqrt{2}$ **59.** ≅1.555724

71. $(\cos x \sin y - \cos y \sin x)\backslash(\sin x \sin y + \cos x \cos y)$
73. $\cos^2 x/\sin x$ **75.** $\cos \phi$

77.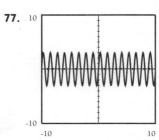

79.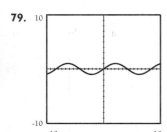

/1.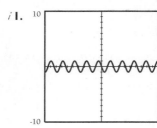

83.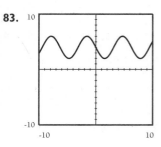

85.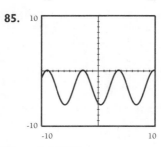

87. ≅ 0.647070 m

89.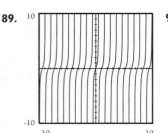

91.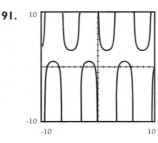

93.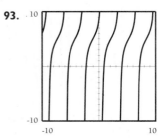

95.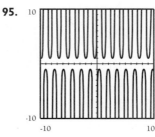

97.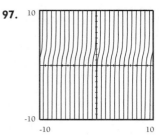

99. (b)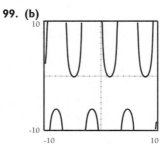

101. (b)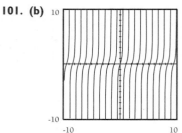

107. $\dfrac{\sqrt{6} - \sqrt{2}}{4}$ **109.** $\dfrac{\sqrt{2} + \sqrt{6}}{4}$ **111.** $-\dfrac{\sqrt{2}}{2}$

113. $\sqrt{0.84} \cong 0.916515$ **115.** $\sqrt{0.4284} + 0.28 \cong 0.934523$
123. $2 \sin a \cos B$ **125.** $\dfrac{\sin(A + B) + \sin(A - B)}{2}$

127. $\dfrac{\sqrt{2}}{2}$ **129.** $\dfrac{\sqrt{3} + 1}{4}$ **131.** $x - \dfrac{x^3}{3!}$

CHAPTER 7 TEST

1. $4.5\pi''/\text{sec} = 270\pi''/\text{min}$

7. 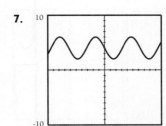 **9.**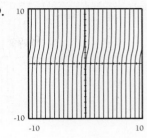

11. $\frac{63}{65} \cong 0.969231$

CHAPTER 8

SECTION 8-1

Exercises 8-1

1. $B = 40°; a = 10 \tan 50° \cong 11.917536; c \cong 15.557238$

3. $B = 70°; b = 5 \tan 70° \cong 13.737387; c \cong 14.619022$

5. $B = 50°; a = 8 \sin 40° \cong 5.142301; b = 8 \cos 40° \cong$ 6.128356

7. $A = 60°; c = 20; a = 10\sqrt{3}$

9. $A = 8°; b = 5 \tan 80° \cong 35.5765849; c = \dfrac{5}{\cos 82°} \cong$ 35.926483

11. $A \cong 27.8°; B \cong 62.2°; b = \sqrt{176} \cong 13.266499$

13. $A \cong 25°; B \cong 65°; c = \sqrt{274} \cong 16.552945$

15. $A \cong 78.5°; B \cong 11.5°; a = \sqrt{96} \cong 9.797959$

17. Not a right triangle.

19. $A \cong 39.5°; B \cong 50.5°; b = \sqrt{72} \cong 8.485281$

21. **(a)** $\theta = 50°; \phi = 20°$

 (b) $h = 10 \sin 40° \cong 6.427876; x = 10 \cos 40° \cong 7.660444$

 (c) $y = \dfrac{10 \sin 40° \cos 70°}{\sin 70°} \cong 2.339556;$

 $a = \dfrac{10 \sin 40°}{\sin 70°} \cong 6.840403$

 (d) $c = 10$

23. $C = 75°; b = \dfrac{5\sqrt{3}}{\sin 45°} \cong 12.247449;$

 $c = 5 + 5\sqrt{3} \cong 13.66625$

25. 2000 meters   **27.** $\frac{1}{2}$ meter   **29.** $\cong 30°$ **31.** 5 km

33. $\cong 87.1557$ meters

35. time = 6.25 seconds; horizontal distance $625\sqrt{3}$ ft $\cong$ 1082.531755 ft.

37. Yes, by at least 8 ft. Airborne distance > 68.0625 ft.

Problem Set 8-1

1. **(a)** infinitely many   **(b)** $(0.201358, 0.2)$
 (c) $(2.940235, 0.2)$ **(d)** $\cong 0.2013579$ **(e)** $\cong 2.9402347$

3. $y = \tan^{-1} x$ appears to be a mirror image of

$$y = \tan x \left(-\frac{\pi}{2} < x < \frac{\pi}{2}\right)$$

about the line $y = x$. The graph of $y = \tan x$ consists of infinitely many piecewise continuous curves whereas the graph of $y = \tan^{-1} x$ consists of a single continuous curve.

5. $[0, \pi]$

SECTION 8-2

Exercise 8-2

1. $\frac{\pi}{6}$ **3.** $\frac{2\pi}{3}$ **5.** $-\frac{\pi}{4}$ **7.** $\frac{\pi}{4}$ **9.** $\frac{\pi}{4}$

11. $\cong 0.304692654$ **13.** $\cong 2.214297436$

15. $\cong 0.3947911197$ **17.** $\cong 0.3947911197$

19. $\cong 1.107148718$

21. $\cot^{-1} x = y$ where $\cot y = x$ and $x \in (0, \pi)$

23. $\sec^{-1} x = y$ where $\sec y = x$ and $x \in \left[0, \dfrac{\pi}{2}\right) \cup \left(\pi, \dfrac{3\pi}{2}\right]$

25. $\frac{\pi}{4}$ **27.** $\frac{3\pi}{4}$ **29.** $\frac{\pi}{3}$

31. $\dfrac{2\pi}{3}$ or $\dfrac{4\pi}{3}$, depending on the definition

33. 0.4   **35.** 5   **37.** $\frac{\pi}{3}$ **39.** $\frac{\pi}{6}$ **41.** $\frac{5\pi}{6}$ **43.** $\frac{\pi}{3}$

45. $\frac{\pi}{3}$ **47.** $\frac{\sqrt{3}}{2}$ **49.** $\frac{1}{2}$ **51.** $\sqrt{1 - x^2}$

53. $\tan^{-1}\left(\dfrac{x}{\sqrt{1 - x^2}}\right)$ **55.** $\tan^{-1}\left(\dfrac{1}{x}\right)$ **57.** $\cong 0.463647609$

59. $\cong -0.2914567945$ **61.** $\cong 0.2013579208$

63. For $x = -0.2$, $\tan^{-1}\left(\dfrac{\sqrt{1 - x^2}}{x}\right) \cong -1.369438406$. Because of the range differences in $\tan^{-1}$ and $\cos^{-1}$, the actual value is $\cos^{-1}(-0.2) \cong 1.772154248$.

65. $\cong 1.04719755$ (Note: Exact value is $\dfrac{\pi}{3}$.)

67. $\cong -1.1.07148718$ [Because of range differences in $\tan^{-1}$ and $\cot^{-1}$, the actual value is $\cot^{-1}(-0.5) \cong 2.034443936$]

69. No answer because 3 is not in the domain of $\sin^{-1}$.

71. No answer because 0.5 is not in the domain of $\csc^{-1}$.

77. $\dfrac{5}{15 + 4x^2}$ **79.** $\dfrac{315 + 161x^2}{945 + 735x^2 + 64x^4}$

81. See the answer to problem 80 and the discussion in problem 82.

83. $y = \sin\left(\dfrac{x}{2}\right) - 1$ **85.** $y = \dfrac{2 - \cot^{-1} x}{3}$

87. $\cong 812{,}000$ miles

89. The actual diameter of the sun is more than 376 times the actual diameter of the moon. However, from the earth, the apparent diameters are about the same.

91. $d = 2000 - 80t$

Problem Set 8-2

1. $e^{i\pi} = \cos \pi + i \sin \pi = -1$ **3.** $e^{ix}e^{-ix} = e^0 = 1$
5. $\cos^2 x + \sin^2 x = 1$ **7.** See problem 5.
9. Since $e^{ix} = \cos x + i \sin x$, then $e^{-x} = e^{i(ix)} = \cos (ix) + i \sin (ix)$.
11. Since $e^{-ix} = \cos x + i \sin x$, then $e^x = e^{-i(ix)} = \cos (ix) - i \sin (ix)$.
13. Since $e^{-x} = \cos(ix) + i \sin(ix)$ and $e^x = \cos(ix) - i \sin(ix)$. Add these to obtain $e^x + e^{-x} = 2 \cos(ix)$. Hence, $\cos(ix) = \dfrac{e^x + e^{-x}}{2}$

15. See the answer to problem 13. This time subtract rather than add.

17. $\sinh (x) = \dfrac{e^x - e^{-x}}{2}$

19. Set the imaginary part of the answer to problem 8 equal to the imaginary part of the answer to problem 6.

21. The domain of f is $\left[-\dfrac{\pi}{2}, \dfrac{\pi}{2}\right]$. The domain of g is $[-\infty, \infty]$.

25. (a) lowest $\cong$ (-0.787234, -1.217478); highest $\cong$ (0, 1.0016946)
  (b) The x-intercepts of $y = 16 \sin(4x) + \cos(4x)$ appear to line up with low points and high points of the graph of $y = e^{-x/4} \cos(4x)$.
  (c) $\cong -0.0156047625 + \dfrac{k\pi}{4}$, where $k = -1, 0, 1, 2, 3, 4$.
  (d) See answer to part b.
27. $\cong$ (0.6, 3); $\cong$ (5.6, 3)
29. $\cong$ (1.2, 2); $\cong$ (1.9, 2); $\cong$ (3.3, 2); $\cong$ (3.9, 2); (5.4, 2); (6.1, 2)
31. $\cong$ (1.6, 2); $\cong$ (3.5, -0.7); $\cong$ (5.9, -0.7)
33. infinitely many   **35.** no solutions   **37.** no solutions

SECTION 8-3

Exercise 8-3

1. (a) $-\dfrac{\pi}{2}$ (b) $\dfrac{3\pi}{2}$ (c) $\dfrac{3\pi}{2} + 2k\pi$

3. (a) $\dfrac{\pi}{3}$ (b) $\dfrac{\pi}{3}, \dfrac{5\pi}{3}$ (c) $\dfrac{\pi}{3} + 2k\pi, \dfrac{5\pi}{3} + 2k\pi$

5. (a) $\dfrac{\pi}{3}$ (b) $\dfrac{\pi}{3}, \dfrac{2\pi}{3}$ (c) $\dfrac{\pi}{3} + 2k\pi, \dfrac{2\pi}{3} + 2k\pi$

7. (a) $\dfrac{\pi}{4}$ (b) $\dfrac{\pi}{4}, \dfrac{5\pi}{4}$ (c) $\dfrac{\pi}{4} + k\pi$

9. (a) $\dfrac{\pi}{4}$ (b) $\dfrac{\pi}{4}, \dfrac{5\pi}{4}$ (c) $\dfrac{\pi}{4} + k\pi$

11. (a) $0, \pi$ (b) $0, \pi$ (c) $k\pi$

13. (a) $\dfrac{\pi}{2}, 0$ (b) $\dfrac{\pi}{2}, \dfrac{3\pi}{2}, 0, \pi$ (c) $\dfrac{k\pi}{2}$

15. (a) $\dfrac{\pi}{2}, 0$ (b) $\dfrac{\pi}{2}, \dfrac{3\pi}{2}, 0$ (c) $\dfrac{(2k+1)\pi}{2}, 2k\pi$

17. (a) $-\dfrac{\pi}{3}, \dfrac{\pi}{3}$ (b) $\dfrac{\pi}{3}, \dfrac{2\pi}{3}, \dfrac{4\pi}{3}, \dfrac{5\pi}{3}$
  (c) $\dfrac{\pi}{3} + k\pi, \dfrac{2\pi}{3} + k\pi$

19. (a) $\dfrac{2\pi}{3}, 0$ (b) $\dfrac{2\pi}{3}, \dfrac{4\pi}{3}, 0$ (c) $\dfrac{2k\pi}{3}$

21. (a) $\dfrac{3\pi}{4}$ (b) $\dfrac{3\pi}{4}, \dfrac{7\pi}{4}$ (c) $\dfrac{3\pi}{4} + k\pi$

23. (a) $\dfrac{\pi}{9}$ (b) $\dfrac{\pi}{9}, \dfrac{5\pi}{9}, \dfrac{7\pi}{9}, \dfrac{11\pi}{9}, \dfrac{13\pi}{9}$ (c) $(6k+1)\dfrac{\pi}{9}$

25. (a) $\dfrac{\pi}{6}$ (b) $\dfrac{\pi}{6}, \dfrac{\pi}{3}, \dfrac{7\pi}{6}, \dfrac{4\pi}{3}$ (c) $\dfrac{\pi}{6} + k\pi, \dfrac{\pi}{3} + k\pi$

27. (a) $\dfrac{\pi}{2}$ (b) $\dfrac{\pi}{2}$ (c) $\dfrac{\pi}{2} + 2k\pi$

29. (a) $\dfrac{3\pi}{4}$ (b) $\dfrac{3\pi}{4}$ (c) $\dfrac{3\pi}{4} + 3k\pi$

31. (a) $0, \dfrac{\pi}{3}$ (b) $0, \dfrac{\pi}{3}, \dfrac{2\pi}{3}, \dfrac{3\pi}{3}, \dfrac{4\pi}{3}, \dfrac{5\pi}{3}$ (c) $\dfrac{2k\pi}{3}$

33. (a) $\dfrac{\pi}{2}, 0$ (b) $\dfrac{\pi}{2}, \dfrac{3\pi}{2}, 0, \pi$ (c) $\dfrac{k\pi}{2}$

35. (a) $0, -\dfrac{\pi}{2}$ (b) $0, \pi, \dfrac{3\pi}{2}$
  (c) $k\pi, \dfrac{3\pi}{2} + 2k\pi = \dfrac{(4k + 3)\pi}{2}$

37. (a) $\dfrac{\pi}{6}, \dfrac{\pi}{3}$ (b) $\dfrac{\pi}{6}, \dfrac{5\pi}{6}, \dfrac{\pi}{3}, \dfrac{2\pi}{3}$
  (c) $\dfrac{(6k + 1)\pi}{6}, \dfrac{(6k + 5)\pi}{6}, \dfrac{(3k + 1)\pi}{3}, \dfrac{(3k + 2)\pi}{3}$

39. (a) $\dfrac{2\pi}{3}, 0$ (b) $\dfrac{2\pi}{3}, \dfrac{4\pi}{3}, 0$ (c) $\dfrac{2k\pi}{3}$

41. $\varnothing$ (no answer) **43.** (a) $\dfrac{\pi}{2}$ (b) $\dfrac{\pi}{2}$ (c) $\dfrac{(2k+1)\pi}{2}$

45. (a) $\dfrac{2\pi}{3}$ (b) $\dfrac{2\pi}{3}, \dfrac{4\pi}{3}$ (c) $\dfrac{(6k+2)\pi}{3}, \dfrac{(6k+4)\pi}{3}$

47. $\varnothing$ (no answer)

49. (a) $0, \dfrac{\pi}{4}, \dfrac{\pi}{2}$ (b) $0, \dfrac{\pi}{2}, \pi, \dfrac{3\pi}{2}, \dfrac{\pi}{4}, \dfrac{5\pi}{4}$

(c) $k\pi, \dfrac{(2k+1)\pi}{2}, \dfrac{(4k+1)\pi}{4}$

Problem Set 8-3

1. model $y = 2\sin\left(\dfrac{2\pi t}{15} + \dfrac{\pi}{2}\right)$; time 12.5 seconds

3. distance $x = 10$ ft; total distance $30\sqrt{10} \cong 94.868$ ft; $\theta \cong$ 1.249 radians (or 71.565°)

5. $x \cong -0.737$ **7.** $x \cong 0.335$ **9.** $x = 0, x \cong 0.875$

11. $x = 0, x \cong 1.24, 3.15, 5.05$ **13.** $x \cong 0.8, 3.95$

15. $x = 0, x \cong 1.6$ **17.** $x \cong 0.59, 3.1, 6.28$ **19.** $x \cong 1.3$

21. $x \cong 0.737$ **23.** $x \cong 0.78, 2.36, 3.93, 5.50$

25. $x \cong -0.7375069$ **27.** $x \cong 0.3354188$

29. $x \cong 0.8767262$

SECTION 8-4

Exercise 8-4

1. AAS; $b = \dfrac{5\sqrt{3}}{\sin 30°} \cong 11.305159$

3. SSA; $B = 90°$ **5.** SSA; $A \cong 53.41°$, or $A \cong 126.54°$

7. AAS; $a = 14\sin 50° \cong 10.724672$

9. SSA; $\sin A = 1.25$: not a triangle

11. SAS; $a = \sqrt{21} \cong 4.582596$

13. SAS; $b = \sqrt{58 - 42\cos 50°} \cong 5.568027$

15. SSS; $A = \cos^{-1}\left(\dfrac{38}{70}\right) \cong 57.12°$

17. SSS; $B = \cos^{-1}\left(\dfrac{29}{36}\right) \cong 36.34°$

19. SAS; $c = \sqrt{41 - 40\cos 40°} \cong 3.218419$

21. SSS; $B = \cos^{-1}\left(\dfrac{65}{70}\right) \cong 21.79°$, $C = 120°$, $A \cong 38.21°$

23. SAS; $a = \sqrt{221 - 220\cos 35°} \cong 4.648902$, $B \cong 65.38°$, $C \cong 89.62°$

25. ASA; $A = 70°$, $c = 10$, $b = \dfrac{10\sin 40°}{\sin 70°} \cong 6.84$

27. SSA; not a triangle

29. SSS; $A \cong 22.33°$, $B \cong 108.21°$, $C \cong 49.46°$

35. 3.75 **37.** $10\sqrt{3}$ **39.** $\cong 92.541658$

41. Shortest distance $\cong 7.7924$ km. Longest distance $\cong$ 13.7579 km

43. $\cong 29.93°$ **45.** $\cong 239800$ miles **49.** $\cos^4 21.47° \cong 0.75$

51. A shorter tube requires a larger deflection angle, hence the intensity at the same distance from the center screen will be less.

Problem Set 8-4

3. 30 **5.** 0 **7.** not triangles **9.** ellipse **11.** ellipse

13. ellipse **15.** size of ellipse **17.** 3 leaf clover

19. 5 leaf clover **21.** 12 leaf clover **23.** $2k$ leaf clover

25. $2k$ leaf clover

27. If k is odd, the clover has k leaves. If k is even the clover has $2k$ leaves.

SECTION 8-5

Exercise 8-5

11. $(5, 0)$ **13.** $(0, 2)$ **15.** $(2\sqrt{2}, 2\sqrt{2})$ **17.** $(3, 3\sqrt{3})$

19. $(5\sqrt{3}, 5)$ **21.** $(2, 60°)$ **23.** $(2, 45°)$ **25.** $(5, 90°)$

27. $(6, 30°)$ **29.** $(2, 0°)$ **31.** $\left(-\dfrac{5\sqrt{2}}{2}, -\dfrac{5\sqrt{2}}{2}\right)$

33. $(0, -4)$ **35.** $(3, -3\sqrt{3})$ **37.** $\left(-\dfrac{5\sqrt{3}}{2}, \dfrac{5}{2}\right)$

39. $\left(-\dfrac{5\sqrt{3}}{2}, \dfrac{5}{2}\right)$ **41.** $\left(5\sqrt{2}, \dfrac{\pi}{4}\right)$ **43.** $\left(5\sqrt{2}, \dfrac{3\pi}{4}\right)$

45. $\left(5\sqrt{2}, \dfrac{5\pi}{4}\right)$ **47.** $\left(6, \dfrac{5\pi}{3}\right)$ **49.** $\left(6, \dfrac{4\pi}{3}\right)$

61. $\tan\theta = 2$ **63.** $\tan\theta = -4$ **65.** $\tan\theta = \dfrac{1}{2}$

67. $r = 2$ **69.** $r^2\cos(2\theta) = 25$

71. 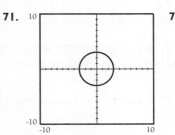 **73.**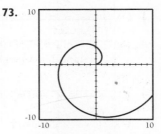

75. 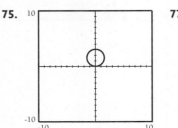 **77.**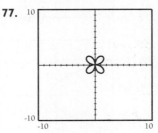

79. 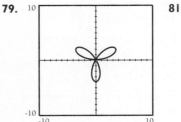 **81.**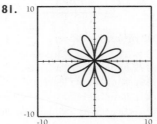

83.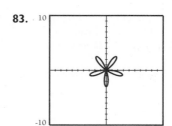

85.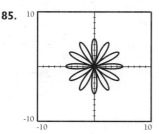

87.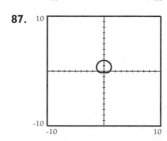

89.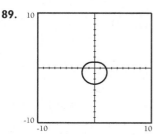

91.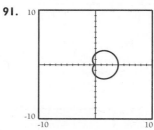

93.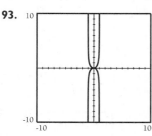

95.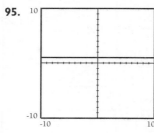

97.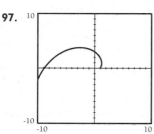

99.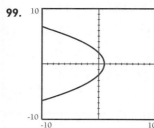

Problem Set 8-5

13. 180°   **15.** $(-|r|, \theta) = (|r|, \theta + 180°)$

17. Number of "leaves" is $2k$.

19. Number of "leaves" is k.

21. Number of "leaves" is $2k$ when k is even and number of "leaves" is k when k is odd.

23. k affects the "length of leaves."   **27.** spiral

SECTION 8-6

Exercise 8-6

1. $5; \left\langle \frac{3}{5}, -\frac{4}{5} \right\rangle; \theta \cong 53.13°, \theta \cong 143.13°$

3. $1; \langle 1, 0 \rangle; \theta = 0°, \phi = 90°$

5. $13; \left\langle -\frac{12}{13}, \frac{5}{13} \right\rangle; \theta = 157.38°, \phi = 67.38°$

7. $\sqrt{2}; \left\langle \frac{\sqrt{2}}{2}, \frac{\sqrt{2}}{2} \right\rangle; \theta \cong 45°, \phi \cong 45°$

9. $1; \left\langle \frac{1}{2}, \frac{3}{2} \right\rangle; \theta = 60°, \phi = 30°$

11. $1; \left\langle \frac{\sqrt{2}}{2}, -\frac{\sqrt{2}}{2} \right\rangle; \theta = 45°, \phi = 135°$

13. $6\sqrt{2}; \left\langle -\frac{\sqrt{2}}{2}, \frac{\sqrt{2}}{2} \right\rangle; \theta = 135°, \phi = 45°$

15. $1; \langle 0, -1 \rangle; \theta = 90°, \phi = 180°$

17. 0; direction cosine is undefined

19. $10; \left\langle -\frac{\sqrt{2}}{2}, \frac{\sqrt{2}}{2} \right\rangle; \theta = 135°, \phi = 45°$

21. $\langle 2, 4 \rangle$ **23.** $\langle -8, 0 \rangle$ **25.** $\langle -4, 8 \rangle$ **27.** $\langle 30, -4 \rangle$

29. $\langle -2, 12 \rangle$ **31.** $\langle 1, 1 \rangle$ **33.** $\langle 3, 4 \rangle$ **35.** $\langle 4, 3 \rangle$ **37.** 2

39. $\langle x, y \rangle$ **41.** 90°   **43.** 45°   **45.** 75°

47. $\cong 36.87°$ **49.** 90°   **51.** 1   **53.** $\dfrac{\sqrt{2}}{2}$

55. $\dfrac{3(\sqrt{6} + \sqrt{2})}{2}$ **57.** $\dfrac{4}{5}$ **59.** 0   **63.** $k \langle a, -b \rangle, k \neq 0$

Problem Set 8-6

1. $\langle 5, 10 \rangle$ lbs   **3.** $\langle 5, 0 \rangle; \left\langle \frac{3}{2}, \frac{3\sqrt{3}}{2} \right\rangle, \left\langle \frac{13}{2}, \frac{3\sqrt{3}}{2} \right\rangle, 7$mph

SECTION 8-7

Exercise 8-7

1. cis 330°   **3.** 2 cis 135°

5. $r \text{ cis} \left(\cos^{-1}(-\frac{3}{5}) \right) \cong 5 \text{ cis } 126.87°$ **7.** 2 cis 240°

9. 2 cis 90°   **11.** $\dfrac{5\sqrt{3}}{2} + \dfrac{5}{2}i$ **13.** $-\dfrac{3\sqrt{2}}{2} + \dfrac{3\sqrt{2}}{2}i$

15. $-1 - \sqrt{3}i$

17. $4 \cos 70° + 4i \sin 70° \cong 1.368081 + 3.758770i$

19. $-\dfrac{5\sqrt{3}}{2} - \dfrac{5}{2}i$ **21.** 72 cis 100°

23. $\left(\frac{9}{8} \right) \text{ cis } (-20°) = \left(\frac{9}{8} \right) \text{ cis } 340°$ **25.** 72 cis 170°

27. 9^3 cis 120°   **29.** 3 cis 20°   **41.** $\sqrt[q]{r^p} \text{ cis} \left(\dfrac{p\theta}{q} \right)$

43. $r^\pi \text{ cis } (\pi\theta)$

Problem Set 8-8

1. cis 0°, cis 72°, cis 144°, cis 216°, cis 288°

3. cis 30°, cis 90°, cis 150°, cis 210°, cis 270°, cis 330°

5. cis 30°, cis 150°, cis 270°   **7.** cis 15°, cis 135°, cis 255°

9. $\sqrt[3]{4}$ cis 100°, $\sqrt[3]{4}$ cis 220°, $\sqrt[3]{4}$ cis 340°

CHAPTER 8 REVIEW EXERCISES

1. (a) $A = 50°$, $a = 7 \cos 40°$, $b = 7 \sin 40°$

1. (b) $A = 10°$, $a = \cot 80°$, $c = \csc 80°$

3. $B = 36°$, $a = 10 \sin 54°$, $b = 10 \cos 54°$

5. $b = 2\sqrt{10} \cong 6.324555$, $A = \sin^{-1}\left(\dfrac{3}{7}\right) \cong 25.377°$,

$B = \sin^{-1}\left(\dfrac{2\sqrt{10}}{7}\right) \cong 64.623°$

7. $b = \sqrt{21} \cong 4.582576$, $A = \sin^{-1}\left(\dfrac{2}{5}\right) \cong 23.578°$,

$B = \sin^{-1}\left(\dfrac{\sqrt{21}}{5}\right) \cong 66.422°$

9. $b = 40\sqrt{2} \cong 56.568542$, $A = \sin^{-1}\left(\dfrac{1}{3}\right) \cong 19.47°$,

$B = \sin^{-1}\left(\dfrac{2\sqrt{2}}{3}\right) \cong 70.53°$

11. $h = 5\sqrt{3}$, $x = 5$

13. $AB = 5 + \dfrac{5\sqrt{3}}{\tan 70°} \cong 8.1520$

15. $12\sqrt{2} \cong 17.0$ cm   **17.** $3\sqrt{2} \cong 4.2$ km   **19.** $\dfrac{-\pi}{3}$

21. $-\dfrac{\pi}{6}$ **23.** $-\dfrac{\pi}{4}$ **25.** $\cong 0.927295$ **27.** domain error

29. $\csc^{-1} x = y$ means $\csc y = x$ where $-\dfrac{\pi}{2} \le y \le \dfrac{\pi}{2}$.

31. π **33.** 0.7 **35.** $\dfrac{\pi}{4}$ **37.** $\dfrac{5\pi}{6}$ **39.** $\dfrac{\sqrt{3}}{2}$

41. $\sqrt{1 - z^2}$ **43.** $\tan^{-1}\sqrt{x^2 - 2}$ **45.** $\cong 0.982794$

47. $\cong -0.206464$ **49.** domain error   **51.** domain error

55. $y = \left(\dfrac{1}{\pi}\right)\cos^{-1}\left(\dfrac{x}{5}\right)$

57. $\theta = 2 \tan^{-1}\left(\dfrac{1}{D}\right)$, $\theta \cong 4.58°$

59. $-\dfrac{\pi}{6}$; $\dfrac{7\pi}{6} + 2k\pi$, $\dfrac{11\pi}{6} + 2k\pi$

61. $\dfrac{\pi}{4}$; $\dfrac{\pi}{4} + k\pi$

63. $\dfrac{\pi}{4}, \dfrac{-\pi}{4}$; $\dfrac{\pi}{4} + \dfrac{k\pi}{2}$ **65.** 0; $k\pi$

67. $0, \dfrac{-\pi}{2}, \dfrac{3\pi}{2}$; $k\pi, \dfrac{(4k + 3)\pi}{2}$

69. $-\dfrac{\pi}{6}, \dfrac{\pi}{2}$; $\dfrac{\pi}{2} + 2k\pi, \dfrac{11\pi}{6} + 2k\pi, \dfrac{7\pi}{6} + 2k\pi$

71. $\dfrac{2\pi}{9}$; $\dfrac{2\pi}{9} + \dfrac{2k\pi}{3}, \dfrac{4\pi}{9} + \dfrac{2k\pi}{3}$

73. $-\dfrac{\pi}{2}$; $-\dfrac{\pi}{2} + 2k\pi$

75. $-\dfrac{\pi}{6}, \dfrac{\pi}{6}$; $\dfrac{(4k + 1)\pi}{6}, \dfrac{(4k - 1)\pi}{6}$

77. $\dfrac{\pi}{6}, -\dfrac{\pi}{3}$; $\dfrac{\pi}{6} + 2k\pi, \dfrac{7\pi}{6} + 2k\pi, \dfrac{5\pi}{3} + 2k\pi$

79. $\dfrac{\pi}{2}$; $\dfrac{\pi}{2} + 2k\pi$

81. $\dfrac{\pi}{6}, -\dfrac{\pi}{6}$; $\dfrac{(12k + 1)\pi}{6}, \dfrac{(12k + 5)\pi}{6}, \dfrac{(12k + 7)\pi}{6}$,

$\dfrac{(12k + 11)\pi}{6}$

83. $0, \dfrac{\pi}{2}$; $2k\pi, \dfrac{(4k + 1)\pi}{2}$

85. SSA; $B \cong 89°$ or $91°$ **87.** SSA; $A \cong 25.55°$ or $154.45°$

89. SAS; $a = 4\sqrt{3}$ **91.** SSS; $B = \cos^{-1}(-0.25) \cong 104.48°$

93. SAS; $c \cong 3.91$

95. ASA; $C = 70°$, $b = \dfrac{10 \sin 60°}{\sin 70°} \cong 9.21605$, $a = \dfrac{10 \sin 50°}{\sin 70°}$
$\cong 8.15207$

97. SAS; $b \cong 9.817387$, $C \cong 67.39°$, $A \cong 47.61°$

99. SAS; $a = 4\sqrt{5 - 2\sqrt{3}} \cong 4.957255$, $B \cong 23.79°$,
$C \cong 126.21°$

101. $\cong 487.3851$ km   **107.** $\left(\dfrac{5\sqrt{2}}{2}, -\dfrac{5\sqrt{2}}{2}\right)$

109. $(-4\sqrt{3}, -4)$ **111.** $(2, 2\sqrt{3})$ **113.** $(2, 60°)$

115. $(5, 90°)$ **117.** $(2, 0°)$ **119.** $\left(6\sqrt{2}, \dfrac{\pi}{4}\right)$

121. $\left(5\sqrt{2}, \dfrac{3\pi}{4}\right)$ **123.** $\left(4, \dfrac{\pi}{3}\right)$ **125.** $(-3, -3\sqrt{2})$

127. $\left(\dfrac{5\sqrt{3}}{2}, \dfrac{5}{2}\right)$

129. 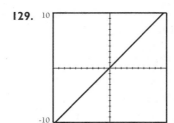   **131.**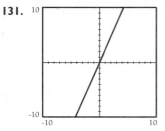

133.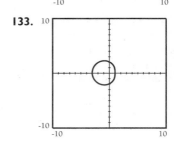

135. $\tan \theta = 3$ **137.** $\tan \theta = -5$ **139.** $r = \dfrac{12}{\sqrt{9 + 7 \sin^2 \theta}}$

141. $13; \left\langle \dfrac{5}{13}, -\dfrac{12}{13} \right\rangle; 67.38°$ **143.** $1; \left\langle \dfrac{\sqrt{2}}{2}, \dfrac{\sqrt{2}}{2} \right\rangle; 45°$

145. $8; \left\langle \dfrac{\sqrt{3}}{2}, -\dfrac{1}{2} \right\rangle; 30°$ **147.** $\sqrt{2}; \left\langle \dfrac{\sqrt{2}}{2}, \dfrac{\sqrt{2}}{2} \right\rangle; 45°$

149. $\langle 13, 15 \rangle$ **151.** $\langle 25, 23 \rangle$ **153.** $\langle -28, -6 \rangle$
155. $\langle 5, -12 \rangle$ **157.** $\langle 2, 7 \rangle$ **159.** $\langle t, u \rangle$ **161.** $45°$
163. $58.11°$
165. $\langle -3, 0 \rangle; \left\langle \dfrac{5\sqrt{2}}{2}, \dfrac{5\sqrt{2}}{2} \right\rangle; \left\langle \dfrac{5\sqrt{2}}{2} - 3, \dfrac{5\sqrt{2}}{2} \right\rangle;$ upstream

167. $6 \operatorname{cis} 300°$ **169.** $10 \operatorname{cis} 135°$ **171.** $1 + i$
173. $-5 - 5\sqrt{3} i$ **175.** $-\dfrac{5\sqrt{3}}{2} - \dfrac{5}{2} i$ **177.** $7 \operatorname{cis} 350°$

179. $625 \operatorname{cis} 200°$ **181.** $2 \operatorname{cis} 310°$
183. $\operatorname{cis} 67.5°, \operatorname{cis} 157.5°, \operatorname{cis} 247.5°, \operatorname{cis} 337.5°$
185. $\sqrt[3]{4} \operatorname{cis} 15°, \sqrt[3]{4} \operatorname{cis} 135°, \sqrt[3]{4} \operatorname{cis} 255°$

CHAPTER 8 TEST
1. $-\sqrt{2}$ **2.** $\sqrt{3}$ **3.** $\dfrac{1}{2}$ **4.** $-\dfrac{1}{2}$

5. $A = 30°, c = \dfrac{40\sqrt{3}}{3} \cong 23.094, a = \dfrac{20\sqrt{3}}{3} \cong 11.547$

8. AAS; $a \cong 2.266816$ **9.** SSS; $B \cong 29.93°$
10. SSA; $\cong 14.7355$ nautical miles
12. (a) $\sqrt{13} \operatorname{cis} (\tan^{-1} 1.5) \cong \sqrt{13} \operatorname{cis} 56.31°$
   (b) $\left(\dfrac{7\sqrt{2}}{2}, \dfrac{7\sqrt{2}}{2} \right)$

13.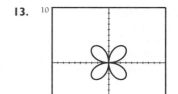

14. $576 \operatorname{cis} 260°$ **15.** $\left(\dfrac{64}{9} \right) \operatorname{cis} 100°$ **16.** $2\sqrt{2} \operatorname{cis} 40°$
17. $\sqrt{5}$ **18.** -3 **19.** $\langle 4, 4 \rangle$ **20.** $\theta \cong 126.87°$

CHAPTER 9

SECTION 9-1

Exercise 9-1
1. $k = 3$ **3.** $m = -1, b = 5$ **5.** $a = 5, b = 3$
7. $a = 3, b = -5$ **9.** $a = \dfrac{1}{25}, b = \dfrac{1}{9}$ **11.** $a = 1, b = -\dfrac{1}{4}$
13. yes   **15.** yes   **17.** no   **19.** yes

Problem Set 9-1
1. parabola   **3.** parabola   **5.** parabola   **7.** circle
9. ellipse

SECTION 9-2

Exercise 9-2
1. vertex: $(0, 0)$;   **3.** vertex: $(0, 0)$;
  axis of symmetry: $y = 0$;   axis of symmetry: $x = 0$;
  focus: $\left(\dfrac{1}{4}, 0 \right)$;   focus: $\left(0, -\dfrac{1}{4} \right)$;
  directrix: $x = -\dfrac{1}{4}$ directrix: $y = \dfrac{1}{4}$

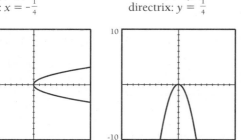

5. vertex: $(0, 0)$;
    axis of symmetry: $y = 0$;
    focus: $(1, 0)$;
    directrix: $x = -1$

7. vertex: $(0, 0)$;
    axis of symmetry: $x = 0$;
    focus: $(0, -1)$;
    directrix: $y = 1$

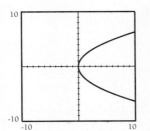 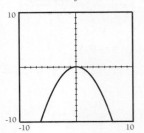

9. vertex: $(0, 0)$;
    axis of symmetry: $y = 0$;
    focus: $(-4, 0)$;
    directrix: $x = 4$

11. vertex: $(-2, 1)$;
    axis of symmetry: $y = 1$;
    focus: $(-\frac{7}{4}, 1)$;
    directrix: $x = -\frac{9}{4}$

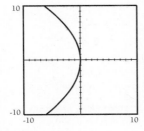 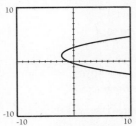

13. vertex: $(1, -2)$;
    axis of symmetry: $x = 1$;
    focus: $(1, -\frac{9}{4})$;
    directrix: $y = -\frac{7}{4}$

15. vertex: $(0, -5)$;
    axis of symmetry: $y = -5$;
    focus: $(1, -5)$;
    directrix: $x = -1$

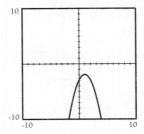 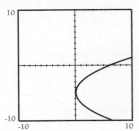

17. vertex: $(0, 2)$;
    axis of symmetry: $x = 0$;
    focus: $(0, 1)$;
    directrix: $y = 3$

19. vertex: $(-2, 3)$;
    axis of symmetry: $y = 3$;
    focus: $(-6, 3)$;
    directrix: $x = 2$

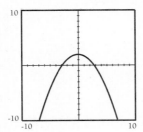 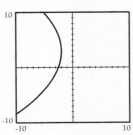

21. $12x = y^2$ **23.** $-12y = x^2$ **25.** $-8y = x^2$ **27.** $8x = y^2$
29. $16x = y^2$ **31.** $8x = y^2$ **33.** $8(x - 8) = (y - 2)^2$
35. $-4(x - 2) = (y - 3)^2$ **37.** $8(y - 1) = (x - 1)^2$
39. $8x = y^2$ or $-8(x - 4) = y^2$ **41.** $4x = y^2$

Problem Set 9-2

1. Focus is 1.125 meters from the vertex.
3. $\dfrac{s^2 - r^2}{s - r}$ **5.** $s + r$ **7.** $2r$ **9.** $y = 2rx - r^2$

11. $y = 6x - 9$ **13.** $y = 6x - 4$ **15.** parallel lines
17. slope is $6r$ **19.** $y = 6rx - 3r^2 + 7$ **21.** right; not
23. down; function **25.** right; not **27.** down; function
29. left; not **31.** right; not **33.** down; function
35. right; not **37.** down; function **39.** left; not

SECTION 9-3

Exercise 9-3

1. vertices: $(0, 3)$, $(0, -3)$,
    $(2, 0)$, $(-2, 0)$;
    focii: $(0, \sqrt{5})$, $(0, -\sqrt{5})$

3. vertices: $(0, 5)$, $(0, -5)$,
    $(4, 0)$, $(-4, 0)$;
    focii: $(0, 3)$, $(0, -3)$

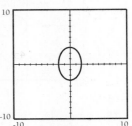 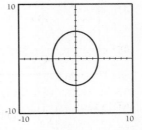

5. vertices: $(0, 2)$, $(0, -2)$,
    $(3, 0)$, $(-3, 0)$;
    focii: $(\sqrt{5}, 0)$, $(-\sqrt{5}, 0)$

7. vertices: $(0, 1)$, $(0, -1)$,
    $(3, 0)$, $(-3, 0)$;
    focii: $(2\sqrt{2}, 0)$, $(-2\sqrt{2}, 0)$

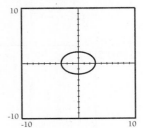 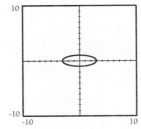

9. vertices: $(0, 1)$, $(0, -1)$, $\left(\frac{1}{3}, 0\right)$, $\left(-\frac{1}{3}, 0\right)$;

focii: $\left(0, \frac{2\sqrt{2}}{3}\right)$, $\left(0, -\frac{2\sqrt{2}}{3}\right)$

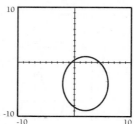

11. vertices: $(0, 0)$, $(4, 0)$, $(2, 3)$, $(2, -3)$;
focii: $(2, \sqrt{5})$, $(2, -\sqrt{5})$

31. $\dfrac{x^2}{25} + \dfrac{y^2}{16} = 1$ **33.** $\dfrac{(x - 2)^2}{9} + \dfrac{(y - 3)^2}{16} = 1$

37. $h = 8\sqrt{14} \cong 29.9$ yds   **39.** $\cong 13200$ miles

41. $\dfrac{x^2}{240000^2} + \dfrac{y^2}{239637^2} = 1$ **45.** $\dfrac{2a}{2c} = \dfrac{a}{c} = e$

Problem Set 9-3

1. $y^2 = -6\left(x - \frac{5}{2}\right)$ **3.** $\dfrac{x^2}{4} + \dfrac{y^2}{3} = 1$

5. (c) $f(t) = \dfrac{b}{a}\sqrt{a^2 - t^2}$,   **(d)** $\dfrac{b(\sqrt{a^2 - r^2} - \sqrt{a^2 - t^2})}{a(r - t)}$

(f) $\dfrac{-b(r + t)}{a(\sqrt{a^2 - r^2} + \sqrt{a^2 - t^2})}$ **(g)** $\dfrac{-br}{a\sqrt{a^2 - r^2}}$

(h) $y - \dfrac{4\sqrt{5}}{3} = \dfrac{-8}{3\sqrt{5}}(x - 2)$

13. vertices: $(6, -4)$, $(-2, -4)$, $(2, 1)$, $(2, -9)$;
focii: $(2, -1)$, $(2, -7)$

15. vertices: $(0, 0)$, $(6, 0)$, $(3, 2)$, $(3, -2)$;
focii: $(3 + \sqrt{5}, 0)$, $(3 - \sqrt{5}, 0)$

SECTION 9-4

Exercise 9-4

1. center:; $(0, 0)$;
asymptote: $y = \pm\left(\frac{3}{2}\right)x$;
transverse: $y = 0$;
conjugate: $x = 0$;
focii: $(\pm\sqrt{13}, 0)$;
e: $\dfrac{\sqrt{13}}{2}$

3. center: $(0, 0)$;
asymptote: $y = \pm\left(\frac{5}{4}\right)x$;
transverse: $x = 0$;
conjugate: $y = 0$;
focii: $(0, \pm\sqrt{41})$;
e: $\dfrac{\sqrt{41}}{5}$

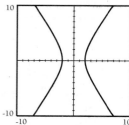

17. vertices: $(-1, 1)$, $(-1, -1)$, $(2, 0)$, $(-4, 0)$;
focii: $(-1 + 2\sqrt{2}, 0)$, $(-1 - 2\sqrt{2}, 0)$

19. vertices: $\left(\frac{1}{3}, 3\right)$, $\left(-\frac{1}{3}, 3\right)$, $(0, 4)$, $(0, 2)$;
focii: $\left(0, 3 + \dfrac{2\sqrt{2}}{3}\right)$, $\left(0, 3 - \dfrac{2\sqrt{2}}{3}\right)$

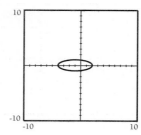

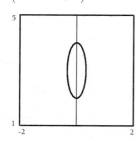

21. $\dfrac{x^2}{25} + \dfrac{y^2}{9} = 1$ **23.** $\dfrac{x^2}{25} + \dfrac{y^2}{9} = 1$ **25.** $\dfrac{x^2}{16} + \dfrac{y^2}{25} = 1$

27. $\dfrac{x^2}{25} + \dfrac{y^2}{9} = 1$ or $\dfrac{x^2}{9} + \dfrac{y^2}{25} = 1$ **29.** $\dfrac{x^2}{25} + \dfrac{y^2}{9} = 1$

5. center: $(0, 0)$;
asymptote: $y = \pm x$;
transverse: $y = 0$;
conjugate: $x = 0$;
focii: $(\pm 10\sqrt{2}, 0)$;
e: $\sqrt{2}$

7. center: $(3, -2)$;
asymptote: $y + 2 = \pm(x - 3)$;
transverse: $y = -2$;
conjugate: $x = 3$;
focii: $(3 \pm \sqrt{5}, -2)$;
e: $\sqrt{2}$

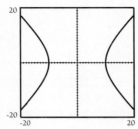

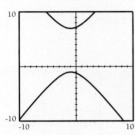

9. center: $(-1, 3)$;
asymptote: $y - 3 = \pm\frac{4}{3}(x + 1)$;
transverse: $x = -1$;
conjugate: $y = 3$;
focii: $(-1, 3 \pm 5)$;
e: $\dfrac{5}{3}$

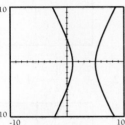

11. center: $(3, -0.5)$;
asymptote: $y + 0.5 = \pm 2(x - 3)$;
transverse: $y = -0.5$;
conjugate: $x = 3$;
focii: $(3 \pm 2\sqrt{5}, -0.5)$
e: $\sqrt{5}$

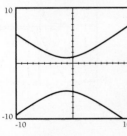

13. center: $(-1, -2)$;
asymptote: $y + 2 = \pm\frac{3}{4}(x - 3)$;
transverse: $x = -1$;
conjugate: $y = -2$;
focii: $(-1, -2 \pm 5)$;
e: $\dfrac{5}{3}$

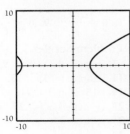

15. center: $(-3, 0)$;
asymptote: $y = \pm\frac{1}{2}(x + 3)$;
transverse: $y = 0$;
conjugate: $x = -3$;
focii: $(-3 \pm 3\sqrt{5}, 0)$;
e: $\dfrac{\sqrt{5}}{2}$

17. center: $(1, -2)$;
asymptote: $y + 2 = \pm 3(x - 1)$;
transverse: $x = 1$;
conjugate: $y = -2$;
focii: $(1, -2 \pm \sqrt{10})$
e: $\dfrac{\sqrt{10}}{3}$

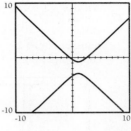

19. center: $(0, 0)$;
asymptote: $y = \pm\sqrt{\frac{3}{5}}x$;
transverse: $x = 0$;
conjugate: $y = 0$;
focii: $(0, \pm 2\sqrt{2})$;
e: $\dfrac{2\sqrt{6}}{3}$

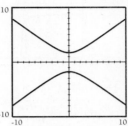

21. center: $(0, 0)$;
asymptote: $y = 0, x = 0$;
transverse: $y = x$;
conjugate: $y = -x$;

23. center: $(0, 0)$;
asymptote: $y = 0, x = 0$;
transverse: $y = -x$;
conjugate: $y = x$;

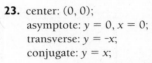

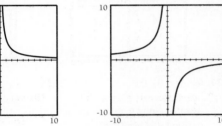

25. center: $(0, 0)$;
asymptote: $y = 0, x = 0$;
transverse: $y = x$;
conjugate: $y = -x$;

27. degenerate

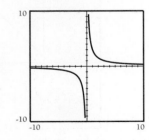

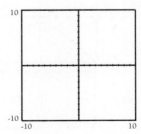

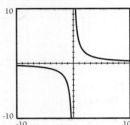

29. center: $(1, -2)$;
asymptote: $y = -2, x = 1$;
transverse: $y + 2 = x - 1$;
conjugate: $y + 2 = 1 - x$;

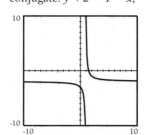

31. degenerate

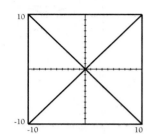

33. center: $(0, 0)$;
asymptote: $y = \pm x$;
transverse: $y = 0$;
conjugate: $x = 0$;

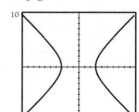

35. parabola

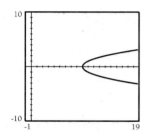

37. center: $(0, 0)$
asymptote: $y = \pm x$;
transverse: $x = 0$;
conjugate: $y = 0$;

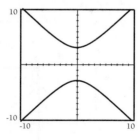

39. line

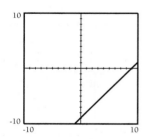

41. $\dfrac{y^2}{9} - \dfrac{x^2}{16} = 1$ **43.** $\dfrac{x^2}{1} - \dfrac{y^2}{8} = 1$ **45.** $\dfrac{x^2}{4} - \dfrac{y^2}{9} = 1$

47. $\dfrac{x^2}{1} - \dfrac{y^2}{9} = 1$ **49.** $\dfrac{x^2}{9} - \dfrac{y^2}{16} = 1$ **51.** $\dfrac{8}{3}$ cm

53. $\dfrac{500}{14} \cong 35.7$ mm

61.

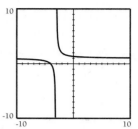

Problem Set 9-4

7. $\dfrac{rx}{a^2} - \dfrac{sy}{b^2} = 1$ or $y = \dfrac{rb^2}{sa^2}x + \dfrac{s^2a^2 - r^2b^2}{sa^2}$

23. As k increases so does the eccentricity and the graph "widens,"

Exercise 9-5

1. $x' = \dfrac{5\sqrt{2}}{2}$

a. 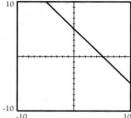   **b.**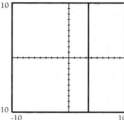

3. $\dfrac{y'^2}{2} - \dfrac{x'^2}{2} = 1$

a. 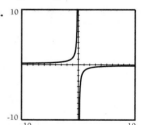   **b.**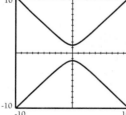

5. $\theta = \dfrac{\pi}{8}$

7. $\theta = \dfrac{3\pi}{8}$ **9.** $\theta = \dfrac{\pi}{6}$ **13.** $x' = x \cos \theta + y \sin \theta$

Problem Set 9-5

7. hyperbola: (49)   **9.** hyperbola: (48)
11. ellipse: (-75)   **13.** hyperbola: (109)   **17.** -7; ellipse
19. -15; ellipse   **21.** -21; ellipse   **23.** 0, parabola
25. 1; hyperbola   **27.** parabola   **29.** hyperbola
31. ellipse
33. $(1 - e^2)x^2 - 2ex + y^2 = e^2$; $e = 1$, parabola; $e < 1$, ellipse; $e > 1$, hyperbola

35. $x^2 + (1 - e^2)y^2 - 2\,dey = d^2e^2$; $e = 1$, parabola; $e < 1$,
ellipse; $e > 1$, hyperbola

37.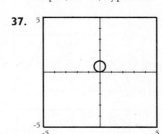

39.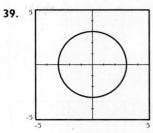

41.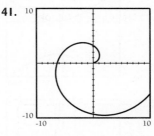

43.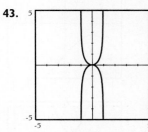

45.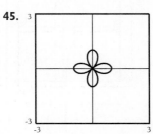

47.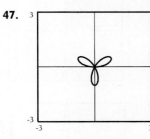

49.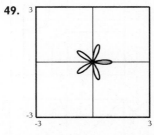

51.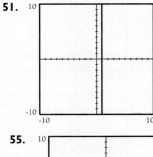

53.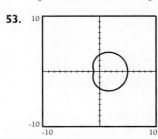

55.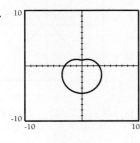

57.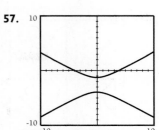

59.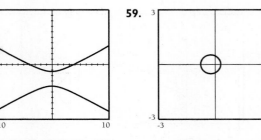

61.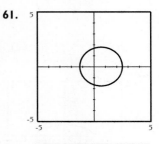

63.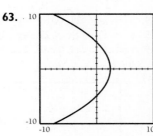

65.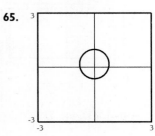

69. $e = \frac{1}{4}, d = 2$ **71.** up/down   **73.** ellipse

REVIEW EXERCISE 9

1. $k = -15$ **3.** $a = \frac{12}{5}, b = -6$ **5.** $a = \frac{1}{25}, b = \frac{1}{9}$

7.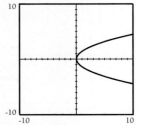

9.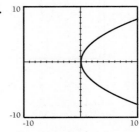

11.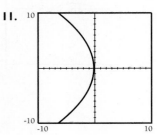

13.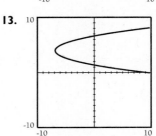

15. 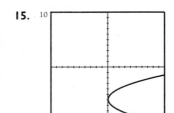   **17.**

19. $8y = x^2$ **21.** $-16x = y^2$ **23.** $x^2 = -4y$

25. $(x - 1)^2 = 8y$ **27.** $4x = (y - 3)^2$

29. vertices: $(\pm 3, 0), (0, \pm 2)$;   **31.** vertices: $(\pm 5, 0), (0, \pm 2)$;
focii: $(\pm \sqrt{5}, 0)$ focii: $(\pm \sqrt{21}, 0)$

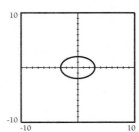   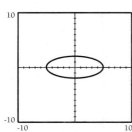

33. vertices: $(\pm \frac{1}{3}, 0)$,   **35.** vertices: $(-1, -2), (7, -2)$,
$(0, \pm 1)$; $(3, 3), (3, -7)$;
focii: $\left(0, \pm \dfrac{2\sqrt{2}}{3}\right)$ focii: $(3, 1), (3, -5)$

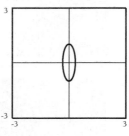   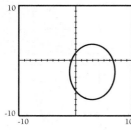

37. vertices: $(-5, 0), (1, 0)$,
$(-2, -1), (-2, 1)$;
focii: $(-2 \pm 2\sqrt{2}, 0)$

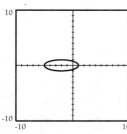

39. $\dfrac{x^2}{9} + \dfrac{y^2}{25} = 1$ **41.** $\dfrac{x^2}{25} + \dfrac{y^2}{16} = 1$

43. $\dfrac{x^2}{21} + \dfrac{y^2}{25} = 1$

45. 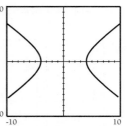   **47.**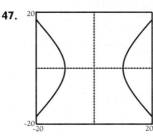

49. 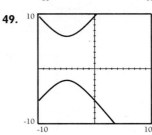   **51.**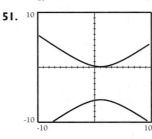

53. 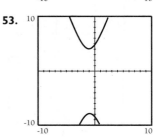   **55.**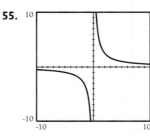

57. 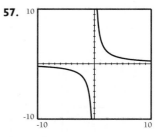   **59.**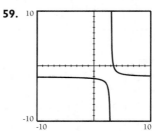

61. 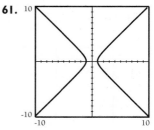   **63.**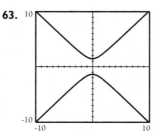

65. $\dfrac{y^2}{16} - \dfrac{x^2}{9} = 1$ **67.** $\dfrac{y^2}{1} - \dfrac{x^2}{9} = 1$ **69.** $x' = \dfrac{\sqrt{2}}{2}$

71. $\theta = \dfrac{\pi}{8}$ **73.** $\theta = \dfrac{\pi}{6}$

CHAPTER TEST 9

1. $a = 2, b = 2$

3.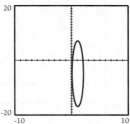

5.

7. $\dfrac{x^2}{16} - \dfrac{y^2}{9} = 1$

9. vertex $(0, 0)$; focus $(1, 0)$; directrix: $y = -1$; axis of symmetry: $x = 0$; latus rectum from $(-2, 1)$ to $(2, 1)$.

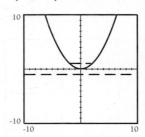

CHAPTER 10

SECTION 10-1

Exercise 10-1

1. $\{(4, 1)\}$ **3.** $\{(3, -2)\}$ **5.** $\varnothing$ **7.** $\{(2, -1)\}$
9. $\{(x, y); x + 2y = 4\}$ **11.** $\{(2, 5)\}$ **13.** $\{(4, 5)\}$
15. $\{(-1.5, 3.8)\}$ **17.** $\{(\frac{7}{2}, -\frac{1}{2})\}$ **19.** $\{(1, 1)\}$

21. $\{(7, 16)\}$ **23.** $\{(2, -\frac{1}{2})\}$ **25.** $\{(\frac{21}{2}, 12)\}$

27. $\{(-\frac{30}{7}, \frac{1}{7})\}$ **29.** $\{(\frac{17}{19}, -\frac{22}{19})\}$ **31.** $(r_1, r_2) = (8, 2)$

33. $(a, b) = (1, 40)$ **35.** $\frac{10}{3}$ liters of 20% and $\frac{20}{3}$ liters of 50%

37. Task A: 38 hours; Task B: 2 hours
39. Price is 100.   **41.** $\{(1, 1, 1)\}$ **43.** $\{(1, -1, 2)\}$
45. $\{(\frac{8}{3}, \frac{4}{3}, 0)\}$ **47.** $\{(0, 1, 3)\}$ **49.** $\{(-1, 3, 1)\}$

51. $\{(w, x, y, z)\} = \{-3, 1, 1, 4)\}$

53. $x = \dfrac{ce - bf}{ae - bd}; y = \dfrac{af - cd}{ae - bd}$

Problem Set 10-1

1. $\dfrac{2}{x - 3} + \dfrac{1}{x + 1}$ **3.** $\dfrac{-1}{x + 4} + \dfrac{1}{x - 5}$

5. $\dfrac{2}{x + 1} - \dfrac{1}{(x + 1)^2}$ **7.** $\dfrac{1}{x - 1} - \dfrac{1}{(x + 1)^2}$

9. $\dfrac{2}{x} + \dfrac{x - 1}{x^2 + 1}$

SECTION 10-2

Exercise 10-2

1. $\{(0, 0), (-1, -1)\}$ **3.** $\{(1, -2), (-3, 6)\}$ **5.** $\{(-1, 1), (2, 4)\}$
7. $\{(1, 1), (-3, -7)\}$ **9.** $\{(-\frac{1}{2}, \frac{3}{4}), (1, \frac{3}{2})\}$
11. $\{(0, 0), (-1, 1)\}$ **13.** $\{(1, 1), (-1, 1)\}$
15. $\{(1, 2), (-3, -6)\}$ **17.** $\{(-1 + \sqrt{3}, \sqrt{3}), (-1 - \sqrt{3}, -\sqrt{3})\}$
19. $\varnothing$ **21.** $\{(1, -1)\}$ **23.** $\varnothing$ **25.** $\{(2, 0), (-2, 0)\}$
27. $\left\{(0, -2), \left(-\dfrac{4\sqrt{2}}{3}, \dfrac{14}{9}\right), \left(\dfrac{4\sqrt{2}}{3}, \dfrac{14}{9}\right)\right\}$
29. $\{(-1, -1), (1, 1)\}$ **31.** $\{(1, 3)\}$
33. $\{(\frac{1}{3}, \frac{1}{2}), (\frac{1}{3}, -\frac{1}{2}), (-\frac{1}{3}, \frac{1}{2}), (-\frac{1}{3}, -\frac{1}{2})\}$

35. height $= \frac{84}{5}$ inches; width $= \frac{63}{5}$ inches

37. $\frac{3}{4}$ **39.** 264 ft²; length $= 20$ feet; width $= 5$ feet

SECTION 10-3

Exercise 10-3

1. 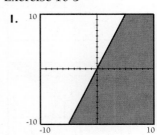   **3.**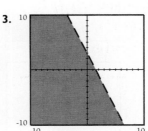

5. 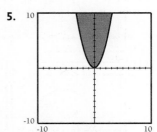   **7.**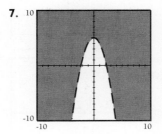

9. 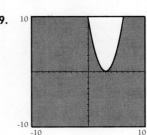   **11.**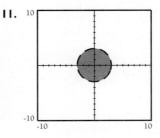

13.
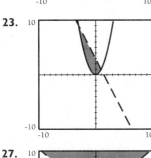

15.

17.

19.

21.

23.

25.
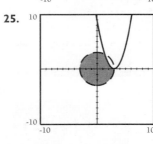

27.

29.
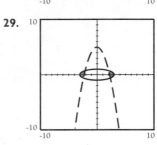

31.
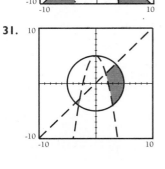

33.

35.

Problem Set 10-3

1. (b) feasible solutions: $(0, 0)$, $(0, 2)$, $\left(6, \frac{4}{5}\right)$, $(5, 0)$

   (d) $C(0, 0) = 0$; smallest $C(0, 2) = -20$; $C\left(6, \frac{4}{5}\right) = 52$; largest $C(5, 0) = 50$

3. maximum $C(8, 0) = 120$

CHAPTER REVIEW EXERCISES 10

1. $\{(5, -2)\}$ **3.** $\varnothing$ **5.** $\{(4, 0)\}$ **7.** $\{(-4, 7)\}$ **9.** $\{(9, -2)\}$

11. $\{(0, -4)\}$ **13.** $\left\{\left(\frac{11}{10}, \frac{12}{25}\right)\right\}$ **15.** $\left\{\left(\frac{17}{23}, -\frac{15}{23}\right)\right\}$

17. $\{(4, 7)\}$ **19.** $\{(5, 1, -2)\}$ **21.** $\{(-3, 1, 8)\}$

23. $\left\{\left(\frac{5}{2}, \frac{17}{2}, -\frac{17}{2}\right)\right\}$ **25.** $\{(1, -1, 2, 3)\}$

27. $\dfrac{1/2}{(x - 5)} + \dfrac{1/2}{(x + 1)}$ **29.** $\{(0, 0), (6, 6)\}$

31. $\{(1, 1), (2, 4)\}$

33. $\left\{\left(\dfrac{3 + \sqrt{105}}{12}, \dfrac{27 + \sqrt{105}}{24}\right),\right.$
$\left.\left(\dfrac{3 - \sqrt{105}}{12}, \dfrac{27 - \sqrt{105}}{24}\right)\right\}$

35. $\{(1, 2), (-1, 2)\}$

37. $\{(-2 + \sqrt{10}, 3 + \sqrt{10}), (-2 - \sqrt{10}, 3 - \sqrt{10})\}$

39. $\{(-1, 2), (3, 6)\}$ **41.** $\{(3, -2), (-3, -2), (3, 2), (-3, 2)\}$

43. $\{(3, 1), (-3, -1), (1, 3), (-1, -3)\}$

45. $\{(\log_3 6, 24)\}$

47.
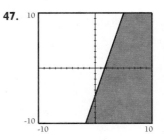

49.
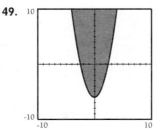

51.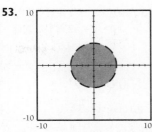

53.

55.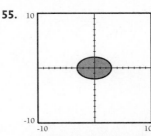

57.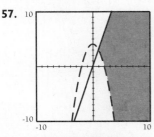

59.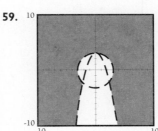

61.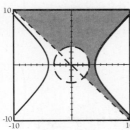

63.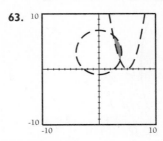

CHAPTER TEST 10

1. $\{(\frac{14}{13}, -\frac{5}{13})\}$ **3.** $\varnothing$

5. Task A: 25 hours, Task B: 15 hours.

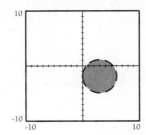

CHAPTER 11

SECTION 11-1

Exercise 11-1

1. $A_{2,4} = 6$ **3.** $A_{1,2} = 2$ **5.** $\begin{pmatrix} 4 \\ 6 \end{pmatrix}$ **7.** $(9\ 8\ 7\ 6\ 5)$

9. $\begin{pmatrix} 1 & 9 \\ 2 & 8 \\ 3 & 7 \\ 4 & 6 \\ 5 & 5 \end{pmatrix}$ **11.** $B_{3,2} = 8$ **13.** $\begin{pmatrix} 3 \\ 6 \\ 9 \end{pmatrix}$ **15.** $(1\ 2\ 3)$

17. $\begin{pmatrix} 1 & 2 & 3 \\ 4 & 5 & 6 \\ 7 & 8 & 9 \end{pmatrix}$ **19.** $\begin{pmatrix} 1 \\ 5 \\ 9 \end{pmatrix}$ **21.** $\begin{pmatrix} 1 & 0 \\ 0 & 1 \end{pmatrix}^T = \begin{pmatrix} 1 & 0 \\ 0 & 1 \end{pmatrix}$

23. $\begin{pmatrix} 5 & 2 \\ 3 & 4 \end{pmatrix}^T = \begin{pmatrix} 5 & 3 \\ 2 & 4 \end{pmatrix}$ **25.** $\begin{pmatrix} 1 & 3 & 6 \\ -3 & 7 & 8 \end{pmatrix}^T = \begin{pmatrix} 1 & -3 \\ 3 & 7 \\ 6 & 8 \end{pmatrix}$

27. $\begin{pmatrix} 1 & 5 \\ 4 & -2 \\ 1 & 3 \end{pmatrix}^T = \begin{pmatrix} 1 & 4 & 1 \\ 5 & -2 & 3 \end{pmatrix}$ **29.** $\begin{pmatrix} 1 & 2 & 1 \\ -1 & 4 & 5 \\ 3 & 0 & 4 \end{pmatrix}^T = \begin{pmatrix} 1 & -1 & 3 \\ 2 & 4 & 0 \\ 1 & 5 & 4 \end{pmatrix}$

33. $\begin{pmatrix} 2 & 3 & 4 \\ 3 & 4 & 5 \\ 4 & 5 & 6 \end{pmatrix}$ **37.** $\begin{pmatrix} 1 & 2 & 3 & 4 \\ 2 & 4 & 6 & 8 \\ 3 & 6 & 9 & 12 \\ 4 & 8 & 12 & 16 \end{pmatrix}$ **39.** $\begin{pmatrix} x \\ y \end{pmatrix} = \begin{pmatrix} 3 \\ 5 \end{pmatrix}$

41. The entry in row 7 column 3 of A^T is also 3.
43. The entry in the third row, third column does not change.

47. $\begin{pmatrix} 1 & 0 & 0 \\ 0 & 1 & 0 \\ 0 & 0 & 1 \end{pmatrix}$ **49.** $\begin{pmatrix} 5 & 3 \\ 8 & 6 \end{pmatrix}$ **51.** $A_{i,j} = A_{j,i}$ implies $i = j$

53. no   **55.** $\begin{pmatrix} 5 & 1 & 1 \\ 1 & 5 & 1 \\ 1 & 1 & 5 \end{pmatrix}$ **57.** $A_{1,2} = 3$ **59.** $\begin{pmatrix} 2 & 3 & 4 \\ 3 & 4 & 5 \\ 4 & 5 & 6 \end{pmatrix}$

SECTION 11-2

Exercise 11-2

1. $\begin{pmatrix} 6 & 1 \\ 5 & 3 \end{pmatrix}$ **3.** $\begin{pmatrix} 2 & 8 \\ 2 & 6 \end{pmatrix}$ **5.** $\begin{pmatrix} 2 & 8 \\ 2 & 6 \end{pmatrix}$

7. $\begin{pmatrix} 3 & 7 & 3 \\ 0 & 7 & 6 \end{pmatrix}$ **9.** $\begin{pmatrix} 4 & 0 & -3 \\ 3 & 4 & 3 \\ 0 & 0 & 4 \end{pmatrix}$ **11.** $\begin{pmatrix} 2 & -1 \\ 1 & 1 \end{pmatrix}$

13. $\begin{pmatrix} 0 & 1 \\ 17 & 3 \end{pmatrix}$ **15.** 11   **17.** -13   **19.** -1   **21.** $\frac{6}{5}$

25. $A \cdot E$ **27.** 103.5P

Problem Set 11-2

1. $\begin{pmatrix} 6 & 0 \\ 4 & 2 \end{pmatrix}$ **3.** $\begin{pmatrix} 18 & 8 \\ 4 & 6 \end{pmatrix}$ **5.** $\begin{pmatrix} 15 & 20 \\ -10 & 40 \end{pmatrix}$

7. $\begin{pmatrix} 3 & 9 & 15 \\ 0 & 7 & 6 \end{pmatrix}$ **9.** $\begin{pmatrix} 5 & 1 & -4 \\ 1 & 6 & 5 \\ 5 & 0 & 5 \end{pmatrix}$ **11.** $\begin{pmatrix} 12 & -1 \\ 4 & 2 \end{pmatrix}$

SECTION 11-3

Exercise 11-3

1. 3   **3.** $\begin{pmatrix} 2 & 6 & 4 \\ 1 & 3 & 2 \\ -1 & -3 & -2 \end{pmatrix}$ **5.** $\begin{pmatrix} 11 & 0 \\ 13 & 14 \end{pmatrix}$ **7.** $\begin{pmatrix} 16 & -1 \\ 10 & 9 \end{pmatrix}$

9. $\begin{pmatrix} -1 & 3 & 5 \\ 0 & 3 & 2 \end{pmatrix}$ **19.** $BA = (700 \ 1100 \ 600)$

Problem Set 11-3

1. 9   **3.** $\begin{pmatrix} 9 & 6 & -3 \\ 6 & 4 & -2 \\ -3 & -2 & 1 \end{pmatrix}$ **5.** $\begin{pmatrix} 15 & -28 \\ 15 & 25 \end{pmatrix}$ **7.** $\begin{pmatrix} 86 & -51 \\ 8 & 31 \end{pmatrix}$

9. $\begin{pmatrix} 93 & 226 & -35 \\ 65 & 13 & -63 \end{pmatrix}$

SECTION 11-4

Exercise 11-4

1. $\{(2, 1)\}$ **3.** $\{(1, -3)\}$ **5.** $\{(2, 1, -1)\}$ **7.** $\{(3, 1, 1)\}$
9. $\{(-1, 1, 1, 1)\}$ **11.** $\{(6 - t, -1, t)\}$ **13.** $\{(2, t, t)\}$
15. $\left\{ \left(\dfrac{14 - t}{3}, \dfrac{7 + t}{3}, \dfrac{-7 + 2t}{3}, t \right) \right\}$ **17.** $\{(0, 0)\}$

19. $\{(\frac{2t}{3}, t)\}$ **21.** $\{(0, 0, 0)\}$ **23.** $\left\{ \left(-\dfrac{2t}{3}, -\dfrac{t}{3}, t \right) \right\}$

25. $\left(\begin{array}{cc|cc} 1 & 0 & \frac{1}{7} & \frac{2}{7} \\ 0 & 1 & -\frac{2}{7} & \frac{3}{7} \end{array} \right)$ **27.** $\left(\begin{array}{ccc|ccc} 1 & 0 & 0 & 0 & \frac{1}{3} & \frac{1}{3} \\ 0 & 1 & 0 & \frac{1}{2} & -\frac{1}{2} & 0 \\ 0 & 0 & 1 & \frac{1}{2} & \frac{1}{6} & -\frac{1}{3} \end{array} \right)$

29. $\left(\begin{array}{cccc|cccc} 1 & 0 & 0 & 0 & -0.5 & 0.5 & 0.5 & 0.5 \\ 0 & 1 & 0 & 0 & 0.5 & 0 & 0 & -0.5 \\ 0 & 0 & 1 & 0 & 0.5 & 0 & -0.5 & 0 \\ 0 & 0 & 0 & 1 & 0.5 & -0.5 & 0 & 0 \end{array} \right)$

Problem Set 11-4

1. $\left(\begin{array}{ccc|c} 1 & 0 & 0 & 3.8 \\ 0 & 1 & 0 & -3.2 \\ 0 & 0 & 1 & -0.6 \end{array} \right)$ **3.** $\left(\begin{array}{ccc|ccc} 1 & 0 & 0 & -0.8 & 0.6 & 1.4 \\ 0 & 1 & 0 & 2.2 & -1.4 & -2.6 \\ 0 & 0 & 1 & 0.6 & -0.2 & -0.8 \end{array} \right)$

5. $\begin{pmatrix} -0.8 & 0.6 & 1.4 \\ 2.2 & -1.4 & -2.6 \\ 0.6 & -0.2 & -0.8 \end{pmatrix}$ **7.** $\begin{pmatrix} 3.8 \\ -3.2 \\ -0.6 \end{pmatrix}$

SECTION 11-5

Exercise 11-5

1. $|A| = -11, A^{-1} = \begin{pmatrix} -\frac{1}{11} & -\frac{4}{11} \\ -\frac{2}{11} & -\frac{3}{11} \end{pmatrix}$

3. $|A| = 14, A^{-1} = \begin{pmatrix} \frac{3}{14} & \frac{1}{14} \\ -\frac{2}{14} & \frac{4}{14} \end{pmatrix}$

5. $|A| = 13, A^{-1} = \begin{pmatrix} -\frac{5}{13} & \frac{4}{13} & \frac{7}{13} \\ \frac{7}{13} & -\frac{3}{13} & -\frac{2}{13} \\ \frac{4}{13} & \frac{2}{13} & -\frac{3}{13} \end{pmatrix}$

7. $|A| = -3, A^{-1} = \begin{pmatrix} 0 & \frac{1}{3} & \frac{1}{3} \\ -2 & \frac{5}{3} & -\frac{4}{3} \\ 1 & -\frac{2}{3} & \frac{1}{3} \end{pmatrix}$

9. $|A| = 4, A^{-1} = \begin{pmatrix} 1 & -1 & 1 & 0 \\ -\frac{7}{4} & \frac{5}{4} & -\frac{1}{4} & \frac{3}{4} \\ \frac{5}{4} & -\frac{3}{4} & -\frac{1}{4} & -\frac{1}{4} \\ \frac{1}{4} & \frac{1}{4} & -\frac{1}{4} & -\frac{1}{4} \end{pmatrix}$

11. $\{(2, 1)\}$ **13.** $\{(\frac{10}{7}, -\frac{9}{7})\}$ **15.** $\{(1, 1, 1)\}$

17. $\{(2, -1, 1)\}$ **19.** $\{(1, \frac{3}{2}, \frac{1}{2}, \frac{3}{2})\}$

21. $\{(20 \text{ wallets}, 10 \text{ purses}, 30 \text{ key chains})\}$
49. $y = -2x + 4$ **51.** $y = x^2 + 5x - 1$ **53.** $y = x^2 - 1$
57. $\langle 1, -5, 2 \rangle$ **61.** $x = -3$ or $x = 2$ **63.** $x = 5$ or $x = -1$
65. $x \cong 4.464$ or $x \cong -2.464$

Problem Set 11-5
1. $\{(2, 1)\}$ **3.** $\{(1.429, -1.286)\}$ **5.** $\{(1, 1, 1)\}$
7. $\{(1, 1, 1)\}$ **9.** $\{(1, 1.5, 0.5, 1.5)\}$

CHAPTER 11 REVIEW EXERCISES

1. 8   **3.** $\begin{pmatrix} 3 \\ 8 \\ 3 \end{pmatrix}$ **5.** $\begin{pmatrix} 1 & 5 & 6 \\ 2 & 9 & 5 \\ 3 & 8 & 3 \\ 4 & 7 & -2 \end{pmatrix}$ **7.** 7   **9.** $(1 \ 2 \ 3 \ 4)$

11. $\langle 2\ 0\ 8\ 1\rangle$ **13.** $\begin{pmatrix} 2 & 0 \\ 1 & 3 \end{pmatrix}$ **15.** $\begin{pmatrix} 4 & -1 \\ 3 & 5 \\ 8 & 2 \end{pmatrix}$

17. $\begin{pmatrix} 1 & 2 & 3 \\ 2 & 4 & 6 \\ 3 & 6 & 9 \end{pmatrix}$ **19.** $\begin{pmatrix} 0.1 & 0.2 & 0.3 & 0.4 & 0.5 \\ 0.2 & 0.4 & 0.6 & 0.8 & 1 \\ 0.3 & 0.6 & 0.9 & 1.2 & 1.5 \end{pmatrix}$

21. $q = 5, t = 7, x = -2, y = 8, z = 10$

23. $a_{3,5} = 10$ **25.** $\begin{pmatrix} 1 & 0 & 0 \\ 0 & 1 & 0 \end{pmatrix}$ **27.** $\begin{pmatrix} 7 & 14 \\ 21 & 28 \end{pmatrix}$

29. $\begin{pmatrix} 5 & 4 \\ 10 & 6 \end{pmatrix}$ **31.** $\begin{pmatrix} 5 & 20 \\ 5 & 15 \end{pmatrix}$ **33.** $\begin{pmatrix} 4 & 8 & 5 \\ 2 & -1 & 4 \end{pmatrix}$

35. $\begin{pmatrix} 3 & 0 & -5 \\ 5 & 3 & 5 \\ 0 & 0 & 3 \end{pmatrix}$ **37.** (0) **39.** no answer

41. $\begin{pmatrix} -8 & 22 \\ 22 & 13 \end{pmatrix}$ **45.** $\{(3, -1)\}$ **47.** $\{(3, -2, 1)\}$

49. $\{(3, 0, 1, -1)\}$

51. $x = \dfrac{10}{3} - \dfrac{t}{3}, y = \dfrac{1}{3} - \dfrac{t}{3}, z = -\dfrac{1}{3} - \dfrac{2t}{3}, w = t$

53. $\{(0, 0)\}$ **55.** $\{(0, 0, 0)\}$ **57.** $\left(\begin{array}{cc|cc} 1 & 0 & \frac{1}{11} & \frac{2}{11} \\ 0 & 1 & -\frac{3}{11} & \frac{5}{11} \end{array} \right)$

59. $\left(\begin{array}{ccc|ccc} 1 & 0 & 0 & \frac{1}{2} & -\frac{1}{2} & \frac{1}{2} \\ 0 & 1 & 0 & -\frac{3}{2} & \frac{5}{2} & -\frac{1}{2} \\ 0 & 0 & 1 & -\frac{1}{2} & \frac{3}{2} & -\frac{1}{2} \end{array} \right)$ **61.** $\begin{pmatrix} \frac{3}{19} & \frac{2}{19} \\ -\frac{2}{19} & \frac{5}{19} \end{pmatrix}$

63. $\begin{pmatrix} \frac{1}{5} & \frac{2}{5} & \frac{1}{5} \\ \frac{1}{5} & -\frac{3}{5} & \frac{1}{5} \\ \frac{4}{5} & -\frac{2}{5} & -\frac{1}{5} \end{pmatrix}$

65. $\{(\frac{36}{19}, -\frac{5}{19})\}$ **67.** $\{(\frac{13}{5}, -\frac{2}{5}, \frac{7}{5})\}$

CHAPTER 11 TEST

1. $w = 5, x = 4, y = 1, z = 3$ **3.** $\begin{pmatrix} 4 & -7 \\ -3 & 6 \end{pmatrix}$ **5.** $\begin{pmatrix} 7 & 5 \\ 7 & 18 \end{pmatrix}$

7. $\left(\begin{array}{cc|cc} 1 & -\frac{1}{2} & \frac{1}{4} & 0 \\ 0 & 0 & \frac{1}{2} & 1 \end{array} \right)$ **9. (a)** 0 **(b)** no inverse

CHAPTER 12

SECTION 12-1

Exercise 12-1

1. $3, 5, 7, 9, 11$ **3.** $1, 8, 27, 64, 125$

5. $\dfrac{2}{3}, \dfrac{3}{4}, \dfrac{4}{5}, \dfrac{5}{6}, \dfrac{6}{7}$ **7.** $5, 10, 20, 40, 80$

9. $-2, 4, -6, 8, -10$ **11.** $1, 3, 4, 7, 11$ **13.** $2, 4, 6, 8, 10$

15. $4, 9, 14, 19, 24$ **17.** $1, 2, 6, 24, 120$

19. $1, 2, 6, 24, 120$

21. $a_n + b_n = n^2 + 2n + 1, 4, 9, 16; a_nb_n = 2n^3 + n^2, 3, 20, 63$

23. $a_n + b_n = n^2, 1, 4, 9; a_nb_n = -2n^3 - 5n^2 - 4n - 1, -12, -45,$ -112

25. $a_n + b_n = (-1)^2 + 3n, 2, 7, 8; a_nb_n = (-1)^n3n, -3, 6, -9$

27. $a_n + b_n = 2^n + \sqrt{n}, 3, 4 + \sqrt{2}, 8 + \sqrt{3}; a_nb_n = 2^n \sqrt{n}, 2,$ $4\sqrt{2}, 8\sqrt{3}$

29. $a_n + b_n = \dfrac{1}{n + 1} + 3^{n-1}, \dfrac{1}{2}, \dfrac{10}{3}, \dfrac{37}{4}; a_n + b_n = \dfrac{3^{n-1}}{n + 1},$ $\dfrac{1}{2}, 1, \dfrac{9}{4}$

31. 15 **33.** 30 **35.** $\frac{11}{6}$ **37.** 441 **39.** -1

41. $S_3 = 6; S_5 = 15$ **43.** $S_3 = 12; S_5 = 30$

45. $S_3 = 39; S_5 = 230$ **47.** $S_3 = \frac{13}{12}; S_5 = \frac{29}{20}$

49. $S_3 = -3; S_5 = -4$

51. $a_n = 2n, a_n = 2n + (n - 1)(n - 2)(n - 3)(n - 4)b_n$, where b_n is any sequence.

53. $a_n = (-1)^n, a_n = \begin{cases} (-1)^n, n \leq 4 \\ (-1)^n + n, n > 4 \end{cases}$

55. $a_n = (-1)^{n+1}n^2, a_n = (-1)^{n+1}n^2 + (n - 1)(n - 2)(n - 3)$ $(n - 4)$

57. $a_n = \dfrac{n}{n + 1}, a_n = \begin{cases} \dfrac{n}{n + 1}, n \leq 4 \\ \dfrac{n + 1}{n}, n > 4 \end{cases}$

59. $a_n = 2 + 3(n - 1),$ $a_n = 2 + 3(n - 1)(n^3 - 9n^2 + 26n - 23)$

65. $\displaystyle\sum_{k=1}^{5} \dfrac{1}{2^k} = \dfrac{31}{32}$. It will never overflow.

Problem Set 12-1

3. $a_n = 2n$ **5.** $a_n = \dfrac{1}{n(n + 1)}$ **7.** $a_n = 3n - 1$

9. $a_n = n^3$ **11.** $a_n = \begin{cases} 1, \text{if } n = 1 \\ 2^n, \text{if } n > 1 \end{cases}$

13. $a_n = \begin{cases} -6, \text{ if } n = 1 \\ 2^n, \text{ if } n > 1 \end{cases}$ **15.** 5, 9, 13, 17, 21

17. 15, 10, 5, 0, -5   **19.** 2, 6, 18, 54, 162

21. 3, -6, 12, -24, 48

23. 8, 13, 18, 23, 28; $a_n = \begin{cases} 8, & \text{if } n = 1 \\ a_{n-1} + 5, & \text{if } n > 1 \end{cases}$

25. 1, -1, -3, -5, -7; $a_n = \begin{cases} 1, & \text{if } n = 1 \\ a_{n-1} - 2, & \text{if } n > 1 \end{cases}$

27. 3, 9, 27, 81, 243; $a_n = \begin{cases} 3, & \text{if } n = 1 \\ 3a_{n-1}, & \text{if } n > 1 \end{cases}$

29. 0.5, 0.25, 0.125, 0.0625, 0.03125; $a_n = \begin{cases} 0.5, & \text{if } n = 1 \\ 0.5a_{n-1}, & \text{if } n > 1 \end{cases}$

Exercise 12-2

1. $a_n = 2 + 3(n - 1)$ **3.** $a_n = -1 + \dfrac{(n - 1)}{2}$
5. $a_n = -3(n - 1)$

7. $a_n = \begin{cases} 7, & \text{if } n = 1 \\ a_{n-1} + 2, & \text{if } n > 1 \end{cases}$ **9.** $a_n = \begin{cases} 3, & \text{if } n = 1 \\ 3 + a_{n-1}, & \text{if } n > 1 \end{cases}$

11. $a_n = 2(3^{n-1})$ **13.** $a_n = 5(-2)^{n-1}$ **15.** $a_n = 2\left(\frac{1}{2}\right)^{n-1}$

17. $a_n = \begin{cases} 4, & \text{if } n = 1 \\ \frac{1}{2}a_{n-1}, & \text{if } n > 1 \end{cases}$ **19.** $a_n\, a_n = \begin{cases} -3, & \text{if } n = 1 \\ -2a_{n-1}, & \text{if } n > 1 \end{cases}$

21. arithmetic: $a_n = 3 + 5(n - 1)$
23. neither   **25.** arithmetic: $a_n = 2(n - 1)$
27. geometric: $a_n = (-2)^{n-1}$ **29.** geometric: $a_n = (0.1)^{n-1}$

31. 120   **33.** -90   **35.** 14,950   **37.** $\dfrac{93}{16}$ **39.** 86

41. 630   **43.** 62   **45.** 46   **47.** 156   **49.** 0.5555555555
51. arithmetic: $10,500, $82,500   **53.** $4,500
59. $1093.44   **61.** $2627.90   **63.** 3 years

Problem Set 12-2

1. $3 - 3^{n+1} = 3(1 - 3^n)$ **3.** $4 - 4^{n+1} = 4(4 - 4^n)$
5. $r - r^{n+1} = r(1 - r^n)$ **7.** $-n^2 - 2n$ **9.** $-n$
13. 2, 3, 4, 5, 6   **15.** 7, 11, 15, 19, 23   **17.** 0, -1, -2, -3, -4
19. 2, 1, $\frac{1}{2}$, $\frac{1}{4}$, $\frac{1}{8}$ **21.** 1, -1, 1, -1, 1

Exercise 12-3

1. rational   **3.** rational   **5.** irrational   **7.** rational
9. irrational   **11.** $0.\overline{3}$ **13.** 0.2 **15.** $0.\overline{16}$ **17.** $0.\overline{27}$
19. $0.\overline{714285}$ **21.** $\frac{1}{3}$ **23.** $\frac{2}{9}$ **25.** $\frac{7}{10}$ **27.** $\frac{3}{11}$ **29.** 2

31. $\frac{13}{100}$ **33.** $\frac{1}{3}$ **35.** $\frac{1}{63}$ **37.** 1   **39.** 1

41. converges: $S = 6$ **43.** diverges   **45.** converges: $S = \frac{20}{3}$
47. converges: $S = -3$ **49.** converges: $S = \frac{5}{4}$

51. converges: $S = -\frac{1}{7}$ **53.** converges: $S = -\frac{1}{11}$ **55.** $3.\overline{3}$

57. 3   **59.** 3, $\frac{1}{3}$ **61.** $0.\overline{9}$, 1

Exercise 12-4

13. 100   **15.** $\dfrac{31}{32}$ **17.** 72   **19.** $\dfrac{10}{11}$ **21.** 225

Exercise 12-5

1. $35a^3b^4$ **3.** $35a^4b^3$ **5.** $1215x^8y^2$ **7.** $-80x^2y^9$ **9.** $\dfrac{8}{27}$

11. $a^6 + 6a^5b + 15a^4b^2 + 20a^3b^3 + 15a^2b^4 + 6ab^5 + b^6$
13. $x^7 - 7x^6y + 21x^5y^2 - 35x^4y^3 + 35x^3y^4 - 21x^2y^5 + 7xy^6 - y^7$
15. $x^5 + 10x^4 + 40x^3 + 80x^2 + 80x + 32$
17. $x^4 - 8x^3y + 24x^2y^2 - 32xy^3 + 16y^4$
19. $32x^{15} + 240x^{12}y^2 + 720x^9y^4 + 1080x^6y^6 + 810x^3y^8 + 243y^{10}$
21. $x^2 + 12x\sqrt{x} + 54x + 108\sqrt{x} + 81$

23. $x^{-3} + 3x^{-2}y + 3x^{-1}y^2 + y^3 = \dfrac{x^3y^3 + 3x^2y^2 + 3xy + 1}{x^3}$

31. 6   **33.** 2   **35.** 3   **37.** 30

Exercise 12-6

1. 60   **3.** 10   **5.** 10   **7.** 20   **9.** 210   **11.** 495
13. 495   **15.** 125,970   **17.** 56   **19.** 220   **21.** 12,320
23. 100,000   **25.** 120   **27.** 60   **29.** 990 (permutation)

33. 210   **37.** $\displaystyle\prod_{k=1}^{r} a_k$ **39.** $\dfrac{\left(\sum_{k=1}^{n} a_k\right)!}{\prod_{k=1}^{n} a_k!}$

CHAPTER 12 REVIEW EXERCISES

1. 1, 4, 7, 10, 13   **3.** $\frac{3}{2}, \frac{5}{4}, \frac{7}{6}, \frac{9}{8}, \frac{11}{10}$ **5.** 3, 7, -4, 11, -15

7. 5, 10, 15, 20, 25   **9.** 3, 6, 18, 72, 360
11. $a_n + b_n = n^2 + 3n + 2, 6, 12, 20$; $a_nb_n = 3n^3 + 2n^2, 5, 32, 99$
13. $a_n + b_n = (-1)^n + n + 1, 1, 4, 3$;
$a_nb_n = (-1)^n (n + 1), -2, 3, -4$
15. $a_n + b_n = 2^{n+1} + \sqrt{n}, 5, 8 + \sqrt{2}, 16 + \sqrt{3}$; $a_nb_n = 2^{n+1}\sqrt{n}, 5, 8\sqrt{2}, 16\sqrt{3}$

17. 55   **19.** $\dfrac{23}{15}$ **21.** -22   **23.** $S_3 = 12; S_5 = 30$

25. $S_3 = 17; S_5 = 60$ **27.** $a_n = 4n, a_n = \begin{cases} 4n, \text{ if } n \leqslant 4 \\ 8n, \text{ if } n > 4 \end{cases}$

29. $a_n = (-1)^{n+1}2^n$,
$a_n = (-1)^{n+1}2^{n+(n-1)(n-2)(n-3)(n-4)}$

31. $a_n = \dfrac{n + 1}{n + 2}; a_n = \dfrac{n + 1 + (n - 1)(n - 2)(n - 3)(n - 4)}{n + 2}$

35. $a_n = 5 - 3(n - 1)$ **37.** $a_n = 5(n - 1)$
39. $a_n = 3 + 4(n - 1)$ **41.** $a_n = (0.5)^{n-1}$
43. $a_n = -16(-\frac{1}{2})^{n-1}$ **45.** $a_n = 3(-2)^{n-1}$

47. $a_n = 2 + 5(n - 1)$ **49.** neither
51. geometric: $a_n = 10(0.1)^{n-1}$
53. 185   **55.** 24,950   **57.** 430   **59.** 1050   **61.** 22
63. 4.444444444   **65.** geometric: $81,000, $121,000
67. rational   **69.** irrational   **71.** irrational   **73.** $0.\overline{6}$

75. $0.\overline{6}$ **77.** $0.\overline{857142}$ **79.** $\frac{7}{3}$ **81.** $\frac{57}{10}$ **83.** $\frac{314}{99}$

85. $\frac{3}{20}$ **87.** 1   **89.** $\frac{1}{3}$ **91.** converges: 10   **93.** diverges

95. -6   **97.** $\frac{30}{17}$ **103.** 110   **105.** 70   **107.** $126a^5b^4$

109. $20412x^{12}y^2$ **111.** $\frac{40}{243}$

113. $a^7 + 7a^6b + 21a^5b^2 + 35a^4b^3 + 35a^3b^4 + 21a^2b^5 +$
   $7ab^6 + b^7$
115. $x^5 + 15x^4 + 90x^3 + 270x^2 + 405x + 243$
117. $x^4 - 12x^3y + 54x^2y^2 - 108xy^3 + 81y^4$
119. 336   **121.** 84   **123.** 21   **125.** 30   **127.** 504
129. 1287   **131.** 116,280   **133.** 39,600   **135.** 480

CHAPTER 12 TEST

1. $S_1 = 1, S_2 = 5, S_3 = 14$ **3. (a)** 155   **(b)** $\frac{1023}{256} \cong 3.996$

5. $1024x^5 - 640x^4y + 160x^3y^2 - 20x^2y^3 + \frac{5}{4}xy^4 - \frac{1}{32}y^5$

9. (a) $a_n = 1 + 2(n - 1)$; $a_n = \begin{cases} 1 + 2(n - 1), n \leqslant 4 \\ 1 - 2(n - 1), n > 4 \end{cases}$

(b) $a_n = 8(\frac{-1}{2})^{n-1}$;

   $a_n = 8(\frac{-1}{2})^{n-1} - (n - 1)(n - 2)(n - 3)(n - 4)$

INDEX

SPECIAL VALUES OF TRIGONOMETRIC FUNCTIONS

θ degrees	$\sin \theta$ $\sin x$	$\cos \theta$ $\cos x$	$\tan \theta$ $\tan x$	radians x
0°	0	1	0	0
30°	$\dfrac{1}{2}$	$\dfrac{\sqrt{3}}{2}$	$\dfrac{\sqrt{3}}{3}$	$\dfrac{\pi}{6}$
45°	$\dfrac{\sqrt{2}}{2}$	$\dfrac{\sqrt{2}}{2}$	1	$\dfrac{\pi}{4}$
60°	$\dfrac{\sqrt{3}}{2}$	$\dfrac{1}{2}$	$\sqrt{3}$	$\dfrac{\pi}{3}$
90°	1	0	undefined ∞	$\dfrac{\pi}{2}$

FUNDAMENTAL IDENTITIES

$$\tan x \equiv \frac{\sin x}{\cos x}, \qquad \sec x \equiv \frac{1}{\cos x}, \qquad \csc x \equiv \frac{1}{\sin x}$$

$$\cot x \equiv \frac{\cos x}{\sin x}, \qquad \sin^2 x + \cos^2 x \equiv 1, \qquad 1 + \tan^2 x \equiv \sec^2 x$$

COMPLEMENTARY FUNCTIONS

$$\sin x \equiv \cos\left(\frac{\pi}{2} - x\right), \quad \sec x \equiv \csc\left(\frac{\pi}{2} - x\right), \quad \csc x \equiv \sec\left(\frac{\pi}{2} - x\right)$$

$$\cos x \equiv \sin\left(\frac{\pi}{2} - x\right), \quad \tan x \equiv \cot\left(\frac{\pi}{2} - x\right), \quad \cot x \equiv \left(\frac{\pi}{2} - x\right)$$

RECIPROCAL FUNCTIONS:

$$\sin x \equiv \frac{1}{\csc x}, \qquad \csc x \equiv \frac{1}{\sin x}, \qquad \cos x \equiv \frac{1}{\sec x}$$

$$\sec x \equiv \frac{1}{\cos x}, \qquad \tan x \equiv \frac{1}{\cot x}, \qquad \cot x \equiv \frac{1}{\tan x}$$

ODD FUNCTIONS

$$\sin(-x) \equiv -\sin x, \qquad \csc(-x) \equiv -\csc x$$
$$\tan(-x) \equiv -\tan x, \qquad \cot(-x) \equiv -\cot x$$

EVEN FUNCTIONS

$$\cos(-x) \equiv \cos x, \qquad \sec(-x) \equiv \sec x$$

SUM FORMULAS

$$\cos(a + b) \equiv \cos a \cos b - \sin a \sin b$$
$$\sin(a + b) \equiv \sin a \cos b + \cos a \sin b$$
$$\tan(a + b) \equiv \frac{\tan a + \tan b}{1 - \tan a \tan b}$$